Lecture Notes in Networks and Systems

Volume 139

The series "Lecture Notes in Networks and Systems" publishes the latest developments in Networks and Systems—quickly, informally and with high quality. Original research reported in proceedings and post-proceedings represents the core of LNNS.

Volumes published in LNNS embrace all aspects and subfields of, as well as new challenges in, Networks and Systems.

The series contains proceedings and edited volumes in systems and networks, spanning the areas of Cyber-Physical Systems, Autonomous Systems, Sensor Networks, Control Systems, Energy Systems, Automotive Systems, Biological Systems, Vehicular Networking and Connected Vehicles, Aerospace Systems, Automation, Manufacturing, Smart Grids, Nonlinear Systems, Power Systems, Robotics, Social Systems, Economic Systems and other. Of particular value to both the contributors and the readership are the short publication timeframe and the world-wide distribution and exposure which enable both a wide and rapid dissemination of research output.

The series covers the theory, applications, and perspectives on the state of the art and future developments relevant to systems and networks, decision making, control, complex processes and related areas, as embedded in the fields of interdisciplinary and applied sciences, engineering, computer science, physics, economics, social, and life sciences, as well as the paradigms and methodologies behind them.

**** Indexing: The books of this series are submitted to ISI Proceedings, SCOPUS, Google Scholar and Springerlink ****

More information about this series at http://www.springer.com/series/15179

Svetlana Igorevna Ashmarina ·
Valentina Vyacheslavovna Mantulenko ·
Marek Vochozka
Editors

Engineering Economics: Decisions and Solutions from Eurasian Perspective

Springer

Editors
Svetlana Igorevna Ashmarina
Applied Management Department
Samara State University of Economics
Samara, Russia

Valentina Vyacheslavovna Mantulenko
Department of Applied Management
Samara State University of Economics
Samara, Russia

Marek Vochozka
Institute of Technology and Business
in České Budějovice
České Budějovice, Czech Republic

ISSN 2367-3370 ISSN 2367-3389 (electronic)
Lecture Notes in Networks and Systems
ISBN 978-3-030-53276-5 ISBN 978-3-030-53277-2 (eBook)
https://doi.org/10.1007/978-3-030-53277-2

This Springer imprint is published by the registered company Springer Nature Switzerland AG
The registered company address is: Gewerbestrasse 11, 6330 Cham, Switzerland

Contents

Decisions in Production and Project Management in the Context of the Digital Economy

Globalized Economy and Russian Industrial Enterprises: Development Specifics and International Co-operation

Specificity of Sustainability Assessment for Industrial Enterprise Functioning in the Digital Economy

E. M. Pimenova(✉)

Samara State University of Economics, Samara, Russia
pimenova-elena@rambler.ru

Abstract. The purpose of this article is to consider the main provisions of assessing the stability of enterprise functioning in the conditions of digitalization of the Russian economy. The paper investigates the essence of the digital economy and proves its relevance at the current development stage of the Russian market. The key indicator of a high level of digitalization is the adoption of management decisions based on objective business analysis data, as well as the use of technological tools to improve the stability of enterprise functioning. Therefore, the author examines the Russian and foreign methods of assessing sustainability, identifying similarities and differences between them. The author also considers the need for a comprehensive assessment of sustainability not only from the financial point of view, but also from the point of view of the environmental and social significance of this issue. This is a time-consuming, painstaking process, so you cannot do without developing appropriate information programs. It can be argued that with the help of digitalization, an industrial enterprise is able to strengthen the sustainability of its functioning in the market, and increase its work efficiency.

Keywords: Digital economy · Industrial enterprise functioning · Sustainability

1 Introduction

The topic of the digital economy is extremely relevant nowadays. The term "digital economy" itself has moved from scientific journalism to the mass media relatively recently, in the last 5–6 years. The maximum attention to it arose in 2017 during the rise of the cryptocurrency profitability, after which various structures, businesses, and individuals became interested in this problem. Accordingly, at the state level, they began to deal with digitalization issues: the national program "Digital economy of the Russian Federation" was developed [15]. It stated that the Russian state policy in the digital economy is focused on creating conditions for its full development. So now "digital economy" is a common term. It is common in many sectors of the national economy of our country, and even in the sphere of state management (for example, the Ministry of Digital Economic Development was created in the structure of the Government of the Russian Federation in May 2018).

The digital economy is based on a qualitatively new type of digital information and communication technologies (ICT), which cover and transform all areas of the

S. I. Ashmarina et al. (Eds.): ENGINEERING ECONOMICS WEEK 2020, LNNS 139, pp. 3–10, 2021.
https://doi.org/10.1007/978-3-030-53277-2_1

industrial and social life of the world economy in general and the national economy of Russia in particular. This economy type is still in the process of formation, but already now, with its powerful potential, it offers businesses and countries a chance to take leading positions in the main areas of the socio-economic development (if it is successfully implemented).

The financial component of digitalization plays a crucial role in the efficient allocation of resources, economic growth and job creation. Having an efficient enterprise to stimulate and maintain the growth of the state economy is a requirement for all countries [12]. In Russia, the construction of a digital model of the economy is based on an industrial and production model. Therefore, the main leverage of the digital breakthrough for Russian enterprises is to obtain sufficient finance for the development and acquisition of digital assets by increasing the efficiency of industrial enterprises and strengthening the stability of their functioning in the market.

The basis of the digital economy is the digitalization of production. It ensures the optimization of the workflow by making software and hardware decisions. In turn, this implies, in addition to replacing production tools, the application of analytical systems that can make production as efficient as possible. Key indicators of a high level of digitalization are: making management decisions based on objective business analysis data, as well as using technological tools to improve the labor productivity and, ultimately, the sustainability of enterprise's functioning.

At present, the company's stability is one of the main factors of its activity in modern conditions, so the importance and role of its assessment, as well as the search for improvement factors, has significantly increased both for the company itself and for its contractors. In turn, instability can have a negative impact on the state of the Russian economy in general, and can also cause non-payments and a slowdown in the business activity of enterprises in particular. That is why it is so important to prevent an unstable financial state of an industrial enterprise at an early stage. This can be achieved through timely identification of factors that affect the sustainability level, as well as through a continuous analysis and development of specific measures based on it to restore stable functioning of the enterprise, which in turn guarantees its survival.

Sustainable development has become a fundamental issue in the modern world, a task that businesses can no longer ignore. But here the main drawback of the existing methods of assessing sustainability in the Russian economic literature comes to the fore – they consider only financial stability, since the main goal of the enterprise's functioning in the market is fairly considered to be profit maximization. In other words, entrepreneurship is mainly focused on the economic development and welfare growth, while environmental and social issues are mostly avoided [13]. This is extremely narrow consideration of the problem. Through digitalization, businesses can maintain their sustainability by acting responsible or more precisely: financially, environmentally and socially sustainable [4].

The term "sustainable development" was first used in 1987 when the report "Our common future" (the World Commission on Environment and Development, WCED) was published [3]. According to this report, sustainability is aimed at ensuring the equality between generations. The sustainability principles expressed in this way are indisputable. Most people want to live like their parents, and they want their children to enjoy the same opportunities. The same logic applies in business – most managers want their business to be at least as profitable as it was in the past, and, ideally, to grow profits. Thus, time is a central factor in the sustainability concept [1].

The growing importance of environmental, social issues and sustainable enterprise development has given rise to the concept of sustainable entrepreneurship, which can be considered as an opportunity to assess financial sustainability at the intersection of environmental and social aspects. Digitalization helps to solve this problem and industrial enterprises are increasingly strengthening their competitive positions through innovative actions in the field of sustainable development, considering issues of financial stability in conjunction with the social and environmental sustainability. Based on this, it can be stated that sustainable entrepreneurship opens up new opportunities for the development of corporate responsibility, taking into account environmental and social aspects [16].

2 Methodology

Historically, the success of the enterprise was explained almost exclusively by its economic efficiency. Many authors emphasize the importance of "planning and efficiency as the main elements of integrating sustainability into enterprise operations" [9]. The research purpose in the field of entrepreneurship was to generate economic benefits or, in some cases, create sources of employment. These factors traditionally determine the contribution of sustainable entrepreneurship to the development of the national economy as a whole. Therefore, value creation was usually measured in economic and financial terms by indicators such as sales, profit, or return on investment, and was always understood solely as maximizing individual profits.

Generalization and comparative analysis of methodological provisions for assessing the sustainability of the enterprise allows identifying the most advanced methods of managing sustainability, so that they can be used in the practical activities of analytical services of Russian enterprises. For this purpose, it is necessary to describe typical procedures for assessing sustainability in Russian and foreign economic literature and identify similarities and differences between them.

First of all, it should be noted that both in the Russian Federation and abroad, qualitative (non-financial) aspects that reflect current trends in the sustainable business development are increasingly supplemented by quantitative (financial) indicators. Here, we are talking about the process of creating new concepts for measuring, accounting and evaluating non-financial aspects that are related to sustainable business, as well as how the classical model for assessing performance and management affects this process [2].

Considering differences between domestic and foreign methods for assessing sustainability, we can note a fairly broad system of sustainability indicators in the foreign literature, when it includes, for example, the return on equity (a variable widely used in research related to finance and accounting); the return on assets (the growth of this indicator in dynamics is considered as a necessary condition for the stability growth of the enterprise's functioning) [7]. In the domestic economic literature, profitability indicators clearly relate to criteria for evaluating the efficiency of the enterprise performance, and the Russian economists usually do not mix sustainability indicators and profitability indicators.

The main specific feature of the Russian approach is that business sustainability can be assessed using absolute and relative indicators. To determine a sustainability type using a three-component indicator (depending on the coverage degree of reserves with various financial sources, four types are distinguished: absolute, normal, unstable, and crisis), it is proposed to calculate three absolute indicators, and then compare them with the size of reserves. Considerations of this method by different authors may not coincide in detail, but in general its meaning is unchanged. Abroad, this method is not used at all (the use of absolute indicators implies the impossibility of their spatial and temporal comparisons). As for relative indicators of business sustainability, there are a significant number of them in the Russian economic literature, and they are most often divided into two groups:

1. Indicators of the capital structure (independence coefficients, loan-to-equity ratios, etc.) – in fact, they represent the ratio of various elements of the enterprise's capital;
2. Indicators of asset finance (coefficients of working capital mobility, maneuverability of equity capital, and others) – the ratio of current assets or own working capital to various elements of an asset or liability.

A comparison of Russian and foreign methods of assessing the business sustainability using relative indicators allows us to conclude: in the foreign economic literature, there are no analogues of the second group of coefficients at all. However, it is replaced by a group of indicators (for example, the ratio of profit before taxation and interest paid for the use of borrowed funds or "times interest earned ratio") that characterize the ability to service debt – this group of indicators, in turn, is not used in the Russian methodology for assessing sustainability [10].

As another difference, we can highlight the fact that in the foreign economic literature, the normative values of financial stability coefficients are most often absent, and in the domestic one – almost all coefficients have regulatory restrictions. However, their use is not effective enough, because they are not differentiated by industries and therefore the very meaning of establishing such strict criteria is lost. In addition, the existing regulatory restrictions have not been revised for a long time, and during this period, the changes in the business environment have made this revision very relevant.

As a result of this research, we can conclude that the main content of the methodology for assessing the stability of the enterprise's functioning does not differ in the domestic and foreign literature. At the same time, directions of this assessment and the set of indicators for each of them are quite stable abroad, so results can be comparable in the spatial and temporal aspect, and are effective for their practical use. In the Russian economic literature, we can note a variety of approaches, but there is still no unity: different authors have their own definition of key concepts, their own set of sustainability indicators, their own methods of calculating and analyzing them. Therefore, it is necessary to expand research on the problem of selecting indicators for assessing the enterprise sustainability (financial, environmental and social – in a complex) in the conditions of modern Russian economy and develop a unified methodology for this assessment. This extension is necessary to improve the analysis quality of the enterprise sustainability and improve the reliability results obtained. But this is painstaking and extremely time-consuming work that requires statistical processing of a lot of data, and for this you need appropriate software. Consequently, the relevance of the production digitalization in modern economic conditions is confirmed.

3 Results

The main goals of enterprise development in the modern conditions of the Russian market are: profit maximization, growth of the share, diversification, etc. These goals make business entities to develop new approaches to achieve them. It is necessary that these approaches allow you to create and more effectively use the existing resource potential of the enterprise, are comprehensive and focused on the application of innovations in organizational, economic, economic and managerial issues.

The economic development process is uneven, and traditional management methods are extremely weak in linking goals of an industrial enterprise with such a dynamic market situation. Therefore, traditional management methods should be replaced with more advanced ones, which would be aimed at adapting the enterprise to modern market conditions. The constantly changing market situation determines the main goal of Russian enterprises – to maintain the stability of their activities.

In modern market conditions, the basis for sustainable operation of the enterprise is the stability of its functioning. A sustainable enterprise has many advantages over other enterprises of the same sphere in attracting investments, loans, selecting suppliers and qualified personnel [8]. The lack of stability is fraught with a lack of funds to finance current and investment activities and can lead to insolvency, and if the financial condition deteriorates – to bankruptcy. In turn, an overabundance of sustainability puts the company on a difficult development path, burdening it with unnecessary expenses.

The concept of enterprise sustainability is attracting more and more attention from the point of view of the theory for studying this issue and its practical application. It becomes especially important for Russian enterprises because of the sanctions (i.e. restrictions on the purchase and sale of goods from foreign countries), the deterioration of foreign trade relations, and the depreciation of the Russian national currency. However, this has also become a stimulating factor for activities of industrial enterprises at the regional level in Russia. In addition, the constantly changing situation in the field of accounting and reporting has contributed to the adaptation of economic entities to dynamic environmental conditions. All this together contributes to a comprehensive analysis of sustainability issues. Thus, we can say that the basis for the stability of the company's position and the key to success is its sustainability. Without its strengthening, no enterprise can present itself on the market as a competitive organization, operate atlist full capacity, and is able to move forward in its activities. Assessment of the company's sustainability helps to identify the financial condition at a certain point in time, evaluate the dynamics of its changes, disclose reasons for ist deterioration or improvement, and develop a plan of actions to increase the liquidity and solvency of the organization.

4 Discussion

Innovative development of an industrial enterprise using digital technologies is always aimed at increasing the efficiency of its activities and forms appropriate reserves for this growth. But in practice, different types of innovations have different effects on the economic efficiency, i.e. they determine different ways to increase (reserves) the effectiveness of forming an enterprise's innovation and technological potential. Innovations provide for an increase in the production efficiency in some cases by increasing demand for products, improving their quality, expanding production of products with new properties; in other cases – by reducing production costs as a result of using new, more advanced and cost-effective technologies. There are many types of innovations in the activities of industrial enterprises at present. And here, it is most appropriate to identify and analyze those innovations that have a direct impact on the efficiency of an industrial enterprise and the stability of its functioning through new technologies, organizational and managerial structures [6].

The review of the Russian economic literature on the research issues allows us to conclude that sustainability is directly related to the financial condition (solvency and liquidity) of the business entity and persists for a fairly long period of time. Therefore, as sustainability in relation to an industrial enterprise can be considered the creation of such an internal system of organization for its production and financial activities, which could guarantee the operation of the enterprise in the market through effective management of its assets. At the same time, in conditions of instability of external factors, own and attracted sources of capital are used, so without effective use and distribution of the enterprise's resources, as well as cost-effective management of the sources of their formation and financial risks, it is impossible to achieve success in the market. In other words, if an enterprise has a system of organizing production and financial activities that ensures its financial sustainability for a long time, then it has financial and economic stability.

A well-functioning enterprise is capable of self-regulation. In other words, it is adapted to changing external and internal conditions. In this regard, we can highlight an important feature of the term sustainability of an industrial enterprise as an economic system: it reflects the ability of the enterprise to maintain its integrity as a system, and at the same time to develop (progress), despite the impact of external and internal factors. The company has responsibilities and obligations to many different internal and external stakeholders in the market and society. This underscores the need for organizations not only to maximize their revenue (value), but also to do so in a sustainable and socially responsible manner [14].

At the request of global investors and regulators, enterprises disclose internal information about sustainability indicators, on the basis of which scientists have begun to conduct research to assess reporting of enterprises and the financial stability of their functioning. The goal of creating value for an organization can be achieved if its management takes into account the interests of all stakeholders and integrates the five dimensions of sustainability (economic, managerial, social, ethical, and environmental) into management strategies, actions, and reporting [11]. This is more relevant in modern conditions, when financial management implements tactical measures to the detriment of the strategic interests of the company's development [17]. During the crisis, most Russian organizations revised their strategies under the influence of factors such as the devaluation of the ruble, difficult access to capital (primarily credit resources because of the lack of liquid collateral), lower oil prices, and consequences of the introduction of sectoral sanctions by the United States and the European Union against Russian companies.

The enterprise management is associated with the delegation of managerial goals and tasks. The concentration of financial leverages (rewards and restrictions) in the hands of a manager increases the importance of these communications for the sustainable development of the enterprise. The controlling influence from above rises from below as a signal related to the surrounding circumstances and processes. Digitalization allows management personnel to maximize the efficiency of the enterprise and strengthen the sustainability of its functioning.

5 Conclusion

The issue of the economy digitalization in modern conditions of the Russian market (and the world economy as a whole) is one of the priorities. With the help of digitalization, an industrial enterprise can strengthen the sustainability of its functioning in the market. To achieve maximum success in the market and increase the efficiency of its activities by several percent, the company needs to put digitalization as the basis for working out its development strategy [5].

Our country has all the prerequisites for accelerating the pace of the economy digitalization. The Russian Federation has all the necessary intellectual and scientific base. In addition, it has original organizational and technological solutions for creating an effective infrastructure for the digital economy. It is essential for the implementation of the program "Digital economy of the Russian Federation" [15], not only R&D funding was increased, but also training in the field of IT was improved, because in our country there is a serious shortage of qualified personnel in the ICT sector. It is necessary to solve this problem at all levels: at the level of high educational institutions, at the level of industrial enterprises (during corporate training), and at the state level (during the implementation of the state program to support the development of education in the field of ICT).

References

1. Bansal, P., DesJardine, M.R.: Business sustainability: it is about time. Strateg. Organ. **12**(1), 70–78 (2014). https://doi.org/10.1177/1476127013520265
2. Dvořáková, L., Zborková, J.: Integration of sustainable development at enterprise level. Procedia Eng. **69**, 686–695 (2014). https://doi.org/10.1016/j.proeng.2014.03.043
3. Environment & Society Portal: UN world commission on environment and development, ed., report of the world commission on environment and development: Our common future. http://www.environmentandsociety.org/mml/un-world-commission-environment-and-development-ed-report-world-commission-environment-and. Accessed 08 Apr 2020 (1987)

4. Flores-Hernández, J.A., Cambra-Fierro, J.J., Vázquez-Carrasco, R.: Sustainability, brand image, reputation and financial value: manager perceptions in an emerging economy context. Sustain. Dev. **28**(1) (2020). https://doi.org/10.1002/sd.2047
5. Istomina, E.A.: Methodology assessment of trends in the digital economy of industry. Bull. Chelyabinsk State Univ. **12**(422), 108–116 (2018). https://doi.org/10.24411/1994-2796-2018-11212
6. Khalabuda, Y., Nikolaev, M.: Increase of efficiency of industrial enterprises activity on the basis of innovations of various types. Procedia Econ. Finance **16**, 299–302 (2014). https://doi.org/10.1016/S2212-5671(14)00805-3
7. Lassala, C., Apetrei, A., Sapena, J.: Sustainability matter and financial performance of companies. Sustainability **9**(9), 1498 (2017). https://doi.org/10.3390/su9091498
8. Mokeev, V.V., Bunova, E.V., Perevedentceva, A.V.: Analysing the economic stability of an enterprise with the help of Eigenstate method. Procedia Eng. **129**, 681–689 (2015). https://doi.org/10.1016/j.proeng.2015.12.091
9. Phusavat, K., Rassameethes, B., Lesjak, D., Chindavijak, C.: Developing model for integrating sustainability into enterprise operations. Int. J. Manage. Enterp. Dev. **17**(4), 329–346 (2018). https://doi.org/10.1504/IJMED.2018.096237
10. Pimenova, E.M.: The main directions and indicators of assessing the financial condition of the enterprise in the domestic and foreign literature. In: Khasaev, G.B., Ashmarina, S.I. (eds.), Proceedings of the 17th International Scientific and Practical Conference "Problems of Enterprise Development: Theory and Practice, pp. 237–242. SSEU Publishing House, Samara (2018)
11. Rezaee, Z.: Business sustainability research: a theoretical and integrated perspective. J. Acc. Lit. **36**, 48–64 (2016). https://doi.org/10.1016/j.acclit.2016.05.003
12. Roghaniana, P., Raslia, A., Gheysari, H.: Productivity through effectiveness and efficiency in the banking industry. Soc. Behav. Sci. **40**, 550–556 (2012). https://doi.org/10.1016/j.sbspro.2012.03.229
13. Sarango-Lalangui, P., Santos, J.L.S., Hormiga, E.: The development of sustainable entrepreneurship research field. Sustainability **10**(6), 2018 (2005). https://doi.org/10.3390/su10062005
14. Svensson, G., Wood, G., Callaghan, M.: A corporate model of sustainable business practices: an ethical perspective. J. World Bus. **45**(4), 336–345 (2010). https://doi.org/10.1016/j.jwb.2009.08.005
15. The National Program: Digital economy of the Russian Federation (2018). http://government.ru/rugovclassifier/614/events/. Accessed 08 Apr 2020
16. Urbaniec, M.: Sustainable entrepreneurship: innovation-related activities in European enterprises. Polish J. Environ. Stud. **27**(4), 1773–1779 (2018). https://doi.org/10.15244/pjoes/78155
17. Vovchenko, N.G., Holina, M.G., Orobinskiy, A.S., Sichev, R.A.: Ensuring financial stability of companies on the basis of international experience in construction of risks maps, internal control and audit. Euro. Res. Stud., **XX**(1), 350–368 (2017). https://doi.org/10.35808/ersj/623

Formation of Industrial Policy Under Uncertainty Growth

G. I. Yakovlev(✉) and A. V. Streltsov

Samara State University of Economics, Samara, Russia
dmms7@rambler.ru, oisrpp@mail.ru

Abstract. With the growing global uncertainty index caused by unforeseen shock scenarios from viral exposure, as well as the rapid development of new digital technologies, it is necessary to find effective ways to manage production activities. The relevance of the study is to justify the need to institutionalize the forms and methods of directive management of the economy to increase its innovativeness and competitiveness, in accordance with the characteristics of each country. The research purpose is to reveal the nature of modern discussions regarding the methods regulating economic processes and to justify the increasing role of industrial policy of states in combination with ensuring freedom of enterprises. The methods include a retrospective analysis of the forms and methods regulating economic activity, a comparative analysis justifying the scenarios of industrial revival on a new technological and social base under centralized state planning and management. As a result, incentives for production and investment activities in high-value added industries are justified using industrial policy tools (investment, tax, financial, administrative regulation, etc.).

Keywords: Competitiveness · Crisis · Development · Industrial policy · Institutionalization · Uncertainty

1 Introduction

At present, profound transformational shifts are taking place in the scale and models of reproductive activity of business entities at the level of world and national economies. Representatives of economic authorities and business are to understand the essence of the changes to develop new lines and directions of development, competencies, methods of organizing production. It is no coincidence that in the scientific and business environment there is increasing interest in the search for effective methods and ways to develop the potential of manufacturing sectors that are adequate to modern challenges of the development of society, the world economy, science and technology, and natural factors.

The exhaustion of purely market mechanisms and models for the priority organization of high value-added industries requires the identification of new ways to boost the business activity of industrialists and entrepreneurs based on focused policy that ensures clear and stable operating conditions on the planning horizon of up to three to five years. In the global economy, the interdependence of firms in individual countries

S. I. Ashmarina et al. (Eds.): ENGINEERING ECONOMICS WEEK 2020, LNNS 139, pp. 11–18, 2021.
https://doi.org/10.1007/978-3-030-53277-2_2

has become strong. The links of a single cross-border value chain, the jumps of uncertainty for them are more synchronized due to the general dynamics of production and consumption, regulatory instruments. In such large integration entities as the European Community, NAFTA, Mercosur, as well as the Eurasian Economic Union (EAEU), close trade and economic relations between the participants lead to more rigorous synchronization of the uncertainty factor caused by unforeseen shock phenomena, such as a modern oil crisis or social stagnation due to the coronavirus pandemic.

The argument has become significant regarding the growing role of unforeseen large-scale factors provoking uncertainty around the world, because of which both domestic production and global value chains will suffer, weakening production links especially in less developed countries of the world that do not have high scientific and technical potential. To avoid a particularly negative scenario of a fall in GDP due to a loss of a place in global value chains, national economies need to switch to production of higher value-added goods, import-substituting policies and increase domestic consumption, which will promote a new economic order using state policy decisions.

2 Methodology

In the course of the study, the authors applied the retrospective analysis of the forms and methods of managing countries with different types of economic activity, attitudes toward freedom of enterprises, a meaningful economic interpretation of the implemented industrial policy, and the characteristics of the economic practice of leading world powers. Changes in distribution of centers of economic power in world economic geography, mainly caused by Industry 4.0, as well as by significant social and humanitarian shocks in the form of a deadly virus pandemic, an oil market crisis, local wars, and West-East sanctions, are considered.

In the business agenda and specialized literature, it has become a habitual place to exploit the terminology of national economic egoism, protectionism, promote various restrictive measures in sanction and counter-control regimes for political purposes, decline in living standards under the influence of the global crisis, etc. The increased macroeconomic uncertainty in the ways and methods of economic development compel even developed capitalist countries to introduce and strengthen directive methods regulating economic activity. The implemented measures of direct and indirect support of private enterprises are more inherent in doctrines of socialism than the concept of "pure capitalism" in the version of the Washington Consensus, and contribute to the revival of industry on a new technological and social base under centralized state planning and management.

3 Results

Since 2011, many researchers note that international value chains have stopped expanding, their volume growth has almost stopped today. According to Ahir, Bloom and Furseri [1], this change along with the digital transformation of industry has been caused by the growth of the uncertainty factor. They showed that from 2008 to 2011, the World Index of Uncertainty (VIN) grew by 200%, and in today's environment of deep geo-economic shock caused by the coronavirus pandemic, the suspension of industrial production and limited human mobility, it has grown even more. Under a high degree of uncertainty regarding the favorable business environment and the deteriorating business environment, the willingness of companies to invest in the development of production and to hire workers, as well as to carry out industrial research and development (R&D), is declining. In the consumer market, people are objectively intensifying their tendency to save, they are reluctant to spend money since there is lack of understanding the favorable nature of ongoing changes and a clear increase in unemployment.

The experience of countries successfully implementing industrial policy, especially in the face of growing uncertainties, shows that all of them used import substitution, but at the same time they focused on exporting products to the world market. They are characterized by systemic measures to attract foreign investors and to form the global competitiveness of their enterprises and their products. The most successful of them were embedded in global value chains, making extensive use of various forms of cooperation, strategic alliances, while maintaining their independence and economic security. The main approaches to the formation of industrial policy are presented in Table 1.

Table 1. The content of industrial policy in scientific works of specialists

No.	Author	Definition	Kea features
1.	Tolkachev and Teplyakov	Russia needs reindustrialization policy, not only import substitution or export-oriented development [16]	We need new industrialization
2.	Tatarkin	The system of relations between state and municipal authorities, business entities, scientific organizations and civil institutions regarding the formation of a structurally balanced, competitive industry	We need institutionalization of the competitive industry on the innovative basis, the intellectual core of which corresponds to the latest technological structure [15]
3.	Rubenstein	The synthesis of relations between the state and business entities aimed at the formation of the competitive industry [10]	We need collaboration policy to improve the competitiveness of industrial enterprises
4.	Simachev, Kuzin, Kuznetsov, and Pogrebnyak	Allocate separate types: – industrial policy in the open economy, – compensatory industrial policy, – technological industrial policy [12]	We need structure industrial support measures in the categories of sustainability, innovation and international competitiveness

(*continued*)

Table 1. (*continued*)

No.	Author	Definition	Kea features
5.	Larionova	A system of measures aimed at developing the national economy, the latest technologies and products with a high degree of processing, modern information and other services, human capital [8]	Industrial policy is considered as part of socio-economic policy of the state
6.	Spring, Hughes, Mason, and McCaffrey	Industrial policies can prevent disruptions to production systems as technology innovations become commercially viable products	Industrial policy should expand its conceptual scope, contribute to a change in the institutional structure of the economy, and organize active interaction with applied research organizations and universities [14]
7.	Shyu, Joseph, Ding	Super-efficient technologies and intensified global competition have generated the desire for industrial revival through relevant active policies in many countries	It is authorized by the state to use certain acceleration instruments of industrial development that are most suitable for the conditions of individual countries [11]

Source: authors.

With the increasing role of public administration in the specialized literature, one can come across various definitions of industrial policy. Some interpret it as a condition for modernization and revitalization of economic growth, assistance in carrying out structural reforms. In many ways, this approach identifies it with economic policy, depriving industrial policy of its subject of study. Others define it as a separate component of general economic policy, along with financial, tax, etc. In Russia, with the beginning of market reforms, following liberal concepts prevailing in the leading capitalist countries, until recently, measures to support the structural transformations of high-tech types of economic activity were ignored by officials and scientists. However, in recent years such measures of "industrial policy" have received increasing support in developed economies, especially in the manufacturing sector, which has been heavily influenced by technological innovations. A comparative analysis of innovation policy undert revitalization of the industrial sector showed differentiation of specific tools in relation to different countries: in the USA, preference is given to demand-side policies and public services; in Germany, scientific and technological development is being added to this; in China, this is additional environmental and regulatory policy.

According to the definitions in Table 1, the tools of industrial policy include methods of various functional links of state regulation (investment, tax, financial, administrative regulation, etc.). Industrial policy has no separate, own tools. This erodes to some extent the boundaries and essential characteristics of industrial policy. In addition, while characterizing the experience of applying industrial policy, it is important to justify the necessity, the direction of its use and the political factor. Despite a wide difference in approaches, all of them have common characteristics:

- need for state influence is determined in one form or another,
- certain prioritization system is singled out either in a sectoral context, or for individual functions of the economic mechanism,
- orientation to the long-term (strategic) perspective,
- comprehensiveness, that is, coverage of a wide range of measures and areas of activity.

In macroeconomic terms, government policy to support the economic activity of industrialists and entrepreneurs will help reduce risks for large corporations in developed countries by stimulating re-sharing processes (returning production capacities to the territory of parent companies). On the other hand, the "reverse pulling" of production and technological lines of transnational corporations into their national borders will negatively affect the industrial appearance of regions – their former locations, mainly in countries with developing economies. In addition to this, in the context of widespread digital transformation, the previous policy of moving production of added value to countries with cheap labor (Asia, Latin America, Russia recently) is currently changing to organizing its own production in countries with developed economies which have fully automated production lines with the prospect of lowest paid workers - robots.

Under the crisis and a high level of uncertainty, the role of industrial policy of the state as an organizing and guiding principle in the structural modernization of high value-added industries will increase in the interests of increasing the competitiveness of domestic enterprises. For this, we introduce various traditional measures, methods of stimulating an increase in the organizational and economic level of the activity of economic structures in the interests of the national economy: financial, regulatory, administrative, technical, foreign trade, labor, social, etc. Objectively, industrial policy is designed to reduce the level of uncertainty for enterprises in the spontaneous market environment and to increase their adaptive abilities to conditions of high turbulence in the business environment, to provide flexibility in modifying successful tools and methods of production and business activities.

It is no coincidence that industrial capital, evaluating future events with a high degree of probability as extremely negative for successful entrepreneurial activity, like the oil crisis of the 70s of the last century or the global financial and economic crisis of a decade ago, very carefully proceeds with new investments, being not sure of state support. For example, capital would never have gone into our domestic agricultural production if it had not outlined the possibility of successful investment and making big profits (the state program for leasing agricultural machinery, the program for supporting food export) gave impetus.

4 Discussion

In the studies of specialists, much attention is paid to the coordinating role of the state, as well as to the problems of improving the efficiency of manufacturing industries, which generate a high level of development in accordance with socio-economic interests of the country in the given period. In the conditions of economic crisis of both in our country and on a global scale, there are many uncertainties regarding forecasts for the global economic development, the goals of scientific and technological development in the face of challenges of Industry 4.0. We need an adequate design of industrial policy based on differentiation of approaches, accounting interests of specific actors of industrial and entrepreneurial activity, industrial complexes.

Andreoni and Chang [2] confirm the debatable nature of modern concepts of industrial policy. Its main theories, which were formed in the analysis of consistent development over the past few decades, revolve around three fundamental provisions regarding:

- roles and structural interdependencies of economic sectors, the conjugation of the processes of actors arising under industrialization,
- need to harmonize economic policy of a set of institutions promoting the industrial agenda,
- recognition of the role of the state in managing economic policy, its policy measures in combination with its entrepreneurial function.

These authors talk about the need for strategic coordination of interactive measures of industrial policy, coordinated interaction of management structures at the national economy level [2].

The work of Andreoni, Chang and Scazzieri [3] points out the need to update the discussion agenda on the content and form of industrial policy, focused on establishing the relationship of structures, institutions and policies. It should consider historical traditions of industrial relations, aspects of industrial, technical and market relations, as well as political economy. It is proposed to consider industrial policy in a broader aspect, within the framework of a holistic structure of socio-economic and political relations of society.

Landesmann and Stöllinger [7] note that the world economy is undergoing rapid structural changes. There are steady changes in the position of countries in global value chains, and they suggest discussing the importance of creating appropriate industrial policies for countries at different stages of development. In the context of financial and economic crises, middle-income countries are especially vulnerable to "structural external imbalances", uncertainties, and therefore the policy of supporting their productive forces is of great importance.

Eikhoff [6] emphasizes the so-called new industrial policy (declared in the European Union), aimed at "increasing the importance of the service sector (thermalization) on a global scale and at preventing the possible consequences of the form of de-industrialization and the associated reduction of jobs". It should fulfill the following two tasks: to restore lost ground in traditional areas of the domestic market (automotive, steel production, shipbuilding), to ensure the entry and conquest of new foreign markets (genetic engineering, biochemical production, etc.) At the same time, considering the characteristics of these types of regulation in more detail, they can be defined as one or another modification of the vertical or horizontal industrial policy.

In turn, Bianchi and Labory [4], analyzing the nature and dynamics of four industrial revolutions experienced by capitalism, the characteristics of the introduced technological and technological innovations, show how radical structural changes occur in industrial policy in the medium term and in production organization methods.

Wu, Zhu, and Groenewold [17] based on a review of China's industrial policy, carried out through central planning for a five-year perspective, note its rigidly prescriptive nature of carrying out the levels of management of national economies and regions. This approach significantly increases the level of production.

A special study by Meckling and Nahm [9] is devoted to green industrial policy for alternative transport technologies, especially electric vehicles, which solve the problem of reducing greenhouse gas emissions. The authors rightly note that such declarations for environmental friendliness of production, the corresponding modernization of industry will promote export opportunities of auto-producing countries and will successfully participate in the competition.

Cheah and Ho [5], while studying 153 public-private technology transfer projects in the context of the Singapore economy, found that project financing through industrial policy and public funding for research and development significantly affects the results of innovation activities.

Ślusarczyk [13] emphasizes the importance of discussions in favor of new industrial policies to develop economies in the crisis. On the materials of industrial enterprises of the European Union, the author confirms the need to formulate new goals of industrial policy aimed at increasing the competitiveness of enterprises of the EU countries in the global economy, exploiting mainly the results of high-tech industry.

5 Conclusion

The industry of the developed country (for example, Russia of the Soviet period) has a highly differentiated complex of industries with both powerful mining and developed manufacturing industries. However, over the years of market reforms, due to various reasons, the state of the manufacturing industry has deteriorated significantly and, even though government authorities are trying to increase the role and share of manufacturing in the industrial structure of the industry, the share of manufacturing does not grow, and the share of extractive industries in gross value added has upward trend. In view of this, the formation and implementation of industrial policy should solve the systemic problem of restoring the proper technological level of manufacturing industries, especially machine-building economic activities. It is necessary to bring the level of contiguity of the technological structure of manufacturing industries to the level of advanced developed countries. At the same time, to neutralize the negative features of catch-up development, it is necessary to consider current geo- and macroeconomic conditions and make extensive use of modern tools of innovation, namely, focus on breakthrough innovations, ensure faster growth of investments in the middle and final stages of technological chains, and facilitate the participation of domestic enterprises in global value chains.

References

1. Ahir, H., Bloom, N., Furseri, D.: 60 years of uncertainty. Finance Dev. **57**(1), 59–60 (2020)
2. Andreoni, A., Chang, H.-J.: The political economy of industrial policy: structural interdependencies, policy alignment and conflict management. Struct. Change Econ. Dyn. **48**, 136–150 (2019)
3. Andreoni, A., Chang, H.-J., Scazzieri, R.: Industrial policy in context: building blocks for an integrated and comparative political economy agenda. Struct. Change Econ. Dyn. **48**, 1–6 (2019)
4. Bianchi, P., Labory, S.: Manufacturing regimes and transitional paths: lessons for industrial policy. Struct. Change Econ. Dyn. **48**, 24–31 (2019)
5. Cheah, S., Ho, Y.: Effective industrial policy implementation for open innovation: the role of government resources and capabilities. Technol. Forecast. Soc. Chang. **151**, 11845 (2020)
6. Eikhoff, N.: Competition support policy or industrial policy. Bull. St. Petersburg Univ. Econ. Finance **3**(69), 8 (2011)
7. Landesmann, M.A., Stöllinger, R.: Structural change, trade and global production networks: an ‘appropriate industrial policy’ for peripheral and catching-up economies. Struct. Change Econ. Dyn. **48**, 7–23 (2019)
8. Larionova, I.V.: The role of innovation in industrial policy. Bull. Yakutsk State Univ. **3**(2), 59 (2006)
9. Meckling, J., Nahm, J.: The politics of technology bans: Industrial policy competition and green goals for the auto industry. Energy Policy **126**, 470–479 (2019)
10. Rubinstein, E.I.: Fundamentals of Industrial Policy. Surgut State Univ., Surgut (2000)
11. Shyu, C., Joseph, Z., Ding, K.: Industrial revitalization via industry 4.0 – a comparative policy analysis among China, Germany and the USA. Global Trans. **1**, 3–14 (2019)
12. Simachev, Yu., Kuzin, M., Kuznetsov, B., Pogrebnyak, E.: Russia on the way to a new technological industrial policy: among alluring prospects and fatal traps. Foresight **8**(4), 8–12 (2014)
13. Ślusarczyk, B.: The requirement of new industrial policy to support Polish economy. Procedia Econ. Finance **27**, 93–101 (2015)
14. Spring, M., Hughes, A., Mason, K., McCaffrey, P.: Creating the competitive edge: a new relationship between operations management and industrial policy. J. Oper. Manage. **49–51**, 6–19 (2017)
15. Tatarkin, A.I.: Industrial policy as the basis for modernization of the Russian economy. Bull. Chelyabinsk State Univ. **19**, 5–17 (2008)
16. Tolkachev, S.A., Teplyakov, EYu.: The law “On industrial policy” and the real industrial policy in the Russian Federation. The World New Econ. **2**, 113 (2015)
17. Wu, Y., Zhu, X., Groenewold, N.: The determinants and effectiveness of industrial policy in China: a study based on five-year plans. China Econ. Rev. **53**, 225–242 (2019)

Local Energy Strategies: A Digital Challenge

E. N. Koroleva(✉)

Samara State University of Economics, Samara, Russia
korol388@mail.ru

Abstract. The paper addresses the problem of ensuring the sustainable development of local territories in the context of increasing demands for energy efficiency in regional development and new technological possibilities of energy due to digitization. The aim of the study is to carry out a typological characterization of energy strategies of the municipalities of the region (a Russian subject) on the basis of an analytical study of the strategic objectives adopted and implemented by them. The author's methodology has been developed and tested, and result in a set of energy strategies of municipalities of a specific region; their groupings and differentiation of all municipalities into various groups of innovative energy technologies. The author has done conclusions on readiness of local territories to digital challenges in the energy sphere, emphasized the role of prospective projects of Smart City, and identified barriers to Smart Energy Systems implementation.

Keywords: Energy Internet · Energy strategy · Municipal entity · Strategic planning · Smart City · Smart energy systems

1 Introduction

The strategic objective of the regional energy policy is to create a sustainable and self-regulating system for regional energy security which requires optimizing the territorial structure of fuel and energy production and consumption [1]. The strategic initiatives in the field of energy are of particular importance among regional energy policy instruments: development and territorial diversification of energy infrastructure; promotion of new renewable energy sources and energy carriers; development of digital energy conservation technologies. These initiatives are being decomposed at different hierarchical levels in relatively autonomous territorial subsystems (country-regions-municipalities-organizations) [12].

The regional dimension is important in Russia because of the existing economic specialization of certain regions, as well as the growing regional and subregional imbalances in the development of the Russian regions [4, 13]. The importance of the regional dimension is also determined by the rapid change in scientific approaches to energy management, influenced by digital technological trends: Smart Grids – Smart Energy – Smart Energy Systems.

The Smart Energy System concept represents a scientific shift in paradigms from single-sector thinking to a coherent understanding of energy systems that integrate all sectors and infrastructures on a digital technology basis (electricity, heating, cooling,

S. I. Ashmarina et al. (Eds.): ENGINEERING ECONOMICS WEEK 2020, LNNS 139, pp. 19–25, 2021.
https://doi.org/10.1007/978-3-030-53277-2_3

industry, construction, transport, etc.) [6]. This will require researching into energy strategies at the subregional level, taking into account: (i) advanced energy technologies and (ii) the long-term development priorities of each local territory.

The research is aimed at carrying out a typological characterization of energy strategies of the municipalities of a region (a constituent entity of the Russian Federation) on the basis of an analytical study of the strategic planning documents adopted and implemented by them.

2 Methodology

The study is based on the author's methodology, which makes it possible to perform a typological characterization of the energy strategies of the municipalities of a Russia subject based upon an analytical study of the strategic planning documents adopted and implemented by them (primarily targeting).

The methodology consists of the following steps:

1. The formation of a body of strategic planning documents of first-level municipalities (urban districts and municipal raions) of the Russian subject.
   For each municipality, the following documents posted on the official site of the local self-government body (as of 31 December 2019) are used:
   - municipal regulations on strategic planning,
   - the current strategy for the socio-economic development of a municipal entity (hereinafter referred to as the municipal strategy), which was approved not earlier than 2017 or an up-to-date municipal strategy approved earlier in 2017, plus an updated plan of activities to implement the municipal strategy.
2. The content analysis of key legal acts and strategic documents which resulted in a set of municipal energy strategies.
   The energy strategy of a municipality in this study refers to the set of priorities, goals, objectives and policies and project for the long-term development of various elements of the energy sphere of the municipality which are presented in the municipal strategy. The municipal energy sphere covers the fuel and energy complex as well as all energy-related facilities and processes. This is part of the heating sector owned by non-energy agencies; direct fuel installations; energy management of direct fuel and energy consumers (energy consuming industries and households) located at the territory.
3. The construction of groupings that allow all municipalities to be placed within various groups of energy technologies reflected in their energy strategies.
4. The implementation of the typological characterization energy strategies of the municipalities of the Russian subject.
   The methodology involved the following methods to ensure the reliability of the research and the scientific validity of the conclusions: theoretical methods (scientific abstraction, induction and deduction, analogies, graphic methods); empirical methods (classifications and typologies, structuring, expert estimates).

3 Results

Within the framework of the study, the methodology was applied to 37 municipalities of the Samara region consisting of 10 urban districts and 27 municipal raions.

A list of promising energy technologies has been compiled in five groups:

1. Electrical and heating networks.
2. Technologies of the oil industry.
3. New materials and coatings.
4. MicroGrid and autonomous renewable energy (RES).
5. Intelligent systems (smart technologies).

The results of the study Volkova, Burda, Gavrikova, and Konev [17] were used to mark and characterize the identified groups. Based on the results of the content analysis of the key legal acts and strategic documents of the municipalities of the Samara Region, a set of energy strategies was developed and an analytical table for urban districts and municipal raions was constructed (Table 1). This has enabled all municipalities to be positioned within the different groups of energy technologies reflected in their energy strategies.

Table 1. The availability of advanced energy technologies in energy strategies of municipal entities of the Samara region

Advanced energy technologies	Urban districts	Municipal raions
Electrical and heating networks	Samara, Togliatti, Syzran, Novokuybyshevsk, Chapaevsk, Zhigulevsk, Otradny, Kinel, Oktyabrsky, Pohvistnevo	Alekseevsky, Bezenchuksky, Bogatovsky, Bolsheglushitsky, Bolshechernigovsky, Borsky, Volzhsky, Elkhovsky, Isaklinsky, Kamyshlinsky, Kinel-Cherkassky, Kinelsky, Klyavlinsky, Koshkinsky, Krasnoarmeysky, Krasnoyarsky, Neftegorsky, Privolzhsky, Pokhvistnevsky, Sergiyevsky, Stavropolsky, Syzran, Khvorostyansky, Chelno-Vershinsky, Shentalinsky, Shigonsky
Technologies of the oil industry	Samara, Novokuybyshevsk	Neftegorsky
MicroGrid and autonomous renewable energy (RES)	Samara, Zhigulevsk	Borsky, Shigonsky
New materials and coatings	Samara, Togliatti, Pohvistnevo	Borsky, Pokhvistnevsky, Sergievsky, Shigonsky
Intelligent systems and smart technologies (smart city)	Samara, Togliatti, Syzran, Novokuybyshevsk, Zhigulevsk, Otradny, Kinel, Oktyabrsky, Pohvistnevo	Bezenchuksky, Elkhovsky, Isaklinsky, Neftegorsky, Pestravsky, Pokhvistnevsky jh21

Source: author.

The most popular advanced energy technology is electric and heating grid technology, which is used by all urban districts and municipal districts. As can be seen from the table, the energy strategies of the largest agglomeration centres – the urban districts of Samara and Tolyatti – incorporate promising energy technologies from most of these groups.

Among the small towns, we highlight the strategic initiatives of the city of Zhigulevsk within the framework of the project «The creation of an innovative solar generation complex and renewable energy sources aimed at the green economy» [16] with the implementation date of 2020–2030. Alternative energy as a new specialization of the urban district of Zhigulevsk is also reinforced at the regional level in the Strategy for socio-economic development of the Samara Region until 2030 [15].

It is noteworthy that all city districts propose to introduce the technologies of «smart city», along with technologies from the group «Electric and heat networks». Three cities of Samara, Tolyatti and Novokuybyshevsk are participants of the federal pilot project «Smart City» [14]. The project is being implemented in Russia within the framework of the national project «Housing and Urban Environment» [9] and the national program «Digital Economy» [8]. The analysis of the data presented at Fig. 1 shows a lack of innovation in municipal energy strategies.

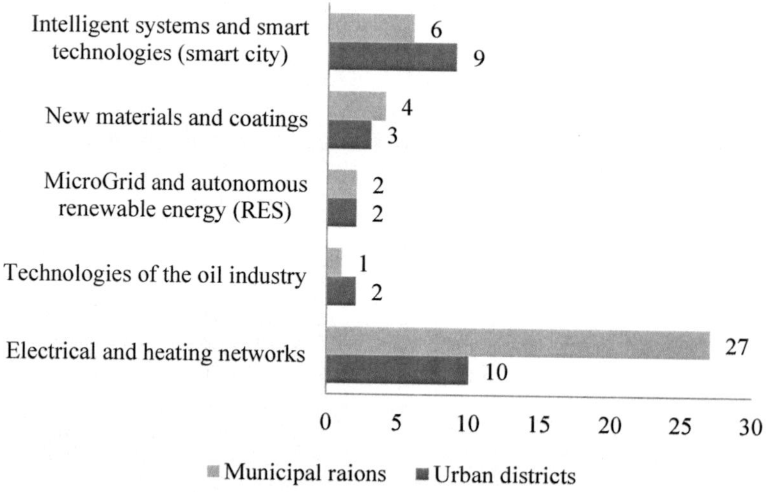

Fig. 1. The distribution of energy strategies of the municipal entities of the Samara Region according to the groups of advanced energy technologies, units . (Source: author)

Only 7 municipal raions out of 27 (25.9%) use advanced energy technologies of more than two groups. Less than a quarter of municipal raions (6 out of 27) plan to use intelligent systems and smart technologies (smart city). The Borsky, Neftegorsky, Pokhvistnevsky and Shigonsky municipal raions stand out positively in the overall array of municipalities. As an organizational innovation we shall highlight the strategic project envisaged by the municipal strategy of the Syzransky raion aimed at the creation of «Innovative center of thermal energy of small towns in the urban settlement of

Mezhdurechensk» . Thus, urban district energy strategies can be characterized as innovative-oriented, responding to digital challenges. The conclusion about municipal energy strategies is the opposite - their strategies do not meet current digital challenges.

4 Discussion

The results of the study can be interpreted in two ways. First of all, let us consider the digital dimension of the innovative technologies. The main technological trend that is rapidly transforming the energy sphere is the transition from Smart Grids to Smart Energy. Smart Grid technology is emerging as a result of the transformation of an electrical network based on rapidly evolving technologies: Microgrid, smart sensors, two-way communication, the digitization and automation of energy management processes. Smart Grid significantly changes the operating principles of the power supply chain, offering new principles of active and decentralized interaction between producers and consumers.

The integration of Artificial Intelligence (AI), Internet of Things (IoT) and blockchain technologies into the energy sphere ensures the digital response of the electrical network to rapidly changing real-time electricity needs. Thanks to solar panels, wind turbines and other energy sources, consumers are becoming energy producers, so the energy flow goes in two directions. The emerging Smart Energy has many advantages - from efficient power transmission to rapid energy recovery, energy efficiency and tariff reduction, and increased integration of various energy sources.

The driver for Smart Energy is Energy Internet (IoE). According to Zhou, Yang, and Shao [18], the Energy Internet is a new form of energy system development that implements the integration of energy flow, information flow and business flow. Energy Internet is an innovative presentation of energy systems at the fourth stage of development (smart & connected energy system). Energy Internet is a new solution for green development of smart cities [5]. Its wide application contributes to the improvement of the quality of life of the population and sustainable energy-efficient urban development [7].

In this context, the projects of the «smart city» within the municipal strategies mentioned above seem to be promising. We agree with the findings of O'Dwye, Pan, Acha, and Shah that digital solutions for systems with a wide range of characteristics and temporal scales require overcoming new barriers, related to the lack of proven methods and algorithms for dealing with complex networks of actors, often with competing objectives [10]. This leads us to the important question: are the regions ready for the development of their energy infrastructure, which is localized in their municipal entities? The existence of an innovation infrastructure, in turn, is a source and condition for the growth of innovation and, as a consequence, the overall competitiveness of each Russian region [2].

The second aspect relates to the implementation of the institutional conditions for strategic planning and management of the development of regional socio-economic systems. The relevant question is whether municipal strategies incorporating energy strategies meet current digital challenges. Note that the digital technologies that we consider in a broad sense include innovative management technologies [3]. This view of the problem is consistent with the conclusions about the prospects of introducing the concept of Smart Energy Systems. In this regard, a possible regional initiative in this area could be the intersectoral integration of Smart Energy Systems at the territory of Samara-Togliatti agglomeration (STA) which comprises 8 urban districts and 9 municipal raions of the Samara Region [11].

5 Conclusion

Thus, the study defines the essence of municipal energy strategies. Based on the results of the application of the author's methodology, the characterization of the energy strategies of the municipal entities of the Samara Region on the basis of an analytical study of the strategic planning documents adopted and implemented by them for the period up to 2030. The main points and conclusions of the presented study develop the theory of spatial socio-economic systems from the perspective of methods of identification and expert assessment of factors and conditions of digital transformation of the regional energy sphere. The practical significance of the research is determined by the possibility of the use of the results obtained by business structures, state and municipal governments of Russian regions in decision-making on the implementation of Smart Energy Systems technologies.

Acknowledgments. The study was carried out as part of the state assignment of the Ministry of Science and Higher Education of the Russian Federation, project No. 26.4131.2017/4.6.

References

1. Energy Strategy of Russia for the period up to 2030. Adopted by the Regulation of the Government of the Russian Federation No. 1715-r of 13 November 2009. http://www.consultant.ru/document/cons_doc_LAW_94054/. Accessed 01 Apr 2020
2. Khmeleva, G.A., Semenychev, V.K., Koroleva, E.N., Agaeva, L.K., Korobetskaya, A.A., Zvereva, I.V., Zenina, K.S.: Innovative development of Russian regions under sanctions. SamNZ RAS Publishing House, Samara (2017)
3. Kurnikova, M.V., Koroleva, E.N., Timofeev, A.G.: The model of municipal strategic planning in the context of digital economy. In: Mantuenko, V. (ed.) Proceedings of the International Scientific Conference "Global Challenges and Prospects of the Modern Economic Development". The European Proceedings of Social & Behavioural Sciences, vol. 57, pp. 1469–1482. Future Academy, London (2019). https://doi.org/10.15405/epsbs.2019.03.149
4. Lapin, A.V.: Socio-economic Territorial Differentiation: Theoretical and Methodological Aspects, Assessment and Analysis. Perm State National Research University, Perm (2018)
5. Li, L., Zheng, Y., Zheng, S., Ke, H.: The new smart city programme: evaluating the effect of the internet of energy on air quality in China. Sci. Total Environ. **714**, 136380 (2020)
6. Lund, H., Østergaard, A., Connolly, D., Mathiesen, B.V.: Smart energy and smart energy systems. Energy **137**, 556–565 (2017)
7. Mahapatra, C., Moharana, A.K., Leung, V.C.M.: Energy management in smart cities based on internet of things: peak demand reduction and energy savings. Sensors **17**(12), 2812 (2017). https://doi.org/10.3390/s17122812

8. National Program: Digital Economy. http://government.ru/rugovclassifier/614/events/. Accessed 01 Apr 2020
9. National Project: Housing and urban environment. http://government.ru/rugovclassifier/846/events/. Accessed 01 Apr 2020
10. O'Dwyer, E., Pan, I., Acha, S., Shah, N.: Smart energy systems for sustainable smart cities: current developments, trends and future directions. Appl. Energy **237**, 581–597 (2019)
11. Pavlov, Y.V.: Development of urban agglomerations: problems and solutions. Cent. Russ. J. Soc. Sci. **14**(5), 112–140 (2019)
12. Salina, T.K., Tchaikovskaya, D.D.: Nature and content of fuel and energy complex as an economic system. Probl. Modern Econ. **2**(42), 316–321 (2012)
13. Skufyina, T.P.: Socio-economic differentiation of space: inconsistencies between the theory and regulation practice. Econ. Soc. Changes Facts Trends Forecast **6**(30), 60–68 (2013)
14. The list of pilot municipalities of the Departmental project for the digitalization of urban economy "Smart City". https://russiasmartcity.ru/documents/22. Accessed 01 Apr 2020
15. The Strategy of the social and economic development of the Samara Oblast until 2030. https://economy.samregion.ru/upload/iblock/0ca/1_proekt-strategii-so_2030.pdf. Accessed 01 Apr 2020
16. The Strategy for the social and economic development of the Zhigulevsk urban district of the Samara Oblast until 2030. https://www.zhigulevsk.org/strategiya_2030/. Accessed 01 Apr 2020
17. Volkova, I.O., Burda, Y.D., Gavrikova, E.V., Konev, A.V.: Assessing conditions for the development of advanced energy technologies on the regional level in Russian Federation. Manage. Sci. **9**(1), 47–67 (2019)
18. Zhou, K., Yang, S., Shao, Z.: Energy Internet: the business perspective. Appl. Energy **178**, 212–222 (2016). https://doi.org/10.1016/j.apenergy.2016.06.052

Multidimensional Statistical Analysis of the Construction Cluster of the Samara Region

S. I. Makarov(✉)

Samara State University of Economics, Samara, Russia
matmaksi@yandex.ru

Abstract. The article is devoted to the study of the interaction of the participants in the construction cluster, which determines the dynamics of the volume and pace of construction in the region. The main subjects of this process are construction organizations and manufacturers of building materials, products and structures. The basic requirements of regional construction companies to suppliers of building materials are determined. Factors are revealed that reflect the dependence of the results of the activities of construction companies on the production and economic activities of manufacturers of building materials. These factors comprehensively affect the growth rate of construction, the development of infrastructure facilities and the introduction of innovations in production processes. Based on the available observations for 7 years, a variance analysis of the samples was performed and regression dynamics models were built monthly and by year, and the significance and adequacy of the constructed models were estimated. The constructed mathematical models reflect the dependence of the volume of work performed in construction on the production indicators of manufacturers of building materials and supplying organizations.

Keywords: Building cluster · Forecasting · Regression models

1 Introduction

The capital-intensive branch of the economy, capable of increasing the country's economic growth rate, creating new jobs, and solving the sore social problems of the population, is construction [3]. To understand the reasons for the slowdown in this economic sphere, it is necessary to conduct a comprehensive analysis of the factors that comprehensively affect the growth rate of construction in the Russian Federation and the Samara region [1, 4].

This study examines the industrial and economic relations between construction organizations and manufacturers of building materials, products and structures. Developing organizations have a need to identify factors affecting the formation of industrial and economic relations and contractual relations in the field of construction in the region [2], the formation of optimal criteria for choosing sources of material and technical support for the construction process [6, 10].

S. I. Ashmarina et al. (Eds.): ENGINEERING ECONOMICS WEEK 2020, LNNS 139, pp. 26–36, 2021.
https://doi.org/10.1007/978-3-030-53277-2_4

2 Methodology

Formalized forecasting methods are based on the mathematical processing of statistical data; their accuracy largely depends on how accurate the information received used in the calculations [15]. If we consider the methods of extrapolation, then a great deal of complexity is the study and understanding of the dynamics of changes in a process or phenomenon in the previous period. Moreover, it is necessary to identify patterns and transfer them to the next period. Of course, modeling can be considered an effective means of forecasting [7]. One of the most acceptable forecasting methods is the regression analysis method. When studying the object of modeling, it is necessary to establish the relationship between the parameters included in the model and the effective attribute. In addition, the regression method is universal, there is the possibility of a wide selection of functional dependencies. But the most important thing to note is the use of information technology.

The regression models include only those factors that are most associated with the effective indicator. In addition, factors that duplicate the same information about an effective indicator should not be included in the model in order to exclude the phenomenon of multicollinearity.

3 Results

A survey of regional construction companies revealed a list of their requirements for suppliers of building materials. The significance of certain criteria for the selection of sources of material and technical support for the construction process is reflected in Fig. 1.

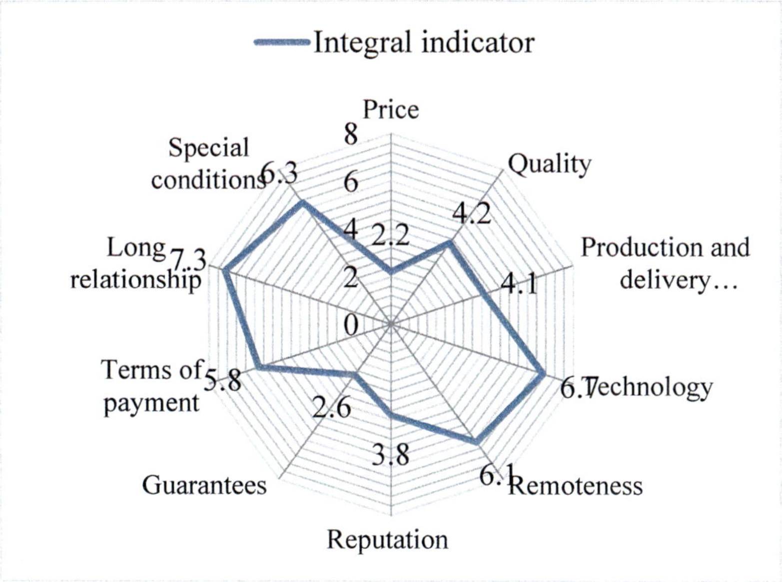

Fig. 1. An average assessment of the significance of the selection criteria by construction organizations of the Samara region for suppliers of material resources (on a 10-point scale). (Source: author)

According to the results of the survey, the following significant selection criteria for suppliers of building materials can be identified that affect the nature of industrial and economic relations and contractual relations with both existing and potential contractors:

1. The duration of industrial and economic relations, estimated by regional builders at 7.3 points, determines the choice of customers in favor of reliable suppliers, cooperation with which is time-tested, which ultimately reduces the risks of partners not fulfilling obligations under the transaction, observing the logistics principles of organizing goods supply.
2. The requirement to use modern technologies for the production of building materials (6.7 points) is justified by the desire of builders to maintain a high level of energy efficiency of the erected buildings and structures, their internal environmental friendliness, which ultimately affects the reduction of material consumption and increased reliability and durability of construction objects.
3. Special conditions determining the terms acceptable for the production process and the minimum batch size are estimated at 6.3 points. The delivery times set by manufacturers of building materials for their products significantly affect the duration of technological processes in construction. The requirement of a minimum supply lot size often does not correspond to the financial capabilities of the customer, who in the conditions of fierce competition in the building materials market will give preference to the supplier whose conditions the builder satisfies.
4. The territorial remoteness of the supplier (6.1 points) is of no small importance. Construction organizations give preference to suppliers located in relative transport accessibility, helping to reduce a number of logistics costs associated with the supply and storage of material resources, as well as increasing the efficiency of deliveries.
5. The terms of payment are traditionally significant in the selection of the supplier (5.8 points), which is also due to the financial capabilities of the construction company, for which full prepayment or a large down payment is undesirable [11].
6. The quality of building materials, as a factor in the selection of a supplier of construction by the organization, is estimated at 4.2 points. As noted earlier, in the analysis of factors affecting the production and economic activities of developers, the quality of the materials purchased should satisfy the need for the construction of energy-efficient and environmentally friendly buildings and structures.
7. The schedule of production and delivery (4.1 points) as the basis for planning the production and economic activities of the construction organization forms its choice in favor of those suppliers whose schedule allows for uninterrupted technological processes in construction.
8. The supplier's reputation as a criterion for his choice is assessed by the construction companies of the region at 3.8 points, since this parameter of the manufacturer of building materials is largely ensured by the fulfillment of all obligations under the contractual relationship. The smaller, in comparison with the above criteria, significance of the supplier's reputation is explained by the emergence of new manufacturers in the construction materials market of the Samara region, whose duration of work does not yet allow to draw adequate conclusions regarding its reliability and commitment.

9. Guarantees of production (supply), estimated at 2.6 points, respectively, having less significance, in contrast to the criteria listed above, are nevertheless important in the conditions of instability of the construction market. Failure to fulfill obligations for the production and supply of material resources leads to the builder's unpreparedness to search and select a new supplier, to search for financial resources at a new price, taking into account changes in market conditions.
10. The least significant (2.2 points) when choosing a supplier is the cost of the supplied building materials, due to the competitors achieving the same level of price offer while differentiating other competitive advantages that determine the nature of production and business relations and contractual relations of partners in the construction industry.

According to a survey of construction organizations in the Samara region, the main problems determining the nature and prospects of further cooperation with manufacturers of building materials are the following (Fig. 2):

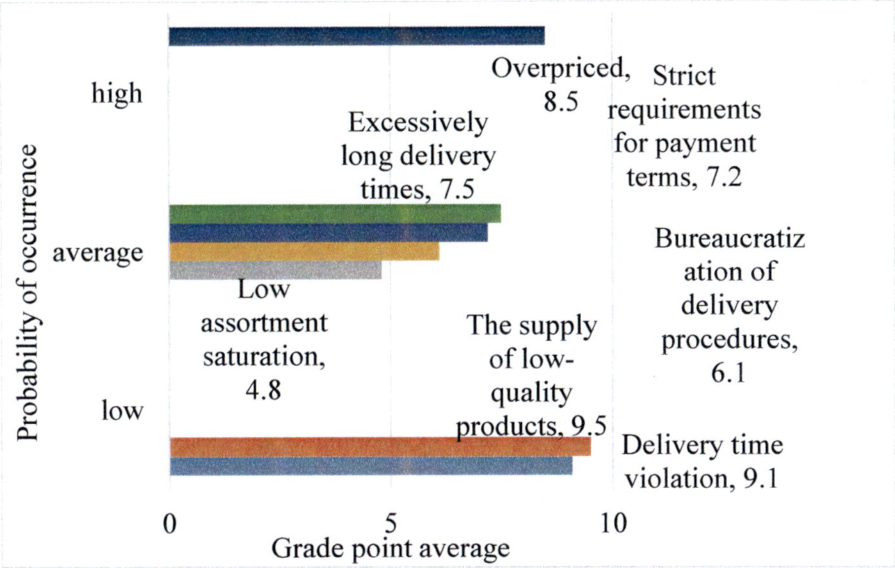

Fig. 2. Distribution of problems of developers in the organization of industrial and economic relations with suppliers of building materials. (Source: author)

1. With a high probability of occurrence:
   - overpricing - 8.5 points. The significance of the influence of the factor of the high cost of building materials, products and structures is described above, in the analysis of factors affecting the production and economic activities of construction organizations.

2. With an average probability of occurrence:

   – low saturation of the assortment - 4.8 points, which expresses the builders' need for integrated supplies of material resources for the construction of various kinds of objects,
   – bureaucratization of delivery procedures - 6.1 points, which increases the delivery time and violates the established construction schedule,
   – stringent requirements on terms of payment - 7.2 points, which, given the limited financial resources of builders, determines their choice in favor of more loyal suppliers,
   – excessively long delivery times - 7.5 points, along with bureaucratization of delivery procedures, leading to disruption of production and technological processes of construction organizations.

3. With a low probability of occurrence:

   – violation of the delivery time - 9.1 points, which has the same consequences as the previous problems of interaction between builders and suppliers,
   – delivery of low-quality products - 9.5 points, which fits into the influence of the factor of lack of materials highlighted by the builders of the Samara region.

According to the results of surveys, research of scientific works, regulatory legal acts and materials of discussion platforms relating to sources of information on the development of the construction industry in the country and the region, the list of factors affecting the formation of industrial and economic relations and contractual relations in the field of construction and production building materials in the Samara region should include such as:

1. Good faith (timely payment of supplies, adherence to a delivery schedule, delivery of products of the required quantity and quality) of contractors, determining the prospects for long-term cooperation of specific developers with specific manufacturers of building materials, and which can be expressed by indicators of their accounts payable.
2. The price level of material resources, also determining the choice of a particular supplier of building materials.
3. The territorial concentration of the production of building materials, narrowing the choice of suppliers of material resources for developers, in connection with an increase in the cost of their transportation.
4. The requirements of suppliers regarding the minimum size of the supply lot, which is reflected in the dynamics of costs for both the transportation of products and the organization of their storage.
5. The manifestation of the conditions for the development of the digital economy, which creates the possibility of inclusion in a single information and logistics space, to update the information about the existing needs for building materials or work and promptly generate the appropriate proposal [11].
6. State support for resolving issues of providing construction organizations with innovative materials produced in the Samara region [13, 14].

Analysis of the conditions of industrial and economic activity of enterprises in the field of construction and production of building materials in the Samara region allows us to confirm the above list of factors affecting the formation of industrial and economic relations and contractual relations between the subjects studied.

All of the above factors of the formation of industrial and economic relations and contractual relations between the subjects under study are reflected in statistical indicators that demonstrate the level of socio-economic development of the Samara region as a whole, business activity in the field of construction and production of building materials, in particular [8, 9].

To simulate the interconnections of the enterprises of the construction cluster and build forecasts, we selected the dynamic series of 2011–2017 on a monthly and yearly basis by the main indicators of the construction industry. Methods of multivariate statistical analysis were applied to the obtained data.

Using the correlation analysis, a close connection was established between the indicators X22 - the commissioning of residential buildings in total, thousand square meters. m and X23 - commissioning of residential buildings by individual developers, thousand square meters. m with indicators of production of building materials in the Samara Region with other factors. It was found that these indicators are significantly affected by the following production indicators:

X5 - Production of large wall blocks (including basement wall blocks) from concrete, one million conventional bricks,
X10 - Production of frame structures for buildings and prefabricated reinforced concrete structures, thousand cubic meters,
X12 - Production of slabs, panels and flooring of floors and coatings of reinforced concrete, thousand cubic meters,
X18 - Production of concrete, ready for pouring (ready-mixed concrete), thousand cubic meters,
X20 - Production of mixtures of asphalt concrete road, airfield and asphalt concrete (hot and warm), t.

In addition, the indicator X23 - the commissioning of residential buildings by individual developers, thousand square meters. m significantly affect indicators:

X3 - Production of ceramic non-refractory building bricks, million conventional bricks,
X9 - Prefabricated reinforced concrete foundation structures, thousand cubic meters,
X15 - Production of structural and architectural-building buildings and prefabricated reinforced concrete elements, thousand cubic meters,
X16 - Manufacture of prefabricated structures for reinforced concrete other, thousand cubic meters.

The presence of this deviation suggests that individual construction is more focused on local manufacturers of building materials.

Since the number of observations in the time series is not large (12 observations), it is advisable to select no more than 2 factors as an argument. In addition, some factors should be excluded from consideration for substantive reasons and interdependent argument factors.

Upon further analysis of the selected factors for building models, the following were left:

Resulting factor X52 (Y1) - Volume of work performed (million rubles) and factor-arguments:

- X9- Price dynamics of finished metal products (%),
- X12- Price dynamics transmission and distribution of electricity (%),
- X38- Price dynamics of cement, lime production (%),
- X47- Dynamics of prices for the production of asphalt mixtures (%);

The following regression models were obtained for the resulting factor X52 (Y1) - the volume of work performed (million rubles):

$$Y_1 = 5512750 - 38452,1X_9 - 1606,75X_{12},$$
$$Y_1 = 152787,9 - 826,52X_{38} - 0,43X_{47}$$

We turn to the statistical analysis of the obtained regression equations: checking the significance of the equations and its coefficients. The tightness of the combined influence of factors on the result is estimated by the coefficient of multiple correlation. In the first model, the coefficient of multiple correlation R = 0.75. Determination coefficient R2 = 0.622. those. in 62.2% of cases, changes in X lead to a change in Y1. The variation in the indicator, the volume of work performed by 62.2% is due to a variation of factors included in the model, and by 0.8% due to a variation of other factors not included in this model. Similar statistical indicators of adequacy have the other models presented in this paper. In other words, the accuracy of the selection of regression equations is high. The above indicates the good quality and adequacy of the constructed models.

The regression coefficient bi shows how many units on average the value of the productive attribute Y changes when the attribute Xi changes by 1 unit is presented in the Table 1.

Table 1. The change in the value of an effective attribute X52 (Y1) when the attribute Xi changes by 1 unit

Variable	Factor	Factor change (%)	The increment of the resulting factor (million rubles)
X9	Price dynamics finished metal products	1	−38452,1
X12	Price dynamics power transmission and distribution	1	−1606,75
X38	Cement, lime production price dynamics	1	−826,5
X47	Dynamics of production prices of asphalt mixtures	1	−0,43

Source: author.

Similarly, the resulting factor X54 (Y2) was considered:

X54 (Y2) - The volume of work performed (thousand square meters) significantly depends on:
X9- Price dynamics of finished metal products (%).
X12- Price dynamics transmission and distribution of electricity (%).
X38- Price dynamics of cement, lime production (%).
X40- Production volume silicate blocks (mln. pcs.).
X45- Dynamics of production of concrete for pouring (%).

The following regression models were obtained for the resulting factor X54 (Y2) - the volume of work performed (million rubles):

$$Y_2 = 7289,98 - 49,36X_9 - 21,68X_{12}$$
$$Y_2 = 101,04 - 1,01X_{38} + 14,29X_{40} + 1,25X_{45}$$

The regression coefficient bi shows how many units on average the value of the productive attribute Y changes when the attribute Xi changes by 1 unit is presented in the Table 2.

Table 2 The change in the value of an effective attribute X54 (Y2) when the attribute Xi changes by 1 unit

Variable	Factor	Factor change (%)	The increment of the resulting factor (thousand sq. M.)
X9	Price dynamics finished metal products	1	49,36
X12	Price dynamics power transmission and distribution	1	−21,68
X38	Cement, lime production price dynamics	1	−1,01
X40	Silicate production volume	1(million pieces)	14,29
X45	Dynamics of production of concrete for pouring	1(%)	1,25

Source: author.

At the third stage of the study, 23 indicators were selected. Based on these factors, the dynamic series for six years 2012–2017 were selected. As the resulting indicators were considered: X1 - the volume of work performed by type of activity "Construction" (million rubles), X2 - the commissioning of residential buildings, total area (thousand square meters).

Some factors were excluded from consideration from substantive considerations and interdependent factors-arguments were eliminated. The following trend was identified is presented in the Table 3.

Table 3. The change in the value of an effective attribute X2 (T2) when the attribute X21 changes by 1 unit

Variable	Factor	Factor change	The increment of the resulting factor (thousand sq. M.)
X21	Producer price indices for construction products	1(%)	−28,02

Source: author.

X2 (T2) - Commissioning of residential buildings, thousand square meters. m of total area significantly depends on:
X21 - Producer price indices for construction products, %.

$$T_2 = 4731,05 - 28,02X_{21}$$

4 Discussion

The criteria established by construction organizations for the selection of suppliers of building materials, products and structures are the result of an individual policy in the field of procurement logistics, however, taking them into account when organizing production and business relations with contractors can minimize the negative impact of the above factors on counterparty insolvency, lack of financing, and high cost materials, structures and products.

The importance of a multi-criteria evaluation of suppliers of material resources for construction needs is emphasized by the fact that in industrial and construction activities construction companies of the Samara region often encounter problems in relations with regional suppliers, differentiated by frequency and significance for the customer (Fig. 2).

An analysis of the problems of developers in organizing industrial and economic relations with suppliers of building materials allows us to supplement the idea of the factors affecting the industrial and economic activities of construction organizations in the Samara Region, since most of them fit into the lists highlighted above and are also reflected in a number of statistical indicators of the state of construction industries of the region [12].

5 Conclusion

Summing up the above, we note that in many respects the nature of the relationship between the subjects of the construction industry is determined by price factors. The low degree of provision of machinery and mechanisms, the localization of the raw material base, the lack of industrial products of innovative building materials, the high cost of energy and transportation services, limit their ability to meet the full range of needs of construction organizations at competitive prices.

In turn, the volume of orders from construction companies is determined by the pace of construction, the dynamics of which, first of all, are limited by declining consumer demand amid inflation in the country and the region. The affordability of housing is also determined by the high costs of various types of resources of the construction organization, including the tax burden and state credit policy [5].

Nevertheless, a study of the markets for construction and production of building materials in the Samara region showed the presence of prospects for its growth. The growing investment attractiveness of the region, state support for the introduction of innovative technologies and equipment in the production processes of industry entities contribute to the development of industrial and economic relations and contractual relations in the field of construction and production of building materials in the Samara region and beyond.

Significantly reduce the negative impact of a number of the described factors, the implementation of the option system for ordering the supply of building materials at the request of developers can significantly reduce the existing imbalance of supply and demand in the construction industry of the Samara region. The positive influence of the described factors must be used in the formation of the organizational and economic mechanism of cluster interaction between builders and manufacturers.

References

1. Decree of the Government of the Russian Federation of 07.02.2011 N 165-r: On approval of the Strategy for socio-economic development of the Volga Federal District until 2020. https://bazanpa.ru/pravitelstvo-rf-rasporiazhenie-n165-r-ot07022011-h1612615/strategiia/5/6/. Accessed 11 Feb 2020
2. Decree of the Government of the Samara Region dated 03.06.2014 No. 315: On approval of the concept for the development of the cluster of the building materials industry and industrial housing in the Samara region until 2020 and an action plan for the development of the cluster of the industry of building materials and industrial housing in the Samara Region until 2020. http://docs.cntd.ru/document/464011116. Accessed 11 Feb 2020
3. Federal Law: On participation in the shared construction of apartment buildings and other real estate objects and on amending certain legislative acts of the Russian Federation, dated December 30, 2004 N 214-FZ. http://www.consultant.ru/document/cons_doc_LAW_51038/. Accessed 11 Feb 2020
4. Federal State Statistics Service: Construction in the Russian Federation (2018). https://www.gks.ru/folder/14458. Accessed 11 Feb 2020
5. Federal Tax Service: Tax analytics (2018). https://analytic.nalog.ru/portal/index.ru-RU.htm. Accessed 11 Feb 2020
6. Government of the Samara Region: Raw material base (2018). https://www.samregion.ru/economy/source/. Accessed 11 Feb 2020
7. Ivanushkin, N.S., Kirshin, I.A.: Features of financing construction projects in Russia. Actual Dir. Sci. Res. Theory Pract. **2**(4), 198–200 (2015)
8. Ministry of Economic Development and Investments of the Samara Region: RAEX rating agency (Expert RA) upgrades investment climate of Samara region (2017). http://economy.samregion.ru/press-center/sobytiya/reytingovoe-agentstvo-raex-ekspert-ra-povysilo-reyting-investitsionnogo-klimata-samarskoy-oblasti-/. Accessed 11 Feb 2020

9. National Rating Agency: Rating of investment attractiveness of Russian regions by the end of 2018 (2018). http://www.ra-national.ru/sites/default/files/analitic_article/IPR-6-06112018.pdf. Accessed 11 Feb 2020
10. Ru-Stat: Import to the Samara Region for 2013–2018 (2018). http://ru-stat.com/date-Y2013-2018/RU36000/import/world/0525. Accessed 11 Mar 2018
11. Samarastat: Finance. Organizations (2018). https://samarastat.gks.ru/finance. Accessed 11 Feb 2020
12. Samarastat: The population of the Samara region (2018). https://samarastat.gks.ru/population/. Accessed 11 Feb 2020
13. The state program of the Samara region dated 09.12.2015 No. 822 "Development of the infrastructure of urban development in the territory of the Samara Region" for 2016 – 2019 years". http://docs.cntd.ru/document/434605300. Accessed 11 Feb 2020
14. The state program of the Samara region dated November 27, 2013 No. 684 "Development of housing construction in the Samara Region" until 2020. http://docs.cntd.ru/document/464012122. Accessed 11 Feb 2020
15. Tikhomirov, N.P., Tikhomirova, T.M., Ushmaev, O.S.: Methods of econometrics and multivariate statistical analysis. Economics, Moscow (2011)

Foreign Experience in Organizing and Evaluating the Internal Control and Audit System

T. Tarasova(✉)

Samara State University of Economics, Samara, Russia
tarasova2004@inbox.ru

Abstract. For Russian enterprises, the formation of effective internal control and audit systems is not just a matter of efficiency or ensuring market competitiveness, but also national security, which determines the relevance and increases the significance of this study. The aim of the study is to study international experience in the formation and assessment of the reliability of the internal control and audit system. During the study, such methods as analysis, observation, formalization, classification, modeling and others were used. The analysis allows us to conclude that very often in Russian companies there is a situation where the same unit combines the functions of internal control, internal audit and risk management. On the other hand, little time has passed since the introduction of the organization's best practices, internal control and internal audit at domestic enterprises. These processes can become successful only in conditions of a developed corporate mentality, which the Russian business environment cannot boast of so far.

Keywords: Internal control · Internal audit · Risk management · Risk-based approach

1 Introduction

The current processes of globalization against the backdrop of the global financial crisis make states not only focus on strategically important sectors of the economy, but also introduce the most effective tools for their accelerated development in order to ensure a high level of international competitiveness. Strengthening the role of internal control and audit as effective tools for the development of entities of various levels has become a strict trend in almost all leading economies of the world. A key factor in the development of this trend is the ability of internal control and audit systems to act as a barrier, providing a barrier to the inefficient use of financial, material and other resources, as well as a powerful factor in mobilizing on-farm reserves. In the current geopolitical conditions of the development of the Russian economy, the organization of reliable internal control systems is of fundamental importance in the practical plane of the formation and improvement of enterprise development strategies as a locomotive of the country's economic development in the near future.

S. I. Ashmarina et al. (Eds.): ENGINEERING ECONOMICS WEEK 2020, LNNS 139, pp. 37–44, 2021.
https://doi.org/10.1007/978-3-030-53277-2_5

2 Methodology

The theoretical basis of the study was formed by scientific results presented in the fundamental works of Russian and foreign experts on the problems of organizing effective internal control and audit systems [2–4, 11, 14, 15].

The author considers the methodology for constructing internal control systems in large German companies. According to this methodology, internal audit system is formed taking into account the principles and standards of the Institute of Internal Auditors, although there are no legislatively established requirements for the application of these particular standards.

3 Results

Consider the foreign experience of organizing and assessing the reliability of the internal control and audit system. We would like to start the study of the current practical models of organizing and assessing the reliability of the internal control and internal audit system from foreign experience, and it is from the experience of European countries, which is often taken by Russian companies as the basis for the implementation of certain innovations, such as, for example, COSO standards and etc. First of all, among European countries, the experience of Germany as one of the most economically developed countries is of interest.

The methodology for constructing internal control systems in large German companies is based on the well-known and widely used COSO document [5]. The internal audit system is formed taking into account the principles and standards of the Institute of Internal Auditors, although there are no legislatively established requirements for the application of these particular standards.

In addition to the well-known international rules and principles, Germany has developed a system of standards for the National Institute of Internal Audits. At the corporate control system level, all national and international standards are supplemented by their own internal rules. The experience of the functioning of the internal control system at German enterprises indicates that the internal audit in these systems is considered as a central independent link (including relying on the Law on Accounting Reporting Reform (BilMoG)).

Let us consider one example of creating an effective system of internal control and internal audit using the example of Konzern revision Deutsche Bahn AG (DB AG) (Konzern revision Deutsche Bahn AG official website: http://www.deutschebahn.com/en/start/). A subsidiary company responsible for scientific and innovative development has been identified in the structure of this holding, therefore it is useful to analyze the experience of this corporation from the point of view of application in Russian organizations [3].

Following the COSO model, in DB AG, the processes of internal control and internal audit are integrated into the system of functioning of all processes of the enterprise, and management is personally responsible for the quality construction and maintenance of an effective internal control system. The personal interest and responsibility of top management determines a very high level of involvement of employees of all levels in the processes of internal control and internal audit. The internal control system actively interacts with all divisions of DB AG: accounting, human resources, IT, security, etc., as well as members of the Management Board and the Supervisory Board of the enterprise [6].

On the basis of a developed corporate culture that offers a positive attitude to internal control procedures, a control environment is formed in the company that defines a management style focused on the complete suppression of illegal actions.

In many German companies involved in intellectual products, including the concern in question, internal control is based on corporate standards that provide for the use of control procedures at all stages of existing business processes. The risk-based approach based on the processes of internal control and internal audit in DB AG defines a risk management system in which risks are analyzed and assessed taking into account the probability of their occurrence by studying the causes, consequences and ensuring system control over each type of risk.

In addition, risks are necessarily taken into account in the process of planning the activities of the enterprise [5], as a result of which the objects with the worst risk indicators are selected and prioritized. For a subsidiary engaged in the promotion of intellectual products, regardless of the level of risks identified, audits are carried out at a specified frequency (2 times a year), and in addition, a risk self-assessment system has been built as an effective element of internal control. This was done in order to protect an effective development strategy of the concern and prevent the diversion of strategically significant developments. For a subsidiary of DB AG, as well as other companies with intellectual products, loss of copyright in a particular product may lead to a complete loss of competitiveness in local as well as global markets.

Given that in the current business environment, the management of any company is personally responsible for violations in subordinates and problems in controlled processes, the management itself is focused on improving the effectiveness of the internal control system and its reliability, which is assessed through internal audit.

Thus, the quality of internal control and internal audit in German companies is determined by:

- the synergy of generally recognized and intra-company standards and rules; integrated involvement of the processes of the internal control system and internal audit in all business processes,
- independence and free access to information,
- professional training of employees of the internal control system and internal audit,
- the presence of a high corporate culture; systematic solution to the problems of internal control and internal audit.

Next, we consider the experience of Japanese companies in organizing an internal control system and internal audit and, in some aspects, compare it with German. This experience is interesting in that a significant number of companies here are high-tech and produce intellectual property products. Japanese corporations have an exceptionally high level of development of the internal control system as an integral component of corporate governance. Internal control functions are included in the responsibilities of all employees of any company from top management to employees in initial positions [7]. Practical experience in the functioning of Japanese companies suggests that special committees work on a regular basis to improve the corporate internal control system.

If we analyze in general, the principles of building an internal control system in Japanese companies are similar to German ones and are based on generally accepted principles in the world according to: Sarbanes-Oxley law, INTOSAI International Standards for Internal Auditing and the Institute of Internal Auditors (all these institutions were mentioned above). However, at the same time, unlike German companies, which although they have the freedom to choose the model of internal control, but often prefer the already established approaches, Japanese corporations, taking into account the national mentality and peculiarities of doing business, build internal control and internal audit based on the national legal framework. First of all, these are the laws "On financial instruments and exchanges" (J-SOX law, developed on the basis of the Sarbanes-Oxley law) and "On corporations" [9].

In Japan, the law provides for the submission of an internal control report to the Ministry of Finance (for companies whose shares are traded in accordance with J-SOX on the Tokyo Stock Exchange). The responsibility for its preparation lies with the management of the company, which controls the conduct of an internal subjective assessment of the effectiveness of internal control and the preparation of a report. Internal control in Japanese companies is defined as a process with a view to: effective economic activity; the reliability of financial statements; strict compliance with laws and others. Internal audit as an element of internal control and the corporate governance system provides an assessment of the effectiveness of internal control [7].

By analogy with the German experience in implementing internal control systems in large corporations in Japan, its procedures are integrated into all processes of companies, are included in the duties of employees, ranging from ordinary posts to top management. In addition to independent internal control and internal audit, self-assessment procedures and tools are actively used (i.e., activities for self-monitoring and self-assessment of their responsibilities in accordance with established internal documents). Self-assessment as a process includes: identification of risks, the formation of control measures, monitoring of their implementation, a continuous process of improving their activities.

It should be emphasized that the extremely important attention in Japanese corporations is paid to the risks of internal fraud, bribery and corruption. In Germany, increased attention is also paid to the prevention of such situations, but nevertheless, in German companies this issue is not posed as harshly as in Japanese corporations.

Based on the requirements of the standards and principles of internal control and internal audit, Japanese enterprises organize and integrate anti-corruption and fraud procedures in the corporate governance system with the maximum degree of transparency of this mechanism from the level of top management to ordinary employees [10].

In conclusion of the analysis of the features of foreign experience in organizing the internal control system and internal audit, let us dwell on the practice of organizations of Canada as a country of the American continent. According to tradition with European practice, in Canada a risk-based approach to organizing internal control is also quite common for doing business, but first of all, the Sarbanes-Oxley law is the leading guide [12].

In contrast to the components of the organization of the internal control system in companies in Germany and Japan, here are just a few aspects of the internal control assessment system, which, in our opinion, is suitable for any organization, and some of its components can be used in Russian practice.

In Canadian corporations, the quality assessment of the internal control system is carried out by the Board of Directors, which evaluates the actions of top management as a whole. Moreover, the assessment itself is qualitative; as a rule, several levels are established (strong, medium, low) [13].

The presented procedures related to the development of the internal control system in Canadian corporations are focused on a high level of corporate culture and discipline. At the moment, in our opinion, not all of them are applicable in Russian practice, but at the same time they can serve as a guide for the development of the internal control system in national companies.

4 Discussion

It is important to note that the state imposes certain requirements on the organization of internal control, internal audit and risk management, and the requirements in many aspects comply with international best practices (in particular, COSO standards). In this regard, the experience of Germany in terms of implementing COSO standards, as well as the accumulated corporate experience of Germany, Japan and Canada, will be able to become a guideline for Russian companies with regard to the adaptation of world best practices.

If we draw a parallel between current practices abroad and in Russia, we note that it is recommended for Russian enterprises to create independent structural divisions of internal audit and separate divisions for risk management and internal control.

An analysis of the practices of some Russian and foreign companies allows us to conclude that in almost all of them internal control and internal audit are carried out by separate structural units. However, there are cases when all these units are subordinate to one top manager, which reduces the effectiveness of these processes and leads to a conflict of interest.

The analysis allows us to conclude that very often in Russian companies there is a situation where the same unit combines the functions of internal control, internal audit and risk management (according to the research of O. Ivanov, this situation is repeated in every fifth company [8]). In our opinion, the reason for this situation may be: a misunderstanding of the key principles of internal control and internal audit, a lack of readiness to bear the costs of their correct implementation, a lack of understanding of the appropriateness of implementing internal control and internal audit tools, an undeveloped corporate culture (this factor is manifested in a significant number of Russian enterprises of different sectors, but recently the situation has begun to change for the better).

Over the past five years from 2011 to 2016. Internal control and internal audit in Russian organizations began to develop actively, although if we pay attention to organizations with state participation, approximately 75% of them introduced a risk-based model of internal control only in 2014–2015 [1].

With regard to internal audit, statistics show that only 30% of organizations of the internal audit service have existed for more than 10 years, and in about 35% of the analyzed companies only 1–2 years [8]. At the same time, one must understand that it is not the period of existence that matters, but the quality of the processes built, which, in our opinion, is determined not only by the number of inspections per year. The fact is that some authors suggest that the number of inspections per period (for example, a year) be considered the leading indicator of the effectiveness of internal audit at the enterprise, citing statistics to confirm their position: almost 75% the studied companies carry out checks at least 50 times a year; almost 20% of companies – 90–100 inspections per year; 22% of companies - 10 inspections per year [8]. Of the total number of inspections, about 25% are unscheduled (approximately for each of the selected groups of respondents).

If you look at the survey data, then almost 96% of company executives are sure that the number of inspections within the framework of internal audit is the most significant indicator in terms of efficiency (Fig. 1).

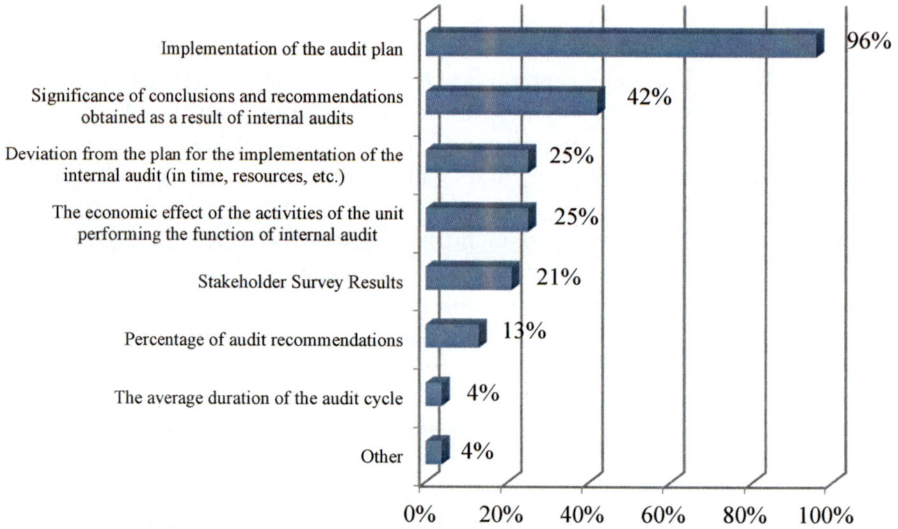

Fig. 1. Key performance indicators of internal audit in Russian companies with state participation (Source: author).

In our opinion, this indicator is important, but perhaps not the main one in the list of indicators of the effectiveness of internal control.

The significance of the number of inspections is manifested to a greater extent in preventing the development of various negative situations (process failures, inconsistencies with established criteria for product quality, fraud, etc.), but if the internal control system is effective, then other indicators come to the fore, for example, the significance of conclusions on inspection results (Fig. 1), etc.

At the same time, it is fair to take into account the fact that the legal framework for the formation of the internal control and internal audit system in companies with state participation is quite young, is in the process of formation, and some common standards and performance indicators are needed for the transition period, allowing accumulate information on the implementation of certain internal control and internal audit procedures at Russian enterprises in order to take it into account in order to improve the already adopted and develop s legal provisions.

From the point of view of identifying the main problems for the effective development of best practices of internal control and internal audit for companies with state participation, most organizations highlight such difficulties as a lack of qualified personnel, a weak interest of senior management in implementing internal control and internal audit, and an insufficiently effective legal framework regulation of these procedures in the Russian Federation, weak interaction between internal control, internal audit services and other Thousands separator companies (especially those who are most responsible for the financial results of the business), the lack of a positive attitude of the overwhelming number of employees to the services of the internal control, internal audit, and others.

5 Conclusion

The considered problems show that there are urgent tasks of developing internal control, internal audit and risk management for Russian organizations. Among their priority ones can be attributed: increasing the level of integration of procedures and functions of internal control processes, internal audit, risk management in the main business processes of the organization; increase in the number of controlled/audited processes; increasing the level of interaction between all departments of the organization within the framework of the internal control and risk management system; intensification of the process of developing the principles of a risk-based VK model at Russian enterprises increasing the level of interest of senior management in implementing the best practices of internal control, internal audit and risk management; automation of internal control procedures, internal audit and risk management [8]).

As one of the significant directions of increasing the effectiveness of internal control and internal audit, we can single out automation of such procedures as: data collection, information exchange, documentation, reporting, archiving, etc. At the same time, the presented functions can be significantly expanded. In addition, we note that most enterprises with state participation prefer to use their own software.

In conclusion, we note that, on the one hand, it is necessary to change the idea of the top management of Russian companies that internal control and internal audit are unnecessary processes and/or do not have (have partially) economic efficiency. To do this, it is necessary to build a new corporate culture (perhaps at the initial stage of policy). On the other hand, little time has passed since the introduction of the best practices of organizing internal control and internal audit at Russian enterprises. These processes can become successful only in the conditions of a developed corporate mentality, which the Russian business environment cannot boast of yet, but it has already chosen these landmarks as the goals of the near future.

References

1. Bondina, N.N., Zubkova, T.V., Lavrina, O.V.: The financial mechanism and the assessment of its impact on improving the efficiency of agricultural production. Bull. Samara State Agric. Acad. **2**, 29–35 (2014)
2. Borovikov, A., Smolyakov, O.: Principles of organization and structure of the management system of a holding type company. Upravlenie **7**(3), 54–62 (2019). https://doi.org/10.26425/2309-3633-2019-3-54-62
3. Dmitrieva, I., Kharakoz, Yu., Zhakan, T.: Development of the state financial control and audit system in the EAEU member countries. Auditor **5**(12), 10–15 (2019). https://doi.org/10.12737/1998-0701-2019-10-15
4. Huang, W.: Built on value. Palgrave Macmillan, Singapore (2019). https://doi.org/10.1007/978-981-13-7507-1_15
5. COSO: Enterprise risk management - Integrating with strategy and performance (2017). https://www.aicpastore.com/AST/Main/CPA2BIZ_Primary/PRDOVR ~ PC-ACOSOERM/PC-ACOSOERM.jsp. Accessed 11 Feb 2020
6. Ivanov, O.B.: Organization of internal audit in a German company. STAGE Econ. Theory Anal. Pract. **5**, 129–142 (2013)
7. Ivanov, O.B.: Organization of internal control in the corporate governance system of Japan. STAGE Econ. Theory Anal. Pract. **5**, 125–146 (2014)
8. Ivanov, O.B., Egorova, E.A.: Status and development directions of internal audit systems, internal control and risk management in companies with state participation. ETAP Econ. Theory Anal. Pract. **6**, 7–28 (2015)
9. Liker, J.: The Toyota Way: 14 Management Principles from the World's Greatest Manufacturer. Alpina Business Books, Moscow (2005)
10. Messier, W., Reynolds, J., Simon, C., Wood, D.: The effect of using the internal audit function as management training ground on the external auditor's reliance decision. Account. Rev. **86**, 2131–2154 (2011)
11. Nusa, I.B.S.: Quality of audit system information for internal control effectiveness. In: Warlina, L., (ed.) Proceedings of the International Conference on Business, Economic, Social Science, and Humanities – Economics, Business and Management Track (ICOBEST-EBM 2019), pp. 198–202. Atlantis Press, Paris (2020). https://doi.org/10.2991/aebmr.k.200108.046
12. Peshkova, T.Yu.: Risk management, internal control and internal audit of insurance organizations in the Canadian insurance supervision system. Financ. Analytics Probl. Solutions **42**, 48–56 (2013)
13. Tsyganov, A.A., Bumbaji, Ya.V.: International experience of integrated financial supervision. Financ. Analytics Probl. Solutions, **43**, 45–52 (2012)
14. Vahdani, M., Rashedi, R., Mehr, J.M.: Investigating the effect of business strategy and internal control on audit quality. Humanit. Soc. Sci. Rev. **7**(6), 787–798 (2019). https://doi.org/10.18510/hssr.2019.76119
15. Younas, A., Md Kassim, A.A.: Essentiality of internal control in audit process. Int. J. Bus. Appl. Soc. Sci. **5**(11), 1–6 (2019). https://doi.org/10.33642/ijbass.v5n11p1

Institutional Measures to Support Industrial Enterprises of the Samara Region

E. S. Morozova(✉)

Samara State University of Economics, Samara, Russia
dhrsseu@gmail.com

Abstract. The article contains an analysis of the main industrial development trends in the Russian Federation and the Samara region in particular, identified on the basis of aggregated statistical data. The author considered the industrial potential of the Samara region and its structure. The analysis of the main programs of support and development of the industry until 2018 and after (for the period of the implementation of national projects 2018–2024) at the level of the federal and regional authorities is presented, the main results of this support in the Samara region are presented. The author conducted a survey of 70 representatives of business and scientific communities, one way or another connected with the industry of Russia and the Samara region. Based on the results of the survey, conclusions were drawn on the conformity of the measures being implemented to the needs of the business and market conditions, a number of recommendations were proposed.

Keywords: Industry · Information support · Institutional support · Support measures

1 Introduction

Due to geographical and historical preconditions, a number of objective advantages of the Samara region allows the region to occupy positions in the top 10 regional economies of Russia. The region is located in the European part of the Russian Federation, as part of the Volga Federal District, the total area is 53 565 km 2 (0.3% of the territory of the Russian Federation; 5.1% of the territory of the Volga Federal District (VFD)), the population is 3 205 975 people (2.2% of the population of the Russian Federation; 10.94% of the population of the Volga Federal District) according to regional authorities. The region owes much of its rich industrial potential, an extensive research, development, scientific and cultural base to the decisions of the State Defense Committee (1941–1945) on the transfer to Kuibyshev region (until 1991, later the Samara region) of a significant part of the aviation, bearing, and radio-electronic enterprises, machine-building, plaster, petrochemical, peat, gas industries, as well as decisions on the transfer to Kuibyshev (until 1991, later Samara) of foreign embassies, broadcasting, arts and cultural organizations.

In the postwar years, the rise of the industrial base was provided mainly due to rocket science and aircraft construction, as well as the production of components. The article provides an analysis of the current state of the industry in the Samara region,

S. I. Ashmarina et al. (Eds.): ENGINEERING ECONOMICS WEEK 2020, LNNS 139, pp. 45–53, 2021.
https://doi.org/10.1007/978-3-030-53277-2_6

analyses support measures created by authorities at various levels, and also, through a survey, the conclusion is made about the information security of the measures taken.

2 Methodology

To assess the level of information security of the implementation of industrial production support programs at the federal, regional and municipal levels, it is necessary to identify the economic institutions through which their development is carried out. Table 1 presents data on the main industrial support programs, the timing of their implementation, as well as institutions at different levels.

Table 1. Institutional support for industry

Name of the program (Purpose of creation)	Implementation level	Implementation dates	Economic Institutions
State program of the Russian Federation “Development of industry and increasing its competitiveness”	Federal	2014–p.t.	Industry Development Fund (IDF), State Industry Information System (GISP)
National project “International Cooperation and Export” (As part of the federal project “Export of Services”)	Federal	2018–2024	Ministry of Economic Development of the Russian Federation, Ministry of Industry and Trade of the Russian Federation
Federal project “Implementation of the best available technologies” (In terms of improving the environmental friendliness of industrial production)	Federal	2018–2024	Ministry of Economic Development of the Russian Federation, Ministry of Industry and Trade of the Russian Federation
The project to create a favorable investment climate in the Samara region	Regional	2016–p.t.	Ministry of Economic Development, Investments and Trade of Samara Region, Agency for Investment Attraction of Samara Region

Source: author.

Based on the above programs, it is worth noting that most of the measures taken are at the federal level, in addition, these measures were developed as an economic tool designed to minimize geopolitical threats. However, the industry did not become a real driver of economic development (2015–2018) and a way out of the crisis situation (2018–p.t.). Meanwhile, it is worth analyzing the actual indicators characterizing the state of industrial production in the Russian Federation (Fig. 1). Hereinafter, according to the Federal State Statistics Service, the analytical center under the Government of the Russian Federation and the Ministry of Economic Development of the Russian Federation.

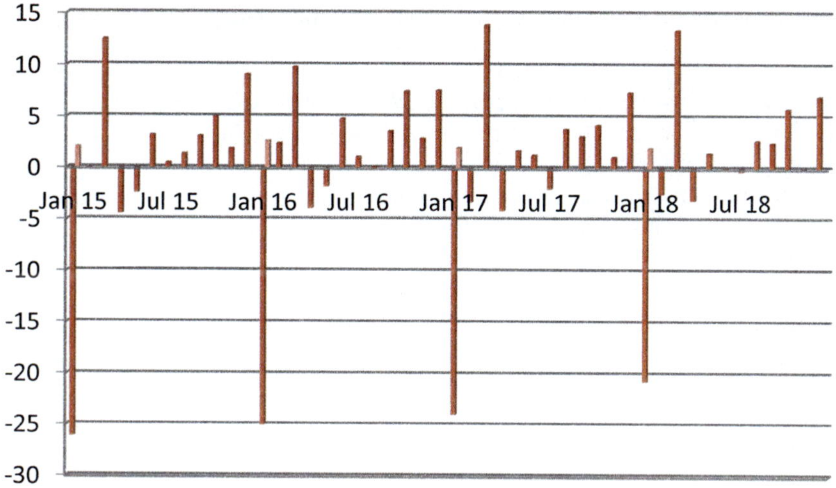

Fig. 1. The increase rate of industrial production compared to the previous month in the Russian Federation in 2015–2018, % (Source: author)

Getting rid of the influence of periods of an intermittent drop in the increase rate of industry in January of each year (due to long holidays), it should be said that the average annual increase rate of this indicator fluctuates in the range of 2–3%. So, according to the analytical center under the Government of the Russian Federation, in 2018, GDP growth amounted to 2.3%, and industrial output increased by 2.9%. And in the first quarter of 2019, the increase rate of industrial production slowed to 2.1% in annual terms, and GDP - to 0.5%.

However, general indicators of industrial production growth cannot objectively reflect a number of negative trends:

1. The growth of industrial production was primarily ensured by sustainable long-term growth of the extractive industry (4% by 2017, 14% by 2014), that in the conditions of an already uneven structure of the Russian economy and falling energy prices, could lead to budgetary insecurity (the share of oil and gas production in the production structure in the Russian Federation is about 70%).
2. Uneven dynamics of industrial production in the constituent entities of the Russian Federation (industry growth in 2018 was observed in 66 constituent entities of Russia (–10% compared to 2017).
3. The outstripping growth in industrial output in 2018 was recorded in the city of Sevastopol (+29.2%), the Yamal-Nenets Autonomous District (+17.0%), Astrakhan (+16.2%) and Tambov (+14.4%) areas. The analyzed Samara region (+1%).

4. In 2018 and in the first half of 2019, the output of high-tech manufacturing activities decreased (–11.5% by 2018).
5. According to a study by the Analytical Center under Government of Russian Federation, the factors limiting production growth in mining and manufacturing are insufficient demand for the products of enterprises in the domestic market, a high level of taxation and the uncertainty of the economic situation as a whole [2].

Based on international conditions (regionalization, sanctions confrontation, emergencies, such as closing borders and production due to COVID-19) and all of the above, it is worth noting the importance of transforming industry in the direction, primarily of the processing high-tech industry. To implement the tasks set by the President of the Russian Federation V.V. Putin and ensuring GDP growth of 3–4% annually require serious institutional changes.

The Russian model, based on the super centralization of economic and political institutions in Moscow and St. Petersburg, practically does not allow regions that are not involved in oil and gas production and processing without serious transfers from the Federation to execute budgets. In turn, this inevitably leads to an outflow of economic entities (or their parent enterprises) to capitals and regions offering preferential conditions for the functioning of the business. The decrease in the number of large taxpayers affects the investment climate and even more leads to the unsecured budgets of the entities. Samara region (a region with a diversified economy) in the ranking of regions is included in the second decile of the economies of the constituent entities. Largely because it takes 2nd place in Russia in terms of oil refining (19.5 million tons per year).

Regional authorities are making efforts to create a favorable investment climate in the Samara region. On the electronic site (online platform) of the Agency for attracting investments, basic information is presented that may interest key stakeholders: investors, project authors, regional authorities. Key attention is paid to the creation of one of the largest agglomerations in the country, Samara-Tolyatti, which is an attempt to adopt the experience of agglomerations in Moscow, St. Petersburg, Tyumen and so on. The portal provides funding for socially important projects both in the field of public-private partnership (PPP), and by attracting third-party investors.

Conditionally, the activities of regional authorities can be divided into three key areas:

1. Regulatory support of the field of PPP (approval of the list of objects the property right to which belongs or will belong to the Samara region, regarding which it is planned to conclude concession agreements).
2. Development of the institutional environment in the field of PPP (at the end of 2018, the Samara Region became one of the 6 leading regions of the Russian Federation in terms of the development of public-private partnerships along with Moscow, St. Petersburg, the Moscow Region, the Republic of Bashkortostan and the Khanty-Mansi Autonomous District scoring 90 points).
3. Accumulation, aggregation and transfer of experience in implementing PPP projects (22 projects are currently under implementation (implemented) for a total amount of

extrabudgetary investments attracted 25.86 billion rubles, of which 14 projects amounting to 6.22 billion extrabudgetary funds raised rubles have already been implemented and 1345 jobs have been created, 25 projects are under consideration and structuring).

Among the main projects it is worth highlighting:

1. The construction of the highway “Central”.
2. Construction of a bridge over the Volga River with a detour of Togliatti and access to the M-5 Ural highway.
3. Construction and exploitation of railways to ensure the operation of the special economic zone of the industrial-production type “Togliatti”.
4. Organization of high-speed railway communication “Samara-Kurumoch airport-Togliatti”.

3 Results

According to the results of an electronic questionnaire, a representatives of the business and scientific field of the city of Samara and the Samara region conducted by the authors of the article, it is worth noting the similarity of indicators of statistical bodies and the empirical assessment of respondents. The survey involved 70 people, of whom 52 identified themselves as being associated with industrial production. Distribution of respondents by gender: men - 61% (82% of those related to industry), the most frequent range value of age: 30–40 years. Respondents were asked the following questions:

1. “How do you assess the general state of industrial production in the Russian Federation on a scale from 1 to 5?”, Where 1-unsatisfactory, 2-poor, 3-lowered, 4-corresponds to the average world level, 5-exceeds the average world level.
2. “How do you assess the general state of industrial potential in the Russian Federation on a scale from 1 to 5?”, Where 1-unsatisfactory, 2-poor, 3-lowered, 4-corresponds to the average world level, 5-exceeds the average world level.
3. “Refer the Russian Federation to one of the following groups”: 1-driver countries of economic development, 2-developed countries, 3-developing countries, 4-lagging countries, 5-backward countries.
4. “How do you assess the state of extractive industrial production in the Russian Federation on a scale from 1 to 5?”, Where 1-unsatisfactory, 2-poor, 3-lowered, 4-corresponds to the average level of developing countries, 5-exceeds the average level of developing countries.
5. “How do you rate the state of manufacturing in Russia on a scale from 1 to 5?”, Where 1-unsatisfactory, 2-poor, 3-lowered, 4-corresponds to the average level of developing countries, 5-exceeds the average level of developing countries.
6. “What drivers of industrial development in the Russian Federation do you consider the most important?”: Technology adoption, technology development, planning system, strengthening state regulation, increasing competition, militarization, environmental friendliness, I find it difficult to answer.

7. "Do the current methods of industrial management in Russia correspond to the scale of the new economic paradigm?": Yes, no, I find it difficult to answer.
8. "Is the existing scientific and technological base capable of ensuring the development of industry in the Russian Federation in the medium term?": Yes, no, I find it difficult to answer.

According to the results of the survey, 48% of respondents assess the state of Russian industry as lowered (while 69% believe that the development of extractive industries is above average among developing countries, and 59% consider the development of the manufacturing sector as lowered). At the same time, the respondents positively assess the industrial potential of the Russian economy (47% corresponds to the world average, 37% exceeds the world average).

Among the other survey results, it is worth highlighting the respondents' hopes that technology development and a well-planned planning system will be able to become drivers of the social and economic development of the Russian Federation. More than half (56%) believe that the current methods of industrial management in the Russian Federation do not correspond to the scale of the new economic paradigm, however 60% are convinced that the existing scientific and technological base is capable of ensuring the development of industry in the medium term. The second block of the survey questions concerned the information security of the measures taken to support the industry and their effectiveness.

So, the following questions were asked:

9. "How do you assess institutional measures taken by federal and regional authorities to stimulate industrial production on a scale from 1 to 5?", Where 1-unsatisfactory, 2-poor, 3-average, 4-good, 5-high.
10. "How do you assess the measures taken by federal and regional authorities to stimulate industrial production on a scale of 1 to 5?", Where 1-unsatisfactory, 2-poor, 3-average, 4-good, 5-high.
11. "How do you assess the ability of small and medium-sized businesses to take part in entrepreneurship support programs on a scale from 1 to 5?", Where 1-unsatisfactory, 2-poor, 3-average, 4-good, 5-high.

The survey results are presented in Figs. 2, 3 and 4.

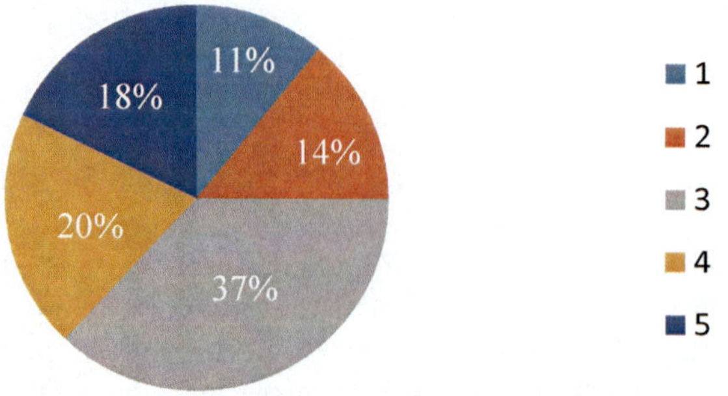

Fig. 2. The results of the survey "How do you assess the institutional measures undertaken by federal and regional authorities to stimulate industrial production?" (Source: author)

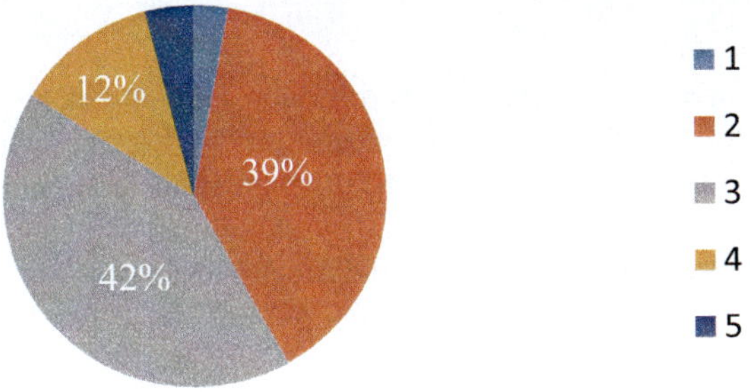

Fig. 3. The results of the survey "How do you assess the information measures taken by the federal and regional authorities to stimulate industrial production?" (Source: author)

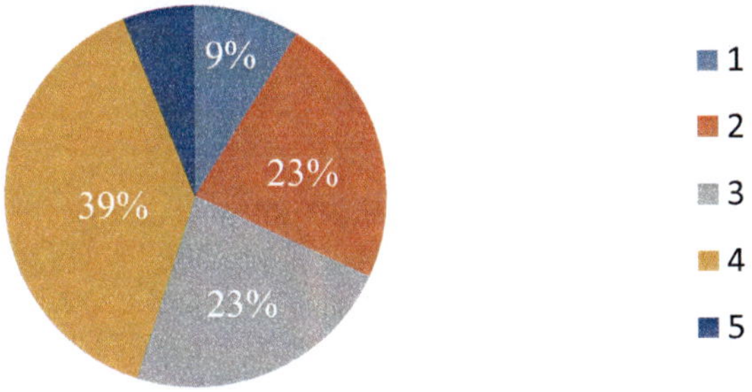

Fig. 4. The results of the survey "How do you assess the capabilities of small and medium-sized enterprises to take part in entrepreneurship support programs?" (Source: author)

Thus, it is possible to assess the degree of satisfaction of the business and scientific sectors in ongoing programs and their information support as average.

4 Discussion

The development of the domestic industrial field is constantly discussed in the scientific and business communities. It is no coincidence that industry is at the forefront of discussions: on the one hand, the presence of a strong industry can ensure national security and hedge the risks of international instability, and on the other hand, the industry must act as a driver of sustainable economic growth, which, unfortunately, failed to implement in Russia. Significant studies of institutional measures, including those applicable in the context of this issue, were published by Aganbegyan, Glazyev, Ivanter, Zubarevich, and Safronov [1, 5, 7, 9].

However, despite a number of objective negative aspects: a high degree of centralization in Moscow, the Moscow region and St. Petersburg, a high degree of imbalance in the economies of constituent entities, varying degrees of industrial development, the share of transfers, lack of funding for science and technology, the outflow of knowledge capital, etc., about which experts say industry has a high potential for development.

Foreign researchers, in turn, are considering such a development of the industry that can allow them to go completely to a green economy [6, 8]. The work also reveals the relationship and interdependence of social and industrial (economic) growth [3]. In addition, the researchers noted the advantage identified by the authors to create agglomerations, the basis for which is an industrial cluster [4].

5 Conclusion

The author analyzed the current state of industry in the Russian Federation, examined the industrial potential of the Samara region and its structure. Based on the analysis of statistical data, a conclusion is drawn about industrial development trends in the Russian Federation.

The article provides an analysis of the main programs of support and development of industry at the level of federal and regional authorities, the main results of this support are presented.

The authors conducted a survey of representatives of business and scientific communities, one way or another connected with the industry of Russia, based on analytical data and the results of the survey, it is possible to draw a number of conclusions and recommendations.

The main findings include:

1. Industry is a prerequisite for the formation of national security, an adequate response to economic and political challenges.
2. Industry can and should act as a driver of the socio-economic development of the Russian Federation, as in the cases of developed countries manufacturing high-tech industry.
3. The measures to support domestic industry that existed before the launch of national projects in 2018 were not enough to at least somehow significantly increase the GDP (GRP) and living standards.
4. Industrial support programs in the Russian Federation in 2018–2024 are mainly built on the implementation of national projects (especially after the change of the Government of the Russian Federation).
5. Implemented measures to support industry do not fully meet the expectations of the real sector.
6. Information measures to support industry development programs were rated by 81% of respondents as poor and medium.
7. It is recommended to develop at the federal level a methodology for balancing regional inequality, a transparent system of distribution of transfers, tax exemptions, the goals of which will be to attract business in general and industry in particular to the regions.
8. Development of regional programs to attract significant financial resources to the Samara-Tolyatti agglomeration on favorable terms is recommended.

References

1. Aganbegyan, A.: Lessons from the crisis: Russia needs modernization and an innovative economy. ECO **1**(427), 34–60 (2010)
2. Analytical Center under Government of Russian Federation: Industrial dynamics production in Russia: Faster growth extractive sector (2019). https://ac.gov.ru/files/publication/a/23445.pdf. Accessed 02 Apr 2020
3. Basakha, M., Hossein, S., Kamal, M.: Industrial development and social welfare: a case study of Iran. Socio-Econ. Plann. Sci. **68**, 100661 (2019). https://doi.org/10.1016/j.seps.2018.10.012
4. Gan, L., Shi, H., Hu, Y., Lev, B., Lan, H.: Coupling coordination degree for urbanization city-industry integration level: Sichuan case. Sustain. Cities Soc. (2020). https://doi.org/10.1016/j.scs.2020.102136. Accessed 26 Mar 2020
5. Glazyev, S.: A Leap into the Future. Russia in the New Technological and World Economic Order. Book world, Moscow (2018)
6. Guo, Y., Tong, L., Mei, L.: The effect of industrial agglomeration on green development efficiency in Northeast China since the revitalization. J. Clean. Prod. **258**, 120584 (2020). https://doi.org/10.1016/j.jclepro.2020.120584
7. Ivanter, V.: The potential for economic growth in Russia. Sci. Works Free Econ. Soc. Russia **218**, 70–80 (2019)
8. Xu, L., Tan, J.: Financial development, industrial structure and natural resource utilization efficiency in China. Res. Pol. **66**, 101642 (2020). https://doi.org/10.1016/j.resourpol.2020.101642
9. Zubarevich, N., Safronov, S.: Inequality in the socio-economic development of Russian regions and cities in the 2000s: growth or decline? Soc. Sci. Modern. **6**, 15–26 (2013)

The Rational Enterprise Strategy Development: Marketing Aspect

I. A. Toymentseva[1(✉)], O. N. Denisova[2], and V. D. Chichkina[2]

[1] Samara State University of Economics, Samara, Russia
tia67@rambler.ru
[2] Syzran Branch, Samara State Technical University, Syzran, Russia
phdenisova@gmail.com, vd_00@mail.ru

Abstract. Developing and implementing a rational enterprise strategy issues are becoming especially important for domestic organizations in today's unstable economy. The relevance of the article is: an enterprise engaged in the production, the sale of goods or services will not be able to withstand competition, act effectively in accordance with market requirements without the rational strategy based on a marketing approach. At the same time, special attention should be paid to the developing strategy principles, methods and tools reflecting the industry features of enterprises functioning. Consumer behavior research in the pharmaceutical industry has shown a potential customer has a significant impact on pharmacy retails and pharmaceutical manufacturers marketing communications. Domestic market restructurisation of the pharmacy retails increases competition for consumers. The problem is the following: the pharmaceutical market is actively entering the network business, which often exceeds the local business amount. The effective strategy development by the Federal departments top management leads to local and regional enterprises displace from the pharmaceutical market. At the same time, this can open up new offers of the most affordable and high-quality products in the pharmaceutical industry to the consumers.

Keywords: Competition · Consumer · Factors · Marketing · Pharmaceutical industry · Strategy

1 Introduction

Today, there is a difficult situation in the field of Russian public health due to population ageing, increase in the prevalence rate of chronic diseases, and existence of socio-economic inequality in obtaining access to health services. The main state socio-economic value is to preserve and strengthen public health. Therefore, building an effective, open to public, and safe health care system and a rational medicine assistance system is crucial for the country's prospective success to achieve its long-term strategic goal, that is to say, a higher level of economic and social development. Nowadays, any form enterprises and any activity sphere face increasing competition at the domestic and international commodity markets. Therefore, in order to adapt to the constantly changing market conditions, as well as to the increasing customer preferences, the company needs to develop an effective strategy using marketing tools which allow meeting consumer demand fully by providing high-quality goods and services.

S. I. Ashmarina et al. (Eds.): ENGINEERING ECONOMICS WEEK 2020, LNNS 139, pp. 54–65, 2021.
https://doi.org/10.1007/978-3-030-53277-2_7

In order to be successful, the organization must have a clear focus, in other words, the enterprise must have a specific strategy that will determine the enterprise management system. The enterprise effectiveness depends on the achieved strategic goals, so the research relevance is undoubtful. The strategy should be subjected to changes, depending on the external and internal environment state. The organization strategy should have its own uniqueness. Concerning the latter point, it is important to emphasize if the organization uses standard technologies, have already been tested by someone, the company cannot expect a high result, since it had been achieved by the leader in this field.

2 Methodology

Modern ideas about the role and importance of strategic management in the economic entities activities make us conclude that there is an urgent need for its use in overcoming the economic crisis, domestic enterprises competitiveness increase on the marketing tools use basis. A strategy is an organization pre-planned reaction to the external environment change, a policy chosen for achieving the desired result. For example, Chandler believes that the strategy should "define the long-term goals of the enterprise, as well as plan actions and allocate the necessary resources to solve the tasks" [2, p.122]. Ansoff defines strategy as "one of several sets of rules for making decisions about the organization behavior" [3, p.45]. Mintsberg, Ghoshal, and Quinn note that the strategy is not just a plan, but a set of actions to solve specific tasks [9]. Ackoff considers strategy as the category of planning. He believes that it is "the process of making and evaluating a number of interdependent decisions predicting a certain activity…" determining "… the desired future state…" [1, p.62]. Porter expresses his approach to defining a strategy: "Strategy is the creation of a unique and profitable position providing a certain choice of activities" [10, p.95].

Vikhansky and Naumov [14] emphasize that strategic management shifts the focus of managers attention to the company surroundings in order to respond to external changes properly. Strategic management tools – Balanced Scorecard, BSC [13], developed by Kaplan and Norton, [7] make the enterprise management process rational.

Hofer [6] defines strategic management as a special process that results in an enterprise interaction with its external environment. Meskon, Albert, and Heduri [8] emphasize the strategy is a comprehensive plan, the goal of which is to achieve the enterprise mission, and its main goals. The researchers Toymentseva, et al. [12] also state the strategic planning leads to the individual strategy development for each individual enterprise and helps the company to achieve its goals.

To define management strategic guidelines the following market tools were used in the Vita pharmaceutical enterprise: PEST and SNW and SWOT analyses, as well as Porter Competitive Forces model. In modern conditions the methodological basis for development strategy forming is both Western and Russian specialist theories and concepts. Evaluation of the general level of satisfaction with the quality of services provided by Vita pharmacy chain was based on the economic and mathematical modeling methods. Through the use of the sociological research method, it became possible to identify the main problems of the pharmaceutical industry and determine strategic development benchmarks.

3 Results

Modern requirements for the economic growth quality improving are accompanied by significant structural changes in the national economy. Today, there is a market share decrease of material production industries and service intensive development: educational, medical, financial, tourism, infrastructure, etc. The original structuring of social services involves the social-economic activities marking directly related to the production of health, education and culture capital, as well as social-economic activities providing conditions for its development (Fig. 1).

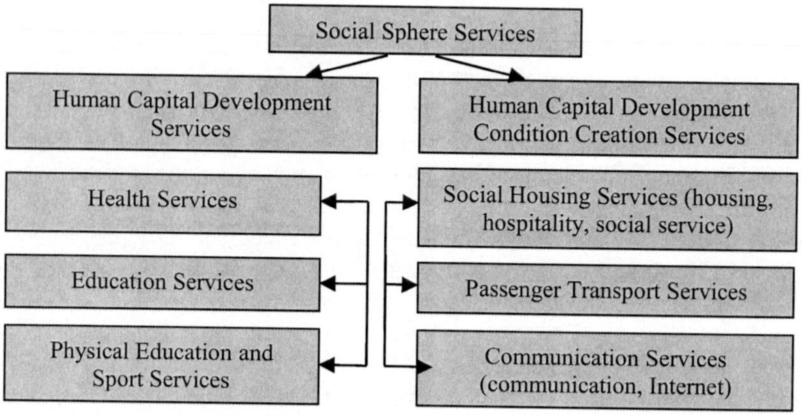

Fig. 1. Social services structure (Source: authors).

The economic value of this type of service is determined by the number of circumstances. Firstly, saving time in the human capital reproduction cycle, that is, increasing the time of its useful functioning, associated with health capital, education and culture increase. Secondly, increasing its mobility, that is, the speed and deadlines of implementation methods and development guidelines.

Methodological approach to determine the integrated index of health services development, including the pharmaceutical industry, comprises the following algorithm for calculating the index. At the first stage, the expert evaluation method allowed to determine the list of the most significant quantitative and qualitative indicators to the population that characterize the consumer properties of the service. Consequently, the main indicators include availability, quality, safety, and price. Based on the fuzzy sets theory, the per unit indicator of each parameter is calculated at the second stage. The performance evaluation of service development indicators in the pharmaceutical industry in 2010–2015, as well as the projected rate for the period up to 2030, are presented in Table 1.

Table 1. Calculation of the integrated index of customer satisfaction

Indicator name	2010	2015	2020 (projection)	2030 (projection)
	Performance evaluation			
Availability, weight coefficient 0.283	0.692	0.692	0.873	0.951
Quality, weight coefficient 0.209	0.692	0.873	0.873	0.951
Safety, weight coefficient 0.349	0.368	0.873	0.873	0.951
Price, weight coefficient 0.159	0.368	0.692	0.873	0.951
Integrated index	0.755	0.792	0.873	0.951

Source: authors.

The calculations allowed determining the integrated index of the service development in the form of the sum of productions of expert evaluations of the indicators under consideration by the corresponding weight coefficients.

The economic and mathematical model of the interpolation problem includes a transposable row-vector of annual time period valuations (X), a row-vector of valuations of the corresponding integrated indexes of service development (Y), as well as cubic spline interpolation operators (cspline, interp).

The calculation of the integrated index of service development within a dynamic series through a software product is carried out below:

$X = (2010\ 2015\ 2020\ 2030)^T$

$Y = (0.523\ 0.781\ 0.873\ 0.951)^T$

s := cspline (X, Y)

A (t) := interp (s, X, Y, t)

Therefore, the values of integrated targeted indexes of long term service development are the following:

A (2020) = 0.885 A (2025) = 0.897 A (2027) = 0.91 A (2029) = 0.933

According to the results of the survey on the issues faced by pharmaceutical and healthcare companies in Russia over the period of 2019, the significance of insufficient solvency of the population (by 0.30 pt.) as well as the lack of state funding for health programs and Russian manufacturers support (by 0.24 pt.) has increased dramatically.

At the same time, the significance of the problems related to imperfect legislative regulation (by 0.32 pt.), corruption (by 0.18 pt.), and geopolitical risks (by 0.32 pt.) has slightly decreased. The difficulties associated with the marking system implementation rank fourth. The results are shown in Fig. 2.

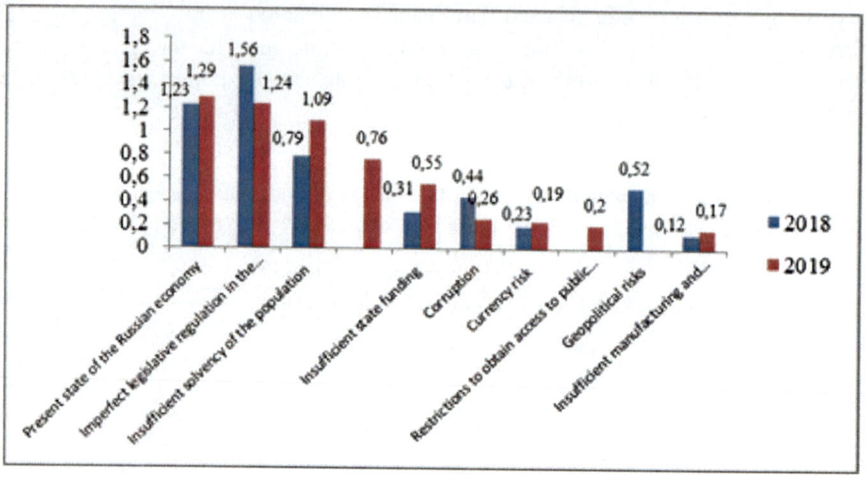

Fig. 2. Dynamics of problems of pharmaceutical companies (Source: authors).

By the end of 2019, 67.2 thousand pharmacies are successfully working in Russia, which is 2% higher than in the previous period. The reported period revenue of 2019 was 1,299 billion rubles, which is also 8% higher than in 2018. The analytical company RNC Pharma has ranked pharmacy chains by their share in the retail market for 2019 [11]. The results are shown in Table 2.

Table 2. TOP 5 pharmacy chains rating in Russia by the end of 2019

№	Pharmacy chain	Central office location	Retail number for 01.01. 2020	Volume, bln. rub	Market share, %
1	Rigla	Moscow	2846	57,9	6,6
2	CC Erkafarm	Moscow	1168	46,4	5,2
3	Pharmacy chain “36 и 6”	Moscow	1403	38,0	4,2
4	Neo-Farm	Moscow	736	31,6	3,8
5	Vita	Samara	1742	30,9	3,4

Source: authors based on [11].

By the end of 2019 leader in the pharmaceutical market share is pharmacy chain “Rigla” (6, 6%), the second place is the company group “Erkafarm” (5, 2%), the third is pharmacy chain “36, 6” (4, 2%). The pharmacy chain “Vita” ranks fifth with a market share of 3.4% and sales volume of 30.9 billion rubles.

The Vita pharmacy chain management pays great attention to customer structure analyzing, the average purchase price and the customer revenue volume. According to the survey results in Vita pharmacies, 82% of customers are women aged 29–75 years. The highest-income client group are people aged 36–45 years (21%), and 20% of revenue coming from the 56–65 age group. Low-income categories include the elderly over 76 (5%) and young people under 22 (2%).

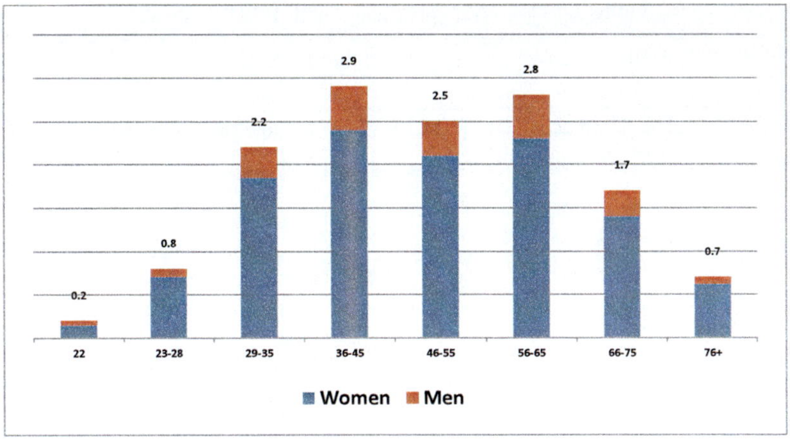

Fig. 3. The distribution of the proceeds of the company "Vita" depending on the gender and age of customers, rub (Source: authors).

Men make purchases 4 times less, and the men receipt average cost is 11% higher than the average receipt cost. The distribution of the proceeds of the company "Vita" depending on the gender and age of customers is shown in Fig. 3. According to research, the main turnover of the company – 87% - is formed by men and women aged 29 to 65 years. To analyze the "Vita" pharmacy retail external environment a PEST analysis was performed, the results of which are presented in Table 3.

One can clearly see the technological advances provide a good opportunity to develop the trade in medicines and medicinal products from the Table 2 data. The main threat is the state prices regulation for VEIMP (vital essential and important medical product) and the over-the-counter medical product sale legalization in retail chains. SNW analysis was performed to identify the company's strengths and weaknesses. The results of the analysis are presented in Table 4.

The following conclusions can be made based on the results of the analysis:

- the weak points of the "Vita" pharmacy chain are a small number of outlets, compared to the main competitor, and a smaller market share,
- the strong points – the loyalty program, the own brand development, the wide stock.

SWOT analysis was performed to compare external and internal environment factors, as well as opportunities and threats. The results are presented in Table 5.

The following conclusions can be made, based on the results of the SWOT analysis:

- the company key competence: new form of trade pharmacies (the "supermarket" format), with a wide stock, low prices and the possibility of expert advice on any type of product offered,
- the main threat to the company: possible competition from the retails and insufficient use of promotion channels contribute to the customers outflow, reducing purchasing power.

Table 3. The "Vita" pharmacy retail PEST analysis

Characteristic	Possibility	Weight	Threat	Weight
1	2	3	4	5
P	1. The pharmaceutical industry development state program	5	1. State price regulation for VEIMP (vital essential and important medical product)	5
			2. The government legislation changes related to allowing over-the-counter medical product sale drugs in retail chains	5
			3. The state spending increase on pensioners maintenance and support in medical establishments	4
E	1. The demand growth for PLP (Private Label Products)	4	1. The medical product fraud increase	3
	2. The adult paying capacity increase due to the retirement age raising	5	2. The population income decrease, the unemployment rate growth	4
			3. Currency instability	4
			4. The non-medical product demand reduce	4
S	1. The population growth	4	1. The alternative medicine promotion	4
	2. The medicine demand increase during the population increase morbidity period	5		
	3. The aging population proportion increase	3		
	4. The government acts development to healthy lifestyle goods and services increase demand	3		
T	1. The automated sales technologies development in the sales area	5	1. The new hardware treatment methods in medical establishments substituting the medicine consumption	4
	2. The e-trade development	5		

Source: authors.

Table 4. The "Vita" pharmacy retail SNW analysis

Rating system	S	N	W	Comment
1	2	3	4	5
Sales system, organization of retail outlets				
Quality opening of new retails	1			665 pharmacies were opened in 2018 which increased revenue by 54%
Pharmacies with open display	1			Open retail equipment
Merchandising	1			Own layout standards
Double quality control	1			The delivered product control
The clients communication standards	1			Communication checklists, equipment, complaint hotline
Organizational system				
Planning Prediction		1		Planning processes do not always work on time
Warehouse and transport logistics			1	Product delay, the warehouse insufficient work organization, lack of warehouses in other regions
Divisions and departments communications		1		Some divisions communications are not effective enough
Financial system				
Profitability achievement	1			Decent pricing, quality work with marginal goods
Financial control	1			Tracking expenses, accounting
Financial stability	1			Company scale
Marketing				
Brand image	1			The company perfoms marketing events to improve its image
Marketing campaigns advertising and promotion			1	Not full coverage of promotion channels, the problem of their understanding by the staff
Staff				
Salary level		1		High competition compared to other pharmacies
Motivation system		1		Motivation is registered for all positions
Training system	1			The company has a professional training base

Source: authors.

The authors also made a strategic analysis of the "Vita" pharmacy retail based on the M. Porter model of competitive forces. The results are shown in Fig. 4. Based on the study, the most rational strategy for the Vita pharmacy retail will be to optimize costs (new pharmacies led to a significant increase in costs in 2018–2019). Today, the costs incurred must be minimized. Thus, we can conclude, generally, the main factors

Table 5. SWOT-analysis

Strong points	Weak points
1	2
1. A rapidly growing pharmacy chain; 2. Highly qualified personnel; 3. Open layout form; 4. Merchandising standards availability	1. Low level of stock management; 2. Weak organization of transport logistics; 3. High costs for training and staff development; 4. Insufficient use of advertising and marketing promotion channels
Capabilities	Threats
1. Private Label Products demand increase based on attractive prices for customers; 2. Preventive agents demand increase connected with the healthy lifestyle promotion; 3. Information technologies development, including the e-trade	1. VEIMP prices state regulation; 2. National currency instability; 3. Increased competition due to the possible sale of parapharmaceutical products in retail chains

Source: authors.

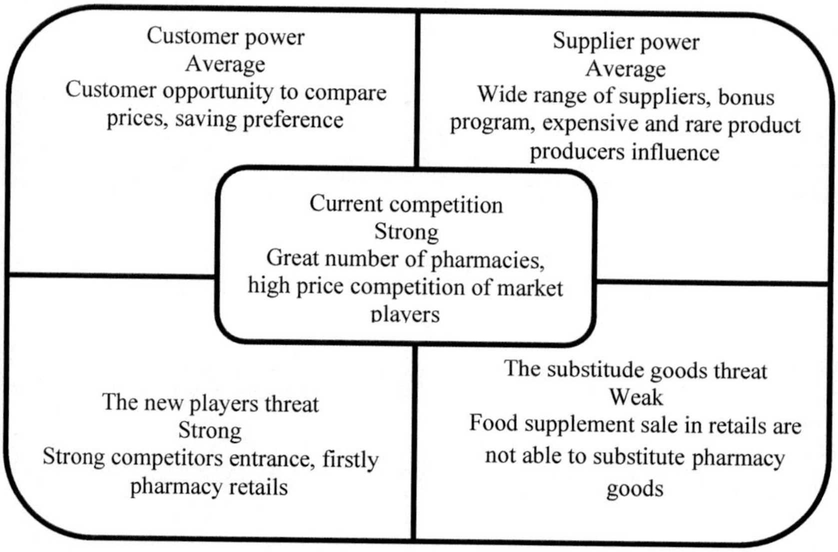

Fig. 4. Model of competitive forces (Source: authors).

having the most significant impact on the pharmaceutical market are income decrease and dietary supplement consumer demand increase. It is also noticed the traditional retail to an online system transition. The "Vita" pharmacy retail is actively developing, and many awards for the company's results and the profit increase in its core business show a positive and sustainable development.

4 Discussion

Promotion of any product or service is not possible without advertising tools. Medicine, medical services, in particular prevention, diagnosis, treatment and medical rehabilitation methods, medical devices advertising should be accompanied by side effects warning in their use, the instructions acquaintance or the expert advice. These requirements do not apply to ads that are exclusively consumed by medical and pharmaceutical workers.

In Russian legislation, medicinal products advertising is regulated by two acts: the General Federal law "On Advertising" [4] and article 44 of the Federal Law "On Medical Products Circulation" [5]. Existing restrictions do not allow many companies to spend money on direct advertising of their products.

The profile Committee of the State Duma on Health Protection has prepared amendments to the second reading of the bill allowing the retail sale of medicines on the Internet. Currently, a draft law on the legalization of remote trade in medicines is being discussed. So far, we are talking about over-the-counter medical products, which can only be sold by pharmacies. This can increase the over-the-counter products demand due to the availability increase and ability to compete with cheaper prescription products offline. In this regard, one of the most important tasks for the "Vita" pharmacy retail is to increase activities in digital. Currently, there is an increase in requests for the topic "online pharmacy".

According to the respondents, the most strategic way to promote pharmaceutical products over the next two years (+40%) is interaction with medical personnel. Pharmacy chains stimulation by paying bonuses, as well as by entering into service contracts (the balance is +36% and +34%, respectively) ranks second. As for online sales through the own online store, the opinions are divided: 46% of respondents believe that it is a priority, while 54% of respondents do not share this opinion.

5 Conclusion

The "Vita" pharmacy retail takes one of the leading positions in the pharmaceutical market. A lot of promotions with discounts, reduced prices, and gifts are held for customers every month. The main qualities the company develops in employees are leadership and partnership. The "Vita" company was one of the first companies to introduce new technologies in retail (IT-solutions for stock management, products addressing and displaying, effective promotions and offers for customers), in warehouse logistics, in personnel training and management.

The need to shape the state policy in the field of medicine assistance is determined by the role of the state in public health protection. In the context of implementation of "Development strategy of the pharmaceutical industry in the Russian Federation for the period up to 2030", every second respondent (50%) expects new medicines and medical devices to be launched on the market. The respondents also forecast a reduction in product imports (31%) and an increase in country-specific production/extension of local manufacturing content (25%).

Two-thirds of respondents (67%) from foreign companies without any local manufacturing content in Russia declare to reduce the volume of imports as a result of the strategy implementation.

The advantages of the state support measures to further business development are:

- guaranteed public procurement (+60%),
- preferential access to public procurement (+52%),
- tax benefits (+36%).

To enhance the pharmacies image and increase advertising contacts with customers, the cross-marketing is recommended to use. In this case, two companies that run their own business combine and exchange their target customers. That is, if customers trust one company, they become potential consumers of the second company. At the same time, cross-marketing campaigns can be aimed at attracting absolutely new customers. Partner marketing tools should be used for the "Vita" pharmacy retail as a part of strategic advertising actions:

Free tickets offer to the children's entertainment centers. The pharmacy will give a free ticket to visit children's entertainment for a certain amount check. This is an opportunity to increase customer loyalty and the check average cost for a pharmacy, and an additional ticket purchase for a children's entertainment center.

Thus, we can conclude that the implementation of a rational strategy will allow "Vita" to develop new products, improve the services quality, find new customers, enter new markets, develop successfully, strengthen its competitive position in the pharmaceutical market and improve its activity efficiency.

References

1. Ackoff, R.L.: The Democratic Corporation: A Radical Prescription for Recreating Corporate America and Rediscovering Success. Oxford University Press, New York (1994)
2. Chandler Jr., A.D.: Strategy and Structure: Chapters in the History of the American Industrial Enterprise. MIT Press, Cambridge (1962)
3. Ansoff, H.I.: Strategic Management. Palgrave Macmillan, London (2007)
4. Federal Law from 13.03.2006 N 38-FZ (edited from 01.05.2019) "On Advertising" (from changes and additions, valid from 12.05.2019). www.consultant.ru/document/cons_doc_LAW_58968. Accessed 07 May 2020
5. Federal Law from 12 April 2010 г. № 61-FZ "On Medical Products Circulation". www.consultant.ru/document/cons_doc_LAW_99350/. Accessed 07 May 2020
6. Hofer, C.W.: Conceptual Construct for Formulating Corporate and Business Strategies. Intercollegiate Case Clearing House, Boston (1977)
7. Kaplan, R.S., Norton, D.P.: The Balanced Scorecard: Translating Strategy into Action. Harvard Business Press, New York (1996)
8. Mescon, M.H., Albert, M., Heduri, F.: Fundamentals of Management. Business, Moscow (1992)

9. Mintzberg, G., Ghoshal, S., Quinn, J.B.: Strategic Process: Concepts, Issues, Decisions. Piter, St. Petersburg (2001)
10. Porter, M.E.: Competitive Advantage of Nations. Free Press, New York (1990)
11. RNC Pharma: Rating of Russian pharmacy chains. *KatrenStyle* (2020). https://www.katrenstyle.ru/articles/journal/news/opublikovan_reyting_aptechnykh_setey_po_itogam_2019_goda_tablitsa. Accessed 07 May 2020
12. Toymentseva, I.A., Karpova, N.P., Toymentseva, A.A., Chichkina, V.D., Efanov, A.V.: Methods of the development strategy of service companies: logistical approach. Int. J. Environ. Sci. Educ. **11**(14), 6820–6836 (2016)
13. Van Vliet, V.: Balanced scorecard (2010). https://www.toolshero.com/strategy/balanced-scorecard/. Accessed 18 Sep 2019
14. Vikhansky, O.S., Naumov, A.I.: Management. Economist, Moscow (1998)

Priority Forms of Building Communications in Organizing International Cooperation

A. L. Beloborodova(✉), N. A. Ilina, O. V. Martynova, and N. G. Antonchenko

Kazan Federal University, Kazan, Russia
a-beloborodova@mail.ru, olgavl982@mail.ru, nadezh.iljina@yandex.ru, anton4enkonataly@gmail.com

Abstract. Finding partners in the market of another country is a complicated task for any enterprise. This article is an example of the Serbian company Elitas, engaged in the production of advertising structures, and it describes the process of forming a partnership base of Russian companies. The following communication channels were selected to search for potential partners: direct e-mail, search for business partners on Internet sites (such as Sberbank partner, business partner, Partnersearch), participation in international exhibitions and fairs, participation in tenders for the supply, personal contacts and relations. The result of the work makes it possible to assess the effectiveness of each communication channel.

Keywords: International market · Partners · Marketing communications · Advertising constructions · Advertising signs · Marketing research

1 Introduction

Russia has an active international cooperation policy with many countries, including Serbia, which has very close trade relations with the Russian Federation. The features of the development of the Serbian economy are described by Dudic, Smolen, Dubic, and Mirkovic [5] and Iskenderov [6], which emphasize the stages of growth and decline, as well as ways of integrating the country's economy. At present, more than 20 intergovernmental agreements and protocols have been signed between Russia and Serbia, the most important of which is the free trade agreement, which has been in force since 2000 and the declaration on strategic partnership between Russia and Serbia (in force since 2013). According to the document, 95% of the goods that have joint Russian-Serbian production were exempted from customs duties.

According to Ru-stat [14], trade between countries is developing in leaps and bounds (Fig. 1). Trade turnover between Russia and Serbia amounted to $8.71 billion for the period 2014–2018. The lowest rate for this time was $1.26 billion, the highest is $2.12 billion. Import to Russia amounted to $4.55 billion for the period 2014–2018. The indicated dynamics leads to the fact that the development of relations between countries is highly dependent on the political situation. It is worth noting the advantages of economic agreements attracted Russian business, but not everyone is fully acquainted with them.

S. I. Ashmarina et al. (Eds.): ENGINEERING ECONOMICS WEEK 2020, LNNS 139, pp. 66–77, 2021.
https://doi.org/10.1007/978-3-030-53277-2_8

One of the companies set the goal of developing partnerships with Russia was Elitas, a Serbian company engaged in the production of advertising signs for domestic and foreign markets.

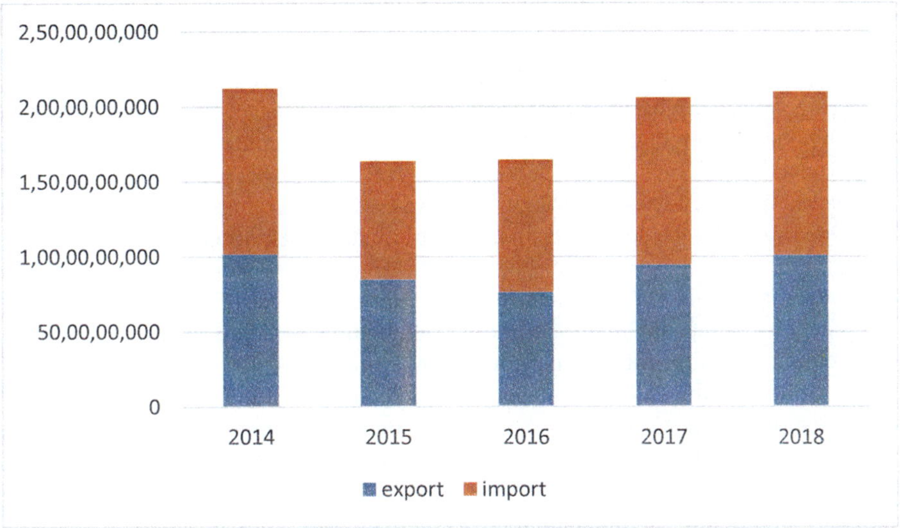

Fig. 1. Trade between Russia and Serbia in the "All Products" section, USD (Source: authors)

It is possible to find work on decorating gas stations, including LED screens, the appearance of large retailers and shopping centers, the internal and external equipment of the Belgrade airport, and other works in the company's portfolio. The company exports its products to 24 countries, mainly to European countries. The main competitive advantages of Elitas:

- favourable customs transport agreements,
- an available labour force,
- international quality standards: ISO 9001 [8], ISO 14001 [7] and OHSAS 18001 [13].

It allows the company to take a leading position the ratio "price-quality" on the market. Consequently, Elitas is attractive for the Russian advertising market, which today is experiencing clear difficulties.

The following negative phenomena inherent in the modern Russian advertising market are identified and reflected in the article by A.V. Shokhorov "Study of the characteristics of the advertising market in Russia" [16]:

1. The increase in the cost of advertising products due to the growth of the foreign component of materials for production.
2. The interruptions in the supply of materials for production associated with sharp surges in the exchange rate of the national currency.
3. The strong reduction in credit lines from suppliers, which leads to non-compliance with the deadlines for the execution of orders.

The above data demonstrate and prove the interest of both Serbian and Russian advertising industry enterprises in organizing mutually beneficial cooperation. The question remains open of choosing the form for the Serbian company Elitas to enter the Russian market.

The following classical forms of companies entering international markets are highlighted in works on foreign economic activity and international marketing by the authors Verbeke, Ciravegna, Lopez, and Kundu [18], Wirtz, Pistoia, Ullrich, and Göttel [19], Meyer and Peng [12], Mellahi, Frynas, Sun, and Siegel [11]: export, joint ventures, auction trading, international trading, international trading in licenses, international renting (leasing), franchising, foreign investment.

The works identified the advantages and disadvantages of the company's entry into the international market and then the authors express an opinion on which form is the most relevant and effective. Some of them give preference to such forms as franchising and joint ventures more often than others. So, for example, Korolev [10] notes that it is practically impossible to compete alone with small companies in highly competitive international markets, therefore, in his opinion, the most promising form of organization for cooperation is joint ventures. Thus, Elitas chose such a form of organization of interaction with Russian partners as the creation of joint ventures. An obligatory step in the step-by-step algorithm of organizing foreign economic activity is the stage of searching for partners. This stage is devoted to our research. The purpose of the present study is to search for business partners in the Russian market to establish joint ventures with the Serbian company Elitas.

2 Methodology

Initially, out of 86 constituent entities of the Russian Federation, regions with millionaire cities were selected. Then, the quantitative composition of manufacturers of advertising signs was determined in each of the selected regions according to AllCompanies. The data are presented in Table 1.

Table 1. Number of companies producing signs in different regions in 2018

Regions	Region number on the map (Fig. 2)	Companies producing advertising signs in the region (in pieces)	Million-plus cities in the region (population)
Moscow region	1	875	Moscow – (12.5 million people)
Leningrad region	2	342	St. Petersburg – (5.3 million people)
Novosibirsk region	3	65	Novosibirsk – (1.6 million people)
Sverdlovsk region	4	148	Yekaterinburg – (1.46 million people)
Novgorod region	5	72	Nizhny Novgorod – (1.25 million people)
Republic of Tatarstan	6	119	Kazan – (1.24 million people)
Chelyabinsk region	7	77	Chelyabinsk – (1.2 million people)
Omsk region	8	19	Omsk – (1.17 million people)
Samara Region	9	62	Samara – (1.17 million people)
Rostov region	10	104	Rostov-on-Don – (1.13 million people)
Republic of Bashkortostan	11	71	Ufa – (1.12 million people)
Krasnoyarsk region	12	34	Krasnoyarsk – (1.09 million people)
Perm region	13	40	Perm – (1.05 million people)
Voronezh region	14	43	Voronezh – (1.04 million people)
Volgograd region	15	53	Volgograd – (1, 013 million people),
Krasnodar region	16	142	Krasnodar – (1 million people)

Source: authors.

Logistics costs are high for joint ventures. In this connection, we have selected regions as close as possible to the Serbian border, as well as regions with the highest standard of living: Moscow region, Leningrad region, Sverdlovsk region, Rostov region, Krasnodar region and the Republic of Tatarstan. Although these regions are characterized by a high concentration of companies producing advertising signs (Table 1), which shows a high level of competition, we have chosen to do so. The high logistics costs will not allow the joint venture to establish competitive prices in regions with low levels of competition, which makes cooperation irrelevant.

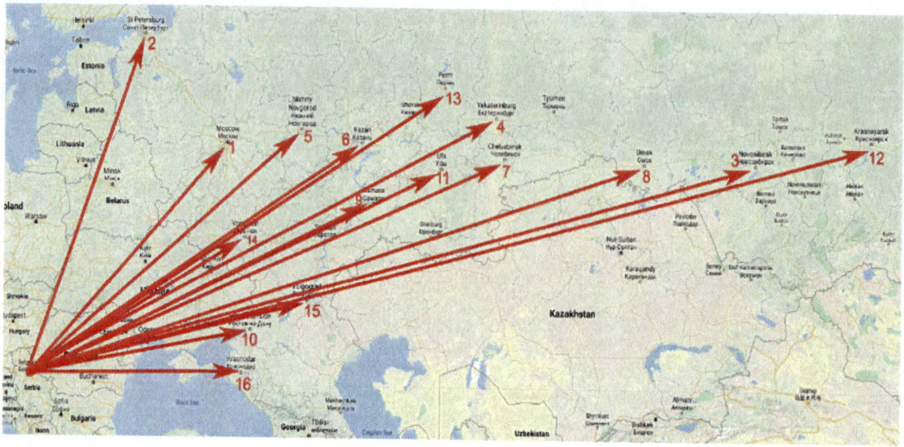

Fig. 2. Selected regions for possible cooperation on the map (Source: authors).

It is necessary to choose cities where large international and Russian networks are represented so that logistics costs are justified. This is directly related to population density and living standards (Fig. 2). Accordingly, such a solution in the future may become not a single supply from country to country with expensive logistics, but a planned order to signs more than one network point. In this case, companies take the selection of a contractor seriously. European quality certificates and a company portfolio with market leaders can be decisive factors. In this case, Elitas has a significant advantage over Russian contractors in the manufacture of various types of signs both indoor and outdoor.

The main competitors cannot provide the signs for all possible customer's orders in the regions, especially when it is necessary to execute an individual and varied request. Elitas can offer technical novelties and a large volume of production on a turn-key basis. If the customer needs different kinds of works, then the Serbian company can process everything in a «single window» and one term. This allows the customer to reduce the risk of the number of contractors for different types of work. One contractor - all types of work.

The contact stage with potential partners comes after the regions have been identified. To begin with, a content strategy was developed to familiarize potential partners with Elitas, its capabilities and its offer of cooperation. The content was compiled from the following parts:

- history of the creation and development of the company,
- main competitive advantages of the company,
- description scales of production activity,
- description of production technologies and capabilities,
- key customer (Gazprom, Metro, Subaru, etc.),
- large completed projects (design of the entire Belgrade airport, etc.),
- possible ways of cooperation between Russian partners with Elitas,
- basic offers for potential partners and possible points of contact (Russian company's entry into the international market).

The site has become a platform for posting content. The next task for Elitas was to identify the channels for contacting potential partners and demonstrate them the created content on the site and gather contacts.

We have reached the conclusion that the following channels for the search for business partners abroad are being prioritized by all listed authors (Karagulian [9], Korolev [10], Vasilieva [17], Balakireva [3], Dorn, Schweiger, and Albers [4]): use of informal connections, use of the Internet, written appeal by mail (fax), participation in conferences, tenders, exhibitions, fairs and participation in international economic forums. We used almost all of them: sending of e-mails to potential partners for cooperation; search for business partners on Internet platforms (such as Sberbank partner, Business partner, Partnersearch); participation in international exhibitions and fairs; participation in tenders; personal relations. Such a channel as participation in international economic forums was not used, since the programs did not fit on the dates.

3 Results

3.1 Official E-Mail to Potential Partners

A database of 48 signage companies based on analysis of the regions of the Russian Federation, which could become potential partners of Elitas for further joint realization of large orders, was created. Table 2 presents the number of possible partner companies by region.

Table 2. Number of possible partner companies relative to the city

City	Number of possible partner companies
Moscow	8
Kazan	4
Tomsk	2
Novosibirsk	2
Nizhny Novgorod	2
Krasnodar	4
Anapa	2
Goryachiy klyuch	2
Surgut	9
Tyumen	2
Ekaterinburg	11

Source: authors.

The following information was collected on the websites of potential partner companies:

- information about the company itself,
- location, provided products and services of the company,
- general contact details,
- contact details of the director, assistant directors or development department.

The next step was the preparation of an official letter summarizing the company Elitas, the offer of cooperation and the enclosed file with a presentation and a link to the site containing the above-mentioned content.

Table 3 presents the statistics of the feedback received.

Table 3. Statistics of feedback received from potential partners of Elitas during the organization of e-mail newsletters

Target action	Number of actions
Number of companies	48
Number of emails	52
Number of replies to emails	5
Number of proposal reminder letters	47
Number of responses since last letter	1
Total number of companies responding to letters	6
Number of companies that have stopped communication through correspondence	3
Number of companies with which further contact is agreed	2
Number of companies with a Skype conference was held to discuss details	1

Source: authors.

The following conclusions can be made:

1. The e-mail distribution of letters brought the company 6 interested contacts out of 52, which is 13.63%.
2. Agreements were reached on cooperation in the process of business correspondence with 3 of 6 potential partners (6.81%). More details:

- the first company is ready to take joint large orders in Russia,
- the second company is ready to supply part of its products to Europe,
- the third company is ready to consider Elitas as a supplier of signage for federal orders.

The result of the communication channel - official emails, Elitas acquired 3 partners out of 48 possible, which is quite good for the results of cold mailing. The proportion of letters read to letters that received replies is 92.3% to 13.63%. A growth point for this communication channel is improving the content of the presentation and site.

3.2 Internet Platforms for Business Partner Search

The way to find business partners on specialized Internet platforms was chosen as one of the channels. The cooperation information was added in each announcement: a link to the Elitas partnership site and presentation materials detailing the activities of the Serbian company and its vision of partnership with Russian companies.

The following sites were selected for testing the effectiveness of this channel: Sberbank partner (https://bankofpartners.com/en); Business partner (http://www.businesspartner.ru/) and Partnersearch (https://www.partnersearch.ru).

The following results were obtained after 2 weeks:

Table 4. The results of the effectiveness of online platforms for finding business partners from 02.15.19 to 03.01.19

Source	Feedback	Target feedback	Negotiating
Sberbank partner	4	0	0
Business partner	7	0	1
Partnersearch	2	0	0
Total	13	0	1

Source: authors.

The business platforms are active today (Table 4). The main participants in such resources are companies seeking sponsors for their own projects. A potential non-target partner was found with an attractive interaction option for Elitas - the SvetoShow project. The Russian side proposed to produce patented products at Elitas facilities. In the future, such developments can be used as a creative form of advertising. SvetoShow's patent portfolio includes test works for Sberbank and several other companies. In conclusion, the effectiveness of business platforms as a communication tool has been proven. Such a method is worth using when prospective sponsors are being sought, but it is not to be expected that the response will be targeted.

3.3 Participation in International Exhibitions and Fairs

It was suggested to participate in professional fairs as one of the options for finding potential partners. This is an opportunity for Elitas to present and showcase their products, as well as to find targeted contacts. First of all, the Russian exhibitions were analyzed, and international exhibitions with similar topics were also found.

Potential exhibitions are presented below in Table 5 (Date of access to the site-February 2019, the information is current on this date).

Table 5. Relevant professional fairs for Elitas

№	Name of the exhibition	Country	Description	Site
1	European Sign Expo (ESE)	Germany (Munich)	Europe's largest exhibition of printed signage and visual communications	https://ese.fespa.com/
2	Design and advertising Next	Russia (Moscow)	Exhibition of marketing communications "Design & Reklama NEXT" has established itself as one of the most important and interesting specialized b2b-forums in Russia	http://www.design-reklama.ru/eng/about.html
3	Reklama, Polygraf, Obaly	Czech Republic (Prague)	26th International Exhibition of Advertising, Printing, Packaging and Innovative Technologies	http://reklama-fair.cz/en/
4	Central Asia Reklam	Kazakhstan (Alma-Ata)	5th International Exhibition of Advertising, Printing, Signage Technologies and Materials	http://reklamexpo.kz/?lang=en
5	Store Promotion Expo Summer	Japan (Tokyo)	Japan's largest exhibition of marketing and advertising products and solutions	https://www.reedexpo.co.jp/en/Exhibitors/207292/Store-Promotion-EXPO-July
6	Exposign	Argentina (Buenos Aires)	18th International Exhibition of Visual Communication	https://www.exposign.com.ar/en/
7	Reklama	Russia (Moscow)	Top industry event for manufacturers and customers.	http://www.reklama-expo.ru/en/

Source: authors.

3.4 Tendering

Tendering is an option to find potential partners and direct clients. It is possible to get a large order by participating in tenders, which supposedly could recoup Elitas logistics and production costs. But these expectations did not materialize during the analysis. Antonov and Kiseleva [2], Akhmetova [1],Shinkevich, Misbakhova, Bashkirtseva, Fedorova, Martynova, and Beloborodova [15] wrote about the public procurement system, and this issue is also regulated by the Federal Law of 2013. So that the international company Elitas could take part in tenders of the Russian Federation, we contacted the Interregional Tender Center (ITC) and consulted with them. This organization has provided us with the following data:

1. Current tenders related to the manufacture of signs. The most attractive options for the sum of a million rubles for the order (3 tenders out of 15) were highlighted in this list.
2. Elitas has two ways to participate in Russian tenders: directly (as a Serbian company with a registered office and registration not in Russia) and through a counterparty. The second option is the most attractive, as the circulation of documents will be easier.
3. However, the market is oversaturated by companies engaged in the production of signs, so participation in tenders gives no more than 10–15% of the net profit from the order. Therefore, taking logistics into account, the cost of the final product will not be as high as expected.

After examining all the tender information, it was concluded that this is a possible but rather complicated way of obtaining orders in Russia. In order to simplify the procedure, it was decided to enter into a partnership with Russian companies (similar to Elitas) that participate in the tenders. And in the future to realize a large order jointly.

3.5 Personal Relations

Personal acquaintances are a widespread tool of promotion in Russia.

The following can be attributed to personal interaction channels:

- personal account on Instagram/Vkontakte or other social networks,
- personal connections and contacts with people who may be interested in the Elitas offer.

Personal Instagram accounts showed the following results.

This communication tool can be considered relevant and effective. We speak about quick response, trust and careful examination of the conditions of the partnership. So, this channel is also targeted (Table 6).

Table 6. Efficiency of Personal Contacts channel

First step	Feedback	Process	Conclusion
2 stories posted on two Instagram personal accounts with Russian audience	2 contacts of Decision Makers have been received	Correspondence with potential partners	1 target contact received

Source: authors.

4 Discussion

The company's entry into the international market is a very complex issue. There are many ways of finding partners and organizing cooperation. The choice depends on the industry in which the company operates, the participating countries and the specific political and economic relations between them. In this regard, the forms of international cooperation described in the literature demonstrate varying effectiveness in each case. The example of Elitas company tested some of the methods in this article. In practice, the most effective methods have been those described more frequently in the scientific periodical literature. Thus, the set of methods proposed in the article is not universal, but the most effective in most scenarios of organizing international cooperation.

5 Conclusion

In conclusion, we can draw the following conclusion:

1. The Serbian market is attractive to Russia. International partnerships are mutually beneficial.
2. Targeted cold mailing of official emails to potential partners gives a result and can be considered a working tool. Improving the content of the presentation and the site is the main condition for using such a channel.
3. The performance of business platforms as a communication tool has been proven. The drawback and at the same time the advantage of this channel is the unearmarked feedback can be potentially useful.
4. Participation in professional fairs can be seen as a channel of communication. The main condition is the convergence of topics and interests of the fair.
5. Participation in Russian tenders is a difficult way for international companies, which makes this channel of communication time-consuming and costly.
6. Personal connections can be considered the most targeted communication channel. It's about trusting relationships, understanding personal relationships, and reacting quickly.

References

1. Akhmetova, L.Ya.: Public procurement system (2010). http://www.audit-it.ru/news/finance/207237.html. Accessed 03 Mar 2019
2. Antonov, V.I., Kiseleva, O.V.: Foreign experience in regulating the placement of state orders and the possibility of its use in Russian practice. Mod. Probl. Sci. Educ. **3** (2013). http://www.science-education.ru/ru/article/view?id=9274. Accessed: 03.03.2019
3. Balakireva, S.M.: Principles and channels of searching for new foreign trade partners in a crisis. Russ. Foreign Econ. Bull. **4**, 63–72 (2016)
4. Dorn, S., Schweiger, B., Albers, S.: Levels, phases and themes of coopetition: a systematic literature review and research agenda. Eur. Manage. J. **34**(5), 484–500 (2016). https://doi.org/10.1016/j.emj.2016.02.009
5. Dudic, B., Smolen, J., Dubic, Z., Mirkovic, V.: Support for foreign direct investment inflows in Serbia. Econ. Ann. XXI **169**(1–2), 4–11 (2018). https://doi.org/10.21003/ea.V169-01
6. Iskenderov, P.: The role of diplomacy in the history of the Balkans. Hist. Questions **4**, 114–126 (2018)
7. ISO 14001. https://www.iso.org/iso-14001-environmental-management.html. Accessed 15 June 2019
8. ISO 9001. https://www.iso.org/iso-9001-quality-management.html. Accessed 15 June 2019

9. Karagulyan, E.A.: The work of the enterprise to enter the foreign market. Marketing research. Strategies for the company to enter the foreign market. Progress Group, Moscow (2015)
10. Korolev, V.I.: Problems of organizing a company's entry to the foreign market. Russ. Foreign Econ. Bull. **11**, 70–77 (2016)
11. Mellahi, K., Frynas, J., Sun, P., Siegel, D.: A review of the nonmarket strategy literature: toward a multi-theoretical integration. J. Manage. **42**(1), 143–173 (2016). https://doi.org/10.1177/0149206315617241
12. Meyer, K., Peng, M.: Theoretical foundations of emerging economy business research. J. Int. Bus. Stud. **47**, 3–22 (2016). https://doi.org/10.1057/jibs.2015.34
13. OHSAS 18001. https://www.lr.org/en-gb/ohsas-18001/?AspxAutoDetectCookieSupport=1. Accessed 15 June 2019
14. Ru-stat: Trade between Russia and Serbia « all products » (2018). https://en.ru-stat.com/date-Y2014-2018/RU/trade/RS. Accessed 15 June 2019
15. Shinkevich, M., Misbakhova, C., Bashkirtseva, S., Fedorova, T., Martynova, O., Beloborodova, A.: Institutional factors of micro, mezzo and macro systems' innovative development. J. Adv. Res. Law Econ. **8**(1), 229–236 (2017). https://doi.org/10.14505/jarle.v8.1(23).26
16. Shokhorov, A.V.: Study of the Characteristics of the Advertising Market in Russia. Bulletin of the Pskov State University. Series: Economics, Law, Management, vol. 1, pp. 92–100 (2015)
17. Vasilieva, A.: Search for foreign trade partners in the context of geopolitical tensions. Actual Probl. Innov. Econ. **13**, 213–218 (2016)
18. Verbeke, A., Ciravegna, L., Lopez, L.E., Kundu, S.: Five configurations of opportunism in international market entry. J. Manage. Stud. **56**(7), 1287–1313 (2019). https://doi.org/10.1111/joms.12355
19. Wirtz, B., Pistoia, A., Ullrich, S., Göttel, V.: Business models: origin, development and future research perspectives. Long Range Plann. **49**, 36–54 (2016)

Importance of International Entrepreneurship Skills Among MBA Students: Global Comparative Study

E. Velinov[1,2], S. I. Ashmarina[3], and A. S. Zotova[3(✉)]

[1] Skoda Auto University, Mlada Boleslav, Czech Republic
emil.velinov@savs.cz
[2] RISEBA University of Applied Science, Riga, Latvia
[3] Samara State University of Economics, Samara, Russia
asisamara@mail.ru, azotova@mail.ru

Abstract. The paper tries to investigate what are the most important skills among MBA students, who are starting transnational entrepreneurship. In the context of globalization, the borders of the world for the professional career of students are open. Therefore, it is important to develop students' skills of multicultural communication, which will be promoted by the development of international relations through exchange programs, summer school programs, internships, and cooperation with international companies. Another important factor is economy development and the facts prove that both national and world economy development basically based on international entrepreneurial activities. The paper focuses on online technologies and blended learning methods as the ways of intensifying the educational practices in the issue. For the purpose of the paper, there have been conducted more than 400 questionnaires with MBA students in selected countries with emerging economies for the period of 2017-2018. Paper results indicate that flexible approach to international on-line project teams design helps students understand regulation ways of their on-line networking procedures and coordinate their activities with other participants of their distant team.

Keywords: Entrepreneurship · International business · Knowledge transfer · Networking

1 Introduction

Entrepreneurship has a social potential because it encourages to take into account the needs of potential customers (thereby improving their conditions, at least in principle), and also because doing business requires the creation and management of a team of people and the creation of jobs. Entrepreneurship can also promote social cohesion by increasing employment, economic rewards and job satisfaction. However, entrepreneurial education is still relatively immature and rarely adequately considered at the strategic level by universities or national policies. Kyrö [5] focuses on such features of entrepreneurship as the challenge of finding and grabbing the opportunities, setting

S. I. Ashmarina et al. (Eds.): ENGINEERING ECONOMICS WEEK 2020, LNNS 139, pp. 78–84, 2021.
https://doi.org/10.1007/978-3-030-53277-2_9

start-ups and their development, adopting risks and allocation of resources in order to find your segment at the market.

Many countries around the world are making efforts to modernize their human resources development systems in order to bring them into line with national and global economic requirements. Education and training systems are usually managed by government agencies, and most countries provide a framework for education and business cooperation in their legislation or strategies. Examples of educational and business cooperation in the field of higher education are in many countries. Factors potentially stimulating such cooperation: Presence of multinational companies; Technological advances in industry; Educational programs.

The concept of the "triple helix" focuses on the distinctive role of the university in economic and innovation development through more close cooperation of three main powers: university, business and government.

What are the relationships between the components of the triple helix?

- technology transfer,
- collaboration, leadership and moderation of conflicts,
- network interchangeability.

Universities are increasingly generating and transferring technology. Being previously considered as the source of humanresources and knowledge now they have got the status of key stakeholders of innovation, with constantly growing internal organizational mechanisms and resources. The World Economic Forum global competitiveness report highlights such future competences as design thinking, communication (especially crosscultural one) and entrepreneurial skills to be the core competences for the competitive professional [9]. The issues of development of the international entrepreneurship education concept in Central and Eastern Europe inspired the authors for making the research.

2 Methodology

For the purpose of this applied research the learning cycle of Kolb as a method in entrepreneurship studying was used, also some methods of statistical analysis and expert analysis were implemented. We suggest that entrepreneurship education for cross-border entrepreneurship can apply several modifications of the experiential learning cycle. Stages of the research work:

- the first stage included the theoretical analysis of various approaches to international entrepreneurial education at the targeted countries,
- the second stage was performed through monitoring and statistical analysis of master students interviews about their entrepreneurship skills,
- the third stage was focused on blended learning technology as the instrument of international entrepreneurial education intensification.

3 Results

Cooperation between universities and business environment in the sphere of entrepreneurship might be performed in several functions:

- training of entrepreneurs in order to acquire skills in the field of digital marketing, innovative, financial, operational management, business planning, legal regulation in the field of copyright protection, etc.,
- generation of ideas and team building, preparation for participation in relevant contests,
- providing a variety of services with the assistance of experts in business planning, product design, marketing research, motivation system, legal support,
- organization of interaction with investors, including venture funds, business angels, strategic partners, participation in international events, interaction with funds.

The university participates in the formation of the business environment and interacts with the business, creating open innovation programs for various sectors of the economy or certain business representatives. The result of international cooperation in the entrepreneurship education is the European project EU-XCEL for startups. During the week, intensive entrepreneurship training is conducted for participants in business incubators of six European universities in Ireland, Denmark, Germany, Greece, Poland, Spain. This is a summer school for startups. In accordance with the terms of the project, 30 students are sent to foreign universities, 30 students come to Athens. Each team must include at least three countries. Teams work for 12 weeks under the supervision of one coach. Every second week, results and plans are discussed. The program was funded by the European Union and the program was completed in 2017.

Various approaches to international entrepreneurship education summarize main competences which are necessary for successful entrepreneurship at the global market. Morris, Neumeyer, & Kuratko [8] explain that potential entrepreneurs should possess primarily business competencies, since the functional responsibilities of specialists are associated, first of all, with the development of the university's affiliate network, negotiating and concluding transactions, forming value offers that are competitive in the eyes of business partners. Ramoglou and Zyglidopoulus [10] continue discussion of professional readiness to recognize your business opportunity and develop it into the efficient and effective business. Learners also need reflective learning based on gamification methods or diversionary analysis in order to recognize critical events [6] that reflect risk and uncertainty related to transnational business. Kwong & Thompson [4] have pointed out that students focused often prefer to learn from experience of successful entrepreneurs, seek for future financing of their ideas and set the network before starting their own business in practice. The global comparative study in the paper was based on the collected 405 answers from the master students from the selected countries with the applied regression analysis as follows:

- Number of strata = 1
- Number of obs = 405
- Number of PSUs = 8

- Population size = 405
- Design df = 7
- F(7,1) = .
- Prob > F =
- R-squared = 0.0828

Then, after applying the correlation analysis in the accumulated countries analysis (Table 1) we have come up with interesting results that the master students in Estonia posses more comprehensive entrepreneurship skills than the rest of the countries.

Table 1. Regression analysis of master students entrepreneurship skills (Linear regression)

Country	Param.	Std. err	.t	P > |t|	[95% Conf. interval]
Estonia	.0226114	.1219853	.0.19	0.858	−.2658381 .3110608
Poland	−.1394425	−.1394425	.−1.50	0.176	−.3587661 .0798811
France	.09363	.09363	0.90	0.400	−.153361 .340621
India	.1285541	.1285541	0.95	0.372	−.1899028 .447011
Israel	−.2127932	−.2127932	−1.78	0.118	−.4947808 .0691944
S.Korea	−.0934428	−.0934428	−0.81	0.445	−.3667347 .1798491
Russia	.1437891	.1288485	1.12	0.301	−.1608892 .4484674
Czech Republic	.0666806	.3081904	0.22	0.835	−.6620739 .7954351
_cons	3.971759	. 3.12789	1.27	0.245	−3.424527 11.36804

Source: authors.

Liu [7] has demonstrated that network-learning has to deal with experience, dialogue, articulation of needs of learners, and pollination, meaning combination of diversified knowledge sources. Online networking offers opportunities for cognitive diversity that enables innovative learning, but it can also lead to communities where like-minded persons repeat and reinforce the narrow-minded views of each other. This danger is relevant for closed communities that do not have people in the role of "gatekeepers" looking for ideas from other communities. Widespread peripheral participation helps "gatekeepers" to transfer information between different communities. Face-to-face learning in classroom can be interactive knowledge sharing and team building experience for students involved but convergence of e-learning ad face-to-face learning is an essential tool for increasing international networking and knowledge sharing opportunities with students from different cultures [2]. Blended learning has mainly been defined as systems combining face-to-face and on-line educational approaches. Blended learning can be arranged as a multi-stage process rather than a single time and place bounded event in order to encourage students to be independent learners outside the classroom Bidder et al. [1] stress that blended learning is a flexible approach that overcomes time, location and cultural constraints. Blended learning involves combination of face-to-face classroom learning activities with asynchronous and/or synchronous online learning. In the context of cross-border entrepreneurship readiness it is important that blended learning helps students to develop their networking skills and get familiar with students in other countries. There are several questions to answer, when searching for blended learning options that enhance cross-border learning readiness:

1. Is blended learning focused on conceptual knowledge sharing or does it also involve learning by doing and reflection of joint experience?
2. How learner-centred is the learning process – how independent are students to search online knowledge sources and to involve other learners to virtual or class-room teams?

In 2019–2023 global EdTech will grow at more than 5% per year. Today it is measured in the amount of about $ 165 billion. Eastern Europe is in among the fastest growing regional markets in online education. In turn, the driver of Eastern Europe is Russia. In the most conservative scenario, the average annual growth of the Russian

Online education in the next five years will be at 20%. The fastest growing companies involved in digital educational solutions are those based on game mechanics (+22.4% per year until 2021) and on the simulation of real processes (+17%). But our X-Culture virtual teamwork practice has also indicated that experience of using online collaboration platforms is not similar in different countries and universities. And blended - learning is the best variant for some educational areas. In Russia blended learning is characteristic primarily for long-term, comprehensive programs involving the acquisition of complex professional skills.

Another way of implementing blended-learning approach and it's especially essential for entrepreneurial education is performing it in mixed aged groups with mature professionals in some spheres and young students. This way young people being digital natives might share their experience of online networking and social media activities with older generation and older generation might share their offline experience of communicating with investors, developing business in real economic environment.

Globalisation trends also influencing greatly in composing national approaches to entrepreneurial education. It is not self-evident any more that all students will start an export-oriented business in the country of their birth and/or education without involving to their start-up teams founders from other countries or transfer his/her enterprise to a new location outside the country, where the business was at first launched. Digital entrepreneurship through innovative business models was greatly influenced by the global trend of sharing economy and they changed greatly the established rules at many modern markets [3, 11].

4 Discussion

Several researches studying the experience of project-based work-integrated learning for international students presented evidence that such projects are valuable both for developing conceptual and problem-solving skills and for self-management in a new business environment. Erasmus student exchange in Europe has created opportunities for educational practices that support learning about cross-border entrepreneurship and for entrepreneurship, although involvement of international students in the longer learning cycle than one semester is more suitable for learning through a student enterprise practice in order to contribute to more stages of enterprise creation and development processes. Cross-border networking skills increase readiness for entrepreneurial project work in international teams and ability to understand added value of team members from other cultures and business environments. Understanding knowledge sharing and project work style differences in face-to-face and online communities is also an important learning outcome of networking practices. Networking is a tool for creating knowledge base for ambitious entrepreneurship initiatives that are driven by diversity of creative ideas and may create scalable and sustainable start-up opportunities that are not limited to local markets and locally available knowledge, human and financial capital.

5 Conclusion

The current goal of any university if it's going to stay competitive at the educational market is to identify and develop young leaders based on experience in the development and implementation of real student entrepreneurial projects, including social, innovative, educational ones. Main activities which can be performed are:

- attracting active students to enter international entrepreneurial programs,
- development of students' leadership skills, business communication skills, ability to work in an independently created team,
- development of entrepreneurial and critical thinking,
- gaining experience in creating and managing real projects,
- creation of a mechanism for the replication of successful projects in the social, innovative and educational fields,
- creation of a platform for the exchange of experience between beginners and established entrepreneurs,
- formation of career opportunities for students: access to the best employer companies interested in leading students,
- creation of a student communication platform to discuss the possibilities of implementing entrepreneurial initiatives.

University students develop and implement projects to solve pressing social, innovative, educational problems, improve well-being and living standards of the population in the regions. During the implementation of their projects, students develop project, social and entrepreneurial competencies and among them the ability to address to adequate online resources or social media resources and networks as well as intercultural communication and transnational cooperation might be the most essential competences. Vakkayil and Chatterjee [12] describe four globalization routes of business schools: moving from local distinctiveness to global patterns, infiltration of global norms, global expansion and moving from global patterns to pursue distinctiveness. Implementing entrepreneurial education in regional communities of Russian Federation or in small economies like Czech or Estonian ones does not allow to implement the global expansion strategy but the development of EdTech or MOOCs and concentration on trans border contacts might open new business opportunities for students in the areas. For example, the Volga region is traditionally a contact zone of

Indo-European Slavic, Finno-Ugric and Turkic-speaking peoples, which represents miniature model of Eurasia. The region is multi-confessional character, which allows us to consider its experience as an example of successful integration of various crosscultural practices. The significant feature of the Samara region in the Volga region is the role of transfer region - the territory of economic, legal, cultural, linguistic communications between the West and the East. The cross boarder educational program, aimed at studying Eurasian cultural and economic integration, will allow accumulating and broadcasting the experience of intercultural interaction in the region in the educational space of the countries of Eurasian continent.

Acknowledgments. The publication was inspired by the project 574060-EPP-1-2016-KZ-EPPKA2-CBHE-SP "Enhancement of Higher Education and Corporate Sectors Integration in Accordance with New Social Environment" (ENINEDU) co-funded by the Erasmus + Program.

References

1. Bidder, C., Mogindol, S.H., Saibin, T.C., Andrew, S.A., Naharu, N.: Students' perceptions of blended learning and achievement. In: Luaran, J.E., Sardi, J., Aziz, A., Alias, N.A. (eds.) Envisioning the Future of Online Learning, pp. 213–225. Springer, Singapore (2016)
2. Bleicher, J., Forrester, P., Honal, A., Velinov, E.: Virtual business projects in the classroom: enhancing intercultural and business skills of students. In: Gonzalez-Perez, M., Lynden, K., Taras, V. (eds.) The Palgrave Handbook of Learning and Teaching International Business and Management, pp. 747–768. Palgrave Macmillan, Cyam (2019)
3. Kelley, D., Singer, S., Herrington, M.: Global entrepreneurship monitor. 2015/16 global report (2016). http://www.gemconsortium.org/report/49480. Accessed 03 Jan 2020
4. Kwong, C., Thompson, P.: The when and why: student entrepreneurial aspirations. J. Small Bus. Manage. **54**(1), 299–318 (2016)
5. Kyrö, P.: The conceptual contribution of education to research on entrepreneurship education. Entrepreneurship Reg. Dev. **27**(9–10), 599–618 (2015)
6. Lindh, I., Thorgen, S.: Critical event recognition: an extended view of reflective learning. Manage. Learn. **47**(5), 525–542 (2016)
7. Liu, R.: Management learning in business networks: the process and the effects. Manage. Learn. **46**(3), 337–360 (2015)
8. Morris, M.H., Neumeyer, X., Kuratko, D.F.: A portfolio perspective on entrepreneurship and economic development. Small Bus. Econ. **45**(4), 713–728 (2015)
9. Penaluna, K., Penaluna, A., Usei, C., Griffiths, D.: Enterprise education needs enterprising educators: a case study on teacher training provision. Educ. Training **57**(8/9), 948–963 (2015)
10. Ramoglou, S., Zyglidopoulos, S.C.: The constructivist view of entrepreneurial opportunities: a critical analysis. Small Bus. Econ. **44**(1), 71–78 (2015)
11. Richter, C., Kraus, S., Syrjä, P.: The shareconomy as a precursor for digital entrepreneurship business models. Int. J. Entrepreneurship Small Bus. **25**(1), 18–35 (2015)
12. Vakkayil, J., Chatterjee, D.: Globalization routes: the pursuit of conformity and distinctiveness by top business schools in India. Manage. Learn. **48**(3), 328–344 (2016)

The Process Approach to Cost Management in Project-Oriented Enterprises of Engine Building

I. Naugolnova(✉)

Samara State University of Economics, Samara, Russia
naugolnovaia@mail.ru

Abstract. Improving the competitiveness of domestic enterprises of engine building is an urgent task in the country. The key factors of competitiveness are the cost and quality of products. The purpose of the article is to justify the possibility of applying a process approach to cost management in project-oriented enterprises of engine building. This approach allows you to distribute direct and indirect costs per unit of output, this allows you to correctly justify its cost and market price; optimize the value chain, eliminate unnecessary time and other resources; to increase the qualitative and quantitative indicators of products manufactured by engine building enterprises. The proposed project and process approach to cost management at the enterprise is based on modern systems and information technology for managing quality and business processes. The article proposes a scheme for implementing the project-process approach to cost management at the enterprise, describes the key rules for its implementation, and identifies its key advantages.

Keywords: Cost management · Information technology of process management · Process management · Process approach · Project management · Project and process approach

1 Introduction

Engine building in Russia is one of the leading manufacturing industries. A significant share of engine manufacturing enterprises is high-tech. High-tech products (innovative products, the production of which uses the results of research and development work corresponding to the priority areas of development of science, technology and technology of the Russian Federation) [2] are closely associated with a significant level of costs for their production.

This fact necessitates increasing control over the level of costs throughout the production cycle of products (works, services), improving and expanding the methods and tools used to manage costs. Proper management of innovation costs, taking into account the particular development of industries, will reduce the risks of innovation; improve the competitiveness of domestic enterprises.

An important tool for cost management in the enterprise engine is a competent system for accounting and cost allocation by product. Traditional approaches to cost management in industrial enterprises have ceased to meet modern requirements.

S. I. Ashmarina et al. (Eds.): ENGINEERING ECONOMICS WEEK 2020, LNNS 139, pp. 85–91, 2021.
https://doi.org/10.1007/978-3-030-53277-2_10

It happened because has changed the ratio of overhead costs. In the days of Henry Ford, it was the interest, and now can exceed half the cost.

As in the allocation of direct and indirect costs, and the application of the method direct-costing unit costs per unit of output are distorted. It doesn't justify the cost of the final product and establish a competitive price on the market. Large industrial companies around the world are currently increasingly widespread method of cost control based on process-oriented accounting [1, 6].

The essence of the process approach to cost management is that the product (product) is considered as a complex of interrelated processes (functions) from the point of view: usefulness, effectiveness, necessity, etc. The costs of the enterprise are distributed by function, process, and then by type of product according to the principle of involvement.

The basis of this approach rests the method of Activity-Based Costing. In the United States rapid spread of this method became in the 1980's thanks to the work of Johnson [4], Kaplan [7]. This method for the 1990-th years used about 10% of large companies, including USA, UK, continental Europe, Australia, started using it in Japan [10]. This wide acceptance of this method is obtained thanks to the possibility of its application not only in industrial but also in the trade and service industries [12].

In Russia, a partially similar method, a functional-cost analysis (hereinafter referred to as FCA), has developed almost simultaneously in parallel. This method fits perfectly into the concept of process cost management in industrial enterprises, in particular in the production of innovative products. It allows you to:

- increase the number of functions and usefulness of an innovative product, improve its quality,
- identify the correspondence between the significance of the function performed by the product and the costs of its creation,
- eliminate unnecessary functions, therefore, operations or individual processes,
- rationally and cost-effectively set the selling price for this product, taking into account the functions included in it,
- activate the creative thinking of the team,
- upgrade production technologies,
- increase the competitiveness of products and the enterprise as a whole.

The use of functional-cost analysis is aimed at minimizing costs at the stages of project, production and operation of the product. Practice shows that companies using this method of analysis were able to reduce production costs by 20–25% [3].

The result of the FCA should be a reduction in costs per unit of beneficial effect, which is achieved by: reducing costs while improving the consumer properties of products; improving product quality while maintaining the level of costs; cost reduction while maintaining the level of quality; cost reduction with a reasonable reduction in technical parameters to their functionally necessary level.

The introduction of a process approach to cost management is fraught with significant difficulties. For an engine company, project management is an urgent need, as these enterprises belong to the so-called "project-oriented" organizations, whose activities are closely associated with the implementation of complex innovative projects for the development and organization of production of new engines and components for aircraft. At the same time, the application of the project approach is often insufficient, because project documentation describes the goals, but it does not pay enough attention to reflect changes in the business processes of the organization. Given the project focus of the engine building enterprises, as well as the proven effectiveness of the process approach to cost management and the enterprise as a whole, the task is to justify the possibility and effectiveness of the project and process approach to cost management in the engine building enterprises.

2 Methodology

Modern process control systems are based on the following main approaches:

- TQM (Total Quality Management) – a system of universal quality management,
- PIQS (Process Integrated Quality System) – a quality management system integrated with business processes,
- ISO standards 9000 series - standards governing the requirements for quality management systems [3],
- BPMS (Business Process Management System) – a business process management system,
- ERP (Enterprise Resource Planning) – a comprehensive system for planning and managing the resources of the organization.

Structural and functional modeling, which allows to visually describe a static system of business processes, is performed using modeling standards (SADT, IDEF, DFD) and software tools (BPWWin, Business Studio, ARIS).

Over the past decade, information technology has developed significantly, a number of tools for process simulation have been developed: tools for discrete-event simulation (Service Model, SimProcess, etc.); dynamic modeling tools (ReThink, PowerSim, ItThink, etc.); simulation tools based on flowcharts (Process Charter, Optima, ARENA, etc.).

3 Results

The main differences of the project and process approach to cost management are:

- main and auxiliary processes are allocated for each project separately, owners of these processes are appointed,
- the resources, parameters and consumers of processes are determined, process inputs and outputs,
- formalization of process control rules is carried out within the project in order to maximize customer satisfaction at the exit.

The advantages of the project-process approach to cost management at the enterprises of engine building include the following:

- within the project it is easier to ensure the coherence of the actions of process performers, to optimize the value stream,
- the project-process approach to cost management at the enterprise harmoniously fits and is easier to implement in the already existing matrix project management system at the engine-building enterprise.

An important factor in the application of the project and process approach to cost management in the enterprise is the determination of the sequence of stages of its implementation. The author proposes the following scheme for the implementation of the project-process approach to cost management at the enterprise (Fig. 1).

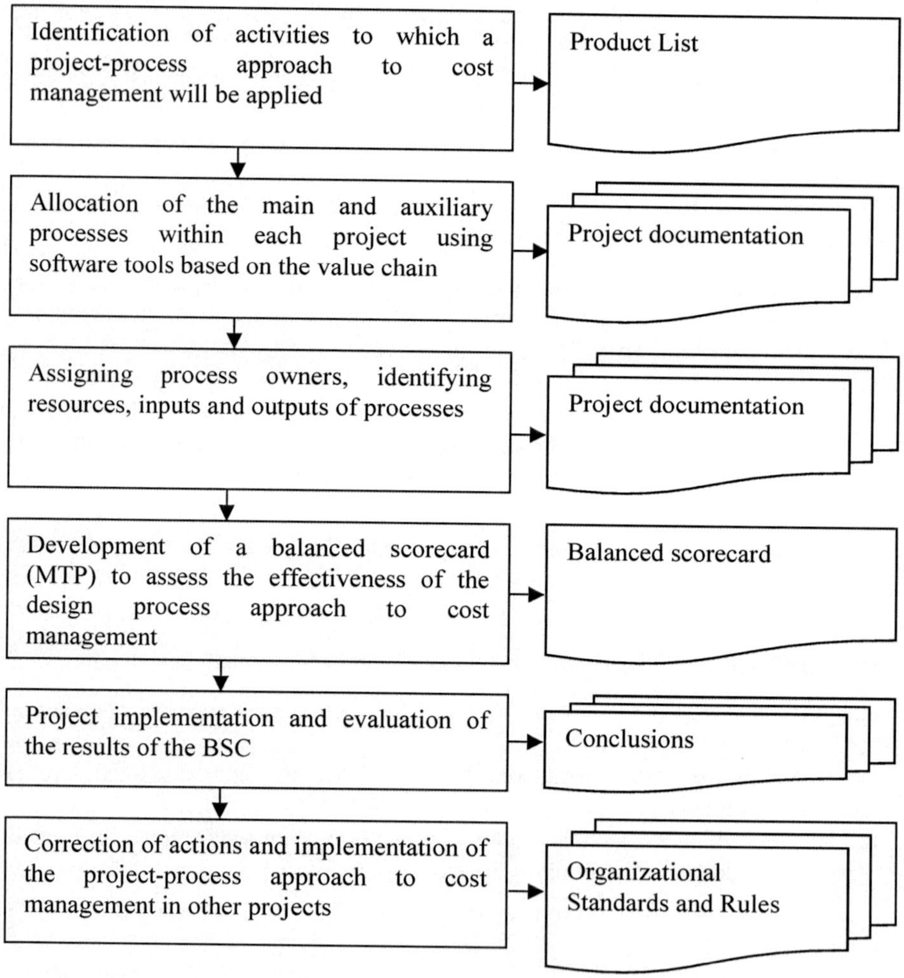

Fig. 1. Organization of the project and process approach to cost management in the engine building enterprise (Source: author).

It is advisable to apply the project and process approach to cost management to all ongoing projects at the enterprise, testing the approach on one or two relatively short-term projects. When selecting processes, it should be borne in mind that the main processes should be no more than 7 ± 2. This figure comes from the fact that the top leader, like any person, cannot effectively manage and perceive information from more main areas of activity. The number of auxiliary processes should not be more than 5 ± 2. Otherwise, the top manager loses control of the organization for the same reason.

The main business processes should be represented as a value chain, highlighting the main stages of creating results. It is important to start at the end of the chain: from the result provided to the consumer. Results can be created by one or different value chains, while chains at different stages can be combined and disconnected.

A general example of a value chain on the example of an industrial enterprise, including an engine manufacturing enterprise producing innovative products, is presented in Fig. 2. The consolidation of process owners should be documented, which will improve the efficiency of process management within the project. The balanced scorecard should include economic and non-economic indicators. To assess the effectiveness of the cost management system, economic indicators must include, for example, the cost of products, the cost of a ruble of marketable products, the share and amount of direct and indirect costs. Non-economic – the duration of the production cycle, tact time, indicators of product quality.

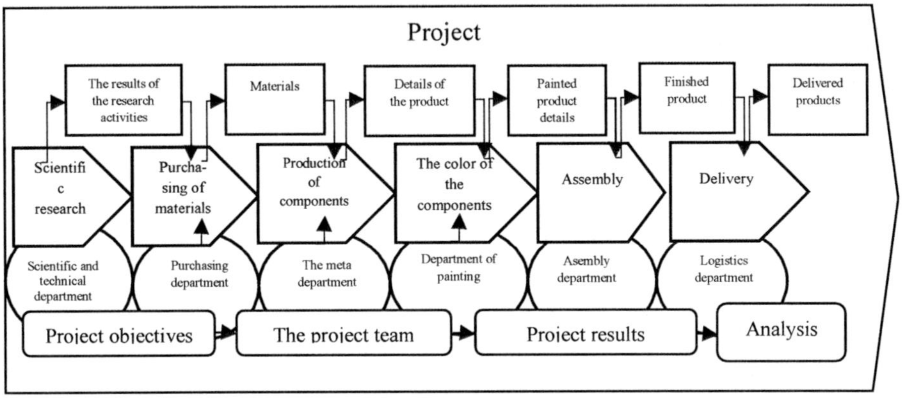

Fig. 2. The value chain of the project on the example of an industrial enterprise (Source: author).

According to the results of the project, conclusions should be drawn based on a comparison of planned and actual indicators for a reasonable system of balanced indicators. Based on the findings, adjustments are made with the further implementation of the project-process approach to managing costs on other projects. Process modeling is effectively carried out using modern information technologies. The most popular in Russia were the ARIS software product, mainly due to the ease of use.

4 Discussion

The introduction of a project and process approach to cost management is justified and feasible in project-oriented enterprises, with an already established and not amenable to adjustment in the short term organizational structure. The project approach to the implementation of core business at the engine-building enterprises has shown its effectiveness. It allows you to significantly improve the qualitative and quantitative indicators of the project (produced products) by securing responsibility for specific performers. The project approach has been widely used for several decades both at domestic and foreign enterprises [11].

However, it does not allow solving the problem of calculating the cost of a unit of production with a significant share of overhead indirect costs. Justify the involvement of overhead costs to specific types of products just allows the process approach to cost management. That is why the symbiosis of the project and process approach in such cases is justified and effective.

These judgments are confirmed by the work of Mikhailova & Sbitnev [9]. The main strengths of the project process approach to cost management in the enterprise are:

- interest of project managers in its high efficiency,
- simplification of internal relationships between departments and individual performers due to documenting process owners and concretizing performers, which will increase the flexibility of the project as a whole [8],
- acceleration of decision-making (for the same reasons),
- improving project performance by limiting its scope, setting specific goals, and concretizing the list of participatory processes,
- increased responsibility for the performance of the project not only of its leader, but also of the owners of all involved processes,
- control of the project budget, reduction of costs for its implementation [5],
- the correct distribution of indirect costs per unit of production, which allows you to correctly justify the cost and market price of the product, as well as develop a marketing policy of the enterprise.

5 Conclusion

The project and process approach to cost management in the enterprise proposed in the article is relatively similar to the process approach, but in contrast, it is used within individual projects of the enterprise. This greatly simplifies the implementation and application of this approach, does not require a significant reorganization of internal corporate relations and organizational structure. It harmoniously fits into the matrix system of project management at the engine-building enterprise.

This solves the key problem of the process approach to cost management in the enterprise - the problem of its implementation. The company has the opportunity to test the process approach to managing costs for individual projects, bring it to a high level and then scale it to other projects. The proposed project and process approach to cost management at engine building enterprises is copyrighted and is currently being implemented at one of the leading engine building enterprises in Russia (PJSC Kuznetsov). The main advantages of the project-process approach to cost management at the enterprise is that it allows you to correctly distribute direct and indirect costs per unit of production, which allows you to correctly justify its cost and market price; optimize the value chain, eliminate unnecessary costs; to increase the qualitative and quantitative indicators of the project as a whole.

References

1. Duran, O., Pereira Afonso, P.S.L.: An activity based costing decision model for life cycle economic assessment in spare parts logistic management. Int. J. Prod. Econ. **222**, 107499 (2019)
2. Iskoskov, M.O.: Problems of cost management and methods of their reduction in the conditions of innovative economy. Kazan Sci. **9**, 60–64 (2011)
3. ISO 9000. Quality management principles and guidelines on their application. Fort Myers, Florida. https://docplayer.net/9314898-Iso-9000-quality-management-principles-and-guidelines-on-their-application.html. Accessed 12 Feb 2020
4. Johnson, T.: Activity-based information: a blueprint for world class management accounting. Manage. Account. (June), 23–30 (1988)
5. Jørgensen, L., Åsgård, T.: Trust and control in project management. Procedia Comput. Sci. **164**, 397–406 (2019)
6. Kaiser, M.J.: Activity-based costing. In: Kaiser, M.J. (ed.) Decommissioning Forecasting and Operating Cost Estimation, pp. 489–522. Gulf Professional Publishing, Houston (2019)
7. Kaplan, R.S.: One cost system isn't enough. Harvard Business Review (1988). https://hbr.org/1988/01/one-cost-system-isnt-enough&cm_sp=Article-_-Links-_-Top%20of%20Page%20Recirculation. Accessed 12 Feb 2020
8. Loiro, C., Castro, H., Ávila, P., Cruz-Cunha, M.M., Putnik, G.D., Ferreira, L.: Agile project management: a communicational workflow proposal. Procedia Comput. Sci. **164**, 485–490 (2019)
9. Mikhailova, E.O., Sbitnev, S.N.: Project and process approach to management at an aircraft engine company. Proceedings of MAI **67**, 23–37 (2013)
10. Tibor, L.C., Schultz, S.R., Menaker, R., Weber, B.D., Young, P.M.: Improving efficiency using time-driven activity-based costing. J. Am. College Radiol. **14**(3), 353–358 (2017)
11. Toledo, R.F., Junior, H.L.M., Filho, J.R.F., Costa, H.G.: A scientometric review of global research on sustainability and project management dataset. Data Brief **25**, 104312 (2019)
12. Tsai, W., Jhong, S.: Production decision model with carbon tax for the knitted footwear industry under activity-based. J. Clean. Prod. **207**, 1150–1162 (2019)

Logistic Foundations of Conducting Transnational Business in the Regional Market

L. K. Kirillova(✉)

Samara State University of Economics, Samara, Russia
kirillova.sseu@mail.ru

Abstract. The article views the adaptation of transnational corporation logistics solutions to the conditions of the national market. Methodological approaches to revealing of problems in the sphere of logistics are presented and indicators of logistics problems are proposed on the example of Coca-Cola Company functioning in the Russian market. Methods of product range analysis from the standpoint of demand and cost of production stability are presented. On the basis of the proposed assortment analysis methodology recommendations are given on stock management in order to minimize storage costs. It is suggested carrying out stock regulation in the retail sales system on the basis of sales promotion measures. A recommendation is considered to reveal the correlation and regression dependence of sales on price as a tool for demand forecasting and stock level maintaining.

Keywords: International business · International logistics · Logistics · Logistic expenses · Markets · Stock

1 Introduction

International business is an integral part of the development of national entrepreneurial structures and is an essential driver of growth of both national and global economies. The most prominent role in the global economy is played by transnational corporations which in total create up to 40% of the world trade. An essential part of this type of activity is logistics implemented on the international scale. The issues of organizing of logistics on the international scale are the subject of scientific research of a number of specialists [3, 5–10]. At the same time, the problems of adapting of transnational companies logistics solutions to the conditions of the countries where production units are located remain unresolved. It is important to take into account the specific characteristics of functioning of transnational corporations that use production units abroad and function to serve local markets. In this regard, international logistics may be of a limited nature because of the need to supply specific units and components for organizing of completed process of consumer value creating. Logistics is a key factor in the development of international trade; it forms a new level of competitiveness [7]. As an object of the research we took the company Coca-Cola HBC Eurasia Samara which has been working in the Russian market since 1994. In total, Coca-Cola Company, having the ownership of the recipe of drinks, forms the marketing policy of the entire company, and the partners of the company located throughout the world, including ten

S. I. Ashmarina et al. (Eds.): ENGINEERING ECONOMICS WEEK 2020, LNNS 139, pp. 92–99, 2021.
https://doi.org/10.1007/978-3-030-53277-2_11

regions of the Russian Federation, are engaged in the production, bottling of drinks, logistics and sales. The specific features of the organization of Coca-Cola Company divisions logistics activities, the assessment of the state and functioning of the logistics system are of particular scientific interest. For the analysis we used only open sources of information from the Internet concerning the performance of Coca-Cola HBC Eurasia Samara LLC [4].

2 Methodology

The penetration into the Russian market was carried out by the method of gradual construction of plants for bottling of soft drinks, first in Moscow, then in a number of regions, including the Samara region, bringing the number of such enterprises to 13. According to the reports of the company, the Samara region provides about 11% of the total sales in Russia [2]. The structure of the products manufactured by Coca-Cola HBC Eurasia Samara LLC includes a variety of products, such as: carbonated drinks (42%) bottled water (29%), juices (25%), tea drinks (2%) and energy drinks (1%). Such sales structure was reported in 2019. The analysis of the current management structure showed the presence of all departments dealing with the regional market research and the emerging demand satisfaction. That is, in the management structure of the company, marketing and logistics services functioning within the regional market with a certain degree of independence are highlighted. The assessment of the financial situation and the development trends of the company was carried out for the period of 2017–2019 [11] and its results are presented in Table 1.

Table 1. The dynamics of relative economic indicators of Coca-Cola HBC Eurasia Samara LLC

Commercial performance indicators	Indicator growth rate 2018/2017, %	Indicator growth rate 2019/2017, %
Revenues from product sales	96.53	94.07
Production prime cost	92.15	91.83
Gross profit	102.98	97.37
Commercial expenses	112.87	115.39
Management expenses	202.63	193.68
Sales profit (loss)	36.24	2.14
Stock	88.16	143.26

Source: author based on [2].

The presented dynamics of changes in commercial indicators shows deterioration in the situation in the company, judging by the indicator of a sharp decline in profits. The profit of 2018 was only 36. 24% of the indicator of 2017 and became even less by 2019. This fact is associated with the decrease in the revenues from sales of products, while the growth rate of the prime cost is slightly behind the parameters of revenues decreasing. Therefore, it is necessary to find out the reasons for such a situation in the field of sales and logistics. The need to find problems in the field of logistics is stipulated by the growth of the indicator of the state of the stock. During the period from 2017 to 2018, the stock decreased by 12%, and by 2019 it increased by 43%. As you know, stock is a key indicator of management in the field of logistics and optimization of stock level is a source of increasing of management efficiency in the company. Logistics management methods are aimed at optimizing costs when performing any process, taking into account the given parameters [8]. An important addition to the analysis of the situation was the study of profitability indicators of Coca-Cola HBC Eurasia Samara LLC. In particular, according to the company's reports, sales profitability for the period from 2017 to 2019 decreased from 7.5% to 1.3%. That is why a detailed analysis of the state of stock and an assessment of the functioning of the company's logistics system are necessary.

3 Results

Taking into account the profile of the researched organization specializing in the production and sales of soft drinks, commodity circulation, stock turnover and the duration of one turnover were studied in detail. The identified parameters of stock status became the basis for management decisions in the logistics of Coca-Cola HBC Eurasia Samara LLC. The general analysis of the state of stock showed an increase in the share of stock in the total turnover from 6.2% in 2017 to 8.0% in 2018 and 9.5% in 2019. This indicates a slowdown in the sales and an increase in the stock, and, as a result, an increase in the cost of the stock maintaining. For a more accurate assessment of the situation during the research, additional calculations were carried out, the results of which are presented in Table 2.

Table 2. Indicators of stock state of Coca-Cola HBC Eurasia Samara LLC product groups

Product group	Stock structure, %			Turnover coefficient		Turnover duration, days	
	2017	2018	2019	2018	2019	2018	2019
Fizzy drinks	35.8	49.0	55.5	22.9	15.0	16	24
Juice	34.9	27.7	23.1	8.5	9.0	43	41
Bottled water	23.5	17.7	15.1	4.2	5.0	87	73
Energy drinks	2.3	2.5	3.0	7.9	3.6	46	101
Tea drinks	3.5	3.2	3.4	7.9	5.8	46	63
Total	100	100	100	13.8	11.3	26	32

Source: author based on [11].

The table allows making a number of conclusions. Firstly, fizzy drinks occupy the largest share in the stock structure; their share is gradually increasing from 35 to 55%. This is a traditional range of products that is most popular with customers. Juice occupies the second position, but their stock share began to decline from 34.9% to 23% by 2019. Bottled water closes the top three in the structure of the company's range, having a decrease in the stock level from 23.5% to 15.1%. It should be noted that the increase in the stock level indicates a decrease in demand, while the reverse trend reflects a favourable situation. Thus, an increase in the share of fizzy drinks stock reflects sales problems and increased storage costs.

Secondly, the analysis of the indicator of stock turnover coefficient confirms the noted trend of decline in sales of fizzy drinks, which is confirmed by the decrease in its values from 22.9 in 2018 to 15 in 2019. Thus, the number of production turnovers decreased markedly and the production volumes remain at the same level. A similar trend is noted for energy drinks, where the turnover ratio fell almost 2 times. The sales of tea drinks also became worse, though not so markedly, and the turnover coefficient decreased from 7.9 to 3.6 turnovers per year.

Thirdly, the duration of the products turnover in general for all items was 26 days in 2018 and 32 days in 2019. These figures are acceptable for a manufacturing company. However, there are three product groups (bottled water, energy drinks and tea drinks) where the indicators for the duration of one turnover are significantly higher and reach parameters of 60–100 days. These values are a signal for making decisions on stock management and eliminating of wasteful costs. The current situation with the level of stock is stipulated by the complexity of the functioning of the company's management structure and the presence of centralized management on the part of the regional branch of the multinational company. The general scheme of production and sales planning for Coca-Cola HBC Eurasia LLC includes a number of successive steps:

1. Regional subdivisions make a sales forecast for each month, in the context of range groups.
2. Regional subdivisions carry out planned calculations of required stock, production plans and other indicators for organizing of logistics activities of a regional subdivision.
3. The range production is organized in volume and quantity, according to the proposed forecasts.
4. Operational management of the logistics system for the organization of supplies and control over the level of stocks by regional subdivisions.

The sales process is carried out through various sales channels, including: modern trade, traditional trade, immediate consumption and wholesale. The organization of interaction between partners is carried out through the transfer of the order to the manufacturer, taking into account the demand for a particular market participant. At the same time, it is important to take into account changes in consumer preferences to make adjustments to production plans, avoiding the creation of excess stock. For these purposes, ABC and XYZ analysis was used in the research, which allows identifying the most popular range positions and positions which are not in demand. As initial data, the annual sales volumes of drinks in 2019 of various brands in packages from 0.33 to 2 L were used. The results of the comprehensive analysis are presented in Table 3.

The obtained results make it possible to give recommendations on the company's product lines, in dependence with the position in the quadrant of the matrix. The products in the AH quadrant include the products that are in steady demand and do not need reserve stock creation.

Table 3. ABC + XYZ analysis matrix of the main brands of Coca-Cola HBC Eurasia Samara LLC for the year 2019

Category	X	Y	Z
A	Coca-Cola 1.0 Coca-Cola 2.0 Coca-Cola 1.5 Coca-Cola 0.5	Coca-Cola Zero 1.0 Coca-Cola Zero 0.5 Dobryi 1.0 Dobryi 2.0 Bonaqua Still 0.5	
B	Rich 1.0	Coca-Cola Zero 0.33 Sprite 1.0 Sprite 0.5 Fanta 1.0 Bonaqua Highly Carbonated 1.0 Bonaqua Still 2.0 Bonaqua Still 1.0 Bonaqua Viva 0.5	Bonaqua Still 5.0 Sprite 1.5
C	Shweppes Lemon 1.5	Nestea Lemon 0.5 Fanta 2.0 Fanta 1.5 Fanta 0.5 Burn 0.5 Burn 0.33 Bonaqua Highly Carbonated 2.0 Bonaqua Highly Carbonated 0.5	Sprite 2.0 Nestea Wildberry 0.5 Dobryi 0.33 Nestea Green Tea 0.5 Nestea Wildberry 0,5 Shweppes Pomegranate 1.5 Fanta Strawberry 1.5

Source: author.

Products in the quadrant AY have low reliability of consumption and an average price level. These products require stock because of the possible fluctuations in demand. Products of the BX category have high price and unstable demand. The group of BY category products forms reliable revenues for the company, but is characterized by unstable demand. For a guaranteed level of sales stock should be created and maintained. Products of the BZ category are in steady demand but have seasonal fluctuations in demand, which requires special stock management during certain periods of the year. The CY category products are characterized by stable demand, which allows planning production and sales volumes clearly without creating reserve stock. The main disadvantage of this product group is its low profitability. The CZ category includes the products that should be produced in a minimum quantity to maintain the breadth of the range. The revenue from this product is very low and the sales are unpredictable. The presented range analysis methodology and recommendations for managing production items are a reliable tool for reducing unproductive stock and the corresponding costs for logistics processes.

4 Discussion

Logistics management is aimed at optimizing of the total costs stipulated by the movement of material flows. The key logistics tools are stocks and their movement to maintain sales and meet consumer demand. If we talk about the production of soft drinks, which the organization under the research specializes in, the optimization possibilities are connected with the postponement of the production process and the creation of goods in anticipation of future sales. That is, maintaining stocks is one of conditions for the stable functioning of the logistics system. Any improvement of the entrepreneurial structure and the processes of its functioning should be aimed at improving the efficiency and, above all, the efficiency of economic activities [1].

To overcome possible problems concerning the reduction of the profitability of business, you can use a number of recommendations:

- increasing of control over the level of stocks of finished products in retail trade and establishing a monitoring system,
- introducing of ABC-XYZ stock analysis into practice on ongoing basis,
- the development of measures to stimulate demand for products with unstable demand, often of a seasonal character.

The products of Coca-Cola HBC Eurasia Samara LLC are sold to the final consumer more often through the retail sales channel. Therefore, it is important to include measures to stimulate demand in the programme of activities, using a set of tools, such as outdoor advertising, mobile billboards, advertising modules in magazines and the Internet, banner and contextual advertising, mass media, television and promotion activities. In addition, it is necessary to take into account the peculiarities of purchasing behaviour of Russian customers as well as the low level of income caused by the depreciation of the national currency. In this situation, it is important to understand the relationship between price and sales. The study of this issue was carried out on the basis of actual observations of sales at different prices, with the use of correlation and regression analysis on the example of one of the brands of products on the basis of Microsoft Excel programme. According to the calculation results, the values of regression indicators, presented in Table 4, were obtained.

Table 4. Regression statistics of sales and price level of the sprite brand, 1 L

Indicators	Value
Multiple R	0.662
R-square	0.439
Normalized R-square	0.383
Standard error	7.464
Observations	12
Regression level	$y = -331.9x + 45899.8$

Source: author.

This table presents the results of the study of the linear dependence of sales on price. For estimation, the least squares method was used, and the statistical significance of the parameters of the equation was checked with the use of the coefficient of determination. The calculations make it possible to conclude that in the situation under research 43.9% of sales variability is due to the impact of prices. The obtained calculated indicators confirm the statistical significance of the model parameters. The regression equation can be used to predict sales when planning incentive programs. This will allow estimating the necessary volumes of stocks and satisfying the demand with optimal logistics costs.

5 Conclusion

The research considers measures for demand management based on the assessment of the impact of price factors. At the same time, it is necessary to take into account non-price demand factors, which include: tastes of consumers, total number of consumers, customer income, availability of substitute products, seasonal character of consumption of goods, the influence of fashion trends on consumption and general expectations of customers. The recommendations on stock management proposed in the work, aimed at logistics costs reduction, will generally increase the use of stock of finished products for stable satisfaction of demand; improve range conformity of goods in retail trade and optimize the volume of stocks maintained for stable functioning of the logistics system. The development of international trade in the context of globalization asks for the creation of an effective logistics system in the countries that has a set of competitive advantages: the volume of storage space, terms of cargo handling and transportation costs [6]. The functioning of multinational corporations in the regional markets requires serious work, on the one hand, to maintain accepted standards of customer service and, on the other hand, to adapt them to the conditions prevailing in the regional markets.

References

1. Ashmarina, S.I., Kandrashina, E.A., Streltsov, A.V., Yahneeva, I.V.: The peculiarities of the organizational changes in entrepreneurial structures. Econ. Bus. **7**(84), 1085–1089 (2017)
2. Audit-It: Coca-Cola HBC Eurasia LLC: Accounting reports and financial analysis (2019). https://www.audit-it.ru/buh_otchet/7701215046_ooo-koka-kola-eychbisi-evraziya. Accessed 27 Apr 2020
3. Blyde, J., Molina, D.: Logistic infrastructure and the international location of fragmented production. J. Int. Econ. **95**(2), 319–332 (2015)
4. Coca-Cola Company: Our company (2019). https://www.coca-colacompany.com/company. Accessed 27 Apr 2020
5. Hallberg, N.L.: The micro-foundations of pricing strategy in industrial markets: a case study in the European packaging industry. J. Bus. Res. **76**, 179–188 (2017)
6. Kirillova, L.K.: International logistics. Samara State University of Economics, Samara (2016)

7. Kirillova, L.K.: Globalization factors and their influence on logistics development. In: Khasaev, G.R., Ashmarina, S.I. (eds.) Proceedings of the 16th Scientific and Practical Conference Problems of Enterprises Development: Theory and Practice, pp. 176–178. Samara State University of Economics, Samara (2017)
8. Kirillova, L.K., Evtodieva, T.E.: Logistics principles of international multi-modal transportation of dangerous goods. Bull. South Ural State Univ. Ser.: Econ. Manag. **13**(3), 184–192 (2019)
9. Lyapin, S., Rizaeva, Y., Kadasev, D., Sysoev, A.: Methods to analyze traffic demand formation in intelligent transportation and logistic regional network. Transp. Res. Proc. **45**, 522–529 (2020)
10. Pečený, L., Meško, P., Kampf, R., Gašparík, J.: Optimisation in transport and logistic processes. Transp. Res. Proc. **44**, 15–22 (2020)
11. Test Firm: The financial state of Coca-Cola HBC Eurasia LLC (2019). https://www.testfirm.ru/result/7701215046_ooo-koka-kola-eychbisi-evraziya. Accessed 27 Apr 2020

Latin America Face Global Threats: Movement of Capital and Investment Attractiveness

O. N. Fomina(✉)

Moscow State Institute of International Relations (University) of the Ministry of Foreign Affairs, Moscow, Russia
o.fomina@inno.mgimo.ru

Abstract. In modern conditions, investment attractiveness determines the level of the countries' participation in the movement of capital, the rate of economic growth, the position of national economies in the structure of the world economic system. The article presents theoretical and methodological foundations of the process of forming the investment attractiveness of Latin American countries as the main factor in attracting capital under conditions of action and development of global threats. It is proposed to use systematic and structural-logical approaches to develop and justify stages and factors of the process of forming investment attractiveness of Latin American countries in the context of the growing impact of non-economic processes on the nature and level of trends in the world economic development. These stages and factors allowed the author to create model for the development of competitive advantages of Latin American countries to foster the participation of countries in the international capital movement. That will determine the competitiveness of these countries on the world stage.

Keywords: Global threats · Investment attractiveness · Latin America

1 Introduction

To date, there is quite a large number of studies on international capital flows, investment attractiveness of countries and regions of the world in the context of traditional overproduction crises, financial crises, and other types of crisis related to economic and political factors, demographic, social, and environmental processes. Modern challenges are not just international, country-specific, regional factors of global crises, but also global factors and threats that affect all countries in one way or another, regardless of country differences, differences in the level and quality of life, socio-economic development and potential. On the agenda of many economic studies is the research of the impact of modern types of threats and possible counteraction to them. Despite the initial predictions of possible changes in the global architecture of the international capital movement, the state of the capital markets, changes in the investment attractiveness of countries and regions of the world under pandemic conditions (coronavirus) there are not still enough informed and structured

S. I. Ashmarina et al. (Eds.): ENGINEERING ECONOMICS WEEK 2020, LNNS 139, pp. 100–111, 2021.
https://doi.org/10.1007/978-3-030-53277-2_12

recommendations on the formation of investment attractiveness of regions in the world, including Latin America.

Therefore, research on the process of forming investment attractiveness of Latin American countries in the context of the action and development of global threats is an extremely relevant area for analysis of international capital flows at the present stage.

We conducted a review of research on features and trends in capital flows and investment attractiveness of country groupings and individual national economies. The assessment of the development level of the problem of the capital movement and investment attractiveness of economic systems in the face of global threats of both economic and non-economic nature was carried out. It is revealed that the development genesis of international economic relations in the context of regional economic groupings on a global scale are influenced by general trends caused by the action of global threats, regardless of the economic development level, investment attractiveness, technology development, or the financial system of a particular state. Therefore, general principles and recommendations for creating and improving the investment attractiveness of country groupings and individual national economies simply do not work in these new conditions. As a result, at the second research stage, general approaches and principles were studied to understand the category of investment attractiveness, factors and conditions for its formation in order to identify stages and factors of forming and increasing the investment attractiveness of national economies in the face of new global threats. The critical analysis revealed that there is practically no research on stages and factors of the process of forming the investment attractiveness of Latin American countries in the context of the action and development of global threats.

2 Methodology

The main issues raised in this study were solved using such methods as: system and institutional approach, method of structural and logical analysis, analysis and synthesis, and evolutionary analysis. The use of systemic and institutional approaches allowed us to reveal the nature and essence of the category «investment attractiveness». The use of structural and logical analysis helped to identify and justify stages and factors in the process of forming investment attractiveness of Latin American countries as the main factor in attracting capital.

General scientific methods of analysis and synthesis, elements of evolutionary analysis, allowed us to identify and describe the main stages and factors in the process of forming the investment attractiveness of Latin American countries as the main factor in attracting capital in the face of the action and development of global threats.

The analysis of components and trends of forming the investment attractiveness in Latin America countries has allowed to determine key competitive advantages of the countries in this region under conditions of action and the development of global threats and to shape the author's position on the development of a model to counter global threats in the Latin America countries, increasing their investment attractiveness and their participation in the international capital movement.

3 Results

Investment attractiveness is a complex category, generalizing a set of economic and non-economic factors, conditions, principles, methods, and tasks for attracting investment to a certain territory. According to Grishina, Shakhnazarov, Roizman the concept of investment attractiveness at the regional level is a set of objective features, means, opportunities and restrictions that determine the intensity of attracting investment to a certain territory [7]. Tumusov determines the investment attractiveness of a region as a set of investment resources that make up a part of accumulated capital that is represented in the investment market in the form of potential investment demand and has the ability to turn into a real investment demand meeting material, financial and intellectual needs for capital reproduction on a certain territory [16].

Developing this point of view, Tumusov identifies potential investment demand and specific investment demand as the real supply of capital, the first of them occurs in the absence of the intention of a subject of investment demand by the presence of available income or profit to direct it to the accumulation [16]. Calling this approach «formal», researchers define the investment attractiveness of a region as a source of future investment in a certain territory [3, 4, 11].

Based on scientific positions of Tumusov, the investment attractiveness of a certain national-state territory is a set of investment resources that characterize a part of the capital in the form of potential investment demand, ready and able to turn into a real investment offer [16]. According to Golovikhin, Nezhivenko, the investment attractiveness of a national-state territory is considered from the perspective of supply: it is a set of characteristics that determine the potential offer of a socio-economic system for the development of investment resources and providing conditions for their development [6]. Elsukov, Mayevsky, Cheberko [5] identify the investment attractiveness of a national-state territory with the investment capacity of the region. For example, according to Konyagina [9], investment attractiveness of a region or investment capacity of a territory is a sum of objective prerequisites of a national-state territory for investment, depending both on the availability and diversity of areas and objects of investment, and on their economic efficiency.

The third approach to determining the investment attractiveness of a national-state territory is based on the theory of comparative and absolute advantages [18], and the starting point of this theory is a condition of limited resources for functioning of national-state territory, which leads to the fact that each of the regions has certain advantages. Thus, the level of investment attractiveness of a national-state territory is a constant dynamic process for distributing absolute and relative advantages [1, 2]. Anticipating such changes opens up additional opportunities for any national-state territory to attract investments to the territory. The reviewed scientific sources have a common point: the concept of investment attractiveness of a region is often considered as «a set of conditions and resources, factors and advantages».

The definition of investment attractiveness of a national-state territory through a set of conditions and resources, factors and benefits is appropriate, therefore, for purposes of further research, the following interpretation of the term is suggested: investment attractiveness of a region is a collection of investment resources located in a territory and having high importance for attracting investment into the economic system of the national-state territory.

Investment attractiveness of a national-state territory, based on the semantic content of the word "attractiveness", is a subjective concept, in other words, it reflects a subjective attitude of a potential investor to the object of investment. Investment attractiveness of a region includes a specific set of characteristic properties and conditions that promote investment, which reflects the objective state of the investment object.

Thus, the investment attractiveness of a national-state territory in Latin American countries is a view of a specific investor on the potential object of investment. In contrast, the category «investment potential of a region» does not depend on characteristics of the subject of investment, which excludes the subjectivity of opinions. As a result, we can conclude that the investment attractiveness of Latin American countries is different for different categories of investors, and the investment potential remains unchanged for all investors.

The problem of attracting investment is the main issue of the socio-economic development of the Latin American countries, territories, industrial and economic entities in the current market conditions and an increasingly urgent task. The system of investment attractiveness of a region as a whole consists of a set of interrelated subsystems, processes and components. As in any general system, the system of investment attractiveness of the Latin American countries contains a subject, an object of management, a mechanism for implementing the process of forming the investment attractiveness of a region, the input and output of the investment attractiveness system in the Latin American countries, and at the moment – global threats.

Figure 1 shows a general diagram of the process of forming the investment attractiveness of Latin American countries in the face of global threats. The external environment for the investment attractiveness formation of Latin American countries in the face of global threats includes resources, conditions, markets and technologies, political, economic, social, and environmental factors. The development strategy for the formation of investment attractiveness in these countries should reflect the main criteria, goals and objectives of the investment attractiveness formation. The main goal of developing the investment attractiveness in the Latin American countries is to increase the level of investment attractiveness in the region in comparison with other regions of the world and the share of global capital flows, which are the main driver of the economic development. Kostyunina, Baronov notes that in some developing countries investors are guaranteed protection of their rights and interests, including protection from nationalization or expropriation (direct or indirect), which is possible in the interests of the state or citizens of the country [10].

There are some tasks of the investment attractiveness formation at the regional level in the context of global threats:

- to improve regional legislation in the field of investment attractiveness of a region;
- to improve the efficiency of investment projects in a region;
- to reduce administrative barriers and procedures in the implementation of investment projects in a region;
- to increase the level of counteraction to global threats.

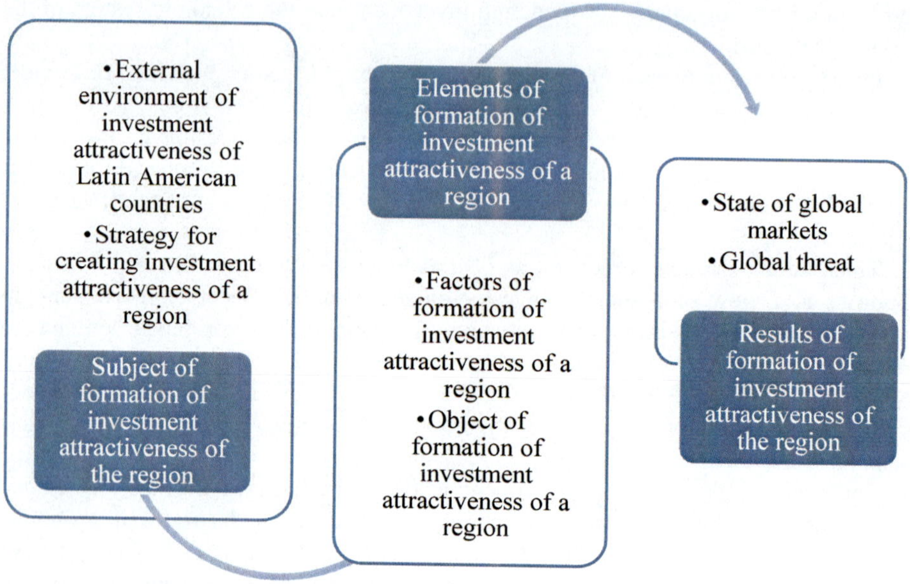

Fig. 1. General scheme of the formation process of investment attractiveness in the Latin American countries in the context of global threats. (Source: author)

Subjects of forming the investment attractiveness of the Latin American countries in the face of global threats are all economic entities and processes whose tasks are to develop and implement measures to improve the investment climate in the region, attract capital, and develop business initiatives. The stages of the process of forming and managing the investment attractiveness in these countries in the context of global threats are shown in Fig. 2.

Let's consider in more detail stages of the formation process of the investment attractiveness in the region under conditions of global threats. The external environment, and the market conditions (for which measures are being developed to create investment attractiveness of Latin American countries) are the starting point in terms of countering global threats. It is the current situation which includes trends and prospects for the development of the investment market that serve as the main criteria for selecting an investment object, the basis for formulating goals and objectives to create investment attractiveness of the region in the face of global threats. The main subject of forming the investment attractiveness of the region together with the political leadership is big business.

Depending on a type of investment project, market type, and product, the investment market is analyzed and evaluated in the context of global threats – the market is very dynamic and unpredictable – and it is necessary to constantly monitor needs and target consumers of this market, its capacity and geography. Assessment of the region's competitive advantages in the face of global threats is an important component of forming the regional investment attractiveness, since it is difficult to predict demand

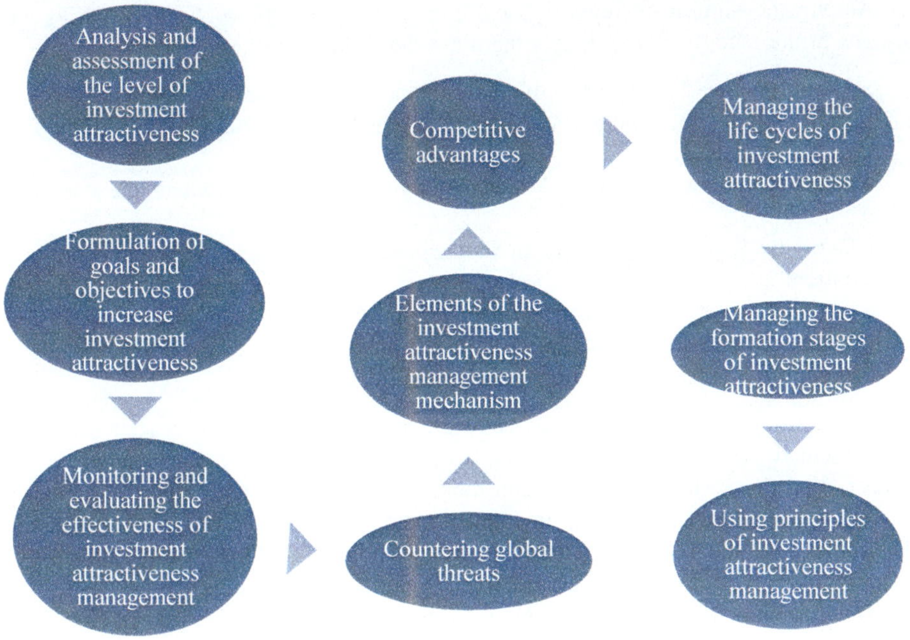

Fig. 2. Stages of the formation process of the investment attractiveness in Latin American countries in the context of global threats. (Source: author)

forecasts for a particular product, which is the basis for making decisions on investment in a region and choosing priorities in the development of industries and economy sectors.

There are objective prerequisites for the movement of capital on a global scale. First of all, the reasons for this are the excess of capital and the lack of acceptable investment objects in one region and the opposite situation in another: the lack of capital and a favorable investment climate in another region.

Among the significant reasons for the capital movement between countries and regions in the face of global threats there are the following ones:

- lack of a balance between supply and demand of capital in the region;
- formation of new commodity markets and regression of existing markets in the face of global threats;
- availability of cheap labor and raw materials in the region where the capital is exported to;
- a favorable investment climate in the region, accompanied by a stable political situation;
- availability of a preferential investment regime in special economic zones;
- low environmental standards in the host region;
- a strategy for expanding participation in new markets.

Among the significant reasons for the capital movement between countries and regions in the face of global threats there are the following ones:

- lack of a balance between supply and demand of capital in the region;
- formation of new commodity markets and regression of existing markets in the face of global threats;
- availability of cheap labor and raw materials in the region where the capital is exported to;
- a favorable investment climate in the region, accompanied by a stable political situation;
- availability of a preferential investment regime in special economic zones;
- low environmental standards in the host region;
- a strategy for expanding participation in new markets.

And in modern conditions, a factor of global threats is also the coronavirus pandemic, which has brought down the markets for energy resources and many types of industrial products. Therefore, the main motive for the export and import of capital is now the epidemiological situation in countries and regions of the world. Global threats determine the global demand, dramatically destabilizing global capital flows, and changing the level of investment attractiveness in different regions.

The formation of the investment attractiveness in Latin American countries is always associated with risks and factors that shape it. In the process of forming the investment attractiveness in this region, the least predictable and often the most important elements are identification, assessment and management of countering global threats that are currently forming the level of investment attractiveness.

At the moment, we can distinguish two groups of factors that form the level of investment attractiveness of Latin American countries: financial factors that shape the level of investment attractiveness in the region; strategic factors that shape the level of investment attractiveness of the region as a whole. Let's consider these factors that form the level of investment attractiveness of Latin American countries in more detail. Figure 3 shows factors that form the level of investment attractiveness of these countries, which can be included into the group of financial factors.

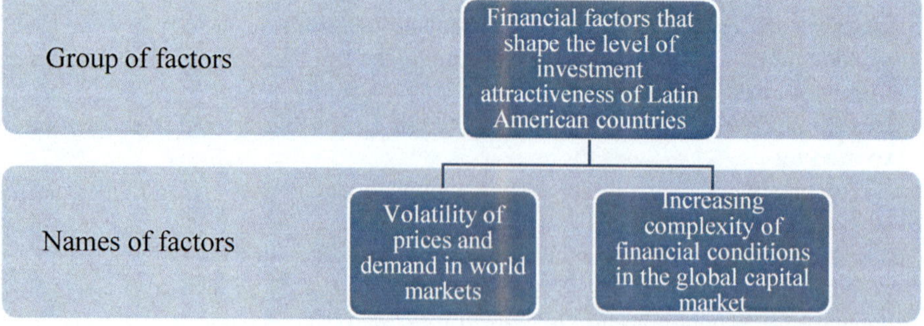

Fig. 3. Group of financial factors that form the level of investment attractiveness of Latin American countries. (Source: author)

The instability of prices and demand on international markets because of global threats is perhaps one of the most significant factors shaping the level of investment attractiveness of Latin American countries in modern conditions. Falling prices for oil and gas products and other industrial products on international markets not only leads to a reduction in revenues, but also reduces the ability of regions to provide additional financing. Financial conditions in the global capital market are becoming more complex. This factor is related to the first one, since a sharp decline in demand on world commodity markets inevitably leads to a similar deterioration of the capital market.

The second group of factors that form the level of investment attractiveness in Latin American countries is presented in Fig. 4. The group of strategic factors that form the level of investment attractiveness in the considered region includes two types of factors: factors of a political nature and global threats. Political factors that have traditionally been the main limiting factor for increasing the investment attractiveness in these countries for several decades are added to the effect of global threats that reduce demand for products, raw materials and materials which are used in the production and exported from the region.

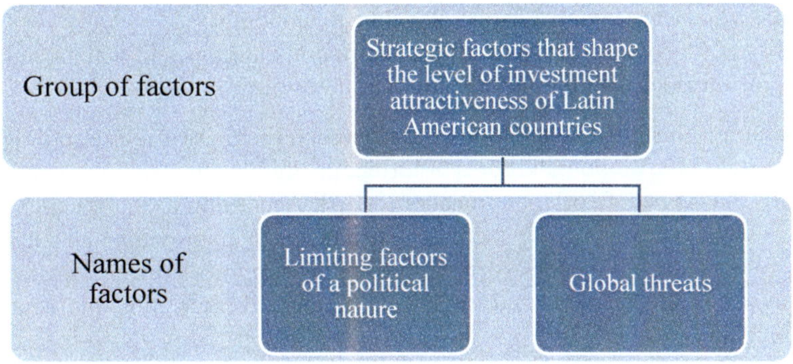

Fig. 4. Group of strategic factors that form the level of investment attractiveness of Latin American countries. (Source: author)

In these conditions, it is necessary to use other competitive advantages, in addition to those traditionally used in Latin American countries (the availability of extensive energy resources and cheap labor forces). When the demand for industrial products and energy resources decreases, it is necessary to re-focus on the production increase of food products taking into account that there are resources and conditions for this in the region. The demand for food products in the context of a pandemic, as any global threat, will be stable and may even grow. Therefore, in the face of global threats, the main factor in the investment attractiveness of Latin American countries should be the maximum use of the agricultural potential of countries in this region to increase production and export of food products to other countries and regions of the world.

The use of agricultural potential of the considered countries is impossible without entrepreneurial initiative and support from state authorities, without attracting additional financing for the production and export of food products to other countries and regions of the world. Therefore, the development of this competitive advantage of Latin American countries in the face of global threats, the coronavirus pandemic, instability and a sharp drop in activity in the financial markets and industrial goods markets is possible with the development of entrepreneurial initiative, the search and use of sources to attract investments and capital. This is possible thanks to the use of digital FINTECH technologies, access to markets not only for large private and public capital, but also for small-scale capital, funds of individuals, and small businesses.

Thus, Santos, Fernandes and Ferreira [14] determined that the financial crisis and recession returned the intermediary effect between socio-demographic characteristics and informal entrepreneurship at the current stage of economic relations development. Williams and Kedir [17] note that in the World Bank's survey of 67515 enterprises in 142 countries, it was found that one in five (19.9%) of the official enterprises started their activities without registration, although this varies in some countries (for example, in Pakistan, Slovakia). Reasons for these trends are:

- insufficient economic development and lower quality of management (modernization theory),
- too much state intervention (neoliberal theory),
- too little state intervention (theory of political economy),
- inconsistency between laws and rules of formal institutions and beliefs, values and norms of informal institutions (institutional theory).

In addition to insufficient economic development, the lack of registration is related to too little state intervention and rules of formal institutions that are incompatible with the socially shared beliefs of entrepreneurs [8, 13]. According to the research by Mac an Bhaird, Owen, Freel [12], to increase the efficiency of providing financing from alternative sources, it is necessary to improve the supply of finance in the private investment and equity markets and to provide greater diversification and depth in the financial markets.

Taking into account these aspects, the following model of forming the investment attractiveness of Latin American countries in the context of global threats is proposed (Fig. 5).

According to this model, the global scale and volume of global threats' impact on processes of capital movement and investment attraction negatively affect the level of investment attractiveness in Latin American countries, their participation in the capital movement. In these conditions, it will not be possible to simply wait for changes in the global market environment due to the high uncertainty in terms of timing and recovery parameters for goods and capital markets, possible changes in the trends of the world economy, the increasing role of digital technologies in processes of the capital movement and the implementation of business activities themselves.

It is necessary to search for and implement other competitive advantages that will work in new difficult economic conditions. That determines the expediency of choosing resources, conditions, markets and technologies for implementing these competitive advantages. For the Latin American countries, such resources, conditions, markets and technologies for implementing competitive advantages in the production and export of

Fig. 5. Model of investment attractiveness formation for Latin American countries in the context of global threats. (Source: author)

food products to other countries and regions of the world will be natural resources of plant and animal origin, climate conditions, small and medium-sized capital markets, and modern digital FINTECH technologies, which are already actively developing in most countries of this region.

Nowadays it is needed to move from such institutional forms of attracting and developing investment in Latin American countries as large integrated business structures with broad involvement of state and foreign banking capital, TNC capital, to more flexible and mobile institutional structures, such as small and medium-sized businesses. So, Tsigelkova, Platonova, Frolova note that in China, working closely with large firms in the structure of global value chains, small businesses can be credited by a specially created financial institution or take advantage of cross-border investments [15]. Such structures may well attract investment in the markets of individuals' funds both within the Latin American countries and abroad, especially since modern digital FINTECH technologies allow this to be done with lower costs and terms of their attraction without losing economic independence and simultaneously forming pre-orders for their products. The systemic effect of actively using new sources of investment attraction without deviating from the traditional schemes of forming investment attractiveness will allow the countries of the region to overcome the next larger-scale global crisis in the face of global threats, but also use their own competitive advantages, to improve the level and quality of life in the countries of the considered region.

4 Discussion

Research on the development of the world economy and the international movement of capital in the works of domestic researchers, such as Kostyunina, Baronov [10].

Studies of regional aspects of investment attractiveness, institutional aspects of attracting investment and financial entrepreneurship are marked in the works of such scientists as Buzyrev, Polyakov [2], Zhukov [18], Santos, Fernandes, Ferreira [14], Tumusov [16], Chub [4], Mac an Bhaird, Owen, Freel [12] and other authors.

The analysis of components and trends of the investment attractiveness formation of a region as a whole has allowed to identify the key competitive advantages of the countries in the considered region in the conditions of action and the development of global threats and to shape the author's position on the development of a model for countering global threats, increasing the investment attractiveness of Latin American countries and increase their participation in the international capital movement.

5 Conclusions

The author's model of the investment attractiveness formation for Latina Americans countries in the conditions of global threats is proposed. In general, the obtained research results are necessary and in demand in terms of finding and implementing other competitive advantages of the considered countries that will work in the new difficult economic conditions.

Directions of further research are related to the justification of the choice of resources, conditions, markets and technologies for the formation of investment attractiveness in Latin Americans countries and will concern the detailed development of institutional factors for the use of new sources of investment attraction based on modern digital technologies in the face of new global threats.

References

1. Abykaev, N.A.: Investment potential and economic growth (methodological aspect). Economist **6**, 58–66 (2000)
2. Buzyrev, V.V., Polyakov, V.G.: Identification of the degree of influence of functional factors of the territorial organization of the region's economy on its investment attractiveness. Bull. TOGU **2**(17), 147–156 (2010)
3. Cheberko, E.F., Mayevsky, A.V.: Development of recreation and tourism as a factor of increasing the investment attractiveness of the region. Manag. Consult. **9**(105), 82–91 (2017)
4. Chub, B.A.: Assessment of the investment potential of Russian economic entities at the meso-level. Betony, Moscow (2001)
5. Elsukov, M.Yu., Mayevsky, A.V., Cheberko, E.F.: Investment climate and investment attractiveness of Russian macro-regions. Manag. Consult. **12**(132), 70–89 (2019)
6. Golovikhin, S.A., Nezhivenko, E.A.: Competitiveness of Russian regions: monograph. Classical University, Chelyabinsk (2017)

7. Grishina, I., Shakhnazarov, A., Roizman, I.: Comprehensive assessment of investment attractiveness and investment activity of Russian regions: methods for determining and analyzing relationships. Invest. Russia **4**, 5–16 (2001)
8. Kistanov, V.V., Kopylov, N.V.: Regional economy of Russia. Finance and Statistics, Moscow (2009)
9. Konyagina, M.N.: Russian investor's portfolio: current management issues. Mod. Sci.: Actual Probl. Theory Pract. **9**, 56–61 (2018)
10. Kostyunina, G.M., Baronov, V.I.: Investment regulation in Myanmar. Southeast Asia: Curr. Dev. Issues **3**(40), 122–141 (2018)
11. Litvinova, V.V.: Investment attractiveness and investment climate of the region: monograph. Financial University, Moscow (2013)
12. Mac an Bhaird, C., Owen, R., Freel, M.: The evolution of entrepreneurial finance – 10 years after the global financial crisis. Int. J. Entrep. Innov. **20**(4), 235–238 (2019)
13. Pierrakis, Y.: Peer-to-peer lending to businesses: investors' characteristics, investment criteria and motivation. Int. J. Entrep. Innov. **20**(4), 239–251 (2019)
14. Santos, E., Fernandes, C.I., Ferreira, J.J.: The driving motives behind informal entrepreneurship: the effects of economic-financial crisis, recession and inequality. Int. J. Entrep. Innov. (2020). https://doi.org/10.1177/1465750320914788. Accessed 07 May 2020
15. Tsigelkova, E., Platonova, I.N., Frolova, E.D.: Comparative analysis of support for small and medium-sized enterprises in the EU and China in order to increase their added value. Econ. Region **15**(1), 256–269 (2019)
16. Tumusov, F.S.: Investment potential of the region: theory, problems, practice. Ekonomika, Moscow (1999)
17. Williams, C.C., Kedir, A.: Evaluating competing theories of informal sector entrepreneurship: a study of the determinants of cross-country variations in enterprises starting-up unregistered. Int. J. Entrep. Innov. **19**(3), 155–165 (2018). https://doi.org/10.1177/1465750318782766
18. Zhukov, L.M.: The role of regions in shaping the investment climate in Russia. Russian Foreign Econ. Bull. **5**, 44–50 (2006)

International Organizations' Approaches to Digital Assets Legalization (Monetary Policy and AML/CFT)

V. E. Ponamorenko(✉)

Russian Foreign Trade Academy, Moscow, Russia
vladpon@inbox.ru

Abstract. The article investigates doctrinal approaches of international organizations to the legalization of digital assets in the light of monetary policy and Anti-Money Laundering/Combating the Financing of Terrorism (AML/CFT). The purpose of the research is to study current aspects of digital assets legalization at the international level in the context of the monetary policy and AML/CFT. This study is predominantly based on analytic papers and documents issued by Global Financial Regulators – IMF, FATF, BIS. The author pays attention to the regulation of crypto-assets by the EU authorities, especially the European Banking Authority (EBA), develops a concept of international legal personality of the FATF as a vivid example of an informal intergovernmental organization. The analysis of legal and other documents of the EAEU bodies plays a very important role. The study also focuses on the position of the FATF in relation to virtual assets and virtual asset service providers (VASP), as well as the monetary taxonomy of the IMF and the BIS, the relationship between the FATF and the FATF-style regional bodies, the contribution of such organizations to the regional economic integration in comparison with formal intergovernmental organizations. The methodology of the research comprises historical, comparative, formal-legal and functional methods, systemic approach.

Keywords: Crypto-assets · Digital assets · Eurasian economic union · Monetary policy · Stablecoin · Virtual currencies

1 Introduction

The world community emphasizes the issue of developing optimal (that is, balancing risks and opportunities) models of legalization of new digital entities both via global financial regulators and the main analytical centers: primarily crypto assets (digital assets, virtual currencies), and activities related to them (mining, exchange, trading, custody, etc.). At the international level, the Twenty Group, the FATF, the Bank for International Settlements, the IMF, the Financial Stability Board, the IOSCO (International Organization of Securities Commissions) and European regulators (European Commission, EBA, ESMA, ECB) focus on the regulation of digital assets. Key institutions that determine trends in the development and regulation of digital technologies are the summits of the Group of Twenty and the Group of Seven.

S. I. Ashmarina et al. (Eds.): ENGINEERING ECONOMICS WEEK 2020, LNNS 139, pp. 112–121, 2021.
https://doi.org/10.1007/978-3-030-53277-2_13

In the final Declaration of the G20 Summit (Osaka, June 28–29, 2019), at the highest level of interstate interaction, it was stated: "Technological innovations can provide significant benefits for the financial system and, more broadly, for the economy. While crypto assets do not jeopardize global financial stability, we remain sensitive to existing and emerging risks. We reaffirm our commitment to apply the FATF Recommendations, recently supplemented with provisions on virtual assets (VA) and virtual asset service providers (VASP), to combat ML/FT (money laundering and terrorism financing). We welcome the adoption by the FATF of the Interpretive Note and Guidelines [26].

The finance ministers and central bank governors of G7 countries in July 2019 agreed to develop and disseminate cryptocurrency control rules throughout the world [29]. By the Osaka Summit of G20, the FATF prepared a new Guidelines for a Risk-Based Approach to Virtual Assets and Virtual Asset Service Providers [20], which, together with the amendments to Recommendation 15 ("New Technologies") and the FATF Glossary, as well as the Interpretive Note to Recommendation 15 were approved at the FATF Plenary Week, held in Orlando (USA) on June 16–21, 2019. Two new terms were included in the FATF Glossary: "virtual assets" (VA) and "virtual asset service providers" (VASP). In the Recommendation 15, as amended, the new requirements were established which state that VASP should be regulated for AML/CFT (anti-money laundering and countering the financing of terrorism), licensing or registration, and should be subject to effective control or surveillance systems.

The Interpretive Note to Recommendation 15 further explains how the FATF requirements should apply to VA and VASP. In particular, it deals with a risk-based approach to activities and operations related to VA, and VASP; monitoring or control of VASP for AML/CFT purposes; licensing or registration; preventive measures, including customer due diligence, storage of records and documents, and reporting of suspicious transactions; sanctions and other enforcement measures; international cooperation. The above Guidelines clarifies that countries and VASPs, as well as other entities involved in activities with VA have to understand the ML/FT risks associated with their activities and take appropriate measures to mitigate them. In particular, the Guidelines provides examples of risk signs that should be especially taken into account in virtual assets, and also focuses on factors that can be additional disguise for operations or limit the ability of VASP to identify customers [20].

2 Methodology

In this article, the author uses the concepts of "monetary policy" and "monetary sphere" as the basic ones ("monetary sphere" is used in the works of Abramova [1], Golovnin [23], Guseva [24] and other authors). The "monetary sphere" is an artificially constructed structure (like other economic spheres: monetary, credit, financial, etc.), which aims to expand the object of legal regulation beyond one of the traditional systems (in our case, monetary). In this case, "monetary sphere" includes elements of banking, payment systems, financial market.

Moreover, the set of tools that influences monetary relations fits into the toolbox of monetary policy, but the author goes further beyond the research of traditional target package of this policy. The author believes that today the relations between the traditional types of stability targeted by monetary policy (price, exchange rate, stability of monetary aggregates/money supply) with financial stability and financial security are essential, to which monetary policy makes a significant contribution. This is more important aspect among digital challenges, as well as fintech lending, cryptocurrencies, cryptocurrency crowdfunding, digital payments, etc.

The key standard setters in the monetary sphere are BIS, IMF, FSB and FATF. The first two organizations influence monetary policy and banking supervision, as the politicians that have the greatest impact on the monetary sphere. The FATF addresses the issues of financial security in the monetary sphere regarding combating the laundering of illicit incomes using various types of money and monetary instruments. The Financial Stability Board, which implements macroprudential policies and is in close collaboration with the IMF and BIS, plays a crucial role in ensuring global financial stability.

3 Results

The vigorous activity of American and European regulators resulted in the publication of several consulting documents, reports, strategies, in which the emphasis is the taxonomy in the field of crypto assets, as well as their classification.

As a basic definition can be used the one given by the European Banking Authority (EBA) in its Report [18] of January 2019: "Crypto-asset is an asset that:

a) is determined mainly by cryptography and distributed ledger technology (DLT) or similar technology;
b) is not issued or guaranteed by the central bank;
c) can be used as a mean of exchange and/or for investment purposes and/or for access to a product or service".

As the basic (accepted by many regulators) classification, the following one can be specified:

1. Exchange (payment) tokens, they are also cryptocurrencies.
2. Security tokens.
3. Utility tokens.

The IMF also has some findings in the field of taxonomy of digital money. In July 2019, it launched a series of reports entitled FinTech Notes, the first report of which was "The Rise of Digital Money" from IMF experts Adrian and Mancini-Griffoli [3], which, however, studies mainly how cash and bankmoney compete with electronic money. The document also highlights the risks associated with electronic money: new monopolies; threats to weak currencies; consumer protection and financial stability; and the risk of encouraging illegal activities. The essential element of the report is the taxonomic "money trees" proposed by the authors to understand the types of modern money, their relationship to each other (including CBDC, cryptocurrencies, investment money, etc.). The issue of stablecoins has to be studied in detail, they are not so risky in comparison with ordinary cryptocurrencies.

On October 18, 2019, the Financial Stability Board issued the report, "Regulatory issues of stablecoins", which gave a definition of stablecoins and global stablecoins. The Board refers stablecoins to a crypto asset that maintains its stable value by linking to another asset or basket of assets. They can be provided with fiat currency, goods or supported by algorithms. The "global stablecoins", are referred to stablecoins with potential global reach and the ability to quickly scale from the point of view of users/holders of crypto assets [21].

The Board makes a point that the nature of stablecoins requires careful research, as well as the legal mechanisms to minimize the risks of stablecoins.

In October 2019, the G7 issued the Report on stablecoins [22], which indicated that:

- global stablecoins, although not risk free, can solve the problem of slow and expensive money transfers,
- on the other hand, stablecoins also do not yet have global regulation and can be unstable, on the conditions of rules' observance, "stable coins" can be convenient as a mean of payment and savings,
- conventional cryptocurrencies cannot be an effective solution, since they are "extremely unstable", difficult to use, have scalability limitations and problems with management and regulation.

On October 18, 2019, the FATF also gave its opinion about stablecoins. According to the regulator, new assets, so-called "global stablecoins", and offered global networks and platforms can potentially cause a shift in the ecosystem of VA and affect the risks of ML/FT. However, the FATF defines two problems: the massive use of such VA and their transfer from person to person without a regulated intermediary. The FATF emphasizes that global stablecoin providers will comply with the FATF standards either as VASP or as traditional financial asset providers. The FATF will continue to study the characteristics and risks of stablecoins and provide further clarification on how the FATF standards are applied to global stablecoins and their service providers.

It is significant to understand that stablecoins can be not only private (LIBRA, Gram, JPM Coin projects, etc.), that is, issued by private corporations, but also state-owned, that is, issued by central banks. The state-owned ones are investigated by the Bank for International Settlements (BIS) under the name "Central Bank Digital Currencies (CBDC)". CBDC, as defined by the BIS, is the digital form of central bank money that differs from balances in its reserves or in current accounts [9]. At the same time, BIS sticks to the point that central banks should take into account the problems of ML/FT when issuing CBDC. It is still unclear how AML/CFT requirements can be implemented for anonymous forms of CBDC, which are easily cross-border, but can meet significant challenges in this regard.

BIS taxonomy distinguishes between three forms of CBDC. Two forms are based on a token (token-based), and the other is based on an account (account-based). The two token-based versions differ primarily in those objects that have access to them, which depends on the potential use of CBDC. The first is a widely available payment instrument, which is primarily intended for retail transactions. The other is limited in access and is intended for wholesale (interbank) payments. The account-based version assumes that central bank opens general-purpose accounts for all agents in its jurisdiction. Currently, China, Sweden and some other countries are about to create their own CBDC. The Bank of Russia launched testing stablecoins.

The analysis of CBDC by the BIS and the IMF is carried out and it investigates how the new form of money issue will transform the banking business, and will also affect the monetary policy (its channels, transmission mechanism, tools).

In a press release published on January 21, 2020 [11], BIS announces that the Bank of Canada, the Bank of England, the Bank of Japan, the European Central Bank, the Swedish Central Bank and the Swiss National Bank, together with the BIS created the group to exchange experiences in the area of assessment of potential cases for CBDC in their jurisdictions. The group will assess the cases of CBDC using, cost-effectiveness, functionality and their design for manufacturability, including cross-border compatibility, and organize the exchange of knowledge about new technologies.

In an IMF blog that is aimed to express the views of IMF staff on current economic and political issues, December 12, 2019, the IMF experts Adrian and Mancini-Griffoli spoke about the future of CBDC [2]. The experts believe that CBDC is a complex and multidisciplinary issue that requires active analysis and discussion. It raises the issues related to monetary policy, functioning of payment systems, financial stability. The IMF can assist in this matter in three ways: informing about political discussions, organizing platforms for discussing policy variations, and helping countries develop policies. Since CBDC is a new topic, the IMF was active in two areas, but is gradually moving into the third area, as member countries are considering options of CBDC implementation and are seeking the advice of the IMF.

Central banks highlight a number of potential benefits of implementing CBDC:

- reducing the cost of cash emissions,
- improving financial inclusion,
- improving the stability of the payment system,
- opposing to new digital currencies,
- support for distributed ledger technology (DLT),
- possible increase in transmission mechanism's efficiency of monetary policy.

At the same time, attention is drawn to the problems and risks of implementing CBDC:

- deprivation of banks' interest income on deposits,
- growth in the balance sheet of the central bank,
- dollarization in the countries with high inflation and an unstable exchange rate,
- high implementation costs for central banks.

One of the trade-offs proposed by experts is the "synthetic CBDC", where private companies issue digital money to the general public (which can be either account-based or represent DLT-based tokens), and the central bank will lend credibility to their system and require the coins to be fully secured by the reserves of the central bank, as well as control the money' issuers.

4 Discussion

The digitalization of corporate compliance instruments and supervision in the financial market (including the banking sector) is a kind of digital response to fintech challenges of economic, information and other types of security, which, along with innovative advantages, presents new technologies.

The main "tangle" of regulatory innovations in the financial market are SupTech/RegTech [7]. In the future the effectiveness of the regulatory influence of central banks will be largely determined how the regulators mastered the ideology and tools of digital supervisory technologies (SupTech), also it will be determined by the introduction of fintech solutions in the systems of internal (compliance) control of credit organizations and other participants in the financial market (RegTech).

We consider these directions in more detail.

Supervisory Technology (SupTech). SupTech is a new concept in the vocabulary of global financial regulators and the banking community. It means the use of digital technological innovation by supervisory authorities in order to increase the effectiveness of supervisory activities. In their core work, "Innovative Technology in Financial Supervision (SupTech) – The Experience of Early Users" Broeders and Prenio define SupTech as supervisory technologies based on innovative technologies used by supervisory authorities to improve their activities [13]. However, in the work "The suptech generations" [16], experts from the Institute for Financial Stability of the BIS makes a point that it remains unclear what innovative technologies fell under the conceptual list "SupTech". Different levels of technological progress in states resulted in differences to interpret the term SupTech. Therefore, experts clarify this concept as follows: the term "innovative technology" refers to the application of Big Data or artificial intelligence (AI) to the tools used by financial authorities, while "financial authorities" are both supervisory and non-supervisory authorities [25]. The second ones are primarily related with financial intelligence units, which in some countries are not authorized with supervisory powers.

Digitalization of supervisory tools helps regulators minimize costs and increase efficiency in the following areas:

1. Data collection (reporting automation, data management, virtual assistants – chat bots).
2. Data analysis (market research, analysis of discipline violations, microprudential supervision and macroprudential supervision) [13].

The role of SupTech in AML/CFT functions is specially emphasized. In the BIS Report [15], on this particular aspect is analyzed based on the experience of nine jurisdictions (including Russia). Taking into account the advantages of new tools (in particular, they can help AML/CFT supervisors and financial intelligence units analyze large amounts of information coming from primary financial monitoring entities, first of all banks), experts at the same time put forward some problems in this field: ensuring high computing power, ensuring confidentiality of information, the difficulty of evaluating the effectiveness of these tools.

The Basel Committee on Banking Supervision (BCBS) made several recommendations for central banks and other banking supervisory authorities (hereinafter referred to as the supervisory authorities) on introducing SupTech into supervisory practice and correlating this tool with financial innovations (FinTech) in the financial market [7].

Not only big data algorithms and artificial intelligence can improve the supervisory activity of regulators, but also the blockchain. Thus, Rafael Auer, the economist of the BIS, in his report "Embedded supervision: How to build regulation into blockchain finance" [10] suggests monitoring tokenized markets using blockchain. Blockchain technology effectively encourages to decentralize trading of tokenized assets, however, it is possible to track such markets using smart contracts, machine learning and artificial intelligence.

The expert underlines that systems based on distributed ledger technology (DLT) can potentially help implement Embedded Supervision as a set of regulatory requirements that allow authorities to automatically track tokenized markets, so the companies don't have to actively collect, verify and provide data.

Regulatory (compliance) technology (Regulatory Technology, RegTech). For the first time, the definition of RegTech was given by Deloitte: it is a technology designed to provide fast, adaptable, easy to integrate, reliable, safe and economic solutions that meet the standards of the regulator.

The FATF emphasizes the development of RegTech tools, according to which this tool will optimize AML/CFT in the primary financial monitoring entities, which, in turn, will contribute to the reduction of ML/FT risks and facilitate communication with financial intelligence units, Central Bank and other supervisory bodies in the field of AML/CFT.

By creating the compliance system (internal control unit), the credit institution implements a compliance function, that is, a function for managing compliance risk. Initially, the concept of compliance risk originates from the documents of the BCBS [12], it is considered only in the context of banking risks and, first of all, in connection with the ML/FT risk. However, gradually the concept of "compliance risk" expanded to conform external and internal rules, norms and standards [27]. RegTech is designed to facilitate, through automation, the procedures for ensuring compliance of an organization (for example, a bank) with the requirements of the regulator and other (in particular, industry and corporate rules and standards).

The BCBS emphasizes that innovative technologies can help financial institutions comply with regulatory requirements (prudential requirements, reporting, consumer protection, AML/CFT). In this context, RegTech can provide banks with more effective ways of risk management.

The Bank of Russia responded in its documents to the pressing issues of digitalization of supervision and compliance in timely way. So, it defines these areas as relevant in its Guidelines for financial market development for the period 2018–2020 [4], and in October 2018 publishes the Report for public consultations "Issues and trends for the development of regulatory and supervisory technologies (RegTech and SupTech) on the financial market in Russia" [6]. The Report analyzes international experience, justifies the necessity to introduce these technologies and suggests a questionnaire for market participants to choose the optimal model for such introduction.

To develop the approach, on October 9, 2019, the Bank of Russia publishes the "Action Plan" (roadmap) for SupTech and RegTech in the Bank of Russia" [5]. The implementation of this action plan should allow: to reduce the regulatory burden on the supervised organizations; improve the quality of Bank of Russia activities in the field of control and supervision; optimize compliance with the requirements of the Bank of Russia by financial market participants.

The issue of international coordination of monetary policy in the post-crisis period of world economic development is one of the most acute and debatable. Experts disagree on the advantages of formal coordination of central banks about the issues of monetary policy implementation, although the unconventional monetary policy implemented by the monetary authorities of the leading countries during crisis resulted in negative effects for the countries with emerging markets, largely due to the lack of such coordination.

The BIS brought into question the benefits of monetary cooperation. However later, it marked that achieving more effective result would require enhanced cooperation, including special joint actions, and perhaps even an agreement on the rules of the game that limit domestic policy. Today, BIS proposes to go beyond the national mandates of central banks and consider the internationalization of financial regulation as an example for monetary policy [8]. Many people reject the global monetary policy outlook. Accordingly, national mandates require large central banks issuing reserve currencies to establish policies for a smaller economic area than that one covered by their currencies. This interpretation of internal mandates contrasts sharply with successful international cooperation in financial regulation and supervision. National mandates there do not impede widespread international cooperation and the development of global rules.

Bruni and his colleagues, relying on the reports of the BIS, believes that the heads of central banks should regularly and officially discuss and agree on some strategic provisions of their monetary policy, and communicate their decisions to the markets. She believes that this discussion is not in conflict with national mandates; in addition, at the international level, coordinated decisions are already the norm in micro- and macroprudential regulations, which are increasingly connected with the principles of monetary policy [14].

As for the European Central Bank, it usually combines orthodox statements about the inevitability of an internal orientation of monetary policy with the liberal recognition that global general shocks and international flows justify the search for some informal and implicit coordination of monetary policy [17].

Much greater consensus is reached among experts when it comes to coordinating monetary policy to contribute to the financial stability and financial security of the global economy. Benua Cure, a member of the Executive Board of the ECB and head of the BIS Innovation Hub, draws attention to this, which focus that a higher level of globalization requires a higher degree of policy coordination, with more clear and mandatory forms of coordination: "In particular, we must strive to strengthen global financial security systems so that we can better cope with global or regional liquidity crises" [19]. Ed Sibley of the Central Bank of Ireland writes about the necessity to strengthen coordination of central banks in the context of ensuring financial stability [28].

The fintech challenges facing the global economy bear the risks of financial stability and financial security (the core of which is the AML/CFT regime), as mentioned above. Their effective minimization is just possible with the interaction not only of regulators in the financial market, but also of monetary authorities.

5 Conclusion

We can conclude that the focus of global monetary regulators to the fintech challenges of our time is reflected in increasingly formalized documents and a deep research of the issue. The risks of financial security and financial stability are pushing the international community to realize the importance of international coordination of various policies, including monetary policy.

The level of reflection of the international banking community regarding the digitalization of banking supervision and compliance reached a new level: global financial regulators publish more and more detailed documents on this issue and summarize more and more good practices. Based on the above stated, in the nearest future innovative regulatory and compliance technologies can be expected to have a significant influence on key parameters of financial market and, in particular, the banking sector: such as stability, security, efficiency, and consumer protection.

References

1. Abramova, M.A.: Concept of development of the Russian monetary system in the context of modernization of the national economy. Ph.D. thesis. Financial University under the Government of the Russian Federation, Moscow (2014)
2. Adrian, T., Mancini-Griffoli, T.: Central bank digital currencies: 4 questions and answers (2019). URL: https://blogs.imf.org/2019/12/12/central-bank-digital-currencies-4-questions-and-answers/. Accessed 25 Oct 2019
3. Adrian, T., Mancini-Griffoli, T.: The rise of digital money. FinTech Notes No. 19/001 (2019). https://www.imf.org/en/Publications/fintech-notes/Issues/2019/07/12/The-Rise-of-Digital-Money-47097. Accessed 25 Oct 2019
4. Bank of Russia: Main directions of financial market development for the period 2018–2020 (2017). http://cbr.ru/StaticHtml/File/36231/ON_FinTex_2017.pdf. Accessed 25 Oct 2019
5. Bank of Russia: The action plan ("road map") in the SupTech and RegTech sphere in the Bank of Russia (2017). https://cbr.ru/fintech/reg_sup/map/. Accessed 25 Oct 2019
6. Bank of Russia: Report of public consultations. Issues and directions of development of regulatory and Supervisory technologies (RegTech and SupTech) in the financial market in Russia (2018). https://cbr.ru/Content/Document/File/50667/Consultation_Paper_181016.pdf. Accessed 25 Oct 2019
7. BCBS: Sound practices: Implications of fintech developments for banks and bank supervisors. Consultative document (2017). http://www.bis.org/bcbs/publ/d415.pdf. Accessed 25 Oct 2019
8. BIS: 85th annual report (2015). https://www.bis.org/publ/arpdf/ar2015_ec.pdf. Accessed 25 Oct 2019
9. BIS: Committee on payments and market infrastructures, markets committee. Central bank digital currencies (2018). https://www.bis.org/cpmi/publ/d174.pdf. Accessed 25 Oct 2019

10. BIS: Working Papers No 811. Embedded supervision: How to build regulation into blockchain finance by Raphael Auer (2019). https://www.bis.org/publ/work811.pdf. Accessed 25 Oct 2019
11. BIS: Central bank group to assess potential cases for central bank digital currencies. Press release (2020). https://www.bis.org/press/p200121.htm. Accessed 25 Oct 2019
12. BIS/BCBS: Implementation of the compliance principles (2008). http://www.bis.org/publ/bcbs142.pdf. Accessed 25 Oct 2019
13. Broeders, D., Prenio, J.: Innovative technology in financial supervision (suptech) – The experience of early users (2018). https://www.bis.org/fsi/publ/insights9.pdf. Accessed 25 Oct 2019
14. Bruni, F., Serrate, J.S., Villafranca, A.: The quest for global monetary policy coordination. Econ.: Open-Access Open-Assess. E-J. **13**(2019-5), 1–16 (2019). http://dx.doi.org/10.5018/economics-ejournal.ja.2019-5. Accessed 25 Oct 2019
15. Coelho, R., De Simoni, M., Prenio, J.: Suptech applications for anti-money laundering (2019). https://www.bis.org/fsi/publ/insights18.pdf. Accessed 25 Oct 2019
16. Di Castri, S., Hohl, S., Kulenkampff, A., Prenio, J.: The suptech generations (2019). https://www.bis.org/fsi/publ/insights19.pdf. Accessed 25 Oct 2019
17. Draghi, M.: The international dimension of monetary policy (2016). https://www.ecb.europa.eu/press/key/date/2016/html/sp160628.en.html. Accessed 25 Oct 2019
18. EBA: Report with advice for European commission on crypto-assets (2019). https://eba.europa.eu/documents/10180/2545547/EBA+Report+on+crypto+assets.pdf. Accessed 25 Oct 2019
19. European Central Bank: The internationalisation of monetary policy (2014). https://www.ecb.europa.eu/press/key/date/2014/html/sp140430.en.html. Accessed 25 Oct 2019
20. FATF: The FATF guidance on the application of the risk-oriented approach "Virtual assets and virtual asset service providers" (2019). https://mumcfm.ru/repository/6c5b5be48c59f9ce7ee3bc73b23fbc95b61e92bf26b6a7259a45ff5df10e2ec9. Accessed 25 Oct 2019
21. FSB: Regulatory issues of stablecoins (2019). https://www.fsb.org/wp-content/uploads/P181019.pdf. Accessed 25 Oct 2019
22. G7 Working Group on Stablecoins: Investigating the impact of global stablecoins (2019). https://www.bis.org/cpmi/publ/d187.pdf. Accessed 25 Oct 2019
23. Golovnin, M.Yu.: Monetary sphere in transition economies: Comparison of the experience of Russia and the countries of Central and Eastern Europe. Ph.D. thesis. Institute of International Economic and Political Research of the Russian Academy of Sciences, Moscow (2003)
24. Guseva, M.V.: Regulation of monetary and credit spheres in Russia at the present stage. Ph. D. thesis. RUDN, Moscow (2004)
25. He, D., Leckow, R.B., Haksar, V., Mancini-Griffoli, T., Jenkinson, N., Kashima, M., Khiaonarong, T., Rochon, C., Tourpe, H.: Fintech and financial services: initial considerations. Staff Discussion Notes No. 17/05 (2017). http://www.imf.org/en/Publications/Staff-Discussion-Notes/Issues/2017/06/16/Fintech-and-Financial-Services-Initial-Considerations-44985. Accessed 25 Oct 2019
26. Official Website of the President of Russia: Osaka Declaration of the leaders of the Group of twenty countries (2019). http://www.kremlin.ru/supplement/5425. Accessed 25 Oct 2019
27. Ponamorenko, V.E., Korovyakovsky, D.G.: Compliance as an actual concept. Custom. Bus. **1**, 28–30 (2017)
28. Sibley, E.: Monetary and financial stability: the implications for prudential supervision (2019). https://www.bis.org/review/r190503b.pdf. Accessed 25 Oct 2019
29. TASS: The G7 Finance Ministers agreed to develop an approach to regulating cryptocurrencies (2019). https://tass.ru/ekonomika/6698293. Accessed 25 Oct 2019

Taxonomy and Typology of Crypto-Assets: Approaches of International Organizations

M. I. Inozemtsev(✉)

Moscow State Institute of International Relations (University) of the Ministry of Foreign Affairs, Moscow, Russia
inozemtsev@inno.mgimo.ru

Abstract. Subject of research: current doctrinal and legal-technical issues of taxonomy and typology of crypto-assets by international organizations. In the paper the author analyzes approaches to taxonomy and typology of crypto-assets by standard-settings bodies such as IOSCO, FSB, FATF, IMF. Besides the global crypto-asset regulatory landscape, the author pays attention to the regulation of crypto-assets at the EU level, especially ESMA policy activities. Special attention is paid to legal regulation of investment tokens and regulatory approaches to a new type of intermediaries in the crypto-market such as crypto-asset trading platforms. The first and foremost issue for regulators to consider is the legal status of crypto-assets and, as a result, it is determined whether financial services rules are likely to apply and, if so, which piece of legislation provides the adequate level of protection to investors. Therefore, the author also concentrates on the legal qualification of crypto-assets. The results of the study of problematic issues of crypto-assets' regulation at the global level and in the EU form the ground for developing the position on the opportunities to adopt the international regulatory experience to facilitate the development of Russian civil law.

Keywords: Digital assets · Digital law · Standard-setting bodies · Virtual currencies · Token · Financial instrument

1 Introduction

Over the past two years, the world has come to understand that without global, even soft-legal regulation of new digital entities in the form of crypto assets, it is very difficult to track and minimize the risks associated with their circulation. The lack of a common understanding between global regulators creates a field for regulatory arbitration, as well as the emergence, maintenance and translation of systemic risk.

Today, we are experiencing a period of intense and productive reflection of the scientific community, global and national regulators, representatives of the crypto community over the problems of digitalization of the economy and the release and circulation of crypto assets.

The need for legal regulation of crypto assets at both the supranational and national levels is caused by such risks of their circulation as violation of financial integrity (financial integrity), protection of the rights of consumers and investors (investor and consumer protection), laundering of criminal proceeds and financing of terrorism (AML/CFT).

S. I. Ashmarina et al. (Eds.): ENGINEERING ECONOMICS WEEK 2020, LNNS 139, pp. 122–133, 2021.
https://doi.org/10.1007/978-3-030-53277-2_14

This study focuses on summarizing the positions of global regulators serving the securities market, since global and national regulation of digital assets is shifting from cryptocurrencies (payment tokens) to securities tokens (investment tokens, asset tokens, digital securities) and crypto derivatives and smart contracts.

2 Methodology

When analyzing and generalizing the positions of global regulators in the context of the topic under discussion, the formal legal and comparative legal research methods were used. When discussing the problems of taxonomy and typology of digital assets, the following methods were used: historical and legal, the method of interpretation of legal norms, as well as some logical methods: analysis, synthesis, induction, deduction and analogy (traduction). Based on the data obtained, key conclusions are formulated that made it possible to reasonably apply the terms and establish the need for systemic improvement, primarily, of Russian legislation.

3 Results

3.1 Evolution of the Problems of International Regulation of Digital Assets in Documents of International Organizations

One of the first international regulators to issue an official document on virtual currencies back in 2014 was FATF. The same organization received an informal mandate to formulate a global approach to understanding the essence of a virtual asset and requirements for service providers of virtual assets from the Group of Twenty in 2028. And since June 2019, the updated FATF virtual assets guide [14], together with the new updated Recommendation 15 of its standard [13], are both reference points for other global standard-setting bodies and basic elements of the virtualization model for virtual assets at the national level.

The Financial Stability Board (hereinafter referred to as FSB), established in 2009, is focused on monitoring the vulnerabilities of the global financial system for the generation and translation of systemic risk. Acting as an aggregator of global financial standards and a conductor of G20 initiatives, the Financial Stability Board has compiled a compendium of 15 key global financial standards. In recent years, the focus of FSB has been on digitalization of the financial sector. Initially, the attention of FSB was drawn to fintech credit and possible risks to global financial stability from this type of activity [20, 21].

Today, FSB carries out important work on the aggregation of information collected at the level of supranational and national regulation of crypto assets, and its conceptualization, considering the initiatives and priorities of the Group of Twenty. In this regard, two documents issued by the Council should be mentioned: "Crypto-assets: Work underway, regulatory approaches and potential gaps" [23, 25], which summarizes the work of global standard setters in the cryptocurrency market, and "Crypto-assets regulators directory" [24], which summarizes information on national authorities with competence in the crypto market. According to the FSB work programme for 2020 [26], the Council will pay special attention to the FinTech sector, global stablecoins, cross-border payment systems. However, the Council will continue to work to ensure the security of the derivatives market.

One of the international organizations, which is a key global standard-setter in the field of the securities market, is the International Organization of Securities Commissions (hereinafter – IOSCO). In November 2017, IOSCO made a statement in which it informed its member national regulators about the risks of initial coin offering (ICO) and analyzed various approaches to ICOs' members. After that, in January 2018, IOSCO issued a statement to the general public, which outlined its concerns regarding ICO [28]. The IOCCO's important initiatives are to create the ICO Consultation Network first, through which participants discuss events and share their concerns, and then the ICO Support Framework as an educational resource for participants to help deal with ICO risks in their own jurisdiction. One of the most important documents issued by IOSCO is the Consultation Report of May 2019 [29], in which the regulator examined the problems and risks associated with trading crypto assets on specialized platforms (Crypto-Asset Trading Platforms, CTPs).

Next, we pay attention to the development of cryptocurrency market regulation issues in the European Union, drawing on the experience of Switzerland, the UK and the USA.

The EU crypto market regulation, as in many other jurisdictions, started with warnings from supranational regulators about the risks rooted in crypto assets and actions for their sale, conversion, initial placement. Among the European financial services market regulators, there are the European Securities Market Supervision Authority (ESMA) and the European Banking Supervision Authority (EBA) which are actively involved in communication with market participants regarding the risks arising in the crypto asset trading and ICO. So, in November 2017, ESMA made a Statement [7] and informed investors about the potential threats contained in ICO and reminded the companies involved in ICO that their actions could be subject to current EU legislation. In February 2018, after a request from the European Commission, all three European supervisory authorities in the EU financial market (ESMA, EBA, EIOPA) issued a warning to investors and users about the risks associated with the purchase of crypto assets [9].

Entering a new level of reflection regarding not only risks, but also the innovative potential of the cryptocurrency market, ESMA [10] and EBA [6] published two Reports on January 9, 2019 assessing the regulatory coverage of crypto assets in the EU countries, as well as made recommendations for the European Commission regarding potential regulatory initiatives.

These reports emphasize, among other things, the need for a technologically neutral approach to the regulation of crypto assets, suitable for homogeneous activities and types of assets. It is also noted that most of the crypto assets are not regulated by the current EU legislation on financial services or there are gaps and shortcomings in the legislation, and in different EU countries crypto assets are classified differently as financial instruments, which creates problems for the control and regulation of this sector. ESMA recognizes that the existing regulatory framework for the functioning of the EU securities market is developed without considering the crypto assets market.

Thus, if crypto assets are considered as financial instruments of The Markets in Financial Instruments Directive (MiFID), then there are significant gaps and problems with the application of the regulatory framework to this sector.

One of the priorities in activities not only of ESMA, but also of other European financial regulators is "promoting supervisory convergence", based on the principle of universal, template management of similar risks ("one-size fits all" approach), which should contribute to the creation of a single supervisory field in the EU in relation to both the securities market as a whole and the crypto market, as well as its parts.

Currently, only part of the crypto assets will qualify as financial instruments in the understanding of MiFID. Therefore, many crypto assets will not be subject to the provisions of the EU financial services legislation. According to ESMA, this means that there is virtually no protection for investors and consumers from the risks of fraud, cyber-attacks, money laundering and market manipulation.

As for the European Commission, it gradually came to understand the importance of the innovative potential of new digital technologies, including blockchain technology, which was manifested in several documents of a strategic and conceptual nature. So, first, attention should be paid to the FinTech Action Plan approved by the European Commission in March 2018 [11]. In it, the European Commission indicates that at the EU level, some measures to counter the risks of virtual currency circulation have already been taken (5AMLD, GDPR, statements by ESAs, etc.). However, new financial services do not always fully fall within the existing EU regulatory framework; moreover, the existing national regulation in the EU countries is incompatible with each other. The Commission is in favor of assessing the applicability of the current EU regulatory framework with respect to ICOs and, more broadly, the turnover of crypto assets. The Commission advocates a reasonable balance between measures to minimize the risks associated with crypto assets and the use of their high potential for the development of the EU economy. This is especially true for blockchain technology in general.

The development of these ideas took place in the blockchain report published by the European Commission's Joint Research Center (JRC) in July 2019 [12]. The report is optimistic about the transformational potential of distributed ledger technology. The document notes a significant influx of investment in the industry. However, the position regarding the prospects of cryptocurrencies is more restrained.

3.2 Key Taxonomy

The main taxonomic issues in the field of legal regulation of digital technologies are concentrated around the basic product of distributed registry technology – the "digital asset"/"crypto asset" in its different interpretation. The vocabulary of international and national financial regulators migrated from the concepts of "cryptocurrency" and "virtual currency" to a wider class – "assets" ("digital", "crypto", "virtual"). The Cambridge Center for Alternative Finance draws attention to this trend in its report [2].

The concept of "asset" can be understood in the categories of domestic civil law and, in particular, in the context of "objects of civil rights". An increasing number of researchers and regulators (for example, CFTCs) agree that they are dealing with a "new asset active" [3]. This asset may be tangible or intangible, representing a thing or property right. It can be recognized that the concepts of "digital asset" and "crypto-active" are used today as identical, since the digital form of the existence of such an asset is based on the technology of a distributed registry, such as blockchain.

A narrower concept in relation to the concepts of digital asset/crypto-asset is the concept of token, which departs from the concept of coin. The term "crypto assets" is used as the base term by the Council on Financial Stability and EU regulators [22]. Having abandoned the term "virtual currencies", the term "virtual assets", as already mentioned, now it is used by FATF. The term "digital assets" is popular in the vocabulary of US regulators and professional communities.

First, you need to pay attention to the ratio of the concepts of "digital asset", "crypto-active" and "token". According to the well-established opinion of the crypto community, they are correlated in volume from a wider to a narrower one. This understanding is most detailed, although somewhat outdated, reflected in one of the early ESMA documents [8]. It uses the term "crypto assets" as a generic term for cryptocurrencies, virtual currencies, virtual assets, and digital tokens. The term "token" is declared more neutral, since it does not carry the claim to the legitimacy inherent in the "currency". It is a broad term that encompasses various virtual assets, and which can be defined by contrasting its assets based on accounts. The abbreviation ICO is used in the ESMA document for the initial offer of any crypto asset. Thus, the concept of "crypto assets" is the broadest for ECMA in this document, the concept of "virtual currency" is narrower, the concept of "cryptocurrency" is even narrower.

In a later Report [10], ESMA uses the term "crypto assets" as a base term for itself, more clearly reducing it to distributed registry technology: "crypto asset" is a type of private asset that is mainly based on cryptography and distributed ledger technology (DLT) or similar technology as part of their perceived or inherent value. The regulator stipulates that unless otherwise specified, ESMA uses this term to mean both the so-called "virtual currencies" and "digital tokens".

At the same time, the "digital token" here is any digital representation of interest, which may be cost, the right to receive benefits or perform certain functions, or may not have a specific purpose for use. Moreover, it seems that in the ESMA taxonomy the adjective "digital" is redundant, tautological. It is not supported by any major national or other supranational regulators.

The British Crypto Asset Task Force, consisting of the Treasury, the Financial Conduct Authority (FCA) and the Bank of England, refers to "crypto assets" as a cryptographically secure digital representation of value or obligations created using distributed ledger technology that can be transferred, stored or traded in electronic format. FATF uses the term "virtual asset" to refer to a digital representation of value that can be traded digitally and can be used for payment or investment purposes, including a medium of exchange, unit of account and/or storage of value.

The most difficult is taxonomy of American regulators. American regulators and the crypto community are very pluralistic about naming new digital entities. For example, the terms "convertible virtual currencies" (FinCEN terminology), "virtual currencies" (IRS and CFTC terminology), "digital assets" (SEC and primarily American expert community terminology), and "cryptocurrency" (IRS terminology) are used in parallel. However, in the 2019 Joint Document, CFTC, FinCEN and SEC use the most universal term "digital assets" in the American expert environment. According to this document, digital assets include instruments that can be qualified in accordance with the current legislation of the United States as securities, commodities or security-or commodity-based instruments, such as futures or swaps [32].

This definition for the American cryptocurrency market should be recognized as consensus, basic. It should be noted that American regulators use an approach focused on the economic meaning and purpose of a financial instrument, regardless of its name ("label").

3.3 Actual Questions of Typology of Tokens

Typology of tokens allows regulators to customize/fine-tune legal regulation tools to the specifics of crypto assets and the types of activities associated with crypto assets. First, it is about determining the features of the civil law regime of tokens-securities and service tokens [4, 30].

One of the main criteria for classifying tokens is functional. In this regard, the basic classification of tokens from international and national regulators is their division into 3 categories: exchange or payment tokens, they are also cryptocurrencies; utility tokens; security tokens. This classification is adhered to by the British regulators FCA [15], HMRC [27], European Union regulators EBA [6], ESMA [10].

At the same time, quite unexpectedly, in its final guide to crypto assets [16], the UK Financial Conduct Authority (FCA) introduces, along with the aforementioned, "electronic money tokens", allowing itself a mixture of fundamentally different in nature types of money: electronic and digital.

The Swiss regulator FINMA classifies tokens with small nuances: along with payment tokens and utility tokens, it allocates asset tokens.

Some American authors offer a wider, alternative classification of digital assets: "pure cryptocurrencies" – bitcoin, lightcoin (decentralized storage of value); privacy-focused coins – monero, Zcash; general-purpose digital currencies, platform currencies, (platform) – Etherium, NEO, RavenCoin. It is the latter that allow you to create new digital assets called service tokens and security tokens. American experts are conducting a deeper classification of tokens. So, in their report, experts from the U.S. Digital Chamber of Commerce draw attention to combinations of different types of tokens that are carried out in a single transaction. In this regard, they distinguish three specific types of tokens: placeholder tokens, mutable tokens, and dividend-paying tokens [31, 34].

A separate legal issue is hybrid tokens. For example, in the United States, for a security token, which is also a payment token, regulators and courts may provide for various types of legal regulation.

The American Bar Association draws attention to hybrid tokens, citing service tokens as an example, which can act simultaneously as a means of payment (for example, Ether) [1].

There are theoretical questions regarding payment tokens, which experts divide into functioning in account-based payment systems and token-based payment systems. This is especially evident in the implementation of "central bank digital currencies" (CBDC).

A separate legal problem is security tokens: first, the issue is being resolved regarding whether to extend existing legislation to a new class of objects or to accept new ones. EBA and ECMA in their documents refer to these tokens as "investment-type crypto-assets", British and American regulators as "security tokens", FINMA as "asset tokens". The documents of the American SEC and FINRA also mention "digital asset securities" as products of tokenization of existing securities. A separate group of assets is occupied by crypto derivatives, which we consider below.

European regulators EBA, ESMA describe the "investment-type crypto asset" as a type of crypto asset similar to a financial instrument. From FINMA's perspective, "asset tokens" are debt instruments or equity claims on the issuer. However, they can be both digital and digital.

Based on this aspect, today one of the most pressing issues requiring scientific reflection and legal registration is the issue of asset tokenization. First, two types of tokens can receive a different legal regime: primary digital assets, i.e. assets that exist in digital form only within the boundaries of the information system creating them), as well as tokens issued by tokenization of existing assets, including rights.

According to the FINMA definition, asset tokens representing intangible assets are digital assets because they exist exclusively in a computer system. Asset tokens that allow you to trade physical assets on the blockchain are digital representations of physical assets. Therefore, they are digitized assets [18].

The division of tokens into digital assets and digitized assets is supported in large part by the American crypto community. So, experts indicate that a "digital asset" is an electronic record in which an individual has a right or interest. A term does not include an underlying asset or liability, unless the asset or liability is an electronic record. That is, the digital asset is just a code. In this case, the "digitized asset" is the asset (which may be a security or a physical asset), the ownership of which is presented in an electronic record. An example of the digitized asset is an electronic record of ownership of real estate held in the distributed registry [1]. In this regard, the focus of both the market and regulators is gradually shifting from the Initial Coin Offering (ICO) model to the Security Token Offering (STO) model.

According to widespread belief, STO is the next evolutionary step after the boom Initial Coin Offering (ICO), which defines the vector of industry development towards a more regulated and transparent market. A distinctive feature of STO is that this type of token placement is supposed to be in full compliance with the requirements of the securities legislation. This should provide more protection of investor rights and lower regulatory risks for token issuers. In addition, STOs are guided by a different target audience – only professional (accredited) investors can participate in such a placement.

Another current focus of attention of regulators and the crypto community is derivatives of securities based on crypto assets, or crypto derivatives. In this case, the main risk stimulating the development of regulations in this area is the risk of violation of the rights of consumers and investors.

So, in its Consultative Document [15], published on July 3, 2019, the FCA announced that it will begin a consultation process to ban the sale, advertising and distribution to retail consumers of derivatives and exchange-traded bonds (ETNs) that mention certain types of crypto assets. The American regulator CFTC allows listing new virtual currency derivatives contract and details their regulation [3]. In the document, the regulator indicates that it sees it necessary to promote innovations arising from virtual assets. However, it intends to do so within the framework of the basic principles of trade professed by the Commission.

3.4 The Issue of Adjustable Perimeter

The key issue of typology of tokens according to their functional purpose is the solution of questions: a) whether the tokens fall into the adjustable perimeter; b) whether it is necessary to create new legislation for them or to apply the existing one. For regulators, it is important whether a new tool is included under the adjustable perimeter. These issues are decided by US SEC and CFTC in relation to the Securities Law (SEA) and the Commodity Exchange Act (CEA), EU regulators – in relation to MiFID II, other directives and regulations. The legal regime of crypto assets is the main problematic issue, since "crypto assets" are not defined in the EU legal acts on securities. Clarity in legal regulation is required both at the EU level and at the level of individual states.

Currently, ESMA considers the provisions of MiFID II as the main document for crypto assets. In the event that cryptocurrencies are regarded as financial instruments, ESMA indicates that the following legal provisions apply: Prospectus Regulation (PR3); Transparency Directive; The Market in Financial Instruments Directive framework (MiFID II/MiFIR); Market Abuse Regulation (MAR); Short Selling Regulation; The Central Securities Depositories Regulation (CSDR); Settlement Finality Directive; Alternative Investment Fund Managers Directive (AIFMD); Investor-compensation Schemes Directive; Anti-Money Laundering Directive V (AMLD5). It is expected that the EMIR and GDPR directives will also be adapted to the cryptocurrency market.

The situation is similar in the UK. So, in the already mentioned FCA Guide, it is explained that the regulated perimeter includes: Security tokens, which are regulated for the reason that these tokens provide rights and obligations similar to those specified in the Regulated Activities Order, RAO; E-money tokens, which meet the definition of electronic money according to the UK Electronic Money Regulations; These tokens include some form of stablecoins. Outside the adjustable perimeter are Exchange tokens, Service tokens. FCA notes that, however, some activities that use unregulated tokens can be regulated, for example, when they are used to facilitate regulated payments. In addition, the Anti-Money Laundering Directive V (AMLD5) introduces AML/CFT for exchange tokens.

Although the FCA Guide is more a political statement in form, market participants should use it as a basis for understanding how FCA will treat certain crypto assets. The FCA's position in the Guide, while not binding, may be a convincing factor in litigation.

4 Discussion

As you know, in the first of the "digital" laws adopted in Russia, namely, Law 34-FL [17], by applying the civil law debate concept "right to right" in Article 128 of the Civil Code of the Russian Federation, the concept of "digital rights" was introduced as a new object of civil rights, attributed by the legislator – along with cashless funds and non-documentary securities – to property rights, and disclosed in the new article 141.1 as follows: "Obligations specified in such capacity in the law are recognized as digital rights and other rights, the contents and conditions of which are determined in accordance with the rules of the information system that meets the statutory criteria" [33].

At the same time, "digital financial assets", as the second of the basic "digital" concepts, are not conceived by the legislator, as could be assumed, the object of digital rights, but they are defined in paragraph 2 of Art. 1 of the Draft Federal Law "On Digital Financial Assets" [5]: "Digital rights are recognized as digital financial assets ....", which poses serious challenges for lawyers to logically interpret the laws envisaged for adoption.

In this regard, we need a complex taxonomic structure, taken as a basis for creating a domestic model of legalization of a new class of assets. If there were a more thorough study of international and foreign experience in legalization of new digital entities in Art. 128 of the Civil Code of the Russian Federation, than "digital rights" could sound "digital assets" (or "digital financial assets"). The same as "electronic rights" are not applied to cashless funds and non-documentary securities in Article 128 of the Civil Code of the Russian Federation [33].

In August 2019, the second package of "digital" laws was adopted – the so-called "The Law on Crowdfunding" [16], which introduced the concept of "utilitarian digital rights", which includes: the right to demand the transfer of things; the right to demand the transfer of exclusive rights to the results of intellectual activity and (or) the rights to use the results of intellectual activity; the right to demand the performance of work and (or) the provision of services. It should be noted that this law does not provide for the possibility of investing through cryptocurrencies. In addition, the law does not allow the exercise of utilitarian digital rights outside of Russian specialized sites.

The concept of "utilitarian digital rights" is not supported by the vocabulary of the new "digital" articles of the Civil Code of the Russian Federation, as well as the bill on digital financial assets, which will raise the problem of the correlation of basic concepts and the construction of a "taxonomic tree" of new digital legislation.

In October 2019, the Ministry of Finance proposed dividing "cryptocurrencies" into three types: technical tokens, virtual assets and digital financial assets [19]. The Deputy Minister of Finance of the Russian Federation A. Moiseev explained that tokens are needed exclusively for the functioning of certain systems; "virtual assets" are bitcoin and similar tokens; "digital financial assets" are tokens that appear as a result of ICOs.

Considering the proposal of the Ministry of Finance through the prism of the international and foreign regulatory vocabulary considered above, we can assume that in this classification, the Ministry of Finance means "utility tokens", "virtual assets" – payment tokens", and "digital financial assets" – "security tokens". It should also be said that this classification is problematic with the concepts of "utilitarian digital rights" given in the Crowdfunding Law, which, of course, does not contribute to the consistency of the legislation regulating the cryptocurrency market. In addition, this position does not correspond to the definition of "virtual assets" by FATF, because the latter - being wide in scope – includes all three types of tokens known to the world (payment, service, token-securities), while in the position of the Ministry of Finance "virtual assets" – and this is nonsense – turn out to be a kind of "cryptocurrency".

5 Conclusion

Thus, we can conclude that conceptualization and legalization of the basic elements of the crypto market regulation - taxonomy and typology of crypto assets - has reached a new level of maturity and formal certainty. Global regulators infiltrate their standards with cryptographic issues, publish specialized guidelines, form a common understanding and a uniform legal field. Soon, one should expect a high level of perception of the approaches of global regulators by national legal systems. At the same time, transformation of civil legislation and modernization of civilistic doctrine will play a significant role in this process. It seems that it is advisable for the domestic legislator to pay serious attention to international approaches and foreign experience in conceptualization and legalization of crypto assets when finalizing/adopting "digital" legislation.

References

1. American Bar Association Derivatives, Derivatives and Futures Law Committee, Innovative Digital Products and Processes Subcommittee, & Jurisdiction Working Group: Digital and digitized assets: Federal and State jurisdictional issues (2019). https://www.americanbar.org/content/dam/aba/administrative/business_law/buslaw/committees/CL620000pub/digital_assets.pdf. Accessed 30 Apr 2020
2. Cambridge Centre for Alternative Finance: The global cryptoasset regulatory landscape study (2019). https://www.jbs.cam.ac.uk/fileadmin/user_upload/research/centres/alternative-finance/downloads/2019-04-ccaf-global-cryptoasset-regulatory-landscape-study.pdf. Accessed 30 Apr 2020
3. CFTC: Staff Advisory No. 18-14. Division of market oversight, division of clearing and risk (2018). https://www.cftc.gov/csl/18-14/download. Accessed 30 Apr 2020
4. Cook, J., Cohen, R., Denisenko, J.: Initial coin offering: a comparative overview of securities regulatory environments in the US, UK and Asia Pacific. In: Dewey, J. (ed.) Blockchain & Cryptocurrency Regulation, pp. 34–47. CPI Group, Croydon (2019)
5. Draft Federal Law N 419059-7 "On Digital Financial Assets and on Amending Certain Legislative Acts of the Russian Federation (on Digital Financial Assets)" (text for second reading). http://cmokhv.ru/news/2019-03-25-news20190325/. Accessed 30 Apr 2020
6. EBA: Report with Advice for European Commission on Crypto-Assets. 9 January (2019). https://eba.europa.eu/documents/10180/2545547/EBA+Report+on+crypto+assets.pdf. Accessed 30 Apr 2020
7. ESMA: ESMA highlights ICO risks for investors and firms (2017). https://www.esma.europa.eu/press-news/esma-news/esma-highlights-ico-risks-investors-and-firms. Accessed 30 Apr 2020

8. ESMA: Advice. Own initiative report on initial coin offerings and crypto-assets (2018). https://www.esma.europa.eu/system/files_force/library/esma22-106-1338_smsg_advice_-_report_on_icos_and_crypto-assets.pdf?download=1. Accessed 30 Apr 2020
9. ESMA: ESMA, EBA and EIOPA warn consumers on the risks of virtual currencies (2018). https://www.esma.europa.eu/file/47283/download?token=FJvg9rX7. Accessed 30 Apr 2020
10. ESMA: Advice. Initial Coin Offering and Crypto-Assets (2019). January, 9. URL: https://www.esma.europa.eu/file/49978/download?token=56LqdNMN. Accessed 30 Apr 2020
11. European Commission: Communication from the commission. FinTech Action plan: For a more competitive and innovative European financial sector (2018). https://eur-lex.europa.eu/legal-content/EN/TXT/?uri=CELEX:52018DC0109. Accessed 30 Apr 2020
12. European Commission: Blockchain now and tomorrow. Assessing multidimensional impacts of distributed ledger technologies (2019). https://ec.europa.eu/jrc/en/facts4eufuture/blockchain-now-and-tomorrow. Accessed 30 Apr 2020
13. FATF: International standards for combating money laundering, the financing of terrorism and the financing of the proliferation of weapons of mass destruction. FATF Recommendations. Veche, Moscow (2012)
14. FATF: Guidance on the application of the risk-oriented approach “Virtual assets and virtual asset service providers” (2019). https://mumcfm.ru/repository/6c5b5be48c59f9ce7ee3bc73b-23fbc95b61e92bf26b6a7259a45ff5df10e2ec9. Accessed 25 Oct 2019
15. FCA: Prohibiting the sale to retail clients of investment products that reference cryptoassets (2019). https://www.fca.org.uk/publication/consultation/cp19-22.pdf. Accessed 30 Apr 2020
16. Federal Law N 259-FL dated 02.08.2019 “On Attracting Investments Using Investment Platforms and on Amendments to Certain Legislative Acts of the Russian Federation”. https://www.garant.ru/hotlaw/federal/1286343/. Accessed 30 Apr 2020
17. Federal Law N 34-FL dated 18.04.2019 “On Amendments to Parts One, Two and Article 1124 of Part Three of the Civil Code of the Russian Federation”. http://www.kremlin.ru/acts/bank/44088. Accessed 30 Apr 2020
18. FINMA: Guidelines for enquiries regarding the regulatory framework for initial coin offerings (ICOs) (2018). https://www.finma.ch/en/~/media/finma/dokumente/dokumentencenter/myfinma/1bewilligung/fintech/wegleitung-ico.pdf?la=en. Accessed 30 Apr 2020
19. Forklog: The Ministry of Finance of the Russian Federation proposed a new approach to the regulation of cryptocurrencies (2019). https://forklog.com/minfin-rf-predlozhil-novyj-podhod-k-regulirovaniyu-kriptovalyut/. Accessed 25 Oct 2019
20. FSB: Financial stability implications from fintech (2017). https://www.fsb.org/wp-content/uploads/R270617.pdf. Accessed 30 Apr 2020
21. FSB: FinTech credit: Market structure, business models and financial stability implications (2017). https://www.fsb.org/wp-content/uploads/CGFS-FSB-Report-on-FinTech-Credit.pdf. Accessed 30 Apr 2020
22. FSB: Crypto-asset markets potential channels for future financial stability implications (2018). http://www.fsb.org/wp-content/uploads/P101018.pdf. Accessed 30 Apr 2020
23. FSB: Crypto-assets report to the G20 on work by the FSB and standard-setting bodies (2018). http://www.fsb.org/wp-content/uploads/P160718-1.pdf. Accessed 30 Apr 2020
24. FSB: Crypto-assets regulators directory (2019a). https://www.fsb.org/wp-content/uploads/P050419.pdf. Accessed 30 Apr 2020
25. FSB: Crypto-assets: Work underway, regulatory approaches and potential gaps (2019b). https://www.fsb.org/wp-content/uploads/P310519.pdf. Accessed 30 Apr 2020
26. FSB: FSB work programme for 2020 (2019c). https://www.fsb.org/wp-content/uploads/P171219.pdf. Accessed 30 Apr 2020

27. HMRC: Policy Paper. Cryptoassets for Individuals. HM Revenue and Customs. 19 December (2018). https://www.gov.uk/government/publications/tax-on-cryptoassets/cryptoassets-for-individuals. Accessed 21 May 2020
28. IOSCO: IOSCO board communication on concerns related to initial coin offerings (ICOs) (2018). https://www.iosco.org/news/pdf/IOSCONEWS485.pdf. Accessed 30 Apr 2020
29. IOSCO: Issues, risks and regulatory considerations relating to crypto-asset trading platforms. Consultation report (2019). https://www.iosco.org/library/pubdocs/pdf/IOSCOPD627.pdf. Accessed 30 Apr 2020
30. Klayman, J.A.: Mutually assured disruption: the rise of the security token. In: Dewey, J. (ed.) Blockchain & Cryptocurrency Regulation, pp. 60–90. CPI Group, Croydon (2019)
31. Rowland, G.S., Kiviat, T.I.: Cryptocurrency and other digital assets for asset menedger. In: Dewey, J. (ed.) Blockchain & Cryptocurrency Regulation, pp. 90–101. CPI Group, Croydon (2019)
32. Tarbet, H., Blanco, K., Clayton, J.: Leaders of CFTC, FinCEN, and SEC Issue joint statement on activities involving digital assets CFTC, FinCEN and SEC joint statement on activities involving digital assets (2019). https://www.sec.gov/news/public-statement/cftc-fincen-secjointstatementdigitalassets. Accessed 19 May 2020
33. The Civil Code of the Russian Federation. http://www.consultant.ru/document/cons_doc_LAW_5142/. Accessed 19 May 2020
34. Token Alliance: Considerations and guidelines for securities and non-securities tokens (2019). https://digitalchamber.org/wp-content/uploads/2019/08/Security-Non-Security-Tokens.pdf. Accessed 30 Apr 2020

Reforming the Russian Oil and Gas Sector (Comparison with Norwegian Practice)

M. Yu. Belyakova(✉)

Moscow State Institute of International Relations (University) of the Ministry of Foreign Affairs, Moscow, Russia
ma-arina@mail.ru

Abstract. The objective of this paper is to analyze the basic taxation principles in the Russian petroleum industry, identify problems that impede its effective functioning and consider the possibility of reforming the industry's regulation. The current Russian taxation system, based on the hydrocarbons' volumes extracted and not on the financial performance of oil and gas companies, does not fully take into consideration the economics of oil and gas projects. Various measures taken by the state in this area, such as tax relieves and the introduction of lowering ratios, have a positive effect on the efficiency of resource development, but do not affect the very foundations of the system. The current situation in the Russian oil and gas industry makes it extremely relevant to review its taxation policy. In this regard, the paper further considers the Norwegian experience in the taxation of oil and gas industry. The author uses methods of comparative analysis and generalization, as well as statistical and systematic methods of scientific analysis. In conclusion the author analyzes the possibility to apply basic principles of Norwegian oil and gas industry taxation to the Russian petroleum sector.

Keywords: Hard-to-recover reserves · Mineral extraction tax · Oil and gas industry in Norway and Russia · State direct financial interest · Tax on financial result

1 Introduction

At the present time there is an acute problem with the quality of explored and developed reserves in Russia:

- existing fields have already been significantly depleted, the average industry depletion rate is 55% [4],
- most of the explored and unexplored reserves are classified as hard to recover – their share in the total structure of reserves is more than 70% [13],
- due to the insufficient volume of geological exploration, the reproduction of the mineral resource base is not properly ensured. It can become a serious threat to the energy and economic security of the country in the future, especially in terms of rapid growth in oil and gas production [8, 12, 17].

S. I. Ashmarina et al. (Eds.): ENGINEERING ECONOMICS WEEK 2020, LNNS 139, pp. 134–144, 2021.
https://doi.org/10.1007/978-3-030-53277-2_15

In order to solve all the above-mentioned problems, it is necessary to invest in new expensive technologies, equipment, conduct research and development, work on the reproduction of the mineral resource base. However, the current tax system does not take these factors into account and thus does not contribute to the solution of these problems. At the same time, taxes from oil and gas activities account for almost half of state budget revenues.

2 Methodology

The methodological basis of the research includes various methods of scientific research: statistical and system methods; comparative analysis and generalization, which allowed to study thoroughly various aspects of taxation in the oil and gas industry of Russia, identify the main problems in Russian experience and develop recommendations for improving the tax system in Russia, taking into consideration the Norwegian experience.

3 Results

The basis of the tax burden on Russian oil and gas companies engaged in production of hydrocarbons is the mineral extraction tax (hereinafter – MET). In 2018, its share in the structure of all taxes paid by oil and gas companies accounted for 66.6% [14]. The MET rate is constantly growing. The sharp increase of the MET rate was associated with the introduction of the "tax maneuver". It implies reduction of export duties on oil from 59% in 2014 to 30% in 2017 with a simultaneous increase in the MET rate for oil from 493 rubles per ton in 2015 to 919 rubles in 2017 (Table 1). It also provides for a reduction in the level of export customs duty on light petroleum products (including gasoline) and an increase of export duties on dark petroleum products, with a reference to the export duty on oil [14].

On January 1, 2019, the process of completing the "tax maneuver" began. Under this process it is planned to make the export duty equal to zero by 2024 (in 2019 its rate is 30%) with a simultaneous increase of MET rate.

MET was introduced on January 1, 2002 (Chapter 26 of the "Tax Code of the Russian Federation" [15]) to solve the problem of using transfer pricing by the companies within one structure. The three previously used taxes calculated on the basis of gross indicators could not solve this problem.

Initially, MET had a flat rate, which considerably contributed to the increase in budget revenues, but later it turned out that some companies received very high profits and others incurred losses due to the fact that the current rate did not take into consideration the particularities of the developed fields. In this regard, it was decided from January 1, 2014 to introduce a differentiated MET, which rate would depend on several production factors and production conditions.

Table 1. The evolution of the mechanism for calculating the MET rate for oil by periods of its application

Period	MET rate calculation formula	Grounds
01.01.2002–31.12.2003	340 · price ratio	Federal Law No. 126-FL of 08.08.2001
01.01.2004–31.12.2004	347 · price ratio	Federal Law No. 126-FL of 08.08.2001 (as amended No. 117-FL on 07.07.2003)
01.01.2005–31.12.2006	419 · price ratio	Federal Law No. 126-FL of 08.08.2001 (as amended No. 102-FL on 18.08.2004)
01.01.2007–31.12.2011	419 · price ratio · Ids	Federal Law No. 151-FL of 27.07.2006
01.01.2012–31.12.2012	446 · price ratio · Ids · Ir	Federal Law No. 258-FL of 21.07.2011
01.01.2013–22.08.2013	470 · price ratio · Ids · Ir	
01.09.2013–31.12.2013	470 · price ratio · Ids · Ir · Ic · Idh	Federal Law No. 213-FL of 23.07.2013
01.01.2014–31.12.2014	493 · price ratio · Ids · Ir · Ic · Idh	Federal Law No. 263-FL of 30.09.2013
01.01.2015–31.12.2015	766 · price ratio · Ids · Ir · Ic · Idh	Federal Law No. 366-FL of 24.11.2014
From 01.01.2016	857 · price ratio · Ids · Ir · Ic · Idh	
From 01.01.2017	919· price ratio · Ids · Ir · Ic · Idh	

Source: author.

Currently, MET on oil (R) is calculated according to the following formula [16]:

$$R = 919 * \text{price ratio} * \text{Ids} * \text{Ir} * \text{Ic} * \text{Idh},$$

where: Price ratio is the coefficient characterizing the dynamics of world oil prices. It is determined by the following formula: Price ratio = (AP - 15) * AER/261, where AP – the average monthly price of Urals on the Rotterdam and Mediterranean exchanges (dollars/ barrel) and AER is the average monthly exchange rate of the ruble to the US dollar.

Ids – indicator characterizing the degree of depletion of reserves of a subsoil plot. The indicator provides for a reduction in the MET rate for oil for fields with a high degree of depletion.

Ir – indicator characterizing the size of reserves of a subsoil plot. This indicator provides for a reduction in the MET rate for small deposits [15].

Ic – indicator characterizing the degree of difficulty of extracting oil. It varies from 0 to 1 depending on the complexity of oil production from a specific reservoir [16].

Idh – indicator characterizing the degree of depletion of a particular raw hydrocarbon deposit. If the degree of depletion of the reserves of the deposit exceeds "1", Idh is equal to "0.3". In other cases, Idh is equal to "1". For deposits containing hard-to-recover oil reserves, Idh is equal to "1" [15].

The formula for calculating MET is established not only for oil, but also for gas and gas condensate (in accordance with the amendments introduced by Federal Law No. 263-FL of 30.09.2013). The formula for calculating MET for gas is as follows:

$$METgas = 35 * Ufe * Ic + It,$$

where Ufe – basic unit of fuel equivalent; Cc – indicator characterizing the degree of complexity of gas production of combustible natural and (or) gas condensate from a hydrocarbon reservoir; It – indicator characterizing the costs of transporting natural gas. And the formula for calculating MET for gas condensate is as follows:

$$METgas\ condensate = 42 * Ufe * Ic * Icf_{,}$$

where Ufe – basic unit of fuel equivalent;

Ic – coefficient characterizing the degree of difficulty of natural gas production and (or) gas condensate from a hydrocarbon reservoir; Icf – adjusting factor;

The introduction of MET, taking into consideration the peculiarities of field development, in general, had a positive impact on improving the situation in the oil and gas industry. The application of reduction indicators which depend on the conditions of field development to the basic MET rate for oil, gas and condensate has a significant impact on the final performance of projects and relieves the tax burden for oil and gas companies. At the same time, the problem of developing hard-to-recover and significantly depleted reserves is not settled, especially taking into consideration the constantly increasing MET rate and the ratio characterizing the world oil price.

The tax system does not contribute to the development of mature production fields either. Such actions may lead to premature termination of their development. As a result, a significant part of the hard-to-recover hydrocarbon reserves may be lost. Oil and gas companies have to complete production when operating costs do not pay off and do not bring profit. Such actions lead to a loss of reserves and a reduction in recoverable resources.

Since it is unprofitable to continue developing new or significantly depleted deposits companies apply to the state for tax relieves. Tax incentives to the MET were provided for hydrocarbon fields in certain regions of Eastern Siberia, the Nenets Autonomous Okrug, Yamalo-Nenets Autonomous Okrug and offshore fields for a certain number of years or after reaching a certain level of production. Also, the zero MET rate was applied to oil production with a viscosity of more than 200 MPa. On the one hand, the further growth of tax incentives threatens the efficiency of the entire tax system. On the other hand companies will not be interested in further development of hard-to-recover hydrocarbon reserves without incentives.

In order to attract investment in the development of new deposits, and at the same time to get away from the "manual setting" of the taxation system, the state introduced a new revenue added tax (hereinafter – RAT) with a 50% rate [7]. The tax base is the revenue from oil sale reduced by operating and capital expenses, export duties and reduced MET. The performance of this tax will be monitored until 2024 and after that it will be possible to make conclusions about its efficiency and possibilities of further application to other fields [13].

Thus, it is impossible to say that the Government does not take any measures to improve the tax system in the most important for the Russian economy industry. The introduction of a new revenue added tax, the use of lowering ratios in calculating MET, and the provision of incentives for the development of hard-to-recover reserves prove this fact. But at the same time, the measures taken by the Government, except for RAT, do not affect the fundamentals of the entire taxation system. That is why it is necessary to work out new approaches to the taxation policy.

In this regard the Norwegian experience in the taxation of oil and gas sector is a very good example. The current taxation policy in petroleum sector of Norway on the one hand allows the Government to maximize its economic rent from the realization of oil and gas projects on the territory of Norway (78%), but on the other hand makes the development of the shelf commercially viable for investors due to the taxation of net profit, but not the gross revenue.

The basic principle of taxation of oil and gas activities in Norway is the taxation of profits [6], but not the revenues – as it is in Russia (Table 3). This approach contributes to the development of both new fields, where no infrastructure has been created for production, and mature fields, where complex, expensive technologies are required to maintain production at the current level.

Corporate income tax (at the rate of 22%) is levied on profits from the sale of hydrocarbons formed after the deduction from the revenue of exploration, research and development costs, depreciation, CO2 fees, license fees and operating expenses (these types of taxes are presented in Table 2) [3, 16].

Special petroleum tax (at the rate of 56%) is levied on the same basis as the corporate income tax with an exception of the uplift deducted from the profit. The uplift is an additional depreciation on equipment for the extraction of hydrocarbons on the shelf, pipelines through which oil and gas is delivered onshore [7]. The uplift rate is 20.8% of capital expenditure in offshore infrastructure and it is recognized for four years, starting from the first year of depreciation of operating assets (that is, 5.2% annually) [7].

Through the uplift, the state enables the investor to carry out an accelerated return on investment by providing him with the right to account for more expenses than expenses incurred by the company.

Table 2. The main types of taxes and fees for oil and gas production in Norway and their rates

Tax	Rate
Corporate income tax	22%
Special petroleum tax	56%
Carbon dioxide (CO2) tax	0.98 krone per 1,000 cubic meters m of burned or released gas, or 1 L of burned oil
N2O emission tax	16.14 kroner per 1 kg of nitric oxide
VAT	25%
License fee (license fee paid at the end of the 10 year license period)	1st year – 34,000 kroner per sq. km 2nd year – 68,000 kroner per sq. km further – 137 000 kroner per sq. km

Source: author based on [1].

Table 3. Calculation of the tax base of oil and gas activities in Norway

Income from the sale of hydrocarbons:
- Operating expenses
- Depreciation (straight-line method of depreciation for investments – 6 years)
- Exploration, research and development costs - Incurred decommissioning costs
- Fee for CO2 emissions and license fee
- Net financial costs (limited by the rule of thin capitalization: 20% of stakeholder equity)
= Tax base of a legal entity (standard corporate income tax rate is 22%)
- Uplift (additional depreciation on installations located on the shelf, that is, interest on capital expenditure (20.8% for 4 years – 5.2% from the volume of investments annually)
= Special tax base (tax rate - 56%)

Source: author based on [2].

The Norwegian taxation policy in oil and gas sector, based on corporate income tax and a special petroleum tax, is easy to administer and it allows the government to maximize the economic rent from oil and gas activities, while taking into consideration the interests and capabilities of oil and gas companies. It stimulates the application of highly efficient technologies to the development of fields, activates exploration activities and encourages the development of infrastructure around fields developed by subsoil users.

4 Discussion

In Russia, the possibility of introducing the tax on financial result [5] was repeatedly considered, however, each time the solution of this issue encountered certain difficulties. The fact is that when switching to a new tax system based on the tax on financial result in Russia, the state fears that it will face the problem of insufficient receipt of tax revenues to the federal budget, primarily because the amount of tax paid depends on the profitability of the project, which at the initial stage, is usually low. In this regard, the system is certainly more favorable to oil and gas companies. Secondly, the calculation of the tax base based on cost indicators offers opportunities for oil and gas companies to lower their tax base due to overstatement of costs. Therefore, the Government does not dare to radically restructure the tax system, but introduces changes "surgically". At the same time, the current tax system based on the MET provides the federal budget with guarantees of stable tax revenues regardless of the profitability of the project at all stages of development and, being tied to the volume of extracted raw materials, exclude any possibility of tax evasion by reducing the tax base.

In this regard, before reforming the taxation system in the Russian oil and gas industry, it is necessary to consider the possibility of introducing a monitoring system for field development, which provides for the establishment of a hydrocarbon production control system in terms of quality, production capacity of wells and other parameters that determine the amount of costs and will be further taken into account when calculating the taxable base. Such a monitoring system could be ensured through the state participation in the development of deposits following the example of the Norwegian State Direct Financial Interest (hereinafter – SDFI).

From the economic point of view, SDFI means the possibility of direct participation in the management of field development and in distribution of profit [11]. In order to manage the SDFI portfolio [11] the state-owned company Petoro was established in summer 2001. At that time Petoro had shares in 80 licenses [9]. Petoro is not an oil company. It acts as the trustee of SDFI in licenses on the Norwegian continental shelf. The state is legally liable for all Petoro's obligations under concluded agreements and other obligations [9, 10].

Currently, Petoro owns almost one third of all Norwegian oil and gas reserves. As of the end of December 2018, Petoro is participating in 208 licenses, 34 developed producing fields and 18 joint ventures (JVs) that own pipelines and oil and gas processing facilities in Norway (Tables 4, 5, 6 and 7) [9, 10].

Table 4. Participation of Petoro in field development

Group of fields	Share of SDFI (as of December 31, 2018)	Remaining production period	License duration period
Flyndre Unit	0,354%		–
Fram H-Nord Unit	11, 2%	2030	2024
Gimle Unit	24,19%	2032	2023
Grane Unit	28, 90%	2043	2030
Gullfaks Unit	30%	2034	2036
Haltenbanken Vest Unit (Kristin)	19,57%	2034	2027
Heidrun Unit	57,79%	2044	2024
Johan Sverdrup Unit	17,36%	2058	2036
Martin Linge Unit	30%	2031	2027
Norne Inside	54%	2035	2026

(continued)

Table 4. (*continued*)

Group of fields	Share of SDFI (as of December 31, 2018)	Remaining production period	License duration period
Ormen Lange Unit	36,49%	2042	2040
Oseberg Area Unit	33,60%	2040	2031
Sindre Unit	27,09%	2032	2023
Snorre Unit	30%	2040	2040
Snøhvit Unit	30%	2051	2035
Statfjord Øst Unit	30%	2025	2026
Sygna Unit	30%	2025	2026
Troll Unit	56%	2054	2030
Valemon unit	30%	2026	2031
Vega Unit	28,32%	2030	2024
Visund Inside	30%	2040	2034
Åsgard Unit	35,69%	2032	2027
Field			
Alta	30%	2019	2025
Draugen	47,88%	2035	2024
Dvalin	35%	2032	2041
Ekofisk	5%	2050	2028
Eldfisk	5%	2050	2028
Embla	5%	2050	2028
Gjøa	30%	2027	2028
Heidmal	20%	2021	2021
Johan Castberg	20%	2052	2019
Kvitebjørn	30%	2035	2031
Maria	30%	2033	2036
Rev	30%	2019	2021
Skirne	30%	2020	2025
Skuld	24,55%	2035	2026
Statfjord Nord	30%	2025	2026
Svalin	30%	2043	2030
Tordis	30%	2030	2040
Tune	40%	2021	2020
Urd	24,55%	2035	2026
Veslefrikk	37%	2025	2020
Vigdis	30%	2040	2040

Source: author based on [9, 10].

Table 5. The share of Petoro in the ownership of pipelines

Pipeline	Share of participation (as of December 31, 2018)	License duration period
Oseberg transport system	48,38%	2031
Troll I + II oil pipeline	55,76%	2023
Grane oil pipeline	42,06%	2030
Kvitebjørn oil pipeline	30%	2020
Norpipe pipe line	5%	2028

Source: author based on [9, 10].

Table 6. The share of Petoro in the ownership of onshore infrastructure for oil production

Pipeline terminal	Share of participation (as of December 31, 2018)	License duration period
Mongstad DA	35%	Unlimited

Source: author based on [9, 10].

Table 7. The share of Petoro in the ownership of gas pipelines

Gas pipeline	Share of participation (as of December 31, 2018)	License validity period
Gassled	46,69%	2028
Haltenpipe Gassled	57,81%	2020
Mongstad Gas Pipeline	56%	2030
Nyhamna	26,14%	2041
Polarled (NSGI)	11,94%	2041
Valemon Rikgassror	30%	2031
Dunkerque Terminal DA	30,35%	2028
Zeepipe Terminal J.V.	22,88%	2028
Vestprosess DA	41%	Unlimited
Norsea Gas AS	40,06%	Unlimited
Ormen Lange Eindom	36,49%	Unlimited

Source: author based on [9, 10].

SDFI serves as a support tool for the realization of the state strategy in the development of oil and gas sector in Norway. Due to this mechanism the state remains the largest owner in the industry, Petoro's revenues are accumulated in the Norwegian Oil Fund. Moreover Petoro administers the taxation policy during the development of hydrocarbon fields. Thus, Petoro is an efficient tool for monitoring license performance, which, on the one hand, allows to increase the efficiency of field development, and on the other hand, provides the state with additional income from oil and gas activities.

5 Conclusion

The Russian authorities responsible for the development of the oil and gas industry in Russia could consider creating a mechanism similar to the Norwegian SDFI, which would allow the state to control directly oil and gas production in the fields under license agreements, as well as provide the state budget with stable additional income from oil and gas activities.

After ensuring state control over the development of deposits, following the example of the Norwegian SDFI, it would be possible to create a new tax system for the oil and gas industry in Russia, based on the financial results of companies. A new taxation policy could provide oil and gas companies with a differentiation of the tax burden depending on specific conditions of oil and gas production and would facilitate the development of hard-to-recover reserves.

References

1. BAHR: Oil and gas in Norway (2019). https://bahr.no/wp-content/uploads/2019/01/Oil-and-Gas-in-Norway-January-2019-1.pdf. Accessed 19 Apr 2020
2. Delloitte: Oil and gas taxation in Norway (2014). https://www2.deloitte.com/content/dam/Deloitte/global/Documents/Energy-and-Resources/gx-er-oil-and-gas-taxguide-norway.pdf. Accessed 19 Apr 2020
3. Goncharenko, I.A.: Legal regulation of taxation of oil and gas fields development in the Russian Federation and foreign countries. Statute, Moscow (2009)
4. Mazurina, E.V.: Hydrocarbon production – a source of incomes of the state and subsoil users. Petrol. Geol. Theory Pract. **9**(3), 1–14 (2014)
5. Nedbaylik, I.O.: Mineral extraction tax as a source of state revenue. Int. J. Appl. Basic Res. **9**(2), 276–279 (2016)
6. Noreng, O.: The Oil Industry and Government Strategy in the North Sea. Groom Helm, London (1980)
7. Norwegian Petroleum: The Petroleum tax system (2019). http://www.norskpetroleum.no/en/economy/petroleum-tax/. Accessed 19 Apr 2020
8. Oleynov, A.G.: Fuel and energy complex of the world. MGIMO, Moscow (2008)
9. Petoro: SDFI and Petoro annual report (2012). https://www.petoro.no/petoro-annual-report/2012-/accounts/accounts-sdfi/notes-11-21. Accessed 19 Apr 2020
10. Petoro: SDFI facts (2019). https://www.petoro.no/about-petoro/sdfi-facts. Accessed 19 Apr 2020
11. Roginsky, S.V.: Formation of a model of state regulation of the fuel and energy complex (Using the example of the Norwegian fuel and energy complex). Ph.D. thesis. RSHU, Moscow (2001)
12. Sidorenko, E.L., Bartsits, I.N., Khisamova, Z.I.: The efficiency of digital public administration assessing: Theoretical and applied aspects. Public Adm. Issues **2**, 93–114 (2019)
13. The draft energy strategy of the Russian Federation for the period up to 2035 (version dated 18.12.2019). https://minenergo.gov.ru/node/1920. Accessed 19 Apr 2020
14. The Ministry of Finance of the Russian Federation: Execution of the federal budget and budgets of the budget system of the Russian Federation for 2018 (preliminary results) (2019). https://www.minfin.ru/common/upload/library/2019/04/main/02_Ispolnenie_federalnogo_budzheta_za_2018_god.pdf. Accessed 19 Apr 2020

15. The Tax Code of the Russian Federation (part two) No. 117-FL dated 05.08.2000 (as amended on 07.29.2017) with amendments and additions that entered into force on 01 September 2017. http://www.consultant.ru/document/cons_doc_LAW_28165/992cab53d45857cbe7b61b9a688ed54275c0dbb4/. Accessed 19 Apr 2020
16. Tormodsgard, Y. (Ed.): Facts 2014. The Norwegian petroleum sector (2014). https://www.regjeringen.no/globalassets/upload/oed/pdf_filer_2/faktaheftet/fakta2014og/facts_2014_nett_.pdf. Accessed 19 Apr 2020
17. Vygon Consulting: The tax reform of the oil industry (2015). https://vygon.consulting/upload/iblock/6b7/vygon_consulting_upstreamtaxreform.pdf. Accessed 19 Apr 2020

Approaches to the Formation of an Industrial Enterprise Management System

R. S. Ibragimova(✉) and D. S. Golovkin

Ivanovo State University, Ivanovo, Russia
irozalia@hotmail.com, dm-golovkin@yandex.ru

Abstract. The authors consider the problem of increasing the efficiency of industrial enterprise management in an unstable and volatile economic environment. The article aims to justify and develop a methodological approach to the formation of an industrial enterprise management system, as well as improving methods and tools for managing an industrial enterprise. The authors conducted critical analysis of the advantages and limitations of modern management systems described in the economic literature and used in the practice of managing industrial enterprises. The study proposes a solution to the problems of managing an industrial enterprise based on the integration of the enterprise economic potential concept and modern management tools: balanced scorecard, the concept of the industry life cycle, foresight, benchmarking and value-based management.

Keywords: Balanced scorecard (BSC) · Benchmarking · Enterprise economic potential · Foresight · Industry life cycle · Value-based management

1 Introduction

At present, the unstable and volatile economic environment creates barriers for business success and increases the risks of enterprise development. New management methods are needed to ensure the flexibility of the enterprise management system that would contribute to successful and long-term development of the enterprise. The increasing complexity of business functioning entities requires permanent improvement of management tools.

Issues of increasing the industrial enterprise management efficiency were resolved in the works of Brierley, Cowton & Drury [1], Chapman [2], Drucker [3], Feigenbaum [4], Horváth [5], Kaplan & Norton [9], Knight [10], Martin [11], Parmenter [12], Pearson & Clair [13], Porter [14], Povey [15] and others.

Modern management are characterized by a wide variety of techniques, methods and tools that allows to improve the management efficiency and make effective management decisions, for example: Balanced Scorecard, Total Quality Management, Controlling, Budgeting, Industry life cycle concept, Foresight, Benchmarking, Key performance indicators, crisis management, Value-Based Management, Management by Objectives and others.

A comprehensive system of assessment, planning, organization and control that takes into account the current situation of the enterprise, its material resources,

S. I. Ashmarina et al. (Eds.): ENGINEERING ECONOMICS WEEK 2020, LNNS 139, pp. 145–154, 2021.
https://doi.org/10.1007/978-3-030-53277-2_16

competencies and capabilities in the global environment is required to determine the strategic directions of increasing the competitiveness, efficiency and sustainability of enterprises in any industry. We believe that such system can be formed on the basis of the concept of «enterprise economic potential» (EEP).

2 Methodology

Based on a critical analysis of domestic and foreign literature, we propose definition of the economic phenomena «enterprise economic potential» (EEP) [7]. EEP is the ability of an enterprise to ensure its long-term functioning and achievement strategic goals based on the use of a resources system, taking into account the opportunities and its competencies in order to satisfy consumer demand for goods and services in the volume and quality that is determined by his individual needs.

The purpose of our research is to justify and develop methodological approach to formation of enterprise management system based on the concept of enterprise economic potential and to improve methods and tools of industrial enterprise management. The specific objectives of the study are:

- critical analysis of the advantages and limitations of modern management systems described in the economic literature and used in the practice of industrial enterprises management;
- management system development on the basis of enterprise economic potential concept.

In the research, economic phenomena were investigated from the standpoint of the theory of systems. We used the methods of exploratory and descriptive studies such as literature review, case-study, trends analysis, benchmarking, expert interviews and others.

3 Results

In the study the advantages and limitations of existing enterprise management systems were identified. Comparative analysis allowed us to develop a holistic and comprehensive approach to the formation of a strategic management system based on the integration of the concept of the enterprise economic potential and effective management methods and tools.

Table 1 shows the results of a comparative analysis of modern industrial enterprise management systems described in the economic literature and applied in practice.

An industrial enterprise is a multidimensional and complex system with many interconnected subsystems and functions, the effectiveness growth of which can be achieved by applying a system methods, that consider the organization as a holistic hierarchical structure consisting of various interconnected and interdependent elements, and are used to achieve a qualitatively new state of organization. Such systematic methods that would be applicable in most enterprises are controlling, total quality management, crisis management and management by objectives. The other systems

Table 1. Comparative analysis of enterprise management systems

Advantages	Limitations
Controlling [5]	
– aims to coordinate management activities and to achieve the goals of the enterprise; – allows for operational management	– short-term oriented; – strategic controlling is seen as a means of ensuring survival, not of development
Budgeting [1]	
– provides transparency of financial results due to the system of indicators; – reduces production costs	– short-term oriented; – implementation depends on the beliefs and support of managers; – lack of coordination of operational budgets with enterprise strategy
Balanced scorecard (BSC) [9]	
– an integrated approach; – able to translate vision and strategy into specific indicators; – customer and market oriented; – focuses on a limited number of key indicators	– consumer-oriented, ignores other audiences; – there is no final benchmark for measuring effectiveness; – lack of trends consideration; – there is no threshold values
Industry life cycle concept [8, 14]	
– decomposition allows to define a business strategy for each phase that characterizes certain conditions and features of the functioning of the business	– there is no methodology for determining the moments of transition between phases; – it is necessary to obtain a large amount of information for analysis
Foresight [6, 11]	
– focuses on a comprehensive assessment of opportunities; – involves a large number of experts from various fields of activity; – designed for practical activities	– suggests subjective influence of individual participants; – possible lack of experts from related fields; – possible insufficient information base
Benchmarking [15]	
– allows to get an information of company's positioning in the industry market; – allows to assess global industry's trends	– any experience, including foreign, is necessary to adapt to Russian realities; – most effective methods are costly
Total quality management [4]	
– suggests setting long-term goals for the enterprise; – quality is created at all phases of the product life cycle, all departments of the enterprise are involved in process	– there is a lack of a quantifiable indicator of the management system; – focuses on marketing management tools and moves away from the financial and economic processes
Key performance indicators [12]	
– allows to formulate a strategy and convey the goals of the enterprise to employees; – closely related to staff motivation; – suggests constant feedback from employee	– difficulties with the assessment of non-standardized processes; – focuses only on financial indicators of the enterprise

(*continued*)

Table 1. (*continued*)

Advantages	Limitations
Crisis management [13]	
– allows to anticipate crises, minimize losses, turn their negative consequences into positive changes; – transparency of enterprise management	– high degree of individualization in management process; – high requirements for the qualifications of managers
Value-based management [10]	
– focus on long-term strategic prospects for the development of the enterprise and strategic analysis and choice of market strategy; – takes into account the interests of the investor and the level of risk; – based on market information and reflects actual practice of transactions; – takes into account the influence of industry	– variety of assessment approaches to the value of enterprise creates confusion in the assessment and decision making process; – enterprise valuation often does not take into account certain types of intangible assets; – owners are not always interested in increasing the value of their business
Risk management system [2]	
– focused on long-term sustainable development; – allows to minimize certain types of specific (systematic) risks; – the ability to account, analyze and evaluate various options of scripts, the ability to take into account risk factors	– some risks cannot be avoided or the costs of their management exceed the damage from their occurrence; – the enterprise is deprived of additional sources of profit due to the formation of a large reserve fund and other costs associated with risk management
Management by Objectives [3]	
– provides a holistic vector of enterprise management system; – allows to make decisions based on management objectives, to ensure the effectiveness of management actions according to the criteria of necessity and sufficiency; – promotes centralization processes; – every employee understands his role in business	– the introduction of the entire system in the complex is quite time-consuming and often accompanied by natural resistance of the staff; – the development and implementation of the system requires highly qualified management personnel; – the system is not recommended to be implemented if the organization is in a crisis state

Source: authors.

described in literature are used to achieve a particular goal of enterprise management (assessment of performance on the basis of BSC and value-based management; construction of a motivation and control system based on KPI; organization and control of financial and economic activities based on budgeting; risk management; strategic planning based on benchmarking, Foresight and industry life cycle), which can be used as a local tool in solving specific tasks.

Crisis management shows all these advantages in crisis conditions (enterprise bankruptcy, external shocks and others), but does not reflect effectiveness in a relatively stable environment. Controlling is more focused on managing short-term performance indicators (revenue and profit) rather than achieving the strategic goals (strategic controlling aims at «survival» of the business). In market conditions short-term oriented controlling does not allow achieving long-term effectiveness of the management system. Total quality management is primarily aimed at improving quality indicators and is far from the financial parameters and processes of the enterprise. Drucker's Management by Objectives concept is the most comprehensive and systematic management system, but does not provide clear implementation tools [3]. Thereby, the system methods and applied management tools discussed above complement each other, and their complex application will contribute to leveling the limitations.

4 Discussion

As has been noted earlier, the enterprise is a complex system with many interrelated parts. It is necessary to use a comprehensive management system that takes into account the current and future situation of the enterprise – strengths and weaknesses, opportunities and threats of external environment. Such system allows to form a comprehensive assessment of the current state of the enterprise, to analyze interdependencies within the enterprise, to develop timely and economically feasible strategic decisions, to make forecasts and plans.

In our opinion, such a comprehensive system for increasing the enterprise efficiency can be formed on the basis of the concept of «enterprise economic potential», and the economic potential itself as well as its individual elements can be subject of management. The concept of the EEP implies an industrial enterprise management system that considers enterprise as a complex hierarchical structure and reflects its ability to ensure long-term functioning, development and achievement of strategic development goals. EPP management is based on a systematic analysis of an industrial enterprise in order to cover the entire financial and economic cycle of the organization and the capital cycle. The strategic directions of the socio-economic and innovative development of an industrial enterprise can be assessed by the structure of economic potential – elements that reflect key points in the formation of enterprise efficiency (production, personnel, finance, innovation, marketing, organization and management) and resources (material and intangible resources of the enterprise participating in the commercial activities of the company and the formation of its income) competencies (reflecting the efficiency of using resources) and opportunities (reflecting internal and external effects – prospects for the enterprise development) [7].

Considering the category «enterprise economic potential» from the position of 6 elements (subsystems) and 3 aspects of management (resources, competencies and opportunities) allows us to consider the economic potential of the enterprise as a concept of industrial enterprise management. An industrial enterprise, like any complex system of elements united by close causal relationships that are in continuous interaction, is a complex of subsystems that perform certain functions. The enterprise economic potential, as a complex category, reflects the basic subsystems and functions of the enterprise, as well as the relationships and patterns that arise between them, which allows creating a management system based on the concept of EPP.

In the framework of this study, obtaining holistic and systematic management tools is possible through the integration of the concept of the enterprise economic potential and modern methods of enterprise assessment and analysis, organizing, control and planning. Such complex system reduces the influence of the limitations of individual tools and maximizes their positive effect. Figure 1 reflects our methodological approach to building a management system based on the concept of the EEP and proposes management tools that are used to implement management functions in the process of making management decisions – planning, evaluation and measurement, organization and motivation, control.

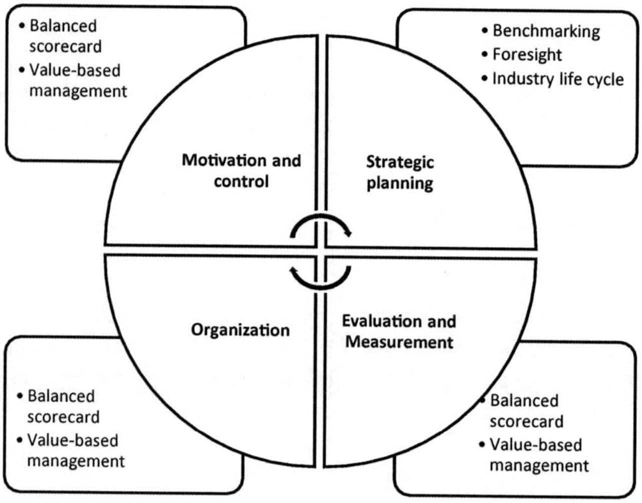

Fig. 1. Approach to the management system of industrial enterprise economic potential (Source: authors)

Planning is the process of determining the goals of the organization, the means to achieve these goals, as well as the required resources, which allows to develop activities aimed at achieving the goals. Effective planning of the enterprise's activity, which means the growth of its economic potential, allows studying the phases of the industry's life cycle, forming priorities in the field of economy, science, technology and society, as well as searching for the best industry experience.

The conclusions of the analysis of industry cycles allow to formulate a common system of key factors for each highlighted stage for the growth of the enterprise economic potential according to its individual elements – production, personnel, finance, innovation, marketing, organization and management. The results of the foresight provide an opportunity to consolidate the opinions of the expert community and identify the current state, problems, opportunities and risks of the industry, formulate strategic development goals and a holistic image of the future, and identify priority areas for industrial development. Benchmarking in the planning system allows the process of determining and adapting existing examples of effective functioning of companies to improve their own work.

The evaluation function is expressed in measuring the level of individual elements of the EPP and the integral level of economic potential. The literature has very deeply and thoroughly studied the methodology for assessing the resource component (production capital, financial and investment resources, certain types of natural resources involved in economic turnover, as well as human resources), but insufficient attention has been paid to assessing competencies and capabilities.

The paper proposes a methodological approach to the formation of a system for assessing the enterprise economic potential, which consists in the analysis of EEP's six elements and obtaining an integrated assessment of EEP with the allocation of indicators in the following aspects: resources, competencies and opportunities (Fig. 2).

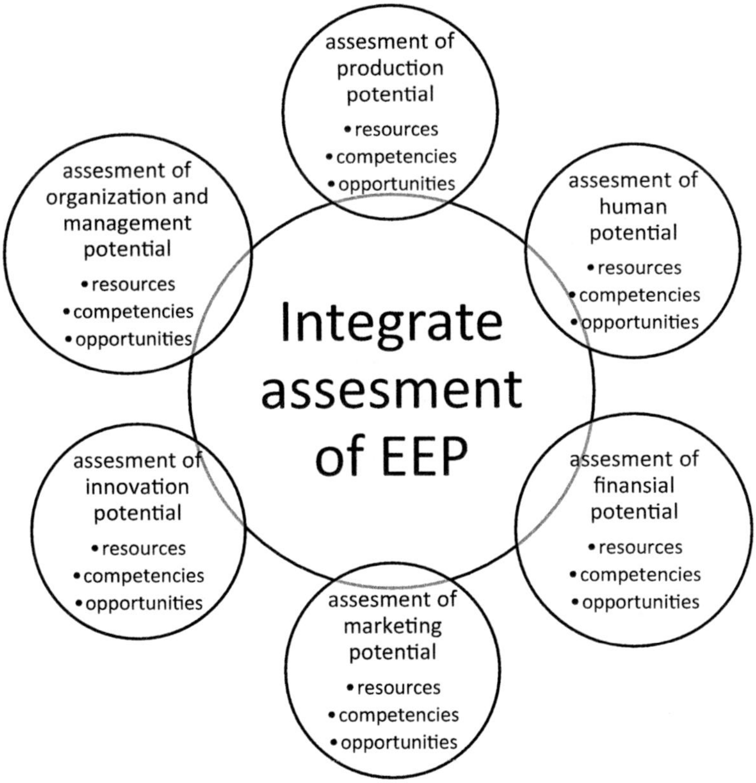

Fig. 2. Approach to the formation of a system for assessing the levels of EEP (Source: authors)

The system for assessing the economic potential of the enterprise, as well as its individual elements, is based on quantitative and qualitative characteristics. The idea of balance, complementarity of financial and non-financial indicators is reflected in the

Kaplan and Norton concept of a balanced scorecard (BSC), which most closely meets modern requirements for assessing the enterprise activities.

Organization is a combination of resources and coordination the actions of individual elements of the system. The functions of motivation and control are to achieve the organization's goals. The balanced scorecards concept and value-based management are the basic tools in the functions of the organization, motivation and control.

Schematically, the industrial enterprise management system based on the concept of the enterprise economic potential and a set of management tools that includes a balanced scorecard, industry life cycle concept, Foresight technologies, benchmarking and value-based management is presented in Fig. 3.

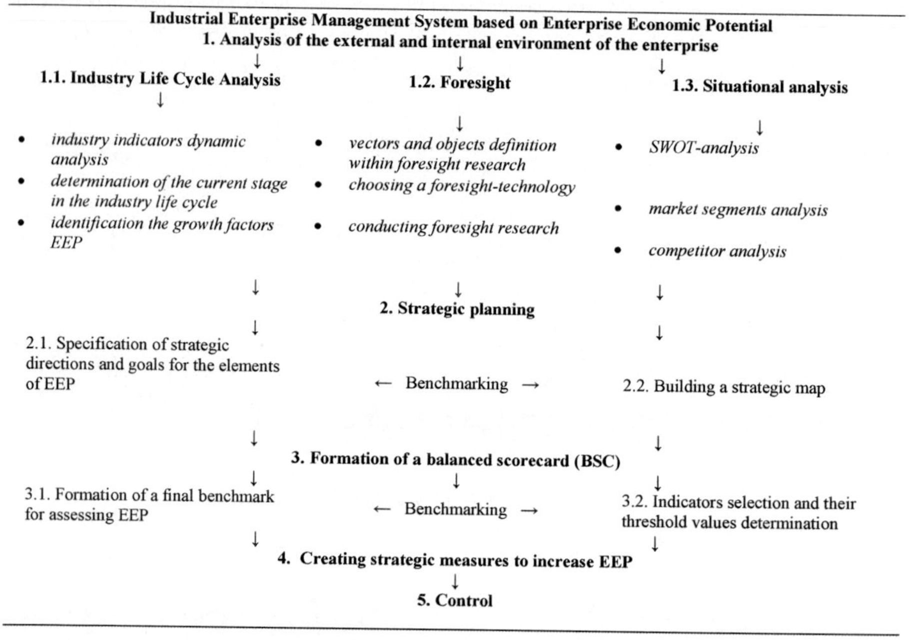

Fig. 3. Industrial enterprise management system based on enterprise economic potential (Source: authors)

Thus, based on a comparative analysis of industrial enterprise management systems reflected in the scientific literature, our study developed and substantiated a management system based on the concept of the enterprise economic potential. The EPP management concept is based on a systematic approach and covers the entire cycle of financial and economic activities and the capital cycle. The approach provides integrated management and makes it possible to identify the impact of various aspects of activity on the enterprise effectiveness.

5 Conclusion

Analysis of systems and tools for industrial enterprises management made it possible to identify their main shortcomings: limited conditions and situations in which the system shows efficiency, orientation toward solving only a few management tasks, insufficient integrity of the approach to enterprise management, focus on short-term performance indicators rather than achieving strategic goals.

The study proposes a solution to the problems in an industrial enterprise management based on the integration of the concept of the enterprise economic potential and modern management tools, such as a balanced scorecard, the concept of the industry's life cycle, Foresight technology, benchmarking, and the value-based management. Such integrated management system makes it possible to minimize the shortcomings of individual tools and maximize their positive effect in the implementation of management functions – planning, evaluation and measurement, organization and motivation, control, which reflect the management decision-making process.

Application of the authors' approach to the formation of the economic potential management system allows:

- to form an understanding of the current state and development trends of the industry as a whole and specific enterprises separately;
- to formulate strategic goals and a holistic image of the future, to identify priority areas for the enterprise development;
- to highlight the strengths and weaknesses of the organization, the problems, threats and opportunities facing the enterprise;
- to create strategic management decisions to ensure the growth of economic potential and capitalization of the enterprise;
- to coordinate strategic goals with operational actions in the implementation of the enterprise strategy;
- to choose a reasonable integral indicator for assessing the economic potential of the enterprise, which allows to compare enterprises with each other.

Acknowledgments. The study was carried out with the financial support of RFFR, project № 18-410-370004 "Forecasting and evaluation of strategic directions for socio-economic and innovative development of the textile industry on the basis of foresight studies".

References

1. Brierley, J.A., Cowton, C., Drury, C.: Investigating activity-based costing in manufacturing industry. Cyprus Int. J. Manag. **13**(1), 21–33 (2008)
2. Chapman, R.J.: Simple Tools and Techniques for Enterprise Risk Management. Wiley, Chichester (2011)
3. Drucker, P.F.: Managing for the Future: The 1990s and Beyond. Routledge, New York (1992)
4. Feigenbaum, A.V.: Total quality management. In: Encyclopedia of Software Engineering. Wiley (2002). https://doi.org/10.1002/0471028959.sof359

5. Horváth, P.: Controlling in der Plattformökonomie. Controlling **31**(6), 1 (2019). https://doi.org/10.15358/0935-0381-2019-6-1
6. Ibragimova, R., Golovkin, D.: Estimation of the textile industry economic potential on the basis of foresight concept. Modern High Technol. Reg. Appl. **4**(56), 128–140 (2016)
7. Ibragimova, R., Golovkin, D.: Methodological substation of the estimation of economic potential of the enterprise. Modern High Technol. Reg. Appl. **3**(47), 64–74 (2018)
8. Ibragimova, R., Golovkin, D.: The industry life cycle concept as a tool for economic potential management at textile enterprises. Perm Univ. Bull. Econ. **12**(1), 124–135 (2017). https://doi.org/10.17072/1994-9960-2017-1-124-135
9. Kaplan, R.S., Norton, D.P.: The strategy map: guide to aligning intangible assets. Strat. Leader. **32**(5), 10–17 (2004). https://doi.org/10.1108/10878570410699825
10. Knight, J.A.: Value Based Management: Developing a Systematic Approach to Creating Shareholder Value. McGraw Hill, New York (1997)
11. Martin, B.R.: The origins of the concept of "foresight" in science and technology: an insider's perspective. Technol. Forecast. Soc. Change **77**(9), 1438–1447 (2010). https://doi.org/10.1016/j.techfore.2010.06.009
12. Parmenter, D.: Key Performance Indicators: Developing, Implementing, and Using Winning KPIs. Wiley, New Jersey (2015). https://doi.org/10.1002/9781119019855
13. Pearson, C.M., Clair, J.A.: Reframing crisis management. Acad. Manag. Rev. **23**(1), 59–76 (1998). https://doi.org/10.2307/259099
14. Porter, M.E.: Competitive Strategy: Techniques for Analyzing Industries and Competitors. The Free Press, New York (1998)
15. Povey, B.: Continuous improvements. TQM Mag. **5**(6). (1993). https://doi.org/10.1108/eum0000000003111

State Support for the Real Sector of the Economy

Impact of Government Funding of R&D Development in Industrial Sector of Russia

A. R. Salkina(✉)

Samara State University of Economics, Samara, Russia
Alya-0508@yandex.ru

Abstract. Specific features of the national R&D sector are "regional", cluster location of research and design work, its transfer and low level of integration at the federal level. According to the survey on the implementation of R&D results in innovative projects in Russia's regions more than 90% of intellectual property objects are addressed to regional industry. This is due to the spatial development of production forces and relations in Russia, historical and current trends. The purpose of the article is to determine the impact of government funding on the development of the entire R&D sector in industrial projects. Despite its weak position in the global comparison, it is possible to notice the preservation of "growth points" in the national research and development sector. When analyzing this problem, one can state a general progressive dynamics of technology exports, as well as growth in certain areas, in particular, manufacturing industry. The study assesses the effectiveness of the development of the industrial sector of R&D in the Russian Federation.

Keywords: Industrial sector · Innovation development · Government funding · R&D · R&D budget

1 Introduction

When comparing the competitiveness of world economies [8], the development of the national innovation system is determined primarily by R&D sector indicators, and secondly by institutional indicators. Over the period from 2012 to 2017 there was a slight positive dynamics in the sector (changes below one point on the rating scale), which can be defined as stagnation. Several constraints can be analyzed, the main of which is the low level of industry investment in technological innovation (3.3 out of 7) and the lack of government mechanisms for the procurement of innovative goods and services (3.3) [10].

The intensive path of development of the sector, due to the implementation of a relatively high (but reducing) intellectual potential, is not closed as an opportunity, but requires [4] changes in the organizational model of the innovation process, R&D cycle integrated into it, and the institutional development of the national intellectual property market. Moreover, it requires not only a change in the organizational mechanisms of R&D, but also the activation of the "transfer" of intellectual property rights into industrial production practice. For the last decade there has been an objective trend of a decline in the volume of transfer of intellectual property into the national economy,

S. I. Ashmarina et al. (Eds.): ENGINEERING ECONOMICS WEEK 2020, LNNS 139, pp. 157–162, 2021.
https://doi.org/10.1007/978-3-030-53277-2_17

expressed by a decrease in the turnover of license rights in absolute terms. In 2017 Russia accounted for 1.41% of the world intellectual property turnover [10] against the background of a breakthrough growth in the number of intellectual property objects in the world, due to the dominance of the "information" and "digital" economy in the structure of social production in economically developed countries. Moreover, technology transfer is not only a key mechanism for the innovative growth of industrial production, but also the basis for integrating the achievements of economically advanced countries into international infrastructure.

In the "Forecast of scientific and technological development of Russia" the Russian Academy of Sciences identified areas where technologies are at the world (or higher) level of development [6]. In particular, global parity - leadership remains in software production; the creation of biocompatible materials, membranes and catalytic systems; bioengineering, bio catalytic, biosynthetic and biosensor solutions; monitoring the state of the atmosphere, hydrosphere, lithosphere and biosphere; energy and energy conservation. These areas are identified by Decree of the president "On the strategy of scientific and technological development of the Russian Federation" as priority areas for R&D investment [1].

2 Methodology

The author analyzes the development trends of the research and development sector of the national industry. The study presents economic characteristics of the national research and development sector, medium-term development trends and comparison with global trends. The process of research, development, experimental and technological work, intellectual property objects created in it are the "core" of added value of technological innovations.

The main research questions are:

- factors of R&D development in the Russian Federation,
- assessment of the development of the industrial sector of the Russian Federation,
- changes in the structure of subjects, interaction processes and the economics of R&D processes,
- allocation of resources for R&D development in most industrial enterprises of the Russian Federation.

The level of R&D costs in the added value of innovative production determines its knowledge intensity, the level of technological efficiency of industries.

3 Results

A key institutional factor in the development of the national research sector is the preservation of a significant amount of basic science funding [4]. The increase in the rate of transfer of fundamental discoveries into applied developments over the period of 10 years (from 2007 to 2017), observed by economists, allows to consider the academic science sector as a resource for innovative activity, a source of breakthrough scientific and technical development [9]. In particular, the President of the Russian Academy of Sciences in 2016 Fortov highlighted number of breakthrough Russian fundamental developments of world level: proton therapy, femtosecond laser, ultrahigh pressure systems, photon computers based on diamonds with germanium defect centers, etc. Moreover, Fortov designated both the horizon for transferring discoveries to applied development and the markets for innovative products [6]. Thus, in the period from 2000 to 2015, R&D cost in the Russian Federation increased from 100 million rubles to 750 million rubles per year; the cost of basic research to 500 million rubles; and the cost of experimental development from 80 million rubles to 190 million rubles per year.

4 Discussion

Based on the development of the R&D sector as a whole, the author analyzed and highlighted a counter positive trend in the growth of applied research funding (Table 1), which creates the potential for innovative developments, both on the basis of fundamental breakthroughs and on global platforms of "open innovation", acquired franchises on intellectual property. The distribution structure of the R&D budget, funded by state programs of the Russian Federation is given in Table 1.

Table 1. R&D budget, funded by state programs of the Russian Federation in 2014–2016 and vertical analysis (VA) for 2016

State programs	2014, bln. rubles	2015, bln. rubles	2016, bln. rubles	VA, %
Development of science and technology	102.73	111.82	119.61	71
Aviation industry development	28.62	27.87	35.22	20
Development of industry and increasing its competitiveness	7.27	8.55	12.80	7
Development of the nuclear power	0.89	0.20	0.01	0
Energy efficiency and energy development	1.49	1.37	0.92	1
Reproduction and use of natural resources	1.50	1.20	1.19	1
Agricultural development	0.04	0.01	0.00	0
Total	142.54	151.03	169.75	100

Source: author based on [3]

Let's consider sectoral distribution of state R&D funding. After analyzing a number of data, we can conclude about cross-sectoral nature of the funding, 71% of which is concentrated in multidisciplinary research and development. The strategy of state R&D financing is not based on breakthroughs in individual industries, typical for the EU countries, [10] but on the creation of a common innovative scientific and technological platform, that provides the potential for technological breakthroughs in priority sectors of industry and infrastructure. It is on these principles that the national program "National technological initiative" was created, which replicates general R&D solutions ("Technet") in industry projections ("AeroNet", "MariNet", etc.).

The formed funding program is based on the promising institutional structure of R&D entities. Over the past 20 years, the global innovation system has undergone significant changes: from clusters led by manufacturing corporations to global networks based on an open technology solutions platform. The new and changing functions of traditional R&D institutions appear, and a new organizational model of research and development is being formed. The research on innovation leadership factors of the United States (and other countries) has identified a key factor in creating new operational models for internal R&D organization and cooperation mechanisms [5]. The Russian sector of research and development, the institutional transformation of which is inseparable from global trends, has undergone significant changes (Table 2). The adoption of the "triple helix" concept at the turn of the 2010s led to a change in the role of Universities in innovation processes. Since 2015 the volume of research and development of scientific and educational organizations in Russia has increased by 15% [2].

Table 2. Dynamics and vertical distribution (VD) of the number of organizations performing R&D by sectors of activity and types of organizations in the Russian Federation in 2000 and 2015

Indicators	2000	VD, %	2015	VD, %
Total number of organizations	4099	100%	4175	100%
Distribution by sectors				
State organizations	1247	30%	1560	37%
Entrepreneurial organizations	2278	56%	1400	34%
Universities	526	13%	1124	27%
Nonprofit organizations	48	1%	91	2%
Distribution by types of organizations				
Research organizations	2686	66%	1708	41%
Design organizations	318	8%	322	8%
Design and survey organizations	85	2%	29	1%
Pilot plants	33	1%	61	1%
Universities	390	10%	1040	25%
Industry divisions	284	7%	371	9%
Others	303	7%	644	15%

Source: author based on [7]

The transfer of applied research to universities reduces the role of specialized research institutes (−25%) in innovative industry programs. Educational centers demonstrate clear leadership; due to their activity in innovative R&D processes [5]. The role of the state and its institutions in the implementation of R&D projects (+7%) is increasing, which is in line with world practice. On the one hand, state funding makes up for the deficit of investments in R&D, smooths out the investment gap in R&D

expenses and profits from the introduction of technological innovations, on the other hand, the government financial initiative is holding back R&D investments in the private capital sector (−22%). The increase in the participation of internal scientific divisions of the industry in applied research and development (+2%) is largely due to the state financing programs, rather than an investment initiative of private entrepreneurship. In other words, there is a change in the structure of subjects, interaction processes, and the economics of R&D processes.

5 Conclusion

Thus, the development prospects of the national R&D sector are determined by the following key medium-term trends:

1. Lagging in R&D funding from world parity.
2. Reduction of intellectual resources in scientific and technical sphere.
3. Insufficient implementation of R&D potential.
4. Institutional transformation.

Summing up the analysis, we can confidently conclude that the key direction of intensification of the national sector is the improvement of managerial approaches to the organization of research and development work in innovative industrial projects. One can also speak with confidence about the clear potential for intensifying the R&D sector of industry in terms of the objectivity of "growth points" and institutional prerequisites.

Huge changes in the share of R&D financing in industry can be observed in dynamics over the past 10 years, and this is due to the enormous demand for the development of this particular national sector. Most Russian and Western economists agree with this view. This development allows further progress of the industrial sector in Russia, which will bring additional revenues to the state budget. In this regard, the current trend of active government funding is more than justified.

References

1. Decree of the President "On the strategy of scientific and technological development of the Russian Federation" (2016). http://kremlin.ru/acts/bank/41449. Accessed 02 Apr 2020
2. Etzkovitz, H., Leydcsdorff, L.: The dynamic of innovations: from National System and "Mode 2" to a Triple Helix of university-industry government relations. Res. Policy **29**(2), 109–123 (2000). https://doi.org/10.1016/S0048-7333(99)00055-4
3. Government of the Russian Federation: Annual monitoring of funds allocated from federal funding budget R&D (2016). https://ac.gov.ru/uploads/_Projects/Presentations/4879.pdf. Accessed 02 April 2020
4. Khlebnikov, K.V.: The effects of investment in human capital for the benefit of in-innovative development of high-tech enterprises. Sci. Inf. J. **10**, 84–85 (2016)
5. Popova, I.N.: Analysis of models for the formation of innovative territorial clusters. Econ. Entrepr. **8**(4), 258–262 (2017)

6. RAS: Forecast of scientific and technological development of the Russian Federation for the long term (until 2030). Conceptual approaches, directions, forecast estimates and conditions of implementation. RAS, Moscow (2008)
7. Rosstat: Russia 2015. Statistical reference book (2015). https://www.gks.ru/free_doc/doc_2015/rus15.pdf. Accessed 02 Apr 2020
8. Schwab, K.: The global competitiveness report 2016–2017 (2017). http://www3.weforum.org/docs/GCR2016-2017/05FullReport/TheGlobalCompetitivenessReport2016-2017_FINAL.pdf. Accessed 02 Apr 2020
9. UNIDO: Industrial development report 2016: The role of technology and innovation in inclusive and sustainable industrial development (2016). https://europa.eu/capacity4dev/unido/document/industrial-development-report-2016-role-technology-and-innovation-inclusive-and-sustainable. Accessed 02 Apr 2020
10. World Bank Group: Digital dividends (2016). http://documents.worldbank.org/curated/en/896971468194972881/pdf/102725-PUB-Replacement-PUBLIC.pdf. Accessed 02 Apr 2020

Corruption Counteraction: Theoretical and Practical Aspects

V. P. Markov[1(✉)] and S. I. Velezev[2]

[1] Samara State University of Economics, Samara, Russia
ladimirmarkov@gmail.com
[2] Samara Law Institute of the Federal Penitentiary Service, Samara, Russia
veleshevsi@mail.ru

Abstract. The article reveals the essence of the phenomenon of corruption. The analysis of legal statistics on corruption-related crimes is given. System value for research of problems of counteraction of corruption in the era of digitization has institutional approach, characterized by effective and comprehensive analysis of the mechanism of participation of the state as a whole and its individual institutions in the development and implementation of the security policy and the position of this sphere of legal regulation in the system of powers exercised by the authorities in the field of state internal and foreign policy and its relationship to individual activities. Within the publication, the main purpose of the article is to reveal the essential characteristics of corruption as a socio-steam phenomenon and analyze statistical information that characterizes trends in the sphere of legal counteraction to corruption in the Russian Federation at the present stage. Work is based on the provisions of the theory of national security (namely, the provisions of prevention in its organization) and modern concepts of state policy in the field of combating corruption.

Keywords: Anti-corruption legislation · Bribery · Corruption · Fraud · Theft · State and municipal procurement

1 Introduction

Unfortunately, the formation of a new model of communication through digital technologies does not allow avoiding corruption in the sphere of public and state activities. Moreover, the use of new technologies helps to make this type of relationship even more latent, which makes this problem one of the most acute socio-economic and legal problems facing scientists who are experts in the field of legal science and practitioners – representatives of law enforcement agencies. This kind of economic deviation, especially at the time of formation of the system of digitalization of communication processes not only undermine the authority of state power bodies and business community, but also have an objective impact on slowing down the pace of social development of the state, retard the realization of economic reforms, reduce the profitability of the core activities of enterprises of all forms of ownership and, ultimately, interfere with the normal work of the authorities and structures of civil society. Despite the introduction of a system of bidding on electronic platforms with the

S. I. Ashmarina et al. (Eds.): ENGINEERING ECONOMICS WEEK 2020, LNNS 139, pp. 163–168, 2021.
https://doi.org/10.1007/978-3-030-53277-2_18

possibility of permanent control by Supervisory and law enforcement agencies, the sphere of procurement of goods services for state and municipal needs remains the most corruption-intensive areas of the domestic economy. To build a reliable and secure procedure for the implementation of state and municipal procurement, it is necessary to develop legal and economic mechanisms of a strategic nature, which will be based not only on domestic scientific and practical research, but also on the experience of anti-corruption activities of developed countries. This will not only help to reduce transaction costs in the form and illegal content of state and municipal contracts, but also lead to real savings in budget funds while improving the quality of goods and services provided.

Within the publication, the main purpose of the article is to reveal the essential characteristics of corruption as a socio-steam phenomenon and analyze statistical information that characterizes trends in the sphere of legal counteraction to corruption in the Russian Federation at the present stage.

2 Methodology

The methodological basis of the research is represented a wide range of scientific approaches: comparative, problematic, systemic, structural, analytical, and a number of others, and a significant number of methods of modern legal science, political science and other related sciences. The complex and multifaceted nature of the problem under study determined the priority use of the principles of unity, comprehensiveness, integrity and movement from the abstract to the concrete.

System value for research of problems of counteraction of corruption in the era of digitization has institutional approach, characterized by effective and comprehensive analysis of the mechanism of participation of the state as a whole and its individual institutions in the development and implementation of the security policy and the position of this sphere of legal regulation in the system of powers exercised by the authorities in the field of state internal and foreign policy and its relationship to individual activities.

The formal legal approach, which is used in the study, determines normative-legal framework of the study and the conditions for the implementation of powers by public authorities. This helps to strengthen the institutional approach and allows you to identify the most priority areas of activity and specify the tasks facing the state-power institutions. The focus on the methodology of economic research is of great importance for the study. This allows a much deeper understanding of the nature and essence of corruption processes and their perniciousness in the modern world.

It should be noted that the tools of psychology, cultural studies, and sociology are important for the study of economic security problems. Besides, the work is based on the provisions of the theory of national security (namely, the provisions of prevention in its organization) and modern concepts of state policy in the field of combating corruption.

3 Results

The current system of views associates corruption crimes exclusively with the political and economic sphere. However, the use of broader and more universal methods, which were mentioned above, allows us to consider corruption not as a universal phenomenon, but as an interdisciplinary, complex object of research: here we see the manifestation of not only economic and legal components – political science, history, philosophy, cultural studies, and even theology cannot be left out. Political science view on this problem is most fully and comprehensively presented in the works "Political corruption: readings in comparative analysis" by Heidenheimer [4] and "The Pathology of Politics – Violence, Betrayal, Corruption, Secrecy, and Propaganda" by Friedrich [3], published in the last quarter of the twentieth century and laid the foundation for the controversy about the essence of corruption, presenting the first terminological arguments. In addition, for the first time, the question was raised about the causes and global consequences of corruption, its impact not only on the economic component, but also on the political life of society. I must say that the issues raised have not lost their relevance today, but we should note a certain shift in the debate towards the formation of concepts for the emergence of corruption manifestations and the development of methods to struggle it. At the same time, there is practically no comprehensive interdisciplinary research on the essence of the phenomenon of corruption [1].

The analyzed phenomenon is a complex formation with several ambiguous views. According to Borzenkov, corruption is an exclusively legal concept, which is a combination of several official crimes (namely theft, bribery, abuse of authority, etc.) that fall under the Criminal code of the Russian Federation [2]. The failure of this kind of view in modern conditions is beyond doubt. Komissarov, criticizing this point of view, raises the question of the need for a broader view of the problem, analyzing cases in which an official acts based on the desire to satisfy an intangible interest. An example of such manifestations can be considered the employment of their own relatives or relatives of "the right people", the desire to satisfy the interests of higher management, etc. [5]. However, this kind of manifestation, characteristic of the late 80's and early 90's of the last century, has undergone significant changes, which also allows us to speak about the imperfection of the described point of view in the modern world. The most successful is the definition of the term "corruption", presented in the works of Kostenko. The author proposes to consider it as a multidimensional socio-economic, political, and moral-ethical phenomenon, which is based on a number of anti-social actions (both illegal and immoral), covering a complex of offenses that are interrelated [6]. An interesting feature of corruption in their works is pointed out by Nesterov and Sukhovarova, who note cases when corruption activities are exclusively immoral in nature, which does not allow to attract the violator under the rules of criminal or administrative law [7]. This allows us to look at corruption not only from the legal point of view, but also from the position of assessing morality. In this case, bribery in state and local government bodies and commercial bribery in the activities of non-state economic entities will be just one of the most striking forms of corruption. The use of such a broad and diverse position suggests that corruption should be considered from the point of view of criminal law, tort relations and morality. This approach allows us to attribute to corruption manifestations not only the extraction of material benefits by an official, but also various kinds of violations of ethics and morals.

Sources indicate an extremely high level of corruption in the Russian Federation. According to international anti-corruption organizations, we are ranked 136th out of 174, which points the inefficiency of existing methods and mechanisms for fighting corruption and the need to develop completely new approaches to the formation of principles for monitoring activities to meet state and municipal needs and conditions for preventing the considered criminal manifestations. One of the most effective methods is a preliminary assessment of corruption risk in the system of procurement of goods and services for state needs. Experts of the transparency international center, having conducted a number of studies on the problem of combating corruption in the Russian Federation for a long time, point to a number of gaps in the legislation that allow satisfying the illegal interests of interested persons [10]:

- there are no mechanisms for public consideration and discussion of state and municipal needs,
- the law does not establish a mechanism for providing mandatory guarantees of compliance of goods or services with the conditions specified in the tender documentation,
- there are no or veiled conditions for disclosure of information about the characteristics of purchased goods,
- the criteria for evaluating bidders and the conditions for their admission are not objectively justified,
- there are no real guarantees of the independence of persons who must control the course and conditions of procurement,
- lack of guarantees of protection for persons who reported unfair bidding,
- the inability of civil society structures to monitor the progress and results of the bidding process.

4 Discussion

Further, it is necessary to analyze open data reflecting the current state and trends of anti-corruption activities in the Russian Federation. The source base for the analysis was legal statistics (Fig. 1; Table 1). These data mean that the number of crimes under article 290 "Bribe reception" of the Criminal Code of the Russian Federation dated 13.06.1996 No. 63-FZ [8] remained fairly stable during the period under review. A sharp turnaround was observed in 2017, when the indicators of the perfect number of crimes have more than halved (from 6,495 to 3,187 registered crimes). Meanwhile it cannot please the fact of further reducing the number of crimes committed in 2018 and 2019. It appears that this trend is a consequence of introduced in 2016 changes to existing legislation, to strengthen the responsibility for taking bribes, and increase the effectiveness of oversight bodies.

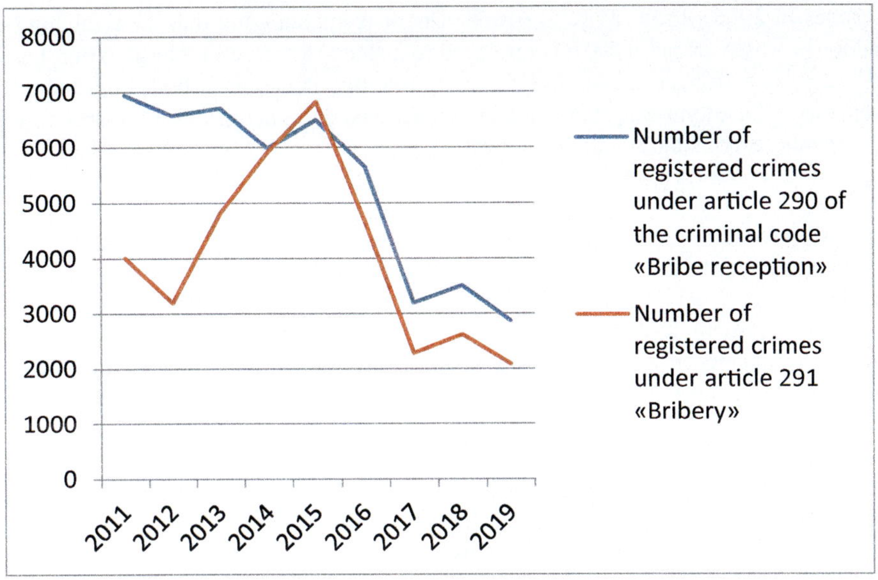

Fig. 1. Indicators of registered corruption crimes for the period from 2011 to 2019 in the Russian Federation (Source: authors based on [9]).

Table 1. Number of registered corruption crimes for the period from 2011 to 2019 in the Russian Federation

Statistics	2011	2012	2013	2014	2015	2016	2017	2018	2019
Crimes under article 290 of the criminal code of the Russian Federation	6947	6576	6710	5980	6495	5644	3187	3499	2866
Crimes under article 291 of the criminal code of the Russian Federation	4005	3182	4811	5913	6816	4640	2272	2612	2077

Source: authors based on [10].

A different scene is shown by data on the number of crimes committed under article 291 of the criminal code of the Russian Federation “Bribery”. The year 2011 is marked by a marked decrease in the number of registered crimes, but since 2012 there has been an annual 30% increase, which led to more than double the total increase in the number of crimes in 2015 – from 3182 to 6186 cases. It is encouraging that the tightening of sanctions for this crime in 2016 contributed to a sharp decrease in their number, but in 2018 there is a slight increase. Let's assume that this is based on both improving the mechanisms for monitoring the conduct of auctions, and changes in the worldview of the officials responsible for their conduct.

5 Conclusion

All the above ideas about the phenomenon of corruption, which have received theoretical development in the scientific and legal field, allow us to conclude that the domestic legal science still lags behind the needs formed by practice. The improvement of technologies, the development of communication tools, and the modernization of society and the economy are taking place much faster than new scientific concepts and ideas defining the concept of corruption. Statistical data indicate that the mechanism for improving anti-corruption legislation is slow and, unfortunately, the mechanism for committing certain crimes is improving. In this regard, it seems that the most promising activity is not only the scientific substantiation of changes in legal acts that establish responsibility for corruption, but also the analysis of the corruption risk of various types of decisions and actions of public authorities and local self-government.

References

1. Bekyashev, K.A.: International Public Law. Prospect, Moscow (2014)
2. Borzenkov, G.N.: Corruption in the system of criminal offences. Bull. Moscow Univ. **1**, 21–26 (1993)
3. Friedrich, C.J.: The Pathology of Politics – Violence, Betrayal, Corruption, Secrecy, and Propaganda. Harper and Row, New York (1972)
4. Heidenheimer, A.J.: Political Corruption: Readings in Comparative Analysis. Holt Rinehart and Winston, New York (1970)
5. Komissarov, V.S.: Criminal-legal aspects of the fight against corruption. Bull. Moscow Univ. **1**, 26–30 (1993)
6. Kostenko, N.I.: International Criminal Law at the Present Stage: Trends and Problems of Development. Yurlitinform, Moscow (2014)
7. Nesterov, A.V., Sukhovarova, E.L.: Transformation of corruption in the conditions of transition to the market. In: Materials of the Scientific-Practical Conference Corruption in Russia: State and Problems, 26–27 March 1996, pp. 59–63. Institute of MIA of Russia, Moscow (1996)
8. The Criminal Code of the Russian Federation dated 13.06.1996 No. 63-FZ (1969). http://base.garant.ru/10108000/. Accessed 25 Sept 2019
9. The Portal of Legal Statistics: Number of registered crimes under articles 290 and 291 of the criminal code "receiving a bribe" (2019). http://crimestat.ru/regions_chart_total. Accessed 25 Sept 2019
10. Transparency International Russia: Russia in the corruption perception index 2019: 28 points and 137th place (2019). https://transparency.org.ru/research/indeks-vospriyatiya-korruptsii/rossiya-v-indekse-vospriyatiya-korruptsii-2019-28-ballov-i-137-mesto.html. Accessed 25 Sept 2019

Legal Regulation of International Payment Systems in the Russian Federation

M. N. Zubkova(✉)

Samara State University of Economics, Samara, Russia
zubkova_maria@mail.ru

Abstract. The dynamic development of payment systems, the formation of new mechanisms for payment infrastructure functioning, the emergence of new types of payment instruments and forms, which are taking place recently, have a significant impact on all spheres of our life, including the sphere of economic security of the state and, as a result, require a special approach and careful legal regulation. Due to the historical development of the monetary system, when paper, credit and electronic money was included into the monetary circulation, there is a need to develop special rules for their circulation and transfer procedures, which should be followed by all participants in payments. As a result, a payment system for transmitting payment information is formed. The purpose of the article is to determine the place and role of international payment systems in the legal regulation of this sphere in the Russian Federation. The main research method is the complex analysis, since legal regulation of payment systems cannot be considered in isolation from the economic content of these relations. Effective regulation and control of payment systems enables national banks to develop a payment system infrastructure.

Keywords: Foreign payment systems · Money transfers

1 Introduction

In the last decade, there has been a rapid development of the payment infrastructure: the emergence of new entities that provide payment services and apply modern payment schemes for goods, works and services, and a radical change in the approach to making payments. Such changes are caused by global factors: scientific and technological progress, the emergence of a worldwide system of unified computer networks for storing and transmitting information (the Internet) and its penetration into almost all spheres of the human activity, structural changes in the economy – globalization and integration. Economic relations between countries are becoming more complex [15]. With the increasing globalization of economic and political processes, the role and importance of payment systems is significantly growing. Rapid and, most importantly, secure implementation of cross-border payments is a basic element for effective functioning and development of the world economy. Among the fundamental institutions that should ensure the continuity and security of cross-border payments are international payment systems, which have developed rapidly in recent decades. In the scientific literature on this topic, it is noted that even if today customers are able to pay

S. I. Ashmarina et al. (Eds.): ENGINEERING ECONOMICS WEEK 2020, LNNS 139, pp. 169–174, 2021.
https://doi.org/10.1007/978-3-030-53277-2_19

for purchases by cash on delivery, which seems safe and reliable, there is still a need for electronic payments that can only be provided through banks [4]. Currently, banks have special systems such as Internet banking, electronic transfers, plastic money (credit and debit cards) and mobile banking [5, 9, 10, 13].

As follows from the norms of domestic legislation, a foreign payment system is a set of organizations that have joined the rules of the payment system organized in accordance with the foreign legislation, and interact under the rules of a payment system (participants of the international payment system), according to which a foreign bank (a foreign credit organization) can act as a payer and recipient of funds for transferring funds of participants in the foreign payment system [2]. This concept was included into the specified regulatory act in 2014 in order to ensure the smooth functioning of payment systems, as well as the legal regulation of the national payment card system.

Initially, as a result of changes made to Federal law No. 161 regarding the definition of a foreign payment system, it was used only for the purpose of legal protection of operations of a foreign central payment clearing counterparty that has a bank account in the Bank of Russia [2]. Since, in accordance with paragraph 11 of Art. 29 of Federal law No. 161-FZ, a transfer of funds made by a foreign central payment clearing counterparty that has a bank account opened in the Bank of Russia, or with its participation, within a foreign payment system, cannot be invalidated on the grounds provided by the legislation of the Russian Federation [2]. However, the economic and political events that happened in the following years predetermined and necessitated clearer legal regulation for functioning of foreign payment systems on the territory of the Russian Federation. The result was changes made to Federal law No. 161 by the Federal law No. 264-FZ of 02.08.2019 [3].

2 Methodology

In the course of research on this topic, the following methods were used. Firstly, it is a method of comparative law which is based on the study and use of the legal regulation of similar relations. In cases where this method is used, not only the content of a particular legal institution is subject to research, but first of all, reasons for its origin in a particular national legal system, as well as development forms of this institution are studied, which makes it possible to establish both general, universal features and legally specific features in the considered institute. This, in turn, allows us to establish a certain basis for purposes of a new research in this area and the formation of proposals for further improvement of the domestic legislation, taking into account national specifics. Secondly, the method of complex analysis was used. Studying a specific legal phenomenon, other tools that are used in other scientific branches are also used together with legal instruments.

In particular, as for this study, it is impossible to consider the issue of legal regulation of payment systems functioning in isolation from the economic science, and from processes that it studies. After all, the evolution of the legal formalization of economic relations will not take place if lawyers avoid analyzing the economic nature of these relations. As for the general formal-logical methods, analysis, synthesis and generalization were applied (in particular, when determining specific features of legal regulation of payment systems and identifying the subject composition of these relations).

3 Results

A foreign organization, which is an operator of a foreign payment system, where cross-border transfers of funds from individuals are made, has to send to the Bank of Russia a statement in the prescribed form through a separate division of a foreign operator payment system on the territory of the Russian Federation which enables to include information about it into the register of foreign operators of the payment systems. The application of the operator of a foreign payment system, among other documents provided by law, should be accompanied by rules of the foreign payment system in Russian.

The rules of a foreign payment system regarding cross-border money transfer with the participation of money transfer operators determine the procedure for payment clearing and payment itself in a foreign payment system. It is also mandatory to specify in the rules the risk management system in the foreign payment system, including the procedure for ensuring the fulfillment of obligations related to the settlement. The scientific literature also repeatedly emphasizes the need to analyze a credit risk, which is an important criterion for the stability of the banking sector [14].

In addition, it is necessary to provide requirements for ensuring the protection of information in a foreign payment system. This is because the development of digital cashless payments also raises various issues that national banks should take into account. In particular, these payments can create problems regarding payment stability and cybersecurity. Security and privacy are becoming increasingly important as operators accumulate huge amounts of customer data through digital payment platforms [17].

It is also important to indicate the order of interaction between the foreign operator of the payment system and the operators for transferring money in disputed and emergency situations, as well as requirements for combating legalization (laundering) of proceeds from crime, financing of terrorism and financing of proliferation of mass destruction weapons.

In addition, the current version of the Federal law No. 161-FZ grants the Bank of Russia the right to establish additional requirements regarding the content of the rules related to the procedure for conducting cross-border money transfers with the participation of money transfer operators, as well as requirements for ensuring information protection when conducting cross-border money transfers with the participation of money transfer operators [2].

In the course of its activities, a foreign payment system operator must: ensure public availability of the rules of the foreign payment system on the territory of the Russian Federation, as well as their observance in the sphere of cross-border money transfer with the participation of money transfer operators; attract service operators from the payment infrastructure that provide payment infrastructure services within payment systems whose operators are registered by the Bank of Russia.

The operator of a foreign payment system and operators of payment infrastructure services attracted by it may not suspend or terminate unilaterally the provision of services within the framework of a foreign payment system to other operators of money transfer and their clients.

As for money transfer operators, they are entitled to participate in a foreign payment system only if the Bank of Russia includes information about the operator of a foreign payment system in the register of foreign payment system operators. At the same time, money transfer operators are required to send a notification to the Bank of Russia about participation or termination of participation in a foreign payment system.

It should be noted that all these requirements do not apply to foreign organizations that have subsidiaries on the territory of the Russian Federation that are registered by the Bank of Russia as payment system operators, as well as to international financial organizations and foreign central (national) banks.

The Bank of Russia is responsible for monitoring and supervising activities of foreign payment system operators. For this purpose, the Bank of Russia carries out some specific functions: check whether the of a foreign payment system operator follows the rules of the foreign payment systems to cross-border transfer of funds with the participation of operators for transferring funds, as well as ensure the public accessibility of the rules of foreign payment systems on the territory of the Russian Federation; verification of compliance by the operator of a foreign payment system with additional requirements established by the Bank of Russia regarding the operation of a foreign payment system; assessment of measures taken by the operator of a foreign payment system to eliminate identified violations of legal requirements; verification of the operator's observance of the obligation to attract payment infrastructure service operators from among the operators providing payment infrastructure services within payment systems whose operators are registered by the Bank of Russia [11].

4 Discussions

At the moment, the world is striving to create a single financial market, and modern technologies allow us to move capital in real time not only from city to city, but also from country to country. In modern conditions, the financial sphere has acquired an independent character, has become "a thing in itself and for itself", having broken away from the material production and exchanges of goods and services. The latter account for only 30% of global cash flows, and 70% serves the financial sector itself [1]. In the context of modern globalization, payment systems are a key element of the global financial system, providing payments both at the national and interstate level, which contributes to the effective implementation of monetary policy and ensuring financial stability of the economy. Due to the increased integration and economic development processes, there is a constant increase in the turnover of existing payment systems [8].

The Federal law "On the national payment system" has laid the basis for a legal framework that is adequate to modern economic realities. The legislator takes serious steps aimed at creating and establishing effective functioning of the payment system as an important institution of the financial system [2].

In the scientific literature, the payment system is understood, for example, as "a set of economic relations that ensure the movement of interrelated cash flows between subjects of the total payment turnover in the process of functioning and integration of the company's finances" [12, p. 42]. At the same time, other authors consider the payment system in a narrower sense, meaning only the system of plastic cards. At the same time, this approach does not take into account money transfers to bank accounts that are carried out by legal entities and individuals, money transfers without opening a bank account, and other operations that are performed without using a bank plastic card. In the legal science, there is a slightly different approach to the definition of a payment system. For example, Tarasenko interprets the payment system as a set of legal norms, institutions, software and other means that ensure interbank settlements [16]. A somewhat broader view of the payment system is proposed by Karchevsky. He identifies a number of qualifying features of the payment system, based on which the author understands the payment system as a set of interrelated elements of an institutional and infrastructural nature that ensure the implementation of settlements and settlement of debt obligations between participants in the economic turnover [7].

5 Conclusion

The development of payment systems helps countries to make financial operations in the international transactions, and citizens to make purchases and pay for trips abroad. The main tool for performing these operations is a payment card that is served by one of the payment systems. In addition, payment systems help countries to control risks that may arise in the course of financial transactions, which indicates their dominant role in the payment system of countries as a whole. Effective regulation and control of payment systems enables national banks to develop a payment system infrastructure. The world experience in the organization of non-cash payments indicates that cash is inevitably eliminated from the payment turnover. At the same time, frequent changes in rules and norms that regulate the sphere of non-cash payments and functioning of payment systems can lead to legal conflicts in practice. Analysis of modern research on functioning of foreign payment systems, in particular, shows that the financial systems of national economies are very sensitive to any changes in the payment space and its legal regulation. Due to the importance of payment systems and their impact on the banking sector, as well as the fact that a perfect payment system has not yet been found, this topic remains complex and relevant for further research on payment systems [6].

References

1. Alekseeva, D.G.: Banking security: legal issues: Ph.D. thesis, Peoples' Friendship University of Russia, Moscow (2011)
2. Federal law of 27.06.2011 N 161-FZ (ed. from 02.08.2019). On the national payment system (2019). http://www.consultant.ru/cons/cgi/online.cgi?req=doc&ts=80359556709219016542332128&cacheid=B716A701C4860B534AFF6B34A455964F&mode=splus&base=LAW&n=331203&rnd=0.12130959861038093#2b49a9lufby. Accessed 28 Feb 2020

3. Federal law of 02.08.2019 N 264-FZ. On amendments to the Federal law "on the national payment system" and the Federal law "On the Central Bank of the Russian Federation (Bank of Russia)" (2019). http://www.consultant.ru/document/cons_doc_LAW_330671/. Accessed 28 Feb 2020
4. Gubenko, E.S.: Financial and Legal Regulation of Payment and Settlement Systems. Prometheus, Moscow (2017)
5. Hemmati, M., Yaghoub, S.: The effect of the execution of the real time gross settlement system (SATNA) on electronic banking (central bank of Islamic Republic of Iran). Adv. Environ. Biol. **7**(13), 3946–3950 (2014)
6. Issa, F., Ahmad, M., Alyaseen, J., Mourad, M.: Towards decreasing the associated risks with payment settlement systems. Int. Rev. Comput. Softw. **9**(10), 1767–1775 (2014)
7. Karchevsky, S.P.: Payment system: Concept, structure, typology, and construction principles. Payments Oper. Work Commer. Bank **4** (2007). https://www.lawmix.ru/bux/62843. Accessed 28 Feb 2020
8. Karmin, I.A.: Formation and development of payment systems in the context of globalization of world finance. Ph.D. dissertation, Saint Petersburg University of Engineering and Economics, Saint Petersburg (2006)
9. Khomenko, E.G.: Financial and legal regulation of the creation and functioning of the national payment system of Russia. Ph.D. thesis, Moscow State Law University, Moscow (2017)
10. Milkau, U.: International payments: current alternatives and their drivers. J. Payments Strat. Syst. **13**(3), 201–216 (2019)
11. Regulations of the Bank of Russia On the procedure for supervision by the Bank of Russia of compliance by non-credit organizations with payment system operators, payment infrastructure service operators, and foreign payment system operators with the requirements of Federal law No. 161-FZ of June 27, 2011 "on the national payment system" and Bank of Russia regulations adopted in accordance with it (2011). http://www.consultant.ru/cons/cgi/online.cgi?req=doc&ts=8035955670921901654233212 8&cacheid=29ADC56C9464F198D9A993B13C8D264A&mode=splus&base=LAW&n=344213&rnd=0.12130959861038093#5d0itasgjv. Accessed 28 Feb 2020
12. Savinova, N.G.: Functioning of the national payment system of the Russian Federation in modern conditions. Ph.D. dissertation, Samara State University of Economics, Samara (2012)
13. Sha, N., Shariq, M.: Virtual banking and online business. Banks Bank Syst. **12**(1), 75–81 (2017)
14. Singh, E.R., Sivasankar, E.: Risk analysis in electronic payments and settlement system using dimensionality reduction techniques. In: Bansal, A. (ed.) Proceedings of the 8th International Conference Confluence 2018 on Cloud Computing, Data Science and Engineering, pp. 14–19. IEEE, New Jersey (2018)
15. Surodeev, Yu.V.: Financial and legal regulation of the national payment system of the Russian Federation: Ph.D. dissertation, Moscow State Law University, Moscow (2016)
16. Tarasenko, O.A.: Payment system of the Russian Federation. Actual Prob. Russ. Law **1**, 65–70 (2008)
17. Yamaoka, H.: The future of central banking. Accounting, Economics, and Law: A Convivium (2019). https://www.degruyter.com/view/journals/ael/ahead-of-print/article-10.1515-ael-2019-0003/article-10.1515-ael-2019-0003.xml. Accessed 28 Feb 2020

State Support for Investment Policy in the Real Economy Sector

A. V. Kozlovsky and O. E. Astafieva(✉)

State University of Management, Moscow, Russia
{av_kozlovskiy, oa_astafyeva}@guu.ru

Abstract. In modern economic conditions, it is urgent and necessary to apply new technologies, innovative technical and technological solutions. The most important market for innovations is construction and related industries ensuring commissioning of facilities and production capacities in the real economy sector. One of the segments of this market is housing construction, which also solves important social problems. In terms of capital investments, housing construction is a leader. The demand for housing facilities together with other social infrastructure facilities is constantly growing. Taking into account the high level of deterioration of the existing housing stock and living conditions that do not meet modern requirements, the need for new construction of residential facilities is obvious. The article investigates the main directions of housing construction development. The authors evaluate advantages of mortgage lending, the application of the housing renovation program, and changes in the structure of construction projects in favor of low buildings in small and medium-sized cities. Considerable attention is paid to the growth prospects of individual housing construction. The relations of developers with the financial and credit system and the state are analyzed. Specific proposals aimed at improving the efficiency of housing construction are given.

Keywords: Housing policy · Loans · Mortgage · Net cost · Renovation

1 Introduction

Construction, as a sphere of engineering and economic activities, plays a significant role in the formation of the gross domestic product and the development of related industries – the construction materials industry, metallurgy, chemistry, and others. The industry is developing constantly and stable, facing quite ambitious goals – to bring the volume of housing construction to 120 million square meters per year, against 80 million square meters at present. Solving this problem will require a significant increase in the production capacity of contractors, structural changes, and many other tasks [3]. The government of the Russian Federation has set goals not only to ensure economic growth rates of 3% by 2020, but also to increase investment in fixed assets to 25% of the total volume of all financing sources. The focus of the investment policy on the real sector of the economy will ensure an increase in production capacity in various industries and, first of all, in capital-intensive areas of economic activity, including capital construction. Temporarily available funds of both legal entities and, most

S. I. Ashmarina et al. (Eds.): ENGINEERING ECONOMICS WEEK 2020, LNNS 139, pp. 175–182, 2021.
https://doi.org/10.1007/978-3-030-53277-2_20

importantly, individuals should be directed to investment, not speculative purposes. It is obvious that the state should create appropriate conditions: legal, tax, tariff and other [3].

2 Methodology

The concept of state support for investment policy should primarily concern the creation and further support of national projects and investment programs at all governmental levels: from the federal to the municipal ones. These are usually projects aimed at implementing state infrastructure initiatives of social orientation. The essence of the concept is to create a single organizational and economic mechanism for managing the investment process at all stages of the life cycle of a program or project. First of all, it is the improvement of the current legislation, which should take into account interests of all participants in the investment process, including private investors; new tools for the development of construction and reconstruction of infrastructure facilities; regulations for managing the development of federal and regional innovation systems, which should include a system of indicators and methods for the development, implementation and evaluation of innovative development programs, as the most investment-intensive, on the one hand, and long-term related to the return on investment, on the other.

The regulations should cover a procedure for approval of federal interests and interests of specific territories, while respecting principles of the supremacy of national interests over private and regional ones; a factor of interactions with other public policies, programs and measures undertaken by the Government of the Russian Federation. They should also take into account peculiarities of specific territories in the sphere of influence of separate mechanisms for the regional policy implementation; the possibility of creating intra-regional or inter-regional clusters in various industries or activity areas, allowing to activate the innovation and investment potential, which in turn will make it possible to optimize the internal innovation and investment resources of the whole system. The adoption of such regulations will enable an internal assessment of the state of the regional innovation system, the availability and balance of resources, the readiness of the regional infrastructure to accept innovations and implement simple and expanded reproduction, on the basis of which federal and regional strategic programs for innovative development can be formed with a sufficient degree of justification.

It is obvious that for successful functioning of innovation systems under such regulations, a clear balance of actions of all participants in the process is necessary, and this is possible only when constant monitoring is carried out, not only of investment revenues, but also of regulatory measures that ensure effective interaction of the system elements.

Currently, the most active investors in the Russian Federation are federal and regional budgets, centralized investment funds (at the disposal of the government), large companies, joint ventures that are being built and operating in the territory of the Russian Federation, and non-resident external investors. The state's participation in the development of the economy should not be limited to providing targeted, gratuitous and irrevocable assistance to commercial structures, even with a share of participation similar to the budget financing of investments. It is necessary to create a large state bank that will finance innovative and investment national projects and programs on the return basis. Interest rates should be set depending on the profitability level of projects being implemented. The bank branches should be created at the regional level, and their funds can be supplemented from regional budgets.

3 Results

A number of prerequisites for the growth of investment activity are already visible. First of all, this is a stable growth in federal budget revenues, which allows implementing large investment projects in the real sector of the economy. At the regional level, structural innovation projects are being actively implemented within the framework of the construction of medium-sized enterprises in the processing industries, the construction materials industry, the food industry and many other sectors of the economy [10–12]. At the municipal level and especially regional levels, investments should primarily address social issues through the construction of housing, social services, engineering infrastructure, road construction, and communications upgrades.

The investment climate formed at the level of the state policy should encourage enterprises to make long-term investments. Such a source as an investment tax credit is almost forgotten. Bank loans, especially long-term loans, continue to be expensive.

Mortgage lending is slowly developing, which in turn hinders the construction and implementation of residential facilities development. Project financing at the infrastructure level is practically not used.

Project financing assumes that investments directed within specific projects are necessarily profitable, and the level of profitability should ensure continued financing (at least 25%). The funds raised for these purposes should ensure not only production and current investment costs, but also the debt obligations of investors. The real viability (stability, profitability, creditworthiness) of projects must be ensured. Any long-term investment involves the return of these funds in a mandatory period. Cash flow, as a result of investments, assumes the receipt of income sufficient for the implementation of strategic development goals of organizations and enterprises [9]. In real practice, the cash flows generated at the operating enterprise do not coincide with the volume and structure of capital investments for production facilities under construction.

It should be noted that debts of individuals to banks on consumer loans, which practically do not bring a real income, reach almost 15 trillion rubles, and investment lending stands still. To solve the problem, project financing should become an integral part of business plans of investment projects. Huge funds, about 8 trillion rubles, without taking into account mortgage lending, are diverted from the financial turnover. Banks cannot invest, investments in household lending are growing, and the banking system is moving further and further away from the real sector of the economy. If the state does not take measures aimed at reorienting the banking system to the investment environment, the most severe consequences are possible – bankruptcy of individuals, mutual non-payments of legal entities, a return to barter schemes, and other adverse consequences. Such measures may include: the creation of a state mortgage bank, further administrative regulation of interest rates, guarantees in obtaining loans (household loans) in the form of property pledge, guarantees from employers or other third parties. Currently, there is a decrease in investment in fixed assets.

Another problem is the development of network (engineering) infrastructure, without which the increase in construction volumes is hopeless. According to various estimates, the development of this network requires about 300 billion rubles a year, compared to 100 billion rubles at present. The question remains "who and how?" should invest these significant funds. There are several options: public-private partnership, concession, project financing and, in particular, refinancing, bank lending.

4 Discussion

Public-private partnership is limited to the participation of private investors in connecting constructed facilities to various networks and minimal reconstruction of communications, which is associated with the allocation of land for construction by public authorities. Private investors not only do not want to be responsible for the replacement of 75% or more worn out engineering networks, but also cannot. If they are required to do so, the cost of new housing and other facilities will more than double.

As practice shows, the use of concession agreements in the sphere of housing and communal services is also unlikely. Land plots and various natural objects in the extractive industries are among the attractive objects for transferring them to a concession. But it is extremely unattractive to take almost completely worn-out networks into the concession, which are not yet subject to depreciation.

You can consider project financing, but not as a tool to attract investment in the implementation of new projects, but as a tool for re-crediting (refinancing) [5, 6]. However, this option does not provide for repayment of investment debts. In this case, the main project agreement is concluded, which provides for the conclusion of contracts for the purchase of various types of products: electricity, water, pipes, etc., this requires long-term price agreements. The activities of the project company participating in the transaction must be licensed. The advantages of this scheme are the lack of own funds, resource costs, wide access to external sources of financing, the use of savings on the project, the presence of the state among the owners of the project, and more.

Issues of lending in the investment and construction sector have long been discussed in the economic literature, but a final, effective solution has not been found yet [4]. If you follow the accepted classification, loans are divided into long-term and short-term. If current short-term loans are easy, long-term loans are not. The fact is that construction products have a long production cycle, and contractors have their own working capital sufficient for 1–2 months of construction and installation work. Then the customers pay for the previously completed works in stages. If we take into account that these works are paid for at the estimated cost, then the estimated profit gradually enters the economic turnover, which causes hidden inflation, since the finished object is put into operation in 1.5–2 years. Lending contractors in the construction process is possible, but this will inevitably leads to an increase in the cost of building products, and in housing construction it increases prices of 1 sq. m. of the property value, which now can hardly be acceptable for a buyer.

The transition to mortgage lending does not solve this problem either. Mortgages are expensive (about 8–10% per), and in 2018–2019 there was a trend towards an increase in bank interest (up to 9–11% per annum). Taking into account that citizens take mortgages for a period of 10 years or more, then square meters purchased on credit cost are twice or more times more expensive for buyers. Government regulation is needed, not just state assistance in mortgage lending. Such a regulator can be a state mortgage bank with a dominant share of state capital, a single state policy, common principles and terms of investment risk insurance, a single mechanism for maintaining accounts of creditors (citizens) and borrowers (general contracting construction organizations). Funds from regional budgets can also be accumulated, which will be transferred to developers after the objects are put into operation. If commercial banks reduce their interest in mortgage lending, and this will necessarily be for at least two reasons: first, it is a long-term and stable income (the guarantee is an apartment under construction), and second, the decline in household income (and this trend is confirmed by surveys of the population) will affect the flow of citizens' funds to the banking system in the downward direction. But such participation must pass through the state mortgage bank and opening of special accounts, which will significantly strengthen the state control over the movement and expenditure of investment resources. It will also be appropriate to note that the debts of individuals to banks exceeded 15 trillion rubles.

The development of the construction industry is hindered by the presence of large volumes of unfinished construction. According to Rosstat, the volume of construction in progress is estimated at 2.5 trillion rubles [8]. These funds are withdrawn from the economic circulation and practically do not bring any real benefit. More than 85 thousand objects are either preserved or temporarily suspended from construction. The consequences of this situation are obvious, it not only negatively affects the efficiency of investments, but also the speed of turnover of funds, return on assets, aging of fixed assets, increasing costs in the process of the construction completion [2].

In Moscow, where the minimum number of dilapidated housing is located, a solution has been found in the form of renovation of the housing stock. But Moscow is an exception in terms of the regional budget, land value, and financial capabilities of the population. Demolition of mentally and physically worn-out five-story houses solves several problems at once: improving the housing conditions of the population, improving the comfort of housing and social infrastructure, and improving the architectural appearance of cities. In the Soviet era, in the early 60s, the strategy of large-panel construction was chosen in the field of housing construction and more than 600 plants for the production of panels and other large-size structures were operating in the country. In all cities and towns of urban type and even in rural areas, a huge number of five-story buildings were built with a standard operating life of 25 years. These deadlines have long expired, but this does not mean that the example of Moscow is being applied in other cities and regions. In Moscow, industrial zones are used for construction, the most significant example is the development of the territory of the former ZIL automobile plant.

In 2019, 79.2 million square meters of residential premises were built in Russia, which is slightly less than in 2014–2018 [1]. The estimated cost of construction continues to grow. It should be noted that these figures include individual, cooperative construction and even self-construction projects, i.e. houses built without permits, land acquisition and other violations. The main developers are organizations based on private ownership, which have very little working capital of their own. Investment and construction companies have been leaving the market in recent years, and new ones are coming much slower, namely, they can ensure the completion of the full construction cycle and put ready-made houses into operation.

In modern conditions, renovation is the most effective mechanism for solving the housing problem, but how to extend the experience of Moscow to the territory of Russia. Obviously, it is impossible without the support of the federal budget. In addition, the renovation program is long-term (15 years) and very capital-intensive. It would be rash to rely only on the state, and we are once again returning to private investors and bank lending. Private investors are interested in investing in housing construction because the difference between the estimated cost of construction, for example, 1 m^2, and the selling price (market value) is two or more times. Part of this revenue is used to cover additional infrastructure costs, but the profit of investors is several times higher than the estimated profit of contractors. If you move residents from five-story buildings to high-rise buildings, some of the space can be transferred to investors for subsequent sale on the housing market. In addition, the investor saves on engineering networks and social facilities, since renovation programs are implemented within walking distance of demolished residential buildings that have long been integrated into the existing urban infrastructure around them.

When we evaluate certain investment programs in terms of their value, in millions or in billions of rubles, these figures often lose the idea about the quality level and other consumer characteristics, which is interpreted by us as the final result of the investment activity. Construction differs from other sectors of the economy in that the final result will be visible in 2–3 or more years after the first investment. The renovation program has a natural-real expression. Currently, 15 construction grounds have been identified, where it is planned to build 500 thousand square meters of residential space. The demolition of five-story buildings, which may soon fall into the category of dilapidated or emergency housing, solve the problem of settlement. Construction of high-rise houses on the site of demolished five-story buildings significantly increases the load on infrastructure and engineering facilities. The consumption of water, electricity, and the length of pipelines, sewer networks, and other infrastructure facilities is increasing. There is an increasing need for inter-house driveways, car parking spaces, recreation areas, playgrounds, and much more. Therefore, the renovation program is accompanied by a significant restructuring of the environment, which requires significant additional investment costs.

The amount of housing construction is significantly affected by the debt of individuals to banks, which exceeds 15 trillion rubles. These funds could be used by banks as credit resources. The practice of issuing second and subsequent loans should be limited (individuals take these loans to pay off debts on previously issued loans).

As it was stated at the investment forum in Sochi, in 2019, to solve the housing problem, it is necessary to build 1.5 million apartments annually, in parallel, it is necessary to change the structure of housing construction, encourage the construction of individual homes, develop mortgages and other areas of financing [7].

5 Conclusion

In order to attract investment in housing construction, enterprises and organizations should intensify their activities for the construction of departmental housing stock. For this purpose, potential developers can be exempt from income tax, in terms of deductions to regional budgets, for the period of construction of departmental housing. For small and medium-sized towns, the transition to the construction of low-rise buildings is relevant. Reducing the cost of building such houses is possible by saving on load-bearing structures, foundations, and elevators. In total, this can save up to 30–32% of the estimated cost of 1 m^2 of living space. If such houses are combined into a structure like cottage settlements, then you can simultaneously solve infrastructure and social problems. Experience of such construction already exists on the territory of "new Moscow", where even the market price of 1 m^2 does not exceed 60–70 thousand rubles. It is obvious that such a solution to the housing problem for megacities is hardly possible because of the acute shortage of land plots. Saving land in relation to the renovation program is another advantage.

The development of individual housing construction is hindered by the lack or high cost of land around urban areas. Regional and local authorities should not only take care of land acquisition issues, but also finance costs of connecting engineering networks and other infrastructure facilities to such areas. As incomes grow in the country's economy, the possibilities of budgets at all levels expand. From the point of view of this approach, subsidized regions remain problematic. Taking into account that these regions receive federal assistance, it is possible to allocate a separate line of targeted funding for the development of housing construction as part of the total amount of subsidies. The problem remains in relation to self-built houses that have already been built and are already inhabited. It is obvious that demolishing these buildings means aggravating the housing problem. The authorities can limit to penalties against developers who violate the law.

Thus, the coordinated work of all financial institutions is necessary to activate housing construction. Funds of the population – citizens, buyers – should be accumulated on the accounts of the state mortgage bank, which will lend to developers before completing residential buildings and social cultural facilities at a minimum fixed percentage. A guarantee can be the volume of construction and installation work performed, linked to specific physical meters: section, floor, apartment. We need to make greater use of mortgage holidays and preferential lending during the transition period, gradually change the current system of payments for work performed, and activate the insurance system in the investment and construction sector. These measures will work if the volume of construction in progress, the population's debt on all types of loans and cross-lending are reduced. Such steps will negatively affect the cost of finished construction products. To reduce or completely eliminate the impact of these factors, it is necessary to develop an effective mechanism for reducing the cost of construction and installation work, including through innovative engineering and technological solutions, at all stages of production and sale of construction products.

References

1. Decree of the Government of Moscow N 497-PP dated August 1, 2017 to the "Program for the renovation of the housing stock in the city of Moscow" (2017). https://www.mos.ru/programmarenovacii.pdf. Accessed 20 Mar 2019
2. Decree of the Government of Moscow of September 27, 2011 N 454-PP on the approval of the State Program of the city of Moscow "Housing" (2011). https://www.mos.ru/dgi/documents/normativnye-dokumenty/view/209518220/. Accessed 15 Mar 2019
3. Decree of the Government of the Russian Federation of April 15, 2014 N 316 on the approval of the state program of the Russian Federation "Economic development and innovative economy" (2014). http://www.consultant.ru/document/cons_doc_LAW_162191/. Accessed 15 Mar 2019
4. Fleischer, G., Bensoussan, B.: Business and Competitive Analysis: Effective Application of New and Classic Methods. FT Press, Moscow (2007)
5. Germaschewski, Y.: Stabilization policy, infrastructure investment, and welfare in a small open economy. Econ. Model. **84**, 322–339 (2020)
6. Keeley, A.R., Matsumoto, K.: Investors' perspective on determinants of foreign direct investment in wind and solar energy in developing economies – review and expert opinions. J. Clean. Prod. **179**, 132–142 (2018)
7. Roscongress: Sochi Russian investment forum 2019. Forum results (2019). https://roscongress.org/news/rossijskij-investitsionnyj-forum-sochi-2019-itogi-raboty-foruma/. Accessed 15 Mar 2020
8. Rosstat: Construction (2020). https://www.gks.ru/folder/14458. Accessed 15 Mar 2020
9. Sharafutdinova, N.N., Persteneva, N.P.: Basic human resource management models. Bull. Young Sci. Samara State Univ. Econ. **2**(30), 228–233 (2014)
10. Waldron, R.: Capitalizing on the state: the political economy of real estate investment trusts and the 'resolution' of the crisis. Geoforum **90**, 206–218 (2018)
11. Wesseh Jr., P.K., Kin, B.: Environmental policy and 'double dividend' in a transitional economy. Energy Policy **134**, 110947 (2019)
12. Zhang, C., Zheng, N.: Monetary policy and financial investments of nonfinancial firms: new evidence from China. China Econ. Rev. **60**, 101420 (2020)

Features of State Regulation of Small Business Financing in the Russian Federation

N. A. Petrov(✉) and A. M. Mikhaylov

Samara State University of Economics, Samara, Russia
petrovnkt@gmail.com, 2427994@yandex.ru

Abstract. Under current economic conditions, the existing problem of the impact of foreign sanctions on reducing the operational efficiency of domestic business representatives is especially relevant from the point of view of maintaining the country's economic security. In such conditions, the state development assistance of small business as a future alternative in the form of replacing external counterparty suppliers for national enterprises is particularly relevant. In the institutional environment of the Russian economy that has formed now, the state is the most important participant performing regulatory functions of the funding process that are encouraging investments in real business. The purpose of this study is to structure the existing mechanisms of state control over the funding process for small businesses. Descriptive, systemic and statistical methods will be used to solve this purpose. The result of the study will be the formed spectrum of existing state methods for regulating financing of enterprises in the real sector of the economy and an assessment of its effectiveness, which can be further used by both the scientific and business environment in forming scientific approaches to the study of the problem identified above, as well as in the development of firms' own investment strategies.

Keywords: Business environment · Financing · Industry · Subsidizing

1 Introduction

Under current economic realities, small businesses are steadily increasing their share in the production of gross domestic product of the Russian Federation. Further development of small business in the domestic economy is difficult without improving the regulatory framework that regulates the provision of small business loans available from the federal budget, as well as investment resources of other market participants. Currently, the development of small businesses, including certain aspects of its financing by state authorities, is regulated by the Federal Law "On the Development of Small and Medium Enterprises in the Russian Federation" (hereinafter - the Law) [5].

Among new directions in the legislative sphere that regulates the activities of small businesses in 2018, there has been a reform of the Civil Code of the Russian Federation (hereinafter - the Civil Code [4]). From 01.06.2018, amendments to the Civil Code began to apply in terms of financial transactions, bank deposits, accounts and settlements. The most significant change for the business was the opportunity to conclude a loan agreement according to the consensual model: the agreement will now be valid for the parties from the moment of signing, and not from the moment the money is transferred.

S. I. Ashmarina et al. (Eds.): ENGINEERING ECONOMICS WEEK 2020, LNNS 139, pp. 183–190, 2021.
https://doi.org/10.1007/978-3-030-53277-2_21

Lack of attention to the specifics of small business activities in the development and implementation of regulatory decisions reduces the level of confidence of entrepreneurs in the state, creates additional incentives for businesses to leave in the shadow sector of the economy. Taking into account socio-economic and foreign economic conditions, the Government of the Russian Federation has developed a long-term positive program of activities in the field of small business development (hereinafter – the Strategy) [10].

The Strategy involves the implementation of the following principles: when designing and implementing government decisions in regulating entrepreneurial activity, the interests of representatives of small and medium enterprises are primarily taken into account. The second principle is legalization of business. It will be profitable for an entrepreneur to work legally if his interaction with regulatory authorities is minimized and a trusting relationship is established between business and the state. The third principle of the Strategy is to promote accelerated development. The fourth principle is to form a system of incentives for the involvement of state authorities and local governments in the development of small and medium-sized enterprises. The fifth principle is to ensure guaranteed and stable "rules of the game". In addition to the initiatives indicated in this Strategy, a multichannel system of financial support for small businesses at various stages of development has been formed at the federal level, involving the use of both repayable and non-repayable financing instruments. At the same time, financial support for small business will be provided on the basis of differentiation of small business entities according to selected target sectors - mass and high-tech. Commercial banks are encouraged to expand lending to enterprises in the real sector of the economy, microfinance and long-term financing are developed by improving syndicated lending to small and medium-sized enterprises, and the development of credit securitization, leasing, factoring, and direct financing instruments.

2 Methodology

The following methods were used in the study: a system method is to systematize the key factors mediating the process of attracting real capital in the small business segment. This study also places great emphasis on the comparison method. The results of studies obtained by other authors are analyzed and conclusions are consolidated. Based on them, a short-term forecast on the provision of investments is compiled. Often, the studies of investing problems represent theoretical justification for investing as a mediation between investors (who have money to invest) and corporations (which need capital to grow and run their business). The method of analysis and synthesis was used. Expert assessments of domestic and foreign researchers were analyzed.

3 Results

We believe that the prepared legislative environment regulating the activities of small businesses is not complete for the broad development of financing operations for small businesses. A significant variety of legislative norms under consideration, including internal standards of financial organizations to control financing operations of small businesses, leads to heterogeneity of the legislative framework in the country's economy [1].

Studying historical dynamics of legislation in developing and financing small business in Russia, the following stages can be distinguished, and they are shown in Table 1 [13].

Table 1. Stages of development of legislation in supporting small business

Stage	Regulatory Acts
Policy formation	1. Law of the USSR "On the State Enterprise"; 2. "Regulation on Small Enterprises" (Minutes No. 6 of the Meeting of the Commission on Improving the Economic Mechanism at The Council of Ministers of the USSR); 3. Law of the USSS "On Cooperation in the USSR"; 4. Law of the USSR "On the Enterprise in the USSR"; 5. Law of the RSFSR No. 445-1 "On Enterprises and Entrepreneurial Activity"; 6. Decree of the Council of Ministers of the RSFSR No. 406 "On Measures to Maintain and Develop Small Enterprises"; 7. Law of the RSFSR "On Corporate Income Tax"; 8. Decree of the Council of Ministers of the Government of the Russian Federation No. 446 "On Priority Measures for the Development and State Support of Small Business in the Russian Federation"; 9. Decree of the President of the Russian Federation No. 2270 "On Some Changes in Taxation and Relationship of Budgets of Various Levels"; 10. Decree of the Government of the Russian Federation No. 409 "On Measures for State Support of Small Business in the Russian Federation for 1994–1995"
Legal foundations	1. Civil Code of the Russian Federation; 2. Tax code of the Russian Federation; 3. Federal Law No. 88-FL "On State Support of Small Business of the Russian Federation"; 4. Decree of the Government of the Russian Federation No. 1256 "On Federal State Program Support of Small Business in the Russian Federation in 1996–1997"; 5. Decree of the Government of the Russian Federation No. 697 "On Federal State Program Support of Small Business in the Russian Federation in 1998–1999"; 6. Federal Law No. 14-FL "On Limited Liability Companies"

(*continued*)

Table 1. (*continued*)

Legal acts in the context of economic crises	1. Resolution of the Government of the Russian Federation No. 121 "On Federal State Program Support of Small Business in the Russian Federation in 2000–2001"; 2. Federal Law No. 128-FL "On Licensing of Certain Types of Activities"; 3. Federal Law No. 129-Fl "On State Registration of Legal Entities and Individual Entrepreneurs"; 4. Federal Law No. 94-FL "On Placing Orders for Supply of Goods, Works, Provision of Services for State and Municipal Needs"; 5. Federal Law No. 209-FL "On Development of Small and Medium-Sized Enterprises in the Russian Federation"; 6. Federal Law No. 223-FL "On Procurement of Goods, Works, Services by Certain Types of Legal Entities"; 7. Federal Law dated 05.07.2013 No. 44-FL "On Contract System in Procurement of Goods, Works, Services to Ensure State and Municipal Needs"; 8. Decree of the Government of the Russian Federation No. 98-r "On Approval of the Priority Action Plan to Ensure Sustainable Economic Development and Social Stability in 2015"; 9. Government Decree No. 265 "On Establishment of Income Limits for SMEs"; 10. Federal Law N 356-FL "On Amending Articles 251 and 252 of the Federal Law "On the Development of Small and Medium-Sized Enterprises in the Russian Federation"

Source: authors based on [13].

Legislation in this direction should be carried out as part of liberalization and amount of income tax payment, especially enterprises working in the innovation sphere, since there is the growth in production of value added in this sphere [3].

4 Discussion

To correctly determine the category of small businesses and analyze the regulation of their financing, it is necessary to have a correct understanding of the specific activity of the segment "Small business". There are a number of industry characteristics for enterprises in this market segment. In Table 2, we indicated the main industries in which small businesses operate.

Table 2. Classification of industries in the segment "Small business"

Industry	Description
Wholesale	- Inventories account for more than 15% of assets; - Significant (more than 15% of assets) volumes of receivables and payables; - Insignificant (less than 15% of assets) share of fixed assets
Retail	- Small proportion of fixed assets (mainly retail and office equipment); - Minimum amount of equity is possible; - Sufficiently high level of liquidity ratios; - Short financial cycle
Services	- Significant proportion of fixed assets through which enterprises operate; - High proportion of receivables and a small amount of accounts payable; - Minimum amount of equity is possible for enterprises in this industry

Source: authors.

Below we analyze the main trends in small business. Figure 1 shows the territorial structure of the number of small businesses in the Russian Federation [9].

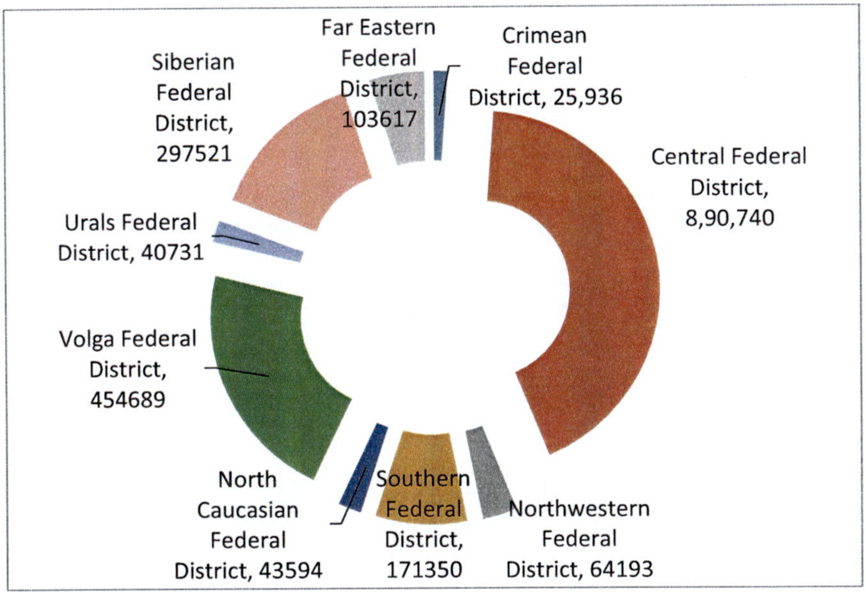

Fig. 1. The territorial structure of small businesses in the Russian Federation, units. (Source: authors)

The highest growth for the noted period is in the Urals Federal District (an increase of 9.7% or 13.9 small enterprises for every 100 thousand inhabitants), the Southern Federal District (an increase of 7.7% or 8.9 small enterprises for every 100 thousand inhabitants), the Central Federal District (an increase of 2.8% or 4.8 small enterprises for every 100 thousand inhabitants) [9].

We analyzed the relationship between the dynamics of institutional indicators prepared by the World Bank specialists and the dynamics of overdue debts of small businesses in the Russian economy that is a sign of the undeveloped system of financing enterprises in the segment we are considering. One of the indicators of the business environment development in the country's economy is the Doing Business Index. Its components are indicators "Time required to start a business" and "Percentage of contract execution" [7] (Fig. 2). We took the data for indicators from the World Bank DATA [11].

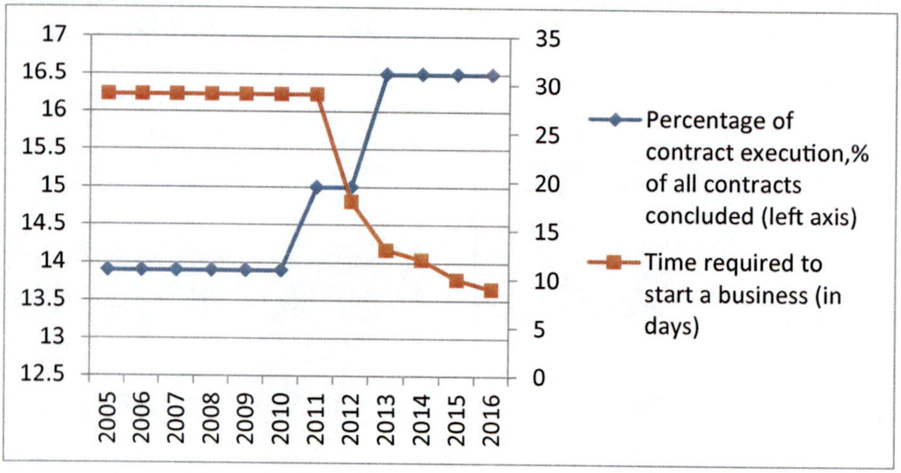

Fig. 2. Institutional indicators in the Russian Federation. (Source: authors)

According to the dynamics of these indicators, it has become easier to organize a business in the Russian economy since 2010. Business organization attracts external financing. However, the dynamics of overdue debt in the Russian economy shows us a negative trend (Fig. 3). «A significant moderator in the development of small businesses is the presence of corrupt patterns of interaction with government bodies. There is a so-called corruption risk» [8, p. 117].

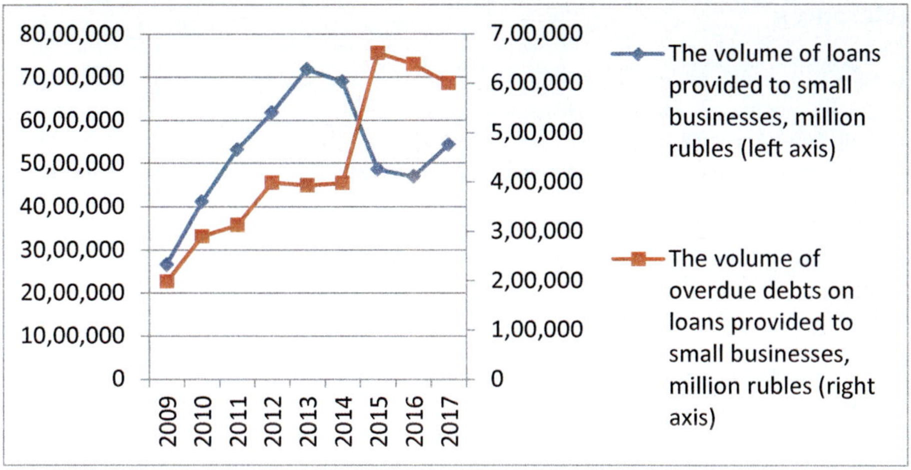

Fig. 3. Indicators of lending and overdue debt of small businesses. (Source: authors)

According to the results of eight months of 2016, the volume of loans provided to small and medium-sized businesses decreased by 11%, to 5.3 trillion rubles. More than half of enterprises (54.4%) had reduced financial stability, of which almost 10% - significantly. Thus, a graphical analysis of overdue loans for small businesses shows us that problems in the legislative environment aimed at regularly improving business conditions in the country affect the quality of loan debt of small businesses [6]. Support in the development and promotion of financing of small businesses is one of key factors in the formation and expansion of activities of small businesses. Among the types of assistance can be identified the creation of funds to support entrepreneurship, simplification of the tax base, the development of business training programs, financial support in the form of subsidies, loans [12].

5 Conclusion

To solve structural problems in the economy, especially the formation of a system for financing enterprises in the real sector, it is necessary to consider the structural features of our national economy, as well as internal and external challenges and restrictions in conducting monetary policy. At the same time, the role of mechanisms aimed at combining the efforts of the state and business in solving problems such as the use of public-private partnership tools in priority projects and programs for economic development [2] should increase. It is also important to maintain the stability of the economy and prevent the accumulation of imbalances in financial and real sectors, to exclude excessive growth in bank lending. Thus, state support of business, in all its forms, whether it is educational, consulting services or direct financial assistance, is the most important factor in the formation of entrepreneurship in the country.

References

1. Berezina, E.: TOP-6 laws that will change the life of business in 2018. Russian Newspaper (2018). https://rg.ru/2018/01/03/top-6-zakonov-kotorye-izmeniat-zhizn-biznesa-v-2018-godu.html. Accessed 01 Feb 2020
2. Cole, A.R., Sokolyk, T.: Debt financing, survival, and growth of start-up firms. J. Corp. Financ. **50**, 609–625 (2018). https://doi.org/10.1016/j.jcorpfin.2017.10.013
3. Ding, S., Jia, C., Wu, Z., Yuan, W.: Limited attention by lenders and small business debt financing: Advertising as attention grabber. Int. Rev. Financ. Anal. **49**, 69–82 (2017). https://doi.org/10.1016/j.irfa.2016.12.003
4. Federal Law dated 26.01.1996 No. 14 "Civil Code of the Russian Federation (part two)" (1996). http://www.consultant.ru/document/cons_doc_LAW_9027/d73c103b664fa1b30b63f8fa944a2e0f33cd67f0/. Accessed 01 Feb 2020
5. Federal Law of July 24, 2007 No. 209-FL "On the Development of Small and Medium Enterprises in the Russian Federation" (2007). http://www.consultant.ru/document/cons_doc_LAW_52144/. Accessed 01 Feb 2020
6. Haaker, T., Bouwmann, H., Janssen, W., Reuver, M.: Business model stress testing: a practical approach to test the robustness of a business model. Futures **89**, 14–25 (2017). https://doi.org/10.1016/j.futures.2017.04.003
7. Jindal, V., Seth, R.: A new order of financing investments: evidence from acquisitions by India's listed firms. J. Corp. Financ. **58**, 307–328 (2019). https://doi.org/10.1016/j.jcorpfin.2019.04.007
8. Makhmudov, T., Konovalova, M., Kuzmina, O., Persteneva, N.: The impact of the institutional environment on the shadow economy. Acta Oeconomica **68**(1), 115–133 (2018). https://doi.org/10.1556/032.2018.68.1.6
9. Mikhailov, A.M., Petrov, N.A., Nedorezova, E.S., Belyakova, O.V.: Influence of factors of production interaction on processes of world economy globalisation. In: Mantulenko, V., (ed.) International Scientific Conference Global Challenges and Prospects of the Modern Economic Development. European Proceedings of Social and Behavioural Sciences, vol. 57, pp. 747–756. London: Future Academy (2019). https://doi.org/10.15405/epsbs.2019.03.74
10. The Order of the Government of the Russian Federation of June 2, 2016 No. 1083-r on the "Strategy for the development of small and medium-sized enterprises in the Russian Federation for the period until 2030, and the action plan ("road map") for its implementation" (2016). http://base.garant.ru/71418202/. Accessed 01 Feb 2020
11. The World Bank: Ease of doing business index (1 = most business-friendly regulations) – Russian Federation (2019). https://data.worldbank.org/indicator/IC.BUS.EASE.XQ?locations=RU. Accessed 01 Feb 2020
12. Troshani, I., Janssen, M., Lymer, A., Parker, D.L.: Digital transformation of business-to-government reporting: an institutional work perspective. Int. J. Account. Inf. Syst. **31**, 17–36 (2018). https://doi.org/10.1016/j.accinf.2018.09.002
13. Vershinina, A.P.: Development and improvement of regional policies to support small business. Internet J. Sci. of Sci. **7**(1), 2–15 (2015)

Innovations in Financial Sustainability Management of Public Health Care Organizations

E. N. Valieva[1(✉)], R. R. Yarullin[2], N. N. Zhelonkin[3], and A. S. Mayorskaya[3]

[1] Samara State University of Economics, Samara, Russia
rad8063@yandex.ru
[2] Financial University under the Government of the Russian Federation, Ufa Branch, Ufa, Russia
jrr61@mail.ru
[3] Samara State Medical University, Samara, Russia
33221111@mail.ru, as-mayorskaya@rambler.ru

Abstract. Healthcare organizations operating under the program of state guarantees of free medical care in Russia are not financially stable because of underfunding. In this regard, the development of methods for analyzing the financial stability and break-even management is an urgent task of the theory and practice of financial management for medical organizations. The aim of the research is to develop new methodological approaches to financial management of public health organizations to ensure a balance between funding and costs. The work is based on the methodology of financial management, namely the concept of cash flow and the concept of alternative costs. The study showed that on average, the plan for the number of visits to clinics is not fulfilled by 4%, and for the amount of funding spent – by 7%. Financial performance and efficiency indicators are excluded from the planning, management, and analysis process. Based on the operational analysis, the authors obtained parameters for adjusting the structure of medical care, the volume of variable and fixed costs that can be implemented in the framework of strategic planning.

Keywords: Break-even · Cost management · Finance of medical organizations

1 Introduction

Permanent health care reform in the Russian Federation since the early 90's of the 20th century aims to provide the population with affordable medical care at the appropriate quality level that meets modern requirements through the effective use of targeted financial resources of the budget system. The most acute problems of the first years of market transformations were solved by creating a system of compulsory medical insurance (CMI). At the same time, difficulties remain in providing free medical care financially.

S. I. Ashmarina et al. (Eds.): ENGINEERING ECONOMICS WEEK 2020, LNNS 139, pp. 191–200, 2021.
https://doi.org/10.1007/978-3-030-53277-2_22

State medical organizations (MOs) are not able to provide the full amount of medical services ordered by the state every year, because their funding is not enough to fulfill the planned tasks. Modern problems of health care finances and medical organizations, in particular, are caused by the transition from financing in accordance with the approved cost estimates to paying for services at the expense of the CMI and subsidizing socially significant medical services ordered by the state. The declared advantages of such a system of financing were the provision of targeted and economical use of state financial resources. However, with a low share of value added in the gross domestic income of the Russian Federation, on which premiums for CMI and budget security depend, it has not been possible yet to realize these advantages. Thus, an urgent task of today is to organize an effective financial management mechanism in medical organizations (MOs), which would be integrated into the financial system of the industry as a whole and would increase their financial stability.

The purpose of this research is to develop new methodological approaches to financial management of public medical organizations to ensure a balance of funding and costs. In particular, it was assumed:

- to identify shortcomings of financial management practices in MOs,
- to formulate innovative methods for substantiating management decisions to ensure the financial stability of MOs,
- to quantify parameters of scenario conditions for achieving break-even of a multi-profile MO.

2 Methodology

In addition to general scientific methods of cognition, the authors based their research on the methodology of financial management, namely the concept of cash flow and the concept of alternative costs. These concepts imply to take into account in the financial management of the organization, on the one hand, estimates of positive and negative cash flows and its determinants, on the other hand, the analysis of alternative scenarios for a possible way of investing financial resources, determination of the cost structure and creation of a system of management accounting and control. In some scientific works, researchers justified the use of appropriate indicators of operational analysis in the construction of effective models of financial management in medicine, as well as the validity of the analysis of the MO's efficiency based on the definition of a financial risk zone and a financial security zone, focusing on the break-even point and solvency indicators [12].

In this research, the authors applied direct costing, break-even analysis (margin analysis) and operating analysis of indicators of economic-financial activities of MOs, which allowed us to determine compliance of available resources with volumes of medical care commissioned by the founder and to suggest ways to optimize the cash flows. The purpose of using these methods was to assess the optimal volume of medical services, their structure, volume of funding, variable and fixed costs to achieve break-even. Among the existing set of methods for calculating the break-even point, we selected the graphical method.

To calculate the level of conditional fixed costs, the following formula was used:

$$FCi = Qi * (Pi - SVCi) \quad (1)$$

where:

FCi – conditional fixed costs;
Qi – number of medical services (units);
Pi – price per unit of service;
SVCi – specific variable costs.

To calculate the correction coefficient for conditional fixed costs, the following formula was used:

$$K = 1 - D/FC \quad (2)$$

where K – reduction coefficient of conditional fixed costs; D – lack of funding.

The standard of variable costs per unit of medical services was calculated in the context of services types (ambulance services, polyclinic services, hospital services, day hospital services) in accordance with the cost structure provided in the tariff agreement for compulsory medical insurance, using the formula:

$$nSVCi = \sum_{i=1,n}^{j=1,m} Fi * dj/Qi \quad (3)$$

where:

Fi – the volume of funding for the i-th service,
dj – the stipulated share of j-th expenses

3 Results

The lack of a positive effect of optimizing healthcare is largely because of its underfunding over the past three decades. Estimates of government spending on medical care as a percentage of the gross domestic income (GDI) published in open sources range from 3% to 6% and, in our opinion, are not correct enough. According to our estimates, at the beginning of market reforms, this indicator was at the level of 2% of the GDI, by the end of the 90 s – it increased to 3% and remained at this level until the early 2000s [9]. Then the growth rate in budget spending on health care slowed down. According to the Accounting Chamber of the Russian Federation, in 2018, the indicator was 2.6% of the GDI (about 2.6 trillion rubles). The implementation of budget expenditures on medicine in 2019 was approximately at the same level – 2.7% of the GDI. The budgets for 2020 were adopted taking into account the growth of medical expenses up to 3% of the GDI. The state program for the development of healthcare in Russia, adopted in 2014, intended to bring state funding for healthcare (including CMI, the share of which accounts for about 40%) to 11% of the GDI by 2020, was not fulfilled. Based on information published by the World Health Organization (WHO) in the global health expenditure database, we have obtained the following comparative estimates: as a percentage of the national income, Russia spends 7 times less on medicine than in the United States, 4 times less than in France and Norway, and 3 times less than in Israel and Estonia. In terms of health care expenditures in US dollars per capita in 2016, Russian indicators were 19, 9, 16, 6, and 2 times less, respectively [11].

External financial conditions make MOs to pay special attention to the development of a mechanism for operational and strategic management of financial resources aimed at maintaining the necessary level of financial stability and break-even. Modern rules for financing the activities of MOs in accordance with the state task do not imply its correlation with the provided quantitative indicators of their activity. The planned task reflects the number of medical services, as well as the maximum amount of funding, which is determined based on low (compared to the objective need) standards of costs per unit of service. The corresponding share of planned financial resources is transferred to MOs on a quarterly basis. Payment for services as they are rendered is not provided. Ensuring that actual costs correspond to the amount of funding within the year is not provided for in the planning task. This approach can be considered as a reduced version of estimated funding.

In the course of performing a planned task, the medical organization may exceed or under-perform planned indicators of the provision of medical services. This will not affect the amount of government funding. The study showed that on average, the plan for the number of visits to clinics is not fulfilled by 4%, and the amount of funding disbursed by 7%. Financial performance and efficiency indicators are excluded from the planning, management, and analysis process. The financial result of a medical organization is not taken into account when planning the volume of services, does not appear as an indicator when forming the cost of services, and is not displayed in the organization's annual reports.

When developing the financial management system of a medical organization, a number of factors determined by the legislation and external economic conditions should be taken into account, namely:

- single-channel principle of financing within the CMI,
- subsidizing medical services provided by the order of state authorities,
- embedding MO's finances in the budget process,
- specific features of financial relations with subjects (participants) of the CMI,
- the state regulation of business activity of MOs,
- ensuring the financial sustainability of the medical organization in conditions of insufficient funding,
- the need to ensure social effectiveness, financial efficiency of the MO and compliance with the legislation.

The analysis of the financial management system of a multi-disciplinary MO serving a large administrative-territorial entity of the Samara region revealed shortcomings inherent in the current management practice:

1. The MO does not independently plan indicators of the volume and structure of medical care. It is satisfied with the indicators provided by the regional commission for the development of the program of state guarantees, in terms of CMI services,

and the indicators provided by the regional Ministry of Health in terms of socially significant and high-tech services. Thus, the planned performance indicators are not linked to the financial condition of the MO, including break-even.
2. Financial analysis, practiced by the MO, does not involve comparing the timing and volume of inflows (payment for medical services by patients, payment by insurance medical organizations for CMI services, budget financing, payment for services by insurance companies for the voluntary medical insurance VMI) and outflows (payment of wages, taxes, payment for contractors' services) of funds. There is no necessary management accounting for this purpose.
3. The organization of operational management of financial and economic activity of the MO is characterized by insufficient control over compliance of the actual volume of medical care to planned indicators, which could lead to the failure of the approved planned task. This, in turn, may lead to non-receipt of planned revenues from the CMI.
4. The current financial management procedure does not imply the development of leverages that would ensure break-even or profitability of the MO's activities in the current and strategic period. In particular, the profitability of a type of medical care or service, as well as the structural division or the MO as a whole, is not evaluated.
5. Reporting is limited to three areas: the number of medical services provided at the expense of the CMI and budget subsidies, accounting reports, and compliance of the quality of services provided with established indicators.

Developments carried out in the course of this research are an attempt to overcome the above-mentioned shortcomings by integrating into the existing management system tools that can be considered as innovative for the Russian medical organizations. The development of methodological recommendations for using the capabilities of operational analysis and economic justification of management decisions was carried out on the basis of data from a multidisciplinary MO for three years. Financial flows in connection with the provision of medical services under the CMI were analyzed.

The CMI rules are so that the MO cannot set prices for medical services in the framework of the program of state guarantees or change them. However, there is a legitimate opportunity to manipulate the cost of providing services in the context of types of medical care if the planned volume of services is met.

To enable the use of operational analysis tools, it is necessary to develop an internal cost accounting system that allows you to assess the cost volume in the context of assistance types or structural divisions of the MO. In addition, for each direction, the division of costs into variables and constants is mandatory.

For the research purpose, the variables included the cost of medicines and other medical expendable materials, the cost of feeding patients in hospitals, and the cost of fuel and lubricants for emergency medical stations. Analysis of the structure of actual costs for providing medical services of the CMI allowed us to draw the following conclusions:

- the share of variable expenses in the structure of actual expenses is generally 12% of the CMI,
- the share of variable costs in the context of types of medical care is significantly differentiated (from 4% in polyclinics to 25% in day hospitals).

Depending on the specific weight of variable costs in the cost of the corresponding assistance type, the break-even point will be more or less sensitive to them. Accordingly, it is possible to ensure the break-even of a multidisciplinary MO by manipulating not only the cost structure, but also the structure of types of medical care provided.

An analysis of existing cost accounting practices shows that material costs are attributed to a division that is randomly selected for this purpose. Expenses are charged directly to the services of a polyclinic, hospital, day hospital, or ambulance only if this is explicitly defined in the regulatory documents. Accrual and accounting of expenses for wages and social insurance contributions in accordance with the directive automated system occurs on the whole for the entire MO. Such databases are useless for cost and break-even management purposes.

We offer two solutions to the problem:

- to reflect the corresponding principle in the accounting policy and keep records of all costs in the context of constant and variable types of medical care,
- variable costs are accounted by types of medical care, and fixed costs are attributed to conditional divisions, so that they are distributed in proportion to variable costs or the number of staff when analyzing the implementation of the plan for expenses, break-even, etc.

In the process of break-even analysis, which usually involves determining the sales volume that covers the variable and fixed costs, a definition of such a cost structure (and corresponding amount) is required which would, on the one hand, allow to provide the volume of services ordered by the state, on the other, ensure the equality of costs and public funding at the expense of CMI means. Calculations of the necessary initial data and the break-even point of the MO are carried out in accordance with the authors' methodological approach.

The break-even point was estimated using the following three scenario conditions:

1. Provision of a planned volume of medical services at the expense of planned financing while maintaining the amount of costs and their structure at the level of the base year.
2. Provision of a planned volume of medical services at the expense of planned financing, while maintaining the basic fixed costs and bringing variable costs to the standard level.
3. Provision of a planned volume of medical services at the expense of planned financing when bringing variable costs to standard ones and reducing fixed costs to the optimal level (expert assessment).

The first scenario option showed that the break-even point will be reached at the volume of 101353 bed days (with the planned value of 90553 bed days). The amount of financing should be 196,746 thousand rubles, with the planned volume of 175781 thousand rubles within the framework of the program of state guarantees, i.e. increased by 12%.

Scenario parameters of the second option ensure that the break-even point is reached with the volume of services of 106,153 bed days. The amount of financing, respectively, should be equal to 2,064 thousand rubles, i.e. it should be increased by 17% in comparison with the planned task for the program of state guarantees.

Below there is a graph illustrating the third scenario option for achieving a break-even for the MO, when providing medical care in hospital conditions (Fig. 1).

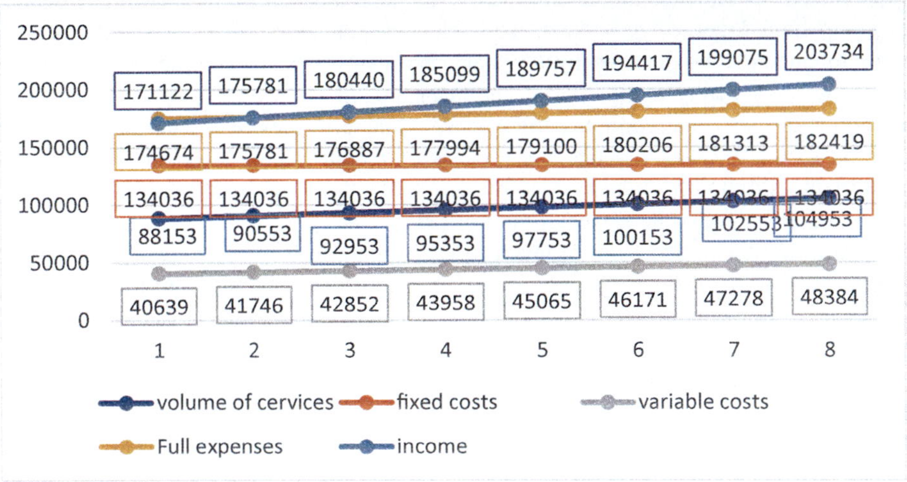

Fig. 1. Analysis of break-even of inpatient care when bringing variable costs up to standard ones and reducing fixed costs to the optimal level. (Source: authors)

According to the third option, the break-even point should be reached if both the variable cost standards and the planned volume of services and financing are met, as a result of a reduction in fixed costs from 155269 thousand rubles to 134036 thousand rubles, i.e. by 14%. Thus, the balance of financial resources and costs in the provision of inpatient medical care in modern conditions of financial security is possible when optimizing fixed costs, consisting mainly of labor and social insurance costs, as well as utility costs. The preferred option, in our opinion, is to increase the state order for medical services (according to our estimates by 17%) while maintaining the number of staff and other factors that affect fixed costs. However, in conditions of insufficient funding, management may choose the latter option.

4 Discussion

The problems considered in this study are widely presented in scientific publications. The variety of interpretations and methods used does not negate the common position of economists, which is to recognize the importance of managing finances of public health organizations in order to ensure, at least, a balance of income and expenses, including on the basis of models and methods originally intended for business. The objective background of this consensus, in our opinion, is the widespread rise in the cost of medical services, which is increasingly difficult for national health insurance systems and budget systems to cope with. Let's focus on the points of view that are closest to us on the issues under study.

The Russian authors draw attention to the need to modernize mechanisms for the formation and use of financial resources of medical organizations that provide socially important medical services and services for CMI. Options for the economic justification of quantitative parameters that determine the financial mechanism in the primary health care outpatient level are proposed. The relevant indicators (funding standards and prices for medical services set by the state, payment rates for contractors, insurance premiums) are the most mobile part of the management system and should be modified depending on external economic conditions, social priorities, etc. The authors draw attention to the need to regulate the process of redistribution of financial resources between the structural divisions of clinics in order to provide financial support for expenses and organize the financial control [7].

A significantly higher level of healthcare financing in the economically developed countries of Europe does not reduce the relevance of the discussion on improving the source structure of financial resources in the industry and the efficiency of their use. In the Netherlands, for the past ten years, the cost of capital construction and technical re-equipment has been funded by insurance premiums but from the mandatory medical insurance system. The analysis of this relatively new economic practice in terms of its efficiency and future prospects is the focus of attention of our Dutch colleagues [6].

Russian scientists pay much attention to issues of financial control in medical organizations, which is particularly relevant in the context of a shortage of financial resources. Attention is drawn to the lack of a unified approach (methods of federal or regional agencies) to the organization of control over the use of both state financial resources and entrepreneurial income. The generally accepted system of budget accounting and commercial accounting does not allow tracking the direction of using funds of MOs. The existing system of financial accounting and reporting does not allow us to quickly assess the implementation degree of the planned activity parameters. Options for improving the financial management tools of medical organizations, including their internal control, are proposed based on the creation of a management accounting system [8].

In a number of works, it is proposed to analyze the economic efficiency of the MO, determining the risk zone, the financial security zone and its actual financial position on the basis of quantitative parameters, and make management decisions based on these conclusions. Parameter estimation is implemented using operational analysis techniques. It is recommended that zones are characterized by break-even indicators, as well as by indicators of solvency and economic potential. It is assumed that financial management organized in this way will contribute to the achievement of strategic goals, namely, economic efficiency and social performance of operating activities [1].

Risk assessment in healthcare organizations is a popular issue. In particular, it is proposed to use a balance mathematical model for managing economic risks in municipal health organizations that provide emergency care. The model has a modification for organizations that carry out business activities. Its use allows you to forecast basic financial indicators. The ease of implementation of the methodology is emphasized. The basis of predictive estimates is the average values of accounting data for a number of reporting periods [5].

Italian colleagues are investigating the revenue management process in public health trusts. The directive condition for functioning of the latter is balancing of incomes and expenses. The studied state medical organizations in some cases showed insignificant losses in reporting, without any distortion of accounting data on income. It was also noted that trusts allow manipulations of certain types of expenses in order to prevent a positive financial result. It is emphasized that state incentives to achieve a budget balance in healthcare can lead to a decrease in the quality of medical services [3].

Some researchers note that in the newly liberated states of Eastern Europe, in most of the observed cases, cost accounting and analysis is carried out for the entire medical organization. The distribution of costs by type of medical care or by completed cases in public health institutions is not common. It is noted that the application of managerial cost accounting would increase the added value in this sector of social services. This conclusion is similar to the one we obtained in the course of a study of medical organizations in the Russian Federation [4].

Published research works show that one of the weak elements of the management system is planning costs and prices for paid medical services. A variant of the operational accounting development and cost analysis is proposed, which will improve pricing and competitiveness of a MO, on the one hand, and increase its financial stability, on the other [10]. Brazilian scientists are also concerned about the problems of risk management in healthcare organizations. They note that healthcare is a complex subject from the point of view of financial management. In this regard, it is insufficiently studied, and the risk assessment methods used are not without disadvantages. One of the reasons for this state of affairs is the lack of sufficient transparency of financial activities of medical organizations. The possibility of using the ERM model, which assumes an assessment of the riskiness of cash flows and variability of costs, is justified. All MOs that tested this risk assessment model recognized the effectiveness of using the information obtained from it to justify management decisions [2].

5 Conclusion

In a standard situation, an organization should increase sales or reduce expenses in order to ensure its break-even. Subsidizing the state order of medical services and funding from the CMI are in short supply. In most cases, MOs follow the first path and increase the volume of services paid by patients within the framework of permitted entrepreneurial activities. But here they face restrictions related to both the population's ability to pay for them and the own "production capacity."

In a situation where it is impossible to influence the incoming financial resources, as it is the case in healthcare organizations, the structure of services and optimization of expenditures become the leverages of the financial stability regulation. The most difficult task is to determine specific optimization parameters that would not lead to a decrease in the quality of medical care. Operational analysis, including the determination of the profitability threshold, can become a methodological tool for justifying the relevant management decisions.

The parameters for adjusting the structure of medical care, the volume of variable and fixed costs obtained on the basis of operational analysis can be implemented within the framework of strategic planning for at least three years. In this regard, financial management can only be used as an element of the MO's general strategic management system. The financial strategy should take into account what changes are expected in the profile of the MO and its organizational structure, personnel composition, volume of services by assistance types, etc. Financial strategy should be based on the following indicators for the whole MO in the context of the structural divisions specializing in the provision of a particular type of medical care: scope of services, fixed costs, the ratio of specific variable costs, prices for services, the amount of state subsidy, funding of the CMI, the financial result.

References

1. Efremova, Yu.S.: Application of strategic management tools in healthcare. Humanit. Res. **1** (65). http://human.snauka.ru/2017/01/18947. Accessed 29 Mar 2020 (2017)
2. Etges, A.P.B.S., de Souza, J.S., Neto, F.J.K., Felix, E.A.: A proposed enterprise risk management model for health organizations. J. Risk Res. **22**(4), 513–531 (2016)
3. Ibrahim, S., Noikokyris, E., Fabiano, G., Favato, G.: Manipulation of profits in Italian publicly-funded healthcare trusts. Public Money Manage. **39**(6), 428–435 (2019)
4. Jovanović, T., Dražić-Lutilsky, I., Vašiček, D.: Implementation of cost accounting as the economic pillar of management accounting systems in public hospitals – the case of Slovenia and Croatia. Ekonomska Istraživanja **32**(1), 3754–3772 (2019)
5. Kraftová, I., Kašparová, L.: Assessment of the financial health of regional emergency medical services in the Czech Republic using the BAMF model. J. Bus. Econ. Manage **18** (2), 340–353 (2015)
6. Kroneman, M., Boerma, W., van den Berg, M., Groenewegen, P., de Jong, J., van Ginneken, E.: Netherlands: health system review. Health Syst. Trans. **18**(2), 1–24 (2016)
7. Razumovskaya, E.A., Lugovtsov, R.Yu., Kozlov, A.V., Fomenko, V.V.: Economic justification for the need to reallocate financial obligations in healthcare. Bull. VSGUTU **2**, 70–76 (2016)
8. Vakhrushina, M.A.: Management accounting of medical (municipal) institutions: perspective or reality? Account. Budgetary Non-Profit Organ. **6**, 17–29 (2016)
9. Valieva, E.N.: The structure and features of the financial mechanism of Russian non-profit organizations that provide public services. Bull. Samara State Univ. Econ. **10**(120), 59–64 (2014)
10. Voronina, L.P.: Financial management of healthcare institutions. Municipality Econ. Manage. **3**(58), 57–67 (2019)
11. World Health Organization: Global health expenditure database. http://apps.who.int/nha/database. Accessed 29 Mar 2020 (2017)
12. Yashina, N.I., Yashin, K.S., Artamonycheva, E.V.: Problems and approaches to assessing the financial condition of medical institutions to determine the types of organizations. Econ. Anal. Theory Pract. **37**, 44–57 (2014)

State Support as Incentive for Developing Enterprises and Increasing Competitiveness

G. P. Gagarinskaya[1], A. V. Gagarinskii[1], E. S. Potokina[2], and T. N. Obuschenko[1](✉)

[1] Samara State Technical University, Samara, Russia
eyo080505@mail.ru, tatyanaobushenko@mail.ru, cantstoped@gmail.com

[2] Samara State University of Economics, Samara, Russia
vev.sseu@gmail.com

Abstract. The study represents the features of state support for enterprises in the state control system of the agricultural sector. The importance of support is associated with the need of enterprises for supportive and stimulating development mechanisms that meet the effective functioning of the economy. The authors analyze the mechanism of state support, focus on the goals of support, internal and external factors of support, and methods of influence on factors. Analyzing the practice of subsidizing, the authors reveal the features of using this form in technological, investment, and marketing processes of products. In conclusion, approaches to assess the relevance and effectiveness of various forms of state support for enterprises are summarized. The authors conclude that at the current level of the agro-industrial complex, support types aimed at increasing the purchasing power of farmers by investing in production of high value-added products using competitive technologies and long-term strengthening of their positions in dynamically changing global markets are becoming relevant.

Keywords: Compensation · Export · Insurance · Investment · Public procurement · Stimulation

1 Introduction

State support is of great importance for agricultural enterprises growing plants and animals in difficult climatic conditions. Natural disasters can provoke the complete death of crops and animals, and the authors examined the state's willingness to support farmers in such situations. In 2019, 303 billion rubles were allocated to support agriculture and it was up by 62 billion rubles in funding in 2018. The agro-industrial complex is evolving, so the state supports for the competitiveness of products and enterprises, develops an efficiently functioning market, creates a favorable investment climate and investment growth, maintaining parity of price indices for agricultural and industrial products, which are also reflected in government support policies through a wide range of ongoing programs.

S. I. Ashmarina et al. (Eds.): ENGINEERING ECONOMICS WEEK 2020, LNNS 139, pp. 201–211, 2021.
https://doi.org/10.1007/978-3-030-53277-2_23

Klimenko and Minchenko analyze approaches to optimize state control, considering market shortcomings and possibilities for improving it, and conclude that the opposition of state control to market control is unjustified [6]. Lenchuk notes that public policy should ensure the formation of harmonious proportions in the economy through the development and implementation of a set of state control measures at the macro, meso and micro levels [7]. A few authors consider state support measures as a tool for the development of small and medium-sized businesses [4]. The purpose of this study is to analyze the mechanisms of assistance to the agro-industrial complex and identify the causes that affect the demand for individual support measures. Many of the applied measures to support agricultural enterprises remained behind the analysis, which represents an independent research topic.

2 Methodology

A way to organize this study is to consider state support as a system which is to create favorable conditions for the development of the agricultural sector. The authors revealed the elemental structure of the state support system, its internal organization, determine qualitative properties that ensure its integrity and peculiarity, determine external relations of state support with other systems. Historical consideration made it possible to find out the conditions for the emergence of state support, the stages passed, the current state, as well as possible development prospects.

3 Results

3.1 Support for the Cultivation of Crop and Livestock Products Under Unsustainable Agriculture

The production process in agriculture has its own characteristics associated with the fact that biological processes, the prevailing weather and climate conditions, especially during the growing season, often influence the value of the produced products, affect the quality, and make it impossible to produce the planned volume products and, as a result, get the necessary profit.

Compensatory support is provided for a range of agrotechnological work in crop production, support for dairy farming, livestock breeding, elite seed farming, meat livestock farming, the development of traditional crop and livestock farming subsectors, and insurance premiums.

Analyzing regional requirements, we can conclude that subsidies allocated are not determined from the actual costs incurred by the farmer, but they are set at the rates – per 1 hectare of sown area, per 1 hectare of low-productivity arable land, per 1 kilogram of milk, per 1 head of farm animal, per 1 ton of farm animal weight. The approach of the lower border is applied - the required livestock is established, the productivity of animals is achieved by the farmer at which subsidies will be allocated to him. Restrictions on the lower limit of purchase (at least 100 heads) of pedigree cattle are allowed. The amount of the annual rate is limited, they can split the annual rate by quarters. Information on regional subsidies is available on the Internet.

The mechanism of any control system should contain a motivational part. The increasing coefficient is introduced when the yield exceeds the average for the area. In regions, the task is to put into circulation unused land, including irrigated land, here farmers can compensate up to 80% of the costs incurred. Regarding elite seed production, the acquisition of seeds of the first reproduction, elite, superelite (depending on what has been achieved in the regions) is motivated, regional authorities apply demotivation, which reduces the coefficient if elite seeds are acquired outside the region.

The amount of support is formed based on the share of each region in the total value of indicators for the country, while if support is distributed among several indicators, a significance coefficient is established for each indicator. The stimulating approach involves supporting accelerated development, some sectors need to be given an impetus for development, while for other sectors, support is directed to strengthen their activities.

In accordance with the Budget Code of the Russian Federation [10], the State Program for the Development of Agriculture and Control of Agricultural Products, Raw Materials and Food Markets [9], priority sub-sectors of the agro-industrial complex are defined: production of grain and leguminous plants, oilseeds, open-ground vegetables, viticulture, fruit and berry plantations, stimulation of production of flax, milk, development of specialized meat cattle breeding, development of sheep husbandry – the financing of priority sectors is provided with state support.

In the regions of Russia, state control is being developed and it explains the procedure for providing incentive subsidies to farmers to help achieve the goals of regional programs for the development of agriculture and control of agricultural products, raw materials and food markets. In breakthrough areas, indicators and enterprises that will ensure this growth to regions until 2024 have been identified. A stimulating subsidy is provided based on the share of each region in the total value of indicators for the country. In the calculation of the incentive subsidy, significance factors are applied both for each priority sub-sector of the agro-industrial complex and for each of indicators involved in the calculation.

Farmers can receive short-term loans at a preferential rate for Russia of no more than 5% on working capital: the purchase of fuels and lubricants, chemical and biological plant protection products, fertilizers, seeds, payment for the repair of agricultural machinery and equipment. Soft working loans cover 70% of the need for financing the sowing campaign. Soft loans are provided for the purchase of young farm animals, fish stock, and for the payment of insurance premiums. At the same time, subsidization of soft short-term loans is increasing from year to year and the list of industries that can receive soft loans is expanding, now the state is ready to assist processing enterprises that have not been practiced before, now they are supporting those processing industries that are nearby and will cause agricultural growth.

3.2 Stimulating Modernization and Innovation

A measure of support for investment projects is the reimbursement of 10–25% of the direct costs incurred, such as monetary costs of agricultural producers to create and modernize facilities equal to the actual value of the facility. It is considered one of the well-proven measures that really affect the development of agriculture, contributes to the influx of new investors in the industry since it significantly reduces the payback period of projects and improves the attractiveness of certain areas [8]. It is especially evident in the example of the greenhouse industry, where, many projects were supported and implemented, and the saturation of the market was achieved.

Assistance is currently being given to investment projects aimed at construction and modernization of storage facilities, dairy complexes, sheep breeding complexes, breeding centers, flax processing, production of dry dairy products for baby food, which indicates the coordination of the state support program, the agricultural development program, and the scientific and technical agricultural development program.

The state gives signals which projects can count on winning the competitive selection: modern technologies, high-performance jobs that provide yield growth in crop production and productivity and livestock. At the design stage of the facility, the total amount of financing, the estimated amount of state support, the financial results of the project, the budget effect, and the payback period are determined.

For projects of dairy farms, investment support will be provided if modernization of both equipment for animal husbandry systems and equipment for feed production and manure removal is provided. Seed centers are supported if elite and superelite seeds are grown. To receive support, projects of dry infant formula of baby food need to be provided with their own raw materials by 70%–90%.

Support will be given to storages with a one-time storage capacity of at least 30,000 tons of agricultural products, and the state indicates what storage it needs: meat and meat products, milk and dairy products, fruits and berries, vegetables and food melons, potatoes, fish and fish products. These should be premises with controlled temperature conditions, the availability of equipment for sorting, calibration, washing, packaging, part-time work, freezing, re-freezing, ripening, the presence of a veterinary and phytosanitary control point, and an automated information management system.

Another very important and successful measure of support is soft investment lending. Like any borrower, farmers have important goals for which they can take a soft loan, the terms of the loan, the opportunity to prolong it. Regions distribute soft loans based on priority areas and can set a limit for one borrower on the territory of the subject. The priorities of regions are also small business forms. There are new opportunities for soft loans for construction of complexes for growing berry and mushroom products. As for individual equipment, the range of support is expanding, for example, to expensive milking installations, highly specialized agricultural machinery. The provision of credit funds for reconstruction and modernization of oil and fat industry enterprises is expanding, they began to expand access to soft loans for flour mills and bakeries, and facilities and equipment for the development of fish farming were added to the list of targeted areas for soft loans, which will enable aquaculture to develop.

Soft investment loans are granted for a period of 2–5 years – to purchase agricultural machinery and equipment for animal husbandry, for a period of 2–8 years to construct vegetable stores, to create facilities for deep processing of products, to construct livestock and reconstruct poultry farms, for up to 15 years – to purchase cattle breeding products, to construct breeding and genetic centers. Farmers faced the need to increase the terms of soft loans to construct slaughter shops from 8 to 12 years, as profitability in the meat industry began to decline.

The issue of subsidizing engineering plants remains debatable. Subsidies reimbursed the manufacturer from 15 to 30% of the price of agricultural machinery sold to agricultural enterprises or leasing companies, the manufacturer of agricultural machinery must be a tax resident of Russia, carry out the established mandatory list of technological operations in Russia, have service organizations to maintain and repair agricultural machinery. Such support gave stability to machine builders and the opportunity to buy agricultural equipment cheaper for farmers.

The support mechanism accepts a new vector and is replaced by a leasing scheme. The state believes that, firstly, under the new conditions, small farms will be able to renew the equipment fleet, earlier they could not do it even with a discount. Secondly, the new mechanism allows farmers not to withdraw funds from circulation, but to direct them to further business development, which makes it possible to update more equipment at equal costs. Thirdly, the state is more likely to stimulate solvent demand of farmers, rather than directly subsidize the engineering industry.

3.3 State Insurance Support

In world practice, Multi-Risk Crop Insurance or Index-Based Crop Insurance is used. Index-based insurance based on easily observable external phenomena is advertised as an inexpensive insurance product [5]. In Russia, traditional insurance products are common. The insurance theory has always considered the mandatory form of insurance as a tool to achieve certain state goals in economic policy. The purpose of state insurance support is to protect property interests of agricultural producers from possible damage associated with natural climatic factors. Under state insurance, all crops in crop production, all livestock of animals (from a certain age) are insured.

The insurance period is one year. Insurance risks are natural disasters and contagious diseases. High insurance rates force the government to provide state support by reimbursing 50% of the insurance premium under the concluded agricultural insurance contracts with state support. The state subsidy for insurance premiums is needed to stimulate the participation of farmers and to prevent insurance failures from entering the market [11].

Considering the state support measure at this stage remains not very popular among farmers. Historically, the legislative base of the state insurance support system has been improved in stages, but after next innovations, insurance did not become in demand, and even despite growth expectations, reduction indicators were recorded.

Low demand indicates the imperfection of insurance mechanisms:

- high cost of insurance policy,
- flexible line of insurance products,
- insurance subsidies were the unprotected budget item, as a result of which they were redistributed to other purposes,
- in case of emergency, support is also provided to uninsured enterprises, which reduces the interest in insurance,
- funds are not allocated within the period specified by the contract.

To solve this problem, the state shows a flexible policy as well as coercion. Motivation is through optimization of insurance objects and optimization of insurance risks. Historically, only crop production was insured, then livestock and aquaculture were introduced into insurance objects.

The Ministry of Agriculture is optimizing the list of insured natural phenomena, according to statistics, the main risk for crops is phenomena associated with a lack of moisture, drought, and dry wind [1]. Due to the geographical extent of Russia, these risks are not the main ones for some territories. Here the crop dies from prolonged rains during harvesting, freezing of the upper soil layer and early snow cover.

Compensation for animal insurance accounts for the damage caused by contagious diseases. If damage was previously compensated for distemper, now the insurance system has been amended and farmers will be able to receive insurance compensation for forced slaughter of livestock during an outbreak of the disease. The violation of electricity, heat, water supply because of natural hazards and natural disasters were included in risks, which is important for a business specializing in growing products in sheltered soil, on reclaimed land, for pig and poultry complexes.

Following the principle of openness of insurance, the Ministry of Agriculture of the country annually develops, approves and places a property insurance plan by regions, groups and types of crops, by types, sex and age composition of farm animals, species and age composition in the Internet [1]. The basic marginal insurance rates are differentiated for each region, for each risk group considering the participation of the insurant (franchise). The flexibility of insurance with state support gave legal opportunity to sign the agreement to one or more risks, more specific to the region and thereby reduce the overall cost of insurance protection.

Before, farmers could receive insurance compensation only if they lost more than 20% of the planned crop, now this threshold has been canceled. However, a mandatory unconditional deductible has been introduced, the minimum amount of which cannot be less than 10% of the sum insured. This will allow for insurance compensation payments of minimum loss of crops. In 2018, only 16% of livestock and 1.5% of the sown area were insured with state support in Russia.

3.4 State Support for Price Stabilization

The maximum level of minimum grain prices for purchasing interventions is a regulator of falling market prices and protects the interests of producers. The maximum level of maximum grain prices for commodity interventions is a regulator of the growth of market prices and protects the interests of processing enterprises. The marginal tariff for storage and insurance is a cost regulator and protects the interests of the budget.

Public procurement interventions are carried out with a strong reduction in the market for product prices. Thanks to this program, farmers do not need a loan and interest payments. The access of small enterprises to public procurement will solve other problems, such as employment, the growth of tax revenues from small businesses in the budget system and will also increase competition [4]. Commodity interventions are conducted when prices rise above the maximum estimated prices through the sale of purchased agricultural products.

The public procurement mechanism has a buyback instrument, farmers who previously sold grain to the state can return it, i.e. purchase it from the state fund at the purchase price, having paid the costs of insurance, storage and paying taxes. Then farmers can sell the grain on the free market, and the state can get rid of the costs of storing and earning grain.

The mechanism of interventions has been determined, which provides farmers with participation in this procedure, as well as control and compliance with the rules. An important indicator is the target volume of the intervention fund, which provides an opportunity to quickly carry out commodity interventions in the future. The following methods and rules are developed:

- methods for calculating the limit levels of minimum and maximum prices;
- methods for calculating the maximum size of fees for services for the storage of purchased products;
- rules for competitive selection of insurers participating in public procurement;
- rules for competitive selection of custodians of reserves of the intervention fund – other documents.

Agent – an enterprise that organizes storage and insurance of reserves of the intervention fund, servicing loans taken by the agent to form reserves of the intervention fund. Public spending on interventions:

- purchase of products from agricultural producers and processing organizations is carried out through loans from authorized banks;
- agent remuneration expenses;
- elevator storage costs;
- stock insurance costs;
- in case of sale at a price lower than the purchase price, the losses are repaid from the funds provided for in the federal budget for financial support of measures in conducting public procurement interventions.

Revenue received by the state from the sale of inventories at high prices is allocated to the federal budget. Many national governments are faced with the need to regulate the industry market, firstly, due to overproduction or insufficient production of agricultural products, and secondly, under the influence of external factors - externalities. Externalities include geographic and demographic factors in Russia due to transport isolation, as well as limited domestic sales due to population outflows and low birth rates in the Asian part of the country.

The value of this type of state support is highly valued as it is the only type of support that directly affects the amount of revenue for grown products. Researchers are considering, along with traditional forms of public procurement, public-private partnerships [2]. Public procurement creates a guaranteed managed market for suppliers where market volumes are known in advance, sources of payment from the budget and a guaranteed level of purchase prices are provided. Optimistic estimates are now changing to realistic: if production and trade are guided by distorted prices and will not be effective, there are price risks for the state that make the realization of reserves acquired in previous years unprofitable; intervention purchases relieve regional authorities and farmers of responsibility for planning and maintaining the balance of production and consumption of agricultural products.

3.5 Support for Export of Products

A new paradigm in the state support system is support for export of products, giving export orientation to cultivation, storage, processing, logistics, creation of economic growth points in the agricultural sector. State approaches to this state support – today it is necessary to help remove trade barriers so that enterprises continue to increase the share of their presence in markets of respective countries.

The federal project "Export of Agricultural Products" was developed in two scenarios: inertial scenario (target export volume of 25 billion dollars), breakthrough scenario (increasing foreign deliveries of agricultural products to $ 45 billion in 2024) will require additional allocations from the federal budget [3].

Support for export:

- create a new commodity mass of agricultural products;
- create an export-oriented commodity distribution infrastructure;
- remove barriers;
- promote and position agricultural products.

The implementation of the federal project provides for a change in the structure of agricultural exports by increasing the supply of higher value-added products. The countries where Russian products will be in demand are identified – China, India, countries of Southeast Asia, Africa, and the Persian Gulf.

Export sums up all the finances that will be required for investments, including government subsidies. The amount of project financing is 696,092 million rubles, more than 53% is soft lending (including construction and modernization of seaports), the remaining 47% is in other areas of support, such as additional capitalization of Rosselkhozbank JSC, reimbursement of part of costs of agricultural machinery, reimbursement of part of direct costs for construction and modernization of export-oriented objects, land reclamation, reimbursement of costs of renting warehouses in target countries, part of costs of transporting products.

The state of the transport infrastructure is a significant obstacle to the development of exports – insufficient throughput of transport routes, insufficient port and transshipment capacities, and uncompetitive logistics. Investments are needed to create transport corridors: railway, sea, automobile. Support is provided for investments in additional port transshipment facilities. The organization of regular route shipments is also in the field of state control.

Another filter of business access to the market is the approval of permits and certificates of veterinary state supervision, sanitary-and-hygienic (epidemiological) control. Permission has been obtained to subsidize certification of agricultural products, comprehensive measures in the field of veterinary and phytosanitary control, monitoring activities and ensuring epizootic safety. There is a wide range of subsidized activities for the creation and promotion of brands and regional sub-brands, the creation of a network of agricultural representatives abroad, the functioning of the Agroexport Center.

The enterprises of the southern agricultural regions do not experience problems with the transportation of goods to sea ports for transshipment of goods. Enterprises of the Siberian, Ural and Central regions record high transport costs. The adoption of the breakthrough scenario caused an adjustment in the amount of compensation for part of costs of transporting products through the expansion of subsidies for export transportation. 50% of costs of exporters for transportation of products by rail (railway tariff + rolling stock rental + loading/unloading) are compensated.

Farmers showed a lack of competencies to promote the export of products, especially regional enterprises and small and medium-sized businesses. To acquire competencies, the government authorized costs of the export assistance program, these subsidies were collected for consultations on foreign economic activity in the target country, on local control and participation in tenders, costs of establishing contacts with foreign companies, and the organization of business meetings, negotiations, examination of foreign trade contracts. As part of the product promotion program, up to 90% of costs of assessing the conformity of products to the requirements of the external market are compensated; up to 50% of costs of R&D performed for these purposes are compensated; up to 80% of costs of congress and exhibition events are compensated to small and medium-sized enterprises; 100% of costs of business missions are compensated to all organizations. As part of solving the problem of lack of competencies, an educational online platform has been launched for small and medium-sized enterprises interested in export activities.

4 Discussion

1. State control currently has a twofold effect on the agro-industrial complex. This is manifested in the fact that, on the one hand, there is an annual increase in the amount of state support, on the other hand, these investments contribute to the development of a market mechanism and suggest a fading interest in separate measures of state support in the future.
2. The process of production in agriculture is taking place under conditions of uncertainty about biological processes, the prevailing weather and climate conditions, especially during the growing season of plant growth. The need for support is also generated by eternals, such as demographic processes, population outflow from agricultural production territories, and geographical factors of remoteness from seaports.
3. In connection with the achieved profitability and saturation of the market with products, some investment projects are withdrawn from state support. At the same time, support is given to wholesale distribution centers and projects that appeared in response to changing preferences in the nutrition of the population. The stimulation of agricultural engineering turned into stimulation of leasing, stimulation of greenhouses through subsidies went into stimulation through soft loans. The range of equipment subject to lending is expanding, credit extension is allowed, some of them are revised due to the deteriorating situation of the sub-sector.

4. The challenge was to determine the relevant insurance risks for farmers, ensure the variability of risks for regions, and determine the level of franchising. Optimization of the insurance system with state support required the formation of the insurance plan, flexibility in the selection of insurance risks, allowed farmers to reduce their own costs and provided access to insurance tools. The state makes it clear that it intends to shift many risks to farmers, as it is customary in world practice and the state is taking coercive measures to do it.
5. The most popular measure of state support is state procurement interventions and state trade interventions, a transparent market with known volumes, demanded types of products, guaranteed prices. Budget expenses are associated with costs of storage and insurance of grain, possible losses from the sale at prices below the purchase price.
6. Export-oriented state support with tasks of financial and non-financial assistance in creating transport corridors, modernizing the reclamation system, epizootic monitoring, umbrella brands, will require large amounts of state and municipal subsidies.

5 Conclusion

Russia's example shows that state support can be given a vector of compensation for losses and a vector of export promotion in developing sectors of the economy. Export facilitation will bring surplus crops to the foreign market, and not just to domestic intervention funds. The increase in the investment component in the long term will affect the pace of development and qualitative transformations of the agro-industrial complex.

The state believes that, under the new conditions, small farms will be able to renew the equipment fleet, earlier they could not do it even with a discount. The new mechanism allows farmers not to withdraw funds from circulation, but to direct them to further business development, which makes it possible to update more equipment at equal costs. The state is more likely to stimulate solvent demand of farmers, rather than directly subsidize the engineering industry.

References

1. Agricultural insurance plan for 2020. Approved by order of the Ministry of Agriculture of Russia N 667 dated December 2, 2019 (2019). https://www.garant.ru/products/ipo/prime/doc/73264647/. Accessed 20 Apr 2020
2. Buso, M., Marty, F., Tra Tran, P.: Public-private partnerships from budget constraints: looking for debt hiding? Int. J. Ind. Organ. **51**(C), 56–84 (2016). https://doi.org/10.1016/j.ijindorg.2016.12.002
3. Federal project "Export of agricultural products" dated February 7, 2020 (2020). http://mcx.ru/upload/iblock/013/013f266cee8d39bce5ca867381ff0da1.pdf. Accessed 20 Apr 2020
4. Gafurova, G.T., Notfullina, G.N., Fukina, S.P.: Government procurement as a tool for the development of small and medium business in Russia. Econ. Region **12**(4), 1233–1243 (2016). https://doi.org/10.15826/recon.2016.2.4.046

5. Jensen, N., Barret, C.: Agricultural index insurance for development. Appl. Econ. Perspectives Policy **39**(2), 199–219 (2017)
6. Klimenko, A., Minchenko, O.: State control of the economy: theory and best practices. Public Adm. Issues **3**, 7–30 (2016)
7. Lenchuk, E.B.: Formation of industrial policy of Russia in the context of the new industrialization. J. New Econ. Assoc. **3**(39), 138–145 (2018)
8. Razvadovskaya, Yu., Shevchenko, I., Grezina, M.: The role of the sectoral concentration, foreign investment and government support in the innovation strategy of the Russian enterprises of the ICT industry. Terra Econ. **13**(4), 65–82 (2015). https://doi.org/10.18522/2073-6606-2015-4-65-82
9. State Program for the Development of Agriculture and Control of Agricultural Products, Raw Materials and Food Markets N 717 dated July 14, 2012 (2012). https://base.garant.ru/70210644/. Accessed 20 Apr 2020
10. The Budget Code of the Russian Federation N 145-FL dated July 31, 1998 (1998). http://www.consultant.ru/document/cons_doc_LAW_19702/. Accessed 20 Apr 2020
11. Yanuarti, R., Aji, J.M.M., Rondhi, M.: Risk aversion level influence on farmer's decision to participate in crop insurance: a review. Agric. Econ. **65**, 481–489 (2019)

Program-Target Management of Russian Aircraft Industry Development

N. M. Tyukavkin[1], V. Yu. Anisimova[1(✉)], and T. E. Gorgodze[2]

[1] Samara University, Samara, Russia
tnm-samara@mail.ru, ipanisimova@yandex.ru
[2] Samara State University of Economics, Samara, Russia
gorgodze_t@mail.ru

Abstract. The purpose of this study is to manage national special-purpose programs distinguished by limited time frames for implementation and available resources. The authors consider the problem of program-target state financing of development programs for aircraft construction enterprises; the essence of program-target management at aircraft construction enterprises is to develop promising changes to be implemented in the form of a project with certain goals and a team, with certain resource and time costs. The current state of the Russian economy distinguished by no competitive management system at aircraft construction enterprises leads to dependence on imports, which significantly affects the country's security, therefore, it is necessary to evaluate the effectiveness of state financing of special purpose aircraft construction programs based on the calculation of investment project indicators. The efficiency of import substitution policy implementation and provision of national aircraft industry technological security is based on the appropriate physical infrastructure and organization of scientific sector activities in the aircraft industry.

Keywords: Aircraft construction · Innovative activities · Innovation project · Program-target state financing · Science · Technology

1 Introduction

The main method of managing programs with limited time frames and available resources is program-target management. Its object is a program, its intended purpose. The target program is a single set of scientific, economic, technical, organizational and production activities united by a general goal achieved under single management.

The essence of program-target management at aircraft construction enterprises is the functioning of promising changes in the form of a project with certain goals and a team, the implementation of which is associated with resource and time costs. A peculiar feature of program-target management is the unity and integrity of taken measures combined into an innovation cycle produced by a special team of performers and managers. It is important to introduce a set of estimated figures that are calculated with a high degree of accuracy and should be used to analyze enterprises' effectiveness [4]. The study of program-target management was carried out by many scientists [3, 7, 8, 12, 13].

S. I. Ashmarina et al. (Eds.): ENGINEERING ECONOMICS WEEK 2020, LNNS 139, pp. 212–223, 2021.
https://doi.org/10.1007/978-3-030-53277-2_24

Innovative aircraft construction projects are characterized by a significant degree of uncertainty and complex implementation associated with the management of a significant number of components in a dynamically changing external environment. When assessing the level of project complexity, it is required to highlight its main parameters: the time frame for implementation; final result; analysis of information; the degree of team interaction; business needs [6].

Elements of the target program are as follows: goals and objectives; the functional and management structure (scientific and research activities, work on the implementation and introduction of innovations); organizational structure (implementation mechanism, rights and responsibilities of the customer and contractors, the procedure of financing, etc.); resource justification and feasibility study of the program to be implemented.

The category "innovation project" in program-target management is defined as the process of innovation, which represents the implementation of scientific, production, technological, organizational, commercial and financial activities in a given sequence, as a result of which we obtain innovations; as a form of targeted management of innovative activity of a business entity, which is a system of measures interrelated and interdependent in terms of time, use of resources and performers to achieve goals in the areas of priority development of science and technology; as a set of organizational and planning, technical and financial documentation necessary to achieve the objectives of the draft documentation.

The prevailing feature of managing the aircraft industry is requirements for using a program-target approach having no alternative, which determines the use of certain tools, mechanisms and methods determined by the vector of increasing innovative activity of the entire complex of industry enterprises.

The development of innovative activity of aircraft construction enterprises within the framework of the existing strategy and special-purpose programs for the development of the aircraft industry has the following features:

1. It is necessary to rank and identify priority areas of R&D to increase aircraft industry efficiency.
2. The industry needs a unified strategy for R&D created and implemented in the form of special-purpose programs.
3. Special attention should be paid to projects, which implementation target is the development of competitive innovative technologies.
4. Organizational innovations in aircraft industrial production shall be carried out on the basis of modernization through the introduction of innovative technologies that increase efficiency and reduce costs for the introduction of innovative products, increase labor productivity using modern management technologies, information systems for quality and business processes management.
5. Expansion of inter-branch networking, relations with scientific organizations, universities and innovative organizations.
6. Creation of a new talent pool in the aviation industry.

In the circumstances concerned, today it is required to make the scientific and industrial complex of the industry ready for the creation and implementation of innovative projects through the introduction of innovative technologies for program-target management in order to increase the competitiveness of domestic aircraft construction enterprises. Considering problems that cause the use of the program-target control method in the aircraft industry we can distinguish the following ones: economic; technical and technological; financial; process (or organizational and managerial); informational; environmental; social, etc., the main of which are technical, technological and process problems [2]. Technical and technological problems related to an increase in the innovative activity of the aircraft industry lie in the design and production potential of the industry that requires a strong research, experimental and production base and highly qualified personnel.

2 Methodology

Organizational, managerial and process innovations that develop the innovative activity of aviation industry enterprises consist in the use of new technologies for managing business processes, both at individual enterprises and in the complex as a whole. Here, it is worth noting that the special relevance of process innovations lies in organizing program-target management based on planning and implementing individual projects related to designing innovative products, reengineering of technological processes within the complex, improving product quality, designing and implementing integrated informatization systems at enterprises. A drawback of innovative activity development is the distribution of resources used in linear-functional structures by units, but not by targeted programs. These problems can be eliminated through the transition to new forms of management organization – the implementation of special-purpose programs.

In terms of the structure a special-purpose program includes the following elements: tasks and objectives; functional structure (research and development, personnel training, introduction, mastering and implementation of innovations, technology transfer, import and export of licenses) with the provision of accurate information about performers; resource support and a feasibility study (costs, timing, results); organizational structure (customers, performers, procedure for delivery, acceptance of works, financing, sanctions and fines, bonuses, etc.).

The organization of the system of target-oriented management at aircraft construction enterprises is aimed at solving the following problems:

1. Creation and implementation of effective interaction of all structural units involved in the introduction of innovations in production processes.
2. Increasing the level of operational efficiency of current management by the development of innovative activity through the approximation of management bodies to performers and the organization of a direct relationship between them.
3. Provision and implementation of functions of effective control over the implementation of works creating the basis for administrative regulation, as well as motivation to achieve the highest results of works in terms of the time frame and quality.
4. Increasing the degree of responsibility of governing bodies and performers involved in the development and introduction of innovations for the deadlines and results of works [1].

The main objective of program-target management in the aircraft industry is to eliminate the divergence of interests of functional services of enterprises in the process of implementing a particular program, which provides the most optimal way to implement it. Innovative programs at aviation enterprises are associated with the development and implementation of new types of products and technologies and, therefore, with the implementation of new tasks and objectives, so the arrangement of program-target management in terms of the organization is the most effective area allowing to achieve the objective quickly. The state priorities of the Strategy for the Development of the Aviation Industry for 2013–2025 of the Russian Federation consist in the creation and implementation of large-scale program-target projects for the civil and military aircraft industry [9]. The program-target approach to managing the aircraft industry requires all managerial and industrial subsystems to be ready for the organization of activities based on innovation.

3 Results

Program-target development of the aircraft industry of the Russian Federation is carried out on the basis of Federal Law of the Russian Federation dated June 28, 2014 No. 172-FZ "On Strategic Planning in the Russian Federation" through the development of documents related to special-purpose programs by the state [5].

In 2018, the planned amount of state subsidies for the industry under State Program of the Russian Federation "Development of the Aviation Industry for 2013–2025" was 59,105 million rubles (in 2016 – 53 billion rubles; in 2017 – 45.4 billion rubles) [11]. In 2019, 135 aircraft and 234 military and civilian helicopters were supplied to customers according to federal programs. According to the Ministry of Industry and Trade of the Russian Federation, an increase in the volumes of industrial production in the industry in 2019 will be 0.6%.

According to Decree of the Government of the Russian Federation dated June 26, 2018 No. 733, the priority areas of R&D covered by subsidies include works carried out under import substitution programs regarding the onboard equipment of the Sukhoi Superjet 100 aircraft [2]. 1,705 billion rubles in 2018 and 1,491 billion rubles in 2019 were provided for these purposes in the state budget for 2018 and for the period of 2019–2020. State subsidization of research and development in the aircraft industry will stimulate the production of competitive domestic aviation products.

The Ministry of Industry and Trade of Russia proposes changes in the financial support of State Program "Development of the Aviation Industry for 2013–2025" based on the federal budget for the period from 2019 to 2021. It is provided for to increase the amount of budget funds sent to for program implementation: [11].

- 2019 – by 19,588 billion rubles,
- 2020 – by 46,235 billion rubles,
- 2021 – by 43,309 billion rubles.

The total amount of funds sent to for this program will be:

- 2019 – 59,93 billion rubles,
- 2020 – 86,809 billion rubles,
- 2021 – 83,85 billion rubles.

This adjustment of budget allocations is justified by the need for temporary restriction on the use of credit resources in 2008–2018 and putting the MS-21 aircraft into serial operation. In addition, the program provides for budget financing of manufacturers of aircraft and their individual components for financial provision of costs associated with the production, sale and maintenance of the MS-21 series aircraft in the following amounts:

- 2019 – 1,579 billion rubles,
- 2020 – 4,109 billion rubles,
- 2021 – 4,811 billion rubles.

As for the scientific research and carrying out of design and experimental measures for the development of the MS-21 aircraft, there are no changes in budget financing for 2019 and 2020 – 7,378 billion rubles and 5,443 billion rubles, respectively. As for 2021, according to the clarification of the needs for financial resources for research and development, budgetary funds were reduced by 5,248 billion rubles, which is 2,53 billion rubles. Additional budget investments are planned for the Russian-Chinese project for the creation of a long-range wide-body CR929 aircraft:

- 2019 – 4 billion rubles,
- 2020 – 6 billion rubles,
- 2021 – 7 billion rubles.

Additional financing from the federal budget is provided for the upgrading of the Il-114 aircraft up to its new version – the Il-114-300:

- 2019 – 967 million 840 thousand rubles,
- 2020 – 2,35 billion rubles,
- 2021 – 616 million rubles.

In 2019, additional budgetary funds in the amount of 2,221 billion rubles were provided for the industrialization of the Il-114-300 aircraft. Budget financing is not provided for 2020–2021 due to the need in more precise indicators and deadlines for the implementation of this program. In 2019, financing of the fatigue tests of the Il-76MD-90A aircraft remained unchanged (135 million rubles). Budget financing of these activities in 2020–2021 is not provided for. R&D of the Sukhoi Superjet program (modification – 75 seats) provides for additional budget funds:

- 2019 – 9,92 billion rubles,
- 2020 – 17,33 billion rubles,
- 2021 – 14,1 billion rubles.

No budget funding is provided for in 2019–2021 for the modernization of the Il-96 aircraft and production infrastructure facilities for the Il-96-400M aircraft. Additional financing of state program "Development of the Aviation Industry for 2013–2025" will be provided for at the expense of secondary and subsequent issuance of shares of UAC PJSC [12]. Based on the above information, it is planned to increase the budget financing of the industry in 2019–2021 by 109,1 billion rubles (by 38%), of which about 41,35 billion rubles is provided for R&D on the creation of a regional SSJ75 aircraft [11]. Financing of MS-21 aircraft research and development remained unchanged due to the completion of the program. So, budget financing in the following amounts is planned under the project for the development of the fleet of domestic aircraft for 2019–2021:

- the SSJ75 – 41,35 billion rubles (75 seats),
- the CR929 – 17 billion rubles,
- the MS-21 – 10,5 billion rubles,
- the Il-114-300 – about 3,934 billion rubles.

E = 0.18 – discount coefficient.

4 Discussion

Now let us determine the effectiveness of state financing of aircraft construction programs by calculating investment project indicators. Al-Invest software was used to carry out calculations. First, let us determine the net profit (NP) that is planned to be obtained from investing. The planned values from the implementation of the investment project are presented in Table 1.

$$NP = R - C \tag{1}$$

Table 1. Generated income from project implementation

Projects/PV	Step of the calculation period	SSJ75	CR929	MS-21	Il-114-300	Discount coefficient
2019	0 Investments	−13,35	−6,0	−3,5	−1,3	1
2020	Investments	13,0	−6,0	−2,5	−1,3	0,85
2020	1	2,9	2,9	1,1	0,4	0,85
2021	Investments	−15,0	−5,0	−3,5	−1,33	0,72
2021	2	9,8	4,3	2,8	0,8	0,72
2022	3	1,2	1,5	2,7	0,3	0,61
2023	4	4,6	3,6	1,2	0,9	0,52
2024	5	8,6	9,2	1,5	0,8	0,44
2025	6	13,5	13,4	1,1	0,8	0,37
2026	7	18,3	16,8	1,8	0,9	0,31
2027	8	23,7	20,2	2,3	1,0	0,27
2028	9	28,5	24,5	2,5	1,0	0,23
2029	10	36,8	27,1	2,8	1,2	0,19
–	In total	41,35	17,0	10,5	3,93	–

Source: authors.

R is the result, profit.
C is project costs.

$$a = 1/(1 + E)^n \tag{2}$$

n is the step of the calculation period.
E = 0.18 – discount coefficient.

A year, six months or a month can be chosen as the step depending on the preliminary feasibility study of the project. For the convenience of calculations we chose a step of one year, since if we choose one month as the step, it will significantly complicate our calculations due to a large amount of processed information. Now let us determine net present values (NPV) of the projects.

$$\text{NPV} = \sum\nolimits_{n=0}^{T} NP \times a - K \times a \tag{3}$$

m is the calculation step number.
T is the number of periods.
N = 0 – the moment of starting investments ("zero" year).
K is (capital) investments.

Future cash flows are adjusted to the current point in time in Table 2.

We can observe from Table 2 that the MS-21 and Il-114-300 projects have a negative discounted income, therefore, there will be no profit on the project and the investments will not pay off.

Now let us determine the project profitability index (PI). If it is greater than 1, then the project is effective (Table 3).

$$PI = NPV/K \tag{4}$$

We can observe from profitability index values that the indices of MS-21 and Il-114-300 projects are less than zero, therefore, the projects are not effective. Now let us determine the discounted payback period (PP). It shows the number of years from the start of the project to the period when the *NPV* with accumulation changes its value from negative to positive taking into account the coefficient *X* (Table 4).

$$X = S - m/(S - m + S + m) \tag{5}$$

S − m is a negative value of the *NPV* on a cumulative total at the step of changing the sign from negative to positive.
S + m is a positive value of the *NPV* on a cumulative total at the next step.

We can observe from Table 4 that *X* is part of the discounted payback period. It means that the payback period in the first project is not 9 years, but 9.08 years. The value of 0.08 is the value of *X*. It follows from Table 4 that it is inexpedient to use the MS-21 and Il-114-300 projects.

Table 2. Discounted income from project implementation for the entire calculation period

Projects/PV	Step of the calculation period	SSJ75	CR929	MS-21	Il-114-300
2019	0 Investments	−13,35	−6	−3,5	−1,3
2020	Investments	−11,017	−5,085	−2,966	−1,102
2020	1	2,458	2,458	0,932	0,339
2021	Investments	−10,773	−3,591	−2,514	−0,955
2021	2	7,038	3,088	2,011	0,596
2022	3	1,576	1,523	0,628	0,513
2023	4	2,373	1,857	0,619	0,464
2024	5	3,759	4,021	0,656	0,35
2025	6	5,001	4,964	0,407	0,296
2026	7	5,745	5,274	0,565	0,283
2027	8	6,305	5,401	0,612	0,225
2028	9	6,425	5,524	0,564	0,225
2029	10	7,031	5,178	0,533	0,229
NPV	In total	11.726	24,001	−0,436	−0,099

Source: authors.

Table 3. Profitability index of planned projects

Projects	PI
SSJ75	1,33
CR929	2,64
MS-21	0,95
Il-114-300	0,97

Source: authors.

Table 4. Discounted payback periods for the projects

Projects	X	Название	PP
SSJ75	0,08	Plus 9	9,08
CR929	0,4	Plus 6	6,4
MS-21	0,98	Not payback	Not payback
Il-114-300	1,0	Not payback	Not payback

Source: authors.

Now let us determine the internal rate of return *(IRR)* generated by a particular project, provided that the costs of the project are fully covered by income (Table 5).

$$IRR = E' + NPV' \times (E'' - E')/(NPV' - NPV'') \qquad (6)$$

E′ is the rate of discount at which the value of NPV is positive (or negative) and closest to zero.
E″ is a discount rate increased (or decreased) by one point as compared to the value of E″, at which the value of NPV is negative (or positive) and closest to zero.
NPV′ – corresponds to the value of E′.
NPV″ – corresponds to the value of E″.

Table 5. Internal rate of return on investment generated by the project

Projects	NPV	IRR, %
SSJ75	Between E′ = 0,24 and E″ = 0,25. At E′ = 0,24 NPV′ = 1,039, at E″ = 0,25 NPV″ = −0,312	24,77
CR929	Between E′ = 0,43 and E″ = 0,44. At E′ = 0,43 NPV′ = 0,242, at E″ = 0,44 NPV″ = −0,128	43,65
MS-21	Between E′ = 0,16 and E″ = 0,17. At E′ = 0,16 NPV′ = 0,1, at E″ = 0,17 NPV″ = −0,178	16,360
Il-114-300	Between E′ = 0,17 and E″ = 0,18. At E′ = 0,17 NPV′ = 0,016, at E″ = 0,18 NPV″ = −0,099	17,139

Source: authors.

IRR shows the level of the rate of discount *E*, above which it is not profitable to implement the project, since *NPV* will become negative.

Now let us determine the average rate of return (*ARR*) for the projects (Table 6).

$$ARR = (NIavg/I) \times 100\% \tag{7}$$

NIavg – the average project income.
I – total investments.

ARR takes into account undiscounted data, i.e. it is the usual profitability of project implementation.

Table 6. Calculation of the average rate of return (*ARR*) for the projects under consideration

Projects	*ARR, %*
SSJ75	35,768
CR929	72,706
MS-21	18,857
Il-114-300	20,941

Source: authors.

We can observe from Table 6 that the MS-21 and Il-114-300 projects show low profitability.

General conclusion: state investment in the SSJ75 and CR929 projects is effective, since the *NPV* of the projects is positive, *PI* is greater than 1, and they will pay off. Therefore, it is expedient to invest in these projects, and it is not expedient to invest in the MS-21 and Il-114-300 projects.

Then the authors carried out an analysis of the efficiency of using the funds allocated according to federal special-purpose programs for the enterprises under consideration (Table 7).

Table 7. Analysis of the efficiency of using the funds allocated according to the federal special-purpose programs for aircraft construction enterprises

Indicators	2012	2013	2014	2015	2016	2017	2018
JSC Aviaagregat							
Revenue, thousand rubles	2 382 880,0	2 561 186,3	2 632 922,1	1 988 942,4	2 746 274,0	4 454 530,0	3 447 320,0
Revenue from the Federal target program, thousand rubles	77 341,0	73 090,0	101 829,0	1 659 934, 5	521 106,0	334 564,0	399 722,0
Share of products under the Federal target program in total sales,%	10,6	3,9	3,9	6,8	7,5	20,6	12,6
JSC Sukhoi							
Revenue, million rubles	6 481 283,0	8 864 487,2	9 153 821,3	10 482 143,1	11 315 451,2	12 551 311,2	14 123 419,5
Funds for innovation, thousand rubles	14,0	18,0	35,0	23,0	15,0	28,0	14,0
Revenue from the Federal target program, thousand rubles	12,0	20,0	35,0	21,0	16,0	27,0	14,0
Share of products under the Federal target program in total sales %	5,2	7,3	8,6	9,4	11,3	12,1	12,8
JSC «GidroMash»							
Revenue, million rubles	3 812,0	4 132,0	4 751,0	5 409,0	7 415,0	8 135,0	7 226,0
Funds for innovation, thousand rubles	–	459 246	459 943	866 130	641 821	480 321	556 739
Revenue from the Federal target program, thousand rubles	–	18	35	23	15	28	14
Share of products under the Federal target program in total sales %	–	11,1	9,7	16,0	8,7	5,9	7,7

Source: authors.

Based on the values presented in Table 7 we see that federal funds allocated for the development of innovative activities of aircraft construction enterprises have a small share in the total volume of activities of these enterprises.

5 Conclusion

Today the following program-target projects are used in the aircraft industry:

1. Completion of static tests of the MS-21-300 aircraft at the Central Institute of Aerohydrodynamics, assembly and installation of aircraft resource model units, the start of flight tests of the second prototype of the aircraft.
2. Continuation of the development of design and technical documentation for the components and parts of the Il-96-400M aircraft, reconstruction and technical re-equipment of the production of this airliner.
3. Continuation of development of design and technical documentation for components and parts of the Il-114-300 turboprop aircraft, start of their production.
4. Organization and carrying out of works on technical re-equipment and reconstruction of production at Rostvel MiG aircraft plant for the purposes of serial production of aircraft.
5. Certification of the Mi-38-2 helicopter for IFR flights at high altitudes and high temperatures.
6. Completion of the assembly of a standard transmission of the Ka-62 helicopter and its fatigue life expenditure of at least 50 h.
7. Approval of the certification test program for the Ka-62 helicopter.
8. Certification of the extended temperature range (up to +40 °C) for the Ansat light helicopter and an increase in its life up to 10 thousand hours.
9. In the engine industry, the development of technical specifications and outline designs of gas generator assemblies for the demonstrator of PD-35 engine technologies.
10. Development of “critical” technologies” of PD-35 engine.
11. Carrying out tests using an in-flight simulator and ensuring certification of PD-14 engine.
12. Production of TV7-117ST-01 automatically controlled engine intended for the installation on the Il-112V and Il-114-300 aircraft to carry out preliminary bench tests.

In this study the authors analyzed target financing for the implementation of aircraft construction projects and concluded: state investment in the SSJ75 and CR929 projects is effective, the NPV of the projects is positive, the PI is greater than 1, the projects will pay off, and it is proposed not to invest in the MS-21 and Il-114-300 projects [10]. The efficiency of the use of budgetary funds at aircraft construction enterprises was determined. It showed that federal funds for the development of innovative activities of aircraft construction enterprises have a small share in the total volume of activities of the enterprises.

References

1. Aleshin, B.S., Manturov, D.V.: Strategic planning for the development of the aviation complex at the national level. In: Kleiner, G.B. (ed.) Scientific Reports of the Thirteenth All-Russian Scientific Symposium "Strategic Planning and Enterprise Development", vol. 4, pp. 8–11. Department of Social Sciences of the Russian Academy of Sciences, CEMI RAS, Moscow (2012)
2. Decree of the Government of the Russian Federation of June 26, 2018 N 733 "On approval of the Rules for the provision of subsidies from the Federal budget to Russian organizations for the reimbursement of part of the costs of research and development work in priority areas of the development of the aviation industry" (2018). https://www.garant.ru/products/ipo/prime/doc/71878910/. Accessed 09 Jan 2020
3. Efimova, N.P., Mokretsova, O.I.: Theoretical assumptions of the program-target management of the budget expenditures in Russia. Econ. Entrep. **12**(4), 810–818 (2015)
4. Efimova, N.S., Kalachanov, V.D.: Simulation of production processes at aircraft building enterprises. Amazonia Investiga **8**(24), 335–345 (2019)
5. Federal Law of the Russian Federation dated June 28, 2014 No. 172-FZ "On Strategic Planning in the Russian Federation" (2014). http://base.garant.ru/70684666/. Accessed 09 Jan 2020
6. Gafiatullin, V.A.: Features of project management of innovative development of a complex of aircraft manufacturing enterprises: parameters and models. Fund. Res. **5–3**, 559–563 (2016)
7. Gurova, E., Larinina, T., Zakorin, N., Andreeva, L., Manukhina, L.: Program-target methods of management small business. In: Murgul, V. (ed.) Proceedings of the International Science Conference SPbWOSCE-2016 "SMART City". MATEC Web of Conferences, vol. 106, p. 08085. EDP Science, Les Ulis (2017)
8. Munns, A.K., Bjeirmi, B.F.: The role of project management in achieving project success. Int. J. Project Manag. **14**(2), 81–87 (1996)
9. Strategy for the development of the aviation industry for 2013–2025. Approved by Decree of the Government of the Russian Federation of April 15, 2014 No. 303 (2014). http://fundamental-research.ru/ru/article/view?id=40341. Accessed 09 Jan 2020
10. Technological Platform Air Mobility and Aviation Technologies: Government funding of the aviation industry (2017). https://aviatp.ru/statefunding#01012020. Accessed 15 Jan 2020
11. The state program of the Russian Federation "Development of the aviation industry for 2013–2025." It is approved by the order of the Government dated December 24, 2012 No. 2509-r(2012). http://government.ru/rugovclassifier/849/events/. Accessed 09 Jan 2019
12. Vasetskaya, N.O.: Program-target management as a tool of financial implementation of special federal programs. Econ. Sci. **5**(3), 160–169 (2019)
13. Xiao, M., Yang, B.: Optimization study of earned value method in construction project management. In: Cheung, E., Yongsheng, Y. (eds.) Proceedings of the International Conference on Information Management, Innovation Management and Industrial Engineering, pp. 201–204. IEEE, New Jersey (2012)

Extending the Confidentiality Regime to Tax Information: Comparative Legal Research

E. Kalashnikova[1(✉)], A. Dashin[2], and L. Podgornova[3]

[1] Samara State University of Economics, Samara, Russia
kalashnikova-helen@yandex.ru
[2] Kuban State Agrarian University named after I.T. Trubilin, Krasnodar, Russia
avdashin@mail.ru
[3] Migration Department of the Police Body 2 (Industrial District) of the Department of the Ministry of Internal Affairs of Russia in Samara, Samara, Russia
lpodgornova2@mvd.ru

Abstract. The research goal is to conduct a comparative analysis of norms of national and foreign law regulating the extension of the confidentiality regime to tax-relevant information and to identify problematic issues in this area. In this article, the author classifies tax information as confidential and considers the problem of extending the confidentiality regime to the tax secrecy. The ratio between tax information (tax secret) and other data that is subject to a similar regime is analyzed. The problem here lies in the ambiguity of the definition of "tax secret". This makes it difficult to determine the volume and content of tax-relevant information that can be subject to such a confidentiality regime, leads to an increase in the number of entities that have the right to access such information and, as a result, the possibility of violating the constitutional rights of taxpayers. In this research, the author used both general scientific and private scientific methods. Based on the results of a comparative legal study of Russian and foreign legislation, the author outlines a direction of reforming the national legislation system regulating the procedure for extending the confidentiality regime to tax-relevant information.

Keywords: Access mode · Confidential information · Information · Regime · Tax secrecy

1 Introduction

In the modern realities of the democratic society development, issues of ensuring the security of the state and society as a whole, especially economic security, are of particular importance [19]. At present, information is crucial for almost all spheres of public life. In these conditions, the role of the information sphere increases. This sphere is a set of different data, subjects that collect, form, distribute and use it, as well as the system of legal regulation of public relations arising in this case. The era of digitalization, which began at the end of the XX century, creates a need to ensure the information security of relations between various economic entities. In this regard, it is

S. I. Ashmarina et al. (Eds.): ENGINEERING ECONOMICS WEEK 2020, LNNS 139, pp. 224–230, 2021.
https://doi.org/10.1007/978-3-030-53277-2_25

particularly important to protect data in the economic sphere, and especially those whose access is restricted by their confidential nature.

The growth of informatization and digitalization of the modern society, the presence of legal conflicts in the regulation sphere of relations related to the confidentiality of tax-relevant data as part of economic information, and the increase in data turnover in the public administration highlights the need to increase the protection degree of individuals who have the right to the tax secrecy [9].

2 Methodology

In the research process, the following methods were used: general scientific methods (abstraction, analysis and synthesis, deductive, ascent from the abstract to the concrete), private scientific methods (special legal research methods): a systematic approach, formal dogmatic (normative), comparative legal, formal logical method. In order to carry out a large-scale study, the synthesis method was used, that is, the use of the above methods in combination. This allowed us to reflect the real level of knowledge of this subject area.

3 Results

Information in the taxation field should be defined as a type of financial information, and therefore part of economically significant information. In the Russian legislation and the legislation of a number of former Soviet republics, the definition of "tax secret" is formulated by listing data that cannot be subject to such a regime. Confidential tax-relevant information (tax secret) is a type of official secret, since such information can be obtained by a subject only when performing official duties in the field of taxation. As for confidential information with the access mode "personal data" and "tax secret", the first one is primary. However, the volume of such information types is different.

Today, the concept of tax secrecy is enshrined in part 1 of Article 102 of the Tax Code of the Russian Federation (part one) of 31.07.1998 No. 146-FZ, according to which it includes any information received by the tax authority, internal affairs agencies, investigative bodies, bodies of the state extra-budgetary fund and the customs authority about a taxpayer, a payer of insurance premiums [25] indicating a number of exceptions stipulated in the same article. However, tax secrecy as a separate institution is not included in the list of information endowed with the regime of confidential information, which includes information constituting, for example, commercial secrets, official secrets, private information, etc. [8]. It can be concluded that the institute of "tax secrecy" is an integral part of another and/or other institutions of confidential information, such as official secrets and/or commercial ones. But despite this, tax secrecy is defined as one of the objects of the criminal law protection together with commercial and banking secrets (Art. 183 of the Criminal Code of the Russian Federation of 13.06.1996 N 63-FZ) [6].

Therefore, the legislator does not include it as an integral part of either commercial or banking secrecy. Tax secrecy as a separate information institution is a subject to the confidential information regime is also mentioned in Federal law of 07.08.2001No. 115-FZ (as amended on 16.12.2019) "On countering the legalization (laundering) of income from crime and financing of terrorism" [10] altogether with official, commercial, bank and secret secrecy. The concept of a "tax secret" is formulated in the form of a list of information included in it through the enumeration of data not included in it. And this information contains data of restricted access, that is, it has characteristics typical for the legal regime of information with a restricted access. The legal regime of information resources is defined in the legal literature as the possibility of performing or not performing certain actions with the object of law that entails a certain legal result, including the procedure for documenting information, provisions on access to information resources, depending on their categories, taking measures to protect information [20].

Scientists distinguish several regimes of confidential information: state secret, official secret [17], professional or commercial secret [14], banking secret [22], and privacy secret. Based on understanding official secrets as information about individuals and legal entities that become known to various officials according to the nature of their official activities, but due to their special nature cannot be freely distributed, the secret information that we are investigating can be attributed with full confidence to one of its types. A number of scientists consider it as a type of professional secret, including information about the "hidden" official activities of state and municipal bodies (for example, information related to the organization and conduct of operational activities of tax authorities in order to perform tasks assigned to them by law) [27] together with such secrets as lawyer, medical, investigative secrets, etc.

In our opinion, this approach limits the volume of the tax secrecy to information obtained by tax authorities and their officials with control measures. It is advisable to define tax secrecy as one of the types of official secrets, since it includes information that is obtained by official subjects in the performance of their official duties in the state or municipal service. Considering a relation between the tax secrecy and personal data, we note that personal data in the Russian legislation is defined as any information relating directly or indirectly to a specific or identifiable individual (subject of personal data) [12]. The list of such data is not fixed at the legislative level and it can include: surname, first name, patronymic, address of residence, profession, social status, as well as any information containing data directly or indirectly mediating a certain or defined individual. Based on this wording "personal data", we can say that the legislator does not associate personal data with a restricted access regime, but in fact most of them may be subject to the regime of various types of confidential information: medical, banking secrets, etc. Although this information is included in the list of confidential information and Federal law of 27.07.2006 No. 149-FZ (as amended on 02.12.2019) "On information, information technologies and information protection" [11] indicates limited access to these data, without calling them confidential or restricted access information, specifying a special procedure for accessing it, which is provided by a special law. But this legal act does not guarantee the confidential nature of such information. Based on this state of affairs, scientists suggest that personal data should not be considered as restricted access information, but as "confidential personal data" [2], which includes personal data to which the legislator has extended the confidentiality requirement. You can define personal data as a "primary" secret, and tax data as a "secondary" secret, but the tax secrecy regime and the personal data regime do not coincide [26].

4 Discussion

Tax secrecy may also include data that is considered as personal. In this case, they simultaneously fall under the personal data regime, in addition to the tax secrecy regime, and then we can talk about the transition of confidential information from one regime to another and, as a result, the need to meet special requirements for the protection of such data. And as it is correctly noted in the legal literature, the provisions of the Federal law of 27.07.2006 № 152-FZ (as amended on 31.12.2017) "On personal data" will also apply to activities of tax authorities and credit institutions, since the law applies to all relations related to processing of personal data using (or without) automation tools, if processing of data without the use of such tools corresponds to the nature of actions (operations) performed with data using automation tools [12].

We agree with the opinion expressed in the scientific literature about the tax secrecy regime that applies to both economically significant information and personal data. The totality of such information is usually defined as tax-relevant information, which is a special type of information intended solely for tax purposes, containing information about the taxpayer's economic activities and personal data. Based on this, we can define the purpose of the "tax secrecy" regime, which is to ensure the protection of the rights and legitimate interests of taxpayers in relation to information classified by the tax legislation as protected objects.

In our opinion, to eliminate discrepancies, it is advisable to include the tax secrecy as a separate institution of confidential information in the list of information of a similar nature approved by the decree of the President of the Russian Federation. So, as we can see, the volume of information covered by the "tax secret" regime is not precisely defined by the legislator, and the consequence of this state of affairs is ambiguity in determining the volume of its content, which causes difficulties in the law enforcement practice. This is also complicated by the increase in the number of entities that have the right to access information related to the tax secrecy. Together with the Tax Code of the Russian Federation, in the Russian legislation there are other normative legal acts containing norms on the tax secret regime (part 2 of article 11 Code of Administrative Procedure of the Russian Federation, article 11 of the Arbitration Procedure Code of the Russian Federation, article 10 of the Code of Civil Procedure, article 241 of the Criminal Procedure Code of the Russian Federation) [1, 3, 4, 7]. "Disclosure" in Russian tax law means the use or transfer to another person of information constituting a commercial secret (secret of production) of a taxpayer or payer of insurance premiums [25] by an official who became aware of it in the course of performing his official duties.

In this regard, we note the inaccuracy of the definition of this concept, enshrined in article 102 of the Tax Code of the Russian Federation [25]. The range of information constituting a tax secret is much broader and includes, in addition to information constituting commercial and industrial secrets, also other data about the taxpayer. In addition, disclosure can be expressed not only in the use and transmission of information, but also in making it public, for example, by publishing it in the media. We agree with Shekhovtseva, who suggests that disclosure should be understood as a guilty illegal violation of the tax secrecy regime, expressed in the action (inaction) of an official of a tax authority, internal affairs body or customs body, as a result of which it became available to third parties [21].

It should be noted that the tax secrecy (Steuergeheimnis) in the German law is defined as the obligation to keep secret by tax officials or experts of information about taxpayers that became known to them in connection with the performance of their official duties [18]. The tax secrecy regime applies to information received by an official in the course of taxation [15]. At the same time, the law stipulates cases when such a regime can be removed, for example, if it is necessary for the tax process; disclosure is permitted by law; the consent of the owner of the secret is obtained; this serves to conduct a criminal process that is not related to a tax crime, and if the information was obtained because of a tax crime or violation of the procedure for collecting taxes and fees; there is a forced public interest for disclosure [13].

According to the Tax Code of the Republic of Belarus 166-Z of 19.12.2002, the regime of the tax secrecy applies to information held by tax, customs and financial bodies, state control bodies, republican bodies of the state administration, bodies of local government and self-government, authorized bodies, organizations and officials who are participants of tax relations, any information about payers in the presence of exceptions, which are enshrined in Tax Code of the Republic of Belarus (Art. 29) [24]. Disclosure of such information is recognized as an offense that entails legal liability.

The concept of tax secrecy in the Code of the Republic of Kazakhstan of December 25, 2017 No. 120-VI "On taxes and other mandatory payments to the budget (Tax code)" is similar to the Russian tax legislation regulating this regime of confidential information (article 30) [5]. Violation of these rules under the legislation of the Republic is recognized as an offense, and information with the "tax secret" regime cannot be disclosed by persons who became aware of them either during their employment or after. It should be noted that this provision differs from the provisions of the Russian legislation. In our opinion, it is necessary to clarify the obligation of non-disclosure of tax secrets by persons who became aware of it during the performance of their official duties and after being removed from such a service.

According to Art. 30 of the Tax Code of the Republic of Azerbaijan No. 905-IQ of July 11, 2000, any information received by the tax authority and its officials about a taxpayer is considered a tax secret and personal life information, with the exception of a certain kind of information [23]. Separately, the legislator refers to commercial secrets, referring to the norms of the Law of the Republic of Azerbaijan No. 224-IIQ of December 4, 2001 "On commercial secrets" [16].

Tax officials of the tax authorities are required not to disclose tax secrets both while serving and after being dismissed from service. Another distinctive feature of the Azerbaijani tax legislation is the indication that the violation of the "tax secret" regime can be carried out both in the form of the information loss and its disclosure which is not limited only to the use or transfer of it to another person. It is worth mentioning the principle of "tax publicity" implemented in a number of European countries: Norway, France, Italy, Sweden. In these countries, tax authorities are required by law to maintain the tax secrecy, while the legislator also provides for the possibility to publish a certain list of strictly defined tax data. This principle is based on the constitutionally established right of citizens to get acquainted with documents and materials of state authorities that directly affect their rights and freedoms, and since the duty to pay taxes is public, a number of tax data are recognized as publicly available, which allows each taxpayer to control the performance of this duty by other citizens.

5 Conclusion

The legislator creates the concept of tax secrecy by listing information that is not covered by the tax secrecy regime. Similarly, the concept of tax secrecy is formulated in the legislation of the CIS countries. The tax secrecy regime applies to economic information and personal information about a taxpayer from the moment of its registration with the tax authorities, and is therefore regulated by the Federal law of 27.07.2006 № 152-FZ (as amended on 31.12.2017) “On personal data” [12]. The extension of the tax secrecy regime to information is aimed at ensuring the protection of rights and legitimate interests of taxpayers in relation to information classified by the tax legislation as protected objects.

References

1. Arbitration Procedure Code of the Russian Federation dated 24.07.2002 N 95-FZ (2002). http://www.consultant.ru/document/cons_doc_LAW_37800/. Accessed 15 Feb 2020
2. Bobrova, N.A.: The paradoxes of application of the legislation on personal data. Law Enforc. Rev. **2**(1), 54–62 (2018). https://doi.org/10.24147/2542-1514.2018.2(1).54-62
3. Code of Administrative Procedure of the Russian Federation dated 08.03.2015 N 21-FZ (2015). http://www.consultant.ru/document/cons_doc_LAW_176147/. Accessed 15 Feb 2020
4. Code of Civil Procedure of the Russian Federation dated 14.11.2002 N 138-FZ (2002). http://www.consultant.ru/document/cons_doc_LAW_39570/. Accessed 15 Feb 2020
5. Code of the Republic of Kazakhstan of December 25, 2017 No. 120-VI “On taxes and other mandatory payments to the budget (Tax code)” (2017). https://online.zakon.kz/document/?doc_id=36148637. Accessed 15 Feb 2020
6. Criminal Code of the Russian Federation of 13.06.1996 N 63-FZ (as amended on 27.12.2019) (2019). http://www.consultant.ru/cons/cgi/online.cgi?req=doc&base=LAW&n=341913&fld=134&dst=1000000001,0&rnd=0.5549880538831127#0935947766169215. Accessed 15 Feb 2020
7. Criminal Procedure Code of the Russian Federation dated 18.12.2001 N 174-FZ (2001). http://www.consultant.ru/document/cons_doc_LAW_34481/. Accessed 15 Feb 2020
8. Decree of the President of the Russian Federation of 06.03.1997 No. 188 (ed. from 13.07.2015) “On approval of the list of confidential information” (2015). http://www.consultant.ru/cons/cgi/online.cgi?req=doc&base=LAW&n=182734&fld=134&dst=1000000001,0&rnd=0.10986309365482638#04019919055324 0194. Accessed 15 Feb 2020
9. Dvinskikh, D.Yu., Talapina, E.V.: Risks of data turnover development in public administration. Public Adm. Issue **3**, 7–30 (2019)
10. Federal law of 07.08.2001 № 115-FZ (as amended on 16.12.2019) “On countering the legalization (laundering) of income from crime and financing of terrorism” (2019). http://www.consultant.ru/cons/cgi/online.cgi?from=32834-0&rnd=70E67DDC6B8910637807798E65978293&req=doc&base=LAW&n=330804&REFDOC=32834&REFBASE=LAW#2ksgd00zfow. Accessed 15 Feb 2020
11. Federal law of 27.07.2006 № 149-FZ (as amended on 02.12.2019) “On information, information technologies and information protection” (2006). http://www.consultant.ru/cons/cgi/online.cgi?req=doc&base=LAW&n=339396&fld=134&dst=1000000001,0&rnd=0.5068544565616508#0851124464429981 6. Accessed 15 Feb 2020

12. Federal law of 27.07.2006 № 152-FZ (as amended on 31.12.2017) "On personal data" (2017). http://www.consultant.ru/cons/cgi/online.cgi?req=doc&base=LAW&n=286959&fld=134&dst=1000000001,0&rnd=0.537934809419724#09084059841657302. Accessed 15 Feb 2020
13. Geiger, H., Mürbe, M., Lederer, S., Obenaus, W.: Beck'sches Rechtslexikon: Rund 1.800 Rechtsbegriffe für Beruf und Alltag. Deutscher Taschenbuch Verlag, Auflage (1992)
14. Halt, G.B., Donch, J.C., Stiles, A.R., VanLuvanee, L.J., Theiss, B.R., Blue, D.L.: Trade Secret. Springer, Cham (2019)
15. Köbler, G.: Juristisches Wörterbuch. Vahlen Studienreihe Jura, 14 (2004). http://www.agmiw.org/wp-content/uploads/2016/05/K%C3%B6bler-Juristisches-W%C3%B6rterbuch-.Auflage.pdf. Accessed 15 Feb 2020
16. Law of the Republic of Azerbaijan No. 224-IIQ of December 4, 2001 "On commercial secrets" (2001). https://online.zakon.kz/Document/?doc_id=30612904. Accessed 15 Feb 2020
17. Purdon, J.: Secret Agents, Official Secrets. Oxford Scholarship, Oxford (2016)
18. Regulations on the procedure for collecting taxes and fees in Germany (2017). https://www.gesetze-im-internet.de/englisch_ao/englisch_ao.html. Accessed 15 Feb 2020
19. Revina, S.N., Sidorova, A.V., Zakharov, A.L., Tselniker, G.F.: Economic relations and economic systems. In: Popkova, E. (ed.) The Impact of Information on Modern Humans, HOSMC 2017. Advances in Intelligent Systems and Computing, vol. 622, pp. 549–554. Springer, Cham (2018)
20. Sellheim, N.: Legal regimes. Seal Hunt **3**, 111–216 (2018)
21. Shekhovtseva, E.V.: Tax secrecy: legal protection regime. Leningr. Law J. **1**(31), 38–42 (2013)
22. Sidorova, A.V., Cherevichenko, T.S., Efremova, E.A., Medentseva, E.V., Azarkhin, A.V.: The institute of banking secrecy in the CIS: rather-legal analysis. J. Adv. Res. Law Econ. **9** (3), 1110–1115 (2018)
23. Tax Code of the Republic of Azerbaijan No. 905-IQ of July 11, 2000 (2000). https://online.zakon.kz/Document/?doc_id=30414629. Accessed 15 Feb 2020
24. Tax Code of the Republic of Belarus 166-Z of 19.12.2002 (2002). http://www.nalog.gov.by/uploads/documents/NK_ot_2018-12-30_N_159-z.pdf. Accessed 15 Feb 2020
25. Tax Code of the Russian Federation (part one) of 31.07.1998 No. 146-FZ (ed. from 27.12.2019) (2019). http://www.consultant.ru/cons/cgi/online.cgi?req=doc&base=LAW&n=327825&fld=134&dst=1000000001,0&rnd=0.07002359312554685#014932158127365236. Accessed 15 Feb 2020
26. Tikhomirov, Yu.A.: Vectors in the focus of law. Public Adm. Issues **1**, 136–159 (2019)
27. Tumanova, L.V., Snytnikov, A.A.: The Promotion and Protection of the Right to Information. Urite, Moscow (2001)

Railway Vehicles Enhancement in the Course of Rail Freight

N. P. Karpova[1(✉)], V. A. Haitbaev[2], and A. A. Kremnev[2]

[1] Samara State University of Economics, Samara, Russia
natk08@rambler.ru
[2] Samara State Transport University, Samara, Russia
vhaitbaev21@mail.ru, a.kremnyov@samgups.ru

Abstract. The rail transportation market in Russia is monopolized by PJSC "Russian Railways". In this regard, the operator dictates the interaction terms to cargo owners, which results in searching optimal ways to do business. The issue of interaction between cargo owners and PJSC "Russian Railways" is to determine the sources of efficiency to deliver cargo to consignees. The purpose of the paper is to find ways to enhance the railway vehicles use in the course of rail freight. The author carried out a comparative study of working efficiency indicators of cargo owners' fleet of cars, own and leased, through the example of PJSC "SG-Trans". The main difficulties and sources of cost in the course of rail freight were identified, and the ways to neutralize and minimize these problems were offered.

Keywords: Carrier · Cargo · Cargo owner · Efficiency · Enhancement · Railway vehicles

1 Introduction

Railway vehicles in Russia play an important role in the country's economy. They provide high mobility and speed for goods and passengers delivery, develop markets, international trade, supply chains, etc. [1, 2] Investments in the rail transport infrastructure development bring numerous benefits to the state in the context of the country's region development, which should not be limited only to the physical aspect in terms of new railway tracks increase [4]. In general, investments should contribute to improve the quality of cargo owners service and speed of freight traffic and passenger trains, and to increase the rail freight volume, the capacity growth of transportation hubs, etc. [5, 6, 8–10].

In fact, PJSC "Russian Railways" is a monopoly at the market of freight rail transportation in Russia. The company dictates the terms how to use public tracks, fleet of freight cars, etc. Therefore, the issues to coordinate the carriers and cargo owners' actions are relevant. First of all, this concerns issues of railway vehicles enhancement in the course of rail freight. The study subject is PJSC "SG-Trans", which is a leader in the oil and gas and petrochemical cargoes transportation and provides transportation of half cumulative volume of all liquefied petroleum gases in Russia. To achieve this purpose, the author analyzed the working efficiency indicators of cargo owners' fleet of

S. I. Ashmarina et al. (Eds.): ENGINEERING ECONOMICS WEEK 2020, LNNS 139, pp. 231–239, 2021.
https://doi.org/10.1007/978-3-030-53277-2_26

cars (own and leased), justified the need to reduce delay time of cars at train stations, identified the main cost items in the course of rail freight, and proposed ways to minimize them.

2 Methodology

The research draws on the mathematical logic methods based on such principles as scientific approach, system approach, completeness and reliability of the infobase, and practical feasibility of the research results. To justify on logical grounds the decisions of railway vehicles enhancement in the course of rail freight, quantitative and qualitative methods of economic research were used. The largest Russian operator of the rail transportation of liquefied gases was the experimental base for making optimal decisions in the course of rail freight.

3 Results

The beginning of 2020 can be characterized by an increase in the presence of PJSC "Russian Railways" at the container transportation market. According to January 2020 data, the growth rate of container rail traffic was 118.2%, compared to the same period in 2019. As compared to 2016, the container traffic volume in the country increased by 98.3%. The total PJSC "Russian Railways" fleet of cars increased in January 2020 by 67 thousand cars, while the fleet of empty cars increased by 98.7 thousand units, and loaded cars decreased by 31 thousand units. As for the rail freight structure, it should be noted that according to January 2020 data, the largest relative share in freight traffic is occupied by coal (29.3%), oil and petroleum product (19.9%), construction cargo (8.4%), and ferrous metals (6.4%) [7].

Under current conditions of traffic intensity increase at the rail freight market, coordination of actions between carriers, consignors and consignees is becoming relevant. First of all, this is important to PJSC "SG-Trans", a railway operator of the oil and gas and petrochemical cargoes transportation, which provides transportation of half cumulative volume of all liquefied petroleum gases in Russia. It should be noted that the interaction policy at such an activity scale requires competent coordination of actions between PJSC "SG-Trans" and PJSC "Russian Railways". Therefore, the companies' cooperation in the course of rail freight should be developed in order to optimize the railway vehicles use in the course of rail freight. This implementation requires an economic justification of the problem how to use own or leased railway vehicles. In this regard, the following input data are the conditions to solve the problem of PJSC "SG-Trans" own or leased railway vehicles enhancement in the course of rail freight: analyzed period is 1 year with monthly detailed elaboration; considered fleet of cars is 30 units (both for own and leased fleet); costing rate for repair and steaming of the own fleet of cars is 238 rubles; costing rate for repair and steaming of the leased fleet of cars is 208 rubles. The calculations are presented in Tables 1 and 2.

In the case of operating the own fleet of cars, the costs of PJSC "SG-Trans" will consist of the costs for cars repair and steaming, entry lines lease, and the company's

Table 1. Calculation of operating efficiency indicators of the own fleet of cars

Indicator/Month	01	02	03	04	05	06	07	08	09	10	11	12	Total
Car turnover coefficient	1.6	1.5	1.8	1.7	1.9	1.7	1.7	1.6	1.7	1.6	1.5	1.6	–
Static load, tons per car	56	57	57	58	57	57	57	56	57	57	58	56	–
Quantity of shipments, tons, in thousands	1.68	1.71	1.71	1.74	1.71	1.71	1.71	1.68	1.71	1.71	1.74	1.68	–
Own fleet of cars, units	30	30	30	30	30	30	30	30	30	30	30	30	–
Operating revenue (at a rate of 256 rubles per ton), RUB, in thousands	688	657	788	757	832	744	744	688	744	700	668	688	8,699
Costs for repair and steaming, RUB, in thousands	640	610	733	704	773	692	692	640	692	651	621	640	8,087
Entry lines lease at oil refineries and gas processing plants, RUB, in thousands	1.45	1.45	1.45	1.45	1.46	1.46	1.46	1.46	1.46	1.46	1.46	1.46	17.53
General economic expenses (depreciation, communication service, wages, etc.), RUB, in thousands	7.03	7.03	7.03	7.03	7.04	7.04	7.04	7.04	7.04	7.04	7.04	7.04	84.5
Total costs, RUB, in thousands	648.2	619.0	741.1	712.5	781.8	700.4	700.4	648.3	700.4	659.7	629.7	648.3	8,190
Profit, RUB, in thousands	39.9	37.7	46.9	44.7	50.0	43.8	43.8	39.9	43.8	40.7	38.5	39.9	509.6

Source: authors.

Table 2. Calculation of operating efficiency indicators of the leased fleet of cars

Indicator/Month	01	02	03	04	05	06	07	08	09	10	11	12	Total
Car turnover coefficient	1.6	1.5	1.8	1.7	1.9	1.7	1.7	1.6	1.7	1.6	1.5	1.6	–
Static load, tons per car	56	57	57	58	57	57	57	56	57	57	58	56	–
Quantity of shipments, tons, in thousands	1.68	1.71	1.71	1.74	1.71	1.71	1.71	1.68	1.71	1.71	1.74	1.68	–
Leased fleet of cars, units	30	30	30	30	30	30	30	30	30	30	30	30	–
Operating revenue (at a rate of 256 rubles per ton), RUB, in thousands	688	657	788	757	832	744	744	688	744	700	668	688.	8,699
Costs for repair and steaming, RUB, in thousands	559	534	640	615	676	605	605	559	605	569	543	559	7,068
Cars leasing, RUB, in thousands	97.7	94.5	97.7	100.8	104.0	100.8	97.7	104.0	104.0	97.7	107.1	104.0	1,210
Entry lines lease at oil refineries and gas processing plants, RUB, in thousands	1.45	1.45	1.45	1.45	1.46	1.46	1.46	1.46	1.46	1.46	1.46	1.46	17.53
General economic expenses, RUB, in thousands	6.03	6.03	6.03	6.03	6.03	6.03	6.03	6.03	6.03	6.03	6.03	6.03	72.44
Total costs, RUB, in thousands	664	636	745	724	787	713	710	671	716	674	657	671	8,368
Profit, RUB, in thousands	23.9	21.1	42.6	33.7	44.5	31.2	34.4	17.6	28.1	26.2	10. 7	17.6	331.5

Source: authors.

general economic expenses (including depreciation). In the case of operating the leased fleet of cars, the costs of PJSC "SG-Trans" will consist of the cars leasing costs, cars repairing and steaming, entry lines lease, and general economic expenses. Figure 1 shows a comparative histogram of the cost and profit dynamics of PJSC "SG-Trans", including two versions of the railway vehicles use.

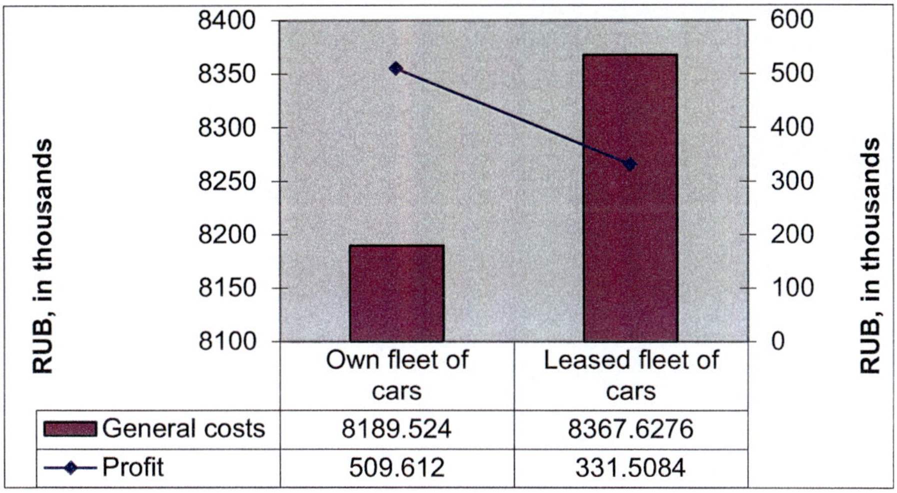

Fig. 1. Histogram of the cost and profit dynamics of PJSC "SG-Trans", including two versions of the railway vehicles use (Source: authors).

According to the calculations, the total cost of PJSC "SG-Trans", while in operation of 30 own fleet of cars, will amount to 8,189.52 thousand rubles in a year, and while in operation of the leased fleet of cars, it will be 8,367.62 thousand rubles.

The company's profit, based on the rate of 256 rubles per ton, will amount to 509.612 thousand rubles in a year while operating the own cars, and 331.5084 thousand rubles while operating the leased cars.

The efficiency of the rail vehicles use in the company for two options were calculated:

1. Operating efficiency of the own fleet of cars will be: 509.612/8,189.52 * 100% = 6.22%.
2. Operating efficiency of the leased fleet of cars will be: 331.5084/8,367.62 * 100% = 3.96%.

Thus, the operating efficiency of the own fleet of cars is higher than the efficiency of the leased fleet of cars. The operating efficiency of the railway vehicles in Samara gas transportation branch of PJSC "SG-Trans" was evaluated. The initial data for analysis is presented in Table 3.

According to Table 4, the empty mileage coefficient of the company's cars is very high and continues to grow. This is primarily due to the type of cargo being transported. For comparison, it should be noted that this coefficient is lower for building

Table 3. Performance indicators of railway vehicles use in Samara branch of PJSC "SG-Trans"

Indicator	2017	2018	2019
Loaded run (km)	250	277	300
Empty run (km)	210	220	230
Schedule speed (km/h)	42	42	42
Average number of train stations a car does at turn-around time	2	2	2
Average car detention at one train station (h)	36	51	60
Average car detention during load/discharge (h)	63	70	80
Average car static load (amount of cargo loaded) (tn)	20	20	20
Average length of haul (km)	400	380	400
Ratio of operating ton-km and tariff ton-km	1.045	1.3	1.3
Amount of transit cars with shunting service	20	28	35
Amount of transit cars without shunting service	24	35	50
Local operations coefficient	0.5	0.5	0.7

Source: authors.

materials transportation, since cars (low-sided cars, dumping cars, etc.) do not require such careful preparation for transportation. And rail tanks for liquefied gases transportation can be loaded only with gas but nothing else.

Table 4. Indicators of operating efficiency of railway vehiclesin Samara branch of PJSC "SG-Trans"

Indicator	2017	2018	2019
Average car dynamic force (tn)	33.44	35.67	34.67
Empty run coefficient	0.76	0.74	0.77
Coefficient of cars shunting service at train stations	0.83	0.8	0.7
Car turn around (24-h day)	4.77	6.2	7.86
Average daily car mileage (km)	96.46	80.14	67.44
Car efficiency (ton-km net)	20,160	13,892	10,019

Source: authors.

The table also shows that car detention at train stations is increasing. This is due to the low train-handling capacity of the stations. Recently, there have been difficulties in developing large train stations that are part of railway hubs. Although there were some difficulties with this development in earlier times, it was quite simple to increase the arrangement of tracks (adding and lengthening of tracks, etc.). At present, in most of the hubs, the territorial possibilities for expanding the railway economy have already been exhausted, and new equipment (electric interlocking, catenary system, mechanical and automated gravity humps, handling equipment on the tracks, etc.) leads to a significant cost increase in reconstruction, and it is necessary to find new ways to develop the hubs.

Car detention during load/discharge also tends to increase. According to the standards, delay time should not exceed 6 h during the load and 24 h during the discharge. But in fact, the cars delay for several days, and PJSC "Russian Railways" fines for it. The author believes that this is primarily due to the fact that the factories, when sending a request for loading, indicate only the amount of cargo and the terms to provide it, without attaching a shipping schedule. Scheduling is done by a specialist in PJSC "SG-Trans" logistics department, who does not know the plant's production schedules and makes a shipping schedule evenly. For example, the factory sends a request for 1,200 tons, which must be shipped within 30 days (from 1 to 30 June). The shipping schedule will look like this: 200 tons on 1 July, 200 tons on 5 July, 200 tons on 10 July, 200 tons on 15 July, 200 tons on 20 July, 200 tons on 25 July. As a result, loading is not performed at all on some days, but on others 500 or more tons are loaded at once, and PJSC "Russian Railways» fines for deadline violation.

Also, in the course of the study, it was found that the most expensive item of expenditure in the company is repair services of railway vehicles. This is due to the strong depreciation of the car fleet. The shelf life of 1,375 oil cars will have expired by 2020, which is 9% of the rail tank fleet of PJSC "SG-Trans" for the transportation of liquefied gases and light hydrocarbon raw materials. 48% of the specialized cars fleet of PJSC "SG-Trans" to transport liquefied gases and light hydrocarbon raw materials consist of oil cars of outdated design with a boiler volume of 54 m^3. These models have the following main disadvantages: increased specific metal content of the oil car compared to other types of cars (high empty weight-to-carrying capacity ratio), due to the large thickness of the boiler wall; underutilization of the allowable axial load (23.5 tn/axle); high operation and maintenance costs per unit.

As part of the update program, the company has already made the following changes to the car fleet:

- 37 oil cars with a boiler volume of 54 m^3 and 5 oil cars with a boiler volume of 54 m^3 were discarded because of their technical conditions and service life,
- in accordance with the Innovative Development Program, the outdated oil cars were replaced with the new large-capacity ones with a boiler volume of 75 m^3 or more. In 2019, PJSC "SG-Trans" acquired 225 oil cars from domestic manufacturers PJSC "Ruzhimmash" and PJSC "Salavatneftemash", including:
- 200 units of railway vehicles with a boiler capacity of 75 m^3, which were delivered under leasing agreements (121 units were put into operation in 2018 and 79 units in the first quarter of 2019),
- 25 oil cars with a boiler capacity of 83.9 m^3, which were delivered under the sales contract and put into operation in 2018.

The author believes that updating the fleet of rail tanks and using the models with improved technical and economic parameters should be a priority in the company's development.

4 Discussion

Railway transport occupies a leading position in the Russian transport system. PJSC "Russian Railways" is one of the world's largest companies at the rail transportation market. It ensures the safety and accessibility of Russian transport and reduces the traffic load in the economy [6]. PJSC "Russian Railways" also covers rail freight in the country using the capacity of its logistics infrastructure: railways, train stations, cars, containers, oil cars, etc. Cargo owners are forced to adapt to the monopolist's activity and look for ways to minimize the rail freight costs under current circumstances.

The study revealed a number of problems of a major railway operator, which ships oil and gas and petrochemical cargoes in Russia. First, freight owners have increasing coefficient of cars empty mileage. Secondly, the car detention at train stations increases, which is due to their low train-handling capacity. Third, the oil cars detention during load/discharge also tends to increase, which results in penalties increase for deadline violation on the part of PJSC "Russian Railways". Fourth, the most expensive item of expenditure in the company is repair services of rail vehicles. According to the experts, the potential for strengthening the coordination relationships between a carrier and a client is based on awareness of the clients' needs and increasing the cargo handling speed on the part of PJSC "Russian Railways", and on increasing the shipment planning quality on the part of the client. The most significant potential to improve the efficiency and quality of customer service is based on improving the interaction between the commercial services of cargo owners and the management of PJSC "Russian Railways", as well as increasing the speed of handling cargo, cars and locomotives directly at the holding's stations. In exchange, the potential for increasing cargo owners' satisfaction is also associated with a deeper understanding of the different cargo owners' needs on the part of the subdivisions of PJSC "Russian Railways". On the one hand, these needs have some common features, such as cargo safety and provision with immediate information about its location. On the other hand, they may differ significantly. For example, for the largest consignors with a continuous production cycle (oil-processing plants, chemical and metallurgical enterprises), the key values are the rhythm and accomplishing supplies for rail vehicles, while in the segment of container and perishable goods transportation, speed and delivery terms guarantee is more valuable [3].

5 Conclusion

The research has shown that the optimal solution in the course of rail freight is to use the own fleet of cars. Thus, the operating efficiency of the own fleet of cars in the course of rail freight is almost twice as high as the operating efficiency of the fleet of cars leased for this purpose from PJSC "Russian Railways". Also, the study revealed that the freight owner has an increasing coefficient of cars empty mileage. This is primarily due to the type of cargo being transported, that is liquefied gas. To transport it, rail tanks are used, and only liquefied gas can be poured into them but nothing else. This results in their empty mileage growth. In addition, the car detention at train stations is increased, which is due to their low train-handling capacity. This is caused by the fact that there are some technical difficulties in territorial expansion of railway hubs. The rail tanks detention during load/discharge also tends to increase, which leads to an increase in penalties for terms violation on the part of PJSC "Russian Railways". This is primarily due to the fact that the factories, when sending a request for loading, indicate only the amount of cargo and the terms to provide it, but they do not attach a shipping schedule. The most expensive expenditure item in the company is repair services of railway vehicles. This is due to the strong depreciation of car fleet in PJSC "SG-Trans". The author believes that updating the fleet of rail tanks and using the models with improved technical and economic parameters should be a priority in the company's development.

References

1. Butko, T., Prokhorov, V., Kalashnikova, T., Riabushka, Y.: Organization of railway freight short-haul transportation on the basis of logistic approaches. Procedia Comput. Sci. **149**, 102–109 (2019)
2. Dolgopolov, P., Konstantinov, D., Rybalchenko, L., Muhitovs, R.: Optimization of train routes based on neuro-fuzzy modeling and genetic algorithms. Procedia Comput. Sci. **149**, 11–18 (2019)
3. Ishmukhammetov, F., Skorlygina, N.: Tandem of the carrier and the consignor. Kommersant (2019). https://www.kommersant.ru/doc/3969767. Accessed 11 Mar 2020
4. Jarocka, M., Glińska, E.: The state and prospects for development of railway transport infrastructure in Eastern Poland – secondary data analysis. Procedia Eng. **18**, 299–305 (2017)
5. Karpova, N., Noskov, S., Toymentseva, I., Shvetsova, E.: Strategically supplier oriented management in the procurement process of JSC "Russian Railways". Problems Perspect. Manag. **16**, 14–27 (2018)
6. Karpova, N., Noskov, S., Toymentseva, I., Shvetsova, E.: Sources of rail transport efficiency in supply chains. Helix **8**(5), 3607–3614 (2018)
7. PJSC "Russian Railways": RR in figures (2020). http://www.rzd.ru/static/public/ru?STRUCTURE_ID=5232. Accessed 11 Mar 2020
8. Romanova, A., Luzhnov, Y., Vygnanov, A., Sokolova, E., Ivashkovska, N.: Resource-saving management on sections of high-speed railway lines. Procedia Comput. Sci. **149**, 297–300 (2019)
9. Štefancová, V., Nedeliaková, E., López-Escolano, C.: Connection of dynamic quality modeling and total service management in railway transport operation. Procedia Eng. **19**, 834–839 (2017)
10. Yan, Z., Li, X., Han, B.: Collaborative optimisation of resource capacity allocation and fare rate for high-speed railway passenger transport. J. Rail Transp. Plann. Manag. **10**, 23–33 (2019)

Finance of Public-Private Partnership in the Russian Federation

T. M. Kovaleva(✉), L. N. Milova, O. A. Khvostenko, and E. V. Popova

Samara State University of Economics, Samara, Russia
fikr@bk.ru, larisamilova2009@yandex.ru, olegkhvostenko@yandex.ru, katirinna@mail.ru

Abstract. The paper deals with the role of finance in public-private partnership and the procurement system in the development of the real sector of economy. Under the conditions of market relations, the economy is becoming the determining factor of social development, and the state is intended to harmonize its interests with the interests of economic entities. One of the means of such harmonization is a public-private partnership (PPP) mechanism, which has been rapidly developed in the Russian Federation in recent years and is used primarily to implement significant state infrastructure and innovative projects. At the same time, the federal contractual system in procurement is aimed to ensure the public commitments of the Russian Federation. Under the contractual system, the state is interested in timely and complete performance of contracts concluded on behalf of the Russian Federation to supply products, works and services, as well as to save budget funds. In the current context, the achievement of these goals is due to such new factors as integration of public and private finance.

Keywords: Public finance · Private finance · Public-private partnership · Real sector of economy · State procurement

1 Introduction

Public investment in infrastructure and innovation projects can be considered as one of the measures to improve the Russian economy. However, this implementation is associated with many problems, and the most serious one is the lack of budget funds. The investment process allows to earn income in the future, but these days, searching for resources may not give the desired result. Gratuitous financial assistance is not always available, and bank loans are too expensive, when interest rates are high. As for private businesses, their interest in the project is often limited to very short terms, due to the rapid capital cycle and the short life cycle of individual enterprises.

Therefore, it is advisable to build an integrated model of the state financial policy to exploit advantages of public finance: large volumes, tax preferences, social benefits, financial control, and private investment. To implement such a policy, it is necessary not only to integrate managerial efforts of the state authority and commercial companies, but also build interaction between public and private finances. Such interaction creates approaches to define PPP finance.

S. I. Ashmarina et al. (Eds.): ENGINEERING ECONOMICS WEEK 2020, LNNS 139, pp. 240–246, 2021.
https://doi.org/10.1007/978-3-030-53277-2_27

One of the mechanisms to unify private business and state activities is the sphere of procurement that requires a state contractual system. It involves the state and municipal customers' activities in interaction with one another, including a commercial sector: suppliers (contractors, performers) and banks. There is a convergence of public and private finance in this interaction.

2 Methodology

The methodological and theoretical ground of the research is based on the fundamental propositions of economic theory, and public and corporate finance. The authors of the study examined the leading foreign and national researchers' works on public-private partnership and the public procurement system. The methodology used in this study combines general scientific methods (synthesis, analysis and collation, analogy, and historical and logical methods) and methods of empirical knowledge (classification, observation, description, and comparison).

3 Results

One of the relatively new forms of interaction in Russian practice is public-private partnership. Interaction between public and private finance can be carried out through the use of the following instruments: budget subsidies, lending, shared and joint participation, tax and other allowances, and contractual system in the field of procurement. PPP is broadly defined in the national literature as an institutional and organizational alliance between the state and business in order to implement national, international, large-scale and local, but always socially significant, projects in a wide range of areas.

It is important to mention that public-private partnership is a specific relationship that has a number of features which affect the finance organization as:

- *PPP subject of is state and municipal property, as well as the services provided by the state and local authorities,*
- *PPP is formed by a special contract, and its life cycle includes not only the stages to build and place the unit in operation, but also the process to make a profit for the investor by meeting the society needs in the project products,*
- *a mandatory condition for PPP is the state and private companies' co-financing and risk sharing between them (sometimes the project can be 100% funded by the private sector),*
- *PPP project is carried out within a certain period established under the contract, but the partnership can be continued on the basis of a new contract,*
- *distribution of financial risks and costs, as well as results achieved, which takes place among the parties within predetermined proportions.*

Basically, the state is the introducer of PPP, that is to say, it invites businesses to participate in the implementation of socially significant projects, but not vice versa. The main forms of PPP in the economy and management of public and private finance include: mutually beneficial forms of interaction between the state and business, state contracts, state assets lease, financial leasing, establishing and operation of joint public-private enterprises, implementation of production sharing agreements, and conclusion of concessionaire agreements.

The scope of public-private partnership is very diverse, from road building to hospitals and universities construction. Improving the PPP system makes it possible to increase efficiency and competitiveness of a country in the world market. Consequently, PPP finance is a form to build money relations where the participants are the state and specific economic entities. PPP finance is a set of economic relations concerning the formation and distribution of financial resources, which are necessary to the state and economic entities for implementing socially significant projects. The need to form PPP is due to its profitability and effectiveness for both participants: public authorities get additional opportunities to solve their specific tasks, and business gets favorable economic management conditions and guarantees for profit. At the same time, PPP is a way to avoid the extreme forms of relations between business, government and society, when they range from privatization to nationalization, depending on the economic cycle. A significant state contribution to the PPP project is granting of a monopoly right to its implementation. However, in exchange for it, the state demands the right to interfere in affairs. Such intervention takes various forms: first of all, tariff and price regulation. Under the conditions of PPP, business is not free to determine the price of its services, since monopoly privileges and protection from competitors provided by the state are the ground to obtain the natural monopolist status. Public-private partnership has a number of advantages that make it possible to effectively implement infrastructure projects, taking into account the strengths of each partner.

The main advantages of PPP finance in Russia are:

- *budget savings,*
- *overcoming problems related to insufficient budget funding,*
- *use of* private investors' *experience in project management,*
- *flexible and balanced risk sharing between the parties of the PPP agreement,*
- *improvement of* services *quality to the public.*

The PPP *advantage is that the state uses the* private sector *efficiency in the construction process. It is achieved through a system of payments that are linked to the phased implementation of work within the time specified. If these terms are not respected, the agreement includes fines.* PPP contracts typically define rights and obligations regarding the design, build, operation and maintenance of infrastructures and the mechanisms for their supervision but also may include provisions about tariffs, access and interconnection rights or levels of service, conditions that are conceptually considered as economic regulation [4].

Another advantage is costs improvement during the entire life cycle of the project. A private investor is responsible for operating costs *in well prepared PPP contracts. This leads to the fact that the investor is interested in the highest quality of construction in contrast to conventional projects where different companies carry out building and operation. The advantages also include the provision of services focused on qualitative results. Since in infrastructure PPP projects, the investor depends on payments by the population or the concessor (related to the results of PPP object operation), the private sector is directly interested in the high quality of service provision. On the other hand, there is a number of risks that participants of public-private partnerships have to face.*

These risks include a lower quality of control and management compared to conventional projects. This is due to the long terms of PPP contracts and complex project management, which does not have enough mobility to adapt to rapidly changing external conditions. Financial risks are an integral part of public-private partnership projects, since financial flows in such projects depend on a number of factors which are very difficult to predict. Depending upon the national legislation, there may also be problems with changing the essential terms of agreements, which are required to be included due to the force majeure circumstances or unethical practices at the hands of one of the parties. The practice to use PPP in a number of foreign countries shows that this mechanism is applied when the state and business have intercomplementary interests, but they are not able to perform all alone and independently of each other.

Concessionary partnerships are successfully developed in transportation (roads, railways, airports, ports, and pipeline service), social infrastructure (health, education, entertainment, and tourism), housing and utilities (water, electricity and gas supply, water cleaning, etc.) and other areas (prisons, defense, and military facilities). At the same time, the leading position in the total number is occupied by projects implemented in transportation and social infrastructure.

As for increasing the role of the state to develop the real sector of the Russian economy, as a system of public procurement, we can mention the following. Under the conditions of the contractual system, the integration of public and private finances is observed. It is clearly evident in the expansion of the bank guarantees to ensure the performance of government contracts.

An application for participation in a competition or auction may be made either through a cash deposit or a bank guarantee, that is to say, at the option of the participants. Starting from January 1, 2014 to ensure tender security and performance of contracts, participants of procuring activity can provide bank guarantees issued exclusively by those banks that are included in the list of banks, according to Article 176.1 and Article 74.1 in Tax Code of the Russian Federation [11, 12].

Generally, it can be noted that commercial companies, participating in the contractual system as suppliers, seek to obtain budget funds in order to produce goods and perform labour. The state, under the conditions of the contractual system, should not only have budget savings, which is achieved by selecting a supplier on a competitive basis, but also timely execution of contracts. Therefore, to ensure their performance, funds of commercial banks, which are guarantors of the work performed by contractors, are attracted and used [8]. As a result, there is an active interaction and integration of public and private economic finance.

A bank guarantee, offered by a participant of procurement activity to ensure tender and closed auction security or contract performance must be included in the bank guarantees register [5]. According to Decree of the Government of the Russian Federation of November 8, 2013, No 1005 "On bank guarantees used for the purposes of the Federal Law "On the contractual system in procurement of goods, works, and services to ensure the state and municipal needs", the function of maintaining the register of bank guarantees is executed by the Federal Treasury on the official website of the Russian Federation [3].

As the federal executive authority, the Treasury of Russia is currently performing the function of maintaining the register of bank guarantees, and the process of close interaction between public and private finances also clearly demonstrates it.

It should be noted that under the current conditions of the contractual system, the issues to increase participants' responsibility for the contract execution, intended use of payments in advance, and bringing budget funds to the real sector of the economy are becoming urgent. Therefore, it is important to shift in emphasis from follow-up financial control to preliminary and current monitor, intense control of the Russian Treasury over state contracts execution, and building the legal and organizational basis for the mechanism to support state contracts by the Treasury.

The practice of Treasury support for contracts of corporate bodies' that are not participants of the budgeting process provides for payment under the state contracts only with the authorization of Federal Treasury through the opened checking accounts. This ensures maximum control over the budget funds that are granted with confirmation of their use in accordance with the conditions and purposes they were provided to performers (contractors). This is another important aspect of the interaction between public and private finance that makes it possible to identify and assess possible risks, prevent financial misconduct before it occurs, and avoid non-execution of the government contracts.

4 Discussion

The problems of public-private partnership development in the current economy, its role in forming the innovative component of the country's economic system and improving this country's competitiveness are constantly in the focus of many scientists' attention.

The issues of how to develop public-private partnership in the innovation sphere are considered in the works of Cheng, Wang, Xiong, Zhu, Cheng [2]. The authors, using the example of different countries, proved that the mechanism to form PPP resources differently depends on the state sustainable development. Along with bank lending and tax preferences, public-private partnership is an effective form to finance small and medium-sized businesses. Such points are reflected in the works of the following authors: Luo, Chen and Liu [9]. Hueskes, Verhoest and Block consider the sustainability of projects implemented under PPP. This study identifies varios options for governing for sustainability [6]. Korab-Karpowicz considers PPP as a tool to deal with international issues. PPPs can lead to the increase of both the effectiveness and the legitimacy of global governance in terms of democratic participation and accountability [7].

Some aspects of the state contractual system are presented in the works of foreign authors such as Buchanan [1] and Williamson [13]. Tammi, Saastamoinen and Reijonen consider public procurement as a means of innovation. In their research, the scientists determine the features of the relationship between competition and the innovative activity of small and medium-sized businesses [10]. This paper contributes to the literature on the role of public procurement in encouraging innovations in the economy by empirically examining how the in verted-U relationship between competition and innovation is related to SME behavior in public procurement [10].

5 Conclusion

The lack of national practice in interaction between public and private finances is a significant obstacle to form and develop an effective model of the state financial policy through the use of current market methods and tools. In turn, this development is impossible without studying the world experience and adapting it to the Russian conditions. The successful solution of this task will allow to find new sources of financial resources, increase the investment attractiveness of the Russian Federation, and ensure the expansion of the innovation sector. Development of this process is aimed at increasing the profit of commercial companies, which is the basis to provide money to the budget at all levels of the country's budget system, social insurance funds, and national welfare. Consequently, creating optimal conditions for investment activity is an important factor for economic growth. In this regard, the issue of investor protection and risk management of investment projects is becoming more and more urgent.

The main factors hindering the development of PPP finance in Russia include:

- changing the "rules" in the project implementation process,
- low level of competitive selection of projects and lack of actual competition,
- insufficient transparency of project management procedures and poor qualification of state and municipal employees in the field of project management,
- low borrowing capacity of regions and municipalities,
- unclear division of possible risks and responsibilities when they occur.

The limiting factors can often include difficulties to define the roles and tasks that private institutions should assume, as well as the associated responsibilities for results. In the near future in Russia, the development of PPP finance will be carried out at the expense of projects which:

- have regional and interregional significance (building a modern infrastructure of the national innovation system); in particular, they are the projects that are aimed at developing a network of technological parks, business incubators, and technology transfer centers,
- are related to improving competitiveness of the basic industries such as energy, oil and gas, and chemical sector,
- develop the mineral resource base and increase efficiency of resource exploitation including projects to create high-tech enterprises for extraction and processing of natural raw materials,
- are aimed at restructuring the economy of non-diversified communities, which can be financed by attracting budget allocations from the Investment Fund of the Russian Federation,
- are focused on developing competitive agriculture and transport infrastructure, in particular, high-speed rail passenger traffic and regional airport networks.

As a result, the trends in the state financial policy of the Russian Federation noted above reflect beneficial transformation in the relations between public and private finance and the emergence of PPP finance. This all enriches the essence and form of appearance of these relations in the public finance system. It can also be noted that the interaction between state and municipal customers and suppliers (contractors), banks, and Federal Treasury bodies generates integration of public and private finance which ensures contractual system effectiveness.

References

1. Buchanan, J.: Is economics the science of choice? In Streissler, E. (ed.) Roads to Freedom: Essays in Honor of F.A. Hayek, pp. 47–64. Routledge&Kegan, London (1964)
2. Cheng, Z., Wang, H., Xiong, W., Zhu, D., Cheng, L.: Public-private partnership as a driver of sustainable development: toward a conceptual framework of sustainability-oriented PPP. Environ. Dev. Sustain. (2020). https://doi.org/10.1007/s10668-019-00576-1. Accessed 20 Mar 2020
3. Decree of the Government of the Russian Federation of November 8, 2013, No 1005 "On bank guarantees used for the purposes of the Federal Law" "On the contractual system in procurement of goods, works, and services to ensure the state and municipal needs" (2013). http://base.garant.ru/70502258/. Accessed 20 Mar 2020
4. Diaz, G.R.: The contractual and administrative regulation of public-private partnership. Util. Policy **48**, 109–121 (2017). https://doi.org/10.1016/j.jup.2016.04.011
5. Federal Law of April 5, 2013 No 44-FL «On the contractual system in procurement of goods, works, and services to ensure the state and municipal needs» (2013). http://www.consultant.ru/document/cons_doc_LAW_144624/. Accessed 20 Mar 2020
6. Hueskes, M., Verhoest, K., Block, T.: Governing public-private partnerships for sustainability: an analysis of procurement and governance practices of PPP infrastructure projects. Int. J. Project Manag. **35**(6), 1184–1195 (2017)
7. Korab-Karpowicz, W.J.: The united citizens organization: Public-private partnerships in global governance. Research in Globalization (2020, in Press). https://www.sciencedirect.com/science/article/pii/S2590051X20300010. Accessed 20 Mar 2020
8. Kovaleva, T.N., Milova, L.N.: Integration of public and private finance in the field of public procurement. In: Sotova, A.S. (ed.) Proceedings of the International Scientific and Practical Conference «Development and Interaction of the Real and Financial Sectors of the Economy in the Context of Globalization», pp. 343–345. Orenburg State University, Orenburg (2019)
9. Luo, P., Chen, B., Liu, F.: Investment and financing for SMEs with bank-tax interaction and public-private partnerships. Int. Rev. Econ. Finance **65**, 163–172 (2020)
10. Tammi, T., Saastamoinen, J., Reijonen, H.: Public procurement as a vehicle of innovation – what does the inverted-U relationship between competition and innovativeness tell us? Technol. For. Soc. Change **153**, 119922 (2020)
11. Tax Code of the Russian Federation (Part 1), July 31, 1998, No 146-FZ (1998). http://www.consultant.ru/document/cons_doc_LAW_19671/. Accessed 20 Mar 2020
12. Tax Code of the Russian Federation (Part 2), August 5, 2000, No 117- FZ (2000). http://www.consultant.ru/document/cons_doc_LAW_28165/. Accessed 20 Mar 2020
13. Williamson, O.: The theory of the firm as governance structure: from choice to contract. J. Econ. Perspect. **16**(3), 172–173 (2002)

Economic Security Theory: State Support for Real Sector and New Political Economy

V. A. Noskov(✉)

Samara State University of Economics, Samara, Russian Federation
noskov1962@inbox.ru

Abstract. This article investigates various views on the problems of ensuring the economic security, presented in both Russian and international scientific literature. These views are often contradictory and do not cover the entire problem as a whole. Concepts and categories used by researchers are often not fully formed. This paper presents the political and economic analysis of the theory of economic security. It is proposed to use a conceptual spatial approach to overcome the existing contradictions. This approach is theoretically justified, since the research objects are not only economic entities (individuals, households, firms, municipalities, federal subjects, sovereign states), but also economic objects (phenomena and processes). It is proposed to apply this approach to the analysis of state support for the real sector of the economy. The methodological basis of the research includes methods of new political economics. The author takes into account the problem of state support development for subjects of the real reproductive process.

Keywords: Methodology of economic science · New political economy · Real sector of the economy · Theory of economic security · Spatial approach · State support

1 Introduction

The turbulence of maintaining and developing the economic security of the Russian macroeconomic system determines the economic sovereignty of our country. The question of whether our economy is sovereign was raised at the economic department of Moscow State University 10 years ago. A monograph was published under the editorship of Osipov, Sinelnikov and Zotova [11]. In general, the country's economic security can be considered as a condition for self-sufficient development of the national economy. But what is the essence of the category of economic security as an economic category?

Economic security is often defined as guaranteed protection of national interests and socially oriented development of a country [1]. Let us, however, question this thesis. If the economy is a system, then statics and dynamics are not friends in it. Scientific articles also tend to avoid realities of the public life. But the problem of ensuring the national economic security and its essence cannot be fully understood without reference to real facts. The identification of socio-political factors of the society instability, their recognition and not only economic threats and risks that are traditionally considered by scientists, will form the theory of economic security in close

S. I. Ashmarina et al. (Eds.): ENGINEERING ECONOMICS WEEK 2020, LNNS 139, pp. 247–252, 2021.
https://doi.org/10.1007/978-3-030-53277-2_28

continuity of the theory and socio-political practice, taking into account interests of different subjects of economic relations [6], and the priority of the state support of the real sector of the economy. Otherwise, "economic security" as a category of the economic science is not the security of economic entities, countries, regions, enterprises, organizations, but the protection of nature and the human from the operational processes (business processes).

2 Methodology

Creating a theory of economic security within the framework of maintaining and developing the national security is an attempt to understand the system of economic sovereignty. It should be emphasized that while economic sovereignty has been studied by many foreign and domestic scientists, economic security abroad is not studied. It may also be so because the economic security is not an independent problem, but a dependent part of the overall security of the existence and functioning of an institutionalized society as a state.

Apparently, the question of creating a theory of economic security is caused by the presence of a mass of near-scientific conversations about types of industrial security (energy, food, etc.), similar to conversations about types of capital (technological, technical, cultural, organizational, etc.). There are even different methods for determining threshold values, for example, food security or energy security, which are based on production indicators (not consumption or consumption prices, etc.).

This leads to the possibility of "pressure" on the population in the form of a pension policy, restrictions on energy consumption, restrictions on housing, taxes on buildings in suburban areas, etc. And why is the consumption of water and air not regulated by social norms?

Is there no theory of economic security today? What then is presented in the content of textbooks on economic security, for example, in the textbooks prepared by Senchagov's team from the Institute of Economics of the Russian Academy of Sciences [13], as well as in the mass of articles on economic security issues [2, 4, 5, 9, 14].

So, the question is whether the theory of economic security is necessary. Who is it useful for? How does it relate to the modern role of the state in the economy, and the state support for its real sector?

3 Results

The possible answers are based on a list of subjects (hence the types of security) and a list of threats. Paradoxically, today there is not the slightest factual reason to assert that the problem of creating a theory of economic security has been satisfactorily solved. We need a theory of economic security protecting people from internal threats. We will name the most significant threats. These include the development of distorted economic relations, the "vulgarization" of pensions, the "optimization" of healthcare and education, the growth of managerial personnel as a commodity with a load on the people's trust in the president, and the productive impotence caused, in particular, by the lack of the state support for the real sector of the economy.

Why may some types of economic relations be considered as a threat to the Russian economic security? Because the political economy defines economic relations as one side of the production way of material products and conditions for the human life [10].

Structuring economic relations into technical-economic, socio-economic, and organizational-economic ones [8], plus an assessment of the pace, rhythm, and stages of the development of organizational-economic relations allows us to identify very productively some of them (with appropriate characteristics) as threats to the economic security. This methodological approach is especially important when analyzing deformations of economic relations, which allows distinguishing such types of deformed economic relations "arvarny", "refouler" ones (refouler is a type of dredging machine that transfers selected soil immediately to the shore, through a special sleeve with a conveyor inside).

Disclosure of their content as a threat to the maintenance and development of the country's economic security can become part of the theory of economic security. Here there are just a few examples that illustrate the proposed approach. The first example is from the sector "energy security" outside of its traditional interpretation. Here (as in housing and utilities), there are "refouler" economic relations, when there are intermediaries between producers and consumers (sales companies).

We remember the statements made by the Russian Prime Minister Vladimir Putin to the state-owned company JSC "RusHydro" after a terrible tragedy that occurred in 2009 at the Sayano-Shushenskaya hydroelectric power station. Vladimir Putin said that from 352 people of the management staff of the energy complex checked after the accident, 169 officials have affiliation with 385 commercial organizations. There are far from isolated cases when the energy complexes of entire regions of the Russian Federation are controlled by family clans. From this example, there is considerable material for the development of the theory of economic security. Now let's give another example from the "food security" sector.

We called deformed economic relation with the term "arvarny" economic relations. Arvarny economic relations arise in the destruction process of a created food product or its withdrawal from a specific reproductive system. The mass media provides a lot of illustrations of arvarny relations: the destruction of black caviar obtained by poachers; the destruction of meat products, cheese, and apples imported to Russia illegally.

Here are the events of October 2018. Kamchatka is littered with tons of rotting salmon. What comments did the Deputy Prime Minister of the Association of fish market enterprises Alexander Fomin give on this situation? He explained that fishermen worked in the sea according to the usual scheme for the required number of days. It became necessary to finish the catch earlier, realizing that they had already obtained enough raw materials for processing. As a result, the catch is 540 thousand tons, which is 2 times more than usual [12]. It seems that the problem is a fault of the fish itself. Probably, some of the problems could be solved by changing the role of the state in the modern economy.

4 Discussion

A theory of economic security is needed, but it should be structured according to the types of threats and economic activities. This position was voiced by Kuznetsova, doctor of economics, professor of economic security of Moscow University of the Ministry of the Interior Affaires named after V. Kikotya in the report "Theory of economic security in the evolutionary development of modern science" on Senchagovsky readings in April 2019 [3].

Are food, information, financial, and energy securities types of the economic security? No, they are not! Because these are types of security, but not types of the economic security.

Economic security is the ability to function for achieving goals and objectives of a system without disrupting its functioning and deforming its goals. This is risk management and prevention (protection) from threats, elimination of consequences from the aftereffect of threats, i.e. from any deviations in functioning, return to a specified state. Protection as a warning (in humans – prevention). Protection as a return (in humans – treatment). Protection as (quasi-strength of functioning) an ability to ignore threats (in humans – immunity). Hence the importance of increasing the modern role of the state and developing the state support for the real sector of the economy.

5 Conclusions

The definition of the content for the concept "economic security", and especially the essence of the category "economic security" (as a category of economic science) "branched out" into a huge number of areas. In particular, the question of whether the word "economic" is related to the economy (the goal is to protect economic interactions) and/or to nature (the goal is to protect the nature from consequences of the economic interactions development). That is, the economic activity of people in terms of inconsistency with the Nature laws is a threat to the Nature (and to humans as part of the Nature) and, consequently, economic security (Nature and Society) is considered as an appropriate economic activity of people.

We will try to determine approximate elements of the theory of economic security that, in our opinion, require priority development:

- the concept of economic security of education systems of different levels,
- the analysis of the efficiency of the state support for the real economy,
- the role of the educational system in the economic security structure of municipalities, regions and the country as a whole,
- the relation between the formation of a personality, the prospects for its socialization and the economic security of the Russian society and the state,
- socio-cultural and value grounds for maintaining and developing the economic security,
- economic security as a factor for the formation of a new management paradigm for the socio-economic development (within the framework of the conceptual system of economic institutionalism),

- economic security in the context of interaction between a formal economy (in the legal field of the state), a "shadow" one (outside the boundaries of the legal and economic regulators, economic development from the state side) and an "informal" economy,
- the place and role of economic security in the structure of ensuring and implementing strategies for the development of economic systems of various ranks – the Russian economy, regional economy, and enterprise economy,
- maintaining the economic security in the context of innovative development of the Russian economy,
- economic security measure as a system of economic risks,
- prospects for reindustrialization of the Russian economy in the context of ensuring the country's economic security [7].

So, is it possible to create a theory of economic security? Yes, it is possible if the collective creativity is under the patronage of the Russian Academy of Sciences.

References

1. Bogomolov, V.A.: The Scale of the Shadow Economy and Economic Security of the Russian Federation. ROAT, Moscow (2010)
2. Eminova, N.E.: Investment security of Russia in the context of globalization. Russian Entrep. **16**(20), 3377–3386 (2015). https://doi.org/10.18334/rp.16.20.2013
3. Kuznetsova, E.I.: Theory of economic security in the evolutionary development of modern science. Ensuring Russ. Econ. Secur. Mod. Conditions **2**, 253–258 (2019)
4. Kuznetsova, Yu.L., Ipatieva, I.A.: Provision of the logistic system safety: issues and solutions. Russ. J. Retail Manag. **3**(2), 163–172 (2016). https://doi.org/10.18334/tezh.3.2.35407
5. Manyaeva, V.A., Piskunov, V.A., Fomin, V.P., Noskov, V.A.: Accounting policy for purposes of ensuring economic security of a commercial organization. Bull. Samara State Univ. Econ. **8**(154), 76–84 (2017)
6. Mikhalenok, N.O., Noskov, V.A.: Problems of Managing the Development of Formal and Informal Institutions in the Education System. SNC RAN, Samara (2007)
7. Noskov, V.A., Chekmarev, V.V.: The issue of ensuring economic security in the process of re-industrialization. Innov. Dev. Econ. **2**(38), 340–346 (2017)
8. Noskov, V.A., Chekmarev, V.V.: Political economy assessment of the educational system promotion in Russia in the conditions of digital economy formation. In: Ashmarina, S., Mesquita, A., Vochozka, M. (eds.) Digital Transformation of the Economy: Challenges, Trends and New Opportunities. Advances in Intelligent Systems and Computing, vol. 908, pp. 518–528. Springer, Cham (2020). https://doi.org/10.1007/978-3-030-11367-4_52
9. Noskov, V.A., Nesterov, O.V.: Contradictions of the Russian economy in the context of economic security theory. In: Mantulenko, V.V. (ed.) Proceedings of the 18th International Scientific Conference "Problems of Enterprise Development: Theory and Practice", European Proceedings of Social and Behavioural Sciences, vol. 82, pp. 266–274. European Proceedings, London (2020). https://doi.org/10.15405/epsbs.2020.04.34

10. Noskov, V.A., Nesterov, O.V., Abramov, D.V.: A critical infrastructure background of the economic security of regional development. In: Ashamrina, S.I., Mantulenko, V.V. (eds.) Proceedings of the 2nd International Scientific Conference GCPMED 2019 "Global Challenges and Prospects of the Modern Economic Development", European Proceedings of Social and Behavioural Sciences, vol. 79, pp. 1039–1047. European Proceedings, London (2020). https://doi.org/10.15405/epsbs.2020.03.150
11. Osipov, Yu.M., Sinelnikov, Yu.S., Zotova, E.S.: Economic Sovereignty and Sovereign Economy. MAKS Press, Moscow (2007)
12. Putin, V.V.: Be strong: Guarantees of National Security for Russia. Editorial URSS, Moscow (2012)
13. Senchagov, V.K.: Economic Security of Russia: General Course. Delo, Moscow (2005)
14. Skipin, D.L., Bystrova, A.N., Kutyreva, E.V., Trufanova, K.N.: Corporate fraud: the essence, risks and impact on economic security of business. Russ. J. Entrep. **18**(22), 3605–3614 (2017)

Business Group Contradictions as a Factor of Integration into the Global Economy

E. S. Gordeeva[1] and Z. P. Antipina[2(✉)]

[1] Saint-Petersburg State University of Economics, St. Petersburg, Russia
antipina_kate@yahoo.com
[2] Samara State University of Economics, Samara, Russia
zhanna-antipina@yandex.ru

Abstract. The purpose of the article is to identify contradictions of business groups as a development source of the real economy sector and theoretical justification of tools for constructive conflict resolution in the context of globalization. To achieve this goal, the authors analyzed economic interests of stakeholders of Russian and foreign business groups, which allowed them to identify and reveal the economic essence of general, special and specific contradictions that determines the development of affiliated industrial enterprises and the dynamics of transnationalization of the real sector of the economy. The research work provides a theoretical justification for integrated planning and market tools for constructive resolution of business groups' contradictions, as well as suggests ways to improve planning tools for coordinating interests of Russian business group stakeholders in order to develop and transnationalize the real sector of the domestic economy.

Keywords: Business groups · Economic contradictions · Economic interest · Stakeholder

1 Introduction

In a globalized market infrastructure, under conditions of production and capital transnationalization, and the expansion of global value chains, business groups have a decisive influence on the development of real-sector enterprises. Institute of business groups concentrates enormous investment of resources, but the contradictions that exist between stakeholders undermine the meaningful use of these funds. The model of interaction of economic interests (that has developed in Russian business groups) blocks the development of industrial enterprises and hinders the transnationalization of the real sector of the domestic economy. Taking into account these problems, the research on contradictions of economic interests that develop in the course of business groups' economic activity, as well as the search for an optimal combination of market and planning mechanisms for their constructive resolution in the context of globalization, become particularly relevant. The development of business groups is given considerable attention in the works of modern economists. This range of research includes: factor analysis of the limit of expanding boundaries of the hierarchy [11, 15], problems of placing property rights [13, 14], changes in information flows and financial

S. I. Ashmarina et al. (Eds.): ENGINEERING ECONOMICS WEEK 2020, LNNS 139, pp. 253–261, 2021.
https://doi.org/10.1007/978-3-030-53277-2_29

markets in the context of globalization and their impact on the integration trends of business groups [6, 12], the relation of innovations with trends of changing firm boundaries [2, 3, 5, 8], analysis of the business groups' development in the process of structural transformations in the economy in the context of globalization [1, 9, 10]. Under-researched are issues associated with achieving constructive conflict resolution of business groups in the context of globalization based on a combination of planning and market instruments for the coordination of economic interests.

2 Methodology

In the context of globalization, the instability and heterogeneity of the institutional and economic environment is increasing, while factors of increasing the business competitiveness are innovations, development of human capital and increase in social activities. The activities of business groups in the process of adapting to these changes are contradictory. The contradiction of a business group is understood as an interaction of opposite, mutually exclusive economic interests of its stakeholders, concerning production, distribution, exchange and consumption, which at the same time are in the internal unity and interpenetration, being a source of self-movement and development of this business group. Business groups' contradictions arise as a result of the interaction of economic interests in several axes: "shareholders – experts", "shareholders – managers", "shareholders – shareholders", "professionals – managers" "shareholders – external stakeholders", "management – external stakeholders." Resolving contradictions of a business group from the perspective of the Eurasian perspective requires the formation of an integrated mechanism for coordinating economic interests – a planning and market one. The combination of planning and market tools will allow, on the one hand, to more effectively support functional trends in the development of business groups (the practice of socially responsible business, increasing competitiveness through investment in human capital and innovations) and, on the other hand, to counteract strengthening of dysfunctional trends – the disproportionate redistribution of income in favor of the parent holding, financial speculation, and opportunism of managers and shareholders. In general, the solution of a complex of contradictions consists in creating such conditions for the interaction of subjects of economic relations, in which it is possible to freely coordinate their interests, and its effectiveness largely depends on achieving the right balance or combination of the market and the plan.

3 Results

Based on the analysis of the specifics of interaction between stakeholders of Russian and foreign groups, it is concluded that in the context of globalization, business groups are characterized by the following general contradictions: between formal and real economic growth; between current and investment consumption. However, changes in the external environment during the period of financial globalization and, in particular, the global economic crisis allowed us to identify special and specific contradictions.

In the first case, these are contradictions between the social and economic performance of the business group; in the second case – between public and private forms of ownership (Fig. 1).

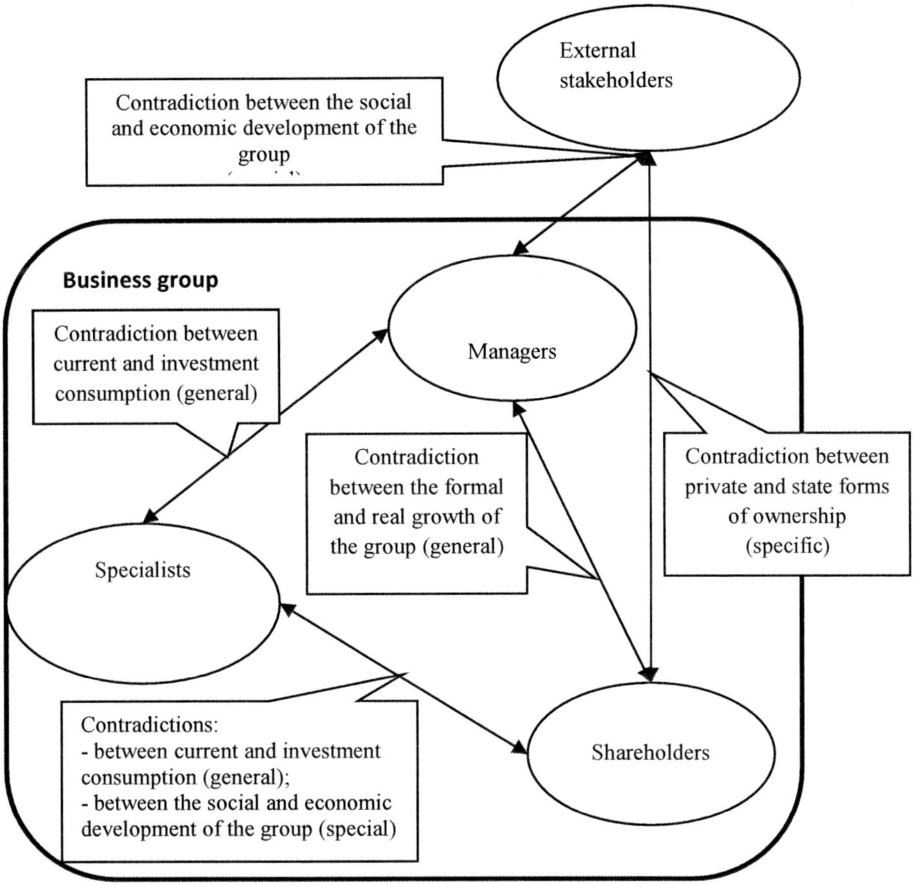

Fig. 1. Contradictions of a business group (Source: authors).

The Contradiction Between the Formal and Real Economic Growth of the Group. In the context of globalization, business groups are increasing investment in innovative development, but as a result of the faster pace of development and implementation of financial innovations, there is a relative independence of the movement of financial capital in relation to real capital. The growth of financial capital dominates. The development of this contradiction is facilitated by the tendency to dilute shares of large shareholders as a result of the business alliances formation. It is common for several groups to control individual key firms, which increases the likelihood of shareholders' conflicts and in some cases hinders the development of other firms in the industry chain. Another consequence is a reduction of the time horizon of large shareholders' interests and strengthening of the managers' positions against this background. Strengthening the power of management, first, exacerbates a contradiction between income growth and investments in the real capital of the business group, and, second, expands the conflict field between the management levels in the business group hierarchy. As a result, the dynamics and structure of the group's growth are mainly influenced by the balance of managers' power factor, while the factor of economic efficiency becomes secondary.

The Contradiction Between Current and Investment Consumption. The economic practice of business groups indicates an increase in investments in human capital, however, this trend is accompanied by a growing disparity in the distribution of income of the group's firms. The multinational structure of the group is designed in such a way that income is redistributed in favor of controlling shareholders and top managers of business units, while the lost profit of specialists is not compensated by investments in human capital, which reduces incentives to work.

The Contradiction Between Social and Economic Efficiency. As a result of the institutionalization of the concept of corporate social responsibility, some business groups (transnational corporations), and later a number of sub-holdings, began to adapt the practice of socially responsible behavior, which allowed resolving contradictions between their internal and external stakeholders. However, social responsibility is characterized by a dual nature – as a significant tool for coordinating interests, it requires investments which causes a decrease in the competitiveness level of firms. As a result, there is a dilemma between the social and economic performance of the business group.

The Contradiction Between Public and Private Forms of Ownership. In the period 2008–2019, positions of state and quasi-state structures as shareholders and creditors of business groups in both developing and established markets have strengthened. Moreover, transnational state ownership has expanded at an unprecedented rate. Investments of sovereign funds from developing countries in the corporate control market in the amount of more than $ 91.5 billion led to the redistribution of part of the capital of multinational banks, investment funds, exchanges and transnational corporations from private business groups in the Western economy to public investors in developing countries. At the same time, the expansion of the financial base due to the state participation is accompanied by a decrease in incentives to increase the economic efficiency of investments. The business group, as a mechanism for coordinating the economic activity, is a meso-level planning and administrative mechanism. At the macro level, planning is implemented through the state regulation, while the market permeates the entire economic system.

Internal mechanisms of the business group aimed at resolving its contradictions are implemented in the field of corporate governance. Planning and administrative coordination of companies' activities within the group allows for a long time to smooth out contradictions of economic interests, preventing their development. Cross-meeting in the director boards is widespread among the group's firms, so that the implementation of the personal union mechanism helps prevent conflicts of interests between firms at different levels of the hierarchy, but from the point of view of the economic relations development, this factor plays a rather negative role. The main mechanisms of self-regulation of the group in the field of conflict resolution should include ensuring the proportional participation of stakeholders in the income of firms, and in a broader sense – the group's adherence to principles of social responsibility. However, the corporate governance mechanisms that have developed in business groups are primarily aimed at protecting the contractual rights of management and the property rights of shareholders of firms located at the highest levels of the group hierarchy, while the economic interests of the management of ordinary firms, employees and external parties to conflicts are implemented on a residual basis. In this regard, the internal mechanism for resolving contradictions in some cases is ineffective.

Prerequisites for the transition of stakeholder relations to a new level are created when the state, institutions and mechanisms of global markets exert a disciplining influence on the group. In general, the role of the state is to create a framework for the constructive coordination of interests, institutionalization of incentives for the development of functional trends, and regulation of destructive trends in the development of business groups. The main instruments are corporate and antitrust laws and regulations that determine the movement of financial capital. The market ensures the participation of groups in the competition, however, given the actual status of business groups as a meso-level institution, the full implementation of the market mechanism in the process of resolving contradictions is possible only at the global level. In this case, the function of a "regulator" is performed by the labor markets and the global capital market, in particular the corporate control market. The incentive role belongs to institutions: exchanges, rating agencies, audit companies and institutional investors. The regulatory influence of the market mechanism in domestic markets is limited with the increasing market power of business groups, but the lack of competition in local markets is compensated by the openness of groups to forces of the global competition. According to the specifics of the Russia's economic development, we believe that the state's capacity as a regulator and its role in growing institutions should be taken into account. The optimal balance of planned and market mechanisms for coordinating interests varies depending on how the subject composition of the group's contradictions changes. The planned-market tools for resolving contradictions are differentiated for one or another axis of interaction between stakeholders.

Integrated Planning and Market Tools for Constructive Conflict Resolution of Business Groups. The analysis of contradictions of business groups allowed us to reveal some conditions for their constructive resolution, which are to ensure: proportional participation of subjects of contradictions in the income of the group's firms; medium-term balance of income and investments; implementation of principles of corporate social responsibility for the purpose of private redistribution of public goods. The creation of these conditions is based on the integrated impact of corporate governance instruments (dividend policy, incentive system for managers, staff pension programs, social investments), state macroeconomic regulation, self - regulation of global markets (stock indexes as an instrument) and institutional impact (ratings, audit control, etc.). Further, we present the implementation of these conditions in the framework of separate conflict areas of economic interests of Russian business groups.

Axis "Shareholders-Specialists". Coordination of interests in general is determined by the state regulation of labor relations. Corporate governance mechanisms are based on the staff participation in profits, management and ownership, as well as human capital development based on the group's investments in professional development, improving knowledge sharing mechanisms, social support programs for staff, and improving the group's corporate culture. The authors found that the development of functional trends is facilitated by the long-term ownership of fictitious capital in the form of staff pension if the shares of key firms are used for stimulation. However, this mechanism is negatively affected by the cyclical nature of financial markets.

Axis "Shareholders – Managers". When forming a group structure, a number of contradiction sources are excluded due to the division of ownership and management at the level of sub-holdings, but such a structure becomes effective when the incentive system for managers focuses on criteria of economic rather than financial efficiency, i.e. the productivity of factors and the structure of economic growth of the group are taken into account. A tool for coordinating interests is the inclusion of managers in the shareholders' list. Contradictions that have entered the conflict stage are resolved through the labor market. If a significant share of joint-stock ownership is concentrated by the management, a market for corporate control starts operating.

Axis "Shareholders – Shareholders". The resolution of contradictions depends on the effectiveness of the law in the field of dividend policy, but in case of a business group's collapse, the mechanisms of the corporate control market will be activated. The development of the real sector depends largely on the reallocation direction for resources of the group's firms. The regulatory role in this case is played, first, by the high level of legal protection tolls for investors' rights, which in the case of a group takeover of individual firms prevents the redistribution of cash flows of the absorbed firm in favor of the parent holding. Second, important is the state regulation of transfer pricing in order to bring intra-company and market prices into line.

Coordination of interests of shareholders, managers and external stakeholders of the business group is based on the integration of socially responsible behavior tools into one economic mechanism. The planned basis for this process is laid within the framework of the social budget and the corporate code of conduct of the business group, however, increasing the level of social responsibility of the group is facilitated by: legislative consolidation of environmental and social responsibility standards, tax incentives, public-private partnership in the field of socially significant projects. Russian business groups are adapting their social responsibility practices to meet demands of the global financial market to its participants. Stock indexes respond to information from social reports. That increases reputational risks in the context of globalization and provides stimulating pressure from institutional investors, audit companies and rating agencies.

The Main Directions for Improving Planning Tools and Coordinating Stakeholders' Interests in Russian Business Groups. Analysis of the dynamics of contradictions between Russian business groups in the process of transnationalization revealed that the ongoing constructive transformations of economic relations in Russian business groups as a whole are determined by the impact of global markets. The main factor hindering the development of the real sector is the low potential of planning instruments for coordinating economic interests. Several key issues are highlighted. First, there are no trends in the formation of an employee-owner in Russian business groups. Second, there is an inefficient mechanism for transferring knowledge. Third, it is the orientation of the incentive system for managers to increase in the fictitious capital. Fourth, there is inefficient coordination of shareholders' interests within the framework of transnational alliances of business groups. Fifth, there is a low level of activity in implementing corporate social responsibility practices.

Axis "Shareholders – Specialists". It is necessary to increase the volume of social investments; encourage employees of the group's firms, and not only top managers of sub-holdings, to acquire equity shares of key transnational companies; and improve the efficiency of knowledge transfer mechanism based on the staff rotation between Russian and foreign firms of the group.

Axis "Shareholders – Managers". The growth of the group's real capital should be encouraged by introducing additional criteria for evaluating the performance of managers. These criteria should be based on integrated indicators of the economic efficiency that take into account changes in the labor productivity and the structure of investment in long-term assets and innovations. It is necessary to extend the time interval of remuneration programs up to 5 years.

Axis "Shareholders – Shareholders". The economic development of the real sector requires the expansion of transnational strategic alliances between business groups. Preliminary cooperation of groups in the framework of "soft" forms of integration will help to develop a mechanism for reconciling conflict interests, the implementation of which will stimulate the mutual development of partners in the conditions of the subsequent transformation of alliances into unions of groups.

Axis "Shareholders – Managers – External Stakeholders". Increasing the level of the information disclosure concerning the content of non-commercial activities of business group firms is required not only at the level of transnational corporations, but also at the level of firms serving them, as well as business groups in general. It is necessary to introduce the practice of making social and human capital investments by the head holdings of Russian business groups.

4 Discussion

Further research on the constructive resolution of contradictions in the development of business groups seems to be productive from the perspective of the concepts of integration of neoclassical and neoinstitutional schools, and the theory of financial capital. Scientific analysis of the dynamics of contradictions in the process of business group diversification is of particular interest [13]. Additional results will be obtained by introducing the risk management optimization factor into the analysis of economic interests of business group stakeholders when entering the corporate control market [14]. A number of researchers consider the transformation of the ownership structure in business groups, namely, the change in the ratio of private and state forms of ownership

[1, 10]. There is a tendency to strengthen the position of the state ownership by investing national wealth funds in fictitious capital of private corporations and banks. Of interest is the analysis of this trend in the context of general structural transformations of business groups from the Eurasian perspective [15]. The main trend in the development of the real sector in the era of globalization is the expansion of global value chains. Business groups are the main participants and coordinators of this process [9]. In recent years, one of the main goals pursued by Russian business groups through cross-border mergers and acquisitions is to control value chains based on the Russian natural resource potential. Thus, the issue on constructive resolution of contradictions of business groups in the conditions of forming new factors for the technological development of the real economy sector requires further research [4, 7].

5 Conclusion

The analysis expands scientific understanding of contradictions that determine the development of the real sector of the economy within business groups, as well as mechanisms for resolving these contradictions and development trends in the context of globalization. The article investigates tools for resolving conflicts of economic interests of business group stakeholders, which will allow this institute to ensure the development of the real sector of the Russian economy in the context of globalization. The formulated provisions, conclusions and recommendations can be used in practical activities of state bodies and corporate structures aimed at developing and improving tools for coordinating the economic interests of group stakeholders and creating economic incentives for the development of the real sector of the Russian economy.

References

1. Antipina, Z.P., Gordeeva, E.S.: Business groups, innovation and structural transformations in developing economies. In: Mantulenko, V. (ed.) Proceedings of the International Scientific Conference "Global Challenges and Prospects of the Modern Economic Development". The European Proceedings of Social & Behavioural Sciences, vol. 57, pp, 102–112. Future Academy, London (2019)
2. Bessonova, E., Gonchar, K.: Incentives to innovate in response to competition: the role of agency costs. Econ. Syst. **41**(1), 26–40 (2017)
3. Chepurenko, A.: Innovation entrepreneurship in transition economies: problems and outlook. Foresight STI Gov. **11**(3), 6–9 (2017)
4. Chetverikova, A.S.: Cross-border value chains in ferrous metallurgy: Russian participation. World Econ. Int. Relat. **62**(8), 97–103 (2018)
5. Golichenko, O.: Public policy and national innovation system failures. Econ. Issue **2**, 97–108 (2017)
6. Gupta, G., Mahakud, J.: Alternative measure of financial development and investment-cash flow sensitivity: evidence from an emerging economy. Financ. Innov. **5**, 1 (2019)
7. Kondratev, V.B., Popov, V.V., Kedrova, G.V.: Transformation of global value chains: experience of three industries. World Econ. Int. Relat. **64**(3), 68–79 (2020)

8. Lee, C.Y., Lee, J.H., Gaur, A.S.: Are large business groups conducive to industry innovation? The moderating role of technological appropriability. Asia Pac. J. Manag. **34**, 313–337 (2017)
9. McKinsey: Globalization in transition: the future of trade and value chains (2019). https://www.mckinsey.com/featured-insights/innovation-and-growth/globalization-in-transition-the-future-of-tradeand-value-chains. Accessed 15 Oct 2019
10. Morck, R., Nakamura, M.: Japan's ultimately unaccursed natural resources financed industrialization. J. Jpn. Int. Econ. **47**, 32–54 (2018)
11. Oh, W.-Y., Chang, Y.K., Lee, G., Seo, J.: Intragroup transactions, corporate governance, and corporate philanthropy in Korean business groups. J. Bus. Ethics **153**, 1031–1049 (2018)
12. Purkayastha, S., Kumar, V., Lu, J.W.: Business group heterogeneity and the internationalization-performance relationship: evidence from Indian business groups. Asia Pac. J. Manag. **34**, 247–279 (2017)
13. Singh, D., Pattnaik, C., Gaur, A.S., Ketencioglu, E.: Corporate expansion during pro-market reforms in emerging markets: the contingent value of group affiliation and diversification. J. Bus. Res. **82**, 220–229 (2018)
14. Yang, K., Schwarz, G.M., Multilevel, A.: A multilevel analysis of the performance implications of excess control in business groups. Organ. Sci. **27**, 1219–1236 (2016)
15. Zhang, L., Sjogren, H., Kishida, M.: The emergence and organizational persistence of business groups in China, Japan, and Sweden. Ind. Corp. Change **25**, 885–902 (2016)

Check for updates

Government Contract as State's Economic Support Instrument: Legal Regulation, Enforcement and Dispute

T. V. Chugurova(✉)

Samara State University of Economics, Samara, Russia
chugurovaTV@yandex.ru

Abstract. The article is devoted to the problems of the state contract as a tool of support to the real sector of the economy in the context of the analysis of certain issues of legal regulation, conclusion, execution and dispute of such contracts. The author, using the example of the Federal Law "On the Contract System in the Sphere of Procurement of Goods, Work, and Services for the Provision of State and Municipal Needs" (hereinafter referred to as the Law on the Contract System), analyzes the law enforcement practice on certain issues of concluding, executing, monitoring the execution of a state contract, its legal nature as a tool of support to the real sector of the economy. The research uses both general scientific research methods, and a formal logical, systematic method of cognition, a method of analyzing judicial practice. As a result of the work, the author singles out the characteristic features of the state contract as a tool of support to the real sector of the economy.

Keywords: Government contract · Public interest · State support · State customer

1 Introduction

Interest in this research topic shows in various scientific circles of the world community [12]. An analysis of foreign experience can be used to develop new approaches to organizing public procurement in Russia, presenting the issue of researching this experience as relevant [3]. An analysis of world experience shows that government contracts are an important tool for developing the market and supporting economic and social stability, as well as for ensuring growth in the real sector of the economy. State contract is one of the contract types to which the general provisions of the Civil Code of the Russian Federation [2] apply, as the question of its legal regulation, conclusion, dispute, invalidation is rather important. The conclusion of a state contract is a complex legal structure, which consists of procurement planning, documentation preparation, including of a draft contract, placing an order in a single information system, identifying the winner that will subsequently be the winner of the contract. All the conditions created by law are aimed at state support of the real sector of the economy, they are enforced to prevent the creation of competitive advantages in public procurement for some participants over others, and violation of competition rules.

S. I. Ashmarina et al. (Eds.): ENGINEERING ECONOMICS WEEK 2020, LNNS 139, pp. 262–268, 2021.
https://doi.org/10.1007/978-3-030-53277-2_30

2 Methodology

The methodological basis of the work is a combination of such general scientific research methods as the analysis of the phenomena studied and the synthesis of the research results, as well as induction and deduction. In the process of developing various aspects of the research topic, formal-logical, systemic methods of cognition were used. To define the concepts of public interest and explicit prohibition, the comparative legal method was used when comparing with related categories, the method of analysis of judicial practice.

3 Results

The subject of the study is the process of improving state contract as a tool of support to the real sector of the economy, and certain issues of the conclusion, execution, control of the state contract.

The law on the contract system regulates relations aimed at realizing state needs in order to support the real sector of the economy, increase the efficiency, effectiveness of procurement of goods, work, services, ensure transparency of state purchases, and prevent corruption and other abuses in the field of such purchases.

When making purchases, customers use competitive methods to identify suppliers (contractors) or purchase from a single supplier (contractor). Competitive methods of determining suppliers (contractors) are tenders (open tender, tender with limited participation, two-stage tender, closed tender, closed tender with limited participation, closed two-stage tender), auctions (electronic auction, closed auction), request of quotations, request of proposals. Paragraph 5 of Article 24 of the Law on the Contract System states that the customer selects the method for determining the supplier (contractor) in accordance with the provisions of Chapter 3 of this law. Moreover, the customer is not entitled to perform actions that entail an unreasonable reduction in the number of participants in the procurement. Purchasing from a single supplier does not apply to competitive procurement methods, the use of such a procurement method can only be carried out in cases established by Article 93 of the Federal Law № 44-FZ "On the Contract System in the Sphere of Procurement of Goods, Works, and Services to Ensure State and Municipal Needs" [5].

At the same time, the content of the specified legal provision provides for the customer the opportunity to conclude low cost procurements in cases where competitive selection procedures are impractical due to the inconsistency of organizational costs for the procurement of the purchase price itself. Based on law enforcement practice, the artificial "splitting" of a single procurement into many procurements in order to avoid public procedures does not meet the goals of introducing such an opportunity to conclude a contract without conducting competitive procedures. In accordance with paragraph 9 of part 1 of Article 93 of the Law on the contract system, the customer has the right to make purchases from a single supplier due to an accident, other emergency situations of natural or man-made nature, force majeure, in case of need for emergency medical assistance or medical assistance in urgent form, when the use of other methods for determining the supplier (contractor) requires much time and thus is impractical. Part 2 of article 93 of the Law on the contract system states that in such a case placement of a notice on the purchase by the customer in a unified information system no later than five days before the date of conclusion of the contract is not required [5]. However, the customer is obliged to notify the procurement control body no less than one business day from the date of conclusion of such contract and justify the impossibility or inappropriateness of using other ways to determine the supplier (contractor), as well as the price of the contract and other essential terms of the contract.

In part 2 of Article 8 of the Federal Law № 44-FZ "On the Contract System in the Sphere of Procurement of Goods, Work, and Services to Ensure State and Municipal Needs", the prohibition set for customers, specialized organizations, their officials, procurement commissions, members of such commissions, and procurement participants to perform any actions that contradict the requirements of the Contract system law, including limiting competition, in particular, unreasonably limiting the number of procurement participants [5]. Failure to comply with the procurement procedure violates the rights of third parties - procurement participants with whom no contracts have been concluded as the advantage is given to a party that does not meet the requirements of the Law on the contract system. Based on the legal position of judges of the arbitration courts of the Russian Federation, a systematic interpretation of the provisions of the Law on the contract system indicates that state and municipal contracts pursue public interest and are aimed at meeting public needs through the use of budgetary funds. The law protects the public interests and the interests of other persons who are entitled to participate in the conclusion of state contracts on the basis of competitive procedures. In addition, in the preparation and execution of budgets, participants in the budget process, within the framework of the budgetary powers established by them, must proceed from the need to achieve specified results using the smallest amount of funds (economy). In accordance with paragraph 2 of Art. 168 of the Civil Code of the Russian Federation [1], a transaction that violates the requirements of the law or other legal act and, at the same time, infringes on public interests or the rights and interests of third parties protected by law is void if the law does not imply that such a transaction is disputable or other consequences of violation are not related to the invalidity of the transaction. The nullity of the transaction connects the literal meaning of the rule of law not with any violation of the law, but only with that which encroaches on the public interests or rights of the interests of third parties protected by law. According to paragraph 75 of the Resolution of the Plenum of the Supreme Court of the Russian Federation № 25 of June 23, 2015 "On the application of certain provisions of Sect. 1 of the Part One of the Civil Code of the Russian Federation by courts" [10], in relation to Articles 166 and 168 of the Civil Code of the Russian Federation, under the public interest, in particular, the interests of the uncertain circle of persons, ensuring the safety of life and health of citizens, as well as the defense and security of the state, and the protection of the environment. A transaction that violates explicit prohibition established by law is void as it harms public interests, for example, a pledge transaction or assignment of claims inextricably linked with the creditor's identity (clause 1 of article 336, article 383 of the Civil Code), insurance of illegal interests (Article 928 of the Civil Code of the Russian Federation) [2].

According to paragraph 18 of the Review of the Presidium of the Supreme Court of Russia of June 28, 2017, "Review of the judicial practice of applying the legislation of the Russian Federation on the contract system in the field of procurement of goods, works, services to meet state and municipal needs" [11], violation of prohibition (concluding transactions for an amount exceeding 100 thousand rubles with a single supplier and concluding an agreement with a party that does not comply with the requirements of the law) indicates the nullity of the transaction.

Currently, article 93 of the Federal Law № 44-FZ "On the Contract System in the Sphere of Procurement of Goods, Work, and Services to Ensure State and Municipal Needs" regulates cases of procurement from a single supplier (contractor), according to which procurement from a single contractor can be carried out by the customer in the event of procurement of goods, work or services for an amount not exceeding 300 thousand rubles (clause 4 of part 1 of this article). In our opinion, the explicit prohibition is set by paragraph 4 of part 1 of Article 93 of the Federal Law № 44-FZ "On the Contract System in the Sphere of Procurement of Goods, Work, and Services to Ensure State and Municipal Needs": the annual volume of purchases that the customer is entitled to carry out on the basis of this paragraph should not exceed two million rubles or should not exceed five percent of the total annual volume of purchases customer and should not be more than fifty million rubles [5].

We believe that the part 4 of paragraph 1 of Article 93 of Federal Law № 44-FZ "On the Contract System in the Sphere of Procurement of Goods, Work, and Services to Ensure State and Municipal Needs" does not establish any other explicit prohibition on the conclusion of a transaction not exceeding 300 thousand rubles [5]. The specified position is confirmed by the executive authority, the competence of which includes state regulation in the field of the contract system – the Ministry of Economic Development of the Russian Federation in the explanations of the Ministry of Finance of the Russian Federation. It seems that the interpretation of the "small volume" procurement as a single lot is an assessment of the activities of the parties to the transaction by bodies and persons who oversee and monitor compliance with the legislation on the contract system. The concept of "artificial split of procurement" is absent in the legislation, it is vague, it is formulated by law enforcement bodies independently for each case under consideration, and therefore cannot be perceived as an explicit prohibition.

4 Discussion

In Russia, public procurement operates a shorter period compared to other countries, so it is relevant to search for effective ways to address the participation of suppliers of goods, works, services in public procurement [13]. Public procurement and the subsequent conclusion of state contracts fulfill several important functions in regulating the economy as reproduction, stimulating, and the function of price regulation, as well as strategic and innovative functions [9]. So, in Sweden, government procurement of organic food has led to a significant effect in agricultural policy [6]. In China, the effectiveness of public procurement is determined not only by price, but also by the time and labor aspects of efficiency [15]. In Russia, every year there is an increase in the number of government contracts, but problems with their use continue to be important [7]. One of the most pressing problems associated with the use of state contracts in different countries [8], including Russia, is the assessment of the actions of customers selecting public procurement participants for the interests of society and the state [14]. We believe that government contracts cannot be considered as those violating public interests if their conclusion does not violate the principles of openness, transparency, competition on them is unlimited, and if work was carried out in accordance with a publicly accepted, openly posted regulatory act, all information about the contracts was openly placed in the public procurement information system.

Also, government contracts cannot be assessed as violating public interests, if the conclusion of the contracts in question did not lead to restriction of competition, that is, to the emergence of facts and circumstances specified in the Federal Law № 135-FZ "On Protection of Competition" [4]: a reduction in the number of business entities that are not included in one a group of persons in the product market, an increase or decrease in the price of a product that is not related to relevant changes in other general conditions for the circulation of goods in the product market, refusal of business entities, not included into one group of persons, from independent actions on the commodity market, determination of the general conditions for the circulation of goods on the commodity market by agreement between business entities or in accordance with binding instructions of another person or as a result of coordination by business entities not included in the same group of persons, their actions on the commodity market, other circumstances that create the possibility for an economic entity or several economic entities to unilaterally influence the general conditions Commodity circulation in the commodity market, and the establishment of public authorities, local authorities, organizations involved in the provision of public and municipal services, with the participation in the provision of such services requirements for goods or business entities not covered by the legislation of the Russian Federation. The conclusion on absence of signs of restriction of competition may follow if the antimonopoly body does not find signs of violation of the law on protection of competition when considering the circumstances associated with the conclusion of state contract in the process of conducting inspection. Also, a similar conclusion can follow if the body authorized to monitor compliance with the requirements of the legislation on the contract system does not reveal the fact of violations in determining the price of the contract when considering an administrative case and there is no proof of the presence of signs of violation of the principle of efficient spending of budget funds in the actions of the customer, and therefore there is no reason to conclude there was violation of public interest.

5 Conclusion

The state contract as an instrument of support to the real sector of the economy has a special place in the system of support tools, it is an independent type of civil contract with special features: the presence of special entities; the presence of the special nature of legal relations arising in the process of its conclusion and execution; the special purpose of the conclusion is to satisfy state needs; special conditions for concluding and executing a state contract. By increasing the requirements of state customers for suppliers' products, work, services of public procurement participants, improving the legal regulation of state contracts, the state contributes to the formation of demand for modern products. State contracts form the economic and legal space in which there are opportunities for the development of all economic entities, including the real sector of the economy.

References

1. Civil Code of the Russian Federation (Part One) 51-FZ (1994). http://www.consultant.ru/document/cons_doc_LAW_5142/. Accessed 29 Apr 2020
2. Civil Code of the Russian Federation (part two) №. 14-FZ (1996). http://www.consultant.ru/document/cons_doc_LAW_9027/. Accessed 29 Apr 2020
3. Esqueda, O., Ngo, T., Susnjara, J.: The effect of government contracts on corporate valuation. J. Bank. Financ. **106**, 305–322 (2019)
4. Federal Law № 135-FZ "On Protection of Competition" (2006). http://www.consultant.ru/document/cons_doc_LAW_61763/. Accessed 29 Apr 2020
5. Federal Law № 44-FZ "On the Contract System in the Sphere of Procurement of Goods, Works, and Services to Ensure State and Municipal Needs" (2013). http://www.consultant.ru/document/cons_doc_LAW_144624/. Accessed 29 Apr 2020
6. Lindström, H., Lundberg, S., Marklund, P.: How green public procurement can drive conversion of farmland: an empirical analysis of an organic food policy. Ecol. Econ. **172**, 106622 (2020)
7. Melnikov, V., Lukashenko, O.: Developing the public procurement system in the post-soviet and modern Russia: challenges and opportunities. Terra Economicus **15**(4), 109–126 (2017). https://doi.org/10.23683/2073-6606-2017-15-4-109-126
8. Nesterova, N., Hans, Q., Streng, J., van Dijk, L.: Public procurement as a strategic instrument to meet sustainable policy goals: the experience of Rotterdam. Transp. Res. Proc. **46**, 285–292 (2020). https://doi.org/10.1016/j.trpro.2020.03.192
9. Popova, O.: Public procurement as an effective tool for regulating the development of the regional economy. Econ. Enterp. Reg. Ind. **6**(90), 61–63 (2014)
10. Resolution of the Plenum of the Supreme Court of the Russian Federation № 25 "On the application of certain provisions of Section I of Part One of the Civil Code of the Russian Federation by courts" (2015). http://www.consultant.ru/document/cons_doc_LAW_181602/2015/. Accessed 29 Apr 2020
11. Review of the judicial practice of applying the legislation of the Russian Federation on the contract system in the field of procurement of goods, works, services to meet state and municipal needs "(approved by the Presidium of the Supreme Court of the Russian Federation) (2017). http://www.consultant.ru/document/cons_doc_LAW_218867/. Accessed 29 Apr 2020
12. Swensson, L., Tartanac, F.: Public food procurement for sustainable diets and food systems: the role of the regulatory framework. Glob. Food Secur. **25**, 100366 (2020)
13. Tedeeva, Z.: Analysis of public procurement mechanism when implementing financial policy of state. In: Bataev, D.K.-S., (ed.) SCTCGM 2018 - Social and Cultural Transformations in the Context of Modern Globalism. The European Proceedings of Social and Behavioural Sciences, vol. 58, pp. 1166–1174. Future Academy, London (2019)

14. Voronin, B., Chupina, I., Pustuev, A., Semin, A., Potekhin, N., Lysenko, Yu., Rubaeva, O., Bukhtiyarova, T.: State procurement mechanism for agricultural products in modern Russia. Int. J. Adv. Biotechnol. Res. **8**(4), 1363–1369 (2017)
15. Wang, Q., Zhang, R., Liu, J.: Price/time/intellectual efficiency of procurement: uncovering the related factors in Chinese public authorities. J. Purch. Supply Manage. **26**, 100622 (2020)

Subsidiary Liability of Controlling Persons of Industrial Enterprises: Main Legal Conflicts

Y. E. Monastyrsky(✉)

Moscow State Institute of International Relations, MGIMO University, Moscow, Russia
monastyrsky@mzs.ru

Abstract. This work describes specific features of applying the institute of subsidiary liability of controlling persons of the enterprise. Creditors who have claims of non-fulfillment of obligations have been given the opportunity to pursue unlimited civil law actions against persons that are not directly related to them, which creates a certain imbalance and hinders the legitimate interests of other participants in the legal relations under consideration, in particular, the interests of managers. This research is aimed at studying the state and practice of applying the institute of subsidiary liability of controlling persons at enterprises and the doctrine of "removing the corporate veil" in Russia using methods of system and comparative analysis. The conducted research allowed us to establish a significant discrepancy between the Russian and international practices of applying the doctrine of "removing the corporate veil", and the presence of a clear imbalance in favor of creditors. As a result, it is concluded that there is a need for detailed regulation in the relevant legislation of terms and procedures for applying subsidiary liability of controlling persons at enterprises. The methodology of the research comprises historical, comparative, formal-legal and functional methods, systemic approach.

Keywords: Bankruptcy · Controlling persons · Doctrine of "removing the corporate veil" · Liability of controlling persons · Subsidiary liability

1 Introduction

One of the most effective methods of protecting rights of participants in economic relations is to create an effective consistent legal framework that excludes the possibility of double interpretation or abuse of definitions used in the law. Subsidiary liability is a key institution that ensures a balance between rights and obligations of creditors and a debtor. Recently, there has been quite a lively debate about the responsibility of controlling persons of the debtor, which include the parent companies, directorate, top management, members of the management board, majority shareholders, as well as group members who actually manage the company [7–9, 14, 15].

In the foreign scientific literature and practice, the doctrine of "removing the corporate veil" is widely used, which assumes the existence of "alterego" of the enterprise in the face of shareholders or other controlling persons at the enterprise as interested

S. I. Ashmarina et al. (Eds.): ENGINEERING ECONOMICS WEEK 2020, LNNS 139, pp. 269–274, 2021.
https://doi.org/10.1007/978-3-030-53277-2_31

ones in the implementation and satisfaction of their personal self-interest through the enterprise activities and avoiding responsibility [2, 17].

The doctrine of "removing the corporate veil" was practically inapplicable to the corporate legal landscape of Russia in the 1990s and 2000s. Representatives of the business community learned about the owners of entire sectors of the economy through the press to the extent that they themselves desired it. Tens of thousands of offshore companies formally owned businesses in the Russian Federation. Many of them had quite an "overseas facade", were headed by Western managers, but de facto and de jure control belonged through trusts to domestic businessmen. The largest jurisdiction from which investments were directed to the Russian economy was small Cyprus with a population of 790 thousand people. The legal directors signed all the title documents. The most popular legal product of foreign consultants was the support of transactions, business purchases, structured under English law and subject to the jurisdiction of a foreign arbitration tribunal. Such constantly cultivated constructions provided two fundamental things: invulnerability of investors, which implied the confidentiality of asset ownership; the best opportunities for raising funds through banks, issuing shares, bonds, and large-scale long-term development.

The designated system was an adaptation to the hegemony of large international banking groups (from the United Kingdom and the United States) with unlimited financial resources, which is the personification of the movement of world money. Credit institutions participated in major projects and determined the contractual structures and names of the consultants involved. Accordingly, the market for legal services was built with a dominant presence of Anglo-Saxon firms in all countries – recipient of investment. It was not realistic to bring controlling persons to justice, in the case of bankruptcy because of ruinous operations that reduce the real value of the participation interest. Rules on the liability of actual managers and owners of legal entities existed in the 1990s, but in Russian legal practice, this institution, despite its existence, was applied only at the end of 2000. For many years, these institutions did not work, and the requirements were not met, although they were repeatedly reviewed. The reason for this is the passivity of courts requiring proof of causation, actual influence, control, etc.

However, the active use of this institution has created a number of new legal problems. Despite repeated assertions in the Russian legal literature that subsidiary liability should be exceptional and extraordinary, it has become a major segment of dispute resolution practice [10].

This legal trend seems to lead gradually or immediately to rethinking the institution of a legal entity as independently responsible for all its assets, as a separate unit in the sense that it does not shift the property burden on the founders and other persons, including its managers. But today, the existing institution of subsidiary liability of controlling persons of enterprises cannot be considered as satisfactory and providing a balance of interests of all participants, because of the imperfection and outright contradiction of legislative acts regulating the bankruptcy procedure.

2 Methodology

The research methodology includes historical, comparative, formal-legal and functional methods, and a systematic approach. When processing the actual material, traditional scientific methods such as dialectical, logical, scientific generalization, content analysis, comparative analysis, synthesis, source studies, etc. were used. The use of these methods allowed us to ensure the validity of the research results, theoretical and practical conclusions, and developed proposals.

3 Results

The origins of problems with subsidiary liability were legal barriers laid down in Paragraph 2 of Article 56 of the Civil Code of the Russian Federation, which presupposes the exclusion of mutual responsibility of the founder (participant) of a legal entity or the owner of its property and the legal entity itself for obligations [3].

This important principle allows you to take risks without incurring liability if financial investments were unsuccessful so much that, having assumed obligations, the created entity could not pay or repay them at the expense of its profits and current earnings. This is captured in the Roman formula *si quid universitati debetur singulis non debetur, nec quod debet universitas singuli debent* (what belongs to the Corporation doesn't belong to its individual members, or: what the Corporation owes, its individual members do not owe).

This canon has been based since the 16th century on the desire of the business community, which has always had levers of influence over the government and can create organizations characterized as "associations of capital", put into circulation surplus funds while being in formal financial security [1, 6].

Reputation, trust in the organization, and its "credit" were measured to a greater extent by two factors: ownership, as well as the prospects of the business idea for which it was created; and to a lesser extent by people behind it, since it is clear that if the project fails, its creators will not be responsible [16].

Paragraph 3 of Article 15 of the civil legislation of the USSR and the republics provided that in case of bankruptcy (insolvency) of a legal entity as a result of unlawful actions of the owner, insufficient funds and its property to meet the creditors' claims, the owner bears responsibility for obligations of the legal entity.

Nowadays, similar provisions are contained in Article 53.1 of the Civil Code of the Russian Federation, Paragraph 1 of which establishes the obligation to compensate for losses caused to an organization by requiring (1) a person authorized to act on its behalf, (2) members of collegial management bodies, (3) a person that had the actual ability to determine actions of the legal entity, as well as to give instructions [3].

It is important to note that now the responsibility of controlling persons comes for bad faith and unreasonableness (indīligentia et insapientia), while most authors firmly believe that these important legal categories replace guilt [11, 12]. For example, a real manager, having issued loans to non-operating organizations on the eve of bankruptcy proceedings, goes abroad in order to obtain permanent residence there. As a result, middle managers are also subject to criminal prosecution: members of the credit committee, accountants, lawyers who make contracts and are not at the top of the hierarchy of labor relations, receive salaries, and so on. The inevitable plot in such a situation is to open a bankruptcy case and direct a group of individuals to full repayment of all creditors' claims. There is a curious phenomenon of group responsibility. The involved circle of allegedly involved citizens begins to testify or give evidence against each other in order to protect themselves from strict liability. The prescribed procedure for establishing a real imbalance between the assets forming the bankruptcy estate and the liabilities provides an incentive for a throwaway sale of assets, the implementation of which is realistic and not time-consuming. Therefore, in Russia, debts to creditors are never repaid in full, or only in a percentage.

4 Discussion

It is argued that failure to satisfy the claims of the company's creditors because of harm to their rights should result in tort liability in accordance with Chapter 59 of the Civil Code of the Russian Federation, as well as for violation of corporate rights [3, 4]. There is, however, another point of view that considers subsidiary liability as non-contractual, but not tort [13]. The defendants compensating for losses, in this case – causing harm and damage to property, in the true sense of the word, did not commit it. So far, the prevailing view is of direct tort liability in indirect claims and special subsidiary liability when controlling persons have caused the organization to lose its full functioning.

We assert that these legal institutions are an expression of the consistent implementation of the principle of full compensation and restoration of subjective rights, and in terms of the imposition of damages, legal claims are doctrinally unified.

Creditors in so-called subsidiary liability and participants in indirect claims lose not so much property as the value of a subjective right. Controlling persons become obligated to restore it only if they are involved and behave culpably. This situation should be regulated by Article 15 of the Civil Code, but in any case not by Chapter 59 of the same document on "direct" tort liability [3]. The tort liability construction is not appropriate for several reasons. Classic harm is a legal action that creates an obligation to replace and correct things that are experiencing physical damage. If it is not present, the need to perform certain actions does not arise, because in the first place, it is unclear what to do. A creditor in bankruptcy has claims that do not change because of the insolvency of an insolvent legal entity. It is not the property that suffers, or so much the property in the form of things, etc., it is a subjective civil right, which is based on expectations of creditors that some actions will occur or will be performed, the claim is repaid, etc. Controlling person, and not the bankrupt, through the outward manifestation of guilt becomes a violator of the civil right that can give effect to Article 15 of the Civil Code and not the Chapter on torts, which should not be applied because the nature of the harm is not so [3].

5 Conclusions

In general, it should be noted that the doctrine of "removing the corporate veil" in Russia, because of the imperfection of the legal technology and the lack of established law enforcement practice, not only has not fully formed, but has transformed into something completely different from the basic idea underlying the Western doctrine. The doctrine, which has an extraordinary character, has become a tool for "influencing" the controlling persons of enterprises.

The Civil Code of the Russian Federation has secured mechanisms for insurance (in the broad sense of this word) of creditors, allowing them to request awards not only from debtors, offenders, but also from other persons, primarily accomplices, obligated to debts, liable by law, named alternative, subsidiary, joint addressees of claims, etc. [3]. These are tools that are increasingly introduced to strengthen the stability of turnover, the fair distribution of the burden of property and losses of persons who relied on law-abiding behavior, the imposition of the risk of loss on other subjects, the preservation of property and material status. The fundamental rule that the causer and no one else is responsible for causing damage has undergone a significant transformation. Since the 19th century, it has been supplemented by a provision on compensation for persons specifically named in the law: representatives, guardians, employers, etc.

But these cases do not end with the imposition of responsibility on an innocent person. Needs called to life contractual structures of insurance, the role of the employer and others, in which, regardless of the fame of the main debtor, who may be a secret, not found criminal, losses are compensated. For this reason, tort liability has a too narrow scope to cover and give a legal foundation to modern legal configurations of liability, for example, in bankruptcy, etc., since in the mass of episodes, the person who caused the damage should not be held as responsible one.

Just the identification of the guilty activities of controlling persons with the offence creates a destructive basis for collective responsibility, making it impossible to file a claim to each other because the extent of their unlawful roles is always different, and requirements of the controlling person, showed slight negligence to the deliberate offender, are possible only with full compensation for the harm, but not earlier (Article 1081 of the Civil Code).

According to Article 15 of the Civil Code of the Russian Federation, some individuals may have losses in the form of "expenses" that they "will have to bear". Article 15 of the Civil Code allows you to recover damages from the guilty party outside of contractual and tort relationships [3]. Unfortunately, the key laws already have adopted concepts that prevent the approval of this simple and understandable logic. For example, the following definition is used: "the transaction caused harm", etc. The result of this terminology is a provision that can make recovery of losses in the field of bankruptcy and in the corporate area unworkable.

The main purpose of the insolvency procedure is to create a bankruptcy estate by returning assets and challenging transactions. In practice, there is legal uncertainty and inconsistency in the actions of courts, as a result, legal activity in the field of insolvency has become the most unpredictable according to many practicing lawyers. The reality is that money transfers are disputed as transactions for the purpose of restitution, and payments are awarded to non-participating individuals who may be part of "groups of controlling persons" and therefore jointly liable.

In our opinion, a detailed study and unified judicial enforcement practice of applying the mechanism of "removing the corporate veil" are necessary, following the example of foreign jurisdictions [5, 18], which will ultimately allow us to maintain a balance between applying new rules and preserving, without distorting the essence and form, the principle of limited liability of participants and shareholders for the obligations of legal entities.

References

1. Asai, K.: Limited liability and corporate efficiency. Int. Rev. Law Econ. **62**, 105886 (2020)
2. Boubaker, S., Chebbi, K., Grira, J.: Top management insider debt and corporate social responsibility? Evidence from the US. The Quarterly Review of Economics and Finance (2019). https://doi.org/10.1016/j.qref.2019.12.001. Accessed 12 Apr 2020. in Press
3. Civil Code of the Russian Federation dated 30.11.1994 N 51-FZ. http://www.consultant.ru/document/cons_doc_LAW_5142/. Accessed 20 Apr 2020
4. Egorov, A.V., Usacheva, K.A.: Subsidiary liability for bringing to bankruptcy – an unsuccessful equivalent of the Western doctrine of removing corporate cover. Vestnik VAS RF **12**, 20–65 (2013)
5. Erhemjamts, O., Huang, K.: Institutional ownership horizon, corporate social responsibility and shareholder value. J. Bus. Res. **105**, 61–79 (2019)
6. García-Granero, A., Fernández-Mesa, A., Jansen, J.J.P., Vega-Jurado, J.: Top management team diversity and ambidexterity: the contingent role of shared responsibility and CEO cognitive trust. Long Range Plan. **51**(6), 881–893 (2017)
7. Kamyshansky, V.P., Tarasenko, A.N.: Some problems of the institute of subsidiary liability of persons controlling the debtor. Civ. Law **1**, 40–42 (2016)
8. Littlewood, D.: Corporate responsibilities. Int. Encycl. Hum. Geogr. **2**, 403–410 (2020)
9. Lomakin, D., Gintovt, O.: Responsibility of controlling persons: the legal nature and mechanism of attraction to it. Econ. Law **1**, 12–39 (2016)
10. Miftakhutdinov, R.: The plenary clearly indicated that subsidiary liability applies in exceptional cases: interview with. Law **7**, 12 (2018)
11. Novak, D.V.: Integrity in corporate law. Bull. Civ. Law **2**, 13–23 (2017)
12. Pletnev, D., Barkhatov, V.: Business success of small and medium sized enterprises in Russia and social responsibility of managers. Proc. Soc. Behav. Sci. **221**, 185–193 (2016)
13. Shitkina, I.S.: Responsibility in corporate legal relations. Econ. Law **6**, 51–117 (2015)
14. Slonevskaya, A.Yu.: Institute of subsidiary liability of persons controlling a bankrupt debtor. Bull. Econ. Justice **1**, 103–115 (2017)
15. Suvorov, E.D.: Subsidiary liability for the debtor's obligations in the bankruptcy process: law enforcement issues. Law **12**, 148–155 (2013)
16. Tarasov, I.T.: The Doctrine of Joint-Stock Companies. Statut, Moscow (2000)
17. Wormser, I.M.: Piercing the veil of corporate entity. Columbia Law Rev. **12**(6), 496–518 (1912)
18. Zhao, T., Xiao, X.: The impact of corporate social responsibility on financial constraints: does the life cycle stage of a firm matter? Int. Rev. Econ. Financ. **63**, 76–93 (2018)

State Support of the Real Sector of the Economy: Legal Regulation

A. V. Fadeev(✉)

Samara State University of Economics, Samara, Russia
Alexfad1960@mail.ru

Abstract. The problem of state support for business is the most important in today's conditions. The cataclysms that have occurred recently have affected the world economy. The world has ceased to be balanced. Business froze in fear of inevitable bankruptcy, unemployment and ruin. Political changes in Russia, which began in January 2020, the drop in oil prices, the collapse of the ruble, which followed the relative stability of the Russian national currency and reached the pandemic country, negatively affected business in the country. The development of urgent state measures to support small and medium-sized enterprises is the only way to maintain its viability. These measures should be based on a clear regulatory framework that does not allow vague interpretations and legal incidents. Using the comparative legal method, it is necessary to draw the experience of those countries in which there is a state policy in relation to small and medium-sized businesses, which is a system of socio-economic actions built on the principle of ensuring the favorable development of this sector of the economy and to develop measures that can preserve it in conditions of prevailing force majeure. The aim of this work is a comparative analysis of measures taken by the world community and Russia to support small and medium-sized businesses during the economic crisis caused by Covid19. Particular attention is paid to the legislative justification of measures taken in Russia and the identification of shortcomings of the newly adopted normative legal acts.

Keywords: Economic crisis · Government support · Pandemic · Small and medium business

1 Introduction

The economy of any modern state is divided into two interconnected sectors – financial and real. The financial sector deals with monetary relations and through the banking system, with its credit, financial and exchange instruments, has an impact on the demand and supply of goods and services. The real sector of the economy is engaged in their production. The real sector of the economy is engaged in the production of tangible and intangible products, since, along with industry and agriculture, it includes scientific enterprises and trade organizations. In each country, the real sector of the economy has its own specifics, which is formed on the basis of various factors – the geopolitical location of the state, the possession of natural resources and minerals, climatic conditions, etc.

S. I. Ashmarina et al. (Eds.): ENGINEERING ECONOMICS WEEK 2020, LNNS 139, pp. 275–282, 2021.
https://doi.org/10.1007/978-3-030-53277-2_32

Almost all states, even with very developed economy, subsidize certain sectors of the real sector of the economy in different ways to maintain economic balance, regulate supply and demand, and set prices for goods and services. This need exists in industry and agriculture, which often due to low labor productivity and technological backwardness feel the need for financial support from the state. With an increase in government spending to support small and medium-sized enterprises, an increase in aggregate demand in the economy is ensured [13]. And the point is not so much in budget financing, but in off-budget investments, subsidies, subventions, loans.

Of course, the situation of the industry often depends on the region where it is located. Here, both the specifics of the climate and the availability of labor are important. It is important to understand that government support for business should balance the financial capabilities of all regions. Such an inter-regional redistribution was characteristic of the post-war period in Italy [16]. The measures necessary to support the economy can be both long-term [2], and urgent, caused by emergency. Those events that occur in the world today must be considered taking into account the experience of past economic crises. But at the same time, one cannot ignore the fact that modern world economic crises have peculiarities. These features are manifested in their causes, not only financial, political, economic or demographic. In recent years, the modern world has undergone tangible disasters, caused not only by globalization processes, but also by environmental disasters.

The research and analysis of factors leading to global crises is not the purpose of writing this scientific article. Our aim is to analyze the legal regulation of the measures that the state has provided in support of small and medium-sized enterprises in the emergency situation that has arisen today. The problems caused by the pandemic are not only momentary in nature. The consequences of a viral attack on the world are long-term and the likelihood that many entrepreneurs will not return to duty is quite high. Only about 5 months a new virus spreads all over the world, but the events following it pose a real menace to the relationships that exist in society [15].

2 Methodology

The research was conducted on the basis of general scientific methods of cognition. The systematic approach made it possible to generalize the existing problems in the legal regulation of small and medium-sized businesses. The comparative method made it possible to find common and identify individual problems in the legal regulation of state support for business in different countries.

The study of a large amount of material made it possible to draw a conclusion about the possible forecasting of economic crises caused by various reasons. An analysis of the legislative framework made it possible to hypothesize the advisability of developing legal norms providing for measures of state support for business in crisis periods "for the future" [17].

3 Results

The author of the well-known theory of the "Black Swan" Nassim Taleb in his work "On the Secrets of Sustainability" says that excessive globalization and interdependence, whether it is biological, cultural or economic, is alien to mother nature. The idea expressed by him is that a person is not able to prevent plagues and epidemics. Errors and accidents will always be present, despite the construction of various forecast models. It is necessary, following the laws of mother nature, to prevent the possible spread of these mistakes and fortuities throughout the system, isolate them in small zones [23].

Covid-19 overtaking the world turns its gaze to this theory and makes us think about the possible forecasting of such events. Crisis predictability was discussed after the global financial crisis in 2008. Due to the presence of clearly identifiable signs in this crisis, it was recognized as an extreme, but quite predictable event. Regardless of the name of the virus, the consequence of which was the concept of economic "sudden stop", it is necessary to develop emergency measures to contain it [5].

The financial and economic crises following the coronavirus pandemic indicate the necessity of serious lessons for all countries [12]. Was it possible to predict a situation that entailed the need for radical measures such as closing state borders, shutting down enterprises, stopping communication between countries, restricting freedom of movement of citizens? The answer to this question will most likely be positive.

Accordingly, it was necessary to be ready for such situations in advance: at the legislative level, develop state measures to support the population and business, create special funds, the funds from which can be used to support citizens and businesses. Clear rules and regulations governing the actions of authorities at all levels, legal entities, and citizens during the arrival of the "black swan", enshrined at the legislative level, would eliminate the need to create them in a hurry and contradict constitutional norms. What measures are being taken in states to prevent the economic consequences of a pandemic? The European Union has allocated 37 billion euros to protect the economies of the block. Of these, 8 billion euros to enterprises and companies that may suffer significant losses. The European Commission allowed the block countries to take loans without restrictions to help business.

In Germany, a package of measures has been adopted to assist private entrepreneurs and companies that could be significantly affected by the pandemic. The total amount of state support will be 750 billion euros. Small firms and private entrepreneurs are able to receive direct subsidies of up to 15 thousand euros, 50 billion euros have been allocated for this purpose. Hospitals will receive 3 billion euros, in all, Germany will spend up to 37% of the country's GDP on the fight against the pandemic.

The Ministry of Finance of England will provide companies with loan guarantees of 330 billion pounds, which is about 15% of the country's GDP, while interest can be paid for 6 months. Small enterprises are exempted from paying business tax for the duration of the epidemic. The Italian Council of Ministers has allocated 25 billion euros to support families and companies in emergency situations. Payments on loans and credits to businesses and households are suspended under state guarantees for banks. In total, 10 billion euros have been reserved for supporting families and entrepreneurs.

The Spanish government has allocated 200 billion euros or 20% of the country's budget to combat the pandemic and its consequences. Of these, 117 billion euros are public funds, the rest is private resources. 100 billion are presented in the form of state loan guarantees. The Polish government has allocated 46.2 billion euros. Part of this money will be provided to entrepreneurs. In addition, credit guarantees, microloans, and operating leasing are provided. For companies that will not lay off employees, the state will cover 40% of the payroll. 1.6 billion has been allocated for healthcare. In the United States, the law "On the provision of financial assistance to casualties of the epidemic" allocated $ 100 billion. The Senate approved a package of measures stimulating the economy in the amount of 2 trillion. dollars, in particular, 500 billion dollars. transferred to the support fund for enterprises, cities, and states, of which 25 billion to airlines, 17 billion to strategically important companies. The government will pay 500 billion dollars to Americans with incomes up to 75 thousand a year – $ 1,200 per adult and $ 500 each. per child. To support small businesses for measures on tax holidays allocated 350 billion dollars. In France, the enterprises that suffered losses were allocated 45 billion euros. Of the same amount, benefits are paid to employees who are transferred to the partially unemployed regime. A state guarantee was provided for bank loans of the enterprise in the amount of 300 billion euros. Small businesses will receive assistance in the amount of 1,500 billion euros. €500billion has been allocated for airlines that incurred losses.

In the world, the main measures taken in support of small and medium-sized businesses are to provide tax and credit holidays, lower interest rates on loans and income taxes, up to full exemption for an emergency, and direct cash support. It is the expediency of measures in the form of direct cash injections, both in business and provided to citizens, that causes the most controversy. The negative impact on the economic efficiency of investments and scientific and technological innovations was expressed even before the pandemic [16]. In the current situation in Russia, the possibility of direct cash assistance to both citizens and businesses finds opponents. The measures chosen by each country depend not only on the economic structure, but also on domestic opportunities. According to many economists, the difficult situation in Russia, caused by the decline in oil prices and the depreciation of the ruble, does not allow "distributing" money to citizens, as in America. For the rest, Russia takes actions similar to those of other states - introduces tax and credit benefits, provides a lease on the lease of municipal and state property for business, and exempts from a number of mandatory contributions until the end of the year.

4 Discussion

What measures and for whom were developed in Russia in the context of the spread of coronavirus to save business?

The measures taken by the Russian Federation are aimed at supporting small and medium-sized enterprises, which, in the opinion of the government, have fallen into a difficult economic situation. The list of such sectors of the economy is approved by Resolution of the Government of the Russian Federation of 03.04.2020 N 434 [18]. Currently, this list is not closed, it is periodically updated.

In April 2020, in the Russian Federation in the direction of economic support for small and medium-sized enterprises, federal laws were adopted, resolutions of the Government of the Russian Federation were issued, numerous letters were published, information was posted. As support measures, the government provides for credit benefits, tax and rental holidays, and a moratorium on bankruptcy.

Of the laws, it is necessary to name the Federal Law of 01.04.2020 N 103 "On Amendments to the Federal Law" "On the Suspension of Certain Provisions of the Budget Code of the Russian Federation and the Establishment of Specific Features of the Execution of the Federal Budget in 2020" [6], Federal Law of 03.04.2020 N 106 "On Amendments to the Federal Law" On the Central Bank of the Russian Federation (Bank of Russia) "and certain legislative acts of the Russian Federation regarding the peculiarities of changing the terms of a loan agreement" [7], Federal Law of April 22, 2020 N 121 "On Amendments to Part Two of the Tax Code of the Russian Federation" [8], Federal Law of April 1, 2020 N 98 "On Amendments to Certain Legislative Acts of the Russian Federation Regarding Emergency Prevention and Response" [9].

A number of government regulations is aimed at supporting the economy. Resolution of the Government of the Russian Federation of 02.04.2020 N 422 approved the "Rules for the provision of subsidies from the federal budget to Russian credit organizations for the reimbursement of shortfalls in their income on loans issued to legal entities and individual entrepreneurs in 2020 for urgent needs to support and maintain employment" [19], Resolution of the Government of the Russian Federation of 04.04.2020 N 466 approved the "Rules for the provision in 2020 from the federal budget of subsidies to Russian air transport organizations for reimbursement of expenses incurred the implementation of measures for the removal of citizens from foreign countries in which there is an unfavorable situation associated with the spread of a new coronavirus infection" [20].

The forced speed of adoption of regulatory legal acts has led to the need to supplement and clarify them with all kinds of letters, information, instructions. The adoption of such serious regulatory documents in a hurry led to the difficulty of their enforcement. Clarifications of the relevant departments, as well as of the Supreme Court of the Russian Federation were required to clarify the concepts introduced by the new laws and decrees. The concept regarding which there is a need for clarification includes, for example, the term "non-working day". Decree of the President of the Russian Federation of March 25, 2020 N 206 "On the announcement of non-working days in the Russian Federation" [3], introduced a new concept that does not apply to a "day off" or to a "holiday". The introduction of the new term caused a number of difficulties, such as payroll, granting leave and dismissal during the "non-working" days. In the aspect of labor relations, clarifications were given by the Ministry of Labor [10], and the meaning that he put into the new term radically differed from the meaning given to the concept of "non-working days" by labor legislation.

An assessment of the term in the civil law sense was given by the "Review of Certain Judicial Practice Issues Related to the Application of Legislation and Measures to Counteract the Spread of the New Coronavirus Infection (COVID-19) No. 1 on April 21, 2020", approved by the Presidium of the Supreme Court of the Russian Federation [21]. According to the Presidium, these days cannot be considered non-working in the sense that gives this concept to the Civil Code of the Russian Federation. Accordingly, they are not grounds for postponement of the fulfillment of obligations under Article 193 of the Russian Civil Code [2].

Many questions raise such a concept, used in regional restrictive acts, such as “high alert mode”, expressions “temporarily suspend”, “restrict”. Firstly, they are not defined by law, and secondly, they do not allow the use of civil law to terminate obligations. Economic crisis is not only financial difficulties, unemployment and impoverishment of the population. It is also a manifestation of cruelty and violence, the criminalization of society. Studies have confirmed an increase in domestic violence from the economic downturn [22], and the experience of the crisis caused by the pandemic has shown an increase in the number of cases of child abuse [11]. Overcoming global crises should occur in interaction and mutual assistance, outside national interests. It is unacceptable, hiding behind the fight against the crisis, to forget about respect for human rights, justice and equality [25].

5 Conclusion

The number of risks that can cause global disasters is only increasing. Whether it is a problem with refugees, cyberattacks, environmental disasters or epidemics, all this affects the economic situation. The world is not always ready for these risks. In addition, factors that threaten our security are often related to each other and reinforce each other. The danger of events lies in their destructive distribution, as our world is interpenetrated, permeated by common networks [14]. It is this problem that mankind faced in the case of coronavirus. Such events should be predicted, and governments and businesses should be prepared for such events.

Small and medium-sized enterprises are included in the economic system of any country. Research on the identification of factors that have an effective impact on their work is given great importance [4]. The most important factors affecting the development of this sector of the economy and identified as a result of the study include government regulation and business support, which are based on a clear and understandable legal framework. Not all states can boast of such, so, according to entrepreneurs from the Czech and Slovak Republics, the level of legislation governing business and its state support is not very high [1]. The need for the functioning and development of small and medium-sized enterprises in accordance with laws and regulations, the solution to the problem of financing them on the basis of laws and regulations, say Chinese entrepreneurs [24].

It is necessary to improve the legal standards governing the support of small and medium-sized enterprises, providing for their effect, both in the normal course of events and in emergency situations. This will provide an opportunity to save time for solving super-urgent tasks in the event of a disaster. In addition, it will be possible to avoid those errors that usually arise during hasty rulemaking. Poor legal regulation leads to problems of law enforcement, free interpretation of legal norms and, accordingly, corruption loopholes. The ambiguity of the proclaimed norms often leads to the fact that they remain declared, but not enforceable. Small and medium-sized businesses always need state support, in a crisis situation, especially. The development of state measures to support entrepreneurs, based on a clear regulatory framework, is the only guarantee of their viability.

References

1. Cepel, M., Belas, J., Dvorsky, J.: The impact of significant political factors on starting a new business. In: Liargovas, P., (ed.) Proceedings of the European Conference on Innovation and Entrepreneurship, ECIE, pp. 220–228. Sonning Common: Academic Conferences and Publishing International Limited Reading (2019)
2. The Civil Code of the Russian Federation (book 1). "Meeting of the legislation of the Russian Federation", 05.12.1994, N 32, Art. 3301 (1994). http://www.consultant.ru/document/cons_doc_LAW_5142/. Accessed 15 Apr 2020
3. Decree of the President of the Russian Federation of March 25, 2020 N 206 "On the announcement of non-working days in the Russian Federation" (2020). https://base.garant.ru/73793643/. Accessed 20 Apr 2020
4. Elver, H.: New constitutionalism and the environment. In: Gill, S., Cutler, A.C. (eds.) New Constitutionalism and World Order, pp. 261–276. Cambridge University Press, Cambridge (2012)
5. Faggini, M., Bruno, B., Parziale, A.: Crises in economic complex networks: black swans or dragon kings? Econ. Anal. Policy **62**, 105–115 (2019)
6. The Federal Law of 01.04.2020 N 103 "On amendments to the Federal Law "On the suspension of certain provisions of the budget code of the Russian Federation and the establishment of specific features of the execution of the federal budget in 2020" (2020). http://www.consultant.ru/document/cons_doc_LAW_349085/. Accessed 15 Apr 2020
7. The Federal Law of 03.04.2020 N 106 "On amendments to the Federal Law" On the central bank of the Russian Federation (Bank of Russia) "and certain legislative acts of the Russian Federation regarding the peculiarities of changing the terms of a loan agreement" (2020). http://base.garant.ru/73842090/. Accessed 15 Apr 2020
8. The Federal Law of April 22, 2020 N 121 "On amendments to part two of the tax code of the Russian Federation" (2020). http://base.garant.ru/73929695/. Accessed 15 Apr 2020
9. The Federal Law of April 1, 2020 N 98 "On amendments to certain legislative acts of the Russian Federation regarding emergency prevention and response" (2020). https://base.garant.ru/73828134/. Accessed 15 Apr 2020
10. Information of the ministry of labor and social protection of the Russian Federation of April 2, 2020 "Questions and answers on the organization of work and respect for the rights of workers during the non-working week" (2020). https://www.garant.ru/products/ipo/prime/doc/73734356/. Accessed 15 Apr 2020
11. John, N., Casey, S.E., Carino, G., McGovern, T.: Lessons never learned: crisis and gender-based violence (2020). https://onlinelibrary.wiley.com/doi/abs/10.1111/dewb.12261. Accessed 15 Apr 2020
12. Lucchese, M., Pianta, M.: The coming coronavirus crisis: what can we learn? Intereconomics **55**(2), 98–104 (2020)
13. Makrelov, K., Arndt, C., Davies, R., Harris, L.: Balance sheet changes and the impact of financial sector risk-taking on fiscal multipliers. Econ. Model. **87**, 322–343 (2020)
14. Masys, A.J.: Complexity and security: new ways of thinking and seeing. In: Masys, A.J., Lin, L.S.F., (eds) Asia-Pacific Security Challenges. Advanced Sciences and Technologies for Security Applications, pp. 1–17. Springer, Cham (2018)

15. McCullough, P.A., Eidt, J., Rangaswami, J., Lerma, E., Tumlin, J., Wheelan, K., Ronco, C.: Urgent need for individual mobile phone and institutional reporting of at home, hospitalized, and intensive care unit cases of SARS-CoV-2 (COVID-19) infection. Rev. Cardiovasc. Med. **21**(1), 1–7 (2020)
16. Öncü, T.S.: Triggering a global financial crisis: COVID-19 as the last straw. Econ. Polit. Wkly. **55**(11), 10–12 (2020)
17. Petraglia, C., Pierucci, E., Scalera, D.: Interregional redistribution and risk sharing through public budget. the case of Italy in times of crisis (2000–2016). Struct. Change Econ. Dyn. **53,** 162–169 (2020)
18. Resolution of the Government of the Russian Federation of 03.04.2020 N 434 (Rev. 18.04.2020), "On approval of the list of branches of the Russian economy most affected by the worsening situation caused by the spread of new coronavirus infection" (2020). http://www.consultant.ru/document/cons_doc_LAW_349344/. Accessed 15 Apr 2020
19. Resolution of the Government of the Russian Federation of 02.04.2020 N 422 (red. of 24.04.2020) "Rules for the provision of subsidies from the federal budget to Russian credit organizations for the reimbursement of shortfalls in their income on loans issued to legal entities and individual entrepreneurs in 2020 for urgent needs to support and maintain employment" (2020). https://www.garant.ru/products/ipo/prime/doc/73741776/. Accessed 16 Apr 2020
20. Resolution of the Government of the Russian Federation of 09.04.2020 N 466 "Rules for the provision in 2020 from the federal budget of subsidies to Russian air transport organizations for reimbursement of expenses incurred the implementation of measures for the removal of citizens from foreign countries in which there is an unfavorable situation associated with the spread of a new coronavirus infection" (2020). http://www.consultant.ru/document/cons_doc_LAW_349996/. Accessed 19 Apr 2020
21. Review of certain judicial practice issues related to the application of legislation and measures to counteract the spread of the new coronavirus infection (COVID-19) No. 1 on April 21, 2020 (approved by the Presidium of the supreme court of the Russian Federation) (2020). https://www.vsrf.ru/press_center/news/28855/. Accessed 15 Apr 2020
22. Schenck-Fontaine, A., Gassman-Pines, A.: Income inequality and child maltreatment risk during economic recession. Child. Youth Serv. Rev. **112**(C) (2020). https://ideas.repec.org/a/eee/cysrev/v112y2020ics019074091931014x.html. Accessed 15 Apr 2020
23. Taleb, N.: On the secrets of sustainability. In: Tyunkina, M., (ed.) The Black Swan. The Impact of the Highly Improbable, pp. 212–246. KoLibry, Moscow (2007)
24. Xi, W.: The modeling and simulation of the cloud financing model for technology-based small and medium-sized enterprises. In: Abawajy, J.H., Choo, K.-K.R., Islam, R., Xu, Z., Atiquzzaman, M., (eds) International Conference on Applications and Techniques in Cyber Intelligence ATCI 2019. Aplications and Techniques in Cyber Intelligence, vol. 1017, pp. 1057–1062. Springer, Cham (2020)
25. Xuan, V.N., Thu, N.T.P., Anh, N.T.: Factors affecting the business performance of enterprises: evidence at Vietnam small and medium-sized enterprises. Manage. Sci. Lett. **10** (4), 865–870 (2020)

State Support for the Russian Economy Affected by Covid-19 Consequences (Legal Aspects)

A. A. Malinovsky[1(✉)], D. M. Osina[1], and E. N. Trikoz[1,2]

[1] Moscow State Institute of International Relations, MGIMO University, Moscow, Russia
Dr.malinovsky@yandex.ru, osina.dina@yandex.ru, alena_trikoz@mail.ru
[2] Peoples' Friendship, University of Russia, RUDN University, Moscow, Russia

Abstract. Measures aimed at combating the coronavirus have already negatively affected the Russian economy. In the future, macroeconomic indicators will deteriorate. The study considers the analysis of relevant government measures to protect the Russian economy in the new realities of the pandemic. For the first time the team of authors (using the method of economic analysis of law) studied the regulatory and analytical material on supporting the Russian economy in the crisis. As a result of the study, the authors come to the following conclusions: 1) the Russian authorities promptly adopted a package of legislative acts to protect the Russian economy; 2) most government measures can potentially be effective and capable of supporting the Russian economy in the context of the coronavirus epidemic, however, it is too early to talk about the final effectiveness and sufficiency of these measures; 3) some measures, despite the fact that they will accumulate funds in the budget, could potentially have a negative effect on the Russian economy in the future.

Keywords: COVID-19 · Economic measures · Financial crisis · State support · Russian economy

1 Introduction

On December 31, 2019, the whole world became aware of the outbreak of a new coronavirus infection in the Chinese city of Wuhan. The Russian authorities promptly responded to the new challenges of the pandemic by developing a kind of "rulemaking vaccine" – a package of legislative acts providing for a set of measures to combat the economic consequences of the coronavirus. These measures were developed considering the views of business representatives, for example the Chamber of Commerce and Industry of Russia presented its proposals for supporting the business, including some e-commerce initiatives [4].

Give to him that asketh thee, and from him that would borrow of thee turn not thou away. (Sermon on the Mount).

S. I. Ashmarina et al. (Eds.): ENGINEERING ECONOMICS WEEK 2020, LNNS 139, pp. 283–287, 2021.
https://doi.org/10.1007/978-3-030-53277-2_33

2 Methodology

The basis of this study is an extensive regulatory and analytical material on the regulatory impact on the Russian economy in the pandemic. The source and empirical base are the appeal of V.V. Putin, the President of the Russian Federation, a package of federal anti-crisis laws and regulations [10].

3 Results

To coordinate the activities of executive authorities during the pandemic, in accordance with Decree of the Government of the Russian Federation No. 236 dated 05.03.2020, the Government Commission was formed to increase sustainable development of the Russian economy (hereinafter referred to as the Commission) and the regulation on it was approved [7]. One of the tasks of the Commission is to develop measures to support the real sector of the economy, including reducing costs for businesses and systemically important enterprises.

On 25.03.2020 V. V. Putin, the President of the Russian Federation, addressed the citizens of Russia, within the framework of which he announced a list of measures to support the Russian economy. On 29.03.2020 the list of instructions of the President of the Russian Federation based on the results of this appeal was published [8].

On 01.04.2020 V.V. Putin signed a package of federal laws to amend certain legislative acts of the Russian Federation to support the Russian economy, which were published and entered into force on the same day. On 02.04.2020 the President of the Russian Federation signed a decree on the extension of "non-working days" in Russia for the whole of April this year with the suspension of the work of companies, banks and enterprises regardless of their form of ownership, which caused additional concern of a few economists. According to estimates by analysts of the Center for Macroeconomic Analysis and Short-Term Forecasting, the direct instant effect of the "non-working day regime" will be up to 20% of GDP losses, or almost 18 trillion rubles [12]. Now it is advisable to consider those measures that have been taken by the Russian authorities to support the Russian economy.

At the federal level, the Law came into force, according to which individual entrepreneurs whose average monthly income will decrease after March 1, 2020 by more than 30% due to the coronavirus are entitled to apply for the so-called credit holidays for up to 6 months, during which they cannot be charged fines and penalties, be foreclosed on collateral or mortgages etc. Federal Law dated 01.04.2020 introduced a new concept of "national guarantee system for supporting small and medium-sized enterprises", which means the system of providing entities with sureties and (or) independent guarantees for their obligations when accessing credit and other financial resources [6].

Other measures were promptly taken at the legislative level, e.g., a temporary moratorium on bankruptcy at the request of creditors was established, insurance contributions were reduced, rent payments were delayed (on municipal and state property), and tax, customs and other inspections of small and medium-sized enterprises (hereinafter referred to as SMEs) were suspended (from April 1 and until the end of 2020).

The Law on the Protection and Promotion of Investment in the Russian Federation was signed, which established a ban on actions that worsen the conditions for doing business related to the implementation of private and public investment projects.

- The system-wide measures to support the economy described in the Government's "Plan of Priority Measures" dated 17.03.2020 can be divided into three blocks: (1) financial and budgetary measures – create a financial reserve of up to 300 billion rubles; create a guarantee fund for the restructuring of company loans; expand measures to support the budgets of the Russian Federation, faced with a fall in tax revenues, by exempting regions from repayment of budget loans in 2020; accelerate state capital investments by simplifying procedures and reducing the composition of documents related to amendments to the federal targeted investment program; (2) preferential and security measures – not apply penalties; extend terms and price adjustments under state and municipal contracts in case of violation of obligations by the contractor due to the consequences of the coronavirus; inform residents about necessary actions in case of non-fulfillment (to prevent the occurrence of non-fulfillment) of the delivery (payment) of goods by foreign counterparties under foreign trade contracts and non-return of funds previously paid by residents if such non-fulfillment is caused by force majeure; give the Government of the Russian Federation the right to determine the procedures for obtaining a deferral (installment plan) for the payment of taxes, fees and contributions, including with regard to the procedure, grounds and deadlines for their submission, as well as the right to extend the deadline for submitting tax returns (calculations); (3) provide operational monitoring of the financial and economic state of backbone organizations of the Russian economy (over 600 companies) [1].

In addition to system-wide support measures, two separate blocks were identified: (4) measures to support SMEs, and (5) priority measures to support sectors of the economy that are at risk (loan borrowers; transport organizations; cultural and physical education and sports institutions; tour operators; leasing companies; construction industry). The same example was followed by some regional authorities [3]. In relation to the subjects in the so-called "Risk areas of the economy" there are the following measures: restructuring loans in connection with the spread of the coronavirus; providing financial support for transport organizations; subsidizing interest rates on loans to developers in the framework of project financing; supporting leasing companies, including their additional capitalization; paying partial compensation for losses to national sports federations; granting a deferral of tax payments (for 3 months); extending labor permits to foreign workers, etc.

4 Discussion

State support measures for SMEs are the following: grant a deferral on loans and all taxes, with the exception of VAT for the next 6 months, and for microenterprises – also provide additional deferrals on insurance contributions to social funds; decrease in half (up to 15%) the amount of insurance contributions by the amount of the salary exceeding the minimum wage; introduce a moratorium on inspections of SMEs, including tax audits, with the exception of issues that pose risks to the life and health of citizens; expand the program for subsidizing the access of SMEs to borrowed funds;

restructure loans of the borrower – a subject of SMEs; subsidy (up to 1/2 the rate of the agreement, but no more than the key one) a part of the interest on loans of SMEs subjects to the postponement of the payment of interest without charging penalties to credit organizations; provide temporary postponement (or moratorium) on the payment of rental payments by subjects of SME-tenants of state or municipal property; provide coverage of preferential microloans by SMEs; expand the capabilities of SMEs to obtain soft loans in the absence of collateral; reduce requirements for securing contracts during public procurement from SMEs; not apply penalties, as well as extend terms and price adjustments in 2020 in case of violation of obligations by the contractor (within the framework of 223-FL) due to the consequences of the coronavirus [5].

In addition to the above, the President of the Russian Federation announced the following measures to support the Russian economy: ensure sustainable lending to the real sector, including the provision of state guarantees and subsidizing interest rates on loans; protect enterprises, that are in a difficult situation, from bankruptcy by introducing a moratorium on filing creditors' applications for bankruptcy of companies and collecting debts and fines for a period of 6 months; introduce a tax on interest income of 13% for citizens whose total amount of bank deposits or investments in debt securities exceeds 1 million rubles in 2021; increase withholding tax rate on dividends to 15% if they are remitted to foreign accounts in countries through which significant resources of Russian origin pass, and therefore the Government decided to review the network of bilateral double taxation agreements [10], starting with the Agreement with the Republic of Cyprus, Article 10 of which provides for the possibility of taxation of dividends at preferential rates of 5 and 10%, provided that they are paid to a company that is a beneficial owner of such dividends and complies with the investment criterion [2]. On 27.03.2020 in pursuance of instructions of the President of the Russian Federation, the Commission identified 22 sectors of the economy that were most affected by the spread of the coronavirus, to provide priority targeted state support [9]. By Order of the Government of the Russian Federation No. 767-r of 28 March 2020 1.5 billion rubles were allocated from the reserve fund of the Government of the Russian Federation to the Federal Air Transport Agency to reimburse the costs incurred by air transport organizations due to the return of Russian citizens from foreign countries with Covid-19 epidemic [11].

5 Conclusion

As follows from the analysis, the Russian authorities are already taking measures to protect the Russian economy and support enterprises (organizations). The measures introduced are undoubtedly intended to alleviate the situation of business entities by mitigating as much as possible the negative consequences of the outflow of visitors and clients against the background of quarantine measures. However, several measures can be ambiguously construed. E.g., the introduction of the above-mentioned tax on interest on deposits does not directly contribute to improving the economic situation, since in fact the state will receive the first revenue from such a measure only in 2022. At the same time, this measure will automatically reduce the attractiveness of deposits as an investment tool, which in fact does not contribute to the attraction of private funds by banks. A potential 15% withholding tax for dividends paid by Russian companies is fraught with a decrease in the attractiveness of Russian jurisdiction for potential investors, which may have a negative cumulative economic effect in the future. In general, the "freezing of the Russian economy" currently being legalized will have the following result – a certain part of SMEs will simply not survive this suspension and will not pay off their obligations. Restructuring of loans to companies and citizens affected by the coronavirus will lead to a significant deterioration in the quality of the loan portfolio of many large banks in Russia.

References

1. Adamchuk, O., Trifonova, P.: The government has 3 times increased the list of backbone companies. Vedomosti (2020). https://www.vedomosti.ru/economics/articles/2020/03/23/825894-pravitelstvo-raza-sistemoobrazuyuschih. Accessed 01 May 2020
2. Agreement between the government of the Russian Federation and the government of the Republic of Cyprus for the avoidance of double taxation with respect to taxes on income and capital (1998). http://www.consultant.ru/document/cons_doc_LAW_23908/. Accessed 01 May 2020
3. Decree of the Governor of the Pskov Region dated March 19, 2020 No. 33-UG "On priority measures to support small and medium-sized enterprises operating in economic sectors which are at risk in connection with the implementation of measures to counteract the spread of the new coronavirus infection in the Pskov Region (2019-nCoV)". https://rg.ru/2020/04/08/pskov-ukaz42-reg-dok.html. Accessed 01 May 2020
4. Economy and Life: RF CCI sent Mishustin a second package of proposals on business support (2020). https://www.eg-online.ru/news/417215/. Accessed 01 May 2020
5. Federal Law No. 223-FL of July 18, 2011 "On procurement of goods, works, services by certain types of legal entities". http://base.garant.ru/12188083/. Accessed 01 May 2020
6. Federal Law No. 83-FL of 01.04.2020 "On amendments to the Federal Law "On the development of small and medium-sized enterprises in the Russian Federation" with a view to developing a national guarantee system for supporting small and medium-sized enterprises". http://base.garant.ru/73827394/. Accessed 01 May 2020
7. Government of the Russian Federation: government commission for improving the sustainability of the Russian economy (2019). http://government.ru/department/549/about. Accessed 01 May 2020
8. List of instructions based on the appeal of the President of the Russian Federation in connection with the spread of coronavirus infection in the country (2020). http://kremlin.ru/events/president/news/63080. Accessed 01 May 2020
9. List of selected areas of activity most affected by the deterioration of the situation due to the spread of the new coronavirus infection to provide priority targeted support (2020). https://www.economy.gov.ru/material/file/be6cf04db17a6caed6bfa9fd97684d7b/26032020.pdf. Accessed 01 May 2020
10. Official website of the President of Russia: meeting with members of the government of the Russian Federation (2020). http://kremlin.ru/events/president/news/63123. Accessed 01 May 2020
11. Order of the government of the Russian Federation No. 767-r of 28 March 2020. http://government.ru/docs/39308/. Accessed 01 May 2020
12. Yastrebova, S., Mereminskaya, E.: Extension of non-working days threatens with reductions and bankruptcies. Vedomosti (2020). https://www.vedomosti.ru/economics/articles/2020/04/02/827003-sokrascheniyami-i-bankrotstvami. Accessed 01 May 2020

State Support Mechanisms for Business in the Context of Coronavirus Pandemia

N. N. Belanova[1(✉)] and A. M. Kuznetsova[2]

[1] Samara State University of Economics, Samara, Russia
bnn371@yandex.ru
[2] Kazan Federal University, Kazan, Russia
Alsu@legenda.travel

Abstract. The spread of coronavirus infection and the subsequent introduction of restrictive measures have brought the economy into a new reality that has no examples of implementation in the past. The Russian government is faced with the acute problem of developing and implementing optimal management solutions. The article investigates the main support measures developed by the government of the Russian Federation. The authors compare these measures with measures implemented in other countries. Based on the comparison of the scale of the state participation, it is concluded that it is necessary to expand the package of anti-crisis measures and their preferential financing at the expense of the national welfare fund (NWF).

Keywords: Consequences of pandemic · Methods and tools of state influence · State support

1 Introduction

The research on the state influence on various spheres of the economy and society is a key focus of many studies. Some researchers analyze development patterns of the state regulation, the others try to establish a relation between the economic growth and the scale of governmental intervention into economy [10]. Researchers do this based on the analysis of empirical data, trying to establish the optimal limits of intervention, but do not see a clear correlation [13]. Some experts identify positive results of interaction between the state and private business [11]. Much attention is paid to the regulation of certain spheres and branches of the economy [2, 7].

Supporters of the state regulation highlight the key limitation for the amount of the state support – this is the budget capacity. Therefore, some authors analyze costs of such a regulation [12], investigate problems of public debt and default [8, 9]. Despite a wide variety of works on the considered subject, issues of the scope and specific instruments of the state regulation need to be refined and updated, since the spread of coronavirus infection has put the economy in a new reality that has no examples of implementation in the past.

S. I. Ashmarina et al. (Eds.): ENGINEERING ECONOMICS WEEK 2020, LNNS 139, pp. 288–295, 2021.
https://doi.org/10.1007/978-3-030-53277-2_34

2 Methodology

The methodological basis of this research is formed by fundamental provisions of the economic theory, scientific works of Russian and international scientists in the field of national and world economy. To date, certain methods and tools of the state regulation have been developed in the economic science. However, a comprehensive approach is needed to identify the main areas of support for the national economy and business. In this research, the authors used dialectical, system-functional, economic-statistical, formal-logical and scientific abstraction methods.

3 Results

The coronavirus pandemic has a significant impact both on the world and the Russian economies. There is no common position of experts on the pace of the economic decline and the GDP decline. Thus, experts of the state development corporation "VEB" of the Russian Federation ("Vnesheconombank") predict that the GDP will decrease by an average of 3.8%. Specialists of the Institute for strategic analysis ("FBK Grant Thorton") believe that the world economy will decline by 2–3% compared to the last year, and the Russian GDP – by 10–20% [1]. According to the forecast of the Central Bank analysts, the GDP decline in the second quarter will be 8% (because of quarantine restrictions) with a rapid recovery and a year-end decline of 4–6% [3].

Based on the macroeconomic theory, it can be noted that the GDP by expenditure is calculated using the formula:

$$\mathrm{GDP} = \mathrm{C} + \mathrm{Ig} + \mathrm{G} + \mathrm{Xn} \tag{1}$$

C – consumer expenditure,
Ig – gross investment expenditure,
G – government expenditure,
Xn – net exports (the difference between exports and imports).
Changes in the GDP occur as a result of changes in these components:

$$\Delta\mathrm{GDP} = \Delta\mathrm{C} + \Delta\mathrm{Ig} + \Delta\mathrm{G} + \Delta\mathrm{Xn} \tag{2}$$

According to the forecast of the Central Bank specialists, investments in fixed assets will decrease by 6–8% this year, and exports will fall by 10–15%. It is difficult to estimate the fall in consumer expenditure, but given the increase in the unemployment rate to 7% on average over the year and the fall in disposable income by 6.5% (17.5% in the second quarter), we can assume that even an optimistic forecast will make it 8–10%.

Government expenditure (G) can be considered as the only limiting factor for the fall in the GDP from the analyzed components. In crisis development periods, when most macroeconomic indicators are declining and the market economy does not provide optimal development rates, the state actively intervenes, increasing expenditures, and thus partially compensating for the fall in the remaining components of the

GDP. However, according to the Central Bank chairman E.S. Nabiullina, declared and implemented by the current government measures will support the GDP by only 2% [3]. According to the head of the Accounting Chamber A. Kudrin, the overall package of the state support measures should be at least 7% of the Russian GDP.

Further, we consider measures that the Russian Government has developed to support the economic development [4]:

1. Loans and grants for business (interest-free credit for payment of wages, grants for payment of salaries and other urgent issues, if at least 90% of the staff remain the same, and 2/3 of interest payments on new loans are covered by the state and banks).
2. Deferral and reduction of taxes (except VAT) and insurance payments for 6 months for small and medium-sized businesses in the most affected industries. For other organizations, the insurance premiums will be reduced by half for the amount of wages exceeding the minimum wage.
3. Deferred lease payments for state and municipal property.
4. Moratorium on the initiation of bankruptcy cases on applications from creditors for system-forming enterprises and organizations of the most affected industries. The most affected industries include road transport, air transport, tourism, exhibition activities, entertainment and leisure, culture and sports, public catering, and non-food retail. This list was approved by a government decree and can be expanded [5].

2. Reducing the administrative burden by:

- automatic renewal of 15 types of licenses and permits that expire in 2020;
- simplifying the conclusion of public contracts, easier requirements for public procurement;
- providing a "green corridor" for imported essential goods, introducing a simplified procedure for certification of imported products and abolishing customs duties on medicinal and medical products that serve to prevent the spread of coronavirus;
- reducing unproductive business costs by introducing a moratorium on inspections, control is planned to be carried out remotely: in the format of audio and video communications;
- cancellation of fines and penalties for contracts executed in 2020.

Support measures for the backbone enterprises include:

- granting preferential loans for 1 year to maintain working capital and save jobs;
- weekly monitoring of enterprises, stress testing and, if necessary, providing targeted support.

Let's consider the practice of applying various forms of the state support during the coronavirus pandemic by individual countries (Table 1).

The conducted analysis shows that he state support measures are implemented mainly in the credit sphere (preferential lending, provision of state guarantees, etc.) and in the tax sphere (reduction of tax rates, provision of deferred tax payments). We have conducted a comparative analysis of the amount of the state support in different countries.

Table 1. Government support for business during the coronavirus pandemic in various countries

Support direction	China	Great Britain	Sweden	Canada
Credit area	Loans for tax debts, simplification of the procedure for issuing loans	Interest-free loans for up to 1 year, credit and financial guarantees	Increased credit resources and state guarantees, special support for export industries	Affordable business loans and interest-free loans to cover operating expenses, issuing preferential loans to the lessee for reducing rental rates
Tax area	Reduction of taxes (VAT, income tax, cancellation of insurance and social benefits for employers)	Tax benefits, deferred VAT payments, tax holidays until the end of 2021 for certain industries	Deferred payment of social insurance contributions, payroll tax, VAT	Deferred income tax, federal goods and services tax, and sales tax
Gratuitous financial support, grants and state investment	Investment in industries that create jobs	Grants to cover current business costs, government payment of 80% of employees' wages for 3 months (maximum amount of 2500 lb per month)	SEK 1 million is allocated to support culture and sports	Subsidizing 75% of employees' wages (up to $ 874 per week), for individual enterprises – 100% compensation of employees' pension and social insurance contributions
Monetary policy	Reducing the mandatory reserve ratio for small and medium-sized banks, increasing the availability of funds	Ensuring financial stability and credit availability	Ensuring financial stability and credit availability	Reducing interest rates, supporting the financial market and liquidity of financial institutions

Source: authors.

Figure 1 shows that, in Russia, the scale of funding for state anti-crisis measures is the lowest (about 1.2% of the GDP). Almost all developed countries have undertaken significant anti-crisis programs, including the use of various monetary and fiscal policy instruments. This is because of the high risk of losses on the part of the economy if SMEs leave the market not having any reserves. A reduction in the production, a sharp increase in unemployment, in poverty, and a reduction in real income will reduce the demand. For a clear demonstration of this development of negative consequences, let's consider the macroeconomic equilibrium graph (Fig. 2).

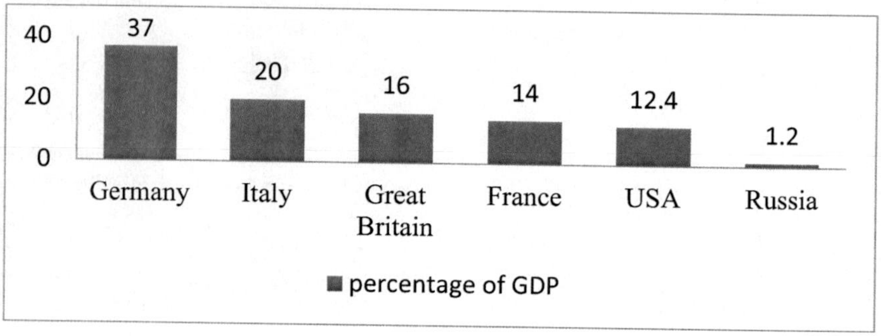

Fig. 1. Comparative scale of anti-crisis measures in Russia and other countries (as a percentage of the GDP). (Source)

Forced self-isolation during the pandemic led to the closure of most businesses. Disruption of production and economic chains and lack of working capital create a high risk of falling production volumes in the economy, which shifts the supply curve from AS1 to AS2. If aggregate demand remains constant, this leads to a decrease in real output (from GDP_{E1} to GDP_2) and an increase in the price level from r1 to r2. However, under current conditions, aggregate demand is also decreasing (as a result of increased unemployment and lower real incomes), which further reduces the real output.

Graphically, this is reflected by the shift of the curve from AD_1 to AD_2, the equilibrium point from E_1 to E_2, and a significant decrease in production volumes from GDP_{E1} to GDP_{E2}. The negative impact on macroeconomic processes occurs both from the factors of aggregate demand and from the factors of aggregate supply.

The high probability of threats being realized and significant losses being incurred forces countries to incur significant financial costs, since it is strategically more profitable to stop the main mechanisms of the crisis spreading at an early development stage than to deal with its negative consequences in the future.

Developing countries are taking more modest anti-crisis measures, but this is explained not so much by the required amount of financial support, but rather by the budgetary capabilities of a state.

For the Russian Government, the priority is to determine needs of the market and its individual subjects for the state support. The development of crisis phenomena in the Russian economy can be much more serious, as the situation is aggravated by the raw material orientation of the economy; that in turn, because of a sharp decline in demand

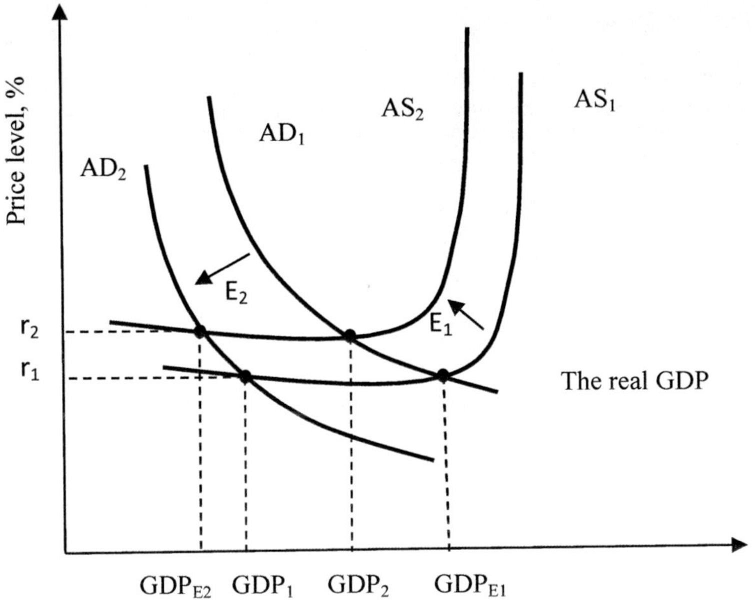

Fig. 2. Macroeconomic equilibrium (Source: authors)

for hydrocarbons, falling oil prices on the world market, will lead to a decrease in the revenue part of the budget.

In our view, the package of anti-crisis measures should be expanded and the scale of financial support from the state should be increased. This is possible as due to the use of funds from the National Welfare Fund (NWF), as well as various monetary and fiscal policy instruments.

In recent years, a strict budget policy has been implemented with the formation of a budget surplus, which will now cover some of the unexpected budget expenditures and budget deficits resulting from the crisis. In our opinion, the funds of the NWF should be used primarily for theses purposes. By April 1, 2020, the federal budget amounted to 12 trillion 855.75 billion rubles, or 11.3% of the GDP. The volume of liquid assets of the fund (funds on bank accounts of the Bank of Russia) amounted to 11,095 trillion rubles (9.8% of the GDP) [6]. Besides, additional budget expenditures can be financed by additional issue of government securities.

Analyzing the package of the Russian anti-crisis measures, it can be noted that the state has mainly limited itself to measures not related to direct expenditures (tax deferrals, state guarantees on loans, etc.). In our view, the state should more actively provide fiscal support to businesses by providing direct subsidies and providing free financial resources to strategically important and system-forming enterprises in case of crisis situations.

4 Discussion

Currently, the economy of Russia and the world economy are going through a crisis. In order to avoid the stage of a prolonged depression, government actions are necessary to revive the markets, maintain the business activity level by increasing public spending, that is, it is necessary to implement a policy of fiscal expansion. The development and implementation of budget policy are determined by goals and objectives of the society. At the same time, a balance should be maintained between public needs and the financial capacity of the state budget. A balanced state budget is an ideal development option, but it is often unattainable. The state's decision on the size of the state budget deficit should be based on an analysis of short- and medium-term development prospects. The choice of specific ways to cover the budget deficit should be based on the analysis of comparative advantages and possible consequences of alternatives.

5 Conclusion

The development of crisis phenomena in the world and Russian economies as a result of the spread of coronavirus infection and the application of restrictive measures puts states in need of rapid anti-crisis decisions. The analysis of methods and tools of the state regulation conducted in this work has shown that the main areas of influence are the tax and credit spheres. The Russian government has mostly limited itself to measures that are not related to direct expenditures (tax deferrals, state guarantees on loans, etc.). In our view, it should more actively implement fiscal expansion, providing direct subsidies and free financial resources to strategically important and system-forming enterprises. Currently, the share of anti-crisis government spending in the GDP is 1.2%, and it is planned to increase it to 4–5%. However, in our view, this is not enough. The state should more actively use fiscal policy tools, covering additional expenses at the base of the National Welfare Fund or placing government bonds on the open market. This policy may lead to a state budget deficit, but it will help to stop the main mechanisms of crisis spreading at an early stage of its development, prevent the development of a crisis phenomena spiral, which will allow us to achieve sustainable growth rates in the long term.

References

1. Bank of Russia: Bulletin of the Department for Research and Forecasting (2020). https://www.cbr.ru/Collection/Collection/File/27780/bulletin_20-02.pdf. Accessed 10 May 2020
2. Domrachev, T.B.: State and non-state support for the development of construction enterprises. Eng. Pers. Future Russia's Innov. Econ. **6**, 43–45 (2017)
3. Starostina, Yu.: Official Central Bank forecast for the Russian GDP (2020). https://www.rbc.ru/economics/24/04/2020/5ea19aff9a7947282c785981. Accessed 10 May 2020
4. Plan to overcome the economic consequences of a new coronavirus infection (prepared by the Government of the Russian Federation). https://www.economy.gov.ru/material/dokumenty/plan_preodoleniya_ekonomicheskih_posledstviy_novoy_koronavirusnoy_infekcii.html. Accessed 10 May 2020 (2020)

5. Resolution of the Government of the Russian Federation of 03.04.2020 N 434 (ed. from 18.04.2020) "On approval of the list of sectors of the Russian economy that are most affected by the deterioration of the situation as a result of the spread of a new coronavirus infection" (2020). https://www.minfin.ru/ru/document/?id_4=117699-svedeniya_o_razmeshchenii_sredstv_fnb_na_depozitakh_veb.rf_po_sostoyaniyu_na_1_aprelya_2020_goda. Accessed 10 May 2020
6. Teryutina, M.M., Tarasov, M.E., Sleptsov, I.I.: State regulation and state support of the agricultural sector of the economy. Econ. Entrepreneurship **9**(86), 365–367 (2017)
7. Cuadrado-Ballesteros, B.: Do progressive governments undertake different debt burdens? partisan vs. electoral cycles¿Se endeudan más los partidos progresistas? ciclo partidista vs. ciclo electoral. Revista de Contabilidad **14**(1), 29–57 (2011). https://doi.org/10.1016/s1138-4891(11)70021-8
8. Giordano, R., Tommasino, P.: What determines debt intolerance? The role of political and monetary institutions. Euro. J. Polit. Econ. **27**(3), 471–484 (2016). https://doi.org/10.1016/j.ejpoleco.2011.02.001
9. Heckelman, J., Wilson, B.: The growth-maximizing level of regulation: evidence from a panel international data. Euro. J. Polit. Econ. **59**, 354–368 (2019). https://doi.org/10.1016/j.ejpoleco.2019.05.001
10. Huanyong, J., Miao, Z.: Corporate social responsibility and collaborative innovation: the role of government support. J. Cleaner Produc. **260**, 121028 (2020). https://doi.org/10.1016/j.jclepro.2020.121028
11. Kearl, J.R.: Rules, rule intermediaries and the complexity and stability of regulation. J. Public Econ. **22**(2), 215–226 (1983). https://doi.org/10.1016/0047-2727(83)90066-x
12. Scuotto, V., Del Giudice, M., Tarba, S., Messeni Petruzzelli, A., Chang, V.: International social SMEs in emerging countries: do governments support their international growth? J. World Bus. (2019, in press). https://www.sciencedirect.com/science/article/pii/S1090951618301585. Accessed 10 May 2020
13. Polat, Z.A., Alkan, M.: The role of government in land registry and cadastre service in Turkey: towards a government 3.0 perspective. Land Use Policy **92**, 104500 (2020). https://doi.org/10.1016/j.landusepol.2020.104500

Decisions in Production and Project Management in the Context of the Digital Economy

Improving Credit Scoring Technology and Stability of a Commercial Bank

V. Ya. Vishnever(✉)

Samara State University of Economics, Samara, Russia
ab3535@mail.ru

Abstract. The technological evolution of credit scoring systems in the Russian consumer lending market deserves close attention in the context of the digital economy. This study investigates the works of leading experts in the development and application of scoring systems. An overview of the most applicable foreign software products in Russian practice is given. Some practical recommendations on the application of advanced technologies in credit scoring of individual borrowers are proposed. The conclusion is drawn about the main directions of credit scoring development in Russia.

Keywords: Consumer lending · Credit risk · Credit scoring · Machine Learning · Stability of a commercial bank · Unsecured consumer loan

1 Introduction

The banking sector in Russia is one of the key segments that supports the stability of the economic system. When making decisions on granting loans to individuals, commercial banks in Russia are faced with the need to minimize credit risks and increase the stability of their organization. In the field of credit risk management, some commercial banks rely on the experience of loan officers in the loan approval process. In assessing the credit risk of the applicant, commercial banks receive the applicant's profile through the application form and from the credit bureau, which checks the applicant's past credit history. The credit rating of the individual borrower is a numerical expression in points for assessing potential risk based on the analysis of the person's credit history. It characterizes the creditworthiness of the person. On this basis, classification of acceptable and unacceptable loan applications is carried out.

One of the most effective tools used by commercial banks to manage credit risks is the credit scoring model. These models allow analyzing a large amount of historical customer data, revealing patterns and characteristics of credit risk to assess new or existing individual borrowers.

When constructing them, traditional statistical and linear programming methods are used, as well as complex methods such as artificial intelligence and Machine Learning. Based on these models, the probability of default - the insolvency of the debtor-individual is determined.

With the rapid development of the credit card business, the credit scoring model has increased its value and prospects. At the same time, some existing scoring models

S. I. Ashmarina et al. (Eds.): ENGINEERING ECONOMICS WEEK 2020, LNNS 139, pp. 299–304, 2021.
https://doi.org/10.1007/978-3-030-53277-2_35

cannot effectively evaluate potentially dangerous customers due to the asymmetry of information in the field of credit card circulation. The introduction of new variable factors into the scoring model may cause their operational failure.

2 Methodology

The research materials are regulatory acts in the field of consumer lending, statistics from the Bank of Russia, the Association of Russian Banks, and research results of Russian and foreign scientists. In the theoretical part of the work, formal logical methods are widely used. In studying the evolution of bank credit scoring programs, system analysis is widely used, which makes it possible to comprehensively consider the transition to new information technologies that can increase the stability of a commercial bank.

Credit scoring analysis is based on a review of expert, statistical and hybrid methods. The expert method is to evaluate a scoring mode, based on subjective judgments of a loan specialist. The statistical method includes the use of logistic regression, decision trees, neural systems. The hybrid method is the most advanced method, based on a systematic analysis of expert and statistical methods, is a procedure fos evaluating all financial and non-financial indicators to build the final model for assessing the borrower's risk.

3 Results

In Russia, for several years until 2018, there was positive dynamics in the growth of consumer lending, especially in the provision of unsecured consumer loans (UCL). This has had a positive effect on maintaining economic growth. In 2019, there was a decrease in the growth rate of UCL. Over the 11 months, the UCL portfolio grew by 20.1%, compared with 22.8% in 2018 [4]. In October 2019, the growth rate of the volume of loans granted to individuals was minimal since February 2018. The share of UCL with overdue debts over 90 days decreased from 9.1 to 7.7% in 2019 [4].

The downward trend in consumer lending is primarily associated with increased regulation of the Bank of Russia. Among the measures taken, the introduction of premiums on the risk factor for high-margin UCL from April 1, 2019, and from October 1, 2019, additional premiums on risk factors for UCL depending on the borrower's debt load (BDL) and the total loan cost (TLC). Loans to borrowers with high debt burden (BDL > 50%) fell under restrictions. The introduction of premiums reduced the possibility of the commercial bank lending due to the need to increase the capital reserve [4].

The second factor influencing the decrease in the rate of growth of UCL was the decision of some of the largest Russian banks independent of the Bank of Russia to tighten lending conditions for individuals to increase their stability. Moreover, the growth of UCL still exceeds the growth of nominal cash incomes of the population, which indicates an increase in the debt burden of the population.

High risks in the field of granting UCL are associated with attempts of banks to increase the terms of loans granted to reduce monthly payments of individuals and, as a result, to reduce personal income tax. This approach can increase risks in the long term, especially in the conditions of economic recession.

The Bank of Russia policy to reduce the key rate, bringing it up to 6.00% does not have a short-term effect on reducing the cost of consumer loans. Current consumer credit rates are largely determined by market conditions, and they depend on the quarterly establishment by the Bank of Russia of the full cost of the loan. The reduction in the key rate has a more pronounced effect on stimulating refinancing in the consumer lending segment.

The consumer lending market in Russia is highly competitive, which increases the degree of risks. Therefore, commercial banks in Russia are striving to apply new methods and banking tools to maintain their competitiveness, increase return on equity, minimize risks, and increase financial stability. In modern conditions, special attention is paid to new information technologies.

The risk management system in Russian commercial banks includes an effective mechanism for assessing the creditworthiness of borrowers, which allows improving the quality of the loan portfolio, which affects profit growth and reduction of credit risk. The observed downward trend in the growth of arrears on loans to individuals is largely determined by modern methods of credit risk management. Among these methods, a special place is occupied by the most technologically advanced, efficient and rapidly developing way of risk assessment in consumer lending - scoring.

Credit scoring involves the use of mathematical and statistical modeling technology to assess the level of credit risk associated with a particular borrower. Credit scoring is used to classify and quantify risk factors for loan repayments by an individual. Commercial credit scoring was first developed in the 1950s in the United States. Currently it is used by most of the world's largest banks. In Russia, credit scoring began to develop with the advent of retail express lending. The high popularity of scoring systems is explained by their advantages due to the computerization of decisions made, which are as follows:

- increase operational efficiency,
- save time,
- increase the accuracy of credit decisions,
- improve the culture of making credit decisions through objective and standardized analysis,
- reduce human errors or bias,
- assign a credit limit based on risk,
- set the price of a loan product based on the analysis of the risk level.

At the same time, modern scoring models have a high cost, difficulties in use. A certain drawback of these models is that analysis is carried out based on historical data, it is assumed that the future period will be similar to the previous period.

On the Russian banking market, both Western and domestic scoring systems are used. The world leader in the development of scoring systems is the American company FICO. Its systems are used by more than 300 Russian banks. It is used in the practice of the Russian National Bureau of Credit Histories (NBCH). All potential borrowers are rated in the range of 250 to 850 points. Borrowers with scores of more than 700 points almost automatically receive a consumer loan. Borrowers with scores of less than 500 points are considered insolvent. Thus, all customers are divided into groups, each of which has its own risk of default and, accordingly, the cost of the loan.

Russian scoring systems used by Russian design bureaus are less functional than foreign systems, but they have lower cost and use the development of Russian applied mathematicians. The largest banks in Russia use their scoring models, which improves the accuracy of credit risk assessment. Medium and small Russian banks, as a rule, use outsourcing services in this market.

The system of retail consumer lending uses 4 main types of scoring:

1. Application scoring. It is an assessment of the creditworthiness of the borrower when granting a loan based on the information available about it. This is the most active type of scoring in Russia.
2. Collection scoring is used for daily effective receivables management.
3. Behavioral scoring. It allows you to predict changes in the creditworthiness of the borrower based on the analysis of the existing history of his relationship with the bank to adjust the established credit limits or generate individual offers.
4. Fraud scoring. Used to statistically assess the possibility of fraud by the borrower.

The application of each type of scoring is carried out using scoring cards. The most famous and widespread model for assessing the creditworthiness of the borrower is compiling a credit card as a set of rating indicators, ranked by points. As a rule, design bureaus develop several variants of scoring cards depending on the type of scoring used.

Scoring cards are an integral part of the scoring model, which is the main tool for credit scoring. Design Bureau follow BASEL recommendations, according to which the assessment and management of credit risk should be based on internal ratings [2]. To do this, their own rating assessment models are created. Each design bureau has its own scores for assessing borrowers who group customers into different risk groups.

The credit rating system makes extensive use of the logistic regression model. As a rule, credit rating systems have ratings of up to 10 points from AAA to D. The algorithm for using the main scoring models in consumer lending is based on classical methods of statistical analysis. At the same time, modern information technologies allow us to expand the range of data being analyzed by studying information about customer activity on social networks, payment behavior, etc.

Currently, there is a global trend in the world for the total automation of banking activities and the widespread introduction of Big Data and Machine Learning financial instruments. The need to further improve the specialization of analytical software training to increase the speed and efficiency of processing large amounts of data has led to the introduction of a new class of technologies and new approaches to information processing, Machine Learning, in the credit scoring process. So, the largest American banks have developed automated Machine Learning models for analyzing documents and aggregating data from them, which has reduced a significant number of loan officers.

Big Data and Machine Learning tools are part of the group of artificial intelligence tools and are widely used in the practice of Sberbank. An important aspect of Machine

Learning is the ability to adapt and retrain on newly arrived data to obtain reliable and representative results.

4 Discussion

Risk is an integral part of the banking market. It cannot be eliminated, but it can be reduced by using appropriate methods. In international practice, there are different approaches to assess the cretitwortiness of the client and the risk of loan default (North American, English and European). The credit scoring system has been deeply and comprehensively studied. In existing works, credit scoring is considered as a process of determining a numerical score based on analysis of the borrower's profile. Many authors emphasize that credit scoring a current assessment of credit risks, the ability and willingness of a borrower to fulfill his financial obligations.

The modern credit scoring system involves scoring at all stages of the movement of the loan product and loan repayment. In credit schemes, a mathematical assessment is usually used in the analysis of historical experience working with the debtor, because of which quantitative models for classification of acceptable and unacceptable loan applications are created.

With the rapid development of bank credit cards, credit scoring models have strengthened their importance [5]. Logistic regression with a step-by-step method of selection is used to assess the parameters of the borrower's credit rating model [3].

The credit scoring method is the basis of the classification method for credit ratings of customers. Credit rating - the process of modeling the creditworthiness of the applicant. The hybrid model improves classification efficiency. The need for control and effectiveness of credit risk management have led financial institutions to improve information technologies and methods developed for this purpose [6].

In recent years, several new credit scoring models have been developed in addition to traditional methods of credit risk analysis. So, new algorithms were proposed and started to be used to increase the accuracy of credit scoring modeling. Among them are: decision trees, neural networks, nearest neighbor coefficient, intelligent Machine Learning, etc. In the short term, considering the pace of development of modern information technologies, self-learning systems will replace banking specialists. Switching to Machine Learning means a breakthrough in credit scoring [1]. Noteworthy is the proposed method for assessing the creditworthiness of the borrower on a continuous scale, in contrast to the binary scale. This method allows assessing each specific borrower, and not a group of "bad" or "good" customers. The results show the feasibility of such a method [7].

5 Conclusion

The study examined the features of the Russian consumer lending market. It is noted that unreasonably high growth rates of consumer lending increase systemic risks of the banking sector, affect the stability of commercial banks, can be an impetus to a recession in the economy, and create an opportunity for a "bubble" in this market.

At the same time, risks of the consumer lending market remain moderate, as the largest banks have enough capitalization and profitability and they can withstand a slight decrease in the quality of their assets.

The financial stability of Russian design bureaus is supported by accumulated reserves. Over the past year, the Bank of Russia has taken a few cardinal measures to cool this market. Among them, one can note a change in the approach of the Bank of Russia to the regulation of unsecured loans granted by commercial banks, which has already led to a decrease in the growth rate of the loan portfolio. These measures forced commercial banks to create a capital buffer that ensures their stability.

The Russian credit risk management system in the consumer lending market includes one of the most progressive, knowledge-intensive and effective methods – credit scoring of the borrower. In our country, this credit risk assessment system appeared much later than in the USA and Europe, where it has been observed the experience of using traditional and latest scoring methods for more than half a century. The largest Russian banks widely use both Western developments and their own models. Medium and small banks use these models on the outsourced basis.

References

1. Agafonova, V., Vishnever, V., Frumusaki, S, Tatishcheva, V.N.: Technological evolution of credit scoring in the banking consumer credit system. Bull. Volga Univ. **2**(3), 21–26 (2018)
2. Investopedia: BASEL accord (2019). https://www.investopedia.com/terms/b/basel_accord.asp. Accessed 12 Feb 2020
3. Bilginci, M.R., Kaya, G.O., Turkyilmaz, A.: Decision support system for credit risk management: an empirical study. Int. J. Inf. Syst. Serv. Sect. **11**(2), 14 (2019). https://doi.org/10.4018/IJISSS.2019040102
4. CBR: About banking's development sectors of the Russian Federation in 2019. Information and analytical material (2020). https://cbr.ru/Collection/Collection/File/25854/razv_bs_19_12.pdf. Accessed 12 Feb 2020
5. Hongmei, C., Yaoxin, X.: The study of credit scoring model based on group lasso. Proc. Comput. Sci. **122**, 677–684 (2017). https://doi.org/10.116/j.procs.2017.11.423
6. Louzada, F., Ara, A., Fernandes, G.B.: Classification methods applied to credit scoring: systematic review and overall comparison. Surv. Oper. Res. Manage. Sci. **21**(2), 117–134 (2016). https://doi.org/10.116/j.sorms.2016.10.001
7. Romanyak, K.: Credit scoring based on a continuous scale for on-line credit quality control. In: Sayed-Mouchaweh, M., Fleury, A., Angelov, P., Lughofer, E., Iglesias, J.A., (eds.) International Conference on Evolving and Adaptive Intelligent Systems IEEE, pp. 158–162. IEEE, New Jersey (2015). https://doi.org/10.1109/eais.2015.7368796

Digital Household Financial Behavior

E. S. Nedorezova(✉), K. N. Ermolaev, and F. F. Salamov

Samara State University of Economics, Samara, Russia
nedlen63@yandex.ru, ermolaevkn@yandex.ru,
farrux_sies@mail.ru

Abstract. In modern-day Russia a holistic perception of the country's economy is impossible without the development of households. According to recent censuses, there are approximately 53 million households in the Russian Federation. This suggests that households play a large role in the country's development and prove the need to study their place and role in the modern economy. A properly organized and efficiently operating financial system of a country has a positive impact on economic development and the growth of national wealth, especially in the context of digitalization. The structure of financial capital of the economic system is heterogeneous and diverse. The condition and development prospects of each financial source are of great importance for the functioning of the financial system as a whole. Household finances occupy a special place among the resources of the national financial market, characterizing not only the standard of living associated with household consumption, income and expenses, but also representing a significant investment resource in Russian economic conditions.

Keywords: Financial behavior · Households · Investment strategy

1 Introduction

At present, Russia faces an important task of transition from the resource-oriented economy to the information one. The experience of developed countries shows that they widely use market-based mechanisms to stimulate investment activity, form programs to attract domestic and foreign investments and a system of investment funds of various types is a standard form of economic behavior.

For the effective use of investments in Russia, it is necessary to attract "long money" into the economy, which has been accumulated, first of all, by households. However, this money plays a modest role in the Russian investment market due to issues of distrust of the population, low financial literacy, problems of using and developing financial institutions. The pace of development of mass investment processes, which Russian households are connected to in order to increase the return on their investments, is significantly behind the similar processes in other countries.

The formation and updating of the financial behavior of households, the financial strategies used by them, is quite complicated and contradictory, and requires a correct understanding of the motivation of potential investors. In this regard, theoretical and empirical studies of the problems of the financial behavior, aggravated in the context of the global financial and economic crisis, and geopolitical contradictions are becoming especially relevant.

S. I. Ashmarina et al. (Eds.): ENGINEERING ECONOMICS WEEK 2020, LNNS 139, pp. 305–310, 2021.
https://doi.org/10.1007/978-3-030-53277-2_36

The study of one of the most important components of the financial behavior of the population – investment – is carried out by scientists using a range of methods to identify the most significant motivating factors. The relevance of such studies is due to socio-economic problems of Russia, which do not allow our country to adequately develop in the context of digitalization. In this case, this is a weak level of investment activity, which affects the formation of the financial and resource base of the investment process, which characterizes a low effectiveness of the financial system. This system is not able to meet the needs of the national economy with sustainable sources of financing. This leads to difficulties in forming the resource base of commercial banks, increasing the dependence of the domestic stock market on foreign capital, and the lack of sustainable long-term sources of financing for the Russian economy.

A promising way to solve the problem may be household savings as one of the possible significant sources of investment resources for the effective restructuring and recovery of the domestic economy in the context of digitalization. In countries with a stable market economy, the investment-saving behavior of the population plays a special role, since savings are a valuable resource for economic development, an internal source of lending to the economy, the basis of the potential for domestic investment, which is a source of free cash that is resistant to fluctuations in the global capital market [1]. In Russia, mechanisms that would allow the effective implementation of these processes and incentives encouraging the population to make their choice in favor of saving and subsequent investment, rather than consuming durable goods, have not been finalized. The investment and savings behavior of the population should be aimed at ensuring the long-term financial needs of the individual or household and earning income from allocated funds.

2 Methodology

As the basic tools in the work, the principles of reproductive, functional, systemic and institutional approaches were used, as well as general scientific methods of cognition (analysis and synthesis, induction and deduction, abstraction), which made it possible to systematize and supplement knowledge in the field of studying the financial behavior of households. The methodological basis of the study was classical tools of formal logic: analysis, synthesis, induction, deduction, abstraction, comparison. Among the widely used methods, we can distinguish mathematical, statistical and graphical methods used to determine the development trend and dynamics of the volumes of the Russian sector. They made it possible to ensure the complexity of scientific research and the reliability of conclusions.

3 Results

Households as one of the most important subjects of the economy have significant financial resources. In modern society, the financial behavior of the population is studied very actively. In the context of geopolitical and economic tensions, household finances have become a vulnerable link in the financial system. At the same time, one should not forget that financial resources of the population are a necessary source of the investment process in the country's economy and a condition for expanded reproduction.

The ability of the economy to overcome negative trends in the context of digitalization to a large extent depends on the financial choice and financial behavior of households. For example, the savings of the population are a powerful investment tool, and the issues of using various financial instruments and the financial literacy of the population show the degree of development of financial institutions in the country.

The financial behavior of the population is the subject of study of various social sciences, in particular economic theory, sociology, psychology. The financial behavior is an important component of the socio-economic behavior of people, an integral element of the life of a modern person. The financial behavior of the population in a broad sense is characterized as the activity of mobilizing and using financial assets. In our opinion, the financial behavior is a set of consumer, savings, and investment behaviors implemented by households aimed at mobilizing, redistributing, and investing money, built to meet their needs.

The financial behavior is determined by a financial strategy for using cash. In modern conditions, financial behavior strategies can be presented in pure form, as well as combined with each other. In our opinion, the key is an important circumstance – household savings are the basis for the formation of a savings and investment strategy. The savings – investment behavior is a guarantee of security of the individual household and one of sources that finance the state system. The financial well-being of the family and people's confidence in their future depend on how rationally and actively households manage their savings. Household cash savings represent a certain portion of cash income that does not go towards consumption and taxes. This part of the income must necessarily have a target structure, otherwise, savings turn into a simple accumulation of money. Household savings can be seed capital for business development or a source of investment [3].

As soon as household savings pass into investment, it should be said that the household carries out the investment behavior. In order to understand the mechanism of transformation of household savings into investments, that is, the formation and implementation of the investment behavior, it is necessary to determine the economic nature of these categories and factors affecting this transformation.

The savings behavior is a structural element of the financial behavior, the basis of which is the category of "savings". The term "savings" is extremely ambiguous and has many interpretations. In the narrow sense of the word, savings are only deposits in bank accounts. An expanded interpretation assumes that the subject has, in addition to all types of cash accumulations, various kinds of shares, liquid real estate goods. Savings are sometimes defined as the balance of household income after making mandatory payments and purchasing necessary goods and services, used to generate future income or ensure future consumption.

In our opinion, savings are the material basis of the saving behavior, the source through which this behavior is realized via the activities of economic agents. The savings behavior is a part of the financial behavior that reveals the activities of households in the formation and use of savings. Now, we examine the category of the investment behavior of households. The investment behavior is considered as actions of individuals and legal entities associated with the investment of their capital (values, money and other liquid scarce resources) in various objects of economic life in order to obtain economic or social effect (profit) in the context of taking into account individual choice from a variety of socio-economic alternatives [4].

The investment behavior should be understood not only as a purely economic process of investing capital (money). This process is associated with the development of economic relations, in which many actors are involved: those who accumulate investment resources, those who carry out the functions of their redistribution, and those who market them. Thus, the essence of investment can be formulated as the investment of values in all various objects of the economic life of society in order to achieve a social and economic effect.

4 Discussion

The financial behavior of households is a complex category that includes various types of behavior that should be considered in a single bundle, synthesizing them into a multi-aspect continuous process. An example of such a symbiosis is the study of the savings and investment behavior. Today, in the context of digitalization, there is an acute problem to develop and implement the savings and investment behavior, rather than consumer activity of households. A weak level of investment activity of households affects the financial and resource base of the investment process, characterizing a stably low indicator of the financial system, which is not able to meet the needs of the national economy with sustainable sources of financing. This leads to difficulties in forming the resource base of commercial banks, increasing the dependence of the domestic stock market on foreign capital, and the lack of sustainable long-term sources of financing for the Russian economy. A promising way to solve the problem can be the development of the saving behavior of households as one of the possible significant sources of investment resources for effective restructuring and recovery of the domestic economy. In countries with a stable market economy, the saving and investment behavior of households plays a special role, since savings are a valuable resource for economic development, an internal source of lending to the economy, the potential for domestic investment, which is a source of free cash that is resistant to fluctuations in the global capital market [2].

In Russia, mechanisms that would allow the efficient implementation of these processes and incentives encouraging the population to make their choice in favor of saving and subsequent investment, rather than consuming durable goods, have not been finalized. The investment and savings behavior of the population should be aimed at ensuring the long-term financial needs of the individual or household and earning income from allocated funds. A multilateral analysis of factors affecting the financial behavior of households in the digital economy and the types of strategies they use in managing money resources is very relevant and requires an integrated approach to the study of the motivation of the population in the field of savings and investments [5].

The basis of the saving and investment behavior of households is the decision-making process, which depends on a whole range of external and internal, subjective and objective factors and motives. It should be emphasized that there are circumstances that bring together and at the same time distance the forming elements of the savings and investment behavior. The logic of cause and effect ties brings together, since savings are the basis for investment. The level of savings and investment are not always determined by the same factors.

5 Conclusion

The research is focused on the issue of placing savings in various financial instruments. This process reflects the transformation of savings into national investment. Tools for investing, as well as how to use them, should be accessible and understandable to the public. Placing the savings of the population in the financial system should bring a positive financial result for the owners of these savings.

In recent decades, the role of household finance in the economy has been growing rapidly: the importance of personal consumption and personal investment is for the banking sector is growing for GDP. Against this background, the financial behavior of households in general and its specific elements (savings, investment, etc.) is attracting attention of scientists and researchers. Of particular relevance is the study of the saving behavior of the population in unstable socio-economic conditions, during periods of crisis. The way the population manages ist money in many respects depends on how the economy as a whole will survive the crisis.

Issues related to identifying the characteristics of the financial behavior of the population and determining the amount of savings accumulated by it are actively studied and discussed both in foreign and in Russian scientific and expert communities. There are several reasons for this. The structure and dynamics of savings are integral indicators of the material well-being of citizens and economic development of the state. They have a significant impact on development conditions of the banking sector and indicate the level of trust in financial institutions and government bodies that have developed in society. But, most importantly, the finances of the population are a significant investment resource, the use of which promises considerable benefits to the state.

References

1. Jünger, M., Mietzner, M.: Banking goes digital: the adoption of FinTech services by German households. Finance Research Letters (2019). https://www.sciencedirect.com/science/article/abs/pii/S154461231930296X. Accessed 11 Feb 2020. in Press
2. Kettlewell, N., Morris, R.W., Ho, N., Cobb-Clark, D.A., Cripps, S., Glozier. N.: The differential impact of major life events on cognitive and affective wellbeing. SSM Popul. Health, **10,** 100533 (2020)
3. Li, J., Meyer-Cirkel, A.: Promoting financial literacy through a digital platform: a pilot study in Luxembourg. Int. J. Financ. Econ. (2019). https://onlinelibrary.wiley.com/doi/abs/10.1002/ijfe.1777. Accessed 11 Feb 2020. in Press

4. Palakvangsa-Na-Ayudhya, S., Pongchandaj, S., Kriangsakdachai, S., Sunthornwutthikrai, K.: KeptAom: savings management system to increase long term savings behavior of children. In: Bartleson, K., (ed.) IEEE Region 10 Annual International Conference, pp. 2247–2252. IEEE, New Jersey (2017)
5. Savinova, V.A., Zhegalova, E.V., Semernina, J.V., Kozlova, A.S.: Problems of the development of the digital economy at the regional level. In: Ashmarina, S., Mequita, A., Vochozka, M., (eds.) Digital Transformation of the Economy: Challenges, Trends and New Opportunities. Advances in Intelligent Systems and Computing, vol. 908, pp. 260–268. Springer, Cham (2020). https://doi.org/10.1007/978-3-030-11367-4_25

Specific Features of Taxation for Foreign Companies Receiving Income from Russian Sources

K. S. Pavlova[1] and N. V. Knyazeva[2(✉)]

[1] Samara State University of Economics, Samara, Russia
mrs.pavlova@list.ru
[2] Association of Professional Forensic Experts, Moscow, Russia
natali_plus@mail.ru

Abstract. In the current model of the economy global digitalization, the state is required to ensure the investment attractiveness of the country for foreign investors, since the involvement of foreign organizations in the economy directly depends on the current system of taxation of foreign organizations. The purpose of the contribution is to assess the current tax legislation determining the tax base and structure of accrued income tax for foreign organizations that receive income from sources in Russia. When receiving income from Russian companies, foreign organizations may form permanent establishments, or pay tax at the payment source of such income by a Russian firm that acts as a tax agent. The paper investigates tax accruals of permanent establishments and income received by foreign beneficiaries. The authors consider some specific features and actual taxation indicators of controlled foreign companies whose activities are controlled by Russian tax residents.

Keywords: Controlled foreign company · Dividends · Permanent establishment · Tax agent · Income tax

1 Introduction

Taxation of foreign organizations is a complex and multifaceted concept, which is especially relevant today in connection with the expansion of international relations, the globalization processes of the world economy, and the direction to the development of the innovative and digital economy in Russia. In recent years, there have been some changes in Russian tax policy related to foreign economic transactions. The application of the deoffshorization policy for the Russian capital in 2015 is already producing results. In 2018, 3,362 controlled foreign companies were registered by tax authorities, which indicates the effectiveness of the capital amnesty conducted in three stages, the last of which is completed in 2020. A lot of questions are raised about criteria for establishing a permanent establishment and the role of a tax agent – Russian individuals who are the source of income for foreign companies in Russia. The preparatory and auxiliary activities for the establishment of a permanent establishment are controversial.

S. I. Ashmarina et al. (Eds.): ENGINEERING ECONOMICS WEEK 2020, LNNS 139, pp. 311–317, 2021.
https://doi.org/10.1007/978-3-030-53277-2_37

In Russia, together with the domestic legislation, provisions of international treaties are applied too, which gives them priority over provisions of the Russian legislation. An example of the operation of international agreements is the common economic customs space of the Eurasian Economic Union, which allows using benefits and a simplified form of customs control and payment of customs duties and taxes for the Russian side and all member countries. Within the framework of international agreements, Russia applies 84 avoidance of double taxation agreements (hereinafter DTAs), which allow not to pay taxes twice under the laws of different countries on the same income, if it was received by a resident of one country from sources in another country. Agreements are also used as a guarantor of the exchange of tax information with other countries, although this provision is made separately, so many agreements are currently subject to changes.

2 Methodology

Different methods were used in the research. These include: theoretical (dialectical logic, rational cognition, etc.); diagnostic (diagnostic analysis of the state and causes); empirical (description of facts, measurement and generalization of research results). In Russia, foreign economic activity is primarily based on the legislative regulation by internal (Russian) legal acts, as well as international agreements. The main federal documents regulating foreign trade include:

1. The Constitution of the Russian Federation.
   Article 15, Paragraph 4, of the Constitution of the Russian Federation states that international treaties are part of the Russian legislation. If an international treaty that is in force (ratified or recognized as valid) in Russia has adopted other terms or rules than those provided by domestic legislation, the rules of the international treaty will be applied [2].
2. The Civil Code of the Russian Federation.
   The Civil Code establishes a rule stated in the Constitution on international treaties (article 7 of the Civil Code [1]), the same rule is repeated in article 7 of the Tax Code [12]. Article 127 states that the liability of legal entities and individuals who conduct activities with the participation of foreign persons is determined by the law on the immunity of the state and its property. Article 317 regulates the use of the currency, monetary obligations in the Russian Federation, currency transactions are more broadly described in the federal law No. 173-FZ “On currency regulation and currency control” of 10.12.2003 [4].
3. The Tax Code of the Russian Federation (part one) and the Tax Code of the Russian Federation (part two).
   The Tax Code regulates issues related to the elimination of double taxation in relation to individuals and organizations; determines the taxation procedure for controlled foreign companies; tax rates and specifics of taxation of foreign persons’ income, including organizations and individuals from sources in Russia; determines the procedure for paying VAT from activities of foreign organizations at the sale place of goods, works and services; the procedure for reimbursement of export

VAT, the specifics of paying excise taxes; exemption from taxes when applying certain customs regimes; defines the taxation procedure for income of foreign companies that pay tax on profits from sources in the Russian Federation through tax agents when carrying out activities that do not lead to the formation of permanent establishments; the specifics of taxation of permanent establishments of the main and dependent types; the specifics of calculating and paying personal income tax and insurance premiums for foreign individuals [12].

4. Federal law No. 164-FZ "On the basis of the state regulation of foreign trade activities" of 08.12.2003.
   This law contains the main provisions on the foreign economic activity regulation; reveals definitions of the most important concepts in the contax of foreign trade. International treaties are an important part of the Russian legal framework. A contract can be concluded between two countries or between a large number of countries. Treaties reflect decisions of top officials of countries and their governments who agree on a specific issue related to geopolitics or economy. Of particular interest for tax purposes are international agreements for the avoidance of double taxation (DTAs), which take into account norms and differentiation of rights for taxation of income of organizations and individuals for each of the countries. This approach allows not to pay tax in full in two different states on the same income, but to either set off the tax, or pay the difference or calculate the tax at the rate specified in the agreement. Accordingly, in the Russian legislation, DTAs address tax issues that relate to personal income tax, corporate income tax, and property taxes of individuals and legal entities [3].

To apply the DTAs rules, a foreign person should provide the tax agent who is the source of income payment in a free form with a document confirming the permanent location of the foreign person in a foreign country. This certificate is usually issued either by foreign tax authorities or by departments of the Ministry of Finance. The tax agent takes this document into account when calculating the tax and applies the provisions of DTAs. If such a document was not provided, or the tax in a foreign country was not paid, the income tax is calculated by the tax agent in accordance with the legislation on taxes and fees.

It should be noted that the regulation of foreign economic activity is provided with an integrated approach to its management. Through tax reports and customs declarations, statistical data is collected on operations related to the conduct of foreign trade, control in this case is assigned to the tax and customs authorities; the analysis is conducted, recommendations for improving the legislation are directed by various ministries, especially the Ministry of Economic Development of Russia, some functions for the foreign economic activity regulation are highlighted for the President and the Government of the Russian Federation.

3 Results

According to the Federal Tax Service, in 2017, dividend payments to foreign organizations amounted to 1.5 trillion rubles, or 17% of all dividends paid to shareholders. This figure was 32% in the structure of tax revenues from income on dividends paid in Russia for 2018, or 2.1 trillion rubles [13]. The growth rate of tax revenues of foreign organizations from a source in the Russian Federation in the form of dividends was 43% in 2018, and 40% in 2017. Among the countries whose resident companies have the most significant shares of investment in the Russian economy and business, there are organizations-residents of such countries as the United States, the Republic of Cyprus and the Kingdom of the Netherlands, which in total own more than half of the shares of Russian companies among foreign shareholders. According to this data, the Russian economy consistently attracts foreign capital, the share of which in Russian companies is slightly less than 20%, and budget revenues in the form of taxes on dividends received by foreign companies are increasing every year and make up a large part of the structure of tax revenues from the profits of foreign organizations.

Much attention is paid in Russia to monitoring the activities of permanent establishments of foreign organizations. According to the Federal Tax Service (FTS) in 2018, the amount of revenue from the profit tax of foreign organizations from their permanent establishments amounted to 29.6 billion rubles (4.9 billion rubles to the federal budget and 24.7 billion rubles to regional budgets), the annual growth rate of tax revenues was 6% in 2018. In 2017, the increase in income tax receipts from the activities of permanent establishments was 15% [13]. However, based on the statistical data of income tax reporting, which reflects the income of permanent establishments of foreign organizations, there is a decrease in the total income of permanent establishments, for example, in 2017, this indicator decreased by 11% (1.1 trillion rubles) compared to 2016 data (1.3 trillion rubles). In 2018, this indicator increased by 20% and amounted to 1.4 trillion rubles [13]. Over the past three years, there has been a general decline in the number of permanent establishments of foreign organizations, with 2,290 companies registered in 2018, and there were 2,500 of them in 2016. It indicates a reduction of the number of companies by 8% for two years. This form of activity in the Russian Federation is not so popular by foreign organizations, since these companies prefer to pay taxes on income from sources in the Russian Federation through tax agents. A permanent establishment is subject to the state registration by the tax authorities as a taxpayer. This procedure leads to obligations to maintain tax and accounting records within the framework established by the legislation of the Russian Federation, i.e. to calculate and pay taxes and submit tax reports. If you pay tax through a tax agent, he takes all the obligations on calculation and payment of taxes.

The participation of Russian individuals in controlled foreign companies (CFC) remains a difficult issue. The Base Erosion and Profit Shifting plan (BEPS) calls for the state regulation of CFC. The application of CFC legislation in Russia involves the introduction of a procedure for the voluntary declaration of foreign accounts and other assets, and the state registration of CFC through the procedure of redomiciling companies. The concept of redomicilation means changing the official address of registration of a legal entity and changing the country of actual residence, according to which the organization will be subject to the tax law of the country where it moved to. In 2018, the amount of tax payable on CFC profits was 2.9 billion rubles (3362 CFC were registered in 2018 according to the Federal Tax Service [13]); in 2017, it was 1 billion rubles (4820 registered CFC in 2017); in 2016, this figure was 2.7 billion rubles (3596 CFC registered in 2016). The decrease in the indicator by almost 3 times in 2016 was because of an increase in the amount of loss accounted for CFC tax purposes in 2017. For three years, the largest amount of corporate income tax payable from CFC activities was in 2018, which generally indicates the fruitful results of the program for deoffshorization of foreign capital in Russia.

4 Discussion

In 2017, the Russian Federation signed a multilateral convention on the implementation of measures related to tax agreements to counteract the erosion of the tax base and to eliminate the withdrawal of profits from taxation (in international practice: "Multilateral Instrument" or «MLI»). Russia has chosen a path of implementing a full package of requirements for limiting benefits applied in the DTAs. We are talking about introducing restrictions on benefits for certain types of passive income, which include dividends, income from the sale of shares and shares in companies; on the development of provisions for organizations to avoid the status of "permanent establishment". Since 2019 Russia has been actively cooperating with the International Convention MLI concerning the implementation of measures on the DTAs, it is planned to make changes in 70 bilateral DTAs, for example, in the DTAs with the United States, France, Italy, Cyprus and others.

International information exchange is a basic condition for improving international taxation, which is the topic of Action 1 of the BEPS plan "Addressing the Tax Challenges of the Digital Economy" [8]. On the official BEPS website, the exchange of data on beneficiaries and tax information is a key criterion for cooperation within the OECD. The main goal of the OECD is to establish an automatic exchange of information between member countries, which will ensure a rapid and targeted information exchange at the request of government agencies that control the taxation and tax evasion. For this purpose, the OECD hosts a global forum on transparency and information exchange of tax issues, whose main objective is to ensure high standards of transparency in tax accounting and information exchange. According to the OECD, since 2014, the transfer of data on income (including dividends, royalties, interest paid from participation in activities, wages, gross income, pensions, etc.) has already taken place between more than 90 countries, thanks to which more than 95 billion Euros have been returned to budgets [9]. Through the exchange of information, facts on 47 million accounts in offshore jurisdictions, whose total value was about 4.9 trillion Euros, were disclosed to the regulatory authorities.

The issue of digitalization is very relevant today. In particular, the role of digitalization of economic processes and its impact on international taxation are considered in the work of Turina [14]. The assessment of the OECD measures aimed at reducing the erosion of the tax base for foreign and international companies is considered in the work of Ismer and Jescheck [6].

Another important issue on the taxation of foreign companies, which is considered in the BEPS plan, is the choice of countries and jurisdictions with lower tax rates made by foreign organizations. This problem is covered in the article by Garner-Beuerle, Mucciarelli, Schuster, & Siems [5]. The authors found that the UK is of the greatest interest for taxation purposes of large corporations in the European Union. The prerequisites and reasons for moving activities of multinational companies in the United States for tax purposes to offshore zones – territories with a low tax burden, are disclosed in the article by Richardson & Taylor [11].

In the framework of Action 7, the OECD clarified requirements for dependent agents of foreign organizations and raised a question of recognizing commission structures whose activities in the Russian tax legislation do not lead to the formation of permanent establishments as dependent agents of foreign companies. The OECD, in Action n 7 of the BEPS plan, states that the use of commission agreements attracts a lot of attention from tax authorities in many countries, and this is also confirmed by the significant number of court proceedings related to such structures in the OECD member countries [7]. In most of these cases, the tax authorities' claims were rejected by the courts. According to the OECD, an independent agent should not be considered as a person who conducts activities for several organizations that are interdependent with the agent. The OECD encourages other countries to be guided in determining the dependent or independent status of an agent of a foreign company not only by the criteria formally established by law, but also by the economic essence of such an agent's activity. Research on this topic was conducted in the work of Petruzzi & Holzinger [10].

5 Conclusion

A significant problem with the taxation of foreign organizations is primarily the imperfection of the Russian legislation in the field of foreign trade taxation. Active cooperation within the framework of indirect participation in OECD programs allows us to expand boundaries of international standards on taxation and build the Russian tax policy in such a way as to increase the trust of foreign partners and arouse interest in cooperation. Active development and application of "digital economy" technology in the scope of taxation is already bringing results: taxpayers use FTS e-services. These services are also convenient for foreign traders, allowing them to address issues on taxation remotely and accelerated. The problem of deoffshorization of Russian capital remains urgent, but for this purpose active preparations are being made ensuring the information exchange which is already used with some countries and helps to control the outflow of foreign and domestic capital from Russia to countries with preferential tax rates, especially in offshore jurisdictions. The introduction of the concept of taxation of controlled foreign companies and the development of an effective mechanism for taxing their income made it possible to obtain additional tax revenues through the "amnesty of foreign capital" procedure and to keep records of transactions of Russian beneficiaries abroad.

Another important area of the Russian tax policy development is to support the OECD principles in modernizing the concepts of dependent and independent agents.

The possibility of excluding the concept of "commission agent" from the list of persons who do not form a permanent establishment of a foreign organization should be considered. This will allow expanding the influence area of the tax legislation, avoiding legal disputes and proceedings on this issue (at the moment there is already such judicial practice in Russia), as well as implementing the principle of fair taxation.

References

1. Civil code of the Russian Federation (1994). http://www.consultant.ru/document/cons_doc_LAW_5142/. Accessed: 20 Feb 2020
2. Constitution of the Russian Federation (1993). http://www.constitution.ru/. Accessed 27 Mar 2020
3. Federal law No. 164-FZ "On the basis of the state regulation of foreign trade activities" of 08.12.2003 (2003). http://www.consultant.ru/document/cons_doc_LAW_45397/. Accessed 13 Mar 2020
4. Federal law No. 173-FZ "On currency regulation and currency control" of 10.12.2003 (2003). http://www.consultant.ru/document/cons_doc_LAW_45458/. Accessed 12 Mar 2020
5. Gerner-Beuerle, C., Mucciarelli, F.M., Schuster, E., Siems, M.: Why do businesses incorporate in other EU member states? An empirical analysis of the role of conflict of laws rules. Int. Rev. Law Econ. **56**, 14–27 (2018). https://doi.org/10.1016/j.irle.2018.05.002
6. Ismer, R., Jescheck, C.: The substantive scope of tax treaties in a post-BEPS world: article 2 OECD MC (Taxes Covered) and the rise of new taxes. Intertax **45**(5), 382–390 (2017)
7. OECD: preventing the artificial avoidance of permanent establishment status, Action 7 - 2015 final report. (2015). https://doi.org/10.1787/9789264241220-enhttps://read.oecd-ilibrary.org/taxation/preventing-the-artificial-avoidance-of-permanent-establishment-status-action-7-2015-final-report_9789264241220-en#page5. Accessed 11 Feb 2020
8. OECD: BEPS actions (2019). http://www.oecd.org/tax/beps/beps-actions/. Accessed 25 Mar 2020
9. OECD: What is BEPS? (2019). http://www.oecd.org/tax/beps/about/. Accessed 25 Mar 2020
10. Petruzzi, R., Holzinger, R.: Profit attribution to dependent agent permanent establishments in a post-BEPS era. World Tax J. **9**(2), 263–298 (2017)
11. Richardson, G., Taylor, G.: Income shifting incentives and tax haven utilization: evidence from multinational U.S. firms. Int. J. Account. **50**(4), 458–485 (2015). https://doi.org/10.1016/j.intacc.2015.10.001
12. Tax code of the Russian Federation from 31.07.1998 (1998). http://www.consultant.ru/document/cons_doc_LAW_19671/. Accessed 22 Feb 2020
13. The federal tax service: statistics and analytics (2020). https://www.nalog.ru/rn77/related_activities/statistics_and_analytics/. Accessed 22 Feb 2020
14. Turina, A.: The progressive policy shift in the debate on the international tax challenges of the digital economy: a "Pretext" for overhaul of the international tax regime? Comput. Law Secur. Rev. **36**, 105382 (2019). https://doi.org/10.1016/j.clsr.2019.105382

Optimization of Production and Sales Program of an Oil Company

A. P. Sizikov(✉)

Samara State University of Economics, Samara, Russia
apsizikov@mail.ru

Abstract. The results of a study carried out within the scope of an oil company management digitalization program are given. Influence of the seasonal fluctuation of demand on production volume and stock of petroleum products has been analyzed. A complex task of determining the optimal levels of stock and coordinating production and sales plans with regard to this stock is discussed. The problem is aimed at establishing the yearly production and sales cycle at which the maximal profit of the company is achieved. The problem is solved using a systemic approach. A two-level optimization algorithm monotonically converging for a finite number of iterations is suggested. Respective software has been developed. Experimental calculations have been completed. Recommendations concerning the intra-year dynamics of production and stock by the main groups of commercial petroleum products are given.

Keywords: Management digitalization · Oil refinery · Systemic approach · Production and sales strategy · Stock optimization

1 Introduction

The demand for petroleum products has a prominent seasonal nature and, as a rule, does not match the current production volume. For example, in winter, the demand for gasoline and diesel fells, but, on the contrary, grows for black oil. During this period, companies have to produce excessive volumes of gasoline and diesel to meet the seasonal demand for black oil.

The discrepancy continuously arising between the current volumes of production and sales is attenuated by stock. Availability of stock makes production and sales relatively independent on each other, rises the degree of operational management freedom. On the other hand, stock means conservation of current assets and direct expenses. A complex problem arises that consists in managing the stock and coordinating production and sales plans in order to achieve the highest general result.

2 Methodology

Let us assume that an oil company manages n businesses producing *m* products. Let us consider the problem of optimizing the company's production and sales operation within a timespan consisting of *T* periods, for example, within a year broken down by months.

S. I. Ashmarina et al. (Eds.): ENGINEERING ECONOMICS WEEK 2020, LNNS 139, pp. 318–323, 2021.
https://doi.org/10.1007/978-3-030-53277-2_38

Let us introduce three groups of variables: x_{jt} is the load of j-th business during the t-th period (it is measured in the volume of oil being refined); y_{it} is the sales of the i-th product during the t-th period; z_{it} is the stock of the i-th product as of the beginning of the t-th period. All variables are nonnegative.

The problem is to maximize the marginal income of the company (the revenue less the expenses on product production and storage):

$$-\sum_{j=1}^{n} p_{jt}x_{jt} - \sum_{i=1}^{m} r_{it}z_{it} - \sum_{i=1}^{m} c_{it}y_{it} \to \max, \tag{1}$$

where p_{jt} is the variable fraction of production expenses falling on one ton of refined oil; r_{it} is specific losses due to stock; c_{it} is the commodity products' prices.

Material balance conditions must be met:

$$z_{it} - z_{i(t-1)} + y_{it} \sum_{j=1}^{n} a_{itj}x_{jt} = 0,\ i = 1, 2, \ldots, m,\ t = 1, 2, \ldots, T, \tag{2}$$

where a_{itj} is the output factor of the commercial product in the current period; z_{i0} is the opening stock of product.

The business load constraints are:

$$U_{jt}^{-} \le x_{jt} \le U_{jt}^{+},\ i = 1, 2, \ldots, m,\ t = 1, 2, \ldots, T, \tag{3}$$

where U_{jt}^{-}, U_{jt}^{+} are the limits of business load in the current period.

The constraints of sales due to the seasonal demand are:

$$D_{it}^{-} \le y_{it} \le D_{it}^{+},\ i = 1, 2, \ldots, m,\ t = 1, 2, \ldots, T, \tag{4}$$

where D_{it}^{-}, D_{it}^{+} are the limits of product sales in the current period.

The stock constraints (for all products in aggregate) are:

$$\Omega_t^{-} \le \sum_{i=1}^{m} z_{it} \le \Omega_t^{+},\ t = 1, 2, \ldots, T, \tag{5}$$

where $\Omega_t^{-}, \Omega_t^{+}$ are the stock limits in the current period.

Since the problem implies investigation of production and sales cycles repeated within a year, it is necessary to take into account that the end of the current year is the beginning of the next year; hence, the opening and closing stock must coincide for all products:

$$z_{i0} - z_{iT} = 0,\ i = 1, 2, \ldots, m. \tag{6}$$

The obtained problem of linear programming would have presented no theoretic interest should columns $\bar{a}_{tj} = (p_{tj}, a_{1tj}, a_{2tj}, \ldots, a_{mtj})^T, j = 1, 2, \ldots, n, t = 1, 2, \ldots, T$, have been constant. In fact, they are variable. Every such column represents the structure of a commodity plan of a business. Production programs of businesses vary depending on the season and in line with external and internal technical and economic conditions. In order to find a really optimal solution, it is necessary, in addition to variables x, y, z, to determine columns $\bar{a}$, i.e. to optimize seasonal production programs of businesses in the context of a general problem. But optimization of a production program of a refinery, even on itself, out of the situation in the system, is a rather intricate problem. That is why it is needless to speak about entering models of businesses into problem (1)–(6) directly.

Let us apply a systemic approach. In our case, it implies application of a two-level model of the company [2, 5]. The first (upper) level is the center represented with model (1)–(6). The second level are models of businesses. The center performs its coordinating function by calculating simplex multipliers π, which are interpreted in this instance as dual evaluations of petroleum products. In this context, we are interested in evaluations corresponding to material balance equations (2).

Taking into account these evaluations, each business represented with a second-level model optimizes its own technology and forms column 'commercial output structure', by which it is represented in the upper-level model. The mechanism of coordinating plans of businesses in order to achieve the optimal general solution follows from representation of model (1)–(6) as a generalized Wolfe problem.

The algorithm is as follows. Problem (1)–(6) is formed using some initial production plans represented by columns $\bar{a}$, and the necessary number of cycles comprising the following actions are realized.

1. We solve the problem using the simplex method with an inverse matrix. We obtain a conditionally optimal solution $x(\bar{a})$, $y(\bar{a})$, $z(\bar{a})$ and respective simplex multipliers $\pi = \pi(\bar{a})$.

2. Each business is trying to increase its own 'profit' in the circumstances offered by the center. To this end, it looks for a respective 'technology' and commodity output structure. In terms of mathematics, this is reduced to determination of vectors $\bar{a}_{tj}$, $j = 1, 2, \ldots, n$, $t = 1, 2, \ldots, T$, which give minimal evaluations of substitution for columns of the current basis of problem (1)–(6). This is completed through solving local problems

$$\Delta_{tj} = \min_{a,p}\left\{\left(\pi, \bar{a}_{tj}\right) - p_{tj} \middle| \bar{a}_{tj} \in M_{tj}\right\}, \; i = 1, 2, \ldots, m, \; t = 1, 2, \ldots, T, \tag{7}$$

where M_{tj} is the set of structural commodity plans of a business in period t.

3. We check condition $\Delta^* = \min\{\Delta_{tj}\} \geq 0$. If it is fulfilled, then the current basis is optimal and the solution has been obtained. Otherwise, we proceed to the next item.

4. We introduce column $\bar{a}$, for which evaluation Δ^* has been obtained, into the basis of problem (1)–(6). We form a new inverse matrix. We delete the column, which has been removed from the basis, from the problem. We go to item 1.

The algorithm has proven monotonous convergence and, if sets M are polyhedral, provides the optimal solution of the problem for a finite number of steps. Local problems represented by formula (7) are explicitly described in [3].

3 Results

Calculations were made using copyrighted software product – SMONPP [4], or, more correctly, one of its components designed to determine a summarized balance depending on production and process parameters of a business and commodity product evaluations. Parameters of Rosneft refineries were used.

Data on the consumption of gasoline, diesel, and black oil in the Samara Region are given in Table 1. Production of petroleum products in the region does not correspond to their consumption either in volumes or structure. The factory yearly output is equal on average to, thousand tons: gasoline – 1200, diesel – 2200, commercial black oil – 1900. The consumption is: gasoline – 640, diesel – 410, black oil – 240. The rest is sold outside the region. We had to assume that consumption of the main commodity groups matches their production on the whole and that seasonal fluctuations of consumption correspond to the presented data in relative rather than absolute units.

Table 1. Intra-Year Dynamics of Petroleum Products Consumption in the Samara Region according to Year 2018 Data, thousand tons

#	Gasoline	Diesel	Black oil	#	Gasoline	Diesel	Black oil
1	52.3	24.2	28.6	7	59.2	36.3	13.3
2	40.2	27.4	25.3	8	63.1	46.3	11.1
3	53.6	23.1	25.2	9	57.0	40.2	8.0
4	47.4	38.8	26.5	10	54.6	29.7	12.7
5	56.8	48.4	29.1	11	57.4	33.5	23.4
6	53.4	40.2	14.7	12	56.9	28.9	26.9

Source: author.

The expenses on petroleum products' storage are equal approximately to 3% of their value [3]. The return on current assets can be assumed equal to 22%. Then it turns out that the expenses (direct and indirect), caused by stock maintenance, amount approximately to 25% of stock value. That was the figure which was input into the model. No constraints on stock levels were imposed. The calculation results (intra-year dynamics of stock, production and sales by groups of products) are shown on Fig. 1.

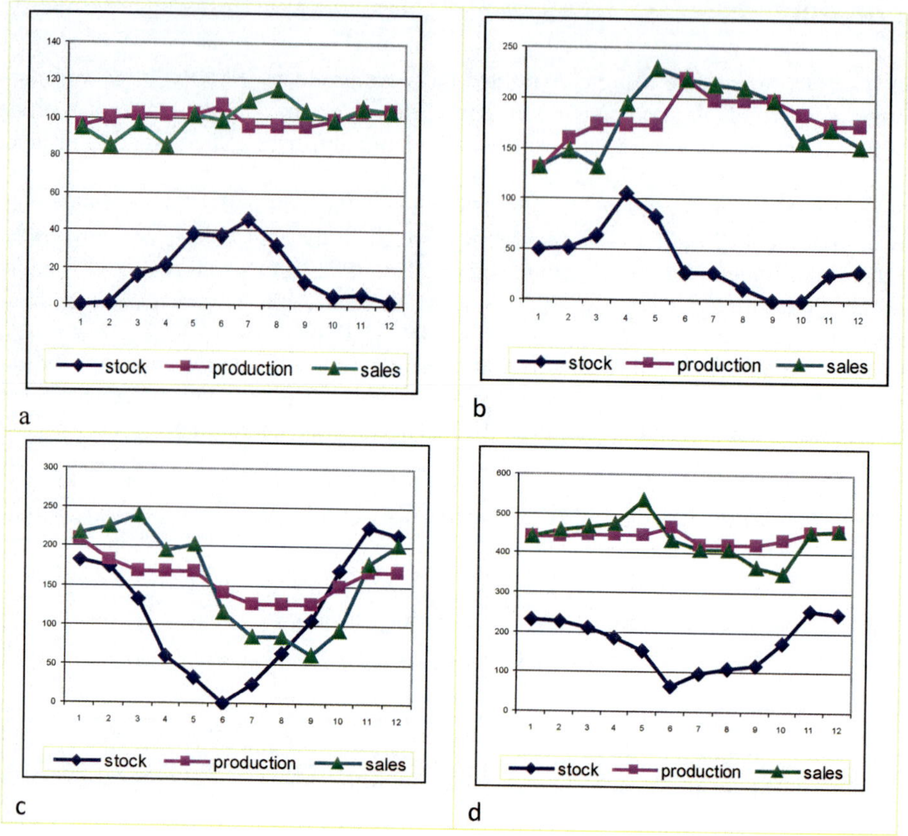

Fig. 1. Calculation results: a) gasoline, b) diesel, c) black oil, d) all three commodity groups (Source: author).

4 Discussion

The results show that in the optimal scenario, production of gasoline during a year does not require any significant changes. Thanks to some decrease of winter demand during the first half-year, gasoline stock is created that compensates the increased summer demand. The result for diesel differs. Here the optimal level of production during a year changes quite significantly. By the middle of a year it increases almost twice relative to the winter minimum, but still is somewhat behind the demand and the difference is made up from stock. The stock is depleted by autumn, then it grows and reaches maximum by spring.

The demand for black oil experiences the highest seasonal fluctuations. The coefficient of variation is equal approximately to 0.6. The dynamics of production is close to the dynamics of demand but is characterized by a substantially lesser variation. Hence, the attenuating role of stock is particularly great in this instance. In summer, black oil production exceeds the demand and the stock grows till November. Later, during the first half-year, it is used to make up the output shortage.

So, the seasonal fluctuation of demand forces the company to maintain a considerable stock. In the example under discussion, the yearly output for the three commodity groups amounts to 5260 thousand tons while the optimal stock level is 2080 thousand tons, i.e. 39%, or 32% in terms of money. This approximately corresponds to findings of other similar investigations [1]. The direct and indirect expenses related to maintenance of such stock amount to 8% of the commodity product value. However, these expenses are compensated through reduction of costs in connection with seasonal restructuring of production.

5 Conclusion

A methodology for optimizing intra-year production and sales cycle of an oil company with regard to the attenuating role of commodity stock has been suggested. A mathematical model and problem solution algorithm have been developed. For algorithm realization, copyrighted software was used. Calculations were made based on Samara Region data. Calculation results allow concluding that in case of optimal stock management it is possible to reduce significantly the production costs related directly or indirectly to seasonal demand fluctuations.

References

1. Chen, H., He, X., Qiu, T., Chen, B.: Refinery inventory management. J. Chem. Ind. Eng. (China) **54**(8), 1118–1121 (2003)
2. Kadambur, R., Kotecha, P.: Optimal production planning in a petrochemical industry using multiple levels. Comput. Ind. Eng. **100**, 133–143 (2016). https://doi.org/10.1016/j.cie.2016.08.008
3. Livshits, M.Yu., Sizikov, A.P.: Multi-criteria optimization of refinery. In: Kuznetsov, G.V., Strizhak, P.A., Zhdanova, A.O., Bulba, E.E. (eds.) Thermophysical Basis of Energy Technologies 2015. Web of Conferences, Les Ulis (2016). EPJ Web of Conferences, vol. 110, no. 01035
4. Sizikov, A.P.: Software product SMONNP. Management of large systems. Collect. Papers **24**, 298–326 (2009)
5. Zhang, L., Jiang, Y., Gao, X., Huang, D., Wang, L.: Efficient two-level hybrid algorithm for the refinery production scheduling problem involving operational transitions. Ind. Eng. Chem. Res. **55**(28), 7768–7781 (2016)

Criteria for Analytical Evaluation and Selection of Investment Programs and Projects

N. A. Moiseenko(✉) and O. A. Gorelova

State University of Management, Moscow, Russia
{na_moiseenko, oa_gorelova}@guu.ru

Abstract. The article investigates problems of evaluating the economic efficiency of investment projects for reconstruction of existing facilities and ways to justify investments in the process of reproduction of fixed assets. Special attention is paid to the selection and justification of criteria that ensure selective selection of an array of investment projects for inclusion in the investment program in the conditions of the economy digitalization. The classification of criteria is analyzed depending on the requirements of various participants in the investment and construction process by the implementation of reconstruction and technical re-equipment projects. The authors focus on the in-depth examination of investment projects for the reconstruction of existing enterprises.

Keywords: Criteria · Investment · Investment efficiency · Investment and construction process · Reconstruction

1 Introduction

When forming an investment program for reconstruction and technical re-equipment, as a rule, there is a situation when several investment projects are proposed that have various innovative solutions for inclusion in the investment program [8], and the total amount of investment is limited, so it is obvious that it is impossible to accept all the proposed investment projects. Therefore, there is a need to develop an algorithm that allows evaluating these projects and then ranking them according to certain criteria [4]. We can offer several approaches to solving this kind of problem in the context of developing digital technologies. Reconstruction is considered effective if the expected return rate exceeds any other investment. However, this should not be sufficient to include the *i*-th object in the investment program of reconstruction and technical re-equipment. In practice, there is a number of restrictions that do not allow carrying out reconstruction in the shortest possible time at all those enterprises where it is effective. In addition, there is a problem of establishing a rational sequence of objects included in the investment program, and the most optimal distribution of reconstructed objects by the implementation periods of the investment program.

S. I. Ashmarina et al. (Eds.): ENGINEERING ECONOMICS WEEK 2020, LNNS 139, pp. 324–332, 2021.
https://doi.org/10.1007/978-3-030-53277-2_39

2 Methodology

Criteria for analytical evaluation of investment proposals in the framework of the formation of investment programs are proposed to be determined depending on the direction of a reconstruction project on a competitive basis. Competitive selection of investment proposals is usually carried out on the basis of expert assessments within the framework of the competition. The selection of offers should not be based on a single criterion, even if it is a complex one. It should relate to all projects, regardless of their complexity or importance. Therefore, it is necessary to conduct a multi-aspect examination [4]. The final decision is formulated taking into account many different, often contradictory characteristics and factors that cover quantitative and qualitative parameters of a project [2]. Some of these factors affect economic, environmental or social consequences of implementing such projects in a country's economy, region, or industry. The other part describes various risks associated with the offer implementation process. Criteria for selecting investment proposals in the framework of the competition can be divided into the following groups:

- target criteria,
- criteria that determine the influence of the external environment,
- criteria that determine the economic efficiency of an investment project for a potential investor,
- criteria that determine environmental aspects,
- criteria of the investor submitting an investment offer,
- criteria that determine scientific and technical prospects,
- criteria that determine commercial prospects,
- criteria that determine the production potential,
- criteria that determine market prospects,
- social performance criteria,
- criteria that take into account various risk factors.

The first group of criteria determines investment directions that require support from the state, i.e. joint projects are formed, so-called public-private partnership, and therefore investment proposals are joint in their nature and have a high certainty degree. The second group forms a specific investment proposal taking into account their innovative component [8]. The criteria for each group are evaluated as mandatory and indicative. Failure to meet the mandatory criteria leads to a deviation of the project from inclusion in the investment program, indicative (recommended) criteria require additional evaluation, depending on the project direction. During in-depth examination of an investment offer, the criteria of each group are evaluated both for the entire offer and for individual participants.

The formation of criteria is based on a socio-economic situation in the country or specific industry in which the investment project is being implemented. Some of the specific criteria may relate to criteria of other groups. This is due to the fact that this type of criteria is taken into account only at the stage of selecting priority investment areas. It is obvious that this is the degree of linking the investment proposal with the goals of an investment program, and then it is clear that the discrepancy of the investment proposal to a certain criterion causes a reasonable management decision of expert commissions to reject a project from inclusion in the investment program [9].

Criteria that determine the impact of the external environment include:

- legal security of the offer, its consistency with the current legislation;
- possible impact of the prospective legislation on the investment offer and results of the investment program implementation;
- a possible reaction of the public opinion to the implementation of a proposal;
- how will parameters in the industry and related industries change as a result of the implementation of the investment proposal,
- impact on the presence of harmful products and production processes in the region (positive, negative, neutral),
- the impact of supply on the employment level in the region.

Criteria that determine the economic efficiency of an investment project for a potential investor should be used as traditional, generally accepted, and related directly to the investment project under consideration, taking into account its features and specific features of its implementation. The composition of these criteria should be subject to in-depth examination, since the optimal indicators of these criteria significantly reduce investors' risks.

The criteria for an investor submitting an investment offer are:

- investor's position in the market;
- competence, experience and management skills of the management team;
- experience and activity of operations in the foreign market, marketing research strategy;
- financial stability and financial sustainability;
- trends in the achieved results of the economic activity,
- data on indicators of change (growth, decline);
- data on changes resulting from diversification and restructuring.

The criteria that determine the scientific and technical prospects affect:

- innovativeness and prospects of innovation implementation;
- patent purity of products and patentability of the used technical solutions;
- the possibility of applying the results obtained in future developments,
- the extent and nature of the impact on other projects of public interest.

These criteria are calculated for an investor who participates in the implementation of an investment offer with their assets.

Criteria for determining commercial prospects contain data on the value of investments, and the implementation of the proposal, the expected annual income, the expected net discounted profit, the value of the index of the internal rate of return that is satisfactory to the investor, the compliance of proposed criteria with indicators of the economic efficiency of capital investments, the payback period and balances of real cash flows, stability of revenues from the proposal implementation, the possibility of using tax incentives, financial investment risks, identified during the evaluation of the proposal.

Criteria that determine the production capacity include:

- opportunity to use available raw materials, materials and necessary additional equipment;
- need to implement technological innovations;
- availability of production personnel;
- possibility of processing and using production waste;
- need for additional production capacity.

These criteria are calculated for an investor who participates in the implementation of an investment offer with their assets.

The criteria that determine the market prospects imply the compliance of the investment offer with market demands, an assessment of the total market capacity in relation to the volume of the products output offered in the investment project, and a probabilistic assessment of the commercial success of the project proposal.

Criteria of social effectiveness:

- possibility of employment growth;
- increase in income of the population;
- improvement of working conditions of employees;
- growth of infrastructure facilities;
- improvement of housing conditions in the region;
- improving the supply of water, gas and electricity to the population of the region.

Criteria that take into account various risk factors:

- likelihood of increasing the cost of construction, reconstruction or technical re-equipment;
- likelihood of increased construction time;
- possibility of reduction of terms of reconstruction and technical re-equipment;
- stable relations between contractors, subcontractors by new construction;
- consistency of primary production and contractors conducting reconstruction works by reconstruction and technical re-equipment;
- political risks;
- environmental risks;
- other risks.

All indicators are accepted taking into account the impact of the investment offer on existing enterprises in the industry. The assessment of each component is taken as a sum of results achieved by the investment proposal under consideration, and changes in the situation at existing enterprises and in the industry or region as a whole [5].

3 Results

After selecting the composition of criteria, the task of selecting investment proposals can be determined: from a set of n objects, the reconstruction of which is effective for a given period of time based on the calculation method given above, it is necessary to include m objects ($m < n$) in the investment program for reconstruction and distribute them according to this investment program.

As a target function, the total economic effect of the reconstruction of objects can be used, calculated according to the described method.

$$y = \sum_{\tau=1}^{t} \sum_{i=1}^{n} x_{i\tau} E_{i\tau} \rightarrow \max$$

where $x_{i\tau}$ = 1, if the *i*-th object is included in the investment program (0, otherwise).
$E_{i\tau}$ – economic effect of reconstruction;

The economic effect of the reconstruction of the *i*-th object in the τ-th year is determined by bringing the economic effect of the reconstruction to the first year of the investment program. Since the reconstruction duration of some objects may exceed a year, $x_{i\tau}$ = I is accepted only for the year of the beginning of reconstruction of the *i*-th object. The duration of reconstruction is assumed in accordance with standards or calculations. Breaks in the reconstruction of an object are not allowed.

It should be noted that some objects may be included in the investment program of reconstruction for non-economic reasons (for example, the emergency state of an industrial enterprise). Such objects are considered as extraordinary, and the task is solved taking into account the corresponding adjustments of capital investments and restrictions on the total capacity [5].

A limited number of capital investments in reconstruction by industry or region $K\tau$ may be a restriction. Restrictions on capital investments should be taken as a cumulative total, since funds not used in the year τ can be realized in the year τ + 1. Thus, restrictions on capital investment can be presented as follows:

$$\sum_{\tau=1}^{t} \sum_{i=1}^{n} x_{i\tau} K_{i(t+1-\tau)} \leq \sum_{\tau=1}^{t} K_{\tau}$$

where *t* is a number of the implementation year of the investment program;
$K_{i\,(t+1-\tau)}$ – capital investment in the reconstruction of the *i*-th object in the (t + 1 − τ) year after the beginning of the reconstruction with a cumulative total.
If (t + 1 − τ) exceeds the standard duration of reconstruction, Ki(t = 1 − τ) is assumed to be equal to the total amount of capital investment in the reconstruction of the *i*-th object;
Kτ – allocated investments for the τ-th implementation year of the investment program.

It should also be taken into account that the reconstruction of existing enterprises is often accompanied by their partial shutdown or partial reduction in production. Therefore it is necessary to take into account the restrictions on the total capacity of industrial enterprises:

$$\sum_{\tau=1}^{t}\sum_{i=1}^{n} x_{i\tau} M_{i(t+1-\tau)} \geq \sum_{\tau=1}^{t} M_{\tau}$$

where *Mi* is the need for the industry or region to increase capacity in the year τ due to the reconstruction;

Mi(t + 1 − τ) – increase (decrease) in the capacity of the i-th object in the (t + 1 − τ) year after the start of the reconstruction. If t + 1 − τ exceeds the reconstruction duration on the i-th object, Mi(t + 1 − τ) is assumed to be equal to the increase in the capacity of the i-th object due to the reconstruction.

Another approach is based on the assumption that if the implementation of all projects is postponed for one year, then based on the resulting index of expected losses of net present value, which acts as one of the main criteria for evaluating any investment project, we can compare and identify projects with minimal losses, the implementation of which can be postponed to the next year of the investment program.

The calculation logic is based on the following: from the business plans of a set of projects, the values of net present value (NPV_0) and the amount of investment ($\sum$ I) required for each reconstruction project are selected; the discount coefficients or return on invested capital are determined when the project implementation is conditionally postponed for one year; the corresponding value of net present value (NPV_1) is determined by multiplying (NPV_0) by the discount coefficient; the absolute value of NPV losses ($NPV_0 - NPV_1$) for each of the projects is determined; then the index of possible losses of the net present value is calculated by dividing the absolute value of losses by the amount of investment in the corresponding reconstruction project. In this way, the priority range of projects is determined by the value of the loss index from a higher value to a lower one. The higher the value of the loss index is, the more feasible it is to implement the project in the first year of the investment program, if there is enough investment.

The same priority series should be built on the basis of a criterion that reflects the efficiency level of the return on investment – the profitability index. This indicator is also calculated when evaluating investment projects. By ranking its values in descending order, you can determine the implementation order of investment projects included in the program. Moreover, practice has shown that the use of these indicators gives, as a rule, the same result.

When forming an investment program for reconstruction and technical re-equipment, it is usually necessary to compare projects with different periods of validity [5]. In this case, the indicators of net present value taken from the business plans of projects are not quite correct. Here you can use the method for calculating the net present value of the reduced flows. The smallest total multiple of the validity periods of the analyzed reconstruction investment projects (γ) is determined, and considering each analyzed project as repeating a certain number of times (n) in the period (γ), the total net present value for each of the pairwise compared projects is determined by the formula:

$$NPV_{(ij)} = NPV_{(i)} \times (1 + \frac{1}{(1+r)^{i}} + \frac{1}{(1+r)^{2i}} + \frac{1}{(1+r)^{3i}} + \ldots n)$$

where: $NPV_{(i)}$ – net present value (reduced effect) of the original project from the business plan;
i – duration of the reconstruction project;
r – interest rate or profitability of the reconstruction project;
$\frac{1}{(1+r)^{i}}$ – a discount coefficient;
n – number of repetitions of the project during the period (γ).

From two projects compared in pairs, the one with the lowest value of the NPV is preferred.

4 Discussion

In the analysis of alternative reconstruction projects, each of which is effective, you can use the incremental indices: here we determined the difference of indicators with the additivity property; in renovation projects, it is the amount of investment, annual income, NPV, etc.; by calculation we determine the internal rate of return and the total return index for a conditional draft from each compared pair of projects with incremental indicators; and the obtained value of the internal rate of return for the conditional project and the analyzed project is compared with the interest rate. If the internal rate of return (IRR) of a conditional draft is more than the profitability of the project, then the project with large capital investments is accepted as the most effective one, and conversely, if the IRR is less than the profitability of the project, the project with less investment is accepted [1, 3]. This is not always legal and appropriate.

In this regard, when faced with certain difficulties in calculating the discount rate, especially in unstable conditions of the economic development, it is necessary to additionally use the option approach to the formation of investment programs. Distinctive features of the real options method are characterized by the complexity and direction of the evaluation process, which enables constant interaction of economists, lawyers, technologists, human resources specialists, financiers, etc. As a result, based on the use of digital technologies, the reasonableness, validity, comprehensiveness and depth of the evaluation results are increased [6, 7, 10]. Evaluation of real investment projects of reconstruction by the optional method consists of several stages. First, the compliance of various factors that affect the implementation of the project in the current conditions is assessed. In parallel, possible measures aimed at minimizing damage and maximizing the use of favorable factors are identified.

As a result of a comprehensive discussion, a general line is worked out with the main development parameters necessary for making future management decisions. In addition, serious attention should be paid to the environmental expertise, so the authors distinguish it as one of the criteria for the analysis of investment proposals. Environmental expertise should always be independent and of high quality. Therefore, it should be carried out by a specialized organization. The investment program itself is pre-evaluated. Compliance with the legislation of the Russian Federation and regional legislation is the most important condition for the implementation of any state program, so if this condition is not met, it is necessary to review and eliminate all the provisions that prevent the implementation of the investment proposal. This is also a qualitative indicator. Further, if necessary, you can conduct an in-depth examination.

The objectives of the in-depth examination are to verify the validity of the economic calculations and forecasts contained in the investment proposal, the correctness of calculations, as well as to verify the estimates of the preliminary examination. In-depth examination is performed by one or more experts from the expert group appointed by the members of the expert council. If necessary, a special examination is carried out with the involvement of external specialists. In the conditions of digitalization, it is possible to build a sufficiently reasonable decision-making algorithm for optimizing the inclusion of the project in the investment program. If the evaluation of individual criteria was adjusted during the in-depth examination, a new offer rating is calculated. Proposals that have passed an in-depth examination are considered at a meeting of the expert council.

5 Conclusion

Based on the available information, it is possible to build an economic and mathematical simulation (or situational) model that will allow you to form an adaptive management plan for the reconstruction investment program. In our opinion, the optional method for evaluating real reconstruction projects is a method of the XX1 century, the use of which in the digital economy may provide specific advantages to industrial enterprises in the future.

The competition for investment proposals can be organized by public or private investment organizations that formulate the main goals of the competition. The competition can be held by an existing expert council or a specially formed competition council. The competition council develops terms of the competition that determine first of all directions of investment proposals and restrictions on their indicators. In accordance with the developed terms of the competition, the system of criteria for evaluating proposals and the procedure for selecting proposals is being clarified. The use of innovative digital technologies allows us to consider and take into account a wide range of criteria, specifics of various reconstruction projects and investment opportunities of customers of such projects. Both existing and newly created expert groups can be used for conducting the examination. To monitor the implementation of the proposal accepted for funding by the expert or competition council, a curator is appointed from the council, who submits a report on the compliance of the actual progress of work compared to the planned ones in the project and makes suggestions for changing the work plan. Funding organizations make a decision on whether to maintain or change the terms of funding based on these reports. Proposals that raise doubts, as well as those related to significant capital investments, can be submitted for in-depth examination.

References

1. García, M.V.R., Krzemień, A., Bárcena, L.C.S., Álvarez, I.D., Fernández, C.C.: Scoping studies of rare earth mining investments: deciding on further project developments. Resour. Policy **64**, 101525 (2019)
2. Karibskii, A.V., Shishorin, Yu.R., Yurchenko, S.S.: Financial and economic analysis and efficiency evaluation of investment projects and programs. Autom. Remote Control **64**, 1205–1224 (2003)
3. Karner, A., Golub, A.: Assessing the equity impacts of a transportation investment program. In: Lucas, K., Martens, K., Di Ciommo, F., Dupont-Kieffer, A. (eds.) Measuring Transport Equity, pp. 277–290. Elsevier, Amsterdam (2019)
4. Makarkin, N.P.: The Effectiveness of Real Investment. INFRA-M, Moscow (2011)
5. Moiseenko, N.A., Kotov, I.A.: Features of managing investment programs for the reconstruction of existing enterprises. Constr. Econ. Manag. **1**(5), 13–27 (2012)
6. Momeni, M.A., Yaghoubi, S., Mohammad Aliha, M.R.: An optimal control model for analyzing quality investment in the project management. Comput. Ind. Eng. **129**, 529–544 (2019)
7. Paquin, J.-P., Gauthier, C., Morin, P.-P.: The downside risk of project portfolios: the impact of capital investment projects and the value of project efficiency and project risk management programmes. Int. J. Project Manag. **34**(8), 1460–1470 (2016)
8. Samosudova, N.V., Cherkas, A.D.: Innovative solutions in modern construction. Modern Innov. **2**(2), 30–32 (2015)
9. Serov, V.M.: On the evaluation of the effectiveness of the functioning and updating of fixed assets. Constr. Econ. Manag. **4**, 7 (2011)
10. Yu, S., Gao, S., Sun, H.: A dynamic programming model for environmental investment decision-making in coal mining. Appl. Energy **166**, 273–281 (2016)

A Modern Tool for Students' Self-organization in the Digital Economy

A. Kutuev, E. Malysheva, and J. Sharikova(✉)

Samara State University of Economics, Samara, Russia
artem-kutuev@mail.ru, sam.malysheva@mail.ru,
sharikovajv@mail.ru

Abstract. The article presents the results of an empirical study of the level of self-organization of students of the Samara State University of Economics (SSEU). The purpose of the study is to determine the level of self-organization of students in the digital economy. It was supposed to diagnose parameters that reflect the degree of self-organization of modern students. The survey was used as a research method. A sample of 120 students. The results of the study demonstrate the degree of application of self-organization tools for students aimed at developing the key competencies of future specialists and the fundamental competencies of future employees.

Keywords: Digital economy · Self-organization · Time-management

1 Introduction

In the new realities of economic and social development, special attention is paid in modern organizations to the sphere of human resources management. The digital age is significantly transforming the labour market, which changes the key competencies that a modern effective employee should have. According to the BCG study, the following skill groups of future employees will be relevant: cognitive, socio-behavioural and digital [12]. Among cognitive skills, experts primarily highlight organization as the degree of self-organization and ability to effectively use such irreparable resource as time.

In today's conditions of accelerating the dynamism of the rhythm of life, many people have a number of questions about the effective distribution of their time, and students are no exception. At present, students' timetable at the higher educational institution is quite dense. It is necessary to visit all lecture and practical classes, to prepare group and individual projects, to make a report, to write a summary, course and graduate qualification works. Many students seek to combine study and work to gain experience and practical skills. At the same time it is impossible to forget about family, friends, hobby, sports. In order to cope with such a rhythm of life, students need to learn to apply time management tools. The ability to select, structure and process information within a short time frame is already a competitive advantage in the labour market under conditions of universal digitalization. In this regard, the aim of the study is to determine the level of self-organization of students as future professionals who are already in global informatization and living in an environment of excessive information

S. I. Ashmarina et al. (Eds.): ENGINEERING ECONOMICS WEEK 2020, LNNS 139, pp. 333–339, 2021.
https://doi.org/10.1007/978-3-030-53277-2_40

flows. The subject of the study is time management as a tool of self-organization. The subject of the study are students of the SSEU. At various times, foreign and domestic scientists have attempted to study various aspects of self-organization and time management [1–4, 6, 11].

Currently, time management is a comprehensive system of managing your time, yourself and your activities. The application of time management principles can help achieve various goals: increasing personal efficiency, more rational use of time, increasing the level of life satisfaction, etc. One obstacle to that is time. Due to the high dynamics of modern life, it is almost impossible to carry out all cases at once. Therefore, when planning, it is necessary to prioritize immediately for purposes, to concentrate their attention on really important cases, and the rest to delegate or abandon them. Sometimes this can greatly improve the efficiency of using your time.

The use of various time management tools allows a modern person to manage more, work more efficiently, focus on the main aspects. But without the internal component of the personality, represented by self-organization, they will remain just tools. In general, self-organization is the ability to organize your activities and your own resources, the ability to organize yourself. Self-organization is a multi-component of the individual that requires the presence and development of the following abilities to:

- self-knowledge – implies the desire of a person to study his personality in order to determine his own strengths and weaknesses;
- self-determination – definition of own desires and formation from them is more whole also arrangement of priorities between them;
- self-monitoring – ability to form constraints for the expected activity with subsequent evaluation after obtaining the result;
- self-realization – use of knowledge of own merits and shortcomings in order to achieve the set goals in the course of their activities;
- self-improvement – ability to learn from the results of their activities and work on themselves, developing their personal qualities for more efficient work in the future [10].

2 Methodology

The methodological basis was the self-organization test questionnaire of Time Structure Questionnaire (TSQ), developed by Fisher and Bond (adaptation in Russian was performed by Madrikova) [8] and modified to meet the needs of the study. In order not to distort the interval of the questionnaire estimates, the scale boundaries were adjusted in proportion to the magnitude of the changes.

The above technique is aimed at systematic analysis of factors of self-organization of the person during the course of their daily activities, which are multi-component components of the person (internal characteristic of self-organization) and assessment of the level of use of auxiliary tools and technologies of time management (external factors of self-organization). All these factors describe the model under study by means of indicators: “regularity,” determination, “perseverance,” fixation, “self-organization,” orientation to the present". The converted questionnaire was distributed to the students of the SSEU. The number of respondents was more than 120 people (1–4 courses).

3 Results

The results of the questionnaire are shown in Table 1.

Table 1. Results of questionnaire of self-organization students

Scale	Degree, number of students		
	Low	Average	High
Regularity	119	55	448
Commitment	22	103	117
Persistence	225	78	119
Fixing	77	111	44
Self-organization	55	72	445
Orientation to the present	66	110	66

Source: authors.

First of all, the involvement of respondents in planning (regularity) of their daily schedule was revealed by determining tasks for the day, as well as the presence of systemicity in building daily cases. It was found that 55 students (45% of the sample) are moderately inclined to develop day plans, a list of cases and try to comply with them. However, elements of spontaneity may be present in their lives.

The next group is slightly smaller than the previous group and is 48 (39%). This group included people who have more advanced tactical planning skills and prefer to move consistently towards their goals. A small group of 19 students (16% of the total sample) prefer to live a spontaneous life and not complicate it with current case planning. It is important to note that such people are evenly distributed in the amount of not more than 5% per course.

Next, a corrective question aimed at determining the degree of personality commitment was used and demonstrates the respondent's ability to concentrate on any goal to fulfill it. The largest group of 103 students, representing 84% of the sample, are able to characterize themselves as sufficiently committed, who are well aware of and understand their own goals and are able to achieve them. However, this is not always the case. Periods of time may appear in their lives when they perform tasks without the necessary clarity and need for them personally.

The next group of 17 respondents (14%) consider themselves purposeful people. They have decided on life goals that have set for themselves, know and go towards their realization. This group prevails among senior students, which may be due to a successful choice of profession and a possible place of future employment.

The smallest group of 2 students are not always inclined to see and set clear goals for themselves. They also describe themselves as people who do not make efforts to achieve their goals. The next stage determined the degree of "Perseverance" of students in carrying out their daily tasks. The ability of respondents to apply will efforts, as well as internal reserves, to bring the cases started to completion was identified. Unlike the two previous scales, this shows a more normal distribution of respondents' responses. A group of 78 students, representing 64% of the total sample, are sufficiently organized and capable of making a will effort to achieve any goal. However, if circumstances so require, they may postpone the current tasks and switch another.

The next group by percentage is 25 people (20%). This group of students has difficulty bringing their cases to the completion stage, which is due to their tendency to be distracted by extraneous cases unrelated to the current one, and may not be meaningful. The smallest group, but not the most important, is the will and organized people. They are able not only to plan goals and objectives, but also to structure them to more effectively complete them. It is not common for them to be distracted by extraneous affairs during their activities.

This group also defined the presence of multitasking and the effectiveness of this form of organization. The results showed that for 20% of the sample it is not typical to take several cases at the same time and not bring them to a logical conclusion, or it happens with a low probability. What can't be said about the others. Half of the students live in multi-tasking conditions, and are often unable to bring all cases to the right time frame. The remainder of the 30% sample does not have internal reserves to bring its cases to a logical conclusion.

A question was then used to determine the degree of student fixation to the pre-planned event structure and the difficulty of adapting to its change. Most respondents consist of flexible people regarding the planning of their activities. This group included 111 people or 91% of the total sample. They can both concentrate on already planned cases and quietly switch to others.

The next largest group included 7 respondents (6%) with a low degree of fixation in a certain occupation. They are flexible and able to switch between cases with ease, as well as to new tasks. A slight group of respondents included 4 people, which is 3% of the total sample. They are characterized by high performance and an obligation to complete their tasks. They try in every way to complete the cases already in place before embarking on new ones. As a result, they lose flexibility when circumstances change.

The tendency of respondents to keep diaries, daily books and other auxiliary means for planning their time was also revealed. It has been determined that most students, namely 72 (59%) may use external sources of self-organization, which may include keeping diaries or diaries and their counterparts. And it is more typical for this group to use either a diary or a daily. At the same time, they can equally rely on their natural self-organization.

Another proportion of 45 respondents (37%) are active users of auxiliary time planning tools, recording their experiences and results in a diary, and using daily and other planning tools. An interesting fact is that this group of students is formed practically at the expense of the 1st year. Their share in the total sample in this group is 43%. The remaining 5 respondents (4%) consist exclusively of 2nd year students. They are not inclined to use external sources of self-organization and do not keep diaries and daily books that can help plan their time more effectively.

The final part of the questionnaire asked about the orientation of students in time, defined the priority of current affairs or the tendency to long-term planning. The largest group of respondents of 110 people (90%) balance well and appreciate what has happened to them, what is going through and happening at present, and what expects them in the future.

The remaining part of the sample was divided into two equal parts of 6 people (5% respectively). One group included students who are characteristic of fixing on their past or future, forgetting the present or ignoring these events. For the other group, it is more preferable to live on the principle of "Here and now" and the past or future cares little about them.

4 Discussion

In order to increase the level of self-organization of modern students, it is advisable to use some time management tools. The use of the Eisenhower matrix can provide substantial assistance to improve planning and scheduling. This time management tool is based on dividing all tasks into two axes – importance and urgency. Before distribution, all targets for the day must be written. In order to determine the degree of importance, it is necessary to assess the possible consequences in case of refusal to perform it. If the consequences are expected to be serious, the goal is important. The urgency criterion is relevance. If the target loses relevance, it is assessed as urgent. This easy-to-implement tool by prioritizing tasks will allow you to plan your day more rationally [5].

For more efficient targeting, it is possible to use the tool of time management "Wheel of life balance". This tool implies the identification of the most important spheres of life for man, in which he wishes to achieve success. Career, financial welfare, family and more, determine the balance [3]. It is necessary to define the main 8–10 directions, to define a key and measurable goal for the long term and to move towards it. The main obstacle to bringing the goal to a logical conclusion may be its incorrect formulation or assessment of significance. Using SMART technology can help solve this problem. It involves several requirements in setting the target:

1. Concreteness implies that the goal must demonstrate the end result in the realization of the goal.
2. Measurable requires that the goal be able to quantify and evaluate its achievement.
3. Reachability is a requirement for an objective that can be achieved while being ambitious. This condition involves evaluating your own resources to achieve your goal, or identifying missing conditions that will be an intermediate step between getting a result and the current state.
4. Consistency is the end result of achieving the goal must be consistent with the longer term and is in one of the directions of the Wheel of Life described above. This will, on the one hand, be a motivating factor, and on the other, make it possible to understand the real significance.
5. Time certainty involves setting time gaps for each target, a deadline for its implementation, after which its execution may lose relevance.

Paper daily is now a fairly traditional and popular tool for structuring and planning your own time, capable of showing an overall picture of your own employment. But the constant compilation of a list of cases and corrections can distort the picture. In the conditions of digitalization, the market of planning applications is actively developing [7]. Based on the fact that students always have a smartphone at hand, such applications can help plan their time more competently and concentrate on really important things.

The introduction of a time management system into its day-to-day operations should begin with the identification of a daily time reserve, which consists of downtime (duration of absence of active activities) and time sinks. Timing technology can help. In the conditions of digitalization, timing becomes possible to transfer from traditional paper media to specialized software products available for installation on a smartphone. This will allow to optimize the time for its maintenance, as well as to build a chart of the daily time balance for further graphoanalytic assessment and increase personal efficiency [9].

There are also alternative methods of timing using Internet services, one of which can be a multifunctional and universal Google form product with good adaptability. By simple transformation and adjustment of formulas in analytical cells within the results table, it is possible to easily determine the duration of activities during the day, which will serve as a basis for monitoring efficiency and further planning of their activities [2]. The main disadvantage of this method is the need to access the Internet for data synchronization and backup. At the moment there is quite a wide selection of applications, performing similar functions of traditional paper daily (external sources of self-organization), but supplemented with additional functionality. As a rule, many of them differ only in the design and variations of key functions (synchronization, backup, ability to project tasks, labels, priorities, etc.), which are available in the free version of the program.

5 Conclusion

In general, respondents who participated in the questionnaire can be described as flexible, able to adapt to changes in the digital economy, while being able to structure their time, having a sufficiently high level of self-organization. Most of them are familiar with the concept of time management and realize the importance of planning their activities in conditions of universal digitalization. However, based on the specifics of the direction of training of students, it is necessary to further inform about existing technologies and tools of tai-management, to develop those competencies that lay the basis for competitiveness in the labor market in the further employment of current students.

References

1. Archangel, G.: Time Management. Alpina, Moscow (2018)

2. Blandford, A.E., Green, T.R.G.: Group and individual time management tools: what you get is not what you need. Pers. Ubiquit. Comput. **4**, 213–230 (2001)
3. Boniwell, I., Osin, E.: Beaond time management: Time use, performance and well-being. Organ. Psychol. **3**, 85–104 (2015)
4. Drucker, P.: The Effective Executive. Eksmo, Moscow (2014)
5. Kochetkova, N.V., Khlebnikova, D.A.: Time management: the art of time management. Modern Socio-Econ. Process. Prob. Patterns, Prospects **3**, 104–106 (2017)
6. Kostricov, S.S., Kerzhentsev, P.M.: Fight in time. Vestnik Univ. **18**, 76–80 (2014)
7. Lekerova, G.Z., Kylyshbayeva, G.B., Karbozova, G.K., Zholdasbekova, K.A., Dlimbetova, B.S., Sikhimbayeva, Z.S., Aymenov, A.Z.: Research of the problem of forming student's motivation. Espacios **25**, 26 (2017)
8. Mandrikova, E.Yu.: Development of the questionnaire of self-organization of activity (OSD). Psychol. Diagn. **2**, 87–111 (2010)
9. Nevolina, V.V., Shcheglova, M.I.: Study of self-organization features as essential characteristic of student time management. Tavrichesky Sci. Obs. **9**(14), 47–49 (2016)
10. Reunova, M.A.: Axiological aspect of self-organization-time student of the university. Vestnik OGU **2**(138), 237–242 (2012)
11. Shmurygina, N., Nikolaeva, N., Bazhenova, N., Bazhenov, R., Tcytcarev, A.: Self-organization activities of college students: challenges and opportunities. Int. J. Environ. Sci. Educ. **11**(17), 9981–9989 (2016)
12. The Boston Consulting Group: Russia 2025: Resetting the talent balance. (2017). https://www.bcg.com/Images/russia-2015-eng_tcm27-187991.pdf. Accessed 15 Feb 2020

Some Basic Mathematical Tools' Principles for Modeling Economic Processes

E. Y. Nuykina(✉)

Samara State University of Economics, Samara, Russia
nuikina1973@mail.ru

Abstract. The purpose of this work is to consider some basic mathematical modeling aspects of economic processes. Economic and mathematical modeling, which is an economic object or process mathematical description, is particularly relevant in management processes. Comprehensive assessment of all important conditions of economic facility operation, management parameters, consequences of decisions made significantly reduces risks and financial losses. The problem is that mathematics does not work with real objects, but with their mathematical models. The problem's mathematical formulation is only half of the success in solving it. The complexity consists in an exception of excess, unnecessary details. Important conditions should be preserved and the problem formulated as a standard model. This makes it possible to identify process patterns and to ensure effective application of this model in practice. By way of illustration, the author presents some economic problems solved by special mathematical techniques means.

Keywords: Economic processes · Mathematical modeling · Optimization

1 Introduction

Modern economics is impossible without the use of mathematical methods as a tool for solving problems of economic content and carrying out various types of financial analysis. Calculation of the goods cost, products and services, costs, profitability of production, arbitrary labor and other calculations is a mandatory function of any economist and accountant. Linear programming, optimization tasks, factor analysis are used to justify economic decisions, identify trends and choose enterprise development strategies.

Economic and mathematical models are very useful in management processes, which are an economic object or process mathematical description [6]. Comprehensive assessment of all important conditions of the economic facility operation, management parameters, decisions consequences made significantly reduces risks and financial losses. A large number of models are distinguished, the selection of which in the real situation depends on specific goals: extrapolation, optimization, balance sheet, expert, network, factor econometric, models of mass service systems and other [7]. In the works of the winner of Prize in Economic Sciences in Memory of Alfred Nobel academician Kantorovich, four groups of economic and mathematical models are distinguished: interaction models of economic subdivisions; macroeconomic models that include demand models and balance sheet methods; optimization models as well as linear models [4].

S. I. Ashmarina et al. (Eds.): ENGINEERING ECONOMICS WEEK 2020, LNNS 139, pp. 340–347, 2021.
https://doi.org/10.1007/978-3-030-53277-2_41

2 Methodology

Linear programming has a very substantial application in economic calculations. Following problem definition.

Function of the purpose:

$$L(\overline{x}) = \sum_{j=1}^{n} c_j x_j \rightarrow extr$$

at restrictions:

$$\sum_{j=1}^{n} a_{ij} x_j \leq b_i, i = \overline{1, k}$$

$$\sum_{j=1}^{n} a_{ij} x_j = b_i, i = \overline{k+1, m}$$

Most often practical tasks use economic-mathematical problems' models about resources optimal use, about diet, about material cutting. These models are so-called "simulators" for compiling and solving a very wide range of economic problems. The current promising direction is economic and mathematical modeling of economic situations taking into account the market participants expectations, as well as using complex variables [3]. This approach is based on statistical study mathematical methods.

Economic and mathematical modeling allows to obtain the most objective idea of the analyzed economic object, to identify the main factors affecting the final result, to determine the force of their impact and the existing relationships [8]. In the case of building mathematical analysis on economic methods and statistics talk about applied science – econometrics. It includes modeling production and financial processes with further calculations [10].

The econometric model is a means of analyzing and predicting the real economic phenomena and processes based on specific statistical information. Classical methods of mathematical analysis include integration, logarithmization and differentiation, which are used both independently and within the framework of mathematical statistics and programming.

The integral method of factor economic analysis, based on the incremented function summation defined as a private derivative multiplied by the increment of the argument at infinitely small intervals, finds use in determining the influence of individual factors using multiplicative, multiple and mixed (multiple-additive) models, regardless of the elements number that are included in the model, as well as regardless of the connection type between them. Using the integral method, the calculation of a certain integral according to a given sub-attribute function and a given integration interval is usually carried out by a program using computer technology.

The integral method solves two main types of problems: statistical and dynamic. In the first case, there is no information on changes in the analyzed factors during the analyzed period, as in the case of the business plans implementation or economic indicators dynamics analysis. Dynamic type of tasks is solved if there is data on analyzed factors dynamics, for example, when studying economic indicators time series [2]. Logarithm is also fairly common, particularly in factor analysis. The essence of the method is to determine logarithmically proportional distribution of factors joint action values in proportion to the share of each of them influence on the total indicator value. Note that in the integral method, this value is equally distributed among the factors. Therefore, the logarithm method makes factor influence calculations more reasonable than the integral method.

Logarithmization does not involve absolute values of economic indicators growth, as in the case of integral method application, but relative - these indicators change indices. In carrying out mathematical analysis, the differential calculation method often finds application, which, formalizing the economic values connections, allows to solve the main economic tasks most effectively: determination of income change directions at taxes increase, the enterprise revenues at the price increase for its products or goods, additional purchased equipment on production and financial indicators influence's calculation, maximum profit search, costs and other minimization.

Each measure is a function of one or more variables, and it is necessary to study their relationships using differential calculus methods. The total change in function, that is, the total key figure, is divided into individual values, the value of which is determined by the product of a certain partial derivative per increment of the variable by which that derivative is defined [2]. For example, if the goal is to maximize profit, as a function of Y from the certain product X sales volume, then at the maximum point, the ratio of the function Y increment to the argument X increment should tend to zero, when the increment X approaches zero.

The Lagrange multipliers method, based on the search for private derivatives of the Lagrange function, is an effective tool for solving economic problems:

$$L(\overline{x}) = f(\overline{x}) + \lambda_1 \phi_1(\overline{x}) + \ldots + \lambda_s \phi_s(\overline{x}).$$

The numbers $\lambda_1, \ldots, \lambda_s$ are called Lagrange multipliers.

The most famous model of the production sphere is the "input-output" method developed by the American economist of Russian origin, Nobel Prize winner V.V. Leontiev, the creator of the intersectoral analysis theory. The method involves the formation of matrix (balance) models built according to a chess scheme and clearly illustrate the relationship between the volumes of final demand for a product or commodity, the total volume and the product industry structure.

If, instead of "product," a more general concept of "resource" is introduced, the balance model is understood to mean a system of equations that satisfy the requirements of matching the resource and its use availability.

$$X_j = \sum_{i=1}^{n} x_{ij} + Z_j, j = \overline{1.n}$$

$$X_i = \sum_{j=1}^{n} x_{ij} + Y_i, i = \overline{1.n}$$

Here

X_i – is the value of total output of industries, i.e. gross product;
Y_i – the industries end products;
Z_j – conditional net production.

Mathematical programming methods are one of the main means of solving problems of the enterprise production and economic activity optimization [9]. They allow to estimate the planned tasks intensity, results deficit, to determine limiting resources, optimal plan of goods transportation and to solve many other optimization tasks. An example of such tasks are transport models, the essence of which is not only to determine the optimal goods transport, the exclusion of too long-distance, counter-transport and repeated transport. With the transport type model, you can solve a wide range of problems, such as optimizing the various parts production or the optimal distribution of company employees [5].

The closed transport task's mathematical model is:

$$L(\overline{x}) = \sum_{i=1}^{m} \sum_{j=1}^{n} c_{ij} x_{ij} \to \min$$

at restrictions

$$\sum_{j=1}^{n} x_{ij} = b_i, \ i = \overline{1, m}$$
$$\sum_{i=1}^{m} x_{ij} = a_j, \ j = \overline{1, n}$$
$$x_{ij} \geq 0$$

In solving economic problems, it is often necessary to examine situations in which the interests of several parties with different objectives compete. Such situations are called conflict situations. In real life and, in particular, in the economy, conflict situations occur very often and are of different nature. Mathematical modeling of conflict situations is carried out by game theory, which allows to properly apply scientific-based methods of conflict situations resolution.

3 Results

Mathematical modeling in economics is carried out in several stages:

- the economic problem presentation and its qualitative analysis. At this stage, the main characteristics of the research object and its structure are identified, as well as the problem and acceptable assumptions are formulated;
- building a mathematical model as a specific mathematical dependency;
- identification of the model and its solutions general properties. An important aspect here is the proof of the existence of solutions in the formulated model;
- preparation of initial information. In this case, methods of probability theory and mathematical statistics can be used to organize selective studies and evaluate reliable data;
- numerical solution of the economic-mathematical model;
- analysis of the results obtained and their application in practice.

Consider the task of determining the amount of capital from known net investments using classical methods of mathematical analysis.

Let us designate capital as a function of time *K(t)*, and net investments –*I(t)*. Then net investment can be determined by how:

$$I(t) = \frac{dK(t)}{dt},$$

that is a derivative of the capital for the time *t*.

It is often required to determine the capital increment over a period of time from t_1 to t_2, i.e. value $\Delta K = K(t_1) - K(t_2)$.

The function *K(t)* is primitive for the function *I(t)*, therefore, we can write:

$$\Delta K = \int_{t_1}^{t_2} I(t)dt = K(t_2) - K(t_1)$$

Let, for example, net investments be given by the function $I(t) = 200t^2$ (cu), it is required to determine the capital increment for 3 years.

Following the primitive formula for net investment, we get

$$\Delta K = \int_0^3 200t^2 dt = \frac{200t^3}{3} = 1800 \text{ (cu)}$$

Given the function of net investment, you can also determine how many years it will take for the capital increment to be a certain amount. These tasks are very much in demand in the field of financial computing. In modern economic conditions, the organization and planning of production in small enterprises is a development promising direction. Lagrange multipliers method allows to find the optimal solution to the distribution and production planning problem.

Let a small enterprise specialize in the production and sale of certain products. The daily production volume is 5 thousand kg. It can sell its products through the store and wholesale through sales agents. When selling x_1 kg of products through the store, sales costs are x_1^2 currency units, and when selling x_2 kg of products through trading agents - x_2^2 currency units. It is necessary to determine how many products should be sold in each way, so that the costs of sales are minimal

In order to solve this problem, it is necessary to draw up a mathematical model.

The objective function is:

$Z(x) = x_1^2 + x_2^2 \rightarrow min$

At restrictions:

$x_1 + x_2 = 5000,$

$x_1 \geq 0, x_2 \geq 0$

Let's make Lagrange's function:

$$L(x_1, x_2, \lambda) = x_1^2 + x_2^2 + \lambda(x_1 + x_2 - 5000)$$

Find private derivatives of Lagrange functions and equate them to zero.

$$\frac{\partial L}{\partial x_1} = 2x_1 + \lambda = 0$$

$$\frac{\partial L}{\partial x_2} = 2x_2 + \lambda = 0$$

$$\frac{\partial L}{\partial \lambda} = x_1 + x_2 - 5000 = 0$$

By solving this equations system, we get

$x_1 = 2500$, $x_2 = 2500$, $Z(x) = 12500000$ currency units.

Thus, in order to obtain minimum costs for products sales it is advisable to sell 2500 kg of products per day through the store and sales agents, at the same time the costs for sales will amount to 12500000 currency units.

The main requirement for any economic-mathematical model is its adequacy, i.e. compliance with the investigated process. Economic and mathematical models are effective only if they reflect only significant characteristics of the studied process, without taking into account secondary parameters [11]. This makes it possible to identify process patterns and to ensure this model effective application in practice.

4 Discussion

The economic and mathematical modeling process can very efficiently select the most optimal solutions to the problem. Today's economic challenges and challenges have a very broad range of issues. The economic-mathematical model analysis can be carried out using various software provisions [5]. For example, using standard, freely available optimizers such as Mathcad and Excel.

In order for the problem to be described qualitatively and quantifiably, it is necessary to carry out an analysis of the situations and objects laid down in this study. Complex objects are broken down into elements, there are relationships, properties, natural relations between them, expressed in the form of mathematical expressions [2]. Step-by-step use of operations research methods and their implementation with various software provisions allow to realize many possibilities that allow to simplify the problem's modeling process and analysis of the obtained result.

System analysis using mathematical methods of economic objects operation conditions, management parameters, decisions consequences significantly reduces risks and financial losses. Mathematics does not work with real objects, but with their mathematical models. The problem's mathematical formulation is only half of the success in solving it. The complexity consists in an exception of excess, unnecessary details. Important conditions should be preserved and the problem formulated as a standard model. To make full use of optimization theory and operation research techniques, one must be convinced of the system mathematical approach usefulness in managing.

5 Conclusion

Mathematical methods and economic and management processes modeling always simplify in one way or another, make more ground-breaking questions to be found. Sometimes nonlinear processes are reflected by linear models, dynamic processes are reflected by static models, and so on. In any case, the researcher should not feel that the mathematical methods application in economic research consists only in the suitable formulas selection, substitution of any numbers in them and obtaining adequate answers to the questions asked. The recommendations of the famous American scientist specialized in the development and application of computational mathematics methods Hemming are known: "The purpose of calculations is understanding, not number" [1, p. 32]. And also: "Before you solve the problem, think about what to do with its solution" [1, p. 68]. It is necessary to possess knowledge and skills in the use of generally accepted and individual techniques, which will allow to achieve the goal faster and easier.

References

1. Hemming, R.: Numerical Method. Nauka, Moscow (1972)
2. Intriligator, M.: Mathematical Optimization Methods and Economic Theory. Airis-Press, Moscow (2002)
3. Kostuchenko, V., Zdanov, A., Rodionov, A.: Mathematical methods of managing economic sustainability of the construction company. IOP Conf. Ser. Earth Environ. Sci. **90**(1) (2017). https://iopscience.iop.org/article/10.1088/1755-1315/90/1/012076. Accessed 09 Mar 2020
4. Kutateladze, S.S., Makarov, V.L., Romanovsky, I.V.: Scientific heritage of L. V. Kantorovich (1912–1986). Siberian J. Ind. Math. **4**(2), 3–17 (2001)

5. Levitin, I.E., Mayboroda, V.P.: Digital economy in management and evaluation of transport and logistics projects and life cycle processes. In: Sofin, A. (ed.) Proceedings of the 2017 International Conference Quality Management, Transport and Information Security, Information Technologies, IT and QM and IS 2017, pp. 240–242. IEEE, Piscataway (2017). https://doi.org/10.1109/itmqis.2017.8085803
6. Okrepilov, V., Chetyrkina, N., Krylov, M.: The role of elements of the economics of quality in mathematical modelling of socio-economic processes. In: Mottaeva, A. (ed.) Proceedings of the International Scientific Conference Construction and Architecture: Theory and Practice for the Innovation Development. EDP Sciences, Paris (2019). https://doi.org/10.1051/e3sconf/201913802008. E3S Web of Conferences, vol. 138 (02008)
7. Podvalny, S.L., Minakova, N.V., Vasiljev, E.M.: Modelling and analysing the dynamic interaction between competing branches of the regional economy as a multiagent control system. In: Korol, E. (ed.) Proceedings of the 1st International Conference on Control Systems, Mathematical Modelling, Automation and Energy Efficiency, pp. 169–174. IEEE, Piscataway (2019). https://doi.org/10.1109/summa48161.2019.8947567
8. Pourjavad, E., Mayorga, R.V.: Optimization of a sustainable closed loop supply chain network design under uncertainty using multi-objective evolutionary algorithms. Adv. Prod. Eng. Manag. **13**(2), 216–228 (2018). https://doi.org/10.14743/apem2018.2.286
9. Voronkova, O.Y., Kundius, V.A.: Economic and mathematical modeling of the process of production of quality agricultural products. Qual. Access Success **20**(173), 116–120 (2019)
10. Wang, Y.: Medium- and long-term predictions for coastal economies based on support vector regression. J. Coast. Res. **98**(sp1), 71–74 (2019). https://doi.org/10.2112/SI98-018.1
11. Zhang, S., Wang, X., Wang, S.: Modeling and computation of energy efficiency management with emission permits trading. J. Ind. Manag. Optim. **14**(4), 1349–1365 (2018). https://doi.org/10.3934/jimo.2018010

Digitalization and Project Management Method in Improving Efficiency of Drilling Wells Construction

O. A. Babordina(✉), M. P. Garanina, P. A. Garanin, and E. K. Chirkunova

Samara State Technical University, Samara, Russia
olgababordina@rambler.ru,
{garaninamarina, ninarag}@ya.ru,
ekchirkunova@gmail.com

Abstract. The authors consider the feasibility of using an innovative approach to solving strategic and current tasks of oil and gas enterprises, which implies optimizing existing processes and integrating modern technologies based on digitalization and project management in these processes. The development of a project implementation schedule can be carried out using software to image the process and create a flowchart of all working phases and decision points, as well as create a Gantt chart using standard programs. An important factor is to determine when to implement technologies that meet the project goals, align with the company's risks, and reduce the cost of the drilling process.

Keywords: Construction of drilling wells · Gantt chart · Project management · Project schedule · Well models

1 Introduction

The oil industry is one of the most important structural components of the Russian economy and one of the key factors in ensuring the country's life. It produces more than a quarter of the Russian industrial output and has a significant influence on the country's budget. Export revenues of the oil industry provide a consistently high trade surplus. Thus, the physical volume of oil exports for 2018 is 236.6 million tons (1.722 billion barrels). The increase, compared to 2017, was only 0.5%, and in value terms, the increase is 38% – 118.42 billion U.S. dollars. In January 2020, Russia's revenues from oil exports increased by 0.6% compared to January 2019 and amounted to $ 9.015 billion [15].

Currently, the labor intensity of oil reserves extraction in Russia is increasing. If innovative technologies and modern equipment are not used, oil production will begin to decline since 2020. The future success requires an innovative approach to solving strategic and current tasks, which implies optimizing existing processes and integrating modern technologies into them [4].

Industry 4.0 technologies such as Big Data analysis, the Internet of Things, and Artificial Intelligence (AI) help to improve the drilling efficiency [8]. However, the lack of serious competition leads to the fact that market participants are not in a hurry to

S. I. Ashmarina et al. (Eds.): ENGINEERING ECONOMICS WEEK 2020, LNNS 139, pp. 348–353, 2021.
https://doi.org/10.1007/978-3-030-53277-2_42

invest significant resources in their development and take on additional risks associated with this. Therefore, the application of new technologies is often an initiative of oil companies themselves [7].

Improving the efficiency of solving current problems at all stages of the life cycle of oil and gas facilities is possible with the complex interaction of all production processes based on the project management method [1].

Project management is a set of measures to coordinate human, financial, information and material resources throughout the project cycle, aimed at achieving the project goals. The entire set of project implementation operations is interconnected in time and space [5].

Oil production from old fields is becoming more and more difficult, its growth is slowing, but the demand on the international market is not decreasing, in particular, China's dependence on oil imports will increase. It is expected that by 2020–2030 it will reach 64–66% of the oil used. In the future, it will be possible to fully automate drilling operations and use solutions based on artificial intelligence, which will make it possible to remotely control the drilling rig.

The urgency of the problem increases in connection with the development of new oil and gas fields lying at great depths with complex mining and geological conditions. One of the most important components of the issue of increasing the drilling operations efficiency is to improve the process of well deepening. A significant amount of scientific and practical work has been devoted to this problem for more than a century in the history of rotary drilling. Currently, these problems need to be solved using a process approach based on the project management method [13], in which the structure and the functioning mechanism of an object or process can be more accurately investigated.

2 Methodology

The use of simulation models and digital technologies in project management is needed not only to optimize processes, to verify the economic efficiency of projects, but also to provide realistic scenarios for the development of the oil industry [6]. According to Russian authors [2], the application of innovations in project management in the oil and petrochemical industry is not only a way to increase the production of raw materials and reduce losses in the search and development of new unconventional fields, but also a tool to increase competitive advantages of companies.

The application of the project management method in the digitization of drilling well construction is a further research of a team of authors [3] related to the management of production processes in the changing internal environment of an oil production enterprise and the use of cloud platforms [9]. The main practical goal of the well construction is to optimize the well design and construction process. At the same time, it should be noted that the well model is a base for analytical and research work of designers, technologists and other specialists.

The well construction model is created from various information modules (geological, ground-based, technical-technological, and economic ones, etc.) that allow calculating, describing, analyzing and predicting technological processes and operations, and changing parameters over time in a certain range of conditions planned during the project implementation.

There is a following classification of well models: conceptual; static; dynamic.

1. A conceptual model of a well is constructed in the process of creating an integrated project for the field development, based on information from a permanent geological-technological model of the field. Based on the conceptual model of the well, the technical specification for the design of its construction (reconstruction) is developed. This well model consists of the following main modules: geological, technical and technological, and economic ones.
2. A static well model describes the initial (original) state of a studied technogenic system – the well before the start of the drilling process. Static modeling is performed to find high-quality and effective engineering and technological solutions used in the project documentation. A static well model is built on the basis of simulation results obtained from engineering and technological solutions formalized in the project documentation and work programs. Engineering (mathematical) calculations are performed with specialized software. Design solutions are a theoretical basis for the construction of wells, the work of professionals (drillers, geologists, geophysicists) in the process of engineering and technological support for the well construction. Design solutions based on modeling of well construction should ensure the variation principle realization, as well as the efficiency and safety of drilling operations. Monitoring of construction (drilling) operations is provided based on the set threshold values (extremes) of the model. A static model consists of a number of modules: ground-based, geological, technical-technological, and so on.
3. A dynamic model of a well describes the state of a technological system – a well in the drilling process. Dynamic well modeling is performed to optimize the well drilling process. Optimization is performed through the operational adjustment of work programs based on a specialized methodology. Primary sources of information on the drilling rig for building a dynamic model are a geological and technological station, a telemetry station, a supervisor's summary, results of a laboratory research. The dynamic well model is based on the processed information during monitoring, namely, technical, technological, geological and geophysical parameters received from the drilling rig in real time. The dynamic model state is time dependent and varies during the drilling process.

Conceptual, static and dynamic models are built in the corporate information system. The results of well modeling are presented in the form of simple forms: time graphs, tables, diagrams, 2D and ZD modules, visualized in the information system [11]. Each of the presented types of wells can be used to optimize the permanent geological and technological model of the field.

According to the presented characteristics of the well models, the most attractive is the dynamic one. On the one hand, it is suitable not only for optimal programming, but also for the operational solution of optimization problems [12]. The latter are especially important when faced with unforeseen constraints, when in the current situation it is necessary to find an optimal solution, i.e. the closest in consequences to the programmed one. If necessary, it can be adjusted in previously accepted program parameters with some deviation from the programmed indicators [10]. On the other hand, dynamic programming allows us to solve problems that are discrete in nature, that is, not transformed from the corresponding continuous problems [14]. Also, this model allows you to predict and prevent sticking of drilling tools based on the selection of optimal technological parameters of the drilling process.

3 Results

The research is determined by the fact that the process of well deepening has not been optimized yet, and the existing reserves are not used for improving the drill bit designs, mechanical drive, technical and technological characteristics of hydraulic downhole motors and drilling modes. The cost of drilling wells can reach up to 75% of all costs for exploration and extraction. For a well to work effectively, the company has to provide $6–10 in revenue in order to spend $1 on drilling. The experience shows that innovative technologies, without using people's technical skills and business processes that ensure their application, will not lead to the desired goals.

Project management can be divided into phases according to the project lifecycle (for example: definition, development, implementation, analysis) [2]. These processes can be at different levels of project detail and implemented in different ways.

The development of a project implementation schedule can be carried out using the following methods:

- using software to image the process and create a flowchart of all working phases and decision points,
- creating a Gantt chart using standard programs.

However, the main thing is to define the project goals, assess technical and commercial risks and opportunities, and initiate technical research. An important factor is also to determine the implementation moment for technologies that meet the project goals and are consistent with the company's risks.

Significant improvements can be made by increasing the efficiency of basic operations: for example, by optimizing equipment maintenance and repair schedules. With the ability to forecast and reallocate work, you can reduce downtime, optimize preventive maintenance, and increase the volume of drilling operations.

Digital projects can be launched in the exploration and production block. The process of well design and construction is now largely digitized. The location of wells, their trajectory, design, and individual operations (lowering casing strings, cementing the trunk, etc.) are calculated in special computer simulators. The software allows you to evaluate the expected return from a reservoir and find the best ways to achieve it. The method application will allow increasing the speed of drilling of the horizontal part of wells in the future, reducing the cost of eliminating errors and improving the quality of sinking, and therefore the efficiency of the future operation.

4 Discussion

The efficiency of drilling wells construction is based on the application of the project management method. The Gantt chart is used when drawing up the work plan. The graphical control system is a result of the construction and operation of drilling wells. Gantt charts provide concise, accurate and visual information about the position of the actual result and help you to make optimal management decisions.

In project management of the construction of complex, communication-rich and equipment-rich technological facilities, a number of problems arise. Most of them are related to errors made at the design stage, which can be eliminated by applying the Gantt method. Its main principles are: the management of future problems is based on past results; the relation of phenomena in time; the relation between work and time. The value of Gantt charts in the well construction is to identify such inconsistencies, as the intersection of structures, the discrepancy between the size of the facility and the installed equipment, overlay of communications, a lack of space to move lifting equipment, barriers for personnel's access and their timely removal while designing.

5 Conclusion

The creation of high-tech drilling rigs is a prospect today for the Russian market, as the situation of falling resource prices on the world market encourages the use of digital technologies to reduce costs. Increasing the efficiency of drilling wells is provided by the optimization process during their design and construction. At the same time, it should be noted that the well model is the main one for analytical and research work of designers, technologists and various specialists. All this will allow switching to full automation of drilling operations and applying solutions based on artificial intelligence, remotely managing the drilling rig based on the flowchart of all working phases and decision points marked in the Gantt chart. The presented research is important in the sphere of drilling directional and horizontal wells in the fields and for the design of well construction in general in the context of project management in solving strategic and tactical tasks in the oil and gas industry.

References

1. Alexandrova, T.V., Prudsky, V.G.: On the conceptual model of oil and gas business transformation in the transitional conditions to the Industry 4.0 (2019). https://hdl.handle.net/10195/74254. Accessed 12 Mar 2020
2. Babordina, O.A., Garanina, M.P., Ilyina, L.A, Garanin, P.A.: Formation of the business model of oil producing enterprises. In: Mantulenko, V. (ed.) Proceedings of the International Scientific Conference "Global Challenges and Prospects of the Modern Economic Development". Future Academy, London (2019). The European Proceedings of Social and Behavioural Sciences, vol. 57, pp. 230–238

3. Bagaeva, I.V., Larionova, G.N., Bagaeva, Y.R.: Comparative analysis of innovative activity of large business in Russia and the United States in terms of digitalization of the petrochemical industry. In: Ilin, I.V. (ed.) Proceedings of the International Conference on Digital Technologies in Logistics and Infrastructure (ICDTLI 2019). Atlantis Press, Paris (2019). Atlantis Highlights in Computer Sciences, vol. 1, pp. 316–321
4. Barney, S., Petersen, K., Svahnberg, M., Aurum, A., Barney, H.: Software quality trade-offs: a systematic map. Inf. Softw. Technol. **54**(7), 651–662 (2012). https://doi.org/10.1016/j.infsof.2012.01.008
5. Barth, C., Koch, S.: Critical success factors in ERP upgrade projects. Ind. Manag. Data Syst. **119**(3), 656–675 (2019). https://doi.org/10.1108/IMDS-01-2018-0016
6. Garvey, M.A., McLeroy, P.G., Weijermars, R.: Reservoir-model-based scenarios for assessing the viability of greenhouse gas mitigation strategies through CO2 enhanced oil recovery. Energy Strat. Rev. **16**, 54–67 (2017)
7. Janocko, J., Ryzhkova, M., Mamzina, T., Bereznev, S.: Digitalization of economic evaluation of mineral deposits. In: Tyulenev, M., Zhironkin, S., Khoreshok, A., Vöth, S., Cehlár, M., Nuray, D., Janocko, J., Anyona, S., Tan, Y., Abay, A., Marasová, D., Stefanek, P. (eds.) Proceedings of the IVth International Innovative Mining Symposium. EDP Science, Les Ulis (2019). E3S Web of Conferences, vol. 105 (04030)
8. Jonak, Ł., Rudnicka, A., Włoch, R.: Digitalization of supply chain transparency: the case of ChainReact. In: Grzybowska, K., Awasthi, A., Sawhney, R. (eds.) Sustainable Logistics and Production in Industry 4.0. EcoProduction, pp. 89–102. Springer, Cham (2020)
9. Meghana, H., Asish, O., Lewlyn, L.: Prioritizing the factors affecting cloud ERP adoption – an analytic hierarchy process approach. Int. J. Emerg. Mark. **13**(6), 1559–1577 (2018). https://doi.org/10.1108/IJoEM-10-2017-0404
10. Nagurney, A., Shukla, S., Nagurney, L., Saberi, S.: A game theory model for freight service provision security investments for high-value cargo. Econ. Transp. **16**, 21–28 (2018). https://doi.org/10.1016/j.ecotra.2018.09.002
11. Somu, N., Kirthivasan, K., Sriram, V.: A computational model for ranking cloud service providers using hypergraph based techniques. Future Gener. Comput. Syst. Int. J. Escience, **68**, 14–30 (2017). https://doi.org/10.1016/j.future.2016.08.014
12. Symons, C.: Exploring software project effort versus duration trade-offs. IEEE Softw. **29**(4), 67–74 (2012). https://doi.org/10.1109/MS.2011.126
13. Syntetos, A.A., Kholidasari, I., Naim, M.M.: The effects of integrating management judgement into OUT levels: in or out of context? Eur. J. Oper. Res. **249**(3), 853–863 (2016). https://doi.org/10.1016/j.ejor.2015.07.021
14. Yakhneeva, I.V., Agafonova, A.N., Fedorenko, R.V., Shvetsova, E.V., Filatova, D.V.: On collaborations between software producer and customer: A kind of two-player strategic game. In: Ashmarina, S., Mesquita, A., Vochozka, M. (eds.) Digital Transformation of the Economy: Challenges, Trends and New Opportunities. Springer, Cham (2020). https://doi.org/10.1007/978-3-030-11367-4_56. Advances in Intelligent Systems and Computing, vol. 908, pp. 570–580
15. RCC: Russia's oil export revenues increased by 0.6% in January (2020). http://rcc.ru/article/dohody-rossii-ot-eksporta-nefti-v-yanvare-vyrosli-na-0-6-72835. Accessed 12 Mar 2020

Regional Market Adaptation to the Demands of Digital Economy

M. S. Guseva(✉)

Samara State University of Economics, Samara, Russia
gusevams@yandex.ru

Abstract. Digitalization of the economy might create conditions for the formation of new disproportion in the national and regional labor markets, the most likely consequences of which will be a further stratification of the population by income level, an increase in structural unemployment, a decrease in both the level of income and the quality of life for citizens with "obsolete" professions. The state has to take measures which will lead to the transformation of the labor market, matching with digital changes and which will contribute not only to economic growth, but to social progress as well. Research is aimed at to identify the socio-economic laws of functioning and transformation in the digitalization of the regional labor market, development of proposals for its adaptation to the demands of the digitalization. The main research methods used were systematic analysis and situational analysis. The article systematizes prominent patterns and tendencies that are common to the labor market as a result of the development of the adaptation to digitalization. Recommendations on the adaptation of the labor market of the Samara region to the demands of the digital economy are proposed.

Keywords: Adaptation · Digitalization · Labor market · Region · Samara region · Transformation

1 Introduction

Digitalization of the economy is a powerful tool for increasing labor productivity and as such, provides developing countries with a unique chance for a non-linear breakthrough to improve in economic growth. For Russia, which suffers technological underdevelopment and demographic constraints, the development of the digital economy can lead to economic growth and state sovereignty. Digitalization is followed by multiple positive economic and social effects, however, it also carries a number of threats and risks to the currently established socio-economic mechanisms, including the labor market. According to experts, in the near future, most traditional companies will be replaced by digital platforms, with 80% of the world trade being conducted through online sales; as such the new labor market will be represented by online employment, with 50% of the workers having a flexible work schedule [4]. The negative effects of digitalization will be the stratification of the population by income level, the growth of unemployment, the decline in income and the quality of life of people who have old professions, and therefore the deterioration of the social well-being of Russian society.

S. I. Ashmarina et al. (Eds.): ENGINEERING ECONOMICS WEEK 2020, LNNS 139, pp. 354–363, 2021.
https://doi.org/10.1007/978-3-030-53277-2_43

The urgent need to minimize the negative effects of digitalization on the pre-existing labor market and provide the digital economy with a workforce that possess appropriate skills, determines the relevance of the chosen research topic. The author aims to identify the socio-economic laws of functioning and transformation in the digitalization of the regional labor market, development of proposals for its adaptation to the demands of the digital economy.

Plan for achieving this aim was to:

1. Study the socio-economic effects of developing the digital economy, as well as employment trends caused by digitalization.
2. Analyze structural, territorial and qualitative characteristics of Samara region labor market.
3. Develop proposals for adaptation of the Samara region labor market to the demands of the economy in a new format – «Industry 4.0».

The object of study is the labor market of the Samara region. The subject of the study is the compilation of socio-economic relations which arise in the regional labor market as a result of the developing digital economy. The significance of the study underlined in the further development of already existing approaches to understanding the specifics, tendencies and issues of the development of the Russian labor market, which manifest and strengthen due to digitalization. The author offers practical ways of adapting regional labor markets to the development factors of the digital economy. The following terminology is utilized in the article. Regional labor market - socio-economic mechanism that operates within a specific economic and geographical space and is influenced by the state and institutional factors, which itself regulates aggregated amount of wage labor implemented in the economy and the employee wage rate. The digital economy implies the transition of all processes and value chains in all spheres of economic production and consumption to digital technologies which allow organizations to work with maximum efficiency. Modern digital technologies consist of: cloud technologies, artificial intelligence, Big Data, the Internet of Things (IoT), the industrial Internet of Things (IIoT), additive technologies, smart services [17].

2 Methodology

The following methods were implemented during the research process: theoretical (analysis and synthesis, comparison, etc.); empirical (measurement and systematization of research results, induction, deduction, grouping, and sampling). The main research methods were system analysis (when determining causal relationship between digitalization and the regional labor market) and situational analysis (when establishing a link between the influence of digitalization factors and trends in regional development of the employment sector). The main experimental base of the study consisted of statistical indicators of the Samara region labor market for the period between 2005 and 2018.

The study of this subject was conducted in four stages:

Stage 1 – collection and processing of data on the state and development of the labor market in the Samara region, VFD and the Russian Federation.
Stage 2 – classification of the main approaches in determining the influence of factors of the digital economy on the formation of an efficient labor market in the region.
Stage 3 – analysis of the main statistical indicators of the Samara region labor market, structural, territorial and qualitative characteristics of the regional labor market.
Stage 4 – diagnostics of vital problems and establishment of possible ways to adapt the labor market under the influence of digitalization factors with the aim of creating an efficient labor market in the region.

3 Results

Labor market in Russia and its regions, operates with the following characteristics and trends:

- satisfactory level of maturity, sustainable long-term relationships between the main market indicators;
- comparatively low level of labor productivity compared with a simultaneous increase in real wages;
- adaptability of real wages to the unemployment rate, which is largely determined by institutional factors: state support for large employers, low minimum wages, and comparatively weak trade unions;
- mechanism of labor market crisis adaptation leads, firstly, to a decrease in real income for the population, an increase in poverty rates and the emergence of the Russian phenomenon of "working poor", and secondly to a decrease in labor productivity as a result of a fall in real GDP and the preservation of employment;
- low territorial and labor mobility of the population;
- high share of employees in the public sector;
- existence of a massive number of unofficial jobs with their own specific mechanisms for determining wages;
- a growing share of highly qualified profession groups in the total number of employees;
- a noticeable differentiation of labor markets in the regions and within regions of the country with a growing conjunction between regional labor; markets and their integration into the general labor market [6].

Samara region with a population of 3.2 million people is a highly urbanized region with Samara-Togliatti agglomeration (STA) established in its territory and has the third largest population in the Russian Federation after Moscow and St. Petersburg. 2.7 million people live in urban districts and municipal districts of STA (over 86% of the region's population). The economy of the region is formed primarily by large and medium-sized businesses. Samara region produces approximately 2% of the gross regional product (GRP) of the constituent entities of the RF. About 29% of industrial production is accounted for by industries traditional for the economy of this particular region: aerospace, metallurgical, and electrical equipment manufacturing. More than 70% of the region's industrial production accounts for motor vehicles, chemicals, foodstuffs, rubber and plastic products, as well as pharmaceutical products (Fig. 1).

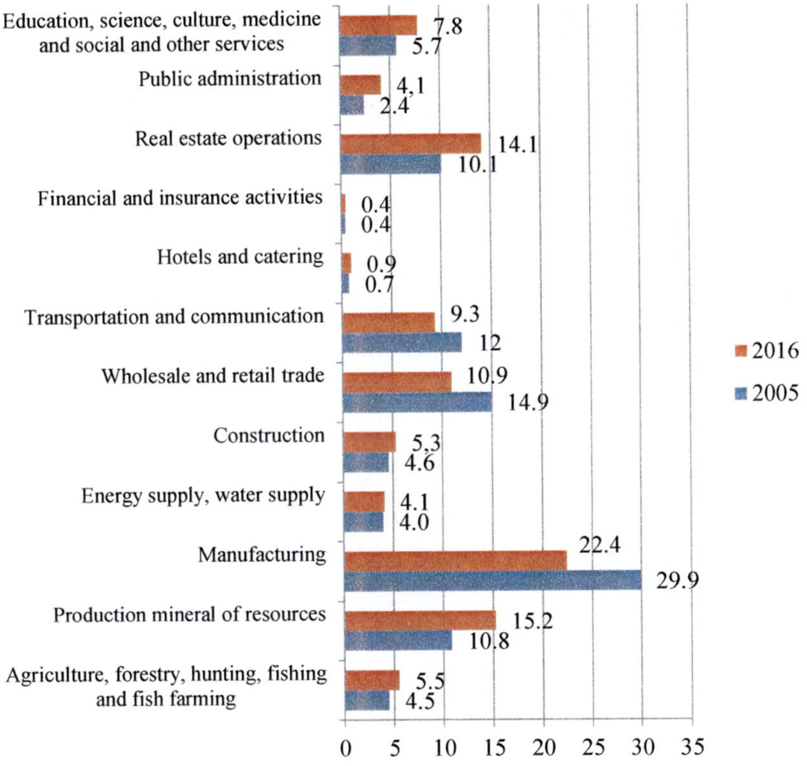

Fig. 1. Sectoral structure of the gross regional product (Source: author).

At present, Russia in the field of digitalization is the leader in the field of telecommunications and communications, IT and software development, the banking sector and financial services. «Catching up» are trade, insurance, education, the auto business, and transport. The «lagging behind» are the sectors of industrial production, oil and gas production, tourism and restaurant services, as well as consulting, publishing and the media. «Beginners» – the construction industry, medicine and the entertainment industry.

Leading industries act as digitalization drivers in related industries. Informatization of the banking sector creates the conditions for the complete computerization of financial processes among its customers, and the development of telecommunications and communications, IT and software development, defined as the IT market, ensures the growth of computerization and informatization in organizations of other industries.

The size of the region's workforce is 2.3%–2.2% (2005–2018) of the country's labor pool (Table 1), and takes the 4th place in the VFD, 3rd place when it comes to the employment rate.

Table 1. Workforce and real wages in the Russian Federation, VFD and Samara region, 2005–2018

Region	2005	2010	2011	2012	2013	2014	2015	2016	2017	2018
Workforce, million people										
Russian Federation	73,8	75,5	75,8	75,7	75,5	75,4	76,6	75,8	76,1	76,2
VFD	15,7	15,9	15,8	15,7	15,6	15,5	15,5	15,5	15,2	15,1
Samara Region	1,7	1,8	1,8	1,7	1,7	1,8	1,8	1,7	1,7	1,7
Participation in workforce of workers aged between 15–72 years, %										
Russian Federation	66,0	67,7	68,3	68,7	68,5	68,9	69,1	69,5	69,1	68,9
VFD	65,8	67,8	68,4	68,6	68,1	68,4	68,9	69,2	68,3	68,0
Samara Region	68,1	68,5	68,8	69,3	69,3	70,4	71,0	71,6	70,5	71,0
The average monthly nominal wage, rubles										
Russian Federation	8555	20952	23369	26629	29792	32495	34030	36709	39167	43724
VFD	6473	15614	17544	20020	22481	24601	25632	27265	29189	31990
Samara Region	7765	16479	18600	20800	23470	25884	26849	28295	30492	33754
Real wages of employees of organizations, compared in % to previous year										
Russian Federation	112,6	105,2	102,8	108,4	104,8	101,2	91,0	100,8	102,9	108,5
VFD	112,0	104,5	103,2	108,8	105,2	101,7	91,0	100,5	103,9	106,8
Samara Region	109,9	104,3	104,4	107,5	106,3	102,4	89,9	98,8	104,6	107,7
Unemployment, in %										
Russian Federation	7,1	7,3	6,5	5,5	5,5	5,2	5,6	5,5	5,2	4,8
VFD	7,4	7,6	6,5	5,3	4,9	4,5	4,8	4,8	4,7	4,4
Samara Region	5,4	5,8	5,1	3,4	3,2	3,0	3,4	4,1	4,2	3,7

Source: author based on [14].

The Samara region is characterized by above the average population workforce participation in both the country and the Volga Federal District, where it securely occupies no lower than 3rd place in that regard. However, from 2010 to 2018, the potential workforce of the region decreased from 20 thousand people down to 14.3 thousand people. The organization's need for workers increased by more than 48% and amounted to 19,066 vacancies. The dominant majority of the able-bodied population of the region are men.

By average monthly nominal wage Samara region ranked 1st in the Volga Federal District in 2005, and the 3rd in 2018 (33,754 rubles). The growth dynamics of real wages for the study period was unstable in the region, as well as in the country and in the VFD. Its growth rate over the general period was lower than both Russian and VFD average. The region is also defined by low labor productivity, which is one of the main constraints for economic growth in the Samara region. The level of registered unemployment in municipalities is characterized by significant unevenness. The following patterns are traced:

1. A higher level of registered unemployment is observed in rural municipal areas; the indicator increases as the municipal district moves away from the STA.
2. The unemployment rate is lower in large cities and higher in small cities.

The dynamics of the Samara region workforce size were negatively affected by the demographic situation. Regional population decreased by 0.67% in the period from 2005–2018, there has been a natural decline in the population since 1991 (−2.9% in 2018). Moreover, since 2015, migration outflow to the region has begun to outpace migration inflow.

The number of foreign citizens entitled to work in the Samara region is growing, but does not have a significant role in the formation of supply in the labor market. Among foreign citizens with a valid work permit, the share of specialists in the field of physical and engineering sciences was 25.8% in 2018 (15.7% in 2016). In the ranking in terms of employment and unemployment in 2018, the Samara region was among the first twenty regions with the highest level of employment and the lowest unemployment. By the share of highly skilled workers in total amount of workers and by the share of the employed population with professional education, Samara takes 5th, 6th and 8th places among the subjects of the Russian Federation. 24% of workers are employed in processing and refinement industry branches, with 5.4% employed in the agricultural sector. Share of people employed in manufacturing sectors is on a slow decline, with construction trade and service being on the rise. In 2018, the number of vacancies in low skill jobs exceeded the supply of labor by 2.1 times, for high skill jobs – 1.7 times.

The Samara region has necessary prerequisites for the creation of high-performance jobs and the development of a digital economy in the region:

1. Over the years, the region has been amongst the leaders of innovative development. According to the Association of Innovative Regions of Russia, in 2018 the region ranks 10th (in 2017 – 9th) in the Rating of Innovative Regions, leading a strong group of mid-tier innovators [3].
2. The Samara region participated in the program of the National Technological Initiative on a competitive basis, and carries out scientific research in the field of EnergyNet (distributed energy), HealthNet (personal medicine), AeroNet (distributed unmanned aerial vehicle systems), AutoNet (distributed network of unmanned vehicles), NeuroNet (distributed artificial components of consciousness and psyche) [11].
3. The innovative development of the region is supported by significant scientific and production potential, the basis of which is university and academic science divisions, as well as scientific departments of industrial enterprises.
4. Innovative developments are then implemented in innovative business. In 2017, Samara region, amongst constituent entities of the Russian federation, took 11th

place in the share of Russian high-tech business and 8th in the concentration of newly created high-tech companies [1].
5. Innovative and high-tech business is developing in a favorable investment climate.

According to the Skolkovo, in 2018 Samara Region reinforced its status of a region with high investment attractiveness. However, according to the results of the "Digital Russia" index for the constituent entities of the Russian Federation in 2018, the Samara region took 17th place (in 2017 – 21st place) [15].

At the same time, the Samara region is ahead of competitors in terms of "Specialized personnel and training programs" sub-index, as well as having high relative indicators in "The existence and development of research competency, technological groundwork and works", but falls short in terms of investment, i.e. the creation of digital jobs, and development of the digital infrastructure.

Digitalization of the Samara region is carried out in partnership between the state and large business and is characterized by the following trends:

- more than 60% of households are connected to broadband Internet,
- about 80% of the region's residents have access to electronic public services,
- all schools in the region have Internet access, and 63% of schools are connected to fiber-optic communication lines,
- actively developing the system of «smart education» and «smart city»,
- as a result of the implementation of the Smart-agro project in the Samara region, up to 100% of arable land was digitized and space monitoring was carried out up to 80% of their area.

The strategic and program documents of the Samara region trace the understanding of the importance of digitalization for the development of a competitive economy, the role of human capital and highly qualified personnel in the creation, development and implementation of digital technologies. Priority is given to education and training, the maintenance and development of research competencies and ICT.

4 Discussion

In modern Russia, interest in the problems of employment and unemployment, the formation and transformation of market relations in the labor sphere, and the specifics of function in regional labor markets has been formed due to development of a market economy. In the works of Aslanova [2], Mokronosov, Matafonov, Chuchkalova, Prudnikov [9], Vaysburd [16], Nedohlebova [10] unique approaches were developed to the definition of the labor market.

Russian scientists have entered the science of modern market problems and are currently working on a study of the staff demands of the digital economy and its impact on employment [2, 12]. Some researchers are united in opinion that digital economy is an entirely new step in development of economic relations, based on digital information trade, which is system-forming, and strongly impacts all aspects of human life [5, 8, 13]. Due to the double-sided effect of digital economy on the labor market such as both a multitude of economic or social benefits and negative effects, scientists are attempting to analyze and predict possible outcomes [4, 7]. However, the regional aspects of the labor market transformation in the context of digitalization have hardly been studied.

5 Conclusion

For the effective formation of the labor market, it is necessary to preserve human capital, support innovation and investment in the digital sphere. Samara region serves as an example that such factors contribute to creation of:

- high-tech jobs that demand new requirements on the qualitative characteristics of workers, generating labor demand for the digital economy;
- conditions and institutions related to the demographic improvement, personal development and development of human capital, the formation of a labor supply that has the characteristics necessary for an economy based on knowledge and digital processes;

The following situation has developed with these factors in the Samara region:

- priority development of education and training, maintenance and development of research competencies and backlogs in the field of ICT;
- infrastructure and investment lag in the digitalization process in the region;
- the lack of a holistic strategic approach to the development of the regional labor market, aimed at increasing its effectiveness, preventing possible imbalances caused by the widespread introduction of high-tech and digital technologies.

Factors in the formation of an effective labor market:

1. Conditions for the development and preservation of human capital.
2. Conditions for maintaining innovation and investment.
3. The role of regional authorities and government agencies in the development of the digital economy.

Taking the specifics of digital economy development in Samara region and the state of the regional labor market into account, its adaptation to the demands of digital economy requires the solution of following imperative issues:

1. Region falls behind in terms of implementation of information and communication technologies whilst having a high scientific and manufacturing potential, scientific developments and groundwork.
2. Lack of IT staff and specialists in various relevant spheres with digital competency.
3. «Maturity» of staff.

Labor market adaptation to the needs of the digital economy should be carried out at the national and regional levels. National level involves the modernization of the education system and the volume of training of IT specialists within the framework of the state program, regulation of migration processes, improving social protection for workers engaged in non-standard forms of labor, improvement to employment promoting policies and increase in efficiency of the public employment service.

On the regional level, measures to adapt the regional labor market to digital transformations must provide am influx of high-tech and digital jobs, jobs in the humanitarian field and areas in which computers and robots cannot replace people (demand generation) and training/retraining of personnel with relevant competencies (labor supply). Specifics of regional labor, its demands and problems, methods of solving such problems must be reflected in a strategy for the development of the labor market and employment of the Samara region, he development of which should be carried out as part of the joint project activity of all involved ministries of the regional government, representatives of unions of workers and employers. Measures to adapt the regional labor market to digital transformations should ensure the creation of high-tech jobs, jobs in the humanitarian field and in areas in which computers and robots cannot replace a person and training (retraining) personnel with relevant competencies.

References

1. Adamaitis, S.A., Barinova, V.A., Zemtsov, S.P., Kidyaeva, V.M., Kotsyubinsky, V.A., Tsareva, Yu.V.: High-tech business in the regions of Russia National report (2019). http://i-regions.org/images/files/ranepa18.pdf. Accessed 28 Feb 2020
2. Aslanova, L.O.: Analysis of the situation on the Russian labor market and forecasting its development in the post-crisis period. Int. Mag. EPOS **4**, 17–23 (2009)
3. Association of Innovative Regions of Russia: Rating of innovative regions (2018). http://www.i-regions.org/reiting/rejting-innovatsionnogo-razvitiya. Accessed 28 Feb 2020
4. Balsmeier, B., Woerter, M.: Is this time different? How digitalization influences job creation and destruction. Res. Policy **48**(8), 103765 (2019). https://doi.org/10.1016/j.respol.2019.03.010
5. Colombo, E., Mercorio, F., Mezzanzanica, M.: AI meets labor market: exploring the link between automation and skills. Inf. Econ. Policy **47**, 27–37 (2019). https://doi.org/10.1016/j.infoecopol.2019.05.003
6. Gurvich, E., Vakulenko, E.: Macroeconomic and structural properties of the Russian labor market: a cross-country comparison Russian. J. Econ. **3**, 411–424 (2017). https://doi.org/10.1016/j.ruje.2017.12.006
7. Hammershøj, L.G.: The new division of labor between human and machine and its educational implications. Technol. Soc. **59**, 101142 (2019). https://doi.org/10.1016/j.techsoc.2019.05.006
8. Kurt, R.: Industry 4.0 in terms of industrial relations and its impacts on labour life. Procedia Comput. Sci. **158**, 590–601 (2019). https://doi.org/10.1016/j.procs.2019.09.093
9. Mokronosov, A.G., Matafonov, M.E., Chuchkalova, Y.I., Prudnikov, D.M.: Forecasting the demands of the regional economy in the training of qualified personnel. Ekaterinburg: RSVPU (2010)
10. Nedohlebova, E.B.: Labor Market. RIC KSERU, Khabarovsk (2018)
11. Program of the National Technological Initiative (2014). https://nti2035.ru/nti/. Accessed 28 Feb 2020
12. Putilov, A.V., Bugaenko, M.V., Timokhin, D.V.: Development of Russian labor market in the context of informatization and computerization of the economy. Procedia Comput. Sci. **145**, 169–176 (2018). https://doi.org/10.1016/j.procs.2018.11.035
13. Richardson, L., Bissell, D.: Geographies of digital skill. Geoforum **99**, 278–286 (2019). https://doi.org/10.1016/j.geoforum.2017.09.014

14. Russian Federal State Statistics Service: Labour market, employment and wages (2020). https://www.gks.ru/labor_market_employment_salaries. Accessed 28 Feb 2020
15. Skolkovo: The digital Russia index (2019). https://finance.skolkovo.ru/downloads/documents/FinChair/Research_Reports/SKOLKOVO_Digital_Russia_Report_Full_2019-04_ru.pdf. Accessed 28 Feb 2020
16. Vaysburd, V.A.: Labor Economy. OMEGA-L, Moscow (2011)
17. Watanabe, C., Tou, Y., Neittaanmäki, P.: A new paradox of the digital economy - structural sources of the limitation of GDP statistics. Technol. Soc. **55**, 9–23 (2018). https://doi.org/10.1016/j.techsoc.2018.05.004

Complexity Theory for Project Management in the Digital Economy

Y. Matveev[1], T. Stepanova[2], and O. Trubetskaya[1](✉)

[1] Samara State University of Economics, Samara, Russia
matveev@gmail.com, olgatrub@gmail.com
[2] Kaliningrad State Technical University, Kaliningrad, Russia
tatyana.stepanova@kgtu.ru

Abstract. The analysis of the digital economy as a global phenomenon involves the use of new methodological approaches that contribute to modern understanding of theoretical and practical aspects of one of the most popular areas in economic science. The use of complexity theory made it possible to justify that the digital economy has all the features of a complex system. The spatial approach revealed heterogeneity of the industry space, the existence of the institutional vacuum in its institutional space, and digital divide, which allow finding new ways to develop both the individual enterprise and the economy. The study of the digital economy in the aspect of the spatial approach revealed a number of problems such as heterogeneity of the industry space, the existence of the institutional vacuum, and digital divide.

Keywords: Complexity theory · Digital divide · Digital economy · Institutional vacuum · Spatial approach

1 Introduction

The adherence to traditional methodological approaches, proven by theory and practice in scientific research, is a natural and legitimate phenomenon. It is a tribute to traditions and a recognized achievement in the methodology of scientific research. However, the desire to go beyond what is usual in the scientific space is insurmountable, although it is risky from the standpoint of adopting new methodological approaches in scientific community. The authors consider that it is necessary and possible to apply complexity theory and methodological approaches, which are relatively new to the interdisciplinary nature of spatial economics, in the study of the digital economy, which is little-known in Russian scientific circles.

Complexity theory came to social sciences, to economics from mathematical theory of complex systems. Complexity theory, in its modern sense, summarizes a whole set of new methodological approaches of a postdisciplinary nature, including interdisciplinary and transdisciplinary approaches aimed at collaboration and integration of various types of scientific thinking (humanitarian and natural science), aggregation of which in the new system paradigm allows us to identify such peculiar complex quality systems like the presence of self-organizing structures: dynamism, emergence, fractality, nonlinear. In the 90s complexity theory took shape institutionally in the

S. I. Ashmarina et al. (Eds.): ENGINEERING ECONOMICS WEEK 2020, LNNS 139, pp. 364–371, 2021.
https://doi.org/10.1007/978-3-030-53277-2_44

framework of the established Santa Fe Institute - known as a center for the study of chaos and complexity. A collaboration of scientists from various fields of knowledge took place at this institute. Complexity theory is multidisciplinary and covers knowledge in areas such as meteorology, biology and chemistry, psychology, sociology and economics. Along with the characteristics noted above, complexity theory operates with such basic concepts as:

- sensitivity to initial conditions,
- strange attractors,
- self-identity,
- self-organization,
- edge of chaos,
- hilly landscape.

2 Methodology

The methodology of the digital economy is presented as a system of principles and methods of organizing and constructing theoretical and practical activities, as well as a doctrine on the progress in changing technological structures, the basis of which is the formation of the network space. This approach necessitates the use of the network management method, including network planning, network organization, network control and accounting. Also in the theory of complexity, analysis is widely used as a way to identify the main problems associated with the development of the digital space, to identify the characteristic features of the digital divide and the existing institutional environment that does not allow the full use of the existing digital opportunities.

3 Results

The digital economy has all the signs of a complex system, namely, dynamism, disequilibrium, self-regulation and self-identity, dissipativity, openness, emergence, attractiveness, fractality. We will reveal all of them.

1. The digital economy is a nonequilibrium system, which means constant dynamic fluctuations in various sectors and levels of management, pendulum oscillations, which each time come into equilibrium in different coordinates. Disequilibrium of the digital economy is closely related to the uneven development, i.e. different speed and direction of changes. The sector of digital technologies - the digital economy is developing rapidly, and the non-digital sector is developing much more slowly. Signs of disequilibrium of the digital economy are a lack of personnel for the digital economy, a lack of knowledge among specialists of various fields, a lack of intellectual resources - inventors, developers, etc., in the field of digitalization, a brain drain and intelligence, a vacuum in modern specialties and profiles in universities. For example, Russian universities still do not train managers for knowledge management, crowdsourcing, and additive technologies.

2. The digital economy has a border of chaos. It can be considered as the boundary between the regulated space and the unregulated space of the digital economy (chaos). Today, there is a rapid decline of traditional institutions, despite the fact that their replacement has not taken shape yet, so the system has signs of randomness, uncertainty between the old and the new. Under conditions of the digital economy, different processes are layered and cogered (from the word coherence), various processes are combined - new institutions are generated that are adequate to a higher level of technical and technological development, old institutions and structures are abolished. The necessary existing institutions are being transformed. All this happens asynchronously, asymmetrically, and does not coincide in time and space. The necessary new institutions and structures are being created slowly. They require significant resources, and, above all, financial, appropriate organizational transformations, which are barriers in the processes of their generation and cause their deficit. The old institutions are no longer functioning and therefore are being eliminated. Objectively unfilled space appears, an unregulated zone - an institutional vacuum, appears, i.e. there is a lack of necessary new institutions, relations between them and the players personifying them.
3. Dissipativity of the digital economy as a nonequilibrium system means a qualitatively different course of various processes at different levels of management. These processes include: the emergence of new spontaneous institutions and institutional structures on a self-regulatory basis, the forms of interconnections between them, the new forms of activity of existing institutions and structures, the implementation of constant dynamic transitions from order to chaos, from chaos to the orderly system with the corresponding order parameters (control parameters). Dissipativity determines "scattering" processes in the formation of single-order structures arising from the conversion of energy scooped from chaos. The consequences of dissipativity of the digital economy are:

   - ability to describe startups based on information theory. A failed startup is a typical dissipative system. A successful startup is an example of generating new information, the value of which significantly exceeds the initial investment in the project,
   - possibility, based on information theory, to describe the process of competition and the filling of market niches in the economy,
   - disappearing postulate of the inevitable maximization of profit of any market actor (company, enterprise). Instead, new concepts are introduced - the development of market niches and the struggle to preserve their conditional information.

   In relation to Russian conditions, dissipativity is associated with the simultaneous existence of more than two technological modes. In reality, there are simultaneously elements of the third, fourth and fifth technological modes in the Russian economy.
4. Strange attractors. From the standpoint of the passing "inheritance" from the industrial economy, one can consider large enterprises, firms, small and medium-sized business structures, within which the able-bodied and economically active population realize their professional competencies. In the digital economy, new strange attractors are also being generated, one of which can be called the Digital

Man. The digital man is one of the new attractors of the digital economy. A digital person must be created even with the highest potential of his inherent self-regulation. Attractors of the digital economy can be called digital enterprises that use the latest, including additive technologies, digital banks, and digital government.

5. Emergence of the digital economy. One of fundamental approaches in complexity theory is Emergence, i.e., emergence of a new quality, quantity, process, relationship in the interaction and interrelationship of various elements, for example, many simple particles or actors, agents that act according to simple rules. In complexity theory as applied to systems, emergence is the presence of a system with special properties that are not inherent in its elements, either individually or in total. A new, more complex substance is formed from this interaction, while the behavior of this new substance is not equivalent to the totality, the sum of the behaviors of its components. Moreover, the behavior of the new substance is logically inexplicable from the standpoint of the behavior of its components. Complexity theory claims that open complex systems, such as the digital economy and the labor market that is inherent in its level of development generate unpredictable patterns of agent behavior due to their inherent emergence and they really resist the effects of direct control measures.

In the digital world, there are several examples of self-organization. For example, there are different types of modern technologies - information, nanotechnology, bio and cognitive technologies, the convergence of which determines technological singularity. Their interaction in the digital economy forms its technical and technological basis and gives new emergence properties - instead of traditional duality - the polarity of being, multidimensionality and autopoiesis come.

The speed of information movement increases, and actors move faster after it. More and more individuals go online (only in 2016–2017 there are more than 2.3 billion of them), and the result is a fundamental and spontaneous restructuring of collective behavior. The upper layer of the developing digital system of planetary scale has absorbed the eternal factors of change - human needs, politics, geography and culture and created new schemes and patterns on their basis. Each of the entities, especially those who manage companies, needs a new coordinate system. This is a basic principle underlying the complex collective behavior of natural systems, which partially explains how computers achieve their useful complexity, despite the fact that their work is based on several simple processes.

In systems theory, emergence is called the presence of a system with special properties that are not inherent in its elements (both individually and in total). If this period is successfully passed, then one can find himself in the historical equivalent of such a moment in the creative process when chaos is replaced by order, and a new model, solution or structure becomes obvious.

The analysis of the digital economy in the aspect of the spatial approach allows us to identify a number of problems that must be solved in the framework of the national project of the same name.

1. Heterogeneity of the industry space of the digital economy in Russia.

In terms of digitalization, Russia occupies a leading position in such areas as education, information and communication technologies, and finance, even at the level of world leaders. However, extractive and manufacturing industries, the transport sector, and agriculture are lagging behind. In metallurgy, the oil and gas industry, mechanical engineering and the electric power industry, there are opportunities for transformation based on digital technologies: robotics, the Internet of Things and augmented reality, efficient capacity utilization. An integrated planning system has been created for Severstal, which includes process management from orders according to customer needs to equipment capabilities and terms of production of goods. An intelligent system for predictive maintenance of equipment has been developed and implemented in industry. Special tags (RFID tags) are superimposed on the units and equipment, which in real time provide data on the state of machines. These indicators are displayed on portable devices, including tablets, smartphones of workers in the workshop, and enter the data warehouse. Then, with the help of big data technologies, an analysis is carried out and a real-time forecast of equipment wear is calculated. Thus, for example, the costs of equipment repair at Novolipetsk Metallurgical Combine are optimized. In the field of transport, logistics services – there is the principle of sharing economy. Yandex taxi is a service for finding a taxi in the Russian market. Another transport project of the sharing economy is a car sharing service. For example, in Moscow these are "Car5", "Anytime", "Delimobil" and "YouDrive". Another example in the field of transport in the Russian market is carpooling, that is, the simultaneous sharing of a car for a trip in one direction - the international carpooling service BlaBlaCar in Russia. There is the use of machine vision technology (smart cameras) to record the movement of motor vehicles in freight yards, taking into account the goods imported or exported, finding vehicles in the parking lot, as well as controlling the access of cars to the parking lot, and registering customers through integration into enterprise systems.

Digitalization of banks in Russia is actively implemented. They store databases in clouds, work using remote desktop technology, reserve information systems on diverse platforms, use hybrid solutions that allow connecting the service provider's infrastructure with the customer's private virtual infrastructure, online banking. The system of electronic payments through mobile devices with biometric user identification is being increasingly developed in the banking sector. At the end of 2016 Sberbank announced the launch of a robot lawyer, launched the Iron Lady artificial intelligence system, which calls debtors [2]. Blockchain technology is still being mastered by Russian banks.

2. The presence of the institutional vacuum in the institutional space of the digital economy of Russia.

Existing state legal institutions are lagging behind the challenges of new digital opportunities, both in matters of Internet regulation and in the application, for example, 3D printers, unmanned vehicles, bioengineering technologies, ensuring cyber security, etc. Cybercrime institutions do not match the number of cybercrimes, which have grown by 75% over the past three years. Russia needs to create a modern information security system for all sectors of the economy. In the Russian Federation, there are no standards for evaluating data processing and storage centers, unlike many other countries, which excludes the objective possibility of assessing the level of services provided to consumers, for example, the possible amount of data for storage. The Institute of Public Control, which is implementing the digital economy program, has not been formed yet. Regarding the Institute of Business, there are a number of proposals in the field of digitalization of management systems and resource planning (ERP-systems) of individual large enterprises. The private sector is experiencing a shortage of relevant knowledge and managerial experience among enterprise managers and employees. There is no competitive pressure due to the high market consolidation in leading sectors, and high barriers for new players to enter the market. In the private sector, the introduction of innovation is in stagnation. This is a consequence of the residual R&D financing of enterprises and existing tax rules that do not stimulate investment in scientific research. Relations with scientific community remain weak, both at the local and international levels. Industry standards for analysis and data integration have not been developed. In Russia there is a shortage of programmers, their training requires financial resources and years of training. As for foreign specialists, they have uncompetitive salaries. According to the data of the Ministry of Digital Development, Telecommunications and Mass Media of the Russian Federation, about 25,000 IT specialists graduate from the country's universities [7]. Only about 400 thousand programmers work in the Russian market, while in the USA – 4 million, in India – 3 million and in China – 2 million. In Germany, about 10% of the population is employed in high-tech industries - in Russia – 2 times less [10].

3. Network space.

In accordance with the Federal Law "On Communication" dated 07.07.2003 N 126-FZ, the creation of a telecommunication network space in the country should be based on a single network [5]. In practice, private capital is building networks spontaneously. An example is fiber-optic temporary huts stretched through the air between buildings, the randomness of architecture and the location of transit nodes of network traffic, the inability to control it even in cases of force majeure. High growth rates of communication networks in the country are due to the absence of administrative barriers and competition in the market. So, over the past 15 years in Russia, more than 50% of households began to have fixed access to the digital environment, which is many times more than the number of phones installed over 120 years. The total coverage of broadband Internet access exceeded 80% of users, and their number is more than 100 million people. However, there is a serious problem: to form a modern infocommunication infrastructure in Russia, it is necessary to develop network-wide regulations and mechanisms; you cannot control the quality and traffic of a single network without it. There is a conglomerate of private fragments of the Internet, which cannot be used as the infrastructure for special networks because of high systemic risks, a low degree of reliability and stability, and security of information exchange.

4. Digital divide in Russia.

This is a direct lack of access to modern information and communication technologies and/or the presence of unequal, selective provision of them in various regions and cities of the country. There are such forms of digital divide in access, in use and in impact of the Internet on the behavior and attitudes of the population. Digital divide bridges the socio-economic gap between entities with and without access to information resources. Regions of the Russian Federation differ in access to the Internet by type of terrain, the availability of infrastructure, the level of education and the mastery of digital skills of the population. The leaders of the ICT rating among the regions in 2018: Moscow, St. Petersburg, Moscow Region, the Republic of Tatarstan, the Republic of Sakha (Yakutia). For 2015–2018 the percentage of the population in Russia using the Internet to order goods and services increased from 19.6 to 29.1% of the total population. Russian users account for 16.6% of Internet users in Europe. At the same time, the number of Facebook users is 12 million people, which is significantly lower than the indicators of developed European countries [6]. At present, the federal project "Eliminating Digital Divide" is being implemented in the Russian Federation, which provides for the creation of access points in settlements with the population of 250 to 500 people and for providing the population access to the Internet at a speed of at least 10 Mbit/s. Russia needs to create its own key niches of digital innovation, which at the lowest cost will allow achieving independence both in the domestic market and the level of recognized world leaders.

4 Discussion

Recently, the economic digitalization has received great attention worldwide. The digital economy is any activity that uses digital data [3]. Economists in their studies consider the transition to digital as a source of new production opportunities for those organizations that can adapt to changing conditions [1], as one way to reduce the costs of firms and consumers in finding and transporting goods and services, as a way to limit price fluctuations for homogeneous goods [12]. Russian researchers study the digital economy as a new reality, which leads to changes in the cultural, social, economic environment [8], to changes in the institutional sphere [4]. Complexity theory allows us to study how individual firms and the country's economy grow, adapt and develop [9]. A number of economists recommend using this theory to model specific results in the strategic management of a firm [11].

5 Conclusion

The digital economy leads to constant fluctuations in economic activity, causes rapid changes in the institutional environment, leads to the emergence of new forms of interaction among economic entities. Digitalization of the economic space of Russia is not uniform, existing institutions lag behind new digital opportunities, leading to the emergence of digital divide both among individual economic entities and among regions of the Russian Federation. The construction of network space is extremely uneven, there are no clearly developed network-wide regulations and mechanisms. The institutional structure also does not correspond to modern digital realities. There is no way to give an objective assessment of the quality of the delivered digital services; the institution of public control has not been formed.

References

1. Anderson, L., Wladawsky-Berger, I.: The 4 things it takes to succeed in the digital economy. Harvard Bus. Rev. (2016). https://hbr.org/2016/03/the-4-things-it-takes-to-succeed-in-the-digital-economy. Accessed 20 Mar 2020
2. Business Online: Sberbank has introduced a robot lawyer that will replace 3 thousand employees (2017). https://www.business-gazeta.ru/news/334226. Accessed 20 Mar 2020
3. Ducharme, L.: Measuring the digital economy international monetary fond (2018). https://www.imf.org/en/Publications/Policy-Papers/Issues/2018/04/03/022818-measuring-the-digital-economy. Accessed 20 Mar 2020
4. Ermolaev, K., Matveev, Y., Trubetskaya, O., Gromova, T.: Institutional changes and digital economy. In: Mantulenko, V. (ed.) Global Challenges and Prospects of the Modern Economic Development. European Proceedings of Social and Behavioural Sciences, vol. 57, pp. 580–589. Future Academy, London (2018)
5. Federal Law "On Communication" dated 07.07.2003 N 126-FZ. http://www.consultant.ru/document/cons_doc_LAW_43224. Accessed 22 Mar 2020
6. Internet World Stats: Internet, facebook and population stats for Europe (2019). http://www.internetworldstats.com/stats4.htm. Accessed 22 Mar 2020
7. Open Innovations 2018: National Program Digital Economy, 16 October 2019. https://digital.ac.gov.ru/upload/iblock/f4b/. Accessed 21 Mar 2020
8. Parker, C., Ramdas, K., Savva, N.: Is it enough? Evidence from a natural experiment in Indias agriculture markets. Manag. Sci. **62**(9), 2481–2503 (2016)
9. Sammut-Bonnici, T.: Complexity theory. Strat. Manag. **12** (2015). https://doi.org/10.1002/9781118785317.weom120210. Accessed 20 Mar 2020
10. TAdviser: The labor market in Russia (2020). http://www.tadviser.ru/index.php/%D0%A1%D1%82%D0%B0%D1%82%D1%8C%D1%8F:%D0%A0%D1%8B%D0%BD%D0%BE%D0%BA_%D1%82%D1%80%D1%83%D0%B4%D0%B0_%D0%B2_%D0%A0%D0%BE%D1%81%D1%81%D0%B8%D0%B8. Accessed 20 Apr 2020
11. Woodsidea, A., Nagyb, G., Megehee, C.: Applying complexity theory: a primer for identifying and modeling firm anomalies. J. Innov. Knowl. **3**(1), 9–25 (2018)
12. Yudina, T.N., Tushkanov, I.M.: Digital economy as a result of the industrial-technological revolution (theoretical and practical aspects) (2017). https://reosh.ru/t-n-yudina-i-m-tushkanov-cifrovaya-ekonomika-kak-rezultat-promyshlenno-texnologicheskoj-revolyucii-teoreticheskie-i-prakticheskie-aspekty.html. Accessed 22 Mar 2020

Methodological Aspects of Credit Portfolio Management in Financing Innovative Projects

A. A. Pomulev(✉) and V. V. Kalmykov

Financial University Under the Government of the Russian Federation, Moscow, Russia
sasha-pomulev@yandex.ru, VVKalmykov@fa.ru

Abstract. There is a high degree of uncertainty and high risks in the financing of innovative projects, which prevents the banking sector from increasing its share in project type lending. Current credit risk assessment methods are unable to assess the change in the borrower's business cycle with exogenous factors in mind. Authors of the research are finalizing and proposing portfolio quality management principles that are applicable for lending to the innovative sector of economy. In particular, a credit risk management system, described in the article below, is modelled using proactive diagnostics based on a free cash flow model calculated using a covariance matrix of factors that determine the dynamics of borrowers revenue. Key indicators used to calculate the probabilistic model, such as estimating the probability of default and evaluating debt exposed to credit risk, are analyzed using Monte Carlo simulations. The borrower's credit risk assessment is modeled using the real options method and using the risk adjusted net present value approach. Authors succeeded in developing a parametric migration matrix according to the multicollinear rNPV model. Article provides an example of a diagnostic card of a loan portfolio for IT borrowers.

Keywords: Credit risk · Default · Innovative economy · Loan portfolio quality · Proactive diagnostics

1 Introduction

Working with the quality of a loan portfolio in a commercial bank is not an easy task. Commercial banks are actively introducing credit risk assessment and management tools. However, according to Banki.ru, the growth in total overdue debt for 2019 amounted to 11.8%, and its share in the total loan portfolio as of January 1, 2020 increased to 5.9% (a year earlier – 5.5%). The share of overdue loans in the corporate portfolio increased from 5.7% to 6.7% [1].

Stagnation of the economy and increased competition in the banking sector may subsequently lead to a decrease in the profitability of interest-bearing lending operations, even in such segments, that provide high investment yields, like innovative economy. In this case to remain profitable, banks need to develop and implement a set of measures aimed to reduce operating and non-operational costs, optimize risk assessment methods, as they directly affect reserves and capital adequacy ratios.

S. I. Ashmarina et al. (Eds.): ENGINEERING ECONOMICS WEEK 2020, LNNS 139, pp. 372–383, 2021.
https://doi.org/10.1007/978-3-030-53277-2_45

One way to increase the marginality of your business is to lend to more marginal, such as innovative projects. For lending efficiency not to be offset by high loan loss provisions, a different approach to working with the loan portfolio is required. In this article, we focused on methods for assessing and managing credit risk, as the most significant form of risk.

Modern approaches to credit risk management (although they are based in large banks on advanced models for assessing the probability of default) do not allow building the management process quite effectively. The reasons for this situation are as follows:

1. Discreteness of factors that are used in the model of assessing the probability of default (financial and non-financial indicators, default indicators), which are identified once a quarter.
2. The model does not take into account the whole variety of exogenous external factors that affect the performance of borrowers.
3. Not transparency of the significance of factors that affect the borrower rating to a greater extent. On the one hand, this helps ensure the independence of the risk management vertical; on the other hand, the business unit does not have the ability to manage portfolio quality.
4. The lack of universal portfolio quality meters, which does not allow them to be managed. The rating of the borrower is static, the fact of delay is a late indicator.
5. The complexity of the management system, expressed in the dichotomy of the goals of internal departments.

Currently existing methods for assessing the credit risk of debt portfolios are not fully able to assess how, when changing the business cycles of one borrowing company, whose loan obligations payments are included in the structure of the loan portfolio, the total loan risk of the portfolio changes. A multivariate analysis of the loan portfolio in the presence of a multicollinearity factor is complicated by the fact that it is not always possible to accurately assess how the selected factors actually affect the dynamics of the total loan risk of the portfolio.

The task is to develop an approach to managing the quality of the loan portfolio of a commercial bank, taking into account the industry specifics of borrowers. The effect is expected from the timely detection of negative signals that may precede the occurrence of overdue debts of 5+ clients of the segment of large and medium-sized businesses in the time horizon of 1–2 months. Solving the problem will improve the quality of the loan portfolio and prevent the occurrence of arrears in the early stages, due to the timely development of a strategy for working with the client, and optimize the reserves for possible losses on loans, to increase the margins of credit operations.

The solution to this problem is to optimize the current process of assessing credit risk and should act as additional tools for assessing and managing the credit risk of a commercial bank on a monthly basis (as opposed to PD models that are calculated once every three months). The information collected as a result of working with this tool should help refine the PD model, make it more elastic due to new factors, and prepare for validation with the regulator.

2 Methodology

The study uses methods of comparison, financial and economic, systemic, statistical, correlation and regression analysis, the Monte Carlo method, the method of principal components, the method of real options. The loan portfolio is a structured system of data on loans and other requirements of the bank in relation to the bank's clients. The loan portfolio is a reflection of the results of the bank's work in credit work, which consists of various types of loans. The quality of the loan portfolio on credit risk is the subject of considerable attention by many authors: Gadzo, Kportorgbi, Gatsi [6]; Samorodov, Azarenkova, Golovko, Miroshnik, Babenko [14]; Ferretti, Gabbi, Ganugi, Sist, Vozzella [5], Qayyum, Riaz [12]. According to GOST ISO 9001-2008 "Quality management systems" [7] quality is a degree of conformity of set of inherent characteristics to requirements.

The quality of the loan portfolio is understood as a property of its structure that has the ability to provide the maximum level of profitability at an acceptable level of credit risk and balance sheet liquidity [9]. The goal of managing a bank's loan portfolio is to ensure maximum profitability at an acceptable level of risk. The main criteria for the quality of the loan portfolio are profitability, degree of credit risk and security. The ratio of these indicators characterizes the effectiveness of credit and total activities of the bank. In the system for evaluating the effectiveness of a credit institution, there is an indicator RAROC Risk-adjusted return on capital. This indicator is characterized by increased sensitivity to borrower risks due to the use of expected losses (EL) and economic capital (EC) in the calculations. The indicator allows you to allocate to any level of customer segmentation in the bank, including the level of the client and the individual transaction. However, in the public reporting of our domestic banking system there is no such indicator. Standard indicators are the value of the prudential reserve, NPL (Non-performing loan).

The NPL indicator is not objective, with an increase in the absolute values of the portfolio, the relative value of the indicator will be adjusted downward, which will not reflect the whole situation. An essential element of the calculation of EL and the subsequent determination of RAROC is the degree of credit risk of the borrower, which is expressed in the rating of the borrower and is determined through PD.

PD is an internal model for calculating the probability of a default of a counterparty and assigning it an internal rating based on selected quantitative and qualitative factors (loan to equity ratio, current liquidity ratio, net debt load, current liquidity ratios, business activity, market position, credit history, ownership share, operational control, state support, fraud of reports etc.).

A practical control problem based on the PD model is the imperfection of the modeling approach itself. For example, in the work of Shatalova and Shatalov [16] noted that often used by a commercial bank methods of calculating credit rating are isolated, not integrated into the sequential process of credit risk management, working independently from this process.

Proactive diagnostics of the loan portfolio of borrowers is based on a fundamental analysis of the external and internal parameters of the borrower, depending on its industry sector. As a rule, financial statements are amenable to substantial adjustments by the client's financial services and are provided with a significant delay. It is not possible to draw up a detailed CF for each borrower due to the different line of loan products.

3 Results

The proposed model is based on the assumption of a high level of direct correlation between the dynamics of the company's free cash flows and the credit risk of the bank's portfolio. The modeling of free cash flows in our model is based on the use of a multi-factor model, among which the key factors are:

- volume of sales,
- customer reliability,
- non-parametric model of stochastic dynamics of interest rates.

The dynamics of these indicators is modeled using a number of stochastic differential equations. Cash flow, which is critical in terms of default risk, is calculated using key variables with a feedback effect. Sales uncertainty is associated with variable product demand. During the growth phase of the global economy, the model can set a high level of sales. Low demand indicates a poor state of the economy, which will be reflected in the reduced creditworthiness of borrowers and the total credit risk of the portfolio. From the above arguments, two key conclusions must be drawn:

- random factors in equations should be positively correlated,
- in order to increase the predictive accuracy of the model, it is necessary to take into account the cyclical nature of the global lending phases (also known as credit cycles).

When modeling free cash flows and calculating a probabilistic model for the phased implementation of the project, it is necessary to take into account that illiquidity problems arise if the debtor company's low income is combined with a significant delay in paying bills (in this case, the company has to sell part of its assets to make a payment, which puts threat to its future prospects). In this sense, when modeling future periods rNPV, it is necessary to carry out additional discounting of the probability coefficient by a correction factor calculated by changing the credit risk of the company.

Key indicators used to calculate the probabilistic model, such as estimating the probability of default and estimating debt exposed to credit risk, are analyzed using Monte Carlo simulations, as well as the likelihood of a forecast cash flow. For the purposes of forecasting discount rates, we use the non-parametric model of stochastic dynamics of Hall-White interest rates (one-factor extended Wasecek model) [2]. To calculate free cash flow, our model refers to the average value of the Ornstein-Uhlenbeck process [4], and the parameters of the free cash flow itself are adjusted by the economic environment model.

Modeling Algorithm (Proactive Portfolio Diagnostics)

a. **Identify a phased structure for the implementation of the project (business operation) for the construction of the rNPV model** [17]

In theory (and in 90% of cases in practice), any project can be divided into specific stages of its implementation. But for companies and projects implemented in a number of sectors (for example, for those, which are operating in the innovative economy framework), the described approach of separating the individual stages of projects implementation has very specific consequences.

b. **Expansion of the correlation matrix using the principal component method**

In the context of increased uncertainty in global capital markets, it is especially important to assess how external systemic risk factors can negatively affect the dynamics of borrowers' quarterly revenue [15]. Since systematic risk is not diversifiable and we consider the entire underlying market for the selected borrowing company as a combination of assets, the systematic risk of an individual borrowing company is the degree to which its risk is associated with market risk. That is why systematic risk of any industry consists of the systematic risks of individual companies, which, by definition, cannot be excluded regardless of how many companies are in the industry.

Among the external factors most often used for constructing covariance matrices, we selected the following by multiple regression analysis: the total financial leverage of the industry companies, the liquidity ratio, the consolidated indicator of the operational efficiency, the reduced profitability ratio of the companies, the variable of the financial crisis. The summary conclusion based on the analysis of data for five sectors (construction companies, banking and insurance companies, food retail companies and airline carriers) is as follows:

- the systematic risk of companies in these sectors is negatively correlated with the combined profitability ratio and this relationship persists for long-term periods of time,
- positive level of correlation between systematic risk and the size of companies in these segments also persists for long-term periods of time,
- financial leverage positively affects the airline's systemic risk, but the correlation coefficient calculated by this argument is not so significant and varies for companies from different industries in the range (0.12–0.17),
- high correlation coefficient also exists between the systematic risk of companies in these segments and the combined liquidity ratio,
- it was not possible to find any significant relationship between the combined indicator of the operational efficiency of companies within sector and systemic risk. The correlation coefficient calculated by this argument varies for companies of different industries in the range (0.23–0.25),
- equally, it was not possible to establish a stable relationship between the systemic risk of a company and the variable of financial crisis. The correlation coefficient calculated by this argument varies for companies of different industries in the range (0.18–0.24).

Below we present the correlation matrices for the given factors for a number of industries (Table 1):

Table 1. Correlation model

	Ratio of profitability	Company size	Financial leverage	Current ratio	Operating efficiency	Variable Fin. crisis
Construction	(0.34)	0.45	0.12	0.76	0.24	0.21
Airlines	(0.12)	0.51	0.16	0.59	0.23	0.19
Banks	(0.29)	0.34	0.15	0.68	0.25	0.24
Insurance	(0.49)	0.29	0.15	0.83	0.24	0.18
Retail	(0.53)	0.68	0.17	0.91	0.23	0.19

Source: authors.

After constructing the correlation matrix, it is required to increase the efficiency of the model to decompose the resulting matrix using the principal component method. But first of all, we need to establish which of the determinants of the correlation matrix have signs of multicollinearity (low level of correlation of indicators with a relatively high coefficient of determination; relatively high level of sensitivity of correlation coefficients to the addition and exclusion of additional determinants from the sample).

In our opinion, the most suitable method to solve the multicollinearity problem is the principal component method (hereinafter: PCA) presented above. PCA is a mathematical procedure that uses orthogonal decomposition and transformation to project a set of observations of possible correlated determinants (i.e., large data) onto a set of values of linearly uncorrelated determinants (i.e. low data), which are global components. The number of major components is less than the number of original determinants. The projection is performed using the least squares method, where large (i.e., main) data variability is recorded, and small variability is ignored. As a result, the first global component will be in the direction of maximum dispersion in the input space. The second global component is orthogonal to the first and coincides with the direction of the second maximum dispersion. Subsequent global components are consistent with the remaining maximum variance up to the last major component, which is the smallest variance of the data.

In the article by Kendryukhov and Tolkachev [8] it is proposed, to conduct a general analysis of components (GCA) and spectral analysis of large empirical covariance matrices based on PCA. So, the first step for conducting PCA analysis is to determine two types of components (global and local) by calculating the coefficients of inverse participation. The second step is to analyze the information embedded in the listed global components. The third stage is the creation of a portfolio of local components and the use of an integrated network approach to identify its correlation structure.

Local components contain information about the correlation of certain assets, and not the entire market as a whole. The most effective method of researching cluster information in the system is a comprehensive network analysis. In addition, portfolios of local components can also be used to establish noise correlations. We use the threshold method to filter noise correlations. We select the 95th percentile of mixed correlations as the threshold in each period. Thus, we filter out most of the noise information contained in local components. Using network analysis, you can discover cluster information contained in portfolios of local components.

As a result, the found noise effects in the correlation matrix we constructed can be removed from the model, thereby increasing its efficiency in the framework of the multifactor analysis.

c. **Modelling probabilities**

The next step in the process of introducing proactive approach to assessing portfolio credit risks is modeling the probabilities of a positive effect on each previous phase of the project (the onset of which is itself a trigger for starting the next phase of the project).

To implement the described stage it is necessary:

- to usereal options method in modelling procedure,
- to build an rNPV model for the purpose of estimations

The methodology of real options is aimed to eliminate the influence of economic uncertainty on the cost of the project by applying the theory of financial options to the project implementation process. This approach considers the process as containing a number of options in the face of unpredictable economic changes. For each phase of the project, an assumption is made that there is a real option for the investor (creditor). If the option holder decides to invest, he will acquire the option to invest in the next phase along with the option for future commercialization of the project. It may also happen that success in the previous phase is accompanied by unfavorable market conditions.

The holder of the option in such circumstances may refuse the project, which limits the negative impact of the strike price of the option. In comparison, conventional NPV-based methods suggest that all investments will occur after making a decision on investing. Projects can include other options, such as extension options, deferrals, and licenses. It is assumed that rational investors value these options, and therefore, the project cost is associated not only with its cash flows, but also with the availability of such options.

After performing the simulation using the real options method, we can begin to build the rNPV, which will serve as the basis for the subsequent construction of the migration matrix and the calculation of the credit risk of the portfolio. Determining the risk-adjusted net present value (rNPV), like NPV, also includes forecasting income (cash inflows), costs (cash outflows), but additionally requires calculating the probability of obtaining forecasted cash flow for each phase of the project. To account for risk, the expected net cash flow for a given period of time is multiplied by the probability of its occurrence [13].

Modeling as part of the process of structuring cash flows by rNPV is carried out taking into account the described conventions:

- cash outflows under Phase 1 represent the outstanding value,
- the described cash outflow occurs regardless of whether the first stage is successful after completion. The cash outflow associated with the costs of Phase 2 is weighted to reflect the likelihood of success of phase 1, because phase 2 can occur only if phase 1 is successful,
- in the same way, the outflow of funds in phase 3 is determined by the cumulative probability of its achievement, which is equal to the success rate of Phase 1 multiplied by the success rate of phase 2, etc.,
- after the net cash flow of each time period has been correctly adjusted for risk, these cash flows are discounted using the appropriate discount rate and the discounted cash flow method.

Moreover, when modelling cash flow using the rNPV method, it makes sense to apply different discount rates at each individual project phase. Discount rates may vary, but typically (for different industries) range from 1.0 to 1.5 standard deviations at different phases of the project.

4 Discussion

For a given basis, operators are represented as square matrices. These estimates are used afterwards to prepare migration matrices estimated by the described methods: cohort for periods T (PI), duration (parametric (D) and nonparametric (AN)), marginal model GLM (M GLM) and transition model GLM (T) GLM). For Russian companies, the average value of the migration matrix for one year (AR) was also calculated. Standard errors were calculated for the matrix elements by the boot method (matrices with the letter B at the end), and the generators for each matrix were determined with the correction calculated for the model described in the described article. Note that transition matrices for duration methods are already continuous time matrices. Then, the transition matrices obtained using the generators (denoted by G) were calculated to make probability estimates possible at any time. A total of 10 matrices were considered.

We separate all the matrices obtained by the factor of the presence/absence of correction for the parametric model of default given in the article. As you know, often large default probabilities are due to the fact that models are extrapolated for a long period of time (in practice, portfolios from the "banking book" are tracked on the horizon up to 10 years). Application of the methodology proposed above for the parametric assessment of the credit risk of portfolios allows us to circumvent this problem. Below we compare the migration matrices obtained without and using the parametric method described above (Table 2, Table 3).

Table 2. Parametric matrix of migration by model Moodys

	Aaa	Aa	A	Baa	Ba	B	Caa_C	D
Aaa	0.8933	0.1018	0.0036		0.0012			
Aa	0.0086	0.8781	0.1065	0.0029	0.0019			0.0019
A		0.0153	0.9027	0.0658	0.0126	0.0027		0.0009
Baa		0.0052	0.0628	0.8356	0.0817	0.0126		0.0021
Ba	0.00		0.0029	0.0391	0.8418	0.1035	0.0039	0.0078
B		0.00	0.0019	0.0038	0.0596	0.8221	0.0298	0.0817
Caa_C				0.0138	0.0276	0.0741	0.6034	0.2800
D								1

Source: authors based on [18]

Table 3. The parametric migration matrix according to the multicollinear rNPV model

	Aaa	Aa	A	Baa	Ba	B	Caa_C	D
Aaa	0.8974	0.1042	0.0071		0.0090			
Aa	0.0093	0.8921	0.1100	0.0109	0.0134			0.0291
A		0.0201	0.9132	0.0801	0.0190	0.0100		0.0081
Baa		0.0112	0.0798	0.8890	0.0922	0.0201		0.0192
Ba	0.0012		0.0189	0.0451	0.8890	0.1101	0.0112	0.0291
B		0.0027	0.0090	0.0111	0.0789	0.8768	0.0671	0.1202
Caa_C				0.0201	0.0450	0.1201	0.8279	0.6728
D								1

Source: authors.

For practical use, the most important are the probabilities of a transition to default. Tables 2 and 3 show the probabilities of transition to default from Ba and Caa_C depending on time. In addition to the average value, confidence intervals were estimated using the boot method.

Models, provided above give different results. Particularly large differences exist for portfolios with lower ratings, such as Caa_C. Migration probabilities provided above are used as opening data for estimating economic capital and, as the result, are directly transposed into financial results. Diversities provided in the process of modeling can lead to significant changes (up to 15%) in portfolio maintenance costs. Based on both methodologies and using retrospective analysis, we were able to conclude that the rNPV parametric model described in this article can significantly reduce the average sampling error, thereby increasing the accuracy of the model.

The major step is to create a diagnostic card for the loan portfolio, taking into account the criticality of the boundaries of financial and other indicators, which allows a static assessment of the level of risk in the portfolio. All indicators are developed individually for each industry. Option scorecard for IT sphere, is given in Table 4, taking into account their features, which are considered in detail in one of the author's articles [11].

Table 4. An example of a diagnostic card of a loan portfolio

Indicator title	Level of risk
Stdevr NPV (1.5)	High
Decrease in contract base by more than 25%	
DSCR in range less than 1	
Failure to comply with LTV indicator (Decrease in estimated value based on annual revaluation results that resulted in loan failure by more than 20%)	
Stdevr NPV (1.2)	Medium
DSCR in the range from 1 to 1, 1	
Failure to comply with LTV indicator (Decrease in estimated value based on annual revaluation results that resulted in loan failure to reach 20%)	
Net assets less than share capital	
Decrease in contract base to 25%	

Source: authors.

The significance of the indicators is determined based on the results of heuristic assessment of experts in the sectors. The construction of the composite index is carried out by a double convolution method [10] or other mathematical weighing methods. The interpretation of the composite index is proposed according to the modified empirical scale of probability of the risk event occurrence (Table 5).

Table 5. Risk level interpretation

Gradation of risk level	Interpretation	Working strategy
High	All deviations are critically exceeded. Probability of default is extremely high	Reduce
Medium	Permissible deviations in the values of indicators are exceeded. Probability of default on obligations average	Hold
Low	There are no deviations in the value of indicators. There is no probability that the liabilities will not be settled	Increase

Source: authors.

For cash flow modelling purposes we use a modified model of adjusted net present value. The theoretical structure of risk-adjusted rNPV calculation consists of a probability tree, which details all likely scenarios and subsequent cash flows, as well as the probability that each forecast scenario will occur. It is also possible to add other qualitative factors, which are significant from the point of view of protection of bank interests.

5 Conclusion

The presented approach to the quality management of the loan portfolio, based on the principles and two-level proactive diagnostics, will help maintain the quality of the bank's corporate loan portfolio and participate in financing innovative projects. The methodology were tested on the loan portfolio of the largest bank in Russia (in Moscow and the Moscow region). The area for further research is the formation of measures to develop and implement a risk management plan, hedging methods, requirements for IT systems, organizational structure and personnel to increase the efficiency of the business process. A special area is the study of the theory of behavioral finance and its impact on borrower behavior [3].

Acknowledgments. We thank our family and friends for their understanding and support.

References

1. Banki.ru: Banking sector in 2019 (2020). https://www.banki.ru/news/research/?id=10919284. Accessed 15 Apr 2020
2. Belsner, O.A., Zhabin, D.N.: Analytical expression of the value of a swap contract in the hall-white model. News Tomsk Polytech. Univ. Eng. Geo-Resour. **309**(3), 15–17 (2006)
3. Bogatyrev, S.Y.: Testing the value function of the prospect theory on Russian M&A market. J. Rev. Glob. Econ. **7**, 731–739 (2018)
4. Dmitriev, D.I.: Analytical expression of swap contract cost in the Hall-White model. Diploma work. Saint Petersburg State University, Saint Petersburg (2016)
5. Ferretti, C., Gabbi, G., Ganugi, P., Sist, F., Vozzella, P.: Credit risk migration and economic cycles. Risks **7**(4), 109 (2019). https://doi.org/10.3390/risks7040109
6. Gadzo, S.G., Kportorgbi, H.K., Gatsi, J.G.: Credit risk and operational risk on financial performance of universal banks in Ghana: a partial least squared structural equation model (PLS SEM) approach. Cogent Econ. Finance **7**(1), 1589406 (2019). https://doi.org/10.1080/23322039.2019.1589406
7. GOST R ISO 9001-2008 "Quality management systems". Requirements (2008). http://docs.cntd.ru/document/1200068732. Accessed 15 Apr 2020
8. Kendryukhov, A.V., Tolkachev, D.O.: The use of the main components method to assess the competitiveness of machine-building enterprises. Mark. Innov. Manag. **4**, 219–227 (2013)
9. Lavrushina, O.I.: Banking risks. KNORUS, Moscow (2007)
10. Nedosekin, A.O., Maksimov, O.B.: A new comprehensive indicator for assessing the financial condition of an enterprise. Quest. Risk Anal. **2–3** (1999). http://masters.donntu.org/2005/kita/sroka/library/art3.htm. Accessed 15 Apr 2020
11. Pomulev, A.A. Selugina, O.N.: Approaches to risk assessment in financing borrowers of the digital economy sphere. In: Ashmarina, S.I., Mantulenko, V.V. (eds.) Proceedings of the II International Scientific Conference Global Challenges and Prospects of the Modern Economic Development. European Publishing, London (2020). The European Proceedings of Social and Behavioural Sciences, vol. 79, pp. 555–562
12. Qayyum, A., Riaz, K.: Incorporating credit quality in bank efficiency measurements: a directional distance function approach. J. Risk Financ. Manag. **11**(4), 78 (2018). https://doi.org/10.3390/jrfm11040078

13. Rambachan, A.: Harvard Economics Math Camp 2018: Econometrics, Probability Review. Harvard University Press, Cambrige (2018)
14. Samorodov, B.V., Azarenkova, G.M., Golovko, O.G., Miroshnik, O.Y., Babenko, M.V.: Credit risk management in the banks financial stability system. Financ. Credit Activity Prob. Theory Pract. **4**(31), 301–310 (2019)
15. Sharpe, W.F.: Capital asset prices: a theory of market equilibrium under conditions of risk. J. Finance **19**(3), 425–442 (1964)
16. Shatalova, E.P., Shatalov, A.N.: Assessment of the Borrowers' Credit Capacity in the Bank Risk Management. KNORUS, Moscow (2020)
17. Svennebring, A., Wikberg, J.E.: Net present value approaches for drug discovery. SpringerPlus **2**, 140 (2013)
18. Yavin, T., Zhang, H., Wang, E., Clayton, M.A.: Transition probability matrix methodology for incremental risk charge (2010). https://arxiv.org/pdf/1102.3857.pdf. Accessed 15 Apr 2020

Implementing Lean Manufacturing and Solving Motivation Problems in Russian Companies

O. P. Maslova(✉), T. A. Ilyina, V. A. Krichmar, and E. G. Safronov

Samara State Technical University, Samara, Russia
ol-masl08@yandex.ru, tanya.ilyina@list.ru,
vera.krich@yandex.ru, ewgenijsafronow@yandex.ru

Abstract. The introduction of lean manufacturing principles in Russian companies often causes dissatisfaction among employees. This is due to the fact that the staff is poorly involved in the process of changes that are imposed on them by the management. The authors suggest paying serious attention to motivation problems when applying principles of lean production by actively teaching employees the basics of the 5S system, which helps to reduce losses in the workplace, as well as involving employees in the process of continuous improvement. A solution to this problem, the authors see in the creation of training classes in companies that model production situations that require managerial influence using the lean production principles.

Keywords: 5S system · Labor productivity · Labor efficiency · Lean manufacturing · Losses · Process factory

1 Introduction

Recently, lean production (Lean system) has become one of the innovative management concepts implemented at Russian enterprises. This concept is based primarily on the production management system developed by the Japanese automobile giant Toyota in the 1950s. It is based on internationally recognized principles of the Toyota Corporation, which is known for its innovative approach to the production organization and strives to continuously improve all processes in the company.

Here it is important to emphasize that the true pioneer in improving the production efficiency is not a Japanese company, but the ideologist of the scientific labor organization Gastev, who in the 20s of the XX century after an active correspondence with Henry Ford implemented his method in the Soviet production, becoming the founder of the Central Institute of Labor (CIL), whose task was the development of productivity and production methods of workers [9].

The Lean system has been widely and successfully implemented in many foreign companies, at enterprises of various activity fields. Although the lean production methods have been most fully implemented at automotive and mechanical engineering enterprises, the successful implementation results are demonstrated by construction

S. I. Ashmarina et al. (Eds.): ENGINEERING ECONOMICS WEEK 2020, LNNS 139, pp. 384–391, 2021.
https://doi.org/10.1007/978-3-030-53277-2_46

companies [10, 15], horticulture [13], furniture industry [1], food industry [3], mining [14], etc.

This concept is based on enterprise management, which allows you to improve the quality of work by reducing losses. Losses are defined as anything that reduces the work efficiency and does not increase the value of a product or service to the consumer. In practice, such losses are extra movement of equipment and employees, which leads to an increase in production time; overproduction which leads to excess inventory and storage costs, etc. In general, losses lead to an increase in the cost of production.

For many Russian enterprises, the issues of high cost price, along with insufficient quality of products are quite acute. At the same time, problems of outdated equipment, high energy consumption, excessive inventory and inconsistent production processes remain also relevant [6]. However, despite the fact that the concept of lean production is generally recognized in the world and its main principles are aimed at solving these problems, its implementation at domestic enterprises is usually painful and it comes not always to the desired result [16].

2 Methodology

The authors used general scientific methods of systematic and integrated approaches, as well as traditional methods of economic analysis (observation, comparison) and non-traditional (heuristic techniques), analysis, synthesis, description, and comparison. Their application is determined by the theoretical nature of this research, which includes the following stages: problem statement, analysis of information on the research topic, comparison and description of various scientific views on the issues under study, synthesis of different approaches to the problem. The results of the study are based on the implementation of lean production and training of personnel according to lean production principles at industrial enterprises of the Samara region over the past three years.

3 Results

The main reason for these problems is the reluctance of employees to accept new ideas and work in accordance with them, as well as the inability of management to motivate staff in the new environment. This is already evident at the very first stages, when implementing one of the lean manufacturing tools of the 5S system (workspace management system). The 5S system implies the implementation of 5 steps (Fig. 1). The effectiveness of the 5S method is very important and has a correlation with the overall performance of production [18], but the resistance of the personnel begins at the stage of "removing unnecessary". However, this state of affairs is typical for all spheres (production, services, education, health, etc.).

As a clear example, let's consider a machine-building enterprise. On the assembly, maintenance, and other grounds of domestic enterprises, there is often a picture when a large number of unnecessary tools are located in the workplaces (locksmiths, turners, assemblers, etc.), which are practically not used or are used in emergency situations.

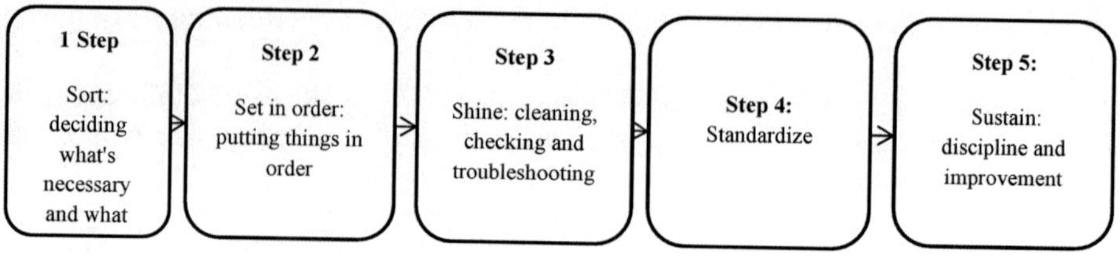

Fig. 1. Basic steps of the 5S concept (Source: authors).

The presence of an extra tool hinders the performance of current work and contributes to temporary losses in the search for a necessary tool.

At the enterprise level, an extra tool at each site leads to higher costs. At the same time, companies solve this problem by purchasing cheaper and, as a rule, low-quality tools. However, this tool wears out faster and a whole chain of unnecessary actions (ordering, delivery) is performed again to replace it, which increases the company's expenses. In addition, poor-quality and constantly changing tools (and sometimes their absence) affect the psychological condition of workers and, as a result, the quality and efficiency of production. So instead of time, effort, and money being spent on developing and improving the quality, it is spent on performing actions that do not add value to the product for the consumer.

The same applies to excess inventory. Many enterprises have a problem of overproduction, resulting in increased stocks of finished products in warehouses, or have the practice of purchasing excessive raw materials, without calculating the optimal volume of these materials for normal functioning of production. As a result, stocks and the size of warehouse space increase, that leads to an increase in the cost of their maintenance and the deadening of working capital, which could also be used for the production development. However, as a rule, enterprises solve the problem of working capital shortage not by using a lean production mechanism, but with additional credit resources, which further aggravates the negative financial situation. At the same time the most difficult part is trying to convince the staff, who are used to working according to the usual way, to change this state of things. Moreover, this applies to the entire hierarchical structure of the enterprise, from the simple worker to the general director.

It is quite difficult, for example, to persuade workers to organize their workplace with the 5S requirements: to remove all unnecessary things from the point of view of technological process, to place working tools rationally, to clean garbage and to wash the equipment to a condition when the smallest damage on it is visible, etc. For most of the staff who tried to implement lean production principles, the first reaction was to reject the proposed measures. People who have been used for years or even decades to work with a certain, familiar way for them, are wary of such innovations and do not see any sense in them. It should take enough time for the employee to feel the effect from the rational organization of the workplace. However, in general, no matter how much management tries to implement lean production system in the workplace of its employees, it will never be realized if the manager himself does not adhere to the "philosophy" of the lean production.

Most managers of domestic enterprises see the success of building a business in the growth of financial indicators and give them priority, which is fundamentally wrong. Only if business managers pay sufficient attention to the staff development, motivational and social activities, the impact of the lean production on the productivity growth will be sustainable. For successful implementation of lean production, first of all, the management itself should firstly be morally ready for changes (for example, it starts implementing the 5S principles in its workplace, and not only requires cleanliness from subordinates), constantly participate in the process of monitoring and improving the lean production processes. Otherwise, the removal and non-participation of management leads to demotivation of employees and no measures (especially of an administrative nature) can affect this. Secondly, the management of enterprises should pay special attention to measures for developing the corporate culture and social support for employees.

It should also be noted that, in general, methods of motivation in the lean production system are mainly non-material. It can be a moral incentive (by providing employees with flexible scheduling and special privileges), regular communication with subordinate managers, maintaining the corporate culture, etc. For unstable conditions of the Russian economy, such measures are essential. In addition, they are consistent with the domestic mentality, based on a propensity to cooperative work.

The success of implementing lean production principles depends more on how the company's management communicates this philosophy to its team. There are many ways for managers and subordinates to interact, which is the essence of management. Administrative and command methods, democratic and liberal – none of them will be successful if the lean philosophy is not accepted by every employee of the company.

At Toyota, the lean production philosophy is based on the principle of continuous training of employees in the framework of these principles in specialized training classes. This experience is considered as advanced and is now used all over the world, including Russia.

In our country, the creation of training classes to teach the principles of lean production at enterprises is a relatively new phenomenon. Its origins date back to the approval of the national project "Labor Productivity and Employment Support" in 2017, during which it is planned to ensure that by 2024 the growth rate of the labor productivity at medium and large enterprises of the basic non-resource sectors of the economy is not less than 5% per year [4, 17]. To implement this project, an operator was created – the "Federal Center of Competence" (FCC), its mission is to improve the quality of goods and services and increase the competitiveness of the Russian economy by creating a culture of high productivity and efficiency among employees of organizations in each region of Russia.

The state is interested in improving the efficiency of enterprises, so the FCC is entrusted with the main task of creating a methodology for training the staff of companies that have applied for participation in this project. Currently, 1513 enterprises from 62 subjects of the Russian Federation have become participants of this national project. It is important to emphasize that in addition to valuable knowledge and experience in implementing the principles of lean production, enterprises receive tangible financial benefits from participating in the national project: up to 300 million rubles for a period of up to 5 years at 1% per annum, under a number of conditions.

A mandatory condition for obtaining a loan under this program is to get a conclusion (certificate) of the FCC on the availability of key elements of the production system and a sufficient level of internal resources use to increase the productivity [7].

In the Samara region, one of the first participants in the national project is the company JSC "Sredne-Volzhsky mechanical plant". More than two years have passed since the company entered the labor productivity program. During this time, impressive results were achieved: revenue increased by 44.6%, the number of employees increased by 10%, and the labor productivity – by 40.6%. This success would not have been possible without the company's employees who deeply understand and support all the management's initiatives.

One of these initiatives was to build the production system of the enterprise. The production system of this plant is a set of tools and methods used to convert all types of resources into finished products and based on improving the efficiency of processes and continuous improvement of the company's activities. At the first stage, the basic guidelines for building a production system were formulated.

JSC "Sredne-Volzhsky mechanical plant" is very sensitive to the best practices in the field of improving the labor productivity, which are used by both foreign and domestic companies. Among the key elements of the production system, the company's management has placed a bet on the following ones:

- search for 7 types of losses,
- applying lean production principles to eliminate losses, such as mapping, 5S, standardized work, and unit product flow,
- continuous improvement of all internal processes by mobilizing unused human potential, increasing the competence of employees and spreading the lean production culture. In order to achieve these goals, it was decided to create a training class "Process Factory" within the enterprise, which plays a serious role in training personnel for the implementation of plans outlined by the plant management to increase the labor productivity.

In the practice of global companies, such training centers solve three tasks at once. The first is to involve employees in the improvement process and motivate people by the preparation for changes. As a rule, the changes themselves have not yet begun at the enterprise, and the management needs to interest employees, teach them the basics of the lean production. Secondly, there is a need to provide employees with such a starting level of knowledge and skills that they could immediately use in their workplace without unnecessary theory, without large time expenditures, on the principle of "take and do". Third, you need to bring them to a certain level of understanding of the lean production system so that they can share their knowledge with their colleagues and involve them in changes. After all, the main idea of lean production is that it is formed, first of all, as a culture, and only then as a skill for using appropriate tools in the workplace [12].

The training class "Process Factory" was created with the support of the "Federal Center of Competence" (FCC), which is the operator of the national project "Labor Productivity and Employment Support" [17]. Among the company's employees, candidates were selected for the position of internal trainers who have received serious training in methods used by the FCC. As a result, they have a high level of competence that helps them train other employees in the sphere of lean production culture. To date, the "Process Factory" of JSC "Sredne-Volzhsky mechanical plant" is a modern training class in which internal trainers train employees in key areas of lean production:

- search for losses by modeling production and office processes,
- process mapping,
- application of the 5S system in production and in the office,
- SMED-quick changeover,
- standardized work.

For each direction, methodological materials have been developed that have a close relation with the real processes taking place at the enterprise. Therefore, employees easily and with great interest are involved in training, learn to see losses, formulate and solve problems, and balance processes. This is a fully interactive learning format that doesn't have boring theory and long slides, but has a lot of practice. Internal trainers set a task for a group of participants: to optimize the production flow with specific, measurable performance indicators that should increase from shift to shift. And in the process of solving this problem, expert trainers teach how to do this using lean production tools. Due to the fact that participants work with real situations and see the effect of their actions in numbers, they come to understanding where they can and should put effort in their enterprise.

4 Discussion

Numerous studies show that innovative management concepts do not work at all or work poorly without building an appropriate organizational culture. So, researchers conducted a study among 295 British manufacturers [2]. Their findings show that the lean practice is positively associated with an organizational culture that is process-oriented, employee-oriented, structurally open, socially free, rule-based, and market-oriented. An even earlier study by a group of German scientists [11] shows that the lean philosophy can increase productivity and efficiency when it is applied in a complex way. According to them, the concept of Learning Factory is already actively used by some companies to train employees in lean production methods.

Some authors pay special attention to the need to revise the approach to the development of human resources that are important for the development of modern industrial systems [19]. They focus on improving, collaborating, communicating and transferring knowledge between industry and universities, facilitating curriculum development, and developing applied knowledge skills and methods. In their opinion, this ensures an effective and efficient transition from the academic to the industrial environment. At the very beginning of the professional career, a person should already have a certain set of knowledge about lean production in order to reduce the time for its adaptation in a company that applies these principles.

Some researchers of lean production consider quality and productivity as priority criteria for the business sustainability [8]. Others speak today about the concept of the Triple bottom line, according to which lean production should be multidimensional and pursue three goals: economic growth, environmental conservation, and social responsibility [5]. But no one denies the importance of such a factor in the implementation of lean production as social responsibility.

5 Conclusion

Nowadays, the application of lean production is an urgent task for many Russian enterprises, since the issues of low labor productivity and product quality, high production cost are quite acute. However, the implementation of lean production in practice does not always bring the expected effect.

The authors use the example of a machine-building enterprise to investigate how losses are formed, what is their reason, and why the application of lean production (namely, one of its tools – the 5S system) does not bring the desired result. One of the main reasons, according to the authors, is the lack of motivation of employees to work in the new conditions (and this applies to the entire hierarchical structure of the enterprise). To solve this problem, it is proposed to use methods of non-material motivation of employees, which are organically consistent with the Russian mentality and are indispensable in unstable economic conditions.

In the article, a special place is given to the mechanism of state support for the introduction of lean production at Russian enterprises, which is implemented through the national project "Labor Productivity and Employment Support" [17]. The operator of this national project was the autonomous non-profit organization "Federal Center of Competence" (FCC), which was established in 2017. The mission of this center is to improve the quality of goods and services, increase the competitiveness of the Russian economy and the labor productivity at enterprises. One of the main tasks is to train the personnel of Russian enterprises in lean production methods.

In addition, the article investigates positive experience of one of the first participants in this national project, JSC "Sredne-Volzhsky mechanical plant". The introduction of lean production has significantly affected the growth of labor productivity and increased revenue of this enterprise. The moral and psychological climate in the team has significantly improved. This company not only uses advanced foreign and domestic methods to improve production processes, but also successfully implements personnel training activities.

In general, we can say that the introduction of lean production at Russian enterprises, although it is associated with a number of difficulties, is still being implemented, as it is evidenced by the positive experience of domestic enterprises.

References

1. Abu, F., Gholami, H., Saman, M.Z.M., Zakuan, N., Streimikiene, D.: The implementation of lean manufacturing in the furniture industry: a review and analysis on the motives, barriers, challenges, and the applications. J. Clean. Prod. **234**, 660–680 (2019)
2. Cadden, T., Millar, K., Treacy, R., Humphreys, P.: The mediating influence of organisational cultural practices in successful lean management implementation. Int. J. Prod. Econ. (2020, in press). https://doi.org/10.1016/j.ijpe.2020.107744. Accessed 20 Apr 2020

3. Costa, L.B.M., Filho, M.G., Fredendall, L.D., Paredes, F.J.G.: Lean, six sigma and lean six sigma in the food industry: a systematic literature review. Trends Food Sci. Technol. **82**, 122–133 (2018)
4. Decree of the President of the Russian Federation No. 204 of May 7, 2018 "On national goals and strategic objectives for the development of the Russian Federation for the period up to 2024" (2020). http://www.consultant.ru/document/cons_doc_LAW_297432/. Accessed 15 Apr 2020
5. Henao, R., Sarache, W., Gómez, I.: Lean manufacturing and sustainable performance: trends and future challenges. J. Clean. Prod. **208**, 99–116 (2019)
6. Ilina, T.A.: The 5S system as the initial stage of implementing lean manufacturing in industrial enterprises. Bull. Samara Municipal Inst. Manag. **4**, 70–77 (2017)
7. Ivanov, O.B., Buchwald, E.M.: National project on labor productivity: the key to "economic breakthrough" for Russia. ETAP Econ. Theory Anal. Pract. **2**, 28–44 (2019)
8. Jimenez, G., Santos, G., Sá, J.C., Ricardo, S., Pulido, J., Pizarro, A., Hernández, H.: Improvement of productivity and quality in the value chain through lean manufacturing – a case study. Procedia Manuf. **41**, 882–889 (2019)
9. Khokhlova, T P., Nazaretian, P.V.: The legacy of Russian Taylor in modern management (to the 135th anniversary of the birth of A. K. Gastev). Manag. Russ. Abroad, **4**, 102–111 (2017)
10. Li, S., Fang, Y., Wu, X.: A systematic review of lean construction in Mainland China. J. Clean. Prod. **257**, 120581 (2020)
11. Makumbe, S., Hattingh, T., Plint, N., Esterhuizen, D.: Effectiveness of using Learning Factories to impart lean principles in mining employees. Procedia Manuf. **23**, 69–74 (2018)
12. Markova, N.A., Markov, D.A.: Motivation, training, and leadership in lean manufacturing. Bull. PNRPU Soc. Econ. Sci. **4**, 329–341 (2016)
13. Pearce, D., Dora, M., Wesana, J., Gellynck, X.: Determining factors driving sustainable performance through the application of lean management practices in horticultural primary production. J. Clean. Prod. **203**, 400–417 (2018)
14. Seifullina, A., Er, A., Nadeem, S.P., Garza-Reyes, J.A., Kumar, V.: A lean implementation framework for the mining industry. Ifac-Papersonline **51**(11), 1149–1154 (2018)
15. Solaimani, S., Sedighi, M.: Toward a holistic view on lean sustainable construction: a literature review. J. Clean. Prod. **248**, 119213 (2020)
16. Suetina, T.A., Rakhimova, G.S.: Improving the company's competitiveness through a lean manufacturing system. Russ. Entrepr. **15**(18), 72–80 (2014)
17. The passport of the national project "Labor productivity and employment support" was approved by the decision of the Presidium of the Council under the President of the Russian Federation for strategic development and national projects on December 24, 2018 (2018). http://government.ru/rugovclassifier/865/events/. Accessed 20 Apr 2020
18. Veres, C., Marian, L., Moica, S., Al-Akel, K.: Case study concerning 5S method impact in an automotive company. Procedia Manuf. **22**, 900–905 (2018)
19. Vukadinovic, S., Macuzic, I., Djapan, M., Milosevic, M.: Early management of human factors in lean industrial systems. Saf. Sci. **119**, 392–398 (2019)

Check for updates

Digitalization of Higher Legal Education in Russia in the Age of Covid-19

D. M. Osina(✉), G. P. Tolstopyatenko, and A. A. Malinovsky

Moscow State Institute of International Relations (MGIMO University), Moscow, Russia
osina.dina@yandex.ru, decanatmp@inno.mgimo.ru, dr.malinovsky@yandex.ru

Abstract. The study covers topical issues of digitalization of higher legal education in Russia. Even though the process of digital transformation of higher education (including law schools) was launched before the Covid-19 pandemic, it was the coronavirus that acted as a catalyst for digitalization of almost all spheres of public life. Universities were faced with the need to create a comfortable and high-quality digital information and educational environment as soon as possible, and many enterprises (including their legal departments) decided to switch for remote work due to the self-isolation regime, which triggered further digitalization of the legal profession. In turn, digitalization of the legal profession can affect the labor market, and, therefore, higher education, as universities must consider the needs of future employers. The authors applied both general methods and methodological techniques (analysis, synthesis, deduction, induction, etc.) and special legal methods (formal legal and comparative legal). While analyzing, the authors conclude that the potential digitalization of legal education is widespread, since it is not only about the use of digital technologies in education, but changing the content of legal education due to digital transformation of the legal profession.

Keywords: Covid-19 · Higher education · Legal education · Digital transformation · Distance education · Legal tech

1 Introduction

Currently, we are actively struggling with the spread of the new coronavirus infection Covid-19 all around the world. Since neither a vaccine nor a specific treatment for the coronavirus has been invented, the most effective measure is social distancing, and even complete self-isolation. In this regard, the governments of many states, including Russia, are taking unprecedented measures, such as the suspension of thousands of enterprises alongside with the transition of millions of workers around the world to remote work. Aware of all the risks of the spread of the infection, both the Ministry of Education of the Russian Federation and the Ministry of Science and Higher Education of the Russian Federation decided almost simultaneously to transfer schools and universities to distance learning [12, 13]. E.g., by the Order of the Ministry of Science and

S. I. Ashmarina et al. (Eds.): ENGINEERING ECONOMICS WEEK 2020, LNNS 139, pp. 392–398, 2021.
https://doi.org/10.1007/978-3-030-53277-2_47

Higher Education for universities (including law schools and faculties) the training should be provided exclusively in the electronic format.

Thus, all Russian universities were urgently required to improve or even create a digital environment to ensure a continuous and high-quality educational process. Even though the course for digitalization of the social sphere (including education) was set long before the coronavirus pandemic [3, 14], the issue of digitalization of Russian education has never been so acute.

Now two trends in digitalization of higher legal education can be outlined: 1) a more obvious one covers the change in the form of education in the Covid-19 era (a greater emphasis is made on distance learning by the use of digital technologies and computer software) and 2) a less noticeable one, but getting more weight in the future – a change in content, since digitalization affects all spheres of public life, including the legal profession, since the needs of the labor market are gradually changing, and hence the requirements for graduates of law schools and faculties will inevitably undergo transformation.

In the age of digitalization accelerated by Covid-19 outbreak, law schools and faculties faced some problems, both of formal (like retaining the familiar form of oral exams) and of substantive nature (changing shape of the legal profession itself and the role of lawyers in particular, including the steps universities should undertake so that graduate lawyers could remain in demand on the labor market). It makes perfect sense that such questions arise as they have already touched the legal community around the world: for example, in the USA there were difficulties with the format of the traditional exam to obtain a lawyer's licence (Bar Exam) [1].

The team of authors attempts to consider the possible digitalization vectors of higher legal education in Russia, considering the general trends towards digitalization of the economy and all spheres of public life (including digitalization of the legal profession and the popularization of legal tech). The authors consider digitalization of legal education in the following key areas: digitalization of the educational process; digitalization of student-university/student-teacher relations; the impact of digitalization of the legal profession on the labor market and, therefore, on legal education. It is important to study both the ways and processes of digitalization and the consequences of digitalization for higher legal education in Russia.

2 Methodology

For the purposes of analysis, the authors have raised the following main questions (related to Russian law schools and faculties):

- What are the features of digitalization of the educational process and what are the significant differences between distance learning and the learning process offline?
- How will digitalization of the legal profession affect the labor market and digitalization of legal education?

The purpose of this study is a comprehensive review of the process and the possible consequences of digitalization of higher legal education in Russia, including digitalization of the legal profession; assessment of the prospects of distance legal education, considering digitalization of all spheres of public life. The results of this study can potentially be used while further studying digitalization of higher legal education in Russia.

The study is based on extensive normative, scientific, analytical and statistical material regarding digitalization of higher legal education as well as all spheres of public life and their impact on the legal profession. Research methods include both general methods and methodological techniques (analysis, synthesis, deduction, induction, abstraction, etc.), as well as special legal methods (formal legal and comparative legal). In the process of preparing the study, a significant amount of material was obtained from different legal research services, such as ConsultantPlus, Garant, Westlaw Academics.

3 Results

Despite the existing trend towards digitalization of practically all spheres of society, including education, in Russia, the coronavirus pandemic acted as a catalyst for digitalization and exposed all the problems in this area. Regarding digitalization of higher legal education and considering the questions posed in the study, the following was established.

3.1 The Features of Digitalization of the Educational Process and the Differences Between Distance Learning and the Learning Process in the Offline

When using digital educational platforms and appropriate technical equipment and software, most academic disciplines can be taught just the same way as in the offline. For example, according to the results of monitoring classes in digital remote format at MGIMO-University, it turned out that more than 99% of the classes were held in accordance with the timetable [8]. In fact, digitalization in this case is carried out through the use of digital technologies in education: the form of presentation of the material changes, but not the content itself (or such changes are not significant). Currently, Russian universities can choose necessary tools and software on their own and most often they use one of the following software products: Webinar, ZOOM, Microsoft Teams, Google Hangouts, Skype, Discord etc. At the same time, the issue of personal data security and confidentiality is especially relevant: for example, a lawsuit has already been filed against the ZOOM developer company in the USA based on an alleged disclosure of users' personal data, including recordings of teleconferences with their participation [2]. Despite the experience of leading universities, the overall situation in Russia shows that the digital infrastructure for science and education is still only at the initial stage of its formation [15].

The features of digital education are also reflected within the relationship between a teacher and students. Firstly, in the context of the digital organization of the educational process, it is getting even more difficult to objectively control the involvement of students within the educational process in class, and therefore the importance of such general cultural competence as the ability to self-organization and self-education inevitably increases [11]. Secondly, the student's perception of the teacher is formed not through personal communication, but through communication within the digital environment. For this reason, the role of the so-called "Personal brand" of the teacher is getting very important, which, in turn, requires a lot of effort from the teacher. A personal brand can be created by using a personal website/an account in a social network, or even by participation in major scientific events, etc. A "brand-teacher" is recognizable and popular. He is trusted by students, the target audience understands his/her "depth" as a specialist, expert [7].

3.2 The Impact of Digitalization of the Legal Profession on the Labor Market and on Digitalization of Legal Education

Nowadays (partially because of the Covid-19 pandemic and the respective need to switch to a remote mode of work), legal technologies (Legal Tech) are actively developing, and it significantly changes the qualitative approach of lawyers to work [6]. In this context, Legal Tech refers to platforms, IT services, and software that first made law firms and lawyers more efficient in performing their activities [22]. Lawyers tend to use more smart contracts instead of regular ones. Smart contracts, being in fact a computer program, are contracts that self-execute once the parties meet agreed-upon conditions [9]. The development of blockchain technology has also affected the legal profession, by both influencing smart contracts based on distributed ledger technology [23] and the development of cryptocurrencies. As of today, the legal status of smart contracts and cryptocurrencies is not legally defined in Russia, but we believe that this is just a matter of time: at present, several bills on digital financial assets are known [18].

Students of the International Law Faculty of MGIMO-University are taught the following legal tech competencies within the framework of the practical course "Legal Technique": 1) be able to work with the State Automated System of the Russian Federation "Justice" (SAS Justice); 2) have the skill to compile legal documents in electronic form and to carry out professional activities through "My Arbiter" tool and other electronic applications and/or official websites of government bodies; 3) be able to use electronic databases providing information about unfair counterparties; 4) be able to work with domestic and foreign legal research services like ConsultantPlus, Garant, Westlaw Academics; 5) know the basics of using Russian and foreign lawbots.

4 Discussion

Some experts believe that the essence of the legal profession has irrevocably changed: most of the work can be automated and performed by robots [17], clients want more services for less money, and legal services cease to be a unique product. In this case, there is only one way out: to learn to provide more services at lower costs, which means that it is necessary to attract new technologies to the profession [20, 21]. In this regard, expert opinions are being actively expressed on the need to adjust the curriculum at law schools in terms of introducing such subjects as, for example, coding for lawyers [4]. Nevertheless, the Russian state educational standards consider digitalization of the legal profession only in its most general form: for example, the already mentioned educational standard for undergraduate studies states that a Bachelor of Laws must be well-aware of the basic methods and means of obtaining, storing, processing information and must be able to work with information in global computer networks. In recent years, the labor market has already begun to require graduates to consider digitalization [10]: it is now important for successful employment to be able to work professionally with legal research services, fill out and file documents in electronic form (for example, with tax authorities and courts), hold client webinars and business conference calls with clients, etc. Thus, it has been established that digitalization of the legal profession directly affects the labor market. For universities, in turn, one of the key performance indicators is the employment of graduates [5]. For this reason, the issue of the timely development of new skills among future lawyers is extremely important. Given the transition to remote work for lawyers and remote training for students, one can assume an increase in the growth rates of digitalization: it is believed that up to one third of office workers can permanently remain on a remote mode of work after the end of the pandemic [16], and digital transformation before the pandemic was considered as a measure to reduce transaction costs by large companies [19]. According to the authors, under such conditions, the need for digitalization of legal education, especially by adjusting the curriculum, will increase.

5 Conclusion

To sum up, there is an increasing trend in digitalization of higher legal education, which has begun before the Covid-19 pandemic. The pandemic is a mere catalyst for the digital transformation process. Now, for the absence of any alternatives to offline learning, law schools are forced to adapt to new conditions as quickly as possible and create/improve a digital educational environment. However, potential digitalization of legal education is much larger, since it is not only about the use of digital technologies in education, but changing the very content of legal education, considering digital transformation of the legal profession. These changes are not on the surface, but new trends in labor market conditions will gradually bring changes into the educational process.

References

1. Angelos, C., Berman, S., Bilek, M.L., Chomsky, C.L., Curcio, A.A., Griggs, M., Howarth, J.W., Kaufman, E.R., Merritt, D.J., Salkin, P., Wegner, J.W.: The bar exam and the COVID-19 pandemic: the need for immediate action. Ohio State Public Law Working Paper, 537 (2020). https://ssrn.com/abstract=3559060. Accessed 20 Apr 2020
2. Blumberg: Zoom sued for fraud over privacy, security flaws (2020). https://www.bloomberg.com/news/articles/2020-04-08/zoom-sued-for-securities-fraud-over-privacy-security-flaws. Accessed 20 Apr 2020
3. Decree of the President of Russia No. 204 dated 05.05/2018 “On National Goals and Strategic Objectives of the Development of the Russian Federation for the Period Until 2024”. http://kremlin.ru/acts/bank/43027. Accessed 20 Apr 2020

4. Fenwick, M., Kaal, W.A., Vermeulen, E.P.M.: Legal education in a digital age. In: Corrales Compagnucci, M., Forgó, N., Kono, T., Teramoto, S., Vermeulen, E. (eds.) Legal Tech and the New Sharing Economy. Perspectives in Law, Business and Innovation, pp. 1–31. Springer, Singapore (2020)
5. Gadzhiev, G.M., Ramazanov, S.A., Maithanova, K.D.: Employment of graduates as an indicator of the effectiveness of the educational environment. News of the Dagestan State Pedagogical University. Psychol. Pedagogical Sci. **1**(30), 51–56 (2015)
6. Law Technology Today: AI trends driving the legal industry (2018). https://www.lawtechnologytoday.org/2018/03/ai-trends/. Accessed 20 Apr 2020
7. Mantulenko, V.V., Yashina, E.Z., Ashmarina, S.I.: Personal brand of university teachers in the digital age. In: Ashmarina, S., Vochozka, M., Mantulenko, V. (eds.) Digital Age: Chances, Challenges and Future. Lecture Notes in Networks and Systems, vol. 84, pp. 62–70. Springer, Cham (2020)
8. MGIMO: Distance learning continues: statistics of classes (2020). https://mgimo.ru/about/news/main/distant-11-04/. Accessed 20 Apr 2020
9. Morgan, N.T.: Blockchain challenges traditional contract law: Just how smart are smart contracts? Wyoming Law Rev. **19**(1), 87–117 (2019)
10. Online Edition of the Clerk: TOP 10 key lawyer skills (2020). https://www.klerk.ru/law/articles/466818/. Accessed 20 Apr 2020
11. Order of the Ministry of Education and Science of Russia No. 1511 dated 01.12.2016 (as amended on 11.01.2018) "On the Approval of the Federal State Educational Standard of Higher Education in Training 40.03.01 Jurisprudence (undergraduate level)". http://base.garant.ru/71579274/. Accessed 20 Apr 2020
12. Order of the Ministry of Education of the Russian Federation No. 104 dated March 17, 2020 "On the Organization of Educational Activities in Organizations Implementing Educational Programs for Primary, Basic, General and Secondary General Education, Educational Programs for Secondary Vocational Education, Corresponding Additional Professional Education, and Additional General Educational Programs, in the Context of the Spread of the New Coronavirus Infection in the Territory of the Russian Federation". https://www.garant.ru/products/ipo/prime/doc/73679183/. Accessed 20 Apr 2020
13. Order of the Ministry of Science and Higher Education of the Russian Federation No. 397 dated March 14, 2020 "On the Organization of Educational Activities in Organizations Implementing Educational Programs of Higher Education and Relevant Additional Professional Programs, in the Event of the Prevention of the Spread of the New Coronavirus Infection in the Territory of the Russian Federation". https://www.garant.ru/products/ipo/prime/doc/73645128/. Accessed 20 Apr 2020
14. Passport of the national project "National Program "Digital Economy of the Russian Federation" (approved by the Presidium of the Presidential Council for Strategic Development and National Projects, Minutes dated 04.06.2019 N 7). https://legalacts.ru/doc/pasport-natsionalnogo-proekta-natsionalnaja-programma-tsifrovaja-ekonomika-rossiiskoi-federatsii/. Accessed 20 Apr 2020
15. Pogorelova, E.V., Efimova, T.B.: Digital transformation of education, science and innovations. In: Ashmarina, S., Vochozka, M., Mantulenko, V. (eds.) Digital Age: Chances, Challenges and Future. Lecture Notes in Networks and Systems, vol. 84, pp. 54–61. Springer, Cham (2020)
16. RBC: The viral revolution: How the pandemic will change our world (2020). https://www.rbc.ru/opinions/society/27/03/2020/5e7cd7799a79471ed230b774. Accessed 20 Apr 2020
17. Remus, D., Levy, F.S.: Can robots be lawyers? Computers, Lawyers, and the Practice of Law (2016). https://ssrn.com/abstract=2701092. Accessed 20 Apr 2020

18. Sidorenko, E.L., Khisamova, Z.I.: Legal regulation of the digital assets market: advantages and possible risks of an alternative bill. Bank. Serv. **11**, 20–23 (2018)
19. Sidorenko, E.L., Khisamova, Z.I.: The readiness of the economy for digitalization: basic methodological approaches. In: Ashmarina, S., Vochozka, M., Mantulenko, V. (eds.) Digital Age: Chances, Challenges and Future. Lecture Notes in Networks and Systems, vol. 84, pp. 308–316. Springer, Cham (2020)
20. Susskind, R.: The End of Lawyers? Rethinking of Nature of Legal Services (Revised Edition). Oxford University Press, Oxford (2010)
21. Susskind, R., Susskind, D.: The Future of the Professions: How Technology will Transform the Work of Human Experts. Oxford University Press, Oxford (2016)
22. Vermeulen, E.P.M.: The future of law, lawyers and law professors…And the exponential growth of disruptive technology. Medium: Chatbots Life (2017). https://chatbotslife.com/the-future-of-law-lawyers-and-law-professors-and-the-exponential-growth-of-disruptive-technology-b5c979608c9c. Accessed 20 Apr 2020
23. Werbach, K., Cornell, N.: Contracts ex machina. Duke Law J. **67**(2), 313–382 (2017)

Digitalization in Activities of Non-state Actors: Example of the Church

E. O. Shebalina[1(✉)] and D. D. Shebalin[2]

[1] Moscow State Institute of International Relations (University) of the Ministry of Foreign Affairs, Moscow, Russia Russian Academy of Sciences, Moscow, Russia
shebalina.e.o@my.mgimo.ru
[2] Federal Agency for the Commonwealth of Independent States Affairs, Compatriots Living Abroad, and International Humanitarian Cooperation (Rossotrudnichestvo), Moscow, Russia
d.shebalin@rs.gov.ru

Abstract. This study considers the role of digitalization in activities of religious institutions using the example of the Roman Catholic Church. The digital innovations undertaken by the Holy See and their results are being studied. The analysis revealed the intense impact of digitalization on approaches of the RCC with the world community, namely the opportunities expanded that allowed the Holy See to strengthen the efficiency of its own ideas. It is noted that digitalization processes have acquired special significance for the church during the coronavirus pandemic in 2020, when traditional formats for the implementation of political aspirations were not available. Thus, digitalization contributes to the fact that even the church, being oriented towards traditionalism and keeping the fundamental principles in its activity, successfully uses digitalization as an instrument of adaptation to modern social changes in society.

Keywords: Digitalization · Holy See · Non-state actors · Roman Catholic Church · Vatican

1 Introduction

Like any social revolution, digitalization has a significant impact on activities of non-state actors in international relations, for example, religious institutions. The authors' choice of the church as an example to consider the impact of digitalization is due to the desire to demonstrate that activities of even the most "conservative" actors in international relations did not escape the impact of the "knowledge economy", but rather gave it new opportunities. If we turn to the history of the issue, the emergence of monotheism would have been unthinkable without the invention of writing, and the birth of Protestantism would have been impossible without the emergence of a set through moving letters. In turn, digitalization can lead to a new religious way of life, possibly even to the formation of new religions [13].

Digitalization leads to a gradual decrease in the distance between people and groups - this leads to a gradual but inexorable devaluation of the material component in many

S. I. Ashmarina et al. (Eds.): ENGINEERING ECONOMICS WEEK 2020, LNNS 139, pp. 399–407, 2021.
https://doi.org/10.1007/978-3-030-53277-2_48

areas of human existence, primarily in the field of information exchange. Now, not only the human voice has received a digital dimension, but the phones through which it is transmitted are becoming less common in the form of an analog gadget. Not only communication devices, but also all traditional means of culture have been digitized. A paper book is becoming increasingly rare, as is a vinyl record for sound reproduction, and in the cinema, where digital technologies are increasingly used, real actors are less featured [8].

But this devaluation of the material and the increasing role of numbers have affected other social areas [7]. Not only books and informational materials are converted into digital form, but such universal human notions, friendship, and even love are gradually being taking into the digital sphere. Thus, the concept of "friendship" from a real relationship between people turns into a banal connection between two accounts registered in a social network. Of course, these trends are widely used by NGOs and public institutions to digitalize their activities in those areas where this is appropriate.

2 Methodology

The authors determined the methods considering the purposes and objectives of the study in the format of the interdisciplinary approach, involving the use of historical and political science research principles in a comprehensive manner. The study is dominated by qualitative and quantitative content analyzes. The qualitative research methods are mostly based on documentary research, scientific articles and analyses. The objective is to understand opinions, perceptions and activities of the Roman Catholic church implementing digitalization. Quantitative methods proved to be especially useful while analyzing the church's activity on online platforms.

3 Results

Active digitalization of its resources is carried out by the Roman Catholic Church, represented in international relations by the state called the Holy See. The Holy See, due to its unique legal personality (a combination of both state and religious components), at the present stage develops foreign policy and diplomacy, making maximum use of all types of influence at the religious level: activates forms of communication alternative to official diplomacy, uses a flexible approach to changing conditions of international relations.

It must be understood that the term "Holy See" combines two entities – religious and political. Religious is the Roman Catholic church, with its extensive network of dioceses around the world. Political is the State Secretariat, which conducts the country's foreign policy through the Second Section on Relations with States and some nunciatures and papal representations abroad. Both institutions are directly subordinate to the Pontiff. The city-state of the Vatican, on the territory of which the administrative buildings of the Secretariat are located, in turn, stands as a separate formation under the management of a special body – the Governorate with the Chairman at the head [6].

The Holy See is actively updating its public diplomacy, making it more inclusive, open to all, rebranding the Roman Catholic Church. The Pope pursues this policy gradually and imperceptibly: using existing methods and internal resources, redistributing functions. Solving the problem of reducing the number of believers and parishioners, the Pope does not tighten politics, but is looking for reasonable ways, meets principles which are contrary to some canonical ones, and incidentally it is confronted by certain religious and public circles.

As part of informatization of its resources, the Holy See launched a site that is visited by more than 0.5 billion people, Pope Francis has launched his pages on social networks Facebook, Twitter and Instagram. More than 1 million users subscribed to the Pontiff's Instagram account within 12 h after its launch. The most popular pontiff account is on Twitter in different languages. The number of subscribers to the English version is 18.3 million, plus 17.6 million subscribers to the Spanish version, 4.9 million – Italian, 4.5 million – Portuguese, 1.4 million – French, almost 1 million – Latin, and about 500 thousand – Arabic. Thus, the total number of followers of the Roman First Hierarch on Twitter reaches about 50 million people. This is almost 10 times more than that of French President E. Macron (4.9 million) and almost 100 times more than that of Italian Prime Minister J. Conte (525.4 thousand). Appeals that the Pope addresses every Sunday flock, broadcast on all information channels of the Holy See. These appeals are an effective tool for soft power.

According to a 3RDPlace study [11], the number of Pope Francis searches on Google reached 1.737.300 per month. In 2013, it was one of the most requested on the network (more than 49 million references) and fell in line with V. Putin, B. Obama and A. Merkel.

Pontiff's Twitter account (in different languages) with publication statistics of 0.79 per day shows an average engagement level of 6.637. For comparison, Barack Obama's account, according to the study, has an average engagement level of 2.309, despite the statistics of daily publication of 7.76 tweets per day. This feature is explained by the fact that Obama's account is focused on delivering his own political idea, and the Pope uses social networks to communicate with believers and to develop the concept of a universal church.

The Vatican has always communicated using the latest technology. In 1989, the Vatican's press service installed 40 of the most modern computers at that time. Since 1991, the Vatican began to actively use programs and services (WORM, SEDOC), which contribute to more active and high-quality dissemination of information in a format convenient for recipients. In 1992, a special DEI system was created through which all the messages of the Pope are transmitted [5]. Thus, throughout the 90 s, the Vatican did everything possible so as not to remain in the "backyard" of new technologies and to "keep up with the faithful". In the mid-90 s, the Vatican did not have its own website. However, all the important events of the Catholic Church (Christmas masses, solemn blessed papal blessings of Urbi et Orbi) were broadcast through the services of the French satellite operator Eutelsat and the British WRN Broadcast. Pope John Paul II addressed believers during the genocide in Rwanda in 1994 and the conflict in Kosovo in 1999. For five and a half months, the Vatican Radio broadcast Albanian programs for the people of Kosovo with calls for peace and solidarity, as well as regularly updated information about the victims of the conflict for people who have lost their loved ones. John Paul II considered the technological development of modern media to be an advantage for humanity [2]. As you know, pontiffs pay special attention to digital diplomacy in the acute phases of crises and conflicts, which often Italian authors call "direct dialogue with the disadvantaged". So, during the Apostolic visit of Pope Francis to the United States in September 2015, the Vatican Television Center, together with the American television channel ABC, released a program in which American and Cuban young people could discuss various topics relevant to them. This was another stage at normalization of public relations between the two countries after the establishment of diplomatic dialogue.

On March 30, 1997 the Vatican launched its own official website www.vatican.va. According to the former head of the Pontifical Council for Mass Communications, Claudio Maria Celli, the Vatican's activities in the Network space have become "a continuation of the Pope's pastoral mission" and, accordingly, his digital diplomacy. The Vatican is closely following new trends in technology and does not stand aside: in 1999, on the eve of the Jubilee Year, Sony and the Vatican City launched a large-scale project, AbbaPater, in which 11 tracks were recorded combining music and the voice of John Paul II, in five languages. The media of the Vatican worked vividly and effectively in 2000, when the Vatican media was tasked with the most intensive and round-the-clock broadcasting of anniversary events, acquainting the public (not only the Catholic, but the entire world community) with the goals of the Holy See in the new millennium.

On January 28, 2013, in Rome, at the conference "The Future of Communication is Responsibility" (Rome, January 28, 2013), a study by Eurisko was presented under the auspices of the Pontifical Council for Mass Communications, which aimed to determine the potential of Catholic media that expand and implement new principles of work, based on the following criteria: seriousness, accuracy, firm position, clarity of criteria and logic. According to the results of the study, being Catholic is becoming popular, which means participating in the universal "right" mission [3].

In the framework of this area, a policy is being developed to reorganize the activities of the Holy See mass media, which resulted in the creation of a unified information network Aleteia. This reform has yielded results – more than 67 million Catholics from the USA, Great Britain and Ireland have already been on the social network [11]. Aleteia is the first worldwide Catholic information network under the auspices of the Pontifical Commission on Mass Communications, which discusses issues of life, faith and society in 6 languages (English, Spanish, French, Portuguese and Arabic). According to the agency, more than 1300 are participating in its work. Catholic partners and more than 800 experts preparing materials. In addition, in the very near future it is planned to launch a new electronic resource, Dot.catholic, which will become the main electronic platform, uniting all the marked churches and dioceses of the Roman Catholic Church.

From the point of view of the Vatican, the church is called to be where there are people. Since now the entire electronic network also consists of people, there should be a church there. The Pope calls not to be afraid to become full-fledged "citizens" of the 21st century information society. It is important to note that Francis puts a special meaning in the concept of "citizen" in his messages, bearing in mind that being so means commitment and the desire to create a common good for society [12]. In his opinion, modern digital technologies have given impetus to the creation of a completely new social space within which it is possible to build such connections that will have an impact on societies and their culture. Thus, the Pontiff raises the question of relations between the church and society, which are regulated at different levels, the most important of which, in his opinion, is the global Internet. He gives him central importance in building a new type of relationship between believers and representatives of secular society. With his call for active participation in digital contacts, he paves the way for building such a dialogue. This model of behavior of the Catholic Church is rooted in history. At the time when society was moving from verbal to written communication, the Church was the first to use a new type of communication. Thus, its written appeals acquired importance, and even holiness. The same attitude was won by their authors. So now the Catholic Church is striving to be the first to apply new forms of communication in the digital space for the effective implementation of its policy.

The rapid development of digitalization of the Roman Catholic Church is due to the orientation of young believers. It is no coincidence that the XV General Assembly of the Synod of Bishops, which was held in the Vatican in October 2018, was dedicated to the theme "Youth, Faith, and Recognition of Callings". The work of the synod was not only accompanied by official comments of the Vatican, but was also tracked and posted on social networks by the faithful Catholics: 1,400 tweets marked with the official hashtag "# Synod2018" in six different languages published by the Vatican News and Vatican Radio channels, with 100 000 reposts on Twitter, 10 million users following the event on Facebook, 650 thousand views on Instagram [14]. In total, five hundred different full-fledged materials were posted on social networks, including publications, articles, images and videos. And 50 million people all over the world noted a message on Twitter by Pope Francis on the Synod and published on October 24, in which the Pontiff explained in nine languages that "this Synod should become a symbol of the Church that can listen, a church that does not always have ready Answers". These statistics are published by the Secretariat for Communications as a report on the activities of the Vatican in social networks and on the activities of the Synod on Youth Issues.

During the Synod, participants were presented a study conducted by the Catholic information network Aleteia in collaboration with St Mary's University in London and Ramon Lull University in Barcelona, which analyzed Facebook and Instagram accounts of 540 million people around the world – representatives of young people aged 18 to 25 years who used social networks to search for information on religious issues. Among the most mentioned queries – the primacy was given to the figure of the Pontiff – Pope Francis.

The results of the study showed that only 4% of young people share content related to Catholicism in social networks. This share is much less than the number of those who use the Internet to search for branded items and consumer goods. However, it was noted that young people who are interested in religion have a higher level of education. The overall picture was approximately as follows: 5 million young people are interested in the issues of faith in Brazil, 2 million are registered in the Philippines, 1 million in India, and 700 thousand young people in Italy, and a smaller number were recorded in other countries.

When considering digitalization of religion and the Roman Catholic Church, the Catholic world was divided into two camps on social networks, as well as in real life – the so-called "liberals" who support the reforms of Pope Francis, and the "conservatives" who they do not approve of his innovations and are grouped around several cardinal oppositionists. One of the most prominent representatives of the latter is Cardinal Raymond Leo Burke, who is considered the leader of the traditionalist wing of the church. However, when typing the hashtag #cardinalburke in any search service, you can easily find the full chronicle of the life of the prelate, published by different users from all over the world, and see that it is very popular.

At the same time, digitalization is not only an instrument of the internal polemics of the church, it is an important element in the spread of dogma around the world. Earlier the Congregation for the Doctrine of the Faith (the current Congregation of Evangelization) sent missionaries to different parts of the globe, now it can be done through online resources [10]. In any case, online sites that unite Catholics on Skype or other instant messengers regularly publish new services for Catholics: for example, a specially created application for the DinDonDan smartphone, which allows you to find the closest mass to the user's location, as well as get the current working hours of temples and service schedules.

The next striking achievement of Catholic digitalization was the development of electronic beads "Click to Pray eRosar"". An interactive cross is placed in the center of the device (the cost of which does not exceed 100 euros), which stores the technological data of the connected application on the smartphone. This device was created to assist believers in prayer [9].

The application that ensures the operation of the gadget is very easy to use: after activation, it allows you to access audio guides, exclusive images and personalized content, depending on the type of rosary chosen (there are thematic collections of prayers for young people, the elderly, for migrants and refugees, etc.). The Click to Pray eRosary application was completely "launched" by the Vatican. According to Frederic Fornos, head of the Pope's Worldwide Prayer Network website (Rete Mondiale di Preghiera del Papa), "the challenge is to offer the world one of the millennia-old spiritual traditions of the Church through the best of modern technology," always with a focus on acquiring new and supporting the youngest believers. After creating the "DinDonDan" application (Android and Apple), the Roman Catholic Church received the "Church 2.0" feature in the media, and the appearance of "Click to Pray eRosary" "updated" the religious institute in Italian media to version 3.0.

4 Discussion

Digitalization of the church played a role in the crisis that took place in the world; it turned out to be prepared for isolation caused by the coronavirus. RCC broadcasts not only the Holy Mass, which is served by Pope Francis every day, but also many other priests. All broadcasts are live thanks to the Tv2000 Vatican TV channel, which completely changed its schedule during the day, offering people, in addition to news about the coronavirus infection, all the necessary programs about spiritual life.

In conditions of self-isolation, the possibilities for leading a spiritual life for believers became even more increased. This is because 10 years ago, the Conference of Bishops of Italy (CEI) began to actively develop a project called the "Digital Shepherd". Between 2009 and 2011, CEI organized four national conferences to push the Italian dioceses and Catholic media to make more active use of the digital platform. The conference data already received the following names: "Church on the Web 2.0" (2009), "Digital Certificate" (2010), "Diocese on the Web" (2010) and "Digital Residents" (2011). At that time, Don Domenico Pompili, who was firmly convinced in the choice of digital language as a new tool for evangelization and dissemination of Catholic culture, was the director of the National Service for Social Communication of CEI. Today, D. Pompili is a bishop and it is no coincidence that he belongs to the circle of proxies of Pope Francis. The site of his diocese is an example of the spread of the Christian faith over the Internet. The materials of the electronic resource called "Faith Clings to the Network" (developed by D. Pompili) contain information on dozens of social initiatives of parish priests, laity and Catholic educators.

Thus, digital investments made by CEI from 2009 to 2011 are bearing fruit. The diocesan sites of the Roman Catholic Church that have arisen in recent years provide opportunities for streaming divine liturgies performed by bishops, as well as many other religious events. In the same way, there is an extensive network of Catholic weekly newspapers that work full time and guarantee online support for believers informing them about the activities of the parish/diocese and enable people to participate in the spiritual life of church communities in a timely manner.

At the same time, the idea of digitalization belongs not only to the Roman Curia and to the staff of the respective church institutions. Digitalization initiatives are supported by simple priests and laity. For example, Don Paolo Padrini (diocese of Torton) is one of the first Italian priests to be engaged in digital shepherding: every day he sends WhatsApp to his endless list of friends, links to the live broadcast of the mass on the Youtube channel from the episcopate's chapel Tortona.

Don Alessandro Di Medio, the parish priest of San Francesco Saverio in Garbatella, is well known in Rome for his "innovative pastoral ways", thanks to which young people from 19 to 29 years old can go through the daily "spiritual route for thought" during Lent through the Sir website (informational CEI agency). Also, in Rome, Father Roberto Cassano, the parish priest of São Paulo della Croce in Corvial, did not want his believers to forget about the liturgy, so he broadcasts the mass via Facebook. He also activated a chat for believers in WhatsApp and prepared an app for the parish, where every parishioner can find comments on gospel, prayers, fundraising announcements for charity, in times of crisis.

Parish priest initiatives are being reported throughout Italy, especially regarding maintaining relations with their communities. "There are many masses on the net, but for each of us it is of great importance," prelates remark, "if we conduct them specifically for our parishioners". In addition, any church, for example the Roman Catholic Church, has a significant archive. The Vatican Archive is a valuable repository of documents of interest to historians from around the world. Following current trends, the Holy See began to translate and archive activities in digital format.

The process of archiving all editorial documents, now in a multimedia format, will allow you to access the documents, as well as make their full digitization and preservation. In addition, the media will also include media inherited from other organizations that are part of the Holy See Communications Department. Starting from 2019, materials are acquired and gradually digitized, to become publicly available to scientific community.

The Vatican Archive is the central archive of the Holy See, where all documents relating to the state and pastoral activities of the Pontiff and the activities of the Holy See are stored. The documents limited to specific periods are available for review by qualified researchers upon written application and submission from the accredited historical research institute [1]. The documentary heritage stored in its vast collections covers a chronological period of about twelve centuries (VIII–XX centuries) and consists of more than 600 archival collections. Although it is not the largest archive in the world in number, it is the largest in the world in geography of coverage – all continents and all states where the Catholic Church is present.

5 Conclusion

The Roman Catholic Church is an example of a traditional religious institution that, despite two major stages of reformation - the First and Second Vatican Councils, has changed many principles, but has not abandoned the fundamental doctrine of Christian doctrine [4]. Traditional religious institutions that adhere to this doctrine cannot fully transfer their activities to online mode. That is why the sacraments of the church, such as confession and communion, will always remain inaccessible to the digital format. However, in cases where this seems possible, the church actively and successfully applies the achievements of digitalization, which is a positive example of adaptation to modern social changes in society. The presence of such an approach allows a religious institution (as, indeed, any other actor in public life) to exist and strengthen its position in the present with a view to the future.

References

1. Albertoni, M.: Vendetta and career: Nunzio Decio Francesco Vitelli and Ferrante Pallavicino. Hypotheses and Documents Originating from the Secret Vatican Archive. Studi Secenteschi **59**, 195–223 (2018)
2. Anyidoho, P.: Pope John Paul II and contemporary media. Integrating understanding of media in culture. Compass Rev. Top. Theol. **50**(2), 6–8 (2016)
3. Ballardini, B.: Leader come Francesco. Piemme, Milano (2014)
4. Barnebeck, T., Peter, A., Jensen, S.: Preaching democracy: the second vatican council and the third wave. J. Comp. Econ. **47**(3), 525–540 (2019)
5. Bea, F., De Carolis, A.: Ottant'anni della Radio del Papa (T. I, II). Città del Vaticano: Libreria Editrice Vaticana (2011)
6. Borrelli, M.: "To show concern": Early coverage of John Paul II, Benedict XVI, and Francis in the American National News Media. Palgrave Studies in Religion Politics and Policy, Las Vegas (2018)
7. Carlton, N.: Digital culture and art therapy. Arts Psychother. **41**(1), 41–45 (2014)

8. Carolyn, H.: The Multimediated Rhetoric of the Internet: Digital Fusion. Routledge, New York (2014)
9. Click to Pray: Click to pray erosary: New smart rosary to pray for peace in the world (2019). https://clicktopray.org/click-to-pray-erosary-the-new-smart-rosary-to-pray-for-peace-in-the-world/. Accessed 02 Apr 2020
10. Gesiak, L.: Reform of the Vatican media, a difficult search for a new path. Person Challenges J. Theol. Educ. Canon Law Soc. Stud. Inspired Pope John Paul II **7**(2), 5–22 (2017)
11. La Vita del Popolo: La rete ama Papa Francesco: Il Pontefice "personaggio dell'anno" anche online (2020). https://www.lavitadelpopolo.it/Chiesa/La-rete-ama-papa-Francesco-il-pontefice-personaggio-dell-anno-anche-online. Accessed 02 Apr 2020
12. Regan, E.: The Bergoglian principles: pope Francis' dialectical approach to political theology. Religions **10**(12), 1–16 (2019)
13. Serres, M.: Hermes: Literature, Science, Philosophy. The Johns Hopkins University Press, Baltimore (1983)
14. Synod18: Conferenza Stampa di presentazione della XV Assemblea Generale Ordinaria del Sinodo dei Vescovi I giovani, la fede e il discernimento vocazionale (2018). https://press.vatican.va/content/salastampa/it/bollettino/pubblico/2018/10/01/0711/01519.html. Accessed 02 Apr 2020

Legal Regulation of Digital Platforms: Reference Points of Modern Legislation

E. L. Sidorenko[1(✉)], I. Sh. Galstyan[2], and A. A. Sitnik[3]

[1] Moscow State Institute of International Relations (University) of the Ministry of Foreign Affairs, Moscow, Russia
12011979@list.ru
[2] Moscow State Regional University, Moscow, Russia
inessagal@mail.ru
[3] Kutafin Moscow State Law University, Moscow, Russia
aasitnik@msal.ru

Abstract. The paper analyzes the current state of Russian and foreign legislation regarding the regulation of digital platforms. The study is based on the differentiation of platforms according to the area of their activities and the main functional tasks. But at the same time, the authors insist on common reference points in the legal regulation of different platforms. Their analysis allows not only to assess the quality and completeness of legal regulation, but also to formulate a number of proposals for improving the current legislation. To the reference points of the rights of digital platforms the authors refer the following ones: the legal status of the platforms, the contractual basis of their activities, the responsibility of the parties for inappropriate goods (services), labor relations with employees of digital platforms, etc. These issues are considered in the article with reference to current legislation and modern judicial practice.

Keywords: Digital platforms · Digital platforms' right · Contract · User agreement · Legal status · Consumers

1 Introduction

The effectiveness of today's digital economy is directly connected with the development of digital platforms. In the Report on Digital Economy at United Nations Conference on Trade (UNCTAD) in 2019, "over the past decade, many digital business models emerged transforming key industries. In 2017, the aggregate value of platform-based companies with a market capitalization of more than $ 100 million exceeded an estimated $ 7 trillion, up 67% from 2015. Some global digital platforms gained very strong market positions in certain segments." As an example, UNCTAD specialists cite examples of giant platforms that became first on search engine market (Google company), online sales (Amazon and Alibaba companies), payments via cellular networks (Alipay), etc.

A set of factors provides the basis for the economic success of platform solutions:

- platforms create a digital market infrastructure, providing the user with necessary digital services and facilitating interaction with contractors,

S. I. Ashmarina et al. (Eds.): ENGINEERING ECONOMICS WEEK 2020, LNNS 139, pp. 408–418, 2021.
https://doi.org/10.1007/978-3-030-53277-2_49

- consumer engagement is carried out through multi-channel marketing, as a rule, in narrow segments and niches,
- if the growth is higher, Metclough's law comes into force (the value of the network for its participants is in direct quadratic dependence on their number),
- platform economy is based on the collection and analysis of big data, that generates forecast demand at first, and then manage it.

The marketing relies on the following principle: as the platform's sales increase, users' costs related to the transition to competing providers increase significantly. Thus, the market monopolizes taking over competing platforms. International organizations highlight the high dynamics of digital platforms' development mainly in the areas of operational and innovative solutions, they express concern about the insufficient elaboration of legal issues for their functioning. In particular, there are disputes regarding the definition of jurisdiction in relation to multinational companies, as well as fixing the legal status of platforms, an algorithm for the responsibility of the company and its customers for goods or services of inadequate quality, etc. Solving these issues requires specialists to give a competent answer to the question if modern law can satisfy current needs of the platform economy or if there should be new legal constructions that reflect economic and technical features of digital platforms.

2 Methodology

The methodological basis of the research is the system of general scientific and special cognition methods of legal phenomena. The authors used historical, comparative legal, formal-logical and system-structural research methods. For complex and reliable analysis, the logical-legal and statistical methods are also used. Conclusions regarding the prospects for the legal regulation of digital platforms are based on forecasting and modeling methods. The methods of analogy and sampling of the maximum similarity assume considering digital platform as a theoretical construction of legal relations and offer solutions in a similar way with related legal phenomena and institutions.

3 Results

During the study, differentiation of platforms is carried out depending on what kind of legal relations are between users; the most important legal issues related to platform's status, applicable law, regulation of labor relations between the platform and its employees are identified. The legal status of digital platforms is researched via corporate regulation and consumer protection.

Applying systematic analysis of general and applied issues of legal regulation of digital platforms, the authors offer a number of recommendations on the development of legal parameters and protection of user rights. Emphasizing the importance of legal regulation of digital platforms, World Bank experts identify four blocks of legal issues:

- antitrust regulation: the World Bank proposes either to recognize them as natural monopolies with a limit on their income, or to establish an obligation to share customer data with new market participants, partially equalizing their chances;
- licensing: there is possibility of introducing licensing or certification requirements which are common for services provided by traditional participants and for platforms (such as taxis or rental housing). At the same time, licensing standards should be specified for people providing payment services on the platform to ensure high speed of payments, and their security;
- taxation. Accounting the specifics of the platforms, several taxation models are proposed: tax on the users' number, on data collected from citizens, treating this information as a valuable resource;
- protection of personal information. The directions of this regulation can be divided into 1) owners rights' of personal data; 2) cross-border movement of data; 3) data protection regulations.

Taking into account the fact that now developed and developing countries use opposite approaches to regulating digital platforms (the first ones restrict data exchange, the second ones, on the contrary, stimulate market openness), the World Bank supports "sensible middle": stimulating transnational exchange through establishing the optimal legal restrictions so that digital platforms can function safely.

In Europe, there is no solution if the framework for legal regulation of digital platforms should be created at the level of the European Union or individual states [32]. Nevertheless, experts of the European Commission identify 5 main areas and ways of possible future regulation of the platform economy at the European level:

- licensing and other restrictions on market entry;
- detailed legislation on the regulation of labor relations at the level of individual countries;
- establishing basic requirements for the protection of consumer rights, protection of copyright and personal data at the European Union level;
- data laws: legal regulation of data transfer between digital platforms is necessary; access mode to authorities' data, use of open data received from authorities; as well establishing new copyright regimes for text and data mining [32].

According to experts of the commission antitrust and tax law, it doesn't impede the development of digital platforms, although they need additional elaboration in new reporting forms. It is essential that the European Union obliges digital platforms to collect VAT on sales, and this practice is recognized by OECD exports as effective [25, 26]. OECD highlights the main activities of digital platforms, they are now related with the services' providing, making payments or searching for information. The Organization considers the protection of consumer rights as the main task of developing the law of digital platforms.

OECD developed the basic principles for protecting users of digital platforms: fair business and advertising, the same as in usual trade; transparency of contracts and transactions; simple and convenient payment methods; clear client's consent to the transaction; ensuring information security and customer confidentiality, product safety; effective dispute resolution mechanism within the platform [24].

Determining principles and priority areas of legal regulation suggested by international organizations seems to be welcomed. However, the proposed program is quite abstract and can be considered as a guide when answering a number of key questions related to the platforms' operation.

First of all, digital platforms differ from each other in their functionality.

UNCTAD experts suggest highlighting operational and innovative platforms [36]. Depending on the activity area, the level of information processing and the main beneficiary, instrumental, infrastructural and applied digital platforms are distinguished [28].

The European Commission suggests distinguishing the following types of crowdfunding digital platforms: donation crowdfunding, reward or pre-payment crowdfunding, crowdfunding lending, equity crowdfunding [15]. In addition, the platform is proposed to be divided by functionality into operational, innovative, aggregated and social, by scale into global, regional and national [17].

Accounting the diversity of digital platforms, proposals for their legal regulation should not cover the individual amendments to laws, but answer to the question: "Should the rules be differentiated depending on the business model, beneficiary or functional task of the digital platform?" However, at present, the answer to this question cannot be unique, not only because of platforms' variety, but also because in modern law the reference points on law scale of digital platforms are still not defined.

The dynamism and flexibility of new digital technologies, as well as obvious differences in characters, interests and needs of participants in economic relations make it difficult to find common parameters for legal regulation of mobile commerce platforms (m-commerce), but they have to be defined [31].

Global digital platforms are the basis for new legal relations, and in this regard, it is fundamentally crucial to highlight key parameters that, on the one hand, will satisfy universal requirement, and on the other hand, will indicate significant differences in the legal structures of digital platforms. We propose to consider among such criteria the legal status of digital platforms and contractual relationship "Platform - Client".

4 Discussion

4.1 The Legal Status of Digital Platforms

Most modern experts are forced to admit that at present the legal status of digital platforms is blurred, but not because of the novelty of this tool, but because most businessmen use imperfect legislation and create hybrid designs to minimize responsibility.

The most detailed issues of the legal status of digital platforms are studied in European law. Here digital platforms are considered mainly as commercial organizations that perform different information and other services. According to the rule proposed by the European Commission, a digital platform should be considered as an enterprise operating in bilateral or multilateral markets and using the Internet to ensure interaction between two or more interdependent groups of users (consumers) [14]. The European Parliament significantly developed this definition by including provision or aggregation of services (content) from service providers (content) to final consumer [16]. In accordance with the requirements of the European Union, most online platforms are considered information society service providers (EU Electronic Commerce Directive (E-Commerce Directive 2000/31/EC)) [11]. However, directive is not the case of interest but judicial decisions taken regarding individual digital platforms in connection with its application. So, on December 20, 2017, the European Court passed the judgement that the Uber service, which via application connects an unprofessional driver using his own car, with a person who wants to use the car, should be considered as a transport service, not an information society service, and this means that Uber does not fall under the regulation of the EU Directive 2000/31/EC [11] and should be considered as a company providing transport services [5]. This decision caused controversy about the platform: if it is possible to limit its functionality only to digital information services, or the platform takes the responsibility about the companies' activities which are placed on it [22].

The next platform that the EU Supreme Court assessed is Airbnb. The key issue for the court was whether its activities could be defined as "information society services" (EU Directives 2000/31/EC) [11]. In its decision of December 2019, the European Court concluded that Airbnb and Uber have different legal status [27]. Refusing to recognize Uber as a company providing only information services, the court emphasized that it offered transport services, without it drivers could not carry passengers, therefore, Uber has a "determining influence" on economically significant aspects of the transport business, and has to bear responsibility as a transport services organization.

As for Airbnb, this company is subject to Directive 2000/31/EC [11], because the Platform: 1) did not offer accommodation services; 2) hosts could offer their hosting services regardless the platform; 3) did not have a "determining influence" on economically significant aspects of service; 4) did not set prices for accommodation, etc. Thus, in European law, a serious step was taken to develop common parameters for determining the legal status of a digital platform as an organization providing information or other services. In fact, the following algorithm was proposed: determination of the legal status of a digital platform actually depends on its control and influence on the services offered. Platforms should control the services that they consider most essential for their business model. It does not matter if the Platform collects fees for the services offered, the important issue is if it participates in setting prices, commissions, etc.

The proposed model exceeded the scope of individual decisions and is now actively used by the courts of various countries in terms of determining the platform's responsibility for people's actions who provide the services. The challenging issue was European Court's decision of Airbnb case (C-390/18) [20] in terms of assessing further development of digital platforms in December 2019. This case brought into sharp focus the problem of determining the company's jurisdiction and the possibility of imposing additional restrictions on it.

Airbnb was accused of unfair competition with hotels. The Tourism Law Violation Association reported that Airbnb acts as an unlicensed real estate agency and violates the 1970 Act, known as the Hoge Act. The European Court was not asked if Airbnb meets the requirements of Directive 2000/31 and if Directive 2006/123/can be applied to it (Directive 2006/123/EC of the European Parliament and of the Council of 12 December 2006 on services in the internal market) [12] as far as domestic markets are concerned.

According to this Directive, a country's law can restrict the freedom of an economic operator to provide services established in another Member State, provided the fulfillment of detailed conditions. This issue is of particular importance due to the fact that the European Union intends to revise Directive 2000/31 and strengthen the principle of country of origin for digital companies [11].

In the court judgement, Member States can take measures that violate the principle of freedom to provide the services of the information society, if two general conditions are observed: it is important to notify the Control Commission and the state on which territory the service provider is located.

Thus, the decision confirmed the ability of the countries to go beyond the framework of European standards and impose additional restrictions. However, this requirement is mitigated by articles 3 (4) and 15 Directives 2000/31, which allow platforms to resist the laws that require them to report data to public authorities. This makes it incredibly complicated to reveal violations for local authorities, creates gray areas and encourages breaking the law. The demarcation of platforms on the basis of their control over the services offered was one more dangerous issue for digital law (European Court decisions C-434/15 Uber and C-320/16 Uber France) [20].

In fact, the issue of the platform's responsibility is dependent on how ready it is to control the price of the service and manage the service. But as practice shows, this criterion is relative. It is not clear which company invests more in optimizing its services: the one that directs the location of the client to the taxi driver or the one that creates an interactive map with the selection of the most suitable hotels. Due to existing variables, the issue is completed and cannot have a unique solution.

The criteria proposed by the European Court led to unexpected results. There is such dependence: the higher the company's control over the quality of services, the higher its responsibility. Digital platforms responded to this situation by looking for options to weaken control over the services.

The alternative option for the platforms' development was proposed to design them as consumer cooperatives or interest platforms. These models are well known to both Russian, European and Asian markets. But they are also linked with a number of difficulties in terms of assessing the cost of services provided by the consumer platform, determining the team of responsible persons, etc. [1, 31].

Russia shows an interesting approach to assess the legal status of digital platforms – aggregators. In paragraph 18 of the Decree of the Plenum of the Supreme Court of the Russian Federation of June 26, 2018 No. 26 "On some issues of the application of the legislation on the contract for goods carriage, passengers and baggage by road and on the contract of transport expedition" [8], it indirectly indicates the responsibility of the person to which the client applies to conclude a contract to carry passengers and baggage. This person has responsibility before a passenger for harm caused during the carriage, if this person entered into a carriage contract on his own behalf or from the circumstances of the contract (included the information on Internet website, correspondence between parties when concluding an agreement, etc.). Fair consumer can have the opinion that the contract of carriage is concluded directly with this person, and the actual carrier is his employee or a third party involved to fulfill transportation obligations. Now this rule is fully applied to digital platforms that provide transport services (Gett, Uber, Yandex.Taxi).

This approach was legally formalized in the Federal Law of the Russian Federation of July 27, 2018 No. 250-FZ «On Amending the Law of the Russian Federation "On Protection of Consumer Rights"» [18]. The law defines the status of the aggregator of information on goods (services) as organizations or individual entrepreneurs who are the owners of the site or program that provide the consumer the opportunity to simultaneously study seller's proposal to conclude a contract for the sale of goods, conclude such an agreement, and also make advance payment by transferring funds to the owner of the aggregator within the applicable forms of cashless payments.

The aggregator is determined to be responsible for: losses incurred by the consumer as a result of providing him with inaccurate or incomplete information about the product or seller; for performing the contract, as well as for the observance of consumers' rights violated as a result of transferring goods of inadequate quality to the consumer and the exchange of non-food goods of appropriate quality for similar goods. Thus, Russian law did not copy the European way which recognizes a person providing the services offered on the platform as an aggregator, the law just indicated information services. And this means that in order to carry out its activities, the aggregator is not obliged to have an appropriate license to provide transport, tourism or other services. The USA uses the similar approach [22, 23, 37].

4.2 Contractual Relations "Platform-Client"

Platforms often represent tripartite business models that obligate the lender and the supplier not only to conclude an agreement with each other, but also to conclude an agreement with the platform operator [13]. It is important to take into account the fact that the platform is primarily a technological base that combines computer programs, a domain name and a website, databases with access, storage, exchange and other processing by sellers and buyers to ensure the network or electronic trading platform functioning [29]. Moreover, the owner does not always combine the functions of an administrator and a platform operator at the same time. Often the type of relations is built in the framework of the agency contract model.

Depending on the platform's business model, the following agreements can be the basis of its activity: an assignment agreement, a commission agreement, a service contract for the transfer of goods and manufacturer recruitment, or mixed models formed within the agreement's freedom. Some authors highlight sui generis, a special type of contractual relationship involving a digital platform [21]. The three-way construction of legal relations (consumer, aggregator and contractor) is also proposed to be applied on the basis of two contracts: a contract for services' providing concluded between the aggregator and the consumer, and a contract for services' providing between the aggregator and the contractor [9].

Increasingly, in the literature there is the idea for the necessity to develop a fundamentally new area of law - transnational consumer law, which protects the interests of platform users regardless their citizenship and company location [2]. However, unresolved issues remain: on what principles these norms will be established, on what will be based and how they relate to national law. In fact, further development of this theory can result in the discussion of law autonomy of the platforms, that is not relevant now [33].

Currently, contractual relationships of digital platforms directly depend on the area and nature of their activities. If they are companies that provide information services, then contracts are concluded for providing services with customers, and agency; mixed contracts are made with companies that perform work or provide services. As our analysis showed, these agreements are concluded 90% of Russian digital platforms. At the same time, modern legislation uses the principle of contract freedom and create various legal structures, provided that they do not establish enslaving conditions and do not violate the rights and legitimate interests of parties to the agreement or third parties.

In this regard, the case of Craig Comb v. PayPal, Inc. is of great interest. The court recognized the PayPal user agreement as unfair under the laws of the State of California on the basis that it was concluded in the form of an adhesion agreement, i.e. "standard contract", drawn up and imposed by the stronger party in the transaction, when the other party can just join it [6].

Currently, there are similar PayPal provisions in user agreements of the largest American (Amazon, eBay, Walmart) and European corporations. And this means that this judicial precedent can become global. The user agreements of digital platforms as new sources of law should be emphasized. Nowadays, a new subject of digital civil circulation is being formed – digital technology platforms (digital corporations). Many of them become transnational, and internal regulatory standards are used to optimize their work [3]. It is presumed that the user, choosing a platform, agrees with the rules described in the user agreement. Platforms moderate user behavior within the "on/off" model. Sanctions are usually applied by system administrators (banishment).

The establishment of rules and responsibility for their non-compliance is a form of rule-making, which is not regulated by national law at all. In fact, there is one more parallel regulatory environment (digital platform law [22]), which forms individual rules of behavior and business traditions [7, 19, 30].

The contradictions in the legal status of digital platforms also affected the determination of the legal parameters of the relations between the platforms and their employees. The courts are proceeding to establish their position, but this position is no longer unified. For example, on March 4, 2020, the Cour de Cassation (the highest court of France) recognized that there is an employee-employer relations between Uber and the driver who operated under the platform [4]. In the future, the labor court will resolve the issue if this former employee is entitled to compensation under the rules of labor law [10]. In 2017, Uber was deprived of a license to work in London, because it did not fulfill the obligation to report crimes and check the convictions of its drivers [18]. A year earlier, an English court recognized that there is a full-fledged labor relations between Uber drivers and Uber, which entitles drivers to paid weekends and a minimum wage [34, 35]. As for other countries, the issue of labor relations on digital platforms remains open. In particular, in Russia, drivers of transport aggregators are mostly attached to taxi companies and only use the platforms' services.

5 Conclusion

The paper considers the main issues in the field of legal regulation of digital platforms. They clearly show the obvious deficit of legal norms in this sphere of relations. Due to the uncertainty of its legal status, digital platforms are forced to lower the threshold for monitoring the work of employees and companies, which inevitably affects the quality of goods and services. The lack of uniform standards of legal regulation have to be taken into account as it leads to chaotic rule-making of companies. User agreements withdraw some of the contradictions in terms of specifying the legal status of aggregators, determining the contractual basis for their work, protecting the rights of users, etc., But there are contradictory issues related to determining the scope of responsible persons and applicable law. The internal law of the platforms undermines the law power, which in the future can adversely affect the development of transnational aggregators. While supporting the proposal of international organizations regarding the legal regulation of digital platforms, the prioritizing tasks have to be emphasized. First of all, it is crucial to develop global criteria for differentiating platforms depending on their legal status and approve common principles – basic requirements for protecting the rights of platforms' consumers.

References

1. Appeal ruling of the Krasnodar Regional Court of December 19, 2017 in the case N 33-41539/2017
2. Calliess, G.-P.: Transnational consumer law; Co-regulation of B2C-e-commerce. In: Dilling, O., Herberg, M., Winter, G. (eds.) Responsible Business: Selfgovernance in Transnational Economic Transactions, pp. 225–258. Hart Publishing, Oxford (2008)
3. Cohen, J.E.: Law for the platform economy. Univ. California **51**(133), 133–204 (2017)
4. Cour de Cassation: Arrêt n°374 du 4 mars 2020/ruling/sentencia (2020). https://www.courdecassation.fr/jurisprudence_2/communiques_presse_8004/prestation_chauffeur_9665/374_4_44528.html. Accessed 23 Apr 2020
5. Court of Justice of the European Union Press release N 136/17 Luxembourg, 20 December 2017. Judgment in Case Asociacion Profesional Elite Taxi v. Uber Systems Spain SL. https://curia.europa.eu/jcms/upload/docs/application/pdf/2017-12/cp170136en.pdf. Accessed 23 Apr 2020
6. Craig Comb, v. PayPal, Inc. Cases N C-02-1227 and C-02-2777 JF (N. D. Cal., August 30, 2002). https://casetext.com/case/comb-v-paypal-inc-2. Accessed 23 Apr 2020
7. Dasteel, J.H.: Consumer click arbitration: a review of online consumer arbitration agreements. Arbitr. Law Rev. **9**(1) (2017). https://elibrary.law.psu.edu/arbitrationlawreview/vol9/iss1/18/. Accessed 23 Apr 2020
8. Decree of the Plenum of the Supreme Court of the Russian Federation of June 26, 2018 No. 26 "On some issues of the application of the legislation on the contract for goods carriage, passengers and baggage by road and on the contract of transport expedition". http://www.consultant.ru/document/cons_doc_LAW_301030/. Accessed 23 Apr 2020
9. Deryugina, T.V.: The legal nature of the contract mediating the emergence of legal relations with the participation of the aggregator. Civil Law **6**(5), 3–6 (2018)
10. Dillet, R.: Uber driver reclassified as employee in France (2020). https://techcrunch.com/2020/03/04/uber-driver-reclassified-as-employee-in-france/. Accessed 23 Apr 2020

11. EU: Directive 2000/31/EC of the European Parliament and of the Council of 8 June 2000 on certain legal aspects of information society services, in particular electronic commerce, in the Internal Market ('Directive on electronic commerce') (2000). https://eur-lex.europa.eu/legal-content/EN/ALL/?uri=CELEX%3A32000L0031. Accessed 23 Apr 2020
12. EU: Directive 2006/123/EC of the European Parliament and of the Council of 12 December 2006 on services in the internal market (2006). https://eur-lex.europa.eu/legal-content/EN/TXT/?uri=celex%3A32006L0123. Accessed 23 Apr 2020
13. EU: Online platforms and the digital single market opportunities and challenges for Europe (2016). https://eur-lex.europa.eu/. Accessed 23 Apr 2020
14. European Commission: Online platforms and the digital single market (2019). https://publications.parliament.uk/pa/ld201516/ldselect/ldeucom/129/129.pdf. Accessed 23 Apr 2020
15. Euronews: What does the European Commission do? (2019). https://www.euronews.com/2019/04/26/what-does-the-european-commission-do. Accessed 23 Apr 2020
16. European Parliament: Digital platforms: An analytical framework. Challenges for competition in a digitalised economy (2015). https://www.europarl.europa.eu/RegData/etudes/IDAN/2015/563432/IPOL_IDA(2015)563432_EN.pdf. Accessed 23 Apr 2020
17. Evans, P.C., Gawer, A.: The rise of the platform enterprise (2016). https://www.thecge.net/app/uploads/2016/01/PDF-WEB-Platform-Survey_01_12.pdf. Accessed 23 Apr 2020
18. Federal Law of the Russian Federation of July 27, 2018 No. 250-FZ «On Amending the Law of the Russian Federation "On Protection of Consumer Rights"». https://rg.ru/2018/08/01/potrebiteli-dok.html. Accessed 23 Apr 2020
19. González, A.G.: eBay law: the legal implications of the C2C electronic commerce model. Comput. Law Secur. Rep. **19**(6), 468–473 (2003)
20. InfoCuria: Judgment of the court. http://curia.europa.eu/juris/document/document.jsf?text=&docid=221791&pageIndex=0&doclang=EN&mode=lst&dir=&occ=first&part=1&cid=4084659. Accessed 23 Apr 2020
21. Ivanov, A.A.: Business aggregators and law. Law **5**, 145–156 (2017)
22. Kartskhia, A.A.: Digital technological (online) platforms: Russian and foreign regulatory experience. Civil Law 3 (2019). https://urfac.ru/?p=2426. Accessed 23 Apr 2020
23. Newman, K.T., Murphy, G.R.: Oberdorf v. Amazon.com, Inc. (2019). https://www.martindale.com/legal-news/article_thomas-thomas-hafer-llp_2519310.htm. Accessed 23 Apr 2020
24. OECD: OECD digital economy outlook 2017 (2017). https://espas.secure.europarl.europa.eu/orbis/sites/default/files/generated/document/en/9317011e.pdf. Accessed 23 Apr 2020
25. OECD: Toolkit for protecting digital consumers (2018). https://www.oecd.org/internet/consumer/toolkit-for-protecting-digital-consumers.pdf. Accessed 23 Apr 2020
26. OECD: The role of digital platforms in the collection of VAT/GST on online sales (2019). www.oecd.org/tax/consumption/the-role-of-digital-platforms-in-the-collection-of-vat-gst-on-online-sales.pdf. Accessed 23 Apr 2020
27. Poyton, D.: What makes Uber and Airbnb different in the eyes of the EU – And why it matters (2019). https://theconversation.com/what-makes-uber-and-airbnb-different-in-the-eyes-of-the-eu-and-why-it-matters-121708. Accessed 23 Apr 2020
28. Rostelecom: Approaches to defining and typifying digital platforms (2019). https://files.data-economy.ru/digital_platforms.pdf. Accessed 23 Apr 2020
29. Rozhkova, M., Kuznetsova, L.: On the responsibility of aggregators in the context of understanding their activities as agency services. Econ. Law **1**, 27–34 (2019)
30. Schiek, D.: Private rule-making and European governance – issues of legitimacy. Eur. Law Rev. **20**(1), 3 (2007)

31. Soifer, T.V.: Civil law regulation of relations on the collective use of goods and services (Sharing economy): Development directions (2019). https://urfac.ru/?p=2752. Accessed 23 Apr 2020
32. Strowel, A., Vergote, W.: Digital platforms: to regulate or not to regulate? Message to regulators: fix the economics first, then focus on the right regulation (2016). https://ec.europa.eu/information_society/newsroom/image/document/2016-7/uclouvain_et_universit_saint_louis_14044.pdf. Accessed 23 Apr 2020
33. Symeonides, S.C.: Party autonomy and private-law making in private international law: The Lex Mercatoria that isn't (2006). http://dx.doi.org/10.2139/ssrn.946007
34. The Guardian: Uber loses appeal over driver employment rights (2018). https://www.theguardian.com/technology/2018/dec/19/uber-loses-appeal-over-driver-employment-rights. Accessed 23 Apr 2020
35. The Guardian Weekly: Factbox: Uber's legal challenges around the world (2016). https://www.theguardian.com/technology/2016/oct/28/uber-uk-tribunal-self-employed-status. Accessed 23 Apr 2020
36. UNCTAD: Value creation and capture: Implications for developing countries (2019). https://unctad.org/en/PublicationsLibrary/der2019_overview_ru.pdf. Accessed 23 Apr 2020
37. Wolfer, M.: And you are not done; The future of products liability claims against Amazon still uncertain (2019). https://www.jdsupra.com/legalnews/and-you-are-not-done-the-future-of-41953/. Accessed 23 Apr 2020

Development of an Innovative Project to Provide Cosmetic Services on the Move

O. V. Astafeva(✉), E. R. Grehova, and I. A. Osipova

Financial University under the Government of the Russian Federation, Moscow, Russia
astafeva86@mail.ru, grehovakate2001@yandex.ru, saynotogmp@yahoo.com

Abstract. Quite often, due to the lack of time, people face the choice between business and beauty. It can be very difficult to make this choice regardless of the status and position of the person. Therefore, it seems relevant to develop an innovative project to provide a unique proposal for providing cosmetic services on the move. In this regard, this article presents some provisions of the market study of cosmetic and transport services in the metropolis. Beauty-taxi is a unique innovative project that allows active people to become more beautiful in the process of moving around the city in the car, and using time rationally. It seems that this proposal can compete with traditional taxis and represent a certain interest for business people, whose activities make it incumbent on them not only to solve a large number of business problems and actively move around the city, but also to look well-groomed, tidy and stylish. In the development of the innovative project, the methodological basis was the study of trends and patterns of the development of the cosmetic market, including the sale of cosmetics and the provision of care services, as well as research of key participants in the taxi market.

Keywords: Beauty-taxi · Cosmetic services · Taxi market

1 Introduction

Each stage of human development is conditioned by certain stereotypes, ideals and standards of beauty to which everyone aspires. Beauty criteria are changing at an unthinkable pace. Beauty and fashion are very popular spheres, so not only women, but men as well watch and follow the modern fashionable trends. Manicure, pedicure, hairdressing services, care products, make-up, cosmetic services, stylish clothes – these are integral elements of modern people. Nowadays, any modern person before any event can make arrangements and sign up for the salon, hire a stylist to create the appropriate image, sign up for manicure and pedicure, but to do all this you need to find time, which in today's world one lacks badly. The cosmetic market is deluged with a large number of studios and salons that are trying to win the attention of customers through various services and offers, but, to a large extent, due to the high competition it is quite difficult to form a unique selling point. A lot of scientific papers are devoted to the study of these questions, including the articles by Gudyreva [3], Laskina [4],

S. I. Ashmarina et al. (Eds.): ENGINEERING ECONOMICS WEEK 2020, LNNS 139, pp. 419–427, 2021.
https://doi.org/10.1007/978-3-030-53277-2_50

Petrova, Rezaikina, & Rotanov [5], Bajenova, Danilova, & Chetoshnikova [1]. However, this article, which is based on an analysis of taxi and cosmetic services markets, attempts to describe the key provisions of an innovative project to provide cosmetic services on the move for the inhabitants of the metropolis, the project is supposed to contribute to the effective solution to many problems of modern business people and the formation of a new market niche.

2 Methodology

The beauty industry is an industry that is rapidly gaining momentum in its development, which includes a wide range of activities. In recent decades, the consumption of cosmetic products has been actively increasing. In turn, the range of goods offered by the market is huge, and manufacturers are increasing production every year. Russia occupies about 2% of the global market for cosmetics and perfumes. Studies show that the average woman in Russia consumes much more decorative cosmetics than the average woman in Europe. Despite the decrease in incomes of the population during the economic downturns, the market of cosmetic products in Russia continues to grow, albeit at a slower pace. Despite the fact that the average receipt amount is getting smaller, the frequency of purchases does not decrease, at least because people get accustomed to the use of a certain product and need it on a permanent basis. The most preferred products for which the consumers are willing to pay, according to Euromonitor International are facial care, perfumes and decorative cosmetics.

A new trend in the beauty industry is the production of organic products, it has become important to buyers that the composition should consist of natural components. Also in today's world, the market benchmarks for product creation are versatility and immediate effect. This is due to the change in the rhythm of life in the society. Nowadays people have a lot to do and to worry about every day and in order to save time they want to acquire such products that would have an instant effect, and would combine several functions and would be compact at the same time. The development of the cosmetics market is also evidenced by the active use of digital technologies, for example, as the ways to promote goods and services manufacturers use websites, mobile applications, advertising in various social networks, which are gaining more and more popularity.

According to many years of practice, in order to increase the customer base and reduce the risk of bankruptcy, it is necessary to offer a set of services designed for the average family, that is, to adhere to the traditional format, without divisions by gender and age factors. Nowadays, the "express" salons are gaining momentum, which are focused on accelerated mode, and services are provided simultaneously by several masters to save the client's time.

Since the most suitable transport for providing cosmetic services on the move is the taxi, it is necessary to investigate the features of the development of the taxi market. This means of transport has a lot of advantages over others. First, if a person is in an unfamiliar place or does not know how to get to the destination, a taxi is the best solution, as you only need to know the address to get around. Secondly, in comparison with any other public transport taxi has more comfortable conditions, there are no crowds of people or stuffiness and suffocation, and you also do not need to make transfers and walk, because taxi will deliver you to the destination. And thirdly, taxi rides allow you to do the necessary things on the way, which brings significant time savings. It is the latter advantage that plays a fundamental role in the choice of transport for the provision of cosmetic services on the move. The Russian taxi market is developing quite actively, as evidenced by the volume growth of the market (Fig. 1).

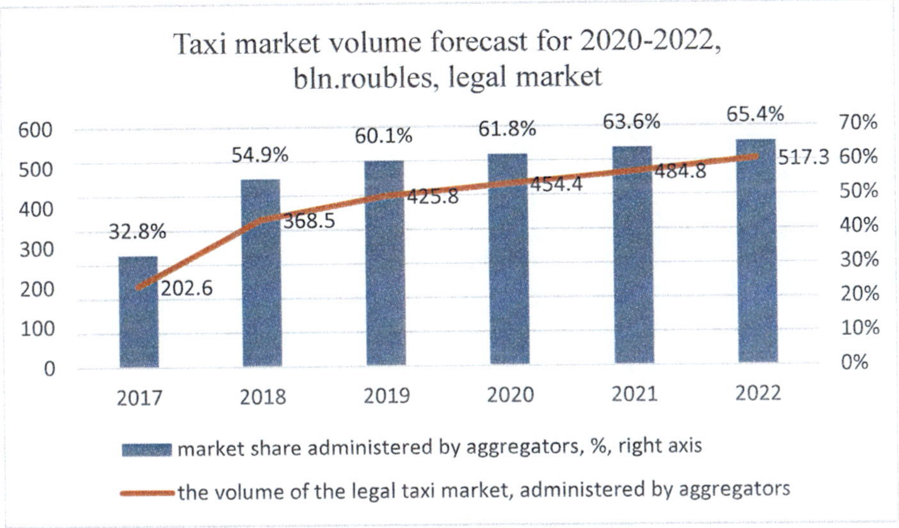

Fig. 1. Taxi market volume forecast for 2020–2022, bln.roubles, legal market (Source: authors based on [7]).

In general, people have started to use taxis more often than before, this is due to the establishment of fair prices and the emergence of new aggregators. The market has become popular. The increase in the number of competitors encourages the search for new advantages, the creation of various bonus systems and convenient online services for the implementation of orders. The main indicators of work quality of the aggregators in the taxi market are digitalization, safety and comfort. In other words, in order to enter and maintain a position in the market, it is necessary to establish acceptable prices for services which are similar to the prices of other aggregators, to develop a unique mobile application for orders, to reduce the time of delivery of the car and to provide a safe and comfortable trip, which will provoke an increase in demand.

In general, according to the report of Discovery Research Group at the moment in the structure of the legal Moscow market the key positions are occupied by Yandex. Taxi, Vezet and Maxim (Fig. 2) [7].

These aggregators occupy the lion's share not only of the capital, but also of the regional market. Taxi Vezet and Maxim are widely known among consumers in cities with a population from 100,000 up to 500,000 people, but in cities with the population of over 500,000 people the leader is Yandex.Taxi.

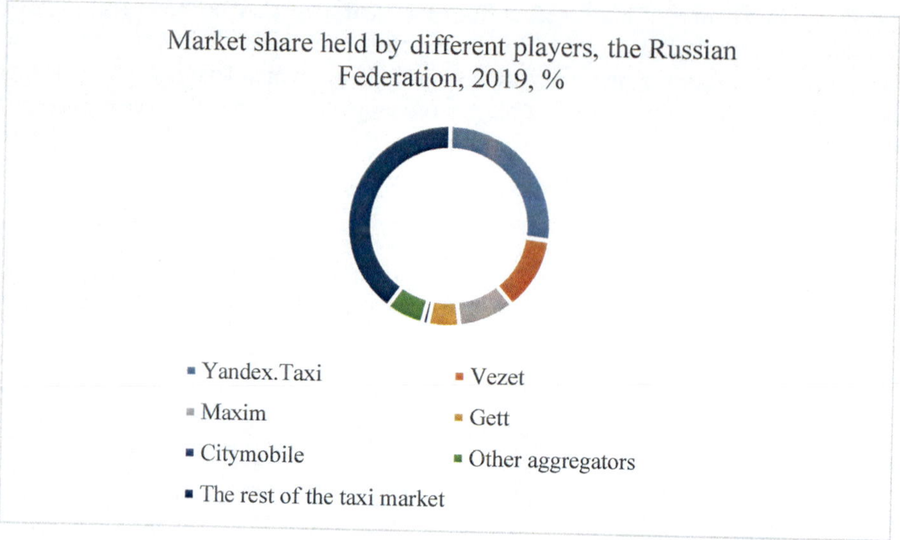

Fig. 2. Market share held by different players, the Russian Federation, 2019, % (Source: authors based on [7]).

Having read the customer reviews of Yandex.Taxi, you can identify the following main positive aspects that must be taken into account when ensuring the quality of the new beauty-taxi:

- fast car delivery;
- democratic prices in comparison with other taxis;
- the ability to call the driver or write in a chat;
- the ability to track the way on the map;
- fast support service;
- availability of additional services.

The main drawback – inadequate taxi drivers. The main claim of the customers – boorish, strange behavior of the drivers, the delivery of another car instead of the promised one. Yandex.Taxi should organize more thorough criteria for selecting drivers not only on high-end tariffs, but also in the economy sector, if they value their customers and want to increase their number.

3 Results

In megacities modern people live in a tight schedule, at a dynamic pace of life people are busy with a lot of things and try to use their time wisely and efficiently, which refers to looking after their appearance as well. As a result, quite a lot of people find it challenging to organise their schedule in such a way that it is possible to combine business and skin and bodycare rationally using time. Therefore, this challenge could be solved with the help of the proposed business idea of providing cosmetic services on the move. The concept of the idea is to develop beauty-taxis in megacities, which will allow not only to get to the desired destination within a fixed period of time, but also to use the provided cosmetic services, which will significantly save time and improve the emotional state of the client.

It is worth noting that despite the non-typical conditions of the cosmetic services provision, beauty-taxis can provide a fairly wide range of care services. The target audience of the beauty-taxis may include the representatives of the middle and upper-middle class. These services can be in demand among business people with tight schedules, women with children, tourist visitors to the capital or flight attendants/conductors, who are forced to go to meetings or events immediately after the airports or stations. Women are the predominant customers of the service, but men can also occasionally call a beauty taxi.

In order to test the business idea of providing cosmetic services on the move regarding its relevance, you need to evaluate it against certain criteria. The authors have presented the customer value proposition (CVP) of the "Voyage de beaute" project using the sketch by Peter Thompson – a model that best allows you to assess the needs of customers and provide the information in a structured way (Fig. 3).

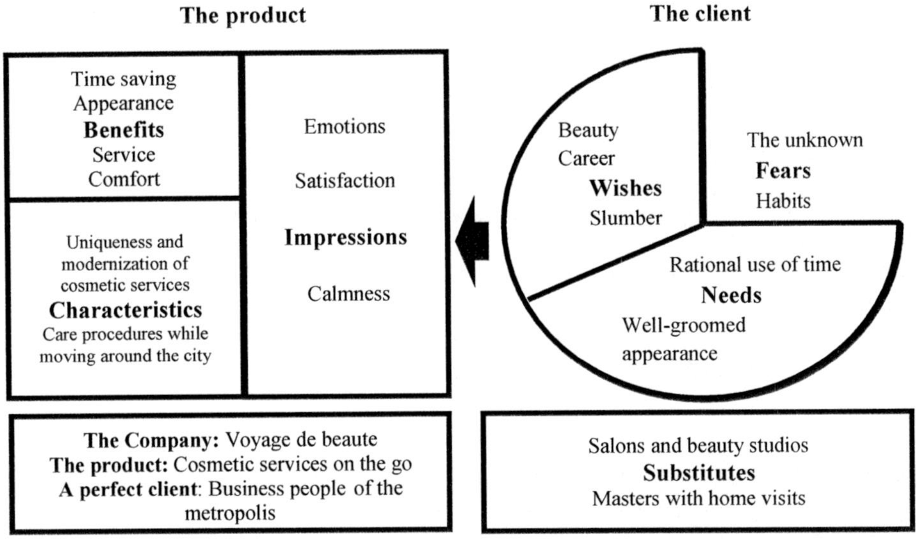

Fig. 3. Value Proposal of the Innovative Project «Voyage de beaute» (Source: authors).

Characteristics

«Voyage de beaute» – is a unique project to provide cosmetic services on the go. The key idea of the project is the synthesis and modernization of the cosmetic and taxi services that are already familiar to everyone. The main task is to solve the current problem of society, namely the answer to the question "How can one look great without compromising the planned business schedule?". Beauty-taxi allows you to save the most valuable resource of today – the time, as the customers can simultaneously engage in at least three activities: moving to the desired destination, getting cosmetic or care services and, for example, doing business with the help of electronic devices, or they can simply relax and enjoy the beauty ride.

Benefits

- Time-saving
- Great looks
- Opportunity to combine several things at the same time
- Ride in a cozy and comfortable car
- Unusual pastime in Moscow traffic jams
- Express service, when there is no time at all
- A great gift to someone
- Relaxing services after a hard day or flights
- A quality service and an individual approach
- Learning games for children

Impressions
The customers who will use the offered services will get unforgettable emotions and impressions, because never before have they beautified themselves while looking at the changing streets of the city from the windows of the car, never before could they transform their appearance during a trip to the events. People will feel a kind of relief and satisfaction, because one of their problems gets solved. They will become calmer, as their tight schedule will provide them with some free time.

Wishes
Modern people want to look 100% every day, achieve success at work and move up the career ladder, at the same time everyone dreams of waking up later in the mornings without spending time on grooming themselves. But due to their schedule not everyone can afford to go to the salon, as it is a very time-consuming process. Therefore, new ways to mix business with pleasure that is the personal care or beauty routine are required nowadays.

Needs

- Looking well-groomed and beautiful in a busy schedule,
- Spending time wisely and efficiently
- Looking after one's appearance not at the expense of everyday business
- Making oneself look presentable after the airport/station

Fears
The fundamental fear of all consumers is the fear of the unknown. Changing any familiar product to a new product is always associated with the so-called "switching pain". In this case, there is a transition not just to a similar product of a different manufacturer, but to something that previously was not presented on the market and no one has heard about it. People are inherent in conservatism, they get accustomed to what they use on a regular basis, and to take a step into the unknown and try something new is sometimes difficult, so it takes time for people to see the benefits of the new offers before they start using them.

Substitutes
At this stage, for the purpose of cosmetic services, people can turn to salons and beauty studios, which abound, they can also invite a home salon master, but these substitutes do not meet the current needs, which are mentioned above, so the project "Voyage de beaute" is a really valuable proposition.

Advertising is a potential advantage of "Voyage de beaute", provided the services are of a high quality. In addition to the fact that the taxi is a new direction in the market of taxis and cosmetic services very few people know about, people are also incredulous about everything new, so they need time to comprehend how this proposal will simplify their lives. To speed up this process, you need a well-built quality advertising that will reflect the whole concept of the idea and will arouse interest in the audience. In order to attract customers, it is advisable to use the following promotion tools:

- websites with modern design and layout technology, where you can book a taxi, see samples of work and compare prices. In order to better understand what a beauty taxi is, you need to attach a commercial to the web page. Since this video is designed to introduce the general idea to the people, it should be more conceptual, dynamic, informative, vivid and memorable;
- endorsement from celebrities is a common way to form an audience, as they have a large number of subscribers and people trust them;
- one's own account in the popular network Instagram.

4 Discussion

In order to assess the potential and threats that affect the success of an innovative project, to formulate the strengths and errors and shortcomings that a company may face, it is advisable to conduct a SWOT analysis – a classic tool of strategic management. It reveals internal and external factors that influence the success of the business. The results of the SWOT analysis of "Voyage de beaute" project are presented in the consolidated summary Table 1.

The final step in the development of a business strategy is to consider the paired ratios of factors previously presented in the matrix.

Strengths + Opportunities. Taken together, these items provide active business development and increase the solvent consumer demand. "Voyage de beaute" helps make people's lives easier while they benefit at the same time.

Weaknesses + Opportunities. Listed opportunities will reduce the number of weaknesses. Fashion and rising incomes will help solve the problem of payback and the inconstancy of cash flows. The absence of a large number of competitors will help to gain a significant share of the market.

Weaknesses + Threats. The combination of these factors makes it difficult to develop the business, leading it to losses. In order to minimize risks, you can use well-built advertising to increase market share, thereby resisting the emergence of new competitors, as well as forming your own competitive advantages.

Strengths + Threats. Thanks to the strengths, the project will be able to minimize the negative impact of the external factors. High-quality promotion and endorsement

Table 1. SWOT-analysis of "Voyage de beaute" project

Strengths	Weaknesses
– Unique offer/selling point – High-quality service – Qualified personnel – Well built marketing – Variety of cosmetic services – Comfortable car with electricity, light music, TV and comfortable chairs – A glass of champagne as a compliment – Qualified masters – Prime high-quality cosmetics – 24/7 work schedule – Celebrity endorsement –Sanification	– Complexity of the car equipment – Weak market representation – The inconstancy of cash flows – Relatively long payback period
Opportunities	**Threats**
– The absence of a large number of competitors – Creating separate vehicle lanes in the city for public transport – Increasing incomes of the population – Fashion for the provision of services in the car	– Lack of solvent demand – The emergence of new competitors and the improvement of the competitor's advertising campaign – Legislative ban on the provision of cosmetic services in the car – Tightening of sanitary standards

Source: authors.

can reduce the threat from competitors, and the continuous improvement in the quality of services will help to comply with all existing sanitary standards, including when they are tightened or new regulations are put in place.

5 Conclusion

The market for cosmetic services, as well as the taxi market in Moscow is actively developing. They are characterized by a huge number of sellers and buyers. With the emergence of new entrants in these areas of activity, we have to come up with new ways to attract and retain the customer's attention, develop quality advertising campaigns, expand services and surprise with unusual offers [2, 6, 8]. Despite the fact that the market of cosmetic services is overflowing with various goods and services, competitors are trying to surprise customers with new beauty services and design solutions, but the modern customer needs the quality of service, so that cosmetic services are mobile and accessible. The main idea of the innovative beauty-taxi project is to modernize the usual format of beauty salons and to implement them in the area of taxis at the convenience and enhancing life standards of business people. All this can generally contribute to the development of a new market niche, which could prove interesting for many residents of the metropolis.

References

1. Bajenova, S.A., Danilova, L.G., Chetoshnikova, L.A.: The marketing of anti-aging services in aesthetic medicine. Problems Soc. Hyg. Health Care Med. Hist. **27**(4), 424–428 (2019). https://doi.org/10.32687/0869-866X-2019-27-4-424-428
2. Faggio, G., Silva, O.: Self-employment and entrepreneurship in urban and rural labour markets. J. Urban Econ. **84**, 67–85 (2014). https://doi.org/10.1016/j.jue.2014.09.001
3. Gudyreva, L.V.: Key aspects of marketing communication policy in shaping a sustainable positive image of the cosmetics market. Eur. Soc. Sci. J. **2–2**, 7–14 (2018)
4. Laskina, O.S.: Strategic marketing of innovations in the cosmetic market. Bull. Voronezh State Univ. Eng. Technol. **1**(67), 277–281 (2016)
5. Petrova, N.P., Rezaikina, A.V., Rotanov, S.V.: System of independent evaluation (confirmation) of correspondence of perfumes and cosmetics in the Russian Federation. Herald Dermatol. Venereol. **6**, 90–96 (2015)
6. Razomasova, E., Romanova, N.: Lack of free time of citizens as an incentive for entrepreneurship in services. Ideas Ideals **11**(3), 303–316 (2019). https://doi.org/10.17212/2075-0862-2019-11.3.2-303-316
7. The Analytical Center under the Russian Government: Study of the taxi market (2019). https://ac.gov.ru/files/content/24166/rynok-taksi-2019-pdf.pdf. Accessed 14 Apr 2020
8. Tripon, A.: Requirements on management practices for the development of entrepreneurial creativity in a global social-ecological system – case study. Procedia Technol. **19**, 483–489 (2015)

Big Data and Big Challenges in Production Networks and Systems

Big Data Technology Application in the Taxation Sphere

O. L. Mikhaleva[1(✉)], D. V. Syradoev[2], and T. A. Terekhova[3]

[1] Samara State University of Economics, Samara, Russia
Mikhaleva2007@yandex.ru
[2] Kazan Federal University, Kazan, Russia
sdv377@mail.ru
[3] Private Educational Institution of Higher Education "Institute of Social and Humanitarian Knowledge", Kazan, Russia
tta377@mail.ru

Abstract. In the context of the economy digitalization, modern technologies penetrate into all spheres of human activity. The sphere of taxation is no exception. Digital technologies allow processing and transmitting large amounts of information, thereby reducing the risk of errors. The use of information technologies contributes to the creation of a transparent system of relations between the state and society. A new form of interaction between tax authorities of different countries is being formed by automatic exchange of tax information. The article investigates the use of Big Data technologies in the field of taxation. Opening up new potential opportunities, the use of digital technologies, at the same time, creates certain difficulties in the process of tax administration. The practical significance of this research work is to assess consequences of the Big Data technology application for tax authorities and taxpayers.

Keywords: Automated control system · Big data · Digital technologies · Digitalization · Tax control

1 Introduction

Today, we can see that digital technologies are being actively implemented in all spheres of activity. Digitalization is making significant changes in the field of taxation. In all countries of the world, there is a process of digital technologies application in tax relations. In the field of taxation, digitalization changes the current legislation, tax authorities use modern software in their activities, there is an increase in tax revenues to the budget, and a decrease in the tax evasion. The use of Big Data processing technology in the tax sphere is pushing for a revision of the existing information interaction between the state and society. In the future, the application of Big Data technologies will expand possibilities of tax administration.

S. I. Ashmarina et al. (Eds.): ENGINEERING ECONOMICS WEEK 2020, LNNS 139, pp. 431–436, 2021.
https://doi.org/10.1007/978-3-030-53277-2_51

2 Methodology

When studying issues in the field of taxation, we pay special attention to the development of empirical research and the application of theory to practical solutions. Applied research is determined by the interest of participants in tax relations. The authors of this research examine the role of such advanced technologies as Big Data. In particular, a comparative analysis is carried out, which allows structuring the information and determining a direction of further research. In the practical aspect, significant problems were identified, further study of which is possible with the help of theoretical tools. The development of applied research, in turn, requires a variety of methods. During this research, the authors used such methods as theoretical (dialectical logic, rational cognition, etc.); diagnostic (diagnostic analysis of the state and causes); empirical ones (description of facts, measurement and generalization of research results).

3 Results

Digitalization makes it necessary to review some approaches to taxation. In modern conditions of a wide use of new digital technologies, the format of interaction between participants in tax relations is changing. The tax authorities are faced with the task of using, processing and analyzing large amounts of information. In this regard, new ways and methods are emerging for processing, storing and analyzing data sets using automated information systems.

The use of Big Data technology has conceptually changed the approach to conducting control checks, minimizing the impact of the human factor. Big Data technologies allow tax services to quickly get the necessary information about a taxpayer, allowing them to have all the relevant data about tax objects. Due to Big Data technologies, the Federal Tax Service of Russia will be able to improve its work tools, which will facilitate the transition to contactless service mode and enable to create personalized services. In the future, digital technologies will increase the level of trust between the state and society, ensure the transparency of the economy and guarantee the stability of state revenues.

4 Discussion

There is no official definition of digitalization, although this term is widely used in various sources to describe transformations in all spheres of public life. The term "digitalization" came into use relatively recently – at the turn of the XX–XXI centuries, although the spread of digital technologies began a little earlier – in the 70 s of the XX century with the development of computer technology [6]. According to Afanasenko, Borisova [1] digitalization represents a process of adopting a new form of data submission, the growth in the use of digital or computer technologies in an individual enterprise, in an industry, or in a country as a whole. Yudina and Tushkanov consider digitalization in a narrow and broad sense of the word:

- in a narrow sense, it is a creation of information and digital platforms and operators at different levels of the economy (global, mega, macro, meso, micro, nano) that allow solving various economic tasks, including strategic ones: the development of medicine, science, education, transport, new industrialization, state regulation of the economy and planning, etc.;
- in a broad sense, it means changing the production nature or economic relations, changing their subject-object orientation. With the help of algorithms, machine-to-machine (M2M) relationships appear, where a person may no longer act as a subject. The productive forces of the society and/or factors of production are also changing. This is how the Internet of Things and, even the Internet of "Everything", arises [15]. Marei considers digitalization as a change in the paradigm of communication and interaction with each other and society [3]. Vartanova, Vyrkovsky, Smirnov, Makeenko clarify the content of this concept – it is not only the transition of information into digital form, but a complex solution of infrastructure, management, behavioral, and cultural character [14].

Digitalization is a modern stage in the development of informatization, characterized by the predominant use of digital technologies for generating, processing, transmitting, storing and visualizing information, which is determined by the emergence and spread of new technical tools and software solutions [11]. Analyzing the various approaches to defining the term "digitalization", it should be noted that this process includes the use of digital technologies. Thus, digitalization can be understood as a process of implementing digital technologies in all areas of human activity, changing the approach to processing, storing and using information.

According to the digitalization level of the Russian tax system, it ranks first in the digital tax administration. The use of Big Data technologies allows tax authorities to get more benefit from existing data, fight tax evasion and fraud. Digitalization also raises new, fundamental questions, such as privacy protection, the future of work, cybersecurity, the market power of digital platforms, who has access to data, what information is reliable, how new technologies can be used ethically, and how to ensure that people and companies can keep up with this transformation [4]. Digitalization and integration of data from all sources of information into a single information space with subsequent analysis based on the use of Big Data technologies will improve and increase the work efficiency of tax authorities.

Big Data refers to the use of heavy-duty computers and high-tech software to collect, process, and analyze huge amounts of data with rapid changeability. The term Big Data was first mentioned by the editor of the journal "Nature", who prepared a special issue of the journal with the theme: how can technologies that open up opportunities to work with large amounts of data affect the future of science? [10]. According to him, new tools and more advanced technologies will help to master the growth of world information. This technology, radically expanding sources of information available to the organization in the online mode, as well as its collection, storage, processing and analysis, turns into a valuable and unique resource in its characteristics. Cloud storage will also be developed further, which will provide easy access to a huge amount of information and contributes to more effective communication between employees [9].

In 2012, IBM published a report called "Big Data Brings Marketing Big Numbers" [8] with a number of indicators that characterize the current state of Big Data: 5 exabytes are generated in the world, every 2 days, and 2.7 zetabytes are in the world in digital form. In the IBM forecast for 2020, there are figures about 35 zetabytes of information accumulated by mankind and stored in digital form.

In 2017, the Federal Tax Service of Russia conducted a research to study the level of digitalization of tax systems in different countries and their success in managing Big Data [2]. 26 tax authorities of countries that are members of the OECD forum on tax administration (FTA) advisory group were interviewed. Most of the countries surveyed understand the importance of Big Data and its management. The more countries are adopting new technologies, the better the taxpayers feel. This increases taxpayer transparency and reduces tax uncertainty and administrative burden.

A system of data processing centers (DPC) has been created in Russia to process and store a huge amount of data. A federal data warehouse, which is formed in the DPC, allows you to compare and analyze any tax data. This greatly complicates attempts to use illegal schemes of evasion from payment of taxes. Big Data technologies allow taking a fresh look at the risk assessment process. If you previously had to use information about past events and conduct a long-term analysis, then using Big Data, risk assessment can be performed automatically in real time using all the incoming information that can affect the situation. In turn, tax authorities can use the constantly received data to improve their activities and interaction with taxpayers.

The advantage of Big Data is its almost unlimited potential for storing information, as well as its speed [12]. As defined by Tapscott and Tapscott, "... blockchain is an eternal digital distributed economic transaction log that can be programmed to record not only financial transactions, but almost everything that has value" [13, p. 201]. The use of Big Data ensures transparency of tax processes, allowing you to track VAT transactions and tax refunds, tax returns. The use of Big Data analytical tools by tax authorities allows controlling VAT accrual and payment. The automated control system of the VAT-2 analyzes a huge array of taxpayer data according to the established risk criteria. This makes it possible to detect and prevent tax evasion or attempts to recover it through illegal schemes. For 2019, the Federal Tax Service reduced the share of suspicious transactions within the system of VAT-2 to 0.47%. This indicator is one of the best in the world [5].

The Federal Tax Service of Russia provides taxpayers with secure access to personal tax information, as well as services, regardless of the taxpayers' location. Now through their personal accounts, taxpayers have the opportunity to communicate contactless with the tax authorities from anywhere in 24/7 mode. Switching to the remote interaction between tax authorities and taxpayers allows abandoning the printed form of documents. The Federal Tax Service of Russia has joined the Common Data Transmission System (CTS) of the OECD, which will technically ensure the automatic exchange of information with foreign partners. The information obtained via the CTS will allow detecting the presence of undeclared foreign financial accounts and assets, as well as comparing information provided by Russian citizens on controlled foreign companies (CFC).

Despite the large-scale and rapid digitalization of the taxation sphere, the most common problem still remains the lack of stable access to channels of information and communication interaction in individual localities. The application of Big Data technologies in the field of taxation determines the tax risks affecting the information security of participants in tax relations. Goncharenko, Malkova, and Advocatova noted the importance of tax risks for individuals and legal entities in the new digital environment [7]. In this regard, it is necessary to improve the system for protecting user's data, completely abandon foreign software, switching to the cooperation with Russian manufacturers.

5 Conclusion

Recently, the Federal Tax Service of Russia has become a source of information for analytical databases. The breakthrough technologies applied in the field of taxation include Big Data and its real-time analytics. Big Data technologies open up new opportunities for tax authorities not only to analyze and improve the efficiency of tax administration, but also to interact with taxpayers. At the same time, there are technological challenges associated with information processing. As a result, there is a need to modernize the software and develop new analytical tools. The use of Big Data technologies will bring a number of significant effects such as: reducing the shadow sector of the economy, increasing transparency of relations between the state and society, comfortable working conditions and reducing the administrative burden. Effective operation of the tax system in the context of a huge increase in the volume of tax information is possible thanks to the use of Big Data technologies for storing, processing and analyzing an information array.

References

1. Afanasenko, I.D., Borisova, V.V.: Digital Logistics. Peter, Saint Petersburg (2018)
2. Bazanova, E.: The Federal tax service has studied the tax technologies of other countries. Vedomosti (2017). https://www.vedomosti.ru/economics/articles/2017/04/25/687343-fns-tehnologii. Accessed 21 Apr 2020
3. BCG: Digitalization as a paradigm change: Interview with A. Marey (2019). https://www.bcg.com/ru-ru/about/bcg-review/digitalization.aspx. Accessed 23 Apr 2020
4. Dutch Digitalisation Strategy. Ministry of economic affairs and climate policy, 06 June 2018 (2018). http://www.government.nl/documents/reports/2018/06/01/dutch-digitalisation-strategy. Accessed 21 Apr 2020
5. Federal Tax Service: Daniil Yegorov summed up the work of the tax authorities in 2019 (2019). https://www.nalog.ru/rn63/news/activities_fts/9611238/. Accessed 21 Apr 2020
6. Gnatyshina, E.V., Salamatov, A.A.: Digitalization and formation of digital culture: social and educational aspects. Bull. Chelyabinsk State Pedagogical Univ. **8**, 19–24 (2017)
7. Goncharenko, L.I., Malkova, Yu.V., Advocatova, A.S.: Current problems of the tax system in the digital economy. Econ. Taxes Pravo **2**, 166–172 (2018)
8. Karr, D.: Report IBM Big Data brings marketing big numbers (2012). https://martech.zone/ibm-big-data-marketing/. Accessed 21 Apr 2020
9. Larkin, J.: HR digital disruption: the biggest wave of transformation in decades. Strateg. HR Rev. **16**(2), 55–59 (2017)

10. Lynch, C.: How can technologies that open up opportunities to work with large amounts of data affect the future of science? Nature **455**(7209) (2008). https://www.nature.com/nature/volumes/455/issues/7209. Accessed 21 Apr 2020
11. Plotnikov, V.A.: Digitalization of production: theoretical essence and prospects of development in the Russian economy. Izvestiya St. Petersburg State Univ. Econ. **4**(112), 16–24 (2018)
12. Reese, H.: Understanding the differences between AI, machine learning, and deep learning. Tech Republic (2017). https://www.techrepublic.com/article/understanding-the-differences-between-ai-machine-learning-and-deep-learning/. Accessed 21 Apr 2020
13. Tapscott, D., Tapscott, A.: Blockchain Revolution: How the Technology Behind Bitcoin is Changing Money, Business, and the World. Portfolio: Penguin, London (2016)
14. Vartanova, E.L., Vyrkovsky, A.V., Smirnov, S.S., Makeenko, M.I.: Industry in the Russian Media: The Future is Digital: Academic Monograph. Mediamir, Moscow (2017)
15. Yudina, T.N., Tushkanov, I.M.: Digital economy through the prism of economic philosophy and political economy. Philos. Manag. **1**(109), 193–201 (2017)

Identification Ramsay Curve Total Instrumental Variables

D. V. Ivanov(✉)

Samara State University of Economics, Samara, Russia
dvi85@list.ru

Abstract. Logistics curves are widely used in various fields of economics, technology, biology, chemistry. Estimating the parameters of logistic trends from the results of observations of the dynamic process in the economic system, with the aim of the reliable analysis of economic indicators and predicting their future behavior, is one of the main tasks in the economy. One of the logistic models is the Ramsay function. The advantage of this function is the ability to use a linear difference equation to estimate its parameters. At the same time, non-linear data transformations are not required for the logistics functions of Ferhulst or Gompertz. A two-step algorithm based on the method total instrumental variables is proposed for estimating the parameters of the Ramsay curve. To improve the numerical stability of the solution, augmented systems of equations are used. Tests showed that the accuracy of the proposed estimate of parameters is higher than that of the estimate obtained using the least squares (LS) method.

Keywords: Instrumental variables · Logistic curve · Ramsay function · Total least square

1 Introduction

In the mathematical description of dynamic processes in various fields of technology, biology, chemistry, economics, logistic functions (logistic curves, logistic trends, logisticians) are widely used [7, 9, 11]. The problem of increasing the accuracy of the parametric identification of dynamic processes described by logistic functions is the most important problem for applied mathematics, an effective solution of which is possible based on statistical methods for processing the results of observations.

Estimating the parameters of logistic trends from the results of observations of the dynamic process in the economic system, with the aim of the reliable analysis of economic indicators and predicting their future behavior, is one of the main tasks in the economy [4, 8]. Various models are used to describe the logistic dynamics, for example, the Ferhulst (Pearl-Reed) model, the Gompertz function, etc.

S. I. Ashmarina et al. (Eds.): ENGINEERING ECONOMICS WEEK 2020, LNNS 139, pp. 437–442, 2021.
https://doi.org/10.1007/978-3-030-53277-2_52

2 Methodology

One of the logistic models is the Ramsay function [5], which depends on two parameters A and b. The advantage of this function is the ability to use a linear difference equation to estimate its parameters. At the same time, non-linear data transformations are not required as for the logistics functions of Ferhulst or Gompertz:

$$y_i = A(1-(1+b\cdot T_i)\exp(-b\cdot T_i))+\xi_i, \tag{1}$$

where $T_i = i\cdot\Delta t$ and Δt is a sampling interval.

An efficient identification technique for short-range samples is expressing Eq. (1) as a linear difference equation based on Z-transform tools [2, 6]. Since expression (1) contains a constant coefficient, we will write a difference equation with first-order differences for

$$z_i - z_{i-1} = c_1(z_{i-1}-z_{i-2}) + c_2(z_{i-2}-z_{i-3}),$$

$$y_i = z_i + \xi_i, \tag{2}$$

where $c_1 = 2\exp(-b\cdot\Delta t)$ and $c_2 = -0.25c_1^2$.

The generalized error for the model (8) can be written as

$$\varepsilon_i = \xi_i - \xi_{i-1} - c_1(\xi_{i-1}-\xi_{i-2}) - c_2(\xi_{i-2}-\xi_{i-3}) \tag{3}$$

The use of the ordinary least squares method does not allow consistent estimates of parameters to be obtained for Eq. (3) because of the autocorrelation in the generalized error ε_i.

We use the extended instrumental variables method [3, 10] to find estimates of parameters vector $\hat{c}$

$$\left(\frac{1}{N}\sum_{i=1}^{N}\psi_i\varphi_i^T\right)c = \left(\frac{1}{N}\sum_{i=1}^{N}\psi_i y_i\right) \Leftrightarrow \hat{R}_{\psi\varphi}c = \hat{r}_{\psi y}, \tag{4}$$

where

$$\hat{R}_{\psi\varphi} = \tfrac{1}{N}\sum_{i=1}^{N}\psi_i\varphi_i^T,\ \hat{r}_{\psi y} = \tfrac{1}{N}\sum_{i=1}^{N}\psi_i y_i,$$
$$\varphi_i = (\,y_{i-1}-y_{i-2} \quad y_{i-2}-y_{i-3}\,)^T,$$
$$\psi_i = (\,y_{i-4}-y_{i-5} \quad y_{i-4}-y_{i-5} \quad \dots \quad y_{i-20}-y_{i-21}\,)^T.$$

Since Eq. (4) contains the covariance estimates, we will seek a solution based on the total least squares method

$$\min_c \frac{\left\|\hat{r}_{\psi y} - \hat{R}_{\psi\varphi}c\right\|_2^2}{1+c^Tc}, \tag{5}$$

The criterion minimization (5) can be based on a solution to a biased normal system.

$$\hat{c} = \left(\hat{R}^T_{\psi\varphi}\hat{R}_{\psi\varphi} - \sigma^2 I\right)^{-1}\hat{R}^T_{\psi\varphi}\hat{r}_{\psi y}, \tag{6}$$

$\sigma = \sigma_{\min}\left(\hat{R}_{\psi\varphi}, \hat{r}_{\psi\varphi}\right)$ is the minimal singular value for the matrix $\left(\hat{R}_{\psi\varphi}, \hat{r}_{\psi\varphi}\right)$.

The summand $(-\sigma^2)$ in expression (6) makes it impossible to calculate $\hat{c}$ with efficient numerically stable methods that do not involve first forming the matrix $\hat{R}^T_{\psi\varphi}\hat{R}_{\psi\varphi} - \sigma^2 I$.

In paper [12] proposes using an augmented system that is equivalent to the biased normal system:

$$\left(\begin{array}{c|c|c} I & 0 & \hat{R}_{\psi\varphi} \\ \hline 0 & I & j\sigma \\ \hline \hat{R}^T_{\psi\varphi} & j\sigma & 0 \end{array}\right)\left(\begin{array}{c} \upsilon \\ \hline \upsilon_{\xi} \\ \hline \hat{c} \end{array}\right) = \left(\begin{array}{c} \hat{r}_{\psi\varphi} \\ \hline 0 \\ \hline 0 \end{array}\right). \tag{7}$$

System (7) is solvable with standard methods for solving linear algebraic equation systems – for example, with LU decomposition.

Using the estimate of $\hat{c}$, we can find an estimate for $\hat{b}$:

$$\hat{b} = -\frac{1}{\Delta t}\ln(0.5c_1) \text{ or} \tag{8}$$

$$\hat{b} = -\frac{1}{\Delta t}\ln(0.25(c_1 + 2\sqrt{-c_2})). \tag{9}$$

Estimates of the parameter $\hat{A}$ can be found as a solution to the normal system of equations:

$$\hat{A} = \left(\Phi_A^T\Phi_A\right)^{-1}\Phi_A^T Y_A,$$

$$\Phi_A = \begin{pmatrix} 1 - \left(1 + \hat{b}\cdot T_1\right)\exp(-\hat{b}\cdot T_1) \\ \vdots \\ 1 - \left(1 + \hat{b}\cdot T_N\right)\exp(-\hat{b}\cdot T_N) \end{pmatrix}, \; Y_A = \begin{pmatrix} y_1 \\ \vdots \\ y_N \end{pmatrix}.$$

Björck [1] proposes using the augmented system that is equivalent to the normal system:

$$\bar{\Phi}_A\bar{A} = \bar{Y}_A, \tag{10}$$

$$\bar{\Phi}_A = \left(\begin{array}{c|c} kI & \Phi_A \\ \hline \bar{\Phi}_A^T & 0 \end{array}\right), \; \bar{A} = \left(\begin{array}{c} k^{-1}r \\ \hline A \end{array}\right), \; \tilde{d} = \left(\begin{array}{c} \hat{r}_{\psi y} \\ \hline 0 \end{array}\right), \; k_2 = \frac{\sqrt{2}}{\mu_{\min}}.$$

$\mu_{\min}$ is the minimum eigenvalue of the matrix $\Phi_A^T\Phi_A$.

3 Results

The proposed identification algorithms of total instrumental variables were run in MATLAB and compared with the algorithms that estimate autoregression parameters in the first step of the least-squares technique [8].

We used the determination coefficient

$$R^2 = 1 - \frac{\sum_{i=1}^{N} (y_i - \hat{z}_i)^2}{\sum_{i=1}^{N} (y_i - E[y_i])^2}$$

and the relative mean-square error of parameter estimation

$$\delta c = \sqrt{\frac{\|\hat{c} - c\|^2}{\|c\|^2}} \cdot 100\%$$

as quality indicators for the model.

The Ramsay curve is given by the equation

$$y_i = 2(1 - (1 + 0.05 \cdot T_i)\exp(-0.05 \cdot T_i)) + \xi_i, \tag{11}$$

Difference equation corresponding to the Ramsay curve (11)

$$z_i - z_{i-1} = 1.9025(z_{i-1} - z_{i-2}) - 0.9048(z_{i-2} - z_{i-3}),$$

$$y_i = z_i + \xi_i.$$

The number of observations is $N = 100$ the sampling interval is $\Delta t = 1$.

$\{\xi_i\}$ is a Gaussian distributed sequence with zero mean and standard deviation $\sigma_\xi = 10^{-4}$.

The results of a computational experiment are presented in Table 1 and Fig. 1.

Table 1. The results of the identification of the model (10).

	c_1	c_2	δc, %	b_1	A	R^2
TRUE	1.9025	−0.9048	–	0.05	2	1.00
LS	1.8333	−0.8357	4.64	0.0870	1.633	0.888
TLS	1.9046	−0.9070	0.15	0.0489	2.022	0.969

Source: author.

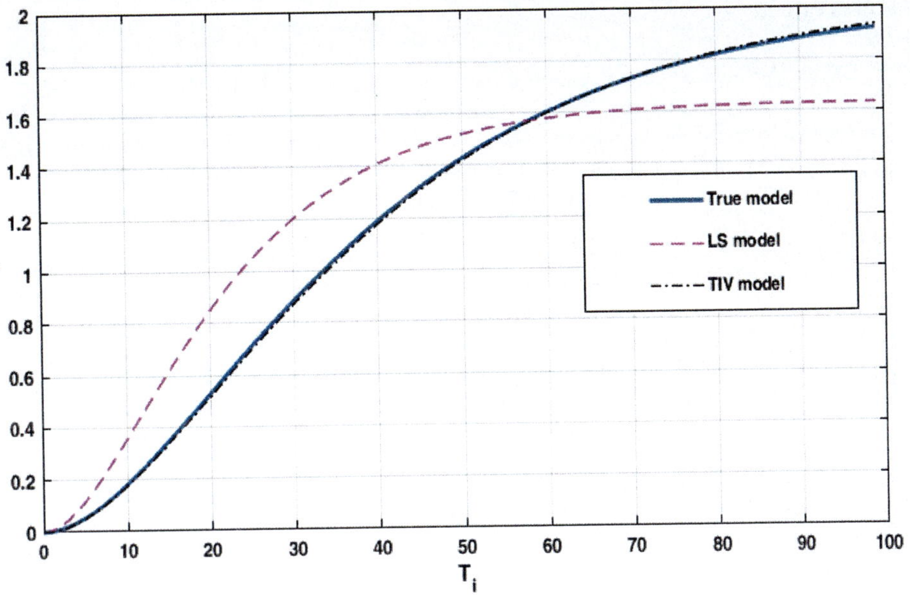

Fig. 1. The results of the identification of the model (10) (Source: author).

4 Discussion

In this paper, we propose an algorithm for estimating the parameters of the Ramsay curve based on the method of total instrumental variables. The advantage of the proposed method is better noise immunity compared to the least squares method [8]. A further area of research may be the construction of an algorithm that takes into account the quadratic relationship of the parameters of the difference equation. The development of regularized algorithms [13, 14] for estimating Ramsay curve parameters is also an urgent task.

5 Conclusion

The method proposed in the article allows one to evaluate the parameters of the Ramsay curve at a higher noise level than the least squares method. The proposed algorithm can improve prediction accuracy in many economic applications using S-curves. The proposed method also shows the best results in the case of an inaccurate model. The use of instrumental variables makes it possible to obtain high accuracy of parameter estimates on short data. This is because instrumental variables do not require knowledge and estimation of the correlation function of the prediction error. The algorithm can be generalized to the case of colored noise by an appropriate choice of the vector of instrumental variables.

References

1. Björck, Å.: Iterative refinement of linear least squares solutions. BIT Numer. Math. **7**, 257–278 (1967). https://doi.org/10.1007/BF01939321
2. Ivanov, D.V., Chertykovtseva, N.V., Terekhova (Zharkova), A.A., Andreeva, E.A.: Identification of exponential trend models with fractional white noise. In: Journal of Physics Conference Series, vol. 1368, p. 042061 (2019). https://doi.org/10.1088/1742-6596/1368/4/042061
3. Ivanov, D.V., Sandler, I.L., Kozlov, E.V.: Identification of fractional linear dynamical systems with autocorrelated errors in variables by generalized instrumental variables. IFAC-Papers-OnLine **51**(32), 580–584 (2018). https://doi.org/10.1016/j.ifacol.2018.11.485
4. Ivanov, D.V., Tikhomirov, E.A., Nazarenko, E.B., Andreeva, E.A., Ilmushkin, A.G., Grygoryants, I.A.: Innovation growth: mathematical modeling and forecasting. In: Vihman, V.V. (ed.) Humanities and Social Sciences: Novations, Problems, Prospects. Advances in Social Science, Education and Humanities Research, vol. 333, pp. 700–704. Atlantis Press, Paris (2019). https://doi.org/10.2991/hssnpp-19.2019.133
5. Ramsay, J.O.: A comparative study of several robust estimates of slope, intercept and scale in linear regression. J. Am. Stat. Assoc. **72**(359), 608–615 (1977). https://doi.org/10.1080/01621459.1977.10480624
6. Refaat, E.A.: Lecture Notes on Z-Transform. Lulu Press, Morrisville (2005)
7. Semenichev, V.K., Kurkin, E.I., Semenichev, E.V., Danilova, A.A., Fisun, G.A., Kasatkina, E.I.: Non-renewable recourses life cycles modeling aspects. Procedia Comput. Sci. **65**, 872–879 (2015). https://doi.org/10.1016/j.procs.2015.09.046
8. Semenychev, V.K., Korobetskaya, A.A., Kozhukhova, V.N.: Proposals of Econometric Tools for Modeling and Forecasting Evolutionary Processes: Monograph. SAGMU, Samara (2015)
9. Shields, L.B., Gertz, T.A., Wilson, K.C., Figg, G.L., Hester, S.T., Honaker, J.T.: The S-curve discontinuity theory predicts the path towards a "well" society and increased longevity. Med. Hypotheses **121**, 99–102 (2018). https://doi.org/10.1016/j.mehy.2018.09.006
10. Söderström, T.: A generalized instrumental variable method for errors-in-variables. Automatica **47**(8), 1656–1666 (2011). https://doi.org/10.1016/j.automatica.2011.05.010
11. Tomić, D.: Empirical evidence of an S-curve in croatia. Ekonomska Istraživanja **32**(1), 2212–2230 (2019). https://doi.org/10.1080/1331677X.2019.1645718
12. Zhdanov, A.I., Shamarov, P.A.: Direct projection method in the problem of complete least squares. Autom. Remote Control **61**(4), 610–620 (2000)
13. Golub, G., Hansen, P., O'Leary, D.: Tikhonov regularization and total least squares. SIAM J. Matrix Anal. Appl. **21**(1), 185–194 (1999). https://doi.org/10.1137/S0895479897326432
14. Zhdanov, A.I.: Regularization of unstable finite-dimensional linear problems based on augmented systems. Comput. Math. Math. Phys. **45**(11), 1845–1853 (2005)

Small Business in the Regions of Russia: Statistical Analysis and Modeling

E. I. Sukhanova(✉)

Samara State University of Economics, Samara, Russia
elsu5463@gmail.com

Abstract. The article is aimed to analyze the dynamics of small business development in the Russian Federation. The study offers an econometric model for small business development in the RF regions based on preliminary distribution of regions into homogeneous groups. The study follows statistical and econometric design. Small enterprises turnover is analyzed as one of the major indices of small business operational activity. The dynamics of the RF small enterprises turnover for 2005–2018 is explored. The econometric *ARIMA* (1, 1, 0) model is constructed and point and interval forecasts are made. The impact of socio-economic indicators on small business development in the RF regions is also studied for 2018. Cluster analysis was applied to break down multidimensional statistical data into homogeneous groups of regions (clusters). Regression models of small business turnover were constructed for every cluster. The hypothesis, that there is a regional differentiation in small business development, is proved. The constructed models demonstrate relevant explanatory power and can be applied to the analysis of small business, e.g. in the sphere of investment and state support issues, and to forecasting.

Keywords: Cluster analysis · Econometric model · Forecast · Small business · Small enterprises turnover · Sustainability

1 Introduction

Small business sector is an important and integral part in economic system of any country. Small business contributes to increasing employment, creating a competitive environment, affects the innovative activity of economic entities, and responds quickly to consumer demand, which affects the well-being of the society at whole. Assessing the sustainability of small businesses [10], various tools and methods of such assessment are always of great relevance.

Updated IT operating model is now a prerequisite for successful operation of a small enterprise. It ensures business's sustainable development and competitiveness when entering the digital economic system. This allows managers to continuously monitor the company's activities, reduces the time for making management decisions, and makes it possible to forecast the economic indicators of the market. Information technologies are used for collecting, processing, analyzing, and systematizing statistical data about an enterprise. IT is also applied to modeling and forecasting the results of business activities [6, 11]. In this study, information technologies for sustainable

S. I. Ashmarina et al. (Eds.): ENGINEERING ECONOMICS WEEK 2020, LNNS 139, pp. 443–453, 2021.
https://doi.org/10.1007/978-3-030-53277-2_53

development of small businesses are defined as an environment for the implementation of mathematical, statistical and econometric methods to study enterprise performance indicators and their relationships.

Researchers have treated small business issues in much detail. They are reflected in a number of modern studies. The papers that consider the statistical-econometric approach to small business problems differ in the statistical methods used [3, 5, 8, 12], in model types [7, 9, 13, 15], and in analyzed indicators [1, 4, 14].

The purpose of this paper is to analyze the dynamics of small business development in the Russian Federation, as well as to econometrically model the level of small business development by regions of the Russian Federation with a preliminary classification of regions to obtain homogeneous groups.

2 Methodology

Small enterprises turnover indicates the efficiency of small business operational activity in the Russian Federation. The results of small business operation depend on the state of the country's economy as a whole, on the values of its socio-economic indicators that reflect the degree of small businesses activity.

In this study the RF small business was analyzed in two stages. First, on the basis of official statistics [2] the dynamics of turnover of small enterprises in Russia from 2005 to 2018 was investigated. The econometric model of turnover time series was constructed and the forecast for 2019–2020 was made.

Next, the RF regions were analyzed separately. The regions were grouped according to certain socio-economic indicators and the econometric model was constructed for each regional group. For this, various factors, that reflect the level of economic development of regions, indicators of industry, construction activity, trade, education and employment, were considered. The turnover of small enterprises, which is seen as one of the major enterprise operation characteristics, was chosen as a effective feature (explained variable). As a result, the set of following indicators was selected:

Y – small enterprises turnover (billion rubles);
X_1 – gross regional product (GRP) (million rubles);
X_2 – monthly average per capita income (rubles);
X_3 – the volume of own production shipped goods, work and services regionally performed (mining, manufacturing, electricity, gas and water production and distribution) (million rubles);
X_4 – retail turnover (million rubles);
X_5 – the number of vocational secondary education institutions (institutions);
X_6 – the proportion of 15–72 years old in labor force (labor force sample study data) (%);
X_7 – labour requirements from employers to employment agencies (people);
X_8 – commissioning of residential and nonresidential buildings (total area) (thousand square meters);
X_9 – the volume of fee based service (million rubles).

80 regions of the Russian Federation were selected for analysis. After Grubbs' test the regions with the surge indicators (discordant observations) were excluded from the sample, namely Moscow, St. Petersburg, Moscow *oblast*, as well as the Nenets and Chukotka Autonomous *okrugs*.

The regional information array of the study was formed on the basis of official data [2]. It represents statistical data on selected indicators for the regions of the Russian Federation for 2018 (the sample represents 80 observations for ten indicators). The methodology used in this study complies methods of multidimensional object classification (cluster analysis) and methods of econometric modeling. Information processing, calculations, and analysis were performed using Excel, Statistica, and Gretl programs.

3 Results

3.1 Dynamics Analysis and Forecast of Small Enterprises Turnover

The dynamics of changes in the turnover of small enterprises in the Russian Federation from 2005 to 2018 is shown in Fig. 1. During this period, on average, there was an increase (although uneven) in the turnover of small enterprises in the Russian Federation. It was especially noticeable in the period from 2015 to 2018. Over these 14 years, the value of this indicator increased by more than 5,5 times.

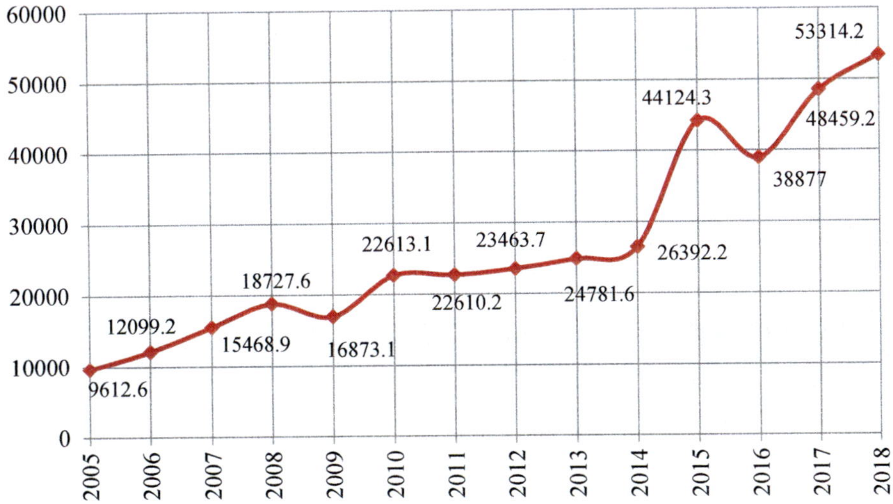

Fig. 1. The dynamics of the RF small enterprises turnover for 2005–2018 (in bln. rub.) (Source: author based on [2]).

To forecast the turnover of small enterprises in the Russian Federation, an econometric model of this indicator (Y) was built based on statistical data from the study period (14 observations). The model of autoregression and integrated moving average of the first order *ARIMA* (1, 1, 0) was optimal for the time series of this indicator (according to Akaike and Schwarz information criteria). The results of the model parameters estimation are presented in Table 1.

Table 1. Model parameters estimation (1)

Variable	Parameter estimate	Standard error	*z*-statistics	*p*-value
const	3345,34	888,241	3,7663	0,0002
ΔY_{t-1}	−0,483635	0,225745	−2,1424	0,0322

Source: author.

Thus, time series of the RF small enterprises turnover takes the form of:

$$\Delta Y_t = 3345,34 - 0,48\Delta Y_{t-1} \quad (1)$$

Estimates of the constructed model parameters are statistically significant on 5%-level of significance. It was found that the model residues (1) are stationary (according to the Dickey–Fuller test, the observed value was $t = -3,28$ at p-value equal to 0,03998). There is no autocorrelation of residues (according to the analysis of the autocorrelation (ACF) and private autocorrelation (PACF) functions of residues of the model). The point and interval forecast of small business turnover for 2020 was calculated with the constructed model. According to the calculations, in 2019, the turnover of small businesses in the Russian Federation averaged 55929,4 billion rubles, and in 2020-might average 59627,9 billion rubles. It can be expected with 95% confidence, that in 2020, the turnover will be from 4941,3 to 69842,5 billion rubles. Thus, the turnover of small enterprises in the Russian Federation will continue to grow.

3.2 The RF Regions Grouping and Econometric Modeling

It is important to take into account that enterprises of different industries operate in different regions of Russia and that small businesses in most cases have a closed cycle in a particular region. Consequently, the impact of socio-economic factors on the results of economic activity of small businesses throughout the Russian Federation should be analyzed at the regional level.

While studying the *Y* dependence (small enterprise turnover) from the factors $X_1,\ldots, X_9$, we hypothesized that the factors under study for different homogeneous regional groups could influence the *Y* differently. That was proved by the results obtained. Therefore, before econometric modeling, first the entire array of data (for ten indicators) was divided with cluster analysis method into homogeneous (in a statistical sense) groups (clusters). Next, econometric models were constructed separately for each cluster.

Hierarchical methods of cluster analysis were used to divide the analyzed set of regions into homogeneous groups, namely the full link method, the unweighted and weighted pairwise average methods, the Ward method, and the iterative clustering method – *k*-means method [14].

The results of regional grouping with different hierarchical methods in dendrograms, allowed estimating the quality and approximate structure of clusters obtained. Based on these results the number of clusters was set in *k*-means method $k = 3$. The regional structure and number of regions by cluster is shown in Table 2.

Table 2. Classification of regions with *k*-means method

Cluster number	Number of regions	Cluster structure (regions)
1	14	*Oblast:* Rostov, Nizhny Novgorod, Samara, Sverdlovsk, Tyumen, Chelyabinsk, Novosibirsk, Kemerovo. *Republic:* Tatarstan, Bashkortostan. *Territory:* Krasnodar, Perm, Krasnoyarsk, Primorsky
2	27	*Oblast:* Belgorod, Vladimir, Voronezh, Ivanovo, Kaluga, Lipetsk, Tula, Yaroslavl, Arkhangelsk, Vologda, Kaliningrad, Leningradskaya, Volgograd, Orenburg, Saratov, Irkutsk, Omsk, Tomsk, Sakhalin. *Republic:* Komi, Dagestan, Udmurtia, Sakha (Yakutia). *Territory:* Altai, Stavropol, Khabarovsk, Khanty-Mansi Autonomous *Okrug*
3	39	*Oblast:* Bryansk, Kostroma, Kursk, Oryol, Ryazan, Smolensk, Tambov, Tver, Murmansk, Novgorod, Pskov, Astrakhan, Kirov, Penza, Ulyanovsk, Kurgan, Amur, Magadan. *Republic:* Karelia, Adygea, Kalmykia, Crimea, Ingushetia, North Ossetia-Alania, Karachay-Circassian, Mari El, Chechen, Mordovia, Chuvash, Kabardino-Balkarian, Altai, Buryatia, Tyva, Khakassia. *Territory:* Transbaikal, Kamchatka. Jewish Autonomous Oblast, Yamalo-Nenets Autonomous *Okrug*. Sevastopol

Source: author.

The average values of indicators throughout three clusters and for comparison throughout Russia (without excluded regions) are demonstrated in Table 3.

As it can be seen from the table, the average values of the regions in the 1st cluster are much higher than the average values of the corresponding indicators in Russia (except X_5, X_6). The regions in the 2nd cluster have average values of indicators closer to the average for Russia (but still exceed them), and the indicators of the regions of the largest 3rd cluster have average values below the average values for Russia. The highest level of economic development is observed in the regions of the first two clusters. Thus, the information array of initial data on the considered indicators was divided into three homogeneous groups of regions. For each group (cluster), econometric models were constructed to determine the impact of socio-economic factors (X_1, …, X_9) on the effective feature – the turnover of small enterprises (Y).

First, the matrices of paired correlation coefficients built in each cluster analyzed. As a result, multicollinearity was eliminated and factors that do not have a significant impact on Y were excluded.

Table 3. Average values of indicators

Indicators	Cluster 1	Cluster 2	Cluster 3	Russia (80 regions)
Y	1023,2	389,7	142,5	384,6
X_1	1337002,7	766428,8	280397,1	636629,2
X_2	29051,2	27980,3	24655,6	27259,7
X_3	1571999,4	836592,4	242638,7	685157,3
X_4	704675,1	300140,1	128839,4	290595,8
X_5	29,3	30,3	18,9	24,7
X_6	68,1	68,3	67	67,7
X_7	35031,6	16300,9	7858,3	15638,2
X_8	3096,8	1366,7	573,3	1292,6
X_9	218349	85792,2	36194,3	85727,9

Source: author.

Different models of paired and multiple linear dependence of small business turnover (Y) on the analyzed explanatory variables were constructed for each cluster. The best model was selected based on the analysis of the model quality characteristics, namely the determination coefficient R^2, the adjusted determination coefficient $\bar{R}^2$, the standard error of the regression residues S_e, and the observed value of the F-statistics. Parameter estimates and statistical characteristics of model quality for three clusters are shown in Tables 4, 5, 6 and 7.

Table 4. Parameters estimates and statistical properties of the model quality (Cluster 1)

Variable	Parameter estimate	Standard error	t-statistics	p-value
Const	−3690,79	1908,56	−1,934	0,0793
X_4	0,000775484	0,000228052	3,400	0,0059
X_6	61,2229	27,8522	2,198	0,0502
$\boldsymbol{R^2} = 0{,}6531$; $\boldsymbol{\bar{R}^2} = 0{,}590$; $\boldsymbol{S_e} = 356{,}6254$; $\boldsymbol{F} = 10{,}3547$				

Source: author.

The models for the dependence of the turnover of small enterprises (Y) on the following explanatory variables: (X_1), (X_1, X_6), (X_4), (X_4, X_6), (X_9), were constructed for the 1st cluster regions. When comparing the quality indicators of the constructed models, the model with variables (X_4, X_6) was selected. The parameters estimates, which were found, are statistically significant (the critical value $t(0,1;11) = 1,7959$), so is the model (the critical value $F\ (0,1;2;11) = 2,8595$) on 10% level of significance (Table 4). The model residues were checked for heteroscedasticity. The White test (p-value = $P(\chi^2(5) > 3{,}4142) = 0{,}6364$) and the Breusch-Pagan test (p-value = $P(\chi^2\ (2) > 1{,}7209) = 0{,}4229$) demonstrated homoscedasticity of the model residues.

Table 5. Parameters estimates and statistical properties of the model quality (Cluster 2)

Variable	Parameter estimate	Standard error	*t*-statistics	*p*-value
Const	84,3100	57,9505	1,455	0,1587
X_4	0,000515328	0,000178612	2,885	0,0081
X_7	0,00924611	0,00242380	3,815	0,0008
$R^2 = 0{,}5903$; $\bar{R}^2 = 0{,}5562$; $S_e = 101{,}8507$; $F = 17{,}2899$				

Source: author.

Table 6. Parameters estimates and statistical properties of the mel quality after heteroscedasticity elimination (Cluster 2)

Variable	Parameter estimate	Standard error	*t*-statistics	*p*-value
Const	86,6117	62,2360	1,392	0,1768
X_4	0,000403965	0,00013679	2,953	0,0012
X_7	0,0113968	0,00269981	4,221	0,0003
$R^2 = 0{,}6344$; $\bar{R}^2 = 0{,}6039$; $S_e = 2{,}5025$; $F = 20{,}8228$				

Source: author.

Table 7. Parameters estimates and statistical properties of the model quality (Cluster 3)

	Parameter estimate	Standard error	*t*-statistics	*p*-value
Const	−7,4941	17,0141	−0,4405	0,6622
X_4	0,00078	0,00016	4,875	<0,0001
X_8	0,0874	0,0253	3,455	0,0014
$R^2 = 0{,}7547$; $\bar{R}^2 = 0{,}7410$; $S_e = 48{,}8346$; $F = 55{,}3659$				

Source: author.

Thus, for Cluster 1 the small enterprises turnover model takes the form of:

$$\hat{Y} = -3690,79 + 0,00078X_4 + 61,22X_6 \qquad (2)$$

The determination coefficient shows that the variation in the turnover of small businesses (Y) by an average of 65,31% is due to the variation of the factors included in the equation (retail trade turnover (X_4) and the level of 15–72 aged participation in the labor force (X_6)).

For Cluster 2 the models with the following explanatory variables were constr (X_4, X_7), (X_4), (X_7), (X_8), (X_8, X_9). The set of (X_4, X_7) provided the best model. The estimates of Cluster 2 model parameters (other than the absolute term of the equation) are statistically significant (the critical value $t(0,1;24) = 1,7109$), the model is also statistically significant (the critical value $F(0,1;2;24) = 2,5383$) on 10%-level of significance (Table 5).

Thus, for Cluster 2 the model takes the form of:

$$\hat{Y} = 84,31 + 0,0005X_4 + 0,009X_7 \quad (3)$$

According to the White test, the model (3) showed heteroscedasticity (p-value = $P(\chi^2$ (5) > 5,3897) = 0,0097), and the opposite result was demonstrated according to the Breusch-Pagan test (p-value = $P(\chi^2$ (2) > 10,2904) = 0,3492). After eliminating heteroscedasticity using generalized least squares estimation, the model (3′) was obtained, which has all the standard errors reduced and the quality characteristics improved (Table 6).

Thus, after the elimination of heteroscedasticity, the following econometric m small enterprises turnover in Cluster 2 was performed:

$$\hat{Y} = 86,61 + 0,0004X_4 + 0,011X_7 \quad (3')$$

The coefficient of determination shows that the variation in the turnover of small enterprises by an average of 63,44% is explained by the variation of the factors included in the equation. The effective feature (Y) in the regions of the second cluster is significantly affected by the retail turnover (X_4), as well as the need for employees declared by employers to the employment agencies (X_7).

The best model for Cluster 3 was selected among models with the following set of explanatory variables: (X_4, X_8), (X_4), (X_5 X_7), (X_5 X_8), (X_7), (X_7, X_9). Among them the model with the variables (X_4, X_8) was selected. The results of its estimation are given in Table 7. The estimates of Cluster 3 model parameters (other than the absolute term of the equation) are statistically significant (the critical value $t(0,05;36) = 2,0281$), the model is also statistically significant (the critical value $F(0,05;2;36) = 3,2595$) on 5%-level of significance. The residues of the model were also tested for heteroscedasticity. Both White and Breusch-Pagan tests confirmed its absence.

Thus, for Cluster 3 the model takes the form of:

$$\hat{Y} = -7,49 + 0,00078X_4 + 0,087X_8 \quad (4)$$

The model 4 determination coefficient (4) shows that the variation in the turnover of small enterprises by an average of 75.5% is explained by the variation of the factors included in the equation. The effective feature in the cluster with the largest number of regions is determined by the retail turnover (X_4) and the total area of residential and non-residential buildings being put into exploitation (X_8).

Based on the models constructed for each cluster, the forecast of small enterprises turnover (Y) was made with the average values of the factors included in the corresponding models. Calculated forecast values in comparison with actual values (calculated in Table 3) are presented in Table 8.

The forecast values, which were obtained, do not differ significantly from the actual average Y values (for each cluster).

Table 8. Actual and forecast values of small enterprises turnover (Y) for each cluster

Cluster number	Model	Actual value	Forecast value
1	(2)	1023,2	1027,94
2	(3′)	389,7	385,98
3	(4)	142,5	143,10

Source: author.

4 Discussion

In Russia the small business sector represents one of the forms of economic activity with good prospects. The researchers pay great attention to this sector and assess its level of development. A number of papers address the problems of small business at the micro level. For example, in order to model the turnover of small enterprises, three-factor production functions of the dependence of turnover on the wages of employees, investments in fixed capital and the number of employees are estimated by Pinkovetskaya and Lebedev [9]. Malesios, Skouloudis, Dey, Abdelaziz, Kantartzis, Evangelinos use Bayesian regression modeling to assess the sustainable development of small and medium-sized businesses [7]. Sukhanova, Shirnaeva and Repina use binary choice models to reflect the financial sustainability of an enterprise [15].

The formation and development of small businesses in the Russian Federation is influenced by social and economic indicators of regional development. That is also reflected in a number of studies. Filonova and Bukreeva analyze the state of Russian regions by indicators of small and medium-sized businesses, using cluster analysis for grouping the regions [3]. Ermolina analyzes factors of innovative activity development in small industrial enterprises [1]. A comprehensive assessment of the state of small businesses in the Russian Federation is carried out using the Maximin method [5], and the principal components analysis is used [8]. Ignatova and Ignatov note a direct correlation between the positive dynamics of small business development and the volume of state subsidies [4].

In this study, the turnover of small businesses is considered as an indicator that reflects the level of development and efficiency and sustainability of small businesses in the regions of the Russian Federation. The study confirmed the hypothesis that socio-economic indicators have different effects on the turnover of small businesses in different groups of regions. This result confirms the need for a differentiated approach to tax policy, investment, and government support for small businesses.

The sustainable development of a small enterprise, along with other factors, is not possible without constant monitoring of its performance indicators, as well as temporary forecasting of crisis situations. Updated information technologies of data processing can be necessary tools here. Special software products for data analysis can be useful (SPSS, Statistica, Stata, Excel, EViews, Gretl, etc.).

5 Conclusion

The paper analyzes the dynamics for 2005–2018 and forecasts the turnover of small enterprises in the Russian Federation for the upcoming periods. Socio-economic indicators that affect the economic activity of small businesses in the regions of the Russian Federation are also identified. The regions under analysis were divided into three homogeneous groups (clusters). For each cluster econometric models were constructed for the indicator of the small business development level – the turnover of small enterprises. It is found that in each group of regions, the turnover of small enterprises is characterized by a different set of factors. Despite the fact that over the past 14 years, almost all regions of the Russian Federation have demonstrated fairly sustainable development of small businesses, the level of development in the regions of the first and second groups is higher. The models have a fairly good explanatory power and can be used for further analysis, e.g. in projects to stimulate small business activity in the regions of the Russian Federation, as well as for forecasting. The use of information technologies allows expanding the range of possible applications of statistical and econometric methods to analyzing and forecasting both the activities of small businesses and other organizations.

References

1. Ermolina, A.A.: Socio-economic characteristics of the region as determinants of innovation activity of small enterprises in Russia. Innov. Dev. Econ. **6**(42), 47–54 (2017)
2. Federal Service of State Statistics: Regions of Russia. Socio-economic indicators (2019). https://gks.ru/folder/210/document/13204. Accessed 19 Dec 2019
3. Filonova, E.S., Bukreeva, Y.V.: Analysis of the Central Federal district regions by indicators of small and medium-sized businesses. Bull. Financ. Univ. **6**(78), 35–47 (2013)
4. Ignatova, I.V., Ignatov, E.S.: Development of entrepreneurship in the region: drivers and problems. Econ. Soc. Changes-Facts Trends Forecast **10**(5), 143–157 (2017). https://doi.org/10.15838/esc/2017.5.53.10
5. Leonov, S.N.: Current state and the features of location of small business in regions of Russia. Econ. Soc. Changes-Facts Trends Forecast **11**(5), 103–119 (2018). https://doi.org/10.15838/esc.2018.5.59.7
6. Lu, H., Pishdad-Bozorgi, P., Wang, G., Xue, Y., Tan, D.: ICT implementation of small- and medium-sized construction enterprises: Organizational characteristics, driving forces, and value perceptions. Sustainability **11**(12), 3441 (2019). https://doi.org/10.3390/su11123441
7. Malesios, C., Skouloudis, A., Dey, P.K., Abdelaziz, F.B., Kantartzis, A., Evangelinos, K.: Impact of small- and medium-sized enterprises sustainability practices and performance on economic growth from a managerial perspective: modeling considerations and empirical analysis results. Bus. Strategy Environ. **27**(7), 960–972 (2018). https://doi.org/10.1002/bse.2045
8. Manukhina, A.V.: Analysis of the impact of socio-economic factors on the formation of small businesses in the regions of the Russian Federation. Bull. Univ. **3**, 172–177 (2012)
9. Pinkovetskaia, I., Lebedev, A.: Entrepreneurial capital and the volume of small enterprises production: Russian regions data. J. Dev. Entrepreneurship **24**(2), 1950013 (2019). https://doi.org/10.1142/S1084946719500134

10. Soto-Acosta, P., Cismaru, D.M., Vatamanescu, E.M., Ciochina, R.S.: Sustainable entrepreneurship in SMEs: a business performance perspective. Sustainability **8**(4), 342 (2016). https://doi.org/10.3390/su8040342
11. Spirina, S.G.: Information technologies for assessing the financial stability of enterprises (corporations). Inf. Resour. Russia **5**(147), 22–26 (2015)
12. Sukhanova, E.I., Shirnaeva, S.Y.: Different approaches to macroeconomic processes simulation and forecasting. Fundam. Res. **12**, 406–411 (2015)
13. Sukhanova, E.I., Shirnaeva, S.Y., Mokronosov, A.G.: Econometric models for forecasting of macroeconomic indices. Int. J. Environ. Sci. Educ. **11**(16), 9191–9205 (2016)
14. Sukhanova, E.I., Shirnaeva, S.Y., Konstantinova, E.O.: Statistical analysis and modeling of the influence of socio-economic factors on the level of small business development in the regions of the Russian Federation. Fundam. Res. **5**, 126–132 (2019). https://doi.org/10.17513/fr.42471
15. Sukhanova, E.I., Shirnaeva, S.Y., Repina, E.G.: Binary choice models-based assessment of company's financial sustainability. In: Mantulenko, V. (ed.) Proceeding of the International Conference «Problems of Enterprise Development: Theory and Practice 2018», SHS Web of Conferences, vol. 62, p. 13002. EDP Science, Les Ulis (2019)

Analytical Support of the Retailer's Purchasing Activity Security

Y. A. Tatarovsky(✉)

Samara State University of Economics, Samara, Russia
tatarovsky.yury@yandex.ru

Abstract. Despite the crisis in the domestic economy, the reduction in the number of small and medium-sized enterprises, it is impossible not to notice the growing level of competition between different manufacturers. According to many experts, the rate of growth in the supply of goods today exceeds the growth of the population's welfare, and the margin of financial strength of most suppliers does not allow an adequate reduction in output, which provokes them to pursue an aggressive sales policy. Aggressive promotion of non-competitive products carries risks not only for the supplier, but also for the retailer. Offering customers products that do not have consumer value is a strategic mistake of any retailer, which is a significant risk to its economic security. The purpose of the study is to develop a methodology for determining the competitiveness of a retailer's suppliers based on the full range of information available from external and internal, accounting and non-accounting sources. The main methods used in the study were analysis, abstraction, modeling, and induction. The result of the study is to obtain a methodology that has a possibility of digitalization and implementation as a software product within the overall management system of the retailer's purchasing activity.

Keywords: Business analysis · Competitiveness analysis · Economic security · Management accounting · Sustainable development · Quality information

1 Introduction

The current state of the domestic retail market is characterized by a high share of large federal retail chains, and the introduction of information systems for which is not only a way to increase their competitiveness and to provide their sustainable development, but also a necessary condition for their survival. In addition, these economic entities have both the resources to develop and implement modern information technologies, and the scale to quickly repay investments in the development of information technologies [1].

Perhaps no management decision is made today without analyzing the activities of competitors. Competitive analysis is carried out both by the organization itself (its marketing and analytical services), and with the involvement of outsourcing consulting, marketing and analytical agencies. Recently, retailers have started to occupy a certain niche in providing competitive analysis services for product suppliers. The economic dictionary defines a retailer as a retail operator, a commercial organization that has a network of retail objects. In the Russian legal field, the concept of "retailer" is absent,

S. I. Ashmarina et al. (Eds.): ENGINEERING ECONOMICS WEEK 2020, LNNS 139, pp. 454–464, 2021.
https://doi.org/10.1007/978-3-030-53277-2_54

but it is appropriate to use the concept of "retail network", which implies a set of two or more retail objects belonging to one economic entity (several interdependent), or using the same means of individualization.

As for the analysis of the retailer's competitiveness, despite the apparent paradoxical nature of this situation, it turns out to be quite logical when examined in depth [8]. Thus, a retailer has a lot of practical experience and competence in the field of product sales, interacting with the end consumer. In addition, only he is the primary owner of an objective information and statistical database containing quantitative and qualitative data related to sales volumes, returns, defects, specifics of transportation and storage, customer loyalty to the brand of a particular manufacturer against the background of its competitors, etc. Close interaction of the retailer with suppliers reveals yet another aspect of the competitive struggle of manufacturers: the flexible sale strategy, the willingness to invest in channels to promote their products, reliability of construction of logistics channels – availability of all the declared products in warehouses to organize their delivery. Thus, the appearance on the market of consulting services of retail companies (or subsidiaries of these companies, which is especially important in the light of future changes in the field of legislation regulating trade activities) can significantly displace the positions held by consulting, marketing and analytical agencies in a number of areas, and is an indicator of increasing the level of domestic retail trade.

2 Methodology

To make adequate management decisions, a company needs to know almost everything about its competitors. Some methods of competitive analysis even provide for the preparation of psychological portraits of managers and senior managers of a competitor company, and the methods of obtaining information vary from customer surveys and analysis of open sources to bribery of the organization's staff, which raises the question of the ethics of the events. However, the scope of information submission should be determined not so much from the point of view of ethics (although this is important), but from the point of view of the legality of information collection and disclosure procedures.

In determining how to gather information for competitive analysis should be guided by the Federal laws "On accounting" and "On commercial secret" [3, 4]. Note that the purpose of this study is not to study the legal side of the issue concerning commercial secrets of an economic entity.

Summarizing the experience described in the periodical literature concerning accounting and record-keeping in organizations, we come to the conclusion that, despite the existence of a legal framework, business entities, in particular small and medium-sized enterprises, rarely take measures to protect their trade secrets (issuance of local acts; definition of a list of information related to trade secrets; conclusion of appropriate contractual relations with staff and third-party organizations), which contributes to its unpunished dissemination by both employees and contractors. But it is the protection of commercial information in the presence of an effective legal framework that should be the basis of information support for civilized competitive analysis conducted by external organizations (including retailers), since, for example, the presence of facts of opening various types of legal proceedings significantly reduces the image component of any economic entity.

Taking all this into account, we have developed a method of competitive analysis recommended for consulting, marketing, analytical agencies, as well as retailers. Since the purpose of competitive analysis is to identify the real and potential advantages of the organization and its competitors, first of all, we will separate the concepts of the organization's competitiveness and product competitiveness. These concepts are interdependent, but we believe that the competitiveness of an organization consists of the competitiveness of its products is incorrect. Organization's competitiveness is based not only on the commercial success of its products, but also for proper formation of the business model, effective management of financial flows and financial condition, quality of management [14], potential development, etc. However, to underestimate the competitiveness of products should not, after all, generated the cash flow (revenues from sales) is the natural basis for the functioning of the business. Product competitiveness is a key, fundamental, but by no means the only factor of competitiveness, because the organization usually works with a wide range of products of different levels of competitiveness. Thus, the task of the proposed method is to form and analyze qualitative and quantitative characteristics of activities that are important for building sales activities. We believe that the information obtained during the analysis can be used as a separate information block when conducting a classic benchmarking analysis, which consists in comparing sales volumes, financial results, financial condition indicators, market capitalization, etc.

3 Results

The first information unit – analysis of organizational and legal information concerning the competitors – includes analysis of data about organizational-legal form, about the size of the share capital, number of founders (as potential possibilities of increasing the financial resources of the economic entity), changes in the governing body, the place of incorporation, types of activities, the presence of litigation, open enforcement proceedings, the availability of debt to budget and extrabudgetary funds. Access to this information is possible with the help of most information services, and their use and publication is not a disclosure of commercial secrets [4].

The analysis of this information block allows us to form assumptions about a number of fundamental aspects of the functioning of a business. In particular, among the important aspects, it is possible to highlight the following:

- number of founders. The presence of several founders may indicate greater or at least diversified financial support for the organization,
- personalities of the founders and the sole management body (head). The question of the emerging trend towards the emergence of the Institute of top management, the differentiation of ownership and disposal functions in business is becoming more and more relevant for Russian organizations, including small and medium-sized businesses. In relation to the methodology under consideration, it is assumed that if

the founder and the Director are the same person, then this organization is more mobile, prone to rapid strategic decision-making, but has less financial support. If there are several founders and a top Manager involved, it is possible to expect more professional tactical management against the background of slow, bureaucratic strategic decision-making. At the same time, personal characteristics of individuals are of great importance, information about which can be obtained from the media;
– "seriousness" of the business: the absence of a "mass" legal address and a "mass" Director; registration of the organization to the actual address of the office (production), and not to the home address of the Director (founder); a logically formed list of activities; the declared amount of authorized capital, etc.

We offer to analyze the information of this block in tabular form using the point system of evaluation (Table 1). The maximum number of points is 8. A positive conclusion on the indicators of this block is given when you set 4 points or more.

Table 1. Tabular representation of organizational and legal information analysis

Criterion	Value	
	0 points	1 point
Number of founders	1	More than 1
The founder and CEO are not the same person	Yes	No
Media coverage of personalities (CEO, founders)	Negative or absent	Positive or neutral
The use of "massive" legal address	Yes	No
Having a "mass" CEO	Yes	No
Compliance of the actual type of activity declared during the registration of a legal entity	No	Yes
Logic of forming a list of economic activities	Absent	Exist
Amount of authorized capital	Up to 50 thousand rubles	More than 50 thousand rubles

Source: author.

The second information block is information generated by the organization. This includes the official website of the organization, its presence on business portals ("Goods and services", "Pulse of prices" and their analogues), activity in social networks, advertising and media coverage.

At this stage, the competitive advantages of the business, its mission and goals, the originality of the idea underlying its operation, the scale of its activities (including in geographical terms) and possible prospects for development are revealed, which in most cases the organization does not hide, publishing information on finding new partners or scaling the business through franchising in its official resources. We suggest analyzing this information as shown in the Table 2.

Here, the maximum number of points is 6 a positive conclusion is issued if the score exceeds 2 points. The result of consideration of the presented information blocks

Table 2. Tabular representation of the information generated by the organization analysis

Criterion	Value	
	0 points	1 point
The adequacy of the mission and achievable goals	No	Yes
Scale*	Small and medium-sized businesses	Major business
Company web-site evaluation	Outdated. Does not meet modern requirements	Meets modern requirements (ease of navigation, timely updates, information, feedback, online payment, etc.)
Presence on business web-portals	No or in free mode	Availability of companies own page and paid services
Promotion in social networks	No	Yes
Regional development (own branches and representative offices or franchisees)	No	Yes

Source: author.
*According to [5].

is the formation of a certain background – qualitative characteristics of the organization's activities that form the basis for interpreting the results of the analysis of quantitative indicators. In addition, we recommend that all interested parties systematically analyze the submitted blocks of information as part of measures to verify the reliability of contractors.

The third block of information is product analysis [11]. Because the retailer gets the product to the end customer, then, as noted earlier, it has the most accurate data about the demand for a product by consumers and their loyalty to the brand and what the retailer knows about the properties and qualities of the product relevant to the shelf life (the actual time stated), etc.

In addition, the competitiveness of a product depends on the ability of its manufacturer to actively respond to emerging challenges, which may include:

- the need to eliminate the shortcomings identified in a particular batch of goods or in the product itself. Disadvantages may arise due to technical shortcomings or obsolescence, the appearance of new requirements from consumers;
- activity of competitors. Deep research of competitors activities is important, but not the only condition for winning the competition; however, the company must take into account the price-quality ratio of competitors products;
- working with the assortment. Updating the product range, eliminating items that are not in demand, and introducing competitive new products is one of the best ways to increase sales. Quality-price matching of the offer to demand increases the probability of purchasing this product, which is why it is important to identify and strengthen the position of a popular product, refusing to support products that are not in demand. The evaluation of the work with the assortment can be made using

the coefficient of updating the assortment series (Cua), based on the SKU (Stock Keeping Unit), that is, the assortment unit entered during the year (period):

$$Cua = SKUb/SKUe \tag{1}$$

where

SKUb – number of SKU at the beginning of the year (period);

SKUe – number of SKU at the end of the year (period).

- the interpretation of this indicator is related to a detailed analysis. So, the value of the Cua close to zero indicates that there are no changes in the range, which can be considered as a lack of work with the range, but it can also indicate that the organization has selected the same range that is most in demand from customers and is competitive against the existing analogues;
- the ratio of the volume of return of the product to the volume of its delivery, estimated by the indicator Cr (return coefficient):

$$Cr = R/D, \tag{2}$$

where

R – return of the product for the year (period) (cost expression);

D – the product delivered to the retailer for the year (period) (cost expression).

This indicator has a common basis with the previous one and is interesting because today the goal of most organizations (mainly dealers and distributors, who play an important role in organizing the supply of regional retailers) is not so much to deliver a demanded product as to "load" large batches. In this case, there is a financial interest: the implementation of sales plans is rewarded with bonuses from manufacturers, and the manufacturer gets the opportunity to save due to the economies of scale, distributing costs to a larger volume of products. The success of such events, as a rule, is not of a long-term nature and is contradicted by increasing the volume of return of unrealized products by the retailer. However, this practice is quite common and reflects the imperfection of the system of motivation of dealers and distributors;

- smoothness of logistics processes. Maintaining an optimal balance level is one of the key tasks of the retailer, which cannot be solved without an established supply process from suppliers. That is why the contracts with the retailer pay great attention to the organizational aspects of the product delivery process, and significant penalties are provided for their violation. It is possible to assess the smoothness of logistics processes through the quality of delivery indicator (Iqd):

$$Iqd = V/D, \tag{3}$$

where

V – number of cases of violation of the terms of delivery for the year (period);

D – number of deliveries per year (period),

- marketing support. The activity of an organization in stimulating the sale of its products can be estimated in relation to the funds aimed at promoting the product

(FP), to the volume of deliveries of this product (D), expressed in terms of the promotion coefficient of the Cpr:

$$\mathrm{Cpr} = \mathrm{FP}/\mathrm{D} \quad (4)$$

Analysis of the third information block is recommended according to the criteria given in the Table 3. Here the maximum number of points is 12. A positive conclusion is issued if you get more than 6 points.

Table 3. A tabular view of the organization's products analysis

Criterion	Value		
	0 points	1 point	2 points
Elimination of product deficiencies	Missing or delayed	Low	Fast
Reaction to competitors' activities	Missing or delayed	Low	Fast
Coefficient of updating the product range*	Lower than competing counterparts	At the level of competing analogues	Higher than competing counterparts
Return rate	Higher than competing counterparts	At the level of competing analogues	Lower than competing counterparts
The rate of shipments quality index	Higher than competing counterparts	At the level of competing analogues	Lower than competing counterparts
Product promotion coefficient	Lower than competing counterparts	At the level of competing analogues	Higher than competing counterparts

Source: author.
*Subject to the identified need to update the product range.

The fourth block of information is a vertical and horizontal analysis of sales of the organization's products and its competitors. Taking into account the richness of benchmarking analysis tools, the possibilities of detailed information and its graphical representation, we consider it necessary to highlight key indicators in this information block:

- sales volume – the number of goods sold in kind for the year (period);
- dynamics of sales volume – the growth rate of product sales over the previous five years (periods);
- dynamics of sales volume of the previous year (period);
- occupied share in sales volume – the share of product sales in physical terms in the total sales of similar products;

- dynamics of the occupied share in sales volume – the growth rate of the share of product sales in physical terms in the total sales of similar products for the previous five years (periods);
- dynamics of the occupied share in the sales volume of the previous year (period).

It is extremely important to analyze the dynamics of changes in sales volume and the occupied share in sales volume both in the long and short-term retrospect. A long-term retrospective allows to determine the trend of change, while a short-term retrospective allows to pay attention to current risks and opportunities. Moreover, it is assumed that the use of natural indicators in the calculations will contribute to a more objective analysis of sales volumes, since the factor of price changes is excluded.

As a base for comparison, it is advisable to use the average values of the listed indicators according to the retailer's data. This is due to the fact that all competitors within the same retail chain follow the same rules and operate under the same conditions (differences in conditions are detected when analyzing the third information block). In Table 4 the analysis of sales of goods of the organization and its competitors is presented [15].

Table 4. A tabular view of the sales analysis

Criterion	Value		
	0 points	1 point	2 points
Sales volume	Below average level	Slightly* below (above) the average level	Above the average level
Dynamics of sales volume	Below average level	Slightly* below (above) the average level	Above the average level
Dynamics of sales volume for the previous year (period)	Below average level	Slightly* below (above) the average level	Above the average level
Occupied share in sales volume	Included in the retailer's TOP 10 sales	Included in the retailer's TOP 5 sales	Included in the retailer's TOP 3 sales
Dynamics of the occupied share in sales volume	Negative (share reduction)	No (or minor* positive change)	Positive (increase in the occupied share)
Dynamics of the occupied share in the sales volume of the previous year (period)	Negative (share reduction)	No (or minor* positive change)	Positive (increase in the occupied share)

Source: author.
*The deviation does not exceed 3%.

The absolute value of the third and fourth information blocks is the exclusivity of the source of information, concentrated in retailer database.

4 Discussion

The issues of increasing the competitiveness of retailers and, consequently, creating conditions for their sustainable development are not only typical for domestic practice. This is a global task that many foreign scientists are engaged in, and it is worth noting the contribution of Asian researchers. So, Yu, Yin, Yang and others considering the retail energy market note a high share of uncertainty, for which they suggest using the fuzzy evaluation method [16]. Also, the scientific interest of Chinese scientists is directed to the study of methods for improving the competitiveness of retail network suppliers using mathematical modeling of key business processes [7]. A study by British scientists Pantanoa, Priporasa, Stylos [13], their research notes not only the need for timely implementation of innovations, but also the rapid withdrawal of techniques and technologies that are losing their relevance. It should also be noted that British scientists pay special attention to the activities of retailers. Such companies not only play a key role in the process of ensuring product turnover, but also note their significant contribution to the development of advanced innovative technologies, including in the field of digitalization. That is why Jones, Comfort [10] identify retailers as the main drivers of sustainable economic development. A joint study by Japanese and American scientists Kim, Takashima and Newell [12]. this paper notes the need for increased attention not only to the clarity of the organization of internal business processes related to procurement, but also the processes of the external environment, as a new source of increasing the competitiveness and sustainable development of the retailer.

An alternative approach to ensuring the economic security of a retailer is presented by Balas, Kaya [2]. According to the presented research, the basis for ensuring the economic security of a retailer is not to debug key business processes or minimize the risks of working with suppliers through the development of digital analytical products. The authors see the solution to the problem of ensuring the economic security of the retailer in the use of protective financial instruments, the market of which was significantly risky during the crisis of 2008.

The analysis of big data and metadata is considered by Grewal, Puccinelli, Monroe, Roggeveena, Rajendra, Nordfält [6, 7]. According to them, the key area of analytical work should be the marketing aspects of the business. In General, the authors' approach presented in this article largely corresponds to the position of the presented authors. Develop the importance of marketing in ensuring economic security and sustainable development of the retailer and scientists Hillebrand, Driessen, Koll, [9]. According to them, the key element is the stakeholders, whose interests should first pay attention to the retailer.

5 Conclusion

The results of the study are largely consistent with the international experience of studying the activities of retailers in the context of their innovative development. The quality of the information basis for making management decisions depends directly on the completeness and detail of the information provided. However, the complexity and financial costs of recording, collecting, systematizing, analyzing and interpreting such a large amount of information make these actions economically impractical. In this case, digitalization and development of modern software products (including those used within CRM platforms) can reduce labor and financial costs, improving not only the quality of management decisions, economic security, but also the profitability of the business.

The advantages of the proposed method are that it can be integrated into other methods of competitive analysis, used both in an integrated form (comparing the amounts of points received) and in a detailed form (suitable for identifying the shortcomings of the activity). In addition, its use helps to improve the quality of sales channels.

References

1. Angle, J.W., Dagogo-Jack, S.W., Forehand, M.R., Perkins, A.W.: Activating stereotypes with brand imagery: the role of viewer. Polit. Identity J. Consum. Psychol. **27**(1), 84–90 (2017). https://doi.org/10.1016/j.jcps.2016.03.004
2. Balas, A.N., Kaya, H.D.: The global economic crisis and retailers' security concerns: the trends. SocioEconomic Challenges **3**(2), 5–14 (2019). https://doi.org/10.21272/sec.3(1).5-14.2019
3. Federal law "On accounting" No. 402-FL of 6 December 2011 (2011). http://www.consultant.ru/document/cons_doc_LAW_122855/. Accessed 07 May 2020
4. Federal law "On commercial secret" No. 98-FL of 29 July 2004 (2004). http://www.consultant.ru/document/cons_doc_LAW_48699/. Accessed 07 May 2020
5. Federal law "On the development of small and medium-sized businesses in the Russian Federation" No. 209. July 24, 2007 (2007). http://www.consultant.ru/document/cons_doc_LAW_52144/. Accessed 07 May 2020
6. Grewal, D., Puccinelli, N., Monroe, K.B.: Meta-analysis: Integrating accumulated knowledge. J. Acad. Mark. Sci. **46**, 9–30 (2018). https://doi.org/10.1007/s11747-017-0570-5
7. Grewal, D., Roggeveena, A.L., Rajendra, S., Nordfält, J.: Enhancing customer engagement through consciousness. J. Retail. **93**(1), 55–64 (2017). https://doi.org/10.1016/j.jretai.2016.12.001
8. Han, K., Jung, J., Mittal, V., Zyung, J.D., Hajo, A.: Political identity and financial risk taking: insights from social dominance orientation. J. Mark. Res. **56**(4), 581–601 (2019). https://doi.org/10.1177/0022243718813331
9. Hillebrand, B., Driessen, P.H., Koll, O.: Stakeholder marketing: theoretical foundations and required capabilities. J. Acad. Mark. Sci. **43**, 411–428 (2015). https://doi.org/10.1007/s11747-015-0424-y
10. Jones, P., Comfort, D.: "Better retail, better world": a commentary on the localisation of the sustainable development goals. J. Public Aff. **19**(2) (2019). https://www.doi.org/10.1002/pa.1910. Accessed 11 Feb 2020
11. Jost, J.T.: The marketplace of ideology: "elective affinities" in political psychology and their implications for consumer behavior. J. Consum. Psychol. **27**(4), 502–520 (2017). https://doi.org/10.1016/j.jcps.2017.07.003
12. Kim, C., Takashima, K., Newell, S.: How do retailers increase the benefits of buyer innovation? An intra- and inter-organization perspective. Asia Pac. J. Mark. Logistics **30**(3), 571–586 (2018). https://doi.org/10.1108/APJML-03-2017-0043

13. Pantanoa, E., Priporasa, C.-V., Stylos, N.: Knowledge push curve (KPC) in retailing: evidence from patented innovations analysis affecting retailers' competitiveness. J. Retail. Consum. Serv. **44**, 150–160 (2018). https://doi.org/10.1016/j.jretconser.2018.06.004
14. Saim, K., Vijay, M.: Values that shape marketing decisions: Influence of chief executive officers' political ideologies on innovation propensity, shareholder value, and risk. J. Mark. Res. **54**(2), 260–278 (2017). https://doi.org/10.1509/jmr.14.0110
15. Yang, H., Guo, N.: Analysis of supplier competitiveness in supply chain profit distribution from the perspective of quantity supplied and price research based on cournot model and Bertrand model. In: Zhang, L. (ed.) International Conference on Strategic Management, pp. 22–27. Web of Proceedings, Milan (2019). https://doi.org/10.25236/icsm.2019.003
16. Yu, X., Yin, S., Yang, M., Wang, L., Zhao, Y., Jiang, Ch.: Competitiveness evaluation and research of electricity retailers based on fuzzy comprehensive evaluation method. In: Qui, X. W. (ed.) Asia Power and Energy Engineering Conference (APEEC), New Jersey, pp. 322–325. IEEE (2019). https://doi.org/10.1109/APEEC.2019.8720686

Intellectual Property Protection in Conditions of Engineering and Digital Economy

O. A. Bulavko(✉) and L. R. Tuktarova

Samara State University of Economics, Samara, Russia
vikigor163@mail.ru, tuktarovalr@rambler.ru

Abstract. One of the main resources of the economy is intellectual property, which currently forms the market for digital, information and engineering technologies. Obtaining a right for an intellectual property object is a particularly relevant aspect during the development of engineering and digital economy. The authors consider a way to protect intellectual property and digital data as an opportunity to ensure the security of an enterprise and any individual. Improving the development of the engineering economy can increase economic indicators, economic growth rates, and the share of the gross national product. This goal can be achieved through the interaction of engineering economy, industrial enterprises' development, and digital and information technologies. The purpose of this research is to develop and implement an algorithm that protects intellectual property.

Keywords: Digital technologies · Digital watermarks · Digital data protection · Engineering economics · Intellectual property

1 Introduction

Currently, the digital economy is developing, information technologies are rapidly improving, computer programs are being updated, and complex engineering tasks are being solved. New methods of using intellectual property in the field of engineering economics are developing, and the efficiency of industrial application of an intellectual product is growing.

In this regard, it is necessary to pay attention to protection methods of intellectual property rights. There are several ways to protect information, such as cryptography and steganography. The main difference between these methods is that cryptographic methods hide the contents of a file by encryption, while steganography methods hide the fact that any information is transmitted.

In this context, there is a close relation between engineering, digital economy, economic growth, intellectual and human capital. In the course of this research, the authors consider the impact of human capital formation and the impact of the Internet on economic growth [4].

At this point in time, the development of information technology is very rapid, and the number of Internet users is growing every day. Any file submitted in the digital form can be copied, modified, and distributed. This is why multimedia products are

S. I. Ashmarina et al. (Eds.): ENGINEERING ECONOMICS WEEK 2020, LNNS 139, pp. 465–472, 2021.
https://doi.org/10.1007/978-3-030-53277-2_55

vulnerable to their illegal use. To solve this problem, various methods of information protection are being developed.

The problem of copyright protection focuses on the protection of intangible property and intangible assets. The main component of intangible property is intellectual property [3]. This issue has become particularly important with the development of digital technologies and the transition to a digital economy. One way to solve this problem is to embed digital watermarks in multimedia files, in particular in digital images and video data.

A digital watermark is usually understood as a special label that is used to authenticate digital data, as well as to protect it from illegal use [10]. An embedded sign can be either visible or invisible to the human eye. The second option, in turn, is divided into fragile, semi-fragile and persistent digital watermarks. Fragile digital watermarks are destroyed after any minor container transformations. In the case of semi-fragile digital watermark systems, the watermark has selective stability: it can be resistant to some transformations and unstable to others. In contrast to fragile digital watermarks, persistent signs must be resistant to various types of attacks: geometric transformations, compression, etc. The main tasks that can be solved with the help of digital watermark systems:

- protection against unauthorized copying;
- copyright protection;
- protection from changes;
- protection against falsification.

The problem of copyright protection can be solved using the "demonstration of ownership" scenario [5]. If the author of a work wants to prove that he is its sole legal owner, then after creating this object, he inserts a persistent watermark that can uniquely identify him as the owner. Due to the specific nature of this application area, the digital watermark has to be as resistant as possible to a wide range of container distortions, such as linear and nonlinear filtering, lossy compression, cropping, and others. In addition, the algorithms for embedding digital watermarks should have a low computational complexity.

The spread of such technologies at enterprises shows the need to spread and develop scientific and technological developments that are a prerequisite for a long-term economic growth of the Russian economy.

Scientific developments in this area are possible with the use of intellectual capital, which, according to some authors, is closely related to the effective implementation of an intellectual product and the development of new technologies [1]. Currently, the problem of research in the area of intellectual capital, copyright and digital technologies in the space of interaction with the engineering economy is not sufficiently studied. In order to study this issue, we propose to consider the interaction between information technology, engineering economics and copyright protection. The problem of information literacy and the development of information and communication technologies have been considered in the works of scientists at the domestic and international levels [2].

2 Methodology

In the course of the research, methods of mathematical analysis were applied, empirical research methods were used, as well as methods of analogy, interpretation, comparison, and generalization. The applied statistical methods helped to obtain quantitative characteristics that allow us to reject minor deviations in the research process. Computer steganography is a branch of steganography that studies systems of hidden information transmission, in which computer hardware or software, digital data act as a container. The popularity of research in the field of steganography is due to the growing need to transmit hidden information through public communication channels, as steganography systems allow you to make changes to the file container unnoticed by third parties.

The main directions of computer steganography are text steganography methods, digital steganography methods, and data format-oriented methods. In digital steganography methods, the container is one-dimensional or multi-dimensional digital signals that have a physical nature.

These signals include images, video data, and audio files. One of the most popular and effective methods in the field of steganography for copyright protection and protection against unauthorized copying is to embed steganographic inserts – tags that carry a certain identifier of the copyright holder in the protected object. These labels are called digital watermarks. It is worth noting that there is no algorithm that is resistant to all types of attacks. Every existing approach to creating a digital watermark embedding system has been designed to counter certain attacks or to counter a specific set of attacks. The purpose of an attack on a system for embedding information in video data may be not only to extract or delete embedded information, but also to distort it in such a way that it becomes impossible for a legitimate copyright holder to detect or extract this information. Attacks of this sort are geometric attack in a shift of each frame to an indefinite number of pixels along the spatial coordinate axes, or rotate each frame for an uncertain corner, or uncertain changes in the magnitude of the frame size.

When shifting by a fixed amount, all points in the spatial area of the image-container are moved according to a certain rule. The position of the container counts after the shift is determined by the following formulas:

$$x' = x + t_x, \quad (1)$$

$$y' = y + t_y, \quad (2)$$

where

x is the initial reference position on the abscissa axis;
y – initial reference position on the ordinate axis;
t_x – shift along the abscissa x-axis;
t_y – shift along the ordinate y-axis;
x' – position of the reference point on the abscissa axis after the shift;
y' – position of the reference point on the ordinate axis after the shift.

If embedding a watermark was performed in a spatial area, then, accordingly, the digital watermark will be transformed according to the same rule as all points in the image-container. In this case, the task of the detector is reduced to restoring synchronization, since the digital watermark does not deteriorate, but moves to unknown points in the spatial area of the container.

When zooming, the corresponding coordinates of points in the spatial area will be multiplied by the zoom factors:

$$x' = a_x \times x, \tag{3}$$

$$y' = a_y \times y, \tag{4}$$

where

a_x is a zoom factor for the abscissa axis;
a_y – a scale factor for the y-axis.

Scaling results in a loss of synchronization between the digital watermark and the detector.

When rotating relative to the origin, the angle φ is used to set the rotation. Let's assume that there is some vector r that specifies a point from the container's spatial area. Then the coordinates of this point can be described by the following formula:

$$x = |r| \times \cos(\alpha), \tag{5}$$

$$y = |r| \times \sin(\alpha), \tag{6}$$

where α is the angle describing the position of the point in the spatial area.

When the container is rotated by an angle φ, the final coordinates of the image can be set as follows:

$$x' = |\mathrm{r}| \times \cos(\alpha + \varphi) = |\mathrm{r}| \times (\cos(\alpha) \times \cos(\varphi) - \sin(\alpha) \times \sin(\varphi)) = \\ = \mathrm{x} \times \cos(\varphi) - \mathrm{y} \times \sin(\varphi), \tag{7}$$

$$y' = |\mathrm{r}| \times \sin(\alpha + \varphi) = |\mathrm{r}| \times (\sin(\alpha) \times \cos(\varphi) + \cos(\alpha) \times \sin(\varphi)) = \\ = \mathrm{x} \times \sin(\varphi) + \mathrm{y} \times \cos(\varphi). \tag{8}$$

After the container is rotated, the digital watermark, as in the case of a shift, moves to the spatial area counts, the coordinates of which are unknown to the detector. There is no loss of information, but the synchronization of the detector and watermark is disrupted [9].

Due to the possibility of carrying out the described attack scenarios, the following requirements should be met for the developed algorithm of embedding persistent digital watermarks:

- resistance to lossy compression;
- resistance to the framing of the container;
- resistance to the rotation of the container.

In order to understand the complexity of ensuring the stability of the algorithm, it is necessary to consider problems of the consequences of distortions and existing approaches to the counteraction.

3 Results

One of the methods to counteract geometric distortions is direct iteration. Since the main problem with geometric distortions is not the loss of information, but rather the desynchronization of the detector and the digital watermark, a possible solution is to search for a watermark by all possible shifts, zoom factors, and rotation angles. A significant disadvantage of this approach is the computational complexity. The detector has to take into account a huge number of geometric configurations, which leads to an extremely high computational load.

Another way to resist geometric attacks is to evaluate the conversion parameters and invert it. To do this, a certain synchronization pattern is inserted at fixed points in the frequency domain. This synchronization pattern can be a set of peaks or have a more complex form. To increase the security, the pattern may depend on a secret key that is known only to authorized users [5].

The same watermark is periodically embedded in the spatial or frequency domain, and the period of repetition of the digital watermark is known. This period is then estimated by the detector by analyzing autocorrelation peaks. Such templates do not contain confidential information, but only serve for synchronization [9].

The third way to solve the problem of geometric distortion is to use self-synchronizing watermarks. This approach is based on the autocorrelation properties of the watermark [9]. By comparing this period with the expected one, the detector can determine zoom factors and the rotation angle that was applied to the data carrier during the conversion process [5]. Using the found coefficients, the geometric transformations are inverted, and the original position of the counts is restored.

Another solution to deal with the problem of geometric transformations is to embed in areas that do not depend on geometric transformations. For example, to ensure the stability to shifts, the most common solution is to embed the digital watermark in the DFT coefficients of the container.

The next type of algorithms aimed at solving the problem of geometric distortions is based on geometric normalization based on features. The idea of this method is to embed and decode or detect a watermark when the container takes some reference geometric position, that is, it has some reference values of zoom factors, and rotation angle. The reference geometric position has to be assumed relative to the coordinate system that is known to the encoder and detector. To do this, the reference geometric position is assumed relative to any container features. These signs can be corners or edges, if the container is an image.

The reliability of this method depends on the stability of features used for the image normalization. Such algorithms can usually be very sensitive to framing, since reference features may be lost during framing. Next, we will consider the existing algorithms for embedding persistent digital watermarks, which are based on some of the listed methods to counteract geometric distortion.

The algorithms for embedding persistent digital watermarks that currently exist can be divided into two classes:

- algorithms for embedding the digital watermarks in the area of transformation invariant to the required types of distortion;
- algorithms that use "sync labels" [8].

4 Discussion

Many existing algorithms for embedding persistent digital watermarks have a significant drawback: embedding in the frequency areas of the image-container. Accordingly, the necessary stages of embedding a watermark are the forward and reverse Discrete Fourier transformation of the image-container. This approach, although it ensures the stability of the digital watermark to geometric transformations, significantly increases the computational complexity of the embedding algorithm [8]. When embedded in the transformation area, the information is embedded in the container's conversion coefficients. The most common embedding is in the frequency domain of the container. Most often, the Fourier transform or cosine transform is applied to the image-container. Usually, methods of embedding a digital watermark in the container transformation area show higher resistance to various attacks than methods of embedding in the spatial area. In particular, these methods are more resistant to image framing.

Also, embedding in the transformation area shows its resistance to other types of geometric transformations, such as shifting and scaling. This is achieved by choosing the transformation in such a way that it is invariant to the required transformations of the container.

The disadvantage of such algorithms is their instability to "watermark template attack" attacks. In such an attack, the attacker, without knowing the embedding key, can detect and delete the "sync tags". In fact, the intruder finds the peaks – "sync marks" and averages them with neighboring values, which makes it impossible to detect the digital watermark further.

In scientific papers [6], a double ratio of four collinear points was used for embedding a digital watermark. The dual relation is preserved for fractional-linear transformations, so it is invariant to affine transformations. During embedding, sets of four collinear peaks are embedded in the DFT conversion coefficients. At the digital watermark extraction stage, the decoder identifies these peaks and searches for suitable sets of four collinear points.

The algorithm for embedding a digital watermark uses the "sync labels" method, which shows that before embedding a digital watermark, distortion-resistant labels are embedded in the image-container, which allow evaluating the degree of distortion made in the container during embedding, as well as compensating for them [7].

5 Conclusion

An important aspect, in our opinion, is the proposed algorithm that helps solve an important problem – copyright protection. The successful implementation of the national project "Digital economy" contributes to the creation of intellectual capital and the efficient use of intellectual property. In many industries called "creative industries", products are created through the development of creative and human capital. The contribution of creative industries is more than 3% of the world's GDP and should increase several times by 2024.

In the near future, the intellectual property market will become highly efficient and reach a new development level. Engineering and technological discoveries will be implemented on the basis of modern digital infrastructure, the development of innovative systems, methods of regulating the information and innovation economy, and the use of human and intellectual capital. It will have a huge impact on the development of industrial production, increasing competitiveness, and regional development. The authors were motivated by their choice to develop an algorithm by the multidimensional nature of this topic, which covers a wide range of issues related to the formation and development of the digital and engineering economy. All of the above problems require a synergistic approach and scientific understanding of all the structural elements of a complex system in the process of this interaction. Thus, the competitiveness and efficiency of industrial production is determined by the choice of an innovative course and development strategy aimed at the implementation in international technological systems while using their key advantages.

References

1. Evans, N., Price, J.: Human capital and information asset management behaviours in law firms. In Tsui, E., Cheung, B. (eds.) Proceedings of the 14th International Conference on Intellectual Capital, Knowledge Management & Organisational Learning (ICICKM 2017), pp. 63–71. Academic Conferences and Publishing International Limited, Sonning Common (2017)
2. Feldvari, K., Varga, M.: Young people as human capital – what type of information literacy education is required? In: Tonković, A.M. (ed.) 6th International Scientific Symposium Economy of Eastern Croatia – Vision and Growth, pp. 318–326. Ekonomski Fakultet Osijeku, Osijek (2017)
3. Gaysin, F.F.: Problem of copyright protection. Bulletin of the Mari State University. Chapter "History. Law" **3**(11), 73–78 (2017)
4. Haini, H.: Internet penetration, human capital and economic growth in the ASEAN economies: evidence from a translog production function. Appl. Econ. Lett. **26**(21), 1774–1778 (2019). https://doi.org/10.1080/13504851.2019.1597250
5. Huang, W., Li, R., Xu, J., Huang, Y., Hosam, O.: Intellectual property protection for FPGA designs using the public key cryptography. Adv. Mech. Eng. **11**(4) (2019). https://doi.org/10.1177/1687814019836838. Accessed 09 Apr 2020
6. Isinkaye, F.O., Aroge, T.: Watermarking techniques for protecting intellectual properties in a digital environment. J. Comput. Sci. Technol. **12**(1), 27–31 (2012)

7. Kur, A., Maunsbach, U.: Choice of law and intellectual property rights. Oslo Law Rev. **6**, 7 (2019)
8. Rizzo, S.G., Bertini, F., Montesi, D.: Fine-grain watermarking for intellectual property protection. EURASIP J. Inf. Secur. **10** (2019). https://doi.org/10.1186/s13635-019-0094-2. Accessed 09 Apr 2020
9. Schellekens, M.: Digital watermarks as legal evidence. Digit. Evid. Electron. Signature Law Rev. **8**, 152–165 (2012)
10. Townsend, L.G.: Should I use a digital watermark on my work? (2018). https://www.lgt-law.com/blog/2018/01/should-i-use-a-digital-watermark-on-my-work/. Accessed 09 Apr 2020

Managing an Organization's Innovative Development: How to Build a Learning Organization

L. I. Fishman[1], A. P. Zhabin[2], O. V. Karsuntseva[3], and Yu. P. Grabozdin[1](✉)

[1] Samara State Social and Pedagogical University, Samara, Russia
fishman@pgsga.ru, yg63@bk.ru
[2] Samara State University of Economics, Samara, Russia
apzhabin@yandex.ru
[3] Samara State Technical University, Syzran Branch, Syzran, Russia
samgtu209@mail.ru

Abstract. The article is aimed at analyzing the problem of transferring decision-making and control powers to lower levels of management subsystems of organizations in the framework of their innovative development. The question is raised about the expediency of using the methodology of quality management in educational services developed by the authors, based on the delegation of feedback as a mechanism for improving the quality of education. It is concluded that the methodology of setting key performance indicators for managers and employees, which is successfully applied in a number of management situations, cannot be effectively used in the process of building control in the management of innovative development of organizations.

Keywords: Delegate feedback · Learning organization · Rationalization of management

1 Introduction

The most important aspects of working out the organization's innovative development strategy are decision-making and control, which are usually focused on the level of top management, while the development and implementation are carried out on the levels below: middle management and specialists of the management subsystem. This imbalance inevitably leads to significant time expenditures, numerous adjustments and corrections in these processes, and often significant defects in the development and implementation of innovative development strategies, in particular, in the economic and social aspects of the organization's development. In this regard, the scientific problem of transferring decision-making and control powers to lower levels of management subsystems in organizations is relevant within the framework of their innovative development. This transfer should not lead to a loss of manageability in this process.

S. I. Ashmarina et al. (Eds.): ENGINEERING ECONOMICS WEEK 2020, LNNS 139, pp. 473–479, 2021.
https://doi.org/10.1007/978-3-030-53277-2_56

2 Methodology

Modern management concepts describe an organization as innovative and developing one. The organization is seen not only as a *learning organization*, where each employee is a builder of a new reality that changes the company [1–4, 8], but also as a *self-learning system*, where employees influence the learning process consciously while building a corporate environment [5–7, 9, 10]. The scientific literature presents a number of features that characterize learning organizations, namely: employee's involvement in decision-making, monitoring of the internal and external environment by all participants, established intra-corporate contacts, the distributed nature of power, the network nature of interaction in the organization, etc. [1, 3, 10]. In such organizations, edges between work and learning are blurred, and management control is primarily aimed at monitoring the external environment, customer requests through personal contacts and a balanced information management system. As a result, in such (often used term "intellectual") organizations, the employees decide where to apply their knowledge, how to build corporate relationships and show their experience and talents, depending on the current situation [1, 3, 10–12]. However, it is important to note that available publications do not provide operational answers to the question: how to ensure these characteristics of organizations.

The use of the concept of a learning organization and the task of its operationalization, in our view, make it appropriate to use the previously developed methodology for quality management of educational services, based on the delegation of feedback as a mechanism for improving the quality of education. Giving an organization characteristics of a "learning", innovative company does not mean refusing the role of its top management in setting development goals. Therefore, top management and middle managers of an organization perform approximately the same role as managers and teachers in an educational organization [2–4]. Finally, to extrapolate the methodology of quality management of educational services to the concept of "learning organization" in order to operationalize the latter, it is advisable to use an information approach to the analysis of management processes [10–12], supplemented by the idea of internal activity of all managers and employees of the organization in collecting and processing information [2–4].

3 Results

In accordance with the methodology of the information approach, management can be considered as a process of collecting, processing and issuing information, and feedback as data carried out precisely by receiving and processing information about activities of a managed subsystem. As the main subjects of management can be considered the head of the organization (top management), middle managers, employees (just like in the educational sphere, the main subjects of management are the head of the organization, teachers and students). This hierarchy of the management structure allows us to consider the management of the organization as a set of ways to interact with the head (top management), with middle management and employees. At the same time, the middle management and employees should be subjects-objects of management, and therefore active users of information. Therefore, information modeling of all subjects' activities should be built as a model of interaction of a "middle manager" as an object with two subjects: with the top management and the middle manager himself, and the latter acts as both an object and a subject of information exchange. The same can be applied to an ordinary employee.

The developed approaches allow us to describe the problems of rationalization of decision-making (with the transformation of the organization into a learning one) at all management levels, namely, the rationalization of feedbacks, i.e., their construction, in which the higher-level subject of management would minimally interfere in functioning, and would be minimally engaged in their correction. This, obviously, can be achieved by a certain delegation of authority – "delegation of feedbacks", i.e. they are temporarily or permanently redirected to lower-level entities, which are now starting to operate in the offline mode.

The problem of transforming an organization in a learning one becomes a problem of reducing the rank of the information user (in some cases with the necessary reduction in the source rank) and organizing reverse and internal links (internal feedback), which is equivalent to the transfer of internal links of the head (through which he receives the basic information for his making management decisions) into internal links of middle management. Similarly, building rational management systems at lower levels involves a similar transfer to employees. Thus, the implementation of feedback is a method of activity that is constantly implemented by all management subjects, and therefore feedback should be considered in management not so much as a way of obtaining information by a higher management subject, but as a way of influencing all lower management levels.

In other words, all subjects (and not just senior managers) should be considered both as active users of feedback channels and sources of information. Thus, it is necessary to refuse the absolute objectivity of information and the assumption of independence of processes occurring in the organization from the content, methods of obtaining, choosing sources and users of information when building feedbacks. On the contrary, in the process of collecting information, it is necessary to communicate goals to all management subjects and ensuring that they accept these goals. At the same time, the idea of the maximum objectivity and comprehensiveness of performance evaluation can be one of the declared principles for selecting evaluation parameters, and not a direct goal.

Any assessment of lower-level subjects should be considered, first of all, as a way to exert managerial influence on their activities. Moreover, the success of building an internal control system should be considered primarily in terms of its impact on certain aspects of the activity and (or) professional consciousness: the evaluation result is not to measure and evaluate, but to influence the situation. Thus, when building a system for monitoring activities of a management subsystem of the evaluation organization, it is necessary to rely on principles that provide:

- consideration of all subjects of the named subsystem as active users of information coming through the feedback channels, and at the same time its sources;

- refusal to idealize the absolute objectivity of information, as well as the independence of decision-making processes at all levels of management from the content, methods of obtaining, selecting sources and users of this information;
- delegation of feedbacks – their temporary or permanent redirection by the top management to lower-level subjects whose performance is evaluated; the necessary conditions for which are the unambiguity of requirements, evaluation parameters, criteria, as well as the availability of technologies and evaluation methods.

Taking into account the specifics of measurements in social systems, we can say that the measurement results and thus the interpretation of these results will be most significantly influenced by the characteristics of decisions made by the subject of assessment: the model of this reality fragment developed by him and the measuring tools used by him. In this case the evaluation results are already embedded in the model and methods of measurement in a certain sense.

The interpretation of the source data is initial in relation to the results interpretation of measuring the quality of decisions made. In the process of evaluating the quality of decisions, the information user is always interested only in certain properties and relations of certain objects in the managed system. Therefore, the source data – a specific system with many complex relations and links within it – is processed into a model in which each object of such a system is considered as a carrier of certain properties, not in the entire set of complex relations, but as a carrier of certain relations. The second (formal) level of input data is the set of measurement results, when certain mathematical values describing their properties are assigned to objects of the evaluated system. Using various methods of analysis or evaluation criteria, the subject works with this formal system, and in the process of final interpretation of the measurement results makes again the transition through the model to a specific system. Therefore, the interpretation of source data in the assessment (control) process is the formalization of a fragment of the organization's reality that is of interest to the user (subject).

Based on any hypotheses, the subject – user of information always assumes the presence of certain relations between the elements of the model created at the stage of data interpretation that reflect properties of the evaluated system (even when there was no formal modeling of these properties for the actual system). Also, the subject (user of information) almost always assumes the uniformity of the initial data, without which it is impossible to provide a reasonable interpretation of measurement results. This forces us to rely on interpretation principles:

- the principle of interpretation matching, which determines the need to align the interpretation of the analysis results with the interpretation of the source data;
- the principle of complementarity of formalism, which determines the need to to reflect those meaningful assumptions that were not used in the interpretation of the original data while interpreting the analysis results.

Thus, it is possible to formulate several basic requirements for the interpretation of source data, i.e., to develop indicators and procedures for obtaining data in the process of monitoring activities of the organization's management subsystem.

1. Compliance of the received information with the management goal. This requirement is equivalent to the requirement to interpret the source data. It involves

building exactly the model of reality, fixing the characteristics of a real organization that interest the subject (user), in this case, the strategy of innovative development.

2. Compliance of subjects (sources of information) and measuring tools with the management goal (the goal of influencing the quality of decisions made). This requirement means creating an algorithm for displaying elements of the evaluated organization, which allows you to guarantee the full compliance of the elements' relations in the real system and the elements' relations in the developed model. In this case, the algorithm should allow you to fully and reliably transform all the relations and properties of the elements of the evaluated system that are of interest to the user into the relations and characteristics embedded in the model.
3. Compliance of the influence produced by the measurement itself with the management goal. This requirement when building control should be implemented through "excessive" interpretation of the raw data, which helps to minimize interpretation of results by measuring the quality of decisions: the selection of indicators, indicators, methods of measurements should be made so that the evaluation grade (high or low, indicating the increase or decrease of some aspect of the decision quality) follows directly from the measurement results, and the object of the results interpretation is just a reason for these values.

4 Discussion

As it was shown above, one of the conditions for building an effective management is its rationalization, involving such a design of feedback that the parent entity of the management hierarchy would minimally interfere in functioning of the system, minimally adjusting its activities. A delegation of feedback may be considered as a way of rationalization, implying that the process of monitoring and evaluating the quality of decisions should encourage the lower subjects in the management hierarchy to improve their activities. Moreover, this improvement should be a result of the impact of the measurement procedure, and not of the subsequent direct management impact.

The fundamental point for building such a mechanism is a clear compliance of control tools with goal setting (in this case, in the field of innovative development of the organization). The more difficult creating such a tool is, the more goals are pursued and the greater the discrepancy between explicit and hidden goals is allowed. It is also important to ensure clear understanding of the innovative development goals by a subject whose performance is being evaluated, and therefore whose goals are being measured or evaluated. This is significantly facilitated by the choice of a control method. Thus, when evaluating the quality of decision-making, it is necessary: - to determine which sides or aspects of the activity could and should be effectively affected in the assessment process; to select the appropriate objects (groups of objects) and evaluation parameters, taking into account their limits; to develop appropriate measurement tools and techniques based on the need to use fairly simple, resource-intensive technologies; to identify adequate subjects (sources of information) for each object and tool used.

Therefore, in the process of developing an innovation development strategy in the organization, it is advisable for the top management to set indicators for middle management and specialists (strategy developers) related, for example, to the absence of patents in the field of innovation being developed or the presence of a certain number and quality of patents; minimum expected economic results, expected social effects, giving freedom to performers to act within these restrictions. In other words, restrictions at the first development stages should not concern the content of the strategy. At later stages of strategy development and implementation, "the leash may be shortened somewhat". However, at all stages, there should not be many restrictions, and they should not be the focus of attention: everything is allowed that is not prohibited.

5 Conclusion

Thus, the use of the concept of a learning organization, coupled with an information approach to the analysis of management processes and extrapolation of the concept of quality management of educational services to the management processes of the organization innovative development, leads to the conclusion that it is necessary to maximize the transfer of decision-making rights and control over the own activities to the lower levels of the management subsystem. The mechanism that ensures this transfer without loss of control is the delegation of feedback from top management through the most accurate recording of the principal results (in particular, economic and social effects), indicators and ways to assess their achievement. In this case, representatives of middle management and specialists can freely "create" (while learning), without going beyond the fixed principal results and effects. The organization thus becomes learning, innovative, but remains manageable. This, in particular, means that the methodology successfully applied in a number of management situations for setting key performance indicators for managers and employees (based on the idea of the most objective assessment) cannot be effectively used in the process of building control in the management of the organization's innovative development. In conclusion, it is necessary to fix a fundamental difference between developing approaches for quality management of "normal" educational services (where all the results are known in advance, standards are set), but a significant similarity with the model of modern higher education, where only the exterior frame is defined, not limiting the student (researcher), and only the master vector of self-study is defined and results often are innovative.

References

1. Annosi, M.C., Martini, A., Brunetta, F., Marchegiani, L.: Learning in an agile setting: a multilevel research study on the evolution of organizational routines. J. Bus. Res. **110**, 554–566 (2020)
2. Braham, B.J.: Creating a Learning Organisation. Elex Media Komputindo, Jakatra (2003)

3. Fishman, L.I.: Methodology and mechanisms for improving the quality of educational services based on delegating feedback. Russian Academy of Sciences, Samara Scientific Center, Samara (2009)
4. Fishman, L.I.: Specific educational services in the context management improve their quality. Samara J. Sci. **2**(3), 61–65 (2013)
5. Fishman, L.I., Sosunova, L.A., Noskov, S.V., Agafonova, V.V., Sivaks, A.N.: Conceptualization of notions and formation of quality indicators of educational services. Helix **8**(5), 3651–3660 (2018)
6. Hughes, J., Liu, J., Liu, J.: Information asymmetry, diversification, and cost of capital. Account. Rev. **82**(3), 705–729 (2007)
7. Pahurkar, R.N., Nair, J., Kolte, A.: Measuring the impact of learning organization on employee retention, competitive advantage and financial performance in the BPO industry. Int. J. Recent Technol. Eng. **8**(3S), 546–553 (2019)
8. Pedler, M., Burgoyne, J., Boydell, T.: The Learning Company: A Strategy for Sustainable Development. McGraw-Hill Book Company, London (1991)
9. Rupčić, N.: Learning-forgetting-unlearning-relearning – the learning organization's learning dynamics. Learn. Organ. **26**(5), 542–548 (2019)
10. Senge, P.M.: The Fifth Discipline: The Art & Practice of the Learning Organization. Bantam Doubleday Dell Publishing Group, New York (2004)
11. Tsvetkov, V.Ya.: Information interaction as a mechanism of semantic gap elimination. Eur. Res. **45**(4–1), 782–786 (2013)
12. Volkodavova, E.V., Zhabin, A.P., Yakovlev, G.I., Khansevyarov, R.I.: Key priorities of business activities under economy digitalization. In: Ashmarina, S.I., Vochozka, M., Mantulenko, V.V. (eds.) Digital Age: Chances, Challenges and Future. Lecture Notes in Networks and Systems, vol. 84, pp. 71–79. Springer, Cham (2019). https://doi.org/10.1007/978-3-030-27015-5_9

Check for updates

Transdisciplinary Strategy of Continuing Engineering Education

F. F. Sharipov(✉), T. Yu. Krotenko, and M. A. Dyakonova

State University of Management, Moscow, Russia
{fanissh, krotenkotatiana}@rambler.ru, marie.d@mail.ru

Abstract. Acceleration of developments in science and technology requires a new approach to training and retraining of personnel working in real economy and first of its elite – engineering personnel. Globalization of world economy presents new claims to the quality of education of contemporary engineers. National engineers are facing a task of modernization of technology in industry, access to new markets with up-to-date knowledge-intensive products. In this connection, present-day engineering education requires comprehension of its cultural traditions, philosophical and historical framework. Starting from his college years integration of a present-day engineer into the current system of knowledge management, accumulation, storage and conveying new knowledge actualized as intellectual property becomes a key factor for the successful engineering training. This requires available modern laboratories, librarian search engines with segmentation function, system of protection and access to necessary information, interaction with the university industrial partner based on a twofold education. The prospect to partake in technology enterprise, in project work, in grant schemes, start-ups and venture funds is a popular area in engineering training. Thereupon the authors establish the development and realization of a new strategy of continuing engineering education on the grounds of transdisciplinarity and association of engineering, economic, linguistic and IT knowledge.

Keywords: Engineering and economics · Engineering education · Innovative development · Knowledge management · Transdisciplinarity · University ecosystem

1 Introduction

Engineering and economics is a field of knowledge emerged at the intersection of natural, technical and economic sciences. It considers engineering capacity to ensure competitiveness of products and services in market conditions by using economic tools. Current tasks in engineering and economics require an individual educational model [19]. To our point of view the concept of engineering and management dominating in the 19th century and which reached its top point in the beginning of the 20th century, when elements of economic science and management naturally entered engineering tuition (and we will talk about it below) is becoming relevant today again.

Tasks which contemporary engineering economy is facing today requires from engineers and managers, and they are the major actors in production operation,

S. I. Ashmarina et al. (Eds.): ENGINEERING ECONOMICS WEEK 2020, LNNS 139, pp. 480–488, 2021.
https://doi.org/10.1007/978-3-030-53277-2_57

transdisciplinary knowledge, comprehensive engineering and managerial training, which includes humanities, mathematical and natural sciences. Articulated priorities for the scientific and technological development of Russia set goals of technological renovation of its industry, development of knowledge-based sectors of economy and promote national innovative products to global markets [21]. Problems of engineering education in view of present-day requirements are obvious.

Transition of the economy to a new economic order sets new tasks in training and retraining of present-day engineering personnel [20]. First, it is necessary to know more than one foreign language. Today the English language is a communicator in global knowledge management system. The second foreign language is required as a key element in cross-cultural communication depending on the region selected by business goals. (In Europe – German, French, Spanish, and Italian to your choice; Chinese and Arabic in Asia). Second, a new type of computer competence for digitalization of production, procurement, sales in new reality of a "smart enterprise", "smart region", "smart country". Now this refers to harmonization of production operation, social infrastructure of all types, and services delivered by the state [9, 15, 17]. Third, the necessity to create a new economic reality creates new occupations like solar panels installer; wind turbine service operator; environmental specialist/scientist; cybersecurity specialist [4, 22]. Fourth, a new model of higher education system is taking place. The development of students' research work in higher education institutions is having a new type of infrastructure. Laboratories of a new type with levels of access and protection of information and librarian search engines are necessary for the effective participation of students in a new global knowledge system. Fifth, in this approach tutors-researchers come to the front of the stage. They involve students in their work; invite experts from certain fields of knowledge or research to the university [6]. Sixth, this model of tuition dictates the necessity of facilities for practical training where students could continue work following the results of practice and qualification tests [18]. Here arises duality of training engineers of a new type, possibility of alignment of theoretical background with practical tasks on their future work place. Seventh, here arise international aspects of engineering training. Globalization naturally brings technology transfer. However, due to the existing system of protection of national intellectual property international research projects are easier to declare than to implement. Even in countries where students go to study and intend to work later.

Therefore, the eighth, there required an up-to-date real time system of authorization of new knowledge, protection of intellectual property in a new environment where school and production enterprise are integrated into one system. Quality of solving the above tasks will determine the efficiency of the nation transition to a new economy.

2 Methodology

Methodological base for the analysis of philosophical, historical and economic aspects in engineering education and drawn conclusions served scientific content of sites of Russian and European research and educational institutions where the headlines and content has such terms like engineering and economics, engineering and economic education, transdisciplinarity, continuity, knowledge management and ecosystem of university. We also used as a source of information views of national and international experts in engineering education on the concept of its transdisciplinarity and continuity which are available in cross references system.

3 Results

To our view transdisciplinarity is one of the important modern trends in engineering education. It is a frequently used term in scientific discourse. Boarder lines separating different fields of knowledge are becoming less tight and more permeable. Flows of information circulating despite rigid disciplinary limits are becoming powerful and acting more and more confidently. Open public discussion of important scientific information is gradually becoming a norm useful for solving vital issues of the society. There is no need today to prove that the world is complicated, multi-dimensional and changeable. It is logical that human consciousness is seeking means to reflect this complexity. Additionally to the forms of cognition, the necessity to combine meanings located outside fields of specific subjects' is becoming a condition for the development of scientific and practical knowledge.

The concept of transdisciplinarity provides broad opportunities for the interaction of heterogeneous information and knowledge environments for solving integrative problems of Nature and Society. The potential for success of the research visible on meeting edges of different substantialities, reasoned moving beyond strict scientific boundaries, new directions of research that connect previously not intersecting fields of knowledge, real achievements of complex experimental teams - all these come from transdisciplinarity. Good perspectives of this direction of development of science and practical experience are proved by the positive results of breakthrough science and technology teams and can be considered as a competitiveness factor of the engineering economy.

It is impossible today to discuss such categories like "man", "knowledge", "consciousness", "information", and "communication" within a sole substantive area especially when worldviews combine [14]. Emerging transdisciplinary theories and concepts become a solid foundation for the innovative engineering designs. The fact is that transcendent considering of strictly disciplinary knowledge roots in the very nature of practices, which are conveyed by cultural tradition, and engineering is exactly such kind of practice. Man's capacity for invention and designing is promoted not only by science. Since the earliest times community services is supported by the people's wisdom, religion, philosophy, art, technique equipment, craft and even by "professional magic". Experience of bright engineering designs in its essence is not limited to the rigid disciplinary boundaries, findings and discoveries occur due to the author's openness to different non-academic means of perception of the world. Examples of successful engineering decisions on complicated tasks implemented in transdisciplinary mode lift many methodological contradictions, which seemed insurmountable before.

We see continuity as another global trend in engineering education. Starting from the 1960s of the 20th century and up today the idea and attempts to create a concept of "continuing education" is the main ideological, organizational and sociocultural concept in many countries. Continuing education emerged to certain extend as an alternative to the existing educational programs subordinate to the industrial tasks and preparing "specific" workers for the vacancies on labour market sees its tasks differently. It must be concerned with the universal education of people. This is the only way to develop individual's natural capabilities and use this limitless potential during all his working life to the benefit of society.

The idea and reasons of this approach are the following. Human being needs continuing realization of individual plans and undertakings, backed up by the required skills and knowledge but not a discreet dividing his life in stages - study, labour self-fulfilment, professional aging. This requires performing unbiased review and re-evaluation of the resources available in the system of education. Education has to be an active non-interrupted process within the whole lifetime of an individual, which means that it cannot be enchained and split by periodization. Knowledge is valuable if obtained in appropriate mode and in good moment, but not when released to be learnt in frame of certain period of life. This perception of the value of education calls for integrated social, political, economic and individual aspects.

French researcher Lengrand attempted to formulate conceptually the idea of continuing lifelong education at UNESCO Forum in 1965 and summarized it in the book "Introduction a l'education permanente" [11]. The world is one whole thing and all its elements are interrelated and interdependent. Human being empowered with the reason, values, strivings and interests is the centre of all world processes. The world is becoming more and more complicated since human being is not only effected by society and nature but is an active subject himself. Human capacity to make quality decisions that are adequate to challenges of time requires new contents and respective means of cognition and hence a consistent review of goals, techniques and technologies of education and tuition. In the middle of the 90 s there was published a book "Areas of Learning Basic to Lifelong Education" edited by Lengrand [12]. In this book, the authors sum up achievements of established models of engineering, economical and managerial education in Europe after the 50s of the 20th century and discuss possibilities of integration of different sciences for lifelong education. In this view a human being is seen as the main object of education and the goal is evolving capacities for self-realization interesting for him and useful for the society.

4 Discussion

We see it appropriate to discuss philosophical and historical foundations of engineering education. Logic of contemporary life patterns leaves the question of concordance between what and how is implemented in educational institutions and what and in which way is going on outside them (while formal, non-formal and informal subsystems constitute continuing engineering and managerial education) is open today but it is not new.

The idea of constant enrichment of inner world is expressed in the texts of great philosophers. In their works on the problems of cognition, they talk about tireless perfection of man. These texts prove that the idea of continuing education did not emerge in present-day technological and managerial universities. The idea of lifelong education is the result of continuing in time polylogue between heterogeneous cultures and manifested in specific educational systems. The ideal originated in antiquity presents an individual who improves his intellectual, physical, emotional well-being by means of education. Learning is a vital necessity for human spirit in the same way as food for the body.

Sense that the idea of continuity and wholeness of education is not new comes not only after reading ancient philosophical texts. Humanitarian views of Enlightenment philosophers probably express the idea of continuing engineering education in a clearer way. Philosophers and educators see the possibility of improvement of man and society through the development of sciences. This logic looks like this: there should be an expansion of knowledge; purposeful education of man who can handle the deficiencies of the society; it is necessary to change social reality in the interest of progress and improvement of civil society; and this is only possible by pursuing knowledge, natural history, humanist ideals, the possibility to fulfil and manage the proposed changes.

Komenský the great Czech reformer said that the only true meaning in life is to set and achieve goals of cognition [7, 8]. Any age suits to open yourself to the new. These views of the genius innovator on continuing education formed the basis of a respective educational concept, which turned into a class-and-lesson conveyor – a powerful didactic machine incomparable by its influence on minds to any later ones.

We can find a fore type of lifelong education in the works of Marie Zhean Antoine Nicolas de Caritat, marquis de Condorcet - a philosopher and political figure in the period of French Revolution of 1789. In his programme of public education, described in the work "Sketch for a Historical Picture of the Progress of the Human Mind" [2], he puts forward the principle of universality that is of total education in different directions of all without any exceptions citizens. Such inclusiveness shall guarantee people the ability to maintain old knowledge and skills at suitable for their occupational activity level and acquire new knowledge and a confident look in future for their self-realization in the changing world.

To develop in future reformer a comprehensive picture of the world European tradition of training an engineer in 19th – beginning 20th century combined two approaches - scientific, technological and moral. The word "engineer" roots back to Latin "ingenium" (specifically in Petronius and Cicero works) and indicated first of all a bright mind, talent, broad outlook and not only the capability to devise and invent. Philosophy from Herder to Schleiermacher and Hegel saw "Wissenschaftliche Bildung" (academic learning) as a combination of scientific, musical, mathematical and engineering education. To the understanding of German classical philosophers, a universal, well-educated and industrious engineer is the man who is building his personality and carrying on a celestial process of Creation in history and culture. Gymnasiums and universities in Germany is quite an earthbound and a concrete implementation of the elevated idea of preparation of Maker, designer, a reformer of life and practice. "Educational province" by Goethe [3] (as a relatively enclosed system, implementing humanistic model of raising and educating young people) became a prototype of Castalia – an unreal country emerged in a few centuries after the age of "epoch of feuilletons" in industrialized Europe. For Hermann Hesse in his famous "Das Glasperlenspiel" (The Glass Bead Game) Castalia is a locality isolated from the rest of the world where intellectuals received moral education and a lifelong tuition [5].

In the 18th century, Mikhail Vasilievitch Lomonosov promoted engineering by his scientific activities, educational views and his life itself given up to science and development of education in Russia. He strongly advocated the democratization including engineering education; he considered it important to advance engineering knowledge by lectures, published materials, libraries and museums. In this period, scientific method becomes the main in higher education. The main objective of "people's school" in addition to cultivating in children love for work, teaching them proper rules of behaviour, becomes setting out the ideas about the world order and causal relations between its phenomena.

The core of Russian engineering school of 18–19 century was a profound mathematical and scientific education with a considerable humanitarian component. Graduates of higher technical institutions received deep scientific knowledge, were knowledgeable, excellently knew philosophy, world history, literature, theology, foreign languages, played musical instruments, and drew. The base part of professional competences of a civil engineer in Russia of the beginning of 20th century consisted of technical and artistic disciplines. It should be noted that Russia was not the only country in this approach: engineering activity in France and Germany was thought of and implemented at the intersection of innovative research and technological practice. Here lies the distinction from the English approach strictly focused on practical training of technicians and artisans.

Before the First World War a future engineer was not guided just to the invention but to a comprehensive realization and a completion of the project (building, ship, bridge, engine etc.). Based on the materials of the design, organization of introduction of innovation students prepared study guides and circulated useful experience. Universities fostered in their graduates sense of everyday diligent work and the desire of a constant increase of its effectiveness. It may look strange today but engineer's work was not only to invent but also to generate more economical technologies and decisions; to design while making the product cheaper, improve productivity. Moreover, an engineer did this in a strong alliance with an entrepreneur [1]. This productive cooperation of an engineer and a business person also fell apart together with the well-developed industry of pre-revolutionary Russia.

It is very important to note that higher engineering institutions in Russia trained not just good technicians, but were consistently preparing their graduates for the career of captains of industry, for military or government career, for serving Tsar and Fatherland. Mendeleev, Ipatyev, Shukhov, Krylov, Vyshnegradsky are bright researchers-engineers, leaders of industry and education and were outstanding statesmen. It is astonishing to see reading works of old school engineers how elaborate they were selecting partners, sources of financing, locations, materials, environment, safety modes in operations, costs controls, organization of transportation, personnel management on locations at construction of architectural installations. Such concentration on details, precision in calculations is easy to understand: it was a direct, first-hand conveyance of non-formal knowledge thanks to the family, dynastic tradition of education; and in addition technical curricula of engineering universities included compulsory social disciplines. It was mandatory for the lead engineering universities to have social psychology departments. Alternatively, there were engineering departments in big commercial institutes in large cities.

Economic development of Russia and the events of the 20th century amassed engineering education on the one hand and destroyed it integrity on the other. It should be noted that in the West there was also a visible movement from universality towards speciality, the development of high technologies was going in large corporations; occupation of engineer-researcher was becoming widespread [16].

USSR rejection of market based commercial production, development of sophisticated, knowledge- intensive technologies only in large state enterprises (heavy, chemical industry, military space sector) resulted in deterioration of engineer's economic and managerial proficiency.

An engineer with a core technical education could hardly count on a position of the manager of a large enterprise – controls there were handed over to the "party leaders", "academic scientists", "experienced managers". General Designers grown in old school also became directors. Old school should be commended: most of the managers of design offices were thinking on a big scale, could see things as a complete whole, understood strategic goals and could manage personnel and scientific research in high-technology enterprises.

Of course, in young big soviet country there should be a genuine training centre of professional managers. Moscow Commercial-Economic Practical Institute (State University of Management today) – is in fact the first higher engineering–economic institute. Alexander Commercial School of Moscow Stock Exchange Society founded in 1885, Nikolay Female Commercial School and Trade School after the emperor Nikolay II name united in 1918 in Moscow Commercial–Economic Practical College. In 1919, it was renamed in institute. For hundred years this higher institution prepared tens of thousands of managers for key sectors of the national industry. But they were trained specifically for the key sectors and unfortunately with the rapid decline in quality of teaching engineering disciplines.

The research process natural for the traditional higher engineering and managerial institutions was redirected into research institutes and design offices (once again per sector principle and for mostly military tasks). Negative consequences of such transformation of the 20 s became visible in fifty years after that. And today return of research in school and university as the main type of their activity is restrained by the fact that institutional framework of current research and development works is concentrated in network communication space between professionals from different fields of knowledge. Current research more and more often is carried out by small breakthrough aimed teams, supported by a large database with the developing possibilities of their aggregation [13].

5 Conclusion

Industrial revolution brings new requirements to human capital and thus to the educational system too. The necessity of continuing retraining of personnel and first of all engineers caused the emergence of corporate educational systems all around the world. The brightest example in Russia is probably the Academy of the state corporation Rosatom. This educational institution is training 3600 reserve personnel without discontinuing their main professional activity and has 900 tutors, 13 of whom are from Top-30 management.

Further, we will see the development of a fusion of formal, non-formal and informal types of education, which will support a dynamic model of competences generated to technological and social requirements that has a long built constant as a base and a flexible, easily renewed variable component. Educational path of a future engineer is a good quality comprehensive engineering-managerial education and a methodical upgrade of his skills with additional professional education. Diploma, which conforms to this model, will also be of dynamic nature and will consist of the base part and a modular tuned for the specific tasks of operation, reflecting soft and hard skills of its holder [10].

As for the educational activity of higher engineering institutions it is very important in our view to develop their business ecosystem where students and tutors temporally brought together by the course named "engineering enterprise" would contact parties in engineering and business sector of the economy, officials assigned responsible for the technological development of a city, region and the country. Such multi-sided format will allow to naturally select the best engineering projects, optimize engineering decisions, promote them in professional society and media, establish socially beneficial contacts, set up work groups and breakthrough teams and develop a culture of engineering business. The current strategy of technological development of Russia formulates a new strategy of continuing education.

References

1. Beiranvand, V., Lucet, Y., Hare, W.: Best practices for comparing optimization algorithms. Optim. Eng. **18**(4), 815–848 (2017). https://doi.org/10.1007/s11081-017-9366-1
2. Condorcet, Zh.: Sketch for a Historical Picture of the Progress of the Human Mind. Librokom, Moscow (2011)
3. Goethe, I.V.: Collected works in ten volumes. Volume 8. The years of Wilhelm Meister's wanderings, or the renouncers. Fiction, Moscow (1979)
4. Hansen, D., Møller, N.: Conceptualizing dynamic capabilities in lean production: what are they and how do they develop? Eng. Manag. J. **28**(4), 194–208 (2016). https://doi.org/10.1080/10429247.2016.1238727
5. Hesse, H.: Das Glasperlenspiel. Suhrkamp Verlag, Frankfurt (2001)
6. Jeffcoat, K.L., Eveleigh, T.J., Tanju, B.: A conceptual framework for increasing innovation through improved selection of specialized professionals. Eng. Manag. J. **31**(1), 22–34 (2019). https://doi.org/10.1080/10429247.2018.1548231
7. Komenský, J.A.: Great didactics. Selected pedagogical works: in two volumes, vol. 1. Pedagogy, Moscow (1982)
8. Komenský, J.A.: The universal council about the correction of human affairs (selected chapters). Part four. Pedagogy, Moscow (1989)
9. Konstantinov, V., Sakulyeva, T., Makeeva, V.: Development of economic tools for managing regional innovation clusters. J. Entrepreneurship Educ. **22**(1S) (2019). https://www.abacademies.org/articles/development-of-economic-tools-for-managing-regional-innovation-clusters-7988.html. Accessed 20 Dec 2019

10. Kulakov, N.Y., Blaset Kastro, A.: Evaluation of nonconventional projects: GIRR and GERR vs. MIRR. Eng. Econ. **60**(3), 183–196 (2015). https://doi.org/10.1080/0013791X.2014.1002053
11. Lengrand, P.: Introduction a l'education permanente. Unesco, Paris (1970)
12. Lengrand, P.: Areas of Learning Basic to Lifelong Education. Pergamon Press, Oxford-New York (1986)
13. Ren, X., Rahman, S.: Stochastic design optimization accounting for structural and distributional design variables. Eng. Comput. **35**(8), 2654–2695 (2018). https://doi.org/10.1108/EC-10-2017-0409
14. Rozin, V.: Reflection of the bases of interdisciplinary study of sociality. Quest. Philos. **1**, 64–73 (2020). https://doi.org/10.21146/0042-8744-2020-1-64-73
15. Ryazanova, G.N., Sazanova, A.A., Sazanova, S.L.: The influence of the processes of digitalization of the economy on the activities of non-financial institutions. Management **2**, 52–56 (2018). https://doi.org/10.26425/2309-3633-2018-2-52-56
16. Savage, G., Franz, A., Wasek, J.S.: Holacratic engineering management and innovation. Eng. Manag. J. **31**(1), 8–21 (2019). https://doi.org/10.1080/10429247.2019.1565467
17. Sellitto, M.A., Hermann, F.F.: Influence of green practices on organizational competitiveness: a study of the electrical and electronics industry. Eng. Manag. J. **31**(2), 98–112 (2019). https://doi.org/10.1080/10429247.2018.1522220
18. Sergeeva, M., Shilova, V., Evdokimova, A., Arseneva, N., Degtyareva, V., Zuykov, A.: Future specialists socialization in the context of competence approach. Revista Praxis Educacional **15**(34), 571–583 (2019). https://doi.org/10.22481/praxisedu.v15i34.5796
19. Tinyakova, V.I., Morozova, N.I.: The searching vector of a new educational model in a knowledge-based economy. Account. Stat. **1**(49), 105–111 (2018)
20. Tinyakova, V.I., Morozova, N.I., Gunin, V.K.: Transformation of the system of professional training of personnel competitive under the conditions of economy, based on knowledge. Econ. Sustain. Dev. **37**(1), 242–245 (2019)
21. Yuryeva, A.A., Kuternin, M.I., Gibadullin, A.A., Romanova, Ju.A.: Formation of mechanisms for the development of innovative activity in the industrial production of the Russian Federation. In: Journal of Physics: Conference Series, vol. 1399, no. 3, p. 033099 (2019). https://doi.org/10.1088/1742-6596/1399/3/033099
22. Zimnukhova, D.I., Zubkova, G.A., Morkovkin, D.E., Stroev, P.V., Gibadullin, A.A.: Management and development of digital technologies in the electric power industry of Russia. In: Journal of Physics: Conference Series, vol. 1399, no. 3, p. 033097 (2019). https://doi.org/10.1088/1742-6596/1399/3/033097

Legal Regulation of Big Data in Industrial Systems: Problems and Development Prospects

E. Kh. Gubaydullina(✉) and A. N. Churakov

Samara State University of Economics, Samara, Russia
elmira_zaripova@mail.ru, Churakov-vvo@mail.ru

Abstract. The article analyzes the concept of "Big Data" in terms of semantic content and legislative regulation. It includes a comparative review of legal regulation and law application of this technology in Russia and foreign countries. In this article problems of legal regulation of the application of this technology are identified. It has been established that the most significant factor hindering further development of Big Data in Russia is, firstly, the lack of "Big Data" legal definition. Yet secondly, there is constant search for a compromise between the need of business entities to ensure the maximum possible availability of data so that the business is cost-effective,, on the one hand, and on the other hand, the urgent need to ensure the proper level of confidentiality and security, especially concerning the protection of personal data. Despite a number of existing problems, the article points to the prospects of using Big Data not just in industry, but also in other areas of society.

Keywords: Big Data · Digital economy · Judicial acts · Objects of civil rights · Profitability

1 Introduction

The introduction of such a phenomenon as Big Data into the public and economic life of Russia is undoudtedly a big step towards the development of "digital economy". In accordance with the big data market development strategy until 2024, developed by the Big Data Association (BDA) together with BCG, the overall economic effect and revenue growth of all Russian industries due the use of big data is expected to result in 1.2% up to 3% of GDP growth. Moreover, such a result will be possible due to the additional revenue of Russian companies, which they will be able to get due to the development of products, services and big data technologies [7].

Basically Big Data means not even the information itself but information and technology (as a combination of huge amounts of data processing methods from various sources). This concept (Big Data) consists of two elements. The first element includes the so-called "raw data", which will include the speed of the appearance of new data, the processing speed of this data, as well as huge physical volumes. The second element of Big Data are special developed technologies and methods that are used in the processing and application of this data for commercial purposes [1]. Sources of Big Data are various containers of information, particularly Internet,

S. I. Ashmarina et al. (Eds.): ENGINEERING ECONOMICS WEEK 2020, LNNS 139, pp. 489–494, 2021.
https://doi.org/10.1007/978-3-030-53277-2_58

commercial data of companies, testimonies of technical devices, databases of public authorities and much more.

It is worth noting that Big Data is a valuable economic resource, and the list of the most valuable IT companies of the world - Microsoft, Facebook, Apple, etc. are the proof of that. Many experts call Big Data "the new oil" [5]. The scope of Big Data is diverse, it is used in e-commerce and retail [2], in the field of financial sphere, telecommunications, state and corporate governance, etc. It is in this case that it is important to structure contracts between companies that accumulate Big Data and/or want to commercialize them and companies that can provide services for the technical analysis of such data. The key aspect for the parties would be to ensure the confidentiality of the data that will be submitted for analysis.

For all its perspectives and significant opportunities provided by the use of Big Data, it should be noted that the issue of legal regulation of this concept remains open to this day. From the side of business, there is an increase in suggestions on the need to reform existing legislation in order to reduce the "backlog" of law from technological progress and to promote the development of the digital economy.

2 Methodology

In the process of writing the article, the author applied various methods of scientific knowledge. The main ones are materialist dialectics, as a universal approach to objective knowledge of reality and a method of system knowledge, which allowed to consider the category of legal responsibility as a system object with its own specific manifestation, including the field of administrative law. Amongst general scientific methods used in the article, we can name: analysis and synthesis; induction and deduction. In the process of research such private-scientific methods of cognition as formal legal, comparative legal and logic were applied. So, for example, the formal legal method turned out to be quite useful in the interpretation and study of regulatory documents.

3 Results

The use of Big Data provides new levers of influence on the competitive situation in the business sphere, but at the same time rise the chances of unauthorized distribution of personal information. The digital economy and the amount of processed data, in particular, develop at an enormous rate. This clearly reflects the lag of state regulation from the realities of the sphere in question and this void is becoming increasingly noticeable. Attempts to regulate the concept of Big Data were made in 2020 by introducing the Big Data Bill. However, this bill did not reach the expected result. The provisions of the bill were criticized by business representatives and government bodies, for example, the rules on the need for user consent to Big Data processing are virtually impossible.

More significant steps are the amendments entered to the Civil Code of the Russian Federation (hereinafter referred to as the Civil Code of the Russian Federation) [10] on "digital rights". As a result a new type of agreement was entered into the Civil Code of the Russian Federation on the provision of services for the provision of information, art. 783.1 of the Civil Code of the Russian Federation. Parties to this agreement may set the condition of confidentiality of the transmitted information, set the obligation not to take actions for the alienation of information or other actions that contribute to its disclosure to third parties during a certain period of time. These innovations are especially important in conditions of low efficiency of implementation of NDA-agreements.

The adopted amendments definitely have a significant impact on the formation of the legal regulation of Big Data, but still do not resolve most of the mentioned issues. For example, it remains unclear whether the operator should obtain the consent of each subject of information to provide data, and also inform the subject about the purpose, methods and types of information that will be collected and processed, since at present the processing of data without the consent of the subject is not legal. Thus, today the question arises of the urgent need to adopt a separate regulatory act aimed at revealing the content of concepts and detailed regulation of the use of information technology, as well as the creation of national standards for processing big data arrays.

4 Discussion

The issue of legal regulation of Big Data is one of the most important not only in Russia, but also in the global legal community [9]. Its significant economic impact on the most vital sectors of the economy requires a quick response. For example, Aeroflot, using Big Data, calculated that out of 32 million passengers, the company had only 10 million unique ones, and only 1.2 million of them provided 50% of the revenue [7]. Based on this information, the company managed to reduce advertising costs. At the same time, the legal side of Big Data allows to outline the vector of legal approach to regulating its application, and it has not been properly studied.

Since 2017 the "Digital Economy" program [7] has been running in the Russian Federation, this program is aimed at developing the country's innovative potential. According to this program, Big Data is "end-to-end technology", in other words - breakthrough digital technology. The Digital Economy program provides users with the right to receive remuneration for the collection and use of data, as well as its exclusion from circulation in the event of failure. Business representatives proposed the concept of the Big Data bill that aimed to attract special intermediary companies that have information about finding Big Data from data operators and transmit the required information for a certain monetary reward. I think this proposal is not advisable, firstly, from an economic point of view, since it leads to a rise in price, secondly, it introduces the risk of market monopolies and loss of pricing flexibility, and thirdly, it creates the risk of losing control of subjects of rights over their personal data, causing harm to both operators and Big Data entities.

It is worth noting that Big Data was also mentioned in the Strategy of the information technology industry for 2014–2020. However, in these programs, as well as in the legislative act that attempts to regulate the use of Big Data, Federal Law of July 27, 2006 No 149-FZ "On Information, Information Technologies and the Protection of

Information" [3] does not even contain a legal definition of the concept of Big Data Based on an analysis of the latest revisions of specialized laws, it can be concluded that the Russian legislator regulates Big Data as personal data and goes in the direction of localizing user data, significantly limiting cross-border data transfer. This position contributes to the withdrawal of European companies from the Russian market and damages the Russian economy.

Of course, the Russian approach has a right to exist. But it raises a lot of controversial issues. In conditions of legal uncertainty and discussion of coherence between the concepts of Big Data and personal data, as a guide, you can turn to the two most striking approaches to the search for an adequate balance between personal rights and business interests in the world - American and European.

The basis of the American approach is the understanding of "big user data" as an asset that brings profit to the business. Such ratio is highly bias to the profitability of the business, providing the maximum availability of information. And issues of confidentiality and data security are actually referred to the competence of the consumer and depend solely on his "cleanliness" in the transmission of data. I would especially like to draw attention to the fact that in the United States there is no single regulatory act aimed at the right regulation of the collection, storage and protection of both personal data and Big Data. The rules governing the above issues are dispersed in various industry and state acts [11].

Unlike the American model, on May 25, 2018, the European Union member states adopted the Pan-European Regulation on Personal Data Protection or the General Data Protection Regulation (hereinafter - GDPR, the "Regulation") [4]. As the main principles of data processing, it indicates legality, fairness, transparency, data minimization, goal limitation, accuracy, storage limitation.

When adopting the Regulation, the provisions of many international legal acts on the fundamental rights and freedoms of man and citizen were accounted for. This consequentially influenced the fact that the Regulation governs big data from the position of priority of the rights and interests of the individual, and not as a digital asset. As a disadvantage of GDPR, it should be noted that Big Data is not consolidated into a separate category. But at the same time, the Regulation reveals the concept of Big Data as a large array of information, the source of which is various channels with a high transmission rate. The data itself can be created by humans or generated by computers." The Regulation also significantly expanded the concept of "Personal Data", namely, personal data refers to any information relating an identified or identifiable person (data subject), by which it can be directly or indirectly determined. Such information includes, but not limited to, the last name and first name, location information, online identifier, factors specific to the physical, physiological, genetic, mental, economic, cultural or social identity of this individual, even IP addresses can be classified as personal data. Thus, Big Data is perceived as personal data in a broader sense. The GDPR has attempted to establish legal grounds for data processing. Processing is considered legal if the data subject has given consent to the processing of his personal data, or if it is necessary to fulfill a contract to which the data subject is a party. The latter wording is thought to be not entirely successful and enables companies to process personal data for no reason at all. But at the same time, enormous fines are imposed on companies for misuse of data (the size of the fine can reach 20 million euros) [6].

In general, the Regulation of the European Union is very logical and harmonious in terms of the regulation of primary data collection, its processing, depersonalization, use and destruction. GDPR basically has a number of legal decisions that can be borrowed to improve domestic regulation in this area. For example, the norms on the requirement for companies to publish information on leaks, payment of compensation and large fines are of great interest. It is believed that only such measures contribute to the responsibility of business and government bodies for the security of personal data.

5 Conclusion

Summing up, we can conclude that the EU approach to Big Data as an object of legal regulation is characterized by centrality, the priority of human rights and freedoms, such as the right to privacy, the right of access to information, etc. The US approach is to provide greater legislative power to individual territories (states). Big data is considered solely as a profitable asset.

If we talk about Russia, it is not yet clear which way the domestic legislator shall take on. Yet one thing is clear, partial reform of legislation in the field of personal data and information technologies regulation drawn by making small amendments and clarifications regarding Big Data does not solve the problem of their legal regulation. Obviously, due to their specificity, economic and social value, Big Data needs special regulation in a separate regulatory act. At the same time, it is important that such act accounts all the pressing issues of using Big Data: data collection, the ability to quickly extract useful information from their large volume and variety, data leakage, the liability of Big Data storage entities.

Russia is currently trying to build its own model of Big Data regulation, focused on total centralized control over its processing and use [8]. But as world experience and legal analysis of the GDPR rules show we should follow by the harmonization of the norms of Russian legislation with the rules of the Regulation. This would certainly serve to increase the level of the rights and legitimate interests of personal data subjects protection and at the same time increase entrepreneurial activity in terms of working with Big Data.

References

1. Celesti, A., Galletta, A., Fazio, M., Villari, M.: Towards hybrid multi-cloud storage systems: understanding how to perform data transfer. Big Data Res. **16**, 1–17 (2019)
2. Elshawi, R., Sakr, S., Talia, D., Trunfio, P.: Big data systems meet machine learning challenges: towards big data science as a service. Big Data Res. **14**, 1–11 (2018)
3. Federal Law of July 27, 2006 No. 149-FZ (as amended on April 3, 2020) "On Information, Information Technologies and the Protection of Information" (2020). http://www.consultant.ru/document/cons_doc_LAW_61798/. Accessed 24 Apr 2020

4. Habr: GDPR – New rules for the processing of personal data in Europe for the international IT market (2017). https://habr.com/company/digitalrightscenter/blog/344064/. Accessed 24 Apr 2020
5. Heuberger-Götsch, O.: The value of data from a legal perspective using the example of profiling. In: Fasel, D., Meier, A. (eds.) Big Data, pp. 83–105. Springer, Cham (2016)
6. Pearce, H.: Big data and the reform of the European data protection framework: an overview of potential concerns associated with proposals for risk management-based approaches to the concept of personal data. Inf. Commun. Technol. Law **26**(3), 312–335 (2017). https://doi.org/10.1080/13600834.2017.1375237
7. Portal of the National Project "Digital Economy": Regulation of big data (2020). https://digital.ac.gov.ru/opinions/4284/. Accessed 24 Apr 2020
8. Savelyev, A.: Russia's new personal data localization regulations: a step forward or a self-imposed sanction? Comput. Law Secur. Rev. **32**(1), 128–145 (2016)
9. Tao, H., Bhuiyan, Z.A., Rahmanc, M.A., Wang, G., Wang, T., Manjur, M.A., Li, J.: Economic perspective analysis of protecting big data security and privacy. Future Gener. Comput. Syst. **98**, 660–671 (2019)
10. The Civil Code of the Russian Federation (part one) from 30.11.1994 № 51-FZ (as amended on 03.07.2016) (1994). http://www.consultant.ru/document/cons_doc_LAW_5142/. Accessed 24 Apr 2020
11. Tucker, D.S., Wellford, H.: Big mistakes regarding big data. American Bar Association (2015). http://papers.ssrn.com/sol3/papers.cfm?abstract_id=2549044. Accessed 24 Feb 2020

Stochastic Models of Energy Audit Organization in Agrarian Enterprises

S. V. Oskin, S. M. Morgun, and N. Yu. Kurchenko(✉)

Kuban State Agrarian University, Krasnodar, Russia
kgauem@yandex.ru, morgun@bk.ru, kalyal389@gmail.com

Abstract. There is a strong need in searching optimal decisions concerning the members of energy audit organizations as well as their possibilities in processing of the maximal number of contracts for carrying of the procedure of energy survey. Using the Waiting Theory the needed quality characteristics for energy audit organization could be derived. So, with the basic intensity of getting and processing of the contracts (λ = 0,01 h–1; μ = 0,002 h–1) there is a possibility for an audit organization to process 19 contracts for energy survey during one year. Mentioned contracts had been waiting for in the queuing system around 790 h, or 0.4 year.

Keywords: Audit · Energy auditor · Energy saving · Management · Queue system

1 Introduction

High level of the competition in the world on items and services demands to look for ways of decrease in product cost. One of such ways is reduction of consumption of energy resources. For rational implementation of this process it is necessary to carry out an energy audit. The energy audit, or power inspection of the enterprises and organizations assumes assessment of communication of activity of the enterprise with costs of fuel, energy (different types) and resources. The energy audit purpose – to estimate efficiency of use of fuel and energy resources and to develop effective measures for cost reduction of the enterprise. In Russia a number of documents [2, 4, 5] which also formalized separate stages were developed for systematization of this process. However specifics of the enterprises, climatic conditions, regional policy, qualification of auditors enter the amendments into the procedure of an energy audit [3]. These and similar features are noted also in other countries [1, 9, 10].

In Russia the lowest rates of carrying out power inspections were observed in agriculture. It is necessary to analyse this situation as we pass to the second stage of energy saving – assessment of efficiency of the actions specified in the power passport and to updating of this document the next five years. A certain rate of carrying out an energy audit is connected with difficulties as in farms and at power accountants. It is necessary to carry to the main problems: a difficult financial condition of a number of the agricultural enterprises of edge that forces heads to postpone the decision on energy saving and certification for uncertain terms; the enterprises of edge are included into various concerns, associations and holdings that slows down signing of the contracts on

S. I. Ashmarina et al. (Eds.): ENGINEERING ECONOMICS WEEK 2020, LNNS 139, pp. 495–505, 2021.
https://doi.org/10.1007/978-3-030-53277-2_59

inspection with each firm separately; there are no developed and approved methods of filling of the power passport of the agricultural consumer of fuel and energy resources; the territorial dissociation of the agrarian and industrial complex enterprises, availability of several technological lines, lack of design technical documentation considerably complicate the procedure of power inspection. The first experience of carrying out power inspections of the agricultural enterprises revealed the following: low qualification of the persons responsible for energy saving at the enterprises; lack of motivation of personnel of assistance to power accountants; lack of interest of heads of farms in search of energy saving potential; in a row a case completely is absent or design and technical and contractual documentation is incorrectly issued that complicates definition of balance accessory of power supply networks, water supply systems and wells; use of the unaccounted energy resources complicating calculation of real indicators of energy consumption; presence of the subsubscribers who aren't equipped with metering devices of energy resources. All this results in need of search of optimal solutions on the structure of personnel of power audit organizations, their opportunities on the maximum quantity of the signed contracts for power inspection.

2 Methodology

From a position of modeling of process of mass service of a situation at an energy audit, it can be presented as follows. Having come to the serving system, the requirement joins turn of earlier arrived applications. The channel of service chooses the requirement from being in turn to start its service which already lasts from 2 to 4 months [6, 8]. After completion of the procedure of service of the next requirement the channel starts processing of the following requirement if that is available in the expectation block. The total amount of works on power inspection is meant by service: tool assessment, analytical processing of the obtained data, filling of tables of the energetic passport, development of recommendations about energy saving drawing up the technical report. The cycle of functioning of a system of mass service of this sort is repeated repeatedly. Our system can be two main types: the system with refusals in which the application which came to a system at the moment when all channels are occupied is refused and at once leaves turn; the system with expectation (turn) in which the application which came to the moment when all channels of service are occupied stands in line and waits won't be released one of channels yet.

The simplest single-channel model with indicative distribution of both dlitelnost of intervals between receipts of requirements, and dlitelnost of service is QS (queuing system) with one power accountant. As flows of applications and service protozoa, distribution density of dlitelnost of intervals between receipts of requirements and distribution density of dlitelnost of service will be following [7, 11]:

$$f_1(x) = \lambda \cdot e^{-\lambda \cdot t}; \quad f_2(x) = \mu \cdot e^{-\mu \cdot t} \tag{1}$$

where λ – intensity of receipt of contracts in a system; μ – intensity of information processing and carrying out energy audit.

Let's present the system of mass service in the form of the count which has two states: 0 – the auditor is free (expectation); 1 – the auditor is busy (there is a service of the application – the contract). On a state graph the system of the differential equations of Kolmogorov is worked out. The solution of such system looks as follows:

$$P_0(t) = \frac{\mu}{\lambda + \mu} + \frac{\lambda}{\lambda + \mu} \exp[-(\mu + \lambda)t] \tag{2}$$

For single-channel QS with refusals the probability of P(t) 0is the relative capacity of system q, that is P(t) 0 = q.

Let's determine probabilities for steady state (t = ∞):

$$q = P_0 = \frac{\mu}{\lambda + \mu} = \frac{1}{1 + \frac{\lambda}{\mu}}, P_{otk} = P_1 = 1 - \frac{\mu}{\lambda + \mu} = \frac{\lambda}{\lambda + \mu} = \frac{1}{1 + \frac{\mu}{\lambda}} \tag{3}$$

Knowing relative capacity, it is possible to find absolute. Absolute capacity (A) – average of applications (contracts) which the system of mass service in unit of time can serve:

$$A = \lambda \cdot q = \frac{\lambda \cdot \mu}{\lambda + \mu} = \frac{1}{\frac{1}{\mu} + \frac{1}{\lambda}}. \tag{4}$$

Size $P_{\text{Quality Department}}$ represents an average share of unserved applications among submitted to a system.

Let's determine absolute capacity, taking into account basic intensivnost of receipt of applications and service:

$$A = \lambda \cdot q = 0,005 \cdot 0,17 = 0,85 \cdot 10^{-3}\,\mathrm{h}^{-1} \tag{5}$$

at 8 h working day and 5 day working week (in a year of 248 working days, 1956 working hours) the value A will be:

$$A = 1956 \cdot 0,85 \cdot 10^{-3} = 1,66\,year^{-1} \tag{6}$$

It means that the system with one auditor is capable to carry out on average 1.66 services of contracts a year.

We will determine the probability of refusal by a formula:

$$P_{otk} = 1 - q = 1 - 0,17 = 0,83 \tag{7}$$

Therefore, about 83% of the arriving contracts will be refused in service that is inadmissible.

Let's determine the nominal capacity of a system:

$$A_{nom} = \frac{1}{t_{obsl}} = \mu = 0,001\,\mathrm{h}^{-1}$$

or

$$A_{nom} = 0,001 \cdot 1956 = 1,96\,year^{-1} \tag{8}$$

Thus, A_{number} is 1.18 times more, than the actual capacity determined taking into account the accidental nature of a flow of applications and holding time. The analysis also shows that at basic intensity of receipt of applications ($\lambda = 0,005$ h^{-1}) for increase in relative capacity it is necessary to carry out search of ways of increase in intensity of service. It should be noted that at reduction of intensity of submission of documents from the enterprises, for example at $\lambda = 0,0025$ h^{-1}, relative capacity increases, but the total number of the processed contracts (absolute capacity) decreases.

Let's consider this system with one power accountant and with a possibility of creation of turn. Let's assume that irrespective of how many contracts arrive on a system entrance (turn + the served clients) can't contain more N-of contracts, t.e. the clients who didn't get to expectation are forced to be served at other auditors. This system can be in the following states: 0 – the auditor is free (expectation); 1 – the auditor is busy (there is no turn); 2 – the auditor is engaged (one contract in turn); n – the auditor is engaged (n − 1 of contracts in turn); N – the auditor is engaged (N − 1 of contracts in turn).

In the stationary mode this system will be described by the following algebraic equations [9–11]:

$$\begin{cases} P_1 - \psi \cdot P_0 = 0, \quad n = 0 \\ \dots\dots\dots\dots\dots\dots\dots\dots\dots\dots\dots \\ P_{n+1} + \psi \cdot P_{n-1} - (1-\psi)P_n = 0, \quad 0 \leq n \angle N \\ \dots\dots\dots\dots\dots\dots\dots\dots\dots\dots\dots \\ \psi \cdot P_{N-1} - P_N = 0, \quad n = N \end{cases} \tag{9}$$

where $\psi = \frac{\lambda}{\mu}$ – the given intensity of an entrance flow of contracts, n – number of a state.

The solution of a system of the Eq. (9) for our QS model will have an appearance:

$$P_n = \begin{cases} P_0 \cdot \psi^n, & \psi \neq 1, \quad n = 0,1,2\dots N \\ \frac{1}{N+1}, & \psi = 1 \end{cases}, \frac{1-\psi}{1-\psi^{N+1}} = P_0 \tag{10}$$

Relative capacity of a system will look as follows:

$$q = 1 - P_{om\kappa} = \begin{cases} 1 - \frac{1-\psi}{1-\psi^{N+1}} \cdot \psi^N, & \psi \neq 1, \quad n = 0,1,2\dots N \\ 1 - \frac{1}{N+1}, & \psi = 1 \end{cases} \tag{11}$$

Absolute capacity is calculated on a formula:

$$A = \lambda \cdot q = \begin{cases} \lambda - \frac{\lambda(1-\psi)}{1-\psi^{N+1}} \cdot \psi^N, & \psi \neq 1, \quad n = 0,1,2\dots N \\ \lambda - \frac{\lambda}{N+1}, & \psi = 1 \end{cases} \tag{12}$$

Let's define average of the contracts which are in a system:

$$L_S = \sum_{n=0}^{N} n \cdot P_n = \begin{cases} \frac{\psi \cdot \left[1-(N+1)\cdot\psi^N + N\cdot\psi^{N+1}\right]}{(1-\psi^{N+1})\cdot(1-\psi)}, & \psi \neq 1 \\ \frac{N}{2}, \ \psi = 1 \end{cases} \quad (13)$$

For determination of average time of stay of the contract in a system it is possible to use a formula:

$$T_S = \frac{L_S}{\lambda(1 - P_N)} \quad (14)$$

The average duration of stay of the contract in turn will be:

$$T_q = T_S - \frac{1}{\mu} \quad (15)$$

Average of contracts in turn (turn length):

$$L_q = \lambda(1 - P_N) \cdot T_q \quad (16)$$

Let's consider QS with one auditor and with a possibility of the organization of turn on the basis of basic intensity of receipt of contracts and their processing. Let's introduce the following restrictions: the number of intermediate places for the arrived contracts expecting carrying out processing is limited and equal 4, [(N − 1) = 4]. If all places are taken – in turn there are already three contracts, then the next contract which arrived to processing on service doesn't stand in line.

Let's define the given intensity of an entrance flow of contracts:

$$\psi = \frac{\lambda}{\mu} = \frac{0,005}{0,001} = 5 \quad (17)$$

Calculate probabilities of finding of a system in separate states on formulas (11–13):

$$\begin{aligned} P_0 &= \frac{1-\psi}{1-\psi^{N+1}} = \frac{1-5}{1-5^6} = 0,00026, \\ P_1 &= \psi \cdot P_0 = 5 \cdot 0,00026 = 0,00128, \\ P_2 &= \psi^2 \cdot P_0 = 25 \cdot 0,00026 = 0,0064, \\ P_3 &= \psi^3 \cdot P_0 = 125 \cdot 0,00026 = 0,032, \\ P_4 &= \psi^4 \cdot P_0 = 625 \cdot 0,00026 = 0,16, \end{aligned} \quad (18)$$

$$P_5 = \psi^5 \cdot P_0 = 3125 \cdot 0,00026 = 0,8$$

Refusal probability in processing of the contract will be $P_{otk} = P_5 = 0,8$.
Let's determine the relative capacity of a power auditor system:

$$q = 1 - P_{otk} = 1 - 0,8 = 0,2 \tag{19}$$

Apparently relative capacity increased from 0.17 up to 0.2 (in comparison with a system without turn).
Absolute capacity of a power auditor system will be defined:

$$A = \lambda \cdot q = 0,005 \cdot 0,2 = 0,001\ \text{h}^{-1} = 1,96\ year^{-1} \tag{20}$$

In comparison with the previous system absolute capacity reached rated value.
Average of the contracts which are on service and in turn:

$$L_S = \frac{\psi \cdot \left[1 - (N+1) \cdot \psi^N + N \cdot \psi^{N+1}\right]}{\left(1 - \psi^{N+1}\right) \cdot (1 - \psi)} = \frac{5\left[1 - (5+1) \cdot 5^5 + 5 \cdot 5^{5+1}\right]}{\left(1 - 5^{5+1}\right) \cdot (1 - 5)} = 4,75 \tag{21}$$

Let's determine the average time of stay of the contract in a system:

$$T_s = \frac{L_S}{\lambda(1 - P_N)} = \frac{4,75}{0,005 \cdot (1 - 0,8)} = 4750\ ч = 2,43\ year \tag{22}$$

The average duration of stay of the contract in waiting list for service will be:

$$T_q = T_S - \frac{1}{\mu} = 4750 - 1000 = 3750\ \text{h}^{-1} = 1,9\ year \tag{23}$$

The average of contracts in turn (turn length) will be defined:

$$L_q = \lambda(1 - P_N) \cdot T_q = 0,005(1 - 0,8) \cdot 3750 = 3,75 \tag{24}$$

Work of such system or the auditor can be accepted unsatisfactory as, despite increase in capacity, the contract stays in turn nearly two years, and the customer will receive ready documents in 2.43. In that case most of customers will refuse services of the auditor.

In most cases the systems of mass service are multichannel (n > 1), i.e. for our case in power audit organization several auditors work. The ultimate goal of use of n in parallel of the working channels consists in increase in speed of service of applications (contracts) due to processing at the same time of n of client contracts. The system can is in the following states: 0 – all auditors are free; 1 – one auditor is busy, the others are free; 2 – 2 auditors are busy, the others are free; k – k of auditors are occupied, the others are free; n – all n of auditors are occupied.

After receiving the equations of Kolmogorov and using formulas of the Erlang [10, 11] it is possible to receive probabilities of states:

$$P_1 = \frac{\psi}{1!} \cdot P_0 = 5 \cdot P_0, P_2 = \frac{\psi^2}{2!} \cdot P_0 = 12,5 \cdot P_0, P_3 = \frac{\psi^3}{3!} \cdot P_0 = 20,8 \cdot P_0,$$

$$P_4 = \frac{\psi^4}{4!} \cdot P_0 = 26,04 \cdot P_0, P_5 = \frac{\psi^5}{5!} \cdot P_0 = 26,04 \cdot P_0, P_6 = \frac{\psi^6}{6!} \cdot P_0 = 21,7 \cdot P_0$$

$$P_0 = \frac{1}{\sum_{k=0}^{6} \frac{\psi^k}{k!}} = \frac{1}{1 + 5 + 12,5 + 20,8 + 26,04 + 26,04 + 21,7} = 0,009 \qquad (25)$$

$$P_1 = 5 \cdot 0{,}009 = 0,045; P_2 = 12,5 \cdot 0,009 = 0,11; P_3 = 20,8 \cdot 0,009 = 0,19;$$
$$P_4 = 26,04 \cdot 0,009 = 0,23 = P_5; P_6 = 21,7 \cdot 0,009 = 0,20 \qquad (26)$$

Refusal probability in service of the application will be equal to stay probability in the 6th state:

$$P_{otk} = P_6 = 0,2 \qquad (27)$$

It is much higher, than in earlier considered options.

Relative and absolute capacities respectively will be defined:

$$A = \lambda \cdot q = 0{,}005 \cdot 0{,}8 = 0{,}004 \, ч^{-1} = 7{,}8 \, year^{-1} \qquad (28)$$

Average of busy auditors:

$$\bar{k} = \psi \cdot (1 - P_{омк}) = 5 \cdot 0,8 = 4 \qquad (29)$$

Thus, at the set QS operating mode 4 auditors from 6 on average will be busy. It is not a really good indicator.

3 Results

On the basis of the received mathematical expressions we will define the necessary number of auditors for decrease in probability of refusal to value less than 10% (Table 1).

Table 1. Values of probabilities of refusal depending on the number of power accountants

n	6	7	8	9
P_0	0.009	0.008	0.007	0.007
$P_{Quality\ Department}$	0.2	0.12	0.07	0.04

Source: authors.

Table 2. Indicators of efficiency of various QS

Indicators	The QS type with basic intensivnost, the Part-1: λ = 0.005; μ = 0.001					New receipts of intensity and service, Part-1					
						λ = 0.005; μ = 0.002; QS type			λ = 0.01; μ = 0.002; QS type		
	1 lecture hall with quality department	1 lecture hall with очеред	6 lecture halls with quality department	8 lecture halls with quality department	6 lecture halls with очеред	1 lecture hall with quality department	1 lecture hall with очеред	6 lecture halls with очеред	1 lecture hall with quality department	1 lecture hall with очеред	6 lecture halls with очеред
q	0.17	0.2	0.8	0.93	0.98	0.29	0.39	0.99	0.17	0.2	0.98
$P_{Quality\ Department}$	0.83	0.8	0.2	0.07	0.03	0.71	0.61	0.01	0.83	0.8	0.03
And, year-1	1.66	1.96	7.8	9.1	9.5	2.8	3.8	9.7	3.3	3.3	19
ψ	–	5	5	5	5	–	2.5	2.5	–	5	5
T_S, h	1200	4753	1200	1200	1588	700	900	744	600	2347	790
T_q, h	–	3716	–	–	588	–	1389	244	–	1878	290
L_S, piece	–	4.75	–	–	7.9	–	1.75	6.22	–	4.75	7.9
L_q, piece		3.75			2.9		2.78	1.22	–	3.75	2.9
N_{audit}	1	1	4	4.65	6	1	1	6	1	1	6

Source: authors.

From Table 1 it is visible that 8 auditors will reduce refusal probability to 0.007. At the same time relative capacity will increase to 0.93, and absolute – to 9.1 contracts a year. In this case 4.65 auditors will be on average busy that speaks about their part-time employment. Let's consider the multichannel system of mass service with expectation in turn. The system has S auditors for service. In the set mode multichannel QS with expectation and unlimited turn can be described also by means of the algebraic equations similar to the previous case without turn.

The analysis of functioning of QS shows that the most highly productive are systems with several auditors and with a possibility of the organization of turn. Let's analyze this type of QS regarding influence of intensity of streams of receipt and processing of contracts for system performance. Let's say that due to application of the software and increase in number of measuring devices, it was succeeded to reduce twice a processing time of contracts [9]. The same effect can be gained if the enterprise has less points of measurement or smaller volume of consumption of energy resources or there are no own boiler houses and transformer substations.

In Table 2 indicators of various QS at the basic and changed intensity of receipt and processing of contracts for audit are given. From the table it is visible that the most effective is QS with 6 auditors, a possibility of the organization of turn and with intensity of receipt and processing respectively $\lambda = 0.01$ the Part-1; $\mu = 0.002$ the Part-1 as this system has an opportunity to process within a year of 19 contracts which will be in QS about 790 h or 0.4 years.

QS depending on their type are submitted. From these drawings it is visually visible that QS of the last type (with 6 auditors, a possibility of the organization of turn and with intensity of receipt and processing respectively $\lambda = 0.01$ the Part-1; $\mu = 0.002$ the Part-1) is the most effective.

4 Discussion

In Russia, after the adoption of the Federal Law No. 261 (on mandatory energy audits), a large number of energy audit companies have emerged that are faced with the problem of efficient operation and maximum profit [4]. Registration of energy passports showed a low pace of conducting surveys, which is associated with difficulties both in the enterprises themselves and in energy auditors. Particularly great difficulties arose during the energy audit of agricultural enterprises. The problems of these enterprises include: the difficult financial condition of a number of farmers, which forces managers to postpone decisions on energy conservation and certification for an indefinite period; individual farms are included in various concerns, associations and holdings, which slows down the conclusion of inspection contracts with each firm separately; there are no developed and approved methods for filling out the energy passport of an agricultural consumer of fuel and energy resources; design-technical and contractual documentation is missing or incorrectly drawn up, which makes it difficult to determine the balance sheet ownership of electric networks, the presence of sub-subscribers not equipped with energy metering devices. All this greatly complicates the energy inspection procedure and ultimately leads to the need to find optimal solutions for the composition of energy audit organizations, their capabilities for the maximum number of energy inspection agreements concluded.

To search for the optimal composition of the energy auditing organization (as well as other audit companies), one can apply the queuing theory. From the position of modeling the process of mass servicing the situation during an energy audit, this process can be represented as follows. Initially, an energy audit organization (EAO) sends out questionnaires to potential customers. These sheets are filled out and sent back to the EAO. After assessing the cost and agreeing on the price of services, energy inspection contracts are concluded, and the final document is the receipt of an energy passport. We will not consider this preliminary stage, since it is very small in time in comparison with other stages of work. After concluding a contract for an energy audit, the EAO issues to the enterprise energy passport tables, which must be filled in by leading experts. This period of time usually already takes from 20 to 60 working days, and it depends on many factors: the availability of data at the enterprise, the qualifications of specialists and their employment, the size of the enterprise, etc. Completed tables and data can come in several stages. We will consider this process as a stream of input requirements to the queuing system. Use of a mathematical apparatus of systems of mass service allows to describe processes of functioning of the power auditor companies and to define the directions of improvement of their work. The developed models of systems of mass service with one auditor showed what increase in frequency of submission of applications (to force the enterprises to process quicker data) doesn't lead to increase in relative capacity, and on the contrary – to reduction. For increase in relative capacity it is necessary to reduce holding time.

5 Conclusion

The analysis of systems of mass service with one auditor and refusals showed that at basic intensivnost of receipt of contracts and their processing ($\lambda = 0,005\ \mathrm{h}^{-1}$, $\mu = 0,001\ \mathrm{h}^{-1}$) for increase in relative capacity it is necessary to look for ways of increase in intensity of service. It is also possible to note that at reduction of intensity of submission of documents from the enterprises (increase in time of filling of tables with services of the enterprise), for example at $\lambda = 0,0025\ \mathrm{h}^{-1}$, relative capacity increases, but the total number of the processed contracts (absolute capacity) decreases.

As a result of the analysis of systems of mass service with one auditor it is established that transition from model with refusals to model with turn increases the annual number of the processed contracts from 1.66 to 1.96; refusal probability in service decreases from 0.17 to 0.2.

The developed models of systems of mass service with several auditors who showed higher efficiency of service. For model with refusals increase from 6 auditors to 8 will reduce refusal probability to 0.007. At the same time relative capacity will increase from 0.8 to 0.93, and absolute – from 7.8 to 9.1 contracts a year. In this case 4.65 auditors will be on average busy that speaks about their part-time employment. The schedule of dependence of number of the served contracts on the number of the working auditors and the line of a trend is received. According to this schedule it is visible that further increase in number of auditors won't lead to significant increase in the served contracts.

The developed model of systems of mass service with 6 auditors and a possibility of the organization of turn showed that the number of the served contracts, in comparison with model with refusals increases from 7.8 to 9.5 and relative capacity increases from 0.8 to 0.98. However the time spent of the contract for service also increases from 1000 to 1660 o'clock.

For model with 6 auditors and a possibility of the organization of turn the given intensity of a flow of contracts has no impact on the relative capacity of audit organization. This indicator has the main impact on length of turn and time of stay of the contract at auditors. Reduction of a processing time of contracts doesn't increase the absolute capacity of a system twice, but reduces the time spent in audit organization twice. Reduction of a processing time of tables of the power passport by the enterprise leads to double increase in absolute capacity by audit organization twice at preservation of the time spent of the contract for auditors. This system at basic intensity of receipt and processing of contracts ($\lambda = 0.01$ the Part-1; $\mu = 0.002$ the Part-1) an opportunity to process within a year of 19 contracts for power inspection which will be in QS about 790 h or 0.4 years has.

References

1. Bo, S., Price, L., Lu, H.: Energy audit practices in China: national and local experiences and issues. Energy Policy **46**, 346–358 (2012)
2. Decree of the Russian President No. 889 of 04.06.2008 "About some measures for increase power and ecology efficiency of the Russian economy" (2008). https://minenergo.gov.ru/node/5195. Accessed 29 Apr 2020
3. Dubinsky, M.Yu.: Energy audit of the industrial enterprises. Energy Secur. Energy Saving **3**, 14–18 (2010)
4. Federal law No. 261-FZ "About energy saving and about increase in power efficiency and about introduction of amendments to separate acts of the Russian Federation" of 23.11.2009 (2009). http://base.garant.ru/12171109/. Accessed 29 Apr 2020
5. GOST P 51379-99. Energy saving. Power passport of the industrial consumer of fuel energy resources. Basic provisions. Standard forms (2018). https://internet-law.ru/gosts/gost/8919/. Accessed 29 Apr 2020
6. Illarionov, E.M.: How to distinguish the professional power accountant from the beginning power accountant. Power Light **2**(27), 47–50 (2013)
7. Ivchenko, G.I., Kashtanov, V.A., Kovalenko, I.N.: Theory of Mass Service. HSE, Moscow (1982)
8. Oskin, S.V., Gromyko, D.V.: Mathematical models of power audit organization during the work with the agricultural enterprises. Works Kuban State Agric. Univ. **3**(42), 156–165 (2013)
9. Singh, M., Singh, G., Singh, H.: Energy audit: case study to reduce lighting cost. Asian J. Comput. Sci. Inf. Technol. **2**(5), 119–122 (2012)
10. Wood, E.: Zero touch energy audit: Will it change the game? (2011). https://www.energy-xprt.com/articles/zero-touch-energy-audit-will-it-change-the-game-265181. Accessed 29 Apr 2020
11. Zaytsev, M.G., Varyukhin, S.E.: Methods of Optimization of Management and Decision-Making: Examples, Tasks, Cases. Delo, Moscow (2008)

Check for updates

Derivatives Trading: Digital Transformation

O. Y. Kuzmina[1], M. E. Konovalova[1](✉), T. E. Stepanova[2], and Yu. N. Lyachenkov[1]

[1] Samara State University of Economics, Samara, Russia
pisakina83@yandex.ru, mkonoval@mail.ru, yl9278915597@mail.ru

[2] Kaliningrad State Technical University, Kaliningrad, Russia
tatyana.stepanova@klgtu.ru

Abstract. Qualitative changes in functioning principles and structure of the derivatives market, as well as its instruments in the context of digitalization, have a significant impact on processes associated with investment activities around the world. Therefore, we need to study the practice of using derivatives, as well as the dynamics of the derivatives market and its instruments. This study considers the functioning of the derivatives market and the process of applying its instruments as a method for reducing investment risks. In the process of studying the dynamics of the derivatives market, based on the analysis of statistical data and factual materials, the main development trends of both the Russian and international derivatives markets are shown. Based on historical and statistical information, the authors simulated the process of analysis, development and implementation of a hedging strategy that reveals the content of the hedging procedure with derivative financial instruments in the context of digital transformation of the economic system.

Keywords: Digital technology · Financial risks · Futures contracts · Hedging · Stock exchange

1 Introduction

In the modern economy, the securities market is one of the main financial markets. Many different subjects of economic activity, as well as the economic system as a whole, depend on its condition. An important element in the structure of the securities market is the derivatives market, in which specific financial instruments, called derivatives, are traded. The main feature of derivative financial instruments is that their existence is based on the circulation of other primary instruments. Consequently, the properties of derivatives directly depend on the qualitative and quantitative characteristics of primary financial instruments. Based on the specifics of derivative financial instruments, their circulation is usually carried out in order to achieve one of two specific effects. One of these effects is hedging of risks, that is, minimization and management of risks associated with economic activities in financial markets. Another is that derivatives are used to carry out speculative operations aimed at economic benefits from changes in the value of the underlying asset [3]. Derivative financial

S. I. Ashmarina et al. (Eds.): ENGINEERING ECONOMICS WEEK 2020, LNNS 139, pp. 506–513, 2021.
https://doi.org/10.1007/978-3-030-53277-2_60

instruments enable various business entities to more actively use highly risky financial instruments while controlling risks and deriving additional economic benefits from this.

Due to high technologies used by exchanges and post-marketing infrastructures, investments are becoming closer, more accessible and more understandable to people, opening up new opportunities. There are more and more convenient and safe mobile banks, applications for investments, personal finance planning. All this contributes to the fact that exchange trading is rapidly developing, improving in software and hardware [6]. This process is especially illustrated in the urgent section, where a high level of informatization helps maximize revenues and minimize risks of players.

2 Methodology

Along with classical tools of formal and dialectical logic, among which we can emphasize the method of abstraction, analysis and synthesis, deduction and induction, the unity of logical and historical, the authors widely used tools of mathematical and statistical analysis, which made it possible to more clearly see the functioning of derivative financial instruments in the context of digital transformation, show the mechanism of minimizing risks when concluding transactions in the derivatives market. The key method used in the study is the system method. It is a systematic approach to the analysis of the derivatives market that allows us to identify its main patterns, show the interaction of its key elements, while reflecting the mechanism of their mutual influence.

3 Results

The derivatives market is a complex system of economic relations that arise when concluding derivatives contracts between market participants; this is a segment where very complex financial instruments are traded. In modern interpretation, a financial instrument can be defined as a contract, at the conclusion of which a financial asset arises for the first organization, and an equity instrument or financial liability - for the second organization at the same time. Thus, a financial instrument refers to any asset, for example, a security or a financial liability, the sale of which ensures cash. A derivative financial instrument is a fixed-term contract under which the parties obtain the right to take certain actions with respect to the underlying asset. It can be concluded that despite the fact that derivative instruments are mainly derivatives contracts that are traded mainly within the securities market, their scope is not limited to a separate sector of the financial market [8]. Various economic entities actively use derivatives as auxiliary instruments in carrying out economic activities in various areas of the economy.

The derivatives market has its own internal structure, which can be classified according to various criteria. Using the general approach, this market can be divided into primary and secondary, exchange and over-the-counter. It is worth noting that the derivatives market has a structure similar to the securities market, however, the stock market is both primary and secondary for derivative financial instruments, while for the securities market this market is primarily secondary. This feature of the derivatives market determines unlimited possibilities for narrowing and expanding the volume of the market, which enables market participants to respond flexibly to changes in the economic situation, which is especially important in the context of the digital transformation of the economic system [4, 14].

The basic principles of circulation of instruments in the derivatives market are similar to the principles the securities market as a whole. However, in addition to new opportunities for speculative operations, the derivatives market offers the investor the opportunity to directly control his own investment risks using derivative financial instruments. The implementation of certain operations aimed at minimizing investment risks is called the hedging process. There are a wide variety of derivative financial instruments. However, at the heart of all this diversity there are several standard and generally accepted contracts. The main exchange derivatives used for hedging are futures contracts and options; forwards and swaps are traded on the OTC market [13].

Carrying out the hedging process in the derivatives market, the investor uses a set of specific derivative financial instruments that activate the mechanisms described above. Thus, the investor, using a set of derivative instruments in a certain sequence, forms his hedging strategy. The effectiveness of the strategy depends on the ability of the investor to predict changes in the economic situation, which, undoubtedly, can be helped by digital technologies, for example, robots, electronic advisers.

Let us analyze the dynamics of the derivatives market in the Russian Federation over the past ten years. Currently, Russia does not publish consolidated statements on the state of the derivatives market. The Bank of Russia periodically publishes reports only on certain types of derivative financial instruments, while detailed reports on derivative financial instruments are not published at all. Information on derivative financial instruments can be found on the exchange website, the largest of which is the Moscow Exchange. Therefore, further analysis of the development of the Russian derivatives market will be based on data from the Moscow Exchange [9].

Considering development features of the Russian derivatives market, first of all, it is necessary to determine the dynamics of the total volume of trade and the place of the derivatives market in the structure of exchange trading. Over the past ten years, the total trading volume on the exchange has grown by almost 4.5 times (from 191,904 billion rubles in 2009 to 860,861 billion rubles in 2018). For almost the entire period, the indicator had an increasing growth trend, especially rapid growth was observed from 2011 to 2016, but then the growth rate decreased significantly, the trend weakened, and at the end of 2018, the growth rate assumed a negative value.

For the period from 2009 to 2018, the share of the derivatives market in the total trading volume of the Moscow Exchange ranged from 7.74% to 19.1%. The peak value was in 2011, and later on the indicator ranged from 10% to 13%. For the period from 2009 to 2018, the volume of the derivatives market grew 6 times (from 14 857 billion rubles to 89 263 billion rubles in 2018). The rapid growth in the derivatives market from 2009 to 2011 (from 14 857 billion rubles to 56 791 billion rubles) was followed by a fall in the next two years (to 48 605 billion rubles in 2013). In the period from 2014 to 2016, there was a rapid growth in the derivatives market (from 48 605 billion rubles to 115 271 billion rubles). Over the next two years, the indicator decreased by 26.7% (to 84,497 billion rubles) in 2017, and an increase of 5.64% was observed (to 89,263 billion rubles) at the end of 2018. We will analyze each of these conditional periods in detail.

The period from 2009 to 2011 can be described as the formation period of the Russian derivatives market. The main volumes of trading in fixed assets and derivatives were concentrated on the MICEX and PTC exchanges, which were merged in 2011 into the Moscow Exchange holding. This fact, together with the transition of the derivatives market from the stage of origin to the stage of formation, predetermined the rapid growth of the derivatives market. If we consider the structure during this period, more than 90% of the derivatives market was occupied by futures contracts, most of which were supplied to currency instruments and indices. At the same time, the greatest demand among options was observed for index option contracts. In the period from 2012 to 2013, two events influenced the development of the Russian derivatives market.

Firstly, on September 17, 2012, the integration of FORTS and MICEX Derivatives Markets was completed. Changes in the organization and functioning of the new platform compared to the integrated ones, as well as the change in the list of traded instruments, quite naturally caused fluctuations in trading volumes in the market because market participants were adapting to the new rules.

Secondly, the consequences of the global economic crisis of 2008 reached Russia and had a strong influence on the decrease in investor interest in stock market instruments, and, as a consequence, financial instruments derived from them. This contributed to the active use of instruments of monetary and foreign exchange markets [11, 12]. Significant changes in the ratio of futures and option contracts in cash terms did not occur. The share of currency futures in the structure of exchange trading on the derivatives market more than doubled (from 14.82% in 2011 to 31.83% in 2013), while the share of index futures decreased (from 67.13% in 2011 to 49.47% in 2013), which is directly related to the decrease in the entire derivatives market in this period.

The period from 2014 to 2016 was under the currency crisis in Russia, which provoked the collapse of the ruble. The strengthening of the previously formed trend towards the transfer of investor resources from the stock section to the currency and money has finally strengthened, and as a result, the devaluation of the ruble against the dollar caused increased interest of market participants in foreign exchange derivatives contracts, which were actively acquired both for hedging and speculation in conditions of market uncertainty. In addition, fluctuations in oil prices attracted the attention of investors to commodity futures. Thus, the volume of trading in currency futures more than doubled over this period (from 29,404 billion rubles in 2014 to 64,561 billion rubles in 2016), and the share of trade in commodity futures in the structure of the derivatives market grew from 1.51% in 2014 to 13.32% in 2016, index futures, on the contrary, reduced their share to 22.68% in 2016.

The period from 2016 to 2018 can be described as stabilization. The currency crisis in Russia ended, and oil prices stabilized due to OPEC intervention, which in aggregate significantly reduced the interest of speculators and hedgers in futures for foreign exchange instruments. This provoked a decrease in the derivatives market and changes in its structure. Thus, the share of currency futures in the structure of the derivatives market decreased from 56% in 2016 to 42.42% in 2018, while the share of commodity futures, on the contrary, increased from 12.32% in 2016 to 23.43% in 2018. The ratio of option contracts to futures in cash terms remained stable rather low.

Trading volume on the derivatives market in 2018 amounted to only 10.37% (89,263 billion rubles) of total exchange trading. The structure of trading on the derivatives market is dominated by futures (92.3% or 82.397 billion rubles), most of which are based on currency, (42.42%) index (21.47%) and commodity (23.43%) instruments. The share of index futures for the first ten years of the analyzed period has significantly decreased and over the past five years it has been steadily fluctuating around 20%, and interest-bearing derivatives are not at all popular on the Russian derivatives market.

As for the dynamics and structure of the stock exchange and the OTC international derivatives market, the following can be noted. The dynamics of the OTC market volumes are quite contradictory. So, from 2009 to 2014, the OTC international derivatives market was actively expanding. In the first half of 2009, the nominal value of unfinished contracts amounted to $ 594,596 billion, and in the first half of 2014 it amounted to $ 691,114 billion, but then the trend weakened, and the nominal value gradually decreased to $ 482,421 billion by the second half of 2016. Over the next two years, the OTC market tended to grow, and by the beginning of 2019, the total nominal value of open positions in the OTC market was $ 640,441 billion [1]. The high volatility of the OTC derivatives market is explained by the fact that in recent decades, state and international organizations have been trying with varying degrees of success to tighten the regulation of the OTC derivatives market, trying to reduce the volume of OTC trading. This is done by standardizing and adapting some OTC derivatives contracts for subsequent circulation on the exchange market, as well as by tightening legislation in this area and introducing various restrictions that impede speculative activity. If we consider the structure of the OTC international derivatives market, we can conclude that over the past ten years, most of the market in monetary terms has consistently been occupied by interest rate and currency derivatives. This fact, together with a decrease in the share of credit derivatives in the market, indicates the restoration of the international derivatives market to its main function, namely, hedging of commercial and financial risks.

The dynamics in the volume of the international derivatives market is structurally similar to the OTC international derivatives market, as the tendency to decrease in market volumes is successively replaced by growth trends. So, for the period from 2009 to 2012, the total nominal value of positions opened on the exchange derivatives market decreased from 73,125 billion dollars in the second half of 2009 to 54,122 billion dollars in the second half of 2012. However, then it subsequently increased to 70,223 billion dollars in 2015. Subsequently, a steady upward trend was formed, and the total nominal value of open positions in currency and interest rate of derivatives alone amounted to $ 68,096 billion by the end of 2018 [1]. The structure of the exchange market is generally similar to the structure of the OTC international derivatives market. However, in the total market for exchange-traded derivative financial instruments, interest-bearing derivatives and derivatives on securities have the greatest weight [10].

The weakest segment of the Russian derivatives market is interest-bearing derivatives, while on the international derivatives market they have the largest share in the market. In addition, it should be noted that such low volumes of transactions under option contracts in comparison with futures are not typical for the international derivatives market. These facts, combined with a high share of foreign exchange derivatives in the Russian derivatives market, suggest that instruments in this market are used primarily for speculative operations, rather than hedging risks. It can be concluded that the fully-formed derivatives market in the Russian Federation still remains structurally underdeveloped.

4 Discussion

Derivative instruments are behind the over-the-counter ones if direct delivery of the underlying asset is required. Such a situation may arise if one of the parties to the contract is the manufacturer, which needs to fix the cost of his products. In addition, the combination of foreign exchange and commodity delivery forwards significantly reduces commercial risks. The flexibility of OTC derivatives contracts is also an advantage. However, the use of OTC derivative instruments is associated with credit risk in relation to the counterparty [2].

The stock derivatives market is much more suitable for active financial investors, since they are most often exposed to stock risk. Exchange contracts have high liquidity and performance guarantees, which makes it possible to quickly and efficiently adapt investment and hedging strategies to changes in market conditions. However, standardization of exchange contracts can impede the full hedging of an asset and as a result only partial hedging occurs [5]. In addition, only foreign exchange and interest-bearing OTC derivatives contracts can be so effective in hedging currency and interest rate risks.

Over-the-counter hedging strategies are in many respects opposed to exchange-based hedging strategies [7]. Unlike exchange-traded instruments, over-the-counter derivative instruments, due to their flexibility, make it possible for the investor to provide full hedging of the asset. The low turnover of forwards, combined with the difficulties of entering the OTC market, predetermines the low adaptability of over-the-counter hedging strategies to changes in the market. Therefore, investors usually use separate over-the-counter derivative instruments to protect themselves from currency and interest rate risks. While most of operations they carry out on the regulated exchange market. This market enables the investor to quickly respond to changes in market conditions and implement the fixed-term contract that has become irrelevant for him. Not always the investor can find a standard exchange contract that fully meets his requirements. Therefore, for the most effective hedging of risks in various situations, strategies have been developed that allow maximum protection of the asset at minimum cost. Modeling the process of reducing investment risks using derivative financial instruments is associated with the development and implementation of the hedging strategy in specific given conditions by the investor. This requires the most complete information about the investment portfolio, the market situation, as well as derivative financial instruments.

5 Conclusion

Investors in the Russian derivatives market can get, not as voluminous as internationally, a fairly wide selection of exchange and OTC derivative instruments to implement a variety of risk hedging strategies. As can be seen from the study, the growth dynamics of the derivatives market is unstable in the Russian Federation, as it is directly related to macroevents occurring in the entire financial market as a whole. The distinctive features of the Russian derivatives market, identified as a result of the analysis, include: 1) high volumes of futures trading in comparison with options; 2) the high interest of market participants in currency futures and its absence in relation to interest-rate derivatives; 3) stable index future contracts in the structure of the Russian derivatives market over last five years.

It can be concluded that in many respects macroeconomic problems that impede the normal functioning and development of the entire economic and financial system of the Russian Federation hinder the development of the Russian derivatives market. However, specific institutional factors are also important. Thus, the imperfection of the legislation in relation to the securities market and derivatives market determines the ambiguity of legal interpretations when concluding contracts, which ultimately leads to legal conflicts. In addition, the lack of practice in regulating relations in the derivatives market directly affects the transparency and efficient protection of bona fide market participants. Only the solution of macroeconomic problems, the development of legislation in the field of derivative financial instruments, as well as increasing market transparency through the use of digital technologies can ensure sustainable development of the Russian derivatives market in the future.

References

1. BIS Statistics: OTC derivatives outstanding (2019). https://www.bis.org/statistics/index.htm. Accessed 25 Dec 2019
2. Chatzis, S.P., Siakoulis, V., Petropoulos, A., Stavroulakis, E., Vlachogiannakis, N.: Forecasting stock market crisis events using deep and statistical machine learning techniques. Expert Syst. Appl. **1121**, 353–371 (2018). https://doi.org/10.1016/j.eswa.2018.06.032
3. Gratton, S., Soualmi, N., Vicente, L.N.: An indicator for the switch from derivative-free to derivative-based optimization. Oper. Res. Lett. **45**(4), 353–361 (2017). https://doi.org/10.1016/j.orl.2017.05.003
4. Hinings, B., Gegenhuber, T., Greenwood, R.: Digital innovation and transformation: an institutional perspective. Inf. Organ. **28**(1), 52–61 (2018). https://doi.org/10.1016/j.infoandorg.2018.02.004
5. Huang, P., Kabir, M.H., Zhang, Y.: Does corporate derivative use reduce stock price exposure? Evidence from UK Firms. Q. Rev. Econ. Finance **65**, 128–136 (2017). https://doi.org/10.1016/j.qref.2017.02.004
6. Kang, S.H., Lee, J.W.: The network connectedness of volatility spillovers across global futures markets. Physica A **52615**, 120756 (2019). https://doi.org/10.1016/j.physa.2019.03.121
7. Lau, C.K.: How corporate derivatives use impact firm performance? Pacific-Basin Finance J. **40**(A), 102–114 (2016). https://doi.org/10.1016/j.pacfin.2016.10.001

8. Mi, H., Xu, L.: Optimal investment with derivatives and pricing in an incomplete market. J. Comput. Appl. Math. **368**, 112522 (2020). https://doi.org/10.1016/j.cam.2019.112522
9. Moscow Exchange: Monthly trading volumes: Annual statistics (2019). https://www.moex.com/en/ir/interactive-analysis.aspx. Accessed 25 Oct 2019
10. Nikitopoulos, C.S., Squires, M., Thorp, S., Yeung, D.: Determinants of the crude oil futures curve: inventory, consumption and volatility. J. Bank. Finance **84**, 53–67 (2017). https://doi.org/10.1016/j.jbankfin.2017.07.006
11. Puchinin, N., Zakharov, A.: Financial and economic crises of recent decades and their impact on the Russian economy. Prime (2019). https://1prime.ru/sience/20190402/829858467.html. Accessed 23 Dec 2019
12. Santana-Gallego, M., Pérez-Rodríguez, J.V.: International trade, exchange rate regimes, and financial crises. North Am. J. Econ. Finance **47**, 85–95 (2019). https://doi.org/10.1016/j.najef.2018.11.009
13. Vervoort, J.M.: New frontiers in futures games: leveraging game sector developments. Futures **105**, 174–186 (2019). https://doi.org/10.1016/j.futures.2018.10.005
14. Vial, G.: Understanding digital transformation: a review and a research agenda. J. Strateg. Inf. Syst. **28**(2), 118–144 (2019). https://doi.org/10.1016/j.jsis.2019.01.003

Digital Assets in the United States: Legal Aspects

M. I. Inozemtsev(✉)

Moscow State Institute of International Relations (University) of the Ministry of Foreign Affairs, Moscow, Russia
inozemtsev@inno.mgimo.ru

Abstract. This research investigates specific features of theoretical comprehension and legal regulation of digital assets in the USA. The main objectives are: to study the scientific discussion development and lawmaking process in the field of legal regulation of digital assets in the USA, to analyze topical issues of allocation of competences among US financial regulators (SEC, CFTC, IRS, FINRA, FinCEN) exercised in respect to digital assets (virtual currency), to develop a position on the possibility of adopting the US experience in regulation of a new asset class for the purposes of development of the Russian civil law. The author analyzes approaches to the taxonomy of digital assets, presents a classification of tokens, distinguishes between digital and digitized assets. Special attention is paid to new forms of activities in the crypto-market that require specific legal regulation. On the basis of the analysis of draft laws and theoretical discussion statements, some problematic aspects of the civil regulation of digital assets are identified, peculiarities of allocation of competences among the US financial regulators are highlighted.

Keywords: Cryptoassets · Custody · Digital assets · Digitized assets · Token · Virtual currencies

1 Introduction

The institutional model of regulation of the US financial market is characterized by a complex segmentation of the regulatory field, resulting in a large number of supervisory agencies that have their own part of the competence. This leads to more complex rules for compliance of market participants' activities with the federal legislation and problems of the corporate compliance adjustment. Regulators have to issue multiple individual and joint documents that clarify for market participants their responsibilities to each of the regulators. At the same time, such documents are often difficult to classify as legal acts, but they are important as acts of interpretation or acts of "soft law" that affect administrative and judicial law enforcement.

These features of the prudential supervision model are overlaid with specific features of the American federalism and its manifestations in the legal system. While at the federal level, legal regulation is primarily aimed at minimizing public risks (such as fraud, risks of legalization (laundering) of proceeds from crime and financing of terrorism, consumer rights violations, etc.), at the states' level, more stimulating

S. I. Ashmarina et al. (Eds.): ENGINEERING ECONOMICS WEEK 2020, LNNS 139, pp. 514–522, 2021.
https://doi.org/10.1007/978-3-030-53277-2_61

legislation is being adopted that creates "windows of opportunities" for innovative activities of market participants.

Among the main regulators that extend their competence to the sphere of digital assets circulation are the U.S. Securities and Exchange Commission (SEC), the U.S. Commodity Futures Trading Commission (CFTC), the Financial Industry Regulatory Authority (FINRA) as a division of the SEC. These agencies, with the task of hedging public risks, also address issues of a private legal nature.

The Financial Crimes Enforcement Network (FinCEN) and the Internal Revenue Service (IRS) also belong to the regulation architecture of the U.S. crypto market as agencies with emphasized public legal competence. Regardless of the federal regulator type, all financial institutions that operate digital assets fall under the legal definition of a security and have to comply with the federal Securities Act (SEA). In addition, the main legal framework supporting the CFTC's competence in the financial market is the Commodity Exchange Act (CEA), and FinCEN relies primarily on the Bank Secrecy Act (BSA).

2 Methodology

The methodology of the study comprises comparative, formal-legal, functional methods, systemic approach. The results of the study of problematic issues of digital assets' regulation in the United States form the ground for creating a position on the opportunities to adopt the U.S. experience in the Russian Federation for facilitating the development of the Russian civil law. The findings of this research can be applied for developing models for legal regulation of digital assets in foreign countries as well as using positive aspects of the considered experience to the Russian legal system.

3 Results

Let's consider the specifics of implementing the competence of the most interesting regulators (SEC, CFTC, FINRA) in the U.S. financial market.

The *SEC*'s statutory mission is to protect investors and maintain fair, orderly and efficient markets. The SEC oversees key participants in the securities market, some of which may be involved in digital asset activities. Key participants in the securities markets include national securities exchanges, securities brokers and dealers, investment consultants, and investment companies.

The *CFTC*'s mission is to promote the integrity, sustainability, and dynamism of the U.S. derivatives markets through prudential regulation. In fulfilling this mission, the CFTC regulates key participants in the derivatives market, including trading advisors, futures commission traders, brokers, swap dealers, large swaps participants, retail currency dealers, commodity pool operators, and commodity trading consultants in accordance with the Commodity Exchange Act (CEA).

FINRA strives to protect investors and ensure the market integrity in a way that encourages the dynamic development of the capital market. Among the key activities related to digital assets that are of interest to FINRA are: purchase, sale or execution of transactions with digital assets; purchase, sale or execution of transactions in a joint fund that invests in digital assets; creation, management or provision of advisory services for a joint fund related to digital assets; purchase, sale or execution of transactions with derivatives (for example, futures, options) linked to digital assets; participation in the primary or secondary placement of digital assets (for example, ICO, pre-ICO); creation or management of a platform for secondary trading of digital assets [11].

Public organizations, such as the Chamber of Digital Commerce and the American Bar Association, play an important role in shaping the expert position and influencing the law enforcement in the United States.

We will highlight *the main aspects* of civil law issues of digital assets turnover in the U.S. financial market:

1. Problems of taxonomy and typology of digital assets.

Here we can highlight the following aspects. *First, the key content issues are a special nature of new digital entities and a need to find for them a separate place among the known objects of civil rights*. So, in American studies, it is suggested that the token is not just a new way of selling assets, but a new type of asset (new asset class) [13], and the CFTC points out that virtual currencies are not like any product that the CFTC has dealt with in the past [4]. At the same time, the SEC uses as the main one the approach according to which most digital assets are similar to securities and they can be subject to the securities legislation.

Second, the issue of terminology is debatable – how to name a new class of objects. American regulators and the crypto community are very pluralistic about naming new digital entities. In particular, the terms "convertible virtual currencies" (FinCEN terminology), "virtual currencies" (IRS and CFTC terminology), "digital assets" (the terminology of the SEC and primarily of the American expert community), and "cryptocurrency" (IRS terminology) are used in parallel.

Since 2015, the CFTC has relied on the IRS definition to provide a "general idea" of what is meant by "virtual currency": a digital representation of a value that functions as a medium of exchange, unit of account, and/or savings and that does not have the status of a legal means of payment [4]. This definition has also been used by the CFTC in lawsuits and other regulatory documents [5].

In 2013, FinCEN released a guide describing virtual currency as a medium of exchange that works as a currency in some environments, but does not have all the attributes of a real currency. In particular, virtual currency does not have the status of a legal tender in any jurisdiction [10].

In the 2019, in a joint document, CFTC, FinCEN and SEC use the most universal term in the American expert community "digital assets": digital assets include instruments that can be qualified under the current U.S. law as securities, commodities, or security- or commodity-based instruments, such as futures or swaps [15].

It is definitely worth recognizing that the vocabulary of international and national financial regulators has migrated from the concepts of "cryptocurrency" and "virtual currency" to a broader class – "assets" ("digital", "crypto", "virtual"). The Cambridge Center for Alternative Finance also draws attention to this trend in its report [3].

With all this variety of terms, it is worth noting that their unification is not so important and fundamental in the American legal model, as American regulators use an approach focused on the meaning, content of the operation, and not on formal features. This is an expansive approach that requires independent reflection by market participants and thorough legal compliance.

So in the above-mentioned joint statement, the heads of the CFTC, FinCEN and the SEC urge companies not to forget about the rules of law governing banking and other financial services, regardless of how they call their digital assets – cryptocurrencies or tokens. Regulators indicate that market participants operate digital assets using a variety of different "labels", but these "labels" may not coincide with how this asset, activity or service is defined by the laws or regulations of the CFTC and SEC. Regulators emphasize that regardless of the labels or terminology that market participants may use, by the asset classification we should take into account the economic nature of the asset, application practices, and underlying technology.

In other words, the nature of the digital asset-related activity that a person is involved in is a key factor in determining whether and how that person should be registered by the CFTC, FinCEN, or SEC. For example, some commodity-related activities may involve registration and other obligations under the Commodity Exchange Act (CEA), while some activities may involve registration and other obligations under the Securities Act (SEA). If a person falls under the definition of "financial institution", his/her activities will be monitored for purposes of the Bank Secrecy Act (BSA) by one or more regulators. For example, in the sphere of countering the legalization (laundering) of proceeds from crime and financing of terrorism, the CFTC, FinCEN, and the National Futures Association (NFA) will monitor activities of the futures commission merchant; the money services business (MSB) companies will be monitored by the FinCEN [15].

Certain obligations under the BSA that apply to a securities broker-dealer, mutual fund, futures commission seller, or introducing broker, for example when developing an AML program or reporting a suspicious activity, apply very widely and regardless of whether the particular transaction includes "securities" or "commodities" as these terms are defined in the federal laws.

Third, the issues of classification of digital assets are legally significant, since such classification allows regulators to customize tools of legal regulation to the specifics of digital assets. The basic classification in both the United States and Europe is that of *payment tokens*, *security tokens*, and *utility tokens*. At the same time, the greatest attention of regulators is paid to security tokens.

Complex issues of classification that have practical legal significance, are:

- differentiation of tokens into digital and digitized ones,
- combination of security tokens and non-security tokens in a single transaction,
- problems of hybrid tokens.

The division of tokens into digital (digital assets) and digitized (digitized assets) is supported by the majority of the American crypto community. Thus, in his fundamental work "Digital and Digitized Assets: Federal and State Jurisdictional Issues" experts point out that a "digital asset" is an electronic record in which an individual has a right or interest. The term does not include the underlying asset or liability, unless the asset or liability itself is an electronic record. In other words, the digital asset itself is just a code. In this case, a "digitized asset" is an asset (which can be a security or a physical asset), the ownership of which is represented in an electronic record. An example of a digitized asset is an electronic record of ownership of real estate stored in a distributed registry [1].

American experts also conduct a deeper classification of tokens.

For example, in their reports [16, 17], experts from the U.S. Chamber of Digital Commerce draw attention to combinations of different types of tokens carried out in a single transaction. In this regard, they distinguish three specific types of tokens.

1. Tokens substitute (*placeholder tokens*).

The token issuer may offer a non-tokenized instrument, the most common of which is the "Simple Agreement for Future Tokens" (SAFT), which refers to service tokens. However, instead of an agreement, the token issuer can provide buyers with a substitute token that will perform the same functions as a SAFT or other non-token contract or instrument. In any case, such a pre-functional document is a security and subject to the provisions of the Securities Act.

2. Changeable tokens (*mutable tokens).*

Instead of providing a substitute token that is later replaced with a service token, the token issuer can choose to sell the pre-functionality token as a security, and then develop a system where the already service token will eventually work and, after deploying the system, allow token holders to use the same token in the system and extract its consumer value in the Blockchain application.

3. Tokens that ensure the payment of dividends (*dividend-paying tokens*).

A tokenized security (a security token) may provide dividends to its holder in the form of a second token, and this second token may be a non-security token (for example, stablecoins).

Hybrid tokens are a separate legal issue. For example, for a security token that is also a payment token, regulators and courts may provide different types of legal regulation, such as:

- integrated: a hybrid token should comply with both securities and payment laws,
- hierarchical: a hybrid token should meet the stricter requirements of the securities law or payment legislation (depending on its prevailing functionality).

The American Bar Association draws attention to hybrid tokens, citing the example of service tokens that can simultaneously act as a means of payment (for example, Ether) [1]. There are also theoretical questions about *payment tokens*, which experts divide into account-based systems and token-based payment systems. A separate legal issue in the United States is *security tokens*, which determine *whether to extend existing legislation to a new class of objects or adopt a new one.*

The SEC's authority to regulate digital assets relies on the approach that these assets are covered by the definition of a security. However, this definition includes a wide range of instruments, largely reducing to the universal concept of "investment contract". Since the Securities Act and the Commodity Exchange Act do not explicitly apply to the circulation of digital assets, whether a digital asset will be within the scope of the "security" concept will depend on whether it can meet the criteria of an "investment contract".

In July 2017, the SEC outlined its legal position regarding the public sale of tokens in the report on the case "The DAO" [14], which implies that the U.S. securities legislation applies to certain operations with cryptocurrencies, including the issue of tokens as part of an ICO (initial coin offering), as well as operations for their exchange. Now all participants in the crypto economy, from issuers to exchanges, as well as other persons involved in any way in the offer and sale of tokens that fall under the definition of securities, and have to comply with the requirements of the U.S. securities laws.

On April 3, 2019, the SEC published a guidance on cryptocurrencies, digital tokens, and similar investments (each of which is a "digital asset") [15], intended to help issuers and others to determine whether the applicable digital asset is a security according to federal securities laws. This guide is a long-awaited document in the field of crypto business in the United States, as well as especially important for those who want to register a company under ICO in the United States and get a license for crypto activity. The guidance also confirms the SEC's position, hinted at by public statements from late 2017, that digital assets are in many (if not most) cases securities. The SEC's guidance focuses on elements of the so-called "Howey test", which is taken from the U. S. Supreme Court case (SEC vs. Howey), and is used by courts to assess whether a transaction or financial instrument is an "investment contract" – a security [15].

While the Howie test consists of three elements, the guide focuses on what is usually seen as a key question for analyzing digital assets: whether investors in a digital asset have a "reasonable expectation of profit derived from the efforts of others." In accordance with the guidelines, the SEC has taken a position that this element is performed in a variety of cases, including when the issuer of a digital asset or any other group: has assumed responsibility for the development, improvement, operation or promotion of a network associated with the digital asset; or supported the market or the price of a digital asset.

Under the SEC's guidance, many (if not most) digital assets issued over the past few years should be treated as securities. Thus, individuals who have already registered ICO firms in the United States and are engaged in issuing tokens had to register their offers or ensure that their offers were eligible for an exemption from the registration. Issuers that have not taken these steps may be subject to the SEC enforcement actions and claims from investors.

American lawyers recommend that if a digital token does not claim to be a security, avoid using terminology related to securities in official documents and marketing materials that may confuse readers about the token nature. Examples of such "unsuccessful" turns: "token holders will own" or "profit" from the Blockchain or distributed network; "token holders will receive interest" or "dividends" on their tokens. It is also not recommended to give references to the "market capitalization", "initial coin offering" (ICO).

So, in December 2018, members of the House of Representatives, Congressmen Warren Davidson and Darren Soto introduced the long-awaited Token Taxonomy Act [7], according to which tokens are excluded from the category of "securities" despite recent bans and fines from the SEC. And the Wyoming Blockchain Coalition in March

2018 promoted the passage of five laws removing “consumer tokens” (“consumptive tokens”) from the scope of federal securities laws under the justification of them as a “new asset class”.

A separate regulation in the USA is assigned to the segment of derivative financial instruments. The CFTC allows new virtual currency derivatives contracts to be listed and details their regulation [6]. In a statement dated May 21, 2018, the regulator indicates it necessary to encourage innovation stemming from virtual assets, but assumes to do so within the framework of the basic trade principles of the Commodity Futures Trading Commission’s LabCFTC [5].

The CFTC document corresponds with a document of the British Financial Conduct Authority (FCA) [9] published on July 3, 2019, the FCA announced that it will begin a consultation process to ban the sale, advertising and distribution to retail consumers of derivatives and exchange-traded bonds (ETN) that mention certain types of crypto assets. The document sets out a number of justifications for the proposed ban on the sale of cryptoasset derivatives to retail customers.

2. Problems of legalization of new types of intermediaries.

A separate issue is the issue of specialized regulation of new (or modernized) *activities* in relation to digital assets.

On the one hand, existing regulations may sometimes apply directly to new intermediaries that perform tasks similar to traditional financial activities. On the other hand, some services on the crypto market may require additional regulation. Experts of the Cambridge Center for Alternative Finance consider only a relatively small number of mediation types in the crypto market to be absolutely new (for example, the issue and placement of tokens, P2P services, crypto exchanges, and blockchain analytics) [3]. Recently, the consulting company Financial Integrity Network recommended the U.S. Congress to create a new category of financial institutions in accordance with the Bank Secrecy Act to regulate activities of cryptocurrency companies [12].

4 Discussion

A separate interest of the expert community is drawn to the specifics of digital asset storage services (custodial activities) [2]. Since crypto assets do not have a physical appearance (they exist solely as digital records in a virtual distributed registry), any movement of the crypto asset (for example, the transfer of rights to it) requires permission in the form of a cryptographically signed message from the initiator. Using a private key enables the transaction to be made. In the context of crypto assets, storage is thus no longer related to direct ownership of assets, but to the secure storage of encryption keys.

On July 8, 2019, the SEC’s trade and markets department and FINRA’s office of general counsel issued a joint statement on the custody of digital assets-securities by broker-dealers [8], which discusses certain provisions of federal securities laws applicable to registered broker-dealers that may be affected when such individuals hold digital asset securities. The document explains the content of this term. Since digital assets may not be securities, the term “digital assets securities” is used for those digital assets that are securities.

The joint statement implies that digital assets securities and related innovative technologies raise new and complex issues related to the regulation and compliance. It also notes that there are many significant differences in the mechanics and risks associated with storing traditional securities and digital assets securities. The joint statement then discusses a number of risks and regulatory requirements for broker-dealers seeking to maintain custody of digital asset securities, as well as those whose business models do not imply any storage.

For broker-dealers whose business models include storing securities with digital assets, the joint statement specifically focuses on the requirements of the SEC's client protection rule, which imposes various requirements on broker-dealers to ensure that clients' securities and cash are easily available for return to clients. The joint statement also discusses some of the unique challenges that may arise when using distributed ledger technology in the context of the SEC's rules in relation to books, documentation and financial reporting.

5 Conclusion

Thus, in order to optimally configure legal tools, experts recommend to recognize that "digital assets" is not a homogeneous, but a generalizing concept that describes many types of tokens; use an approach based on actual data on the activities of crypto market entities to develop detailed understanding of the tokens types, their functions, basic infrastructure and other key features; create a typology based on the nature and functional specific features of tokens; identify potential harm to financial markets, investors and consumers from the unregulated token circulation, as well as cooperate with crypto market entities in order to better understand the specifics of their activities.

References

1. American Bar Association: Derivatives and Futures Law Committee, Innovative Digital Products and Processes Subcommittee, Jurisdiction Working Group: Digital and digitized assets: Federal and state jurisdictional issues (2019). https://www.americanbar.org/content/dam/aba/administrative/business_law/buslaw/committees/CL620000pub/digital_assets.pdf. Accessed 01 Dec 2019
2. Baris, J.G.: The custody of digital assets global legal insights (2019). https://www.globallegalinsights.com/practice-areas/blockchain-laws-and-regulations/06-the-custody-of-digital-assets-2020. Accessed 01 Dec 2019
3. Blandin, A., Cloots, A.S., Hussain, H., Rauchs, M., Saleuddin, R., Allen, J.G., Zhang, B., Cloud, K.: The Global Cryptoasset Regulatory Landscape Study. Cambridge Centre for Alternative Finance, Cambridge (2019)
4. CFTC: A CFTC primer on virtual currencies (2017). https://www.cftc.gov/sites/default/files/idc/groups/public/documents/file/labcftc_primercurrencies100417.pdf. Accessed 01 Dec 2019
5. CFTC: An Introduction to virtual currency (2018). https://www.cftc.gov/sites/default/files/idc/groups/public/@customerprotection/documents/file/oceo_aivc0218.pdf. Accessed 01 Dec 2019

6. CFTC: Staff advisory No. 18–14. Division of market oversight, division of clearing and risk (2018). https://www.cftc.gov/csl/18-14/download. Accessed 01 Dec 2019
7. Davidson, W., Soto, D.: Token taxonomy act of 2018 (2018). https://ru.scribd.com/document/396096529/Token-Taxonomy-Act-of-2018. Accessed 01 Dec 2019
8. Division of Trading and Markets, & Office of General Counsel: Joint staff statement on broker-dealer custody of digital asset securities (2019). https://www.sec.gov/news/public-statement/joint-staff-statement-broker-dealer-custody-digital-asset-securities. Accessed 01 Dec 2019
9. FCA: Prohibiting the sale to retail clients of investment products that reference cryptoassets (2019). https://www.fca.org.uk/publication/consultation/cp19-22.pdf. Accessed 01 Dec 2019
10. FinCEN: Application of FinCEN's regulations to persons administering, exchanging, or using virtual currencies (FIN-2013-G001) (2013). https://www.fincen.gov/resources/statutes-regulations/guidance/applicationfincens-regulations-persons-administering. Accessed 01 Dec 2019
11. FINRA: Regulatory Notice 19–24 FINRA encourages firms to Notify FINRA if they engage in activities related to digital assets (2019). https://www.finra.org/rules-guidance/notices/19-24. Accessed 01 Dec 2019
12. Khatri, Y.: Advisory firm suggests US Congress to regulate crypto sector under the bank secrecy act (2019). https://www.theblockcrypto.com/linked/38191/advisory-firm-suggests-us-congress-to-regulate-crypto-sector-under-the-bank-secrecy-act. Accessed 01 Dec 2019
13. Klayman, J.A.: Mutually assured disruption: the rise of the security token. In: Dewey, J. (ed.) Blockchain & Cryptocurrency Regulation, pp. 60–89. Global Legal Group Ltd., London (2019)
14. Report of Investigation Pursuant to Section 21(a) of the Securities Exchange Act of 1934: The DAO, no. 81207, 25 July 2017. https://www.sec.gov/litigation/investreport/34-81207.pdf. Accessed 01 Dec 2019
15. Tarbert, H., Blanco, K.A., Clayton, J.: Leaders of CFTC, FinCEN, and SEC issue joint statement on activities involving digital assets (2019). https://www.sec.gov/news/public-statement/cftc-fincen-secjointstatementdigitalassets. Accessed 01 Dec 2019
16. Token Alliance: Legal landscapes governing digital tokens in the United States (2018). https://digitalchamber.org/wp-content/uploads/2019/08/CHD-19003-16-06-D1-White-Paper-Countries-United-States.pdf. Accessed 01 Dec 2019
17. Token Alliance: Considerations and guidelines for securities and non-securities tokens (2019). https://digitalchamber.org/wp-content/uploads/2019/08/Security-Non-Security-Tokens.pdf. Accessed 01 Dec 2019

Adaptability and Flexibility of Law in the Context of Digitalization

E. L. Sidorenko[1(✉)], L. L. Arzumanova[2], and O. N. Amvrosova[3]

[1] Moscow State Institute of International Relations (University) of the Ministry of Foreign Affairs, Moscow, Russia
12011979@list.ru

[2] Kutafin Moscow State Law University, Moscow, Russia
llarzumanova@msal.ru

[3] North-Caucasian Federal University, Stavropol, Russia
kilinkarov.77@mail.ru

Abstract. The paper investigates current issues of legal transformation in the context of digitalization. The main attention is paid to directions and stages of adaptation, determination of the legal status of digital technologies, their protection capacity and delinquency. Based on expert assessments, key parameters of legal regulation of digital technologies and their products are determined and the need to supplement civil, financial and information law with new categories and institutions is justified. It is concluded that approaches to assessing the legal nature of digital objects are contradictory and there is insufficient consideration of technical aspects of digital technologies, as well as a need to develop a unified legal strategy for the civil and intellectual rights of digital technologies at the international level. Among the priorities and directions of this strategy, there should be the development of issues on the legal personality of digital technologies and the main mechanisms for protecting products of this technology application. The conclusions formulated in the article have important practical and methodological significance and can be taken into account when reforming the current legislation.

Keywords: Copyright and related rights · Digital economy · Digitalization · Legal reforms · Legislation · Legal relations

1 Introduction

In the context of rapid development of the digital economy, there is a need to develop a science-based approach to the legal regulation of digital technologies and the legitimation of new tangible and intangible assets. The problem, however, is that while declaring the importance of legal transformation, scientists mostly ignore the question in what should be expressed the digital transformation of law, and whether modern legislation has the necessary internal resources for adaptation. In fact, the question of digital transformation of law depends on the question of its readiness to adapt to modern digital challenges and realities. Adaptability in most scientific theories is considered as a property of organic and inorganic systems that determine the success of their existence in a changing environment. Unfortunately, in the theory of law, this

S. I. Ashmarina et al. (Eds.): ENGINEERING ECONOMICS WEEK 2020, LNNS 139, pp. 523–532, 2021.
https://doi.org/10.1007/978-3-030-53277-2_62

term is not widely used because of the obvious reluctance of legal scholars to recognize the fact that the law, being aimed at regulating public relations, often does not keep pace with their development. Meanwhile, it is the adaptability of law that is the key parameter in identifying the relationship between the law adaptation degree to regulated relations and the degree of its autonomy as a regulator and controller of the development of these relations.

Speaking about the adaptability of law to the digital economy, it is important to identify key tasks of adaptability – resolving contradictions in the process of implementing digital technologies, stimulating positive development of the digital economy and minimizing digitalization risks. When solving these tasks at the initial stage, the legislation does not regulate relations but adapts to them. The results of this process are development of law, its self-actualization and self-realization. Considering the adaptation of law in the conditions of digitalization as an inevitable process of its development, we can conditionally distinguish its constituent elements: heuristic (the ability to design and find solutions based on a detailed study of digital processes and phenomena that require regulation); communication (active interaction with participants of new legal relations, search for consensus between new needs and traditional legal institutions); axiological (compliance of new digital challenges with the fundamental principles of law); empirical (search for the most acceptable models of regulation through the formation of a new legal practice).

The ideal adaptation is based on 4 elements at the same time. In this case, it allows not only to solve particular problems of regulation, but also to present them in a systematic unity with the main ideas, principles and structures of law. While recognizing the importance of developing a unified concept of digital law, legal science nevertheless ignores the issue of a development strategy for the law. Because of the lack of a theoretical framework for digital law, experts prefer to address applied issues of regulation for individual digital technologies. In particular, the issues of legal regulation of artificial intelligence (hereinafter – AI) are considered within the scope of its use in certain areas of the legal activity [28]; changes in modern delinquency and contract legislation [15], etc. The legal regulation of the use of drones and other breakthrough technologies is analyzed in the context of reforming special legislation (medical [20], informational [26]), etc. Great importance in the scientific literature is given to the regulation of smart contracts in the context of securities turnover [32], the will of the parties [14], the correct legal formalization of smart contracts [17, 23]. Attention is also paid to the regulation of crypto assets. Most of the works are devoted to the basic principles of financial legislation [12], taxation and regulation of the digital payment instruments market [8, 16]. The legal status of Big Data, its use in legal activities [13], models of legal control [22] and legal risks of using Big Data [17, 24] are also discussed.

2 Methodology

This research work is based on general scientific and private scientific methods. Mainly general logical (analysis, synthesis, induction, deduction, abstraction and classification) and general theoretical methods were used. A large role in the research is given to methods of abstract, analog and simulation modeling, which allow building a construct of legal features of new digital phenomena on the basis of traditional legal institutions. These methods of scientific knowledge were supplemented by empirical methods: statistical (statistical observation, summary and grouping, statistical analysis), sociological (study of documents, surveys in the form of questionnaires and interviews, observation) and socio-psychological (testing). The use of these approaches, together with general theoretical methods, made it possible to conduct a comprehensive study of the development of digital law.

3 Results

The authors of this research identified the main approaches to defining legal characteristics of digital technology, evaluated proposals for restructuring of law under modern conditions, delineated boundaries of the legal formalization of digital phenomena and perspectives of the development of international standards for digital rights. A critical analysis of these aspects allowed the authors to formulate a number of recommendations aimed at the legislation development both in terms of sustainability of the digital economy and in terms of minimization of digitalization risks.

Unfortunately, modern science does not address the question of what parameters should be used to assess the adaptability of law in the era of digitalization. Without claiming to be an indisputable position, we consider it possible to assess the adaptive capabilities of the modern law through two groups of factors:

- legal categories and institutions that set the general tone for the digital transformation of law: the legal nature of digital technologies as objects of civil rights, their "protectability" and delinquency,
- particular issues of regulation of individual technologies: the Internet of things, AI, Big Data, machine learning, unmanned aerial objects, robots, etc., as well as a number of previously identified general issues that are gaining a new meaning in relation to some technologies.

The solution of the first block of questions involves the use of heuristic and axiological tools of adaptation, while the second block is focused on the use of communication and empirical tools. The first reference point for assessing the adaptive capabilities of law in the context of digitalization is the protection capacity of digital technologies and their products. Under the protection capacity of digital technologies is understood their ability to act as an object of civil and intellectual rights. In this regard, the important question is whether emancipation (separation) of real and copyright law from traditional legal institutions is permissible.

Opponents of emancipation insist that artificial intelligence and machine learning cannot be used as the basis of law, and they cannot replace the fundamental principles of law, since these principles themselves have been developed for centuries in a way that is called "anchoring learning" in the field of AI, and human behavior is basically too irrational and inconsistent [10]. Proponents, on the contrary, see emancipation as a natural conflict of public and private interests in the digital world and consider a possibility of recognizing digital technologies as legal entities if they have functional independence, acquire value and utility in the economic sense [19].

In the current Russian legislation, digital rights are considered as objects of civil rights, together with things and other property. They are a type of property rights and are mentioned along with intellectual rights. That gives reason for some researchers to deny the possibility of attributing digital objects to objects of copyright protection. They are defined as binding and other rights, the implementation, disposal and restriction of which are possible only in the information system (Article 141.1 of the Civil Code of the Russian Federation), and therefore have a very limited application [6].

As a result, a whole set of issues related to the use of these rights outside the digital environment remains outside the legal framework, namely, the use of the results of the exercise of rights (generalization and analysis of Big Data, machine learning results, etc.). While in the Russian law, the starting point for regulating digital rights is their connection with the information system, in the Anglo-Saxon law, the "protectability" of digital objects is associated with their relation to property [31]. From these positions, digital technologies acquire a status of objects of civil law, as well as products created based on the use of these technologies.

Thus, in the Anglo-Saxon law, it is much easier to answer questions about: 1) Who owns the new code of the self-learning program, which embodies its "experience" of executing commands; 2) Who has rights to works created by such programs and robots; 3) Who bears property and other responsibility for the negative consequences of the use of digital technologies or their products.

However, the experience of individual countries is not sufficient for the progressive and successful development of digital economy. It is important to develop a unified universal approach to recognizing digital technologies, products and rights as objects of civil legal relations.

We believe that this approach should be based on the following methodological provisions:

- recognition of a digital product or phenomenon as an object of civil rights should not be based on establishing its material nature or economic value. As a rule, digital objects have a complex legal nature, and can be considered simultaneously from the position of intellectual and real (property) rights,
- the assessment of digital technologies should be based on a legal model that is as close as possible to the legal essence of relations and can ensure a balance of private and public interests in digital turnover,
- in developing a universal approach to the regulation of digital technologies as an object of civil rights, it is necessary to proceed from the objective emancipation of digital law from traditional legal institutions. In other words, the adaptive capabilities of traditional law in relation to digital technologies are very limited due to the multifunctional nature of objects, technical saturation, uncertainty of regulatory tasks and possible risks of their introduction into civil circulation.

Copyright Protection Ability. Speaking about the development of intellectual property rights in the digital sphere, it is important to note that one of the most relevant is the choice between granting the right to the results of the digital technologies use to individuals and recognizing them as a public domain. This fundamental question leads to a number of additional issues related to 1) the protection of objects created using digital technologies, 2) the definition of the digital form of existence of a virtual work; 3) the qualification of the content of digital platforms, 4) the possibility of changing the nature of the disposal of the exclusive right to a protected object due to its existence in the digital environment, 5) the joint use of exclusive rights in crowdsourcing, etc. But whatever the decision is, it is obvious that it cannot be made before the legal nature of the rights to the products of digital technologies and, in particular, AI is determined.

Difficulties are also associated with objective characteristics of digital products: a high degree of repeatability obtained from the use of these technologies results, the low contribution of the creative human labor, automation of certain processes, the inability to distinguish between creative and non-creative component, the complexity of differentiation of rights of the author and rights of the originator etc. First of all, this applies to digital platforms and platform solutions, databases, data processing algorithms created in the process of machine learning, etc. The digital economy raises a question of how the protection of intellectual property results should be built and interests of creators, users and investors should be fully respected.

4 Discussion

In science and practice, there are three possible models for the protection of rights to results of intellectual activity:

1. Consideration of digital technology products through the prism of copyright. This approach is based on the European Union Directive 96/9/EC [7] in cases where digital objects meet originality criteria for data selection and processing. But this approach negates the difference between the author and the owner of the technology. In addition, it may be difficult to distinguish authorship in cases where, for example, a digital solution (digital platform, blockchain registry, cloud technology) of one author creates a product of another author, as well as in cases of transferring part of the data from one author's system to another without the consent of the manufacturer of the first database. Besides, the proposed solution contradicts one of the main principles of copyright – the protection of a work that has an original, unchangeable and singular form. Digital solutions, as a rule, have a plurality of presentation formats, coding ways and methods, and this excludes the possibility of considering them as an indivisible original work.
2. Regulation of digital technologies in the format of related rights. This approach is more typical for the Russian law, where the activity of manufacturers of databases or other digital products is defined as organizational and technical one [25]. In this case, creators and developers of a database architecture acquire the exclusive right to results of their intellectual work in general. But in this case, boundaries of related rights are blurred and the risk of infringement of the users' rights of digital products is created.

3. The introduction of an independent institute of intellectual rights to digital objects, which would combine both property and non-property rights, as well as reflect technological specifics of protected objects [9, 19]. The development of a fundamentally new mechanism will allow us to build a correct model for regulating relations in the sphere of using digital technology products (works created by robots, codes of self-learning programs, etc.).

The next important issue is to give digital technologies *a property of legal subjects* (legal personality). This proposal has already gone beyond the theory. It is already suggested to recognize robots and objects of robotics as subjects of obligations and consider them as agents when concluding contracts with third parties who act on behalf of their owner and on their own behalf [29]. In particular, in the Arizona law, robot couriers were granted all the rights and obligations of pedestrians, with the exception of the obligation to give way to pedestrians [1, 21].

In traditional law, legal personality is a combination of legal capacity, capacity and delinquency. Today, there are three types of legal capacity: individuals, corporations, and public legal entities. At the same time, the expansion of this list was carried out gradually as it became necessary to introduce new participants in the turnover. In particular, giving legal personality to legal entities required changing approaches to understanding the will, interest, motives the subject. In this regard, it is hardly appropriate to deny the legal personality of digital technologies because of their lack of a strong-willed component [30].

It is much more important to evaluate technologies through the prism of corporate legal personality, namely, from the perspective of their autonomy and ability to make independent decisions. To these properties, we can also add the ability to self-learn and the ability to change the algorithm of actions on its own. It should be noted, however, that modern digital technologies have different degrees of independence. Can the same rules of legal personality be applied to them, or should they be differentiated? In the modern law, this issue has not found its solution yet. If the legal personality of AI and other digital technologies is recognized, it will also be important to resolve such issues as civil liability insurance, criteria for determining the possible danger of robot activity, establish rules for the delinquency of digital technologies, and determine the procedure for accounting and registration of new legal entities.

In general, the idea of legal personality of digital technology is reasonable, but it should not be equated to the legal personality of individuals, and the legal status of corporations. According to experts, the recognition of legal personality in robots is possible, since the legal personality of international associations, in particular, the UNO, has already been recognized. It is also important to consider the autonomy of digital objects. The degree to which the technology depends on a person will determine its delinquency.

Speaking about the law transformation in the context of digitalization, we should also mention the importance of a narrow approach, when experts eventually come to the solution of general theoretical problems of the law adaptation through the solution of particular regulation issues on individual technologies. In particular, issues related to the use of AI need to be addressed. It is difficult to determine who is responsible for errors made by digital devices. Experts have widely discussed the responsibility of

IBM's Watson AI, which diagnoses cancer in South Korea. They conclude that responsibility should be borne by the creators of "Watson" and its associated medical staff [5]. In addition to regulation issues, there is a wide discussion of creating legal guarantees for the safety of AI and robots. In particular C. Chessman raise the question of applying animals treatment rules to robots, up to the establishment of responsibility for cruelty to robots or AI similar to living creatures, since the contemplation of this abuse can cause mental harm to a person [4].

In connection with the active implementation of AI and machine learning, the task of determining possible boundaries of the use of technology in legal activities arises. As most experts note, legal decision-making cannot be automated due to the fact that there are too many value judgments in it, norms are designed to be applied based on intuition and taking into account the context, as well as justice itself in any form [33].

This approach is seen as extremely categorical, especially taking into account that digital technologies have long been used in the activities of lawyers. The question is how this participation should be framed and how the measure of machine responsibility for the final decision can be allocated. According to R. Catterwell, machine learning and AI can be used for automated interpretation of contracts. However, while some provisions may well be interpreted by the machine, others may not. There are two main limitations: 1) some provisions are relative, intuitive and rather "a matter of impression"; 2) some provisions can only be interpreted in accordance with views of the parties and their circumstances. Therefore, machine learning can only help lawyers, but not replace them when interpreting contracts [2]. In this regard, the separation of human and machine powers in terms of decision-making seems reasonable. AI can take a function of data processing and analysis, and a person can make a decision based on a critical analysis of the information prepared by the AI.

Many questions arise when using the Internet of things. Lawyers emphasize insufficient protection of personal data of users of "smart" things, because devices often collect and exchange information without the owners' knowledge, can exchange information in a cross-border manner, etc. [27]. The problem is that provisions of current legislation, and in particular the EU General Data Protection Regulation (GDPR), do not take into account the use of the Internet of things and smart homes, and the literal application of the rules in some cases may impose an unreasonably high responsibility on the creators of devices, and in other cases – create cybersecurity risks [3]. The issue of personal data protection also arises in connection with the use of cloud technologies. Even the most modern and strict General Data Protection Regulation (GDPR) is not applicable in practice to processing of personal data by cloud technologies, in particular, because it may be impossible to identify data controllers and providers, and, moreover, the very fact of data collection by the program may remain secret from everyone [11].

The exchange of data stored in a cloud storage is no less problematic. Currently, many types of data, including those that may be needed during a criminal investigation in country 1, may be controlled in country 2. Modern international treaties on mutual legal assistance, however, cannot help country 1 to obtain data from the cloud of country 2, and simple international courtesy leaves too much discretion, as evidenced by court practice in the cases of *Microsoft Ireland* or *Yahoo! Belgium*. Therefore, experts suggest that countries start negotiations on the conclusion of international agreements on the exchange of information [34].

Speaking about the legal regulation of digital technologies, we cannot say nothing about the applied aspects of the use of unmanned aerial vehicles (UAVs or drones). Although the legislation of many countries, including Russia, has already introduced regulations restricting the use of drones, the issue of protecting personal life remains open. In particular, no country in the world provides land owners with adequate legal protection against malicious actions of drone owners who may use drones to invade someone else's territory and interfere with their privacy [15]. The issue of information collected by drones and the possibility of its use by owners has not been resolved [18].

5 Conclusion

Summing up this research, it is important to emphasize that modern legislation begins the first stage of adaptation, when only the regularities and challenges of digitalization are being investigated. In fact, a layer of axiological and heuristic tasks is formed through the choice of a law development strategy: stimulating the digital economy development (a strategy of progressive development) or minimizing the risks of its use (a security strategy). Without denying the importance of this direction, however, we should note the need for a more consistent and systematic approach to the development of digital law by addressing two fundamental issues: 1) Can traditional legal structures be adapted to realities of digital economy, or is it necessary to develop fundamentally new laws? 2) How can a model of the law universalization of transnational technologies be built?

The solution of the first question directly depends on how modern jurisprudence will assess the legal personality and protection capacity of digital technologies. According to the study, the marginalization of traditional law in the light of the development of digital technologies is evident, and therefore the use of old structures can only lead to mosaic regulation without reserves for further adaptation.

The progressive development of the digital economy is impossible without the development and adoption of fundamentally new legal structures with a margin of safety for the future. In particular, the issue of legal personality of technologies, copyright and related rights, machine liability and insurance needs to be radically revised. It is also important to note that the development of these models should be conducted not at the level of individual states, but at the level of international community to ensure the universalization of rights.

References

1. An act amending sections 28 101 And 28 627, Arizona Revised Statutes; Amending Title 28, Chapter 3, Article 15, Arizona Revised Statutes, By Adding Section 28 913; Amending Sections 28 2153 And 28 4132, Arizona Revised Statutes; Relating To Transportation Devices. https://www.azleg.gov/legtext/53leg/2R/bills/HB2422P.htm. Accessed 20 Apr 2020

2. Catterwell, R.: Automation in contract interpretation. Law Innov. Technol. **81** (2020). https://ssrn.com/abstract=3542549. Accessed 20 Apr 2020
3. Chen, J., Edwards, L., Urquhart, L., McAuley, D.: Who is responsible for data processing in smart homes? Reconsidering Joint Controllership and the Household Exemption (2019). https://ssrn.com/abstract=3483511. Accessed 20 Apr 2020
4. Chessman, C.: Not quite human: artificial intelligence, animals, and the regulation of sentient property (2018). https://ssrn.com/abstract=3200802. Accessed 20 Apr 2020
5. Chung, J., Zink, A.: Hey Watson, can I sue you for malpractice? Examining the liability of artificial intelligence in medicine. Asia-Pac. J. Health Law Policy Ethics (2017). https://ssrn.com/abstract=3076576. Accessed 20 Apr 2020
6. Civil Code of the Russian Federation dated 30.11.1994 N 51-FZ. http://www.consultant.ru/document/cons_doc_LAW_5142/. Accessed 20 Apr 2020
7. Directive N 96/9/EC of the European Parliament and of the Council of 11 March 1996 On the legal protection of databases. OJ L 77, 27 Mar 1996. https://eur-lex.europa.eu/legal-content/EN/TXT/?uri=celex%3A31996L0009. Accessed 20 Apr 2020
8. Edwards, F.R., Hanley, K., Litan, R., Weil, R.L.: Crypto assets require better regulation: statement of the financial economists roundtable on crypto assets. Financ. Anal. J. **75**(2), 14–19 (2019)
9. Entin, V.L.: Copyright in Virtual Reality. New Opportunities and Challenges in Digital Age. Statut, Moscow (2017)
10. Fernández-Villaverde, J.: Simple rules for a complex world with artificial intelligence. PIER Working Paper 20-010 (2020). https://ssrn.com/abstract=3559378. Accessed 20 Apr 2020
11. Fosch Villaronga, E., Millard, C.: Cloud robotics law and regulation (2018). https://ssrn.com/abstract=3305353. Accessed 20 Apr 2020
12. Giudici, G., Milne, A., Vinogradov, D.: Cryptocurrencies: market analysis and perspectives. J. Ind. Bus. Econ. **47**, 1–18 (2020)
13. Goanta, C.: Big law, big data. Forthcoming, law and method (2017). https://ssrn.com/abstract=3166290. Accessed 20 Apr 2020
14. Gomes, S.: Smart contracts: legal frontiers and insertion into the creative economy. Braz. J. Oper. Prod. Manag. **15**, 376–385 (2018)
15. Hacker, P., Krestel, R., Grundmann, S., Naumann, F.: Explainable AI under contract and tort law: legal incentives and technical challenges. Artif. Intell. Law. (2020). https://doi.org/10.1007/s10506-020-09260-6
16. Huang, R.H., Yang, D., Loo, F.F.Y.: The development and regulation of cryptoassets: Hong Kong experiences and a comparative analysis. Eur. Bus. Organ. Law Rev. **21**, 319–347 (2020)
17. Ivantsov, S.V., Sidorenko, E.L., Spasennikov, B.A., Berezkin, Y.M., Sukhodolov, Y.A.: Cryptocurrency-related crimes: key criminological trends. Russ. J. Criminol. **13**(1), 85–93 (2019). https://doi.org/10.17150/2500-4255.2019.13(1).85-93
18. Kaminski, M.E.: Robots in the home: what will we have agreed to? Ohio State Public Law Working Paper No. 292 (2015). https://ssrn.com/abstract=2592500. Accessed 20 Apr 2020
19. Kharitonova, Yu.S.: The legal regime of the results of artificial intelligence. In: Louts, E.B. (ed.) Modern Information Technology and Law: Monograph, pp. 68–83. Statute, Moscow (2019)
20. Konert, A., Smereka, J., Szarpak, Ł.: The use of drones in emergency medicine: practical and legal aspects. Emerg. Med. Int., 3589792 (2019)
21. Lapastora, C.: Arizona law gives delivery robots same rights as pedestrians – but they must abide by same rules (2018). https://www.foxnews.com/tech/arizona-law-gives-delivery-robots-same-rights-as-pedestrians-but-they-must-abide-by-same-rules. Accessed 20 Apr 2020

22. Lei, C.: Legal control over big data criminal investigation. Soc. Sci. China **40**, 189–204 (2019)
23. Liu, Y., Huang, J.: Legal creation of smart contracts and the legal effects. J. Phys: Conf. Ser. **1345**, 042033 (2019)
24. Low, K.F.K., Mik, E.: Pause the blockchain legal revolution. Int. Comp. Law Q. **69**(1), 135–175 (2019)
25. Maggon, H.: Legal protection of databases: an Indian perspective. J. Intellect. Property Rights **11**, 140–144 (2006)
26. Marquès, M.C.: Recreational drones: legal framework, civil liability and data protection. Revista de Derecho Civil **6**, 297–333 (2019)
27. Mohamed, S., Zulhuda, S.: The concept of internet of things and its challenges to privacy. South East Asia J. Contemp. Bus. Econ. Law **8**(4), 1–6 (2015)
28. Mowbray, A., Chung, P., Greenleaf, G.: Utilising AI in the legal assistance sector – testing a role for legal information institutes (2019). https://papers.ssrn.com/sol3/papers.cfm?abstract_id=3379441. Accessed 20 Apr 2020
29. Neznamov, A.V., Naumov, V.B.: Strategy for the regulation of robotics and cyberphysical systems. Law **2**, 69–89 (2018)
30. Ponkin, I.V., Redkina, A.I.: Artificial intelligence from a legal viewpoint. Peoples' Friendsh. Univ. Russ. Bull. Ser.: Jurid. Sci. **22**(1), 91–109 (2018)
31. Rahmatian, A.: Originality in UK copyright law: the old "skill and labour" doctrine under pressure. Int. Rev. Intellect. Property Compet. Law **44**, 4–34 (2013)
32. Sidorenko, E.L.: Stablecoin as a new financial instrument. In: Ashmarina, S.I., Vochozka, M., Mantulenko, V.V. (eds.) Digital Age: Chances, Challenges and Future. Lecture Notes in Network and System, vol. 84, pp. 630–638. Springer, Cham (2020)
33. Wachter, S., Mittelstadt, B., Russell, C.: Why fairness cannot be automated: bridging the gap between EU non-discrimination law and AI (2020). https://ssrn.com/abstract=3547922. Accessed 20 Apr 2020
34. Yunquera, R.: One cloud in the sky and conflicting laws on the ground: Regulating Law enforcement access to e-evidence in cloud computing in the European Union and the United States. Mayer Brown Legal Business Prize (2018). https://ssrn.com/abstract=3271580. Accessed 20 Apr 2020

Relevant Issues of Service Secret: Theoretical and Legal Aspects

E. Efremova[1(✉)], R. Yusupov[2], and S. Metlin[3]

[1] Samara State University of Economics, Samara, Russia
lenoksamara97@gmail.com

[2] Samara National Research University named after academician S.P. Korolev (Samara University), Samara, Russia
r.yusupov@bk.ru

[3] Samara State University of Railway Engineering, Samara, Russia
metlin-sergey@mail.ru

Abstract. Information has always been a valuable facility. Millions of people use information resources. In the digital society, it has acquired new qualities. Now it is not only an informative resource, but also a strategic, administrative and economic one. These resources now largely determine the scientific and technical potential of the country, the level of development of its national economy, the way of life and human activity. Information plays a significant role in the activities of any modern commercial enterprise and public authority. This is especially true for information that is subject to official secrecy. That is why the relevance of forming effective means of its protection is currently increasing. Service secret as a type of confidential information is its own type. Its legal regulation has developed relatively recently. As an independent species, it began to stand out only in the twentieth century. The content and level of protection of official secrets differ from one state to another. This is due to historical legal traditions.

Keywords: Access to information · Confidential information · Secret · Service information · Trade secret

1 Introduction

In the contemporary world, information is defined differently. Undoubtedly, information is wealth [1]. It is generally accepted that information is related to economic resources [15]. Today, information is the object of criminal actions [3]. The global circulation of information has created a number of problems in the collection, dissemination and access to information [20]. Human culture is progressively becoming digital. A good deal of attention is paid to the confidentiality of information [4]. This raises questions about the relationship between ethical norms and the dissemination of information [5]. A clear definition of the confidentiality regime and how to disclose information will help to resolve these issues correctly. The attitude to official secrets is different. According to Graeber the concept of "service secret" is a specific invention of the bureaucracy [13]. Therefore, it fanatically protects its offspring. Russian scientists

S. I. Ashmarina et al. (Eds.): ENGINEERING ECONOMICS WEEK 2020, LNNS 139, pp. 533–538, 2021.
https://doi.org/10.1007/978-3-030-53277-2_63

consider official secrets as the result of a special legal regime, as a type of confidential information, as an element of trade secrets. Official secrets as a category have almost no precise definition in Russian legislation. However, legal liability is provided for its disclosure. This creates a lot of problems in law enforcement and administrative work.

2 Methodology

The authors used the synthesis method to study the problems of service secrets. The research methodology consists of approaches and principles used for conducting general scientific and juridical research. It is based on dialectical and formal-logical methods. The way of analysis and others were also used. The combination of methods made it possible to fully consider the problematic aspects of service information with a confidentiality regime.

3 Results

Today, the legislation of most countries prohibits employees from disclosing information that has become known to them in connection with their service. Over the past few decades, the principle of public administration openness has become increasingly important in most countries, and consequently, employees are increasingly charged with an opposing duty to ensure access to public documents. This is why the problem of establishing boundaries between official secrets and socially important information is urgent in modern society. Its resolution is not possible without establishing clear boundaries between them and improving the legal regulation in this subject matter. Russian legislation is not perfect. The definition of "service secret" is not fixed in Russian legislation. Only the list of information that cannot be covered by such a regime is fixed. There is no federal law regulating this type of information. Today, the "service secret" regime, the basis for its dissemination and responsibility for violation are regulated by a large number of departmental acts. This contradicts the norms of the Constitution of the Russian Federation and leads to problems in law enforcement and management activities. First of all, it is necessary to adopt a Federal law regulating official secrets. The definition of "official secret" will be fixed, which should be understood as confidential, legally protected information received by a person in the course of professional activities related to the management and organizational and legal foundations of the organization.

4 Discussion

In modern Russian legislation, the concept of "service secret" is not fixed. Nevertheless, this mode of data confidentiality is regulated by a number of legal acts. Thus, in the third paragraph of the Decree of the President of the Russian Federation of 06.03.1997 No. 188 "On approval of the list of confidential information", service secrets are defined as: "official information, access to which is restricted by state authorities in accordance with the Civil Code of the Russian Federation and federal laws" [6, p. 3]. It should be noted that such information is not part of trade or professional secrets.

Until 2008, proprietary and trade secret, are governed by article 139 of the Civil Code of the Russian Federation (Part II) of 30.11.1994 No.51-FZ. According to this document, information is an official or trade secret in case when the information has actual or potential commercial value owing to uncertainty to its third parties, it has no free access on lawful basis [20]. Today, these rules are excluded from the civil code of the Russian Federation.

In the Federal law of 27.07.2006 No.149-FZ "On information, information technologies and information protection", service secrets are mentioned in the third paragraph of the first part of the eighth article and the fourth part of the ninth article [9]. The legislator in this normative legal act did not fix the definition of "service secret", did not define its scope and content. But it contains an important rule: the conditions for the extension of the service secret regime to information and responsibility for its violation can be established exclusively by federal law (the fourth part of the ninth article). However so far, the federal law regulating the regime of service secrets in Russia has not been adopted.

In 2004, the Draft Federal Law No. 124871-4 "On Service Secrets" was developed and submitted for consideration to the state Duma of the Russian Federation. The second article of the draft contained a list of information classified as service secret which included: "confidential information generated in the course of administrative activity of the organ or organization, the distribution of which obstructs the authority or organization of the powers granted to him or otherwise adversely affect their implementation" [8, p. 2]. Confidential information was included in this list obtained by the organ or organization in accordance with their competence in accordance with legislation. Here, the service secret regime was defined as: "a set of legal, organizational, technical and other measures taken by authorized officials of public authorities and organizations that provide restrictions on the dissemination of information constituting official secrets, and on access to this information" [8, p. 2].

Nowadays, the "service secret" regime is regulated by the Decree of the Government of the Russian Federation of 03.11.1994 No. 1233 "On Approval of the Regulation on the Procedure for Handling Limited-Service Official Information in the Federal Executive Bodies, an Authorized Atomic Energy Management Body and an Authorized Space Agency" [7, p. 4]. According to this document, official information of limited distribution is "unclassified information related to the activities of organizations, restrictions on the distribution of which are dictated by official necessity, as well as unclassified information received by organizations, access to which is restricted in accordance with federal laws" [7]. The list of information that cannot be covered by this confidentiality regime is also fixed. According to the resolution, documents containing information classified as service secrets must be marked "for official use" (p. 1.4). The decision to classify individual information as an official secret is made at the discretion of the head of the federal executive posts, the competent authority for managing the use of atomic energy, and the relevant authority for space activities within its competence [7]. This goes against the constitutional right of citizens to access information.

Service secrets are also mentioned in a number of other legal acts. For example, the second part of the fourth article of the Act of the Russian Federation of 20.08.1993 No. 5663-1 "On space activities" [2] establishes the obligation to carry out such activities in compliance with the rules for protecting service secrets. According to the seventh article of the Federal law of 27.07.2004 No. 79-FZ "On the State Civil Service of the Russian Federation", employees are prohibited from disclosing or using official information obtained during the performance of official duties for purposes not related to the civil service [10]. Nonetheless, the concept of "proprietary information" is not disclosed, and the distinction between "confidential information and "proprietary information" is not justified.

In addition, today there are a large number of departmental acts regulating the grounds for the dissemination of the "service secret" regime, the procedure for working with such information, and responsibility for violating this mode. For example, the regulation of the judicial Department under the supreme court of the Russian Federation of February 22, 2007 No. 25 States: "Official information of limited distribution includes unclassified information related to the activities of the judicial Department, restrictions on the distribution of which are dictated by official necessity" [16, p. 12]. The regulation also contains a list of information that cannot be classified as official secrets.

Working with tax authorities with restricted access information is regulated by the federal tax service of 31.12.2009 No. MM-7-6/728. In accordance with this document, confidential information is recognized as: "unclassified information concerning the activities of tax authorities, restrictions on the distribution of which are dictated by official necessity" [17, p. 13]. The rules of this provision may not apply to information that is a state or tax secret. Therefore, information that is recognized as a tax secret is not included in the information with the "service secret" mode.

The seventeenth of the Federal Law of 30.12.2004 No. 218-FZ "On credit histories" obliges employees of the central bank of Russia to keep official secrets of credit history bureaus, sources of credit history formation, subjects of credit histories and users of credit histories [11]. However, employees of the central Bank of the Russian Federation are not government employees. They can be recognized as subjects of official secrecy on the basis of the third article of the Federal Law of 10.07.2002 No. 86-FZ "On the Central Bank of the Russian Federation (bank of Russia)" because they directly participate in the implementation of the unified state monetary policy [12].

The question of defining the concept of "service secret", its scope and content is also debatable in the scientific community. So far, scientists have not come to a consensus on these issues. It is proposed to understand service secret as information created or received by a state or local government body, access to which is restricted by federal law in the interests of other people [18]. The exception will be information constituting a state secret. Kamalova specifies: "the official secret is the legal regime of information entered and protected to ensure the confidentiality of official information obtained or developed by state or municipal authorities, public-law organizations on a legal basis" [14, p. 87]. Restriction of access to such information is introduced for the effective implementation of state functions and ensuring information security of the activities of state administration bodies.

5 Conclusion

A large number of voids and collisions characterizes Russian legislation regulating the circulation of confidential information. This also applies to the regulation of official secrets. The relevance of resolving these issues increases due to the widespread use of artificial intelligence in professional and official activities [19]. Service secrets or information with restricted access are regulated by a large number of laws, regulations, and departmental regulations. However, none of them fully reveals its essence. There is also no clear definition of the term “official secret”. To properly regulate this type of information, it is necessary to determine its specific characteristics. Such data has the following specific features: information has no commercial value; does not satisfy private interests; is subject to protection by state agencies and municipal authorities; the restricted use regime is established by officials who are personally responsible for making such a decision; such information can be obtained only in the performance of job description by officials. In the conditions of universal digitalization, the question of the ratio of professional and official secrets becomes particularly relevant.

References

1. Adlakadi, R.: Information Is Wealth. Booksclinic Publishing, Chhattisgarh (2020)
2. Act of the Russian Federation of 20.08.1993 No.5663-1 (as amended on 15.04.2019) “On space activities” (2019). http://www.consultant.ru/cons/cgi/online.cgi?req=doc&base=LAW&n=322588&fld=134&dst=1000000001,0&rnd=0.4562585409763176#04663205344311909. Accessed 11 Apr 2020
3. Bolgova, V.V, Cherevichenko, T.S., Azarkhin, A.V., Sidorova, A.V., Efremova, E.A.: Internet offenses of young population. In: Popkova, E. (eds.) Growth Poles of the Global Economy: Emergence, Changes and Future Perspectives. Lecture Notes in Networks and Systems, vol. 73, pp. 667–675. Springer, Cham (2020)
4. Cariolato, S.: World Without History? Digital Information is Volatile: With It Our Culture Can Disappear but Its Preservation Can Save Us. Youcanprint, Bartolini (2019)
5. Carroll, J.M.: Confidential Information Sources: Public and Private. Elsevier, Amsterdam (2014)
6. Decree of the President of the Russian Federation of 06.03.1997 No. 188 (ed. from 13.07.2015) “On approval of the list of confidential information” (2015). http://www.consultant.ru/cons/cgi/online.cgi?req=doc&base=LAW&n=182734&fld=134&dst=1000000001,0&rnd=0.10986309365482638#040199190553240194. Accessed 11 Apr 2020
7. Decree of the Government of the Russian Federation of 03.11.1994 No. 1233 (as amended on 18.03.2016) “On Approval of the Regulation on the Procedure for Handling Limited-Service Official Information in the Federal Executive Bodies, an Authorized Atomic Energy Management Body and an Authorized Space Agency” (1994). http://www.consultant.ru/cons/cgi/online.cgi?req=doc&base=LAW&n=195608&fld=134&dst=1000000001,0&rnd=0.3063160621801002#03415806630465419. Accessed 11 Apr 2020
8. Draft Federal Law No.124871-4 “On Service Secrets” (2004). http://www.consultant.ru/cons/cgi/online.cgi?req=doc;base=PRJ;n=39972#02334193309286694. Accessed 11 Apr 2020

9. Federal law of 27.07.2006 No. 149-FZ (as amended on 02.12.2019) "On information, information technologies and information protection" (2019). http://www.consultant.ru/cons/cgi/online.cgi?req=doc&base=LAW&n=339396&fld=134&dst=1000000001,0&rnd=0.5068544565616508#08511244644299816. Accessed 11 Apr 2020
10. Federal law of 27.07.2004 No. 79-FZ (as amended on 01.05.2019) "On the State Civil Service of the Russian Federation" (2019). http://www.consultant.ru/cons/cgi/online.cgi?req=doc&base=LAW&n=340367&fld=134&dst=1000000001,0&rnd=0.47113947222427344#022587680661012932. Accessed 11 Apr 2020
11. Federal law of 30.12.2004 No.218- FZ (as amended on 01.05.2019) "On credit histories" (2019). http://www.consultant.ru/cons/cgi/online.cgi?req=doc&base=LAW&n=349448&fld=134&dst=1000000001,0&rnd=0.7979839807468578#0553053609190844. Accessed 11 Apr 2020
12. Federal law of 10.07.2002 No. 86- FZ (as amended on 02.08.2019) "On the Central Bank of the Russian Federation (Bank of Russia)" (2019). http://www.consultant.ru/cons/cgi/online.cgi?req=doc&base=LAW&n=349172&fld=134&dst=1000000001,0&rnd=0.22474775504210842#09694457069318365. Accessed 11 Apr 2020
13. Graeber, D.: The Utopia of Rules: On Technology, Stupidity, and the Secret Joys of Bureaucracy. Melville House, Brooklyn (2015)
14. Kamalova, G.G.: Ratio of legal regimes of official and professional secrets. Bull. South Ural State Univ. Ser. Law **4**, 87–94 (2016) https://doi.org/10.14529/law160415.s.87-88
15. Matveeva, L.G., Nikitaeva, A.Y., Chernova, O.A.: Information as a strategic resource for regional. Terra Economicus **1**, 134–145 (2018)
16. Order of the Judicial Department attached to the Supreme Court of the Russian Federation of February 22, 2007 No.25 "On approval of the Regulation on the procedure for handling limited-service official information in the apparatus of the Judicial Department under the Supreme Court of the Russian Federation and in the departments (divisions) of the Judicial Department in the constituent entities of the Russian Federation" (2007). http://www.consultant.ru/cons/cgi/online.cgi?req=doc&base=EXP&n=398975#02075996907882376. Accessed 11 Apr 2020
17. Order of the Federal Tax Service of Russia of 31.12.2009 № MM-7-6/728@ (as amended on 21.07.2011) "On approval of the Regulation on the procedure for handling restricted official information with the tax authorities" (2011). http://www.consultant.ru/cons/cgi/online.cgi?req=doc&base=LAW&n=315645&fld=134&dst=1000000001,0&rnd=0.5195774155331949#02870330514406234. Accessed 11 Apr 2020
18. Pavlov, I.Y.: Modern problems of legal regulation of state and professional secrecy in Russia. Leningrad Law J. **1**(31), 29–37 (2013)
19. Sidorova, A.V.: Subject structure of the offense in artificial intelligence (AI) and robotics. In: Ashmarina, S.I., Vochozka, M., Mantulenko, V.V. (eds.) Digital Age: Chances, Challenges and Future. Lecture Notes in Networks and Systems, vol. 84, pp. 541–547. Springer, Cham (2020)
20. The Civil Code of the Russian Federation (part II) of 30.11.1994 No.51- FZ (as amended on 26.06.2007) (2007). http://www.consultant.ru/cons/cgi/online.cgi?req=doc&n=69407&base=LAW&from=340325-0&rnd=0.47508663527192585#0027100471325111286. Accessed 11 Apr 2020

Digital Assistants in Managing Customer Relationships in Modern Companies

E. V. Cherkasova[1](✉) and M. R. Zainullina[2]

[1] Samara State University of Economics, Samara, Russia
1-19732807@yandex.ru
[2] Kazan Federal University, Kazan, Russia
Milyausha-zainul@list.ru

Abstract. The paper aims to determine the impact of digital transformation on the functionality of CRM solutions as digital assistants to managers, as well as the role of CRM systems in business processes of modern business entities. The concept of "CRM (Customer Relationships Management)" implies a business strategy based on using business rules and processes that are designed to engage consumers in a dialogue in order to obtain mutual benefits in trusting and transparent business environment. The authors developed the scheme "How to interest the customer", in which an attempt is made to determine the distinctive features of the business process of customer service in companies that use CRM systems. The results obtained in this scientific study can be useful for developers of CRM systems who want to fully meet the needs of customers.

Keywords: Artificial intelligence · Business processes · Company · Customer needs · Manager · System

1 Introduction

In traditional enterprise resource management methods, the company's client is not considered as an element served by the system. This arrangement of the company's planning and management systems is explained by the fact that attention is focused on optimizing only the internal activities of the company itself. This method is now considered significantly outdated. The lack of a unified approach to working with clients in the company's divisions and offices negatively affects the effectiveness of their work: the company loses opportunities to increase sales and increase customer loyalty. The consumer has become more independent, more educated, more sophisticated. And he wants to manage all aspects of his life. It becomes increasingly difficult to understand the needs and opportunities of the buyer, to try to find a product in the aspect that it will attract the consumer because the modern market is full of various offers, and this is the first obstacle to the realization of a sufficient volume of production and customer loyalty in the current market conditions [4].

The second obstacle is the lack of time that customers are willing to spend searching for a company that can meet their individual request. These problems are the reason for the need to create new modern methods of company management. At the moment, digital technologies have become the priority methods of managing business

S. I. Ashmarina et al. (Eds.): ENGINEERING ECONOMICS WEEK 2020, LNNS 139, pp. 539–546, 2021.
https://doi.org/10.1007/978-3-030-53277-2_64

processes of companies. But this approach involves not only installing modern hardware or software at the enterprise. Today, for many companies, the most urgent tasks are to reduce costs, plan accurately, attract customers, work actively with existing customers and retain them. Not having a unified database of customers, the company is losing significant amounts of money due to inefficient work of the departments of marketing, sales and service. Digital technologies make fundamental changes in approaches to management, corporate culture of the enterprise, become full-fledged assistants on the way of building long-term and most importantly - high-value communications with the consumer [3]. As a result, the productivity of the company's personnel increases, the level of customer satisfaction increases, and the company gains a reputation as a progressive and modern organization, which ultimately ensures the company's profitability.

2 Methodology

The methodological basis for identifying key trends in the development of customer-oriented solutions for modern companies using digital tools is the methods of empirical research: observation method, comparative analysis method, descriptive research method, as well as methods of theoretical research-analysis, synthesis and deduction with which the following tasks are solved: 1) evaluate the role of digitalization in modern business processes of business entities; 2) determine the impact of digital transformation on the functionality of CRM solutions as digital assistant managers; 3) identify the specifics of customer base management systems with artificial intelligence elements.

Digitalization is directly related to the digital economy, the implementation of key aspects of which is carried out using special end-to-end technologies [8]. Among these technologies at the moment, the leading ones are: Artificial intelligence, Robotics, Virtual and augmented reality, Big Data, Quantum technologies, Distributed registry systems, New production technologies, Wireless communication technologies, and Industrial internet [5].

Leading experts in the field of digitalization tend to believe that the digital economy includes markets and industries (Public administration, Smart city, Digital healthcare, Education); platforms and technologies (Research and Development); environment (Legislative sphere, Personnel, Infrastructure, Information security, Digital technologies). The leading digital assistants of modern companies are Big Data and predictive analytics – 68%; Chat bots – 51%; Robotization – 50%; Optical character recognition – 36%; AI-technologies – 28%; Internet of things – 24%; Virtual reality – 21%; Blockchain – 19% [5].

According to the data provided in the article by E. Chernysheva, "Business will be enriched with artificial intelligence" [2], Artificial Intelligence (AI) can have the greatest impact on customer relationship management systems (so-called CRM systems). The data allowed us to assess the expected economic effect of this impact: Customer service – 43%; Client support (calculations, logistics) – 41%; Managing relationships with partners – 41%; Creating customer-oriented products and services – 40%; Electronic commerce – 40%; Corporate marketing, branding and advertising –

39%; Product marketing or service marketing – 39%; Field marketing – 39%; Pricing of goods and services – 39%; The study of consumer behavior – 38% [2].

As you can see, CRM technologies play an important role in the era of "numbers". However, it is logical to ask whether these are the same CRM systems that appeared more than 20 years ago, or a new generation of products that have nothing in common with previously used solutions. We will try to deal with this issue. If we consider the historical aspects of the development and formation of customer-oriented solutions, we can identify three key stages: CRM birth, CRM boom, stagnation, and rebranding. It should be noted that in terms of dates the situation in the West and in Russia is different. In Russia, CRM solutions appeared about 10 years later in relation to the Western market (early 2000s). The most interesting period in the development of customer-oriented solutions is, of course, the CRM boom (in Russia, approximately dates back to 2002–2007). This time period accounts for the largest number of implementations (and, as statistics has shown, the largest number of failures) [1].

The key reasons for the failure of most CRM projects at that time was the misconception that an IT product of this class would be a "pill for all diseases" for the company. Undoubtedly, the client-oriented solution had a useful functionality for representatives of the business environment at that time. In this regard, it should be mentioned that depending on the tasks and functions, there are three classes of CRM systems. These include: operational, analytical, and joint CRM systems. The operational class of client-oriented solutions assumes minimal functionality, which consists of collecting, organizing and storing the history of communications with clients. As for analytical solutions, they have become the most popular all over the world at the moment, since they are focused on extracting data that is practically useful for companies from a large array of data. Joint CRM products combine the capabilities of the first two classes (Table 1).

Table 1. Classes of CRM systems

Type of CRM system	Functionality
Operational	Collection, systematization, centralized storage of data about the customer and the history of interaction with them, automation of the sales, marketing and after-sales service process
Analytical	Analysis of the customer base in accordance with certain criteria, segmentation, evaluation of value for the company
Joint	Interaction and transactions between companies within the B2B (Business to Business) segment, when several enterprises jointly implement the creation of a product/service

Source: authors.

The authors of the study suggest a new type of CRM solutions that arose under the influence of digitalization on the process of interaction with customers. We are talking about Artificial Intelligence CRM products that are focused on supporting decision-making within the framework of communication with the customer and take on part of the Manager's tasks (Table 2).

Table 2. Artificial Intelligence CRM systems (AI CRM systems)

Intellectual	Analysis of the work of sales Department employees, determination of the most effective models of working behavior and sales strategy; analysis of customer behavior; forecast of customer outflow; formation of personalized recommendations for developing a policy of interaction with customers; forecast of changes in the company's revenue

Source: authors.

The goal of Artificial Intelligence is to improve technological processes in various fields of knowledge Artificial intelligence (AI) allows you to change almost every socially significant aspect of human life. With the help of Artificial Intelligence, decisions are made faster, and natural errors caused by the human factor are eliminated, which are considered inexcusable from the client's point of view. Thus, Artificial Intelligence is gradually turning to the sphere of customer and company relationship management, where it creates a more convenient customer service system, while the service process is not depersonalized, but solutions are better and faster going beyond their actual limitations. We can say that Artificial Intelligence has begun to solve the daily problems and difficulties faced by employees [7].

For example, customer service departments are often overloaded with a huge amount of customer information. This information includes not only the purchase history, but also socio-economic and demographic data of customers, and employees should take this data into account. Therefore, using AI will provide a better and more efficient way to store and conveniently use a huge amount of information. AI is becoming incredibly useful in managing "Big Data" and rationalizing labor-intensive activities [9]. Therefore, the work of a CRM system with Artificial Intelligence is more functional than a traditional CRM system, and provides companies with great advantages – the use of modern high-tech methods to improve the relationship between the company and the client allows you to increase the level of customer satisfaction [6].

3 Results

In this paper, the authors proposed the descriptive scheme "How to interest the customer" in which an attempt is made to determine the distinctive features of the business process of customer service in the companies which use technological systems for creating relationships between the company and customers. This scheme "How to interest the customer" describes the business process "As it is" and the business process "As It Should Be".

In the business process "As it is", staff interacts directly with the traditional CRM system, where the sales department is the most important link. On the basis of the system the following happens:

1. Sequential processing of information about the interest of a potential client.
2. The CRM system receives information about the interest of a potential client.
3. The CRM system determines whether the client or this interest exists in the database of the system.

4. The manager from the sales department enters information about the client and checks the information entered in duplicates.
5. The CRM-system generates a notification to the validator in the form of the task "Check the entered record".
6. The validator receives the task.
7. The validator gives a positive conclusion about the entered record.
8. The CRM-system generates a notification to the manager about the decision on validation
9. The manager from the sales department fills in the customer's card.
10. The CRM-system generates a completed customer's card with personal data and contacts.
11. The CRM-system records and register the presale activity.
12. The sales manager according to the recorded data organizes telephone calls or generates reminders.
13. The manager registers interest in the system.
14. The CRM system qualifies the interest.

However technology development gives the move to natural expansion of companies needs and automation of business processes. So the next descriptive scheme "How to interest the customer" describes the business process "As It Should Be". The presented business process is focused on close interaction of personnel with the CRM system, but unlike the business process "As it is", the main role is played by the Artificial Intelligence CRM system (AI CRM system). On the basis of the system there are two ways of interaction.

The first is:

1. The AI CRM system informs the decision manager about the interest of a potential client.
2. The AI CRM system searches for this contact in the database.
3. The client is new.
4. The AI CRM system generates a client card.
5. The AI CRM system checks the entered record (without the participation of the validator).
6. The AI CRM system enters the client card into the database.
7. The AI CRM system searches for customer interest in the database of system interests.
8. If the interest is not known to the AI CRM system the task of registering the interest in the system is transferred to the sales department.
9. After registering the interest in the AI CRM system, the sales manager monitors information about the potential customer's interest.
10. The sales manager receives information about the potential customer's interest.
11. The sales manager sends information about the potential customer's interest to the AI CRM system for qualification.

12. The AI CRM system qualifies information about the potential customer's interest.
13. The AI CRM system selects the answer to the potential customer's interest.
14. The AI CRM system sends the answer to the sales manager.
15. The sales manager sends the answer to the customer.
16. The customer receives the answer.

The second is:

1. The AI CRM system informs the decision manager about the interest of a potential client.
2. The AI CRM system searches for this contact in the database.
3. The client is listed in the existing database of the AI CRM system.
4. The AI CRM system searches for the interest of the potential customer in the database of interests.
5. The AI CRM system selects the answer to the interest. of the potential customer.
6. The AI CRM system informs the sales department about the need to inform the client by means of reminders or telephone calls.
7. The sales manager receives the task of informing the customer.
8. The sales manager contacts the customer using the communication channels offered by the AI CRM system.
9. The customer receives the answer.

4 Discussion

Successful business activity is determined by the company ability to efficiently handle huge amounts of information. The level of modern technological development opens great opportunities to simplify all business processes at any company from small and medium enterprises to large conglomerates. AI technologies are developing and becoming incredibly useful in managing big data and streamlining time consuming activities. AI CRM-systems are efficient due to the following points: specialization - the transition from the development of general purpose tools to problem/subject-specific specialized tools, which ensures a reduction in time development, increases the efficiency of using tools, simplifies and accelerates the work of experts; integration, namely, Artificial Intelligence tools integrate more easily with other information technologies and tools (with computer-aided software engineering (CASE), database management system (DBMS); openness and portability - development is carried out in compliance with standards that ensure these characteristics; client/server architecture - the development of an information system in this architecture allows you to reduce the cost of equipment used in the application, decentralize applications, increase reliability and overall performance, because the amount of information sent between computers is reduced, and each application module is run on adequate equipment.

When using AI CRM-systems the manager can evaluate the degree of involvement of each client in e-mail communication, thereby identify typical consumer groups by analyzing user behavior. AI CRM-systems help the manager find the reasons that influence the level of sales and then he can take measures to improve merchandising and store planning. However, there are certain problems with using AI CRM-systems. Even though many solutions can be automated, not all company managers entirely trust AI recommendations. Managers sometimes adjust the results of Artificial Intelligence decisions [10]. There is often lack of a unified system for organizing the company's activities, which also requires specific methods in using AI CRM-systems [3]. Such method is proposed in the scheme "How to interest the customer" with the use of AI CRM-systems.

5 Conclusion

Thus, we can conclude that the proposed scheme "How to interest the customer" on the basis of the AI CRM-system is efficient due to the following points: it enables enterprises to provide a predictable assessment of potential customers, which further helps employees focus on working with a more profitable customer. It performs an analytical assessment of the opportunities that the company can offer to the client, for example, the company informs the user about the trend of the transaction. This method also allows to register the activity of websites and e-mail in order to analyze the data obtained, to analyze further accounting and forecasting; to use a database of ready-made responses to solve a typical customer problem, and this database reduces the time spent working with the application and increases customer satisfaction. It also makes the process of recommending products to customers personal and thus offers better services that can meet customers' expectations. So the method of the proposed in this article scheme "How to interest the customer" on the basis of the AI CRM-system can be appropriate for a lot of companies.

References

1. Biktimirov, M.R., Yelizarov, A.M., Shcherbakov, A.Y.: Trends in the development of big data processing technologies and tools for storing multiformat data and analytics. Russ. Digit. Libr. J. **19**(5), 390–407 (2016)
2. Chernysheva, E.: Business will be enriched with artificial intelligence. Kommersant (2017). https://www.kommersant.ru/doc/3329890. Accessed 26 Apr 2020
3. Dufva, T., Dufva, M.: Grasping the future of the digital society. Futures **107**, 17–28 (2019). https://doi.org/10.1016/j.futures.2018.11.001
4. Hoyer, J.: War for the Client. Loyalty Once and for All. Poppury, Moscow (2010)
5. KPMG: Digital technologies in Russian companies (2019). https://assets.kpmg/content/dam/kpmg/ru/pdf/2019/01/ru-ru-digital-technologies-in-russian-companies.pdf. Accessed 26 Apr 2020
6. Lammi, M., Pantzar, M.: The data economy: how technological change has altered the role of the citizen-consumer. Technol. Soc. **59**, 101157 (2019). https://doi.org/10.1016/j.techsoc.2019.101157. Accessed 26 Apr 2020
7. Medini, K., Andersen, A.L., Wuest, T., Christensen, B., Wiesner, S., Romero, D., Liu, A., Tao, F.: Highlights in customer-driven operations management research. Procedia CIRP **86**, 12–19 (2019). https://doi.org/10.1016/j.procir.2020.01.026

8. Mitin, V.: Seven definitions of the digital economy (2017). https://www.crn.ru/news/detail.php?ID=116780. Accessed 26 Apr 2020
9. Silahtaroğlu, G., Alayoglu, N.: Using or not using business intelligence and big data for strategic management: an empirical study based on interviews with executives in various sectors. Procedia – Soc. Behav. Sci. **235**, 208–215 (2016). https://doi.org/10.1016/j.sbspro.2016.11.016
10. Syntetos, A.A., Kholidasari, I., Naim, M.M.: The effects of integrating management judgement into OUT levels: in or out of context? Eur. J. Oper. Res. **249**(3), 853–863 (2016). https://doi.org/10.1016/j.ejor.2015.07.021

Russian Food Security on the Market of Milk and Dairy Products

A. B. Malina[1], E. P. Afanaseva[1](✉), M. V. Kagirova[2], and M. R. Gafarov[3]

[1] Samara State University of Economics, Samara, Russia
Kuzdavletova_AB@mail.ru, parus82@mail.ru
[2] Russian State Agrarian University – Moscow Timiryazev Agricultural Academy, Moscow, Russia
kagirovamary@gmail.com
[3] Kazan Federal University, Kazan, Russia
gafarov_m_r@mail.ru

Abstract. The paper deals with the issues of country's food security that is the most important indicator of national security of the state. The authors justified the fact that food security governance is a priority direction of the Russian state policy. The dynamics of the total production of dairy products in Russia was characterized, and the indicators of product exports and imports were analyzed. The level of actual per caput consumption of milk and dairy products was evaluated. The dynamics of the level of use of average annual productive capacity of the companies that produce particular types of dairy products and milk production in the context of farm categories was studied.

Keywords: Dairy industry · Food security · Imports of milk · Milk production · National security

1 Introduction

Food security governance remains one of the most important issues not only for a particular country, but also for the whole world. It is possible to solve this global problem, or rather to reduce its manifestation, but this requires not only the efforts of international organizations and specialized agencies, but also an active participation of the entire world community. A large role here is assigned to agriculture, which is designed to provide food to the population of a country. If we compare Russia in terms of food security with a number of foreign countries, our food security index (GFSI) is 42 points from possible 100. The leaders here are Singapore (87.5), Ireland (84), and the United States (83.7). We have a negative trade balance, with agricultural exports amounting to $21.9 million and imports amounting to $21.9 million. This is caused by the lack of the own production, low labor productivity, inflated cost of production, lack of a distribution network. Today, solutions to all these problems are being worked out at the legislative level.

S. I. Ashmarina et al. (Eds.): ENGINEERING ECONOMICS WEEK 2020, LNNS 139, pp. 547–553, 2021.
https://doi.org/10.1007/978-3-030-53277-2_65

2 Methodology

The following key methods were used in this research: monographic, comparative analysis, economic, statistical, and abstract-logical ones. The research was based on official statistics from the Federal State Statistics Service, regulatory acts of the Government of the Russian Federation, and official data from the Federal Customs Service. The analysis was based on information from the Ministry of Agriculture of the Russian Federation, publications of scientists and practitioners, as well as information from Internet portals dedicated to agro-industrial topics. These methods and resources allowed to establish and characterize the state of the Russian milk market, determine trends of its development, offer and justify measures that are aimed at improving the food security of the Russian Federation in the context of milk and dairy products.

3 Results

Food security governance is a priority area of the state policy, as it covers a wide range of national, economic, social, demographic, and environmental factors [8]. It should be noted that the goal to reduce imports in the agricultural sector was set in 2010, long before the implementation of sanctions, which enabled national producers to develop quickly.

In accordance with the Decree of the President of the Russian Federation No. 20 of January 21, 2020 "On approval of the food security Doctrine of the Russian Federation", food independence is defined as the level of self-sufficiency in percentage terms, and it is calculated as the ratio of domestic production of agricultural products, raw materials, and food to the volume of their domestic consumption [3]. Food independence has threshold requirements: at least 85% of meat and meat products (meat in equivalent), 90% of milk and dairy products (milk in equivalent), 95% of potatoes, and 90% of vegetables and gourds.

According to the Federal State Statistics Service, the self-sufficiency level of basic food in 2018 was the following: 95.7% meat and meat products, 83.9% milk and dairy products, 95.3% potatoes, and 87.2% vegetables and gourds [4]. Based on the above, we can notice that the balance in milk and dairy products provision is more disrupted than in other food groups, and this fact demonstrates the relevance of the research.

Let us analyze the general dynamics of production, imports, and exports of dairy products in Russia.

Table 1. Dynamics of production, imports, and exports of dairy products in Russia, mln tons

Indicator	2000	2005	2010	2015	2016	2017	2018	2019
Production	32.3	30.8	31.8	30.8	30.8	30.2	30.6	31.3
Imports	4.7	7.1	8.2	7.9	7.5	7.0	6.5	6.4
Exports	0.5	0.5	0.5	0.6	0.6	0.6	0.6	0.6

Source: authors based on [4]

According to Table 1, the total production of Russian dairy products in 2019 decreased slightly and amounted up to 96.9% compared to 2000. During the period 2000–2019, the production mostly decreased, but the decline was insignificant.

Owing to the governmental policy of import substitution industrialization, the trade balance of food is steadily improving, and the share of imported food on retail markets is decreasing [15]. It should be noted that according to the data in Table 1, the total imports of milk and dairy products in 2019 increased by 36.2% (or 1.7 mln tons) compared to 2000, and it decreased by 22% (or 1.8 mln tons) compared to 2010. Despite the decrease in imports volumes, its level is still very high, and it does not allow to have the proper level of food security. Summarizing the results of 2019, the imports volume in milk equivalent amounted to about 6.4 mln tons (the minimum for the last 16 years).

Cheese and curd, which makes more than 1/3 of the imports of dairy products, ranks first in the structure of dairy product imports in Russia. About 20% of butter, fats, and oils derived from milk, dairy spread, milk and condensed cream ranks second. About 10% of milk and concentrated cream ranks third. The main suppliers of milk and concentrated cream to Russia in 2017–2018 were Belarus, Kazakhstan, and Finland [11].

The reduction in dairy import volumes in 2018 was primarily due to the introduction of temporary restrictions to dairy products supplies from Belarusian enterprises, as well as to reduction in external supplies from non-CIS countries, as a result of high stock levels and decrease in import attractiveness due to the devaluation of the ruble. The dairy products supply from Belarus (the main external supplier of dairy products to Russia) declined by 10–15%. Consequently, the volume of imports from Belarus became the lowest over the last 15 years.

Despite the growth, exports of dairy products from Russia is very low due to the fact that the national production of milk is not enough for domestic consumption, and only limited volumes of the product may be exported. These are mainly whole-milk and fermented-milk products, ice cream, cheese, and analogue cheese.

Russia exports about 90% of dairy products to the Customs Union and CIS countries [13]. This is a traditional consumer for Russian exporters, which has clear preferences and requirements, and historically built logistics. The growth of demand for Russian products in Southeast Asia (especially in China) and Africa can develop a significant market, and Russian milk producers have the opportunity to take advantage and increase exports significantly [1].

In order to evaluate the Russian self-sufficiency level on the dairy market, it is necessary to analyze the amount of milk and dairy products produced.

In 2018, annual per caput consumption of domestic dairy products accounted to 209 kg that is 46.7% less than the balanced rate of consumption, which is 392 kg [6]. The economic accessibility plays a significant role in the aspect of food security [9].

According to Federal State Statistic Service, the annual per caput consumption of milk and dairy products was 229 kg in 2018. Since domestic producers cannot provide a sufficient production level at this stage, Russia has to import foreign products [4] (Table 2).

Russia has recently known a noticeable increase in dairy products production. This is largely due to the release of a significant market share after import deliveries decrease. Milk production in Russia is concentrated in three categories of farms:

agricultural organizations, households' farms, and peasant farms. The main share of milk production is concentrated in agricultural organizations and households' farms (Table 3).

Table 2. Level of self-sufficiency in milk and dairy products in the Russian Federation, %

Indicators	2014	2015	2016	2017	2018
Level of self-sufficiency in milk and dairy products, %	78.1	79.9	80.7	82.3	83.9
Annual per caput consumption of milk and dairy products, kg	239	233	231	230	229

Source: authors based on [4].

Table 3. Dynamics of the structure of milk production in the Russian Federation by farm categories, %

Farm category	2000	2005	2010	2015	2016	2017	2018
Agricultural organizations	47.3	45.1	44.9	49.3	50.6	51.9	53.0
Households' farms	50.9	51.8	50.4	44.0	42.1	40.2	38.9
Peasant farms and self-employed entrepreneurs	1.8	3.1	4.7	6.7	7.3	7.9	8.1

Source: authors based on [4].

The total cow population in Russian farms of all categories contained 7,942,600 animal units in 2018. The cow population decreased by 5.8% over 5 years (488,300 animal units), by 12.3% (1,117,700 animal units) over 10 years, and by 41% (5,530,600 animal units) over 20 years.

Despite the fact that the cow population in private farm holdings slightly exceeds the indicators in the agricultural organizations, the productivity of cows and milk production in agricultural organizations is higher than in other farm categories. The merchantability level is around 90%, which is significantly higher than in agricultural holdings, and it is not comparable with the merchantability level in households' farms where milk is consumed mainly for personal needs.

According to Table 3, milk production by agricultural organizations was 53% of the total production in 2018, which is 5.7 points higher than in 2000, due to the increase in cow productivity over the studied period of time.

The cow population in agricultural organizations in 2000–2018 decreased by 49.4% and amounted to 3,283,000 animal units, and the productivity of animals increased by more than 2.5 times and amounted to 5,945 kg. The cow population decreased by 44% and amounted to 3,360,800 animal units in the households' farms, while the productivity of cows increased by 28.9% and amounted to 3,463 kg.

As a result, milk production in households' farms in 2018 compared to 2000 decreased by 27.4% or 4,505,300 tons. The significant positive dynamics in milk production was observed in peasant farms (by 4.4 times or 1,925,600 tons) and in agricultural organizations (by 6.3% or 960,000 tons).

TOP-5 the regions that produced milk in 2017 included the Republic of Tatarstan (the volume of milk production is 1,823,800 tons, and the share in total milk output in Russia is 5.8%), the Republic of Bashkortostan (1,718,400 tons and 5.5%, respectively), the Altai Territory (1,401,900 tons and 4,5%), the Krasnodar Territory (1,380,900 tons and 4.4%), and the Rostov Region (1,091,100 tons and 3.5%) [13].

The foreclosure of national market of agricultural products including milk and dairy products by the countries that imposed sanctions against Russia has had a positive impact on Russian agricultural development and increase in uncooked milk production. State investment programs also contribute to production development.

Due to the fact that uncooked milk is not the final product, any increase in its production depends on the possibility of further processing including the product-line expansion dairy products: capacity to process whole milk products or to produce cheese, milk powder, and ice cream [7] (Table 4).

Table 4. The level of use of the average annual production capacity by organizations that produced particular types of products in 1990–2017, %

Type of products	1990	1995	2000	2005	2010	2015	2016	2017
Whole-milk products (in milk equivalent)	76	24	32	48	57	61	59	50
Butter	76	35	25	27	28	36	35	38
Cheese	86	51	49	61	63	66	64	48
Concentrated milk products	79	48	55	61	58	60	59	68

Source: authors based on [4].

The analysis of the dynamics of the use of particular production capacities to produce dairy products shows that Russian capacities, taking into account new construction activity, modernization, and reconstruction, have a prospect for growth. Further increase in both uncooked milk production and its industrial processing will depend on the rate of population's solvency increase, the governmental actions in the Russian Federation to further restrict imported products and expand export markets, and the measures of the state support for the dairy industry as a whole.

4 Discussion

Due to the special significance of the topic, a large number of publications of domestic and foreign scientists are devoted to food security issues. So, Nazarenko believes that the concept of food security includes two main aspects. First, it is the level of food self-sufficiency and the availability of reserves that ensure the stability and independence of the state from changes in the global market and weather conditions. Secondly, this researcher refers to the availability of food for all groups of the population, including its poorest part, considering food to be the most important and integral factor of survival and existence of people [12]. Another domestic economist, Larionov, argues that the country's food security can only be ensured within the framework of an integrated approach, while its basic components should be fixed at the legislative level [10].

Kiselev, Strokov, Zhorova, Belugin believe that the leading Western governments treat the food security problem as the opportunity to provide all the population with food in sufficient quantity and quality, while Russia treats food security as the opportunity to provide food self-sufficiency and import substitution industrialization [9]. Foreign researchers pay attention to the fact that food security depends on the policies of different countries [2]. International experience in ensuring national food security is based on three main approaches: protectionistic approach (Japan, Norway) that protects domestic production, intermediate approach (EU countries) that ensures food security within the EU, and export-oriented approach (USA) that subsidizes exports provided that the domestic market is saturated with products. If there is high-level food security, there is economic growth and human well-being [16]. If there is no food security, demographic changes and urbanization become evident [1].

We should take into account the best of each approach in our policy. We should not forget that the new doctrine was adopted in the context of increasing exports [14]. Therefore, we should increase overall production to provide food for both national and international needs. The issue cannot be resolved only by the state programs. In our opinion, it is much more important to create conditions for attracting private business resources to finance research projects on production programs on long-term and mutually beneficial base. This will allow to introduce modern management and marketing approaches to Russian business.

5 Conclusion

Today, food security is an important criterion of national security, especially in the context of the global crisis and turbulence in the world economy [5]. At the same time, it is important to increase the competitiveness of domestic products on the world market. The authors analyzed the dependence of production of milk and dairy products on imports, and they revealed that milk production decreased, and at the same time the volume of imports increased in 2000–2019 in Russia. Cheese and curd, which makes 1/3 of the imports of dairy products, ranks first in the structure of Russian imports of dairy products. About 20% of butter, fats, and oils rank second. The analysis of the dynamics of use of particular production capacities to produce dairy products shows that Russian capacities have a prospect for growth, which will be the technical basis to increase production. The level of Russian self-sufficiency in milk and dairy products increased from 78.1 to 83.9 over the past five years, but on the other hand, the actual per caput consumption of milk and dairy products decreased by 4.2% and amounted to 229 kg per year in 2018. According to the adopted Food Security Doctrine of the Russian Federation, which takes into account the aspects of socio-economic development of the country, it is necessary to expand access to basic food types for the population. Today, we need to focus on meeting the following conditions: improving production efficiency, reducing production costs, increasing competitiveness at all stages, as well as developing exports and stimulating domestic consumption of dairy products.

References

1. AfterShock: Russian market of milk and dairy products. Agrovestnik (2019). https://agrovesti.net/lib/industries/dairy-farming/rossijskij-rynok-moloka-i-molochnykh-produktov.html. Accessed 27 Apr 2020
2. Bureau, J.-C., Swinnen, J.: EU policies and global food security. Global Food Secur. **16**, 106–115 (2018)
3. Decree of the President of the Russian Federation No. 20 of January 21, 2020 "On approval of the food security Doctrine of the Russian Federation" (2020). https://www.garant.ru/products/ipo/prime/doc/73338425/. Accessed 27 Apr 2020
4. Federal State Statistics Service: Indicators that characterize import substitution in Russia (2020). http://www.gks.ru/folder/11188. Accessed 27 Apr 2020
5. Glinskiy, V., Serga, L., Alekseev, M., Samotoy, N., Simonova, E.: The development of the food industry as a condition for improving Russia's national security. Procedia Manuf. **21**, 838–845 (2018)
6. Gusakova, E.P., Malina, A.B., Myasnikova, V.M.: Regional food self-sufficiency assessment as a factor of national food safety. In: Ashmarina, S.I., Mantulenko, V.V. (eds.) Proceedings of the 2nd International Scientific Conference GCPMED 2019 - Global Challenges and Prospects of the Modern Economic Development. European Proceedings of Social and Behavioural Sciences, vol. 79, pp. 1400–1407. European Proceedings, London (2020)
7. Gusakova, E.P., Shchutskaya, A.V., Afanaseva, E.P.: Innovative development management in agribusiness. In: Mantulenko, V.V. (ed.) Proceedings of the 18th International Scientific Conference "Problems of Enterprise Development: Theory and Practice". European Proceedings of Social and Behavioural Sciences, vol 82, pp. 861–868. European Proceedings, London (2020)
8. Hatab, A.A., Cavinato, M.E.R., Lindemer, A., Lagerkvist, C.-J.: Urban sprawl, food security and agricultural systems in developing countries: a systematic review of the literature. Cities **94**, 129–142 (2019)
9. Kiselev, S., Strokov, A., Zhorova, M., Belugin, A.: Agro-industrial complex of Russia under the conditions of sanctions and the need to ensure food security. AIC: Econ. Manag. **2**, 12–18 (2015)
10. Larionov, V.G.: Food security in Russia. Prodovolstvennaya Politika i Bezopasnost **2**(1), 47–58 (2015)
11. MilkNew:s Review: 5 years under sanctions for the Russian dairy industry (2019). https://milknews.ru/analitika-rinka-moloka/rinok-moloka-v-Rossii/5-let-sankcii.html. Accessed 27 Apr 2020
12. Nazarenko, V.I.: Food security (In the world and in Russia). Monuments of Historical Thought, Moscow (2011)
13. Russian Federal Service for Veterinary and Phytosanitary Surveillance: Import. Export. Transit (2017). http://www.fsvps.ru/fsvps/importExport. Accessed 27 Apr 2020
14. Union "Eastern Chamber of Commerce of the Moscow Region": Food Security 2020 (2020). https://vomo.tpprf.ru/ru/news/349446/. Accessed 27 Apr 2020
15. Uzun, V., Shagaida, N., Lerman, Z.: Russian agriculture: growth and institutional challenges. Land Use Policy **83**, 475–487 (2019)
16. Xu, X., Hu, H., Tan, Y., Yang, G., Jiang, B.: Quantifying the impacts of climate variability and human interventions on crop production and food security in the Yangtze River Basin, China, 1990–2015. Sci. Total Environ. **665**, 379–389 (2019)

Statistical Research of the Economy Informatization in the Russian Federation

N. V. Proskurina(✉) and Yu. A. Tokarev

Samara State University of Economics, Samara, Russia
nvpros@mail.ru, tokarev_ya@mail.ru

Abstract. The paper examines the issues of the economy informatization in the Russian Federation. "The Strategy for the Development of the Information Society in the Russian Federation for 2017–2030" adopted by the Government of the Russian Federation confirms the relevance of the topic. The study aims at a comprehensive analysis of the economy informatization in the Russian Federation. The methodological framework applied in the study included time series analysis, cluster analysis, and structural shift analysis. The findings that emerged from the study indicate the consistency in the structure of expenditures for information and communication technologies. The positive trend for major economic indicators was observed. A multidimensional grouping of Russian regions based on cluster analysis has revealed that the number of regions with a low level of informatization has significantly decreased compared to 2010. The results of the study define the level of informatization in the economy a positive one. In terms of achieving the goals defined in "The Strategy" those findings are of quite a noticeable significance.

Keywords: Statistics · Digital economy · Region · Economy informatization

1 Introduction

The conceptual design of modern world has been internationally changed by the introduction and development of information and communication technologies (ICTs) Digitalization has permeated the entire sphere of human life and government. Mobile devices, computers, and the Internet allow the instantaneous exchange of information among different sources and let people be at the forefront of everything that happens in the world. Despite the accompanying difficulties and risks, informatization brings positive changes in the economy, as it minimizes costs, reduces barriers to entry to the market, and, ultimately, increases labor productivity.

At the same time, there is a need to analyze trends in the informatization of the economy, so that the state has the necessary information for rapid response and management decision-making. The role of national statistical services is important in this regard. Their statistical indicators that characterize the development of the ICT sector can be considered as one of the most important constituents for monitoring the development of the modern information society.

"The Strategy for the Development of the Information Society in the Russian Federation for 2017–2030" [11] reflects the importance of the research topic.

S. I. Ashmarina et al. (Eds.): ENGINEERING ECONOMICS WEEK 2020, LNNS 139, pp. 554–560, 2021.
https://doi.org/10.1007/978-3-030-53277-2_66

This document identifies the aspects of using ICTs aimed at the development of the information society and the formation of the national digital economy, and also defines target indicators, and normative values of statistical indicators that have to be achieved.

The purpose of this study is to analyze statistically the informatization of the economy in the Russian Federation. The primary data for this study were collected using the Federal State Statistic Service database.

To date, the problem has received scant attention in the research literature. The studies are either limited to the analysis of informatization for one territory [7, 10], or are of a sociological nature [1]. The most interesting study from the statistical point of view is [4], which analyzes the use of ICT by the Russian population. Our research focuses on the use of ICT in the country's enterprises.

2 Methodology

The study was conducted in the form of a mixed methodology based on the following statistical methods.

1. The method of time series analysis.

To analyze the dynamics, the corresponding traditional analytical indicators are used, such as growth rates. Their calculation is based on a comparison of levels. Statistically, the trend is represented as an indicator of the average growth rate. Based on the extrapolation, the forecast was made for 2020–2021.

2. The method of cluster analysis.

Cluster analysis includes the classification of multidimensional observations, which are used to determine the distances of objects and to obtain homogeneous groups or clusters. It also allows examining, analyzing, and classifying large information arrays.

Two cluster analysis procedures were applied in this paper. Euclidean distance was used for proximity measure, and *k*-means method was used as a clustering algorithm.

3. The method of structural changes analysis.

The Ryabtsev index was used as an indicator in this study. It evaluates the measure of differences in structural shifts in the economy and takes values from 0 to 1. The closer the index value is to 1, the more the differences among the structures are. The "0" value means that the structures are identical.

3 Results

Not only is the state of the economy crucial for the informatization development. Also it depends on the financial and technological support of enterprises, on their involvement in the use of modern software and digital solutions. Statistics is intended to provide a quantitative assessment of the informatization state. Statistics applies the system of indicators to describe its individual aspects, identify trends, and form an information basis for making management decisions.

ICT costs indicator is seen as the basic one in this paper. It is measured in monetary units and in terms of inflation tends to grow linearly. Therefore, it is important to analyze changes in the cost structure by the type rather than by their volume. It should be noted that since 2014, the Federal State Statistic Service has introduced a new classification of costs, which makes it impossible to compare structural changes with the previous periods (Table 1).

Table 1. The structure of costs for information and communication technologies in Russia for 2014–2018 (%)

Costs types	2014	2015	2016	2017	2018
Purchase of IT and office equipment	22.1	20.3	20.0	20.0	20.0
Purchase of telecommunications equipment	13.1	13.5	11.6	10.9	10.4
Purchase of software	13.8	17.6	22.4	18.9	18.1
Telecommunication charge payments	23.8	22.2	19.3	17.6	17.8
Employees training	1.0	0.6	0.5	0.4	0.3
Payments to IT subcontractors and outsourcing companies	17.1	20.1	20.3	25.3	26.6
Other costs	9.1	5.6	5.9	6.8	6.8

Source: authors based on [5].

With the Ryabtsev index it was revealed that there were no significant changes in the costs structure over that period (0.150 is a low level of structural differences. It is apparent that there is a change in the share of costs for IT subcontractors and outsourcing companies (from 17.1% to 26.6%) and for the purchase of software (the share increased from 3.8% to 22.4%, then fell to 18.1%). There is a tendency to reduce the share of telecommunication costs (from 23.8% to 17.8%). As for the other changes, they do not exceed the statistical margin of error.

The dynamics of the selected indicators was analyzed in order to assess how the level of the economy informatization has developed in the Russian Federation. These indicators are of priority for evaluating the implementation of "The Strategy for the Development of the Information Society in the Russian Federation for 2017–2030" [11].

1. The number of personal computers per 100 employees (units). This indicator implies the availability of equipment for solving telecommunication tasks.
2. The number of organizations with their own website (%). The indicator demonstrates the organizations' activity in promoting their goods and services electronically.
3. The number of organizations that use global networks (%). This characterizes technologies that allow exchanging information, using various software, and information resources (Internet, Extranet, Intranet).

Figure 1 shows the growth rate (in % to the previous year) for each of the indicators in 2011–2018. As it can be seen from this figure, in general, the dynamics of the development is similar, i.e. the values increased almost during the whole period under study. The period of active growth was replaced by the phases of moderate growth, and in some periods the decline can be observed. But it is apparent that there is an overall upward trend.

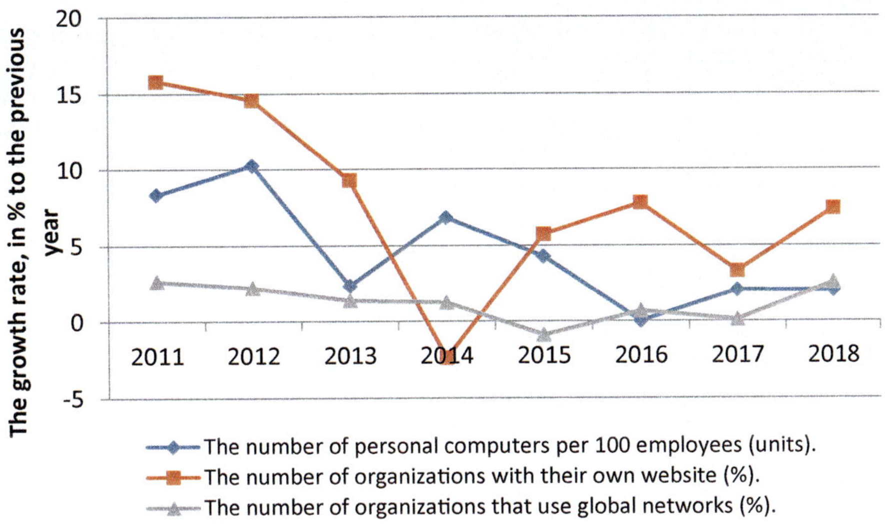

Fig. 1. Dynamics of the main indicators of the Russian economy informatization in 2011–2018 (Source: authors based on [5]).

The number of personal computers (per 100 employees) increased from 36 in 2010 to 51 in 2018, which represents an average annual growth of 4.5%. This growth trend can be explained by the increase in the number of jobs related to intellectual activities, which include the analysis of various data. The reason for this is the expansion of the information database and the complexity of the analysis methods.

The most noticeable growth is observed in the proportion of organizations that have their own website in the Internet (on average 7.5% per year). If in 2010 only one in four organizations had a website, in 2018 the number was one in two. The situation with the use of global information networks in the economy is quite satisfactory, i.e. more than 90% of legal entities use the networks in their activities (average annual growth of 1.2%).

The second task of our research was to forecast the indicators of informatization in the Russian economy. The forecast period is 3 years. The limitation was taken into account, namely, the indicators that were expressed as a percentage could not exceed the value of 100%, since this would contradict their meaning. The results of the forecast calculations and time series models are presented in Table 2.

Table 2. Forecast for indicators of informatization in the Russian economy

Indicators	Time series model	2020	2021
Number of personal computers per 100 employees (units)	$1.833t + 36.167$	56.33	58.17
Number of organizations with websites (%)	$2.505t + 28.331$	55.89	58.39
Number of organizations using global networks (%)	$83.535x^{0.0393}$	91.79	92.10

Source: authors.

Forecast values indicate that the growth of the analyzed indicators is expected. Therefore, the tasks to form a technological basis for the development of the economy and social sphere, as well as the development of information and communication infrastructure, which were announced in "The Strategy for the Development of the Information Society", can be fulfilled.

The cluster analysis methodology was used for a comprehensive multidimensional assessment of the economy informatization level in the regions of Russia.

To describe the level of the economy informatization the following regional indicators were selected:

X1 – The share of internal expenditure on research and development in % to GRP;
X2 – The share of research and development expenditures aimed at economic development in the total volume of internal research and development expenditures, %;
X3 – The share of innovative goods, works, and services in the total volume of goods shipped, works performed, and services provided by industrial companies and service organizations, %;
X4 – The share of innovative goods, works, and services in the total volume of exports of goods, works, and services of industrial companies and service organizations, %;
X5 – The number of patents for inventions issued to Russian applicants, units per 1 mln. people;
X6 – The share of fundamentally new technologies in the total number of developed advanced production technologies, %;
X7 – The share of information and communication technology (ICT) costs in the total volume of products shipped, %.

The entire set is divided into 3 clusters. Thus, the regions of Russia can be divided into the regions with high, medium and low levels of the economy informatization (Table 3). The total number of regions differs by year due to the differences in the mode of providing information by Rosstat.

Table 3. Cluster analysis results on the economy informatization level in the regions of Russia

Level	Cluster number	Number of regions	
		2010	2017
High level of the economy informatization	1	21	18
Medium level of the economy informatization	2	14	42
Low level of the economy informatization	3	48	22

Source: authors.

In comparison with 2010, the situation in 2017 has changed significantly. A noticeable number of regions have changed their rating. For example, the Trans-Baikal *krai* has sharply deteriorated its position, moving from the first cluster to the third. On the contrary, the Voronezh *oblast* has moved from a group with a low level to a group with a high level of the economy informatization.

A positive aspect is the sharp decrease in the number of regions in the third cluster and their mass transition to the second one. This indicates that regional authorities pay attention to the development of informatization of their economies.

4 Discussion

Informatization of the economy is a set of interrelated phenomena and processes that aim to meet the needs of all economic entities in various types of information. Among these entities there can be business units (enterprises) and government agencies, as well as the population (households). Information and communication technologies are a necessary means to achieve this aim.

The period, when the information societies are formed, implies that there is an increasing need for an objective assessment of the impact of information and communication technologies on the regional socio-economic development in the Russian Federation.

Statistical research of the society informatization should be based on the clearly defined category of "information society". The correlation between various indicators of the regional economies' informatization should be taken into account. In many studies one of the important criteria for informatization is represented by the transition of the public sector to the electronic platforms and the ability of citizens to participate in the management of the territory through digital technologies [2, 3, 9].

Moreover, statistical studies should consider the impact of information and communication technologies on the socio-economic development of society, namely the growth of living standards, economic efficiency, labor productivity, and energy efficiency [6, 12]. To solve its problems, statistics can use various methods, including those related to the calculation of multidimensional estimates of the economies' informatization level [8].

The salient feature of the Russian economy is that the development of informatization in the Russian Federation started later than in European countries. This has resulted in the lack of the necessary tools for implementing ICT-related activities provided there is a large amount of knowledge. This has become a catalyst for mass purchases of technical equipment, job restructuring, and training or retraining of personnel to perform new tasks. In turn, that has resulted in increased availability of personal computers, the use of information technology, etc.

5 Conclusion

The importance of the ICT sector development in the transition to a digital economy makes this study ever more relevant in the first place. Next, the findings of this research support "The Strategy of Information Society Development in the Russian Federation for 2017–2030" adopted by the government of Russia [11]. This document defines the target indicators and the guidelines for the development.

To assess the degree of achieving goals, which were set in "The Strategy", the degree of development and trends in the major economy informatization indicators was analyzed. Statistical methods and the Federal Service of State Statistics database were used for this purpose.

The main financial indicator is the organization's costs on information and communication technologies. There have been no noticeable changes in their structure in recent years, which indicates the set balance in this domain. The analysis of time series has shown that in Russia there is an upward trend in the availability of computers, websites and the use of global information networks. However, in some periods the trend was interrupted by the phases of decline. The forecast values indicate the possibility of achieving the standard values of "The Strategy".

A large potential is represented by the regional economy. The number of regions with a low level of informatization in the economy has significantly decreased over the past few years. That was achieved by the measures taken. Overall, information and communication technologies are gradually invading modern management systems in all sectors of the economy, and in public and municipal management.

References

1. Batrakova, L.G.: Development of digital economy in Russian regions. Soc. Polit. Res. **1**, 51–64 (2019). https://doi.org/10.24411/2658-428X-2019-10338
2. Das, A., Singh, H., Joseph, D.: A longitudinal study of e-government maturity. Inf. Manag. **54**(4), 415–426 (2017). https://doi.org/10.1016/j.im.2016.09.006
3. Demushina, O.N.: Influence factors for e-participation. ARS Administrandi **9**(2), 132–151 (2017). https://doi.org/10.17072/2218-9173-2017-2-132-151
4. Dolgikh, E., Efimova, M.: Statistical analysis of information society development in Russian Federation. Vestnik Universiteta **9**, 90–94 (2018). https://doi.org/10.26425/1816-4277-2018-9-90-94
5. Federal State Statistic Service: Monitoring the development of the information society in the Russian Federation (2020). https://gks.ru/folder/14478. Accessed 05 Mar 2020
6. Kumar, R.R., Stauvermann, P.J., Samitas, A.: The effects of ICT on output per worker: a study of the Chinese economy. Telecommun. Policy **40**(2), 102–115 (2016). https://doi.org/10.1016/j.telpol.2015.06.004
7. Liu, C.: Sustainability of rural informatization programs in developing countries: a case study of China's Sichuan Province. Telecommun. Policy **40**(7), 714–724 (2016). https://doi.org/10.1016/j.telpol.2015.08.007
8. Salyutina, T.Y., Kuzovkov, A.D.: Analysis of methods and approaches to measuring processes of informatization and the movement towards the information society. T-Comm **10** (6), 52–57 (2016)
9. Sangki, J.: Vision of future e-government via new e-government maturity model: based on Korea's e-government practices. Telecommun. Policy **42**(10), 860–871 (2018). https://doi.org/10.1016/j.telpol.2017.12.002
10. Shamina, O.V.: Statistical evaluation of informatization of economics of the Nizhny Novgorod region. Vestnik NGIEI **4**(71), 93–100 (2017)
11. The Strategy for the Development of the Information Society in the Russian Federation for 2017–2030 (2017). http://kremlin.ru/acts/bank/41919. Accessed 01 Mar 2020
12. Zhou, X., Zhou, D., Wang, Q.: How does information and communication technology affect China's energy intensity? A three-tier structural decomposition analysis. Energy **151**, 748–759 (2018). https://doi.org/10.1016/j.energy.2018.03.115

Economic and Social Aspects of the Innovation Management: Decision Making and Control

Analysis of Social Risks Impact on Economic Security of Companies

T. A. Korneeva(✉) and V. V. Kozhukhova

Samara State University of Economics, Samara, Russia
korneeva2004@bk.ru, viktoriaart1995@gmail.com

Abstract. An effective system of economic security is one of priority tasks of many companies. One of the most important factors in economic security of companies is their compliance with the principles of corporate social responsibility (CSR). The research challenge is that despite many CSR studies, the regulatory framework in the field of CSR remains unsystematic. In addition, the question of assessing the impact of a company's social responsibility on its economic security has not been analyzed. The purpose of the study is to determine the relationship of social responsibility and economic security of companies. As part of the study, an analysis and systematization of the legislative framework was carried out, a system of indicators for assessing social risks calculated based on non-financial reporting data was developed. According to the reporting data of the largest Russian oil and gas companies, a social risk assessment was carried out.

Keywords: Companies · Corporate social responsibility · Economic security · Non-financial reporting · Social reporting

1 Introduction

In current conditions of instable world economy and multidimensional nature of economic relations, business entities are forced to quickly adapt and seek ways to reduce threats to their functioning and development. Creating an effective system of economic security is one of priority tasks of many companies. The content of "economic security" is characterized by complexity and dynamism. It is constantly updated by researchers in connection with the emergence of new factors affecting it. One of the most important factors of economic security of companies is their observance of the principles of corporate social responsibility, according to which companies voluntarily participate in the social development of society, local territorial communities, take measures to protect the environment, create a favorable social environment through social investment and various social projects and programs.

The organization's performance in relation to society in which it operates, and its environmental impact has become an extremely important part of evaluating its overall performance and ability to continue to function effectively. Awareness of advantages and assessment of organization's activities in the field of social responsibility can affect its competitive advantages, reputation, ability to attract and retain employees, shareholders, investors, customers or users.

S. I. Ashmarina et al. (Eds.): ENGINEERING ECONOMICS WEEK 2020, LNNS 139, pp. 563–569, 2021.
https://doi.org/10.1007/978-3-030-53277-2_67

Certain aspects of corporate social responsibility (CSR) are considered in the scientific works of foreign scientists such as Azzouz & Jack [1], Moneva-Abadia, Gallardo-Vazquez, & Sanchez-Hemandez [4], Han, Li, Lubrano, & Xum [2], Morsing & Spence [5], Nguyen, Kecskés, & Mansi [7], Hsu & Chen [3] and others. Issues of economic security of companies are considered in the works of the following authors: Na, Park, Yu, Kim, & Chang [6] and others. However, issues of the relationship of social responsibility and economic security are insufficiently studied. In addition, the CSR regulatory framework is not systematized. The question of assessing the impact of the company's social responsibility on its economic security is not properly analyzed too.

The main source of information on the effective application of CSR principles, the results of the company's social policy is non-financial reporting, in particular reporting in the field of sustainable development and social reporting. However, the advisory nature of such reporting, as well as the lack of standardized reporting forms, leads to a decrease in the transparency and comparability of non-financial reporting by companies. Thus, the relevance of the chosen topic is due to the need for a detailed study of corporate social responsibility issues in the context of the formation and development of the economic security system of companies. The purpose of the study is to determine the relationship of social responsibility and economic security of companies, as well as the development of indicators to assess the impact of social responsibility of the company on its economic security.

2 Methodology

The study analyzed and systematized the regulatory framework in the field of CSR.

To conduct a social risk assessment, a system of indicators has been developed, one of which is Injury Frequency Rate (Formula 1), which characterizes the number of industrial accidents per 1000 workers

$$IFR = \frac{The\ number\ of\ victims\ of\ accidents}{Average\ number\ of\ employees} \times 1000 \tag{1}$$

However, this indicator does not provide a reliable data on accident rate in the organization. In this regard, it is necessary to calculate additional indicators, such as Lost Time Injury Frequency Rate (LTIFR) (Formula 2), which characterizes the number of injuries resulting in temporary disability, per 1 million hours worked.

$$LTIFR = \frac{The\ number\ of\ victims\ with\ loss\ of\ working\ time}{Total\ hours\ worked\ by\ all\ staff} \times 1000000 \tag{2}$$

Fatal Accident Rate (FAR) shows the number of fatalities due to industrial accidents per 100 million hours worked (Formula 3).

$$FAR = \frac{Number\ of\ fatal\ assidents}{Total\ hours\ worked\ by\ all\ staff} \times 100000000 \tag{3}$$

In connection with the development of infectious and occupational diseases, it is necessary to calculate Occupational Disease Rate (ODR) (Formula 4) and Lost Day Rate (LDR) (Formula 5).

$$ODR = \frac{Number\ of\ cases\ of\ newly\ diagnosed\ occupational\ diseases}{Total\ hours\ worked\ by\ all\ staff} \times 1000000 \quad (4)$$

$$LDR = \frac{Number\ of\ days\ lost\ as\ a\ result\ of\ accidents}{Total\ hours\ worked\ by\ all\ staff} \times 1000000 \quad (5)$$

A decrease in the above indicators illustrates a decrease in social risk and, as a result, increased security of the company.

One of the important indicators of the company's effective HR policy is the turnover rate (Formula 6), as well as dynamics of the number of employees.

$$TR = \frac{The\ number\ of\ retired\ employees\ for\ the\ period}{Average\ number\ of\ employees\ for\ the\ period} \times 100\% \quad (6)$$

Values of the turnover rate in the range of 3%–5% indicate a natural renewal of the team and does not require measures on the part of the personnel department and the company's management. Excessive staff turnover reduces motivation and loyalty of remaining employees, which contributes to an increase in social risk and a decrease in economic security of the company. The calculation of these indicators is carried out according to the reporting in the field of sustainable development of the largest Russian oil and gas companies, published on their official websites.

3 Results

As part of the study, the analysis of the regulatory framework in the field of CSR regulation was carried out. Four main levels of the CSR regulation can be distinguished: international, national, corporate, intracorporate. Table 1 presents international standards used by Russian and foreign companies when preparing reports.

Table 1. Standards/systems used by Russian and international companies

Standards	Russia	World average
Other	10%	29%
UN PRI	3%	6%
Integrated reporting	15%	22%
OECD recommendations	2%	23%
Kyoto protocol to the UNFCCC	26%	37%
CDP	26%	44%
United Nations Global Compact	29%	50%
ISO	76%	63%
GRI	62%	73%

Source: authors based on [14].

The results of statistical studies indicate that more and more companies in Russia and in the world bring their reporting to unified international standards. At the same time, Russia is approaching global trends in the development of the CSR concept and its standardization.

The basis of legal regulation of CSR at the state level is the National Standards of the Russian Federation, Federal Laws, Decrees of the President of the Russian Federation, which regulate companies' advertising policy, charitable activities, guarantee compliance with ethical and social norms, the right of citizens to protection of their health and etc.

The corporate level of regulatory regime of corporate social responsibility relations considers basic international principles, as well as specific requirements of national legislation. Compliance with company certain standards in the field of social responsibility may be a prerequisite for its membership in a particular industry association.

At the corporate level, regulatory documents have a maximum degree of specification. Examples of normative documents of this level are: charter, code of corporate ethics, mission statement, standards of socially responsible behavior, business ethics programs, collective bargaining agreements, etc. Norms prescribed by documents at international, state, and corporate levels form the basis of internal company standards. Thus, from the position of inter-level analysis, the principle of vertical regulation is used here.

To assess the effective application of CSR principles, as well as the degree of influence of the results of the company's social policy on its economic security, a system of indicators has been developed as part of the study. Since economic security is defined as the ability to counteract the negative impact of various factors, it is proposed to assess the impact of CSR on economic security through the assessment of social risks. As part of the study, social risk indicators were calculated according to the social reporting data of the largest Russian oil and gas companies [8–13] (Table 2).

Table 2. Social risk assessment indicators according to financial statements of Russian oil and gas companies for 2018

Company	Average headcount (thousand people)	IFR	LTIFR	FAR	ODR	LDR	TR
PJSC LUKOIL	102,5	–	0,11	0,01	0,06	–	0,078
PJSC Gazprom Neft	70,6	–	0,258	0,89	0,018	9,79	0,129
PJSC Gazprom	466,1	0,28	0,17	0,57	0,03	12,01	0,056
PJSC Rosneft Oil Company	308	–	0,34	2,47	0,06	23,08	0,108
PJSC NOVATEK	13,036	0,79	–	–	–	50	0,07
PJSC Tatneft	48,078	0,14	0,08	0,01	0,027	–	–
PJSC Transneft	115,1	–	–	–	–	–	0,05

Source: authors.

The results of the study indicate that non-financial reports of the studied oil and gas companies have varying degrees of detail and contain different sets of analytical data. The reporting data of PJSC Novatek [10] and PJSC Transneft [13] are insufficient to calculate Injury Frequency Rate, Lost Time Injury Frequency Rate, as well as Occupational Disease Rate. The reports of PJSC Transneft [13] and PJSC Tatneft [12] do not contain data on the dynamics of staff turnover. The reporting of three of the seven companies (43%) [9, 12, 13] included in the sample does not contain information on Lost Day Rate. Four out of seven companies (57%) [8, 9, 11, 13] do not calculate Occupational Disease Rate.

The data in Table 2 indicate low social risks for oil and gas companies. Low values of Injury Frequency Rate indicate that accidents leading to temporary disability or death are rare. Moreover, the dynamics of the given indicators according to the reporting data of the studied companies indicates a reduction in social risk [8–13].

PJSC Gazprom and PJSC Gazprom Neft have the lowest level of social risk. In the reports of these organizations, analytical indicators in relation to the results of social activities are most fully disclosed. The reports of PJSC Tatneft and PJSC Lukoil are less detailed in terms of social policy issues. The least transparent from the point of view of the CSR disclosure is the reporting of PAO Transneft and PAO Novatek. Their reporting contains a small set of analytical indicators. These companies do not calculate Injury Frequency Rate, Occupational Disease Rate and do not disclose data which reporting users could independently calculate using these indicators.

The findings are confirmed by official ratings. The RAEX Rating Review annually compiles ESG rating of Russian companies based on the assessment of a specific set of indicators of the company's impact on the ecological and social environment, as well as the degree of the company's exposure to management risks. The positions of the studied companies in this rating regarding the social component are presented in Table 3.

Table 3. Rating of companies by social factor

Company	Place in the ranking
PJSC Tatneft	2
PJSC Gazprom	3
PJSC LUKOIL	5
PJSC Rosneft Oil Company	7
PJSC Transneft	11
PJSC NOVATEK	14

Source: authors based on [15].

The data in Table 3 indicate that the most transparent from the point of view of social responsibility are the statements of PJSC Tatneft and PJSC Gazprom (PJSC Gazprom Neft does not participate in the rating as a separate company and is considered as part of PJSC Gazprom) [15]. The least transparent from the point of view of the CSR disclosure is the reporting of PAO Transneft and PAO Novatek [15].

4 Discussion

For the most effective application of CSR principles to strengthen the economic security system of companies, it seems necessary to standardize non-financial reporting, such as social report and report on activities in the field of sustainable development. Standardization of the reporting structure will allow identifying significant topics that need to be included in the report, as well as determining a system of non-financial indicators that sufficiently disclose information about key aspects of the company's social policy, risks and opportunities.

The assessment of social risks, as well as analysis of indicators of social risks will allow identifying and eliminating social risks that have the strongest impact on company security. In the past few years, in the practice of exchanging business information between Russian companies and regulatory authorities, there has been a tendency to submit reports in XBRL format, the main task of which is to regulate the exchange of information.

In this regard, it is possible to use the XBRL format in the formation of a social report and a report on activities in the field of sustainable development. World practice shows that the introduction of the XBRL format will reduce the time required to prepare reporting data by almost 90%; reduce the cost of ownership of IT systems by 40%; will provide an opportunity to reduce the cost and speed up the process of exchanging business information. The effective internal control and audit service of companies will allow for monitoring compliance with legal requirements regarding the formation and presentation of social reporting, as well as identifying and eliminating possible problems in a timely manner, thereby increasing the transparency of business reporting.

5 Conclusion

Thus, the condition for the effective functioning of organizations and ensuring economic security of both individual companies and the state is to follow the principles of socially responsible business and the formation of non-financial reporting. The lack of standardization of non-financial, including social, reporting, as well as the optional nature of its use creates difficulties in the preparation and verification of such reports. In addition, a different degree of detail in the reporting of different companies, a different set of indicators used makes reporting less comparable and complicates the assessment of organizations.

References

1. Azzouz, L., Jack, A.: Benchmarking the sustainability reporting of high-speed railways (HSRs): towards a state-of-the-art benchmarking and reporting framework for HSRs. J. Clean. Prod. **250**, 119505 (2020). https://doi.org/10.1016/j.jclepro.2019.119505
2. Han, S., Li, G., Lubrano, M., Xun, Z.: Lie of the weak: inconsistent corporate social responsibility activities of Chinese zombie firms. J. Clean. Prod. **253**, 119858 (2020). https://doi.org/10.1016/j.jclepro.2019.119858
3. Hsu, F.-J., Chen, S.H.: Does corporate social responsibility drive better performance by adopting IFRS? Evidence from emerging market. J. Comput. Appl. Math. **371**, 112631 (2020). https://doi.org/10.1016/j.cam.2019.112631
4. Moneva-Abadia, J.M., Gallardo-Vazquez, D., Sanchez-Hernandez, M.I.: Corporate social responsibility as a strategic opportunity for small firms during economic crises. J. Small Bus. Manag. **57**, 172–199 (2019)
5. Morsing, M., Spence, L.J.: Corporate social responsibility (CSR) communication and small and medium sized enterprises: the governmentality dilemma of explicit and implicit CSR communication. Hum. Relat. **72**(12), 1920–1947 (2019)
6. Na, O., Park, L.W., Yu, H., Kim, Y., Chang, H.: The rating model of corporate information for economic security activities. Secur. J. **32**, 435–456 (2019)
7. Nguyen, P., Kecskés, A., Mansi, S.: Does corporate social responsibility create shareholder value? The importance of long-term investors. J. Bank. Finance **112**, 105217 (2020). https://doi.org/10.1016/j.jbankfin.2017.09.013
8. PJSC Gazprom: Sustainability report 2018 (2019). https://www.gazprom.ru/f/posts/01/851439/sustainability-report-rus-2018.pdf. Accessed 14 Feb 2020
9. PJSC LUKOIL: Sustainability report 2018 (2019). http://www.lukoil.ru/FileSystem/9/347263.pdf. Accessed 14 Feb 2020
10. PJSC NOVATEK: Sustainability report 2018 (2019). http://www.novatek.ru/ru/development/. Accessed 14 Feb 2020
11. PJSC Rosneft Oil Company: Sustainability report 2018 (2019). https://www.rosneft.ru/upload/site1/document_file/Rosneft_CSR18_RU_Book.pdf. Accessed 14 Feb 2020 (2019)
12. PJSC Tatneft: Sustainability report 2018 (2019). http://2018.tatneft.ru/. Accessed 14 Feb 2020
13. PJSC Transneft: Sustainability report 2018 (2019). https://www.transneft.ru/u/section_file/39731/otchet-ob-ustoychivom-razvitii-2018.pdf. Accessed 14 Feb 2020
14. PwC: ESG factors in investing (2019). https://www.pwc.ru/ru/sustainability/assets/pwc-responsible-investment.pdf. Accessed 14 Feb 2020
15. RAEX: ESG-rating (2019). https://raex-rr.com/country/RAEX-600/ESG_rating#table. Accessed 14 Feb 2020

Development Concept for Ski Resorts in the Context of the Economy Digitalization

N. V. Polyanskova(✉)

Samara State University of Economics, Samara, Russia
polynskova@mail.ru

Abstract. The article is aimed at developing a concept for domestic ski resorts under conditions of the economy digitalization. The author considers issues and specifics of the competitiveness of ski resorts with an effective 5A business model used as a foundation that forms an approach to the complex's activities as an ecosystem and allows integration into the digital development process. As a result of marketing research, comparative analysis and economic modeling, it was established that when it comes to skiing and leisure, tourists have a great demand for additional digital services or opportunities to receive "package offers". The availability of such services, in conformity with international trends in the development of the industry, is no longer treated as a competitive advantage, but as a standard for skiing complexes. All stages of the concept of development of the ski resort must ensure implementation of digital technologies in the marketing of services. A digital customer-oriented platform (application) can become an operational foundation of the ski complex and be used in "flexible" pricing. To ensure investment attractiveness, competitiveness and strong social significance, the offered conceptual business model should be aimed at processes of creating "added value", transmitting unique offer to clients and extracting profit.

Keywords: Concept · Digitalization of economy · Ski resorts · Ski sports · Tourism

1 Introduction

Active development of digital technology in Russia and abroad influences efficiency of development for all economic subjects at the modern stage. Tourism industry has begun its implementation into a new digital format as a forerunner. At the current moment, the subject's competitiveness in the tourist industry is mainly determined by the level of its involvement in the digital process.

Information technology is becoming the primary resource in current conditions [1]. In recent years, Russia has been developing in conditions of external challenges that require a significant increment in innovation in all economic spheres. The economic development of the Russian Federation has moved to an advanced format. The established development models of most economic subjects require paramount changes. The development of the tourism industry will be intertwined with investments in innovation and reaching more tourists with mobile channels [9].

S. I. Ashmarina et al. (Eds.): ENGINEERING ECONOMICS WEEK 2020, LNNS 139, pp. 570–577, 2021.
https://doi.org/10.1007/978-3-030-53277-2_68

The leaders in the digitalization of tourism services are major foreign ski resorts. Russian ski resorts have significant development potential. Therefore, the main concept for the company's development in this industry is establishment of an effective business model that allows its integration into the digital development process.

2 Methodology

When conducting research on the advancement of skiing and tourism and establishing a concept in the context of the development of the digital economy, general scientific and applied research methods were used. The author conducted marketing research on the development of ski resorts employing survey methods and content analysis, explored the official sites of the analyzed ski resorts in Russia, and the ski tourism portals of Russia and the world.

The method of comparative analysis was implemented for ski resorts in Russia, the Volga Federal District, as well as the Samara region. An expert analysis of the coordination of interests of subjects in the field of skiing and tourism of the Samara region was carried out. When developing the concept for the development of domestic ski resorts, a prognostic method was used to predict the evolution of the company in the current challenges of the tourism industry. When creating a model for the development of a modern ski complex, the method of economic modeling was used.

3 Results

With each passing year, skiing builds up more and more popularity among Russian population. Climate conditions of the vast majority of Russian regions provide an opportunity to engage in this sport from December to April. The implementation of the Demography national project involves the popularization of sports among Russian citizens and an increase in the share of people engaged in sports from 36.8% in 2018 to 55% by 2024. According to the data for 2018, 125.5 thousand people are engaged in skiing in the Russian Federation (Table 1) [8].

Table 1. Ski athletes training indicators in Russian Federation, 2015–2018

Year	The amount of people engaged in ski sports in Russian Federation, ppl.	The amount of organizations engaged in training ski sportsmen, u.
2015	99262	119
2016	97657	121
2017	107824	114
2018	125509	107

Source: author based on [8].

According to the Ski.ru ski portal, there are approximately 2.5–3 million skiers in Russia, whose annual agenda includes at least one or more trips to the ski resort. On average, a skier performs 1 trip a year for 10 days. Based on this, the market demand can be formulated as 25 million person-days.

In the period between 2015–2019 there is a definite positive trend of an increase in skiing facilities. On current day, there are 325 ski facilities in the country (Table 2).

The majority (23.1%) of ski resorts are located in the Siberian Federal District, which points to a certain sufficiency in the market for ski recreation in this region. The second place in this rating is occupied by the Volga Federal District (21.2%). A positive point is the location of the region in the Volga Uplands grants significant opportunities for the development of this particular type of recreation. The third place in terms of the amount of ski resorts is taken by the Central Federal District - it comprises 17.5% of Russian ski resorts (Table 2).

Table 2. Distribution of ski facilities by federal districts of Russian Federation by October of 2019

Federal district	Ski resorts, u.	Share of ski resorts, %
Siberian FD	75	23,1
Volga FD	**69**	**21,2**
Central FD	57	17,5
Ural FD	39	12,0
Northwestern FD	38	11,7
Far eastern FD	36	11,1
Southern and North Caucasus FD	11	3,4
Total in RF	**325**	**100,0**

Source: author based on [10].

Despite the significant number of ski resorts in the country, only a few dozen of them enjoy considerable popularity.

Below are the 10 best ski resorts in Russia with a total number of visits in the season between 2018–2019 [10]:

Krasnaya Polyana ("Rosa Khutor" & "Gorki Gorod", Sochi), 1,5 mm. ppl.
Sheregesh (Tashtagol, Kemerovo region), 1 mm. ppl.
Sorochany (Moscow region), 0,5 mm. ppl.
Dombai (Karachay-Cherkessia), over 300 k. ppl.
Bobrovy log (Krasnoyarsk), over 300 k. ppl.
Elbrus (Kabardino-Balkaria), 250 k. ppl.
Igora (Priozersky district, Leningrad region), over 200 k. ppl.
Solnechnaya Dolina (Miass, Chelyabinsk region), over 200 k. ppl.
Sviyaga hills (village of Savino, Kazan, Tatarstan), 200 k. ppl.
Gora Belaya (Nizhny Tagil, Sverdlovsk region), 200 k. ppl.

These resorts have a fairly developed infrastructure. They are not inferior to Western "colleagues" in terms of equipment and comfort, and visa-free travel makes them especially attractive to Russian citizens and guests from the CIS countries.

The main competitive advantage of modern ski resorts is their focus on the development of an entirely new support infrastructure. In terms of ski holidays, tourists demand not only the ability to ascend and descend, but also additional digital services and an opportunity to receive special "package deals". Following the mainstream world trends in evolution of the industry, the presence of associated services is no longer seen as a competitive advantage but as a standard practice for ski resorts. Due to this, comparatively small "one day" resorts enjoy sufficient demand for their services on the regional market [7].

On all stages of implementation of the ski resort development project, it is essential to ensure use of digital technology in service marketing. In this case, a digital customer-oriented platform (application) can become an operational foundation of the ski complex and be used in "flexible" pricing. A modern internet application can promptly supply information on weather conditions, traffic load, availability of parking spaces, discounts, ski pass purchase conditions and its status, etc. Partnership with Yandex, Google and other participants in the digital promotion of services is considered as the most promising.

A specific characteristic of ski sport and recreation is defined in segmentation: professional athletes, amateur nonprofessionals and beginners. Using this division whilst creating a marketing strategy in the digital age will allow further increase in efficiency of advertising and other tools.

To organize an efficient ski resort, it is essential to expand the offer and enhance the system of implementation with new offers, conduct SMM campaigns, and offer discounts to certain specific target groups. One of the advantages that can attract regular users of ski resort services is the Ski Pass system. At the same time projects' weaknesses and risks can be counteracted.

Advertising and promotion of ski resort services should be carried out with the ultimate use of the potential of previously established contacts in the B2B (business to business) and B2C (business to consumer) formats - tour operators, media, advertising agencies, partners, web application developers and SMM specialists, etc.

More bleeding edge access systems for ski slopes, such as the Axess system, are relevant as well. Modern standards establish certain demands to the ski resort services, to the training and certification of personnel. Typical basic service package includes 5 blocks: active recreational services; basic services; additional services; supporting services; passive recreational services (Fig. 1).

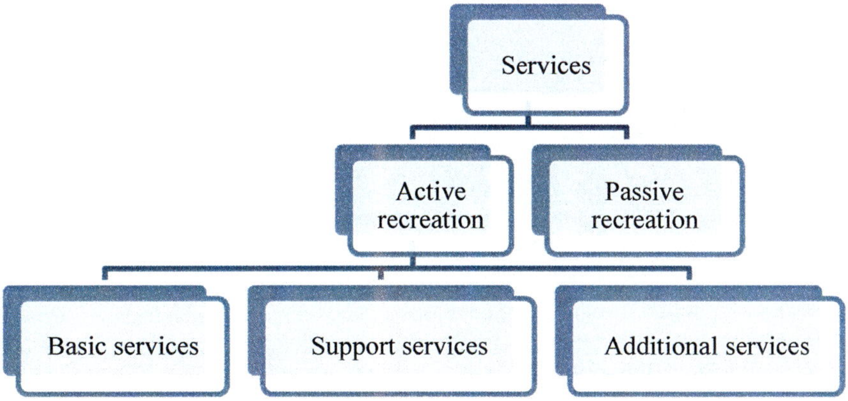

Fig. 1. Ski resort service groups (Source: author).

Foreign countries that rely on development of sports and tourism, the implementation of local projects for the creation (reconstruction) of ski resorts is based on the following basic principles:

- inclusion in the implementation of territory development programs (strategies),
- utilization of public-private partnership mechanism,
- diversification of the range of services (uniform all season).

The best practices are demonstrated by Austria, France, Switzerland, Germany, Italy, Mexico, Turkey (Table 3).

Table 3. Leading countries in ski sports

Country	Main ski resorts
Austria	Innsbruck, Bad Gastein, Zillertal, Mayrhofen, Kitzbühel, Zell am See
France	Les 3 Vallées, Chamonix, Val-d'Isère, Tignes
Italy	Val di Fassa, Val di Fiemme, Breuil-Cervinia, La Thuile
Switzerland	St. Moritz, Zermatt, Grindelwald, Adelboden, Verbier, Saas-Fee, Davos
Andorra	Pas De La Casa/Grau Roig, Soldeu/El Tarter, Arcalís – Ordino, Pal/Arinsal
Turkey	Uludag, Palandöken, Sarikamish
Bulgaria	Pamporovo, Borovets, Vitosha
Slovenia	Kranjska Gora, Bovec, Bohinj
Slovakia	Jasna, Smokovec, Tatranska Lomnica, Strebske Pleso
Germany	Oberstdorf, Garmisch-Partenkirchen
Finland	Pallas, Yllas, Levi
Norway	Geilo, Hemsdal, Lillehammer, Trysil
Czech Republic	Spindleruv Mlyn, Pec pod Snezkou
Poland	Zakopane
Sweden	Öre
Spain	Sierra Nevada
North America (USA)	Aspen, Deer valley, Squaw Valley

Source: author.

For these countries, skiing and tourism is one of the most prioritized directions in economic development, as well as innovative development of resorts with digital technology and bleeding edge equipment used as a foundation. The authorities act as stakeholders in such projects and as the coordinator in all stages of the life cycle and an investor in the infrastructure objects (transport, utilities, social, etc.). After analyzing international practice, domestic and foreign standards it is possible to draw following conclusions essential for drawing up a development concept for domestic ski resorts. The competitive advantage of ski resorts and their sustainable development is determined by the 5A model, which shows an approach to the activities of the resort as an ecosystem.

According to the methodological approach 5A, the business model of the ski resort (or another tourist or sports facility) is based on the following elements (Fig. 2).

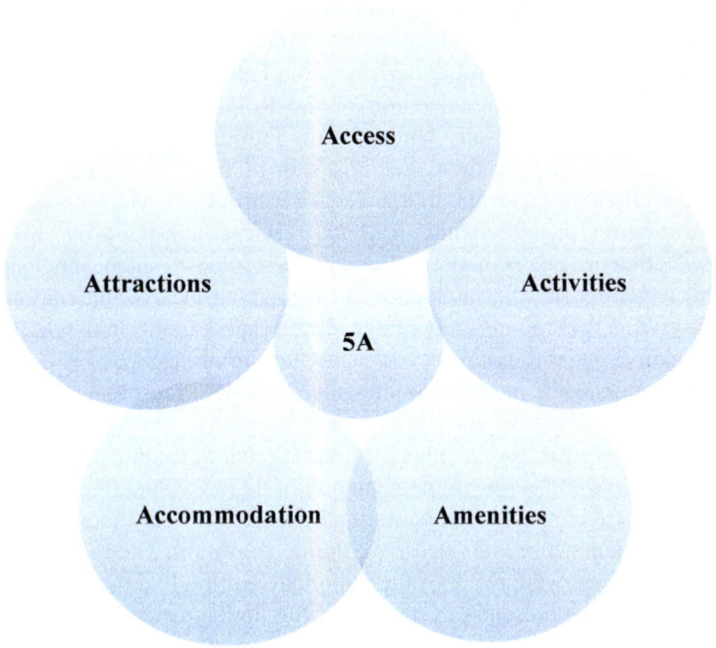

Fig. 2. Elements of a ski resort development business model (Source: author).

More and more resorts question: does their existing business model provide sufficient sales to cover the essential expenses throughout the entire investment cycle? And is it possible to reduce costs by implementing innovative approaches? Aside from the digital transformation of the industry, global trends such as the pursuit of a healthy lifestyle and environmental protection also influence the tourism market.

As such, the business model of a modern ski resort, first and foremost, should prioritize creating "added value", broadcasting its unique offer to the client and making profit. Transfer of tourists to the slope for winter sports, such as skiing and snowboarding, as well as related quality services for associated and additional services, should lie as a foundation for the business model. The offers have to be based on the ability to provide both winter and summer vacations, the originality and innovativeness of the services offered, and the holding of international events. This will provide the ski resort with investment attractiveness, competitiveness and strong social significance in the conditions of modern digital technology development.

4 Discussion

The digital transformation of the tourism industry is being heavily researched and discussed both in the scientific field and at the state level. According to the Decree of the Government of the Russian Federation of 20.09.2019 N 2129-r «On approval of the Strategy for the development of tourism in the Russian Federation for the period until 2035», the maximized digitalization of the Russian tourism sector is heavily implied [4]. The strategy especially notes that the introduction of digital technologies forms the mainstream trends in the development of the tourism industry, increasingly influencing virtually all parts of the tourism service through reducing transaction costs and raising awareness of participants in digital services and platforms. Most studies point to low competitiveness of domestic ski resorts due to an insufficient level of technological development and innovation of resorts and lack of integration into singular digital system. Taking into account new development trends in the digital tourism business, companies have to compete for each client in digital space [5].

One of the factors of increasing the innovativeness of ski resorts is a threat to their economic stability of their development caused by global warming and consequent decrease in snow cover [6]. Researchers also remark that development of ski resorts in face of climate change and active implementation of digital technology provides it with brand new prospect focused on innovation and use of ICT. This should be a channel not solely for communication, but also for commercialization. The results of the study show that ski resorts are well aware of the importance of the Internet. Companies must go beyond the traditional presence, giving priority to digital marketing methods [2]. Ski resorts must provide special attention to their websites. Companies should create efficient customer feedback systems, which will help improve organization management. It is also recommended to introduce more informative content, both about the services of the ski resort and about the aspects of the local environment. This will make the website content relevant for the requirements of the target audience [3].

5 Conclusion

Modern development of the tourism industry should be correlated with development of the international economy and be in line with digital development. This will supply ski tourism facilities with a high level of competitiveness, both in the domestic market and in the field of international tourism.

Promotion of healthy lifestyle on social networks has provided results: the number of tourists who intertwine their travels with various sports has increased. Skiing is one of the popular activities in the field of sports tourism. The dynamics of visits to ski resorts shows growth internationally and in Russia both. At the same time, the infrastructure of the vast majority of domestic ski resorts lags behind modern ski resorts of Europe. And the growing demand for ski services calls for a thorough approach when attracting potential customers in the face of exorbitant competition.

The competitive advantage of modern ski resorts in conditions of developing digital economy is based on certain principles. The most crucial principle is development of innovative supporting infrastructure. Aside from using the routes, modern support infrastructure and the associated services must be provided.

Today, the application of digital technology in ski resort marketing is a primary method of promoting and attracting customers. In this case it is appropriate to implement digital applications and platforms to ensure the operation of ski resorts. As such, the business processes of modern ski resorts should create a specific development concept that allows them to ensure their competitiveness with implementation of innovative technologies based on digital solutions.

Acknowledgments. The study was conducted with financial support from the Department of Economic Development, Investment and Trade of the Samara urban district Administration under municipal contract No. 1289 to provide a range of research work on marketing and development of a concept for the compound territory development of the "Chaika" ski resort located at Samara region, Samara, Krasnoglinsky district, vill. Upravlencheskiy, 65 quarter of the Samara forestry.

References

1. Babkin, A.V., Chistyakova, O.V.: Digital economy and its influence on competitiveness of entrepreneurial structures. Russ. Enterp. **18**(24), 4087–4102 (2017)
2. Cristobal-Fransi, E., Daries, N., Serra-Cantallops, A., Ramón-Cardona, J., Zorzano, M.: Ski tourism and web marketing strategies: the case of ski resorts in France and Spain. Sustainability **10**, 2920 (2018). https://doi.org/10.3390/su10082920
3. Daries-Ramon, N., Cristobal-Fransi, E., Martin-Fuentes, E., Marine-Roig, E.: ICT development in snow tourism: analysis of the online presence of ski resorts in Spain and Andorra. Doc. d'Analisi Geografica **63**(2), 399–426 (2017). https://doi.org/10.5565/rev/dag.348
4. Decree of the Government of the Russian Federation of 20.09.2019 N 2129-r «On approval of the Strategy for the development of tourism in the Russian Federation for the period until 2035». http://government.ru/docs/37906/. Accessed 02 Apr 2020
5. Kornilov, N.Yu.: Improving the business processes of a tourist enterprise in the conditions of digitalization. Bull. Russ. New Univ. Ser. Man Soc. **4**, 85–90 (2019). https://doi.org/10.25586/rnu.v9276.19.04.p.085
6. Moreno-Gené, J., Sánchez-Pulido, L., Cristobal-Fransi, E., Daries, N.: The economic sustainability of snow tourism: the case of ski resorts in Austria, France, and Italy. Sustainability **10**, 3012 (2018). https://doi.org/10.3390/su10093012
7. PWC: Ski resorts of Austria and Switzerland (2018). https://www.pwc.ru/ru/sports-new/assets/pwc-studie-bergbahnen-hoch-hinaus.pdf. Accessed 02 Apr 2020
8. Skiing development program in the Russian Federation for 2018–2022, approved by Order of the Ministry of Sports of the Russian Federation on September 7, 2018 No. 774. https://www.garant.ru/products/ipo/prime/doc/71777346/. Accessed 02 Apr 2020
9. Vintean, A.: Tourism of the future – an on going challenge. Stud. Bus. Econ. **14**(3), 258–272 (2019). https://doi.org/10.2478/sbe-2019-0058
10. Vsnega: Ski resorts of Russia (2019). https://www.vsnega.ru/resorts/russia/. Accessed 04 Apr 2020

Once Again to the Question of Economic Efficiency of Medical Services

A. G. Lukin[1], A. R. Saraev[2], and E. E. Dozhdeva[3](✉)

[1] Samara National Research University, Samara, Russia
Lagufk@mail.ru
[2] Samara State Medical University, Samara, Russia
saraeff10@mail.ru
[3] Samara State University of Economics, Samara, Russia
dozhdeva69@mail.ru

Abstract. The purpose of the article is to develop approaches to assessing the economic efficiency of medical services. The problem of measuring financial performance is very relevant for the financial management of medical care both in the private and in the public sector, especially in the sphere of state management. In modern market conditions, management in the healthcare industry needs to determine the amount of the minimum required financial resources for the successful treatment of incoming patients, in a fairly large set of various methods of treatment of diseases associated with different ways leading to the same clinical effect, a large variety of drugs having the same indications and in the context of sufficiency or deficiency of industry financing.

Keywords: Cost of medical services · Economic efficiency · Financial result · Financial management

1 Introduction

How important is it to evaluate, or to be more precise measure, the efficiency of a medical service? Performance measurement issues are of key importance in conducting research related to management issues, both in scientific, theoretical aspects, and in practical activities. Thus, A. Neely, C. Adams, and M. Kennerley note that in the academic community, research in the field of efficiency measurement is conducted by representatives of functionally different branches of knowledge [11]. Specialists in such fields as accounting, economics, human resource management, marketing, operational management, psychology and sociology, all these spheres study this subject [10].

Moreover, as it is noted by Fedotov [5], despite the fact that a leading position in these studies is the measurement of the efficiency of organizations (organizational efficiency), the components of a subject area studied in economics (productivity and efficiency analysis/measurement) and management (performance management/measurement, business performance management), the evaluation of efficiency is equally important for the management practice of organizations of any type. Neely & Kennerley [7], Krivonozhko et al. [8], and Gumilev [6] noted the great importance of evaluating efficiency in the field of management.

S. I. Ashmarina et al. (Eds.): ENGINEERING ECONOMICS WEEK 2020, LNNS 139, pp. 578–582, 2021.
https://doi.org/10.1007/978-3-030-53277-2_69

It is essential for organizations that provide medical services too. In modern conditions of the Russian reality, when medical care has moved from the category of public goods provided by the state to the category of private sphere, financed from various sources (budgets of the country's budget system, insurance organizations that are part of the system of compulsory medical insurance, and others that provide voluntary medical insurance services, as well as from the pocket of those who need medical care directly, without the above intermediaries), issues of economic and financial management of medical institutions come to the fore.

Without such an important criterion for evaluating management performance as efficiency, which links such important performance indicators as the achieved result and the total costs (expenses) to achieve it, it is impossible to find options for the greatest optimal impact on the management object in order to obtain the best result in certain conditions.

2 Methodology

The starting point of the modern concept of efficiency is considered to be the 50s of the twentieth century, when the works of Debreu [3] and Farrell [4] an approach was proposed to measuring the efficiency based on the ratio of results to costs. The most famous is the second work, in which Farrell identifies the following types of efficiency:

1. Technical efficiency – the state in which the enterprise reaches the maximum output for a given set of resources (performance).
2. Allocative efficiency – the state in which the enterprise reaches the minimum expenditure of resources for a given volume of output (effectiveness).
3. Economic efficiency is a state that combines the two previous types of efficiency (efficiency) [4].

Based on the conclusions of Debreu and Farrell, technologies for evaluating the efficiency of complex systems were developed, for example, the technology developed by American specialists Charnes, Cooper and Rhodes [2] in the seventies of the XX century.

Modern research on ways to measure the cost-effectiveness of medical services focuses on achieving the highest efficiency (allocative efficiency), offering various options for reducing the cost of providing medical services. At the same time, scientific research on the economic efficiency of medical services is conducted, as a rule, by narrow specialists within a particular branch of medicine. So, the team of authors headed by Basinkevich et al. [1], investigating the effectiveness of providing services to patients with coronary heart disease, considers 4 options for evaluating the effectiveness of providing a specific medical service: cost minimization analysis, cost effectiveness analysis, cost utility analysis, cost benefit analysis. At the same time, they chose the first method of minimizing the cost, using as a conclusion the thesis that the most cost-effective service will be one that will require less costs if the result is equal. The same approach to evaluating the efficiency is used by Kukolnikova & Zhukov [9], who offer the "cost/efficiency" ratio as a measurement. At the same time, they consider all four of these methods in detail.

The main problem in applying these methods of evaluating the efficiency is the measurement of the result. In this case, the result is taken into account only in the cost utilitarian analysis and in the cost analysis of profit. When using the cost-based utilitarian analysis, the result of medical treatment is based on the choice of preferred or utilitarian health criteria by patients themselves. The result of this choice is a calculated coefficient, called the "quality-of-life indicator" (QALY), which reflects changes in the life expectancy and quality.

In the methodology of cost analysis of profit, the result, as the name of the method implies, is the profit of a medical institution received from the provision of medical services. But here, as Kukolnikova & Zhukov rightly point out, the difficulty of expressing the clinical effect in monetary terms is obvious [9]. Thus, the stumbling block in measuring the efficiency of medical services is the inability to evaluate its result in monetary units. Therefore, the research task is to bring the main indicators of the efficiency formula to a single unit of measurement:

$$\mathbf{Efficiency} = \mathbf{Result/Costs} \tag{1}$$

If "efficiency" is a relative indicator, then "result" and "cost" should be reduced to a single unit of measurement. The result of the medical treatment of the patient can be expressed in monetary terms only in the amount of the cost of achieving it, but then the effectiveness of any treatment will be equal to 100% or "1", which does not allow us to compare its effectiveness. This conflict has led to the fact that almost all methods of measuring the efficiency are based on the dynamics of changes in costs.

Kukolnikova & Zhukov propose the following solution to this problem. They justify that the method of cost minimization analysis of medical services is applied only when the compared medical actions have "the same clinical effectiveness", i.e. they lead to the same result. Otherwise, they recommend using cost effectiveness analysis and just suggest the "cost/effectiveness" method for this purpose, on the basis of which it is possible to choose a treatment with a desirable clinical outcome [9].

3 Results

It is proposed to develop this methodology for measuring the efficiency based on a reference approach. The essence of this approach is that based on medical observations, the most effective method of treating a specific disease is selected, which makes it possible to achieve the desired clinical result quite quickly, which is taken as a standard. As a postulate (initial message), we assume that the economic efficiency of this treatment method is equal to 1, i.e. the cost of its application expresses its economic result (Formula 2).

$$\mathbf{Result\ 1/Costs\ 1} = \mathbf{1} \tag{2}$$

If it is necessary to measure the efficiency of another treatment method for this disease, then the cost of its implementation is compared with the result of the reference (first) method (formula 3).

$$\textbf{Result 1/Costs 2} = \textbf{Efficiency 2} \quad (3)$$

It is obvious that a more cost-effective treatment method will be measured by a coefficient greater than one, since the desired result is achieved at a lower cost, and vice versa, if the efficiency value is less than one, the reference method is preferable. It is clear that the use of the reference approach is possible only if the same result is achieved.

4 Discussion

As an example, let's consider a typical choice offered to patients in the dental practice. Let's take one common pathology – the absence of one tooth, while two adjacent teeth are healthy. To preserve chewing and aesthetic functions, two methods of treating this pathology are used. That is, the result will be the same, there are only different methods of achieving it. Any dentist, regardless of the specialization, offers the most affordable treatment for the patient. To date, this is the installation of a bridge-like prosthesis on three crowns. The second method is to install an implant in the lower jaw, in place of the missing tooth, followed by the installation of a crown on it.

To solve this problem, we use the proposed method described above. At the first stage, we will take into account costs associated with the installation of a bridge prosthesis. At the second stage, we will calculate the cost of installing the implant and a crown on it. At the third stage, we will compare the economic efficiency of both methods. The first method, widely used now, requires the manufacture of three crowns, depulpation of 2 teeth, preparation of the oral cavity and installation on cement. The average cost of this service is 32 thousand rubles. The cost of the second method is estimated at 26 thousand rubles. If we compare the efficiency of two methods, we need to take the first as a reference, since this is the most common method of treating missing teeth in modern Russia. Thus, its efficiency will be equal 1, and the result expressed in monetary terms, respectively, 32,000 rubles. Now let's calculate the economic efficiency of the second method, based on Formula 3:

$$\textbf{Efficiency 2} = \textbf{32000/26000} = \textbf{1,23}$$

That is, we will get that the efficiency of the second treatment method is higher than the first one and is 1.23. This is a fairly clear illustration of the proposed method of evaluating the economic efficiency of medical services. Thus, for further work it is necessary to consider the number of complications and frequency of visits and many other factors that are associated with the treatment process of other diseases, as in the proposed example, based on the results of the study to date, the complication rate is about the same – 4.2% and 4%, respectively.

But with further research, it may turn out that the number of complications for the second method of treatment may increase the amount of costs and ultimately reduce the economic efficiency of this method to a level less than 1.

5 Conclusion

In our opinion, using the reference approach allows you to:

- express the result in units comparable to the cost (monetary equivalent) and include it in the mathematical model,
- search for a reference method for treating a specific disease makes medical managers to consider more effective ways to treat patients,
- help scientists to evaluate the economic efficiency of the newly developed methods for the treatment of patients,
- measure the efficiency of complex treatment methods by dividing them into simple ones and presenting it as the sum of the efficiency coefficients of the methods or determining the arithmetic mean coefficient,
- develop an ideal model of the most effective way to treat a particular disease, which can serve as a benchmark for the efficient treatment of a particular disease,
- effectively and more realistically plan activities of medical institutions, as well as specific specialists who provide medical services,
- centralize planning of financial provision of medical care based on the approval of reference methods and organize its effective financial control.

References

1. Basinkevich, A.B., Matchin, Y., Bubnov, D.S., Silvestrova, G.A., Shamrina, N.S., Atanesyan, R.V., Vladimirov, S.T., et al.: Clinical and economic efficiency of endovascular treatment of coronary arteries in patients with stable coronary heart disease during hospitalization for "one night". Cardiology **58**(S3), 36–45 (2018)
2. Charnes, A., Cooper, W.W., Rhodes, E.: Measuring the efficiency of decision making units. Eur. J. Oper. Res. **2**, 429–444 (1978)
3. Debreu, G.: The coefficient of resource utilization. Econometrica **19**(3), 273–292 (1951)
4. Farrell, M.J.: The measurement of productive efficiency. J. Roy. Stat. Soc. Ser. A (General) **120**(3), 253–290 (1957)
5. Fedotov, Y.: Measuring the effectiveness of the organization: Features of the DEA method (data convolution analysis). Russ. Manag. J. **10**(2), 51–62 (2012)
6. Gumilev, L.N.: Ethnogenesis and Biosphere of the Earth. AST Publishing House, Moscow (2019)
7. Kennerley, M., Neely, A.: Performance measurement frameworks: a review. In: Neely, A. (ed.) Business Performance Measurement: Unifying Theory and Integrating Practice, pp. 143–162. Cambridge University Press, Cambridge (2007)
8. Krivonozhko, V.E., Utkin, O.B., Senkov, R.V., Antonov, A.V., Volodin, A.V.: Analysis of the Effectiveness of Financial Institutions in the Transition Economy. Nonlinear Dynamics and Control. ISA RAS, Moscow (2000)
9. Kukolnikova, E.L., Zhukov, B.N.: Evaluation of the cost-effectiveness of the treatment of patients with phlebological profile. Pract. Med. **6**(54), 126–129 (2011)
10. Lebas, M., Euske, K.: A conceptual and operational delineation of performance. In: Neely, A. (ed.) Business Performance Measurement: Unifying Theory and Integrating Practice, pp. 125–139. Cambridge University Press, Cambridge (2007)
11. Neely, A., Adams, C., Kennerley, M.: The Performance Prism: The Scorecard for Measuring and Managing Business Success. Prentice Hall, New Jersey (2002)

Internal Control as an Element of Ensuring Sustainable Development of Innovative Enterprises

T. E. Tatarovskaya(✉)

Samara State University of Economics, Samara, Russia
tatarovskaya.tatyana@gmail.com

Abstract. Innovation today is not only a condition for sustainable business development, but also for its survival. However, it is worth noting that innovative enterprises, due to the specificity and dynamism of their activities, have a whole range of risks, which actualizes the need to implement internal control systems. In order to ensure the reliability of reporting and sustainable development of innovative business entities, the issues of organizing the internal control system, approaches to assessing the costs of implementing this system, the main factors limiting the effectiveness of internal control, requirements and recommendations for the formation of a risk report in innovative business are considered. The article describes the main risks inherent to innovative business entities and shows the advantages of implementing the proposed internal control system.

Keywords: Accounting · Economic security · Innovative business entity · Internal control system · Risk appetite · Sustainable development

1 Introduction

Despite the importance of innovation in the current situation, modern conditions are not very favorable for the functioning of innovative businesses. Due to the need of increasing competitiveness and retain a place in the market, subjects of innovative entrepreneurship need to assess possible risks and obtain reliable business information required for making managerial decisions [3]. To ensure the achievement of goals and increase the efficiency of their activities, it is advisable for entities to use a generalized indicator that indicates the success of the business. This can be sustainable development.

Sustainable development is considered by various authors from different points of view: on the one hand, it can be considered as a series of successive stages of economic growth [9], and on the other – as a balanced state of resources that provides stable profitability and optimal conditions for expanded reproduction, economic growth in the long term, taking into account the influence of the external and internal business environment [1]. We will understand sustainable development as a sequence of states that contribute to the growth of economic stability, creating a foundation for maintaining this situation for a long time. Ensuring the sustainable development of an innovative enterprise depends on a variety of factors and tools used. One of them is the internal control system.

S. I. Ashmarina et al. (Eds.): ENGINEERING ECONOMICS WEEK 2020, LNNS 139, pp. 583–592, 2021.
https://doi.org/10.1007/978-3-030-53277-2_70

2 Methodology

In the course of economic activity, an innovative enterprise is influenced by industry, legal and other internal and external factors (significant conditions, events, circumstances, actions). In this regard, there are various risks that can significantly affect the financial position and financial performance of an innovative enterprise. Most of the risks of economic activity have financial consequences and, consequently, have an impact on the accounting statements prepared by an innovative enterprise [12]. Based on this, in order to form a complete picture of the financial position of the innovative enterprise, its financial results and changes in its financial position, the annual financial statements disclose indicators and explanations about potentially significant risks of economic activity to which the innovative enterprise is exposed [8]. Disclosure of this information is one of the components of the internal control system (ICS) of the committed facts of economic life of an innovative enterprise [4].

An innovative enterprise should implement a set of measures to identify risks, minimize the probability of their occurrence, reduce possible negative consequences when they are implemented, and ensure the adoption of rational and effective management decisions [7].

We conducted a survey of the owners of 125 enterprises in the Samara region that actively implement innovations in their activities, about the place of ICS in their organizations. According to the results of the survey 68% of the innovative enterprises stated that their organizations use control, of them in 28% of organizations, there is ICS, the remaining 40% use only certain procedures, while 32% of respondents do not consider appropriate the creation of the ICS, as an innovative company, especially in the early stages of its existence, the need in financial resources for its functioning and creation of additional control procedures, according to representatives of the innovative enterprises, it is only an additional financial burden, as well as a bureaucratic element that reduces mobility, which is necessary for the development of innovative enterprises.

Among the key problems hindering the implementation of the internal control system for innovative enterprises, there were also noted: the lack of financial and labor resources, since the organization and operation of this system involves significant costs for innovative business entities [5]. Therefore, in order to ensure the sustainable development of an innovative enterprise, it is necessary to develop a methodological approach to the implementation of ICS.

The main principle of implementing internal control in an innovative enterprise is the principle of rationality, the essence of which is to assess the ratio of benefits and costs associated with accounting. The concept of International financial reporting standards sees this ratio not as a fundamental limitation, but as a qualitative characteristic, since the benefits on a business scale have both financial and non-financial components.

For an innovative enterprise, we recommend two ways to organize the ICS:

- transfer of responsibilities for the organization and functioning of the ICS to the head of an innovative enterprise,
- the use of certain control procedures in the activities of an innovative enterprise that are sufficient for effective management.

It is assumed that for an innovative enterprise, the most appropriate combination of these methods is when control functions are distributed among several responsible persons (depending on the number of allocated business processes) without creating an internal control Department, and the effectiveness of control and evaluation of the actions of responsible persons is evaluated by the head of the organization.

In addition, due to the small number of employees in an innovative enterprise, it is advisable to introduce a system of collective responsibility for detected violations and a simplified unified document flow to reflect the results of the ICS operation. Optimizing the expenditure of available resources will increase the effect of organizing this system (Table 1).

Table 1. Elements of the internal control system

ICS element	Actions required for the element to function successfully within the system
Control environment	Assessment and optimization of the existing organizational structure Assessment of available labor resources Formation of personnel policy Evaluation of the organization's system of values and ethical principles Assessment of the Manager's competence and work style
Risks evaluation	Identification of risks inherent in the organization's activities Assessment of the significance of identified risks Assessment of the scale of consequences from the occurrence of identified risks Development of measures to eliminate identified risks Development of measures to eliminate the negative consequences of the occurrence of identified risks
Internal control procedures	Development of the optimal number of control procedures that are most appropriate for an innovative enterprise Formation of a list of goals and plans, the implementation and achievement of which should be monitored by the ICS
Information and communication	Assessment of the existing hardware and software necessary for the implementation of the ICS functions, for their viability in the new environment Formation of databases required for employees to perform internal control procedures (list of responsible employees, contractors, etc.) Dissemination of information necessary for effective communication and management decision - making Development of support and simplification procedures implemented by the information system for internal control purposes
Assessment of internal controls	Evaluating the actions of employees involved in the ICS Evaluating the effectiveness of the methods and procedures used

Source: author.

The listed elements and the actions to ensure their functioning must be developed once, standardized and consistently applied. It is advisable to re-evaluate the sufficiency of consistently applied methods and procedures in an innovative enterprise once every five years or more often if significant events occur in the business life of an innovative enterprise.

3 Results

Innovative enterprises, setting themselves the task of improving accounting, analytical and control support in order to achieve its sustainable development, face the need to determine the amount of financial costs for innovations and assess the sufficiency of human resources for their implementation. The proposed method for estimating the estimated costs of implementing the ICS is to compare the risk appetite of the organization and the accepted level of materiality.

The following indicators are elements of this methodology:

- the level of costs for implementing the ICS (CICS), %,
- materiality level (ML) – the maximum amount of error in the formation of accounting and analytical, financial (budget) information, that is, the maximum amount of any negative deviations that the organization is ready to incur, %,
- risk appetite (RA) – the amount of risk that an innovative enterprise is willing to accept to achieve its goals,%.

$$CICS = ML - RA. \tag{1}$$

There is no unified methodological approach to calculate risk appetite for an innovative enterprise, but it is possible to adapt existing methods for certain types of economic activity (banking sector, information technology) to the specifics of innovative business. Based on the results of research by foreign and domestic researchers, the most optimal methods were adapted to meet the requirements of innovative business and two methodological approaches to risk appetite assessment were developed.

1. Calculating the amount of risk appetite as a percentage of the balance currency.

In accordance with International financial reporting standards, the most common assessment of the level of materiality is expressed as a percentage of the balance sheet currency or total revenue, depending on the object being investigated for materiality misstatements.

The balance sheet currency is one of the fundamental indicators for conducting control and analytical procedures. According to the value of the balance sheet currency, they give a primary assessment of the stability of the organization's development, form an opinion about its position in the market. Therefore, it makes sense to consider the indicators of risk appetite and materiality as a percentage of the balance sheet currency. In this context, risk appetite will reflect the size of the share of the balance sheet currency that an innovative enterprise is willing to sacrifice to achieve its goals.

2. Focus on business goals and forecasting future activities.

This method, based on the assessment of risk appetite based on the current and expected future volume of economic transactions and the structure of significant risks, is relevant if there is data on the value of risk appetite in previous periods.

The risk appetite in the future period is calculated by increasing the value of the risk appetite of the previous period by the amount of the predicted increase in market and other risks (changes in interest rates, exchange rates, prices for shares and goods, etc.), identified by research of analytical materials of the largest consulting companies. The calculation of risk appetite in the framework of this approach is made taking into account:

- stage of the life cycle of an innovative enterprise,
- opinions of the main users of accounting (financial) statements,
- complexity of the accounting system, understanding of legislation,
- risk of fraud,
- external economic conditions,
- legal restrictions, etc.

According to the research, the effectiveness of ICS for innovative enterprises may be limited:

- changes in the economic environment or regulatory legal acts, the emergence of new circumstances outside the sphere of influence of the management of the innovative enterprise,
- abuse of official authority by the management or other personnel of an innovative enterprise, including cases of collusion of personnel,
- making incorrect management decisions regarding the facts of economic life, aspects of accounting or management accounting, conducting control and analytical procedures, as well as preparing accounting (financial) statements.

The proposed classification of factors limiting the ICS is expanded and adapted to the features of an innovative enterprise (Table 2).

Table 2. Factors limiting the capabilities of the internal control system subject of innovative entrepreneurship

Factors	Factor characteristics and consequences
Incorrectly set goals	An innovative enterprise is focused on poorly formulated, unrealistic and unacceptable goals, which will make the ICS ineffective
External factor	Environmental factors have a significant impact on the organization's activities, but the ICS can not affect them in any way
Erroneous judgments	The ICS does not guarantee the correctness of judgments formed by employees at all levels and the management of the organization, which can lead to negative consequences
Human factor	Any employee of an innovative enterprise may intentionally or negligently make a mistake, misspell, etc.
Negligence of the management	Managers of an innovative enterprise may neglect the ICS due to the isolation of their person from other employees (due to a high position) or to conceal facts that may negatively affect the organization, both reasons can lead to the formation of incorrect business information, erroneous judgments and the adoption of unacceptable and/or ineffective decisions
Collusion	Collusion of employees with each other or with third parties is difficult to predict and difficult to detect, but its presence can jeopardize the functioning of the entire ICS

Source: author.

Overcoming these factors will increase the effectiveness of the ICS, ensure the sustainable development of an innovative enterprise and achieve its goals. According to the summary of current business practices, the following classification of risks is proposed: financial, legal, country, regional, reputational, etc. According to the current legislation, if when preparing accounting (financial) statements, an innovative enterprise faces the need to disclose additional information about its activities in order to form a complete picture of ensuring sustainable development, its financial position, dynamics and financial results, then additional indicators and required explanations are included in the reporting.

To form a complete and adequate view of the financial position of an innovative enterprise and its main performance indicators in order to assess the achievement of sustainable development, it is advisable to disclose information in the annual accounting (financial) statements, including indicators and necessary explanations about the significant risks of the organization's activities [8]. This includes data on the organization's exposure to risks; on the objects of control and analysis, through which you can get a complete picture of the risks; on measures to identify and manage risks, reduce the negative consequences of their occurrence, as well as changes in this issue compared to previous periods.

Innovative enterprises usually neglect the ability to generate additional forms of reporting, using only a simplified one (balance sheet, report on financial results), although such information would be useful for external users, in particular for investors, venture funds, etc.

It is necessary to formalize the description of risks inherent in the activities of an innovative enterprise in the reporting. Moreover, a number of reporting forms compiled according to foreign standards (IFRS and GAAP) contain indicators that disclose such information [6].

Additional information, including indicators and necessary clarifications about risks, can be provided in the form of a separate section as an explanation to the balance sheet and the statement of financial results. Another way to present this information is to include it in the explanation of the relevant indicators of the accounting (financial) statements. To disclose additional information about risks, a separate report may be provided as part of the accounting (financial) statements.

4 Discussion

The implementation of the ICS is a fundamental element in ensuring the sustainable development of an innovative enterprise due to the quality and effectiveness of management decisions made based on the results of the operation of this system. It is difficult to conduct a full assessment of the effect of the introduction of an innovative business entity ICS. However, as practical experience shows, the introduction of ICS helps to reduce losses ensure the sustainable development of an innovative enterprise [11]. In addition, it is important to understand that the reduction of losses cannot always be measured by financial indicators, as well as the effect of using various management tools; therefore, it is advisable for innovative enterprises to analyze non-financial indicators.

To prove this statement, let's turn to foreign experience. The German consulting company RSM [2] conducted a survey of innovative enterprises in Germany, Spain, Poland, Hungary, Greece and Israel, dividing them into two categories: consumers of innovative products and manufacturers of innovative products [10]. In order to confirm the relevance of the problem of assessing the contribution of the ICS to the sustainable development of an innovative enterprise, we interviewed the heads of 125 organizations in the Samara region on similar issues.

When asked whether you agree that the implementation of the ICS affects the sustainable development of an innovative enterprise, the responses of the heads of consumer organizations were distributed as follows (%):

- fully agree – 32,
- rather agree – 53,
- rather disagree – 6,
- completely disagree – 9,
- I don't know – 0.

As for the producer respondents, their responses are distributed as follows (%):

- fully agree – 48,
- rather agree – 20,
- rather disagree – 8,
- completely disagree – 4,
- I don't know – 20.

The survey results can be summarized as a diagram (Fig. 1).

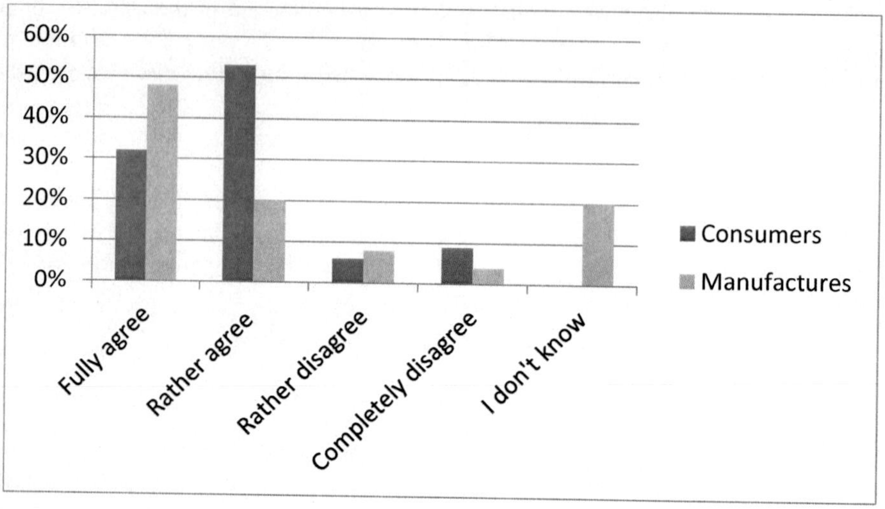

Fig. 1. Distribution of responses from consumers and manufacturers of innovative products on the importance of the role of the ICS in ensuring the sustainable development of their enterprises (Source: author).

From this we can conclude that almost 85% of respondents-consumers agree that the ICS is involved in ensuring the sustainable development of an innovative enterprise. The number of respondents who answered the same way is slightly less – 68%. And only 6% of consumers and 8% of producers disagree with this provision. Consequently, most organizations recognize the contribution of the ICS to ensuring sustainable development, which means that there is a need to determine the most important features of the ICS work for an innovative enterprise.

In response to the question, which aspects of the ICS can be considered the most important, consumers named (%):

- ensuring high financial stability – 42,
- reducing the probability of negative consequences – 50,
- reduction of investment risk – 21,
- high transparency of activity – 50,
- improving long-term performance and future performance – 6,
- increasing the number of efficient and effective business processes – 20,
- improved planning reliability – 5.

Note: it was possible to provide several answers to this question, and the percentage of respondents who marked each question is indicated.

The manufacturers' responses about the advantages of ICS were distributed as follows (%):

- ensuring high financial stability – 38,
- reducing the likelihood of negative consequences – 49,
- reducing investment risk – 11,

- high transparency of activity – 62,
- improving long-term performance and future performance – 7,
- increasing the number of efficient and effective business processes – 37,
- improving the reliability of planning – 3.

Note: it was possible to provide several answers to this question, and the percentage of respondents who marked each question is indicated. The results of the responses of the managers of the surveyed organizations are the same, which means that consumers and producers agree, especially in terms of reducing the likelihood of negative consequences and high transparency of activities.

5 Conclusion

Managers of many innovative enterprises are confident that the functioning of the ICS provides a number of advantages that ensure the achievement of sustainable development. When implementing this system in innovative business entities, it is recommended to clearly define the content of the system elements, develop a set of measures to prevent the occurrence of factors that limit the effectiveness of internal control, develop a report format based on the results of the system, and most importantly, assess the financial capabilities of the organization for the implementation of the internal control system.

Methodology for estimating the estimated costs of implementing an internal control system in an innovative business it is assumed to compare the estimated level of risk appetite of the company and the level of materiality approved in it. The choice of a methodological approach to risk appetite assessment depends on what the attention of management decision makers is focused on: the financial aspect (balance sheet currency) or the strategic aspect (goals, activity forecasting).

The effect of implementing an internal control system in an innovative business entity may be limited by a number of factors. For innovative businesses, the most significant are erroneous goal setting and external factors. The external business environment has a significant impact on the activities of innovative enterprises. Therefore, the internal control system has the task of continuously monitoring changes in the external business environment in terms of identifying risks inherent in the entity's activities, as well as new risks that were not previously considered. In addition, it is advisable to monitor changes in legislation regulating various aspects of financial and economic activities of innovative enterprises.

References

1. Alazzabi, W., Mustafa, H., Karage, A.: Risk management, top management support, internal audit activities and fraud mitigation. J. Financ. Crime (2020, in press). https://doi.org/10.1108/JFC-11-2019-0147. Accessed 30 Apr 2020
2. Bungartz, O., Strobl, G.: Talking points value added by internal control (2012). https://rsmus.com/content/dam/mcgladrey/pdf/talking_points_value_added_by_internal_control.pdf. Accessed 15 Apr 2020

3. Chalmers, K., Hay, D., Khlif, H.: Internal control in accounting research: a review. J. Account. Lit. **42**, 80–103 (2018). https://doi.org/10.1016/j.acclit.2018.03.002
4. Crespo, N., Rodrigues, R., Samagaio, A., Silva, G.: The adoption of management control systems by start-ups: internal factors and context as determinants. J. Bus. Res. **101**, 875–884 (2019). https://doi.org/10.1016/j.jbusres.2018.11.020
5. Kravet, T.D., McVay, S.E., Weber, D.P.: Costs and benefits of internal control audits: evidence from M&A transactions. Rev. Acc. Stud. **23**(4), 1389–1423 (2018). https://doi.org/10.2139/ssrn.2958318
6. Lartey, P., Kong, Y., Bah, F., Santosh, R., Gumah, I.: Determinants of internal control compliance in public organizations; using preventive, detective, corrective and directive controls. Int. J. Public Adm. **43**(8), 711–723 (2019). https://doi.org/10.1080/01900692.2019.1645689
7. Liew, A.: Enhancing and enabling management control systems through information technology: the essential roles of internal transparency and global transparency. Int. J. Account. Inf. Syst. **33**, 16–31 (2019). https://doi.org/10.1016/j.accinf.2019.03.001
8. Loishyn, A., Levchenko, S., Tkach, I., Getmanskii, V.: Assessment of efficiency indicators of internal control system functioning. J. Sci. Pap. «Soc. Dev. Secur.» **9**(4), 120–130 (2019). https://doi.org/10.33445/sds.2019.9.4.9
9. Moskvin, E., Gorodilov, S., Rusyaeva, T., Stotckaia, A.: An approach to overcoming the limitations for launching and functioning the innovative cycle in the scientific and production companies. In: Vorobev, E.N. (eds.) Proceedings of 4th Forum Strategic Partnership of Universities and Enterprises of Hi-Tech Branches (Science. Education. Innovation), pp. 66–68. IEEE, Piscataway (2016). https://doi.org/10.1109/ivforum.2015.7388256
10. Schroeder, J., Shepardson, M.: Do SOX 404 control audits and management assessments improve overall internal control system quality? Account. Rev. **91**(5), 1513–1541 (2016). https://doi.org/10.2308/accr-51360
11. Spanò, R., Allini, A., Caldarelli, A., Zampella, A.: Controlling innovation and innovating control: Insights from a knowledge intensive network. Bus. Process Manag. J. **23**(6), 1359–1384 (2017). https://doi.org/10.1108/BPMJ-02-2017-0036
12. Zhao, Y., Bedard, J.C., Hoitash, R.: Auditor effort and the prevention of financial report misstatements. Audit. J. Pract. Theory **36**(4), 151–177 (2017). https://doi.org/10.2139/ssrn.2693619. (Forthcoming)

Evaluation of the Effectiveness of Personnel Adaptation Process in Small Business

A. B. Shtrikov[1(✉)] and D. B. Shtrikova[2]

[1] Samara State University of Economics, Samara, Russia
ashtrikov@yandex.ru
[2] Samara State Technical University, Samara, Russia
shtrikovdadb@yandex.ru

Abstract. The article is devoted to the discussion of such a personnel management problem as personnel adaptation. The relevance of the work is determined by the fact that poor adaptation always leads to losses: low productivity and rapid dismissal of employees. The aim of the work is to determine the features of evaluating the effectiveness of the adaptation process in small businesses. The components of adaptation and the factors influencing it are considered. The main reasons for the dismissal of employees have been identified. The main indicators for assessing the quality of personnel adaptation are determined and the optimal ones for small businesses are selected. The authors assessed strengths and weaknesses of the adaptation of personnel at a small business – a beauty salon, proposed measures for its improvement and evaluated their effectiveness.

Keywords: Adaptation · Adaptation costs · Dismissal · Methods of adaptation · Personnel

1 Introduction

The relevance of the work is determined by the fact that adaptation of staff is today a process, which is given insufficient attention sometimes. Often, the adaptation of personnel in the organization often occurs spontaneously and the employee himself carries it out. However, it depends much more on the adaptation process than is commonly thought. High-quality staff selection is not enough for excellent work of employees. After the selection process, a period in the life of a new employee necessarily comes when he cannot realize all his potential. This is precisely because he has not yet adapted to the new organization.

Content adaptation is divided into the following components [4]:

1. Organizational adaptation consists of adapting an employee to order and rules, as well as to his new place of work, organization of the work process, place and role of the enterprise in the society. Organizational adaptation solves a number of problems concerning personnel, such as reducing staff turnover.
2. Psycho-physiological adaptation, as a rule, has as its goal to cause the addiction of the human body.

S. I. Ashmarina et al. (Eds.): ENGINEERING ECONOMICS WEEK 2020, LNNS 139, pp. 593–605, 2021.
https://doi.org/10.1007/978-3-030-53277-2_71

3. Socio-psychological adaptation is a process of introducing an individual into the current organizational system of organizational culture, as well as the need for a new employee to get used to the existing moral and psychological climate in the organization.
4. Professional adaptation is the process when a newly hired employee acquires new theoretical and practical knowledge, masters new responsibilities in the professional sphere. It is aimed at ensuring that a newly hired employee in no way can lower his level of success by applying various kinds of skills in his professional field, using objects and the language of his activities.

The process of adaptation management at the enterprise includes the development of a system of measures that can guarantee the success and high speed of this process. Poor adaptation always leads to losses. Moreover, these losses are manifested not only in reducing the productivity of a new employee. Also, the negative consequence of poor-quality adaptation may be the rapid dismissal of an employee from the organization, which, in turn, suggests that again you have to spend money on staff recruitment. However, even if you put up with these costs, it is far from the fact that the new employee will not quit quickly, too. This leads to a meaningless and endless cycle of new employees, which greatly affects the finances of the organization. There are several reasons for the dismissal of employees in the first three to six months of work [3]:

1. Mismatch of his ideas about the organization and the upcoming work and reality, that is, unjustified expectations;
2. An inappropriate employee was initially selected with inappropriate competencies, specialty, etc.;
3. The lack of a clear technological system that would allow a novice to quickly and painlessly join the organization.

The main errors in the adaptation system are:

1. Overload of adaptation;
2. Lack of participation in adaptation of the heads of departments;
3. Disinterest of mentors in training;
4. Inconsistency of adaptation tools to adaptation goals;
5. "Blur" of the responsibility of adaptation performers;
6. Duplication of the content of the adaptation tools used [3].

A high-quality system of adaptation of employees can break the circle and solve this problem. However, it is not enough just to create an adaptation system for employees; it is important to constantly evaluate and improve it to avoid waste of money in adaptation. The aim of the work is to determine the features of evaluating the effectiveness of the adaptation process in small businesses. To achieve this goal, the following tasks should be solved:

- analyze current research on the adaptation of personnel in small businesses,
- determine the methods and indicators for assessing the effectiveness of personnel adaptation in small businesses,

- identify the main problems and errors in the adaptation system,
- suggest measures to improve the effectiveness of the adaptation system using the example of a beauty salon and determine their effectiveness.

2 Methodology

Various persons including the head of the organization, the head of the department, which accepts a new employee, the head of the human resources department, a specialist in the human resources department and the mentor of a new employee, are engaged in the construction, implementation, management and control of the personnel adaptation system in the organization.

Adaptation management tools are understood as a way of transferring information to a new employee undergoing adaptation.

The most common adaptation management methods are [4]:

1. Training – it is implemented within the organization using both corporate training and training in various educational institutions.
2. Staff retraining – it will give an employee the opportunity to master a new position or profession. It is used for employees who have had a long break in work.
3. Cases, trainings, situational tasks – aimed at the successful development of new knowledge through interactive learning, involves interaction with all employees of the organization.
4. Business games – aimed at examining the situation from the inside, searching for a new optimal solution to a particular situation.
5. Instructions – suitable for employees who are more comfortable getting to know their functional activities on their own, using various tools.
6. Mentoring – attaching a more experienced employee with a similar work experience to a newcomer.
7. Budding – infusion of a new employee into the team through the establishment of friendly relations with the team.
8. Staff rotation – implies the movement of personnel within the same organization for different or similar positions, with a change not only in the position, but also in functional responsibilities.
9. Shadowing – a newcomer attaches himself to a more experienced employee of a subdivision or department and observes how he performs his functional duties, how to react in those or other situations and learn to be flexible in this position.
10. Immersion method – developed for leadership positions. Independent and intensive implementation of all their duties. The employee must quickly engage in work, establish communication with the employees of the unit and fulfill their immediate responsibilities.
11. Coaching – individual tuning with a coach that will help to reveal his professional qualities in the employee, achieve a clearly defined professional goal, teach the employee how to make decisions.

The most common adaptation methods in Russian organizations are mentoring and instructions. This is due to the fact that employers are poorly aware of the connection between the process of adaptation of employees and the company's profit. The main indicator of the successful passage of adaptation by a new employee is the fulfillment of his functional duties with maximum labor productivity. The optimal tool for monitoring and evaluating new employees is feedback on the work done, namely the following objective and subjective indicators. The objectives are the reduction of costs for attracting, searching and selecting new staff; and the decrease in the employee turnover rate in the organization for less than a year. Subjective results include:

- the formation of a positive HR brand outside and inside the organization;
- improving the moral and psychological climate in the team;
- the formation of loyalty to the organization of a new employee from the first days of his work;
- satisfaction with the company as a whole.

The following approaches to assess the effectiveness of the adaptation process.

1. Evaluation through staff satisfaction. It is implemented in the company by questioning new employees after the adaptation process. The quality and satisfaction with the process of adaptation of the employee and the difficulties faced by the employee are assessed with use of the questionnaire. Both a new employee and his immediate supervisor fill it. There indicators of "job satisfaction" and "employee satisfaction with the organization." The disadvantage of this approach is the inability to determine the impact of an aspect of adaptation on the success of the whole process.
2. Assessment of the adaptation process through the identification and development of employee performance indicators, as objective: professional (compliance with qualification requirements and place of work), socio-psychological (compliance of the new employee's behavior with the standards adopted in the organization), psychophysiological (fatigue, level of stress) and subjective - determine their own assessment of the employee's attitude to the organization, process and relations with the team and the chief. Both the new employee and his mentor or chief carry the assessment.
3. Integral assessment of the system for evaluating the effectiveness of adaptation programs not only in relation to a new employee and indicators of its effectiveness, but also the impact of this program on internal organizational processes.

The following criteria are used to analyze the level of adaptation:

- costs incurred by the employer for the full adaptation of the new employee, taking into account his profile of activity (necessary printed material and literature, trainings, training courses);
- the cost of working hours of employees responsible for effective adaptation (HR manager, responsible employee in the department and the time of the head of the new employee or specialist who will be the direct mentor);

- expenses for the training of mentors for new employees (measures to improve their professional level, development and creation of narrow-profile training programs, through which mentors will work with new employees);
- number of employees hired and successfully completed the probationary period;
- the percentage of employees who have successfully worked in the organization for at least one year after completing the adaptation;
- the number of employees who left voluntarily during the probationary period.

Qualitative criteria:

- assessment of satisfaction with the process of work in a new place;
- knowledge and understanding of the mission and values of the company, as well as the corporate culture of the organization;
- level of satisfaction with the socio-psychological climate in the company and in its department.

The normative indicators are also compared with the actual results of the new employee, for example, the performance of daily office tasks, namely the speed of their execution, quality and effectiveness, assessment of the sociability of the new employee, for example, if his/her main function is to work with customers, suppliers or other partners. The assessment occurs at regular intervals to evaluate the adaptation system. and professional growth of the employee.

For small businesses with a small number of employees, the following methods for evaluating the effectiveness of the adaptation system can be selected:

- expert – analysis of the presence of adaptation elements in the personnel management;
- economic – a comparison of the labor productivity of a newly hired employee and an employee who has worked for more than one year;
- sociological – a survey of employees on the degree of satisfaction with adaptation methods.

With an improvement in the performance of personnel movement and an increase in the stability of labor resources, the efficiency indicators of personnel use are significantly increased. The performance indicators of employees who work in the team for a long time are significantly higher than those of newcomers. Wages of employees on piecework salary directly depend on production, so, the current situation may affect the decision of a new employee to quit this job, which will negatively affect the efficiency of the enterprise.

The right approach in choosing adaptation tools is to analyze the existing situation in the company and the goals that the company faces and, specifically, the adaptation system in the organization. It is necessary to carefully consider which tools are worth using and which are not (target approach). Unfortunately, often organizations use two approaches while choosing adaptation tools: a problem-oriented approach (the problem is found, it needs to be solved quickly) and a process-oriented approach (as it should be, because others have it).

3 Results

The authors will evaluate the adaptation system for small enterprises using the example of the Chika beauty salon, located in the center of Samara. Chika LLC is a small business enterprise, therefore there is no separate service or personnel management officer, including adaptation processes at the enterprise. The personnel management structure in the cabin is two-level. At the first level, personnel management is carried out by the director of the salon, at the second level, personnel management of the beauty salon is carried out by the administrator, who is the organizer of all labor activities inside the beauty salon.

The most important and priority problems at the enterprise that need to be addressed immediately are:

- inefficiency of the existing mentoring system,
- lack of attention of management to employees,
- ignoring the socio-psychological side of relationships in the team;
- lack of criteria for assessing the effectiveness of the existing system of adaptation of new employees.

The salon is characterized by a fairly high staff turnover - 40% in 2018, despite the fact that in small enterprises in the service sector, turnover, as a rule, rarely exceeds 20%. The reasons for the dismissal were clarified. Dissatisfaction with wages accounts for the largest share in the reasons for dismissal (35%), which negatively characterizes the adaptation system at the enterprise, since in manual processes the productivity of an unadapted employee is much lower than the employee who joined the team. Other reasons for dismissal are directly related to the ineffective adaptation system of personnel in the salon – this is pressure from the team (16.7%) and lack of attention from the management (25%). Let us evaluate the effectiveness of the adaptation system according to a five-point system, and analyze the presence of the necessary types of adaptation (Table 1).

Table 1. Assessment of the personnel adaptation system of Chika LLC

Type of adaptation	Characteristic	Availability	Rating
Organizational	Organizational - familiarization with job responsibilities, the structure of the salon, hierarchy, management mechanisms; acquaintance with the immediate supervisor	Present	4
Socio-psychological	Socio-psychological - acquaintance with the salon and the team, employees, with the traditions that exist in the salon	Partially present	3

(*continued*)

Table 1. (*continued*)

Type of adaptation	Characteristic	Availability	Rating
Labor	Job training	Present	5
Extra education	Improving professional skills to a certain level required in this salon, familiarity with working on certain brands used in the salon	Absent	0
Mentoring	The practice necessary for a newcomer to work at the level of the rest of the masters of the salon, the opportunity to upgrade qualifications "by the chair"	Partially present	2
Financial	The employee must start making money and bring profit to the enterprise, therefore, the help of management is required for the client to develop a base for newcomer	Absent	0
Total			14
Average rating			2,33

Source: authors.

As can be seen from the above calculations, the assessment of the personnel adaptation system in the beauty salon shows very low efficiency – out of the selected six main types of personnel adaptation, the most important in the beauty industry, only two are fully executed, two are partially executed, and two are not performed at all. Evaluation of the adaptation system is 2.33 points, a very low level. To solve these problems, three following directions were proposed to increase the effectiveness of the adaptation system:

1. Development and implementation of a mentoring system.
2. Development and implementation of an adaptation program.
3. Evaluation of the effectiveness of adaptation measures.

Let's consider these areas in more detail.

I. Development and implementation of a mentoring system

In the beauty industry, the introduction of a mentoring system is inevitable in order to get an effective return on the work of salon staff. This is due to the fact that no matter how much and wherever did master study before starting work, he will definitely have questions about a brand new to him and about the features of work in this particular salon, and he will turn to his colleagues with them.

Forms of mentoring in a beauty salon can vary (from the least to the most costly in time and effort):

- tips on working with brand products,
- a novice observes the work of the mentor from the side and asks him questions (assists the mentor),

– mentor-teacher monitors and checks the work of a beginner at the first stage of adaptation in the cabin.

The period of mentoring (in practice) lasts two months or less. For effective feedback from the mentoring system, it is necessary to stimulate mentors, for example: fixed coaching bonus, mentoring bonus and increase of the bonus for services performed by the mentor himself during the period of mentoring.

II. Development and implementation of an adaptation program.

It includes the following steps:

1. Presentation of information about the organization and features of the labor process regulations. Acquaintance of candidates with their upcoming official duties, with the list and price list of services provided by the salon. Identification of financial aspects of future work.

When hiring a time wage employee, salary value during and after the probationary period and payment dates are announced. When hiring piecework employees, it is important to consider that people are afraid to come to a new place of work, realizing that it takes time to generate a large number of clients. During the period of adaptation, such employees must get a fixed salary plus a percentage of the volume of surrendered revenue. It would be logical to pay the adaptation salary until the interest that this master earns would be comparable to the earnings of other colleagues.

2. If the employee is not completely ready to start work, it is necessary to make a schedule for preparing for work, trying to implement it as soon as possible so that he does not lose motivation for a new job and he can survive the financial "downtime".
3. On the eve of the first day of work, familiarization with the organization and the workplace is carried out. It is carried out after registration of the employee for work by the way of excursion and includes instruction on the use of equipment, tools, materials.
4. An explanation of the requirements for conducting professional activities, personal behavior. It includes preparing a workplace, creating a load for beginners and advertising a new specialist among clients. The first 4 stages in total take about 1 week and are carried out until the beginner goes to work.
5. The leader draws up a clear plan of professional and financial development for a newcomer, designs the trainings and seminars that must be completed; describes the revenue plan, the goals of working with clients. Takes 3–4 weeks.
6. The implementation of an individual plan should be constantly monitored by management.

Throughout the entire adaptation period, the leader should include personal communication with a new employee in his weekly work schedule in order to discuss the results of work and ways to further promote the newcomer to the salon.

III. Evaluation of the effectiveness of adaptation measures

The profitability (loss-making) of a newcomer, the attitude of clients to him, collects information about whether the new specialist has taken root in the team, whether it needs some continuation of adaptation (for example, it would be nice for him to undergo additional training or change something in his work, relations with clients and colleagues) is assessed.

The result of the completion of adaptation can be considered:

- the fact that the employee did not quit during the probationary period and is not going to do so,
- consent of employer to keep it working,
- the employee makes a profit and is able to work independently, without a mentor.

The results of the calculation of economic efficiency are shown in Table 2.

Table 2. Cost-effectiveness of measures to adapt staff

Cost item	Number of persons	Expenses for 1 person, rub. in month	Cost, rub.
Expenses			
Salary of mentors for 2 months	3	2500	15000
Contributions to social funds	3	750	4500
Adaptive salary for new employees for 2 months	4	7000	56000
Training costs (duration - 1 month)	4	5000	20000
Unexpected costs			5000
Total costs			**96000**
Labor productivity growth			337000
Saving dismissal costs	1	30198	30198
Saving staff recruitment costs	1	14911	14911
Reception cost savings	1	46	46
Saving adaptation costs	1	9731	9731
Total additional profit			**391886**
Economical effect			**295886**

Source: authors.

The application of these measures from the experience of similar enterprises should give the following results: labor productivity growth is at least 20% of the current, and a decrease in staff turnover is 20%. The results obtained indicate the economic efficiency of the proposed measures.

4 Discussion

During the work, the literature on the issue of staff adaptation was analyzed. Makhmudova, Bikulova, and Eremina discuss the methodology for assessing the effectiveness of adaptation in companies as the main stage that each employee goes through when joining the organization [5]. The main task of working with staff is to retain an employee in the enterprise in order to unleash his/her potential and make a profit in the future. Of course, to achieve this goal, a new employee needs an adaptation period during which he will be able to master the main goals and objectives of the company. However, at the present stage of economic development, the leaders of many companies do not test methods for evaluating personnel adaptation programs, since its quality is difficult to express using quantitative indicators. The existing methodologies for assessing the effectiveness of personnel adaptation and the possibility of testing it in Russian realities are analyzed [5].

Sharapova, Sharapova, and Borisov consider the issue of mentoring as one of the methods of staff training [8]. The main advantages of mentoring for both a new employee and a mentor are considered. The article explains the duration of mentoring and the goals that a mentor must achieve for successful teaching. The importance of financial and non-financial motivation of mentor staff is also identified. The authors take into account indicators such as the number of mentors and the number of trainees. They calculate the workload of one mentor and take into account the number of employees who have been successfully trained through mentoring. The main advantages of mentoring for both a new employee and a mentor are considered. The article explains the duration of mentoring and the goals that a mentor must achieve for successful learning. The main advantages of mentoring both for a new employee and for a mentor are revealed. The importance of material and non-material incentives for mentors is determined. The stages of the mentoring process are discussed. The authors take into account indicators such as the number of mentors and the number of trainees. They calculate the workload of one mentor and take into account the number of employees who have been successfully trained through mentoring. For the training staff, the implementation of the proposed activities will allow to choose a training course within the agreed cost; improve student satisfaction and also motivates employees to further development and career growth. It is proposed to evaluate not only in terms of satisfaction with the mentoring system, but also by calculating the effectiveness of professional responsibilities. It is noted that for future research it will be useful to analyze the impact of the quality of mentors on improving the professional competence of the chief [8].

Bilorus notes that the rating indicator of the competitiveness of the personnel management system of an enterprise is the result of the interaction of such factors: the effectiveness of the motivation and incentive subsystem; the effectiveness of subsystems for the selection, adaptation and assessment of personnel; level of organizational and working conditions, social infrastructure of the enterprise; level of socio-psychological climate in the team; corporate culture and company image; the level of development of the subsystem of training, promotion and personnel development; level of development of the personnel planning and marketing subsystem. Thus, the use of the modified GE/McKinsey matrix model will allow company managers to formulate strategic recommendations for the development of the personnel management system and its competitive advantages intelligently (not intuitively), which would make it possible to remain on the market and increase the attractiveness and image of the company as an employer [1].

Sucre and Chirinos believe that because of constant changes in the environment the staff needs high motivation and qualification, in order to facilitate its adaptation to high variable environment, it is necessary to increase employee productivity by identifying potential that contributes to the formation of career plans and the creation of replacement tables. In the selection of personnel, it is necessary to process large volumes of scattered information that are not agreed upon or not standardized; this can also be interpreted in different ways. To this end, it is proposed to create a structure for managing work profiles, which provides a generalized solution for the exchange of information and services, interacts with business systems related to human talent management [7].

Popova and Shynkarenko consider the issue of maintaining staff competencies and skills that contribute to the process of adaptation to the world of volatility, uncertainty, complexity and ambiguity, is one of the mechanisms for ensuring the development of the enterprise. In accordance with modern trends in personnel management, some specific actions for the adaptation, development and training of personnel were proposed [6].

Blumberga and Treilina note that the competitiveness of modern global business depends on flexibility and adaptation skills. Work abroad has become relevant in recent years, when companies develop their activities at the international level. When moving to work abroad, each expatriate experiences a range of psychological changes in adaptation process. To reduce the number of unsuccessful international appointments, it is necessary to analyze existing processes, understand the weaknesses of the process and find out how to avoid or reduce these shortcomings. The purpose of the study is to find out what are the main shortcomings of the international appointment process in terms of expatriates and returnees in order to improve the processes of the global mobile team. The authors proposed two surveys: one for expatriates and returnees, the other for managers and HR specialists [2].

Klochkov, Didenko, Makov, Zapivahin, Ostapenko, and Volgina believe that professional adaptation of specialists occurs not only after the stages of employment, but also with changes in technology, regulations, and industrial culture. At the same time, forecasting the level of personnel resistance to changes will reduce additional costs, build an adequate system of professional development and prepare personnel for changes by determining the degree of personnel resistance to changes in the requirements for them [3].

Kolesnichenko, Radyukova, and Kolesnichenko note that the problem of staff adaptation is urgent due to a direct impact on the results of employees, but despite this, many organizations do not receive proper attention to staff adaptation, and the process occurs spontaneously. Many managers underestimate the importance of adaptation, believing that it can be spontaneous, do not want to spend available resources, or often partially adapt. Most organizations do not even have basic adaptation programs. However, how well the adaptation process is organized will depend on the employee fulfilling professional duties, the degree of his/her identification with the company, the adoption of its norms and values [4]. In general, the problem of staff adaptation at small enterprises in recent years has not been adequately addressed.

5 Conclusion

The head of enterprise directly carries personnel management functions at small enterprises. The costs associated with the replacement of the employee who quit during the adaptation period, which the enterprise would save when implementing the proposed adaptation system include:1) dismissal costs, 2) staff recruitment costs, 3) adaptation costs, 4) reception costs, 5) the costs associated with lost profits, because an adapted employee works with greater productivity than an unadapted employee. The last point plays the greater role, the greater the proportion of manual labor in the production process. At the same time, in hardware processes, its plays no role.

To implement the proposals for improving the personnel adaptation system, the following costs are required:

1. Salary and contributions to social funds of mentors for first 2 months.
2. Adaptive salary for new employees for 2 months.
3. Training costs.
4. Unexpected costs.

The main economic effect of the proposed measures to improve the process of staff adaptation in small business enterprise is to save money by organizing funds as a result of reducing financial losses associated with staff adaptation. Thus, the introduction of the proposed measures to improve the process of adaptation of employees in the organization will help to increase employee productivity, reduce staff turnover during the adaptation period, as well as costs associated with the adaptation of personnel and increase the stability of the team.

References

1. Bilorus, T.V.: Development strategy formation of the personnel management system of the enterprise based on portfolio analysis. Market. Manage. Innov. **1**, 189–195 (2018). https://doi.org/10.21272/mmi.2018.1-13
2. Blumberga, S., Treilina, D.: International assignment process in organization: expatriates and repatriates perspective. In: Proceedings of the SGEM 2016, BK2: Political Sciences, Law, Finance, Economic and Tourism Conference Proceedings, vol. 5, pp. 401–408. SGEM, New York (2016)
3. Klochkov, Y.S., Didenko, N.I., Makov, K.M., Zapivahin, I.O., Ostapenko, M.S., Volgina, A.D.: Employment and professional adaptation of specialists. In: Shestopalov, M. (ed.), Proceedings of the 6th Forum Strategic Partnership of Universities and Enterprises of Hi-Tech Branches, pp. 162–164. IEEE, New Jersey (2017)
4. Kolesnichenko, E., Radyukova, Y., Kolesnichenko, D.: Improving the management of employees adaptation in regional authorities. In: Ślusarczyk, B. (ed.), Proceedings of the 31St IBIMA Conference Innovation Management and Education Excellence Through Vision, pp. 1481–1487. IBIMA Publishing LLC, King of Prussia (2018)

5. Makhmudova, I.N., Bikulova, P.V., Eremina, I.Y.: Methodology of evaluation of the personnel adaptation effectiveness in companies. In: Mantulenko, V. (ed.) Proceedings of GCPMED 2018 – International Scientific Conference Global Challenges and Prospects of the Modern Economic Development. The European Proceedings of Social and Behavioral Sciences, vol. 57, pp. 1023–1031. Future Academy, London (2019). https://doi.org/10.15405/epsbs.2019.03.102
6. Popova, N., Shynkarenko, V.: Personnel development at enterprises with regard to adaptation to the VUCA world. Econ. Ann. XXI **156**(1–2), 88–91 (2016)
7. Sucre, F.S., Chirinos, D.P.: Ontological framework for the management of job profiles based on labor competencies. GECONTEC-Revista Internacional de Gestion del Conocimento y la Tecnologia **5**(1), 17–41 (2017)
8. Sharapova, N.V., Sharapova, V.M., Borisov, I.A.: Mentoring as a method of staff training. In: Ślusarczyk, B. (ed.) Proceedings of the 31St IBIMA Conference Innovation Management and Education Excellence Through Vision, pp. 3245–3251. IBIMA Publishing LLC, King of Prussia (2018)

New Perspective in Projecting University Graduates' Competencies: Talent Management

I. N. Makhmudova[1(✉)] and N. V. Kozhuhova[2]

[1] Samara National Research University, Samara, Russia
Mahmudova.I@yandex.ru
[2] Samara State University of Economics, Samara, Russia
kuka_55_@mail.ru

Abstract. The paper reveals an innovative approach to knowledge management in the development trend of a Cognitive University. Graduate's competencies are derived from changes in business processes in the organizations that are potential employers, taking into account the digital economy requirements. The authors of the study implemented the talent management approach, not the knowledge management one. The socio-psychological analysis of the components of students' professional and qualification skills was used. The terminological determinacy of the concepts of graduates' loyalty and reliability by the criterion of their participation in the potential business process was established. The approaches to measure the graduates' potential abilities were defined. The differences between highly professional (HiPro) and high-potential (HiPo) graduates were identified. The subject of responsibility for the timely and high-quality formation of new graduates' competencies under the conditions of new reality in employers' organizations was distinguished. The mechanism of HR-technologies, in which a high-potential graduate is not able to reveal their qualification level and reduces the organization's innovative development, was explicated. The role of staff involvement in the personnel reserve formation was determined.

Keywords: Activity · Competencies · High-potential potency · Highly professional potential · Loyalty · Talent management

1 Introduction

The implementation of digital technologies in the Russian economy has also influenced the learning process in higher education. According to the long-term strategy of economic development in the domestic economy, the national objective is to modernize the traditional sectors of economic growth and form the digital economy focused on knowledge and high technologies [3, 13]. Awareness of the importance to evaluate the graduates' potential abilities comes with the formation of new socio-economic relations in the society and the introduction of modern digital technologies in production and training. University graduates' professional competencies are also updated to meet modern development strategies in new organizations and their business ideas.

S. I. Ashmarina et al. (Eds.): ENGINEERING ECONOMICS WEEK 2020, LNNS 139, pp. 606–612, 2021.
https://doi.org/10.1007/978-3-030-53277-2_72

Solving these issues leads to a change in the concept paradigm from knowledge management to talent management, and to the formation and accumulation of students' knowledge capital. An innovative approach to knowledge management in the development trend of a Cognitive University is becoming relevant.

2 Methodology

The process to build the information and innovation economy obliges organizations to keep the most accurate records of the quantity and quality of available resources, as well as to redistribute them with maximum efficiency. However, recent years have been marked by a lack of experienced specialists, a sharp decrease in the efficiency of labour potential, and a large amount of raw data in the work. There is an increase in employees' dissatisfaction with work and turnover of labour and, as a result, a sharp decrease in staff loyalty in the organization.

Such conditions do not contribute to the effective development of organizations. On the contrary, they reduce their effectiveness and profitability. In this regard, business leaders cautiously select applicants for vacant jobs.

Consequently, educational organizations are faced with the task to train a potentially in-demand labour. A new approach to the knowledge component in the formation of graduates' professional and qualification skills is required. The entire training system should be adapted to the graduate's potential demand to work in renewed organizations. Similarly, all the components of the graduate's qualification level, the training content, and technology through the use of advanced cloud solutions become more complicated.

Today, it can be noted that the task to provide enterprises with highly professional personnel has failed. The objective reason is that it is difficult to prepare a specialist for the labour market, who has the necessary qualification corresponding to modern digital technologies. This is due to the fact that the required competencies are not precisely defined, either by the employers or academic staff. Therefore, an innovative approach to build the graduate's qualification level in accordance with the business objectives is required. The main condition to provide an organization with high-end professionals at the right time and in the right place is to create a coherent and logical talent management system. Today, the tasks to motivate employees are shifting towards providing greater opportunities for self-realization, professional growth, and career development within the employing company. According to KPMG research, efficiency improvement of organizations' activity in Russia is becoming a higher priority, which is achieved by increasing employees' engagement (55%) and talent management (51%) [2].

3 Results

Building such a complete and transparent talent management strategy, integrated into all elements of the HR management system, is impossible without terminological determinancy.

Let us try to figure out what kind of employee a modern organization needs. To do this, we need to decide who can be called high-potential employees (HiPo)? How do they differ from highly professional (HiPro) staff? Who has a higher degree of labour activity and/or is more correlative with the strategy of the organization's modern business process? Can they be included in the personnel reserve? How reliable will they be? And what loyalty level should we expect?

The issue of loyalty and reliability is not circumstantial. The level of staff loyalty is closely related to the employee turnover rate. Employee's loyalty is characterized by the staff personal positive attitude to the company's activities, policies, and management. The basis of the loyal employees' behavior is their desire to benefit the company and avoid actions that can damage it. The highest degree of employees' loyalty is when they treat the company's affairs as their own business. Loyal employees, on the one hand, deeply accept some company requirements, but on the other hand, they are ready to put up with others that do not quite suit them [1].

Loyal staff should not be confused with reliable personnel.

Staff reliability reveals the degree of behavior standards in relation to the company. Employees try to respect all rules and regulations established in the company. But in case of unfair treatment by the management or discordance between their personal interests and the interests of the company, they are able to commit an illegal action against the company [6]. In fact, reliable personnel are passive to demonstrate their labour potential. They assume a wait-and-see attitude to evaluate the benefits that are provided at their workplace by the company management. Such an employee can be classified as a highly professional (HiPro) staff.

Under the conditions of technological changes, when we have to look for innovative approaches and operational procedures, deal with raw data, master modern mechanisms and technologies that require an initiative attitude when performing tasks, respect the requirements and standards of labour discipline, and treat the company's policy positively, this means to slow down innovative changes. A highly professional (HiPro) employee uses past experience, which, in the ideal case, may not fit into a new relationship, and in the worst case, it can spoil innovative initiatives.

Loyal staff may not be included in the category of HiPro employees. However, due to high motivation, they are HiPo employees, who are ready for any tasks in new reality. They have inquisitive mind and are ready to search for non-standard solutions, because they are not burdened with experience and imposed solutions (rules and outdated standards). They are proactive to find solutions that results in readiness to actively progress, master and use modern technologies in their work, and get a positive result for the organization.

Respectively, projecting the situation to the educational process, we can conclude that the formation of the students' perspective teaching trajectory in high school requires graduates' enthusiasm in such restricted professional actions as accomplishment of calculation tasks, accurate compliance with the operating procedure requirements, knowledge of professional work algorithms, etc.

At the same time, a high motivation level is an important condition for being a HiPo employee. To stimulate activity and initiative attitude to work, it is necessary to bring the HiPo employee to the partnership and co-management level in the organization, and increase the engagement rate. They should understand their personal benefit not in the form of one-time compensation/bonus, but in the form of the prospects for their personal development in this organization. That is not to become a nominal, but a real personnel reserve in accordance with the performance evaluation, which reflects the achievement level of the employee's business objectives. To achieve this task, the university graduate should gain an active personal responsibility to build their professional competencies, which are necessary to work in a new environment.

With a focus on HiPo employees, the higher school should be guided by an active training in the format of practice-oriented distance e-learning majors courses, in order not to form an "effective" training group. The course may be in demand by individual students who are deeply interested in certain aspects of the field-specific knowledge. Checking the quality of such students training is possible through the use of the testing assignments method. These assignments should be developed in consort with organizations that are potential employers, in which students plan to do their work placement or internship/working activity under contracts. The main requirement is to match two sources: Job success profile and Skill-matrix, which provide a detailed description of the development level of professional skills for a specific job position.

It is necessary to plan student progress development guide (PDG), based on an electronic document (on SAP platform), which allows the student independently and/or with the tutor support conducts self-assessments (according to the competency model developed by the employing company), and elaborate the plan of their future profession-oriented development. When evaluating the professional level of student's competencies, we can follow two criteria: sustained performance (stability of work results) and potential level (from 1 level of non-compliance to 3 maximum). As a result, the graduate will be assigned to a certain category, according to the talent development level.

However, it is important to maintain a balance when involving an employee in self-management and self-assessment processes. The student's activity in the process of getting involved in the profession can decrease dramatically in the case of overcharge, fatigue, lack of assessment of the results achieved, and loss of their own growth prospects. A student may experience occupational burnout and dissatisfaction with the training process and results. Loyalty to their future profession with the loss of their activity transforms the student's qualities in favor of reliability, calmness, and lack of initiative. If the occupational burnout process intensifies, such graduates demonstrate absolute lack of initiative and desire to study. They stop attending classes and, most often change training to work, but not according to their specialty profile. That is, a student either looks for a new place to apply their abilities and the university becomes the primitive mechanism to produce graduates, or goes to the underground (changes the sphere of professional activity), that means that this student leaves the university or changes to another specialty, while still hoping that this change can make a difference in the life.

4 Discussion

The assessment of the required competence level is recommended in accordance with the content of the SHL HiPo formula, presented above, to evaluate potential. The authors [2, 4–12] propose to include only three components, which we support and determine in the study. They are:

1. Desire to achievements (potential to take a managerial position).
2. Abilities (Hogan, Talent Q questionnaires).
3. Engagement (potential assessment, PiF test and Potential in Focus).

At the same time, it is much more important not just to assess the graduate's potential, but, based on the assessment results, bring information about their enrollment in a particular group at the HiPo level when applying for employment. This is an indicator of the formation of students' potential and the graduates' qualification level, recognition of their potential performance in work, and success in implementing the personal development trajectory in the company. Enrollment in a specific HiPo group gives an idea of the graduate's potential enrollment in the personnel reserve of the employing company and the personal advantages that may be available to a particular graduate – job applicant in a potential employing company. In this context, the motivational interest factor determines the level of students' responsibility in the development of their own professional and qualification level.

The main problem in the functioning of this system is that both the university management and the student, due to different circumstances, can accelerate this process. This has a negative impact on the students' well-being. They are overcharged with tasks, power and, sooner or later, they will experience an occupational burnout. This leads to dissatisfaction, disappointment in the chosen profession, and a loyalty decrease in the chosen training area. The student loses the valuable HiPro quality, that is their quality characteristic, even while remaining a high professional.

The developed technology, innovative approaches to build and use the students' potential, and the graduates' qualification level in new environment are based on the analysis and comparison of research results, published for public use in the Internet environment. The authors research to identify specific HiPo and HiPro competencies that correspond to digital technologies [4, 5, 9, 10] was also used as a basis to solve the tasks. The proprietary technology of competencies evaluation to measure the labour activity level and identify new trends and patterns of building competencies of end-to-end technologies in the implementation of labour potential was used [4].

5 Conclusion

For the foregoing reasons, we can highlight several important points. Speaking about the students' potential and university graduates' qualification level, it is necessary to keep in mind not only a set of certain knowledge and skills for performing professional duties, but the level of their active and initiative use by the students in order to solve the tasks.

The activity level is achieved by providing the opportunity to engage in self-development, using advanced cloud technologies (in the format of online e-learning courses). Owing to these courses, a university graduate can be involved in the co-management in the organization after employment. At the same time, the graduates must be extremely goal-oriented, understand their role and the level of participation in the business process in a potential employing company, which is determined by the relevant competencies formation.

The type and list of current competencies is determined and approved by the management of the employer's organization in coordination with leading professors of higher education for each specific business task. Thus, it is important to the management and teaching staff of the university not only to provide a cumulative system for building professional qualities of a HiPo student in accordance with new technologies of personnel management and the business environment. It is also necessary to identify indicators, which allow to timely determine the loss moment, occupational burnout, lower level of loyalty and student's initiative activity to solve problems. All this will result in revealing the graduate's qualification level when using it in a potential employing company.

References

1. Borzunov, A.A.: The importance of the personnel direction to ensure the company's economic security under the conditions of digital transformation. Vector economy. Economics and Enterprise Management 7 (2017). http://www.vectoreconomy.ru/index.php/number7/economicmangment-7-2017. Accessed 09 Mar 2020
2. Business Switzerland: World economic forum (2016). https://business-swiss.ch/mezhdunarodny-e-organizatsii/vsemirny-j-e-konomicheskij-forum-ve-f/. Accessed 12 Dec 2019
3. Federal Law: On independent assessment of qualifications, dated on 3 July 2016, No 238-FL. http://www.consultant.ru/document/cons_doc_LAW_200485/. Accessed 09 Mar 2020
4. Kifa, L.L., Kryuchkova, I.V.: On the issue of staff development trends. Azimuth Sci. Res. Econ. Manage. **4**(17), 175–177 (2016)
5. Laktionova, K.: Potential of employees: Illusiveness and reality. Pro-Personal (2015). https://www.pro-personal.ru/news/1086977-19-m8-zarplatu-nado-perechislyat-po-novym-pravilam-inache-shtraf-100-tysyach. Accessed 09 Mar 2020
6. Mahmudova, I.N., Krivoruchko, Y.: From reliability to staff loyalty to provide personnel security of the organization. Kadrovik **7**, 93–100 (2019)
7. Mahmudova, I.N., Zhuchenko, S.I.: Unclaimed employment in the digital economy. In: Gulyaev, G.U. (ed.), Vth International Scientific and Practical Conference on Economic and Management in the XXI Century: Strategies for Sustainable Development, pp. 84–88. MCNS «Science and Education», Penza (2018)
8. Mahmudova, I.N., Ilyukhina, L.A., Bogatyryova, I.V.: The corporate personnel training system development: the experience of Russian enterprises. In: Mantulenko, V. (ed.) International Scientific Conference on Global Challenges and Prospects of the Modern Economic Development. The European Proceedings of Social & Behavioural Sciences, vol. 57, pp. 654–655. Future Academy, London (2018). https://doi.org/10.15405/epsbs.2019.03.65
9. Makhmudova, I., Belonogova, E.: Using digital technologies in staff training. Slovak Int. Sci. J. **1**(16), 19–22 (2018)

10. Makhmudova, I.N.: Methodology for assessing the staff competence to determine compliance with the professional standard. Sci. Rev. Theory Pract. **1**, 59–74 (2017)
11. Makhmudova, I.N., Novoselova, O.V.: Evaluation of the reliability of a business partner in an updated reality. In: Soloviev, D.B. (ed.) International Multi-Conference on Industrial Engineering and Modern Technologies “Far East Con”, pp. 1200–1204. Atlantis Press, Paris (2019). https://doi.org/10.2991/ISCFEC-18.2019.269
12. Makhmudova, I.N., Bikulova, P.V., Eremina, I.Y.: Methodology of evaluation of the personnel adaptation effectiveness in companies. In: Mantulenko, V. (ed.) Proceedings of the International Scientific Conference on Global Challenges and Prospects of the Modern Economic Development. The European Proceedings of Social & Behavioural Sciences, vol. 57, pp. 1023–1031. Future Academy, London (2019). https://doi.org/10.15405/epsbs.2019.03.102
13. National Program: “Digital economy of the Russian Federation” approved 24 December 2018. http://base.garant.ru/72190282/. Accessed 09 Mar 2020

Inevitability of Company's Corporate Culture Transformation Under Conditions of New Management Technologies

V. M. Svistunov(✉), G. P. Kuzina, and V. V. Lobachev

State University of Management, Moscow, Russia
{svistunov, vvl}@guu.ru, gpkuzina2009@yandex.ru

Abstract. The purpose of this article is to establish the relation between the level of employee's satisfaction and the achieved digitalization level of the company. At the same time, job satisfaction is considered as an important factor in the formation and development of corporate culture. The authors analyze a problem of decreasing the level of job satisfaction in the context of internal organizational changes occurring in the company as part of its digital transformation. The problem is that the creative component and motivational attitudes of employees decrease with the increasing use of modern information technology tools. In the context of digitalization, the development of an effective strategy for interaction between the company's top management and its employees is largely subject to the following chain of criteria: the automation level of business processes – the degree of satisfaction with working conditions by the staff – in the corporate culture of the company.

Keywords: Automation level · Corporate culture · Company employees · Digital technologies · Job satisfaction

1 Introduction

Changes in the economy associated with the transition to a new technological order are invariably accompanied by a serious internal organizational transformation. The active process of digitalization of business processes for most domestic companies is accompanied by a growing number of corporate tasks that are solved using modern information and communication technologies [8]. This leads to the need to form specialized services and divisions within the organizational management structure. While investigating processes of digital transformation of companies, the authors of the article focus on significant changes simultaneously occurring in their corporate culture [4].

Today, hardly anyone doubts that most companies have their own corporate culture. At the same time, corporate culture is not only values and beliefs, rituals and symbols, artifacts and heroes, internal rules and regulations, but also the loyalty of the staff, the employees' satisfaction with the content, characteristics and results of their work. This is something that is valuable to most employees, which determines their behavior, world outlook and motivation to achieve their goals. Corporate culture is a subtle category that does not always lie on the surface. To a greater extent, we observe

S. I. Ashmarina et al. (Eds.): ENGINEERING ECONOMICS WEEK 2020, LNNS 139, pp. 613–623, 2021.
https://doi.org/10.1007/978-3-030-53277-2_73

the manifestation of corporate culture in the behavior and actions of employees, rather than in the declared appeals and internal regulations.

2 Methodology

The authors have developed a model for assessing the current level of employees' satisfaction with the job/work and the level of digital transformation achieved by the company. At the same time, the authors make two assumptions: the level of digital transformation is determined based on the results achieved by the company in automating the most important information, production and technological business processes; the level of job satisfaction is considered as an important component of the corporate culture.

Many factors have a significant impact on the stability and development direction of the corporate culture. Before the beginning of the XXI century, an important element in the formation of corporate culture was the material base of the company. It was the basis of reproduction processes and largely determined the economic efficiency of the company's operation. The more complex and expensive the material base of the company was, the more professionally trained, qualified, responsible and disciplined employees it needed, the more meaningful and creative their work was. High and strict requirements for the personnel selection made it less painful for employees to adapt and their further professional development more effective [2].

In the conditions of transition to the stage of mass digitalization, both at the state and internal levels, there is a change in the basic element of the corporate culture. Researchers increasingly consider the level of automation achieved by the company (Levels of Automation, hereinafter LOA) to be the basic factor in the development of the corporate culture in the conditions of digitalization of technical, technological, financial, economic and information processes implemented at the strategic, tactical or operational levels. LOA is a quantitative and qualitative description of administrative, industrial-technological, information-analytical, documentary and other operations, in whole or in part made by an independent from a person automated complex (during operation), as well as by a dependent on corporate information system (hereinafter CIS).

Today, there is no global standard or unified method for evaluating the achieved level of automation. In practice, specialists most often use quantitative and qualitative indicators and characteristics: the number of computers, in terms of a certain number of employees; the provision level of production, technological and management processes with modern software and hardware; the share of manual, mechanized and automated operations; the degree of reduction (as a percentage) of direct human participation in management, information, production and technological processes; the dependence of automated processes performed on the operator, etc.

Russian researchers use the criterion of the degree of human participation in decision-making processes or in the direct execution of operations to evaluate LOA. According to this approach to the LOA assessment, business processes can be classified as: manual, automated (mechanized) and automatic (robotic).

In the practical work, the Sheridan & Verplank method is often used, based on the assessment of the level of independence from a person of performed production, technological, managerial and other operations. This approach makes it possible to present a formalized description of the ten levels of achieved LOA [7]. Each level is a numeric expression of the automation degree of operations performed by the employee, including the decision whether to perform a particular operation. The minimum level 1 corresponds to a situation when all operations in the company are performed manually by employees without any exception. The maximum level 10 excludes human participation in any production, technological, managerial, information and other operations. The other levels are called intermediate. Table 1 gives a formalized description of all LOA levels.

Table 1. Description of levels of automation (LOA) according to the Sheridan and Verplank method

Levels of automation (LOA)	Formalized description of the automation level
1	Computer assistance is not expected. All operations, including deciding whether to perform each of them separately, are performed by the employee independently
2	The computer is used to develop the maximum number of possible (alternative) solutions. All operations, including deciding whether to perform each of them separately, are performed by the employee independently
3	The computer develops possible alternative solutions or generates a complete list of solutions/actions; selects several most appropriate alternatives. The employee makes a decision about the operations to be performed
4	The computer develops possible alternative solutions or generates a complete list of solutions/actions; selects the only economically (technically, technologically, etc.) feasible alternative. The decision to perform the operation is made by the employee
5	The computer generates one appropriate decision and executes it only after receiving confirmation from the employee
6	The computer generates one appropriate decision and executes it if the employee, within a pre-defined time corridor, does not reject the automated execution of operations
7	The computer generates one appropriate solution and executes it in automated mode, necessarily informing the employee about the progress of the performed task. The employee retains the function of full control over the progress of operations (task solutions)
8	The computer generates one appropriate solution and executes it in automated mode, informing the employee about the progress of the task if the corresponding request is received. The employee partially controls the progress of operations
9	The computer forms an appropriate solution and performs it in automated mode. The employee is informed about deviations in the course of operations (task solutions). The employee performs the final control function, evaluating the result
10	Fully automated task solving cycle that excludes employee participation in any of the stages

Source: authors based on [7].

Analysis of the formalized description of LOA levels in the Sheridan and Weplank methodology shows that tasks assigned to the 6–10 levels of automation can and should be performed without the participation or with the minimum possible participation of company's employees. The more employees have to deal with these LOA levels as part of their job responsibilities, the more clearly two divergent trends become apparent in the company. The first is manifested in a decrease in the creative component of employees' work, reducing the number of communication contacts that they enter into; their professional knowledge and experience become less in demand. In other words, companies with a high level of LOA are characterized by: less meaningful work activities of employees; a lower need for creative employees; not a high need for managers with high leadership qualities. The second trend is characterized by a high need for specialists who are able to make non-standard decisions in critical situations; providing uninterrupted operation of computer systems and complexes, including the development of their own or adaptation of specialized software; having professional knowledge that ensures high-quality expertise of modern information technologies presented on the market. This trend indicates the need of the company's management and HR department to solve a complex structural problem. Its essence is to reduce or retrain personnel whose functions are sufficiently or fully automated. This is a task to search for personnel who can work efficiently in a highly automated environment [1].

Today, the company's achievement of even level 4 of LOA sets fundamentally different tasks for its personnel department in the processes of search, selection or retraining of staff [5]. Achieving 6 or higher LOA levels is not only time-consuming, but also quite expensive and requires a long time to implement changes. This goal will not be achieved without the involvement of specialists with a high level of specialized professional knowledge and extensive practical experience in the development, implementation and operation of multifunctional and multi-user software [11].

What in the conditions of a sharp decrease in the meaningful content of work can "force" an employee to effectively perform an ever-growing volume of low-content administrative and control functions? What can contribute to maintaining a unified corporate communication space? To get answers to these questions, we analyzed the level of employees' satisfaction with their work in companies with different LOA levels. Generalization and theoretical understanding of data collected as a result of the survey of experts allowed: to identify material and non-material factors that significantly affect the formation of the level of labor satisfaction; to determine the direction of influence of various factors; to assess which factors are relevant in the transition to digital management technologies, and which are most effective in the traditional management.

In the study, the proposed evaluation model is applied to employees of two companies that have fundamentally different achieved LOA levels. The authors made a methodological assumption that at the research period the achieved LOA of the first company should not exceed level 3, and the second company should have an advanced LOA level – no less than level 5. Methods used in the research: theoretical, mathematical, methods of comparison and generalization, survey, testing, expert assessments and comparative analysis.

3 Results

The authors analyzed opinions of experts representing two domestic companies that have been operating in Moscow for more than 10 years. The first company, with the conditional name A, has been actively using a specialized multi-user ERP system in its management practice for the past two years. The implementation of this system began in 2015 and took three years. The second company, with the conditional name B, has been using two local standard software platforms for six years, which provide automation of the main accounting and management tasks, taking into account corporate functional and industry specifics. Each software product has limitations for taking into account individual characteristics of the company's business and further development its functional and hardware capabilities. Table 2 provides a description of the personnel structure of companies A and B.

Table 2. Characteristics of the personnel structure of companies A and B

Indicator	Company A		Company B	
	People	%	People	%
Total number of employees	258	100	276	100
Those from them who have:				
Higher education	153	59,30	145	52,54
Secondary vocational education	75	29,06	69	25,00
Secondary general education	30	11,63	62	22,46
Education in the field of development and operation of IT systems and technologies	28	10,85	7	2,54
Number of employees: At the age under 35	73	28,30	53	19,20
At the age from 35 to 60	139	53,88	136	49,28
Over the age of 60	46	17,83	87	31,52
With more than 8 years of experience in the company	160	62,02	194	70,29
With less than 2 years of experience in the company	57	22,09	32	11,59
Average staff turnover over the past 3 years, %		21,86		7,00

Source: authors.

The total number of employees in company B is 18 people more than in company A. Employees with higher education in company A are 6.76% more than in company B; with the secondary vocational education – 4.06% more; with the secondary general education – 10.83% less. This data indicates a higher educational level of employees of company A. Employees of company A are characterized by a younger age. The number of employees who have worked for more than 8 years in company A is 62.02%, and in company B is 70.29%. The data presented correspond to the average turnover rate for the last 3 years (company A – 21.86%, company B – 7.0%).

An important distinctive feature of company A is a significant number of specialists with specialized education in the field of IT (10.85% vs. 2.54% in company B).

Company A is also characterized by the presence in the staff list of the following positions: programmer, system analyst, system administrator, information security specialist, database administrator. Company B's staff includes the following positions: PC operator, programmer, and system administrator.

In the course of the study, 186 people were interviewed, employees of both companies, representing the following categories of experts: senior managers (2% of the total number of experts); 24% – middle managers and 19% – lower managers, 52% – specialists and only 3% – workers. The sample includes deputy directors of companies, heads of divisions, departments and services, line managers, members of project teams and groups, specialists and workers who use corporate information technologies and software products in their daily activities that partially or more automate the work or operations performed. Men in the presented sample made up 64%, women – 36%. By age, the sample structure is as follows: 21 years old – the age of the youngest expert and 67 years old – the age of the oldest one. The average age of the expert group is 53 years old.

The assessment of the LOA level achieved by the companies was based on information collected using the open-question interview method. In their responses, the experts described the level of automation of tasks and operations performed at their workplaces. When preparing their responses, the experts were not influenced by external factors. The responses that were formulated did not depend on the influence of the proposed response options and their sequence. The interview results were subjected to qualitative and quantitative content analysis. The final results were used to calculate the level of LOA using the Sheridan and Verplank methodology. The results of the analysis are presented in Table 3.

Table 3. Characteristics of the achieved LOA level in companies A and B

Indicator	Company A	Company B
Percentage of employees who use computer technology to perform their job duties, %	94	79
Percentage of employees using application software, %	100	100
Percentage of employees using functional software, %	74	48
Percentage of employees using tool software, %	12	7
Percentage of employees using individual software,%	19	48
Percentage of employees using group software, %	17	35
Percentage of employees using the network software, %	64	17
Percentage of employees using general-purpose software, %	14	57
Percentage of employees using method-oriented software, %	24	11
Percentage of employees using problem-oriented software, %	62	32
Percentage of employees who use standard platforms to perform their job duties, %	21	37
Percentage of employees using a ERP-system to perform their job duties, %	62	–
Assessment of the achieved LOA	**6**	**3**

Source: authors.

The evaluation of the achieved LOA in the general and partial assessment was based on a comparative analysis of responses received from experts from both companies. The comparative analysis was carried out within the range of maximum and minimum limits of automation of the main set of functional tasks solved by the company's management in current conditions. Thus, the achieved level 6 (LOA) is confirmed by the survey results, according to which 64% of employees in company A use a specialized multi-user ERP-system at all levels of the corporate hierarchy for: information support of implemented management processes; development and implementation of management decisions; formation and development of a single corporate data bank. At the same time, time corridors are regulated and strictly observed for performing the corresponding computing and other information operations in automated mode, which can only be changed by employees who have the appropriate software permission.

The achieved level 3 (LOA) is confirmed by the survey results, according to which only 37% of company B' employees use standard software platforms to perform their job duties. The unified software and information space is available to 17% of employees. Standard software platforms are used exclusively for informational support of the process of developing possible alternative management solutions at the operational and partially tactical levels of the corporate management. Management decision-making processes are carried out without the active use of digital technologies. The majority of employees (57%) need general-purpose software used to automate the simplest functional tasks. This data is evidence that the company B's employees have little access to modern digital technologies.

The labor satisfaction index was calculated based on the survey results of the same 186 experts who participated in the study of the LOA achieved by companies. The authors used a questionnaire containing a single basic set of indicators for assessing the level of job satisfaction for companies. The experts were asked to answer two questions. First of all, it was necessary to determine the influence degree of each of the basic indicators on the overall satisfaction of experts with the results of their work. The second question was intended to help to assess the expert's satisfaction with each of their indicators. Each basic indicator was evaluated on a five-point scale.

The experts were not able to change the list of basic indicators. The fact is that less important or unimportant (for an expert) indicators that have little effect on the satisfaction index are not able to fully reflect the overall level of their satisfaction with working conditions and results. The satisfaction index of each expert was defined as the arithmetic value of the satisfaction indices for each indicator. The assessment of the employee's overall satisfaction with work was conducted separately for each company and calculated as the arithmetic average of the satisfaction indicators of all experts representing the respective company. The final results of the study are presented in Table 4.

Table 4. The achieved level of job satisfaction in companies A and B

Indicator	Company A (LOA ↑)	Company B (LOA ↓)
Interest in work	2,78	3,14
Satisfaction with work achievements	3,23	4,11
Satisfaction with relationships between employees	2,14	3,96
Satisfaction with the relationship with management	4,19	3,02
The level of demand for professional knowledge	2,94	4,15
The level of demand for professional skills and experience	2,01	4,23
General professional demand	2,11	4,17
Responsibility for the work performed	3,76	4,31
The level of creative component in the work	1,98	3,16
Preference for the work performed to high earnings	1,89	3,45
Satisfaction with the overall content of the work performed	3,67	3,21
Possibility of career growth or promotion	2,33	2,88
Satisfaction with the current state of corporate culture	3,16	3,98
Outlooks (forecasts) on the development of corporate culture	3,43	4,76
The overall level of job satisfaction	**2,83**	**3,76**

Source: authors.

The achieved level of satisfaction is characterized not only by the technical and functional content, but also by the social and psychological comfort of working in the company. Table 5 shows the indicators that have the greatest impact on the employee's satisfaction in companies with fundamentally different levels of automation.

Table 5. The main indicators that affect the level of the employee's satisfaction with the work in companies with different LOA levels

Company A (LOA ↑)	Company B (LOA ↓)
Satisfaction with the relationship with management	Development prospects of the corporate culture
Responsibility for the work performed	The level of demand for professional skills and experience
Satisfaction with the overall content of the work performed	General professional demand

Source: authors.

Table 4 and 5 data show: the multidirectional influence of indicators that determine the level of employee's satisfaction with work in companies with different LOA; a significantly lower overall level of job satisfaction among employees of companies with a significantly higher LOA.

4 Discussion

Automation of business processes, which occurs both within a particular company and in its external environment, is a significant factor that affects the quantitative and qualitative parameters of the corporate culture, regularities of its effective use and development. Serious structural, organizational, content and personnel changes taking place in companies that are on the path of active digitalization are increasingly forcing specialists to discuss what motivational factors in these conditions will be the determining factors for their employees. In this case, it is not only about achieving a higher LOA and increasing production results, but also about maintaining a unified and effective corporate communication space.

Increasingly, researchers are turning to the tools of corporate culture theory in search of an answer to the question of the influence degree of the achieved level of business process automation on the stability and viability of the company's key values system. This question is important and relevant. Often, managers, when making digital changes in the company, encounter a decrease in the level of satisfaction with their work among their employees. Decline processes are observed both in companies that have been characterized by a stable corporate culture for many years, and in companies whose internal culture has not yet been fully formed. This state of affairs is not acceptable. Without the support and understanding of employees, digital changes implemented in the company are doomed to failure [9]. In the context of digital transformation, high production results often depend not on the quality and quantity of employees' labor, not on the number of high-tech technical means used, but on the balanced operation of corporate information systems. At the same time, a large number of employees of the company whose functionality has been successfully automated, demonstrate a serious decrease in the creative component in their professional activities. Employees face: a significant increase in the cost of working time to perform administrative and control functions; reduced diversity in the functions and work performed, independence and responsibility in decisions made. All this negatively affects the level of job satisfaction and staff confidence in the changes being made [7].

Thus, job satisfaction, in the context of total digitalization, should be recognized as an important tool for consolidating the potential of departments and employees in the process of achieving the company's strategic goals. Job satisfaction should be considered as an integral part of: the corporate management philosophy; the processes of formation, functioning and development of an effective corporate social and economic environment. Its main goal is to increase the motivational factors of labor and organizational efficiency of internal corporate processes; formation, preservation and multiplication of basic corporate values. On the one hand, the organization of work to increase the level of job satisfaction assumes and contributes to the maximum return of each employee of the company in accordance with their education, work experience, qualifications, formed personal qualities. On the other hand, job satisfaction should be based as much as possible on the functions that employees are able to perform effectively: they have the necessary professional knowledge and experience, and appropriate personal qualities; sharing not only basic corporate values, beliefs, norms, and rules, but also properly motivated to maximize their own and corporate potential [3, 6, 10].

5 Conclusion

Increasing the company's competitiveness in the Russian and foreign markets today is impossible without the use of modern information and communication technologies. In these circumstances, company management should pay attention to the changing trend of effective motivation of their employees. The previous trend, which was based on the development of corporate culture and demand for an employee as a professional, was replaced by a trend of fairly strict regulation as an integral part of the business process automation.

Labor ceases to carry elements of creativity, its content and attractiveness decrease. All this leads to a significant decrease in the level of employee's satisfaction with their professional activities in companies with a high level of business process automation. Subordination to the power, strict regulations and automation protocols is a way not only to reduce motivation, but also to psychological or social alienation of employees. In these conditions, the readiness and ability of the company's top management to develop and implement a fundamentally new strategy of corporate culture becomes important.

References

1. Astafeva, O.V., Pecherskaya, E.P., Tarasova, T.M., Korobejnikova, E.V.: Digital transformation in the management of contemporary organizations. In: Ashmarina, S.I., Vochozka, M., Mantulenko, V.V. (eds.) Digital Age: Chances, Challenges and Future. Lecture Notes in Networks and Systems, vol. 84, pp. 382–389. Springer, Cham (2020). https://doi.org/10.1007/978-3-030-27015-5_46
2. Kot, M.K., Spanagel, F.F., Belozerova, O.A.: Problems of digital technologies using in employment and employment relations. In: Ashmarina, S., Mesquita, A., Vochozka, M. (eds.) Digital Transformation of the Economy: Challenges, Trends and New Opportunities. Advances in Intelligent Systems and Computing, vol. 908, pp. 227–234. Springer, Cham (2020). https://doi.org/10.1007/978-3-030-11367-4_21
3. Kraev, V.M., Tikhonov, A.I.: Risk management in human resource management. TEM J. Technol. Educ. Manag. Inf. **8**(4), 1185–1190 (2019). https://doi.org/10.18421/TEM84-11
4. Mitrofanova, E.A., Bogatyreva, I.V., Tarasenko, V.V.: Digital reality and perspective of the management of educational organizations. In: Ashmarina, S.I., Vochozka, M., Mantulenko, V.V. (eds.) Digital Age: Chances, Challenges and Future. Lecture Notes in Networks and Systems, vol. 84, pp. 390–397. Springer, Cham (2020). https://doi.org/10.1007/978-3-030-27015-5_47

5. Mitrofanova, E.A., Konovalova, V.G., Mitrofanova, A.E.: Opportunities, problems and limitations of digital transformation of HR management. In: Mantulenko, V. (ed) Proceedings of the GCPMED 2018 – International Scientific Conference on Global Challenges and Prospects of the Modern Economic Development. The European Proceedings of Social & Behavioural Sciences, vol. 57, pp. 1717–1727 (2019). Future Academy, London. https://doi.org/10.15405/epsbs.2019.03.174
6. Schekoldin, V.A., Bogatyreva, I.V., Ilyukhina, L.A.: Digitalization of labor regulation management: new forms and content. In: Ashmarina, S.I., Vochozka, M., Mantulenko, V.V. (eds.) Digital Age: Chances, Challenges and Future. Lecture Notes in Networks and Systems, vol. 84, pp. 137–143. Springer, Cham (2020). https://doi.org/10.1007/978-3-030-27015-5_17
7. Sheridan, T.B., Verplank, W.L.: Human and computer control of undersea teleoperators (1978). https://www.semanticscholar.org/paper/Human-and-Computer-Control-of-Undersea-Sheridan-Verplank/d48b94e6af5093e7cc41e20fa6aca4f3a2d860bb. Accessed 20 Mar 2020
8. Svistunov, V.M., Lobachev, V.V.: Four questions about digital economy. Pers. Intell. Resour. Manag. Russia **40**(1), 5–14 (2019)
9. Svistunov, V., Kuzina, G., Lobachev, V.: The level of trust of the company staff as the factor of the increase of social and economic systems management efficiency. In: Nikitov, S.A. (ed.) Proceedings of *the* XXI International Conference Complex Systems: Control and Modeling Problems, pp. 799–802. IEEE, New York (2019)
10. Svistunov, V.M., Lobachev, V.V., Simonova, M.V.: Staff responsibility as efficiency-driven factor of ERP-systems. In: Ashmarina, S.I., Vochozka, M., Mantulenko, V.V. (eds.) Digital Age: Chances, Challenges and Future. Lecture Notes in Networks and Systems, vol. 84, pp. 480–486. Springer, Cham (2020). https://doi.org/10.1007/978-3-030-27015-5_58
11. Troshina, E.P., Mantulenko, V.V.: Influence of digitalization on motivation techniques in organizations. In: Ashmarina, S.I., Vochozka, M., Mantulenko, V.V. (eds.) Digital Age: Chances, Challenges and Future. Lecture Notes in Networks and Systems, vol. 84, pp. 317–323. Springer, Cham (2020). https://doi.org/10.1007/978-3-030-27015-5_38

Making Decisions in Cyclical Regional Economy: Innovative Import Substitution and Export

V. K. Semenychev, G. A. Khmeleva(✉), S. S. Asanova, and A. A. Korobetskaya

Samara State University of Economics, Samara, Russia
505tot@mail.ru, galina.a.khmeleva@yandex.ru, ssww28@rambler.ru, kornast@yandex.ru

Abstract. The article is aimed to theorize the basis for valuating regional sectoral economic cycles stages and develop an algorithm of decision-making on innovative import substitution and export under regional cyclicity conditions. The methodology of the study is based on the thesis stating that the cycle stage reflects supply and demand rate on production factors and results, describes market expectations and in that regard is important in making decisions on production, investments and regional import substitution and export policy. Current scientific researches evaluating import substitution and export often fail to consider that regional sectoral cycles may significantly vary from the national one which may lead to wrong conclusions and management decisions. When addressing this problem, the authors proposed a "demand-markets-industry cycles" approach. It allows all stakeholders (manufacturers, investors, and regional authorities) to get information on most popular goods groups for innovative import substitution, state preferences, and also to make decisions on production, investments, import substitution and export policy based upon a certain stage of the regional cycles.

Keywords: Economy cyclicity · Export · Import substitution · Region · Sectoral cycle

1 Introduction

In an open economy, the development of innovative import substitution and export is an important part of national and regional economic policies. Import substitution can be considered as a certain stage that allows import-dependent countries to first solve the problem of imports and then move to a competitive production of export products. For instance, that happened in China, where it was possible to provide the development of the machine building industry in just a decade and become one of the world's leading manufacturers in terms of exports [15]. In a dynamic world with a cyclical economy, it is important to make quick decisions about production, investment, and government support for regional import substitution and export. Slow processes of economic reorientation are unacceptable, especially for developing countries where industries have accumulated a certain lag of underdevelopment [11].

S. I. Ashmarina et al. (Eds.): ENGINEERING ECONOMICS WEEK 2020, LNNS 139, pp. 624–632, 2021.
https://doi.org/10.1007/978-3-030-53277-2_74

Litvinova, Talalaeva, and Ledeneva identified the positive impact of innovative import substitution on structural changes in the economy [9]. An increase in the export of innovative goods is accompanied by a decrease in their imports. Weidong and Yeung showed that the introduction of multinational corporations and deregulation by China were the determining factors in the geographical distribution of the automobile industry in China [15]. Researchers note serious difficulties in determining an effective model for assessing import substitution and product competitiveness. Southworth [13] focus on a comparative analysis of imports to determine the effectiveness of the development of domestic production. The research topic is interesting for all countries of the world. A number of papers was devoted to the problems of foreign economic activity participants due to the specifics of production in certain industries. Flores, Villarreal, and Flores, [3] described the necessity to create supply chains in the aerospace industry that could reduce external impacts or deliberate knowledge transfer, as well as create networks with local economies and businesses. The study was aimed to examine the location models of firms and institutions around the aerospace industry throughout Mexico. Kopczewska, Kudła, and Walczyk [8] used spatial secondary effects between taxation and economic growth. The researchers consider the evaluation strategy that is taken into account in applied research of economic policy. The evaluation includes the concept of spatial side effects and their local and global effects, direct and indirect impacts, as well as the role of various spatial weighting schemes, which creates a combined approach to the analysis.

Volkodavova, Zhabin, and Yakovlev proposed a methodological approach to organize the import substituting industry and an index to evaluate the competitiveness of import substituting products [14]. The authors note that for the purpose of effective import substitution, it is necessary to estimate the market dynamics in the framework of strategic analysis, to identify factors affecting the internal and external competitiveness of the national economy and manufacturers of import-substituting products. Scientists have also highlighted the effectiveness of preferential zones. Using the examples of Japan, China and South Korea, Hassink, Hu, Shin, Yamamura, and Gong (2018) note a high impact of government actions on industrial restructuring and differences in approaches [4]. Hence, it is important to have the fullest possible understanding of the state of production processes and export potential in the industries [5]. Researchers point out that regional differences in cyclicity are among significant factors in providing the effectiveness of economic policy decisions [1]. However, the stage of the industry cycle is usually disregarded when making decisions about innovative import substitution and export, which reduces their effectiveness, since market expectations for growth and decline in the industry are not taken into account. The authors of the paper have attempted to fill this gap.

2 Methodology

The authors have worked out an algorithm for making decisions on import substitution and export, based on the study of the most important market factors "demand-markets-industry cycles".

1. The "Preliminary" stage is necessary to understand the dependence of a region on imports of innovative products. For this purpose it is necessary to assess the actual dependence on imports in terms of the volume dynamics and the proportion of exports and imports for at least 10 years.
2. The "Demand" is aimed at determining the groups of innovative products that are most resistant to possible changes in demand. To do this, we use the XYZ analysis method, which gives good results when determining the stability of sales of certain types of goods and fluctuations in their consumption level. The scale of reference values for groups is shown in Table 1.

Table 1. Referent values of the variation coefficient

Group	Variation coefficient	Class characteristics
Group «X»	0.10 < V < 0.20	Absolutely stable demand
Group «Y»	0.20 < V < 0.25	Stable demand
Group «Z»	V > 0.25	Unstable demand

Source: authors.

The level of product groups stability varies from absolutely stable to unstable.

3. The "Markets" stage consists in evaluating the potential domestic market for innovative products and the potential for sales abroad. To assess the domestic market, it is necessary to carry out a content analysis of information portals that place government orders. This will result in the creation of a database of public procurement and region's enterprises that need domestic products instead of imported ones. We estimate the prospects for sales abroad using the previously proposed [7] mutual trade growth potential (P_{ic}):

$$P_{ic} = \frac{X_{ic} - X_{irc}}{X_{ir}} \cdot 100 \quad (1)$$

where X_{ic} is a total sum of exports of the ith product to the cth country; X_{irc} is a sum of exports of the ith product from the rth region to the cth country; X_{ir} is a total sum of exports of the ith product from the rth region.

The main idea of the index is that the region as an economic entity of the country, due to increased competitiveness and active management actions, is able to increase exports of products within the framework of existing agreements between trading countries, thereby replacing the share of exports of other regions. The minimum value of 0 is reached when a region fully meets the country's product needs. The greater the potential is, the greater is the possible exports growth from the region.

4. The "Industry Cycles" stage is necessary to understand the risks of cyclical behavior for producers, investors, and the state and to develop management solutions for implementing the policy of innovative import substitution and export orientation of the economy. At this stage, using the industries selected at the preliminary stage, we construct models of trends, cyclical and seasonal fluctuations of these industries for at least 10 years. This method was first introduced by the

authors in [9]. To evaluate the dynamics models of regional economic indicators, the authors implemented the method as a script in an open-source programming language R.

Each model included four unobservable components, among which are: trend T_t; cyclic component C_t; seasonal fluctuations S_t and stochastic component ε_t, combined in both additive and additive-multiplicative (mixed) structures:

$$Y_t = T_t + C_t + S_t + \varepsilon_t \tag{2}$$

$$Y_t = (T_t + C_t)(1 + S_t) + \varepsilon_t \tag{3}$$

The components are evaluated one by one inside an iterative process.

The authors considered seven trend models: linear trend, growth models (generalized exponent and power functions), logistic curves (generalized sigmoid model and arctangent), symmetric and asymmetric bell-shaped curves. Such a set of trends covers various dynamics' types that differ by direction, speed and symmetricity. The cyclic component was modeled as a sum of three sines with non-proportional frequencies:

$$C_t = \sum_{j=1}^{3} A_j \sin(\omega_j t + \varphi_j), \tag{4}$$

where A_j is the amplitude of the ith sine, $\omega_j > 0$ is its frequency, and $\varphi_j \in [-\pi;\pi)$ is its phase.

The summation of three sines with non-proportional frequencies is a suitable tool to describe complex, not strictly periodical fluctuations which is the most common case for economic cycles. The idea is also responding to Slutsky's hypothesis at the meso-level of economy [12]. To eliminate the seasonal fluctuations, we used the LOESS method which considers seasonality evolution from one year to another.

3 Results

We will show the results of the proposed methodological approach application by the example of one of the Russian regions. The Samara region is among the industrial leaders of Russia, but its economy is dominated by old-industrial sectors. Therefore, the production of innovative products for domestic and foreign markets is an important task for the region.

1. The "Preliminary" stage is necessary to assess the dependence of the economy of the region from imports which is determined on the basis of the indicators: the dynamics of export and import, the structure of foreign trade and commodity structure of import of the Samara region for 2013–2018.

   The calculation results showed that exports and imports are decreasing in absolute values. In 2013–2018, due to sanctions, exports decreased by almost 2 times and amounted to $ 4.52 billion at the end of the analyzed period. The import decreased due to the solvency of Russian enterprises in 2015–2016. During 2017–2018, imports increased and amounted to $ 2.25 billion. A sharp drop in foreign trade indicators is replaced by the increase again in both exports and imports since 2016. This creates additional opportunities for import substitution and expansion of export production.

An assessment of the actual scale of the dependence of the regional economy on imports indicates that the Samara region needs to implement a policy of import substitution primarily for the following groups of goods: machinery, equipment and hardware; transport; metals and their products; food, beverages, & tobacco; chemical products; plastics and rubber; tools and devices, watches.

2. The "Demand" stage. To justify the potential demand for the selected groups of products, a study of sustainable demand is conducted using the XYZ method. The XYZ-analysis calculates the dynamics of the import structure for the groups of products identified at the previous stage, then estimates the yearly variation of import volumes, and after that calculates the standard deviation, variation coefficient, group and level of stability (Table 2).

Table 2. Final comparative data table

Group of products in import of the Samara region	Average growth rate in 2013–2018	Standard deviation	Variation coefficient	Group	Level of demand stability
Food, beverages, tobacco	83.37	16.40	0.17	X	Absolutely stable
Chemical industry products	55.82	43.13	0.25	Y	Stable
Plastics & rubbers	62.93	66.34	0.28	Z	Unstable
Metals and their products	69.06	76.21	0.24	Y	Stable
Machinery, equipment and hardware	72.96	235.77	0.24	Y	Stable
Transport	53.91	180.40	0.32	Z	Unstable
Tools and devices, watches	49.77	25.82	0.27	Z	Unstable

Source: authors.

The results of the analysis show that the following groups of goods are the most resistant to demand for import substitution: food, beverages, & tobacco; chemical industry products; metals and their products; machinery, equipment and hardware.

3. The "Markets" stage allows to provide an indication of potential domestic and foreign markets. This stage and the next one are demonstrated by the example of the group of products "Plastics & Rubber". This choice is justified by the following arguments: First, the Samara region is one of the leaders in the production of this group of products. In 2018, the region was the 4th one among Russian exporting regions with a volume of $318 million. In the structure of chemical industry products exports, the rubber occupies the second position after the fuel and energy

sector. It is necessary to further develop this export direction, since along with the growing global demand the region has all the necessary resources for that. Secondly, this group is characterized by high risk level, since the demand for products is unstable, which makes it particularly necessary to assess the cycle stage of production of chemical products at the method's fourth stage "Industry cycles". Table 3 shows the calculation results of the export potential factor for the item "Plastics and rubber" using the formula (1).

Table 3. Export potential of the Samara region in the sector "Plastics & rubber", units

Country	2018	Interpretation	Country	2018	Interpretation
China	69.84	Medium	Ukraine	90.44	Medium
India	38.42	Medium	Spain	5.39	Low
Germany	53.44	Medium	Belarus	204.76	High
Kazakhstan	177.54	High	South Korea	4.76	Low
Turkey	27.26	Low	Thailand	3.28	Low

Source: authors based on [7].

The results of the calculation have showed a high potential for expanding exports of products of the item "Plastics & rubber" with Kazakhstan and Belarus. Thus, the total export of "Plastics & rubber" from Russia to Kazakhstan is $586 million per year, while supplies from the Samara region are $23.2 million, or 3.9% of the total. Kazakhstan is neighboring the Samara region, and trade agreements provide for a single trade economic space (duty-free trade). In addition, a number of support measures have been adopted at the state level, such as concessional lending, provision of state guarantees, simplification of the customs clearance process, creation of attractive conditions for bringing goods to the world market, information support and expert advice.

4. The "Industry cycles" stage is necessary to improve the efficiency of management decisions regarding the launch of new import-substitution and export-oriented projects.
   To assess the interrelation between the considered industries to the economic cycle stages we first describe the stages graphically (Fig. 1).

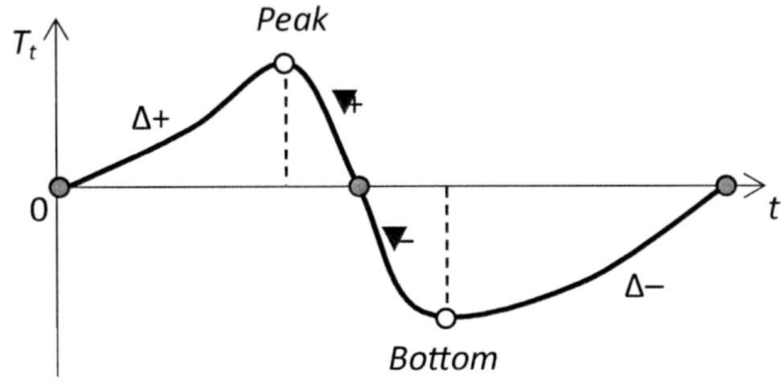

Fig. 1. Cycle stages indication (Source: authors)

Cycles' stages are signed as it is shown on the Fig. 1: symbol Δ means growth, symbol ▼ – decline, plus sign + marks positive area (cycle over trend), minus sign is used for negative areas (cycle below trend).

Table 4 shows the evaluations of the cycle stages performed by the authors for the key industries of Russia and the Samara region.

Table 4. Current cycle stages

Industry	Consumer prices index	Extraction	Manufacturing	Food, beverages. & tabacco	Chemical Industry	Pharmacy	Metals, & their products	Machinery	Cars	Trains
Russian Federation	▼–	▼–	▼+	▼+	Δ –	Δ –	Δ +	Δ –	▼–	▼+
The Samara Region	▼–	Δ –	▼ –	Δ –	Δ –	▼+	Δ –	Δ +	▼–	▼+

Source: authors.

The calculation results show that there is an increase in the economic cycle in the chemical industry both in the country as a whole and in the Samara region in 2019–2020. This leads to the conclusion that exactly now it is necessary to start developing projects to increase the production of import substituting and export-oriented products.

4 Discussion

Public authorities usually pursue several goals at once (in our case, the growth of import substitution and export is assumed), so the process of policy development can be considered as a multi-criteria task. Production development options are possible using innovative or inertial scenarios of the domestic economy [2].

In any case, the success of decisions on import substitution and export is determined by the availability of reliable information about available resources, sales markets and the market situation. To do this, researchers have developed and analyzed methods for evaluating import substitution and production of competitive products in all industries, based on the strategic priorities in the industry and the region. The strategic priorities provide important information from the point of view of goal setting in import substitution, which may be insufficient, as they only provide an understanding of the starting point for developing measures to implement import substitution policies, and this is clearly not enough to develop current investment projects in this area. The information in the "demand-markets - industry cycles" area is an important data frame to make decisions on import substitution and export. The calculation of potential demand indicators allows to identify the most stable product groups over time. Market data is a base to assess the potential of the domestic market for innovative products and their potential for sales abroad.

The picture of the uprisings and downturns of business cycles provides important information to entrepreneurs and authorities about the stage of the industry [10]. This information gives the investors the ability to minimize the risks of investment. The entrepreneurs get the understanding what to do to increase or reduce sales, whether to use a new production method or a new commercial usage approach of an existing product; to create a new benefit or change its quality; develop new market or new source of raw materials; implement organizational innovations. Regional authorities get the opportunity to adjust regional industrial policy.

5 Conclusion

Effective management decisions cannot be made without reliable and complete information about the current situation. However, enterprises and authorities often do not take into account information about the cyclical nature of the economy in general as well as the stage of the industry cycle in particular when making decisions. This increases a rate of cyclical risks. Meanwhile, taking into account the depth of bottom cycles is necessary not only to identify the current stage in the industry, but also for future expectations regarding the cyclical behavior of market participants [6].

The algorithm proposed by the authors complements the existing scientific and methodological basis for the development of import substitution policy, considering the depth and amplitude of cycles. For instance, in a case of fatal failures, it is probably useful to first identify the causes of the decline and thus take targeted import substitution measures addressing these causes. For other example, a deficit or high cost of investment necessitates subsidizing loan rates and government guarantees for investment in import substitution and export projects. If there is a short-term decrease in the cycle, we can expect a positive effect from support measures in a shorter time. From that point of view, it is important to evaluate the duration of economic recovery after economic failures when planning measures in import substitution policy. The amplitude of the cycle indicates the potential response of the industry on economic changes. If the amplitude is wide and sharp, there may be a "swing effect", which causes the risks of an inertial pullback of the economy to negative values. Thus, we can conclude that along with data on markets and potential demand, understanding the cyclical nature of the regional economy significantly complements the information base for making decisions about import substitution and export.

Acknowledgments. The reported study was funded by RFBR, project number 20-010-00549.

References

1. Beraja, M., Hurst, E., Ospina, J.: The aggregate implications of regional business cycles. Econometrica **87**(6), 1789–1833 (2019). https://doi.org/10.3982/ECTA14243
2. Borisov, V.N., Pochukaeva, O.V.: Innovative machine engineering as a factor of developing import substitution. Stud. Russ. Econ. Dev. **26**(3), 225–232 (2015). https://doi.org/10.1134/S1075700715030028

3. Flores, M., Villarreal, A., Flores, S.: Spatial co-location patterns of aerospace industry firms in Mexico. Appl. Spat. Anal. Policy **10**(2), 233–251 (2016). https://doi.org/10.1007/s12061-015-9180-0
4. Hassink, R., Hu, X., Shin, D.-H., Yamamura, S., Gong, H.: The restructuring of old industrial areas in East Asia. Area Dev. Policy **3**(2), 185–202 (2018). https://doi.org/10.1080/23792949.2017.1413405
5. Kale, D., Wield, D.: In search of the missing hand of 'collaborative action': evidence from the Indian medical device industry. Innov. Dev. **9**(1), 1–23 (2019). https://doi.org/10.1080/2157930X.2018.1429807
6. Khmeleva, G.A. (ed.): Russian Regions in the Conditions of Sanctions: The Possibility of Priority Development of the Economy Based on Innovation. Samara State University of Economics, Samara (2019)
7. Khmeleva, G.A., Fedorenko, R.V.: International economic integration of certain regions: the case of the Samara Region. In: Nahlik, A. (ed.) Proceedings from Modern Economy, Smart Development, vol. 19, pp. 516–530. University of Sopron Press, Sopron (2019)
8. Kopczewska, K., Kudła, J., Walczyk, K.: Strategy of spatial panel estimation: spatial spillovers between taxation and economic growth. Appl. Spat. Anal. Policy **10**(1), 77–102 (2015). https://doi.org/10.1007/s12061-015-9170-2
9. Litvinova, A.V., Talalaeva, N.S., Ledeneva, M.V.: Innovative import substitution in Russia and the world: a comparative analysis. In: Popkova, E.G., Sergi, B.S. (eds.) The 21st Century from the Positions of Modern Science: Intellectual, Digital and Innovative Aspects. Lecture Notes in Networks and Systems, vol. 91, pp. 521–533. Springer, Cham (2020)
10. Luo, X., Lu, X., Zhang, Z., Pan, Y.: Regional differences and rural public expenditure cyclicality: evidence from transitory and persistent shocks in China. The Annals of Regional Science (2020). https://link.springer.com/article/10.1007%2Fs00168–020-00985-5. Accessed 17 Apr 2020
11. Schmalz, S.: The Brazilian Economic Policy: From the Crisis of Import Substitution to the *Programa de Aceleração do Crescimento*. In: de la Fontaine, D., Stehnken, T. (eds.) The Political System of Brazil, pp. 265–281. Springer, Heidelberg (2016). https://doi.org/10.1007/978-3-642-40023-0_15
12. Slutsky, E.E.: Addition of random causes as a source of cyclic processes. Market Issues **3**(1), 34–64 (1927)
13. Southworth, F.: Freight flow modeling in the United States. Appl. Spat. Anal. Policy **11**(4), 669–691 (2018). https://doi.org/10.1007/s12061-018-9273-7
14. Volkodavova, E., Zhabin, A., Yakovlev, G.: Dichotomy of import substitution and cooperation strategies in industry. Econ. Ann. XXI **162**(11–12), 48–52 (2016). https://doi.org/10.21003/ea.V162-10
15. Weidong, L., Yeung, H.: China's dynamic industrial sector: the automobile industry. Eurasian Geogr. Econ. **49**(5), 523–548 (2008). https://doi.org/10.2747/1539-7216.49.5.523

Challenges of Entrepreneurship Education in the Republic of Tatarstan

R. M. Akhmetshin(✉) and R. B. Palyakin

Kazan Federal University, Kazan, Russia
renakhmet@mail.ru, roman.palyakin@yandex.ru

Abstract. The relevance of the study falls within the necessity to create conditions for the development of entrepreneurship based on socially-oriented support mechanisms. The purpose of the study is to identify the main directions for the development of educational infrastructure of entrepreneurship in the socio-economic environment of the Republic of Tatarstan. The study used the methodology of institutionalization of infrastructure, based on the theory of social justice and the theory of synergies between the participants in the infrastructure. A mechanism has been developed for institutional infrastructure support of entrepreneurship with implementation of partnership principles. Considering synergies in the use of labor resources, the necessary strategic decisions on infrastructure management are indicated. Challenges facing entrepreneurship in developing educational infrastructure are explored, including key factors causing these problems. The necessity to address issues of developing educational infrastructure through the mechanisms of infrastructural support of entrepreneurship in the Republic of Tatarstan is justified.

Keywords: Business development conditions · Business support · Enterprise infrastructure · Educational infrastructure · Infrastructure

1 Introduction

In the current socio-economic conditions of management accompanied with a drop in oil prices, an increase in the value of foreign currency, a policy of sanctions and restrictions, the competitiveness of domestic enterprises is particularly acute.

On the one hand, protectionism policies – as retaliatory sanctions and countermeasures – help protect domestic households, and on the other, limit healthy competition. The issue of developing the educational infrastructure of entrepreneurship and the implementation of socio-economic programs is especially acute, and is associated with the active work of state ministries and departments.

Creating comfortable conditions for business development in the Republic of Tatarstan addresses several problems at the state level. First of all, it is a solution to the problems of employment and self-employment of the population, when small and medium-sized enterprises create jobs. A separate area is the employment of socially vulnerable part of the population – single mothers, people with disabilities, and persons recently released from prison. Secondly, the effective development of small and medium-sized enterprises increases payments to state budget funds and organizations.

S. I. Ashmarina et al. (Eds.): ENGINEERING ECONOMICS WEEK 2020, LNNS 139, pp. 633–639, 2021.
https://doi.org/10.1007/978-3-030-53277-2_75

A transparent and clean mechanism based on state support and regulated reporting system solves the issue of control and monitoring of receipts in all areas. Thirdly, the widespread development of economic activity increases the welfare of certain groups of the population, and for others makes a wider range of diverse products more accessible: from food products that can be purchased at agricultural fairs in the "Agro-Industrial Park "Kazan" to technological devices and art objects sold by entrepreneurs via the Internet.

The key entities regulating the provision of state subsidies in the region are: the Ministry of Economy of the Republic of Tatarstan and the Ministry of Agriculture of the Republic of Tatarstan. The main areas of state support are [1]

1. The program "Leasing grant"
2. The program "Innovation"
3. The program supporting agricultural producers "Beginner Farmer" and "Family Livestock Farm"
4. Compensation for training costs

The development of the educational infrastructure of entrepreneurship in the Republic of Tatarstan is directly related to the last point, as it sets the trends and patterns of evolution of the studied course. In order to study this area of management, we analyze the concept of "educational infrastructure of entrepreneurial activity". At its core, infrastructure is a system of interconnected and interdependent elements (subjects and objects) that ensure its functioning.

2 Methodology

With regard to enterprise infrastructure development as a system of relations between the participants of infrastructure support in relation to the educational infrastructure of the business, we can distinguish synergies in the use of labor resources. The synergy in the use of labor resources is due to communications between the subjects of demand (enterprises) and the supply of labor resources (objects of social infrastructure), which is reflected in the dynamics of labor market indicators and quality of life [10]. The implementation of labor synergy occurs by bridging the gap between the emerging needs for labor resources of entrepreneurial structures and the supply on the labor market [9]. By putting in place mechanisms of co-creation (co-working), creating special educational programs and coaching (aimed at specific competencies) centers, labor specialization of the population takes place, in line with the priority areas of specific business units. Synergy in the use of labor resources is becoming a catalyst for entrepreneurial activity of the population.

In the study of entrepreneurial activity an important scientific support is given by the concept of Durkheim [4], which solves the problems of solidarity of the social system by considering external factors of regulation, self-organization, freedom and justice. At the same time, entrepreneurial solidarity is determined by the "harmonious interaction of regulation, self-organization, freedom and justice, as well as the quality of life and adaptation, ensuring maximum entrepreneurial effect" [7, p. 74].

Based on the studied ideas about entrepreneurial activity and the institutionalization of entrepreneurial activity, we can propose a mechanism for institutionalizing the infrastructure of entrepreneurship, which includes the actions of agents at the federal and regional levels in order to enhance entrepreneurial potential through targeted impact within the infrastructure (Fig. 1).

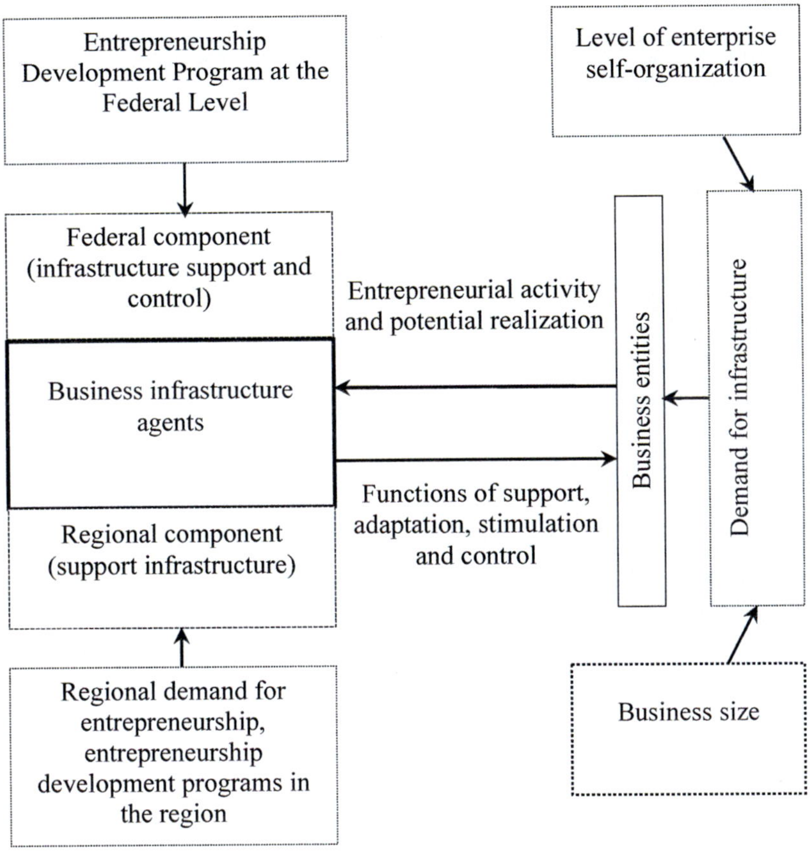

Fig. 1. Implementation of the mechanism of institutional infrastructure support to entrepreneurship. (Source: authors)

Within the mechanism, the infrastructure plays the role of a dynamic open system of agents' interaction in various areas of the management functions of control, adaptation, support and stimulation, with regard to the level of development of self-organization of business structures [8]. Wherein the infrastructure support system acts as an open platform, contributing to the achievement of a high level of entrepreneurial activity based on the regulated conditions for doing business, and at the same time maintaining entrepreneurial freedom and social justice [7]. The main parameters of synergy in the use of labor resources are presented in Table 1.

Table 1. Parameters for building the synergy in the use of labor resources

Parties involved	Objectives of parties	Interaction terms	Interaction indicators	Key interaction structures
Entrepreneurs	To realize the labor and educational potential of employees	Development of employment centers and educational forms in the field of entrepreneurship	– level of educational programs development for SMEs, (points) – development and implementation of co-working projects, (points) – increase in the average salary of SME employees, % – additional jobs at SME enterprises, ppl. – qualification level of SME employees, (points)	– public private partnership – network business structures – objects of social infrastructure and employment, – coach centers, co-working facilities
Infrastructure Owners	To increase workforce competencies			
Government structures	To solve unemployment problems, increase labor activity of the population			

Source: authors.

In the perspective of strategic development of infrastructure participants interaction, the synergy in the use of labor resources is most effective for infrastructure implementation strategy [6]

According to the implementation strategy, achieving synergies in the use of labor resources is permissible, subject to particular tasks

- development of educational services for entrepreneurship
- introduction of educational and training projects in enterprises;
- development of public capital and co-financing tools for educational projects;
- development of mechanisms for open interaction between business, educational institutions and personnel.

3 Results

In our opinion, the educational infrastructure of entrepreneurial activity is a complex of interconnected elements represented by the training and learning subsystems in conjunction with technical and technological means (premises, equipment, the global Internet) and limited to a specific place for projects implementation. The training subsystem is represented by specialists from various sectors of the economy engaged in training and sharing their knowledge, best practices and experience. The learning subsystem is represented by people interested in developing their entrepreneurial skills, acquiring knowledge and solving existing problems in business management [2]. These include students of universities of the Republic of Tatarstan, employees of state and public organizations, commercial enterprises. The technical and technological means include all the necessary audio and video equipment, software, furniture, stationery and others [1].

The problems of developing the educational infrastructure of entrepreneurship in the Republic of Tatarstan are directly related to the aforementioned elements and, by analogy with the theory of constraints by E. Goldratt, the quality of ongoing educational projects will be determined by the quality of the weakest subsystem [3]. Thus, we further identify and formulate the key problems of the educational infrastructure of the region. Currently, a conflict of interests between the learning subsystem represented by the target audience and the training subsystem, represented by various kinds of business trainers, is widespread. This problem can be identified as an information conflict, when the educational material presented is either not properly understood (students cannot understand it), or is insolvent (irrelevant and invaluable).

On the one hand, the target audience of educational events can be described as low-quality, as it is represented by disinterested students. An ineffective advertising campaign, failing to attract the highest quality audience, leads to the fact that the information is not properly absorbed and the effectiveness of the event tends to zero. A separate issue is the lack of a selection, monitoring and control system for those wishing to participate in educational activities. The organizers, upon reaching the required number of participants (as a rule, limited with the capacity of the room where the educational event will be held or with quota according to the statement of work) stop the advertising campaign. In practice, most of the declared participants refuse or cannot attend the event and the customer/organizer does not achieve the goals.

When the organizer is a commercial company, then its goal is to maximize profits, that is why it does not stop advertising campaigning for an event until its inception. However, if the state acts as an initiator, and the public organization as an organizer, then in practice the advertising campaign is ineffective in 80–85% of cases. Effective channels for disseminating information on educational events dedicated to entrepreneurship in the Republic of Tatarstan are: social networks, announcements on websites and digital media, printing materials, mailing lists, calling businessmen and public organizations according to available databases, contextual and targeted advertising.

On the other hand, the quality of the training subsystem is not trustworthy. Most of the so-called business trainers frequently invited to participate in educational events are the same undergraduate and postgraduate students of Kazan universities or regional centers, professors from regional universities, failed entrepreneurs who share theoretical knowledge and lack business management practices. The lack of competencies of a real business trainer complicates the educational process devoted to address challenges facing small and medium-sized enterprises. Such kind of specialists and so called "practitioners" teach a theoretical course by inertia without going into practical aspects; they do not highlight pitfalls and do not provide in-depth materials within an educational event.

The third and most important issue is the discrepancy between the offered educational courses for entrepreneurs in the Republic of Tatarstan and the real needs of the business sector. Quite often, the key principle of marketing is violated - to produce what is needed, and not what is possible. An entrepreneur running business for more than ten years is not interested in wasting his valuable time and listening to a course devoted to "Typical mistakes of a novice entrepreneur" from a person giving a lecture at sight. Businessmen need to consult with more qualified entrepreneurs and get their advice on business development and difficult business cases. Unfortunately, modern business trainers are trying to make their program the most attractive and colorful, but they do not have strong practice and cannot professionally advise heads of business entities.

The fourth problem is the centralization of the educational infrastructure for the development of entrepreneurship in the capital of the Republic of Tatarstan, Kazan. Despite a large number of platforms, industrial parks and other business support entities in other cities and regions of the republic, the main efforts are always happening in the capital, which is a big omission. We cannot say that educational events are not implemented in the districts, however, the question of the competence of the business trainers who implement them, the relevance of the topics and their relevance to the real business problems of a given city or region is especially acute.

4 Discussion

Studies in the field of infrastructure support for entrepreneurship of the Republic of Tatarstan show that republic has a sufficiently developed socio-economic system, but there are certain problems of infrastructure support for small businesses identified in the process of determining the effectiveness of the infrastructure support system, and they form the key problems of the socio-economic development of the Republic Tatarstan. Among these problems, first of all, a weak coordination in the activities of infrastructure elements and the short of qualified personnel in infrastructure sectors. All these problems must be solved within the strategy for developing infrastructure support for entrepreneurship, thereby providing directions for the development of the mechanism of infrastructure support [8].

The system of infrastructure support for entrepreneurship is created to balance the interests of participants in the socio-economic space and to transform the relations between entrepreneurship in the region and state regional bodies from the traditional interaction between the object and the subject (with the participation of infrastructure institutions as intermediaries) to partnership relations [5]. The identified negative trends in educational infrastructure have developed for two main reasons. Firstly, the interest in entrepreneurial activity among the population has increased due to the vigorous activity of state bodies in this direction. As a result, educational events have turned into a specific product sold on a wide market through social networks, because demand creates supply. These products include various kinds of webinars, master classes and speeches by unknown business gurus, without real management experience. Secondly, educational events aimed at supporting and developing small and medium-sized businesses tends to be quickly implemented on their own.

As a result, most qualified trainers are rarely if ever attracted, as the organizers aim not to improve the educational infrastructure of entrepreneurship in the Republic of Tatarstan or to solve the real problems of businessmen by means of professional consulting, but to make a profit. Maximum financial result from such events is possible only through costs optimization or reduction resulting in the quality of implemented events.

5 Conclusion

Achieving sustainable development of educational infrastructure is impossible without providing infrastructure to entrepreneurs. From the position of managerial solutions at the level of developing the enterprise infrastructure in the Republic of Tatarstan, it is necessary to expand the infrastructure support system by creating a unified information and analytical base of entrepreneurial activity, maximizing the interconnection of entrepreneurs with developing infrastructure elements – the development of training and internal communication; establishing a body for the interconnection of entrepreneurs and state authorities as partners, implementing structural and functional relationships between infrastructure elements (by enhancing the self-organization of each element in the overall system and establishing information interaction).

The development of the educational infrastructure of entrepreneurship in the Republic of Tatarstan has a number of weaknesses to work and improve. Undoubtedly, in recent years the situation has significantly changed for the better, but it is necessary to carefully approach the issues of selecting business trainers, presented content, as well as the target audience of such events.

References

1. Akhmetshin, R.M., Narsov, I.A.: Educational entrepreneurial infrastructure development challenges in the Republic of Tatarstan. Kazan Econ. Bull. **5**(25), 73–76 (2016)
2. Antonchenko, N.G., Kalenskaya, N.V.: Developing a methodology for assessing the efficacy of managerial decisions in entrepreneurial establishments. Life Sci. J. **11**(7), 365–369 (2014)
3. Dettmer, W.: Goldratt's Theory of Constraints: A Systems Approach to Continuous Improvement. Alpina Publisher, Moscow (2010)
4. Durkheim, E.: The Division of Labour in Society. The Method of Sociology. Science, Moscow (1991)
5. Kalenskaya, N.V., Akhmetshin, R.M., Grigoryeva, L.L.: The development of state regulation in small entrepreneurship infrastructure provision. Mediterr. J. Soc. Sci. **5**(18), 27–33 (2014)
6. Kalenskaya, N.V., Palyakin, R.B., Tarkaeva, N.A.: Strategic management of regional business infrastructure system. Mediterr. J. Soc. Sci. **18**(5), 311–316 (2014)
7. Palyakin, R.B.: Formation of a Binary Model of the Development of the Entrepreneurial Infrastructure Based on the Asymmetry of the Institutional Environment. Paragraph, Kazan (2018)
8. Palyakin, R.B., Akhmetshin, R.M., Rolbina, E.S.: Modeling of an entrepreneurial infrastructure support system in the Republic of Tatarstan. Issue Multi. Soc. Sci. Manage. **10**(1), 162–167 (2019)
9. Roger, S., Nijkamp, P.: Knowledge spillovers, entrepreneurship and economic development. Ann. Reg. Sci. **43**, 835–838 (2009)
10. Strauss, T.: Growth and government: is there a difference between developed and developing countries. Econ. Governance **2**, 135–157 (2001)

Dispositional Education as a Resource of Innovative Economy

M. O. Suraeva and O. S. Andreev(✉)

Samara State University of Economics, Samara, Russia
marusyasuraeva@mail.ru, andreevoleg1984@mail.ru

Abstract. This article discusses the issues of improving the education management system, since the methodological level of its development does not fully reveal their essence, which indicates a divergence of theoretical positions. The emphasis is on dispositional technologies in education. The main components of scientific novelty is the effective functioning of the higher education system, which is the strategic task of the state at all levels of its development, on the solution of which depend both social stability and economic growth, as well as the country's competitiveness.

Keywords: Innovation · Innovative development · Innovative education · Knowledge management · Innovative economy · Knowledge economy

1 Introduction

The role and impact of higher education on the country's economic growth is increasing every year. As a result of analysis and research of foreign experience in ensuring the economic sustainability of universities, it was noted that a special role is given to both the state and the private sector. The role of the state is focused on financial support for innovation and research, the positioning of universities in the global academic community, while the role of business structures is focused on building cooperative and integration ties with the higher education system. The existing mechanism ensures the economic sustainability of universities, and its main criteria are the development of the export of educational services, the availability of educational services and the integration of academic and scientific activities with business processes in the economy.

The transition to the fourth industrial revolution, aimed at restructuring all spheres of life based on digital technologies, provides a change in the status of a classical university, which becomes a driver of innovative socio-economic development. In modern conditions of economic and social development of society, new management tools for higher education institutions are needed. The search for new models for the integration of innovative higher education will help to identify factors that influence the country's economic growth.

The intensification of the innovative development of Russia should be coupled with the progressive and dynamic development of human capital and the entrepreneurial ability of society, which in world practice are considered as leading state economic

S. I. Ashmarina et al. (Eds.): ENGINEERING ECONOMICS WEEK 2020, LNNS 139, pp. 640–648, 2021.
https://doi.org/10.1007/978-3-030-53277-2_76

resources that ensure the progressive development of industrial production and the most important industrial sectors. The practice of the last 20 years has shown that the high level of use of natural resources and the intensive lending policy of small, medium and large enterprises have not fully contributed to the innovative development of industry, the formation of a new socio-economic structure.

At the present stage, innovative activity in industry does not exceed 2.5–3%, and innovative projects are implemented by an extremely small number of industry entities. A comprehensive change in the current situation should be directly connected by creating an effective system of functioning of the higher education system, taking into account its integration with the real sector of the economy.

For Russia to become one of the five world economies in the world, it is necessary to have an efficiently functioning higher education system in the country, providing fundamental scientific, socio-cultural and practical training of competent specialists, able to adapt to changing situations in the production and social sphere, as well as to create new knowledge necessary for the formation and development of production of the fifth and sixth technological structures. In the light of the above, scientific research aimed at studying the interaction of factors of higher education and other factors of innovative development in terms of economic growth are very relevant.

2 Methodology

To develop the technological sustainability of higher education institutions, the main guideline of the development strategy of Russia is focused on the modernization of teaching methods and the active development of the online education system. Most higher education institutions retain the traditional form of work from potential contractors: enterprises; governing bodies; financial structures; institutional structures of various sectors of the economy [1].

Together, the organizational and economic mechanism for ensuring the sustainable economic development of universities makes it possible to:

- implement in a new quality the existing educational technologies, including distance and dual forms,
- implement at a qualitative level additional subsystems of economic sustainability, such as marketing and branding in the system of university activities.

At the present stage in world practice, in the framework of ensuring technological (production) sustainability, universities have a significant "arsenal" of educational technologies.

In world practice, such innovative teaching technologies as information, communication, and interactive have become widespread.

Interactive technology is based on innovation activity. The organization of interactive learning involves:

- modeling of practical market situations,
- use of variable case studies,
- interaction of information flows and logical methods of thinking.

To ensure a high level of market and technological sustainability, organizational sustainability through the use of managerial and organizational innovations is of particular importance. Based on the scientific generalization of modern theoretical approaches to dispositional education, we believe that the development of universities is the achievement of an equilibrium state between all types of sustainability and the university environment by building adaptive systems (dosing managerial influences and investments in development), and also that the essence of economic sustainability consists of the three backbone categories, such as sustainability, balance and sustainable development.

In the short term, the principles of building adaptive systems should become:

- building an economy based on "knowledge", and not on resource potential;
- integration of academic and scientific activities;
- integration interaction of universities with the manufacturing and non-manufacturing sectors of the economy;
- transition of the education system from an extensive to an intensive path of development. Knowledge management systems belong to the class of information systems used to manage the knowledge of organizations.

Under the management of university knowledge should be understood:

- a set of processes related to the creation, distribution, processing and use of knowledge of the activities of the university;
- a systematic process established by the university to work with information and knowledge resources.

The basis of a new form of managerial technology in education is knowledge management:

1. Management technology for the operational coordination of traditional university values with entrepreneurial ones.
2. Management technology for the construction of extended peripherals, focused on external relations and research units
3. Management technology stimulating the academic structure – the mainstay of an entrepreneurial university.
4. Management technology of the integrated change-oriented entrepreneurial culture.
5. Management technologies for the development of own region and cooperation with other universities.
6. Management technologies for creating a university without borders.

3 Results

Summarizing the conceptual approaches to the definition and essence of dispositional education, we can conclude that the sustainable development of universities is achieved through managerial impact, the availability of financial stability and investment in development, and, therefore, an urgent direction is to consider the principles of building modern mechanisms to ensure this sustainability. For the purposes of studying the increment of new knowledge, directions for their systematization have been developed (Fig. 1).

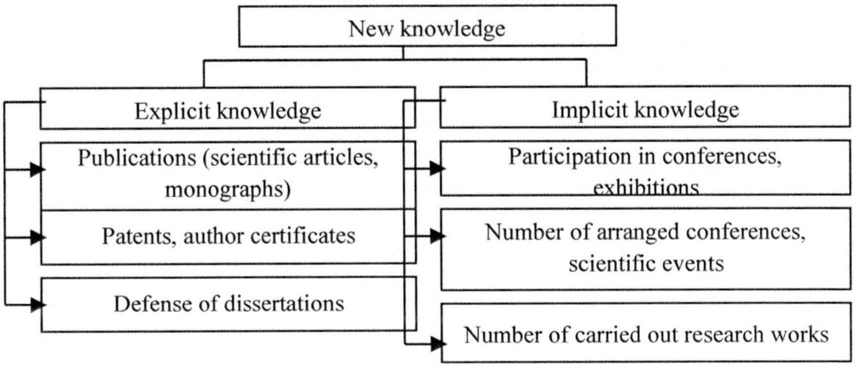

Fig. 1. Systematization of new knowledge. (Source: authors)

The modern concept of education is based on the priority of knowledge, intelligence, and erudition. The specifics of higher education institutions determines the diversity of corporate governance, which includes mechanisms by which the university is managed and controlled. Corporate governance implies a system of relationships between university management, academic council, professional union, board of trustees or supervisory board, shareholders and other interested parties. Corporate governance should be based on a structure that is created to achieve the goals of the university and their control.

Corporate governance is recognized as a model that is designed, on the one hand, to regulate relations within the organization, and on the other hand, to coordinate the goals of the organization with external stakeholders. Corporate management systems of higher educational institutions should be based on the use of innovative organizational and managerial structures. The classical structure of university management is linear-functional. The university management model is a vertical structure. Linear-functional structures in the conditions of innovative development cannot ensure the effective functioning of innovation-oriented autonomous higher education institutions [5].

For innovation-oriented autonomous universities, organizational-managerial structures based on the principle of decentralization are most inherent. In the development and implementation of organizational innovations in higher education institutions, the main provisions of modern development concepts should be taken into account. An assessment of the interaction of components is based on a modified Lotka-Volterra interaction model. This statement is the basis for identifying new management technologies in education for knowledge management. In studies on the management of inter-organizational interaction to identify the technology of the interconnections of the organization, organizational changes and organizational training, the Lotka-Volterra system aroused great interest [6]. The Lotka-Volterra model makes it possible to differentiate the impact of internal interactions and external cooperation on the growth of knowledge in the organization. The following characteristics of knowledge formation are defined in the Lotka-Volterra system:

1. Endogenous growth rate: This is an internal rate of knowledge accumulation (or knowledge destruction) within an organization.
2. Speed of internal interaction.
3. External speed of interaction: This is the speed at which knowledge is created or destroyed in an organization through interaction with another organization. This indicator can be positive if the interaction brings new knowledge to the organization or negative if the organization loses knowledge in the process of interaction with another organization.

Mechanisms that bring new knowledge to the organization:

1. Endogenous growth (positive part a_i) due to the creativity of individuals or groups, the transfer of knowledge from the environment (with the exception of another organization).
2. Constructive internal interactions (the positive part of b_i) that makes it possible to build architectural knowledge when people from different groups in cooperation build new knowledge based on their specific knowledge.
3. Constructive interaction with another organization (the positive part of c_{ij}) in favor of transferring knowledge from one organization to another.

Mechanisms that remove new knowledge from the organization:

1. Endogenous decline (the negative part of the speed a_i) due to the standardization of individuals, the lack of transfer of knowledge from the environment, the outflow of key individuals;
2. Destructive internal interactions (negative part b_i) are often necessary in order to be able to effectively use new knowledge with limited resources;
3. Destructive interactions with another organization (negative part of c_{ij}), where knowledge is stolen or transferred with negative consequences to one of the organizations, for example, with the transfer of key personnel.

The question arises how to interpret the endogenous growth rate a_i, within the population (internal) b_i to interpret the interaction rate, and between the population c_{ij} to interpret the rate of (external) interaction [7]. Thus, the symmetry between both organizations (as is the case for competition or symbiotic models) is not favorable for the creation of new knowledge, as this leads to stability through negative feedback or to instability, through positive feedback.

4 Discussion

In the scientific literature, the problems of education are constantly being raised, but, unfortunately, most often they are unsystematic, fragmented in nature and concern any of its individual sides. In world practice, significant attention of scientists is focused on the economic sustainability of universities. In this aspect scientific works of such scientists as Koryakov [9], Malysheva [10], Mikhalev [11] and others can be noted.

Considered concepts of economic sustainability are associated directly with the activities of industrial enterprises, the organizational and economic mechanisms of which may not be fully associated with the activities of higher education institutions. In recent years, new scientific papers have been devoted to the economic sustainability of the education system. Here can be noted the scientific papers of scientists such as Belyakov, Belyakov, and Klyachko [2], Gusev and Polovova [5], Bely, Bespalova, and Romanova [1], Savenkova and Sovetkina [14], Popov, Vlasov, Orlova [13] and others.

To ensure the sustainable economic development of universities, a series of phased reforms were implemented, such as decentralization of management, the introduction of a credit training system, the development of a three-stage training system, academic freedom and academic mobility. The creation of research universities and the introduction of university autonomy based on corporate governance principles are planned as strategic guidelines. Meanwhile, the complex of unsolved problems in the sustainable economic development of the higher education system remains quite wide.

Despite a large number of fundamental studies by foreign and domestic scientists, the problems of adapting positive global experience in creating a new model of innovative development of the country's economy, based on the integration of higher education, science and industry, have still been poorly studied. Dispositional education involves communicative orientations, personality traits, and individual predispositions.

An analysis of the literature data suggests that the monitoring system for the effects of teaching communicative skills should include those that are directly related to the components of communicative competence. A special place is occupied by studies in which the study of joint (parallel) changes in various types of components of competency is carried out: knowledge and skills, knowledge and attitudes (dispositions), skills and attitudes (dispositions) [2]. The central place in the structure of a person is occupied by the possibilities of its self-realization in communication. The institute is the highest step in the system of continuing business education. In the disciplines, the study of which in accordance with the state educational standard is mandatory, integrated curricula are compiled.

The change in the economic situation in the country posed the institute the problem of providing the educational process with educational and scientific-methodological materials. The accumulated educational literature is now beginning to become obsolete over the years. However, the transition to a personality-oriented education requires a significant change in the information and educational environment. It is about creating an electronic library and video library. And although electronic materials do not replace the teacher as the main carrier of information on the studied subject, the electronic library plays an important role in providing the student with the necessary procedural information in supporting motivation and maintaining interest in learning, setting out common goals and concretizing the objectives of the course, explaining difficult ideas and concepts. The electronic form of the course makes it possible to make quickly changes to the content and structure of the course, due to the development of existing and the emergence of new technologies.

A flexible system of settings makes it possible to adapt the system to the individual characteristics of the university. It reflects the basic data that is used by the entire system during the training process. Among them, the training system, the proportion of average current, rating, exam grades. In the final, the maximum possible percentage change in the individual curriculum. Effective education quality management is the ability to monitor and analyze performance data and make timely decisions [3].

Distance learning with interactive tools, a testing system makes it possible to conduct an interesting and objective learning process. Developing course packages requires the involvement of the most experienced teachers in this creative work. Such an organization of work makes significant changes in the control and assessment of students' knowledge, and contributes to the training of a new level specialist. At all stages of the work, it is proposed to use the included observation method in the educational process of teaching students. In the conditions of the formation of social and moral orientations, the student must develop the ability to independently solve the problems posed, becoming an initiative, business, and creative person. In many universities, new technologies for the provision of services are being tested, distance learning, new information technologies are being introduced. In this aspect, technologies such as credit and distance learning technologies, case technologies can be noted. Technological innovations include new teaching methods and technologies using information technologies, the development of which does not always bring the desired results. Distance learning technologies should be based on distance learning information exchange tools, the strategic focus of which is corporate training [4].

5 Conclusion

In the modern world there is an increase in the economic value of knowledge that:

- the development of progressive forms of organization of production, thereby ensuring an increase in GDP growth,
- increase the competitiveness of domestic goods and services, reduce the return on investment,
- form a healthy social environment, it is better to adapt to its rapidly changing conditions.

We believe that knowledge management plays a key role in shaping intellectual capital. Knowledge is an object (property), while an organization (its employees) is a subject (owner). In modern conditions there is no clearly formed model of higher education aimed at the needs of the market in relevant specialists. But, it should be noted that none can be called innovative. At present, they are leaning toward models of universities directly, since universities themselves determine innovative parameters of education [8].

Nevertheless, several types of modern models of innovative higher education can be distinguished. So, according to the model of network interaction, operate many universities. In market conditions, higher education institutions carry out the production and dissemination of knowledge, the formation of public and professional consciousness, and the development of culture. At the same time, important problems of the underdevelopment of the innovative component of higher education are: ineffective system of professional development of the faculty; low labor motivation, etc. [9].

The domestic higher education system is under the influence of the following threats:

- the lack of an adequate system of performance indicators for universities,
- underdevelopment of the system of strategic partnership between universities and production in solving problems of professional education.

The strength of universities is the development of strategic partnerships between higher education institutions and business structures, state and public organizations. The university development strategy should be innovative, since innovation-oriented organizations are able to best respond to market demands and the actions of competitors. It should satisfy the needs of society in the formation of a professional encyclopedist, able to withstand entropy processes in society, to ensure the achievement of a systematic, comprehensive quality of all aspects of its activities. The functioning and development of the university is carried out by pooling resources with its partners, implementing targeted programs together with the university and participating in its development [10].

The SMART University concept provides students with the opportunity to study at foreign universities on the basis of free access to international educational content; the opportunity to work in the specialty acquired at the university; rational alternation of the process of obtaining knowledge in the classroom with work on-line.

It is advisable to introduce the model of a research university primarily in humanitarian universities, where basic research predominates, and the model of an innovative university in technical universities with a developed material and technical base for applied research and development and commercialization of their results. The introduction of the criteria of economic sustainability in the practice of universities and their widespread use in practice will significantly increase the economic sustainability of the higher education system in Russia [12, 13].

The introduction of innovations should become attractive for business entities themselves, therefore, it is necessary to create free competition in the market. In addition, a mechanism of incentives and tax incentives should be created for domestic innovators. In particular, it is possible to exempt partially from taxes those business entities that are actively introducing innovative products in their activities or it is possible to use a mechanism of partial state funding for the development of innovative products and their implementation in practice.

The theory of system integration of universities with the real sector of the economy is supported by a number of scientists. So, in the works of scientists it is noted that the economy should be futures, that is, work for the future, and the development processes today should not focus on quantity, but on quality and anthroeconomics – investing in a person. Thus, it is necessary to further develop and improve this process. Improving sustainable economic development offers a package of measures that will allow universities to reach the world level of training.

References

1. Bely, E.M., Bespalova, M.P., Romanova, I.B.: On competitiveness and market stability of state universities. High. Educ. Today **6**, 15–17 (2011)
2. Belyakov, S.A., Belyakov, N.S., Klyachko, T.L.: Analysis and Assessment of the Economic Sustainability of Universities. Maks-M, Moscow (2008)
3. Castiaux, A.: Inter-organizational learning Lotka-Volterra modeling of different types of relationships. In: Kelly, G.J., Walker, P. (eds.) International Conference of the System Dynamics Society – Community-Level Systems Thinking, pp. 2–6. Springer, Cham (2004)
4. Fedorova, A.V.: Integration of universities and innovative enterprises: aspects of knowledge management. Econ. Manage. Innov. Tech. **12** (2013). http://ekonomika.snauka.ru/2013/12/3598. Accessed 11 Apr 2020
5. Gusev, Yu.V, Polovova, T.A.: Problems of a stable trajectory of universities functioning. ETAP Econ. Theory Anal. Pract. **1**, 112–121 (2009)
6. How, J.: Crowdsourcing. Collective Intelligence as a Tool for Business Development. Alpina Publisher, Moscow (2012)
7. Ivinskaya, I.V.: Enterprise activity of high school as the factor of economic development of region. Socio-Econ. Phenom. Process. **10**(032), 53–60 (2011)
8. Kalinina, S.V.: Assessment of the competence of the teaching staff in the quality management system of the university. Manage. Russia Abroad **1**, 115–122 (2011)
9. Koryakov, A.G.: Methodological issues of sustainable development of enterprises. Issues Econ. Law **4**, 4–5 (2012)
10. Malysheva, M.A.: Modern Teaching Technologies at the University (Experience of the HSE in St. Petersburg): A Teaching Aid. HSE Research Institute, St. Petersburg (2011)
11. Mikhalev, O.V.: Economic Stability of Economic Systems: Methodology and Practice of Scientific Research and Applied Analysis. Publishing House of the St. Petersburg Academy of Management and Economics, St. Petersburg (2010)
12. Niu, K.-H.: Knowledge management practices and organizational adaptation Evidences from the high technology companies in China. J. Strategy Manage. **3**(4), 325–343 (2010)
13. Popov, E.V., Vlasov, M.V., Orlova, N.V.: Transaction costs of generating explicit knowledge of an academic institute. Manage. Russia Abroad **1**, 107–114 (2011)
14. Savenkova, Y., Sovetkina, A.A.: Management of the competitiveness of a university in modern socio-economic conditions. Educ. Issues **4**, 182–198 (2009)

Innovative Methods of Managing the Company's Financial Results

D. V. Aleshkova[1](✉), A. V. Shepelev[1], and Kh. M. Salikhov[2]

[1] Samara State University of Economics, Samara, Russia
dashajuly343@gmail.com, giz-shepelev@mail.ru
[2] Kazan Federal University, Kazan, Russia
2911379@mail.ru

Abstract. The article investigates the experience of using innovative methods of managing financial results in a joint venture and the possibility of using these methods at a Russian enterprise. The importance of managing financial results for an individual enterprise and for the country's economy as a whole is indicated. The research includes CVP analysis of the enterprise, DuPont model, and profitability analysis of the enterprise. The authors make conclusions about the applicability of classical and innovative methods at each of the considered enterprises. Recommendations for managing financial results within the research object are proposed, their effectiveness is calculated, and the repeated CVP analysis was performed based on the expected results. It is concluded that these measures will be effective for the company in various ways. They can be implemented together or separately. However, their combined effect may differ from the sum of the effects.

Keywords: Financial results of the company · Innovative management methods · Profit · Profitability

1 Introduction

The financial result of an enterprise is an indicator that affects not only the enterprise itself, but also the state as a whole. This indicator allows you to determine the stability of enterprises in a particular industry, as well as to increase the market value of the enterprise. A high-quality system for managing financial results is a fundamental basis for ensuring sustainable growth of an economic entity. The practical significance of this study is to develop specific proposals for improving the organizational and economic system for managing financial results at the enterprise, which can be used by the company's management in developing a comprehensive system for this purpose.

Thus, the study of various methods and techniques for managing financial results of an enterprise is an actual and promising direction of economic research. The enterprise "EPK Samara" is the research object. This enterprise produces more than 4.5 thousand types of radial ball bearings, radial thrust bearings, and roller radial and thrust bearings. The company is part of the European Bearing Corporation, which produces bearings for all branches of mechanical engineering. The foreign representative of this corporation is the joint venture "EPK Brenco" organized with the American partner Amsted

S. I. Ashmarina et al. (Eds.): ENGINEERING ECONOMICS WEEK 2020, LNNS 139, pp. 649–658, 2021.
https://doi.org/10.1007/978-3-030-53277-2_77

Rail. In the framework of this study, the authors try to compare methods used at domestic and foreign enterprises of the same corporation to manage their financial results. Based on this, it will be possible to draw a conclusion about the adjustment of methods used by enterprises to improve the efficiency of their activities.

2 Methodology

First of all, it is necessary to assess the current financial condition of the research object. One of the methods used for strategic analysis of the company's financial results is a CVP-analysis, which provides an opportunity to optimally compare variable and fixed costs, as well as the price and volume of sales [2].

The main elements that are analyzed in this method are margin profit, break-even point (BEP), financial safety margin, and operating leverage. The calculation of indicators is shown in Table 1.

Table 1. CVP-analysis

Indicator	2017	2018
Revenue (thousand rubles)	4 596 856	4 741 589
VC	1 877 535	2 030 560
FC	1 081 378	1 083 262
Marginal profit	2 719 321	2 711 029
Specific margin profit	0,85	0,9
Break-even point (pieces)	1 272 838	1 197 936
Break-even point (rubles)	1 828 008	1 894 625
Financial safety margin (FSM)	2 768 848,58	2 846 964,05
Coefficient of FSM	0,6	0,6
Operating leverage	1,66	1,67

Source: authors.

As we can see, changes in the cost structure in 2018 mostly affected their variable part. Due to the decrease in production volumes in 2018, the share of raw material costs decreased. At the same time, the share of costs associated with paying for the services of third-party organizations involved in the production process has increased almost 3 times. Despite this, margin profit decreased by less than 1%. This change was caused by the revenue growth in 2018, which allowed covering increased variable costs. In 2018, the specific margin profit averaged 0.9 rubles per unit of output.

The value of the break-even point in physical terms has also decreased. At the same time, the break-even point in cost terms has increased. This is due to an increase in the share of the variable part in the enterprise's cost structure. Due to the increase in revenue in 2018, the company's financial safety margin also increased. But, despite these changes, the excess of sales revenue over its threshold value in 2018 compared to 2017 did not change and amounted to 60%. In this regard, the value of the operating leverage has not changed much. That is, the extent of the cost structure impact on the company's revenue has not changed.

Based on the analysis, we can conclude that the increase in revenue is caused by an increase in prices for manufactured products, since the production volume tends to decrease. This method of increasing revenue is the extensive one, which is not always applicable. The "EPK Samara" cost structure is dominated by variable costs, which have increased despite a decrease in production volumes. There is an increase in the cost structure of expenses for third-party services. This leads to the assumption that optimization of this particular cost item can have a positive impact on the company's profit. In this case, it's worth considering the possibility of changing contractors who perform such services to more budget options.

Another way to reduce these costs is to link the cost item to a specific product group. In this case, it is possible to reduce costs by decreasing the production volume of goods that require the involvement of third-party organizations for their production, while increasing the production volume of other product groups. Another approach to optimizing this cost item may be the creation of the own department that would perform functions of third-party organizations. However, it makes sense to evaluate alternatives first, since there are examples where maintaining own divisions is more expensive. The next step was to analyze the profitability using the three-factor DuPont model [1]. The initial data for the analysis of the return on equity (ROE) of "EPK Samara" are presented in Table 2.

Table 2. Initial data on the company's financial results

Indicators	2017	2018	Changes, %
Net profit, *thousand rubles*	1 374 334	1 355 953	−1,3
Revenue, *thousand rubles*	4 596 856	4 741 589	+3,1
Assets, *thousand rubles*	4 953 568	4 487 981	−9,4
Equity capital, *thousand rubles*	3 903 140	3 980 251	+2,0
ROE, %	35,2	34,4	−2,3

Source: authors.

The data presented in the table indicate that the ROE indicator is decreasing in 2018 compared to 2017 by 0.8% points or 2.3%. This is because of a decrease in net profit (−1.3%) by an increase in equity (+2.0) for the period under review. To calculate the measure of influence of each of the factors on the change in the ROE index, an index factor analysis is used, in particular the method of chain substitutions (Table 3).

As it can be seen from the calculations in Table 3, the ROE decreased in 2018 compared to 2017 by 0.8%, which indicates a decrease in the efficiency of the company. ROE decreased by 0.12 p.p. because of a decrease in profitability of sales by 1.3 p.p. or 4.3%. It is possible to increase the profitability of sales (net margin) by reducing the cost per ruble of products and services provided. ROE increased by 0.36 p.p. due to the acceleration of the working capital turnover by 0.111 turns or 12.4%. ROE decreased by 0.32 p.p. due to a decrease in financial leverage by 11.1%.

Table 3. Calculation of indicators using the DuPont model

Indicators	2017	2018	Absolute change	Relative change, %	Influence of the factor
Profitability of sales	29,9%	28,6%	−1,3	−4,3	**−0,12**
Asset turnover ratio	0,893	1,004	0,111	12,4	**0,36**
Financial leverage ratio	1,269	1,128	−0,141	−11,1	**−0,32**
The combined effect of three factors					**−0,8**

Source: authors.

3 Results

3.1 The Modern System of Financial Results Management

Planning and budgeting at "EPK Samara" is carried out by the budget planning and analysis department. When the company's divisions interact, the general budget for the year is drawn up. This is a coordinated plan for all divisions and functions of the enterprise as a whole, which combines private budgets and characterizes the information flow for making management decisions. Monthly budgets are also drawn up for divisions.

The budget planning and analysis department deals with issues related to making settlements with contractors, determining the need for inventory items, drawing up revenue and expenditure budgets, the sales budget taking into account services, the cash flow budget and the balance sheet.

The sales department provides information on sales of the main production, the dynamics of finished products stocks, the project of the annual application for production for the upcoming period, etc. The production department is responsible for preparing the production and procurement budget for production needs. The department of labor and wages is engaged in the formation of the wage fund, the distribution of labor costs and the development of the production calendar for the upcoming period. The HR department is responsible for transferring funds for social programs. The technological department deals with the issues of production reconstruction and modernization, as well as reducing production costs. The production department provides information about work in progress, the accounting department – about the size and structure of tax payments, etc. (Table 4).

After analyzing this system, we can conclude that only one department – the technology management – is indirectly related to financial results management. This department deals with reducing costs, which helps to influence the financial result only from the production point of view.

Based on a comparison of the use of classical methods of managing financial results, it can be concluded that there is no particular difference at the enterprises under consideration. Therefore, classical methods will not be able to give the necessary result

Table 4. Comparison of classical methods used for managing financial results

Method	«EPK Brenco»	«EPK Samara»
Accounting policy	+	+
Information policy	+	+ −
Profit forecasting	+	+
Planning	+	+
Financial structure management	+ −	+ −
Production assets management	+	+
Managing the use of production non-current and current assets	+	+
Depreciation policy	+	+
Price policy	+ −	+ −
Assortment policy	+ −	+ −
Sales policy	+	+
Tax policy	+	+
Dividend policy	+	+
Managing the cost of capital	+ −	+ −
Risk management	+	+
Personnel policy	+ −	+ −

Source: authors.

when managing financial results of the enterprise. Next, we will consider the use of innovative methods for managing financial results at enterprises (Table 5).

Table 5. Comparison of innovative methods used for managing financial results

Method	«EPK Brenco»	«EPK Samara»
CVP-analysis	+	−
Cost-killing	−	−
Kaizen costing	−	−
Benchmarking	+	+
ERP-system	+	−
Budgeting	+	+
BSC	−	−
BRP	−	−
Brainstorming	+	+
5S	+	−
DuPont Model	−	−

Source: authors.

Most of the innovative methods of managing the company's financial results have not been applied at "EPK Samara" yet. At the same time, "EPK Branco" uses some innovative methods for managing financial results that are not used in "EPK Samara". These methods are CVP-analysis, ERP-system and 5S-system. Therefore, it makes sense to consider the possibility of implementing some of these methods in relation to the research object.

3.2 SWOT-Analysis of the Enterprise

SWOT-analysis is one of the most common methods that assess internal and external factors that affect the company's development in a complex. It follows from the SWOT-analysis of the enterprise, that despite the large number of strengths and opportunities of the enterprise, it is necessary to take into account threats and reduce a number of weaknesses (Table 6).

Table 6. SWOT-analysis of the research object

Strengths: 1) availability of unique technologies; 2) high level of training of working personnel; 3) positive reputation of the company; 4) a large share in the domestic market; 5) high level of technical equipment in the production of new types of products; 6) a wide range of products; 7) development of new types of bearing products	**Weaknesses:** 1) lack of staff (workers); 2) uneven distribution of job responsibilities; 3) an insufficiently developed system of interaction between departments; 4) inefficient system of rewarding employees for completing tasks; 5) high degree of depreciation of production assets in the production of basic types of products; 6) insufficient level of profitability
Opportunities: 1) high demand for bearings; 2) new R&D and NTP in the bearing industry; 3) preferential conditions for the production of bearing products; 4) increase in government funding/grants; 5) increasing the state defense order	**Threats:** 1) emergence of new competitors; 2) the economic situation in the country; 3) negative changes in the tax legislation; 4) a fixed state price for bearing products

Source: authors.

SWOT-analysis allows concluding that, despite the development and application of new technologies in the production, the availability of a wide range of products and positive reputation of the enterprise, the weak point limiting its development and, as a consequence, the increase in profits is the lack of staff, lack of clear organization units, and a high degree of the equipment deterioration. State orders build a large share of the company's orders. Therefore, a decrease in their volume can significantly affect the performance of the enterprise. However, a wide range of products can compensate for these trends by entering new markets. The insufficient level of the enterprise profitability indicates the need to improve the mechanisms for managing financial results.

In addition, the results of the SWOT-analysis clearly show that the main weaknesses of the considered enterprise are related to the imperfection of the enterprise work organization. Such features may affect the ability to implement certain measures designed to affect the financial results of the company. Poorly organized management activities will prevent the application of new systems, or the implementation of activities will not be able to give the necessary effect.

3.3 The Improvement Directions of Mechanisms for Managing Financial Results

Based on the results of calculating the ROE indicators, it was noted that the indicators of the enterprise are lower than the values of the main competitors from China (ROE = 42.8) [4]. Previously, CVP analysis and analysis based on the DuPont model were conducted, the results of which allowed us to draw conclusions about the use of asset turnover and increased production volumes as a source for the ROE growth. Focusing on the ROE of Chinese manufacturers, we set the ROE rate at the level of ROE = 42. Based on this, it is possible to determine the target rate of return. So, if ROE is approximately equal to 42, and the average amount of equity is equal for 2018, then, substituting the indicators in the formula, we can say that the target net profit should be equal to 1,655,513 thousand rubles. This value is more than the actual value by about 300,000 thousand rubles.

Based on the results of the CVP-analysis, it should be noted that an extensive way to increase profits by increasing sales prices is not optimal for the company. The risks of reducing revenue while maintaining prices are not so great, which is due to the presence of regular customers, a high level of products quality, as well as a good reputation of the manufacturer. Therefore, it makes sense to use the operating leverage to increase the financial result. In this case, there is a possibility of transferring a part of variable costs to the permanent category. Let's consider the cost structure of the enterprise (Table 7).

Table 7. Cost structure of the company for 2018 (thousand rubles)

Real	Prospective
Variables: 2 030 560 Including: – raw materials: 1 637 041 – third-party services: 393 519	Variables: 1 768 692 Raw materials: 1 768 692
Constant: 713 440 Including: Wages: 431 402 Deductions from wages: 121 724 Depreciation: 111 843 Other expenses: 48 472	Constant: 1 049 810 Including: Wages: 684 227 Deductions from wages: 205 268 Depreciation: 111 843 Other expenses: 48 472
Cost price: 100% Variables: 74% Constant: 26%	Cost price: 100% Variables: 63% Constant: 37%

Source: authors.

As you can see, reallocating a variable cost item such as third-party services to a fixed cost item allows reducing the share of variable costs in their overall structure, which is what the impact of the operating leverage is manifested in. It is also possible to influence the financial result by accelerating the turnover of assets through reducing the duration of the production cycle and the loss of working time. All this can be implemented through the modernization of the company's financial results management system. Based on all calculations and conclusions, we can offer the following recommendations for improving the management of "EPK Samara" in relation to the financial results.

First of all, it is the implementation of the lean production system in the workplaces not only in production, but also in engineering and technical departments. The responsible division for the operation of this system will be the production department and employees who perform the necessary functions at their workplaces. The main costs associated with the implementation of such a system are costs of maintaining cleanliness and order in the production, as well as costs of conducting an audit of the implementation of standards associated with the existing remuneration system. Another activity that can affect the target profit is the influence on variable costs such as third-party services (for example, annealing and rolling out components, as well as their transportation). These costs make up a large share of the total cost structure, and, as a result, have the greatest impact on the production cost.

There are several ways to reduce costs. The first is to change the counterparty. If there is such a possibility, it makes sense to find another supplier with more favorable conditions. If this is not possible, the enterprise management should consider the opportunity of opening the own department that would do similar work. However, first of all, it is necessary to assess whether this will be cost-effective for the company. Costs of creating a workshop, attracting additional employees, purchasing raw materials, etc. should be taken into account.

Initially, this measure requires quite high costs associated with the direct creation of workshops and the purchase of equipment. At the same time, in the future, costs will consist of expenses for raw materials, maintenance of the workshop and wages for additional workers. However, it is assumed that these costs will be significantly lower than when contacting contractors. If such a measure is profitable, the company will be able to reduce the level of variable costs, which will also reduce the production cost. The company will be able to increase production volumes by reducing costs, as well as speed up turnover, since the dependence on contractors will be less. Based on this decision, the authors performed a repeated CVP-analysis after transferring part of the costs from the variable category to the constant one (Table 8).

The results of the repeated CVP analysis show that with unchanged revenue and sales volume, the transfer of a part of variable costs to the category of constant ones enables an increase in the margin profit by about 12%, while the break-even point in physical terms decreases by almost 32%, and the financial safety margin increases by 16.6%. It is a point when the effect of the operating leverage is clearly shown, the value of which has decreased by almost 15%. And the further the company is from the break-even point, the smaller the operating leverage is.

Table 8. Repeated CVP-analysis

Indicator	Expected value	Changes, %
Revenue (thousand rubles)	4 741 589	–
VC	1 697 941	−16,4
FC	892 792	+25,1
Marginal profit	3 043 648	+12,3
Specific margin profit	1,1	+22,2
Break-even point (pieces)	811 629	−32,2
Break-even point (rubles)	1 099 889	−42
Financial safety margin (FSM)	3 319 112	+16,6
Coefficient of FSM	0,7	+16,7
Operating leverage	1,42	−15

Source: authors.

4 Discussion

Considering the management of financial results of an enterprise, a lot of authors note that financial indicators are largely dependent on various management mechanisms used at enterprises [5]. They also describe the impact of innovative methods of managing the company's financial results on the effectiveness of its activities. It is believed that a greater effect is achieved when several innovative mechanisms are used together [7]. Some authors note that the choice of innovative mechanisms for managing financial results for a particular enterprise depends on its accounting policy, financial indicators, risk levels, and the enterprise itself [3]. Other scientists note the inadequacy of classical mechanisms for managing financial results in relation to modern realities. In this regard, the importance of using innovative mechanisms to influence the enterprise financial efficiency is confirmed [6].

5 Conclusion

The study confirmed the need to use recommendations to optimize and improve the mechanism for managing financial results at the enterprise:

- implementation of the lean production system in the workplaces of employees not only in the production, but also in engineering and technical departments,
- creation of additional production workshops that will substitute services of third-party organizations.

Based on the results obtained, we can conclude that the enterprise uses only a few innovative methods for managing financial results, and these methods do not give the result needed. After conducting the CVP-analysis and using the DuPont model, it was found out that the variable part prevails in the cost structure, costs of services from third-party organizations. In order to optimize the cost structure and increase profits, it is proposed to use the 5S system and create the own workshop, which would substitute services of third-party organizations. Summing up, we can conclude that these measures could be effective for the company in various ways. They can be implemented together or separately. However, their combined effect may differ from the sum of the effects.

References

1. Dauphiné, A.: Dupont and systemic models. Bulletin d'Association De Geographes Francais' **95**(3), 361–370 (2018)
2. Kenton, W.: Cost-volume-profit – CVP analysis definition. Corporate Finance & Accounting (2020). https://www.investopedia.com/terms/c/cost-volume-profit-analysis.asp. Accessed 11 Apr 2020
3. Kober, R., Subraamanniam, T., Watson, J.: The impact of total quality management adoption on small and medium enterprises' financial performance. Account. Finance **52**(2), 421–438 (2012)
4. Luo, H.R., Wang, R.: Foreign currency risk hedging and firm value in China. J. Multinational Financ. Manage. **47**, 129–143 (2018)
5. Ma, W., Jin, M., Liu, Y., Xu, X.: Empirical analysis of fractional differential equations model for relationship between enterprise management and financial performance. Chaos Solitons Fractals **125**, 17–23 (2019)
6. Quon, T.K., Zéghal, D., Maingot, M.: Enterprise risk management and business performance Buring the financial and economic crises. Probl. Pers. Manage. **10**(3), 95–103 (2012)
7. Rasid, S.Z.A., Isa, C.R., Ismail, W.K.W.: Management accounting systems, enterprise risk management and organizational performance in financial institutions. Asian Rev. Account. **22** (2), 128–144 (2014)

Analysis of Competitive Strategy Development in an Innovative Company

O. V. Bakanach[1(✉)] and A. V. Komyagin[2]

[1] Samara State University of Economics, Samara, Russia
bakanach@mail.ru
[2] Nonprofit Company "Innovation Fund of Samara Region", Samara, Russia
alexei.komyagin@yandex.ru

Abstract. This paper presents a thorough analysis of the factors that determine the development of competitive strategy in an innovative company. The study is based on the data of "Nanotechnology center of Samara *oblast*" Limited Liability Company for the period up to 2025. The "Strategy for Spatial Development of the Russian Federation" provides for the acceleration of scientific, technological and innovative development of the Russian Federation via the promising large centers of economic growth. This strategy and other national projects in the country require specific instruments for their implementation. Application of modern technologies represents one of such instruments and thus, there is an urgent need to explore them in details. The aim of this study is to analyze the factors that determine the development of competitive strategy in an innovative company. The methodology applied in the paper includes the tools and methods of strategic planning. The strategy development stages of the innovative company (Nanocenter) are investigated. The forecast of the company's key economic indicators is made. The transition to the new "innovation pipeline" model is justified in this study.

Keywords: Forecasts · Innovative company · Startup · Strategic planning

1 Introduction

Central to the constant growth in the RF innovative development is the creation of high-tech companies, implementation of innovative activities bundled up with R&D and aimed at commercialization, alongside with the modernization of existing production facilities and organization of new ones [8]. The concept of technological breakthrough encompasses these and various other tasks. Indeed, it is impossible to create competitive advantages for industries without developing and implementing technological innovations, developing modern infrastructure, and organizing cooperation and collaborations. Along with these, the accompanying tasks of companies are to survive in a crisis, reduce risks, increase sustainability, strengthen the role of Intellectual Property (IP), improve the competence and skills of personnel, invest capital, change the structure of production, etc.

The development of key industries and areas that can bring Russia to a new level of technological development is possible through the formation of a multi-industry

S. I. Ashmarina et al. (Eds.): ENGINEERING ECONOMICS WEEK 2020, LNNS 139, pp. 659–665, 2021.
https://doi.org/10.1007/978-3-030-53277-2_78

complex based on modern and globally competitive technologies, and the activation of the development, implementation and commercialization of innovations. All these activities require the development of methods and tools for the strategic development of both existing traditional industries and modern newly created companies. The trends of the country's economy to move along the innovative path of development determined the setting of strategic planning tasks at the federal and regional levels, as well as at the level of companies and industrial enterprises [10].

This strategic planning process is essential for innovative business. Undoubtedly, the pace of economic growth in this country is largely dependent on innovative companies. The study of the processes of forming competitive strategies in innovative companies is one of the most relevant and paramount both in economic theory and in economic practice. The aim of this study is to analyze the specific factors that determine the development of competitive strategy in nanotechnology center.

It should be noted that most of the research is based on foreign data and evidence and systematizes the concepts that were formed in foreign companies, which, in some cases, makes it difficult to apply it to the domestic business environment [5, 6]. This is caused by the specifics of the organizational structures of large enterprises and limited information about the market [1].

Although the solution to many theoretical, methodological and applied problems in the subject area under study has been found, still the key characteristics of strategic development of innovative companies engaged in serial creation and development of startups, the importance of which for the economy has been empirically proven, nave not been sufficiently studied and justified [3].

2 Methodology

This study uses methodological approach based on the following methods of strategic planning.

1. External audit, which implies collecting information about the macro-and meso-environment in order to identify trends in the development of the industry, especially the relationships of participants in production chains.
2. Internal audit, which implies conducting an analysis of the company's own capabilities for strategy formation, i.e. management analysis, analysis of internal resources and capabilities using the method of expert assessments, economic and statistical forecasting methods.

The investigation is concerned with "Nanotechnology center of Samara *oblast*" Limited Liability Company, which represents a set of business units and business processes aimed at commercializing technologies in the field of nanoindustry, based on the combination of real estate, laboratory and technological equipment, as well as a set of marketing and business support services for innovative companies through technology transfer and the creation of new companies.

3 Results

Nanotechnology centers are among the range of companies that form the framework of the RF innovative ecosystem. Such centers are often defined as venture construction companies and their objectives are determined as follows:

- transfer and commercialization of new technologies – creation of innovative startups,
- identification of advanced technologies,
- support the startups at the initial stage,
- research and development procedures,
- business formation.

To provide venture construction and guarantee an innovative way for development of the economy in the region, "Nanotechnology center of Samara *oblast*" Limited Liability Company (Nano center, NC *SO*) was founded in 2015. Currently, the stage of the regulatory documents pooling alongside with logistics, and partner networks formation has finished and the team building stage has been completed. The development of the company's medium-term strategy for the period up to 2025 is now on the agenda of "Nanotechnology center of Samara *oblast*" Limited Liability Company. As NC *SO* is one of 15 nanocenters, its mission flows organically from the objective of the whole network. The mission of the NC *SO* is to commercialize innovations in nanotechnology and nanomaterials through the serial hatching of high-tech companies and their launch on global markets. Following its mission, NC *SO* aims to become a leading Russian technology center by investing in high-tech projects and bringing them to global markets. NC *SO* strategy 2025 is divided into 2 stages.

STAGE I. Ecosystem V 1.0 in 2015-2019

This stage of NC *SO* development 2015–2019 is characterized by the accumulation of competencies in certain areas and the formation of the first portfolios of small innovative companies. By the end of the stage, 8.86 million rubles were invested. Out of more than 160 projects proposed by the applicants and partners, a total of 22 investment projects (with incorporation) were approved for financing by the NC *SO* Board of Directors in coordination with the Investment Committee.

Based on the results of the V 1.0 ecosystem analysis, the main objective for 2020–2025 was determined, namely to create conditions for the serial launch of startups. Based on this objective, the following tasks were set:

- to create the "perfect" projects under the revised and approved specialization,
- to set up the contract productions of composite nanostructured reinforcing elements for drilling bits (PDC-cutters) and metal dentures with nanocoating,
- to enlist global technology companies' cooperation in the discussion of joint ideas and projects,
- to arrange the designing and launching of a technology platform (TP) in the approved areas that will accumulate competencies, share the risks of participants and can be used as a base for a serial launch of startups in the chosen direction.

Based on the results of solving the tasks set at stage 1, the NC SO ecosystem was upgraded from the version 1.0 to the version 2.0.

STAGE II. Ecosystem V 2.0 in 2020–2025

The ecosystem of NC *SO* in 2020–2025 is characterized by the development of previously created project companies (including building the logic of development in accordance with the new concept of the "innovation pipeline"), as well as the creation of new small innovation company (SIC) and TP based on the accumulated competencies. In contrast to the V1.0 ecosystem, there is a percentage redistribution of the sources for project portfolio generation: 60% of projects are created "inside" the NC *SO*, 30% of projects are created together with the network of nanocenters, and only 10% come from external sources. The difference between the V2.0 ecosystem and V1.0 is the transition to a new level of interaction between projects. This creates the following project relationships:

- "Technological Platform to Startup",
- "Technological Platform to Technological Platform",
- "Startup to Startup".

The stage of 2020–2025 is characterized by the complete formation of NC's *SO* position as one of the elements of the innovation pipeline. There should be a self-sustaining reaction to the creation and development of startups, which, in turn, will provide the basis launching of other startups [2].

The main goal for 2020–2025 is to ensure the launch of the maximum number of startups within the technological platforms, as well as to provide the foundation for the implementation of the nanotechnology center as a serial technological entrepreneur.

Strategic goals and objectives of the company in 2020–2025

1. To ensure constant sequence of businesses creation:
   - launching of five technological platforms until 2025,
     launching of 7 startups in 2020 and subsequent annual launch of not less than 8 startups.
2. To increase the qualitative constituent of startups:
   - increasing the share in the partnership with NC network to 30% to 2025,
   - incorporating technological platforms in partnership with global technological companies,
   - hunting and recruiting partners for international agreements,
   - enlisting international partners in project company's equity participation.
3. To ensure survival and high yield from investments:
   - investment startups at the "Seed" stage – 30% survival, 26% expected return,
   - investment startups at the "Round A" stage– 80% survival, 24% expected return,
   - "Cash cow" startups – 100% survival, 23% expected return.

   Survival desired assessment is an expert one, which is based on venture investment experience and analysis of high tech startups' survival. Expected return is set by NC

SO as the minimum rate of return, which allows positive assessment of investments in the project.

4. To ensure the efficiency of NC's project activity:
   - annual exit with a set return from startups in accordance with the plan shown in the table below.
5. To ensure the integration of NC *SO* in the economy of Samara *oblast*:
   - job creation (not less than 580 jobs to 2025),
   - premises rented (not less than 500 m2 to the end of 2019, not less than 3000 m2 to the end of 2025).

The new key indicators of NC's *SO* efficiency are in compliance with the new approved Strategy of the Fund for infrastructure and educational programs until 2025 (*FIOP*) until 2025 [9] (Table 1).

Table 1. Key economic indicators of NC *SO* for 2020–2025

Indicator	Measuring unit	2020	2021	2022	2023	2024	2025
NC *SO* and SIC revenue	mln. rubles (for the period)	210	250	270	400	550	900
Number of employees in NC *SO* and SIC	people (progressive total at the end of the period)	190	220	285	380	490	580
Performance of contractual R&D	units (for the period)	2	2	2	2	2	2
Creation of intellectual property objects	units (for the period)	8	8	9	9	9	10
Number of legal entities founded within NC *SO* framework	units (progressive total)	44	52	60	68	76	84

Source: authors.

In 2020 the main revenue stream is expected from contract productions organized within the framework of projects in the field of oil development and dental technology. The rest of the revenue in 2020 will be generated by startups CASH COW, PRE SEED and SEED stage.

It is forecasted that in future 10% of revenue will be generated by startups of the PRE-SEED and SEED stage, 20% - by startups of the ROUND A stage. Contract production at full load will generate up to 30% of revenue, and CASH COW projects will account for up to 40% of total revenue (Table 2).

Providing the environment for the serial launch of "perfect" projects should become the main result of the strategic activities of NC SO.

Table 2. NC *SO* revenue forecast in 2020–2025, mln.rubles

Indicator	Total	2020	2021	2022	2023	2024	2025
Investment startup PRE SEED and SEED	311	21	25	27	40	55	90
Investment startup ROUND A	536	42	50	54	80	110	180
CASH COW startup	1 072	84	100	108	160	220	360
Contractual production	804	63	75	81	120	165	270
Total revenue	**2 723**	**210**	**250**	**270**	**400**	**550**	**900**

Source: authors.

4 Discussion

Currently, Russian companies operate in a constantly changing environment, which requires fundamentally new approaches to business management. The instability of the economic environment, tougher competition, increasing consumer demands for the product, and shortening the product life cycle significantly complicate the management process, thus the prospects for development are becoming less predictable. In these conditions, the implementation of strategic management ideas and technologies into the practice of enterprises is of special importance.

The main goal of the company's development strategy is to achieve long-term competitive advantages that will ensure its survival, sustainable operation and development. The scope of strategic solutions is extensive: the choice of areas of activity, justification of priorities in the use of resources, search for long-term partners, organizational forms of management, opportunities to use the strengths of the enterprise, reducing the negative consequences of the weaknesses and threats from the external environment. A strategic approach to solving company management problems makes it possible to form alternatives for its development [11]. Innovation is the main competitive advantage for sustainable development of companies. At the same time, the growth of the market for successful startups can become a driver for the development of priority sectors of the economy [7].

5 Conclusion

According to the Strategy of the Fund for infrastructure and educational programs until 2025 [9], the model "Innovation Pipeline" constitutes the basis for nano centers activities, including NC *SO* among others.

The new proposed model of work for NC *SO* is based on the transition from the "project funnel" (consideration of incoming applications to the nanocenter) to the generation of projects directly "inside" the center. This model involves analyzing the requests and needs of the existing and potential industrial partners, and creating technological solutions in accordance with the analyzed requests and with the possibility of subsequent implementation of these technologies on the market.

Today, currently the activities of NC SO, which are driven by trends described, are aimed not only at creating startups focused on their basic technology. They also strive for creating companies that specialize in popular services in the field of technological entrepreneurship (prototyping, manufacturing, engineering, marketing and sales), thereby forming a block of infrastructure companies. The most important part of such services will be combined within separate contract productions, while the other part will be distributed among other startups. Thus it will open up various opportunities for solving individual tasks by combining elements among themselves. The process of managing, analyzing and prioritizing technological trends and creating components of technological infrastructure will allow startup teams and technology entrepreneurs to focus on the underlying technology [4].

References

1. de Oliveira, L.S., Soares Echeveste, M.E., Cortimiglia, M.N., Gularte, A.C.: Open innovation in regional innovation systems: assessment of critical success factors for implementation in SMEs. J. Knowl. Econ. **10**, 1597–1619 (2019)
2. Eroshenko, E.P.: Development of the business ecosystem in the regions of the Russian Federation as the basis for creating successful startups. Econ. Bus. Law **9**(2), 123–134 (2019)
3. Ershova, A.E., Kuznetsov, A.N.: Strategy competitive development of the company BMW group. Bus. Strategy **5**(61), 26–28 (2019)
4. Feki, C., Mnif, S.: Entrepreneurship, technological innovation, and economic growth: empirical analysis of panel data. J. Knowl. Econ. **7**, 984–999 (2016)
5. Fink, T.M.A., Reeves, M., Palma, R., Farr, R.S.: Serendipity and strategy in rapid innovation. Nat. Commun. **8**, 2002 (2017)
6. Senior, M.: Europe's biotech renaissance. Nat. Biotechnol. **38**, 408–415 (2020)
7. Sergeeva, I.G., Orlova, O.P.: Analysis of high-tech companies formation and development. Russ. Econ. Bull. **3**(1), 156–161 (2020)
8. Sokolov, A.P., Kabannik, E.A.: Economic conditions and priority directions of innovative development of the Russian economy. Vestnik Altai Acad. Econ. Law **1**, 86–94 (2020)
9. Strategy of the Fund for infrastructure and educational programs until 2025. https://fiop.site/o-fonde/strategiya-razvitiya/. Accessed 20 Apr 2020
10. Trofimova, O.E.: Investment of venture capital in Russia as an important element of transitioning to an innovative economy under sanctions. Stud. Russ. Econ. Dev. **28**(4), 398–405 (2017)
11. van Someren, T.C.R., van Someren-Wang, S.: The instrument: strategic innovation as a new foundation for Russian innovation system. In: van Someren, T.C.R., van Someren-Wang, S. (eds.) Strategic Innovation in Russia, pp. 81–161. Springer, Cham (2017)

Demand for Engineering Qualifications in the Transformation Period

M. V. Simonova[1](✉), L. V. Sankova[2], F. I. Mirzabalaeva[3], and E. V. Privorotskaya[4]

[1] Samara State University of Economics, Samara, Russia
m.simonova@mail.ru
[2] Yuri Gagarin State Technical University of Saratov, Saratov, Russia
sankovalv@sstu.ru
[3] Plekhanov Russian University of Economics, Moscow, Russia
faridamir@yandex.ru
[4] Samara State Technical University, Samara, Russia
dmipriv@mail.ru

Abstract. The study analyzes the situation on the labor market in four regions of the Volga Federal District of the Russian Federation in the context of a professional group of engineering and technical specialties. The analysis was performed to identify the conformity of training volumes, the structure of employment by education, the structure and dynamics of active resumes and vacancies. The purpose of the research is to identify trends in demand for engineering qualifications in a strategic perspective and to overcome the consequences of the coronavirus pandemic. The authors revealed the tendency to increase and equalize the expenditures of regional budgets on education, and to reduce output in the studied group of specialties. The structure of supply and demand in labor markets of the Volga Federal District, which is characterized by a highly competitive situation among applicants and a decrease in job offers, is determined. The necessity of advanced training of engineering and technical specialists with basic digital skills is justified.

Keywords: Digital skills · Supply and demand in the labor market · Technical engineers · Transformation · Qualification

1 Introduction

The period of global transformations around the world leads to irreversible consequences in many sectors of the economy, destroys the structure of interconnections in the global economic system. There is a real prospect of mass unemployment, which was recently discussed, but in another aspect – it happens because of automation and robotization of production [9]. The imposition of these processes will have consequences for the entire labor market, including for the specialized labor market oriented to industry [10].

During and after the crisis, it becomes obvious that many formats of the digital economy will not be able to continue to develop or die, such as a bloated financial

S. I. Ashmarina et al. (Eds.): ENGINEERING ECONOMICS WEEK 2020, LNNS 139, pp. 666–675, 2021.
https://doi.org/10.1007/978-3-030-53277-2_79

sector, many types of services built on entertainment and consumer industries. The sphere of real material production of necessities, communications, global infrastructure projects aimed at the domestic consumer in the country will come to the fore. To realize such a strategic perspective, human and labor resources are required that are professionally oriented to fulfill technical tasks in the first place. Basic technical skills become strategic capital, which is not subject to devaluation and has a reproducible base on which new technologies and systems can be created [7]. In this situation, the technical orientation of the labor potential in a particular territory, the regional profile of the correspondence of the labor market and the education system in engineering specialties will be of decisive importance [6].

Consumption, as it has evolved over the past decades, will have to undergo significant changes related to both the forced restriction of consumption and the restructuring of the system of economic relations. In this regard, the structure of the labor market will change significantly. One of the main segments of labor markets in the coming years will most likely be industrial production, by which we mean the production of real consumption products based on digitalization of all processes, these are precisely technical, engineering and economic competencies. One of the trends identified in scientific research in recent years has been the reduction of employment in the sphere of material production and the increase in the services sector, while industrial production has moved into the category of fully automated and almost eliminating human presence [5]. The onset of the global economic crisis provoked by the coronavirus can dramatically change both the structure of the global economy and, accordingly, the labor market. In this situation, industrial production, initiated by regional and national interests, can become the driver that will allow bringing the scale of employment to optimal values. In addition, automation and robotization of routine operations can reduce jobs in the production process, but at the same time creates entire segments of new jobs associated with servicing the technological cycle, designing, programming and organizing labor according to the design principle [1]. In this situation, the basic engineering skills acquired in the vocational training system and the adaptive development and mentoring system at the enterprise are priorities.

2 Methodology

Employment in the region is determined by the development of traditional or priority sectors of decisive national economic importance [15]. Most often, workers live in the region where the production is located. Therefore, the development of the labor potential of the production location is crucial for commercial success of the enterprise. As an example, we consider local relationships in neighboring regions of the Federal District; these are the Samara region, Republic of Tatarstan, Saratov and Ulyanovsk regions. For the analysis, we selected indicators of the education system, demand and supply on the labor market, statistics on employment and unemployment, as well as data presented on one of the most popular online work resource [11].

The indicated regions are in the Volga Federal District. They are comparable in size and population, they are in the same climatic zone near each other and are adjacent to the most important waterway of Russia - the Volga River. However, the regions differ in economic and labor potential, which is characterized, inter alia, by quantitative indicators of the system of professional and higher education. When assessing the region's labor potential by educational characteristics, both quantitative and qualitative indicators are important, for which, in our analysis, the following were taken: the number of organizations and graduates of higher and secondary specialized education in the region, the share of unemployed in the total number of employees, the share of employees with vocational education in the total number of employees and the growth rate of expenditures of the regional budget on education. It should be borne in mind that the costs of local budgets on education have a limitation, since only general and secondary vocational education is financed from them, higher education is provided by the federal budget. The most relevant and dynamic information about vacancies and job seekers is job sites, which have become the most accessible and sought-after employment tool that responds to changes in the situation in almost real time. A joint analysis of statistics and data on the number and structure of resumes and vacancies can show the state of the specialized labor market oriented to industry.

3 Results

The share of specialists with higher and secondary vocational education in the total number of employed and unemployed shows demand for qualified specialists and the possibility of finding a person with vocational training. An assessment of the general situation in regions with the training of specialists can be carried out by graduation dynamics of young specialists with vocational training. For the possibility of comparing data, we took expenditures on education in relative units, taking 2009 as the base value from which we calculated the annual growth rate. However, a comparison of absolute indicators of expenditures of regional budgets is also significant. The comparison was carried out according to the following indicators, indicated in Fig. 1 for each region:

1. Growth rate of regional budget expenditures on education, %.
2. Number of organizations of higher and secondary vocational education, units.
3. Number of graduates of higher and secondary vocational educational organizations, thousand people.
4. Unemployed population with higher and secondary vocational education, in % of the total number of employees.
5. Employed population with higher and secondary vocational education, in % of the total number of employees.

The most significant growth rate of regional budget spending on education is observed in the Saratov and Ulyanovsk regions. However, in absolute terms, the cost of education in these regions remains almost two times lower than in the Samara region and the Republic of Tatarstan. This largely depends on the development of the economy and budget revenues in the whole region, however, we can note the tendency to equalize differences in the cost of education in neighboring regions.

The reduction in the number of educational organizations training workers and specialists, as well as in the number of graduates in all the regions under consideration is indicative, which can only partially be explained by the decline in the demographic wave. An alarming trend may also be associated with the outflow of promising youth in capital regions. This requires additional measures to develop regional education systems and create attractive jobs. An increase in the share of specialists with education in the total number of employees is observed in the Republic of Tatarstan and the Samara region, which shows higher qualification requirements. Accordingly, we can assume that the economic structure of these regions allows us to create more efficient jobs. At the same time, it is also necessary to note an increase in the share of specialists with education among the unemployed, in contrast to the Saratov and Ulyanovsk regions. The situation may be characterized by trends in economic digitalization, when the creation of a new type of jobs increases the requirements for education and qualifications, unskilled labor remains in demand due to low profitability of automation of many processes, and staff with average qualifications are put out of jobs. Thus, development trends of educational components of regions' labor potential can characterize the economic situation and the directions to develop labor markets.

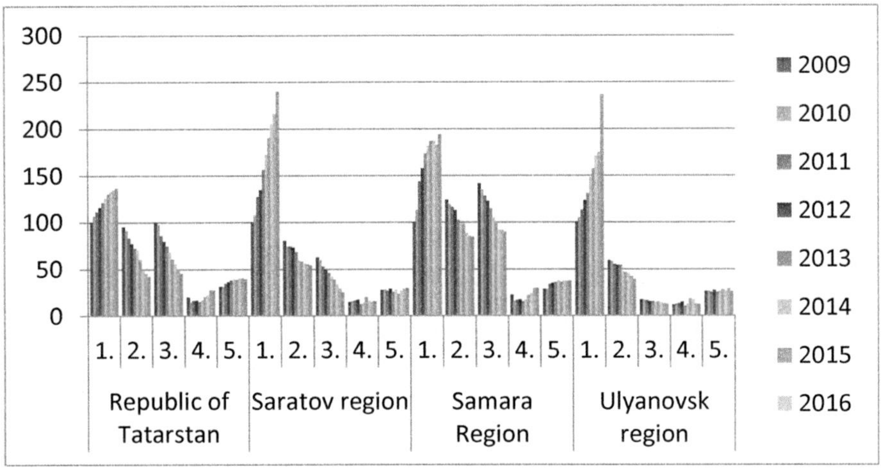

Fig. 1. Analysis of educational components of regions' labor potential of the Volga Federal District (Source: authors).

The growth of qualification requirements is especially noticeable in high-tech sectors of the economy, in which there are high requirements for technical education and special engineering skills, the development of which occurs when receiving basic education in undergraduate, specialty and master's programs. Using the example of the Samara region, we analyzed the region's provision with graduates of several common technical training areas - Heat and Power Engineering, Electric Power and Electrical Engineering, Power Engineering, Construction, as these specialties are the most popular in the region's labor market. To assess the state and identify the trend, we calculated the share of the number of graduates in the total number of graduates of the Samara region and compared with the share of graduates in this specialty in the Russian Federation (Table 1).

Table 1. The proportion of graduates of technical specialties in the Samara region

Name of training (specialty)	2015	2017	2018	2019
Heat power engineering and heat engineering, people	58	116	143	112
Power industry and electrical engineering, people	288	569	529	270
Power engineering, people	31	93	93	62
Construction, people	383	835	825	573
Number of graduates of the Samara region in 4 training programs, people	760	1613	1590	1017
Number of graduates of the Samara region in all training programs, people	11913	19337	19596	19743
Share of graduates of the Samara region in 4 training programs of the total number of graduates of the Samara region, %	6,4	8,3	8,1	5,2
Number of graduates of the Russian Federation in 4 training programs, people	36855	55810	53270	54855
Share of graduates of the Samara region in 4 training programs from the total number of graduates of the Russian Federation, %	2,1	2,9	3	1,9

(Source: authors).

The number of graduates was calculated as the sum of graduates in 3 levels of higher education – undergraduate, specialty, master's degree. In 2019, there was a significant drop in the number of graduates in all the areas of training programs considered, which is characterized by a decrease in both absolute indicators and the share in the total number of graduates. At the same time, in 2017–18 there was an increase in all specialties, after which the number of graduates decreased to a value less than it was in 2015, with a steady increase in the total number of graduates of higher education in the Samara region and in the Russian Federation. The share of graduates of the studied areas in the Samara region is almost three times higher than the share of all graduates in the country, since energy and construction are one of the priority sectors of the Samara region and the education system must correspond to the structure of the local economy. The number of graduates in the Samara region and in the Russian Federation is growing quite proportionally to each other.

The entry of new specialists into the labor market is accompanied by redistribution of labor resources in accordance with economic, professional and personal needs that characterize supply and demand in a territory. Crisis situations are characterized by instantaneous dissonance in the number of resumes and vacancies in the labor market. Reliable distributed data can be estimated at the end of the year, but even now you can evaluate the dynamics and structure of supply on the labor market, which cannot change as quickly as the number of vacancies. Employment in the Russian Federation is regulated by the state to a rather high degree, as shown by previous crises of 2008 and 2014. We analyzed the number of resumes and vacancies in the labor market of the regions of the Volga Federal District to identify labor market conditions and determine employment trends in the industrial sector. The number of resumes and vacancies was estimated over several periods – the total number of active resumes and vacancies not currently filled, including the number of resumes and vacancies posted within the last week (as of 04.25.2020), within the last month, within the last year. Since the professional group with the qualification of Engineer is quite extensive, we compared the number of resumes for all resumes with the mention of the word Engineer in the title and allocated a group of resumes specializing in Industry and Construction (Table 2).

Table 2. Comparative characteristics of engineering specialties resumes in the regions of the Volga Federal District

Number of resumes		Samara		Kazan		Saratov		Ulyanovsk	
		Engineer (all professions of the region), units	Engineer (production, construction), units	Engineer (all professions of the region), units	Engineer (production, construction), units	Engineer (all professions of the region), units	Engineer (production, construction), units	Engineer (all professions of the region), units	Engineer (production, construction), units
Total for the entire period		18271	9999	21868	12628	6884	3742	4588	2885
Number of resumes within the last week		990	588	1028	570	352	191	167	107
Number of resumes within the last month		1999	1124	2110	1179	706	369	380	249
Number of resumes within the last year		6087	3317	6846	3921	2254	1189	1302	836
Expected salary, thousand rubles	Below 20	2171	1033	3640	1952	947	454	751	395
	20–40	7378	4060	9546	5518	2736	1553	2031	1264
	40–60	4333	2647	4801	3104	1203	737	941	674
	60–80	1546	932	1553	998	503	280	271	186
	80 and higher	1298	834	1445	786	594	268	185	105
Work experience, years	More than 6	13328	7639	14225	8651	5084	2901	3443	2247
	3–6	2131	1157	3222	1927	827	434	510	339
	1–3	1193	631	1877	1049	454	239	271	150
	no	1626	574	2553	1001	526	170	363	149
Key skills	PC user	3299	1847	3615	2205	1417	749	861	592
	AutoCAD	1950	1428	2651	1916	661	512	468	383
	Teamwork	1915	1029	2254	1320	792	416	477	321
Age, years	18–30	5522	2537	8008	3987	2013	852	1280	717
	30–40	8422	4653	10340	6195	3039	1682	2126	1359
	40–50	3181	1860	2912	1810	1250	720	752	472
	50–60	1736	1143	1527	1075	829	546	578	413
	More than 60	686	522	593	461	293	230	199	159
Gender	Male	14206	7736	16408	9431	5602	3019	3717	2372
	Female	4012	2231	5402	3155	1277	720	863	509

(Source: authors).

Quantitative characteristics for all periods are comparable and allow dividing resumes into two groups according to quantitative indicators – Samara and Kazan in one group and Saratov and Ulyanovsk in another, in which the number of resumes differs by 3–4 times. Great attractiveness is justified by a significant difference in salary expectations of applicants. The largest number of applicants in all four regions expects to have salary in the range of 20–40 thousand rubles, which shows approximately the same standard of living in the studied regions, but the probability of finding a qualified specialist from employers in Kazan and Samara is much higher. Analysis of the structure of applicants by age shows that the most dynamic age group, prone to changing professional and official position, is a group of 30–40-year-old specialists who have received education, professional experience and have a desire to find more decent working conditions. As age increases, the number of resumes decreases and after 40 years the largest number of resumes, compared to other cities, is posted in Samara, and Kazan is the leader in the number of resumes under the age of 40 years. In general, for all the cities studied, the most active part of the workforce is the age category up to 40 years. The key skills for engineers are the ability to use a computer and specialized programs, the less important is the ability to work in a team, which is determined by the nature of engineering work.

The number of vacancies for engineering positions decreased in all the studied regions (Fig. 2).

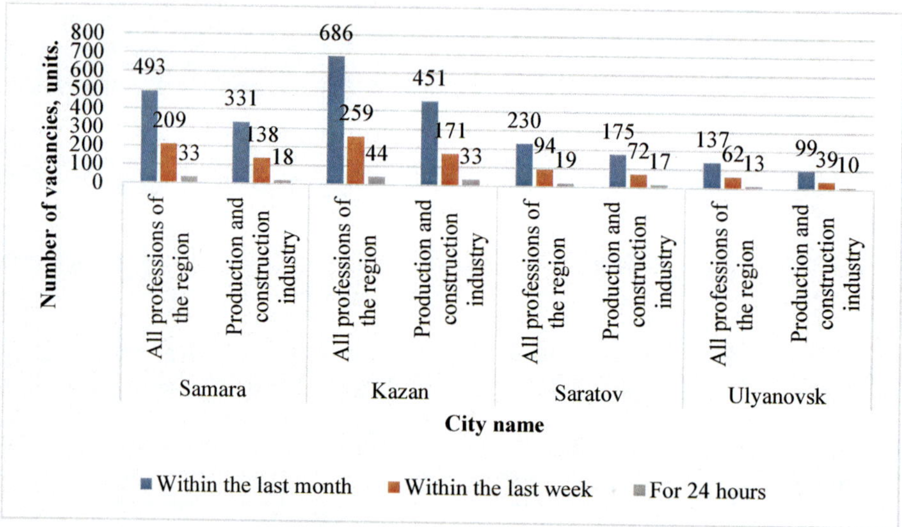

Fig. 2. The dynamics of the number of vacancies posted in the regions of the Federal Volga District (Source: authors).

The increase in vacancies within the last month shows a negative trend, the largest decrease in the number of posted vacancies occurred in Ulyanovsk by 6%, only in Saratov there is a slight increase of 1%. At the same time, in all cities under consideration there is an increase in the number of resumes by 5–13%, the largest in Kazan. The labor market remains highly competitive, 7–9 people per vacancy, this situation in the short term will tend to increase until the end of the pandemic. When implementing the strategy for domestic consumption and national investment projects, the situation may change in the direction of increasing the number of vacancies, competition in the labor market will decrease, and to meet the needs of the new economy in qualified personnel, their preparation should begin at a faster pace.

Salary offers far exceed the expectations of applicants, qualifications and areas of responsibility are of great importance (Fig. 3).

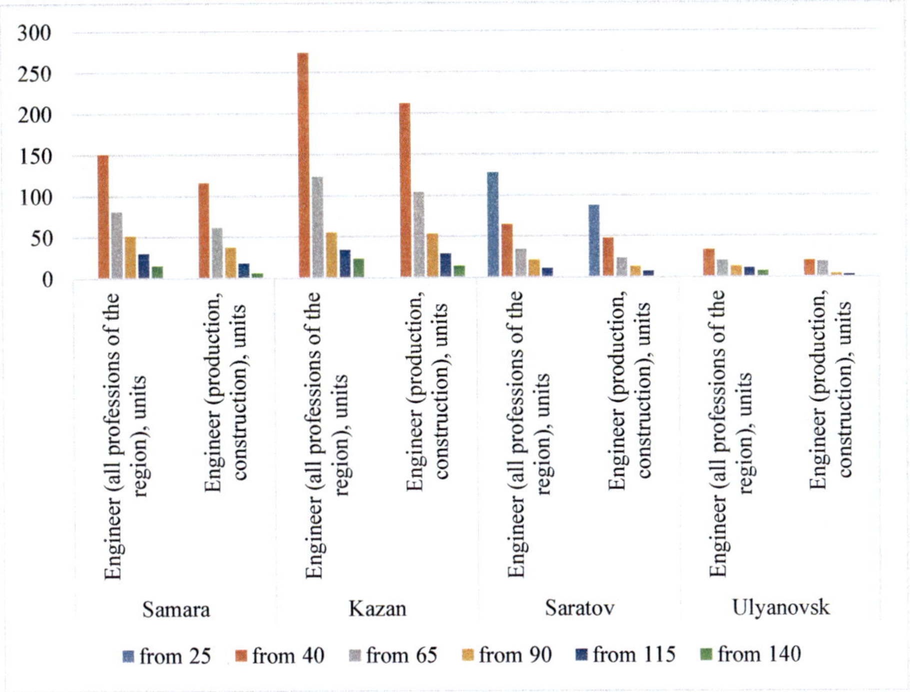

Fig. 3. Salary offers in the posted vacancies of the regions of the Federal Volga District (Source: authors).

There is a significant difference in salaries offered in vacancies in different cities. In Samara, Kazan and Ulyanovsk, there are no vacancies with salary below 25 thousand rubles, such a salary is offered only in Saratov. In Samara and Kazan, the largest number of vacancies with salary of 40–65 thousand rubles, in Saratov from 25 to 40 thousand rubles, in Ulyanovsk with a small total number of vacancies, the proposed level of salary is in the range of 40–65 thousand rubles.

4 Discussion

The relationship between the requirements for applicants, the proposed remuneration and qualifications of applicants has been a debated area of science for many years [8]. Recent years of discussion have been around the impact of competence and qualifications on the level of remuneration, the ratio of "soft" and "hard" skills to form a successful career [3]. Many researchers, in our opinion, unreasonably overestimate the importance of "soft" skills, often to the detriment of basic, fundamental knowledge.

In our opinion, this specificity is more characteristic for leadership, managerial positions, which are crucial for leadership qualities [12]. Our studies show the priority of technical and informational knowledge, which is given by both job seekers in resumes and employers in vacancies. The qualification characteristics and level of job positions are of great importance, when personal qualities practically do not matter in low-skilled positions [4]. Interregional labor market research conducted in our country confirms the trends of a stable technical and technological stratum of specialists with both technical and digital skills in many regions [13].

In recent years, qualification requirements have grown across the entire spectrum of engineering and technical jobs, often in vacancies one can note clearly overstated requirements that do not correspond to the regional labor market. But this situation encourages applicants to increase their competence and to increase competitiveness in the labor market [14]. This primarily concerns the skills associated with the use of artificial intelligence [2]. Such skills become basic for all technical specialists.

5 Conclusion

Studies have shown general trends in engineering qualifications in neighboring regions with a significant difference in absolute terms. The training in the selected group of specialties is carried out in enough volume, but the reduction in output is worrying when the labor market has long-term demand for advanced engineering, technical and digital skills.

The authors analyzed the number of graduates in the group of training areas corresponding to the most demanded qualifications in the regional labor market. They have noted the equalization of regional spending on the formation of local budgets, a gradual increase in the share of employees with vocational education. Based on the identified trend of reducing the number of graduates of the required specialties, it is necessary to develop targeted training of young specialists in educational organizations.

The analysis of the correspondence of the number and structure of resumes and vacancies showed that the labor market in the studied regions of the Federal Volga District is highly competitive, with a significant advantage in favor of the interests of employers and unsatisfied demand of applicants in the most developed regions of the Samara region and the Republic of Tatarstan. The labor market of the Saratov and Ulyanovsk regions is more balanced, although less attractive to applicants.

An alarming situation is when there has been a decrease in the number of vacancies in economically more developed regions within the last month, due to the epidemiological situation. In implementing the strategic goals of the country's economic growth, based on national projects and the development of domestic consumption, we need accelerated growth in the training of engineering and technical specialists with basic digital skills.

References

1. Albuquerque, P.H.M., Saavedra, C.A.P.B., de Morais, R.L., Peng, Y.: The robot from Ipanema goes working: estimating the probability of jobs automation in Brazil. Latin Am. Bus. Rev. **20**(3), 227–248 (2019)
2. Alekseeva, L., Azar, J., Gine, M., Samila, S., Taska, B.: The demand for AI skills in the labor market. SSRN Working Paper (2019). https://papers.ssrn.com/sol3/papers.cfm?abstract_id=3526045. Accessed 29 Apr 2020
3. Alpaydın, Y.: Identifying higher-education level skill needs in labor markets: the main tools usable for Turkey. Kuram ve Uygulamada Egitim Bilimleri **15**(4), 945–967 (2015)
4. Beblavý, M., Mýtna Kureková, L., Haita, C.: The surprisingly exclusive nature of medium- and low-skilled jobs: evidence from a Slovak job portal. Pers. Rev. **45**(2), 255–273 (2016)
5. Börner, K., Scrivner, O., Gallant, M., Ma, S., Liu, X., Chewning, K., Evans, J.A.: Skill discrepancies between research, education, and jobs reveal the critical need to supply soft skills for the data economy. Proc. Natl. Acad. Sci. U.S.A. **115**(50), 12630–12637 (2018)
6. Büth, L., Blume, S., Posselt, G., Herrmann, C.: Training concept for and with digitalization in learning factories: an energy efficiency training case. Proc. Manuf. **23**, 171–176 (2018)
7. Darabi, H., Karim, F.S.M., Harford, S.T., Douzali, E., Nelson, P.C.: Detecting current job market skills and requirements through text mining (2018). https://www.semanticscholar.org/paper/Detecting-current-job-market-skills-and-through-Darabi-Karim/10bd61ee3b72c8003b5f6446e865c9a958c424aa. Accessed 29 Apr 2020
8. Karmaeva, N.: The Oxford handbook of skills and training. J. Econ. Sociol. **20**(4), 114–125 (2019)
9. Maresova, P., Soukal, I., Svobodova, L., Hedvicakova, M., Javanmardi, E., Selamat, A., Krejcar, O.: Consequences of industry 4.0 in business and economics. Economies **6**(3), 46 (2018)
10. Reinhold, M., Thomsen, S.: The changing situation of labor market entrants in Germany. J. Labour Mark. Res. **50**(1), 161–174 (2017)
11. Rosstat: Socio-economic indicators (2019). http://gks.ru/storage/mediabank/Region_Pokaz_2019.pdf. Accessed 29 Apr 2020
12. Sopegina, V.T., Chapaev, N.K., Simonova, M.V.: Integration of pedagogical and technological knowledge in forming meta-competencies of a modern worker. Int. J. Environ. Sci. Educ. **11**(15), 7836–7846 (2016)
13. Simonova, M.V., Sankova, L.V., Mirzabalaeva, F.I., Shchipanova, D.Y., Dorozhkin, V.E.: Assessment problems and ensuring of decent work in the Russian regions. Int. J. Environ. Sci. Educ. **11**(15), 7608–7626 (2016)
14. Vu Anh, T.L., Le Quoc, T.: Development orientation for higher education training programme of mechanical engineering in industrial revolution 4.0: a perspective in Vietnam. J. Mech. Eng. Res. Dev. **42**(1), 71–73 (2019)
15. Yıldız, F.: The role of education in the participation of Turks living in Europe in labor Markets. Proc. Soc. Behav. Sci. **186**, 1179–1187 (2015)

Social and Environmental Competence as a Component of University Graduate's Competitiveness

E. N. Chekanushkina[1], N. A. Timoshchuk[1(✉)], L. A. Kolyvanova[1,2], and E. V. Fakhrutdinova[3]

[1] Samara State Technical University, Samara, Russia
elenacheka@mail.ru, 7.60n@mail.ru, larisaleksandr@yandex.ru
[2] Samara State University of Social Sciences and Education, Samara, Russia
[3] Kazan Federal University, Kazan, Russia
efahr@mail.ru

Abstract. In the modern world of innovative and digital technologies, one of the priorities of the country's strategic development is the transition to environmentally friendly and resource-saving energy, where research of the development of technological systems for climate and ecosystem management is particularly relevant. However, before starting a particular research in this scientific field, it is necessary to solve the main task of organizing professional training of specialists in the field of environmental education. It is important to create a set of measures for the effective formation of socio-environmental competence among students of higher education institutions. Based on our previous research of practical pedagogical experience in developing a competitive specialist with certain professional knowledge and skills in the field of their professional activities, we have developed and implemented in the educational process a model for the formation of socio-environmental competence of students. Within the framework of this model, a set of diagnostic measures was carried out to identify interdisciplinary knowledge in students, including environmental knowledge, as well as their further relationship with the formation of a professionally successful career.

Keywords: Competitiveness · Professional education · Social and environmental competence · University students

1 Introduction

The Ministry of economic development of the Russian Federation created a document of strategic planning of the country's socio-economic development in the long term until 2030. This document provides implementation of strategies and target programs under conditions of dynamic development of the key scientific and technological trends in the field of life sciences, information and telecommunication systems, environmental management, transition to an innovative economy, which determines the relevance of training of highly qualified, competitive specialists to ensure the sustainable economic

S. I. Ashmarina et al. (Eds.): ENGINEERING ECONOMICS WEEK 2020, LNNS 139, pp. 676–681, 2021.
https://doi.org/10.1007/978-3-030-53277-2_80

growth. In the Presidential Decree "On the strategy of scientific and technological development of the Russian Federation" one of the priorities is "the transition to environmentally friendly and resource-saving energy", and "in the long term, the research in the field of nature-like development technologies, man-machine systems, climate and ecosystems management" become particular urgent [3]. Higher education plays a fundamental role in the formation of a competitive graduate, "who is able to demonstrate formed general cultural, professional and meta-subject competencies (including socio-environmental ones), as well as integrative personality qualities (that provide individual, industrial and social needs) the most effectively in unstable working conditions" [11].

2 Methodology

In the works of foreign scientists, it is noted that taking into account principles of sustainable development is an integral part of training specialists in the field of innovative technologies [2], and improving the quality of life of every person in the country is both the goal and responsibility of specialists around the world [1]. According to them, new generations need a different view of the environment, free from technocracy and aimed at preserving and increasing natural resources [4].

Based on our previous research, practical pedagogical experience in environmental training of future specialists, a theoretical model for the formation of socio-environmental competence in students of technical training direction has been designed and is still being implemented. In this sphere, the Samara State Technical University annually conducts diagnostics of socio-environmental competence of graduates, using multi-level tools, one of which is a test to determine the ability to self-management (according to the method of Peisakhov [5].

3 Results

Documentary analysis of the legal base in the sphere of education, the country's socio-economic development, policies, national projects and federal programs, as well as domestic and foreign scientific literature shows that modern enterprises pay great attention to the realization of new approaches, implementation of lean production models, the development of green technologies, green economy, that ensure social and environmental safety, reduce human impact, environment improvement, rational use of resources. This means that there is a need for competitive specialists with environmental training in the labor market.

Thus, the federal state educational standards of higher education, including in the field of training "Engineer in industry and production", indicate the need for students to have knowledge in the sphere of environmental safety and environmental protection.

In this regard, an important task of universities is the formation of the graduates' socio-environmental competence which is understood as the ability (readiness) of individuals to perceive the surrounding reality in the unity of natural and socio-cultural relations, based on the formed knowledge, skills, experience and personal qualities, to organize their professional activities from the position of the environment management and environmental conservation, to adequately assess, analyze, predict and find effective ways to solve tasks and problems arising in a situation of uncertainty by the interaction of our society and the nature [7].

Based on the above, we conducted a survey at the engineering, economic and machine-building faculties of the Samara State Technical University (SamSTU), in order to identify opinions of students of different gender groups about requirements of the modern labor market to them as future professionals who have knowledge and skills not only in the field of their profession (Fig. 1).

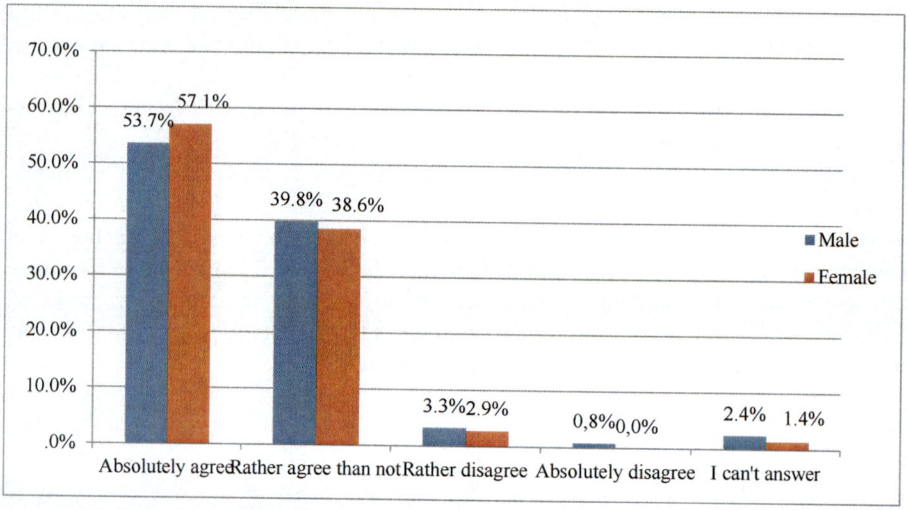

Fig. 1. Indicators of students' agreement on the availability of knowledge not only in the field of their profession (Source: authors).

Figure 1 shows that more than 53% of students absolutely agree with the opinion that it is necessary to have interdisciplinary knowledge, including environmental knowledge. To the question of "how much the possession of professional knowledge and skills, social and environmental competence will help to build a successful career as an engineer" (Fig. 2), the majority of respondents (82.4%) answered positively, and only 1% of university students deny this fact.

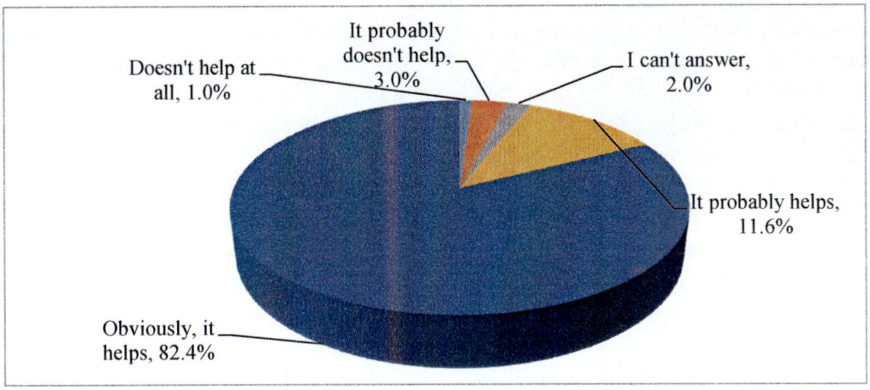

Fig. 2. The opinion of students about competencies that contribute to a successful career (Source: authors).

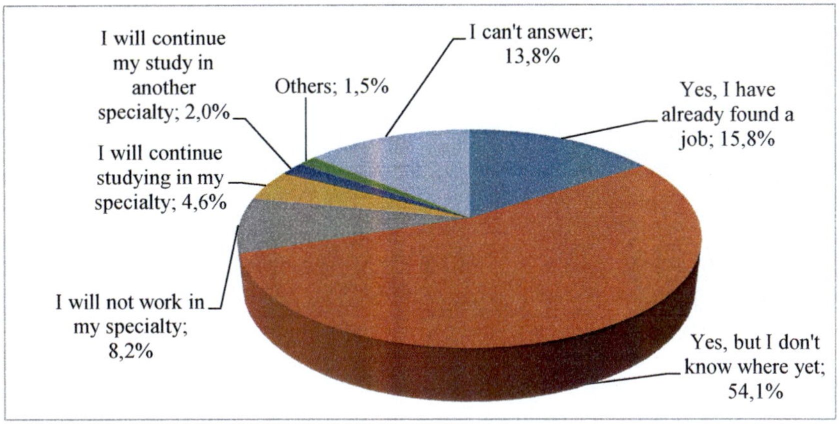

Fig. 3. Distribution of students' answers about their professional future (Source: authors).

An important aspect of the competitiveness of university graduates is that more than half of students (54.1%) expect to work in their specialty after the graduation, but, unfortunately, they do not know where exactly, 15.8% have already found a job and are working. However, 8.2% of respondents do not intend to work in their specialty (Fig. 3).

4 Discussion

Programs developed by Russian universities to improve the competitiveness of future specialists are of particular interest. For example, in the "road map" of the St. Petersburg State Electrotechnical University "LETI" named after V.I. Ulyanov (Lenin), considerable attention is paid to "the implementation of new competitive educational programs at the university together with leading foreign and Russian universities and scientific organizations" [10, p. 8], which in the future will ensure the competitiveness of graduates in the domestic and international labor market. The strategic goal of the Perm National Research Polytechnic University is "to form a world-class engineering research university that trains competitive specialists for high-tech sectors of the regional economy" [6, p. 14]. The mission of the Samara State Aerospace University named after academician S.P. Korolev is to "provide world-class training of personnel with interdisciplinary key competencies" for various sectors of the economy" [8, p. 9]. The effectiveness of the Samara State Technical University in the quality management field is based on such an approach as "an organic combination of research activities with the educational process, which allows to significantly expand opportunities for high-quality training of highly qualified personnel" [9, p. 4]. Based on the above, it can be stated that domestic universities, including higher schools in Samara and the Samara region, consider the formation of a competitive specialist that meets international educational standards to be one of the most significant indicators of their professional activity.

5 Conclusion

In our opinion, the identification of students' ability to control themselves in various life situations is of significant importance in their further professional activities, in ensuring social and environmental safety. Their professional success depends on the organization of the self-management system and the formed abilities of future specialists to forecast, set goals, plan, control, and make decisions.

Thus, in the first year of students' training (83%) at the university, the level of their self-management abilities was lower than the average one, while the graduates' indicators were significantly higher (at the average and high levels), which indicated the effectiveness of our continuous socio-environmental education at the university (Table 1).

Table 1. The formation level of the self-management system among university students

Year	Levels of self-management ability				
	Low <50%	Below the average	Average <70%	Above the average	High >75%
2015	4,5%	83%	9%	3,5%	–
2016	1%	49%	32,5%	16%	1,5%
2017	–	27%	46%	23%	4%
2018	–	11%	50,5%	28%	10,5%
2019	–	–	43%	31%	26%

Source: authors.

Thus, the development of social and environmental competence among high school graduates is one of the most important components of their competitiveness in the labor market, thereby contributing to their successful organization of their professional activities from the position of finding effective ways to solve the tasks set in the field of environmental management and conservation.

References

1. Borkovskaya, V.G., Roe, R., Bardenwerper, W.: Sustainability risk management: the case for using interactive methodologies for teaching, training and practice in environmental engineering and other fields. In: Solovev, D.B. (ed.) Smart Technologies and Innovations in Design for Control of Technological Processes and Objects: Economy and Production. FarEastCon 2018. Smart Innovation, Systems and Technologies, vol. 138, pp. 251–260. Springer, Cham (2020)
2. Chekanushkina, E.N., Mikhelkevich, V.N.: Heuristic methods of developing students' creative potential in the process of forming socio-ecological competence. Acad. Bull. ELPIT **3**, 95–100 (2019)
3. Decree of the President of the Russian Federation No. 642 of December 1, 2016 "On the strategy of scientific and technological development of the Russian Federation". https://www.garant.ru/products/ipo/prime/doc/71451998/. Accessed 25 Apr 2020
4. Kuznetsova, N.V., Soloviev, A.A., Botasheva, L.Kh., Lebedev, I.A., Sarkisyan, K.S., Kurdyumov, V.I., Chistyakov, K.A., Prokopyev, A.I.: Bachelors safety culture formation in the field of environmental protection. EurAsian J. BioSci. **12**(2), 157–166 (2018)
5. Peisakhov, N.M.: The Problems of Individual Differences Psychology. Kazan State University Publishing House, Kazan (1974)
6. Perm National Research Polytechnic University: The program of competitiveness increase for 2016–2025 (2016). https://pstu.ru/files/file/2016_anna/programma_povysheniya_konkurentosposobnosti_pnipu_06_07_16.pdf. Accessed 25 Apr 2020
7. Quadrado, J.C., Ferreira, E.P., Pokholkov, Y., Zaitseva, K.: Higher education institutions as key players to ensure sustainable development goals. In: Petrie, M.M., Rivera, L.F.Z., Aranzazu-Suescun, C. (eds.) Proceedings of the 17th Latin American and Caribbean Conference for Engineering and Technology, pp. 1–5. Latin American and Caribbean Consortium of Engineering Institutions, Montego Bay (2019)
8. Samara State Aerospace University named after academician S.P. Korolev: The program of competitiveness increase (2013). https://ssau.ru/files/info/official_docs/ppk.pdf. Accessed 25 Apr 2020
9. Samara State Technical University: Policy in the field of quality management (2019). http://smk.samgtu.ru/node/1. Accessed 25 Apr 2020
10. St. Petersburg State Electrotechnical University "LETI" named after V.I. Ulyanov (Lenin): The program of competitiveness increase (2019). https://etu.ru/assets/files/university/normativnye-dokumenty/programma_5-100_etu.pdf. Accessed 25 Apr 2020
11. Timoshchuk, N.A.: The concept of forming a competitive engineer as a strategy for the educational policy of technical universities. Doctor thesis. SamSTU, Samara (2019)

Legal Regulation of Remote Work in Russia

M. A. Andrianova(✉)

Moscow State Institute of International Relations (University) of the Ministry of Foreign Affairs, Moscow, Russia
m.andrianova@inno.mgimo.ru

Abstract. Labor law in Russia is expected to change in order to make this sphere more attractive to employers, more accessible to employees, and more effective for the state. The practice meets expectations of optimists for the implementation of progressive reforms, which also affect problems of the labor regulation. Business has long been interested in using this tool, but the lack of skills in new technologies among a wide range of employees and the inertia of the current legislation hindered the process. Currently, there is a clear tendency to overcome all the indicated obstacles to the widespread use of remote work in Russia. This article investigates the current legislation and practice of the remote work application, recent changes in the sphere, as well as development trends in the remote work regulation, taking into account the changing needs of the population.

Keywords: Interaction of labor and civil law in Russia · Labor law reform in Russia · Registration of labor contracts with employees · Remote work · Technology impact on labor law

1 Introduction

At the end of the XVIII century. the increased use of collective labor in the first industrial enterprises led to a paradoxical situation: the state was interested in the industrial development; industry and entrepreneurs needed to use labor; people needed to work. But relations regulated in terms of the long-established principle of freedom of contract (which has always worked well as a natural regulator) stalled.

As a result, all participants were dissatisfied: employees were suffocating from appalling working conditions, employers were dissatisfied with the quality of work and the constant threat of aggravation of relations at the enterprise, the state expected a social explosion.

Labor law arose and formed as an independent branch at that time when all participants in labor management relations came to the conclusion that the situation could not remain so it was. It is interesting that employers and large industrialists initiated the creation of a special regulation of labor relations at enterprises They were the first to realize that the society had reached a new stage of its development.

It seems that labor law is currently experiencing something similar. More and more people talk about the need for reforms in the sphere of regulation of labor relations [9],

S. I. Ashmarina et al. (Eds.): ENGINEERING ECONOMICS WEEK 2020, LNNS 139, pp. 682–687, 2021.
https://doi.org/10.1007/978-3-030-53277-2_81

and there are opinions that the traditional idea of employment – a 40-h working week, mandatory presence in the office – is becoming outdated [8].

It is much more effective to give more freedom to the employer in choosing the form of employment that he is ready to provide, and more freedom to the employee in matters of regulating the labor regime, switching to remote employment. The ongoing technological changes also have a strong impact on the process of performing the labor function. Representatives of many professions can perform their work from anywhere in the world, if they have access to the Internet and a laptop or smartphone.

Even before the coronavirus pandemic, this fact meant that the growth of the number of remote workers around the world, the development of flexible forms of employment re trends of the near future. Currently, the advantages of remote work and the need for it are undeniable. When the peak of the pandemic is passed, the trace effect of the months of work experienced on the remote mode will continue to affect and will inevitably correct the attitude of the legislator to this institution.

2 Methodology

To solve the research tasks, the author used both general scientific methods: dialectical method, methods of analysis and synthesis, and special methods of legal science: formal legal method, comparative legal method, historical legal method, methods of grammatical and logical interpretation of legal norms. A structural and functional method was used in relation to the development of trends in the field of labor law in Russia. These methods allowed us to create a more complete and comprehensive view of the considered problem.

3 Results

At present, we have successfully passed the time when the very possibility of recognizing relations using remote technologies as labor relations was denied. The question was raised by the freedom of the employee from the direct control of the employer arising from the use of remote work. Now we are talking about expanding the scope of labor law to non-standard areas, including remote work and relations in the field of professional sports.

Such positive changes (for the purposes of this study, we consider it appropriate to focus on changes dedicated to remote work), which could be compared with a small revolution in the field of labor law, occurred in 2013, when the text of the Labor Code was added with Chapter 49.1 "Specific features of regulating the work of remote workers" [12]. Similar changes have been made to labor regulations in many foreign countries. In particular, "telework" has been considered in the Labor Code of France, starting from the Article L1222-9 [11].

Part 1 of Article 312.1 of the Labor Code of the Russian Federation provides an official definition of remote work, which is recognized as the performance of a certain function of the employment contract outside the location of the employer [12]. In general, it is not currently possible to say that the issue of remote work is not legally regulated. An employment contract with a remote employee can be concluded by exchanging electronic documents. This situation has significantly facilitated the practice of hiring remote workers, but the situation is far from ideal at the moment.

A mandatory condition for electronic document management, which is clearly established by law, is the use of enhanced qualified electronic signatures of the parties to the employment contract when exchanging electronic documents. The enhanced qualified electronic signature is the strictest type of electronic signature in the opinion of the legislator [5]. Using only such a signature will eliminate problems for the parties with proof of the existence of an employment relationship between them.

Remote work relations that are registered with violations may not give the parties protection under the Labor Code [12]. For example, the Definition of the Moscow City court in 2017 stated that an employment contract concluded between parties, as it follows from the requirements of Article 312.1 of the labor code of the Russian Federation, with the inclusion of a condition for remote work, signed by both the plaintiff and the defendant, is not concluded. However, the presence of electronic correspondence between the parties does not indicate the remote nature of the work [4].

There are obvious difficulties faced by the practice with an overly formal approach to regulating relations within the framework of the labor legislation. In addition to the difficulties that arise at the stage of registration of labor relations, there are issues related to the proper performance of labor duties by the employee.

A lot of researchers and practitioners draw attention to the lack of control available to the employer. It seems that remote mode is not suitable for all forms of employment, but if the employer initially assumes this possibility, the main problem is not the lack of ability to control employees, but ensuring effective communication with them and the ability to receive timely results of high-quality work. This goal can be achieved by more detailed procedures and conditions for interaction between the employee and the employer, including by defining the procedure for obtaining assignments, reporting procedures, technical capabilities for rapid communication between the parties, and so on.

In itself, the possibility of more detailed regulation of labor relations within the framework of individual labor contracts is another trend and urgent need of modern labor law. Judicial practice convinces parties of the need (for security purposes) to use standard labor contracts and other typical documents in the framework of labor relations, however, only individualized labor contracts can give a sense of vividness and effectiveness the labor law, which is so necessary for all the parties.

Continuing to cite examples of unpleasant complications that parties face when using remote work, we will cite several court decisions.

According to the Appeal ruling of the Sverdlovsk regional court of 14.08.2018, if the personnel department is located in one city, and the dismissed employee is in another, the employer should ensure that the employee's work recordbook is issued on the day of dismissal at his place of work [2]. This harmless definition reveals a problem of possible abuse by employees, as well as contradictions between the document management requirements in relation to labor relations in paper form, while the relationship between the parties is based on remote interaction. In the above case, the claim was filed by an employee who considered that the work record book was issued to him untimely and won the dispute. Taking into account the ongoing transition to electronic workbooks, it seems that this practice can also have a positive impact on solving problems of registration of hiring and firing of remote employees.

Another difficult practical problem is a possibility of combining traditional employment and remote work. Based on the analysis of court decisions, we have to conclude that the practice does not allow such a combination [1].

There are problems with determining grounds for dismissal of remote employees. In particular, an employee cannot be dismissed for his absence at the workplace if he performed their work remotely based on the agreement with the employer. This conclusion was reached by the Supreme Court of the Russian Federation in the Definition of 16.09.2019 N 5-KG19-106 [3].

It seems that the Definition was made because of the fact that the agreement between parties did not specify how the parties would carry out the communication necessary in the opinion of the employer for the effective performance of the employment function. In our opinion, an employee's violation of the agreements established in the contract may give a completely different picture in court.

It should be noted that the legislation and legal practice of foreign countries on the considered subject also do not imply that the regulation of remote work has reached its highest development level. The reforms in the sphere of labor law that have been carried out in France in recent years also relate to issues of remote workers. The French legislator proposes collective agreements as the main regulation source of this type of labor relations and establishes in the law a list of issues to be discussed during negotiations. Among them, there are conditions for transferring an employee to "télétravail" and terminating the transfer agreement, provisions concerning methods and the procedure for expressing consent to the transfer, provisions on working hours, frequency of communication and working hours, and the employment regime for persons with disabilities. Thus, two tasks are solved at once. The appropriate method of regulation is proposed, taking into account the specifics of a particular enterprise, which removes the task of the legislator to regulate areas that can not be subject of effective state regulation. A list of issues is also proposed, that are the most difficult from the point of view of the subsequent implementation of relations.

We can conclude that the legal regulation of remote work in Russia currently needs special attention and gentle adjustment. It should also be taken into account that the use of remote work significantly expands the possibility of protecting the employees' rights, is an effective means of combating unemployment and staff turnover, seasonal epidemics and force majeure, and indicates the socially-oriented nature of the regulation norms. However, in such legal relations, the employer is a weak side, which bears both the problem of the inability to implement full control over the employee's compliance with the labor discipline, and all the difficulties associated with the legal uncertainty of registration and notification of employees.

4 Discussion

Currently, there is no shortage of articles on practical issues related to the use of remote work, which give advice to employers on the correct registration of employees, their proper notification, taking into account a significant number of recent bylaws.

However, there are few academic papers on the considered subject. Examples include the work of Lyutov "Remote work: The experience of the European Union and the problematic aspects of its legal status in Russia" [7] and Shuraleva "Distance work: A comparative analysis of the laws of the EAEU member states" [10]. Both works draw attention to the lack of regulation in this sphere.

This position could be accepted if it were not the obvious impossibility to regulate thoroughly the use of remote work at the legislative level. Clarification is required, but in view of the obvious trend away from a centralized regulation. It is necessary to make more active use of local acts and collective agreements in order to create the most effective and flexible labor regulation.

5 Conclusion

The text of this article has collected examples of practical difficulties when using distance labor. In our opinion, it would be wrong to conclude that the considered institute is not viable in Russia. On the contrary, it is precisely such non-standard forms that can give a positive impetus to the effective application of institutions of such an ingenious invention as labor law, which is now completely undeservedly perceived by the parties as an unnecessary burden.

A possible solution to this problem is to abandon the formal approach practiced by the parties in the formation and registration of labor relations in general, and distance labor in particular. It would also be wrong to assume that, in the end, labor law aims at a complete transition to flexible and atypical forms of employment, such as distance work. Flexible forms of employment (temporary work, distance work, and loan work) do not replace traditional employment, but are intended to complement it.

It is only important that these forms are recognized by the legislator and effectively protected in the courts. World practice shows that there is a direct correlation between the effectiveness of legislation on flexible forms of employment and the level of informal employment in the country.

It seems that the state authorities of the Russian Federation also feel the need to choose a more liberal approach to regulating atypical forms of employment in general, and remote work in particular. An example is a letter from the Ministry of Labor dated March 2020 [6]. According to the Letter, a transfer to a remote (distant) job is allowed as a temporary measure without a special reissue of the employment contract. It is necessary to obtain the consent of the employee, the wording of which can be included in the order for temporary transfer to remote work. The employee has to be familiar with the order. It is also needed to make an additional agreement.

We can hardly hope that the procedural nature and algorithmization of labor relations will soon weaken its influence in practice. In the end, these features of the legal technique of labor law were conceived to facilitate the situation for employees in case of violation of their rights. For a similar purpose – more complete realization of the interests of employees, the institute of remote work was also conceived.

References

1. Appeal ruling of the Moscow city court of 04.04.2019 in the case N 33-14969/2019. http://www.consultant.ru/cons/cgi/online.cgi?req=doc&ts=112059073108725661762668853&cacheid=503D9F4BB8CC611F982BF4E35478E9C3&mode=splus&base=SOCN&n=1127700&rnd=60135401FA2DC70E4A13661EF422A9C7#xe7ziah0xq. Accessed 07 May 2020
2. Appeal ruling of the Sverdlovsk regional court of 14.08.2018 in the case N 33-14260/2018. http://www.consultant.ru/cons/cgi/online.cgi?req=doc;base=RASVR;n=160666;dst=100053#011187480356378776. Accessed 07 May 2020
3. Definition of 16.09.2019 N 5-KG19-106. https://legalacts.ru/sud/opredelenie-sudebnoi-kollegii-po-grazhdanskim-delam-verkhovnogo-suda-rossiiskoi-federatsii-ot-16092019-n-5-kg19-106/. Accessed 07 May 2020
4. Definition of the Moscow city court of 14.04.2017 N 4G-2443/2017. http://www.consultant.ru/cons/cgi/online.cgi?req=doc&base=SOCN&n=855332#06766297269837012. Accessed 07 May 2020
5. Federal law of 06.04.2011 N 63-FZ (as amended on 23.06.2016) "About an electronic signature" (from ed. and add., Intro. effective from 31.12.2017). http://www.consultant.ru/document/cons_doc_LAW_112701/. Accessed 07 May 2020
6. Letter of the Ministry of Labor of 27.03.2020 N 14-4/ 10/P-2741. http://www.consultant.ru/document/cons_doc_LAW_348752/. Accessed 07 May 2020
7. Lyutov, N.L.: Remote work: the experience of the European Union and the problematic aspects of its legal status in Russia. Lex Russica **10**(143), 30–39 (2018)
8. Munz, A.: Social innovation and the work of the future. HR Digest **14**, 5–7 (2019)
9. Orlovsky, YuP: Is labor law reform necessary? Law **11**, 39–46 (2019)
10. Shuraleva, S.V.: Distance work: a comparative analysis of the laws of the EAEU member states (part 1). Labor law Russia Abroad **4**, 41–44 (2019)
11. The Labor Code Of France. https://www.legifrance.gouv.fr/affichCode.do?cidTexte=LEGITEXT000006072050. Accessed 22 May 2020
12. The Labor Code of the Russian Federation of 30.12.2001 N 197-FZ (as amended on 16.12.2019). http://www.consultant.ru/document/cons_doc_LAW_34683/. Accessed 07 May 2020

HR Training in Digital Economy: Innovative Investment Projects

Y. V. Veis(✉), O. Y. Eremicheva, I. B. Kostyleva, and V. S. Tikhonov

Samara State Technical University, Samara, Russia
jveis@yandex.ru, oksana-eremicheva@yandex.ru, i.b.kostyleva@gmail.com, leasingagro@mail.ru

Abstract. In the context of the training system transformation and the digital economy development, an important problem is personnel training for a specific requirement from enterprises. Employers are interested in young professionals, but they are not satisfied with the lack of their practical experience. At the same time, universities are not able to take into account all the 'requirements' to employees when developing basic educational programs. One of the solutions to this problem is to invest funds of industrial partners in the process of training students and introduce new forms of practice-oriented project training, taking into account the employers' requirement. The aim of this research is to analyze the practical experience of interaction between universities and industrial partners in the process of personnel training for enterprises. The authors used the five-factor model by J. Phillips and the calculation of ROI indicator to evaluate the investment effectiveness in educational programs, methods of absolute and relative values, retrospective analysis, and a grouping method to analyze statistical and analytical information. The research results show the interest of corporate industrial partners in investing funds for training students at the individual request if they will see the economic effect of this investment.

Keywords: Economic impact of investment · Investment in education · Training

1 Introduction

Digital economy transformation currently meets new challenges to economic relations. The problem of personnel training for enterprises is really crucial now as the personnel has to be flexible, able to adapt to new conditions quickly, to respond to new social tasks. Process of generation of innovative decisions and introduction of innovative elaborations, that facilitate enterprises' investment actualizes a problem of producing new personnel who has creative vision for problems' solving and skills to work in a project team.

The consortium of universities with industrial corporations can become one of the possible solutions. The university in this consortium functions as a fundamental transmitter of competencies and knowledge. Enterprises are stakeholders who are interested in training. An essential aspect in this process is practical orientation in

S. I. Ashmarina et al. (Eds.): ENGINEERING ECONOMICS WEEK 2020, LNNS 139, pp. 688–696, 2021.
https://doi.org/10.1007/978-3-030-53277-2_82

training, which allows the employer to assess not only the knowledge of future personnel, but also practical skills and abilities. Project competencies are really significant and greatly influences successful implementation of innovation projects.

Personnel training is quite cost-intensive process compared to the classical university education in Russia if the HEI (higher educational institution) wants to meet company's challenges and account practical-oriented training based on project approach. Such HEI has such customers as enterprises that have sufficient investment resources and can participate in co-financing.

2 Methodology

Over the years, enterprises highlight a number of major shortcomings in university education: unreasonable theoretical training with a reduced practical component; too general outdated knowledge with an insufficient professional focus; poor research and creativity skills; and low level of teamwork [9]. So now these requirements are new challenges for the university professor and success of his professional tasks' implementation.

Numerous research suggests project-based studying as the most appropriate means to achieve effective competency education in HEI that combines knowledge, skills, and values. Project activities are those ones that create conditions for meta-subject competencies and project teamwork skills, promote practice-oriented learning, and increase motivation [10]. Educational models that integrate project-oriented learning present high practical involvement of students: the transition from a passive knowledge recipient to an active researcher-practitioner.

Project-based learning demonstrates an improvement in teaching and studying quality, supports cognitive development at a higher level, involves students in solving complex, innovative problems, teaches them modern technologies and processes, and generates interest in applied research and self-tuition. The main achievements in project activity in teaching process are the following ones:

- assertive developing the need to acquire new knowledge;
- demand for knowledge and practical skills, based on the need to solve real professional problems;
- mastering professional competences as educational results of main curriculum (hardskills);
- students' professional and personal skills, ensuring developing of universal competencies.

Educational technology aimed at involving business and enterprises of the real economy sector in projects' implementation, training specialists who are ready to introduce technological innovations, becomes the core of the educational model. From the point of view of personnel training technology development, problem-oriented project-based training is a short form of "problem-oriented, interdisciplinary and participant-oriented project work", which creates the basis for the educational approach known as "The Roskilde Model" [1].

The key conceptual aspect here is John Dewey's empirical learning system, which implies the necessity to base education on student's own life experience and specific competencies for each student, the education has to be based on the theory of the American teacher Carl Rogers about "self-targeted education," that differs from Dewey's theory, but comes to similar conclusions [7].

The employers whose requirements are to develop project teamwork skills consider the formation of an innovative university as the basis for HR potential of the economy to be the most successful world practice. The examples of such universities are The Hong Kong University of Science and Technology, The Singapore University of Technology and Design, Ecole Polytechnique Federale de Lausanne, Open University Skolkovo and others [4].

For universities, interaction with corporate customers brings several substantial development factors into the classical education system. Firstly, it is the opportunity for additional funds for the development of individual educational trajectories, taking into account the customer's requirements. Secondly, the involvement of industrial partners into education process provides an opportunity for practical cases and tasks to implement a practice-oriented approach. Thirdly, it is possible to involve employees with unique competences who can level the gap between university theoretical base and practical tasks of the enterprise.

However, industrial partners have the following difficulties. In the conditions of significant reduction of economic growth rate, it is vitally important for enterprises to understand the efficiency of investments in HR development. The clear projection of this effect stimulates stakeholders' activity.

The authors suggest using ROI (return on investment) to show investments efficiency and to evaluate the investments in the process of personnel training taking into account individual corporate needs [8]. The methodology is founded on the four-level model by Kirkpatrick [5], but its main disadvantage is the lack of training effect assessment in money equivalent. In the five-level model of Jack Phillips ROI allows fully assessing the effectiveness of training students within 4–6 years at the university. In addition, this model takes into account an important aspect: the relevance of the educational program to the strategic goals of the enterprise.

3 Results

The result of interaction of university techniques and corporate industrial partners is personnel training system that accounts enterprise's requirements and digital economy development. Samara State Technical University realizes practice-oriented training in three modes: interdisciplinary project communication (IPC), project-educational intensity (PEI) and cross-cutting practice-oriented project-based training (POP), integrated into the main curriculum educational program (CEP) at all levels and degree programs of university preparation [11]. The main stages of assessment methodology and the role of corporate customer, considering training form are presented in Table 1. This table is formed on the basis of investment efficiency evaluation for the industrial partner.

Table 1. Stages of methodology of ROI assessment in Samara State Technical University with the participation of corporate stakeholders

Stages of methodology assessment	IPC	PEI	POP, CEP
Planning:			
Programs goals	Development of a product project team for a specific assignment (MVP of the project)	Creation of a team who can solve specific project tasks (based on the assignment)	Development of competencies for project work on specific cases (cases are provided by stakeholder companies)
Development of an assessment plan and basic data collection	Creation of an assignment, assessment criteria for MVP project	Creation of an assignment	Case studies for project training implementation
Data collection:			
Data collection during studying	Participation in the project work (individual experts and supervisors from the industrial partner), participation of experts in the reporting project analytical sessions, acceptance of project work results	Conducting master-classes in the course of face-to-face and intercontact period of intensity, participation of experts in the reporting project analytical sessions, acceptance of the project work results	Participation of experts in the assessment of case solutions and final defense of the project work
Data collection after the education	Performance assessment on the basis of students' employment	Performance assessment on the basis of students' employment	Performance assessment on the basis of students' employment
Analysis:			
Assessment of material effect	Specific product result in the course of training according to the technical assignment	Specific result in the course of training that solve certain task according to the technical assignment	Multi-optional solution of a case study relevant for an industrial partner
Assessment of nonmaterial effect	Creation of a team oriented for further work on the projects of the industrial partner, promoting industrial partner at scientific events, exhibitions, in	Creation of an effective team oriented for further work in the form of an interdisciplinary project team	Staff recruitment during training that best meets enterprises' requirements

(*continued*)

Table 1. (*continued*)

Stages of methodology assessment	IPC	PEI	POP, CEP
	the public information space during the project work		
Converting of assessment into value indicators	The converting is made accounting total evaluation of economic effect		
Training costs	Training costs include investment costs for project implementation, taking into account the existing material and technical base of the university, as well as compensation of project risks	Training costs include the cost of organizing and conducting master classes based on university's facilities	Training costs include the cost of developing case studies
Calculation of ROI	Calculation and evaluation of the indicator based on the analysis		
Report:			
Individual reports	Reporting on individual projects and forms of training		
Final report	Final report for industrial partner		

(Source: authors)

Such enterprises as PJSC Oil Company Rosneft, PJSC KAMAZ, PJSC T Plus and other companies act as industrial partners investing in personnel training. As project tasks, participants in the educational process are given real innovative project tasks that are relevant for industrial partners. Urgent problems requiring practical solutions are described in case studies. Reports are made in the following way:

1) response to the training program (taking into account reflection on all training stages);
2) relevance of the training program content to the strategic goals and objectives of the industrial partner;
3) practical application and significance for the consumer;
4) extent of results' achievement;
5) evaluation of investments' effectiveness (ROI).

4 Discussion

The success of personnel training process for corporate customers is evaluated not only in the process of studying, but also at the stage of creation of personnel enterprise reserve and the transition of students to permanent employment at the enterprise. The ROI model which is used on the basis of the university by corporate customers to evaluate training effectiveness is not universal. There are alternative options of training efficiency evaluation, such as Mincer method [3], system of integrated evaluation indicators [6], "Human Economic Value Added" (HEVA) indicator [2]. It is crucial to know that none of the methods is universal. Thus, the use of the Mincer method is widespread in evaluating investments in personnel training. This is due to sufficient simplicity of this model. But often mistakes can occur when this method is used due to lack of all information about the volume of investment costs and knowledge applicability. The system of integrated indicators takes into account the cost of investment, but it is directly related only to labor productivity, without evaluating the additional intangible investment effect. Calculation of "Human Economic Value Added" (HEVA) by Fitzenz' method is simple enough, but it doesn't offer the possibility to evaluate the result of the project activity regarding the MVP project result (minimum viable product), as well as the effect from case studies and project tasks and the image component of this training [2].

Certainly, ROI is not a universal indicator. But it gives the chance to evaluate adequately investments' efficiency into personnel training to confirm the efficiency of enterprise and university interaction. Figure 1 shows the process of forming a project funnel taking into account possible forms of training students with active participation of corporate customers.

The number of projects forms the project funnel. At the stage of integrated project training via the courses "Innovative Economics and Technological Entrepreneurship" and "Practical Project" the number of projects related to the solution of project tasks is large. As the investment costs of industrial partners interested in training are not high at this stage due to the integration of these courses into the educational process of the main educational program.

The project-educational intensity includes fewer projects, as it is a rather new form of interaction between the university and the industrial partner. And this form of education is quite expensive in comparison with classical education. The number of projects at the stage of interdisciplinary project teams is not high due to quite large investments from all participants of the educational process. Table 2 presents a comparative analysis of the main indicators characterizing the cost of training for the investor and the effectiveness of investment in forms of training (in comparison with the cost of training in the basic educational program).

Of course, there are risks that students will leave for competing companies, as these forms of training do not imply the conclusion of a contract that requires the student' consent and obligation to work a certain number of years at the enterprise. But, getting the economic effect at the expense of solving project tasks allows compensating investments.

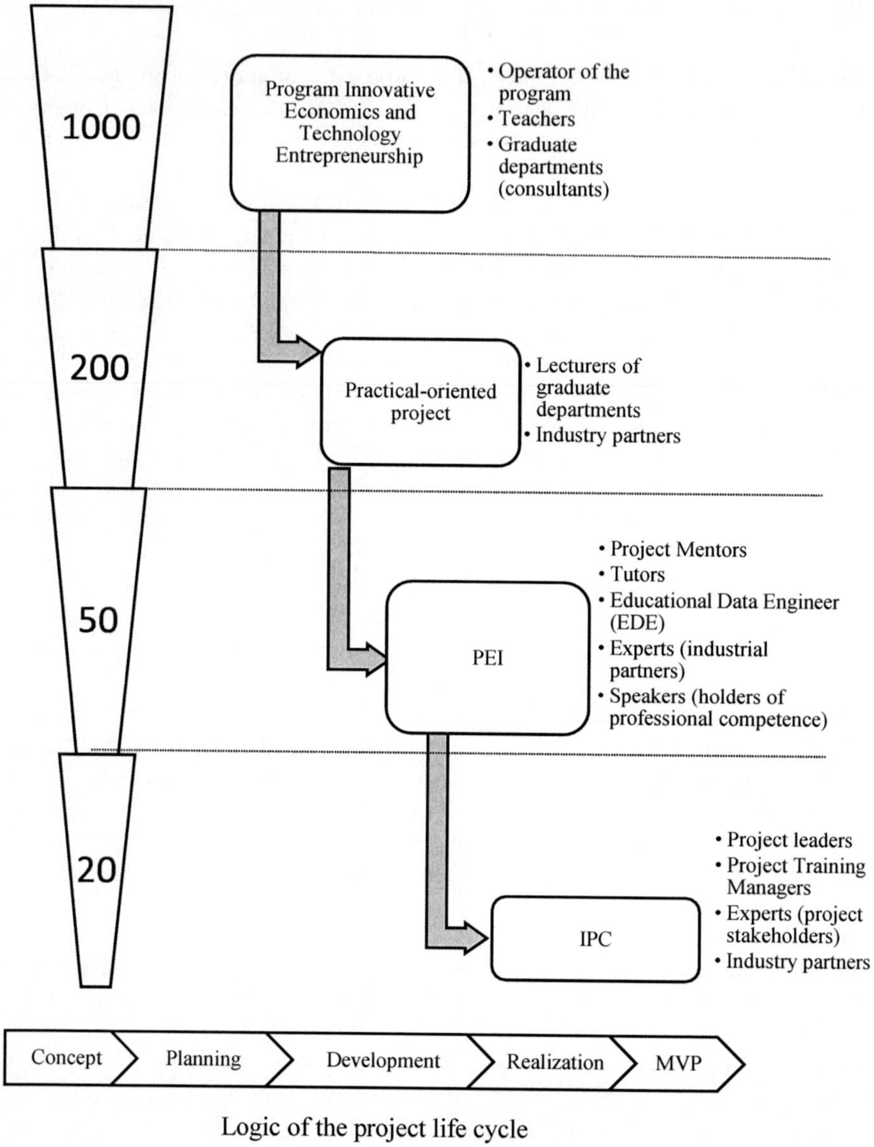

Fig. 1. A funnel of projects in the training implementation with the participation of industrial partners in Samara State Technical University (Source: authors)

Table 2. Stages of methodology of ROI assessment in Samara State Technical University in the implementation of training with the participation of corporate stakeholders (in relative measurement units to the cost of education in the basic education program)

Assessment indicators of investment in personnel training	IPC	PEI	POP, CEP	CEP
Investments	3	1,3	1,1	1
Profit from investing in staff training	4,3	1,6	1,3	1
ROI	43	23	18	0

(Source: authors)

5 Conclusion

Analyzing the practical experience of Samara State Technical University interaction with corporate industrial partners, the following conclusions can be drawn. Corporate customers are interested in investing in training and development of promising students in order to attract them to work for their enterprises. Classical forms of the main educational program do not have time to adapt to constantly changing demands of employers. The process of formation and development of the digital economy sets new challenges for employers and universities. Employers are ready to invest into expensive forms of training if they are clearly aware of the economic benefits. The best way to evaluate investment in training is to calculate the ROI based on Phillips' five-factor model. Practice confirms that industrial partners actively participate not only in investment into training, but also as a carrier of competencies necessary for personnel training [8].

References

1. Andersen, A.S., Kjeldsen, T.H.: Theoretical foundation of PPL at Roskilde university. In: Andersen, A.S., Heilesen, S.B. (eds.) The Roskilde Model: Problem-oriented Learning and Project Work. and Change in Professional Education, vol. 12, pp. 3–16. Springer, Heidelberg (2015)
2. Fitzenz, I.: Profitability of investments in personnel: measuring the economic value of personnel. Vershina, Moscow (2006)
3. Giziene, V., Simanaviciene, Z., Palekiene, O.: Evaluation of investment in human capital economic effectiveness. Econ. Eng. Decis. **23**(2) (2012). https://doi.org/10.5755/j01.ee.23.2.1541
4. Huisman, J., Smolentseva, A., Froumin, I. (eds.): 25 Years of Transformations of Higher Education Systems in Post-Soviet Countries: Reform and Continuity. Palgrave Studies in Global Higher Education, London (2018)
5. Kirkpatrick, D.L., Kirkpatrick, J.D.: Implementing the Four Levels: A Practical Guide for Effective Evaluation of Training Programs. Berrett-Koehler Publishers, Oakland (2007)
6. Milost, F.: Investments in human capital: elements of investments, their valuation, a true and fair view of financial statements compiled by using the classical approach, and the related financial performance ratios. In: Russ, M. (ed.) Management, Valuation, and Risk for Human Capital and Human Assets, pp. 137–159. Palgrave Macmillan, New York (2014)

7. Nielsen, N.M.: Problem-oriented project learning as a first year experience: a transformative pedagogy for entry level PPL. Educ. Sci. **10**, 6 (2019). https://doi.org/10.3390/educsci10010006
8. Phillips, P.P., Phillips, J.J., Stone, R.D., Burkett, H.: The ROI Fieldbook: Strategies for Implementing ROI in HR and Training (Improving Human Performance). Butterworth-Heinemann/Elsevier, Massachusetts (2006)
9. Ríos, I.D.L., Cazorla, A., Díaz-Puente, J.M., Yagüe, J.L.: Project-based learning in engineering higher education: two decades of teaching competences in real environments. Proc. Soc. Behav. Sci. **2**(2), 1368–1378 (2010). https://doi.org/10.1016/j.sbspro.2010.03.202
10. The regulation on the project activities of students at Federal State Budgetary Education Institution of Higher Education "Samara State Technical University". http://cpo.samgtu.ru/sites/cpo.samgtu.ru/files/polozhenie_p_-_405_ot_29.03.2019.pdf. Accessed 18 May 2020
11. Veis, Y.V., Eremicheva, O.Y., Ilyina, L.A., Tikhonov, V.S.: Efficiency evaluating of project training investing as a method of private investors involvement. In: Mantulenko, V. (ed.) Proceedings of the 17th International Scientific Conference "Problems of Enterprise Development: Theory and Practice" 2018, SHS Web of Conferences, vol. 62, p. 04004 (2019). https://doi.org/10.1051/shsconf/20196204004

Formation Mechanism of Industrial Policy for Improving Regional Food Security

A. V. Pavlova[1(✉)], A. I. Sabirova[2], A. R. Safiullin[2], and A. D. Khairullina[2]

[1] Samara State University of Economics, Samara, Russia
930895@list.ru
[2] Kazan Federal University, Kazan, Russia
halbi@mail.ru, Aigylkinyes@mail.ru, safiullin.ar@gmail.com

Abstract. As a result of the pandemic, both the internal needs of the territory (country, region) and the importance of providing humanitarian assistance to the countries with food shortages may grow significantly. The food security assessment and the application of measures to improve it, should be carried out in advance. The reorientation of import substitution policy to export-oriented one should also be based on the competition development in the industry. In the article, the authors suggest to calculate a new type of Herfindahl-Hirschman index, which characterizes the level of producers' concentration at the market by the share of gross profit of economic entities. The research purpose is to provide scientific justification of practical recommendations for the development of food markets in the region, taking into account the level of their concentration. Based on multidimensional statistical calculations, the authors analyze the hypothesis of a positive impact of large economic entities (including monopolies) development on the provision of local food products to the population; consider tools for forming the vector of industrial and competitive policy based on matrix positioning, as well as conduct correlation and regression analysis on the example of the regional beef market.

Keywords: Competition policy · Food security · Industrial policy · Local markets · Regional development

1 Introduction

The current epidemiological situation in 2020 (the pandemic) has shown possible failures in ensuring food security and means of subsistence, preventing a panic reaction that can exacerbate failures and worsen food security. In such conditions, it is particularly important to maintain the functioning of food supply chains both within the country and at the regional level [3]. Russia is the largest country in the world, located in the Eastern part of Europe and Northern Asia, so it is safe to say that its food security is formed based on administrative and territorial division.

The formation of legal foundations in the field of food security in the Russian Federation is rooted in the XIX century [8]. However, in our opinion, the key stage in the history of legislation development, aimed at achieving the goals of providing the

S. I. Ashmarina et al. (Eds.): ENGINEERING ECONOMICS WEEK 2020, LNNS 139, pp. 697–707, 2021.
https://doi.org/10.1007/978-3-030-53277-2_83

population with their own food begins with the Decree of the President of the Russian Federation from 30.01.2010 № 120 "On approval of the Food Security Doctrine of the Russian Federation" [12]. However, taking into account changes in external and internal conditions due to the geopolitical situation, Russia's accession to the WTO (World Trade Organization) and the deepening of integration processes in the EAEU (The Eurasian Economic Union), the Ministry of Agriculture and other departments have developed a project of a new Food Security Doctrine [12], which was adopted in January 2020. The new Doctrine has undergone a number of key changes (for example, in changing the range of a food products set), but the key aspect is still the procedure for calculating of the food independence indicator. It is defined as the level of self-sufficiency of the territory in percentage terms, calculated as the ratio of the volume of domestic production of agricultural products, raw materials and food to the volume of their domestic consumption and has threshold values. Thus, the main principle is the state responsibility for ensuring a guaranteed level of the population food well-being.

In a number of regions of the Russian Federation, the regional authorities, with the help of Federal legislation, have developed a strategy for the food security development, taking into account their natural, climatic and national characteristics. Some of the Russian Federation subjects that can be classified as economically developed regions have legislated their own regulations to ensure food security (Moscow, Saint Petersburg, Tatarstan, Bashkortostan, Nizhny Novgorod, Sverdlovsk, and Ulyanovsk regions).

Moreover, food security in the economic aspect is often considered only at the level of budget allocations amount determining, the dynamics of which is not always based on quantitative and qualitative planning.

The above has led to the purpose of the study, which is to scientifically substantiate the theoretical approach and development of practical recommendations for the development of food markets in the region based on the regulation of their concentration level. Therefore, the analysis of regional food markets is very promising, taking into account the new principles of the formation of the Food Security Doctrine.

2 Methodology

In the economic literature and public policy, usually in all sectors (except for the military and natural monopolies), the preference is given to the competition development. However, the works of the famous economist J. A. Schumpeter indicate that highly competitive markets do not always form the conditions for organizing the more effective resources redistribution. Thus, a positive impact of large farms on the population's food supply due to economies of scale, reduced production costs and the formation of a product with a high degree of processing is possible. To refute or confirm this food market hypothesis, we conducted an empirical analysis using mathematical and statistical analysis tools, we revealed the influence of the Herfindahl-Hirschman index calculated by three parameters (natural indicators, revenue, profit) on social and economic indicators that form the economic and physical availability of food (according to the region local markets).

The study used comparative analysis and statistical methods (the method of regression analysis using the Gretl software product) [5]. For the purpose of economic and statistical analysis, data from the Russian Federal State Statistics Service and data from the Republic of Tatarstan Ministry of Agriculture were used. Together, these methods allowed us to solve the tasks set in the study.

Based on the made in 2020 change in methodology of calculation of food security indicators, in particular the introduction of a new indicator, which reflects the degree of internal market needs meeting through domestic production, the authors calculated the coefficient characterizing the physical security of the population with −i product:

$$F_i = \frac{V \cdot k_{recalculation}}{P \cdot N_{use}} \quad (1)$$

F_i – where is the coefficient of physical supportability of the population with i-food product;

V – the volume of food production in the study area;

$k_{recalculation}$ – coefficient of raw materials conversion into the finished food;

P – average annual population of the study area;

N_{use} – the minimum rational rate of the product consumption according to medical standards (recommendations on rational norms of food consumption that meet modern requirements of healthy nutrition approved by the order of the Ministry of Health of the Russian Federation of 19 August 2016. N 614).

As a result of calculating this coefficient (F), it is possible to compare the market capacity of the studied product according to the established rational medical standards of its consumption, taking into account the population of the territory (on average for all population groups) and the actual production volume of the product in the studied territory.

As a result of calculations there will be two options:

a) F > 1-indicates that the volume of production by commodity producers in the study area is sufficient and fully covers the population demand;
b) F < 1-indicates the lack of production of the product in the study area.

At the same time, it should be understood that economic analysis in the agricultural sector has some features: the raw material does not always fully acquire the form of a finished product for sale, as it can be used for agricultural unit's own consumption. Thus, we agree with the point of view presented in the scientific literature - to evaluate economic entities from the position of full commodity production, therefore, we will consider all manufactured products of the company as a commodity [7]. We will test this approach on the example of the beef market in the regional market of the Republic of Tatarstan - which forms one of the 85 subjects of the Russian Federation, and also conduct a matrix analysis to determine the vector of industrial and competitive policy of the region.

3 Literature Review

The destabilization of the world oil market and large-scale quarantine measures, that made the products delivery complicated, have led to some countries restricting food exports, while others risk of agricultural works disruption. The head of the FAO (Food and Agriculture Organization) notes that strict isolation and movements restriction can disrupt the production, processing, wholesale supplies and sale of food both at the national and international level [1]. For example, due to the borders closure for seasonal agricultural workers, a significant part of the crop may be lost, and difficulties are expected due to labor shortages in a number of industries. Large food suppliers reduce or completely stop supplying the market with certain food products due to a lack of working capital to perform current activities. Such an imbalance in the world market of raw food products leads to prices increasing at both the global and regional levels. In rich countries, the rise in food prices leads to decreasing of the population level and quality of life. In poor countries, it results in an acute shortage or even unavailability of essential food for broad population, which results in socio-political upheavals of massive proportions [2, 9, 10]. According to Gurkov, one of the main causes of the revolutions and uprisings series that broke out in late 2010 in North Africa and the Middle East and is known as the “Arab spring” was the rise in world food prices in 2007–2008 [2].

According to prognosis, by 2050 the world population will increase to 9.6 billion people [14]. In this regard, to satisfy the overall demand, food production should increase by 70–110% [14].

Under these conditions, the economic assessment and implementation of the agro-industrial complex strategic potential is a key goal of state regulation for the last two decades, and has become a central strategy for the governmental and non-governmental programs development [10]. Evaluation of region sustainable development, covering economic, environmental and social aspects of the territory development, is inextricably connected with economic indicators such as food demand and population growth, and increasing the total productivity of production inputs in the region [6, 11].

From an empirical point of view, in the context of agricultural development, the number of studies supporting small business (farming) has grown significantly over the past decade [4]. However, competition and industrial policy directions in the agri-food sector within even one state should take into account regional differences. Differences are manifested both in the population living standard and in economic potential differentiation.

4 Results

The supply assessment in the food market should also include the indicator of the economic availability of food to the population. Such social and economic indicators as the purchasing power of beef by the population, the share of expenditures on beef, the price index, average producer prices, and average consumer prices characterize the economic availability of the analyzed product for the locals. Table 1 shows the values of the above-mentioned indicators in the Republic of Tatarstan.

Table 1. Coefficient of physical accessibility and values of social and economic indicators that form the economic accessibility of beef for the population according to the Republic of Tatarstan records (Territorial body of the Federal State Statistics Service of the Republic of Tatarstan)

Year	Coefficient of physical supportability of beef meat	Consumption per caput "meat and meat products", (on meat basis), kg.	Purchasing power of the population "Meat and meat products", kg.	Share of expenditures on "Meat and meat products", percentage out of 100	Price indices "raw meat of cattle", in %	Average producers prices "live cattle", rouble/tn	Average consumer prices "beef, except boneless meat", rouble/kg	Consumer price indices "Meat products", in %
2013	0.96	73	109	11.8	141.0	49701	171.25	104.4
2014	0.94	74	98	8.9	85.9	59354	223.35	120.9
2015	0.94	78	101	8.4	101.7	69993	240.16	105.4
2016	0.96	80	110	6.9	150.2	71079	240.47	97.8
2017	0.82	80	120	7	103.1	73997	260.51	116.1
2018	0.90	79	104	7.2	100.2	89433	305.1	108.5
2019	0.96	81	109	8	120.9	96447	299.09	100.6

Source: authors.

The regional market of meat products has an internal potential, despite the fact that the level of self-sufficiency is reached almost annually (at least 85%) %. This conclusion can be justified by the fact that the indicators are close to the boundary value set by the Doctrine, and the calculation was made based on statistics that form the indicator value of beef meat production in live weight.

During the study period, the local people' beef supply of agricultural organizations and peasant farms in the republic averaged 92%.

We calculated the correlation coefficients between the concentration level indicators (Herfindahl-Hirschman Index (HHI) and market concentration (CR) on revenue, volume of production in kind and gross profit) and the main social and economic indicators that affect the availability of meat and meat products for the population. Statistically significant correlation coefficients obtained in the course of research on the beef market are shown in Table 2.

Table 2. The revealed relationships between social and economic indicators that form the economic availability of beef meat and the level of producers' concentration according to the Republic of Tatarstan for 2013–2019

No.	Indicators	HHI by natural indicators of Cattle meat	HHI by Cattle meat revenue	HHI by Cattle meat profit	CR3 by natural indicators of Cattle meat	CR3 by Cattle meat revenue	CR3 by Cattle meat profit
1	Share of expenditures on Meat	✓ 0.89	✓ 0.89	0.41	✓ 0.89	✓ 0.86	0.44
2	Average consumer prices of Cattle meat	-0.65	✓ -0.72	-0.13	-0.59	✓ -0.70	-0.18
3	Share of cash expenditures on food	✓ 0.79	✓ 0.74	✓ 0.74	✓ 0.84	✓ 0.74	✓ 0.73

Source: authors.

Since the HHI values for the beginning of 2019 are 0.02–0.04 units and CR3 values are equal to 0.16–0.26, then we can say with confidence that the beef market of the Republic of Tatarstan has a stable dynamics of forming a structure close to the perfect competition. The values of concentration indicators also characterize the fact that in 2018, the three largest producers of the study area accounted for 16% of the total volume of production (or 26% of profit). For further analysis, the indicators "share of expenditure on meat" and "share of money expenditure on food" can be considered identical. Considering the fact that the indicator "share of population expenditure on meat" is more close to the object of analysis, and also has higher values of correlation coefficients, let's analyze the influence of the concentration level directly on it (2).

$$y = 5,91 + 145,34 \cdot x_9 \tag{2}$$

where y is the indicator of the share of population expenditure on meat;

x_9 – Herfindahl-Hirschman index of the beef market (in natural values).

This regression equation is statistically reliable and allows us to draw some economic conclusions. Thus, an increase of producers concentration in the market will negatively affect the availability of meat to the population. A graph of the observed and calculated values confirms the quality of the model (Fig. 1).

The calculations results show that if the largest producers will increase production volumes to 1%, it will immediately lead to an increase in the share of population expenditure on meat. According to the 95% of confidence interval, calculated for the regression coefficient, the increase of the population expenditures share will be [0.6; 2.3] percent and more. Thus, the share of expenditures on meat is not sensitive to changes in the concentration of producers in the beef market, this fact is also confirmed by the elasticity coefficient (0.29%).

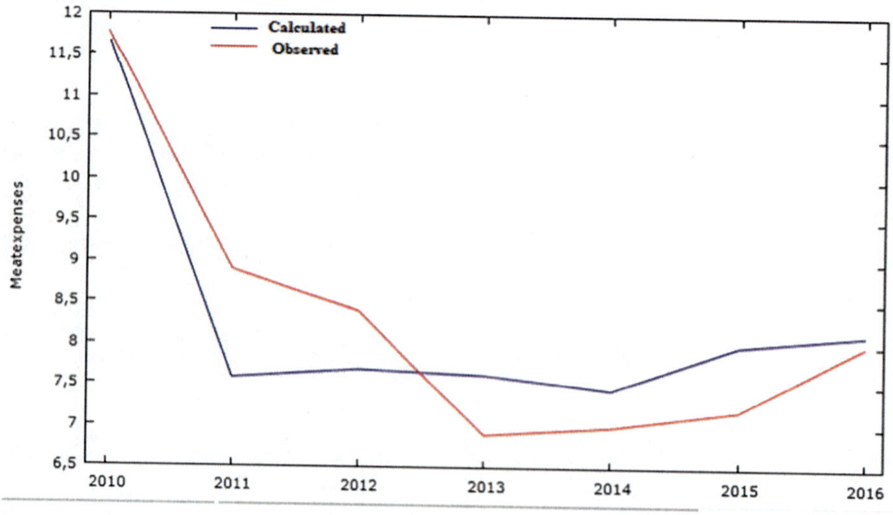

Fig. 1. Dynamics of observed and calculated values of the HHI impact on the beef market on the share of meat expenditures of the population of the studied territory (Republic of Tatarstan), 2013–2019 (Source: authors).

However, the share of population expenditures on meat will increase significantly if the beef market becomes medium-centered (HHI = 0.1). In this situation, the share of population expenditures on meat will reach on average 20.44% or, with a 95% reliability, will be [12.47; 27.92] percent.

If the beef market becomes highly monopolized (HHI = 0.2), then the average expenditures of the population on meat may be 34.98% or, with a 95% reliability, will be [19.20; 50.76] percent. This situation will definitely lead to a significant decrease in the level of well-being of the region's population.

Despite the fact that the results of calculations of correlation coefficients showed a high direct relationship between market monopolization indicators and average consumer prices on beef meat, the Durbin-Watson test characterizes the presence of autocorrelation in time series with a probability of 95% (Table 3).

Table 3. The least square method (T = 7) Average consumer prices for Beef = 287,18–1261,06•HHIrevenueBeef

	Coefficient	*St. error*	*t-statistics*	*P-value*	
const	287.184	21.1343	13.5885	0.00004	***
HHIrevenueBeef	−1261,02	540.264	-2.3341	0.06687	*

Arithmetical dependent mean	248.5614	Arithmetical off. dependent mean	45.90070
Sum of squared errors	6049.661	Standard error of model	34.78408
R-square	0.521435	Cor. R-square	0.425722
F(1, 5)	5.447894	P-value (F)	0.066871
Log likelihood	−33,59904	Criterion of Akaike	71.19807
Criterion of Schwarz	71.08989	Criterion of Hannan-Quinn	69.86099
Rho parameter	0.645082	Durbin–Watson statistic	0.786981

Breusch-Pagan test for heteroscedasticity -

The null hypothesis: heteroskedasticity is missing; Test statistic: LM = 0,615575

p-value = P(Chi-square(1) > 0.615575) = 0.432696

Durbin-Watson statistics = 0.786981; P-value = 0.0131824

t(5, 0,025) = 2,571

Variable	Coefficient	95% confidence interval
const	287.184	(232,856, 341,512)
HHIrevenueBeef	-1261.02	(-2649,81, 127,778)

Source: authors.

The presented data characterize that the model is unreliable and is not subject to economic analysis. In general, the values of all HHI and CR3 indicators show that the beef market is structurally close to the perfect competition market. Our research also revealed the fact that state support in the form of subsidies in the beef market has a positive impact on the formation of their own food security (3).

$$y = 11390,9 \div x_{15} \tag{3}$$

where y is the coefficient of physical security of the population with beef meat; x_{15} – average amount of subsidies allocated to the district;

Despite the fact that only 16% of the dependent variable is explained according to the hyperbolic function in the constructed model, Fischer's test allows us to say that the constructed equation is statistically reliable by more than 95%. In particular, the t-statistic confirms the statistical significance of the regression coefficient by more than 95%. The absence of heteroscedasticity (with reliability of more than 99%) is confirmed by the Breusch-Pagan test. In addition, the remainder of the equation has a normal distribution, p - value of the test statistics = 0.26 [5].

Also, in order to make managerial decisions for the formation of a regional food security strategy on beef meat, it is necessary to take into account the fact that the physical supply of the population with beef meat is not affected by the level of producer prices, the cost of 1 kg of beef meat and the profitability of production.

As a part of the region food security discussion, it should be noted that the development of the regional food market depends entirely on the development of its local markets [7]. At the same time, the economic literature does not clearly distinguish the understanding of this category [4, 7]. Taking into account the administrative territorial division of the Republic of Tatarstan, we have identified municipal districts of the republic as local markets [13]. Thus, this approach takes into account common business conditions for producers, as well as logistic aspects of products implementation [7].

5 Conclusion

Let's analyze the production of beef meat in the region, which is characterized by the following parameters: despite the fact that the market is marked by a loss of production (−25%) the meat and meat products market of the Republic of Tatarstan has a positive dynamics, since the production for 2 years increased by 16% (5 thousand tons), exports to other regions of the Russian Federation increased by 26% (30 thousand tons) on average, 1 ruble of revenue is 1 ruble of budget allocations.

We suggest formulating the definition of industrial and competition policy instruments, using the decision-making matrix. Since the studied territory (the Republic of Tatarstan, which includes many districts-hereinafter local markets), we will reflect in the matrix the ones that produce the beef meat. The diameter of the circle reflects the value of the HHI on the volume of implementation in natural expression, according to the X-axis (x) we can see the values of the local market average production profitability; on the Y-axis (y) we can see the values of product physical security coefficient on the local market. Since in the previously suggested formula of the physical security coefficient calculating (1), food security of the territory will be achieved if its value is higher than 1, the origin of coordinates on the ordinate axis (y) passes through the value equal to "1". We have made recommendations based on which quadrant of the matrix a particular local market will be assigned to (Fig. 2).

Describing the matrix, we should remember that "F" is the coefficient of physical security, and "R" is the profitability of production (calculated as the profit to cost ratio).

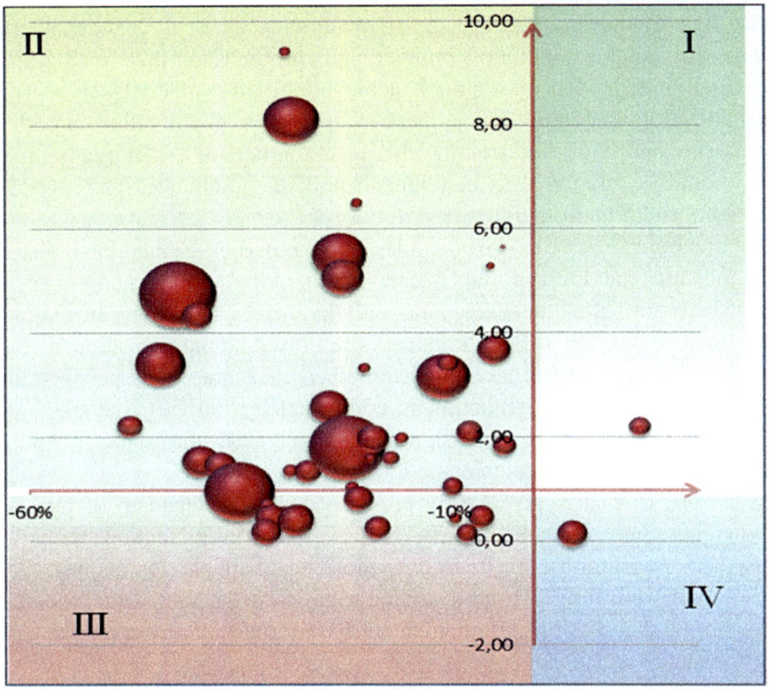

Fig. 2. Matrix for making strategic decisions in order to determine the development vectors of industrial and competitive policy of beef production in the Republic of Tatarstan (Source: authors).

In the region beef production is characterized by indicators of activity loss ratio (−20%). The I quadrant of the matrix (F > 1; R > 0) includes only one local market with 8 producers, so we consider it appropriate to develop competition among them using the tools of industrial and competitive policies.

The second quadrant (F > 1; R < 0) includes 68% of the local markets of the studied territory (255 producers out of represented in the sample 351). In these local markets, the population is provided with products of their own production above rational norms, the necessary level of food security is achieved, so it is necessary to increase the profitability of producers' activities. Producers of all local markets of the II quadrant are unprofitable (17 local markets out of 30 demonstrate a loss of production above 20%). We believe that these indicators were formed due to the sale of products at a below-cost price. To ensure that these companies do not stop their activities, direct and indirect government support should be provided to them, since their closure may increase the risk of food dependence and increase the level of the territory monopolization as a whole. This recommendation is acceptable, since the above-mentioned analysis shows that increasing of the beef market monopolization to the level of medium-concentrated does not significantly increase the share of population expenditures on the product.

Almost 30% of the producers of the III quadrant (F < 1; R < 0) are unprofitable and do not provide the population with enough products of their own production. However, food independence is almost achieved in three districts (F > 70%). We recommend developing competition in these local markets, it will reduce product prices and improve the quality of products. In other local markets of the III quadrant, existing producers should be supported (protectionism policies, guaranteed price system, cost compensation system and others).

The IV quadrant (F < 1; R > 0) includes 1 local market (Zelenodolsk district), the value of F indicates that local producers cater for the population demand in accordance with established rational medical standards only by 15%. The activity of producers are inefficient (R = 5%). Even if 2–3 producers are closed, monopolization in this local market will grow, so it is necessary to strengthen measures of large operating production support. This recommendation is confirmed by the fact that high monopolization in this food market will significantly increase the share of population expenditures on meat (see conclusions after Fig. 1).

Thus, the use of matrix positioning provides a new mechanism of planning soft government intervention in the industry. However, the interpretation of the findings requires special precautions; the reason is the data available for analysis and their spatial resolution were limited by the studied region. At the same time, the suggested approach can be successfully implemented based on data from other territories, including the state level.

By integrating indicators of assessing the competition level in the industry and social and economic indicators, we have developed tools that could be used to improve the level of food security in the region.

References

1. Food and agriculture organization of the United Nations: COVID-19 coronavirus outbreak (2020). http://www.fao.org/2019-ncov/ru/. Accessed 30 Apr 2020
2. Gurkov, A.: COVID-19 pandemic: The world will face the food shortages (2020). https://clck.ru/N6C2D. Accessed 30 Apr 2020
3. InterFax: The UN announced that COVID-19 is the threat for food markets (2020). https://www.interfax.ru/world/701158. Accessed 30 Apr 2020
4. International Fund for Agricultural Development: Smallholders, food security, and the environment (2013). https://www.ifad.org/documents/38714170/39135645/smallholders_report.pdf/133e8903-0204-4e7d-a780-bca847933f2e. Accessed 30 Apr 2020
5. Ismagilov, I.I., Kadochnikova, E.I.: Special models of econometrics in the Gretl environment: a textbook for students, pursuing a bachelor's degree in 38.04.01 "Economics". University of the Kazan, Kazan (2018)
6. Kudryashov, A.: Lean manufacturing. Implementation problems. Bulletin of UNIDO in Russia, 6 (2012). http://www.unido-russia.ru/archive/num6/art6_14/. Accessed 30 Apr 2020
7. Safiullin, A.R., Khairullina, A., Sabirova, A.I.: The impact of competition on the sustainable development of the regional food market. J. Soc. Sci. Res. **20189**(5), 439–443 (2018)
8. Sklyarenko, S.A., Boltaevsky, A.A., Pryadko, I.P., Sharov, V.I.: History and modernity of the food security problem. Agriculture **1**, 27–39 (2016). https://doi.org/10.7256/2453-8809.2016.1.20267

9. Storozh, I.A.: Lean manufacturing implementation algorithms. Stand. Qual. **11**, 58–60 (2016)
10. Tan, H.: A food crisis looms as coronavirus forces farms to stay idle and countries hoard supplies. World Markets (2020). https://www.cnbc.com/2020/03/30/coronavirus-food-crisis-looms-as-farms-idle-countries-hoard-supplies.html. Accessed 30 Apr 2020
11. Teslenko, A., Chernukha, A., Bezuglov, O., Bogatov, O., Kunitsa, E., Kalyna, V., Katunin, A., Kobzin, V., Minka, S.: Construction of an algorithm for building regions of questionable decisions for devices containing gases in a linear multidimensional space of hazardous factors. Eastern-Eur. J. Enterp. Technol. **5**(10(101)), 42–48 (2019)
12. The "Food Security Doctrine of the Russian Federation" was approved by the Decree of the President No. 20 from January 21, 2020 "On approving the Food Security Doctrine of the Russian Federation» (2020). http://kremlin.ru/acts/news/62627. Accessed 30 Apr 2020
13. Turgel, I.D., Bozhko, L.G., Linshi, S.: State support for single-industry towns' development of Russia and Kazakhstan. Financ. Theory Pract. **2**(92), 22–32 (2016)
14. United Nations: World population prospects: The 2017 revision (2017). https://www.un.org/development/desa/publications/world-population-prospects-the-2017-revision.html. Accessed 30 Apr 2020

Author Index

S. I. Ashmarina et al. (Eds.): ENGINEERING ECONOMICS WEEK 2020, LNNS 139, pp. 709–711, 2021.
https://doi.org/10.1007/978-3-030-53277-2

Lecture Notes in Networks and Systems

Volume 798

The series "Lecture Notes in Networks and Systems" publishes the latest developments in Networks and Systems—quickly, informally and with high quality. Original research reported in proceedings and post-proceedings represents the core of LNNS.

Volumes published in LNNS embrace all aspects and subfields of, as well as new challenges in, Networks and Systems.

The series contains proceedings and edited volumes in systems and networks, spanning the areas of Cyber-Physical Systems, Autonomous Systems, Sensor Networks, Control Systems, Energy Systems, Automotive Systems, Biological Systems, Vehicular Networking and Connected Vehicles, Aerospace Systems, Automation, Manufacturing, Smart Grids, Nonlinear Systems, Power Systems, Robotics, Social Systems, Economic Systems and other. Of particular value to both the contributors and the readership are the short publication timeframe and the world-wide distribution and exposure which enable both a wide and rapid dissemination of research output.

The series covers the theory, applications, and perspectives on the state of the art and future developments relevant to systems and networks, decision making, control, complex processes and related areas, as embedded in the fields of interdisciplinary and applied sciences, engineering, computer science, physics, economics, social, and life sciences, as well as the paradigms and methodologies behind them.

Indexed by SCOPUS, INSPEC, WTI Frankfurt eG, zbMATH, SCImago.

All books published in the series are submitted for consideration in Web of Science.

For proposals from Asia please contact Aninda Bose (aninda.bose@springer.com).

Subarna Shakya · João Manuel R. S. Tavares ·
Antonio Fernández-Caballero · George Papakostas
Editors

Fourth International Conference on Image Processing and Capsule Networks

ICIPCN 2023

Springer

Editors
Subarna Shakya
Department of Electronics and Computer Engineering, Pulchowk Campus, Institute of Engineering
Tribhuvan University
Lalitpur, Nepal

Antonio Fernández-Caballero
Universidad de Castilla-La Mancha
Albacete, Spain

João Manuel R. S. Tavares
Faculdade de Engenharia
Universidade do Porto
Porto, Portugal

George Papakostas
Department of Computer Science
International Hellenic University
Kavala, Greece

ISSN 2367-3370 ISSN 2367-3389 (electronic)
Lecture Notes in Networks and Systems
ISBN 978-981-99-7092-6 ISBN 978-981-99-7093-3 (eBook)
https://doi.org/10.1007/978-981-99-7093-3

This Springer imprint is published by the registered company Springer Nature Singapore Pte Ltd.
The registered company address is: 152 Beach Road, #21-01/04 Gateway East, Singapore 189721, Singapore

Paper in this product is recyclable.

We would like to dedicate this proceeding to all members of advisory committee and program committee for providing their excellent guidance. We also dedicate this proceeding to the members of the review committee for their excellent cooperation throughout the conference. We also record our sincere thanks to all the authors and participants.

Preface

It is our pleasure to welcome you to the 4th International Conference on Image Processing and Capsule Networks [ICIPCN 2023] organized on August 10–11, 2023. A major goal and feature of the conference is to bring academia and industries together to share and exchange their significant research experiences and results in the field of imaging science, with a particular interest on the capsule network algorithms and models by discussing the practical challenges encountered and solutions adopted to it.

This conference will deliver a technically productive experience to the budding researchers in the field of image processing and capsule networks by stimulating a good awareness on this emerging research field. ICIPCN promises to provide a produce a bright landscape for the image processing research works, while the response received, and research eagerness witnessed, have truly exceeded our expectations. At the end of the conference event, we are overwhelmed with the high level of satisfaction.

The response for the conference event has been increasing at an unprecedented rate both from Thailand and overseas. Due to the professional expertise of both internal and external reviewers, the papers have been selectively accepted based on their extensive research and publication quality. We have received a total submission of 366 out of which only 48 papers were accepted for publication based on its research effectiveness and applicability. We would also like to extend our thanks to the members of the organizing committee for their hard work in delivering spontaneous response to all the conference participants.

We are now enthusiastic to get the proceedings of the ICIPCN conference event covered by Springer. We also appreciate all the authors of ICIPCN 2023 for their timely response to all the queries raised from the conference.

Finally, we would like to thank the Springer publications for producing this volume.

Lalitpur, Nepal
Porto, Portugal
Albacete, Spain
Kavala, Greece

Prof. Dr. Subarna Shakya
Dr. João Manuel R. S. Tavares
Dr. Antonio Fernández-Caballero
Dr. George Papakostas

Contents

Editors and Contributors

About the Editors

Subarna Shakya is currently a professor of Computer Engineering, Department of Electronics and Computer Engineering, Central Campus, Institute of Engineering, Pulchowk, Tribhuvan University, Coordinator (IOE), LEADER Project (Links in Europe and Asia for Engineering, Education, Enterprise, and Research Exchanges), ERASMUS MUNDUS. He received M.Sc. and Ph.D. degrees in Computer Engineering from the Lviv Polytechnic National University, Ukraine, 1996 and 2000 respectively. His research area includes e-government system, computer systems and simulation, distributed and cloud computing, software engineering and information system, computer architecture, information security for e-government, and multimedia system.

João Manuel R. S. Tavares graduated in Mechanical Engineering at the Universidade do Porto, Portugal, in 1992. He also earned his M.Sc. degree and Ph.D. degree in Electrical and Computer Engineering from the Universidade do Porto in 1995 and 2001 and attained his Habilitation in Mechanical Engineering in 2015. He is a senior researcher at the Instituto de Ciência e Inovação em Engenharia Mecânica e Engenharia Industrial (INEGI) and an associate professor at the Department of Mechanical Engineering (DEMec) of the Faculdade de Engenharia da Universidade do Porto (FEUP).

João Tavares is a co-editor of more than 55 books and co-author of more than 50 book chapters, 650 articles in international and national journals and conferences, and three international and three national patents. He has been a committee member of several international and national journals and conferences, is co-founder and co-editor of the book series *Lecture Notes in Computational Vision and Biomechanics* published by Springer, founder and editor-in-chief of the journal *Computer Methods in Biomechanics and Biomedical Engineering: Imaging and Visualization* published by Taylor & Francis, editor-in-chief of the journal *Computer Methods in Biomechanics and Biomedical Engineering* published by Taylor & Francis, and co-founder and co-chair of the international conference series: CompIMAGE, ECCOMAS VipIMAGE, ICCEBS and BioDental. Additionally, he has been a (co-)supervisor of several M.Sc. and Ph.D. thesis and supervisor of several post-doc projects and has participated in many scientific projects both as researcher and as scientific coordinator.

Antonio Fernández-Caballero received his Master in Computer Science from the School of Computer Science at the Technical University of Madrid, Spain, and he received his Ph.D. from the Department of Artificial Intelligence of the National University for Distance Education, Spain. He is a full professor at the University of Castilla-La Mancha, Albacete, Spain. His interests enforce him to be part of the membership in National Society Networks AERFAI (Spanish Association of Pattern Recognition and Image Analysis), RTNAC (National Natural and Artificial Computation Network), RedAF (Physical Agents Network), AIPO (Association of Human–Computer Interaction) and Spanish Technology Platform on Robotics (Hisparob) and European networks euCognition (The European Network for the Advancement of Artificial Cognitive Systems), and SIMILAR (The European taskforce Creating human–machine interfaces SIMILAR to human–human communication).

George Papakostas received the Diploma degree in electrical and computer engineering in 1999 and the M.Sc. and Ph.D. degrees in electrical and computer engineering in 2002 and 2007, respectively, from the Democritus University of Thrace (DUTH), Greece. From 2007 to 2010, he served as an adjunct lecturer with the Department of Production Engineering and Management, DUTH. He currently serves as an adjunct assistant professor with the Department of Computer and Informatics Engineering, Technological Educational Institution, Eastern Macedonia and Thrace, Greece. In 2012, he was elected as a full professor in the aforementioned Department of Computer and Informatics Engineering. He has co-authored more than 70 publications in indexed journals, international conferences, and book chapters. His research interests include pattern recognition, computer/machine vision, computational intelligence, machine learning, feature extraction, evolutionary optimization, and signal and image processing.

Contributors

Shinu Abhi REVA Academy for Corporate Excellence (RACE) REVA University, Bangalore, India

S. Abhishek Department of Computer Science and Engineering, Amrita School of Computing, Amrita Vishwa Vidyapeetham, Amritapuri, India

Priyanka Acharya Vishwakarma Institute of Technology, Pune, Maharashtra, India

Sheikh Shemanto Afridi Department of Computer Science and Engineering, East West University, Dhaka, Bangladesh

Chashu Agrawal Department of Electronics and Telecommunication Engineering, Vishwakarma Institute of Technology, Pune, India

P. Akhendra Kumar Department of Electronics and Communication Engineering, Faculty of Science and Technology (IcfaiTech), ICFAI Foundation for Higher Education, Hyderabad, India

G. Akila Department of Electronics and Communication Engineering, Meenakshi Sundararajan Engineering College, Chennai, India

Abdullah Al Noman Computer Science and Engineering Discipline, Khulna University, Khulna, Bangladesh

Diana A Andrushia Karunya Institute of Technology and Sciences, Coimbatore, India

T. Anjali Department of Computer Science and Engineering, Amrita School of Computing, Amrita Vishwa Vidyapeetham, Amritapuri, India

M. Antony Freeda Rani Department of Electrical and Electronics Engineering, DMI College of Engineering, Chennai, Tamil Nadu, India

S. M. Arafat Rahman Department of Computer Science and Engineering, East West University, Dhaka, Bangladesh

Abu Shamim Mohammad Arif Computer Science and Engineering Discipline, Khulna University, Khulna, Bangladesh

Enoch Hall Asanda Makerere University, Kampala, Uganda

V. Z. Attar School of Computational Science, COEP Technological University, Pune, India

A. Azeroual Research Center STIS, M2CS, National Graduate School of Arts and Crafts of Rabat, Mohammed V University in Rabat, Rabat, Morocco;
Group of Biomedical Engineering and Pharmaceuticals Sciences, National Graduate School of Arts and Crafts of Rabat (ENSAM), Mohammed V University, Rabat, Morocco

S. Babeetha Department of Computing Technologies, College of Engineering and Technology, SRM Institute of Science and Technology, Chengalpattu, Tamil Nadu, India

M. S. Babitha Department of Electronics and Communication Engineering, Karunya Institute of Technology and Sciences, Coimbatore, India

Y. Baby Kalpana CSE, P.A. College of Engineering and Technology, Pollachi, India

M. Selva Balan Central Water and Power Research Station, Pune, Maharashtra, India

Ashlyn Kim D. Balangcod Department of Mathematics and Computer Science, University of the Philippines Baguio, Baguio, Philippines

Asok Bandyopadhyay ICT and Servics Group, Centre For Development of Advanced Computing, Kolkata, India

Tushar Barai G.H.Raisoni College of Engineering Nagpur, Nagpur, Maharastra, India

Afruza Begum School of Computer Science Engineering, Vellore Institute of Technology, Chennai, India

T. Belhoussine Drissi Laboratory of Electrical and Industrial Engineering, Information Processing, Informatics, and Logistics (GEITIIL), Faculty of Science Ain Chock, University Hassan II—Casablanca, Casablanca, Morocco

Sivaiah Bellamkonda Computer Science and Engineering Department, Indian Institute of Information Technology, Kottayam, Kerala, India

Ilya Belousov Volgograd, Russian Federation

El Benany Mohamed Mahmoud Department of Mathematics and Computer Science, University of Nouakchott AL Aasriya, Nouakchott, Mauritania

Shajulin Benedict Indian Institute of Information Technology Kottayam, Kottayam District, Kerala, India

Chirag Bhandari Department of Electronics and Telecommunication Engineering, Vishwakarma Institute of Technology, Pune, India

Rakhi Bhardwaj Department of Computer Engineering, Vishwakarma Institute of Technology, Pune, India

Utpal Bhattacharjee Department of Computer Science and Engineering, Rajiv Gandhi University, Itanagar, Arunachal Pradesh, India

Nikunj Bhavsar Department of Computer Engineering, School of Engineering, P P Savani University, Kosamba, Surat, Gujarat, India

Shashank Bhosagi JSPM's Rajarshi Shahu College of Engineering, Pune, Maharashtra, India

Ketan Bodhe G.H.Raisoni College of Engineering Nagpur, Nagpur, Maharastra, India

Bedrettin Çetinkaya Computer Engineering, Ted University, Ankara, Turkey

Shantanu Chakraborty Department of Computer Engineering and IT, COEP Technological University, Pune, India

Sai Chandu SR University, Warangal, Telangana, India

Yücel Çimtay Computer Engineering, Ted University, Ankara, Turkey

Mayur Chauhan Department of Computer Engineering, Gujarat Power Engineering and Research Institute (GPERI), Mehsana, Gujarat, India

Shreya Daga Department of Electronics and Telecommunication Engineering, Vishwakarma Institute of Technology, Pune, India

Pranav Dalve Department of Electronics and Telecommunication Engineering, Vishwakarma Institute of Technology, Pune, India

Bivas Dam Department of Instrumentation and Electronics, Jadavpur University, Kolkata, India

Anvay Deshpande Vishwakarma Institute of Technology, Pune, Maharashtra, India

W. Dhafer Research Center STIS, M2CS, National Graduate School of Arts and Crafts of Rabat, Mohammed V University in Rabat, Rabat, Morocco;
Group of Biomedical Engineering and Pharmaceuticals Sciences, National Graduate School of Arts and Crafts of Rabat (ENSAM), Mohammed V University, Rabat, Morocco

S. Dharini CMR Institute of Technology, Bengaluru, India

B. Dhiyanesh CSE, Dr. N.G.P. Institute of Technology, Coimbatore, India

A. Diana Andushia Department of Electronics and Communication Engineering, Karunya Institute of Technology and Sciences, Coimbatore, India

Mohamed El Moustapha El Arby Chrif Department of Mathematics and Computer Science, University of Nouakchott AL Aasriya, Nouakchott, Mauritania

Y. El Ouahabi Laboratory Health and Biology, Faculty of Sciences, Ibn Tofail University, Kenitra, Morocco

M. H. El yousfi Alaoui Research Center STIS, M2CS, National Graduate School of Arts and Crafts of Rabat, Mohammed V University in Rabat, Rabat, Morocco;
Group of Biomedical Engineering and Pharmaceuticals Sciences, National Graduate School of Arts and Crafts of Rabat (ENSAM), Mohammed V University, Rabat, Morocco

Kinjal Gautam Department of Computer Science and Engineering, School of Computer Science Engineering and Technology (SoCSET), ITM (SLS) Baroda University, Vadodara, Gujarat, India

R. Gayathri Department of Electronics and Communication Engineering, Sri Venkateswara College of Engineering, Sriperumbudur, India

Musa Genemo Gumushane University, Gumushane, Turkey

Pratham Ghatkar JSPM's Rajarshi Shahu College of Engineering, Pune, Maharashtra, India

Yiga Gilbert Department of Computer Science, Makerere University, Kampala, Uganda

Pradnya N. Gokhale Sardar Patel College of Engineering, University of Mumbai, Mumbai, Maharashtra, India

Lokesh Gopinath Department of Electronics and Communication Engineering, SRM Institute of Science and Technology, Kattankulathur, Chengalpattu, Tamil Nadu, India

Zameer Gulzar SR University, Warangal, Telangana, India

Fatma Nazli Günay Computer Engineering, Ted University, Ankara, Turkey

Cota Navin Gupta Neural Engineering Laboratory, Department of Biosciences and Bioengineering, Indian Institute of Technology Guwahati, Guwahati, Assam, India

Samia Shameem Haque Department of Computer Science and Engineering, East West University, Dhaka, Bangladesh

P. M. Hareesh Computer Science and Engineering Department, Indian Institute of Information Technology, Kottayam, Kerala, India

N. Herald Anantha Rufus Department of Electronics and Communication Engineering, Vel Tech Rangarajan Dr. Sagunthala R&D Institute of Science and Technology, Chennai, Tamil Nadu, India

Vishwanath R. Hulipalled School of C&IT, REVA University, Bangalore, India

Sanjay Jain CMR Institute of Technology, Bengaluru, India

Sanjeda Sara Jennifer Department of Computer Science and Engineering, East West University, Dhaka, Bangladesh

Rajesh Kumar Jha Department of Electronics and Communication Engineering, Faculty of Science and Technology (IcfaiTech), ICFAI Foundation for Higher Education, Hyderabad, India

Sameer Joshi Miraj Nuclear Medicine AND Molecular Imaging Center, Miraj, Maharashtra, India

Tanmay Kadam Department of Computer Engineering, Vishwakarma Institute of Technology, Pune, India

Joshua Edward Kalungi Department of Computer Science, Makerere University, Kampala, Uganda

G. Kalyani Department of Information Technology, Velagapudi Ramakrishna Siddhartha Engineering College, Vijayawada, India

Bomken Kamdak Department of Computer Science and Engineering, Rajiv Gandhi University, Itanagar, Arunachal Pradesh, India

Al Kawsar Department of Computer Science and Engineering, East West University, Dhaka, Bangladesh

Emmy William Kayanja Department of Computer Science, Makerere University, Kampala, Uganda

Ashfan Khan Vishwakarma Institute of Technology, Pune, Maharashtra, India

Vaishnavi V. Kulkarni School of CSE, REVA University, Bangalore, India; Department of Computer Science and Engineering, Amrita School of Computing Bengaluru, Amrita Vishwa Vidyapeetham, Bengaluru, India

Varun Vinod Kulkarni Department of Computer Science and Engineering, Amrita School of Computing Bengaluru, Amrita Vishwa Vidyapeetham, Bengaluru, India

Puranam Revanth Kumar Department of Electronics and Communication Engineering, Faculty of Science and Technology (IcfaiTech), ICFAI Foundation for Higher Education, Hyderabad, India

Ankita Kumari Vishwakarma Institute of Technology, Pune, Maharashtra, India

Mayuri Kundu School of CSE, REVA University, Bangalore, India

Jonah Mubuuke Kyagaba Department of Computer Science, Makerere University, Kampala, Uganda

Sagar Lahade Pimpri Chinchwad College of Engineering, Pune, India

M. Lasya Department of IT, VR Siddhartha Engineering College, Vijayawada, India

Prabhat Mali Department of Electronics and Computer Engineering, Pulchowk Campus, Institute of Engineering, Tribhuvan University, Lalitpur, Bagmati, Nepal

A. Manasa Department of ECE, G. Narayanamma Institute of Technology and Science (for women), Hyderabad, Telangana, India

K. Manivannan Department of Computer Science and Engineering, Faculty of Engineering and Technology, Jain (Deemed-to-be University), Bengaluru, Karnataka, India

Ggaliwango Marvin Department of Computer Science, Makerere University, Kampala, Uganda

H. Massar Laboratory of Electrical and Industrial Engineering, Information Processing, Informatics, and Logistics (GEITIIL), Faculty of Science Ain Chock, University Hassan II—Casablanca, Casablanca, Morocco;
Research Center STIS, M2CS, National School of Arts and Crafts of Rabat (ENSAM), Mohammed V University in Rabat, Rabat, Morocco

James Mayanja Makerere University, Kampala, Uganda

A. Mehathab Data Scientist with Specialization in Remote Sensing, Expertz Lab Technologies, Kochi, Kerala, India

Pranay Meshram G.H.Raisoni College of Engineering Nagpur, Nagpur, Maharastra, India

M. Miyara Computer Science and Systems Laboratory (LIS), Faculty of Science Ain Chock, University Hassan II—Casablanca, Casablanca, Morocco

Moustapha Mohamed Saleck LAROSERI Laboratory, Department of Computer Sciences, Faculty of Sciences, University of Chouaib Doukkali, El Jadida, Morocco

Monish Mohanty Department of Computer Science and Engineering, Amrita School of Computing Bengaluru, Amrita Vishwa Vidyapeetham, Bengaluru, India

Himanka Sekhar Mondal ICT and Servics Group, Centre For Development of Advanced Computing, Kolkata, India

Vijaya B. Musande Department of Computer Science and Engineering, Jawaharlal Nehru Engineering College, Aurangabad, Maharashtra, India

Joshua Mwesigwa Makerere University, Kampala, Uganda

B. Nsiri Research Center STIS, M2CS, National Graduate School of Arts and Crafts of Rabat (ENSAM), Mohammed V University in Rabat, Rabat, Morocco;
Group of Biomedical Engineering and Pharmaceuticals Sciences, National Graduate School of Arts and Crafts of Rabat (ENSAM), Mohammed V University, Rabat, Morocco

Nagi Ould Taleb Computer Sciences Department, Higher Institute of Digital, Nouakchott, Mauritania

Jaderick P. Pabico Institute of Computer Science, University of the Philippines Los Baños, College, Laguna, Philippines

Barnali Pal ICT and Servics Group, Centre For Development of Advanced Computing, Kolkata, India

Lakshmi Harika Palivela School of Computer Science Engineering, Vellore Institute of Technology, Chennai, India

Sanjeeb Prasad Panday Department of Electronics and Computer Engineering, Pulchowk Campus, Institute of Engineering, Tribhuvan University, Lalitpur, Bagmati, Nepal

Kinjal Patel Faculty of Computer Applications and Information Technology, Gujarat Law Society University, Ahmedabad, Gujarat, India

Peeta Basa Pati Department of Computer Science and Engineering, Amrita School of Computing Bengaluru, Amrita Vishwa Vidyapeetham, Bengaluru, India

Babasaheb R. Patil VIMEET, University of Mumbai, Mumbai, Maharashtra, India

Prathamesh Patil JSPM's Rajarshi Shahu College of Engineering, Pune, Maharashtra, India

Dipak Chandra Patranabis Department of Instrumentation and Electronics, Jadavpur University, Kolkata, India

Kaveri Pawar JSPM's Rajarshi Shahu College of Engineering, Pune, Maharashtra, India

Prathibha Prakash Department of Computer Science and Engineering, Amrita School of Computing, Amrita Vishwa Vidyapeetham, Amritapuri, India

V. Premanand School of Computer Science Engineering, Vellore Institute of Technology, Chennai, India

R. Radha EEE, Study World College of Engineering, Coimbatore, India

S. Rajkumar Sona College of Technology, Salem, India

K. Ravi SR University, Warangal, Telangana, India

Md. Redwan Ahmed Department of Computer Science and Engineering, East West University, Dhaka, Bangladesh

S. M. Renuka Devi Department of ECE, G. Narayanamma Institute of Technology and Science (for women), Hyderabad, Telangana, India

Ahmed Wasif Reza Department of Computer Science and Engineering, East West University, Dhaka, Bangladesh

Sharmin Akther Rima Department of Computer Science and Engineering, East West University, Dhaka, Bangladesh

Aleksey Rogachev Volgograd, Russian Federation

Dmitry Rogachev VNIIGiM named after Kostyakov, Moscow, Russian Federation

A. Ruhan Bevi Department of Electronics and Communication Engineering, SRM Institute of Science and Technology, Kattankulathur, Chengalpattu, Tamil Nadu, India

Md. Sajid Anam Ifti Department of Computer Science and Engineering, East West University, Dhaka, Bangladesh

P. Sajitha Karunya Institute of Technology and Sciences, Coimbatore, India

Priyanka Sakpal JSPM's Rajarshi Shahu College of Engineering, Pune, Maharashtra, India

Tanmayee Samantaray Neural Engineering Laboratory, Department of Biosciences and Bioengineering, Indian Institute of Technology Guwahati, Guwahati, Assam, India

Rushikesh Sanap Department of Computer Engineering and IT, COEP Technological University, Pune, India

Ashwini Sanjay Gaikwad Department of Computer Science and Engineering, Jawaharlal Nehru Engineering College, Aurangabad, Maharashtra, India

Bhakti Sarag Department of Computer Engineering, Vishwakarma Institute of Technology, Pune, India

P. Saraswathi IT, Velammal College of Engineering and Technology, Madurai, India

S. Satish Department of Electronics and Communication Engineering, Vel Tech Rangarajan Dr. Sagunthala R&D Institute of Science and Technology, Chennai, Tamil Nadu, India

Rajesh Saturi Department of CSE, Vignana Bharathi Institute of Technology, Ghatkesar, Hyderabad, India

R. Senthil Rama Department of Electrical and Electronics Engineering, DMI College of Engineering, Nagercoil, Tamil Nadu, India

S. K. Shabeer Department of IT, VR Siddhartha Engineering College, Vijayawada, India

Milind Shah Department of Computer Science and Engineering, Krishna School of Emerging Technology and Applied Research (KSET), Drs. Kiran and Pallavi Patel Global University (KPGU), Vadodara, Gujarat, India

Aman Shakya Department of Electronics and Computer Engineering, Pulchowk Campus, Institute of Engineering, Tribhuvan University, Lalitpur, Bagmati, Nepal

Sahil Shendurkar Department of Computer Engineering, Vishwakarma Institute of Technology, Pune, India

Swati Shinde Pimpri Chinchwad College of Engineering, Pune, India

Aditya Shreeram Neural Engineering Laboratory, Department of Biosciences and Bioengineering, Indian Institute of Technology Guwahati, Guwahati, Assam, India

Jay B. Simha REVA Academy for Corporate Excellence (RACE) REVA University, Bangalore, India

A. Soulaymani Laboratory Health and Biology, Faculty of Sciences, Ibn Tofail University, Kenitra, Morocco

S. S. Sridhar Department of Computing Technologies, College of Engineering and Technology, SRM Institute of Science and Technology, Chengalpattu, Tamil Nadu, India

G. Srikanth Reddy Department of CSE, Vignana Bharathi Institute of Technology, Ghatkesar, Hyderabad, India

D. Srinivas Goud Department of CSE, Vignana Bharathi Institute of Technology, Ghatkesar, Hyderabad, India

Konumuri Kalyan Suhas Department of Information Technology, Velagapudi Ramakrishna Siddhartha Engineering College, Vijayawada, India

S. S. Suni Ilahia College of Engineering and Technology, Ernakulam, India

S. Suresh CSE, KPR Institute of Engineering and Technology, Coimbatore, India

Mandava Venkata Sri Sai Surya Department of Information Technology, Velagapudi Ramakrishna Siddhartha Engineering College, Vijayawada, India

Muddayya Swami Department of Computer Engineering and IT, COEP Technological University, Pune, India

Mohammad Tahir G.H.Raisoni College of Engineering Nagpur, Nagpur, Maharastra, India

Gom Taye Department of Computer Science and Engineering, Rajiv Gandhi University, Itanagar, Arunachal Pradesh, India

Pius Tumwebaze Makerere University, Kampala, Uganda

Adriana Margarita Turriate-Guzman Universidad Privada del Norte, Lima, Peru

Radhesyam Vaddi Department of IT, VR Siddhartha Engineering College, Vijayawada, India

Chandrasekar Venkatachalam Department of Computer Science and Engineering, Faculty of Engineering and Technology, Jain (Deemed-to-be University), Bengaluru, Karnataka, India

Shanmugavalli Venkatachalam Department of Computer Science and Engineering, KSR College of Engineering, Tiruchengode, Tamil Nadu, India

N. Venkatesh School of CS & AI, SR University, Warangal Urban, Telangana, India

M. Venkateshwara Rao Department of CSE, Vignana Bharathi Institute of Technology, Ghatkesar, Hyderabad, India

Rachit Verma Indian Institute of Information Technology Kottayam, Kottayam District, Kerala, India

Sundeep Vijayaraghavan Department of Plastic and Reconstructive Surgery, Amrita School of Medicine, Kochi, India

Sofía Alessandra Villar-Quispe Universidad Privada del Norte, Lima, Peru

S. Vishal Department of Computer Science and Engineering, Amrita School of Computing Bengaluru, Amrita Vishwa Vidyapeetham, Bengaluru, India

Shubham Waghule Department of Computer Engineering, Vishwakarma Institute of Technology, Pune, India

Medha Wyawahare Vishwakarma Institute of Technology, Pune, Maharashtra, India;
Department of Electronics and Telecommunication Engineering, Vishwakarma Institute of Technology, Pune, India

Gökçe Nur Yılmaz Computer Engineering, Ted University, Ankara, Turkey

Meftahul Zannat Department of Computer Science and Engineering, East West University, Dhaka, Bangladesh

Modern Challenges and Limitations in Medical Science Using Capsule Networks: A Comprehensive Review

Milind Shah, Nikunj Bhavsar, Kinjal Patel, Kinjal Gautam, and Mayur Chauhan

Abstract Capsule networks (CapsNet), an emerging neural network architecture, is now used in medical science to develop potential tools and applications. Particularly, in the domain of medical image analysis, CapsNet outperforms the existing CNN models in terms of disease detection and classification tasks, such as identifying abnormalities in retinal images for diabetic retinopathy and tumor detection. Moreover, capsule networks are now used in analyzing the electronic health records (EHRs) such as hospital readmissions and mortality rates. However, the implementation of capsule networks in medical science is still in the nascent stage facing several challenges due to the limited availability of high-quality medical data, lack of interpretability, and ethical considerations. In order to overcome these challenges, more research and collaboration works should be encouraged between medical professionals and artificial intelligence (AI) experts. This research study discusses about the modern challenges faced by medical science and how the challenges can be solved by using capsule networks and algorithms.

M. Shah (✉)
Department of Computer Science and Engineering, Krishna School of Emerging Technology and Applied Research (KSET), Drs. Kiran and Pallavi Patel Global University (KPGU), Vadodara, Gujarat, India
e-mail: milindshahcomputer@gmail.com

N. Bhavsar
Department of Computer Engineering, School of Engineering, P P Savani University, Kosamba, Surat, Gujarat, India

K. Patel
Faculty of Computer Applications and Information Technology, Gujarat Law Society University, Ahmedabad, Gujarat, India

K. Gautam
Department of Computer Science and Engineering, School of Computer Science Engineering and Technology (SoCSET), ITM (SLS) Baroda University, Vadodara, Gujarat, India

M. Chauhan
Department of Computer Engineering, Gujarat Power Engineering and Research Institute (GPERI), Mehsana, Gujarat, India

S. Shakya et al. (eds.), *Fourth International Conference on Image Processing and Capsule Networks*, Lecture Notes in Networks and Systems 798,
https://doi.org/10.1007/978-981-99-7093-3_1

Keywords Neural network · Capsule networks · Machine learning · Medical science · Healthcare · Disease diagnosis · Imaging analysis

1 Introduction

Recently, medical science is facing significant challenges in diagnosing and treating complex diseases. One of the most promising solutions to solve the emering disease detection and classification challenges is the use of capsule networks. Capsule network revolutionizes the process of medical data analysis by enabling a more accurate and efficient data processing and interpretation techniques to analyze complex datasets.

Even though the research and development in the domain of capsule networks is still in nascent stage, it is now being used in medical research and patient care. One of the main advantages of capsule networks is its ability to handle the complex data structures such as 3D images or sequences of data. This makes them well suited for medical imaging applications, where accurate and detailed information of a medical image is required for the purpose of diagnosis and treatment planning [1, 2].

Another potential application of capsule networks in medical science is drug discovery. Capsule networks can be used to analyze the large datasets of molecular structures and predict how different elements will interact with each other. This helps to identify the required drugs more quickly and accurately and leverage new treatments [3].

The purpose of initiating medical research using capsule networks is to analyze its potential in addressing the emerging data processing challenges and enhancing healthcare outcomes. Medical science research intends to advance the application of these models in various areas of health care by improving healthcare diagnostics, treatment methods, patient care, and overall health outcomes [2].

Figure 1 represents some popular machine learning techniques used in medical science research. Each technique plays a unique role in addressing various medical challenges and responsibilities.

These techniques together provide an efficient tool for addressing a variety of medical science challenges, such as diagnosis, prediction, patient stratification, and data analysis.

This paper is divided into the following subsections: Sect. 1 is an introduction that discusses the modern challenges in medical science and how they can be addressed using capsule networks. Section 2 describes the modern challenges in medical science. Section 3 explains how medical science problems can be solved using capsule networks. Section 4 describes how to analyze a large amount of data in drug development using capsule networks. Section 5 includes research questions, which outline the topics discussed in this paper. Section 6 focuses on related work. Section 7 discusses the limitations of existing work, Sect. 8 describes the methodology, and Sect. 9 explains the research motivation. Section 10 presents the problem statement. Section 11 consists of the discussion. Section 12 highlights the existing

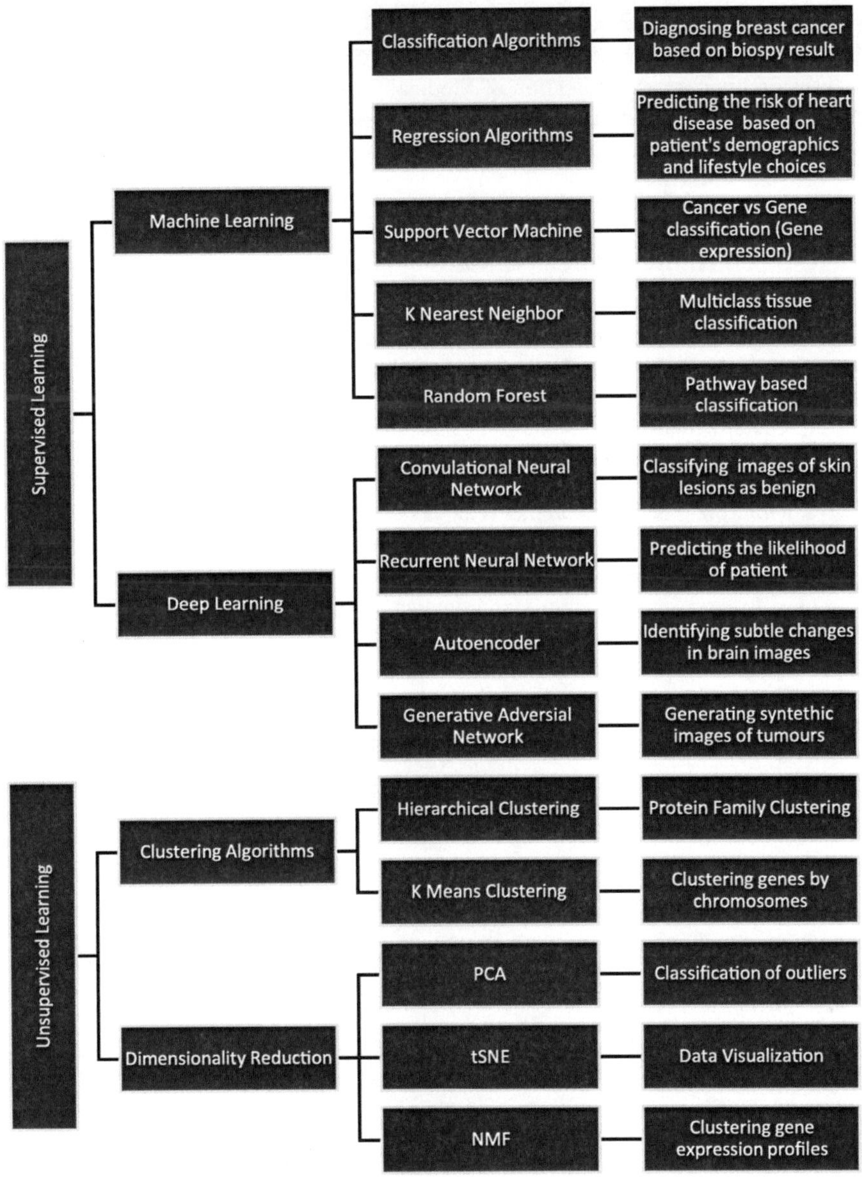

Fig. 1 Various machine learning algorithms used for medical science research

research limitations. Section 13 outlines the identified research gaps. Section 14 discusses the limitations of capsule networks for Medical Science Research, Sect. 15 focuses on the current application of Medical Science, Sect. 16 describes the open challenges and future directions, and finally, Sect. 17 provides the conclusion and future work, which concludes the paper.

2 What Are the Modern Challenges in Medical Science?

Capsule networks, a relatively new type of neural network architecture, have the potential to address several challenges in medical science (Fig. 2).

(1) Medical image classification: With the ability to capture the spatial relationships between different parts of an image, capsule networks are now used to classify the medical images with higher accuracy when compared to the traditional convolutional neural networks (CNNs).
(2) Disease diagnosis: Capsule networks can be used to diagnose diseases by considering a combination of symptoms and medical test results. Here, capsules are used to represent different symptoms and medical test results and further the

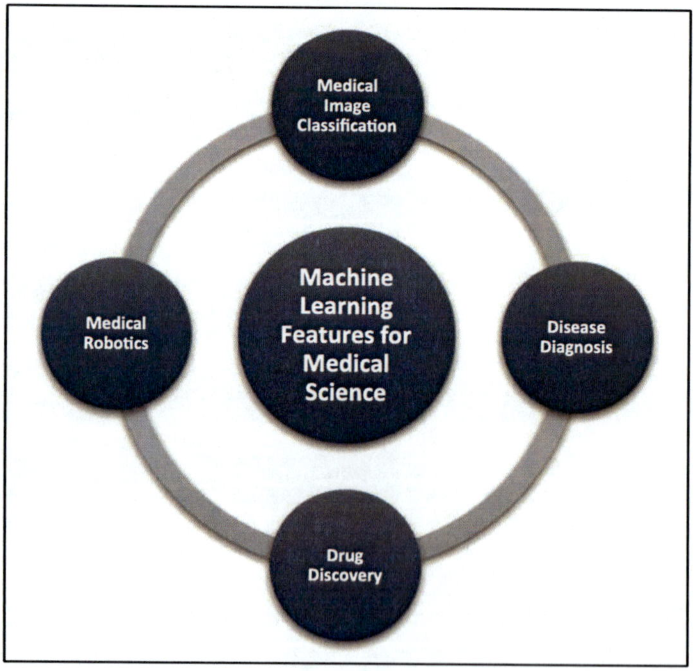

Fig. 2 Modern challenges in medical science

network can capture the complex relationships between them and make more accurate diagnosis.

(3) Drug discovery: Capsule networks can be used to predict the effectiveness of drugs by analyzing the molecular structure of human body. Capsules can be used to represent different chemical properties, and the network helps to learn about the different combinations of properties to find out the effective drug.

(4) Medical robotics: Capsule networks can be used to improve the performance of medical robots by allowing them to better understand and interpret their environment. Capsules can be used to represent different objects present in the environment.

Overall, the capsule networks have the potential to revolutionize many areas of medical science to perform more accurate and efficient analysis of complex data [2–6].

3 How Medical Science Problems Can Be Solved Using Capsule Networks?

Capsule network, a type of neural network, is proposed as a potential solution to overcome some of the limitations faced by traditional convolutional neural networks (CNNs) in image recognition tasks as mentioned below:

(1) Improved medical image recognition: One potential application of capsule networks in medical science is to improve the accuracy of medical image recognition tasks. Capsule networks are designed to be better able to recognize complex patterns in images and to handle variability in image orientation and scale, which are common challenges in medical imaging. This could potentially lead to more accurate diagnosis based on medical images.

(2) Better disease diagnosis and treatment: Capsule networks can be used to identify various medical data patterns that might not be manually diagnosed. For example, capsule networks could be used to analyze medical records and identify patients at risk for a specific disease based on data patterns in their medical history.

(3) Improved drug development: Capsule networks could also be used to analyze large amounts of data on drug interactions and side effects, which may potentially lead to the development of more safe and effective drugs.

(4) Personalized medicine: Finally, capsule networks could be used to develop personalized treatment plans based on a patient's specific medical history and genetic profile. By analyzing large amount of data obtained from patients with similar conditions, capsule networks could identify the most effective treatment strategies for individual patients.

While capsule networks are still an emerging technology in the medical field, there are several ways in which they could be potentially used to solve the emerging problems and improve patient outcomes [2–4, 7–9].

4 How to Analyze a Large Amount of Data in Drug Development Using Capsule Networks?

Capsule networks require a number of important phases to analyze large amounts of data in the process of drug development. Here is an overview of the procedure:

(1) Begin by acquiring and preprocessing the data related to drug development. This can include information from clinical trials, structures of molecules, expression of genes data, patient records, and other sources. Cleanup the data, account for missing values, standardize attributes, and perform any required transformations.
(2) Dataset partitioning: Separate the dataset into training, validation, and testing subsets. To avoid introducing bias, ensure that the partitioning preserves the distribution and characteristics of the data.
(3) Capsule network architecture design: Based on the specific objectives of the drug development task, select a suitable capsule network architecture. Consider variables such as the problem's complexity, the availability of computational resources, and your prior knowledge of the dataset.
(4) Training the capsule network: Use the training dataset to train the capsule network. This requires forward propagation, the computation of capsule activations, and iterative adjustments to the network's parameters using techniques such as gradient descent or backpropagation through time.
(5) Performance evaluation: Using the validation dataset, evaluate the performance of the trained capsule network. Evaluate parameters such as precision, recall, $F1$ score, and the area under the receiver operating characteristic curve (AUC-ROC). Adjust hyperparameters as needed to optimize the efficiency of the model.
(6) Testing and generalization: Finally, evaluate the trained capsule network on an independent testing dataset in order to determine its generalization capability. Validate the model's reliability and efficiency in drug development responsibilities by evaluating its performance on unidentified information.
(7) Interpretability and visualization: Through capsule routing and concept learning, capsule networks offer interpretability advantages. Utilize techniques such as capsule visualization, attention mechanisms, and concept activation mapping to obtain insight into the model's decision-making process and to identify significant data features or patterns.
(8) Iterative refinement: Repeat the processes mentioned above in order to refine and enhance the capsule network model. Experiment with various architectures, hyperparameters, and training techniques to improve the efficiency and robustness of the model.

(9) Consider how the capsule network analysis works within the larger drug development workflow [2, 3, 10].

5 Research Questions

This review paper discusses the following research questions.

Q1: What are the recent challenges in medical science using capsule neural network?
Q2: How medical science problems can be solved using capsule networks?
Q3: Which capsule neural network algorithms are used to solve challenges of medical science?
Q4: How to analyze large amount of data in drug development using capsule networks?
Q5: What are the existing research limitations in medical science using capsule networks?

Through the discussion of these research questions, this research aims to bridge the gap between capsule networks and medical science, clearing the way for effective incorporation into healthcare practices and contributing to improvements in diagnostics, treatment planning, and patient care.

6 Related Work

Till now, a lot of researches has been done to solve challenges of medical science using capsule neural network.

In Quan et al. [7], new coronavirus infections control deep learning analysis of chest X-rays (CXRs). Convolutional neural and capsule networks solve this challenge. DenseCapsNet, a novel deep learning framework, uses the strengths of dense convolutional networks (DenseNet) and capsule neural networks (CapsNet) to reduce data reliance in convolutional neural networks. 750 CXR images of the lungs of healthy people, patients with various diseases, and new coronavirus patients can detect COVID-19 with 90.7% accuracy, 90.9% $F1$ score, and 96% sensitivity. DenseCapsNet can detect novel coronaviruses in CXR radiography. This research uses widely analyzed COVID-19 datasets. 9432 CXR images were obtained in three COVID-19 datasets. Normal, COVID-19, and other pneumonia were classified by COVID-Net. The radiologist sent four CXRs. Radiologists use a computer to randomly choose 200 chest X-rays for diagnosis after getting typical patient lesion characteristics from a public dataset. Radiology specialists agree that the public dataset classifications and pneumonia patient features are typical, justifying its research use. To test our neural network architecture, a computer randomly picked

750 CXR images from the dataset—250 normal, 250 pneumonia, and 250 new coronavirus pneumonia. Randomly split 750 CXR images into training, validation, and testing sets. DenseNet121 extracts features and CapsNet packages capsules. Reduce neural network data reliance. 250 normal chest radiographs, 250 pneumonia radiographs, and 250 COVID-19 radiographs showed good diagnostic findings with 90.7% accuracy and 95%sensitivity.

In Modi et al. [4], a detail-oriented capsule network classifies COVID-19 in this research. In this research, the detailed oriented capsule network classifies chest CT scan images to predict COVID-19 and non-COVID-19 (DOCN). The model evaluation uses accuracy, precision, and sensitivity. The method offers 98% accuracy, 81% sensitivity, and 98.4% precision. This research used Soares et al. and Italian datasets. This collection contains 360 COVID-19 images and 397 CT scans of both sick and healthy individuals. Italy contains a total of one hundred COVID-19 CT images. There is a possibility that the first model pre-training can improve COVID-diagnostic CAPS. It was decided to create and use datasets based on CT scans. Because Image Net's natural images are distinct from the COVID-19 scan dataset, it is not used for the pre-training phase of the process. A previously trained model may be improved using COVID-CAPS to provide more similar X-ray images. First, the entire COVID-CAPS model must be trained using data from external sources. Outer classes determines the capsule amount. Two capsules take the place of the models deep network so that it can accommodate the COVID-19 data. Before beginning weight training, all capsules were easily switched out, and the usual layers were attached in place. TensorFlow, Scikit learn, and a few other similar packages are used in Python 3 for the work proposed. Evaluations are made on the accuracy, specificity, and sensitivity of the model. The research used capsule networks with many convolution layers to outperform earlier models. We will optimize model processing speed.

In Wirawan et al. [11], this research addressed EEG-based emotion identification issues using the continuous capsule network technique. The effective architecture of the first, second, third, and fourth continuous convolution layers with values of 64, 256, and 64, respectively, with kernel sizes of $2 \times 2 \times 4$, $2 \times 2 \times 64$, and $2 \times 2 \times 128$ for the first three layers and $1 \times 1 \times 256$ for the fourth layer achieved this. Layers 1–3 have 64 and 64 values, respectively. Extracting and representing differential entropy and three-dimensional cube characteristics improved the continuous capsule network. These algorithms extracted low-frequency and spatial information from electroencephalogram data. Four sentiments, two arousal classes, and valence are 91.35, 93.67, and 92.82% accurate with DEAP. The proposed approaches have DREAMER dataset accuracies of 94.23, 96.66, and 96.05% for four sentiments, two arousal classes, and valence. The AMIGOS dataset has four sentiment categories, two arousal classes, and two valence classes. The proposed algorithms have accuracies of 96.20, 97.96, and 97.32%. The 3D cube technique and DE methodology for feature extraction enable the continuous capsule network in the classification phase. DE and 3D cube are used to describe low-frequency and spatial data. The continuous capsule network detects four and two emotions more accurately than some previously published methods on three separate datasets. Recent research addresses more parameters than continuous capsules.

In Sanchez et al. [1], explore capsule networks with ConvNets in medical image analysis datasets with minimal labeled data and class imbalance. Our research uses public MNIST, Fashion-MNIST, and medical (histological and retina imaging) datasets. Our method is suitable for medical imaging because capsule networks can be trained with minimal input and resist uneven class distribution. LeNet, baseline ConvNet, and CapsNet were tested for limited training data, class imbalance, and data augmentation usefulness. Three designs, nine data conditions, four repetitions, and four public datasets taught 432 networks. MNIST and Fashion-MNIST, with ten classes, sixty thousand training images, and ten thousand test images, are the first two datasets. CapsNet consistently beats LeNet and baseline. With additional data, the MNIST gap reduces. For DIARETDB1, LeNet with 5% of the data performs similarly to CapsNet and better than baseline with 1%. Structures in this image cause this behavior. All research except TUPAC16 had significance tests with p-values below 0.05.

In Jaber et al. [2], this paper provides an overview of several popular deep learning algorithms, describing their architectural designs and practical implementations. The algorithms discussed include backpropagation, autoencoders, variational autoencoders, restricted Boltzmann machines, deep belief networks, convolutional neural networks, recurrent neural networks, generative adversarial networks, CapsNet, the transformer, embeddings from language models, bidirectional encoder representations from transformers, and absorption. These techniques are widely utilized in the field of artificial intelligence. This research addresses several deep learning issues. Adaptive deep learning, neural architecture search, and AutoML-Zero are these challenges. Examine these algorithms' benefits and drawbacks, healthcare usage, and future. This research defines and inspires new usage of popular algorithms. This paper introduces new deep learning researchers to the benefits, drawbacks applications, and processes for many algorithms. We describe how to utilize many deep learning algorithms in health care, including the COVID-19 pandemic. We discuss deep learning challenges in one part to increase awareness and provide solutions. This may motivate researchers.

In Saif et al. [12], since 2019, COVID-19 has affected the world in unforeseen ways. COVID-19 is extremely dangerous and propagates quickly, killing millions despite our attempts at fighting it. Controlling this deadly disease or sickness requires quick and effective therapy. This research proposes a parallel concatenated convolutional block-based capsule network for COVID-19 diagnosis from multimodal medical images. By concurrently modifying the receptive field, deep convolutional blocks with varying filter sizes can incorporate discriminative spatial information and increase model scalability. By supplying fine-to-coarse information, capsule layers enable the model to represent more complex information. The model was tested on two chest radiograph datasets and one ultrasound imaging dataset. The proposed architecture's impressive COVID-19 detection test results show the models potential. Our deep convolutional capsule network has a deep parallel convolutional neural block. Hence, a larger dataset improves our proposed network performance. COVID-19 has no such medical dataset. US COVID-19 numbers are scarce. We pre-train and fine-tune. Transfer learning involves training a deep neural network using a large source dataset before freezing its early layers. The target dataset retains parts of the networks deeper layers. Because a large dataset can be combined, this pre-training strategy works well for numerous computer vision issues.

In Zhao et al. [8], manual diagnosis of medical images is efficient, and long-term eye exhaustion affects diagnostic precision. To help in the processing and analysis of medical images, a completely automated system is necessary. Machine learning can quickly differentiate COVID-19 from other pneumonia-related diseases or healthy individuals. Nevertheless, the minimal number of labeled images and repetition of the models and data bias the learning outcomes, resulting in an inaccurate additional diagnosis. A deep channel-attention correlative capsule network for spatial, correlative, and fused feature categorization is used to solve these problems and resolve the related concerns. Experiments were confirmed using X-ray and CT imaging datasets, and they outperformed several prior findings that were considered as state-of-the-art research. Using two different datasets, we analyzed the DCACorrCapsNet model for COVID-19 recognition. The computer was equipped with an i5-7300HQ central processing unit, a GTX1050Ti graphics processing unit, and 8 gigabytes of random access memory (RAM). In this section, we will cover the settings of the multifeature extractor, the multi-level capsule, and the classification layer module. The deep architecture is responsible for the generation of correlative data from the nearby capsules, the extraction of attention-based spatial characteristics, and the integration of time- and frequency-based features as a Fisher vector. The feature representation may be obtained by merging the outputs. The DCACorrCapsNet algorithm was validated using X-ray and CT datasets from the medical field. In simulations based on real-world data, the DCACorrCapsNet model performed far better than its competitors.

In Zhang et al. [10], a capsule network model for EMR categorization can extract advanced Chinese medical text features. This model uses LSTM, GRU, and novel routing architecture. This model outperformed baseline techniques on the Chinese electronic medical record dataset with an $F1$ score of 73.51%. It outperformed several baseline models. CHIP2019 task-3 utilized China Clinical Trials Registry screening criteria. Hierarchical clustering and human induction identified seven individuals, 44 semantic categories, descriptive data, and labeling criteria for each category. Capsule + LSTM's average $F1$ is 73.51%. CNN beat GRU and LSTM due to shorter data. GRU outperformed LSTM by 4.5%. The experimental text is shorter than the control text, and therefore, both networks may have poor gating unit output. Longer experimental texts may improve LSTM and GRU network performance and narrow the gap. The capsule network increases $F1$ values by 5.46 and 10.92% in both models when integrated with the LSTM and GRU networks to classify Chinese medical text.

In Zhang et al. [5] presented an improved capsule network for medical image classification. The proposed capsule network adds feature fragmentation and multi-scale feature extraction to the original capsule network. The feature decomposition reduces computations and improves network convergence by extracting richer information. The multiscale feature extraction unit transmits crucial data from low- to high-level capsules. Implementing capsule networks in PCam. Experimental findings show that it can classify medical images well, which might be a model for future image classification applications. Datasets were used to analyze the proposed network. We outperformed the capsule network on MNIST and CIFAR-10. We then used the proposed network to PCam medical imaging data with good results. Deep learning datasets are like MNIST. NIST collects this data. 10,000 testing samples and 60,000 training samples. Alex Krizhevsky and Ilya Sutskever created CIFAR-10 to find universal things. 10 RGB color image categories. 500,000 training images and 10,000 test images are 32 × 32 pixels.

In Afriyie et al. [9] provided a variant of the capsule network that is less complicated but trustworthy and can extract features for the improved classification. For challenging image datasets such as CIFAR 10, Fashion-MNIST, and kvasir-dataset-v2, the proposed model obtains test accuracies of 87.3%, 93.84%, and 93.84%, and 85.50%, respectively. For datasets containing few parameters, the proposed model achieves results comparable to those achieved using state-of-the-art approaches. Squash and amplifier approaches are effectively used in dynamic routing methodology. Our capsule maps onto the primary capsule and provides amplified features. It also contains three convolutional layers and four amplified layers. In order to identify and categorize kvasir-v2, our architecture magnifies photos to a size of 28 × 28. Customized layers with 3 × 3 kernels generate 28 × 28 feature maps from photos. Conv1 uses modified layer 1's 96, 26 × 26 kernels and Rely of 1. Model experimental results are analyzed using performance evaluation methods. We investigate the model accuracy and validation. Our technique-produced results comparable to the baseline model for all test datasets. The proposed PC layer has all necessary classification criteria. This strategy also helps explain and comprehend CapsNets for practical usage in key applications.

In Wang et al. [6], the research demonstrates two contributions of its own. First, in contrast to standard surveys that separate literature on deep learning on medical image segmentation directly into many categories and introduce literature according to a multi-level structure from rough to precise. Second, this research concentrates on supervised and weakly supervised learning techniques and on unsupervised approaches because they have been covered in many older surveys and are now controversial. Regarding supervised learning techniques, we analyze three elements of the literature: the selection of backbone networks, the design of network blocks, and the enhancement of loss functions. Separately, we analyze the literature on poorly supervised learning algorithms according to data augmentation, transfer learning, and interactive segmentation. It has been shown that completely automated deep neural network-based segmentation of medical images is extremely helpful. By analyzing the development of deep learning in the segmentation of medical images, we have highlighted possible challenges. Researchers successfully utilized a range of techniques to increase the segmentation accuracy of medical images. Accuracy improvement by itself can explain algorithm performance, especially in the field of medical image analysis where class imbalance, noise interference, and the severe implications of incomplete tests must be considered as well.

In Akay et al. [3] describe the attempt to apply deep learning to analyze molecular shuttle swarm behavior, highlighting machine learning's advantages and demands in both established and emerging domains. To understand complex events and their functions in the production, maintenance, and management of nano- to macro-scale systems, more data must be collected and processed. There has been progress, but systems lacking the vast quantities of data required by DL algorithms require further development. Complexity of data necessitates additional procedures to validate that the appropriate DL approaches are used and to fully understand how the algorithms arrived at their conclusions. DL algorithms may be utilized to analyze the dynamics of molecular shuttle swarming. We estimate ML's future impact.

In Yu et al. [13], we develop 36,360 question combinations that fulfill Chinese medical questions. This corpus is randomly divided into three sets: 32,360 question pairs for training, 2,000 for development and 2000 for testing. We tested the proposed CapsTM on this corpus and compared it with state-of-the-art methods. CapsTM scores 0.8666. Research shows that CapsTM matches Chinese medical inquiries better than other cutting-edge technologies. Two medical experts interpret 36,360 Chinese medical question pairs. The dataset has 12,610 positive training samples, 19,750 negative training samples, 801 positive development samples, 1199 negative development samples, 798 positive test samples, and 1202 negative test samples.

In Heidarian et al. [14], convolutional neural networks are used in most deep learning-based COVID-19 pneumonia detection algorithms (CNNs). CNNs require large datasets and data augmentation to detect image samples with exact spatial correlations. Earlier CT scan methods used a simple threshold method to extend predictions made at the slice level to those made at the patient level or complex infection segmentation to identify disease. COVID-FACT, a two-stage automated CT-based approach, detects COVID-19-positive patients. Capsule networks gather COVID-FACT spatial data. In the first step, layers with infection are identified, and in the second phase, patients are classified as COVID or non-COVID. COVID-FACT identifies layers with infection and positive cases by using an in-house CT scan dataset consisting of COVID-19, community-acquired pneumonia, and normal people. COVID-FACT requires less supervision and annotation than its rivals, yet it has an accuracy of 90.82%, sensitivity of 94.5%, and specificity of 86.0%, as well as an area under the curve (AUC) of 0.98. From April 2018 to May 2020, 171 COVID-19 patients, 60 CAP patients, and 76 healthy people underwent volumetric chest CT scans for COVID-CTMD. 50–16, 183 males, 124 women. Free dataset with annotations from Figshare. Data analysis COVID-FACT. COVID-19s and 43 non-COVID cases—19 community-acquired pneumonia and 24 normal—were tested. Adam optimizer, 1e 4 initial learning rate, 16 batch size, 100 iterations. The test set model had the lowest validation loss. COVID-FACT had 90.82% accuracy, 94.5% sensitivity, 86.0% specificity, and 0.97 AUC. In another experiment, we trained our model on CT scans without lung segmentation. The model had 0.95 AUC, 90.82 accuracy, 92.72 sensitivity, and 88.37 specificity.

In Tang et al. [15], CapsNet, introduced recently, allows for a broad spatial hierarchy between features and improves linear translation robustness. This research analyzes whole-slide pathological images using CapSurv and survival loss. VGG16 semantic-level features separate discriminative patches from entire slide diseased images for optimal CapSurv training. A public cancer dataset predicts glioblastoma and lung squamous cell carcinoma survival. There is a possibility that CapSurv will improve survival model predictions. Comparisons are made between CapSurv and seven other sample survival models, including DeepConvSurv, Cox, LASSO-Cox, EnCox, PCRM with logistic, exponential, Weibull distributions, and BoostCI. The objective is to enhance the effectiveness of CapSurv. The survival analysis statistic is referred to as the C-index. It conducts model regression analysis. The C-index for CapSurv's GBM and LUSC survival is the highest possible. DeepConvSurv outperformed CapSurv by 5.8% and 7.8%, respectively, in both the TCGA-GBM and TCGA-LUSC datasets. These findings have been confirmed by CapSurv capsule and survival loss. When it came to the identification of survival-related properties, CapSurv fared better than both human design and the convolutional neural network. The AUC values generated by classification tasks for CapSurv and other survival models measure how effectively the models can differentiate between long- and short-term survivors. CapSurv is able to accurately forecast survival for GBM as well as LUSC. In terms of area under the curve (AUC), CapSurv performs better in the TCGA-GBM dataset than DeepConvSurv, Cox, Lasso-Cox, En-Cox, logistic, exponential, Weibull, and BoostCI. Since capsule orientation displays a variety of instantiation factors associated with survival time and survival loss is effective for survival prediction, CapSurv is a reliable method for analyzing survival data. The survival loss presented is compared to loss functions that contain either one or two survival loss elements.

In Monday et al. [16] propose a novel COVID-19 classification. A novel neurowavelet capsule network classifies COVID-19. We use a multi-resolution discrete wavelet transform to decrease noisy and inconsistent CXR data and increase network feature extraction robustness. Subsampling diminishes spatial feature loss in the discrete wavelet transform of multi-resolution analysis, hence enhancing classification performance. We analyzed the proposed approach using COVID-19-confirmed cases and CXR images of healthy individuals. The approach has 99.6% accuracy, 99.2% sensitivity, 99.1% specificity, and 99.7% precision. Our COVID-19 screening approach is current based on experimental results. This innovative approach will assist in COVID-19 and other disease prevention. Keras on NVIDIA GTX1080 creates the model. 104 was the learning rate for a 16-batch size. Thirty Adam epochs train the model. Regularization and batch normalization used a 50% dropout to reduce overfitting (BN). The dataset has three parts—training, validation, and testing. During training, model performance is verified. To evaluate the model performance, the test dataset was used. Our technique is evaluated using accuracy (ACC), sensitivity (SEN), receiver operating characteristic (ROC), and specificity (SPE). According to all facts, our strategy outperformed deep learning algorithms.

7 Existing Work Limitations

See Table 1.

8 Methodology

Capsule networks have shown great promise in various medical science challenges, including medical image analysis, disease diagnosis, and drug discovery. There are various capsule network algorithms that have been used to solve these challenges, including.

8.1 CapsNet

CapsNet stands for dynamic routing between capsules. CapsNet is the original capsule network algorithm proposed by Geoffrey Hinton and his team in 2017. It has been used for various medical image analysis tasks, including injury classification, medical image segmentation, and disease diagnosis. CapsNet is designed to model the part-whole relationships between different features in an image, which makes it well suited for medical image analysis tasks.

There are a few ways in which CapsNet algorithm solves medical science challenges:

1. ***Improved Feature Extraction***: The CapsNet algorithm extracts features via a two-step procedure involving convolutional layers and primary capsules. Typically, the CapsNet algorithm begins with one or more convolutional layers. These layers extract features from the given data. Each convolutional layer consists of a set of filters or kernels that can be learned. The filters are convolved with the input data using a sliding window operation to extract features and capture local patterns. The output of the convolutional layer is a collection of feature maps, each of which represents the activation of a particular learnt feature. After the convolutional layers, the obtained feature maps are used to construct the primary capsules. Each primary capsule represents a collection of neurons that encode the presence of a specific visual entity or component. In the case of image recognition, for instance, primary containers could represent fundamental features such as boundaries or corners. CapsNet uses capsules, which are groups of neurons that represent different features of an image. Capsules allow the network to model the hierarchical relationships between features, which makes it easier to distinguish between different types of medical images. This can improve the accuracy of tasks such as lesion classification and disease diagnosis.
2. ***Robustness to Deformations***: CapsNet is designed to be robust to deformations in an image, which is particularly useful for medical image analysis tasks. For

Table 1 Differential analysis

Author name	Publication with year	Methods/ algorithms used	Dataset used	Accuracy (%)	Technologies used	Limitations
Quan et al. [7]	Elsevier, 2021	DenseCapsNet	NLMMC	90.7	Python and neural network	We will also look into lightweight models, trainable parameters, and model training efficiency
Sanchez [1]	Springer, 2018	LeNet, baseline ConvNet and CapsNet	MNIST and fashion—MNIST	–	Python	This branch's completely linked layers regularize parameter optimization but lose a lot of information. To better understand the learning latent space, we will deconvolve these layers
Saif [12]	IEEE, 2021	CapsConvNet	Point of care ultrasound (POCUS)	100	Python	–
Zhang et al. [10]	MDPI, 2022	Bi-directional LSTM	CHIP2019		Python	We want to improve medical classification network models by expanding the dataset
Zhang et al. [5]	IEEE, 2020		PatchCamelyon (PCam)	88.23	Python	We want to analyze deeper capsule networks and novel routing methods to increase network performance and apply to medical image datasets

example, a lesion in a medical image may appear in different orientations or sizes, but CapsNet can still identify the lesion based on its distinctive features.
3. ***Efficient Routing***: Utilizing dynamic routing allows the CapsNet algorithm to accomplish efficient routing. Dynamic routing enables capsules in various layers to effectively communicate and coordinate their outputs, resulting in enhanced performance and robustness. CapsNet uses a dynamic routing algorithm to calculate the agreement between different capsules in the network. This allows the network to focus on relevant features and ignore irrelevant ones, which can improve the efficiency of tasks such as medical image segmentation.
4. ***Flexibility***: CapsNet is a flexible algorithm that can be adapted to different medical image analysis tasks. For example, researchers have proposed different variations of CapsNet for tasks such as breast cancer diagnosis, brain tumor segmentation, and diabetic retinopathy classification.

Overall, CapsNet is a promising algorithm for medical image analysis tasks, and ongoing research is expected to further improve its performance and applicability to various medical science challenges [1, 7, 14, 15].

8.2 ConvCaps

ConvCaps, which is short for convolutional capsule networks, is a variant of the capsule network (CapsNet) algorithm that incorporates convolutional layers before the capsule layers. ConvCaps is a variant of the CapsNet algorithm that incorporates convolutional layers before the capsule layers. It has been used for medical image analysis tasks such as brain tumor segmentation and classification of breast cancer. It has been shown to achieve state-of-the-art results in several benchmark datasets for image classification and object detection.

There are some ways in which the ConvCaps algorithm can solve medical science challenges:

1. ***Improved Feature Extraction***: ConvCaps' feature extraction process: ConvCaps uses convolutional capsules instead of layers. These capsules collect visual entity and instantiation parameters in input data. ConvCaps starts with main convolutional capsules as feature extractors. Each primary convolutional capsule receives feature mappings from the previous layer. Convolutional filters are used to create these feature maps. ConvCaps explicitly captures spatial correlations in capsules, enabling the network to learn and modify complicated patterns. Convolutional layers extract local features, whereas capsules capture higher-level correlations and agreement. ConvCaps uses convolutional layers to extract features from medical images, which allows it to capture important patterns and structures that are useful for medical image analysis tasks. This can improve the accuracy of tasks such as lesion detection, segmentation, and classification.
2. ***Efficient Routing***: The convolutional capsule (ConvCaps) algorithm achieves efficient routing by utilizing routing-by-agreement and iterative agreement

updates. ConvCaps is a capsule network (CapsNet) extension that integrates convolutional layers. Similar to CapsNet, ConvCaps uses a dynamic routing algorithm to calculate the agreement between different capsules in the network. This allows the network to focus on relevant features and ignore irrelevant ones, which can improve the efficiency of tasks such as medical image segmentation.

3. ***Robustness to Deformations***: ConvCaps is designed to be robust to deformations in an image, which is particularly useful for medical image analysis tasks. Medical images often contain distortions, noise, and artefacts, but ConvCaps can still identify and extract useful features from the images.
4. ***Transfer Learning***: ConvCaps can be pre-trained on large datasets of natural images and then fine-tuned on smaller medical image datasets. This transfer learning approach can help improve the performance of ConvCaps on medical image analysis tasks, even when the medical image datasets are small.

Overall, ConvCaps is a promising algorithm for medical image analysis tasks, and ongoing research is expected to further improve its performance and applicability to various medical science challenges [17].

9 Research Motivation

Capsule networks have recently shown the potential to address some of the challenges faced by the traditional convolutional neural networks (CNNs) to perform image recognition tasks. In the medical field, image recognition plays a crucial role in the areas of disease diagnosis and treatment planning. There are several key motivations to incorporate capsule networks for solving various medical science challenges, they are as follows:

1. ***Improved accuracy***: While performing image recognition tasks, capsule networks can improve the accuracy when compared to traditional CNNs, particularly in situations where a significant variation can be observed in the general appearance of objects (e.g., medical images of tumors or organs).
2. ***Robustness to image variability***: Medical images can often vary in appearance due to the differences in imaging techniques, patient anatomy, and disease progression. Capsule networks have the ability to recognize objects regardless of their position, rotation, or scale, and this ability is particularly useful in medical imaging applications.
3. ***Explainability***: One of the limitations of traditional CNNs is that it is difficult to interpret how the network arrived at a particular decision. Capsule networks explicitly represent the hierarchical structures present within an image, which could potentially make it easier to understand how the network arrived at its decision. This could be particularly useful in medical applications, where it is important to understand the reasoning behind a diagnosis.

4. ***Few-shot learning***: To solve the medical science challenges, it is often difficult to obtain large amounts of labeled data for training machine learning models. Capsule networks have the ability to learn from a small number of examples, which could be particularly useful in medical applications, where the number of available labeled images may be limited.

The use of capsule networks to solve medical science challenges has the potential to improve accuracy, robustness, explainability, and the ability to learn from limited data [1, 2, 4, 7–9, 12].

10 Problem Statement

Despite the vast opportunities of capsule networks in the field of medical science, there are some challenges that prevent their use in solving the issues related to health care. The limited availability of labeled medical data is one of these challenges, as it is the requirement for enhancing interpretability and explainability of capsule network predictions, robustness to variability, and noisy data.

11 Discussion

In [7], a scientific methodology was utilized to collect data, wherein the radiologist was requested to choose four chest X-ray (CXR) images and engage in a comprehensive analysis of the case characteristics. The size of the hilum in both lungs was relatively small, exhibiting an absence of any apparent nodules or protuberances. The dimensions and morphology of the cardiac silhouette exhibited characteristic features. Each of the two diaphragms located along the coast exhibited a planar surface as well as an acute angle. This phenomenon is not different from the standard. The enlargement of the lower left lobe of the lung refers to the process in which the lung tissue in this specific anatomy region becomes more compact and more solid due to various pathological conditions or infections. The establishment of a diagnosis for bacterial influenza holds significant importance. Within the lung region, an increased representation of a fragmented arrangement becomes apparent. The MRI scans revealed the presence of extensive, dense opacities distributed uniformly across the entirety of both lung fields. The testing for viral pneumonia holds significant importance. Radiologists employ a computer-based random selection process to obtain a sample of 200 chest X-ray images for diagnostic purposes. This selection is made following the provision of lesion characteristics associated with typical patients, as documented in a publicly available dataset. This facilitates a more precise determination of the condition. Consensus among radiology professionals exists regarding the reliability of the public dataset's classifications and the shared clinical characteristics observed in various types of pneumonia. These findings highlight the dataset's potential utility for research purposes. To validate our neural network system, a computer randomly selected 750 chest X-ray (CXR) images from the dataset mentioned earlier. Out of these, 250 images represented normal cases, 250 represented other types of pneumonia, and 250 represented cases of pneumonia caused by the new coronavirus. Arbitrary differentiations were established among the 750 CXR images that constituted each set for training, validation, and testing.

In [4], we considered model pre-training for COVID-CAPS diagnosis. CT scan datasets were created and utilized. Image Net is not used for pre-training because its images are natural images and not COVID-19 scans. COVID-CAPS should enhance a pre-trained model for similar X-ray images. Initially, external data must train the full COVID-CAPS model. Outside classes determine the capsule number. Two capsules replace the final layer to modify the model for COVID-19 data. Before weight training, all capsules were transferred effortlessly. The capsule network dataset is label-compatible. COVID-19 is limited to severely imbalanced datasets because it is new. Due to this imbalanced dataset, we increased the loss margin for misclassified positive instances. Pre-trained capsule networks provide the limited dataset. A 5-class external dataset pre-trains five final capsules. The compression of five capsules into two capsules ultimately results in the entire COVID-19 dataset, which contains all capsule layers. In the case of Adam optimizers, we used a learning rate that consisted of 10, 3, 100 epochs, and 16 different batch sizes. For the purposes of model training (80%) and model selection (20%), respectively, we have two sets of data: training sets and validation sets. Evaluation will perform the test of the model using the test set. Accuracy, sensitivity, and specificity are the three measurements used.

In [1], we experimentally show that CapsNets' equivariance features minimize significant data needs, making them suitable for medical image processing. We handle class distribution imbalance and limited annotated data using computer-aided diagnostic (classification) tasks. We run several controlled tests on two visual (MNIST and Fashion-MNIST) and two medical datasets that target mitotic identification (TUPAC16) and diabetic retinopathy detection to verify our claims (DIARETDB1). This is the first study to use capsule networks to solve medical image analysis challenges. We focus on computer-aided diagnostic image classification challenges. We compare capsule networks (CapsNets) to conventional convolutional networks (ConvNets) under biomedical image database constraints such as class imbalance and limited labeled data. CapsNets create hierarchical image representations by passing images through several layers of the network, like ConvNets. Against the trend toward deeper models, the original CapsNet has only two layers: a main caps layer that captures low-level signals and a specialized secondary caps layer that can predict both the existence and position of an object in the image. We propose using CapsNets equivariance features to use structural redundancy in images and minimize the training image number. Image patterns may indicate a good diagnosis. Exudates and hemorrhages might be soft or solid. CapsNet routing algorithms learn to establish relations between features, unlike ConvNets, which use feature detection to make decisions. The routing method collects redundant features instead of repeating them throughout the network to handle invariance. The above advantages directly impact the quantity of data samples needed to train networks.

In [12], an ablation researcher outlined each stage in the proposed process. Preprocessing and pre-training considerably increase the performance of the proposed method. The major preprocessing step, NCLAHE, improves the image resolution by 1.6–2.13%. Technical issues can reduce image quality, which can harm machine learning performance. N-CLAHE improves medical imaging quality and performance. When large-scale labeled data can be aggregated, pre-training improves performance. Pre-training the deep feature extractor block of the proposed technique, which takes a lot of training data, improves its performance. Our technique decreases false positives and false negatives more than current methods. COVID-19 and pneumonia patients share symptoms, making misclassification likely. Our approach handles these scenarios effectively. Most notably, our approach detects COVID-19 patients better from chest X-rays. Six of 1144 COVID-19 chest X-ray patients are misclassified, and only two are pneumonia. Our proposed approach performs well, properly identifying 228 COVID-19 instances and misclassifying only two, one of which is pneumonia. Our proposed approach correctly categorized 645 COVID-19 US frames and misclassified nine, including eight pneumonia frames. AUCs of receiver operating characteristics (ROC) are shown to demonstrate the proposed model's ability to differentiate classes.

In [10], capsule-based Chinese medical text categorization is proposed. Chinese medical text classification enhances the capsule network model in this research. We collected complex medical text characteristics using the capsule network's unique network topology and powerful feature extraction. When the capsule network and the long short-term memory (LSTMs) network handle medical text, $F1$ values are at least 4% higher than other basic models. Capsule network processes. Five methods were tested for screening clear text datasets in clinical trials. Researchers calculated the mean accuracy, recall, and $F1$ scores for 44 categories. According to the research, a shorter experimental text may have decreased the gating unit output in both networks. Longer experimental texts may increase LSTM and GRU network performance, decreasing the result gap. Integrating the capsule network with the LSTM and GRU networks increases $F1$ values by 5.46% and 10.92%, respectively. This shows that Chinese medical literature categorization depends on the capsule network. The research showed that GRU and LSTM with the capsule network model can categorize medical short texts. The accepted information will help categorize medical literature in future research and for the book "Electronics 2022, 11, 2229." 10 of 11 students can create medical knowledge graphs. This research's dataset is small, restricting the model's impact.

12 Existing Research Limitations

Some researchers have utilized capsule networks for lung nodule classification and observed that this approach outperformed traditional neural networks. Nevertheless, the authors acknowledged that the limited size and lack of diversity in the dataset could potentially restrict the generalizability of the findings. Due to the often restricted size of medical datasets, acquiring a sufficient amount of labeled data to effectively train capsule networks can be challenging. This limitation obstructs the generalizability of the results and may necessitate the application of transfer learning or other techniques to address the data shortage [2, 4, 7, 8, 14].

Some articles have summarized the current research on capsule networks in medical image analysis and discussed their potential advantages and limitations. The authors noted that the interpretability of capsule networks could pose a challenge and that there is still a lack of research on their use in many specific medical applications. Capsule networks utilize a complex set of hierarchical representations to make predictions, which can make interpreting the output difficult. This challenge hinders understanding the factors driving the predictions and can limit the ability to diagnose or treat medical conditions effectively. While some studies have investigated the use of capsule networks in specific medical applications, such as medical imaging or diagnosis, there is still a lack of research in many other areas. This limitation restricts the generalizability of the results and necessitates further research to comprehend the potential of capsule networks in various medical contexts [1, 3, 5, 15–17].

Some researchers have also utilized capsule networks for the classification of diabetic retinopathy and observed that this approach outperformed traditional neural networks. However, the authors acknowledged that the limited size of the dataset and the requirement for substantial computing resources could pose challenges. Training capsule networks can be computationally expensive, particularly when working with extensive medical datasets. This limitation can inhibit the utilization of capsule networks in real-time or resource-constrained environments, such as low-resource settings or mobile applications [1, 3].

13 Identified Research Gaps

Labeled data limitations make capsule networks for medical research challenging. Finding large datasets of labeled medical images for capsule network training may be difficult and expensive. Overfitting reduces capsule network performance on unseen data [8].

Computational complexity is another issue with capsule networks in medicine. Convolutional neural networks are computationally simpler than capsule networks. This makes capsule networks challenging to train on large datasets and inappropriate for real-time applications [10, 14].

14 Limitations of the Capsule Networks for Medical Science Research

Capsule networks are another kind of deep learning model that has been proven effective for a variety of tasks, including the classification of medical images. However, capsule networks have some limitations that must be addressed when using them for medical science research.

- Capsule networks have the drawback of requiring an extensive amount of training data. This is because capsule networks learn to precisely represent the spatial relationships between image features, which requires a large quantity of training data. In the field of medical science, it can be challenging to acquire huge amounts of labeled training data, which can restrict the application of capsule networks for certain tasks.
- Data noise can affect capsule networks. Because capsule networks learn to express spatial correlations between features, data noise can break these associations. Imaging artefacts and image acquisition variances cause noise in medical data. Noise inhibits capsule network learning.
- Finally, training capsule networks is computationally costly. Capsule networks contain many parameters to learn, making training tedious and expensive.
- Capsule networks can have difficulty generalize.
- Limited availability of labeled data.
- Capsule networks are difficult to read, making predictions difficult to understand.
- Capsule networks are less popular than other deep learning models, and hence, there is less research regarding how to use them.

Despite these drawbacks, capsule networks are promising for medical research. Capsule networks may become increasingly more powerful for medical science research as research continues to address these limitations [2].

15 Current Applications

Machine learning and deep learning have been widely utilized in both established and emerging domains. The following is a summary of areas that have utilized the current capabilities of machine learning (ML) and deep learning (DL).

15.1 Micro-Robot Adaptation

Certain researchers utilized a sophisticated convolutional neural network (CNN) architecture to enable the real-time adaptation of a minimally-invasive and therapeutic endoscopy robot specifically designed for the gastrointestinal (GI) system.

The methodology that was previously described enabled the robot to determine its spatial location and angular orientation in relation to a specified frame of reference. The combination of the CNN architecture with sophisticated sensor platforms has enabled the realization of precise and instantaneous robot adaptation within the gastrointestinal (GI) system. The integration of molecular robot technologies with CNN (and other) architectures presents an exciting opportunity for advancing research. This combination has the potential to facilitate accurate diagnosis and interventions, as well as enable the incorporation of molecular shuttles within living organisms [3].

15.2 Network Biology

The area of network biology focuses on analyzing the interactions among biomolecules that play a role in the organization and operation of living cells. This includes the analysis and reconstruction of extensive biological networks within cells, as well as the development and implementation of smaller synthetic gene networks. Machine learning algorithms have the capability to utilize and integrate diverse biological datasets in order to construct complex and layered models for analyzing the interactions of different phenomena. These analyses vary from within the cells interactions to intercellular interactions within tissues and organs, encompassing both healthy and diseased conditions. Significant advancements have been made in this field; however, additional research is required to comprehensively understand the mechanisms by which affecting biological process networks can give rise to various diseases. Capsule networks utilize methodologies related to convolutional neural networks (CNNs) in order to visually represent distinct and targeted objects toward specialized capsules or modules. A capsule network incorporates additional layers within each hidden layer, as opposed to simply increasing the number of layers [3].

16 Open Challenges and Future Redirections

Researchers effectively utilized a variety of techniques to improve the segmentation accuracy of medical images. Although accuracy improvement only cannot clarify algorithm performance, especially in the field of medical image analysis where class imbalance, noise interference, and the severe consequences of failed tests must be taken into consideration. Below sections will discuss possible future research directions for medical image segmentation.

16.1 Transfer Learning

Medical imaging is commonly accompanied by considerable interference. Furthermore, the process of highlighting medical images is frequently more costly in comparison with highlighting natural images. Hence, the exploration of utilizing pre-trained deep learning models on natural images for medical image segmentation represents an interesting field for future research. Furthermore, transfer learning plays a significant role in achieving weakly supervised medical image segmentation. Transfer learning refers to the utilization of pre-existing knowledge to acquire new knowledge, with a particular focus on identifying similarities between the existing and new knowledge domains. Transfer learning is a technique that leverages the correlation between data or tasks to facilitate the sharing of model parameters or knowledge acquired by a model. This sharing process enhances the efficiency of model learning by accelerating the learning process. Transfer learning is an effective way to address the issue of limited availability of labeled data [6].

17 Conclusion and Future Work

Capsule networks have shown great potential for addressing various challenges in medical science. These challenges include but are not limited to medical image analysis, disease diagnosis, drug discovery, and personalized medicine. From medical image analysis to personalized medicine, capsule networks have demonstrated improved performance over traditional convolutional neural networks.

In this review paper, we have highlighted what are the modern challenges in medical science, how medical science problems can be solved using capsule networks and the capsule network algorithms used to solve any medical science problems.

While capsule networks show great promise, there is still much work to be done in this field. One area of future research could focus on improving the interpretability of capsule networks, as understanding how these networks arrive at their decisions is crucial for building trust with medical professionals and patients. Furthermore, the efficiency of capsule networks can be improved by exploring novel applications and training methods that leverage their unique capabilities in medical science.

Despite these advantages, there are still several challenges that need to be addressed in order to fully realize the potential of capsule networks in medical science. One of the main challenges is the need for large amounts of labeled training data, which can be difficult to obtain in medical applications. Another challenge is the interpretability of the learned features, which is important for understanding the underlying mechanisms of disease and building trust in the predictions made by the model.

References

1. Jiménez-Sánchez A, Albarqouni S, Mateus D (2018) Capsule networks against medical imaging data challenges. Lecture notes in computer science (including subseries lecture notes in artificial intelligence and lecture notes in bioinformatics), 11043 LNCS, pp 150–160. Available at https://doi.org/10.1007/978-3-030-01364-6_17
2. Abdel-Jaber H et al (2022) A review of deep learning algorithms and their applications in healthcare. Algorithms 15(2). Available at https://doi.org/10.3390/a15020071
3. Akay A, Hess H (2019) Deep learning: current and emerging applications in medicine and technology. IEEE J Biomed Health Inform 23(3):906–920. Available at https://doi.org/10.1109/JBHI.2019.2894713
4. Modi S et al (2021) Detail-oriented capsule network for classification of CT scan images performing the detection of COVID-19. Mater Today Proc [preprint]. Available at https://doi.org/10.1016/j.matpr.2021.07.367
5. Zhang Z et al (2020) Enhanced capsule network for medical image classification. In: Proceedings of the annual international conference of the IEEE engineering in medicine and biology society, EMBS, July 2020, pp 1544–1547. Available at https://doi.org/10.1109/EMBC44109.2020.9175815
6. Wang R et al (2022) Medical image segmentation using deep learning: a survey. IET Image Process 16(5):1243–1267. Available at https://doi.org/10.1049/ipr2.12419
7. Quan H et al (2021) DenseCapsNet: detection of COVID-19 from X-ray images using a capsule neural network. Comput Biol Med 133:104399. Available at https://doi.org/10.1016/j.compbiomed.2021.104399
8. Zhao A et al (2023) DCACorrCapsNet: a deep channel-attention correlative capsule network for COVID-19 detection based on multi-source medical images. IET Image Process 17(4): 988–1000. Available at https://doi.org/10.1049/ipr2.12690
9. Afriyie Y, Weyori BA, Opoku AA (2021) Exploring optimised capsule network on complex images for medical diagnosis. In: IEEE international conference on adaptive science and technology, ICAST, Nov 2021 (Jan 2022). Available at https://doi.org/10.1109/ICAST52759.2021.9682081
10. Chen W et al (2022) Research on medical text classification based on BioBERT-GRU-attention. In: 2022 IEEE international conference on advances in electrical engineering and computer applications, AEECA 2022, pp 213–219. Available at https://doi.org/10.1109/AEECA55500.2022.9919061
11. Wirawan IMA et al (2023) Continuous capsule network method for improving electroencephalogram-based emotion recognition. Emerg Sci J 7(1):116–134. Available at https://doi.org/10.28991/ESJ-2023-07-01-09
12. Saif AFM et al (2021) CapsCovNet: a modified capsule network to diagnose COVID-19 from multimodal medical imaging. IEEE Trans Artif Intell 2(6): 608–617. Available at https://doi.org/10.1109/TAI.2021.3104791
13. Yu X et al (2021) CapsTM: capsule network for Chinese medical text matching. BMC Med Inform Decision Making 21(2):1–9. Available at https://doi.org/10.1186/s12911-021-01442-9
14. Heidarian S et al (2021) COVID-FACT: a fully-automated capsule network-based framework for identification of COVID-19 cases from chest CT scans. Front Artif Intell 4:1–13. Available at https://doi.org/10.3389/frai.2021.598932
15. Tang B et al. (2019) CapSurv: capsule network for survival analysis with whole slide pathological images. IEEE Access 7:26022–26030. Available at https://doi.org/10.1109/ACCESS.2019.2901049
16. Monday HN et al (2022) COVID-19 pneumonia classification based on NeuroWavelet capsule network. Healthcare 10(3):1–18. Available at https://doi.org/10.3390/healthcare10030422
17. Stynes P. Brain age classification from brain MRI using ConvCaps framework Animesh Kumar National College of Ireland Supervisor

Studies on Movie Soundtracks Over the Last Five Years

Sofía Alessandra Villar-Quispe and Adriana Margarita Turriate-Guzman

Abstract In the field of film studies and, especially, music studies, the number of those who dare to go beyond the limits when researching film production is minimal. This is because the soundtrack as a subject has not always been an object of analysis that is investigated as rigorously as it is now. Therefore, this paper seeks to determine the trend of studies on movie soundtracks over the last five years in the Scopus database. This is a systematic review using the PRISMA method. Fifteen articles published between 2017 and 2022 were selected, and, once all the information was reviewed, it was divided into four categories. The main limitations were the lack of information and the number of papers written in English. Finally, it was concluded that the soundtrack fulfills several functions within the narrative of a movie and that the topic needs more academic attention in Ibero-America. Therefore, researchers are invited to continue to problematize this issue in future systematic reviews.

Keywords Soundtrack · Movies · Audiovisual · Film music

1 Introduction

In the field of film studies and, especially, music studies, the number of those who dare to go beyond the limits when researching film production is minimal. This is because the soundtrack as a subject has not always been an object of analysis that is investigated so rigorously [1]. Nowadays, in the research bases, there is diverse information and studies on movie soundtracks. The creation of the soundtrack has a very significant influence on the audiovisual narrative of movies. For example, Serrano [2] indicates that the minimalist music composed by Philip Glass for the soundtrack of the documentary *Koyaanisqatsi* manages to better communicate the

S. A. Villar-Quispe · A. M. Turriate-Guzman (✉)
Universidad Privada del Norte, Lima, Peru
e-mail: adriana.turriate@upn.edu.pe

S. A. Villar-Quispe
e-mail: N00170591@upn.pe

S. Shakya et al. (eds.), *Fourth International Conference on Image Processing and Capsule Networks*, Lecture Notes in Networks and Systems 798,
https://doi.org/10.1007/978-981-99-7093-3_2

projection of the reality in which we live by expressing what is inexpressible about our current state in the movie. Likewise, when associating certain sounds with something specific, such as a city, Grifol-Isely [3] argues that the mnemonics of a soundtrack are essential along with the hues, as this is how Barcelona managed to adopt a brand image. Meanwhile, González [4] states that Daniel Montorio's music for the Spanish Western highlights more the actions than the emotions since he uses musical blocks for them and not for the characters.

Regarding the creative process of the soundtrack for a movie, Fernandez [5] comments that, during the creation of *Carmen, la de Triana*, models of musical quality and coherence were found, such as the balance between cultural, folkloric, and the most popular elements to achieve the distinction from the German version of the movie. In turn, Oliveira [6] states that to achieve this new aesthetics supporting the future of cinema, it is necessary to provoke emotions by adapting the nature of the movie, as evidenced in the movies by Flora Gomes. Furthermore, in the movie *Das Kabinett des Doktor Caligari*, Amorós and Gómez [7] point out that, to provide the movie with greater depth and dynamism, Rainer Viertlböck turned away from classical music, plunged into jazz, and only with that genre created continuous leitmotifs to identify the characters and their actions.

On the other hand, music can also be linked to political communication. This happened in the late 1920s when sound was incorporated into cinema. Instrumentation and rhythm were prioritized as technical resources to represent Latin America through music in movies, and this, in turn, contributed as a precedent for later productions aligned with U.S. foreign policy [8]. Similarly, Farías [9] points out that the character and other timbral features of musical pieces, such as the *Nueva Canción Chilena*, used in the Chilean documentary during the Popular Unity, allowed the audience to connect with the ideas of the Left.

In other studies, varying from movie promotion to education, medicine, psychology, and computer science, the soundtrack serves as a useful tool. For the first category, music videos with movie soundtracks can fulfill their role thanks to commercial communication [10]. In the field of education, Porta and Herrera [11] determined that the soundtrack gives meaning and sense to both the narrative and the multimedia experience of children watching their favorite audiovisuals. Finally, in medicine and psychology, the research aims to obtain results focused on perception, emotions, bodily reactions, and moral judgments. Meanwhile, in computer science, movie genres are investigated according to their soundtrack [12–15]. Therefore, the topic addressed in this research is current and relevant. Therefore, the aim is to determine the trend of studies on movie soundtracks over the last five years in the Scopus database. For this purpose, this systematic review will be carried out, and the PRISMA method will be used. The methodology section is presented below.

2 Methodology

This research addresses the trend of studies on movie soundtracks over the last five years in the Scopus database. It was developed through a systematic review, a study in which information from published articles is managed. The researcher collects all those articles of his/her interest regarding a specific topic, analyzes them, compares them, and, in the end, draws his/her conclusions. This must be developed objectively, either from a qualitative or quantitative perspective [16].

The research was conducted using keywords such as "music AND films", "soundtrack AND movies", and "soundtrack AND films". There were only three searches performed in Scopus, a database of citations and abstracts from neutral sources with more than 1.8 billion referenced citations. This was the only database used due to its strict quality and ethical criteria for the selection of titles, which are selected by the independent Scopus Content Selection and Advisory Board (CSAB), an international group of scientists and researchers with experience in journal publishing, representing the main scientific disciplines [17].

The formulae provided by the system after applying the filters were as follows:

[17] TITLE-ABS-KEY (music AND films) AND (LIMIT-TO (OA, "all")) AND (LIMIT-TO (PUBYEAR, 2022) OR LIMIT-TO (PUBYEAR, 2021) OR LIMIT-TO (PUBYEAR, 2020) OR LIMIT-TO (PUBYEAR, 2019) OR LIMIT-TO (PUBYEAR, 2018) OR LIMIT-TO (PUBYEAR, 2017)) AND (LIMIT-TO (DOCTYPE, "ar")) AND (LIMIT-TO (LANGUAGE, "Spanish")) AND (LIMIT-TO (SRCTYPE, "j")) [17].

[17] TITLE-ABS-KEY (soundtrack AND movies) AND (LIMIT-TO (OA, "all")) AND (LIMIT-TO (PUBYEAR, 2022) OR LIMIT-TO (PUBYEAR, 2021) OR LIMIT-TO (PUBYEAR, 2020) OR LIMIT-TO (PUBYEAR, 2019) OR LIMIT-TO (PUBYEAR, 2018) OR LIMIT-TO (PUBYEAR, 2017)) AND (LIMIT-TO (DOCTYPE, "ar")) AND (LIMIT-TO (LANGUAGE, "English") OR LIMIT-TO (LANGUAGE, "Spanish")) AND (LIMIT-TO (SRCTYPE, "j")) [17].

[17] TITLE-ABS-KEY (soundtrack AND films) AND (LIMIT-TO (OA, "all")) AND (LIMIT-TO (PUBYEAR, 2022) OR LIMIT-TO (PUBYEAR, 2021) OR LIMIT-TO (PUBYEAR, 2020) OR LIMIT-TO (PUBYEAR, 2019) OR LIMIT-TO (PUBYEAR, 2018) OR LIMIT-TO (PUBYEAR, 2017)) AND (LIMIT-TO (DOCTYPE, "ar")) AND (LIMIT-TO (LANGUAGE, "English") OR LIMIT-TO (LANGUAGE, "Spanish")) AND (LIMIT-TO (SRCTYPE, "j")) [17].

Scopus detected a total of 80 articles, and the PRISMA method was applied to refine the search and answer the research question [18]. The 15 articles selected were published between 2017 and 2022, amidst which 2021 stands out with 6 publications. Likewise, it is important to note that Spain has the highest number of publications (i.e., 6) out of the studies collected from the Scopus database. On the other hand, the first reason for excluding some articles, in the end, was because they were essays, and the second was because one of them was an interview. Regarding language, 9 articles were published in Spanish and 6 in English. Finally, out of the keywords of

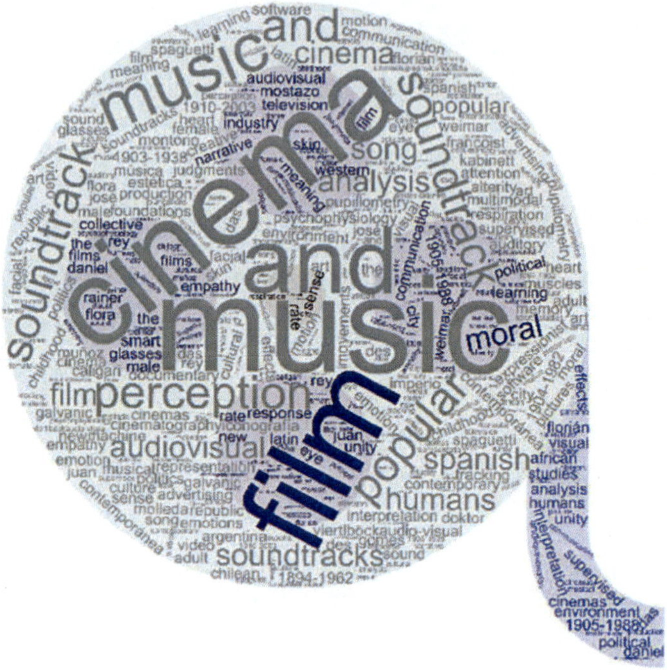

Fig. 1 Word cloud

all the articles, the most outstanding one was "music". Figure 1 illustrates the word cloud.

3 Results

After reviewing the articles, the information was divided into four categories. The list of categories is tabulated in Table 1.

The findings by category are detailed below.

3.1 Soundtrack Influence on the Audiovisual Narrative of Movies

During this research, three articles that reinforce the importance of a movie's soundtrack in determining its narrative were detected by scanning the papers and finding out that what they had in common was their authors pointing out how music, whether by choice of the director or not, need to fit the nature of the film. Firstly, the study

Table 1 List of categories

Category	Articles	Author	Synthesis
Soundtrack influence on the audiovisual narrative of movies	Aullidos de chatarra. Una aproximación a Koyaanisqatsi (1982) a través de la música de Philip Glass y su imbricación ideológica [Scrap howls. An approach to Koyaanisqatsi (1982) through the music of Philip Glass and its ideological interweaving] [2]	Serrano [2]	Minimalism in a documentary soundtrack
	Estrategias narrativas y musicales en el wéstern español de los años sesenta: dos filmes de Ramón Torrado con música de Daniel Montorio [Narrative and musical strategies in the spanish western of the 1960s: two films by Ramón Torrado with music by Daniel Montorio] [4]	Gonzáles [4]	Roles of music in the narrative of two Spanish Westerns
	Diálogos entre imágenes y sonidos en el imaginario urbano. Ritmando Barcelona(s) [Dialogues between images and sounds in the urban imaginary. Barcelona's rhythms] [3]	Grifol-Isely [3]	Importance of sound in the image of the city through audiovisuals
The creative process of a movie soundtrack	La creación musical en el proceso cinematográfico. Una perspectiva crítica [Music creation in the film process. A critical approach] [1]	Lázaro [1]	Problems with the musical creation of a movie
	"I make films to be seen": the narrative issue of Flora Gomes [6]	Oliveira [6]	Elements of narrative cinematography and the direction process
	La música de Carmen, la de Triana (1938): Las fuentes y el proceso creativo [The music of Carmen, la de Triana (1938): Its sources and its creative process] [5]	Fernández [5]	Important implications during the pre-production phase
	La música de Rainer Viertlböck en la versión restaurada del film Das Kabinett des Doktor Caligari, de Robert Wiene (1919) [Rainer Viertlböck's music in the restored version of the film Das Kabinett des Doktor Caligari by Robert Wiene (1919)] [7]	Amorós and Gómez [7]	Creative soundtrack proposal for an expressionist film
Music and political communication	La música en las representaciones de lo latinoamericano en los primeros filmes hollywoodenses con sonido incorporado (1927–1932) [Music in the Representations of Latin America in Hollywood Films of Early Sound Era (1927–1932)] [8]	Poveda [8]	Problems in the musical representation of Latin America

(continued)

Table 1 (continued)

Category	Articles	Author	Synthesis
	Música y comunicación política en el documental durante la Unidad Popular [Music and politic communication in the documentary during Chile's Popular Unity] [9]	Farías [9]	Use of music to achieve a political identity
The soundtrack as a study instrument	Soundtrack Music Videos: The Use of Music Videos as a Tool for Promoting Films [10]	Selva-Ruiz and Fénix-Pina [10]	Features of the soundtrack music video
	La música y sus significados en los audiovisuales preferidos por los niños [Music and its significance in children favourite audiovisuals] [11]	Porta and Herrera [11]	Soundtrack and children's interpretation
	A computational lens into how music characterizes genre in film [12]	Ma et al. [12]	Use of music according to the movie genre
	Locus of emotion influences psychophysiological reactions to music [14]	Merrill et al. [14]	Music listening and its psychophysiological responses
	How soundtracks shape what we see: analyzing the influence of music on visual scenes through self-assessment, eye tracking, and pupillometry [13]	Ansani et al. [13]	Soundtrack effects in the interpretation of visual scenes
	The influence of film music on moral judgments of movie scenes and felt emotions [15]	Steffens [15]	Specific emotions associated with moral judgments

conducted by Serrano [2], in 2018, on the documentary *Koyaanisqatsi*, demonstrates that the music of the whole movie performs an organizing role. In other words, it gives an internal structure to the movie. The beginning contains music that fulfills the introductory, structural, and incidental role; the climax and denouement present the musical notes more prominently. Secondly, Gonzáles [4], in 2021, through the analysis of the soundtrack of two Spanish Westerns by Daniel Montorio, demonstrated how the moral codes and some conventional narrative frameworks of the genre are strategically underlined through the conscious use of leitmotifs. For example, in Los Cuatreros, the villains are associated with the timbres of the horns and trombones, as they usually appear in the dark. Thirdly, Grifol-Isely [3], in a study on the sounds and the image-movement relationship associated with the city of Barcelona, points out how in the movie *Vicky Cristina Barcelona* by Woody Allen, when the first scene begins with the song Barcelona by Giulia y Los Tellarini, it is assigned the image of a labyrinthine place because the rhythm is fast and accentuated by the use of stringed instruments and an accelerated cadence. Thus, these three cases prove the influence of music within the audiovisual narrative of a documentary or fiction feature film.

3.2 The Creative Process of a Movie Soundtrack

Regarding the creation of music during the filmmaking process, four articles coincide. Basically, according to Lázaro [1], the movie director makes all the elements work in a strict film sense. However, to eliminate incompatibilities within the media of expression, the director, as well as the other heads of the filmmaking areas, must be a musician; consequently, the film musician must be a filmmaker. Indeed, Oliveira [6], in a study of the movies by Flora Gomes, in 2019, notes that the soundtrack is a mark and, in turn, it is highly important for the filmmaker. This occurs because music, along with other aesthetic aspects, has a great effect on the viewer: it arouses emotions. Moreover, it is also an adjustable component since it allows for adaptability in production as it can be included when shooting is finished.

Regarding the above, Fernández [5], in his analysis of the movie *Carmen, la de Triana*, in 2021, highlights the creative film process of Florián Rey, the director, and the success of the composers, Muñoz Molleda (soundtrack) and Juan Mostazo (songs). The latter composed a song in situ based on the recitation of a comedy. This, and the fact that the songs were recorded before filming, differentiated them from other productions of that time in Spain. On the other hand, Amorós and Gómez [7] indicate that Rainer Viertlböck innovated in musical composition, using synthesizers and other instruments characteristic of jazz. The result was a modern soundtrack for *Das Kabinett des Doktor Caligari*, an expressionist film from 1919 that, at the time, was not well received for resorting to the music of classical composers. In conclusion, although Lázaro's proposal is far from the current panorama, it is clear, from the examples provided by the other authors, the relevance of music within the creative process in cinema.

3.3 Music and Political Communication

Two articles coincide in the category of music and political communication. On the one hand, Poveda [8], in a study of the musical characterization of Latin America in three Hollywood productions with incorporated sound, from 1927 to 1932, noted three problems in the performances: the use of references to Spanish culture; the sexual connotations through the body, tango and rumba; and the interest in some resources and processes that are used later, during the era of the "Good Neighbor Policy". On the other hand, Farías [9], in 2020, in his analysis of the music of three documentaries produced during the Popular Unity in Chile, found that in *El Derecho al Descanso* certain modern music was used to portray the bourgeoisie and an instrumental huayno piece for the working class (only at the beginning). In contrast, in *Un Verano Feliz* and *Balnearios Populares*, rock-pop is used to symbolize enjoyment and leisure. The latter movie differs from the others because of the absence of the *Nueva Canción Chilena*, and because it was produced by a group of amateurs and not by a government institution. Undoubtedly, this demonstrates its great commitment to Popular Unity.

3.4 Soundtrack as a Study Instrument

Regarding studies in which the soundtrack is used as a study instrument, six articles were found. The first one, by Selva-Ruiz and Fénix-Pina [10], in 2021, focuses on the use of the soundtrack music video as an instrument for movie promotion. The authors analyzed the content of 119 music videos produced over 33 years. Then, according to internal and external characteristics, they determined that 93.3% included images of the artist. In the second study, from the field of education, Porta and Herrera [11] used the soundtrack to know its meaning from the children's interpretation. For this purpose, they showed different clips of 14 audiovisuals edited in 3 versions (sound, image, and everything) to 547 students between 9 and 15 years old. In the end, thanks to an assessment questionnaire, they indicated the superiority of the soundtrack in 3 out of 5 films: The Lion King, Titanic, and Oz, the fantasy world.

Regarding the previous paragraph, now from the computer science area, in the third study, Ma et al. [12] designed neural network models with various clustering mechanisms to predict the genres of 110 movies by listening to their music. With a macro-$F1$ score of 0.62, the models demonstrated higher accuracy than previous work; however, no evidence was found that music of a given genre implies common low-level visual features. Now, in the field of medicine and psychology, the fourth study, by Merrill et al. [14], in 2020, compared the effect of perceiving emotion in music with assessing the emotion induced in the listener by the music, both in psychophysiological responses such as muscle activity, respiration or heart rate, which were measured in 40 participants while listening to a soundtrack excerpt.

Thus, they found that, if more attention is paid to musical expression, bodily reactions improve.

Likewise, regarding the aforementioned, the latest studies are also within the psychology sector. The fifth one, carried out by Ansani et al. [13], in 2020, reports two studies on how music affects the interpretation of visual scenes. In the first one, conducted with an online survey, variables such as empathy toward the character or the perception of the environment were highly influenced by the soundtrack. In the second one, which was the same but added eye-tracking, the results were replicated. Finally, Steffens [15], for his research on the influence of music on perceptions, actions, and moral judgments, selected 81 participants and showed them 2 excerpts from 2 different movies with 4 different soundtrack music pieces. However, in 3 out of 4 cases, the desired emotional induction through film music was not successful. Thus, although in this last case no favorable results were evidenced, the efficiency of the soundtrack as a study instrument is evident through the other examples.

4 Discussion and Conclusions

The soundtrack used for specific purposes for the narrative of a movie has become one of the major sources of musical experience in our daily environment [19]. Currently, Porta and Herrera [11], to some extent, agree with the aforementioned, stating that the importance of the audiovisual pairing—music and media—is because it is part of children's daily life. Meanwhile, José Lázaro [1] indicates that music is infinitely expressive, so it can be used, for example, to accompany the speech of a character and convey both ideas and emotions. All this shows the relevance of the soundtrack for having stayed as an everyday thing for us.

Thus, this study has different scopes. Firstly, at the methodological level, it is identified as a systematic review [16], in which the PRISMA method was used [18]. Secondly, at a practical level, it may be useful to build the theoretical framework for future research on movie soundtracks. Finally, at a theoretical level, it updates the information available on the research topic addressed. Now, regarding the main limitations, the first one was the lack of information to specify the original research problem. Initially, it had been proposed to resort only to soundtrack studies carried out in Ibero-America; however, there were not enough studies to carry out a systematic review. Therefore, researchers in this geographic region must reflect and take an early interest in the subject. On the other hand, the second limitation was the large number of articles in English that are available in the database. Nevertheless, this provides an opportunity for those who are fluent in English to broaden their knowledge.

In summary, this research was carried out using the PRISMA method [18], which facilitated access to 15 documents from the Scopus database [17]. From this information, it was found that the highest publication trend was in the year 2021, that Spain leads in the number of publications, and that out of the keywords of all the articles, and the most outstanding one was "music" (this is depicted in Fig. 1). After a complete review of the material collected, it became evident that the theory can be divided into four categories (for details, see Table 1). Firstly, the influence of the soundtrack on the audiovisual narrative of movies in which music fulfills several roles; secondly, the importance of music during the creative process of a movie; thirdly, the relationship between soundtrack and political communication, in which music is used to connote several elements. Finally, in the fourth category, the effectiveness of the soundtrack as a study instrument for different research projects ranging from the field of education to psychology is highlighted. However, aside from all of the above, the subject still has much to be investigated, especially in Ibero-America, where more academic attention to the subject is needed in the future. Therefore, other collaborators, colleagues, and students are invited to discuss the topic in future systematic reviews.

References

1. Lázaro López J (2021) Music creation in the film process. A critical approach. [La creación musical en el proceso cinematográfico. Una perspectiva crítica] Fonseca J Commun 22:173–188. https://doi.org/10.14201/fjc-v22-23477
2. Serrano J (2018) Scrap howls. An approach to Koyaanisqatsi (1982) through the music of Philip Glass and its ideological interweaving. [Aullidos de chatarra. Una aproximación a Koyaanisqatsi (1982) a través de la música de Philip Glass y su imbricación ideológica] Boletín De Arte 28:479–506. https://doi.org/10.24310/BOLARTE.2007.V0I28.4494
3. Grifol-Isely L (2022) Dialogues between images and sounds in the urban imaginary. Barcelona's rhythms. [Diálogos entre imágenes y sonidos en el imaginario urbano. Ritmando Barcelona(s)] Fotocinema 24:265–289. https://doi.org/10.24310/FOTOCINEMA.2022.VI24.13834
4. González C (2021) Narrative and musical strategies in the Spanish Western of the 1960s: two films by Ramón Torrado with music by Daniel Montorio. [ESTRATEGIAS NARRATIVAS Y MUSICALES EN EL WÉSTERN ESPAÑOL DE LOS AÑOS SESENTA: DOS FILMES DE RAMÓN TORRADO CON MÚSICA DE DANIEL MONTORIO] Anuario Musical 76:181–206. https://doi.org/10.3989/anuariomusical.2021.76.09
5. Fernández R (2021) The music of Carmen, la de Triana (1938): its sources and its creative process. [LA MÚSICA DE CARMEN, LA DE TRIANA (1938): LAS FUENTES Y EL PROCESO CREATIVO1] Anuario Musical 76:151–180. https://doi.org/10.3989/anuariomusical.2021.76.08
6. Oliveira J (2019) I make films to be seen: the narrative issue of Flora Gomes. J Sci Technol Arts 11(1 Special Issue):2-1–2-10. https://doi.org/10.7559/citarj.v11i1.587
7. Amorós A, Gómez N (2017) Rainer Viertlböck's music in the restored version of the film Das Kabinett des Doktor Caligari by Robert Wiene (1919) [A música de rainer viertlböck na versão restabelecida do filme das kabinett des doktor caligari de robert wiene (1919)] Cuadernos De Música, Artes Visuales y Artes Escénicas 12(2):2017. https://doi.org/10.11144/Javeriana.mavae12-2.mrvv
8. Poveda J (2020) Music in the representations of Latin America in hollywood films of early sound era (1927–1932). [La música en las representaciones de lo latinoamericano en los primeros filmes hollywoodenses con sonido incorporado (1927–1932)] Resonancias 24(46):55–77. https://doi.org/10.7764/RES.2020.46.4
9. Farías M (2021) Music and political communication in the documentary during Chile's popular unity. [MÚSICA Y COMUNICACIÓN POLÍTICA EN EL DOCUMENTAL DURANTE LA UNIDAD POPULAR] Universum 36(2):601–621. https://doi.org/10.4067/S0718-23762021000200601

10. Selva-Ruiz D, Fénix-Pina D (2021) Soundtrack music videos: the use of music videos as a tool for promoting films. Commun Soc 34(3):47–60. https://doi.org/10.15581/003.34.3.47-60
11. Porta A, Herrera L (2017) Music and its significance in children's favorite audiovisuals. Comunicar 25(52):83–91. https://doi.org/10.3916/C52-2017-08
12. Ma B, Greer T, Knox D, Narayanan S (2021) A computational lens into how music characterizes genre in film. PLoS One 16. https://doi.org/10.1371/journal.pone.0249957
13. Ansani A, Marini M, D'Errico F, Poggi I (2020) How soundtracks shape what we see: analyzing the influence of music on visual scenes through self-assessment, eye tracking, and pupillometry. Front Psychol 11. https://doi.org/10.3389/fpsyg.2020.02242
14. Merrill J, Omigie D, Wald-Fuhrmann M (2020) Locus of emotion influences psychophysiological reactions to music. PLoS One 15. https://doi.org/10.1371/journal.pone.0237641
15. Steffens J (2020) The influence of film music on moral judgments of movie scenes and felt emotions. Psychol Music 48(1):3–17. https://doi.org/10.1177/0305735618779443
16. Manterola C, Astudillo P, Arias E, Claros N (2013) Systematic reviews of the literature: what should be known about them. [Revisiones sistemáticas de la literatura. Qué se debe saber acerca de ellas] Cirugia Espanola 91(3):149–155. https://doi.org/10.1016/j.ciresp.2011.07.009
17. Elsevier (2022) Coverage you can count on. https://www.elsevier.com/solutions/scopus/how-scopus-works/content
18. Page MJ, McKenzie JE, Bossuyt PM et al (2021) The PRISMA 2020 statement: an updated guideline for reporting systematic reviews. Syst Rev 10:89. https://doi.org/10.1186/s13643-021-01626-4
19. Wingstedt J, Brändström S, Berg J (2010) Narrative music, visuals, and meaning in film. Vis Commun 9(2):193–210. https://doi.org/10.1177/1470357210369886

Blind Source Separation of EEG Signals Using Wavelet and EMD Decomposition

H. Massar, M. Miyara, B. Nsiri, and T. Belhoussine Drissi

Abstract Blind source separation (BSS) is a widely adopted approach for processing biomedical signals, particularly electroencephalography (EEG) signals, which indicate the electrical activities of the brain. Currently, EEG data is heavily polluted by various physiological artifacts such as muscular action, cardiac rate, and blinking, which call for the utilization of BSS techniques. This study intends to evaluate the performance of BSS methods by combining them with two other signal processing techniques, namely the discrete wavelet transform (DWT) and the empirical mode decomposition (EMD). This study proposes two approaches, the first of which involves combining the BSS and EMD algorithms. Here, the intrinsic mode function (IMFs) is extracted from each EEG signal and then separated those using BSS techniques. Based on the results, the ACSOBI-RO method was found to be the most effective. The second approach adds a DWT step to extract the approximation coefficients from the IMFs. The resulting coefficients are then fed into the BSS methods, and the SOBI method is found to be the most effective. In conclusion, the addition of the DWT block has significantly improved the performance of our methods. The obtained research findings demonstrate that the proposed approaches can effectively address the problem of physiological artifacts in EEG data and can be applied to other biomedical signals.

Keyword Blind source separation · Empirical mode decomposition · Discrete wavelet transform · Electroencephalogram

H. Massar (✉) · T. Belhoussine Drissi
Laboratory of Electrical and Industrial Engineering, Information Processing, Informatics, and Logistics (GEITIIL), Faculty of Science Ain Chock, University Hassan II—Casablanca, Casablanca, Morocco
e-mail: Hamzamassar10@gmail.com

M. Miyara
Computer Science and Systems Laboratory (LIS), Faculty of Science Ain Chock, University Hassan II—Casablanca, Casablanca, Morocco

H. Massar · B. Nsiri
Research Center STIS, M2CS, National School of Arts and Crafts of Rabat (ENSAM), Mohammed V University in Rabat, Rabat, Morocco

S. Shakya et al. (eds.), *Fourth International Conference on Image Processing and Capsule Networks*, Lecture Notes in Networks and Systems 798,
https://doi.org/10.1007/978-981-99-7093-3_3

1 Introduction

Blind source separation (BSS) is a set of methods that enables the identification of individual source components from the mixtures collected by numerous instruments. The sources of interest are often not readily accessible due to the sensitivity of the information to various disturbances. BSS utilizes a combination of different sources (observations) to predict the source signals, hence the term "blind" as only the mixtures are recognized as information. BSS has proven to be a solution to many previously unsolved issues in various application areas, including biomedicine, networking, and acoustics. The medical domain is one of the areas where BSS has numerous applications, such as its utilization in electroencephalography (EEG) analysis. EEG is a non-invasive method of observing and logging brain activity, which involves attaching electrodes to the forehead in order to identify the electrical impulses produced by the brain's neurons. It is frequently used to identify and explore various neurological conditions, including seizures, sleep disorders, and brain trauma. Moreover, cognitive functions such as focus, recollection, and language can be studied by using EEG data [1, 2].

Due to the great temporal accuracy of EEG, extraneous noise can cause a variety of artifacts in the data. Artifacts can be caused by measuring equipment as well as human beings, with the former comprising malfunctioning electrodes, line noise, and high electrode impedance. By employing a more accurate recording device and following stringent recording guidelines, these problems may be avoided, although physiological artifacts are more challenging to get rid of. Common physiological aberrations in the EEG data include eye movements, eye blinks, heart activity, and muscle activity [3]. Scientists widely used the BSS methods to separate the artifacts and the neural component to deal with these kinds of artifacts. Other techniques, like the wavelet transform, have been used to cope with artifacts. The wavelet transform (WT) is a technique used for processing and examining signals, enabling a localized breakdown of the spectrum in terms of either space or time. This method involves decomposing single-channel signals, such as EEG signals or other biomedical signals, aiding in their analysis and processing. An additional technique, empirical mode decomposition (EMD), has been used to analyze these data types. The empirical mode decomposition method makes the assumption that any given set of data (or signal), at any given moment, can be deconstructed into a number of coexisting, straightforward natural oscillations, referred to as intrinsic mode function (IMF), which superimpose one another [4].

Several publications have proposed and reported the processing and evaluation of EEG data using BSS, EMD, WT, or a combination of these techniques. Jamil et al. [5] utilized the discrete wavelet transform (DWT) and independent component analysis (ICA) to remove ocular artifacts as an illustration. The authors compared their approach to existing methods and found that it outperformed other techniques in terms of accuracy and reliability. They also discussed potential applications of EEG signal analysis in non-restricted environments, such as brain-computer interfaces and neurofeedback training. In a study by Zhao et al. [6], a multi-step strategy for blind source separation (BSS) is proposed to mitigate ocular, movement, and myogenic artifacts commonly observed in mobile high-density EEG data. This approach effectively addresses various types of artifacts typically encountered in such data, leading to enhanced accuracy and dependability in the subsequent analysis and interpretation of brain activity. Chen et al. [7] introduced an alternate approach using canonical correlation analysis (CCA) in combination with multivariate EMD. Paradeshi et al. [8] employed the wavelet ICA technique to eliminate eye-related artifacts, assessing their method using metrics such as peak signal noise ratio (PSNR), standard deviation (SD), root mean square error (RMSE), and power spectral density (PSD). In another investigation, Mowla et al. [2] utilized a dual-step process involving CCA and SWT to address EMG irregularities initially, followed by the application of the SOBI algorithm and SWT to remove EOG artifacts.

The suggested approach integrates the DWT, EMD, and BSS methodologies. Following the computation of the approximation coefficient and the IMFs produced from the EEG data, this research study aims to evaluate the BSS approaches. Here, the success of the suggested strategy is evaluated using the Spearman correlation coefficient (SCC).

This research study is organized as follows; the methods, approaches, and databases used in this study are covered in Sect. 2; the findings are presented in Sect. 3; and the conclusion is presented in Sect. 4.

2 Material and Methods

2.1 Datasets

A semi-simulated dataset was used in this research study. 27 individuals' EEG data were gathered during an eyes-closed period. Here, 19 EEG sensors are inserted by using the 10–20 International Method. Further, two datasets were gathered for a total of 54 datasets totaling 30 s. The sounds are filtered by using a bandpass filter at 0.5–40 Hz before being notched and filtered at 50 Hz. The data collection occurred at a rate of 200 Hz. Also, by placing four electrodes above and below the left eye, two on each eye, and two more electrodes, they recorded the EOG signals from the same participants with their eyes open. This procedure resulted in the creation of two binary signals: the horizontal EOG (HEOG), obtained by subtracting the recordings from the left and right EOG electrodes, and the vertical EOG (VEOG), obtained by subtracting the recordings from the top and lower EOG electrodes.

The developed semi-simulated EEG dataset was produced by using the contamination model detailed below:

$$\mathrm{EEG}_c = \mathrm{EEG}_N + A \times \mathrm{VEOG} + B \times \mathrm{HEOG} \tag{1}$$

where EEG_N stands for the signals gathered during the session with the eyes closed, and EEG_c for the relevant EEG data that have been purposely polluted. The previously published VEOG and HEOG signals' contamination coefficients are defined by vectors A and B. For more details on this collection, please refer to [9].

2.2 Empirical Mode Decomposition

A potent method for nonlinear and non-stationary data processing is empirical mode decomposition (EMD) [10]. Such signals may be analyzed using its data-adaptive multi-resolution technique by dissecting them into multiple components at different resolutions. EMD divides the signal into intrinsic mode functions (IMFs), which are zero-mean signals with a range of amplitudes. EMD depicts the signal as an expansion of fundamental functions depending on the input signal and approximated by an iterative process known as sifting, in contrast to standard decomposition approaches like wavelets that analyze signals by projecting them onto a collection of specified basis vectors [11].

The decomposition technique primarily relies on the detection of extrema and a screening process. Interpolating the local minima yields a lower envelope, while interpolating the local peaks yields a higher envelope. The first component $C1$ is then derived as the difference between the original data $x(t)$ and the mean $m1$ calculated from the upper and lower envelopes [12]. By using EMD, we can analyze and interpret nonlinear and non-stationary signals more effectively than with traditional techniques. It allows us to uncover patterns and trends in data that may not be visible with other methods, making it an essential tool in fields such as finance, biology, and engineering.

$$C_1 = x(t) - m_1 \tag{2}$$

The C component will be an IMF. If not, new extrema may be formed, and the ones already there can be relocated or magnified. Therefore, to acquire an IMF, this screening process must be used many more once [1]. Each IMF must satisfy two requirements: that the mean value of the envelopes always be zero and that the number of zero crossings and extremes be equal or varied by one. The residue is mapped after the discovery of the first IMF.

$$r(t) = x(t) - \mathrm{IMF}(t) \tag{3}$$

With $r(t)$ representing the residue.

2.3 Wavelet Transform

The wavelet transform can analyze a signal and identify any abrupt shift in the temporal and frequency features of the signal, making it a complicated analytical tool for managing non-stationary data [3]. The wavelet transform breaks down the input signal into a collection of basic building components called ψ_(a, b). These are later produced by dilating and translating the mother wavelet (Eq. 3), a well-localized analysis function with few oscillations [13, 14].

$$\psi_{a,b} = \frac{1}{\sqrt{a}}\psi\left(\frac{t-b}{a}\right) \tag{4}$$

With b being the displacement (translation), associated with the idea of time location, and a being the step size (dilation), related to the idea of frequency (scale).

The continuous wavelet transform (CWT) for a signal $x(t)$ is defined as [15]:

$$C(a, b) = \frac{1}{\sqrt{a}}\int_{-\infty}^{+\infty} x(t)\Psi^*\left(\frac{t-b}{a}\right)\mathrm{d}t \tag{5}$$

where Ψ^* is the mother wavelet's complex conjugate.

In our study, we utilized the discrete wavelet transform (DWT) because of its rapidity and ease of simulation. By consecutively transmitting a signal through low-pass and high-pass filters, we calculated the DWTs, yielding approximation coefficients and details. The application of half-band filters reduces the signal sample's dimension by a factor of 2 with each decomposition stage [16]. Figure 2 shows the corresponding Mallat's decomposition tree (Fig. 1).

In this context, x represents the signal, h stands for the high-pass filters, g represents the low-pass filters, $d1$ and $d2$ correspond to the details at the first and second levels, while $a1$ and $a2$ denote the first and second-level approximation coefficients.

In EEG signal processing, the DWT offers several advantages. Firstly, it enables the identification and extraction of different frequency components present in the signals. EEG signals often consist of a combination of various frequency components that correspond to different brain activities. By decomposing the signal using the DWT, these components can be separated and analyzed individually. This is particularly useful in identifying specific brainwave patterns, such as alpha, beta, theta, and delta waves, which are associated with different cognitive states. Secondly, the DWT allows for efficient denoising and artifact removal from EEG signals. The DWT can help distinguish between the desired brain activity and unwanted artifacts by decomposing the signal into different frequency bands. By discarding or suppressing the wavelet coefficients corresponding to noise, the DWT enables the extraction of cleaner EEG signals for further analysis.

There are several types of wavelets; in our investigation, we used the wavelets that are shown in Table 1. These wavelet functions were selected based on their proven

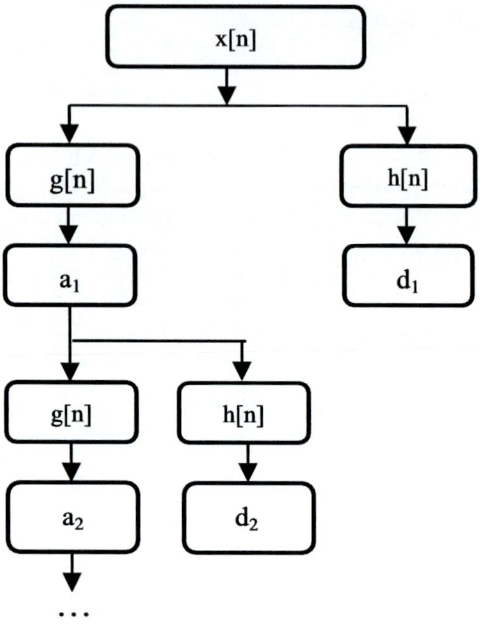

Fig. 1 DWT signal analysis [17]

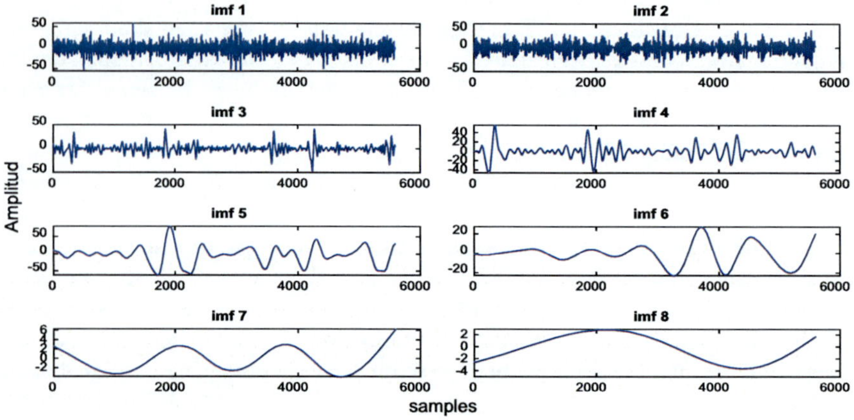

Fig. 2 IMFs component that are extracted from the EEG signals

effectiveness in similar applications and our prior research. These wavelet families possess desirable properties such as orthogonality, compact support, and good time–frequency localization, making them suitable for analyzing and processing EEG signals using the DWT. Their selection was guided by the goal of improving the accuracy and efficiency of the proposed method for EEG signal analysis.

Table 1 Wavelet types used in our research

Wavelets types	Wavelets
Daubechies	db1, db2, db3, db4, db5, db6, db7, db8, db9, db10, db11
Coiflets	coif1, coif2, coif3, coif4, coif5
Symlets	sym1, sym2, sym3, sym4, sym5, sym6, sym7, sym8, sym9, sym10, sym11

2.4 *Blind Source Separation*

A method called blind source separation (BSS) is used to decompose a combination of signals into its individual components. This is achieved by analyzing the statistical characteristics of the mixture, without any prior knowledge of the constituent signals or the mixing process. BSS approaches aim to identify unique statistical structures for each source signal, using statistical independence as a key characteristic. Higher-order statistics, sparsity, and non-Gaussianity are also frequently utilized. BSS can estimate the underlying sources by analyzing the mixture's statistical features. BSS is a powerful signal processing technique used in audio processing to separate mixed audio signals into individual sources, and it has many other applications, including in telecommunications, biological signal processing, and image and video processing.

$$X(P \times M) = A(S(N \times m)) \tag{6}$$

A represents the mixed model, X represents the mixed signals, and S is the source signal.

The most basic variation of the mixed model is the instant linear blend, where the observations are concurrently formed by combining the sources in a linear manner at every instance. In this case, a mixing matrix A with a size of (P, N) serves as the model for mixing.

$$X(P \times M) = A(P \times N)S(N \times m) \tag{7}$$

The BSS's objective is to identify an unmixing Matric B that can separate the input signal.

2.4.1 Independent Component Analysis

To resolve the issue of blind source separation (BSS), independent component analysis (ICA) is a highly effective approach that aims to separate blind sources [18, 19]. ICA works by discovering estimates of unknown sources through the assimilation of these sources to signals with maximal independence. Typically, this method involves two steps: whitening the mixed signals and utilizing higher-order statistics (HOS) to maximize independence.

HOS are a useful tool for achieving statistical independence between signals. By identifying patterns in the data that reflect the degree of independence between variables, HOS can help separate independent sources and reduce the impact of noise and outliers. Several approaches have been developed that use HOS, including maximum likelihood, contrast functions, and quadratic criteria. These approaches can be evaluated and compared using measures such as Kullback–Leibler divergence and mutual information. Overall, HOS provides a powerful framework for identifying and exploiting the complex statistical dependencies that exist in real-world data.

Infomax

Infomax techniques refer to learning algorithms that engage in an optimization procedure with the objective of enhancing entropy to its maximum extent. The underlying rationale is based on the concept that if the components within signal vector Y are self-contained, then the anticipation of combined entropy achieves its highest value [13]. The Infomax methodology is frequently implemented through a neural network architecture known as "feedforward network with lateral inhibition." The subsequent mathematical expression is employed for training the network [3]:

$$J(t+1) = J(t) + n(t)(I - ha(Y)Y^T)B(t) \tag{8}$$

The initial value of J is frequently a random matrix, where h is a function related to the distribution structure and $n(t)$ is a learning rate function.

2.4.2 Second-Order Statistics

Multivariate time series data can have hidden structures that can be found using second-order statistics approaches [8]. Second-order statistics refers to statistical measurements that may be derived from the covariance matrix of the observed data. These techniques are founded on the joint diagonalization of any quantity of random time-delayed covariance matrices calculated from X [20].

SOBI

The SOBI method's fundamental premise is to distinguish the sources in a combination using their temporal structure. It is predicted that there is no correlation between the sources and that each source has a unique time structure. The cross-correlation matrix of the combined signals is first calculated (Eq. 8).

$$Rx(\tau) = E\{x(t)x(t+\tau)^T\} \tag{9}$$

This matrix is then subjected to a process known as eigenvalue decomposition. This results in a collection of eigenvectors that serve as the foundation for the divided sources. In a new coordinate system with uncorrelated sources, the combined signals are then projected using the eigenvectors.

SOBI-RO

The SOBI technique has been modified to perform better in the face of outliers or noise. This update is known as second-order blind identification with robust orthogonalization. It works by jointly diagonalizing a set of randomly selected time-delayed covariance matrices of the recorded signals. Before creating time-delayed covariance matrices for mixed signals, the following formula is a robust orthogonalization formulation:

$$\acute{x} = Ox \tag{10}$$

The next stage is to calculate the eigenvalue decomposition of the covariance matrix of the whitened data in order to determine the mixing matrix.

Amuse

The algorithm for multiple unknown signals extraction (AMUSE) is a widely used second-order statistics blind source separation (BSS) method that was introduced in 1991. The first step in the algorithm involves centering and whitening the mixed signal matrix to obtain $X_b(t)$.

$$X_b'(t) = BX \tag{11}$$

where B is the whitening matrix. Next, the time-lagged covariance matrix is decomposed into its eigenvectors and eigenvalues using an eigenvalue decomposition. Assuming V is the matrix of eigenvectors, the predicted mixing matrix J can be determined as:

$$J'(t) = B^{-1}V \tag{12}$$

This process is useful for estimating the individual sources from the mixed signals without prior knowledge of the sources [21].

2.5 Proposed Method

This study proposes two methods that combine EMD, DWT, and BSS techniques. The first method involves decomposing each EEG channel into a set of IMFs using the EMD algorithm (Fig. 2). Next, we apply the BSS method to each set of IMFs to separate them into independent components (ICs) (Fig. 3). Finally, we use the Spearman correlation coefficient (SCC) to assess the degree of similarity between the ICs and the IMFs retrieved from the VEOG and the HEOG.

For the second method, we used the DWT to extract the approximation coefficient at a specific level from the IMF data (Fig. 4). We then applied the BSS method to these data to decompose them into ICs (Fig. 5). Similar to the first method, we used the SCC to compute the similarity between the created ICs and the approximation coefficient extracted from the HEOG's and VEOG's IMFs.

A statistical metric called the Spearman correlation coefficient evaluates the direction and strength of the relationship between two ranking variables. The value is

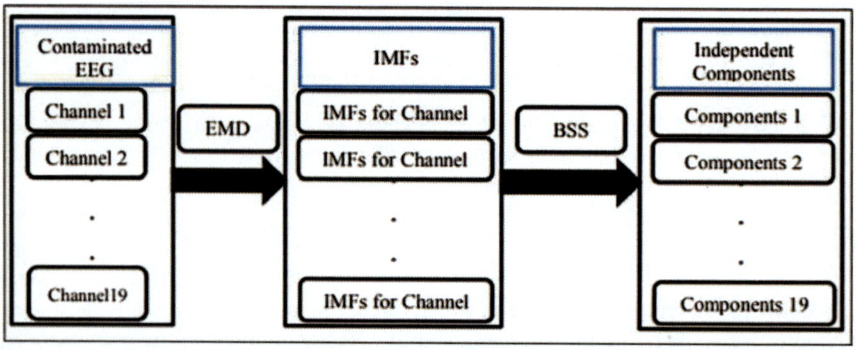

Fig. 3 First method's steps. The illustration includes two steps: (1) EEG decomposition using EMD into IMFs; (2) Application of the five BSS methods to the IMFs, and then computing the correlation coefficient between them and the IMFs extracted from the VEOG and the HEOG

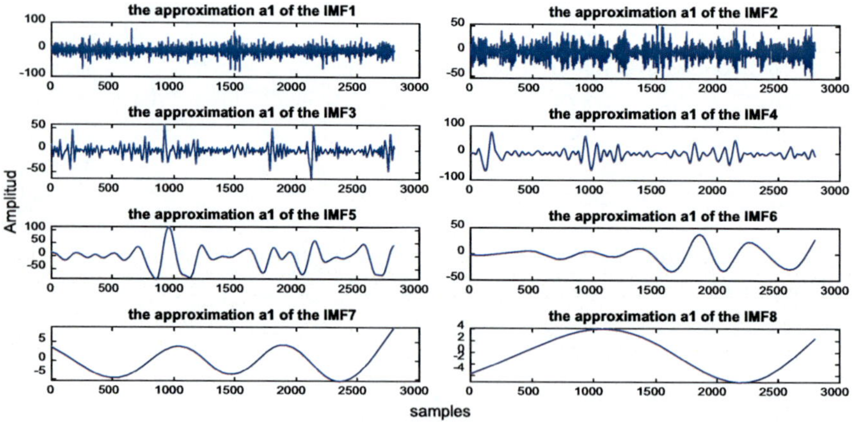

Fig. 4 Approximation coefficient a1 extracted from the EEG's IMFs using the wavelet symlet 1

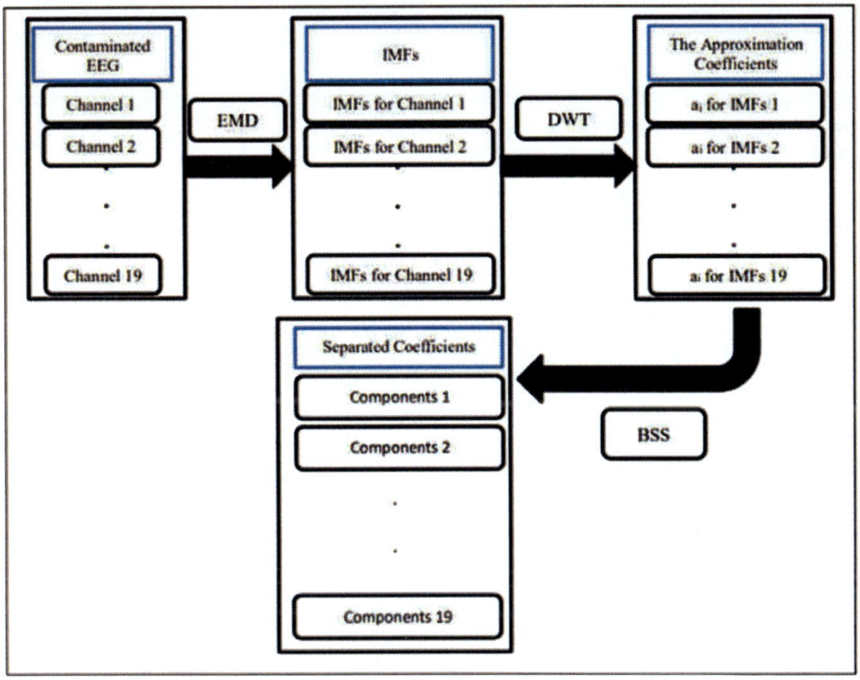

Fig. 5 Second method's step, in this method we add a block of DWT to extract the approximation coefficients from the IMFs and then we try to compute the correlation coefficient using these separated components and the approximation coefficient extracted from the VEOG's and the HEOG's IMFs

between − 1 and + 1. Whereas a number of − 1 denote a perfect negative correlation, a value of + 1 denotes a perfect positive correlation. No correlation exists between the variables, as shown by a value of 0.

The following formula may be used to calculate the Spearman correlation coefficient [22]:

$$\mathrm{SCC} = 1 - 6\sum \frac{r^2}{I\left(I^2 - 1\right)} \tag{13}$$

where r is the square of the difference in the two coordinates' rankings for each data pair, and I is the number of data pairs.

3 Results

In this study, we introduce two techniques that integrate BSS, DWT, and EMD algorithms to segregate the intrinsic mode functions (IMFs) obtained from individual EEG signals, along with the approximated coefficients derived from these IMFs. Our approach involves the assessment of BSS methods applied to both EEG data's IMFs and approximation coefficients, utilizing the previously mentioned SCC metric.

For the initial method, we gauge the SCC values between the extracted components and the IMFs sourced from VEOG and HEOG. Subsequently, we identify the mean values across 54 samples by opting for the larger SCC value.

In the second approach, we calculate the SCC between the isolated components and the approximation coefficients taken from HEOG and VEOG's IMFs, with the inclusion of the DWT block. The proposed technique is implemented using MATLAB 2019a in conjunction with EEGLAB, an interactive toolkit within MATLAB designed for processing EEG data.

EEG signal analysis can benefit from various techniques to extract meaningful information. EMD utilizes a data-driven approach to break down EEG signals into IMFs. These IMFs capture the different oscillatory modes at various scales present in the signal. Conversely, DWT offers a multi-resolution analysis by decomposing EEG signals into different frequency bands. This allows for the identification of specific frequency components associated with different cognitive states or neurological disorders. DWT provides a localized representation of the signal in both the time and frequency domains, aiding in the characterization and classification of EEG patterns. To address the issue of artifacts contaminating EEG signals, BSS techniques such as ICA are employed. These methods aim to separate EEG signals into their underlying source components, isolating and extracting various artifacts like eye blinks, muscle activities, and external interference. The removal of these unwanted components enhances the interpretability of EEG signals.

Combining EMD, DWT, and BSS techniques provides researchers with the complementary strengths of each method. EMD offers flexible and adaptive decomposition, DWT enables frequency-domain analysis and event localization, and BSS aids in signal separation and denoising. This integrated approach facilitates a comprehensive understanding of EEG signals, allowing for accurate interpretation of brain activity, detection of abnormalities, and investigation of cognitive processes. This framework has significant potential for advancing EEG signal processing and its applications in neuroscience, clinical diagnostics, and brain-computer interfaces.

The main objective of our work is to evaluate and assess the effectiveness of BSS methods when applied to the IMFs representing the underlying oscillatory modes of different scales in EEG signals. Specifically, we aim to test these methods by applying them to the approximation coefficients that represent the high-frequency components of the IMFs.

For the first method, the best SCC result is obtained using the ACSOBI-RO method, according to the measurement computed using the VEOG, which is equal to

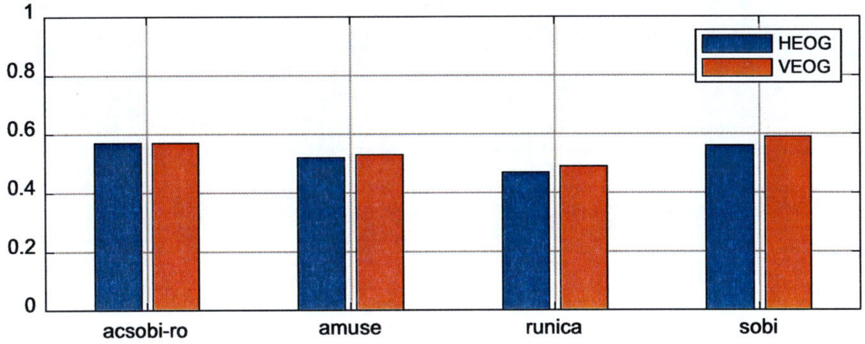

Fig. 6 Results of the first method (EMD-BSS)

0.57 and the higher value obtained using HEOG is given by the SOBI method which is equal to 0.59 (Fig. 6).

Figures 7, 8, 9 and 10 present the SCC results estimated using the HEOG data for each BSS method.

For the ACSOBI-RO method, the higher value of SCC is computed using the wavelet coif4 at level 3 which is equal to 0.59. On the other side, the SOBI method produced a higher value of SCC using the wavelet db1 at level 4 equal to 0.61. The best value for the RUNICA method is equal to 0.56, which is given using the wavelet db8 at level 4. And for the AMUSE method, the greatest value of SCC is 0.54 using sym3 at level4.

For the SCC values calculated using the VEOG signals, Figs. 11, 12, 13 and 14 present our results. The higher value of SCC for the ACSOBI-RO method is calculated using the wavelet coif4 at level 3, which is equivalent to 0.59. The SOBI technique, on the other hand, generated a higher SCC value of 0.60 using the wavelet db1 at level 4. The optimum result for the RUNICA method is 0.57, obtained using

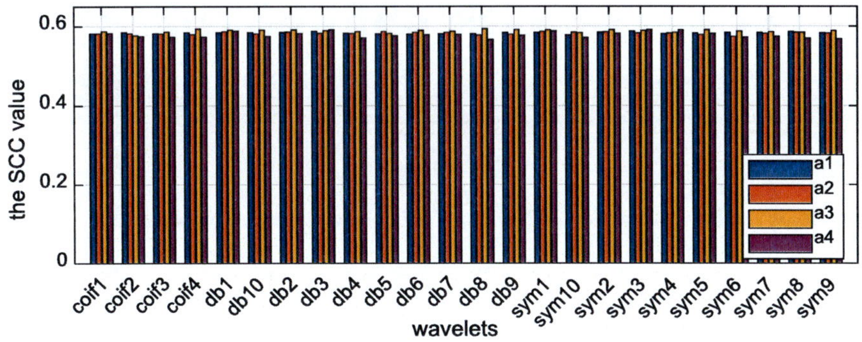

Fig. 7 SCC value calculated using the HEOG data and the separated components using the ACSOBIRO

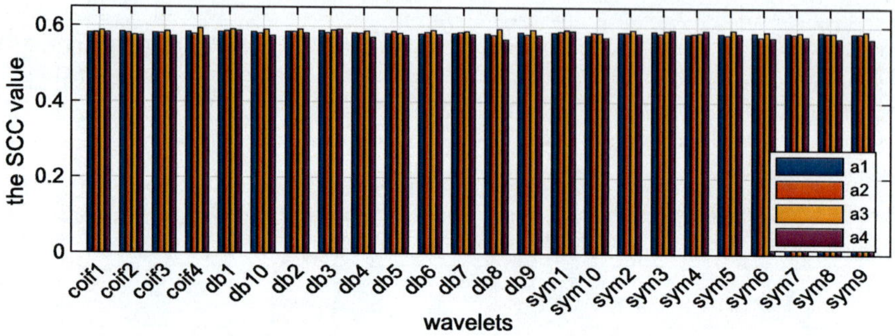

Fig. 8 SCC value derived from HEOG data and the separated components determined from SOBI

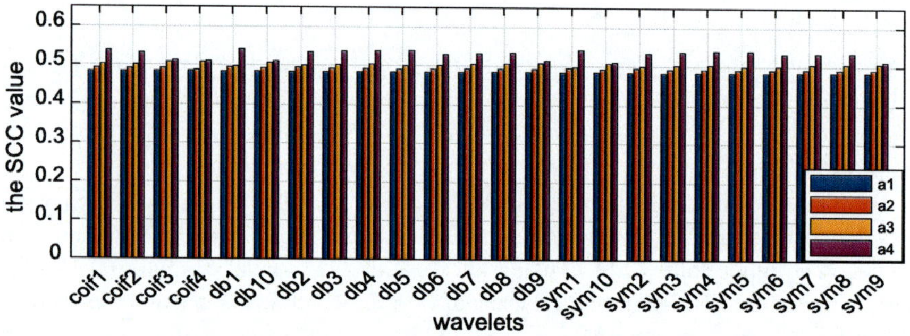

Fig. 9 SCC value obtained from HEOG data and the RUNICA separated components

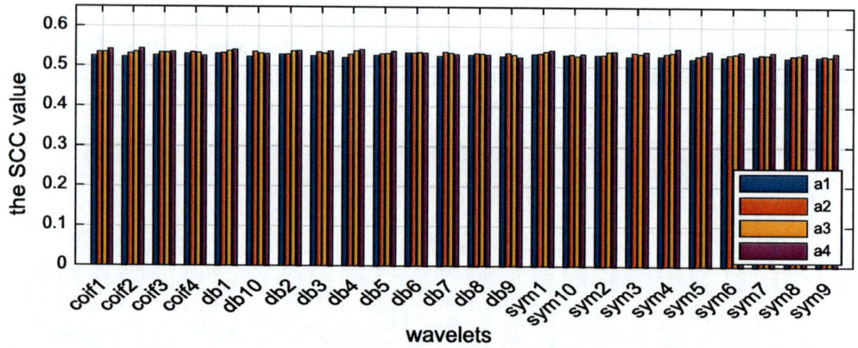

Fig. 10 SCC value obtained from HEOG data and the amuse separated components

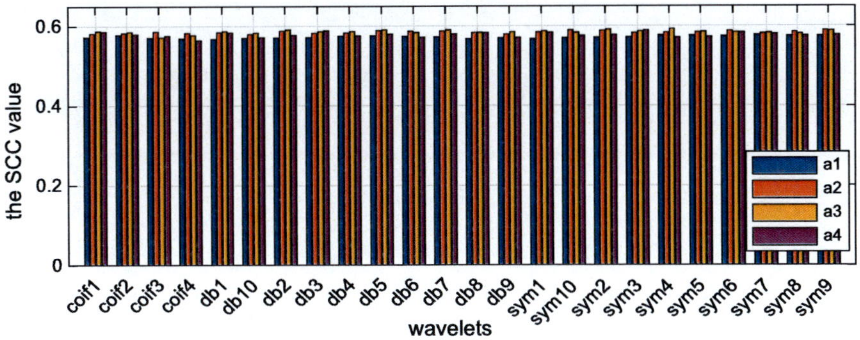

Fig. 11 SCC results obtained using the VEOG signals and the ACSOBIRO results

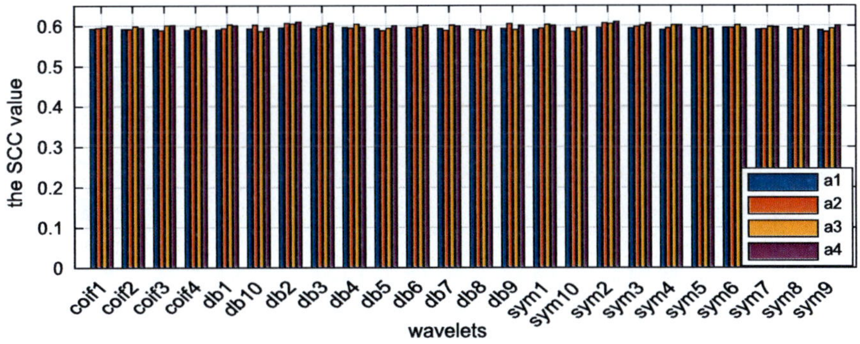

Fig. 12 SOBI results using the VEOG data

the wavelet db8 al level 4. The highest SCC number for the AMUSE method is 0.55 when using sym3 at level4.

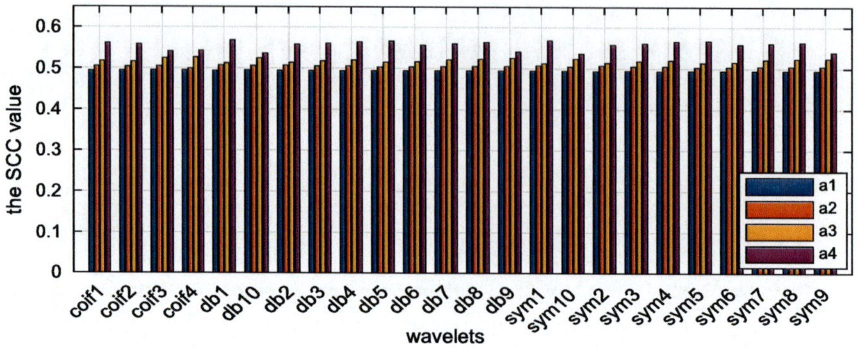

Fig. 13 SCC values calculated using the component extracted using RUNICA and the VEOG signals

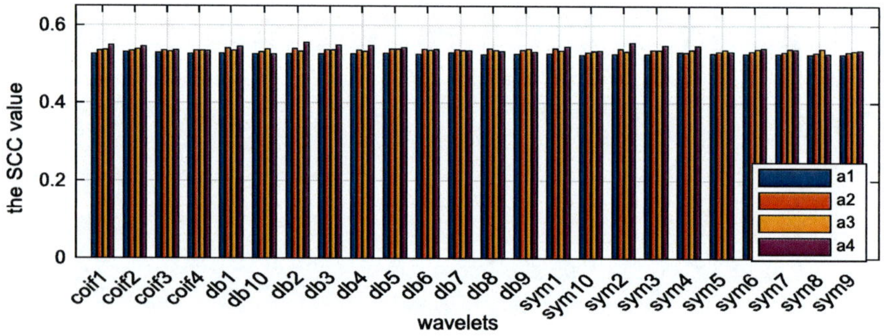

Fig. 14 SCC values derived from the amuse component and the VEOG signals

4 Conclusion

This study presented a novel method by combining the BSS technique with the EMD and DWT method. Further, the performance of the separation method is tested and evaluated after applying them to the data extracted from the EEG signals by using the EMD and the DWT methods.

The EMD method processes a non-stationary signal by decomposing them into IMFs, which are a representation of oscillatory components with different time scales. That makes the idea of separating these kinds of data interesting. As a result of the first approach, the best-performing BSS method is ACSOBI-RO with a value of 0.57. The DWT aims to decompose a signal into two types of coefficients, and one of these coefficients is the approximation coefficient which represents the low-frequency component of the related signal. In terms of this letter, we extracted the approximation coefficient from the EEG's IMFs, and then, we fed them into the

BSS methods, in order to test the separation approaches by applying them to these extracted data.

The best-performing algorithm based on this second method is SOBI, which gives the greatest value by applying the wavelet db1 at level 4, which is equal to 0.61 using the HEOG signals. And by applying the wavelet db2 at level 4, which is equal to 0.60 using the VEOG signal. As a conclusion of the second method, the DWT step has successfully increased the SCC result.

Acknowledgements This paper and the study it supports would not have been possible without the incredible help of my supervisors, Mm. M.MIYARA, Mr. T. Drissi Belhousine, and Mr. B. Nsiri. Their enthusiasm, knowledge, and careful attention to detail have inspired me and kept my work on schedule.

References

1. Mannan MM, Kamran MA, Jeong MY (2018) Identification and removal of physiological artifacts from electroencephalogram signals: a review. IEEE Access 6:30630–30652. https://doi.org/10.1109/ACCESS.2018.2842082
2. Mowla MR, Ng S-C, Zilany MSA et al (2015) Artifacts-matched blind source separation and wavelet transform for multichannel EEG denoising. Biomed Signal Process Control 22:111–118
3. Sahonero-Alvarez G, Calderon H (2017) A comparison of SOBI, FastICA, JADE and Infomax algorithms. In: Proceedings of the 8th international multi-conference on complexity, informatics and cybernetics, 17–22
4. Safie SI, Rahim R (2019) Quality assessment on muscle locations for speech representation. Indonesian J Electr Eng Comput Sci 17:957–967. https://doi.org/10.11591/ijeecs.v17.i2.pp957-967
5. Jamil Z, Jamil A, Majid M (2021) Artifact removal from EEG signals recorded in a non-restricted environment. Biocybern Biomed Eng 41:503–515.https://doi.org/10.1016/j.bbe.2021.03.009
6. Zhao M, Bonassi G, Guarnieri R et al (2021) A multi-step blind source separation approach for the attenuation of artifacts in mobile high-density electroencephalography data. J Neural Eng 18(6):066041
7. Chen X, Xu X, Liu A et al (2017) The use of multivariate EMD and CCA for denoising muscle artifacts from few-channel EEG recordings. IEEE Trans Instrum Measur 67:359–370.https://doi.org/10.1109/TIM.2017.2759398
8. Paradeshi KP, Kolekar UD (2017) Removal of ocular artifacts from multichannel EEG signal using wavelet enhanced ICA. In: 2017 international conference on energy, communication, data analytics and soft computing (ICECDS). IEEE, 383–387. https://doi.org/10.1109/ICECDS.2017.8390150
9. Klados MA, Bamidis PD (2016) A semi-simulated EEG/EOG dataset for the comparison of EOG artifact rejection techniques. Data Brief 8.https://doi.org/10.1016/j.dib.2016.06.032
10. Kopsinis Y, Mclaughlin S (2009) Development of EMD-based denoising methods inspired by wavelet thresholding. IEEE Trans Signal Process 57:1351–1362.https://doi.org/10.1109/TSP.2009.2013885
11. Echeverria JC, Crowe JA, Woolfson MS et al (2001) Application of empirical mode decomposition to heart rate variability analysis. Med Biol Eng Comput 39:471–479.https://doi.org/10.1007/BF02345370

12. Echeverría JC, Crowe JA, Woolfson MS et al (2001) Application of empirical mode decomposition to heart rate variability analysis. Med Biol Eng Comput 39:471–479. https://doi.org/10.1007/BF02345370
13. Toulni Y, Belhoussine Drissi T, Nsiri B (2021) ECG signal diagnosis using discrete wavelet transform and K-nearest neighbor classifier. ACM Int Conf Proc Ser: 3457628. https://doi.org/10.1145/3454127.3457628
14. Toulni Y, Belhoussine Drissi T, Nsiri B (2021) Electrocardiogram signals classification using discrete wavelet transform and support vector machine classifier. IAES Int J Artif Intell 10:960–970.https://doi.org/10.11591/ijai.v10.i4.pp960-970
15. Drissi TB, Zayrit S, Nsiri B et al (2019) Diagnosis of Parkinson's disease based on wavelet transform and Mel frequency cepstral coefficients. Int J Adv Comput Sci Appl 10
16. Chaovalit P, Gangopadhyay A, Karabatis G et al (2011) Discrete wavelet transform-based time series analysis and mining. ACM Comput Surveys (CSUR) 43:1–37.https://doi.org/10.1145/1883612.1883613
17. Daqrouq, K, Hilal TA, Sherif M et al (2009) Speaker identification system using wavelet transform and neural network. In: International conference on advances in computational tools for engineering applications. IEEE, 559–564. https://doi.org/10.1109/ACTEA.2009.5227953
18. Noorbasha SK, Sudha GF (2021) Removal of EOG artifacts and separation of different cerebral activity components from single channel EEG—an efficient approach combining SSA–ICA with wavelet thresholding for BCI applications. Biomed Signal Process Control 63:102168. https://doi.org/10.1016/j.bspc.2020.102168
19. Xi J, Chicharo JF, Tsoi AC et al (2000) On the INFOMAX algorithm for blind signal separation. In: WCC 2000-ICSP 2000. 2000 5th international conference on signal processing proceedings. 16th world computer congress 2000. IEEE, 425–428. https://doi.org/10.1109/ICOSP.2000.894523
20. Langlois D, Chartier S, Gosselin D (2010) An introduction to independent component analysis: InfoMax and FastICA algorithms. Tutor Quantitative Methods Psychol 6:31–38
21. Tong L, Soon VC, Huang YF et al (1990) AMUSE: a new blind identification algorithm. In: IEEE international symposium on circuits and systems. IEEE, pp 1784–1787
22. Myers L, Sirois MJ (2004) Spearman correlation coefficients, differences between. Encyclopedia of statistical sciences, vol 12

Image Extraction Approaches for Density Count Measurement in Obstruction Renography Using Radiotracer 99mTc-DTPA

Pradnya N. Gokhale, Babasaheb R. Patil, and Sameer Joshi

Abstract This research pursued to define quantitative amounts of renal radiotracer's density inside the kidneys in terms of renal obstruction levels by applying image processing transforms FFT, wavelet, and Harr-like extraction and finding correlation among density measured by transforms and radioactive counts by gamma machines. In this retrospective study, we considered 140 renal scintigraphy scans from which 110 were detected with renal obstruction (hydronephrosis) disorder ranging from moderate to severe and 30 cases were found to be normal functioning kidneys. Of 110 cases, 64 cases were diagnosed with left kidney hydronephrosis (LK-HL) with an M/F ratio of 43/21, and 46 cases were diagnosed with right kidney hydronephrosis (RK-HR) with an M/F ratio of 23/23. The mean age of the selected case is 25.65 ± 24.58 years. Three image-enhanced transforms, namely FFT, wavelet, and Harr-like extraction, are applied to renal scans to detect the density of darkness inside the kidney. Correlation among density-measured count and scintigraphy-measured radioactive count by Spearman's correlation method. There was a strong positive correlation between FFT-measured density inside the kidney and dynamic renal scintigraphy using ^{99m}Tc-DTPA measured radioactive counts in the case of both left hydronephrosis kidney and right hydronephrosis kidneys ($p = 0.81$ and $p = 0.80$, respectively). Also found a moderately positive correlation between wavelet

Authors' Contributions The study was designed by Pradnya Gokhale. Material preparation and data collection were performed by Pradnya Gokhale and Prof. Dr. Babasaheb Patil. The data analysis was performed by Pradnya Gokhale. The first draft of the manuscript was written by Pradnya Gokhale, and all authors commented on previous versions of the manuscript. Prof. Dr. Babasaheb Patil played a crucial role in developing the methodology for the study, and Dr. Sameer Joshi provided insightful feedback on the analysis carried out. All authors read and approved the final manuscript.

P. N. Gokhale (✉)
Sardar Patel College of Engineering, University of Mumbai, Mumbai, Maharashtra, India
e-mail: pradnya.gokhale@spce.ac.in

B. R. Patil
VIMEET, University of Mumbai, Mumbai, Maharashtra, India

S. Joshi
Miraj Nuclear Medicine AND Molecular Imaging Center, Miraj, Maharashtra, India

S. Shakya et al. (eds.), *Fourth International Conference on Image Processing and Capsule Networks*, Lecture Notes in Networks and Systems 798,
https://doi.org/10.1007/978-981-99-7093-3_4

and Harr-like transform's measures and scintigraphy measures. In conclusion, this study showed that the application of the FFT object detection technique can be a better option to count the density of renal scan in terms of obstruction level as compared to Harr-like and wavelet transform methods, and these quantitative measures of this density can be considered to define the obstruction/hydronephrosis levels instead of reporting moderate level to severe levels of renal obstruction disorder.

Keywords Renal scintigraphy · Obstruction · Feature extraction · Radiotracer · Hydronephrosis · Correlation coefficient

1 Introduction

Renal scintigraphy images provide information regarding renal radiotracer excretion activity that is regularly categorized by ample noise and low contrast and resolution features while using ^{99m}Tc-DTPA, and it provides observations of the regions of interest in a very complicated way which finds to be difficult to analyze in terms of diagnosing the level of obstruction, i.e., hydronephrosis [1]. Before the scintigraphy procedure, patients were hydrated for 30–60 min. They were then positioned supine to a gamma camera, and the acquisition process was initiated. The radiopharmaceutical DTPA was then injected as a bolus dose of 5–10 mCi for DTPA scintigraphy, depending on the patient's age, gender, weight, medical history, and type of renal parameters. Clinical experts decide the appropriate radionuclide dose based on the patient's characteristics. This renal imaging technique thus provides information on peak time and half time of radiotracer's accumulation inside the kidney, glomerular filtration rate (GFR), and radioactive counts on the bases of the tracer's excretion time activity curve stimulated after drawing the regions of interest around the kidneys [2–4]. In renal scintigraphy, motion detection is restricted due to the high level of noise and fading of organ boundaries, and it represents inherent image characteristics. These restrictions can cause a reduction in the precision of motion detection in cases especially radiotracer's transit process through renal parenchyma to the cortex. Though this imaging provides a diagnosis of renal obstruction in terms of hydronephrosis, it fails to count the levels of renal obstruction in terms of quantity of figures which will directly help the nephrologist to understand the status of obstruction. This research study utilizes 110 renal scan samples which were processed using image edge extraction transforms to quantify the density count of radiotracer inside the kidney.

1.1 Characteristics of DTPA in Renal Imaging

1. **Glomerular filtration**: DTPA is mainly cleared by glomerular filtration which means it is filtered through the glomeruli in the kidneys and evacuated in the

urine. It provides information about the glomerular filtration rate and the overall function of the parenchyma.

2. **Static and dynamic Imaging**: For static and dynamic renal imaging, DTPA can be used, where in the static method images are acquired at specific time points after the injection and in the dynamic method sequential images are acquired over several minutes. In static mode, it allows for the evaluation of renal morphology like renal size, dimensions, and shape while in dynamic mode it provides evaluation of renal blood flow and tubular function [5].

2 Materials and Methods

A. Sample Selection Details

Patients aged less than 10 years and those aged more than 65 years were excluded from the study. Renal scans were selected only if they showed gradual tracer accumulation into the dilated pelvis without significant excretion after diuretic administration. The study did not include cases with horseshoe-shaped kidneys, renal transplants, donors, or single-functioning kidneys. As shown in Table 1, for this study, 64 cases of left kidney hydronephrosis (M/Fe: 43/21) and 46 cases of hydronephrosis (M/Fe: 23/23) were considered. The mean years of age of the selected case are 25.65 ± 24.58 years.

B. Renal Scintigraphy Image Processing

By using suitable color measure, a parametric appearance of each kidney can be acquired; permitting the pixel-by-pixel true mean transit time in the cortical zone to be read immediately and the mean transit time (MTT) image can be accommodated in renal image processing easily [6]. Estimation of absolute renal uptake of radiotracer and its evacuation rate through the kidney can be achieved by using iterative deconvolution and Rutland–Patlak methods in which reported matrix inversion deconvolution values were significantly higher than Rutland–Patlak values [5, 7].

Initial recognition of clinically substantial renal hydronephrosis affected by ureteropelvic obstruction was achieved by using automated signal analysis and a machine learning framework. But the value of the signal investigation depends extremely on the quality of the data obtained by skilled nuclear medicine physicians or technologists' accuracy level of earlier detection of obstructed cases effects [8].

Table 1 Sample selection data

Parameter	Male	Female	Total	Sample types
Left hydronephrosis (HL)	43	21	64	Renal hydronephrosis (renal obstruction) samples
Right hydronephrosis (HR)	23	23	46	
Normal kidney	15	15	30	
Total	81	59	140	

Renal scintigraphy scans are regularly categorized by much noise and low contrast and resolution while using ^{99m}Tc-DTPA, which observes the regions of interest very complicated and difficult to analyze in terms of diagnosing the level of obstruction, i.e., hydronephrosis [9]. Here, each digital image is a set of coordinate values "*nij*" which corresponds to the number of counts in a set of pixels identified by (*i*, *j*). These pixels are signified in matrix form by {(*i*, *j*, *nij*)}. Drawing the defined region of interest (ROI) around the high-density darkened area inside the kidneys and counting the activity on all the selected pixels for *P*-recorded images, not only allows the tracing over radioactivity period but also the evolution of the activity of the structure analogous to the selected pixels. Thus, access to a dynamic study of the functioning of the renal system is achieved. Data is then articulated in a matrix pattern by inserting the time parameter *tk* which designates the development of images: {(*i*, *j*, *nij*), *tk*} [10].

The study incorporates three image enhancement techniques, namely wavelet transforms, FFT transform, and Harr-like extraction, to analyze each sample obtained from renal scans. The objective is to quantify the darkness within the kidney region by measuring the density of the last frame of each scan, specifically the 30th image in the case of SIEMENS gamma scans.

These image enhancement techniques aim to improve the visibility and clarity of the kidney region, allowing for more accurate measurements of darkness. The density of the last frame is considered a representative measure of the darkness within the kidney. To clinically validate these results, they are compared with the amount measured by the Gamma machine's radioactive kidney counts. This comparison ensures that the quantification of darkness obtained through image enhancement techniques aligns with the measurements obtained directly from the Gamma machine, which is a reliable and established method.

Thus, in this study, as shown in Fig. 1, image processing steps are designed and followed to find the best of transform which will approximately match to the clinical report and will provide quantity in terms of density of the obstructed part of the kidney.

C. Significance of image enhancement transforms

In the study, the density counts obtained through the application of FFT, wavelet, and Harr-like extraction transforms indicate the presence of radioactive tracer inside the kidney. Normally, when the renal system is functioning properly, the tracer is

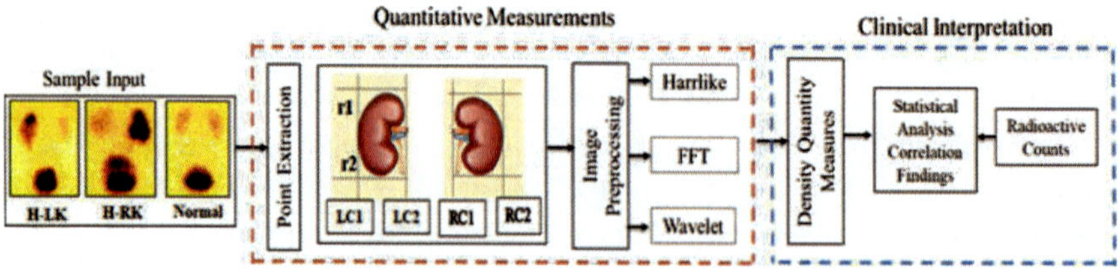

Fig. 1 Renal scintigraphy scan process

able to excrete out of the kidney. However, in the case of hydronephrosis or renal obstruction, the tracer is unable to pass through the obstructed area and remains inside the kidney.

To measure the density of darkness inside the kidney, image processing techniques are applied to enhance and extract the kidney structure from the renal scan images. The transforms, namely FFT, wavelet, and Harr-like extraction, are utilized to improve the visibility and clarity of the kidney region. By enhancing and extracting the kidney structure, the density of darkness, which represents the presence of the radioactive tracer, can be measured.

These image processing techniques provide a quantitative measure of the density of darkness inside the kidney, indicating the extent of renal obstruction or hydronephrosis.

Feature extraction is a process that involves extracting relevant information or characteristics from raw data, in this case, images. The main objective of feature extraction is to represent the raw image in a reduced form while preserving important information. By reducing the dimensionality of the data, feature extraction aims to simplify the representation of the image while retaining its discriminative properties.

One of the primary motivations for feature extraction is to reduce the number of false findings or irrelevant information. By extracting and focusing on the most relevant features, the likelihood of false detections or misleading information can be minimized.

In the context of image feature extraction, the goal is to transform the low-dimensional space of an image sample into a higher-dimensional space where meaningful features can be identified and utilized. These features serve as the fundamental characteristics that can be used to identify and distinguish one image from another [7].

Thus, the better features can entirely recognize the object (here in the case of darkness inside the kidney) which is required to be identified. In this study, three effective feature extraction methods, namely wavelet transform, FFT transform, and Harr-like extraction transform, are applied to renal scan to extract the structures so that the design of classifiers can be specific and more accurate to define the density of darkness [11].

The flowchart mentioned in Fig. 2 represents the steps involved in quantifying the density count of darkness. These steps likely include preprocessing the renal scan images, applying the feature extraction methods, and using the extracted features to train and evaluate classifiers. The flowchart provides a visual representation of the overall process followed in the study.

Renal scan frame of 30 images/sample is selected as an input image. As image sizing for every sample is different, all images were resized with matrix 330 $\times$ 1100 scale. Renal scan is passed through RGB to gray conversion to enhance the required pixels. Point extraction is performed on each of left kidney and right kidney of all cases. Object detection is achieved by applying the mentioned three image enhancement techniques. The dark area inside the kidney is counted in terms of the density of darkness.

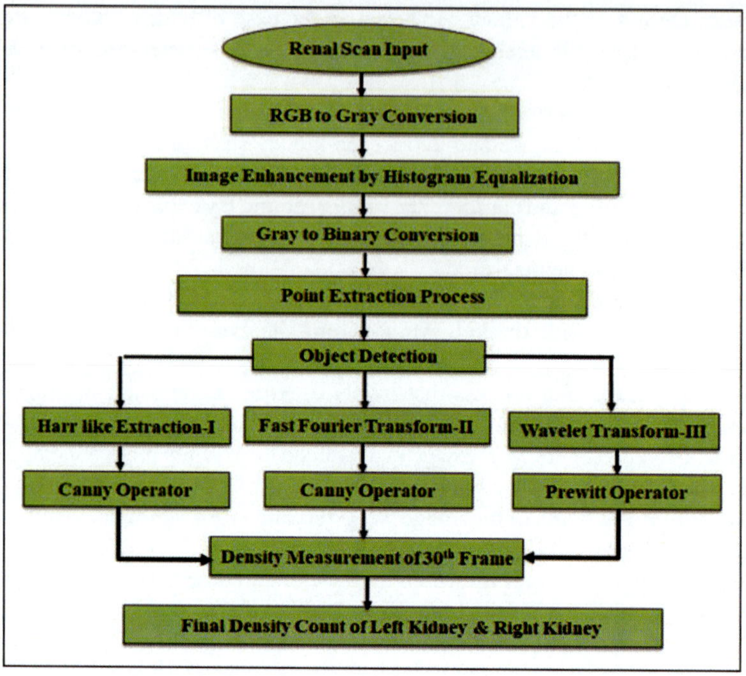

Fig. 2 Renal image processing flowchart

As shown in Fig. 3. for every kidney pair scan edge identification is done using Canny operator for Harr-like and FFT methods and Prewitt operator for wavelet transform by defining edges with row1, row 2 left kidney column LC1 and LC2 and right kidney column RC1, RC2.Three edge detection transforms, namely wavelet, FFT, and Harr-like extraction, were applied on every 30th last frame of every scan, and three density counts were identified using these transform techniques. The wavelet transforms are mathematical tools used for extracting information from images. These transforms provide both frequency and spatial localization of image features, making them useful for analyzing images in both the frequency and spatial domains.

Fig. 3 Point extraction of kidney pair

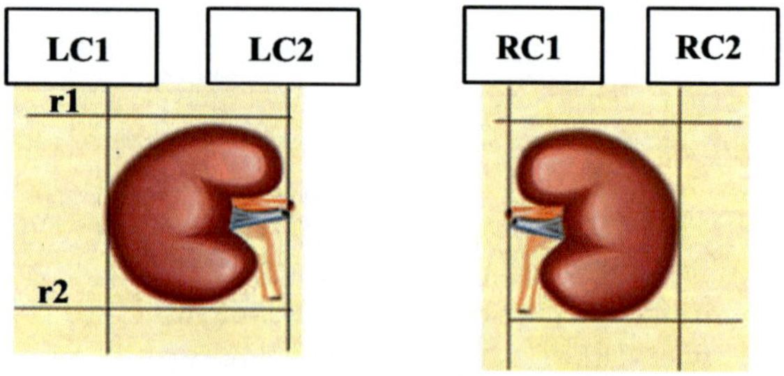

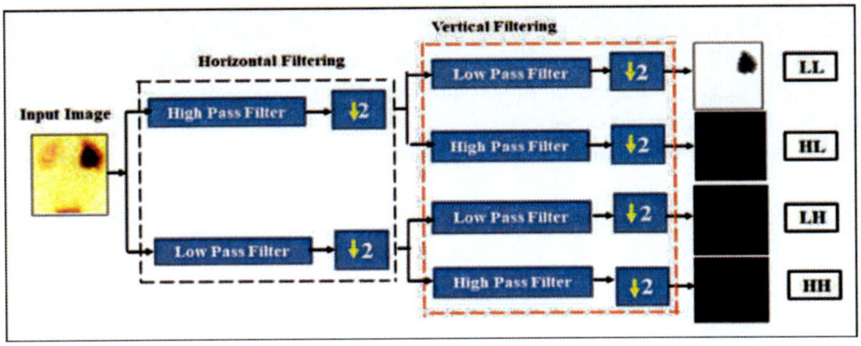

Fig. 4 Image extraction method using wavelet transform

The wavelet transform decomposes an image into different frequency components, allowing the extraction of fine details at different scales. It achieves this by using a set of wavelet functions that are scaled and shifted to analyze different parts of the image. The transform captures both low-frequency components, which represent global image features, and high-frequency components, which represent local image details.

One advantage of wavelet-based transforms is their ability to provide temporal resolution, meaning they can capture changes in frequency and spatial content over time. This temporal resolution makes wavelet transforms suitable for analyzing dynamic or time-varying images, as they can capture variations in both frequency and spatial information [12]. Figure 4 shows wavelets various filters and their concern results.

A Haar-like feature is defined by dividing a rectangular region of an image into sub-rectangles and computing the difference between the sum of pixel intensities in the white sub-rectangles and the sum of pixel intensities in the black sub-rectangles. These features are typically represented as adjacent black and white rectangles. The Haar-like features capture local intensity variations in an image and can be used to detect edges, lines, corners, and other simple patterns. They are computationally efficient to compute and are often used as a preliminary feature extraction step in object detection algorithms, such as the Viola-Jones algorithm for face detection.

By using Haar-like features, the algorithm can learn to distinguish between different patterns or objects based on the variations in intensity captured by these features. These features provide a simple yet effective way to represent and detect visual patterns in images.

Harr-like transform can be stated in matrix form which is expressed as:

$$T = HFH^{\tau}$$

Here, F is an "$N \times N$" image matrix, H is an "$N \times N$" Harr-like transformation matrix, and T is the subsequent "$N \times N$" transform [13].

The fast Fourier transform-low pass (FFT) is a mathematical field. In this study, FFT-HP is used for renal image filtering, reconstruction, and density count measurements. It is an important image processing tool that is used to decompose an image into its sine and cosine components.

While dealing with the derivation of the FFT it is expressed as:

$$F(u) = \sum_{X=0}^{M-1} f(x) W_M^{ux}$$

where, $u = 0, 1, 2, M - 1$, where

$$W_{M=e^{-2j\pi/M}}$$

And M is assumed to be in the form of $M = 2^n$ with "n" positive integer. The transformation output symbolizes the image in the frequency domain, and the input image is the spatial domain equivalent. In the frequency domain image, each point denotes a particular frequency contained in the spatial domain image [14].

The density count measured by these transforms was compared and analyzed to find one of transform performs better in terms of precise and approximate density count close to the radioactive count provided by the renal gamma scan.

3 Results

Sixty-four patients with left kidney hydronephrosis cases (M/Fe: 43/21) and forty-six patients with right kidney hydronephrosis cases (M/Fe: 23/23) were considered in this study. Mean and standard deviation of scintigraphy-measured radioactive count.

3.1 *Mean and Standard Deviation of Transforms*

The mean is a statistical quantity that signifies the average of a set of values. It is calculated by summing up all the values in the set and then dividing the sum by the total count of values. The mean provides a central value that represents the typical value of the dataset.

While standard deviation is a statistical measure that quantifies the amount of variation or dispersion in a set of numbers. It measures how much the individual data points deviate from the mean.

A higher standard deviation indicates that the data points are more spread out from the mean, while a lower standard deviation indicates that the data points are closer to the mean. Both the mean and standard deviation are important statistical

Table 2 Mean ± S.D. of radioactive counts of HL, HR, and normal cases by scintigraphy scan

Type of category		Mean value ± S.D.
DTPA HL	Left kidney	66.76 ± 8.25
	Right kidney	33.24 ± 8.25
DTPA HR	Left kidney	36.66 ± 10.44
	Right kidney	65.34 ± 10.44
Normal	Left kidney	48.43 ± 4.59
	Right kidney	51.57 ± 4.59

Table 3 Mean ± S.D. of transforms detected density counts of HL, HR, and normal cases

Type of category		FFT	Wavelet	Harr
		Mean ± S.D.	Mean ± S.D.	Mean ± S.D.
DTPA HL	LK	68.28 ± 23.11	65.94 ± 22.30	57.06 ± 18.94
	RK	28.59 ± 14.55	27.44 ± 13.76	25.59 ± 13.11
DTPA HR	LK	27.78 ± 18.50	26.87 ± 21.24	22.87 ± 18.50
	RK	70.22 ± 21.22	67.13 ± 20.43	63.43 ± 19.39
Normal	LK	19.13 ± 4.92	18.30 ± 4.60	15.63 ± 4.11
	RK	18.50 ± 4.60	17.53 ± 4.43	16.70 ± 4.28

measures that provide information about the central tendency and spread of a data set, respectively.

Tables 2 and 3 indicate the mean and the standard deviation values of transforms measured density counts including normal functioning kidney cases. As shown in Table 3, hydronephrosis case kidney's mean and standard deviation found to be higher than other two transforms' results.

3.2 *Radioactive Counts Measurement*

Renal scintigraphy counts measurement by gamma machine (Gate's method). The gamma machine records the number of gamma rays detected during a specified time period, typically in counts per minute (CPM). These counts are a measure of the radiotracer activity in the kidneys, reflecting the function and excretion of the radiotracer by the renal system.

It is expressed as:

$$\text{Uptake Kidney Counts} = \frac{(\text{Kidney Counts} - \text{Background Counts})}{e^{(-\mu \times \text{Kidney Depth})}}$$

where μ = Linear attenuation coefficient of radiotracer.

Right kidney depth = 13.3((Wt.)/(Ht.)) + 0.7.

Table 4 Comparison among mean ± S.D. of transforms detected density counts of HL, HR, and radioactive scintigraphy counts

Type of category		Mean ± S.D. of transforms detected density counts			Mean ± S.D. uptake counts
		FFT	Wavelet	Harr-like	
DTPA HL	LK	68.28 ± 23.11	65.94 ± 22.30	57.06 ± 18.94	67.40 ± 7.02
	RK	28.59 ± 14.55	27.44 ± 13.76	25.59 ± 13.11	32.60 ± 7.02
DTPA HR	LK	27.78 ± 18.50	26.87 ± 21.24	22.87 ± 18.50	34.66 ± 10.44
	RK	70.22 ± 21.22	63.13 ± 20.43	63.43 ± 19.39	68.34 ± 10.44

Left kidney depth = 13.2 ((Wt.)/(Ht.)) + 0.7.

Table 4 shows that calculated mean and standard values of all density count measures by three transforms along with radiotracer's uptake counts and we found that, density counts measured by FFT transform are approximately close to that of radioactive scintigraphy uptake counts measured by gamma camera. This identifies that the FFT transform offers more precise uptake count estimation in terms of mean value and variability.

3.3 *Statistical Correlation Findings*

Spearman's rank correlation coefficient, often denoted as ρ (rho), is a statistical measure that quantifies the strength and direction of the monotonic relationship between two ranked variables. It is based on the ranks of the observations rather than the actual values. It is expressed as:

$$\rho = 1 - \frac{6\sum di^2}{n(n^2 - 1)}$$

where di = difference among two ranks of every observation, and n = number of observation

Spearman rank finding is a non-parametric test that is used to quantify the degree of relationship [15]. All statistical evaluations were achieved using the SPSS 0.20 version software.

Findings of Spearman's rank coefficient values help to understand the linear relationship among two continuous variables [15–17]. As shown in Tables 4 and 5, among three techniques FFT techniques were found to be having a strongly positive correlation with radioactive counts (ρ = 0.804:0.801) which indicates that the density count measured by FFT transform is precisely close to gamma machine results.

Table 5 indicates correlation count of the transform counts and scintigraphy counts in case of left kidney hydronephrosis, while Table 6 indicates coefficients in case of right kidney hydronephrosis. In both cases, FFT transform correlation was found to be strongly positive as compared to wavelets and Harr-like's transforms.

Table 5 Correlation of the transform counts and scintigraphy counts in case of LK-HL

Left kidney hydronephrosis		
Type of transforms	Radioactive counts by scintigraphy scan	Correlation interpretation
FFT	$\rho = 0.804$	Strongly positive
Wavelet	$\rho = 0.595$	Moderately positive
Harr-like	$\rho = 0.575$	Moderately positive

Table 6 Correlation of the transform counts and scintigraphy counts in case of RK-HR

Right kidney hydronephrosis		
Type of transforms	Radioactive counts by scintigraphy scan	Correlation interpretation
FFT	$\rho = 0.801$	Strongly positive
Wavelet	$\rho = 0.573$	Moderately positive
Harr-like	$\rho = 0.572$	Moderately positive

4 Discussion

The standard deviation is a summary expression of the differences of each observation from the mean and analysis of these observations indicates the location of the center of the data and the standard deviation is the spread in the data. Comparison among mean and standard deviation values derived from three different transforms and renal scintigraphy gamma machines radioactivity counts indicate that:

- In the case of DTPA-HL, density count values derived by FFT transform are more as compared with remaining transforms, as in the case of the left kidney (68.28 $\pm$ 23.11) is more than the right kidney (28.59 $\pm$ 14.55) and find to be nearer to the radioactive count of the left kidney (67.40 $\pm$ 7.02) and right kidney (32.60 $\pm$ 7.02), respectively, measured by the gamma scanning output.
- Similarly, in the case of DTPA-HR, density count values derived by FFT transform are more than the remaining transforms, as in the case of the right kidney (70.22 $\pm$ 21.22) than the left kidney (27.78 $\pm$ 18.50) and found to be closer to the radioactive count of the right kidney (68.34 $\pm$ 10.44) and left kidney (34.66 $\pm$ 10.44) respectively counted by the Gamma scanning output.
- In the case of the normal functioning kidneys, all three transform density counts are found to be approximately the same as of scintigraphy radioactive counts.

In this study, Spearman's rank value is used to find a correlation between the transform counts results and the scintigraphy count. We found that, in case of the left kidney and right kidney hydronephrosis, there is a strongly positive correlation between FFT's density count and scintigraphy.

radiotracer count, i.e., $p = 0.81$ and 0.80, respectively, while outputs detected from wavelet transform in case of left kidney and right kidney hydronephrosis found to be moderately positive, i.e., $p = 0.59$ and 0.57, respectively. Also, in the case of Harr-like extraction transform results of the left kidney and the right kidney hydronephrosis were found to be moderately positive, i.e., $p = 0.57$ and 0.57, respectively.

4.1 *Statistical Analysis for Clinical Validation*

The accuracy and consistency of a specific method used in clinical studies are proven by the outcomes drawn from clinical validation. This process involves a comparison among the results obtained from the transformations with established clinical standards or reference methods. For clinical validation of these results, regression analysis statistical method is used to assess the relationship between different variables and make predictions or estimations based on this relationship. In this study, it is used to assess the correlation between the density count measured by image processing techniques (such as FFT, wavelet, or Harr-like) and the uptake count of radiotracers in the kidneys.

In relation to the clinical validation of these results, a regression model was used to establish the relationship between the density measurements of normal kidneys from hydronephrotic kidney pair and the density measurements of normal functioning kidney cases. From the graph to assess the goodness of fit of the regression model, the coefficient of determination (R^2) was calculated.

a. **Regression of Coefficient for DTPA_NORMAL Kidneys_HL Cases**

Figure 5 shows a scatter plot indicating the relationship among density counts of DTPA LK-hydronephrosis sample and normal functioning kidney samples measured by three transforms.

As shown in Table 7, the determination coefficient (R^2) value of 0.698 in the case of FFT count and normal kidney counts indicates that approximately 69.8% of the variation in the normal kidney's count can be explained by the FFT transform.

b. **Regression of coefficient for DTPA_NORMAL Kidneys_HR Cases**

Figure 6 shows scatter plot indicating the relationship among density counts of DTPA-RK hydronephrosis sample and normal functioning kidney samples measured by three transforms.

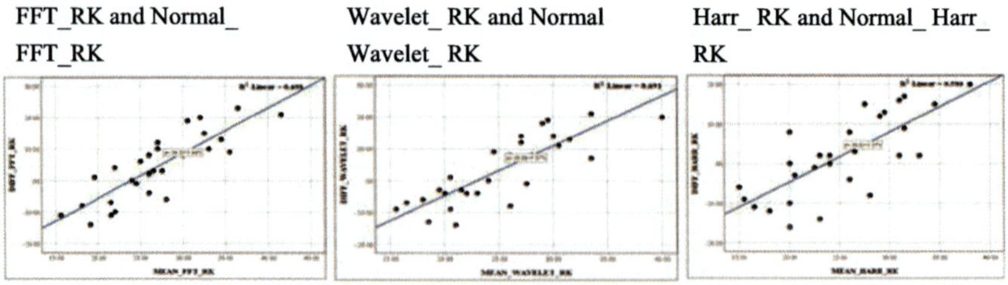

Fig. 5 Regression of coefficient for DTPA_NORMAL kidneys_HL cases

Table 7 Clinical validation analysis-DTPA-HL

Finding parameters	R^2 value	R^2 interpretation
FFT density count and normal kidney density count	0.698	Very strong correlation
Wavelet density count and normal kidney density count	0.691	Strong correlation
Harr-like density count and normal kidney density count	0.580	Strong correlation

FFT_RK and Normal_ FFT_RK   Wavelet_ RK and Normal Wavelet_ RK   Harr_ RK and Normal_ Harr_ RK

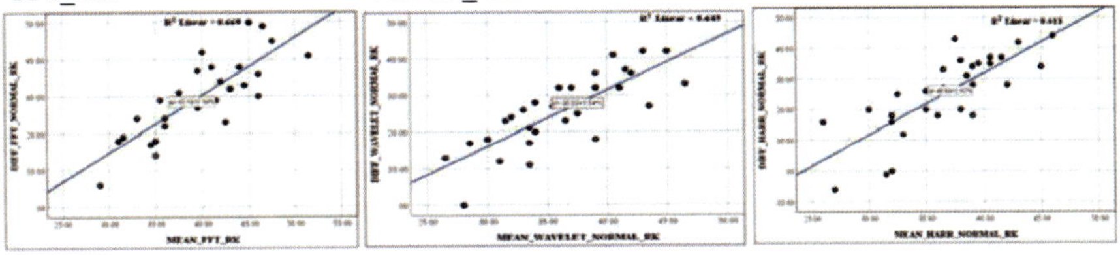

Fig. 6 Regression of coefficient for DTPA_NORMAL kidneys_HR cases

Table 8 Clinical validation analysis-DTPA-HR

Finding parameters	R^2 value	R^2 interpretation
FFT density count and normal kidney density count	0.669	Very strong
Wavelet density count and normal kidney density count	0.654	Strongly
Harr-like density count and normal kidney density count	0.613	Strongly

The R^2 value as shown in Table 8, i.e., coefficient of determination 0.669 shows that approximately 66.9% of the variation in the normal kidney's count can be explained by the FFT transform. This suggests a very strong relationship between the FFT transform's left kidney's density output of DTPA-HR case and the normal left kidney's density count.

5 Conclusion

This study aimed to overcome the limitations of using renal scintigraphy alone for quantitatively assessing the level of hydronephrosis (renal obstruction). By processing the renal obstruction scans using image extraction and enhancement transforms (FFT, wavelet, and Harr-like), you were able to measure the density of radiotracer inside the kidney.

It was found that the density counts obtained from the FFT transform were closely related to the radiotracer radioactive counts measured by the scintigraphy machine. This suggests that the FFT's density count can serve as a supportive measure for quantitatively diagnosing the level of hydronephrosis, rather than relying on subjective levels such as "moderate" or "severe." The direct proportionality between the FFT-measured density counts and radiotracer counts indicates that higher density counts correspond to a greater presence of radiotracer inside the kidney, which can be indicative of a higher level of renal obstruction.

By utilizing the FFT transform and its density count as a quantitative measure, it can be considered as an additional parameter for clinicians to evaluate and report the stage of hydronephrosis in a more objective manner. This can potentially enhance the accuracy and precision of diagnosing and assessing renal disorders, allowing for more targeted and effective treatment decisions.

Conflict of Interest Pradnya Gokhale and Prof. Dr. Babasaheb Patil declare that they have no competing interests.

Ethics Approval NA.

Funding NA.

Acknowledgements We want to thank Dr. Alok Pawasker (SSGCH) and Dr. Sameer Joshi (N.M. & MIC) for their suggestions and assistance in finding data sources through various gazettes of hospital resources.

References

1. Volterrani D, Daniele G, Mazzarri S, Guidoccio F, Dardano A, Giusti L, Del Prato S et al (2019) Annual congress of the European Association of Nuclear Medicine 12–16, 2019 Barcelona, Spain. Eur J Nucl Med Mol Imaging 46:1–952
2. Blaufox MD, De Palma D, Taylor A, Szabo Z, Prigent A, Samal M et al (2018) The SNMMI and EANM practice guideline for renal scintigraphy in adults. European Eur J Nucl Med Mol Imaging 45:2218–2228
3. Hans F, Suzanne D, Ariane B, Greet L, Anne R, Iris Van den H, Campbell M (2009) Dynamic renal imaging in obstructive renal pathology a technologist's guide, ENAM. University Medical Centre Nijmege
4. Andrew T, Taylor DC, Brandon DP, Donald Blaufox M, Emmanuel D, Belkis E, Sandra FG, Andrew JWH (2018) SNMMI procedure standard/EANM practice guideline for diuretic renal scintigraphy in adults with suspected upper urinary tract obstruction 1.0. Seminars Nucl Med 48(4):377–390
5. Piyamas S, Kobchai D, Surapun, Y (2015) The estimation of GFR and ERPF using adaptive edge based active contour for the segmentation of structures in dynamic renal scintigraphy. Int J Innov Comput Inf Control 11(1)
6. Andre AD, Amy P, Hamphrey R (2008) Ham pixel-by-pixel mean transit time without deconvolution. Nuclear Med Commun 29(4):345–348
7. Issa IA (2010) A comparison between the values of renal parenchymal mean transit time by applying two methods, matrix inversion deconvolution and Rutland-Patlak plot. World Appl Sci J 8(10):1211–1219
8. Emily SB, Antonio RP, Elijah B, Pooneh R, Tabrizi RD, Sussman BMS, Eglal S, Massoud M, Hans GP, Marius G (2018) Early detection of ureteropelvic junction obstruction using signal analysis and machine learning: a dynamic solution to a dynamic problem. J Urol 199:847–852

9. Volterrani D, Daniele G, Mazzarri S, Guidoccio F, Dardano A, Giusti L, Del P et al (2019) Annual congress of the European Association of Nuclear Medicine 12–16, 2019 Barcelona, Spain. Eur J Nucl Med Mol Imaging 46:1–952
10. Madsena CJ, Mollerc ML, Zerahnb B, Fynboa C, Jensen JJ (2013) Determination of kidney function with 99mTc-DTPA renography using a dual-head camera. Nucl Med Commun 34(4):322–327
11. Jianhua L, Yanling S (2011) Image feature extraction method based on shape characteristics and its application in medical image analysis. CCIS 224:172–178
12. Van Fleet PJ. Discrete wavelet transformations, 2nd edn.
13. Gonzalez Rafael C. Digital image processing, 3rd edn. Pearson's International, 320–532
14. Oran Brigham E (1988) The fast Fourier transform and its applications. Prentice-Hall, Englewood Cliffs, New Jersey
15. Mukak M (2012) Statistics corner: a guide to appropriate use of correlation coefficient in medical research. Malawi Med J 24(3):69–71
16. Gokhale PN, Patil BR (2022) Correlation between renal tracer parameters derived for recognition of renal obstruction with ^{99m}Tc-DTPA scintigraphy. In: 2022 10th international conference on emerging trends in engineering and technology-signal and information processing (ICETET-SIP-22), 1–6
17. May P, Ehrlich HC, Steinke T (2006) ZIB structure prediction pipeline: composing a complex biological workflow through web services. In: Nagel WE, Walter WV, Lehner W (eds) Euro-Par 2006. LNCS, vol 4128. Springer, Heidelberg, pp 1148–1158

Deep Short-Term Long Memory Technique for Respiratory Lung Disease Prediction

B. Dhiyanesh, Y. Baby Kalpana, S. Rajkumar, P. Saraswathi, R. Radha, and S. Suresh

Abstract Recently, respiratory sampling has been of great help in diagnosing health disorders and diseases by providing valuable information for various medical conditions. Several methods can be used to diagnose respiratory infections, such as collecting blood or sputum samples. It has the potential to predict the risk of respiratory disease-related exacerbations. It is also an essential inflammatory biomarker with which medical professionals can prescribe imaging equipment for respiratory diseases and collect basic patient status information. However, current testing methods can be challenging and time-consuming to diagnose respiratory diseases. Although these techniques are accurate, they increase the risk of contamination and lead to complications when long analysis times are required. To address this issue, the deep short-term long memory method for respiratory lung disease prediction. First, we collected a dataset of respiratory disease sounds, a key indicator of respiratory lung health and disease, from the Kaggle repository. Next, a pre-processing step is performed to ensure a smooth and low-noise signal in the respiratory lung images using Z-score normalization. Next, feature extraction can be performed using the local binary Gabor filter (LBGF) to efficiently detect local spatial patterns and grayscale variation in an image. Finally, the deep short-term long memory (DSTLM)

B. Dhiyanesh (✉)
CSE, Dr. N.G.P. Institute of Technology, Coimbatore, India
e-mail: dhiyanu87@gmail.com

Y. Baby Kalpana
CSE, P.A. College of Engineering and Technology, Pollachi, India

S. Rajkumar
Sona College of Technology, Salem, India
e-mail: rajkumar.s@sonatech.ac.in

P. Saraswathi
IT, Velammal College of Engineering and Technology, Madurai, India

R. Radha
EEE, Study World College of Engineering, Coimbatore, India

S. Suresh
CSE, KPR Institute of Engineering and Technology, Coimbatore, India

S. Shakya et al. (eds.), *Fourth International Conference on Image Processing and Capsule Networks*, Lecture Notes in Networks and Systems 798,
https://doi.org/10.1007/978-981-99-7093-3_5

method can classify airway diseases of step obstruction and non-obstructive localization. The proposed simulation result achieves high efficiency in respiratory disease prediction. In addition, respiratory diseases can be diagnosed by several performance measures such as sensitivity, specificity, accuracy, and $F1$-score.

Keywords Respiratory · Lung disease · Deep learning · DSTLM · Z-score normalization · LBGF · Classification · Accuracy · $F1$-score

1 Introduction

Respiratory diseases are a significant contributor to both illness and death rates. Then, they represent a considerable burden and cause changes globally. Chronic respiratory diseases significantly impact patients worldwide, with 7.4% of the global population affected and 7.0% of all deaths attributed to these conditions [1]. The healthcare sector is a unique and indispensable industry compared to other industries. The profession that demands the highest quality of diagnosis, treatment, and service is the most crucial area where people spend their money. One of the primary factors contributing to the accelerating impact of diseases on human health is the effects of environmental changes, climate change, lifestyle, and other related risks.

The images and sounds produced by different medical devices may be limited due to their subjective nature, complexity, and level of intelligence. Medical image and audio analysis can improve the accuracy of predictive modelling and diagnose diseases immediately. Due to the shortage of trained personnel, healthcare workers welcome technological aids to help handle more patients. The impact of respiratory diseases is increasing, and apart from significant health disorders like cancer and diabetes, it also threatens social life [2, 3].

However, respiratory diseases pose a significant and persistently high burden on global health systems. Diagnosing this condition can be challenging, particularly in remote or resource-constrained settings where the required diagnostic tools may not be available. Respiratory illnesses such as colds and flu can cause mild discomfort and incapacitation. On the other hand, conditions like pneumonia, tuberculosis, and lung cancer pose a serious threat to one's life as they lead to severe respiratory diseases [4].

Instead, millions of people suffer from pneumonia. They affect millions of people and threaten healthcare systems. Globally, respiratory disease significantly contributes to mortality and disability [5].

In this section, we first collected a dataset of respiratory sounds, a key indicator of respiratory health and disease, from the Kaggle dataset. Then, we perform pre-processing steps using wavelet smoothing and Z-score normalization to ensure a smooth and low-noise signal in respiratory lung disease images. Next, we use local binary Gabor filter (LBGF) models to extract features from respiratory lung disease. Finally, obstructive and non-obstructive respiratory diseases can be classified based on airflow limitations using the proposed DSTLM methods. Thus, the proposed

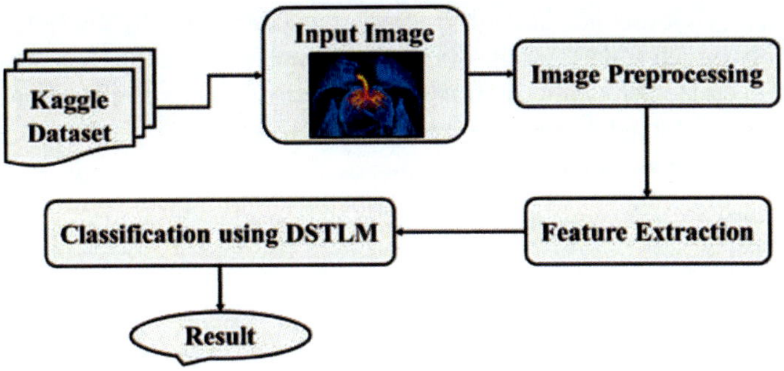

Fig. 1 Basic architecture of respiratory lung disease

method can improve the performance of pulmonary respiratory disease images and increase their application accuracy.

This section explores their applications through the basic structure of respiratory diseases as constructed in Fig. 1. In these, we first collected data from the Kaggle dataset and provided input images. Then, we pre-processed the respiratory disease images and extracted their features. Finally, we can find and obtain the accuracy of obstructive and non-obstructive artefact limits for respiratory disease using the proposed method.

2 Related Work

According to the author, asthma and COPD can be prevented, diagnosed, and treated using decision-support systems in clinical situations. These disorders can be diagnosed by identifying critical factors by conducting empirical breath studies on a representative sample [6].

The author discusses the effectiveness of three DT methods—packing, AdaBoost, and random forest—in predicting the LOS for children with respiratory diseases. The forms can also be used to monitor and treat patients based on data collected from a dataset of 11,206 records from hospital information structures. The capacity expansion technique can be adopted and analysed after pre-processing and computational transformation [7].

The author suggests that Wi-Fi CSI data can be used to track and extract human breathing motion. Through demonstrations, COTS models can be used to implement end-to-end systems composed of hardware platforms, control software, data acquisition algorithms, and processing algorithms [8]. The author proposed that a DL model can be developed and used to evaluate sleep states and respiratory disease events. The AHI score and subsequent calculations can only use the pulse oximetry data set to handle these [9].

The author proposed that the Mel-spectrogram-based deep CNN-RNN model can be applied for breath sound classification. The patient data that has been defined can be utilized to create personalized classification models for screening individuals for respiratory illnesses and identifying any irregularities [10].

The author proposed that new techniques such as OST and deep ResNets can be used to identify breaths, bursts, and natural sounds. In addition, the provided OST raw breath sounds can be processed, and Resnet's OST spectrogram can be rescaled [11]. The author proposed that a DL-based approach could be used in the NLSD to research long-term mortality. A binary classification model can be employed using a 3D ResNet-centric neural network to classify non-lung cancer mortality (cardiovascular and respiratory mortality) [12].

The author proposes using three classification techniques to implement a hybrid CNN model. Among them, the classification technique uses FC layers to classify CXR images and can collect weights with high classification accuracy with a model trained over multiple epochs [13]. The author proposed that non-contact auscultation could handle a DL-based model for lung disease classification. Two pulmonary specialists in respiratory medicine recorded the typical sounds of healthy breathing and five diverse sounds indicative of lung disease [14].

The author proposed that the DL auscultation analysis framework can be used to classify respiratory cycle abnormalities and diagnose disease from respiratory recordings. The framework converts the input speech into a spectrogram representation and starts with front-end feature extraction [15]. The author presents a low-cost PoC method for quantifying active NE in patient sputum using LFA [16]. The author proposed that a lightweight CNN framework can be implemented to utilize map-based composite volumetric lung sound features and classify respiratory diseases based on individual respiratory cycles. This feature offers the capabilities of EMD and CWT [17].

The authors proposed that breath signature analysis from RGB infrared sensors could be used to screen mask wearers' health status for non-contact techniques. In addition, respiratory data collection for mask wearers is functionally saturated in technology [18]. The author aims to obtain the length of the time-regression of the relationship between exposure and response. A novel method for detecting infectious respiratory diseases from human axial sweat samples using an electronic nose (e-nose) is being developed [19].

The author proposed that these can be used to derive regression lengths in exposure–response relationships. Two conventions are commonly used when exploring the connection between disclosure and response: linear and nonlinear. Both confounder- and confounder-free models can be implemented based on these assumptions [20]. The author proposed a model for class 2 and 5 COPD classification using a framework for learning representations of Mel-spectral slices. The method suggested involves pre-processing and data augmentation, creating Mel-spectrogram segments from lung sounds, and refining the pre-existing YAMNet [21].

The author notes that technological advances in USE lung diagnostic studies can be reviewed in detail. It presents USE's rationale and practical strategies and provides information on its various uses in evaluating lung disease [22]. The author suggests exploring the significance and worth of utilizing the FOT scheme when evaluating lung cancer. Different levels of heterogeneity in lung tissue can be recognized through specialized equipment and related methods [23].

2.1 Problem of Statement

1. The risks of contamination and complications with traditional respiratory disease prediction techniques
2. The use of the deep short-term long memory method for improved accuracy in respiratory disease prediction
3. Details about the collection and pre-processing of the dataset used in this study from the Kaggle repository
4. A closer look at feature extraction using the local binary Gabor filter (LBGF) for detecting spatial patterns and grayscale variation in images.
5. Discussion on performance measures such as sensitivity, specificity, and accuracy.

3 Proposed Methodology

In this section, we first used the Kaggle dataset to collect information on pulmonary respiratory disease. We collected data from the Exasens dataset of breath sounds, a key indicator of health and disease in respiratory diseases. Then, wavelet smoothing and Z-score normalization methods can be used to pre-process respiratory disease images to smooth the images and ensure low signal-to-noise. Next, we use LBGF models to extract the features of these images. These methods can detect an image's local spatial patterns and grayscale variation efficiently. Lastly, the proposed E-LSTM method can detect obstructive and non-obstructive respiratory disease artefacts. Furthermore, based on DL methods, the DSTLM proposed method can recover the efficacy and accuracy of respiratory lung disease applications.

As shown in Fig. 2, the architecture diagram for respiratory disease using the proposed method is clearly explained in this section. First, we collect the Exasens dataset of breath sounds from the Kaggle dataset to collect respiratory disease data.

Then, pre-processing can be done using wavelet smoothing and Z-score normalization methods. Next, LBGF methods can be applied to feature extraction to efficiently detect spatial patterns and grey-scale variation in an image efficiently. Then, the proposed DSTLM method can detect the airflow limitations of obstructive and non-obstructive respiratory diseases to improve efficiency and increase accuracy.

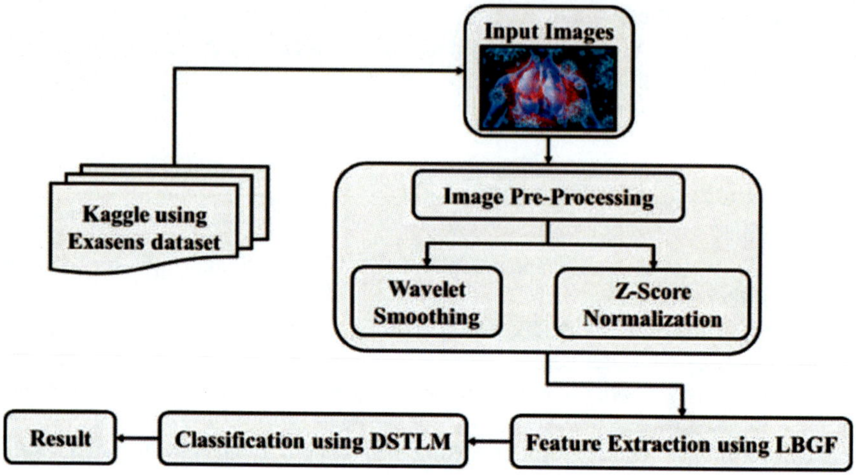

Fig. 2 Proposed architecture diagram (CMSFS-ARNN)

3.1 Dataset Collection

In this section, we collect data for pulmonary respiratory disease from the Kaggle dataset. Additionally, this data has introduced a novel dataset for classifying four respiratory conditions: COPD, asthma, infectious diseases, and healthy controls (HC). The Exasense dataset can collect demographic data from four saliva sampling partners as part of a collaborative project at the Borsdell Research Center at the Northern Biomaterials Bank.

3.2 Image Pre-processing

This section deals with the pre-processing steps after collecting the pulmonary respiratory disease information from the Kaggle dataset. Then, respiratory disease images can be smoothed and used to ensure low signal-to-noise by pre-processing lung diseases using wavelet smoothing and *Z*-score normalization methods on respiratory disease images. These methods may be disturbed by acoustic noise from ambient noise, ambient speech, electronic interference or stethoscope drift, lung sound recordings, and other electronically recorded biosignals. Therefore, it is necessary to ensure the signal is smoothed and pre-processed before the feature extraction process. Following two processes for image smoothing and normalization,

a. Wavelet smoothing
b. *Z*-score normalization.

3.2.1 Wavelet Smoothing

In this respect, wavelet transforms (WT) are widely used to smooth transient signals. Then, instead of using the Fourier transform (FT) to decompose groups of waves, these can be decomposed instead of the signal's complex sinusoids. By converting the time and frequency domains of a signal, useful information can be extracted or noise can be removed. Next, a fundamental frequency can be used as the simulation unit to preserve the original data and decompose a signal into higher and lower frequencies using WT.

This section estimates the signal translated into WT by multiplying the background noise (Eq. 1).

$$\varphi_{uv}(s) = |u|^{-1/2}\varphi\left(\frac{s-v}{u}\right) \tag{1}$$

This category calculates signal smoothing as a time-domain and frequency-domain transformation (Eq. 2).

$$A_m = w_m + z_m \tag{2}$$

Equation 3 is a continuous wavelet transform that computes various frequencies across different sizes and positions.

$$s(u, v) = \frac{1}{\sqrt{u}} \int_{-\infty}^{\infty} A(s)\psi^* \frac{(s-v)}{u} cs \tag{3}$$

Equation 4 is the evaluated discrete wavelet transform that uses finite wavelets at different magnitudes and levels.

$$s_{nm} = \int_{-\infty}^{\infty} A(s)\varphi_{nm}(s)cs \tag{4}$$

Calculate the infinite sequence using high-pass and low-pass filters (Eqs. 5 and 6).

$$\overline{H}_{rs} = \sum_{o=0}^{or-1} \overline{q}_{r,o} A_{s-o} \tag{5}$$

$$\overline{G}_{rs} = \sum_{o=0}^{or-1} \overline{e}_{ro} A_{s-o} \tag{6}$$

Let us assume an s—time instance, u—dilation function, v—translation function, $(\varphi(s))$ wavelet translate and dilate signal, m and n—signal noise, w—real signal, z—noised signal, r—level of the decomposition A_{s-o} infinite sequence, $\overline{H}_{rs}$ wavelet coefficient, $\overline{G}_{rs}$ rs—scaling coefficient, and $\overline{q}_{r,o}$ and $\overline{e}_{ro}$ high-, and low-pass filter. In this area, the signal can be estimated by multiplying the background noise. Then, the signal smoothing can be calculated as a time-domain and frequency-domain transform using continuous wavelet and discrete wavelet transforms.

3.2.2 Z-Score Normalization

In this category, it is necessary to ensure that each signal is normalized by its Z-score after the above signal pre-processing step. Z-score normalization increases the dynamic range between the corresponding values in the time-domain signal. In other words, significant trends in the signal do not necessarily dominate small movements.

This model Eq. 7 calculates the normalization for raw score converting.

$$G = \frac{G - \mu}{\sigma} \tag{7}$$

Equation 8 gives the standard deviation of the score distribution and can be calculated.

$$\sigma = \sqrt{\frac{\sum_{j=1}^{m} (g_j - \overline{g})^2}{m}} \tag{8}$$

Estimate the normalized values according to the attribute's mean and standard deviation (Eq. 9).

$$A_j = \frac{B_j - \text{AVG}(X)}{X} \tag{9}$$

Let us assume the g-raw score and μ—signal exhibits a mean value, σ—standard deviation, m—total number, j—score, $\overline{g}$—average score, X-attribute, A_j—value of the attribute, and B_j—normalized value. To calculate the standard deviation of the distribution, we normalize the raw scores as shown in the equation. A normalized value is also estimated based on the attribute's mean and standard deviation.

3.3 *Local Binary Gabor Filter*

This section uses important features extracted from images to solve specific problems. In addition, these methods can efficiently detect local spatial patterns and grey-scale variation in pulmonary respiratory dog images. Features can be particular structures in an image, such as points, edges, colours, sizes, shapes, and objects. Logically, image type affects feature quality. Although these new features may differ from the original ones, they might exhibit a more vital ability to distinguish in various locations than the current site. Feature extraction aims to enhance the effectiveness of machine learning algorithms for object recognition by generating more valuable features, including through methods such as developing binary Gabor filters. The LBGF represents the size relationship between a central pixel and its neighbouring pixels in a neat form. The LBGF transform is part of the windowed Fourier transform, which uses the LBGF function to extract features related to various sizes and orientations in the frequency domain.

This category, Eq. 10, estimates the response of a spatially-based 2D Gabor filter.

$$A_j(a, b; \lambda, \varphi, \sigma_a, \sigma_b) = \frac{1}{2\pi\sigma_a, \sigma_b} \exp\left\{-\frac{1}{2}\left[\frac{p_1^2}{\sigma_a^2} + \frac{p_2^2}{\sigma_b^2}\right]\right\} \tag{10}$$

Compute the image transformation using the two-dimensional Gabor function expressed in Eqs. (11) and (12).

$$Z(a, b) = \iint \delta J(\alpha, \beta) H(a - \alpha, b - \beta) c\alpha c\beta \tag{11}$$

$$H_{\lambda,\theta,\varphi,\sigma b^{a,b}} = \exp\left[\frac{\left(a^2 + \gamma^2 + b^2\right)}{2\sigma^2}\right] \cos\left[2\pi\frac{a}{\lambda} + \varphi\right] \tag{12}$$

In this section, the imaginary part of the Gabor energy filter can reflect the gray level changes in different directions. (Eq. 13).

$$E(a, b, \theta) = \sqrt{((E^*g)(a, b, \theta, 0))^2((E^*g)(a, b, \theta, \pi))^2} \tag{13}$$

Compute the LBP histogram generated from the smallest values of all micro-patterns (Eq. 14).

$$X_j(s) = \begin{cases} 1, \ j(s) < J\left(s + \Delta r_j\right), \\ 0, \text{ otherwise} \end{cases} \quad (j = 1, \ldots, M_m) \tag{14}$$

Each LBP is transformed into a vector to calculate co-occurrences efficiently in this category (Eq. 15).

$$E_j(s) = \delta_{jl}(y(s)) \tag{15}$$

This category indicates the number of adjacent LBP pixel sizes (Eq. 16).

$$G(a) = \sum_{\delta=J} E(s)E(s + x)^t \tag{16}$$

Let us assume a and b—function, $Z(a, b)$—Gabor residuals, (α, β)—convolution of the image, E—Gabor function frequency of filter centre, δ—set of image points, θ-theta, and filter direction, α, β—alpha and beta, λ—lambda, γ—Gama, 0 or π—phase offset, M_m—no of the neighbor pixel, $y(s)$—label, $G(a)$—element, a—displacement vector. In this section, the size relationship between a central pixel and its neighbouring pixels can be represented in an elegant form, and the LPGF method can be used to extract features related to different sizes and orientations in the wave domain.

3.4 Deep Short-Term Long Memory (DSTLM)

In this section, the proposed DSTLM method can be used to classify obstructive and non-obstructive respiratory diseases to determine the accuracy of pulmonary respiratory illness. A technique utilizing DSTLM has been introduced that can accurately classify pulmonary diseases while requiring minimal training time. During a network's operation, temporary information is managed by cells, while in and outgoing information is controlled by gates. The proposed DSTLM method networks can process data in forward (one-way) or forward and reverse order. In signal analysis, the latter is the preferred option and is crucial for timely learning. A DSTLM network can help predict lesions better by presenting clinical relationships between lesions.

Algorithm
Input: Number of adjacent pixel size G
Output: Prediction result
Start
Step1: Calculate the output of the core function cell
$G_w = E_w G_{w-1} x_w G_w$ (17)
Step 2: Calculate the gate entered in the result central cell
$D_w = Q_w \tan D(G_w)$ (18)
Step 3: The functions of the hidden units selected to operate in the network can be calculated based on the sigmoid function
$x_w = \sigma(z_{Ux} U_w + z_{Dx} D_{w-1} + z_{Gx} G_{w-1} + a_x)$ (19)
$E_w = \sigma(z_{UE} U_w + z_{DE} D_{w-1} + z_{GE} G_{w-1} + a_E)$ (20)
$Q_w = \sigma\left(z_{UQ} U_w + z_{DQ} D_{w-1} + z_{GQ} G_{w-1} + a_Q\right)$ (21)
Step 4: Calculate the limitations of the output's forward and backward direction artifacts
$v_w = z_{\overrightarrow{Dv}} \overrightarrow{D^M} + z_{\overleftarrow{Dv}} \overleftarrow{D^M} + a_v$ (22)
Return v_w
Stop

Let us assume G—functional cell, w—time, E_w—target gate activation, x_w—input gate activation, G_w—input to the central compartment, σ—sigmoid function, Q_w—output gate activation, z—weight, z_U—input gate weight, z_D—hidden to hidden weight, z_G—peephole weight, v_w—bidirectional output. In this category, DSTLM processes pre- and post-respiratory disease information using bilateral artifact thresholding of the production.

4 Analyses and Discussions of Experimental Results

This section enables Python to manipulate the Jupyter toolbox to detect airflow limitations in pulmonary respiratory diseases, enhance image quality, and achieve accurate results. For more information, see the project's Website at https://archive.ics.uci.edu/ml/datasets/Exasens.

Table 1 mentions the data set name, number, training, test time, and languages using their values and categories to achieve the accuracy of respiratory diseases.

4.1 Evaluation Matrix

In this section, we can evaluate the enactment of the developed model using several metrics for analysing classification confusion matrices. Based on the training/classification scheme, a confusion matrix array can be generated after each iteration. Calculating all evaluation metrics is as simple as counting the number of them in the confusion matrix. By identifying true positives, true negatives, false positives, and false negatives, accurate measurements of accuracy, sensitivity, specificity, precision, and $F1$-estimate can be obtained.

$$\text{Sensitivity} = \frac{\text{TP}}{\text{TP} + \text{FN}} \tag{23}$$

$$\text{Specificity} = \frac{\text{TN}}{\text{TN} + \text{FP}} \tag{24}$$

Table 1 Parameter model

Types	Values
Dataset name	Kaggle
No. of dataset	400
Training	180
Testing	75
Language	Python
Tool	Jupyter
Dataset name	Kaggle

$$\text{Precision} = \frac{\text{TP}}{\text{TP} + \text{FP}} \tag{25}$$

$$F1\text{-Score} = \frac{2\text{TP}}{2\text{TP} + \text{FP} + \text{FN}} \tag{26}$$

$$\text{Accuracy} = \frac{\text{TP} + \text{TN}}{\text{TP} + \text{TN} + \text{FP} + \text{FN}} \tag{27}$$

In this section, the performance of pulmonary respiratory disease images can be evaluated using sensitivity analysis to determine their accuracy, as introduced in Fig. 3. Then dealing with the FOT, CNN, and EMD methods presented in the literature review, their accuracy is as low as 49%. Sensitivity analysis of the DSTLM performance of the proposed method with the previously submitted form achieves high accuracy.

Figure 4 suggests that a dataset of pulmonary respiratory disease images was obtained from Kaggle and analysed for specific performance to determine their accuracy. Detailed performance analysis of CNN, FOT, and EMD methods has achieved lower accuracy. Comparing these three methods with the proposed DSTLM method, their accuracy is increased to 70%.

In this section, they can be implemented using algorithmic analysis to determine the accuracy of respiratory disease images, as shown in Fig. 5. Then, the accuracy performance of the proposed DSTLM method increased to 77% over analysis. Then, the three methods EMD, CNN, and FOT, presented in the literature analysis, have 59% lower performance accuracy than the proposed method.

In Fig. 6, $F1$-score performance is used to determine the accuracy of the proposed method for detecting respiratory diseases. A comparison of the EMD, FOT, and CNN

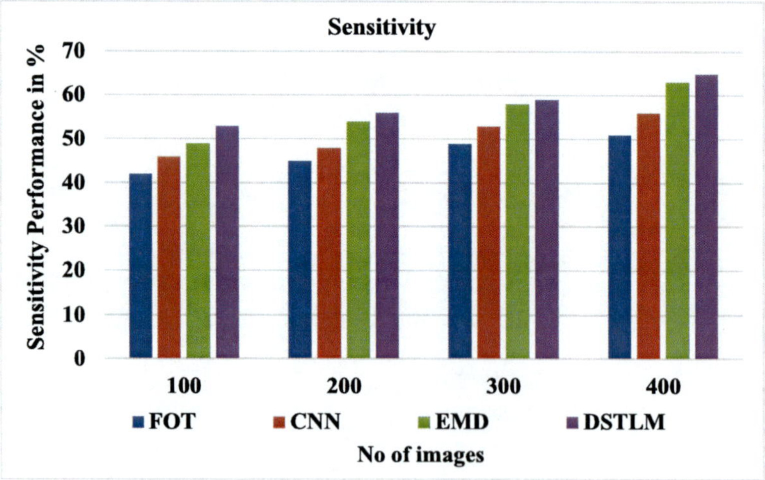

Fig. 3 Analysis of the sensitivity performance

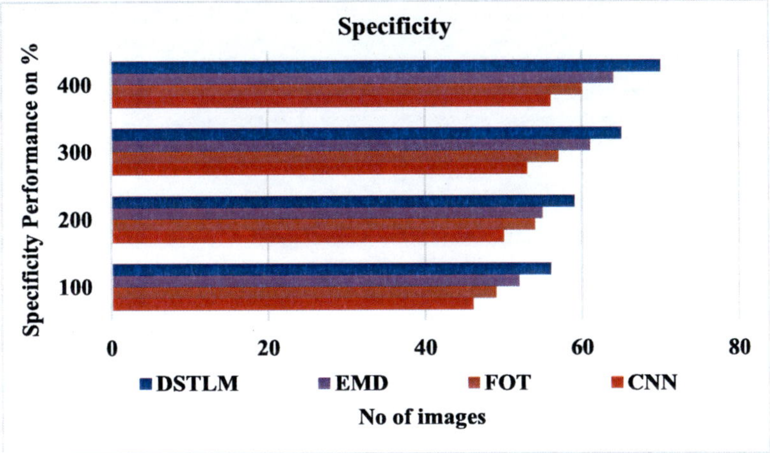

Fig. 4 Analysis of the specificity performance

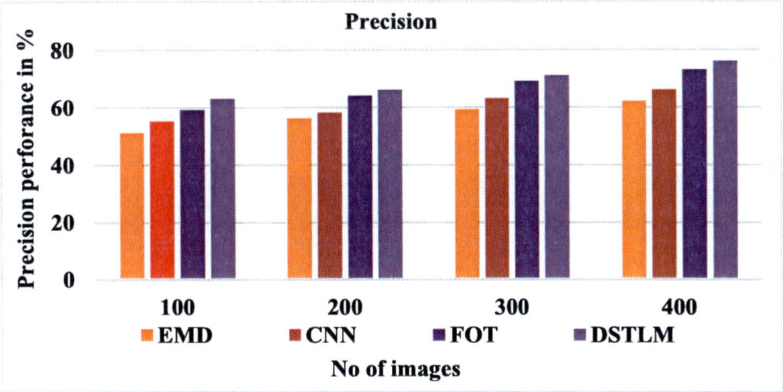

Fig. 5 Performance of the precision

models presented in the literature was done before determining the performance of the proposed method. It is necessary to increase the accuracy of their performance value. Then, the $F1$-score value of the proposed DSTLM method with these three methods increased to 83% with high accuracy.

In Fig. 7, image accuracy is improved by analysing the proposed method and the literature method presented. Thus, the proposed DSTLM and the presented literature methods provide a systematic classification and identify them with high accuracy. The classification accuracy of the DSTLM method is increased to 90% compared to the other three methods.

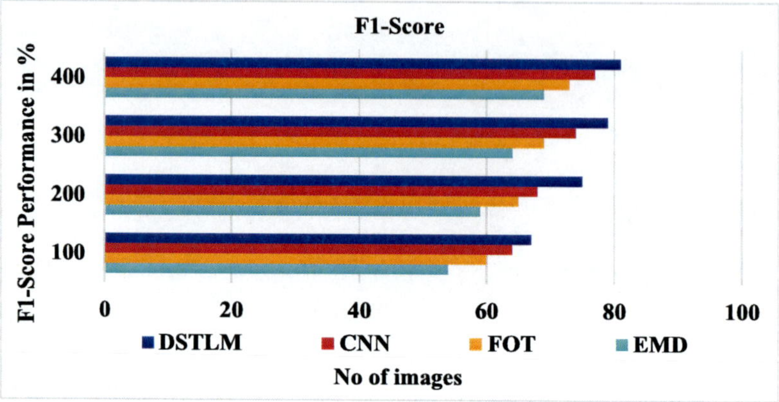

Fig. 6 Analysis of the *F*-score

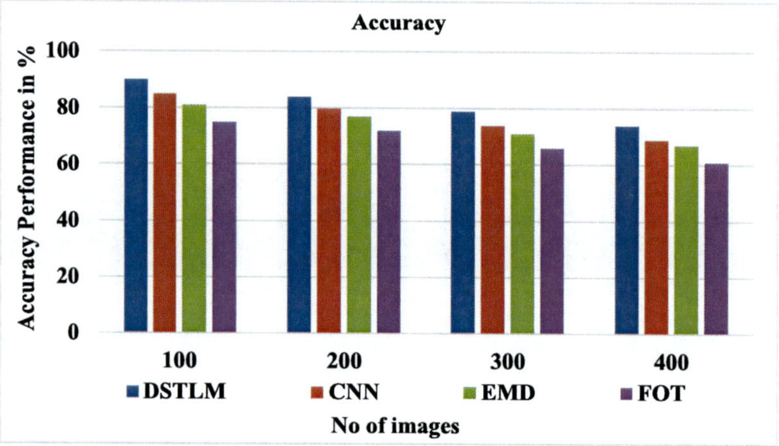

Fig. 7 Performance of the accuracy

5 Conclusion

This section uses a DL model based on DSTLM to classify pulmonary respiratory diseases. Then, based on this finding, a state-of-the-art DL-assisted lung and respiratory diagnosis taxonomy can be developed by summarizing the key concepts and features of current applications of DL in pulmonary respiratory disease diagnosis and their characteristics. This model differs from methods used to diagnose lung and respiratory diseases. Using LBGF feature extraction and wavelet smoothing with *z*-score normalization, only lung-specific features with appropriate image enhancement were considered. We proposed the DL model DSTLM to improve classification performance with minimum computation. We can use the proposed method to analyse the data using sensitivity, specificity, precision, recall, and accuracy measurements. The model was used to classify lung sounds into several pulmonary diseases with an average accuracy of 90%. This improves the dependability of the model being suggested. The performance metrics of classification models are impressive, but there is room for improvement by adjusting the pre-processing and training techniques. Future directions are indicated, and there are new techniques being developed that could potentially improve performance and expand the use of lung respiratory disease applications.

References

1. Wang C, Ma J, Zhang S et al (2022) Development and validation of an abnormality-derived deep-learning diagnostic system for major respiratory diseases. NPJ Digit Med 5:124
2. Bharati S, Podder P, Rubaiyat Hossain Mondal M (2020) Hybrid deep learning for detecting lung diseases from X-ray images. Inform Med Unlock 20:100391. ISSN 2352-9148
3. Srivastava A, Jain S, Miranda R, Patil S, Pandya S, Kotecha K (2021) Deep learning based respiratory sound analysis for detection of chronic obstructive pulmonary disease. PeerJ Comput Sci 11(7):369
4. Jasmine Pemeena M, Priyadarsini KK (2023) Lung diseases detection using various deep learning algorithms. 2023:3563696
5. Sakthivel S (2015) Secure data storage auditing service using third party auditor in cloud computing. Int J Appl Eng Res 10(37)
6. Spathis D, Vlamos P (2019) Diagnosing asthma and chronic obstructive pulmonary disease with machine learning. Health Inf J 25(3):811–827
7. Ma F, Yu L, Ye L, Yao DD, Zhuang W (2020) Length-of-stay prediction for pediatric patients with respiratory diseases using decision tree methods. IEEE J Biomed Health Inform 24(9):2651–2662
8. Mosleh S, Coder JB, Scully CG, Forsyth K, Kalaa MOA (2022) Monitoring respiratory motion with Wi-Fi CSI: characterizing performance and the BreatheSmart algorithm. IEEE Access 10:131932–131951
9. Kiruthiga G et al (2022) Effective DDoS attack detection using deep generative radial neural network in the cloud environment. In: 2022 7th international conference on communication and electronics systems (ICCES), pp 675–681
10. Acharya J, Basu A (2020) Deep neural network for respiratory sound classification in wearable devices enabled by patient-specific model tuning. IEEE Trans Biomed Circuits Syst 14(3):535–544
11. Chen H, Yuan X, Pei Z, Li M, Li J (2019) Triple-classification of respiratory sounds using optimized S-transform and deep residual networks. IEEE Access 7:32845–32852
12. Karthick K et al (2023) Iterative dichotomiser posteriori method-based service attack detection in cloud computing. Comput Syst Sci Eng 44(2):1099–1107
13. Yaseliani M, Hamadani AZ, Maghsoodi AI, Mosavi A (2022) Pneumonia detection proposing a hybrid deep convolutional neural network based on two parallel visual geometry group architectures and machine learning classifiers. IEEE Access 10:62110–62128
14. Choi Y, Choi H, Lee H, Lee S, Lee H (2022) Lightweight skip connections with efficient feature stacking for respiratory sound classification. IEEE Access 10:53027–53042
15. Pham L, Phan H, Palaniappan R, Mertins A, McLoughlin I (2021) CNN-MoE based framework for classification of respiratory anomalies and lung disease detection. IEEE J Biomed Health Inform 25(8):2938–2947

16. Naveenkumar E et al (2022) Detection of lung ultrasound covid-19 disease patients based convolution multifacet analytics using deep learning. In: 2022 second international conference on artificial intelligence and smart energy (ICAIS), pp 185–190
17. Shuvo SB, Ali SN, Swapnil SI, Hasan T, Bhuiyan MIH (2021) A lightweight CNN model for detecting respiratory diseases from lung auscultation sounds using EMD-CWT-based hybrid scalogram. IEEE J Biomed Health Inform 25(7):2595–2603
18. Jiang Z et al (2020) Detection of respiratory infections using RGB-infrared sensors on portable device. IEEE Sensors J 20(22):13674–13681
19. Naveenkumar E et al (2022) Lung Ultrasound COVID-19 detection using deep feature recursive neural network. Intelligent sustainable systems. Lecture notes in networks and systems, vol 458
20. Lu J, Bu P, Xia X, Yao L, Zhang Z, Tan Y (2020) A new deep learning algorithm for detecting the lag effect of fine particles on hospital emergency visits for respiratory diseases. IEEE Access 8:145593–145600
21. Roy A, Satija U (2023) A novel melspectrogram snippet representation learning framework for severity detection of chronic obstructive pulmonary diseases. IEEE Trans Instrum Meas 72:1–11
22. Zhou B, Yang X, Zhang X, Curran WJ, Liu T (2020) Ultrasound elastography for lung disease assessment. IEEE Trans Ultrason Ferroelectr Freq Control 67(11):2249–2257
23. Ghita M, Billiet C, Copot D, Verellen D, Ionescu CM (2023) Parameterisation of respiratory impedance in lung cancer patients from forced oscillation lung function test. IEEE Trans Biomed Eng 70(5):1587–1598

Utilizing Satellite Imagery for Flood Monitoring in Urban Regions

Priyanka Sakpal, Shashank Bhosagi, Kaveri Pawar, Prathamesh Patil, and Pratham Ghatkar

Abstract According to the report, billions of people are affected by floods as it is a natural hazard caused by heavy rainfall or glacier melting. Floods occur when there is an overflow of water bodies like rivers, lakes, or oceans, in which there is a heavy flow of water that enters cities and villages causing lots of damage to homes and businesses. To decrease the impact of floods and decrease the loss of life and property, it is important to use the technology available to us to detect the flood-affected areas. But before using the technologies, extraction of water bodies from remote sensing areas is an important task that is mostly done by using a deep learning network for extraction of water bodies. In this paper, we have surveyed the previous research papers which talk about various techniques to predict the condition of water bodies from satellite-captured images and further use it to detect floods. In this paper, we have written the survey table about the accuracy of the techniques used in the paper and a summary of that paper.

Keywords Flood monitoring · Synthetic aperture radar · Normalized difference water index and Sentinel · Convolutional neural network · Dence local feature compression · VGG-Net · Dense-Net

1 Introduction

In recent years, floods have become increasingly frequent and devastating, causing significant economic and humanitarian crises around the world. Effective monitoring and timely response to flood events are critical to mitigating their impact and protecting vulnerable communities. One promising approach in flood monitoring is

P. Sakpal (✉) · S. Bhosagi · K. Pawar · P. Patil · P. Ghatkar
JSPM's Rajarshi Shahu College of Engineering, Pune, Maharashtra 411033, India
e-mail: pssakpal_comp@jspmrscoe.edu.in

S. Bhosagi
e-mail: srbhosagi@jspmrscoe.edu.in

S. Shakya et al. (eds.), *Fourth International Conference on Image Processing and Capsule Networks*, Lecture Notes in Networks and Systems 798,
https://doi.org/10.1007/978-981-99-7093-3_6

the utilization of satellite images, which offer valuable data for understanding and predicting flood dynamics over large areas.

Satellite imagery provides a comprehensive and synoptic view of the Earth's surface, enabling the detection and mapping of flood-affected regions. By leveraging the advancements in remote sensing technology, flood monitoring using satellite images has witnessed remarkable progress. This survey paper aims to provide a comprehensive overview of the state-of-the-art techniques and methodologies employed in flood monitoring using satellite images.

The objective of this survey is threefold:

First, to review the fundamental principles of remote sensing and satellite image analysis as they pertain to flood monitoring. This includes an examination of different satellite sensors, such as optical, synthetic aperture radar (SAR), and multispectral sensors, and their capabilities in capturing flood-related information. Furthermore, we will explore various image pre-processing techniques, such as radiometric and geometric corrections, and image enhancement methods specific to flood monitoring.

Second, to investigate the different approaches for flood detection and mapping using satellite images. This section will delve into the algorithms and methodologies employed for flood detection, including pixel-based, object-based, and machine learning techniques. Additionally, we will explore the challenges and limitations associated with flood detection, such as cloud cover, image resolution, and image classification accuracy.

Third, to highlight the applications and case studies where flood monitoring using satellite images have been successfully utilized. We will discuss how satellite-based flood monitoring contributes to flood risk assessment, early warning systems, and post-disaster management. Moreover, we will explore the integration of satellite data with other data sources, such as hydrological models and geographic information systems (GIS), to enhance flood monitoring and response capabilities.

By examining the existing literature and advancements in the field, this survey paper aims to provide a comprehensive understanding of flood monitoring using satellite images. The knowledge and insights gained from this survey will not only benefit researchers and practitioners but also contribute to the development of more accurate and efficient flood monitoring systems in the future. Ultimately, the effective utilization of satellite imagery can play a pivotal role in minimizing the impact of floods and safeguarding communities worldwide. Flood causes lots of damage to human life and the economy, according to the report published by the Centre for Research on the Epidemiology of Disasters, flood affected about 1 billion people all over the world from 2001 to 2010, and overall damage is 142 billion dollars. Google Earth Engine provides data for flood mapping which saves time in gathering information and accelerates image processing. High-resolution satellite photos are used to verify the effectiveness of the applied strategy.

Many techniques can help us in the detection of floods by using satellite images so that we can efficiently classify flooded and non-flooded areas. Satellites provide a cost-effective and rapid resource for mapping and monitoring the scope of ongoing flood occurrences in real time. Since the 1970s, numerous studies on the mapping and monitoring of floods using evidence from various triggers aboard different satellites have been published in the scientific literature. "Landsat", "TRMM", "GPM", "Terra", "Aqua", and "GRACE" are a few of the satellites frequently used for flood monitoring and forecasting. The delay between data gathering and picture scattering to the user on the base is short for "RADARSAT" and "ENVISAT."

2 Related Work

In the year 2009, research conducted by Martinis et al. [1] developed an approach based on fuzzy set theory to automatically map flooded areas from multispectral S2 MSI pictures in this work. It derives its evidence scores from two band combinations: (1) Spectral index and (2) Element of hue, saturation, and value color transformation. The results prove that performance is better than earlier single-input feature segmentation algorithms, with more consistent and strong accuracy.

In the year 2013, research conducted by Giustarini et al. [2] hybrid approach integrating backscatter thresholding, region growth, and change detection is detailed in this paper as a method for obtaining flood extent from very high-resolution urban SAR images in an automated and repeatable manner. It is based on a statistical distribution of open-water backscatter values derived from flood images. The proposed methodology was shown to be 81.7% accurate.

Mason et al. [3] 2014 conducted research on the double scattering approach, utilized in this paper to detect flooded metropolitan areas using SAR pictures. To assess double scattering durability, the method uses a SAR image and a high-resolution LADAR map of the metropolitan region. This strategy is 100% accurate in flooded areas and 91% accurate in non-flooded areas.

Verwaeren et al. [4] 2015 researched the speckle uncertainty in a flood mapping approach and the bootstrap method which are employed in this paper. Using an image segmentation method, different speckle patterns were found, and speckle was added to the segment mean backscatter value to create fake flood images. F-values of 0.50 and 0.65 indicate the accuracy of two test instances.

In the year 2015, Kersten et al. [5] studied how to detect the flood crisis in real time, a Terra SAR-X-based flood service is paired with a completely automated processing chain in this research. SAR data pre-processing, computation, and change of overall additional data and unsupervised classification utilizing fuzzy logic-based algorithms are all part of the processing chain. The proposed methodology was tested in two locations: Thailand and Germany, yielding 98.5% flood and 71.9% non-flood areas, and 95.37% flood and 83.52% non-flood areas, respectively.

In the year 2016, Luca et al. [6] created a prototype system and offered a mapping algorithm for real-time flood conditions. To identify the flood, this system was evaluated utilizing COSMO-SkyMed and Sentinel-1 data. The precision was calculated to be 98%.

Renaud et al. [7] in the year 2016 investigated research in which they used the data of synthetic aperture radar and they have also used a probabilistic flood mapping approach. A Bayesian technique was applied in this case. The likelihood of a given pixel was calculated by using the readings taken from a histogram of backscatter.

Chapi et al. [8] 2017 proposed two models for mapping flood susceptibility, bagging-LMT—a combination of bagging ensemble and logistic model tree. Here the bagging-LMT model outperforms the logistic model tree, logistic regression, Bayesian logistic regression, and random forest models. An accuracy of 99.3% was achieved for the training dataset and an accuracy of 95.5% on the dataset for testing the bagging-LMT model.

In the year 2018, Notti et al. [9] present a semi-automated mapping of flood based on the available open-source satellite images. Various techniques are used to detect flooded areas utilizing MODIS, Proba-V, Landsat, and Sentinel-2, multispectral satellite data. To detect places that have been flooded, semi-automatic extraction is used. After that, utilizing auxiliary data web-based data and data such as the digital elevation model by using this map of the flood is manually purified.

Wang et al. [10] in the year 2018 studied the SRFIM-MSI approach using the normalized difference water index to extract spectral information from the multispectral band. Two Landsat 8 OLI multispectral data were used to assess the proposed technique, yielding an overall accuracy of 96.81%.

Murphy et al. [11] 2018 conducted research on deep convolutional and recurrent neural networks for the semantic segmentation of UAV imagery. More correct, expressive, and meaningful division maps were obtained using an end-to-end integration network and also used the combination of RNN and CNN networks.

In the year 2018, Bioresita et al. [12] investigated a completely automated chain of processing for rapid flood and surface water mapping. The main goal is to see if Sentinel-1 data would be used for flood detection and surface water mapping, as well as to develop an automated surface water extraction chain processing system. The accuracy level reached was 98.9%.

Martinis et al. [13] in the year 2018 study on a system for detecting changes automatically based on data from Sentinel-1 SAR. A Jensen–Shannon divergence-based index is used to choose the reference picture from a pool of possible image candidates. Second, using log-ratio data, saliency detection is utilized to determine the prior probability of impacted and unaffected classes.

In the year 2019, Li et al. [14] researched the multitemporal TerraSAR-X data to look at the contribution of the intensity of SAR and interferometric coherence in flood detection in urban areas. In this structure, the multitemporal intensity is essential.

Rudner et al. [15] 2019 proposed the rapid segmentation approach for flooded areas by fuzzy multiresolution, multisensory, and multitemporal satellite images by using multitemporal satellite images. It is very easy to differentiate between flooded areas and other sections. It is fundamental to the larger problem of automated maps from satellite imagery. These maps are useful for tasks like disaster management or locating eligible rooftops for solar panels.

Nallapareddy et al. [16] in the year 2019 studied thresholding and supervised classifications methods to classify flooding areas. The utilization of estimating area data that is considered typical of each rock type to be categorized is referred to as supervised classification.

In the year 2019, Shen et al [17] studied near-real-time flooding maps created using satellite information. Here, NRT system called "Radar Produced Inundation Diary (RAPID)" involved four steps of classification based on multi-threshold and compensation, statistics, machine learning correction, and morphological processing.

Han et al. [18] 2019 researched on "unsupervised detection" of flood by using the "synthetic aperture radar" system for the data. The large data faces the problems like automatic thresholding as some data shows unimodal. For multi-scale chessboard segmentation-based tiles, there are statistical concepts such as bimodal Gaussian distribution and non-parametric histogram-based method.

Huang et al. [19] 2020 investigated how SAR is used to penetrate clouds and heavy rainfall. Sentinel-1 SAR image-set are used to cover the location and backscatter classification with the assistance of Sentinel-2 images using a supervised classifier. By using the same process, SAR pictures categorized the pre-flood and flooding duration to understand "change detection" by uniting them.

In the year 2020, DeVries et al. [20] studied SAR sensors which shows the essential data for flood disaster planning. As "Sentinel-1" SAR satellite data is freely available [GEE]. The authors used very high-resolution optical imagery with an accuracy of $89.8 \pm 2.8\%$ (95%).

Bonafilia et al. [21] in the year 2020 researched "Sen-1" floods 11used to justify the full "convolutional neural network" to parts the flood water. Differentiate the permanent flood and total surface water here using the 4 data divisions learners from training as "FCNN" as this dataset consists of 4831 512 * 512 chips coves 120,406 km^2 so it divides into four parts (1) 446 chips cover water surface from the event of flood (2) 814 chips for the permanent surface of water (3) 4385 chips of water surface classifies the "Sen-1" pictures from the flooding conditions.

In the year 2020, Li et al. [22] conducted research on dynamic water bodies. So to detect such water bodies, old dataset gives some limitations, but the advanced cloud-based "Google Earth Engine" platform with freely available data from "Sentinel-1" is very much useful.

Jain et al. [23] in the year 2020 studied on automatic identification of water from satellite by old methods whose accuracy is not that perfect, so to avoid a deep convolutional neutral network with the addition of water index-based solutions found. Here by combining the normalized difference water index with the blue/NIR spectral index, they found shallow water.

In the year 2020, Ross et al. [24] research on SAR used to find flooded areas. Data provided by the SAR is very effective for flood detection. First, here divide the region into two categories for example (a) monsoon region and (b) not rainy season. Then other steps of the algorithm work.

Hashemi-Beni et al. [25] in the year 2021 researched deep learning-based approaches used in flood detection with the help of a "convolutional-neutral-network" from the optical images. Using a digital electoral model, the region-expanding approach assesses the degree of flooding beneath vegetation that is difficult to observe from an imaginative perspective.

Meysam et al. [26] in the year 2021 studied flood events monitored by "Sentinel-1" images. The Otsu threshold method is used to separate flooded areas. Accuracy is more than 90%.

In the year 2021, Li et al. [27] conducted research on deep learning which used remote sensing images for the separation of water bodies from other areas. So the new network is used which is a dense local feature compression network. The authors created a new dataset called "GaoFen-2". Authors give good results by using "Sentinel-2 ZY-3 images" comparing old methods, and the accuracy of that network is above 90%.

3 Major Techniques Used

3.1 *Ordered Weighted Averaging*

Ordered weighted averaging is an aggregation technique between the logical "or" and "and" which allow for a combination of multiple inputs using weighted averages. Ordered weighted averaging is a step to assign weights to the model updates based on their performance. It is generally used in computational intelligence which is work as an optimized algorithm to train neural networks refers as the federated averaging algorithm. Federated averaging is an optimization algorithm in the field of federated learning, specifically designed for training neural networks. By performing local training on the dataset at the central server extracting features and identifying flood-related patterns. The model updates, in the form of gradients, are then aggregated using federated averaging. This process ensures that the shared model incorporates the collective knowledge and insights from the distributed devices. It contains three major components:

- Central Server: It coordinates the federated learning process. Here initialization of the model is done; it is used to manage the overall training process.
- Client Devices: These may be any smart device like IoT devices. It performs the model training locally using its data.
- Communication Protocol: Rules for communication between the central server and client devices are defined here. It ensures secure communication with data privacy about the model updates.

By using both techniques collaboratively, flood monitoring systems can benefit from the collective intelligence of the distributed satellite devices which may be any smart device like IoT devices while accounting for the varying performance levels or reliability of each device. This collaboration will create a more accurate and comprehensive model for flood detection and prediction.

The following steps are used which are as follows.

a. For each input attribute from estimating data, define data partitioning.

b. Define and choose the ordered weighting average that will be used in the soft integration. And apply federated averaging phases, such as model initialization, local training, model update, and communication and aggregation.
c. Soft-integration builds on pixels to find out the synthetic general evidence degree.
d. Segmentation of synthetic overall verification for creating maps of flooding areas based on the decision-maker mindset.

3.2 *Spectral Indices*

Spectral indices, mostly used in multispectral remote sensing areas, is a mathematical function that is the combination of pixel values from two or more than two spectral bands. It is applied on more than two reflectance bands of the image. It is based on the spectral properties of the object of interest like identifying the properties of the earth's surface.

The steps to create a map of flooding areas using spectral indices:

- Gather satellite data: Gather the data for flooded areas. To spot the changes in land cover and water presence, it is ideal to use photos taken before and after the flood event.
- Pre-process the imagery: To adjust atmospheric conditions, geometric distortions, and radiometric variances, pre-process the satellite images.
- Calculate spectral indices: The methods like normalized difference water index (NDWI), normalized difference vegetation index (NDVI), and modified normalized difference water index (MNDWI) are used to calculate the spectral indices.
- Thresholding and classification: After calculating the spectral indices, we should classify the flooded areas using the appropriate threshold values.
- Create a flooded area map: Construct a binary map labeling flooded areas as "1" and non-flooded areas as "0" to create the flooded area ap by applying the threshold values to the spectral index images.
- Post-processing and visualization: Refine the binary map by using morphological operations to eliminate artifacts and smooth the borders such as erosion and noise reduction of an image. Finally, for visualization and interpretation, place the map of the flooded area over a base map.

3.3 *Region Growing*

It is an image segmentation approach. It is also a pixel-based picture segmentation approach that necessitates seed point selection which is particularly used in image processing to recognize the features of an area of interest. A region grows from this seed point and segments the image. Figure 1 is the image that is used to depict the region's growth.

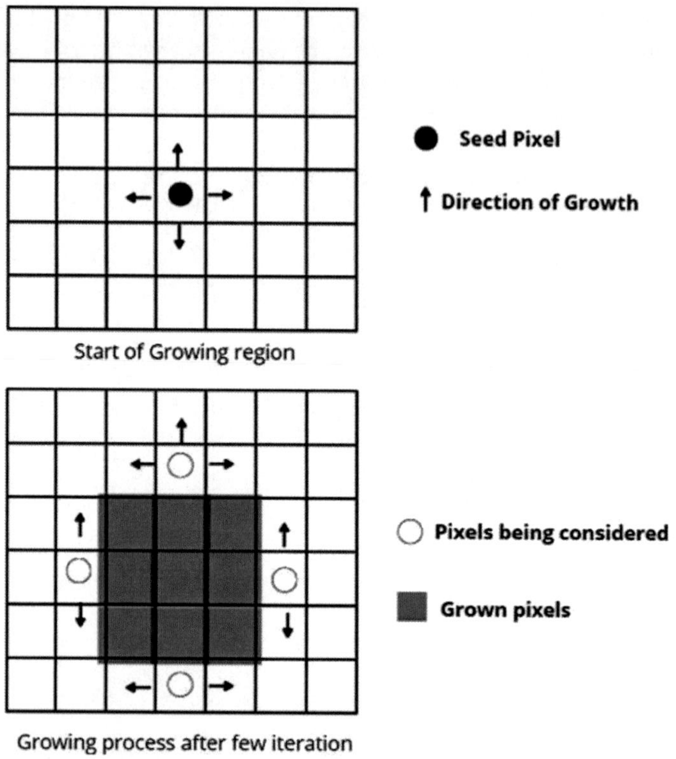

Fig. 1 Region growing [28]

In this algorithm first need to select the seed pixel of the selected region. After that pixel algorithm checks the neighboring pixel and decides whether in the region the pixel should be added or not by examining if those pixels meet certain define criteria or not like pixels with the same intensity or the color values of the pixel. And if those pixels meet that criterion, then by adding those neighbor pixel to the seed pixel the region starts to grow. It can be executed by the plain threshold approach.

3.4 Double Scattering

There are two approaches are used to find the double scattering in the urban flooded areas. The first approach used a single SAR which included flooded areas images. Most of the urban areas in the Terra SAR-X pictures are unflooded, and those areas are called aerial photos which may be flooded or maybe not. SAR images are much more easily available than aerial photos. Even only using synthetic aperture radar images unflooded areas would also detect.

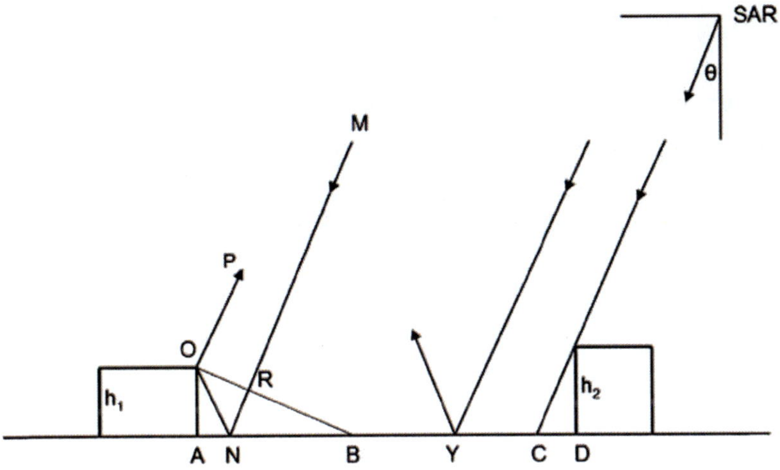

Fig. 2 Double scattering [3]

And the second approach for double scattering used the concept of multiple-scattering theory as detecting the change between the flooded and the unflooded pictures which is used for much complex and heterogeneous media (Fig. 2).

3.5 Bootstrap Method

It is a replacement sampling resampling strategy for estimating statistics. It can be used to compute the mean and standard deviation of the sample data. It is a metric used in applied machine learning to assess how well models perform when they are generating predictions on data that was not part of the training set.

The steps used are as follows:

- Step-1: Pick several bootstrap trials and obtain the sample data of the region of interest.
- Step-2: Choose a select size of test.
- Step-3: Choose several samples from the original sample of the interested region with equal size using the chosen measurement.
- Step-4: Evaluate the sample statistic.
- Step-5: Assess the sample statistics mean using the calculated data and use those distributed samples to calculate errors or execute the hypothesis test on the earliest sample.

3.6 Fuzzy Logic-Based Post Classification

For post-classification fuzzy logic-based algorithm was used with initial labeling. It is a useful method for combining clear data sources by considering their uncertainties to improve the accuracy of classification. Here, first image will convert into the fuzzy set, and according to that set classified result gives the membership value which is nothing but represents the degree of pixels to a different class. By merging this degree with specific information refining of data will be done. And fuzzy logic is used where data is undefined to increase the reliability of the expected result.

The fuzzy threshold helps to determine the degree of similarity which is also called membership value. And it converts images into binary regions.

The fuzzy thresholds are defined as [5],

$$x_{1[h]} = \mu_{h(\text{water})}$$

$$x_{2[h]} = \mu_{h(\text{water})} + f_\sigma * \sigma_{h(\text{water})}$$

where $x_{1[h]}$ and $x_{2[h]}$ are variables, $\mu_{h(\text{water})}$ = mean of the elevation of initially derived water objects and $\sigma_{h(\text{water})}$ = standard deviation of derived water objects.

The factor f_σ is defined by [5]:

$$f_\sigma = \sigma_{h(\text{water})} + 2.5$$

This feature was included to lessen the impact of the digital elevation model in areas of low topography.

3.7 Probabilistic Flood Mapping

The procedure of flood-plain mapping is essential for a number of processes in a probabilistic manner. These procedures include

1. Creating models for flood overflow.
2. Using data from earlier floods, carry out a sensitive inspection of the model.

At each *i*th cell, a weighted total of each Monte Carlo simulation is used to create the probabilistic map.

The flooded state C_j is, therefore [29],

$$C_j = \frac{\sum_k L_k W_{ki}}{\sum_k L_k}$$

where C_j = weighted average flood state for *j*th cell, W_{ki} = a moist cell, and L = a probability weight allocated to each *k*th simulation result and calculated using [29].

$$L_j = \frac{F_k - \min(F_k)}{\max(F_k) - \min(F_k)}$$

F	fit metric that shows how well a replicate flood extent detects the flood extent and
$\max(F)$ and $\min(F)$	minimum and maximum measures of fit metric.

3.8 *Normalized Difference Vegetation Index*

The NDVI is nothing but an unlimited index that measures the thickness of green on a segment of land by describing the distinction in plant cover reflectance linking visible and near-infrared light.

It is calculated as [9],

$$\text{NDVI} = \left(\frac{\text{NIR} - \text{RED}}{\text{NIR} + \text{RED}}\right)$$

where

NIR	reflection in the near-infrared spectrum and
RED	reflection in the red range of the spectrum.

3.9 *Modified Normalized Difference Water Index*

The modified normalized difference water index utilizes green and SWIR bands to amplify open-water features. It is used to lower the characteristics of built-up areas connected to open water in other indexes. This is calculated using the below formula [9].

$$\text{MNDWI} = \left(\frac{\text{Green} - \text{SWIR1}}{\text{Green} + \text{SWIR1}}\right)$$

where

Green	green bands pixel values.
SWIR	Short-wave infrared pixel values and SWIR is defined as band11(1620 nm).

3.10 *Normalized Difference Water Index (NDWI)*

The normalized difference water makes use of open-water features in a satellite image, and it is nothing but the index that is used to identify the scope of water bodies in remote areas and also used to extract the area of land which is covered by water. NDWI index successfully calculates the moisture content. The GREE-NIR mixture is used to enumerate the NDWI, allowing for the identification of minute changes in the water content of water bodies [19].

$$\text{NDWI} = \left(\frac{\text{Green} - \text{NIR}}{\text{Green} + \text{NIR}}\right)$$

where

NIR near-infrared band and
Green green band.

Due to the specific properties of water and the high wavelength of NIR, it penetrates through water easily. So, a combination of near-infrared bands absorbing light causes darker in the NIR band which is high as compared to other bands. Due to those properties, NDWI is used to detect water bodies from the images of remote sensor areas.

3.11 CNN (Convolutional Neural Network)

A CNN which is a deep learning network refers as a convolutional neural network. A CNN takes input as an image and according to weights assigned to each feature in the image differentiates one from the other. It is used for classification purposes for many visual applications like images and pictures. CNN has a multilayer perceptron Fig. 3, which means that all neurons present in one layer are connected to all neurons present in the next layer. A neural network contains minimum input of three layers, a hidden network, and an output. In the input layer, the neurons take the pixel values as input, then adjust the values of input using the activation function. Rectified linear unit (ReLU) is a mostly used activation function due to its simplicity, and it simply set all negative values to zero and keeps positive values unchanged. Using this activation value, network's expressive power enhances. After input it has a hidden network that can contain any number of layers, and then, it has the output layer which has the result.

First, the image is taken, and with the help of a filter or kernel, the feature vector of the image is gathered and then used to generate feature maps. Then those feature maps are fed into a fully connected layer for classification, and then, output is generated.

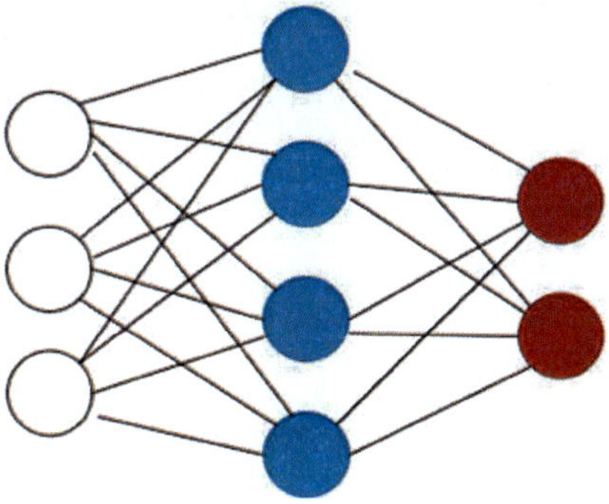

Fig. 3 Multilayer perceptron

4 Literature Survey

See Table 1.

5 Observation on Literature Survey

After the survey of 27 research papers, our paper summarizes techniques for image segmentation. In recent years, researchers have started using "machine learning algorithms" and "neural networks" for segmentation like "CNN", "FCNN", and "RCNN". Some researchers also mixed multiple algorithms to get more accurate results, e.g., "backscatter thresholding", "region growing", and "change detection". We also got some information about the free resource that can be used to get the free satellite images.

Machine learning algorithms and neural networks, such as convolutional neural network (CNN), fully convolutional neural network (FCNN), and region-based convolutional neural network (RCNN), have indeed gained popularity for image segmentation tasks due to their ability to learn complex patterns and features.

Combining multiple algorithms is a common approach to improve segmentation accuracy. Techniques like backscatter thresholding, region growing, and change detection can complement the capabilities of machine learning algorithms and neural networks. These methods often leverage specific characteristics of the satellite images or exploit contextual information to refine the segmentation results.

Here, we use image segmentation methods like Region Growing to create an initial estimate of the flooded areas, Convolutional Neural Network to improve the particular patterns and features connected to flooding and Dense-Net (Densely Connected Neural Network) use to design exact border localization, and the addition of contextual data, improving the segmentation outcomes by precisely defining the flooded areas within the satellite images.

Table 1 Literature survey table

References	Performance matrix	Summary
[1]	*F*-score > 0.90 *F*-score decreased to less than 0.75 in heterogeneous conditions *F*-score is a statistical measure used in binary classification Formula: *F*-score = 2 * (precision * recall)/(precision + recall) The *F*-score ranges from 0 to 1, where a value of 1 indicates perfect precision and recall, while a value of 0 indicates poor performance	This paper studied the single-polarized high-resolution SAR satellite data. With a focus on SAR data, the paper presents an automatic flood detection system that combines histogram thresholding with segmentation-based categorization
[2]	Total good i.e., the accuracy by which the methods could differentiate between flooded and non-flooded regions on SAR image Method M1 = 81.6% Method M2a = 81.8% Method M2b = 81.7%	Synthetic aperture radar sensors are helpful to represent build-up environments where the flood risk is high. This paper has introduced M2b methods which is an advanced version of M2a. Method M1 focuses on SAR images and extracts the water pixels from the image. Method M2a includes the non-flooded image as a reference to detect the water and enhance the performance of the algorithm
[3]	Flooded double scattering detects 100% accuracy Unflooded with 91% accuracy	Here double scattering method is used in urban areas. Using the SAR image data and high-resolution LiDAR height map of the metropolitan region, this approach scattering intensities. The LiDAR data estimates the scattering curves in the images
[4]	*F*-value = 0.50 to 0.65 Where *F*-value is the ratio of two variances	In this paper, two strategies are used which are a non-parametric bootstrap method and another method is flood mapping procedure

(continued)

Table 1 (continued)

References	Performance matrix	Summary
[5]	Thailand Flood accuracy = 98.55% Non-flood accuracy = 71.99% Germany Flood accuracy = 95.37% Non-flood accuracy = 83.52%	Terra-SAR-X SAR using high resolution for real-time flood mapping or detection fully automatic processing chain was provided
[6]	Overall accuracy = 98%	This paper has proposed the algorithm known as flood mapping algorithm that can get SAR data from satellites that are based on the flood forecast. In SAR using dark objects classifier separates the non-flooded pixel and flooded pixel from images
[7]	By applying this method to four images of two rivers, we found error values were ranging from 0.04 to 0.23 It indicates the good performance of the proposed system	The probabilities flood mapping approach is employed in this paper, which is based on SAR data

(continued)

Table 1 (continued)

References	Performance matrix	Summary
[8]	The performance for estimating the dataset of the new hybrid bagging-LMT model was accuracy = 99.3%, kappa coefficient 0.949, which is highest among the other methods The representation for evidence dataset of the new hybrid bagging-LMT model was accuracy = 95.5%, kappa coefficient of 0.863, which is highest among the other methods The kappa coefficient (also known as Cohen's Kappa) is a statistical measure used to assess the agreement between two raters or observers Formula Kappa = (Po − Pe)/(1 − Pe) Where • Po is the observed proportion of agreement • Pe is the proportion of agreement expected by chance	This paper has proposed a "bagging-LMT" model for mapping flood vulnerability. It was validated using data from the Haraz watershed in Iran. LMT, logistic regression, and random forest are compared with the proposed model against soft computing models
[9]	Geomorphic-based flood maps are 95% similar to official flood maps available	This paper has proposed a semi-automatic method for flood mapping using freely available images and software. Multispectral data obtained by MODIS, Landsat, Senitel-2, SAR by Sentil-2, and Proba-V was used to detect flooded areas
[10]	Overall accuracy (traditional method DPSO) = 94.65% overall accuracy (proposed method SRFIM-MSI) = 96.81%	This paper has proposed a novel approach to the SRFIM by providing more spectral information to enhance flood mapping

(continued)

Table 1 (continued)

References	Performance matrix	Summary
[11]	In UAV dataset included the semantic segmentation of flooded areas with an accuracy of 96% and MioU of 92%	In this paper, two dense networks are employed which are CNN and RNN used for correctly segmenting out object boundaries and utilizing it and end learning. The network used for aerial image separation in UAV for flooded areas in Houston, TX
[12]	The overall accuracy of the algorithm when applied on England December 29, 2015 SAR data is 98.86%	Fully autonomous processing was employed in this paper. The procedure is tested on three separate places that have recently been flooded. It has an overall accuracy of above 98% with the value calculated by using F-measure that are between 0.64 and 0.92
[13]	A kappa value of 0.9238 was obtained by the flood crisis at Evros River, Greece A kappa value of 0.8682 was obtained by the flood events at Ouse River and Wharfe River in York, United Kingdom	In this presentation, quick flood mapping uses a two-step automated change detection chain based on Sentinel-1 SAR. A Jensen Shannon uses divergence-based index to choose a reference picture from a pool of probable image choices, and a log-ratio-based saliency detection is used
[14]	Overall Accuracy of supervised CNN $= 90.39 \pm 0.32\%$ A-SL CNN $= 92.82 \pm 0.10$ K statistic of supervised CNN $= 0.614 \pm 0.010\%$ A-SL CNN $= 0.686 \pm 0.004\%$	This study employed multitemporal "TerraSAR-X" data to detect urban floods using SAR intensity. Here, an active self-learning convolutional neural network framework was developed
[15]	Multi3Net has a 93.7% accuracy rate in segmenting flooded structures from non-flooded structures	They suggest a method for rapidly segmenting flooded buildings by combining multiresolution, multisensory, and multitemporal satellite images in a CNN in this paper. Using publicly available medium-resolution data, this program creates highly accurate segmentation maps of flooded buildings. Multi3Net was utilized to segment flooded structures by fusing multiresolution, multisensory, and multitemporal satellite images, and the results showed that full-fusion beat other fusion methods
[16]	–	This study explores thresholding and unsupervised classification approaches for locating areas in the Rapti and Ghaghara rivers in August 2017. Here, optical images and high-resolution multitemporal SAR were employed. It indicates that SAR data can be sufficient for flood water mapping and flood monitoring

(continued)

Table 1 (continued)

References	Performance matrix	Summary
[17]	The automated RAPID system exhibited producer and user accuracies of 93%, 77%, and 75%, respectively	We need real-time maps in case of a flood. Here, they developed an NRT system called Radar produced a diary for processing Sar data and producing reliable inundation maps
[18]	In VV polarization RGA-based LM thresholding gave the highest kappa coefficient of 0.91, with an overall accuracy of 98.82% In VH polarization TRS-based KI thresholding gave the highest kappa coefficient around 0.89, with an accuracy of 98.59%	This paper has introduced a multi-scale chessboard segmentation-based tiles selection method and a non-parametric thresholding algorithm for identifying water areas. Various methods like Otsu thresholding, KI thresholding, LM thresholding, TRS, and RGA were discussed
[19]	Overall accuracy = 80.95% kappa value = 74.1%	Change detection is an aspect, and thresholding selection is the most important part of flood mapping. Here they proposed a novel approach where there will be no need for thresholding selection. Instead of using change detection of thresholding, it uses land cover backscatter classification
[20]	Using high-resolution images they estimated an area-normalized accuracy of 89.8 ± 2.8% (95% c.i.), over Houston, Texas Thessaly, Greece, and Eastern Madagascar in January and March 2018, respectively, receive an overall rate of around 98.5% by comparing the results with the flood map derived by SAR	This paper has proposed an algorithm that takes all the Senitel-1 SAR data and other data required hosted on Google Earth Engine to map the flooding. This algorithm depends upon multitemporal SAR images to identify unexpected floods. The proposed algorithm eliminates the time-consuming data download and pre-processing

(continued)

Table 1 (continued)

References	Performance matrix	Summary
[21]	–	The Sen1 Flood11 dataset was utilized to estimate, verify, and test the neural network which is fully convolutional to segment flood water from the permanent water surface. The deep learning model for radar flood detection performs more accurately with flood water training labels rather than just permanent surface water
[22]	Overall classification accuracy of the HSWDC dataset was 93% and the kappa coefficient of 0.86	In this paper, they used the Google Earth Engine platform and Sentinel-1 imagery to map the surface of water bodies dynamically on a large scale with high spatial and temporal precision. Here uses threshold approach for SAR-based water mapping for the large-scale area with variant land cover types; samples across China are created
[23]	Their system gave an $F1$ score of around 96% and a kappa score of around 92% which good among the other systems	Using the Media Eval 2019 flood dataset, this research paper attempted to combine water index-like characteristics with a deep CNN-based solution for flood identification
[24]	–	In this paper, the approach is to process the SAR pictures of the same location from two different dates and establish flooded and non-flooded differentiation to provide a realistic estimate if the region is flood-prone or not

(continued)

Table 1 (continued)

References	Performance matrix	Summary
[25]	RG (Scenario-1 and no vegetation) = 70.5% RG (Scenario-1 and vegetation) = 81.7% RG (Scenario-1 overall accuracy) = 77.3% RG (Scenario-2 and no vegetation) = 90.7% RG (Scenario-2 and vegetation) = 92.1% RG (Scenario-2 overall accuracy) = 91.8% Integrated method (Scenario-1 and no vegetation) = 98.6% Integrated method (Scenario-1 and vegetation) = 81.7% Integrated method (Scenario-1 overall accuracy) = 88.4% Integrated method (Scenario-2 and no vegetation) = 98.6% Integrated method (Scenario-2 and vegetation) = 92.1% Integrated method (Scenario-2 overall accuracy) = 92.4%	This paper proposed an integrated approach for mapping the flood extent using the concept of FCN deep learning and RG. The digital elevation model was used to find the extent of flood beneath vegetation, and flood surfaces are extracted from high-resolution UAV imagery using the FCN model based on deep learning. Using an integrated strategy, floods can be detected in both visible and hidden places

(continued)

Table 1 (continued)

References	Performance matrix	Summary
[26]	The general reliability was higher than 90%, confirming the pertinence of the Otsu which is the automatic thresholding method in flood mapping	In this research paper, they used SAR images from Senital-1 and studied a flood-affected area in Iran. They have developed an algorithm called the Otsu thresholding algorithm, which will separate flood-affected areas from remaining land cover
[27]	Their overall accuracy was about 98.44% which is among the highest accuracy given by other methods, $F1$-score was about 95.39%, and IoU was 91.25% $F1$-score also known as F-score F-score is a statistical measure used in binary classification Formula F-score = 2 * (precision * recall)/(precision + recall)	Here using dense local feature compression network proposed a novel with the goal of automatically removing water bodies from various remote sensing photos

6 Proposed Architecture

The suggested model will direct the development of a system that will be utilized to minimize any tragedy from the flood by alerting authorities about the state of the water. The two core modules of suggested concepts for real-time flood monitoring systems utilizing satellite imagery are water detection and water body extraction. Our research indicates that CNN designs are the best option, but for real-time applications, CNN's DenseNet architecture is the greatest option. DenseNet is the best option for our system because of its user-friendliness and quicker, more precise results. The suggested model uses DenseNet, which is especially suited for image classification tasks, and it is the backbone for this proposed model. During training, the pre-processed images are fed as input to the DenseNet model, and using DenseNet, features are extracted very effectively throughout the network. DenseNet has dense blocks where every layer is connected to every other layer, helping with feature reuse and addressing the vanishing gradient problem.

7 Methodology

7.1 Training

Given train dataset with images and given a label for those images as below

$$\text{Dataset} = \{(x_1,\ y_1),\ (x_2,\ y_2),\ \ldots,\ (x_n,\ y_n)\}.$$

Pre-processing the image steps include:

- Step-1: Resize the images of the training dataset.
- Step-2: Normalization of images by dividing the pixel value.
- Step-3: To improve or increase the area cover by rotating and transforming the images, data augmentation techniques are mostly used.
- Step-4: Here subtract the mean value for each image.
- Step-5: Standardized the pixel of images.

Training Model

$X = \{x_1, x_2, \ldots, x_n\}$
//Set of images
$P = \{p_1, p_2, \ldots, p_n\}$
//Pre-processed images
for each x of p

p = pre-processed images(x);

end for

Trained Model = CNN (*P*).

7.2 Testing

Taking the sample images after using this dataset as input for the trained model's final output will help to define the meaning of the given image.

8 Conclusion and Future Scope

In this work, we have studied various methods or strategies for software-based systems to detect floods from various types of satellite images. We have studied research papers from reputed journals. In Table 1, we have included the performance of the method used and a summary of the same. Figure 4 has a pie chart that shows the usage percentage of the image segmentation technique. The survey summarizes, we learned many new techniques of extracting the water bodies from the satellite images using image segmentation techniques and machine learning algorithms, and further use it for detection of the flood. The research papers which we referred to for this survey were published between 2009 and 2021. We discovered that there are not many applications like this for alert messages in urban areas, so we are creating one for flood monitoring. We will also try to improve the system's accuracy, which was previously described. In the existing system, processing the image takes a lot of time. However, we will also make improvements to the system's ability to analyze pictures more quickly by utilizing DenseNet architecture, a sort of CNN design (Fig. 5).

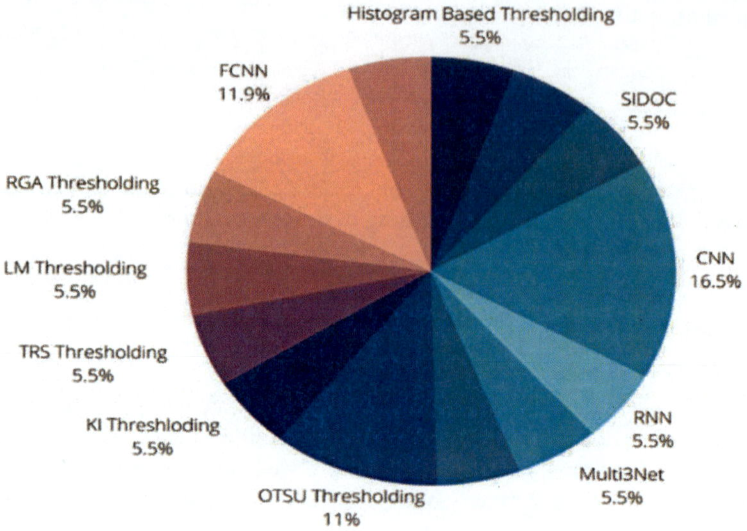

Fig. 4 Usage of various methods used to extract water from the satellite image

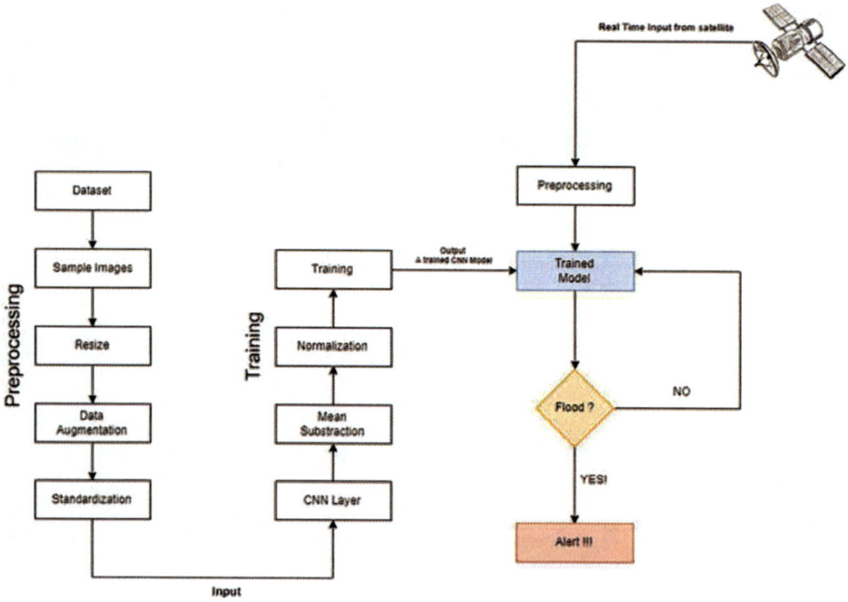

Fig. 5 Flowchart of system architecture

References

1. Martinis S, Twele A, Voigt S (2009) Towards operational near real-time flood detection using a split-based automatic thresholding procedure on high-resolution TerraSAR-X data. Nat Hazard 9:303–314. https://doi.org/10.5194/nhess-9-303-2009
2. Giustarini L, Hostache R, Matgen P, Schumann GJ, Bates PD, Mason DC (2013) A change detection approach to flood mapping in urban areas using TerraSAR-X. IEEE Trans Geosci Remote Sens 51(4):2417–2430. https://doi.org/10.1109/TGRS.2012.2210901
3. Mason DC, Giustarini L, Garcia-Pintado J, Cloke HL (2014) Detection of flooded urban areas in high-resolution Synthetic aperture radar images using double scattering. Int J Appl Earth Observ Geoinf 28:150–159. ISSN 1569-8432. https://doi.org/10.1016/j.jag.2013.12.002, https://www.sciencedirect.com/science/article/pii/S0303243413001700
4. Giustarini L, Vernieuwe H, Verwaeren J, Chini M, Hostache R, Matgen P, Verhoest N, De Baets B (2015) Accounting for image uncertainty in SAR-based flood mapping. Int J Appl Earth Observ Geoinf 34:70–77. https://doi.org/10.1016/j.jag.2014.06.017
5. Martinis S, Jens Kersten, André Twele, A fully automated TerraSAR-X based flood service. ISPRS J Photogrammetry Remote Sensing 104:203–212. ISSN 0924-2716. https://doi.org/10.1016/j.isprsjprs.2014.07.014, https://www.sciencedirect.com/science/article/pii/S0924271614001981
6. Boni G et al (2016) A prototype system for flood monitoring based on flood forecast combined With COSMO-SkyMed and Sentinel-1 data. IEEE J Sel Top Appl Earth Observ Remote Sens 9(6):2794–2805. https://doi.org/10.1109/JSTARS.2016.2514402
7. Giustarini L et al (2016) Probabilistic flood mapping using synthetic aperture radar data. IEEE Trans Geosci Remote Sens 54(12):6958–6969. https://doi.org/10.1109/TGRS.2016.2592951
8. Chapi K, Singh V, Shirzadi A, Shahabi H, Bui D, Pham B, Khosravi K (2017) A novel hybrid artificial intelligence approach for flood susceptibility assessment. Environ Model Softw 95:229–245. https://doi.org/10.1016/j.envsoft.2017.06.012
9. Notti D, Giordan D, Caló F, Pepe A, Zucca F, Galve JP (2018) Potential and limitations of open satellite data for flood mapping. Remote Sens 10:1673. https://doi.org/10.3390/rs10111673
10. Wang P, Zhang G, Leung H (2019) Improving super-resolution flood inundation mapping for multispectral remote sensing image by supplying more spectral information. IEEE Geosci Remote Sens Lett 16(5):771–775. https://doi.org/10.1109/LGRS.2018.2882516
11. Rahnemoonfar M, Murphy R, Miquel MV, Dobbs D, Adams A (2018) Flooded area detection from UAV images based on densely connected recurrent neural networks. In: IGARSS 2018—2018 IEEE international geoscience and remote sensing symposium, pp 1788–1791. https://doi.org/10.1109/IGARSS.2018.8517946
12. Bioresita F, Puissant A, Stumpf A, Malet J-P (2018) A method for automatic and rapid mapping of water surfaces from Sentinel-1 imagery. Remote Sensing 10:217. https://doi.org/10.3390/rs10020217
13. Li Y, Martinis S, Plank S, Ludwig R (2018) An automatic change detection approach for rapid flood mapping in Sentinel-1 SAR data. Int J Appl Earth Observ Geoinf 73:123–135. ISSN 0303-2434. https://doi.org/10.1016/j.jag.2018.05.023, https://www.sciencedirect.com/science/article/pii/S0303243418302782
14. Li Y, Martinis S, Wieland M (2019) Urban flood mapping with an active self-learning convolutional neural network based on TerraSAR-X intensity and interferometric coherence. ISPRS J Photogramm Remote Sens 152:178–191. https://doi.org/10.1016/j.isprsjprs.2019.04.014
15. Rudner T, Rußwurm M, Fil J, Pelich R, Bischke B, Kopackova-Strnadova V, Biliński P (2019) Multi3Net: segmenting flooded buildings via fusion of multiresolution, multisensor, and multitemporal satellite imagery. Proc AAAI Conf Artif Intell 33:702–709. https://doi.org/10.1609/aaai.v33i01.3301702
16. Nallapareddy A, Varadharajulu B (2019) Flood detection and flood mapping using multitemporal synthetic aperture radar and optical data. Egypt J Remote Sens Space Sci 23.https://doi.org/10.1016/j.ejrs.2019.01.001

17. Shen X, Anagnostou E, Allen G, Brakenridge R, Kettner A (2019) Near-real-time non-obstructed flood inundation mapping using synthetic aperture radar. Remote Sens Environ 221:302–315. https://doi.org/10.1016/j.rse.2018.11.008
18. Cao H, Zhang H, Wang C, Zhang B (2019) Operational flood detection using Sentinel-1 SAR data over large areas. Water 11:786. https://doi.org/10.3390/w11040786
19. Huang M, Jin S (2020) Rapid flood mapping and evaluation with a supervised classifier and change detection in Shouguang using Sentinel-1 SAR and Sentinel-2 optical data. Remote Sensing 12:2073. https://doi.org/10.3390/rs12132073
20. DeVries B, Huang C, Armston J, Huang W, Jones JW, Lang MW (2020) Rapid and robust monitoring of flood events using Sentinel-1 and Landsat data on the Google earth engine. Remote Sensing Environ 240:111664. ISSN 0034-4257. https://doi.org/10.1016/j.rse.2020.111664, https://www.sciencedirect.com/science/article/pii/S003442572030033X
21. Bonafilia D, Tellman B, Anderson T, Issenberg E (2020) Sen1Floods11: a georeferenced dataset to train and test deep learning flood algorithms for Sentinel-1. IEEE/CVF Conf Comput Vision Pattern Recogn Works (CVPRW) 2020:835–845. https://doi.org/10.1109/CVPRW50498.2020.00113
22. Li Y, Niu Z, Xu Z, Yan X (2020) Construction of high spatial-temporal water body dataset in China based on Sentinel-1 archives and GEE. Remote Sens 12:2413. https://doi.org/10.3390/rs12152413
23. Jain P, Schoen-Phelan B, Ross R (2020) Automatic flood detection in Sentinel-2 images using deep convolutional neural networks. https://doi.org/10.1145/3341105.3374023
24. Jardosh P, Kanvinde A, Dixit A, Dholay S (2020) Detection of flood prone areas by flood mapping of SAR imagery. Third Int Conf Smart Syst Inventive Technol (ICSSIT) 2020:814–819. https://doi.org/10.1109/ICSSIT48917.2020.9214089
25. Hashemi-Beni L, Gebrehiwot AA (2021) Flood extent mapping: an integrated method using deep learning and region growing using UAV optical data. IEEE J Sel Top Appl Earth Observ Remote Sens 14:2127–2135. https://doi.org/10.1109/JSTARS.2021.3051873
26. Moharrami M, Javanbakht M, Attarchi S (2021) Automatic flood detection using sentinel-1 images on the google earth engine. Environ Monit Assess 193:248. https://doi.org/10.1007/s10661-021-09037-7
27. Li M, Wu P, Wang B, Park H, Yang H, Wu Y (2021) A deep learning method of water body extraction from high-resolution remote sensing images with multisensors. IEEE J Sel Top Appl Earth Observ Remote Sens 14:3120–3132. https://doi.org/10.1109/JSTARS.2021.3060769
28. https://users.cs.cf.ac.uk/Dave.Marshall/Vision_lecture/node35.html
29. Di Baldassarre G, Schumann G, Bates PD, Freer JE, Beven KJ (2010) Flood-plain mapping: a critical discussion of deterministic and probabilistic approaches. Hydrol Sci J 55(3):364–376. https://doi.org/10.1080/02626661003683389
30. https://link.springer.com/article/10.1007/s13198-021-01152-5

Optimizing Permutations in Biclustering Algorithms

Aditya Shreeram, Tanmayee Samantaray, and Cota Navin Gupta

Abstract Data-driven methods used on structural magnetic resonance imaging (sMRI) have been successful in predicting disease subtypes within a population of subjects. Biclustering is a data-mining technique that can cluster rows and columns of a matrix to find submatrices corresponding to subtypes. One such method is N-BiC, which is an N-component biclustering algorithm. The limitation of this algorithm is that it is resource intensive, as it uses recursive Depth-First Search (DFS), which can be less efficient and time-consuming over iterative approaches. The biclusters enlisted in N-BiC are dependent on search order, and it requires a number of permutation operations to explore and merge the possible biclusters. In N-BiC, permutation refers to the order in which the components are sent to the modified DFS algorithm. The task of permutations in N-BiC is to find common subjects between components. For example, a permutation using three components (say 1, 2, and 3) relates to common subjects between components 1, 2, and 3. Similarly, yet another permutation 2, 3, and 1 relate to the same set of subjects between components 1, 2, and 3 in a different order, which is therefore redundant. Our proposed approach eliminates this repetitive process to obtain common subjects in a bicluster. Therefore, Modified N-BiC has optimized permutations and consumes reduced run time. It works on par with N-BiC and has reduced memory requirements. We also suggest that N-BiC algorithm does not find biclusters in the subject by voxel matrix as done by other biclustering studies but only in the decomposed loading matrix (subject by component). We also provide the necessary MATLAB codes for further research and application on other varied datasets.

Aditya Shreeram and Tanmayee Samantaray—Contributed equally.

A. Shreeram · T. Samantaray · C. N. Gupta (✉)
Neural Engineering Laboratory, Department of Biosciences and Bioengineering, Indian Institute of Technology Guwahati, Guwahati, Assam, India
e-mail: cngupta@iitg.ac.in

A. Shreeram
e-mail: sadityaaditya_s@iitg.ac.in

T. Samantaray
e-mail: tanma176106113@iitg.ac.in

S. Shakya et al. (eds.), *Fourth International Conference on Image Processing and Capsule Networks*, Lecture Notes in Networks and Systems 798,
https://doi.org/10.1007/978-981-99-7093-3_7

Keywords Structural magnetic resonance imaging · Depth first search · Biclustering · Data-mining algorithm · Source-based morphometry

1 Introduction

1.1 Literature Survey

Clustering is a popular unsupervised machine-learning technique that has been used to identify clusters within a dataset. Some of the applications of clustering include Image Segmentation and Recognition, Information retrieval, data mining, and so on [1, 2]. Even though it has many uses, clustering does not identify data points that may belong to several clusters over varying conditions (local patterns) [3]. For example, in a gene expression matrix, identifying the local patterns would lead to a greater understanding of genes that up-regulate or down-regulate together and the particular sample/condition in which such expression patterns occur. Biclustering is one such technique that can overcome this limitation of clustering. It is a data-mining technique that clusters rows and columns together. A bicluster refers to a two-dimensional sub-matrix within a larger matrix. They can be classified based on the type of values (constant, scaled variables, etc.), structure (row or column exclusivity and exhaustivity, etc.) or based on numerous evaluation measures like Mean Square Residue (MSR) or Scaled Mean Square Residue, etc., [4].

Biclustering also has applications in disease subtyping [4–9]. Other than biological data analysis, the technique has also been used in text mining [10, 11]. Some of the popular biclustering algorithms have been integrated into the toolbox BIDEAL, developed by [12]. However, it is a comparatively recent approach as applied to complex medical image data [13]. Some of the advancements concerning biclustering with neuroimaging datasets are Biclustered Independent Component Analysis (BICA) [14], N-component biclustering (N-BiC) [15], and Fuzzy non-negative matrix factorization (FNMF) [16]. In BICA, biclustering was performed on two components selected based on previous studies, whereas in N-BiC, multi-component biclustering was performed (in [14] and [15], sMRI data are used). In N-BiC, permutation refers to the order in which components are sent to the modified DFS algorithm. For example, a permutation using three components (say 1, 2, and 3) relates to common subjects between components 1, 2, and 3. Similarly, yet another permutation 2, 3, and 1 relate to the same set of subjects between components 1, 2, and 3 in a different order, which is therefore redundant. Similar FNMF was applied to diffusion tensor imaging (DTI) data, resulting in its decomposition to two different matrices with only positive entries. Overall, the number of studies in this domain is relatively less, and this warrants research for better and faster algorithms.

1.2 *Aim of the Study*

In this study, we attempt to improve upon N-BiC, which has been used for identifying the subtypes in a cohort of patients suffering from Schizophrenia [15]. Although the application of this algorithm onto neuroimaging datasets led to a better understanding of the disease, the method requires high resources due to the $n!$ permutation-based merging step. Our proposed algorithm, Modified N-BiC, which performs biclustering, includes usage of an iterative search method and performing comprehensive merging, thus removing the need for permutations. This optimization of the permutation count brings down the runtime and memory requirements significantly. To illustrate this, we consider two simulated datasets generated using custom MATLAB code and a publicly available multicenter sMRI dataset from Parkinson's Progression Marker's Initiative (PPMI) [17].

2 Materials and Methods

2.1 *Datasets*

Simulated Data and Bicluster Embedding. Two artificial datasets with embedded biclusters to mimic the subject by component matrix were created to compare the performance of N-BiC and Modified N-BiC. These were generated with a normal distribution in size (400 * 10) with the help of a MATLAB function *randn*. The matrix was normalized by using min_max_scaling to the range [0–1]. Once the data are generated, the biclusters were embedded in the data matrix (well separated to enable ease of interpretation) by artificially increasing the value of loading coefficients at particular locations. Dataset 1 has 10 columns, while Dataset 2 (Fig. 1) has 11 columns. Both datasets have three biclusters embedded with the following specifications (Table 1). The datasets have 10 and 11 columns, respectively, to demonstrate the effect of column count on performance.

Patient and Demographic Details. This is a publicly shared large multicenter [17] dataset from which 165 healthy controls (HC) and 216 Parkinson's disease (PD) patients were considered (Table 2). The studies were conducted following the Good Clinical Practice (GCP) principles and Declaration of Helsinki after approval was received from the regional ethics committees of the participating sites.

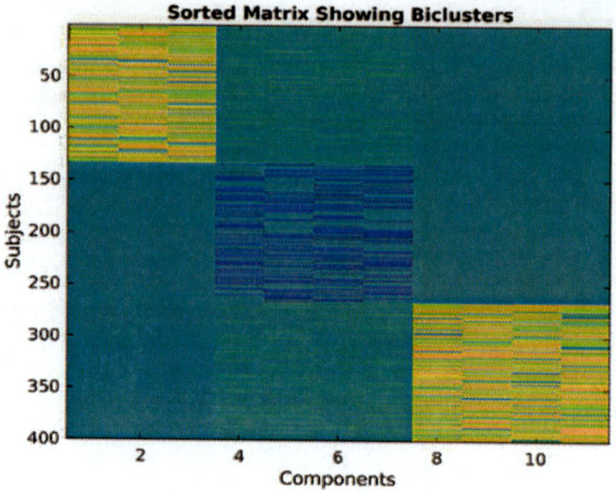

Fig. 1 Sorted eleven component datasets visualized using *imagesc* (MATLAB)

Table 1 Details of the simulated dataset

Dataset number	Bicluster number	Columns	Rows
1	1	1, 2, 3	1–133
	2	4, 5, 6	134–266
	3	7, 8, 9, 10	267–400
2	1	1, 2, 3	1–133
	2	4, 5, 6, 7	134–266
	3	8, 9, 10, 11	267–400

Table 2 Details of the patient demographics for PPMI participants

Measures	HC	PD
Count [total (male)]	165 (107)	216 (130)
Age in years (mean ± std.)	60.50 (11.03)	58.82 (7.39)
Age range	31–83	38–70

2.2 *Device Specifications and Software*

Following are the device specifications under which the experiments were run. CPU: I5-3470S, GPU: GT 710, System memory: 16 GB, Ram Specifications: DDR3 with 1600 MHz. All program codes were written using MATLAB 2022a software.

2.3 Structural Magnetic Resonance Imaging Data

The sMRI images were acquired using multiple scanner models of 1.5 or 3 T with scan acquisition parameters [18]. The sMRI scans were preprocessed according to the methodology followed in [19]. Computational anatomy toolbox, CAT12 [20], was used within Statistical Parametric Mapping (SPM 12) [21] for preprocessing in MATLAB R2021a [22]. The following steps were involved in the preprocessing:

1. Normalization and registration to standard 152 average T1 Montreal Neurological Institute (MNI) template.
2. Segmentation into different tissue types using unified segmentation algorithm [23].
3. Smoothing of gray matter (GM) images using full-width at half maximum (FWHM) Gaussian kernel of 8 mm.

The preprocessed images were regressed with age, gender, site, scanner, and field strength. These scans in the form of a 4D matrix (3D images over subjects) were flattened to a 2D matrix (subjects by voxel). Source-based morphometry (SBM) [24] was used on the subjects by voxel matrix, to obtain the loading coefficient matrix (subjects by components) and component matrix (components by voxel). The Group ICA of the fMRI toolbox (GIFT) was used to perform decomposition [25].

After SBM, a loading matrix of 381 * 30 was used. The number of components used in the current study was followed from previous subtyping studies [14, 15]. Of these 30 components, 10 and 11 number of components were considered separately to provide a balanced comparison between the two algorithms, N-BiC and Modified N-BiC. The N-BiC algorithm requires more than 42 GB to store 12! permutations of 12 components. In MATLAB, this takes up 12! * 12 * 8 bytes, where 8 is the number of bytes allocated by MATLAB to a single number of double data type. This gives around 42.8 GB to be exact.

2.4 Modified N-BiC Algorithm

Overview. A few steps of our algorithm are similar to the N-BiC algorithm [15] where the loading matrix is sorted according to the best methods explained therein (Step 1 of Fig. 2). The loading matrix, sorted using these methods, is further used to identify biclusters, by evaluating every combination of components at the given parameters. The loading coefficients of common subjects between pairs or groups of components are stored in a separate data structure (bicluster search) (Step 2 of Fig. 2). At the end of the search, the biclusters are sent to a local validator, where every bicluster is compared with $(N - 1)$ biclusters in the set. N is the number of biclusters enlisted and biclusters already compared are excluded. The biclusters are merged based on overlap criteria, and thus, final list contains unique biclusters for further analysis. The total number of such comparisons comes out to be $N * (N -$

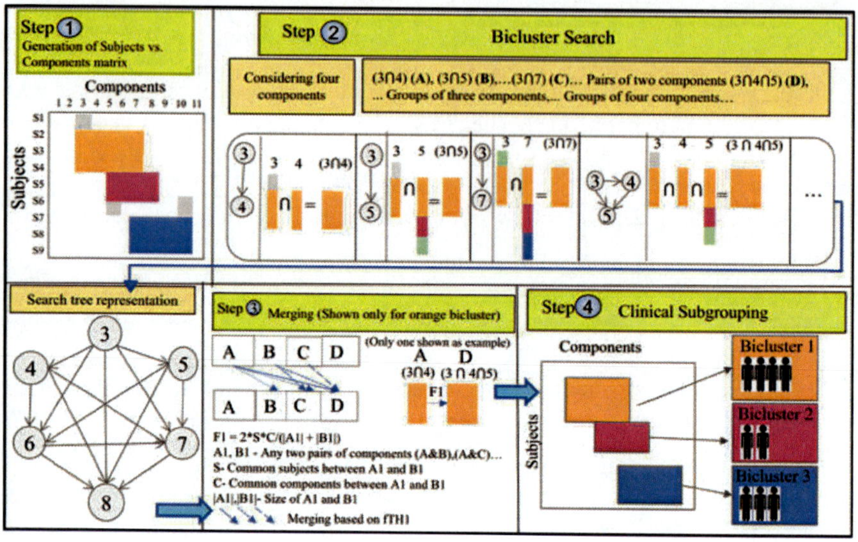

Fig. 2 Graphical abstract depicting methodology behind Modified N-BiC depicting bicluster search, merging, and subgrouping

1)/2. This validation process is shown in Step 3 of Fig. 2 (merging). After Step 3, required subgroups are identified from the bicluster list (Step 4 of Fig. 2).

Sorting. The main idea of sorting is to select only those subjects which show high expression for a particular component. For each component, the individual loading coefficients are compared with the mean loading coefficient and then retained or eliminated based on the threshold. More details on sorting methods could be found in Sect. 3.1 and [15].

Parameters and Search Bic function. The algorithm is a type of Depth-First Search (DFS), called iterative deepening DFS [26], which explores the intersection of all possible subsets of components. It combines the approaches of DFS and Breadth-First Search (BFS) [26] as it looks for the intersection between components at a particular depth before moving to a higher depth. For example, intersection for common subjects is performed between pairs of two components such as (1, 2), (1, 3), (1, 4), and so on, after which intersection between groups of three components (1, 2, 3), (1, 2, 4) happens and so on. This process continues till maximum depth is reached (total number of components) as shown in Step 2 of Fig. 2. The total number of such component pairs is given in Eq. (1).

$$\text{TotalPair(TP)} = \sum\nolimits_{k=2} \binom{n}{k}, \tag{1}$$

where k goes from 2 to n (n is the number of components in the matrix).

We used MINTERSECT [27], a MATLAB helper file, for performing sequential intersection. The parameters (NoS, *K*, *O*) referred to in [15] as (*S*, *K*, *O*) are used for our approach except for *fTH2*, which is the overlap criteria for merging over all the permutations. This is because in Modified N-BiC, the merging takes place with a single validator, requiring a single overlap parameter (*fTH1*). The overlap between two biclusters is determined by the $F1$ index or Dice index, which calculates the similarity between two biclusters [15]. The total number of operations (TO) in the worst-case scenario for Modified N-BiC is given by Eq. (2), while that for N-BiC is given by Eq. (3).

Let *Pair* represent an array of length TP (Eq. 1) containing all the pairs of components, *m* be the number of non-zero elements in a particular column, *N* be the total number of enlisted biclusters, *M, P* be the number of subjects, components, respectively, in bicluster 1, and *Z, Q* be the number of subjects, components, respectively, in bicluster 2 for pairwise comparison.

The total number of operations in our proposed approach, Modified N-BiC, is:

$$\mathrm{TO}_{\mathrm{Modified\text{ - }N\text{ - }BiC}} = \sum_{i=1}^{\mathrm{TP}} \prod_{k=1}^{\mathrm{length(Pair}(i))} m_k + \sum_{j=1}^{\frac{N*N-1}{2}} \left(M_j*Z_j + \left(P_j*Q_j\right)\right) \quad (2)$$

whereas, in N-BiC, the total number of operations is:

$$\mathrm{TO}_{\mathrm{N\text{ - }BiC}} = \sum_{a=1}^{\mathrm{NP}} \left[\sum_{i=1}^{\mathrm{TP}} \prod_{k=1}^{\mathrm{length(Pair}(i))} m_k + \sum_{j=1}^{\mathrm{VN}} \left(M_j * Z_j + \left(P_j * Q_j\right)\right) \right] \quad (3)$$

where VN is $\frac{N*N-1}{2}$ when one bicluster is compared to $N-1$ remaining biclusters.

Here, NP is the number of permutations considered, VN is the variable number of biclusters, and since the bicluster merging is order dependent, merged bicluster pairs are removed. Here the first term in Eqs. (2) and (3) denotes the number of comparison operations involved in a component pair, which is sum of the product of *m* over all the pairs (TP) in a permutation. The second term in Eq. (2) relates to the intersection between two biclusters, which is the computed intersection of the common subjects/indices (*M* and *Z*) and common components (*P* and *Q*) summed over $N * (N-1)/2$. In Eq. (3), this relates to the number of operations in internal validation (internal summation across a variable number of biclusters) and the summation over all the permutation steps. To perform an exhaustive merging across all permutations, NP $= n!$.

2.5 *Evaluation Measures*

Mean Square Residue (MSR). The Mean square residue [4] is a measure reflecting the homogeneity of a bicluster. With MSR, it is possible to identify biclusters with similar values of loading coefficients, allowing one to better understand the dataset.

$$\mathrm{MSR}(B) = \frac{1}{|I| * |J|} \sum_{i=1}^{i=|\mathrm{I}|} \sum_{j=1}^{j=|\mathrm{J}|} (b_{ij} - b_{iJ} - b_{Ij} + b_{IJ})^2, \tag{4}$$

Here, B is a bicluster, and $|I|$, $|J|$, b_{ij}, b_{iJ}, b_{Ij}, b_{IJ} stand for bicluster row size, bicluster column size, element in the ith row and jth column, row means, column means, and total means, respectively.

F1 Index and Consensus Score (CS). The $F1$ index is calculated for each pair of extracted biclusters and original biclusters. Consensus score [28] denotes the consensus between the final list of biclusters (identified/extracted) and the original list of biclusters (ground truth).

Let us suppose there are three ground truth biclusters A, B, and C, and three extracted biclusters $A1$, $B1$, and $C1$. Suppose A has more subjects in common with $A1$, B in common with $B1$, and C in common with $C1$, according to $F1$ indexes. In this case, the $F1$ index is given by Eq. (5), and the consensus score is given by Eq. (6).

$$\text{Number of pairs} = 3\{(A, A1), (B, B1), (C, C1)\}, \tag{5}$$

$$\begin{gathered} \sum F1\,\text{indexes} = F1(A, A1) + F1(B, B1) + F1(C, C1) \\ \mathrm{CS} = \frac{\sum F1\,\text{indexes}}{\text{Number of pairs}} \end{gathered} \tag{6}$$

2.5.1 Comparison with Bimax

We compared the performance of Modified N-BiC with an existing biclustering algorithm called Bimax [29], using the toolbox BIDEAL [12]. Bimax is a divide and conquer algorithm, which divides the whole binarized (1's and 0's) matrix into several partitions and searches for biclusters with only 1's. The algorithm was chosen for its ease of use and interpretation.

3 Results and Discussion

We compare the three algorithms namely, N-BiC, Modified N-BiC, and Bimax, using simulated data and the PPMI dataset. Positive and negative sorting methods as included within the N-BiC toolbox, were the default sorting method for all scenarios.

3.1 Simulated Dataset

The runtime and average MSR on simulated datasets are compared and shown in Table 3. Table 4 lists the biclusters identified by both algorithms on Dataset 1. The comparison of consensus scores of each algorithm over the simulated dataset is shown in Table 5. Table 6 represents biclusters enlisted by both algorithms on dataset 2. Due to space constraints, only the results at 100 permutations are shown in Tables 4 and 6. Parameter settings used for the following experiments: NoS = 125, $K = 2$, $O = 0.8$, where NoS—minimum number of subjects in a bicluster, K—minimum number of components in a bicluster, O—overlap between the biclusters. These parameters were set based on the rationale of the N-BiC algorithm.

Biclusters discovered by Bimax on Dataset 1. Parameters settings used for this experiments: min_rows (minimum rows): 125, min_cols (minimum columns): 2, NumBiClust (no. of biclusters): 20. Details of the biclusters found {Components, Number of Subjects} in Dataset 1: {(2, 3), 128}, {(1, 3), 127}, {(1, 2), 129}, {(1, 2, 3), 126}, {(7, 8), 129}, {(8, 9), 128}, {(7, 9), 131}, {(7, 8, 9), 127},{(7, 10), 131}, {(7, 8, 10), 127}, {(7, 9, 10), 129}, {(7, 8, 9, 10), 125}.

Biclusters discovered by Bimax on Dataset 2. Parameters settings used for this experiments: min_rows (minimum rows): 125, min_cols (minimum columns): 2, NumBiClust (no. of biclusters): 20. Details of the biclusters found by

Table 3 Performance of both algorithms on simulated datasets

Number of components	Number of permutations	N-BiC (RT)	Modified N-BiC (RT)	N-BiC (AvMSR)	Modified N-BiC (AvMSR)
10	1	0.66	0.82	1.04	0.94
10	2	0.74	–	1.07	–
10	5	0.82	–	1.06	–
10	100	3.00	–	1.06	–
11	1	6.61	0.98	0.86	1.01
11	2	6.50	–	0.89	–
11	5	6.68	–	0.90	–
11	100	9.03	–	0.91	–

RT runtime (seconds), *AvMSR* average MSR

Table 4 Biclusters enlisted by both algorithms on dataset 1

S. No.	N-BiC (number of components)	Modified N-BiC (number of components)	N-BiC (number of subjects)	Modified N-BiC (number of subjects)
1	[1, 2]	[1, 2]	125	129
2	[1, 3]	[1, 3]	126	127
3	[2, 3]	[2, 3]	126	128
4	[7–9]	[7–9]	125	127
5	[7, 8, 10]	[7, 8]	125	129
6	[7, 9, 10]	[7, 10]	126	131
7	[8–10]	[8, 9]	125	128
8	–	[8, 10]	–	127
9	–	[9, 10]	–	129
10	–	[1–3]	–	126
11	–	[7, 9]	–	131
12	–	[8, 9]	–	128

Table 5 Consensus score comparison on simulated datasets

Number of components	Number of permutations	N-BiC	Modified N-BiC
10	1	0.806	0.7396
10	2	0.793	–
10	5	0.798	–
10	100	0.802	–
11	1	0.706	0.743
11	2	0.691	–
11	5	0.692	–
11	100	0.682	–

Bimax{Components, Number of Subjects}: {(2, 3), 128}, {(1, 3), 127}, {(1, 2), 129}, {(1, 2, 3), 126}, {(8, 11), 127}, {(10, 11), 128}, {(8, 10), 127}, {(8, 9), 128}.

3.2 *PPMI Dataset*

Parameter settings used for performance comparison using the PPMI dataset are NoS $= 40$, $K = 2$, and $O = 0.8$. The PD-only loadings were considered for biclustering on components: (1:10) and (1:11), where NoS—minimum number of subjects in a bicluster, K—minimum number of components in a bicluster, O—overlap between the biclusters. A comparison of runtime and average MSR on the loading matrix is

Table 6 Biclusters enlisted by both algorithms on dataset 2

S. No.	N-BiC (number of components)	Modified N-BiC (number of components)	N-BiC (number of subjects)	Modified N-BiC (number of subjects)
1	[1, 2]	[1, 2]	126	129
2	[1, 3]	[1, 3]	126	127
3	[2, 3]	[2, 3]	126	128
4	[8, 9]	[8, 9]	125	128
5	[8, 10]	[8, 10]	125	127
6	[8, 11]	[8, 11]	125	127
7	[9, 10]	[9, 10]	125	129
8	[9, 11]	[9, 11]	126	129
9	[10, 11]	[10, 11]	126	128
10	–	[1–3]	–	126
11	–	[9–11]	–	126
12	–	[8–10]	–	125
13	–	[8, 9, 11]	–	125

shown in Table 7. These parameters were set based on the rationale of the N-BiC algorithm, although other configurations can be explored in future studies.

Table 7 Performance on PPMI dataset

Number of components	Number of permutations	N-BiC (RT)	Modified N-BiC (RT)	N-BiC (AvMSR) (* 10^8)	Modified N-BiC (AvMSR) (* 10^8)
10	1	0.61	0.15	3.58	3.15
10	2	0.64	–	3.71	–
10	5	0.67	–	3.33	–
10	100	2.11	–	3.15	–
11	1	6.51	0.19	4.09	3.38
11	2	6.44	–	3.62	–
11	5	6.58	–	3.63	–
11	100	8.46	–	3.38	–

RT runtime (s), *AvMSR* average MSR

3.3 Performance of Modified N-BiC on PPMI Dataset

Parameter settings used for bicluster search using the PPMI dataset: NoS $= 55$, $K = 2$, $O = 0.8$, where NoS—minimum number of subjects in a bicluster, K—minimum number of components in a bicluster, O—overlap between the biclusters. The PD-only loadings were considered for biclustering on all 30 components. Biclusters discovered: Components [3 and 16] with 56 subjects and components [5 and 16], [5 and 17], [25 and 30] with 55 subjects each. Modified N-BiC is capable of finding biclusters from a loading matrix with 30 components, under the same system configuration. Although we have executed the algorithm under various parameters of *NoS*, K, and O, the results have only been presented for one parameter setting as proof of execution. Biclusters common to four or more components were not generated under various parameters, so the algorithm was executed up to the intersection between three components. Modified N-BiC used 5 MB for Dataset 1 and 7.2 MB for Dataset 2 (both approximates). The memory usage was calculated using *whos* command in MATLAB. N-BiC on the other hand used 276.8 MB for Dataset 1 and 3.27 GB for Dataset 2 (just for allocation of the space for permutation possibilities), and it requires 42 GB of system memory for the same (12 components and more). At the same system configuration, N-BiC failed to run, but Modified N-BiC successfully discovered the biclusters efficiently. This indicates the feasibility of execution for Modified N-BiC and a higher number of valid biclusters returned by exhaustive search approaches in a reasonable time. It can be seen that Bimax performed on par with N-BiC as well as Modified N-BiC on Dataset 1 and Dataset 2. Thus, we suggest that Modified N-BiC can work comparably to existing algorithms, although more research on this aspect is required.

It is observed from Tables 3 and 7 that the run time increases for both N-BiC and Modified N-BiC with an increase in the number of components, though the increase in time for N- BiC is more than that for Modified N-BiC. This is because N-BiC is a backtracking algorithm that uses recursive DFS to enumerate search possibilities, and recursion can be less efficient in terms of time and space complexity over iterative approaches, especially with an increase in the number of operations [30].

For the N-BiC, the algorithm has to be run for a higher number of permutations (dependent on the dataset) to get stable results. This is because the arrangement/ permutation plays a role in the enlistment of biclusters. Increasing the number of permutations increases the running time and has a varying influence on the average MSR due to a varying number of biclusters enlisted; every time the algorithm is executed. In the end, it can be seen from Tables 3, 4, 5, 6 and 7 that the Modified N-BiC has an on-par or similar performance with N-BiC.

4 Research Limitations/Implications

The biclusters enlisted by the algorithm are dependent on NoS, K, and O and hence these need to be tuned very properly. These values also affect the runtime of the algorithm (same as in N-BiC [15]). For example, if the overlap parameter (O) is set too high, then too many similar biclusters are listed, making the analysis tedious. In contrast, if the O value is set to be too low, an aggressive merging takes place, and too few biclusters are left for analysis.

5 Originality and Value

This study presents an improvement over N-BiC [15] in terms of reduction in memory requirements and run time while providing on-par performance with the same (bicluster enlistment). The improvement comes in the form of a switch to an iterative deepening DFS over the recursive DFS with backtracking in N-BiC. We have also optimized the merging step by comparison of a bicluster with other biclusters in a single permutation, which reduces the necessity for $n!$ permutations in the worst case down to one permutation. Hence, our approach seems to provide similar results at a lesser computational cost. The value of Modified N-BiC lies in its potential as an alternative biclustering algorithm that can be used for disease subtyping not only from sMRI data but for other varied applications like cancer and gene analysis.

6 Conclusion and Future Research Work

In this study, we suggest improvements for the N-BiC algorithm by using an iterative DFS approach and a comprehensive merging step, removing the need for permutations. The Modified N-BiC approach works on par with the N-BiC approach and uses fewer resources than the latter. We hope that the provided source code and datasets for download will enable further testing and validation of our work. This study also shows the feasibility of execution of exhaustive biclustering algorithms on devices with low specifications, especially for subtyping. This algorithm can still be improved upon, in the future by use of optimizations like dynamic programming approach. It can also be used on other datasets to obtain biologically relevant biclusters.

Code Availability The simulated data and codes used in the current study are available at the below GitHub link: https://github.com/NeuralLabIITGuwahati/Permutations_Biclustering.

Author Contributions CNG proposed the idea to AS and TS. AS did the entire work and had frequent discussions with TS and CNG. The first draft of the paper was written by AS and TS. Numerous discussions then between AS and CNG led to the final submitted manuscript.

Acknowledgements The authors are thankful to PPMI for providing the dataset. Aditya Shreeram was funded by Ministry of Education (MoE) scholarship by the Government of India (GOI). Tanmayee Samantary was funded by MoE doctoral scholarship by GOI. Cota Navin Gupta's time was supported by the Scheme for Promotion of Academic and Research Collaboration (SPARC Grant), Government of India, Project Code: P1073.

References

1. Saxena A, Prasad M, Gupta A, Bharill N, Patel OP, Tiwari A, Er MJ, Ding W, Lin CT (2017) A review of clustering techniques and developments. Neurocomputing 267:664–681
2. Adepu B, Gyani J, Narsimha G (2021) A novel multi-objective differential evolution algorithm for clustering data streams. In: Lecture notes in networks and systems, pp 67–78
3. Orzechowski P, Boryczko K, Moore JH (2019) Scalable biclustering—the future of big data exploration? GigaScience 8(7):giz078
4. Pontes B, Giráldez R, Aguilar-Ruiz JS (2015) Biclustering on expression data: a review. J Biomed Inform 57:163–180
5. Noronha MDM, Henriques R, Madeira SC, Zárate LE (2022) Impact of metrics on biclustering solution and quality: a review. Pattern Recogn 127:108612
6. Ramkumar M, Basker N, Pradeep D et al (2022) Healthcare biclustering-based prediction on Gene expression dataset. Biomed Res Int 2022:1–7
7. Wang YK, Print CG, Crampin EJ (2013) Biclustering reveals breast cancer tumour subgroups with common clinical features and improves prediction of disease recurrence. BMC Genomics 14(1):102
8. Samantaray T, Saini J, Gupta CN (2022) Subgrouping and structural brain connectivity of Parkinson's disease—past studies and future directions. Neurosci Inform 2:100100
9. Sun J, Bi J, Kranzler HR (2013) Multi-view biclustering for genotype-phenotype association studies of complex diseases. In: 2013 IEEE international conference on bioinformatics and biomedicine, 316–321
10. de Castro PA, de França FO, Ferreira HM, Von Zuben FJ (2007) Applying biclustering to text mining: an immune-inspired approach. Lecture notes in computer science, 83–94
11. Dhillon IS, Mallela S, Modha DS (2003) Information-theoretic co-clustering. In: Proceedings of the ninth ACM SIGKDD international conference on knowledge discovery and data mining
12. Verma NK, Sharma T, Dixit S, Agrawal P, Sengupta S, Singh V (2021) Bideal: a toolbox for bicluster analysis—generation, visualization and validation. SN Comput Sci 2(1):24
13. Castanho EN, Aidos H, Madeira SC (2022) Biclustering fMRI time series: a comparative study. BMC Bioinform 23(1):192
14. Gupta CN, Castro E, Rachkonda S et al (2017) Biclustered independent component analysis for complex biomarker and subtype identification from structural magnetic resonance images in Schizophrenia. Front Psychiatry 8
15. Rahaman MA, Mathalon D, Lee HJ et al (2020) N-BIC: a method for multi-component and symptom biclustering of structural MRI data: application to schizophrenia. IEEE Trans Biomed Eng 67(1):110–121
16. Arnedo J, Mamah D, Baranger DA et al (2015) Decomposition of brain diffusion imaging data uncovers latent schizophrenias with distinct patterns of white matter anisotropy. Neuroimage 120:43–54
17. Marek K et al (2018) The Parkinson's progression markers initiative—establishing a PD biomarker cohort. Ann Clin Transl Neurol 5(12):1460–1477
18. Parkinson's Progress Marker's Initiative Scanner Information. https://www.ppmiinfo.org/sites/default/files/docs/archives/PPMI2.0_MRI_TOM_Final_FullyExecuted_v2.0_20200807.pdf

19. Samantaray T, Saini J, Gupta CN (2022) Sparsity dependent metrics depict alteration of brain network connectivity in Parkinson's disease. In: 44th annual international conference of the IEEE engineering in medicine & biology society (EMBC), pp 698–701
20. Computational anatomy toolbox (CAT12). http://www.neuro.uni-jena.de/cat/
21. SPM 12. https://www.fil.ion.ucl.ac.uk/spm/software/spm12
22. MATLAB, Natick, Massachusetts. https://www.mathworks.com/products/matlab.html
23. Ashburner J, Friston KJ (2000) Voxel-based morphometry—the methods. Neuroimage 11(6):805–821
24. Xu L, Groth KM, Pearlson G, Schretlen DJ, Calhoun VD (2009) Source-based morphometry: the use of independent component analysis to identify gray matter differences with application to schizophrenia. Hum Brain Mapp 30:711–724
25. Group ICA of fMRI toolbox. https://github.com/trendscenter/gift
26. Korf RE (1985) Depth-first iterative-deepening: an optimal admissible tree search. Artif Intell 27:97–109
27. MINTERSECT—Multiple set intersection. https://in.mathworks.com/matlabcentral/fileexchange/6144-mintersect-multiple-set-intersection
28. Hochreiter S, Bodenhofer U, Heusel M et al (2010) Fabia: factor analysis for bicluster acquisition. Bioinformatics 26:1520–1527
29. Prelić A, Bleuler S, Zimmermann P, Wille A, Bühlmann P, Gruissem W, Hennig L, Thiele L, Zitzler E (2006) A systematic comparison and evaluation of biclustering methods for gene expression data. Bioinformatics 22:1122–1129
30. Sánchez MR (2018) Basic concepts of recursive programming. Introduction to recursive programming. CRC Press, Taylor & Francis Group, Boca Raton, p 26

Extracting Graphs from Plant Leaf Venations Using Image Processing

Ashlyn Kim D. Balangcod and Jaderick P. Pabico

Abstract Leaf venation is an important characteristic of plant species that can lead to the identification of a plant. The study of leaf venation is laborious if done by manual inspection. This study proposes a leaf venations' extraction process requiring minimal algorithms of image processing. Canny edge detection and Zhang-Suen thinning algorithm were used to extract the venation of the leaf images. Using ground truth tracing, the average values computed are 0.522 for correctness and 0.406 for completeness of the extracted venation. This study also presents the use of graph metrics to describe the leaf venations of plants. Using thirty images of five plant species, leaf venations up to the tertiary veins are compared using graph properties: size, volume, number of claws, number of wedges, and maximum degree. This study proposes these mentioned graph metrics as additional characteristics of leaf venations as results show a possible pattern for the five plant species.

Keywords Leaf venations · Image processing · Graph properties

1 Introduction

The morphological features of leaves are used as a basis for plant identification. Some flowering plants may even be identified by leaves alone. Some characteristics used for identification include the shape of the leaf, texture, margins of leaf blade, and leaf venations [1]. Leaf venations, or the positioning of the veins on the blade, are known to provide the mechanical support of the leaf. It contains the phloem and

A. K. D. Balangcod (✉)
Department of Mathematics and Computer Science, University of the Philippines Baguio, Baguio, Philippines
e-mail: adbalangcod@up.edu.ph

J. P. Pabico
Institute of Computer Science, University of the Philippines Los Baños, College, Laguna, Philippines
e-mail: jppabico@up.edu.ph

S. Shakya et al. (eds.), *Fourth International Conference on Image Processing and Capsule Networks*, Lecture Notes in Networks and Systems 798,
https://doi.org/10.1007/978-981-99-7093-3_8

xylem which transports the nutrients and water, respectively, needed for the plant's processes [2].

There had been studies on the other roles of plant venations aside from plant identification, such as tracing the historical lineage of genus Gunnera using leaf fossil records [3], understanding climatic conditions [2], and predicting season temperature and atmospheric carbon dioxide of an area [4].

Plants can be separated into two categories, and they are the monocots and the dicots. These two categories have different leaf venations. Generally, for monocots, leaves are parallel venations occurring side-by-side without intersecting. Dicots mostly have netted or reticulate leaf venations [1]. An extensive document to classify dicot leaves has been established by Hickey [5] and is still currently in use. An important aspect of leaf veins is the order of venation where the veins are classified according to their width relative to their point of origin. The primary vein of the first degree (1°) is the thickest vein. Second-degree (2°) veins are the next smaller veins and protrude from the primary veins. Higher-order venations such as tertiaries (3°), quarternaries (4°), and quinternaries (5°) are the next finer veins arising from their previous lower-order veins [5]. Aside from the order of venations, some quantitative measures for plant leaf venations are vein density, edges per area, number of areoles, number of nodes, and number of edges [6].

Prior to the computation of these metrics, leaves have to be extracted. There are different methods for extracting the venations, either manually or through image processing. Manual extraction requires chemical clearing of the leaf and staining which, in some cases, requires hours sometimes days to fully clear the leaves to obtain the veins. Another method is the visual inspection of high-resolution leaf photographs, which is another tedious and time-consuming process.

The approach proposed in this study is to use automatic extraction of the leaf images using algorithms of image processing, specifically the use of Canny edge detection and Zhang-Suen thinning algorithm, and a novel graph detection from the extracted venation image. The plants considered for this study are selected dicot plants found in the area.

2 Related Works

Some applications and algorithms have been developed using image processing to extract the leaf venations either automatically or with human intervention. Some of these are the Leaf Extraction Analysis Framework Graphical User Interface (LEAF GUI) by Price et al. [6], PhenoVein by Buhler et al. [7], Leaf Image Analysis Interface (LIMANI) by Dhondt et al. [8], and the Network Extraction from Images (NEFI) by Dirnberger et al. [9]. These approaches used different methods in the extraction of the venations using processes such as image segmentation, skeletonization, and computation of quantitative measures. Figure 1 summarizes the steps defined in LIMANI application. Figure 2 is the summary of the algorithm involved in the Leaf

GUI software, while Fig. 3 is the processes involved in the methodology of the phenoVEIN software.

The methodologies of LIMANI, LEAF GUI, and phenoVEIN produce quantitative data of leaf venations. LIMANI's results are data on lamina area, vascular vein length, length of free-ending veins, areas of areoles, and other data [8]. LEAF GUI has an output of the quantitative data of the vein network such as leaf area, leaf perimeter, edges per area, number of areoles, and areole statistics [6]. The measured

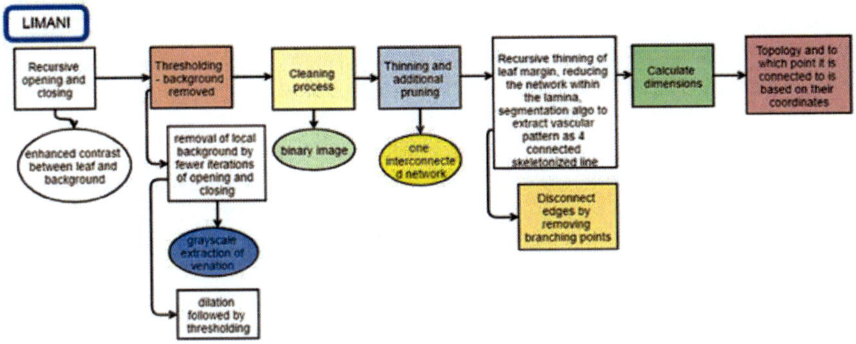

Fig. 1 Summarized methodology of LIMANI [8]

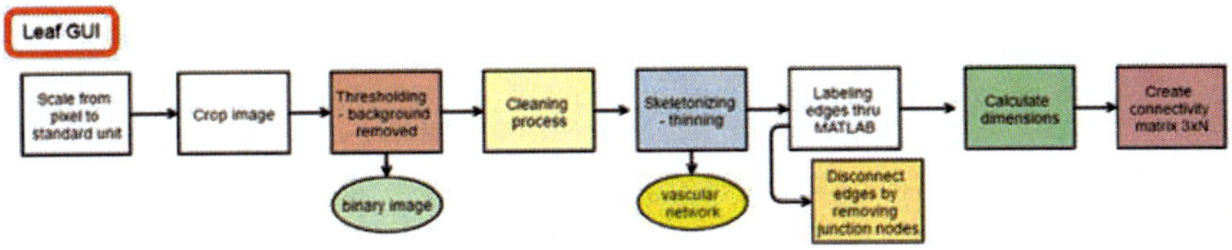

Fig. 2 Summarized methodology of LEAF GUI [6]

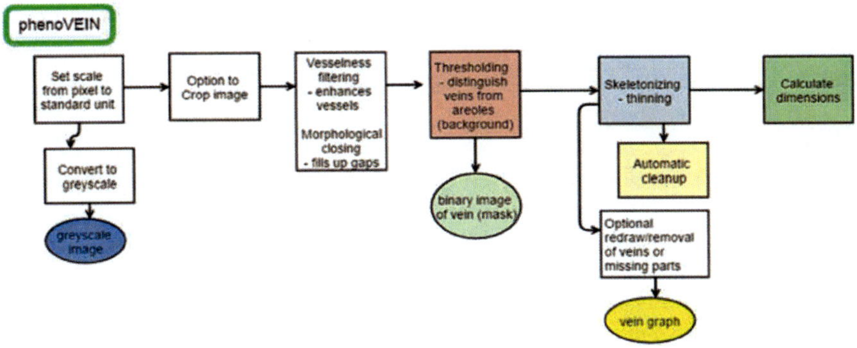

Fig. 3 Summarized methodology of phenoVEIN [7]

traits by PhenoVEIN include skeleton length, projected leaf area, number of veins, and number of areoles [7]. These characteristics are used to describe the plant leaf and venation as the measures used in plant taxonomy. For the goal of this study, which is to extract graphs from the venations, the applications mentioned above lack this step as well as computing metrics from the graphs.

Another study of similar work on extracting graphs from venations uses Canny edge detection, a midrib detection process, vein filling method, and the Zhang-Suen method for skeletonization [10].

The workflow in this study proposes a simplified process wherein, compared to the applications mentioned above, fewer steps are used to extract the venations of the leaves using image processing. This study also presents another metric for leaf venations which is the graph properties of undirected graphs. The venations extracted from the images are converted to a graph $G(V, E)$, where V is a finite set of n vertices $\{v_1, v_2, \ldots, v_n\}$ and E is the set of m edges $\{(v_i, v_j) | v_i, v_j \in V\}$. Metrics for undirected graphs are then computed as another method to describe leaf venations of different species of plants.

3 Methodology

3.1 Image Acquisition

A set of leaves were collected from five plant species selected according to their visibility of leaf venations. A plant taxonomist identified the samples and these plants are namely *Bougainvillea spectabilis* (BS), *Capsicum* sp. (CA), *Cassia* sp. (CS), *Durata erecta* (DE), and *Dahlia* sp. (DS). The leaves were photographed under ample lighting using a DLSR camera with a setting of ISO 200, focal length of 24 mm, and no flash. Using a normal photograph of a leaf under a bright light has proved to hardly capture the details of the venations. Instead, the light had to be supplied underneath the leaf and was achieved using a light pad where the leaves are placed revealing the veins more clearly. The proposed workflow of this study from image segmentation to the extraction of graph metrics is shown in Fig. 4.

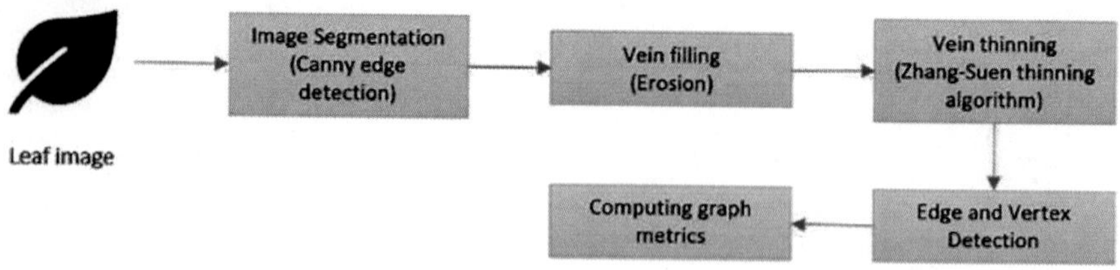

Fig. 4 Summary of the proposed simplified workflow from image segmentation to extraction of graph metrics

There were a total of 30 leaf photographs from five dicot plant species used in this study. Leaf images were selected depending on how well the contrast of the venations is compared to its background. Seven photos each of *Cassia* sp., *Durata erecta*, and *Dahlia* sp. plant species were chosen, while five photos were of *Bougainvillea spectabilis* and four photos were chosen from *Capsicum* sp. leaves. Since the venations of interest are only up to the third degree, the photographs were captured using a low resolution of 72 pixels per inch for faster processing of the algorithms.

3.2 Preprocessing

The raw photos of the leaves had a pixel dimension of 5184 × 3456 pixels (width × height) including largely a white background. The images were cropped to remove some parts of the background reducing the dimension to a width of 800 pixels to an approximate height of 500 pixels. To capture the overall graph, the whole leaf including the boundaries is included in the proposed methodology. Before the image segmentation process of Canny edge detection is applied, the background of the leaf images was first removed, after which the image is converted to a grayscale image.

3.3 Ground Truth (GT) Tracing

The veins of interest were traced manually using a pen tablet (XP-PEN) and an image-editing software (Adobe Photoshop 2023). Using a one-pixel size setting, the digital pen traces the veins of interest which are the primary veins, secondary veins, and tertiary veins. Tracing of one leaf took 30–60 min depending on the complexity of the venation. Another person aside from the tracer reviewed the output for possible editing. The digitized file is saved in a.jpeg format for comparison with the extracted image from the proposed workflow.

3.4 Vein Extraction and Graph Conversion

In the computational workflow for vein extraction, there are two major processes involved. First is the edge detection and second is the skeletonization process. This study utilizes the Canny edge detection algorithm [11] for the first process as it shows superior results best fit for this study [12–14]. Using the OpenCV library and the Java programming language, the photographs had to be converted to grayscale images before subjecting them to the Canny edge detection algorithm. The output image is the boundaries of the leaf and the venations. However, some of the leaf venations are too thick that instead of being detected as one vein, they are detected as two parallel veins. In order to solidify the veins, a morphological operation called erosion was used. In erosion, the darker areas grow in size [15].

The second process is skeletonization where the solid veins including the boundaries of the leaf are converted to one-pixel width. In this step, Zhang-Suen's thinning algorithm [16] was utilized. Zhang-Suen is a thinning algorithm used mostly for character recognition in documents and is also known for its high performance, time-wise [17]. Specifically, this thinning algorithm aids in the clarification of curvature of scripts such as in Arabic texts [18] and Balinese scripts [19]. For this study, Zhang-Suen is used to clearly define the skeleton of the leaf venations to one-pixel size.

From the one-pixel width of the leaf venations including the boundaries, there are two objectives for the graph conversion method. The first is to determine which are vertices, or the intersection of the line segments with different slopes, and the second is to determine the edges which are connected to the vertices. In determining the vertices, each pixel and its immediate surrounding neighbors are compared to a set of 3 × 3 pixel-size patterns marked as vertex templates. If the pixels match any of the patterns, they will be added to the vertex list. Figure 5 shows the detected vertices marked by circles and saved in a text file. The edges are then determined by traversing from one vertex point following the connected lines to the vertex until it reaches another vertex point. The path is then stored as an edge including the pixels in its path. Metrics from graphs are then computed using the vertex and edge lists. Table 1 shows the graph properties used to compute the metrics of the extracted venations.

A limitation of the proposed workflow is that the threshold value of the Canny edge detection must be adjusted to include the needed venations from the extracted images. In this case, the threshold varies from 150 to 180. Also, the quality of the images is crucial such that if the input leaf images are blurry or the venations are not contrasting with its background, the workflow will not be able to extract the venations properly.

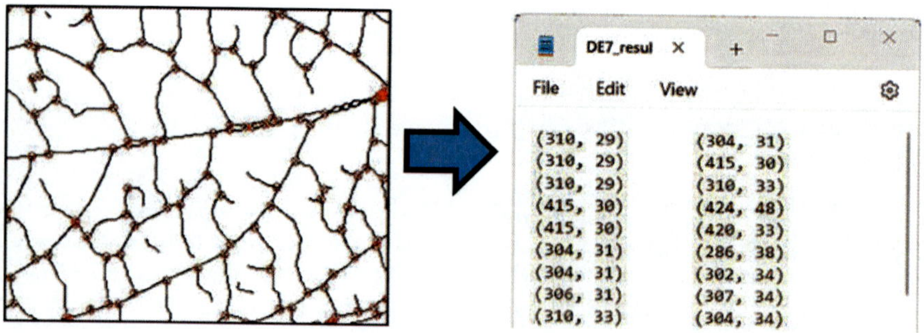

Fig. 5 Detected vertices are marked by circles. The vertices are saved in a text file as pixel points (x, y)

Table 1 Graph properties used as metrics of the converted undirected graphs of leaf venations [20]

Property	Description
Size	$n = \|V\|$
Volume	$m = \|E\|$
Number of wedges, 2-stars or 2-paths	$s = \sum u \in V \binom{d(u)}{2}$, where $d(u)$ is the degree of any vertex u
Number of claws or 3-stars	$z = \sum u \in V \binom{d(u)}{3}$, where $d(u)$ is the degree of any vertex u
Maximum degree	$d_{max} = max_{u \in V} d(u)$

4 Results

The proposed workflow in this study is able to identify the venations of leaves up to the veins of interest using the images used for this experiment. The veins of interest are the order of venations from first (1°) up to the third degree (3°). The parameters for the algorithm for Canny were adjusted accordingly to control the level of detail for the detection process. The range of threshold is from 150 to 180 using the default kernel in OpenCV. The image sizes were of 800 × ~ 500, width by height pixels with 72 ppi. Figure 6 shows the results of the different processes involved in the proposed workflow from Canny edge detection result, eroded image, and the Zhang-Suen thinning result. Venation extraction to vertex detection takes an average of 10 s per leaf running in an Intel Core i7, 16GB RAM 64-bit Windows 11.

4.1 Performance Analysis

The performance of the proposed workflow of venation image extraction can be measured by its correctness and completeness as defined by three criteria, the true positive (TP), false negative (FN), and false positive (FP) [21]. True positive is the pixels that are both in the extracted image and the basis image or ground truth. False positive is the pixels detected in the extracted image but is not in the ground truth image, while false negative is the pixels that are in the ground truth image but is not detected in the segmented image. Correctness (Cor) and completeness (Com) are defined as:

$$\text{Com} = \frac{\text{TP}}{\text{TP} + \text{FN}}, \tag{1}$$

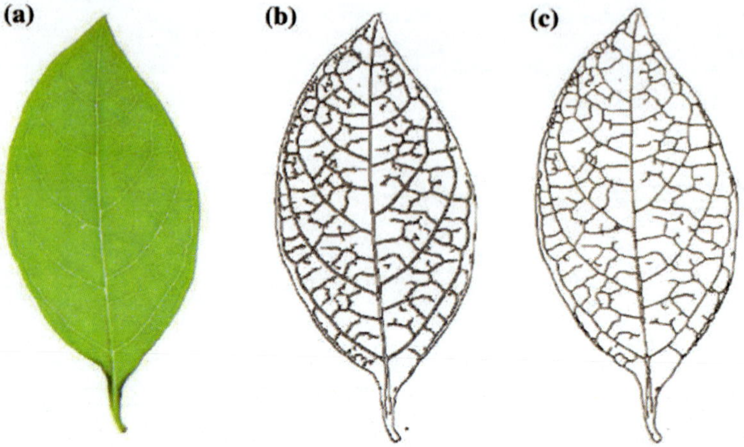

Fig. 6 **a** Photograph of leaf of *Durata erecta*, **b** output of the Canny edge detection, **c** image after the skeletonization using Zhang-Suen thinning algorithm

$$\mathrm{Cor} = \frac{TP}{\mathrm{TP} + \mathrm{FP}}, \tag{2}$$

with values ranging from 0 to 1 and with 1 as the optimum value.

Using the metrics of correctness and completeness, results show that using the proposed workflow, the extracted images revealed a total average of 0.522 for correctness and 0.406 for completeness. The results are not as high given the resolution of the input images of only 800 × ~ 500. The aim of the algorithm was to extract the higher-order veins in order to convert the venations into an undirected graph which also includes the boundaries of the leaf. Gıven also that the lower-order veins were of the vein of interest, the relatively low completeness values are expected. Table 2 shows the results of the completeness and correctness of all leaf samples.

4.2 Graph Metrics Results

Using the formula in Table 1, the values of each metric for each leaf were grouped according to plant species. The result was tabulated and compared. The results show a difference in the computed graph metrics from the different plant leaf venations for each plant species. The graph properties that gave significant differences among the plant species are size (n), volume (m), number of wedges (s), and number of claws (z). The graph metric maximum degree was disregarded as all graphs revealed had a maximum degree of 5. Figures 7 and 8 show the result of the size (n) and volume (m) of graphs taken for each plant species. The results for the number of wedges (s) and number of claws (z) are also presented in the charts in Figs. 9 and 10.

Table 2 Correctness and completeness values of the extracted veins using the proposed workflow for 30 leaf images

Extracted vein image	Correctness	Completeness
BS1	0.513782016	0.433073515
BS2	0.432817885	0.416735799
BS3	0.51902464	0.405329242
BS4	0.548636149	0.406370157
BS5	0.533449879	0.411577396
CA1	0.550867744	0.416515837
CA2	0.532643312	0.405485264
CA3	0.515320031	0.418625251
CA4	0.53071195	0.414908363
CS1	0.575009092	0.395724267
CS2	0.576441763	0.402168918
CS3	0.596319468	0.413250127
CS4	0.572447272	0.396338916
CS5	0.577792286	0.422985717
CS6	0.580717949	0.416684384
CS7	0.570981996	0.411857453
DE1	0.593535011	0.336413653
DE2	0.581963822	0.370073656
DE3	0.588737154	0.362387042
DE4	0.575356265	0.366352883
DE5	0.564749897	0.379045811
DE6	0.587273128	0.384006253
DE7	0.601047469	0.39346225
DS1	0.555439011	0.436750404
DS2	0.537365964	0.418011243
DS3	0.531433765	0.44434346
DS4	0.510484066	0.446662933
DS5	0.515510593	0.427549902
DS6	0.536567181	0.40304408
DS7	0.553917208	0.433935463

There is a consistent pattern for the five plant species using the graph metrics of size (n), volume (m), number of wedges (s), and number of claws (z). DS has the highest range of values in all four metrics with its maximum value as the highest compared to other four plant species. CA has the smallest range in all metrics as well as the lowest values. It is also observed that the same pattern occurs in all metrics with CA getting the lowest values and DS having the highest values. The pattern is consistent when in terms of range of values in increasing order: CA, DE, CS, BS, and DS.

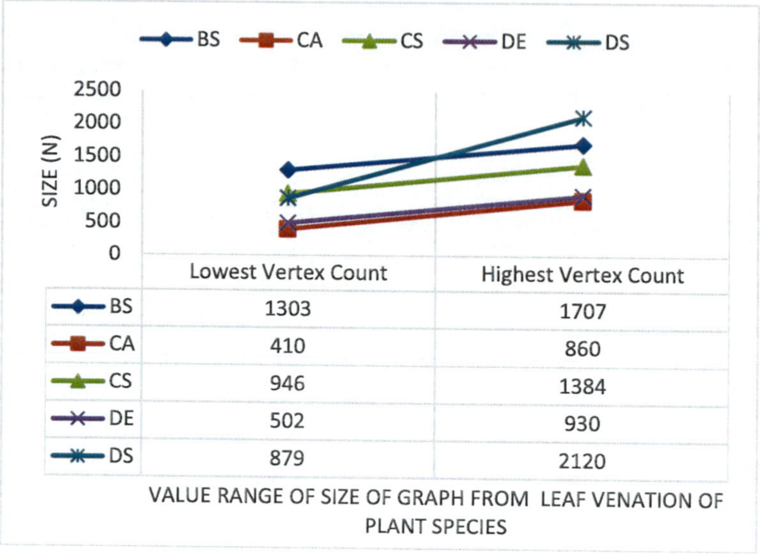

Fig. 7 Range of values of graph sizes (n) of leaves per plant species

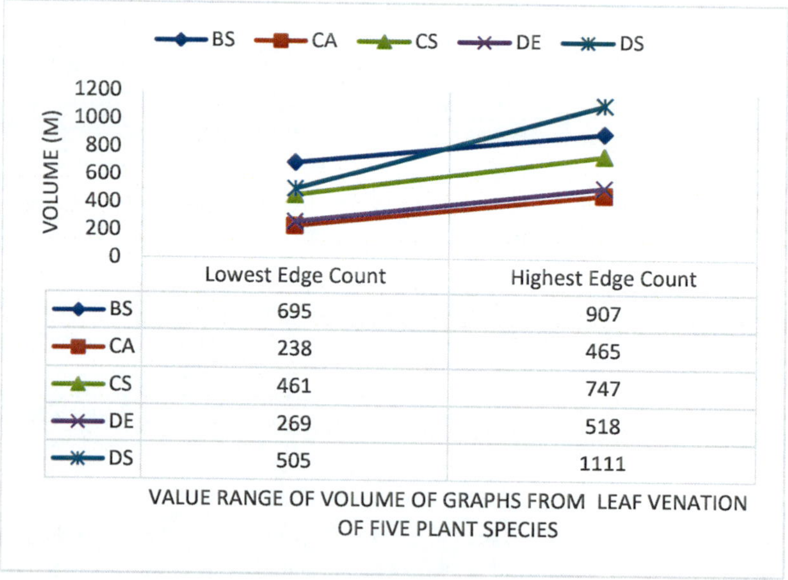

Fig. 8 Range of values of graph volume (m) of leaves per plant species

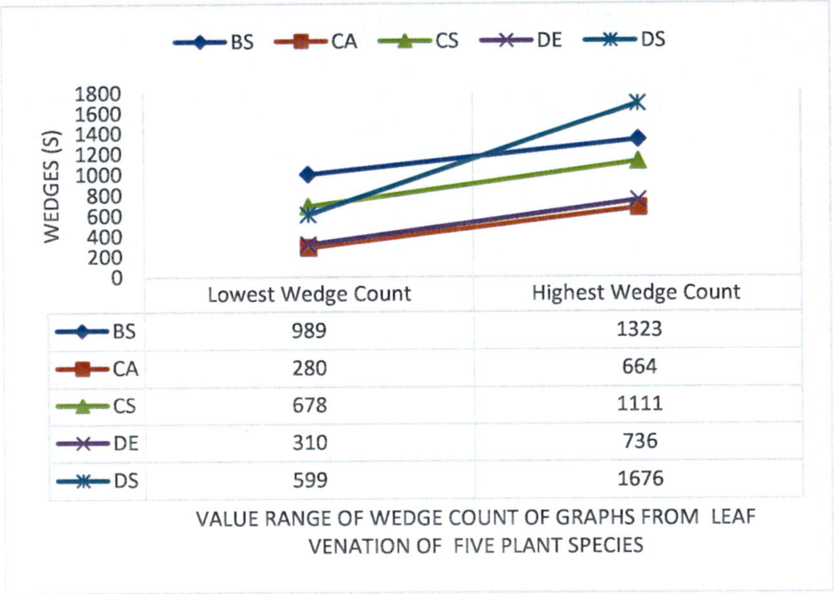

	Lowest Wedge Count	Highest Wedge Count
BS	989	1323
CA	280	664
CS	678	1111
DE	310	736
DS	599	1676

Fig. 9 Range of values of graph wedge count (s) of leaves per plant species

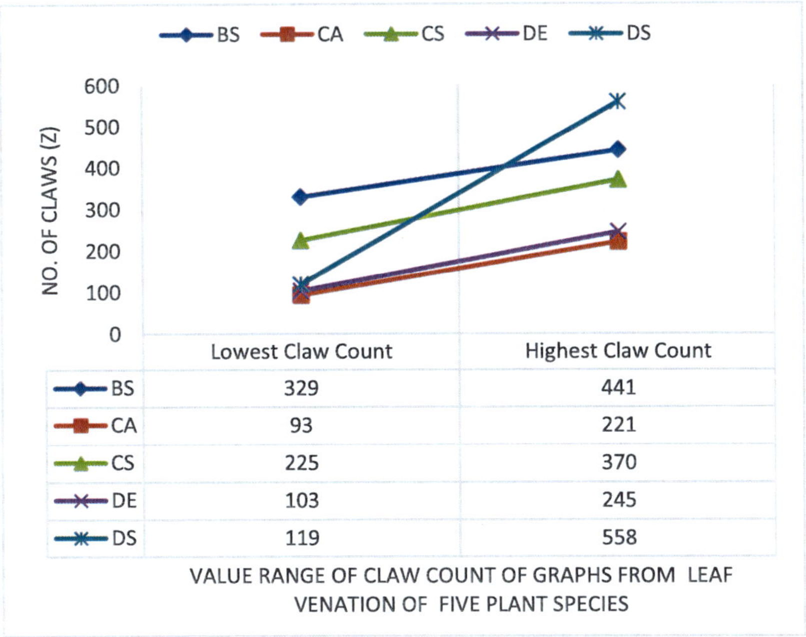

	Lowest Claw Count	Highest Claw Count
BS	329	441
CA	93	221
CS	225	370
DE	103	245
DS	119	558

Fig. 10 Range of values of graph claw count (z) of leaves per plant species

5 Conclusion and Recommendation

The proposed workflow presented in this study to extract the venation of plants has lesser steps involved compared to the other known software, making the computation lesser and hence, faster. The use of the algorithms Canny edge detection and Zhang-Suen thinning algorithm is sufficient to segment the venations up to the venation of interest which is the third degree (3°) from the leaf images. The quality of the input photographs greatly affects the quality of the detection method such that the more distinct the venations are from the background, the better the performance of the proposed workflow. The undirected graph (G) was successfully extracted from the segmented venation images using the proposed workflow. Although the correctness and completeness of the extracted venations can still be improved, this study has shown that patterns can be revealed from the graph properties when converting the leaf venation to undirected graphs. These metrics can be added to the other leaf venations metrics for plant characterization. Other graph metrics can also be used as additional measures for pattern analysis. The proposed workflow can also be utilized in similar vein images where graphs are needed to be extracted. This study can also be improved by adding more sample datasets and with higher image resolutions. With more datasets, the graph metrics can be further analyzed using machine learning to identify patterns.

References

1. Jones Jr S, Luchsinger A (1979) Plant systematics. McGraw-Hill, pp 222–232
2. Sack L, Mckown A, Frole K, Scoffoni C, Rawls M, Havran J, Tran H, Tran T (2012) Developmentally based scaling of leaf venation architecture explains global ecological patterns. Nat Commun 3
3. Fuller DQ, Hickey LJ (2005) Systematics and leaf architecture of the Gunneraceae. Bot Rev 71:295–353
4. Blonder B, Enquist BJ (2014) Inferring climate from angiosperm leaf venation networks. New Phytol 204:116–126
5. Hickey LJ (1973) Classification of the architecture of dicotyledonous leaves. Am J Botany: 17–33
6. Price C, Symonova O, Mileyko Y, Hilley T, Weitz J (2011) Leaf extraction and analysis framework graphical user interface: segmenting and analyzing the structure of leaf veins and areoles. Plant Physiol 155
7. Bühler J, Rishmawi L, Pflugfelder D, Huber G, Scharr H, Hulskamp M, Koorneef M, Schurr U, Jahnke S (2015) PhenoVein—a tool for leaf vein segmentation and analysis. Plant Physiol: 2359–2370
8. Dhondt S, Van Haerenborgh D, Van Cauwenbergh C, Merks RM, Philips W, Beemster GT, Inze D (2012) Quantitative analysis of venation patterns of Arabidopsis leaves by supervised image analysis. Plant J 69:553–563
9. Dirnberger M, Kehl T, Neumann A (2015) Nefi: network extraction from images. Sci Rep 5
10. Balangcod AD, Pabico JP (2020) Automatic identification of selected dicot plant species using graph properties from leaf venations. In: 20th Philippine computing science congress
11. Canny JF (1986) A computation approach to edge detection. IEEE Trans Pattern Anal Mach Intell PAMI-8 6:679–698

12. Kaur S, Singh I (2016) Comparison between edge detection techniques. Int J Comput Appl 145:15–18
13. Acharjya P, Ritaban D, Dibyendu G (2012) Study and comparison of different edge detectors for image segmentation. Glob J Comput Sci Technol
14. Shin M, Goldgof DB, Bowyer K, Nikiforou S (2001) Comparison of edge detection algorithms using a structure from motion task. IEEE Trans Syst Man Cybern Part B (Cybern) 31:589–601
15. Said KAM, Jambek AB (2021) Analysis of image processing using morphological erosion and dilation. J Phys Conf Ser 2071(1)
16. Zhang TY, Suen C (1984) A fast parallel algorithm for thinning digital patterns. Commun ACM 27(3):236–239
17. Widiarti AR (2011) Comparing Hilditch, Rosenfeld, Zhang-Suen, and Nagendraprasad-Wang-Gupta thinning. Int J Comput Inf Eng 5(6):563–567
18. Saudagar AKJ, Mohammed HV (2016) OpenCV based implementation of Zhang-Suen thinning algorithm using Java for Arabic text recognition. Inf Syst Des Intell Appl Proc Third Int Conf India 3:265–271
19. Sudarma M, Sutramiani NP (2014) The thinning Zhang-Suen application method in the image of Balinese scripts on the Papyrus. Int J Comput Appl 91(1):9–13
20. Kunegis J (2013) KONECT — The Koblenz network collection. In: Proceedings of international web observatory workshop, 1343–1350
21. Heipke C, Mayer H, Wiedemann C, Jamet O (1997) Evaluation of automatic road extraction. Int Arch Photogramm Remote Sens 32(3 SECT 4W2):151–160

Multispectral Fusion of Multisensor Image Data Using PCNN for Performance Evaluation in Sensor Networks

S. Dharini and Sanjay Jain

Abstract This research study proposes a Multispectral Fusion of Multisensor Image Data using PCNN by exploring the image visual quality and the object detection process. Initially, the collected IR and VIS images are preprocessed using a sampling, filtering and resizing process. Then, the contrast enhancement is done on the preprocessed images to improve the visual quality of preprocessed image. While performing image fusion, utilizing the less block size image has lowered the functioning of fusion process; thus, image resizing is applied before employing the DCT technique. Here, the pixels of IR and VIS are further improved by PCNN that intentionally performs to increase the quality of the pixels by capturing the local information of fused images. Further, IDCT is also employed to decompose the fused images. Finally, the object presented in the image is detected and the metrics like SNR and PSNR are used to prove the efficiency of the fused image.

Keywords Object detection · Multimodal sensor images · Fusion process · PCNN · PSNR · SNR

1 Introduction

The technological innovations and developments provide numerous benefits in the form of digital applications. Specifically, in the field of object recognition, the complexities in detecting multiple objects are greatly reduced. However, the intrinsic and extrinsic characteristics make the recognition process more challenging. Further to overcome the challenges, existing research works are analyzed in detail. The major

S. Dharini (✉) · S. Jain
CMR Institute of Technology, Bengaluru, India
e-mail: dhariniraghav@gmail.com

S. Jain
e-mail: Principal@cmrit.ac.in

S. Shakya et al. (eds.), *Fourth International Conference on Image Processing and Capsule Networks*, Lecture Notes in Networks and Systems 798,
https://doi.org/10.1007/978-981-99-7093-3_9

characteristics of a mining system are efficiency, and it must be attained by considering the storage features, processing metrics, feature extraction procedures, feature indexing and pattern retrievals [1]. Based on these characteristics, various mining models are evolved in the past decade and are mainly categorized into function-based mining and information-based mining. Further, the function-based mining is widely analyzed by the research community as it provides the important features of data. In this category, numerous clustering and classification models are evolved to attain better image analysis results. Specifically, the classification models are further categorized as unsupervised and supervised models [2]. In the supervised classification process, existing labeled images are used to group the unlabeled images. In case of unsupervised classification process, depending on the content or feature of the images, different groups are created [3]. Various techniques are developed under unsupervised category and supervised categories infer the essential information from the image. Specifically, low-level images are used in the mining process so that the high-level objects can be extracted [4]. Generally, the information will be extracted by four levels, namely, pixel, semantic concepts, object and pattern levels [5]. The features of deep learning algorithms are significantly proved in various classification and pattern recognition applications. Various deep learning models are evolved as recognition models, which imply the increased utilization of human computer interactions. One of the significant aspects of human computer interaction is emotion-related data, which helps to build an effective communication module. Similarly, various image processing applications are evolved based on deep learning algorithms. Deep learning approaches and existing learning models extract information or feature from a single input image. However, to obtain detailed features to enhance the classification or recognition performance, fused images are widely preferred. The process of combining multiple images as single image is termed as image fusion. Rich information can be obtained from the fused image so that it assists all phases involved in image processing and analysis tasks like object detection, image classification, image segmentation and so on. It also helps for better interpretation of processed images that covers the real-time applications like agriculture, head pose tracking system, surveillance applications, weapon detection, etc. [6]. The performance of fused images relies on two factors, namely, image decomposition algorithm and choice of fusion rule. It includes many properties such as redundancy, directional selectivity, shift variance, lossless reconstruction and preservation of edges. The fusion rules must be designed in a way that it should merge the information without any loss and without introducing spectral and spatial distortion in fused images. This research work presents a detailed analysis of image fusion process and validates the efficiency of fusion in object recognition application. This research study is organized as follows. Section 2 presents the related work that describes the merits and demerits of the prior techniques. Section 3 provides the proposed work, and simulation results are presented in Sect. 4. Finally, Sect. 5 concludes the proposed research work.

2 Related Work

In order to gain a broad knowledge on the existing image fusion techniques, a brief literature analysis is presented in this section by considering the methodology and its features. Multiple sensors with multiband and multimode capabilities were studied, and the process of data selection, preprocessing, registration and the fusion methodology are employed in multiple sensory applications and also analyzed. In general, Visible (VIS) and Infrared (IR) images are combined in the fusion process to obtain rich information content. Object location and thermal radiations can also be captured in IR images [7], but it is difficult to obtain more details from IR images due to its low-resolution features. In the case of VIS image, significant details can be obtained with high resolution, but due to environmental factors, the VIS images will lose some features. Thus, IR images are combined with VIS images using image fusion process to obtain the low- and high-level features.

The conventional IR and VIS image modeling involves Sparse Representation (SR)-related methods and the Multiscale Transform (MST)-related methods. The MST-related methods presented the decomposing model to the source images with the multiscale representations with the fusion rule; further by applying inverse transform on fused multiscale representations, fused images are obtained. Some methods like wavelet transform, Pyramid transform, non-subsampled contourlet transform (Hu et al. 2017) and MST methods are employed in image fusion process. [7], have designed a convolution-based Sparse Representation model to decompose the image fusion methods. However, the presented model enhances the fused image accuracy wherein the quality of image is reduced after the fusion process. An activity level measurement is addressed by [8] to generate fusion rules by using deep convolutional neural networks. Here, pixel-based activity information was shared among the users using a weighted map approach [9]. Due to the discontinuity variable, the weighted approach could not match the shared information during the retrieval process. In [10], a deep learning framework was designed to explore the efficiency of content-based medical image retrieval processes. This model takes high computational time to retrieve the relevant content. In continuation with the above study, [11] presented a method to identify the defocused and focused pixels in image and performed pixel-level fusion by using convolutional neural network. Here, the edges of the pixels are not properly detected, which results in creating overlapping issues during the fusion process. Li et al. [12, 13], introduced a fusion layer in deep CNN, wherein each block is connected with the output of the other layers. Though the fusion layer has properly fused the images, the decomposition part of the fused image was not remarkable. Similar study was extended by [14] for engaging more on deep features and zero-phase components to develop a fusion process. Since the components have reduced the sensitivity of the image enhancement, it is not suitable for real-time applications. A multimodal medical image fusion using deep CNN and multiscale transformation-based feature process were studied by [15]. This has proved the efficiency rate of the fused images, however, the interpretation of fused images becomes a complex task.

3 Research Methodology

The proposed Pulse-Coupled Neural Network (PCNN) approach for multispectral fusion of multisensor data is presented in this section. Here, a simultaneous Infrared image and Visible images are employed to fuse the images via Discrete Cosine Transform (DCT)-Pulse-Coupled Neural Network (PCNN). The followings are the proposed steps:

3.1 Image Preprocessing

It is the initial step that describes the details of information collection and the validation of preprocessing techniques based on the collected data. A pixel sampling is performed on the IR and VIS images in which the pixel values are replaced with median value of the neighbor pixels. The medial value is obtained by sorting the pixel values. A window slides over the entire image pixels and replaces it with median pixel value. After sampling, Gaussian filter is applied to remove the noise and enable image smoothening. Generally, convolution operators are termed as Gaussian operators, which smoothens the image. The parameters of Gaussian filters are standard deviation σ and window dimension. If the standard deviation value is high, then it requires high smoothing process.

3.2 Image Enhancement

The quality of the preprocessed image is further enhanced by applying various techniques, namely, contrast enhancement, histogram equalization and adaptive histogram equalization. Initially, the adjustment of the images is done, and then, it is followed by an equalization process.

(a) **Histogram Equalization**

To improve the image contrast, histogram equalization is employed to alter the intensity values to provide contrasted image. Histogram equalization is an effective and convenient image enhancement process. It is the technique used to develop a linear cumulative histogram value of the given input images. Based on the obtained histogram values, the pixel values are redistributed according to the intensity range. The conventional histogram equalization technique is described below:

Let us assume an input image, $I(p, q)$ with an aggregate of n pixels, in which the gray level ranges from $\left[X_0, X_{N-1}\right]$. The probability density function $P(r||k)$ for the image level r_k is given as,

$$P(r_k) = \frac{n_k}{n}, \tag{1}$$

where

n_k count of the r_k in the image.
n aggregate number of pixels with $k = 0, \ldots (N - 1)$.

Finally, the histogram of the image is estimated by the arrangement of n_k against the r_k. Owing to it, the altered intensity is estimated by the cumulative functions which is given as,

$$C(r_k) = \sum_{i=0}^{k} P(r_i). \quad (2)$$

It helps to arrange the image under dynamic range, $[X_0, X_{N-1}]$ which is again elaborated as,

$$f(X) = X_0 + (X_{N-1} - X_0)C(X). \quad (3)$$

The above equation has reduced the unequal histograms that have changed the brightness of the image.

(b) **Adaptive Histogram Equalization (AHE)**

At last, a preprocessed image is fed into Adaptive Histogram Equalization (AHE), which assists for obtaining enhanced image quality. It is a spatial domain model that describes the uniform distribution of pixel intensity for better contrast enhancement. The procedure of AHE is as follows:

(a) Input: preprocessed image.
(b) Obtain the image histogram.
(c) Finding the local minima in the histogram.
(d) Image is divided based on local minima.
(e) Allocating specific gray levels to each divided histogram.

By doing so, the histogram partitioning prevents some parts of histogram being dominated by others.

3.3 *Image Fusion*

To fuse the IR and VIS images, Discrete Cosine Transform (DCT) is applied by combining the pixels in order to provide the fused image. Cosine transform functions combine the VIS and IR images' missing pixels in the fusion process. The cosine functions make use of the Mel frequency range of the pixels and transform it into spatial domain. Further to estimate coefficients, DFT is applied. The final output of DFT provides the Mel frequency cepstral coefficient (MFCC) which is given as follows:

$$C_n = \sum_{k-1}^{k} (\log D_k) \cos[m(k-1)\pi], \tag{4}$$

where $m = 0, 1\ldots k - 1$; C_n represents the Mel frequency cepstral coefficient, and m indicates the number of coefficients. The two-dimensional input signal is then converted into one-dimensional signal using MFCC and then categorized into frames with the condition ($M < N_m$), where N_m indicates the neighbor frames. This condition reduces the pixel information losses. Further by applying FFT, the frequency domain is attained through windowing process. Finally, Mel frequency is used to obtain spectrum magnitude.

3.4 Image Quality Enhancement

To improve the image quality, Pulse-Coupled Neural Network (PCNN) is employed in the proposed work which is familiar network model formulated based on the cat and monkey pulse bursts. In the proposed method, neurons pulse synchronization and global coupling characteristics are employed. The local information of a VIS and IR images is fused using PCNN. The neurons are geared up by the values of a single pixel under the spatial domain. Thus, the spatial frequency with gradient energy is employed in this study. Here, multimodal-PCNN (m-PCNN) is a 2D-linked neuron which is laterally connected. Depending on the pixels' count, the neurons for the network is defined; similarly, the neuron stimulus is selected based on the gray pixel values. In the neuron internal state, data fusion is performed and it is essential that the input images must have identical resolution. Deviation in the resolutions will reduce the fusion performances. PCNN provides better fusion performance over traditional approaches and preserves the texture and edge features in image which makes the fusion suitable for recognition applications.

3.5 Image Reverse-Fusion Process

To enhance the segmentation process in the proposed work, Inverse Discrete Cosine Transform (IDCT) is applied to extract the foreground objects from the images. General applications utilize DCT for compression, but in the proposed work, DCT is used to segment the object. For object segmentation, background and foreground subtractions are performed in DCT considering the decorrelation attributes. This preserves the features, reduces the computation complexity and enhances the retrieval efficiency. Further to enhance the effectiveness of segmentation, user-defined functions are employed. The steps involved in modified DCT are presented as follows:

- Select initial frames.
- Compare each frame with the first frame.
- Employ binary segmentation for frame subtraction and perform morphological operation to select the object from the image.

Background and foreground subtraction-based IDCT provides enhanced performance over conventional DCT. Finally, the output of IDCT is employed to detect the object presented in an image.

4 Results and Discussion

The proposed model performances are discussed in this section from the beginning to final PSNR and SNR measurements. Two types of images like Infrared image and VIS image are used for the fusion process, and finally, the fused image metrics are measured and analyzed.

Figure 1 presents the input images, wherein source image 1 denotes the IR image and the source image 2 presents the VIS images.

Figure 2 presents the preprocessed output. Here, sampling and filtering processes are applied into the images. The irrelevant noise in IR and VIS images is given.

Figure 3 presents the image enhancement that enhances image visual quality. Also, the pixel-level information has been deliberately performed to improve the visual quality.

Results of fused images are presented in Fig. 4, and to attain better performance, DCT functions are applied to resize the images.

Figure 5 presents the output of the PCNN technique. Here, the quality of the pixels is improved layer by layer.

Figure 6 presents the object detection. From the pixel-quality improved images, the objects present in the image are detected.

Figure 7 presents the performance analysis of the DCT technique by using two metrics, PSNR and SNR.

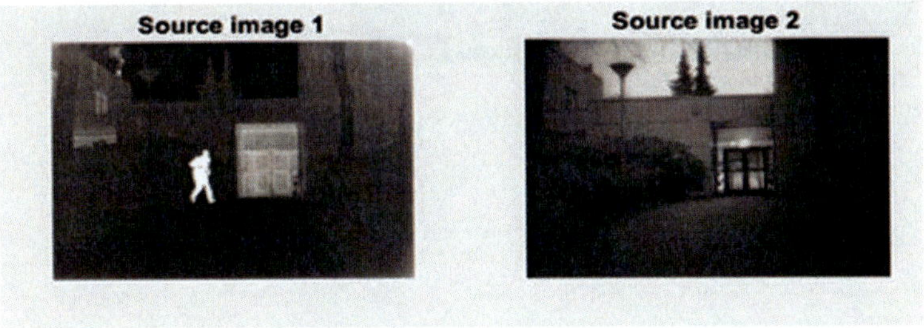

Fig. 1 Input images

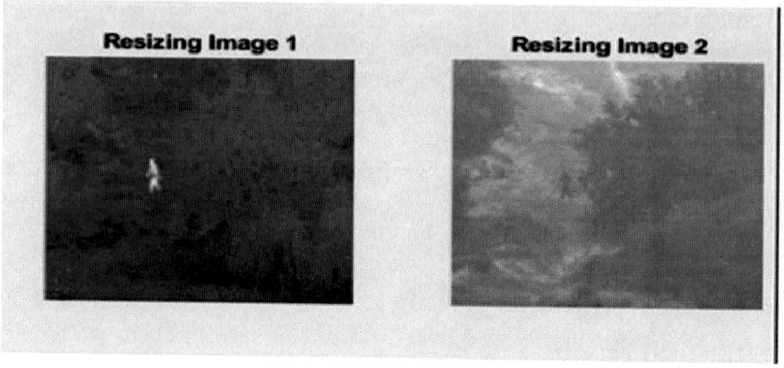

Fig. 2 Preprocessed image

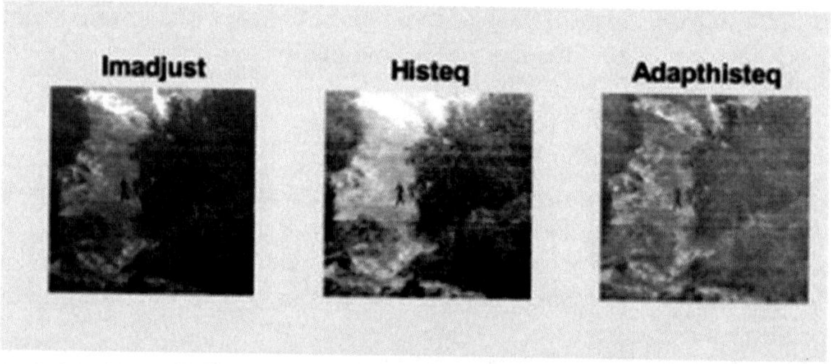

(a)

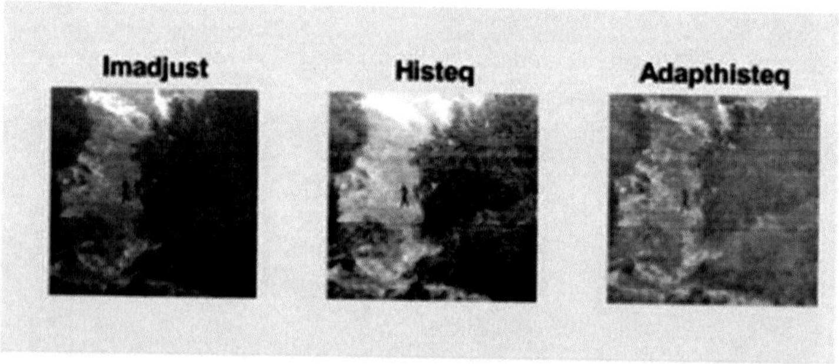

(b)

Fig. 3 Image visual quality enhancement

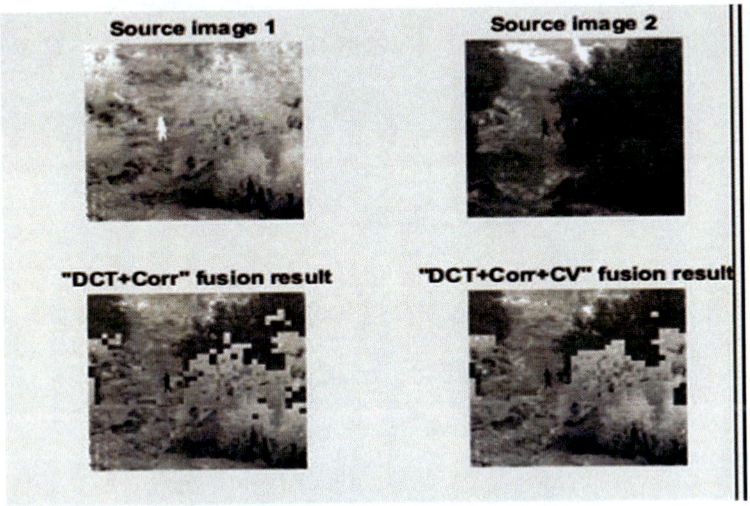

Fig. 4 Image fusion process

Fig. 5 PCNN process

Figure 8 presents the performance analysis of the IDCT-fusion technique. It is estimated to find the efficiency of the image fusion process.

Table 1 shows the error rate analysis of DCT-fusion technique and IDCT-fusion technique.

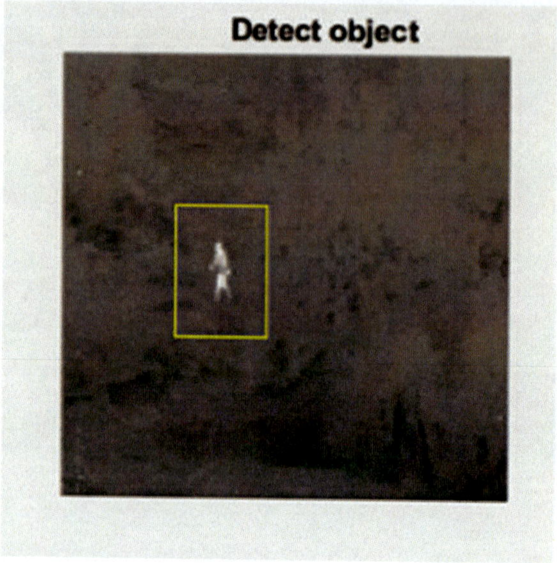

Fig. 6 Detecting the objects

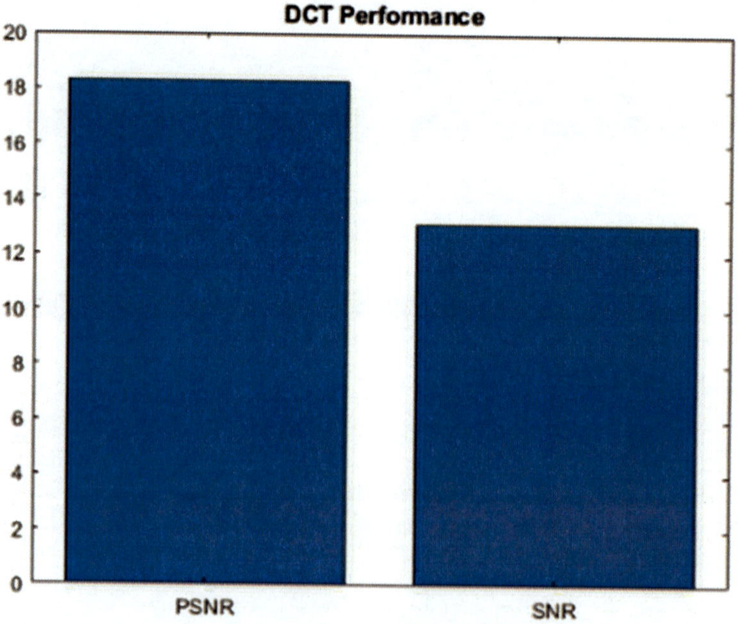

Fig. 7 Performance analysis of the DCT-fusion technique

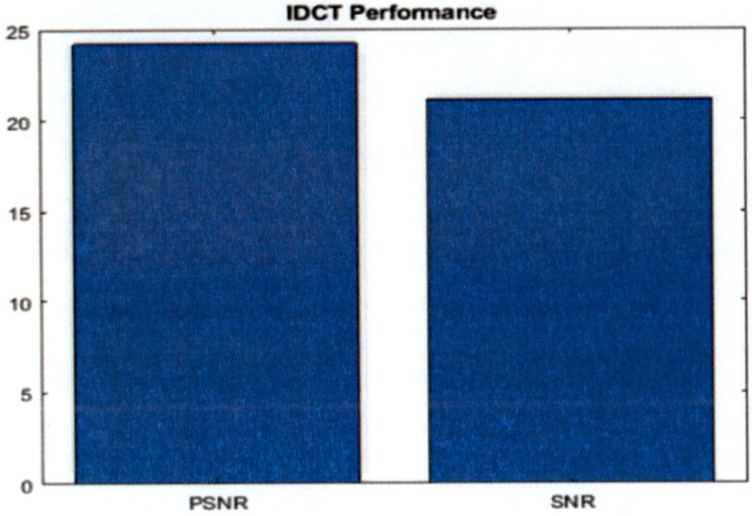

Fig. 8 Performance analysis of the IDCT-fusion technique

Table 1 Error rate analysis

Technique name	Error rate
DCT_PSNR	18.2800
IDCT_PSNR	24.2068
DCT_SNR	13.0999
IDCT_SNR	21.1067

5 Conclusion

In recent times, the analysis of multimodal images has been explored in real-time applications. In specific, object detection in industrial scenes in understanding environments becomes tedious tasks. Image fusion concept is one of the image processing steps focused by several researchers. It accelerates the growth of analyzing multimodal images. Multiple images are combined in the fusion process to provide rich information that assists for all phases involved in image processing and analysis tasks. It also helps in enabling the better interpretation of processed images. The fusion rules are designed to merge the information without any loss and without introducing spectral and spatial distortion in fused images.

References

1. Li X, Orchard MT (2001) New edge-directed interpolation. IEEE Trans Image Process 10(10):1521–1527
2. Zhang X, Wu X (2008) Image interpolation by adaptive 2-d autoregressive modeling and soft-decision estimation. IEEE Trans Image Process 17(6):887–896
3. Liu X, Zhao D, Xiong R, Ma S, Gao W, Sun H (2011) Image interpolation via regularized local linear regression. IEEE Trans Image Process 20(12):3455–3469
4. Guo K, Yang X, Zha H, Lin W, Yu S (2012) Multiscale semilocal interpolation with antialiasing. IEEE Trans Image Process 21(2):615–625
5. Zhong B, Ma KK, Lu Z (2019) Predictor-corrector image interpolation. J Vis Commun Image Represent 61:50–60
6. Bhateja V, Patel H, Krishnan A, Shahu A (2015) Multimodal medical image sensor fusion framework using cascade of wavelet and contourlet transform domains. IEEE Sensor J 25:6783–6790
7. Wang X, Meng J, Liu F (2016) Fusion of infrared and visible image based on compressed sensing and nonsubsampled shearlet transform. Int J Signal Process 9(4):37–46
8. Liu Y, Chen X, Peng H, Wang Z (2017) Multi-focus image fusion with a deep convolutional neural network. Inf Fusion 36:191–207
9. Liu Y, Chen X, Cheng J, Peng H, Wang Z (2018) Infrared and visible image fusion with convolutional neural networks. Int J Wavelets Multiresolution Inf Process 16(03):1850018
10. Qayyum A, Anwar SM, Awais M, Majid M (2017) Medical image retrieval using deep convolutional neural network. Neurocomputing 266:8–20
11. Tang H, Xiao B, Li W, Wang G (2018) Pixel convolutional neural network for multi-focus image fusion. Inform Sci 433:125–141
12. Li J, Song M, Peng Y (2018) Infrared and visible image fusion based on robust principal component analysis and compressed sensing. Infrared Phys Technol 89:129–139
13. Li H, Wu XJ, Kittler J (2018) Infrared and visible image fusion using a deep learning framework. In: 2018 24th international conference on pattern recognition (ICPR). IEEE, pp 2705–2710
14. Li H, Wu XJ, Durrani TS (2019) Infrared and visible image fusion with resnet and zero-phase component analysis. Infrared Phys Technol 102:103039
15. Xia KJ, Yin HS, Wang JQ (2019) A novel improved deep convolutional neural network model for medical image fusion. Cluster Comput 22(1):1515–1527

U-Net-Based Segmentation of Coronary Arteries in Invasive Coronary Angiography

A. Azeroual, Y. El Ouahabi, W. Dhafer, M. H. El yousfi Alaoui, B. Nsiri, and A. Soulaymani

Abstract Coronary artery disease continues to be the primary reason for death, creating a difficult situation for worldwide health strategies. Among the main techniques to diagnose these diseases clinically is invasive coronary angiography. Consequently, the development of a precise deep learning model for coronary artery segmentation becomes imperative for the diagnosis and evaluation of coronary artery disease. This is particularly important to address the challenges arising from issues in coronary angiography, which can adversely affect image quality and hinder cardiologists' medical interpretation. In this study, U-Net model is developed on images from invasive coronary angiography to achieve accurate and effective segmentation. This project's dataset includes 99 patients who underwent invasive coronary angiography systems. The average accuracy of the suggested model was 98% and an intersection over union of 86%. Indeed, the experimental results demonstrated big effectiveness of the U-Net model to help cardiologists in diagnosing this disease.

Keywords U-Net model · Deep learning · Coronary artery disease · Invasive coronary angiography · Binary image segmentation

A. Azeroual (✉) · W. Dhafer · M. H. El yousfi Alaoui · B. Nsiri
Research Center STIS, M2CS, National Graduate School of Arts and Crafts of Rabat, Mohammed V University in Rabat, Rabat, Morocco
e-mail: amal.azeroual@um5r.ac.ma

Group of Biomedical Engineering and Pharmaceuticals Sciences, National Graduate School of Arts and Crafts of Rabat (ENSAM), Mohammed V University, Rabat, Morocco

M. H. El yousfi Alaoui
e-mail: h.elyousfi@um5r.ac.ma

B. Nsiri
e-mail: b.nsiri@um5r.ac.ma

Y. El Ouahabi · A. Soulaymani
Laboratory Health and Biology, Faculty of Sciences, Ibn Tofail University, Kenitra, Morocco
e-mail: soulaymani@uit.ac.ma

S. Shakya et al. (eds.), *Fourth International Conference on Image Processing and Capsule Networks*, Lecture Notes in Networks and Systems 798,
https://doi.org/10.1007/978-981-99-7093-3_10

1 Introduction

As per the guidelines provided by the "World Health Organization", coronary artery diseases (CAD) are classified as a sort of cardiovascular condition that causes death to 17.9 million person each year [1]. Cardiovascular diseases encompass a range of heart and blood vessel conditions, including CAD, cerebrovascular diseases, and rheumatic heart diseases. More than four out of five deaths from cardiovascular diseases are related to myocardial infarctions and strokes, with approximately one-third of these fatalities occurring prematurely in individuals aged below 70 [1].

CAD affects the human heart by a buildup of cholesterol plaques in the coronary arteries that provide the heart by oxygenated blood. The symptoms of CAD vary from being without symptoms to experiencing stable chest pain, acute coronary syndrome, and sudden cardiac death. Arrhythmias and heart failure are frequently observed complications, while myocardial infarction is the prevailing manifestation of CAD [2]. CAD can be clinically diagnosed using three primary techniques: "Invasive Coronary Angiography" (ICA), "myocardial perfusion imaging", and "drug stress echocardiography". These methods are effective in identifying angina symptoms related to myocardial ischemia [3]. The first technique entails the injection of a contrasting substance into the arteries surrounding the heart, allowing for the visualization of coronary arteries under the supervision of a cardiologist. The severity of the narrowing is manually determined by the physician visual assessment (PVA). Meanwhile, quantitative coronary angiography is considered a reliable computer-assisted procedure for measuring the severity of stenosis [4]. However, it is limited by a busy workflow and the measurement of complex lesions. These limitations have motivated the progress of deep learning (DL) and the convolutional neural network (CNN)-based tools that can rapidly and accurately segment and detect lesions (shown in Fig. 1).

Deep learning technology is utilized by Du et al. [5] in the interpretation of diagnostic coronary angiography, which assists cardiologists in detecting and diagnosing the severity and shape of lesions while conducting the intervention. Rjiba et al. [6] introduced a novel method based on a 3D CNN for the automatic extraction of centerlines of coronary arteries in "Coronary Computed Tomography Angiography"

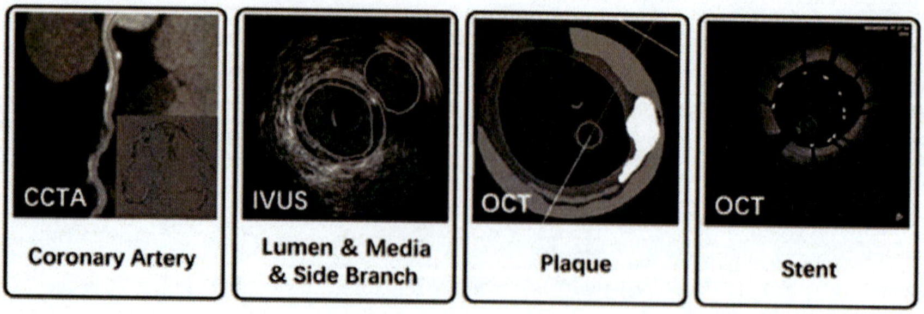

Fig. 1 Example of image segmentation on CAD

(CCTA) data. Gulsun et al. [7] investigated the application of CNN as a postprocessing method to differentiate between genuine CAD and leaks. More recently, Table 1 presents some studies on the utilization of CNN models for the purpose of image segmentation [8].

Also, U-Net model has proven its accurate performance in semantic segmentation of medical images and was applied in various segmentation tasks including retinal segmentation and brain arterial segmentation, that was challenging many aspects of ICA imaging, especially, that they have limited contrast and include blur noise [17]. Furthermore, by leveraging the power of U-Net, the segmentation process becomes more efficient and precise, enhancing the overall interpretation and analysis of ICA images. U-Net model requires a minimal number of annotated images and can easily be applied to various biomedical images' segmentation tasks thanks to the training process, which could be completed within a reasonable timeframe of up to 10 h using an "NVidia Titan GPU" (6 GB).

Long et al. constructed a U-Net network by drawing upon the principles of fully convolutional networks in their research [18]. Dorgham et al. present U-Net Computed Tomography Segmentation, an innovative automated DL segmentation framework tailored for computed tomography images. The primary objective is to merge the DL U-Net approach with computed tomography images, facilitating automatic segmentation within the realm of medical image analysis [19].

The subsequent sections of this paper are structured as follows: Sect. 2 presents a comprehensive review of the pertinent literature in the field. Section 3 of the document details the proposed approach, while in Sect. 4, the outcomes achieved were showcased and examined using the suggested models. Finally, the concluding section summarizes the key findings and implications of the study.

Table 1 Application of CNN models in image segmentation of CAD

Study	Year	Modality	Clinical application
Li et al. [9]	2022	CCTA	Stenosis diagnosis
Du et al. [10]	2022	IVUS	Plaque identification
Lin et al. [11]	2022	CCTA	Prediction of cardiovascular risk
Hong et al. [12]	2022	OCT	Risk stratification
Zhao et al. [13]	2021	XA	Stenosis diagnosis
Tu et al. [14]	2021	XA	FFR computation
Yu et al. [15]	2021	IVUS	FFR computation
Chu et al. [16]	2021	OCT	Plaque characterization

IVUS intravascular ultrasound, *OCT* optical coherence tomography, *XA* X-ray angiography

2 Related Work

2.1 Medical Imaging Works for Coronary Arteries

Medical image analysis involves the identification of specific informative details within an image or the classification of the image itself to address clinical issues. In the existing literature, there have been significant advancements in coronary artery analysis, through the utilization of various image processing techniques, including level-set, graph-cut, Gaussian distribution of lumen, and active contour methods. These methods typically necessitate crucial preprocessing and postprocessing steps [20].

CNN techniques have been extensively utilized in the realm of cardiology for medical imaging purposes, spanning multiple imaging modalities. Moeskops et al. [21] have developed a unified CNN capable of segmenting six different tissues in Magnetic Resonance (MR) images. Their approach extends to segmenting the pectoral muscle in MR breast images as well as the CAD. In other studies, Han et al. [22] carried a study to evaluate the practicality of utilizing DL methods for analyzing CCTA images concerning the detection of coronary artery stenosis. They conclude that the diagnostic accuracy of CCTA-Artificial Intelligence surpasses that of traditional CCTA.

2.2 Image Segmentation with U-Net

In recent years, DL is extensively employed to deliver precise image analysis across diverse medical domains. The key attribute of DL models lies in their ability to extract features, enabling them to provide accurate and distinct descriptions of different types of tissues. Medical image segmentation is a highly demanding undertaking in the realm of medical image analysis, as it endeavors to extract valuable data and enhance the accuracy of clinical diagnosis [23].

CNN techniques have exhibited exceptional levels of accuracy and dependability in numerous applications in computer vision, like image classification, object detection, and semantic segmentation. Previous research utilizing DL methods in the context of medical imaging in cardiology has predominantly concentrated on tasks related to segmentation. Alakwaa et al. have introduced a novel automated approach that utilizes a three-dimensional CNN (3D CNN) for the detection of lung cancer. Their method focuses on segmenting nodules from three-dimensional computed tomography (CT) scans [24].

Similarly, numerous techniques utilizing U-Net and its variations have been suggested to accomplish automated process of segmenting medical images. Recently, Baccouch et al. conducted a comparative analysis between U-Net and standard CNN models for the segmentation of medical images. Their study revealed that U-Net is better suited in the context of medical applications, specifically for segmentation tasks in the field of health care, when compared to other CNN models [23]. Ronneberger et al. have introduced an innovative framework named U-Net, which is a significant architecture for semantic segmentation in CNN. U-Net demonstrates remarkable performance for tasks involving pixel-level predictions [25]. Chen et al. have introduced TransUNet, a novel variation of the U-Net architecture. TransUNet combines transformers with U-Net to address the limitations of U-Net in explicitly capturing long-range dependencies, which can be challenging due to the inherent localized nature of convolution operations in tasks involving medical image segmentation [26]. In other research, Zhou et al. have introduced an enhanced architecture, U-Net++, for the task of segmenting medical images. Unlike the standard U-Net architecture, U-Net++ is a neural network that establishes interconnectedness between the decoder and encoder sub-networks by incorporating a series of nested, dense skip pathways. These pathways are designed to minimize the semantic gap between the feature maps of the encoder and decoder, resulting in a more powerful framework for accurate segmentation of medical images [27]. In summary, U-Net models have been used in many studies for the task of segmenting medical images, and in the rest of the paper, the model based on U-Net was presented.

3 Materials and Methods

3.1 Dataset

The dataset employed in this project was collected from 99 patients' coronary angiography imaging (Github), at the "Jiangsu Province People's Hospital" in China, an interventional angiography system (AXIOM-Artis, Siemens, Munich). The X-ray images were captured at a rate of 15 frames per second and each one with a size of 215×215 pixels. Expert cardiologists have annotated the dataset of left coronary and right coronary artery images. A total of 616 images, along with their corresponding masks, were collected for this study and 512×512 of size divided into 80% for training and 20% for testing.

3.2 Method

The main purpose is to train a U-Net model employed to perform segmentation of coronary arteries in imaging using ICA, which will be used in the next step in the assessment of stenosis and its classification after using oriented methods for the detection of stenosis. The model was applied on both left coronary artery (LCA) in Fig. 2 and right coronary artery (RCA) images in Fig. 3 in order to segment vessel tree. In the overall, the model was implemented utilizing TensorFlow/Keras and took about 15 min in training achieving training and a validation accuracy of 98%.

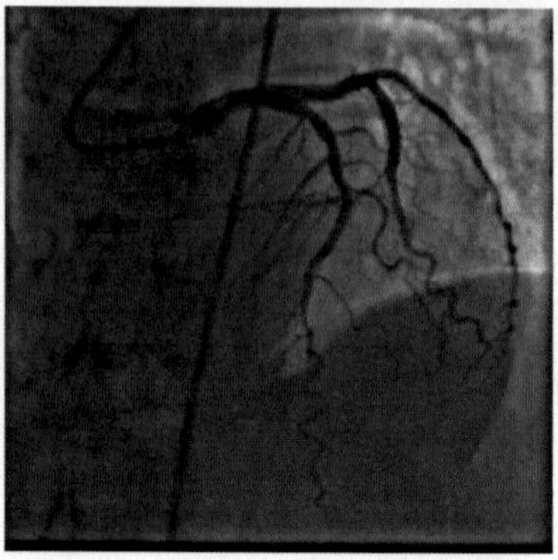

Fig. 2 Left coronary artery (LCA) X-ray image

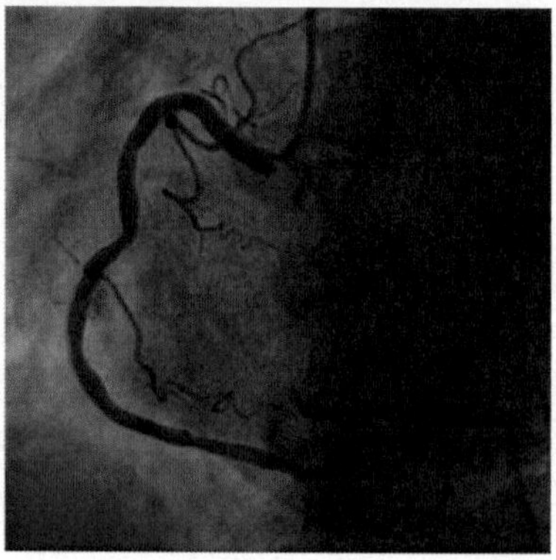

Fig. 3 Right coronary artery (RCA) X-ray image

U-Net Architecture. U-Net model includes an encoder and decoder path. The encoder path adheres to CNN architecture standards by including repetition 3×3 convolution layers followed by max-pooling functions and rectified linear unit (ReLU). A kernel of 3×3 size was applied across the 512×512 input for feature extraction with a total of 16 filters used (shown in Fig. 4). Then, a max-pooling layer got introduced between layers with the aim of down-sampling the input in weight and height directions using a stride of 2 according to the following formula:

$$\mathrm{Floor}((W - F)/S + 1), \quad (1)$$

where W represents the input x or y shape, F the pooling window size, and S stands for the stride. Conversely, the next path of the U-Net model is the decoder one using transposed convolution to up sample the output of the encoder path with both Kernel and stride of size two, according to the following formula:

$$Z = (W - 1)S + K, \quad (2)$$

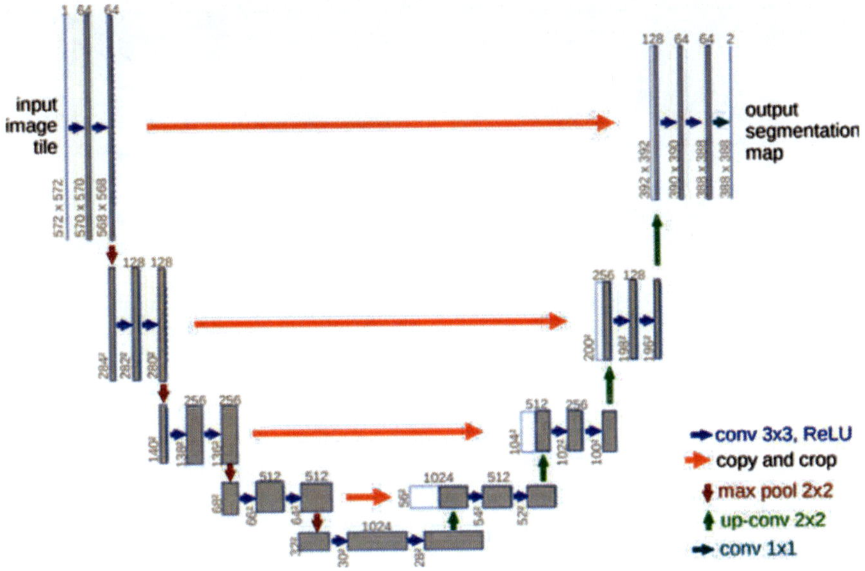

Fig. 4 U-Net architecture (*Source* Navab et al. [28], p. 235)

where Z stands for the output after the transposed convolution operation and K for kernel size.

Training. This U-Net model was applied on 616 images of 512×512 size split into 80% of the data which was allocated for training purposes, while the remaining 20% was reserved for testing. The training endures approximately 15 min for 100 epochs using GPU Colab and TensorFlow/Keras libraries.

4 Results and Discussion

In general, the training has achieved 98% in training and validation accuracy and approximately 86% for mean intersection over union (IoU) which are fairly good results, that could be improved after data augmentation to make it more representative for validation (shown in Figs. 5, 6, 7).

Centerlines' Extraction. In this step, this input is the segmented masks of coronary artery images in BGR color space to finally have binary images with centerline extracted [11]. Figure 8 shows the next process:

- Images are converted from BGR color space to grayscale (meant to be used on reading images using cv2.imread).
- Thresholding to ensure all the images are binary.

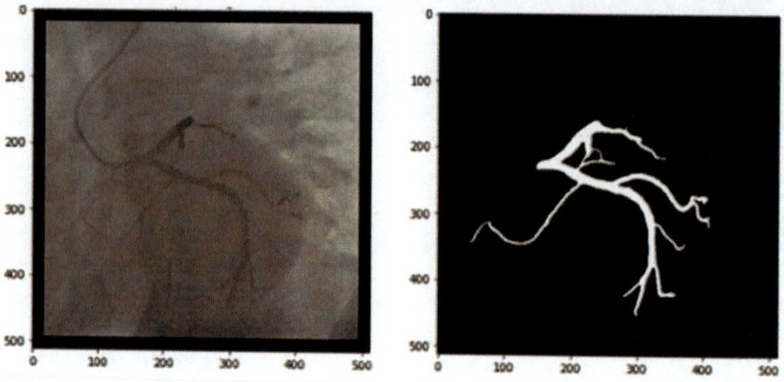

Fig. 5 U-Net segmentation results

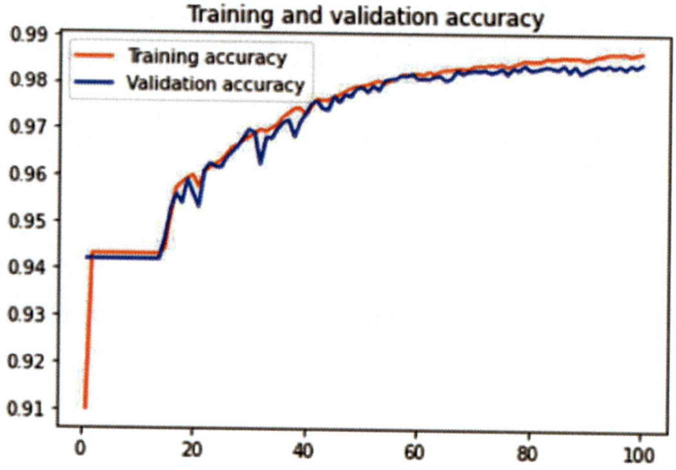

Fig. 6 U-Net learning curve/results of training accuracy

- Normalizing the image, so they only have values 0, 1.
- Closing morphological operation was used to remove any disconnected segments
- Skeletonizing to extract the centerline (it is the process of iteratively removing the border pixels until no more pixels can be removed).

Diameter Calculation. Figure 9 shows the process followed:

- Extracting centerlines.
- Distance transform is used with Euclidean distance; this returns an image in which each pixel represents its distance from the border of the contour.
- Bitwise and operation are performed on the resulting distance transform map with the centerline as a mask, this results in those pixels which are in the centerline containing the distance from the border, and all the remaining pixels are set to 0.

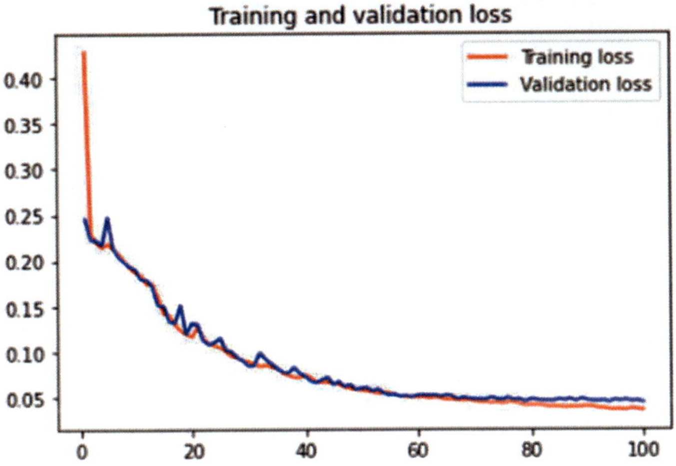

Fig. 7 U-Net learning curve/results of training loss

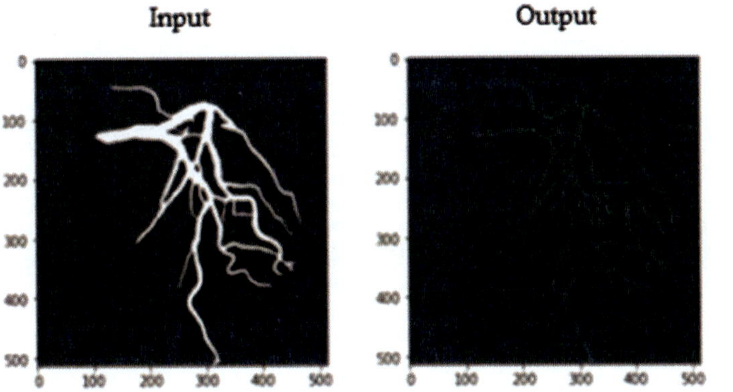

Fig. 8 Centerlines' extraction

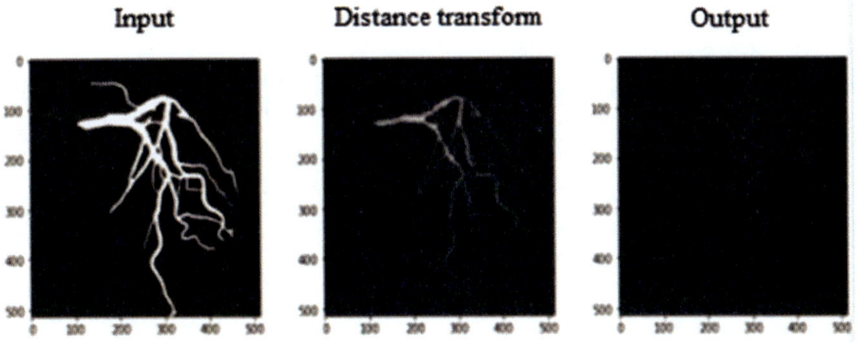

Fig. 9 Diameter calculation

5 Conclusion

In conclusion, having its architecture constantly improved over time, U-Net model is nowadays considered as one of the highly accurate biomedical image segmentation CNN-based models. The basic model, which was utilized throughout this project, reached 98% of accuracy as well as 86% of IoU in the segmentation of coronary arteries from ICA. The subsequent phase aims to further improve the architecture and enhancement of the dataset. This will produce more accurate outcomes that may be exploited in detecting stenosis after measuring arteries' centerlines and diameters.

Acknowledgements The authors would like to extend their heartfelt gratitude to Professor Abdelfettah Daoudy, ESL teacher in Amideast and researcher in the field of Applied Linguistics at Mohammed V University for proofreading this article.

References

1. Cardiovascular Diseases (2023) Available online at: https://www.who.int/health-topics/cardiovascular-diseases#tab=tab_1 (accessed Jun 08, 2023)
2. Hajikhani B et al (2023) COVID-19 and coronary artery disease; a systematic review and meta-analysis. New Microb New Infect 53:101151
3. Zhu S et al (2022) A knowledge graph-based analytical model for mining clinical value of drug stress echocardiography for diagnosis, risk stratification and prognostic evaluation of coronary artery disease. Int. J. Cardiol. 2:109181
4. Alizadehsani R et al (2019) Machine learning-based coronary artery disease diagnosis: a comprehensive review. Comput Biol Med 111(May):103346
5. Du T et al (2021) Training and validation of a deep learning architecture for the automatic analysis of coronary angiography. EuroIntervention 17(1):32–40
6. Rjiba S et al (2020) CenterlineNet: automatic coronary artery centerline extraction for computed tomographic angiographic images using convolutional neural network architectures. In: 2020 10th International Conference on Image Processing Theory, Tools and Applications, IPTA 2020
7. Gülsün MA, Funka-Lea G, Sharma P, Rapaka S, Zheng Y (2016) Coronary centerline extraction via optimal flow paths and CNN path pruning. In: Lecture Notes in Computer Science (including subseries Lecture Notes in Artificial Intelligence and Lecture Notes in Bioinformatics), vol 9902, pp 317–325
8. Chu M, Wu P, Li G, Yang W, Gutiérrez-Chico JL, Tu S (2023) Advances in diagnosis, therapy, and prognosis of coronary artery disease powered by deep learning algorithms. JACC 3(1):1–14
9. Li Y, Wu Y, He J, Jiang W, Wang J, Peng Y, Jia Y, Xiong T, Jia K (2022) Automatic coronary artery segmentation and diagnosis of stenosis by deep learning based on computed tomographic coronary angiography. Eur Radiol 32:6037–6045
10. Du H, Ling L, Yu W, Wu P, Yang Y, Chu M, Yang J, Yang W, Tu S (2022) Convolutional networks for the segmentation of intravascular ultrasound images: evaluation on a multicenter dataset. Comput Methods Programs Biomed 215:106599
11. Lin A et al (2022) Deep learning-enabled coronary CT angiography for plaque and stenosis quantification and cardiac risk prediction: an international multicentre study. Lancet Digital Health 4(4):e256–e265
12. Hong H et al (2022) (2022) Risk stratification in acute coronary syndrome by comprehensive morphofunctional assessment with optical coherence tomography. JACC 2(4):460–472

13. Zhao C et al (2021) Automatic extraction and stenosis evaluation of coronary arteries in invasive coronary angiograms. Comput Biol Med 136(March):104667
14. Tu BXM, Ding D, Chang Y, Li C, Wijns W (2021) Diagnostic accuracy of quantitative flow ratio for assessment of coronary stenosis significance from a single angiographic view: a novel method based on bifurcation fractal law. Catheter Cardiovasc Interv 97(S2):1040–1047
15. Yu W et al (2021) Accuracy of intravascular ultrasound-based fractional flow reserve in identifying hemodynamic significance of coronary stenosis. Circul Cardiovasc Interv 14(2):E009840
16. Chu M et al (2021) Artificial intelligence and optical coherence tomography for the automatic characterisation of human atherosclerotic plaques. EuroIntervention 17(1):41–50
17. Siddique N, Paheding S, Elkin CP, Devabhaktuni V (2021) U-net and its variants for medical image segmentation: a review of theory and applications. IEEE Access 9:82031–82057
18. Long J, Shelhamer E, Darrell T (2015) Fully convolutional networks for semantic segmentation. In: IEEE Conference on Computer Vision and Pattern Recognition (CVPR), pp 3431–3440 (2015)
19. Dorgham O, Naser MA, Ryalat MH, Hyari A, Al-Najdawi N, Mirjalili S (2022) U-NetCTS: U-Net deep neural network for fully automatic segmentation of 3D CT DICOM volume. Smart Health 26(August):100304
20. Huang W, Huang L, Lin Z, Huang S, Chi Y, Zhou J, Zhang J, Tan RS, Zhong L (2018) Coronary artery segmentation by deep learning neural networks on computed tomographic coronary angiographic images. In: Proceedings of the Annual International Conference of the IEEE Engineering in Medicine and Biology Society (EMBS), vol 2018-July(July 2018), pp 608–611
21. Moeskops P, Wolterink JM, Van der Velden BHM et al (2016) Deep learning for multi-task medical image segmentation in multiple modalities. In: Lecture Notes in Computer Science (including subseries Lecture Notes in Artificial Intelligence and Lecture Notes in Bioinformatics), vol 9901 LNCS, no October, pp 478–486
22. Han D, Liu J, Sun Z, Cui Y, He Y, Yang Z (2020) Deep learning analysis in coronary computed tomographic angiography imaging for the assessment of patients with coronary artery stenosis. Comput Methods Prog Biomed 196:105651
23. Baccouch W, Oueslati S, Solaiman B, Labidi S (2023) A comparative study of CNN and U-Net performance segmentation of medical images: application to cardiac MRI. Proc Comput Sci 219(2022):1089–1096
24. Alakwaa W, Nassef M, Badr A (2017) Lung cancer detection and classification with 3D convolutional neural network (3D-CNN). Int J Biol Biomed Eng 11(8):66–73
25. Ronneberger O, Fischer P, Brox T (2015) U-Net : convolutional networks for biomedical image segmentation. In: Medical Image Computing and Computer-Assisted Intervention, pp 234–241
26. Chen J, Lu Y, Yu Q, Luo X, Adeli X, Wang Y, Lu L, Yuille AL, Zhou Y (2021) TransUNet: transformers make strong encoders for medical image segmentation, pp 1–13
27. Zhou Z, Rahman Siddiquee MM, Tajbakhsh N, Liang J (2018) UNet++: a nested U-Net architecture for medical image segmentation BT—deep learning in medical image analysis and multimodal learning for clinical decision support. MICCAI 11045(2018):3–11
28. Navab N, Hornegger J, Wells WM, Frangi AF (2015) Medical image computing and computer-assisted intervention. In: MICCAI 2015: 18th International Conference Munich, Germany, October 5–9, 2015 proceedings, part III, Lecture Notes in Computer Science (including subseries Lecture Notes in Artificial Intelligence and Lecture Notes in Bioinformatics), vol 9351, pp 12–20

Change Detection for Multispectral Remote Sensing Images Using Deep Learning

M. Lasya, Radhesyam Vaddi, and S. K. Shabeer

Abstract Urbanization is a global process. Rapid industrialization and urbanization in developing nations represent a serious threat to the environment. The method of "change detection" examines how a region's features have changed during a span of two or more times. Deep learning methods have outperformed more conventional change detection methods in tests. Artificial neural networks are the foundation of the specialized machine learning method known as deep learning. Comparing the deep learning model to other machine learning techniques reveals significant capability and versatility. Two visuals taken at various timestamps are given as input to a Siamese neural network for change detection. For each of the input images, higher-level feature vectors are produced using a sequence of convolutional layers. This method aids in the extraction of significant features that may be utilized to evaluate and quantify the differences between the images that were viewed. The Euclidean norm is then used to compare the feature vectors to determine the degree of change. Deep Learning Change Detection (DLCD) outperforms traditional change detection due to three factors: better information representation, enhanced change detection techniques, and performance improvements. To encourage better decision-making, it is essential to identify agricultural changes, which we can do with DLCD.

Keywords Deep Learning Change Detection (DLCD) · Computer vision · Multispectral images · Remote sensing · Change detection · Siamese neural network

M. Lasya · R. Vaddi · S. K. Shabeer (✉)
Department of IT, VR Siddhartha Engineering College, Vijayawada, India
e-mail: shaikshabeer2403@gmail.com

R. Vaddi
e-mail: syam.radhe@vrsiddhartha.ac.in

S. Shakya et al. (eds.), *Fourth International Conference on Image Processing and Capsule Networks*, Lecture Notes in Networks and Systems 798,
https://doi.org/10.1007/978-981-99-7093-3_11

1 Introduction

It is essential to comprehend the connections and interactions between human and natural occurrences in order to make better decisions, which calls for rapid and accurate change detection of Earth's surface features. The "change detection" method assesses how the characteristics of a given area have changed over the course of two or more time periods. Change detection can be broadly categorized into two types: supervised change detection and unsupervised change detection [1]. By utilizing change detection techniques [2], it becomes possible to identify alterations in land use and land cover (LULC) [3]. A common approach involves comparing aerial or satellite images captured at different time points to identify and analyze the observed changes in a region.

In order to identify and maintain track of an area's physical characteristics, remote sensing entails monitoring the radiation that an area reflects and emits from a distance (typically from satellite or aircraft).

Remote sensing involves capturing photographs of the Earth from a distance using specialized cameras, which provide scientists with a way to "feel" the Earth and study its features. The cameras are mounted on satellites and aircraft, allowing for large areas of the Earth's surface to be captured in photographs. This provides researchers with a much broader view of the Earth than what can be observed from the ground. With the help of remotely sensed Earth imagery, it is possible to track changes in features such as the growth of cities, alterations in agriculture, and changes to woodlands over periods of years or decades.

1.1 Applications of Remote Sensing

Remote sensing is the practice of using sensors and other instruments to collect data about the Earth from a distance, often from space. The measurement of soil moisture content using active and passive sensors is one of the main applications of remote sensing. Other applications include mapping land use, weather forecasting, environmental analysis, natural hazards analysis, and resource discovery [4].

In addition to these applications, remote sensing can also be used for monitoring large forest fires from space, which allows rangers to observe a much larger area than they could on the ground [5]. This is just one example of the many ways in which remotely sensed photographs of the Earth can be used, including watching erupting volcanoes, tracking clouds to anticipate the weather, and detecting dust storms [6].

2 Proposed Work

2.1 Datasets

The initiative, which is a component of the Copernicus Programme started by the European Union, utilizes Sentinel-2 satellite data. Sentinel-2 is equipped with a multispectral instrument (MSI), which is a multispectral imaging sensor designed to capture high-resolution images of the Earth's surface. It is a global monitoring program with the objective of disseminating reliable and timely information about the climate, land, and oceans of the planet. The European Space Agency (ESA) created the mission, which frequently takes high-resolution optical shots of the Earth's land and coastlines at a distance of between 10 and 60 m. For the years 2012 and 2015, the data was obtained from Swisstopo (Figs. 1, 2). Swisstopo is a federal agency in Switzerland that is responsible for surveying and mapping the country. They provide a diverse range of geospatial datasets that cover the entire country. It is responsible for geodetic, topographic, and cartographic surveys, as well as the production and dissemination of geospatial data and maps. Swisstopo plays a crucial role in maintaining accurate and up-to-date geographical information for various applications, including land management, spatial planning, infrastructure development, and environmental monitoring. The organization collects and maintains data related to topography, elevation, land cover, cadastral information, and geodetic reference systems.

2.2 Architecture

Figure 3 shows the architecture diagram. ResNet34 serves as an encoder in a Siamese neural network. The images are changed to the necessary dimensions. The multispectral image is sent into the convolutional layers of the neural network to produce a feature vector. A feature vector holds all of the features of a picture. The Siamese neural network generates feature vectors, denoted as f(X) and f(Y), for images X and Y captured at different timestamps T1 and T2, respectively. These feature vectors are obtained by passing the respective images through identical convolutional layers that share the same parameters. The resulting feature vectors are encoded with 128 bits. To assess the dissimilarity between the two feature vectors, a difference vector is calculated by the Siamese neural network. This difference vector captures the variations and discrepancies between $f(X)$ and $f(Y)$, enabling the network to quantify the level of change between the two images.

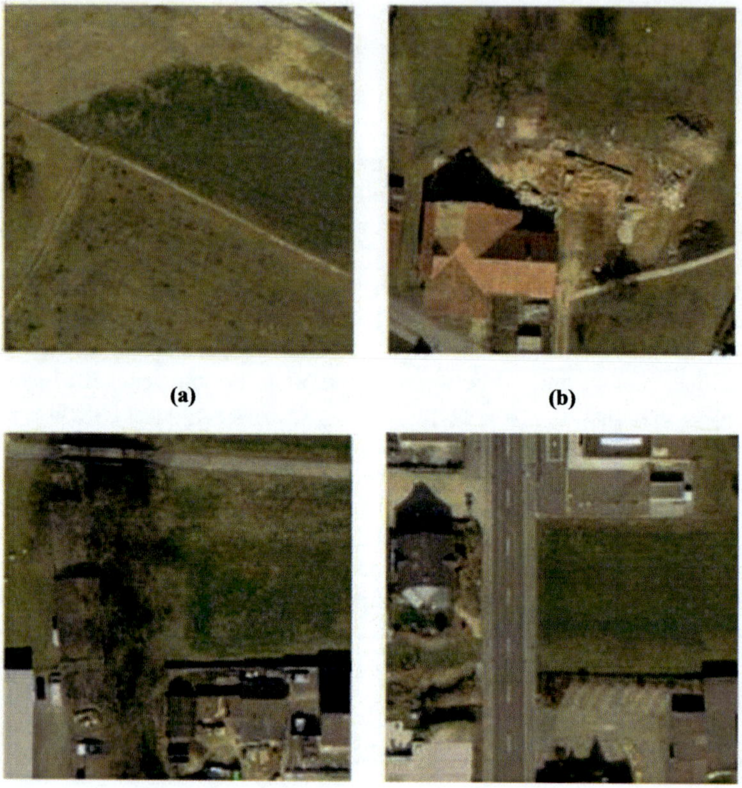

(a) (b) (c) (d)

Fig. 1 **a–d** Images of Switzerland in the year 2012

2.3 Proposed Work

Neural networks are nearly perfect at every task in the deep learning era of today, but they need more data to do so. However, we cannot always rely on obtaining additional data to tackle problems like face recognition and signature verification. To address these kinds of difficulties, we have a new form of neural network architecture called Siamese networks.

In order to obtain better predictions, it only uses a small number of photographs. Siamese networks have been more well-liked in recent years due to their capacity to learn from relatively little data. This post will explain Siamese networks and how to use them to build a signature verification system with Pytorch.

Siamese neural networks are a type of neural network architecture that has gained popularity due to their ability to compare feature vectors and determine the similarity between inputs. The name "Siamese" refers to the fact that these networks have two

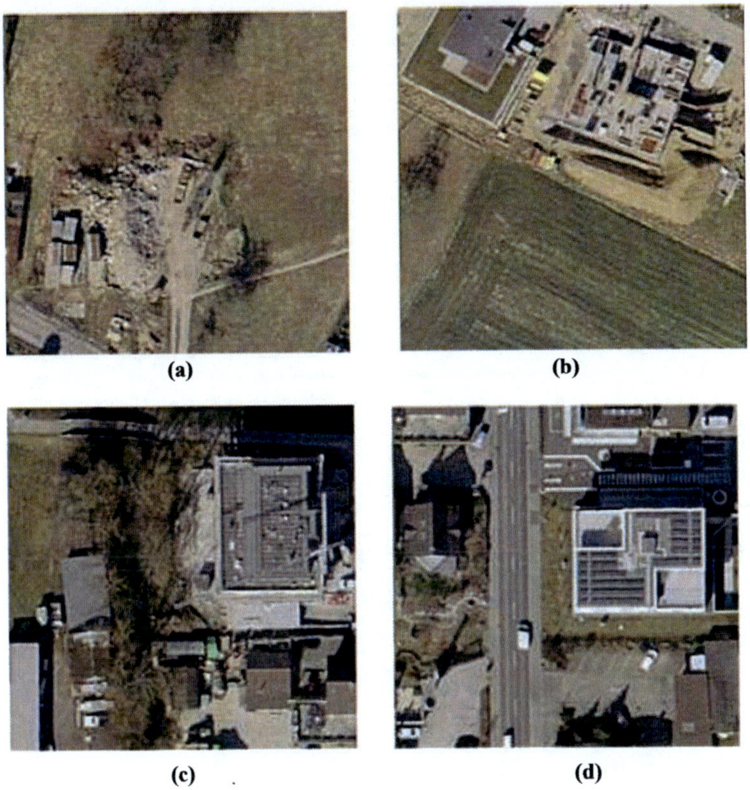

Fig. 2 a–d Images of Switzerland in the year 2015

or more identical subnetworks, which share the same configuration, including their parameters and weights. This means that when the network is trained, the updates to the parameters of one subnetwork are mirrored in the other subnetwork, ensuring that both subnetworks remain identical. The Siamese network takes in two images as input and produces two feature vectors, one for each image. These feature vectors are then compared to determine the degree of similarity between the two images.

Neural networks are capable of predicting different classes, but it might be difficult to add or remove new classes because the network must be retrained using the complete dataset. For training, deep neural networks need a lot of data. When taught with a similarity function, Siamese neural networks (SNNs) can assess whether two images are the same. As a result, SNNs can categorize fresh data without the need for network retraining.

To prepare the input data, we preprocess it. Before converting the data to NumPy arrays, it normalizes the data by taking the mean from each channel's value and dividing it by the standard deviation.

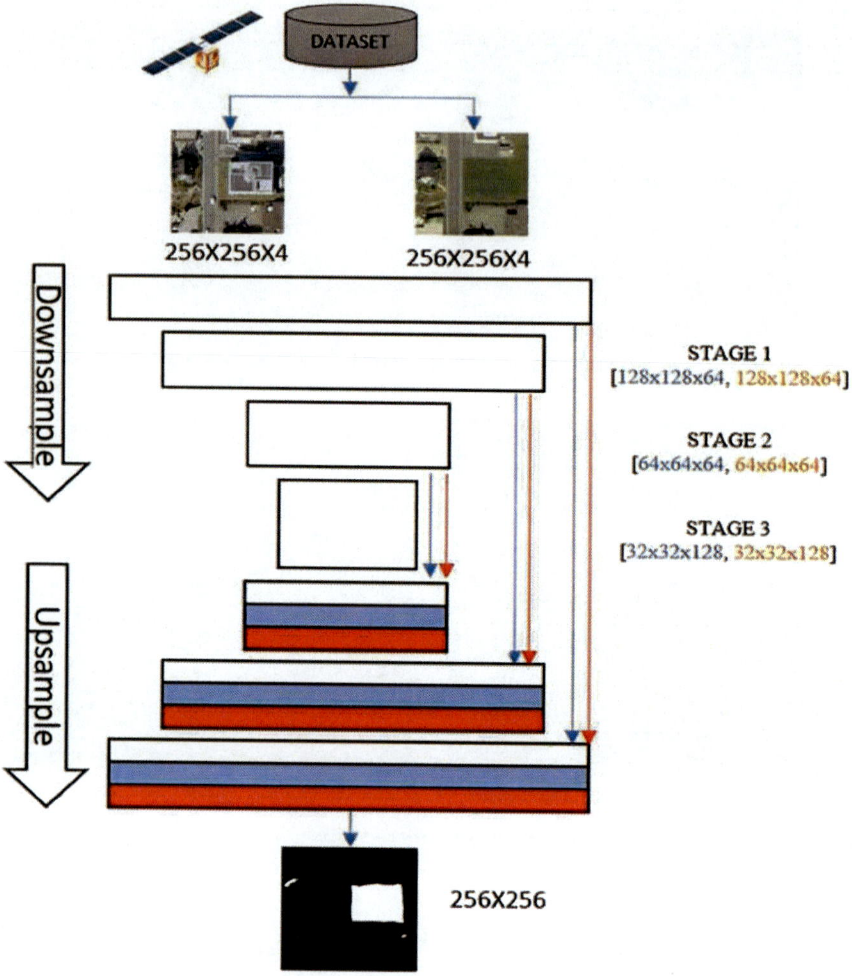

Fig. 3 Architecture diagram

Algorithm: Siamese Neural Network Input: Very high-resolution satellite images.

Output: Change map.

1. Start
2. Consider the two input images as $X = (cX, tX)$ and $Y = (cY, tY)$.
3. Compute the semantic relatedness of the two matrices as the Euclidean distance between their matrix representations:

$$D\phi(X, Y) = \|\{\phi(CX) \oplus \phi(tX)\} - \{\phi(CY) \oplus \phi(tY)\}\|2 \tag{1}$$

$\varphi(cX)$ and $\varphi(cY)$ are caption representations and $\varphi(tX)$ and $\varphi(tY)$ are matrix representations for the query and the candidate matrix, respectively.

4. Define the contrastive loss as a quadratic function of the pairwise distances, aiming for $D\varphi(X, Y)$ to be small (close to zero).
5. If table Y is similar to table X and equal to or larger than the margin m otherwise:

$$L(y, \phi, X, Y) = \frac{1}{2}(1 - y)D_{\phi}^{2}(X, Y) + \frac{1}{2}y\{\max(0, m - D_{\phi}(X, Y)\}^{2} \quad (2)$$

6. y is the true label of the pair.
7. Obtain binary classification labels by thresholding the distance at half of the margin, $m/2$. Pairs with distance scores below this threshold are considered similar, while others are considered dissimilar.
8. Output the difference label of the images $L(X, Y)$.
9. End

Recognizing and quantifying change over time can provide valuable insights in various domains. For example, assessing the progress of a city or town's infrastructure development can help gauge its economic success. However, any model designed for this purpose needs to identify and focus on relevant changes while disregarding unrelated ones. In the case of tracking structural changes, it becomes necessary to filter out variations such as alterations in water bodies, trees, and roadways that may not be of interest. This challenge is particularly evident in data normalization since photographs captured over time can exhibit inconsistencies that are difficult to consistently account for. When using a smartphone to capture facial images, factors like lighting conditions and positioning settings can vary between each capture, further complicating the normalization process. Similarly, satellite images are subject to variations caused by the azimuthal and elevational angles of the satellite, fluctuations in cloud cover, and the reflection of sunlight. These factors introduce additional complexities in analyzing satellite imagery consistently over time.

According to Ref. [7], the traditional method of measuring change by directly comparing pixel differences between images is no longer considered the simplest and most direct approach. While more sophisticated methods exist, they still yield good results. An alternative method involves using a Siamese neural network, which takes two images captured at different timestamps as input. The network applies a series of convolutional layers to each image to produce higher-level feature vectors [8]. The Euclidean norm is then used to evaluate these feature vectors and determine the degree of change. According to the theory, if there is not a clear difference between the two images, the corresponding feature vectors in the dimensional space will look comparable.

To train Siamese networks, one uses a contrastive loss function. If there is no change, it seeks to minimize the distance between the two feature vectors, and vice versa. The picture that follows tries to demonstrate it using a situation with two classes. The actual label would be 1 if neither of the two photos included the modifications we were training for [9–12].

3 Result Analysis

We downloaded satellite images from the Swiss Federal Office of Topography, specifically the Sentinel-2 images, which show views of Switzerland in multiple spectrums. Comparing images with different time stamps is often necessary to identify changes that have occurred in a particular area over time. This approach is useful for monitoring changes in LULC over several years. The process involves comparing images acquired at different times to identify any differences. The Deep Learning Change Detection (DLCD) algorithm generates a binary image that indicates the locations of LULC changes [7]. Outcomes are shown in Figs. 4, 5, 6 and 7.

The suggested method results in the confusion matrix. When assessing a classification model's performance, a confusion matrix is a table that displays the percentage of accurate and inaccurate predictions made. Its four subgroups are True Positives (TP), True Negatives (TN), False Positives (FP), and False Negatives (FN). When both the anticipated and actual values are positive, TP occurs. Negative values of

(a) (b) (c)

Fig. 4 **a** The image of land in Timestamp 1, **b** the image of land in Timestamp 2, and **c** the change map which shows the change that occurred in agricultural lands due to urbanization

(a) (b) (c)

Fig. 5 **a** The image of land in Timestamp 1, **b** the image of land in Timestamp 2, **c** the change map which shows change occurred due to the demolition of the building

(a) (b) (c)

Fig. 6 **a** The image of land in Timestamp 1, **b** the image of land in Timestamp 2, and **c** the change map which shows the change occurred due to urbanization

(a) (b) (c)

Fig. 7 **a** The image of land in Timestamp 1, **b** the image of land in Timestamp 2, **c** the change map which shows the change that occurred due to the construction of a new building in that area

the expected and actual values result in TN. FP, often referred to as a Type 1 error, occurs when the forecast is positive but the actual result is negative. On the other hand, when the actual value is positive, but the forecast is negative, FN, also known as a Type 2 error, takes place.

Table 1 presents the confusion matrix. Metrics like accuracy, precision, and F1-Score are generated by the Siamese neural network to assess its effectiveness in change detection. The percentage of accurate forecasts among all made predictions is known as accuracy. While the F1-score combines precision and recall, taking into account both false positives and false negatives, precision concentrates on the accuracy of positive predictions. A high F1-score indicates accurate positive and negative class prediction, whereas a low score indicates bias toward one class. These metrics shed light on the model's effectiveness and its precision in identifying changes.

Table 2 presents the metrics obtained for the Siamese neural network.

Table 1 Confusion matrix actual values

	Change	No change
Change	90	1
No change	12	106

Table 2 Metrics

Metrics	Formula	Values
Accuracy	(TP + TN)/(Total)	0.938
Precision	(TP)/(TP + FP)	0.989
Recall	(TP)/(TP + FN)	0.882
F1-score	2(Precision*Recall)/(Precision + Recall)	0.932

- The final output of our project is a binary image, where the white area symbolizes the change and the black area symbolizes no change.
- It was observed that the model produces an accuracy of 93.8%.
- Precision is a measure of the proportion of positive class forecasts that actually belong to the positive class. It was discovered that the model produced data with a precision of 98.9%.
- Recall quantifies the number of accurate class predictions generated from the entire dataset's successful cases. Recall for the model was found to be 88.2%.
- A single score provided by F-measure strikes a balance between recall and precision issues.

4 Conclusion

In this project, we used two separate timestamps to look for changes in Switzerland as a particular location. Our investigation utilized multispectral remote sensing photographs taken by the Sentinel-2 satellite to detect changes in the target area. To achieve change detection, we employed a Siamese neural network. Throughout our tests, we discovered that a number of acquisition functions and prediction uncertainty calculation strategies worked effectively. These methods achieved performance comparable to a model trained on the whole dataset with only a few hundred samples, outperforming a naive random baseline by a large margin.

Our project's ultimate objective is to use remote sensing images to track changes in agricultural areas. As a result, we focused on detecting changes in agricultural areas, which may include changes in land use, crop yield, and soil conditions. By identifying these changes, we hope to provide insights that can help improve agricultural practices and increase productivity while minimizing environmental impact. Calculating the percentage change before and after natural disasters, and other features will be a part of our future research. To enhance the performance of a Siamese neural network model for change detection, we can employ techniques such as data augmentation to increase the diversity of training samples, explore different network architectures to capture complex patterns, consider alternative similarity metrics, and consider ensemble methods for combining multiple models. These strategies can collectively improve the model's ability to detect changes accurately and robustly.

References

1. De Jong KL, Bosman AS (2018) Unsupervised change detection in satellite images using convolutional neural networks. Preprint at: arXiv:1812.05815. https://arxiv.org/abs/1812.05815
2. Singh A (1989) Review article digital change detection techniques using remotely-sensed data. Int J Remote Sens 10(6):989–1003
3. Su L, Gong M, Zhang P, Zhang M, Liu J, Yang H (2017) Deep learning and mapping based ternary change detection for information unbalanced images. Pattern Recognit 66:213–228
4. Deng L, Hinton L, Kingsbury B (2013) New types of deep neural network learning for speech recognition and related applications: an overview. In: Proceedings of IEEE International Conference Acoustics, Speech Signal Process, pp 8599–8603
5. Gillanders SN, Coops NC, Wulder MA, Goodwin NR (2008) Application of land-sat satellite imagery to monitor land-cover changes at the Athabasca Oil Sands, Alberta, Canada. Can Geogr 52:466–485. https://doi.org/10.1111/j.1541-0064.2008.00225.x
6. Ma L, Liu Y, Zhang X, Ye Y, Yin G, Johnson BA (2019) Deep learning in remote sensing applications: a meta-analysis and review. ISPRS J Photogramm Remote Sens 152:166–177
7. Daudt RC, Le Saux B, Boulch A, Gousseau Y (2018) Urban change detection for multispectral earth observation using convolutional neural networks. In: International Geoscience and Remote Sensing Symposium (IGARSS). IEEE
8. Hou B, Wang Y, Liu Q (2017) Change detection based on deep features and low rank. IEEE Geosci Remote Sens Lett 14(12):2418–2422
9. Zhang L, Zhang L, Du B (2016) Deep learning for remote sensing data: a technical tutorial on the state of the art. IEEE Geosci Remote Sens Mag 4(2):22–40
10. Cao C, Dragićević S, Li S (2019) Land-use change detection with convolutional neural network methods. Environments 6(2):25
11. Hamdi ZM, Brandmeier M, Straub C (2019) Forest damage assessment using deep learning on high-resolution remote sensing data. Remote Sens 11(17):1976
12. Coppin P, Jonckheere I, Nackaerts K, Muys B, Lambin E (2004) Digital change detection methods in ecosystem monitoring: a review. Int J Remote Sens 25(9):1565–1596

Explainable AI for Black Sigatoka Detection

Yiga Gilbert, Emmy William Kayanja, Joshua Edward Kalungi, Jonah Mubuuke Kyagaba, and Ggaliwango Marvin

Abstract Banana plants are susceptible to the dangerous fungal disease known as Black Sigatoka, which has a negative impact on global economies. Early detection and timely intervention are crucial for preventing the spread of the disease. In recent years, machine learning (ML) has shown great potential for detecting and diagnosing plant diseases, including Black Sigatoka. However, the lack of transparency and interpretability of ML models raises concerns about their use. In this paper, we propose explainable AI approaches for Black Sigatoka detection using Local Interpretable Model-Agnostic Explanations (LIME) and Integrated Gradients. Our methodology involves the utilization of Mobilenet V2 and AlexNet models which are trained on an extensive dataset of banana leaf images and generating explanations to provide a better understanding of the CNN's decision-making process. We demonstrate the effectiveness of our approach through extensive experiments and show that it outperforms existing state-of-the-art methods for Black Sigatoka detection. Our approach not only provides accurate and interpretable results but also promotes responsible AI practices for plant disease diagnosis.

Keywords Black sigatoka · Explainable AI (XAI) · Local interpretable model-agnostic explanations (LIME) · Convolutional neural network (CNN) · Plant disease diagnosis

---

Y. Gilbert · E. W. Kayanja · J. E. Kalungi · J. M. Kyagaba · G. Marvin (✉)
Department of Computer Science, Makerere University, Kampala, Uganda
e-mail: ggaliwango.marvin@mak.ac.ug

Y. Gilbert
e-mail: yiga.gilbert@students.mak.ac.ug

E. W. Kayanja
e-mail: emmy.kayanja@students.mak.ac.ug

J. E. Kalungi
e-mail: joshua.kalungi@students.mak.ac.ug

J. M. Kyagaba
e-mail: jonah.kyagaba@students.mak.ac.ug

S. Shakya et al. (eds.), *Fourth International Conference on Image Processing and Capsule Networks*, Lecture Notes in Networks and Systems 798,
https://doi.org/10.1007/978-981-99-7093-3_12

1 Introduction

1.1 Backgroung and Motivation

Millions of people in Eastern Africa rely heavily on bananas in their diets, making up as much as one-fifth of the region's total calorie consumption per person [4]. Bananas are grown by more than 75% of the smallholders in East Africa [13] due to their special benefits, which include yields that are adequate despite unpredictable rainfall and their longstanding nature combined with their perennial fruiting nature [16]. In terms of consumption per person and total production, as well as the area occupied in Uganda, bananas come in first [15]. These qualities set bananas apart as a desirable crop for domestic food production, nourishment, and steady revenue [1]. Despite being a staple food, the productivity of bananas in Uganda is rapidly declining. This is partially brought on by ailments like Sigatoka leaf disease, which is brought on by fungi belonging to the genus Pseudocercospora [5]. With Black sigatoka, a fruit bunch may develop unevenly and prematurely as a result of severely infected leaves dying, limiting fruit production drastically [2]. In the Ugandan districts of Luwero with a mean Disease Severity Index of 40.4% and Mbarara with a mean DSI = 37.9%, Black Sigatoka disease was equally severe [6]. One potential solution to mitigate the spread of Black Sigatoka is through the use of Artificial Intelligence (AI) systems that can detect the disease early on and enable effective management and control measures [7, 12].

In constrained resource settings, such as smallholder farms, rural communities, and all developing countries at large, there may be limitations on the availability of resources such as computing power, Internet connectivity, and technical expertise [14]. Therefore, any AI system designed for Black Sigatoka detection in these settings must be efficient, cost-effective, and easy to use.

To address these challenges, Explainable AI (XAI) can be utilized to overcome these difficulties by enhancing the AI system's accuracy and dependability while ensuring transparency and interpretability. XAI provides a framework for understanding the reasoning behind the AI system's decision-making process, allowing users to evaluate the system's performance and identify any potential biases or errors.

1.2 Research Contribution

- Creating visual and analytical tools for exploring banana Sigatoka datasets.
- Develop and test models for detecting Black Sigatoka in bananas.
- Compare the performance of different Black Sigatoka detection models and suggest the best one.
- Generate and contrast interpretable explanations for Black Sigatoka classification to enhance model transparency.

2 Research Problem Definition

Banana Black Sigatoka has caused a significant deterioration in the quality and quantity of bananas produced in Uganda, a drop in farmer income, the death of banana trees, and an increase in production costs throughout the control process of BBS. Smallholders unknowingly spread banana Black Sigatoka through plant materials such as suckers locally known as "Ensukusa" from infected gardens and use ineffective techniques to identify the disease; by the time they do, the disease has advanced and begun to manifest symptoms, at which point it is too late.

Farmers require early warning of the disease outbreak. However, all deep learning models used to identify banana Black Sigatoka (BBS) have a black-box nature and are comparably weak in explaining their inference processes and outcomes, making them uninterpretable, non-intuitive, less human-understandable, and providing no explanations for AI predictions.

We opt to develop explainable deep learning models that are credible, comprehensible, and self-explanatory as well as algorithm strategies that will aid in banana Black Sigatoka (BBS) detection. This will enable farmers with limited resources and those without access to extension services to mitigate the economic, social, and environmental impacts of the BBS.

3 Research Approach and Methodology

Figure 1 is the methodology visualization, and the aim was to develop an explainable AI model to detect Black Sigatoka in banana leaves from the image data, and provide explanations for its predictions. The model aimed to help smallholders with limited resources to identify infected plants early. The steps were:

Study existing models for plant disease detection and review their strengths and weaknesses. The main weakness was the lack of explainability in the models.

Develop two models using AlexNet and MobileNet architectures, which are successful in image classification tasks. AlexNet used deep convolutional layers to learn complex features from the images. MobileNet used depth-wise separable convolutions to achieve a trade-off between accuracy and efficiency.

Train both models on a dataset of banana leaf images with different stages of Black Sigatoka infection. Fine-tune and optimize the models for better performance.

Apply Lime and Integrated Gradients techniques to both models to generate explanations for the predictions. These techniques highlighted the image regions that influenced the models' decisions, helping smallholders to understand the model's logic.

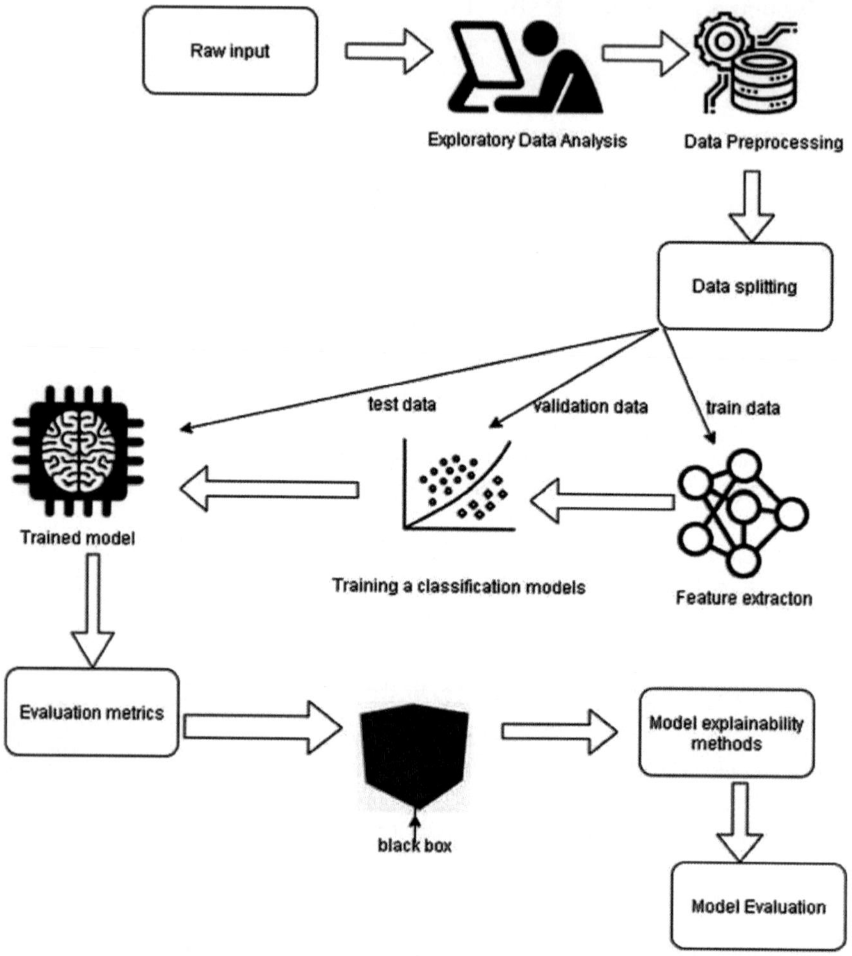

Fig. 1 Methodology visualization

3.1 Data Collection and Preprocessing

We used the Harvard Dataverse banana leaf disease dataset [10], which has 5883 images of healthy leaves and 6147 images of leaves infected with Black Sigatoka. To gain more insights into the disease process and potentially improve the accuracy of our model, we included images of different stages of disease progression in our dataset. We then applied multi-class classification instead of binary classification, dividing the images into four classes: "healthy", "early stage affected", "intermediate stage affected", and "late stage affected" as shown in Fig. 2.

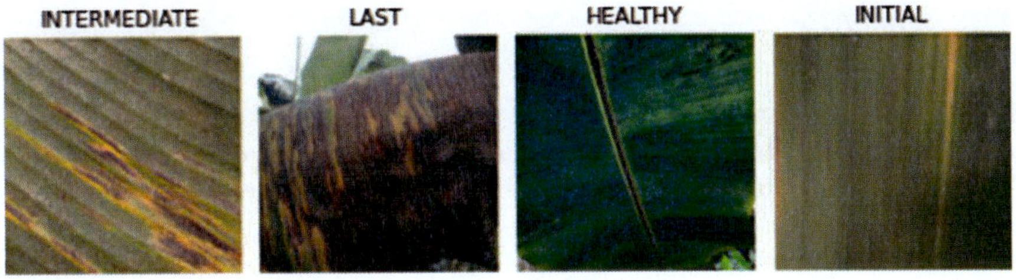

Fig. 2 Image classes

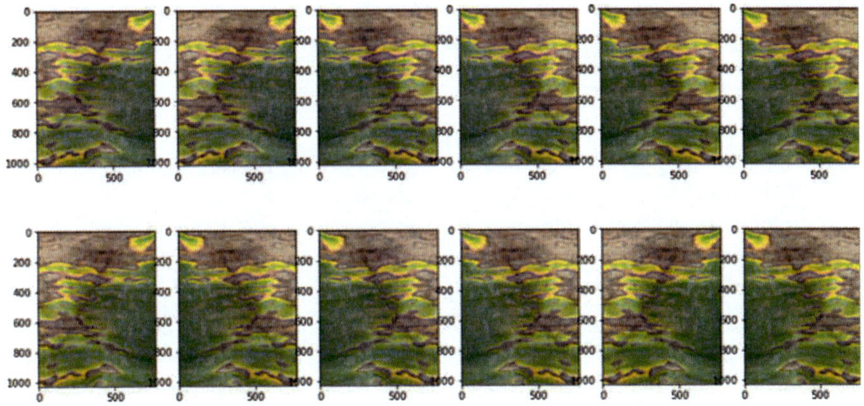

Fig. 3 Augmented banana leaves

We labeled our dataset with the four classes and applied various techniques to optimize our model's performance, such as data augmentation, transfer learning, and hyperparameter tuning. The evaluation criteria we employed for multi-class classification were precision, recall, and $F1$-score for each class. We also visualized the performance of our model in each class using confusion matrices.

As demonstrated in Fig. 3, we employed data augmentation to artificially increase the amount and diversity of training data from the original dataset. We applied slight modifications to each existing data point and generated 16 new data points in the latent space of the original data using machine learning models. We used techniques such as image resizing, rotation, flipping, noise addition, cropping, scaling, and translation. We used the ImageDataGenerator class from the Keras deep learning package to apply data augmentation automatically during model training.

3.2 Model Implementation

AlexNet Architecture and MobileNet

AlexNet is a deep convolutional neural network (CNN) model that consists of eight layers: five convolutional layers and three fully connected layers [3, 11]. The model uses various techniques to enhance its feature extraction and generalization abilities, such as local response normalization, dropout, rectified linear units (ReLU), and max-pooling. The model has over 60 million parameters and is deeper and wider than previous models. The input shape for AlexNet is $227 \times 227 \times 3$. The ReLU activation function is used for all convolutional layers. Dropout with a rate of 0.5 is applied after the fully connected layers to reduce overfitting. Batch normalization is used after each convolutional layer to improve training stability. Max pooling is used to downsample the feature maps after some convolutional layers. The feature maps are flattened before passing through the fully connected layers. The softmax activation function is used in the final layer to output the class probabilities.

The AlexNet Model Architecture is illustrated in Fig. 4.

MobileNet is a CNN model that aims to overcome the difficulties of deploying deep learning models on limited-resource devices like mobile phones [17]. It emphasizes efficiency and compactness without compromising on performance. MobileNet employs depth-wise separable convolutions, a factorized form of standard convolutions, to significantly reduce computational complexity and model size. This approach enables MobileNet to achieve a good balance between accuracy and inference speed. By utilizing depth-wise convolutions followed by point-wise convolutions, the model captures spatial, and channel-wise correlations efficiently. MobileNet's architecture as illustrated in Fig. 5 offers various configurations, allowing trade-offs between model size and accuracy. The input shape for MobileNet is

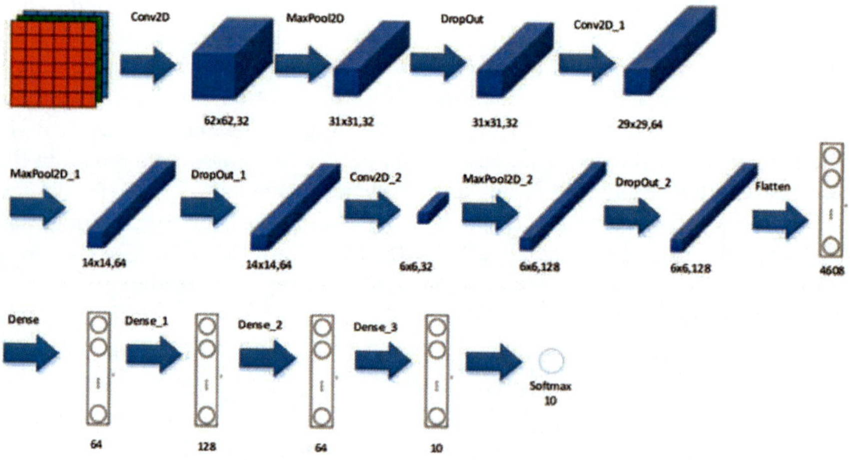

Fig. 4 AlexNet model architecture

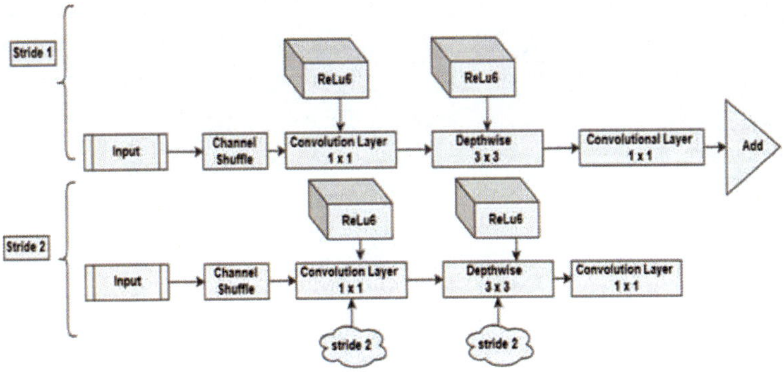

Fig. 5 MobileNet V2 model architecture

$224 \times 224 \times 3$. The activation functions used in MobileNet are ReLU and softmax. Dropout is used in MobileNet to prevent overfitting. The dropout rate is 0.5.

Table 1 shows the AlexNet Architecture that we employed.

Table 2 shows the MobileNet Architecture that we employed.

Model Training
The models underwent training using the Adam optimizer and utilized categorical cross-entropy loss. During the training process, the dataset was iterated over a predetermined number of epochs. Within each epoch, the model received training on mini-batches of images, and the weights were adjusted based on the gradients computed through backpropagation. The training progress was tracked using metrics like loss and accuracy, which offered valuable information about the model's performance throughout the training phase. Once the model training is finished, a separate validation set is employed to assess the performance of the trained models.

4 Major Research Findings

4.1 Model Evaluation

The evaluation process involved a range of performance metrics such as loss, accuracy, precision, recall, and $F1$-score as illustrated in Tabels 3 and 4. To aid in identifying potential misclassifications, a confusion matrix is generated in Fig. 6, visually representing the classification results. These evaluations provide insights into the model's effectiveness in accurately predicting and classifying the validation data.

Figure 7 shows the training and validation accuracy together with the training and validation loss of MobileNet and AlexNet models trained on a dataset of banana leaf images.

Table 1 AlexNet architecture

Layer (type)	Output shape	Parameters #
conv2d (Conv2D)	(None, 55, 55, 96)	34,944
batch_normalization (BatchNormalization)	(None, 55, 55, 96)	384
max_pooling2d (MaxPooling2D)	(None, 27, 27, 96)	0
conv2d_1 (Conv2D)	(None, 27, 27, 256)	614,656
batch_normalization_1 (BatchNormalization)	(None, 27, 27, 256)	1024
max_pooling2d_1 (MaxPooling2D)	(None, 13, 13, 256)	0
conv2d_2 (Conv2D)	(None, 13, 13, 384)	885,120
batch_normalization_2 (BatchNormalization)	(None, 13, 13, 384)	1536
conv2d_3 (Conv2D)	(None, 13, 13, 384)	1,327,488
batch_normalization_3 (BatchNormalization)	(None, 13, 13, 384)	1536
conv2d_4 (Conv2D)	(None, 13, 13, 256)	884,992
batch_normalization_4 (BatchNormalization)	(None, 13, 13, 256)	1024
max_pooling2d_2 (MaxPooling2D)	(None, 6, 6, 256)	0
flatten (Flatten)	(None, 9216)	0
dense (Dense)	(None, 4096)	37,752,832
dropout (Dropout)	(None, 4096)	0
dense_1 (Dense)	(None, 4096)	16,781,312
dropout_1 (Dropout)	(None, 4096)	0
dense_2 (Dense)	(None, 4)	16,388
Total parameters: 58,303,236		
Trainable parameters: 58,300,484		
Non-trainable parameters: 2752		

Considering the overall performance, both models demonstrated high accuracy and achieved competitive results. However, based on the provided evaluation, AlexNet appeared to have a slight advantage in terms of precision, recall, and $F1$-scores across multiple classes. Therefore, the AlexNet model was suggested as the best model for detecting Black Sigatoka in banana leaves.

Table 2 Mobile nets architecture

Layer (type)	Output shape	Parameters #
mobilenetv2_1.00_224 (Functional)	(None, 7, 7, 1280)	2,257,984
average_pooling2d_1 (AveragePooling2D)	(None, 3, 3, 1280)	0
flatten_1 (Flatten)	(None, 11,520)	0
dense_3 (Dense)	(None, 512)	5,898,752
dropout_2 (Dropout)	(None, 512)	0
dense_4 (Dense)	(None, 50)	25,650
dropout_3 (Dropout)	(None, 50)	0
dense_5 (Dense)	(None, 4)	204
Total parameters: 8,182,590		
Trainable parameters: 6,337,406		
Non-trainable parameters: 1,845,184		

Table 3 Comparison of AlexNet and MobileNet results

Metrics	Results	
	AlexNet	MobileNet
Train loss	0.0035	0.0663
Train accuracy	0.9995	0.9892
Row 3, cell 1	0.2164	0.5419
Test accuracy	0.9462	0.9167

4.2 XAI Results

To gain insights into the decision-making processes of AlexNet and MobileNet, we employed LIME and Integrated Gradients techniques [8]. LIME provided local explanations by identifying the salient regions of an image that influenced the model's predictions. By visualizing these explanations, we gained a better understanding of the features considered by each model during classification. Integrated Gradients assigned importance values to individual pixels, allowing us to interpret the contributions of different image regions to the final prediction. These interpretability techniques enhanced our understanding of the inner workings of both models, providing valuable insights into their decision processes.

LIME

We used LIME, or Local Interpretable Model-agnostic Explanations, to comprehend how our black-box classifier model behaves [9]. Permutation of the data input LIME started by making a number of fictitious data closely related to the data point categorized in one class rather than the other. LIME produced a large number of samples

Table 4 Classification results

Class	AlexNet				MobileNet			
	Precision	Recall	$F1$-score	Support	Precision	Recall	$F1$-score	Support
HEALTHY	1.00	1.00	1.00	18	0.97	0.96	0.97	141
INITIAL_STAGE	1.00	1.00	1.00	26	0.92	0.89	0.90	147
INTERMEDIATE_STAGE	0.93	1.00	0.97	28	0.87	0.87	0.87	168
LAST_STAGE	1.00	0.93	0.96	28	0.91	0.96	0.93	134
Accuracy			0.98	100			0.92	590
Macro avg	0.98	0.98	0.98	100	0.92	0.92	0.92	590
Weighted avg	0.98	0.98	0.98	100	0.92	0.92	0.92	590

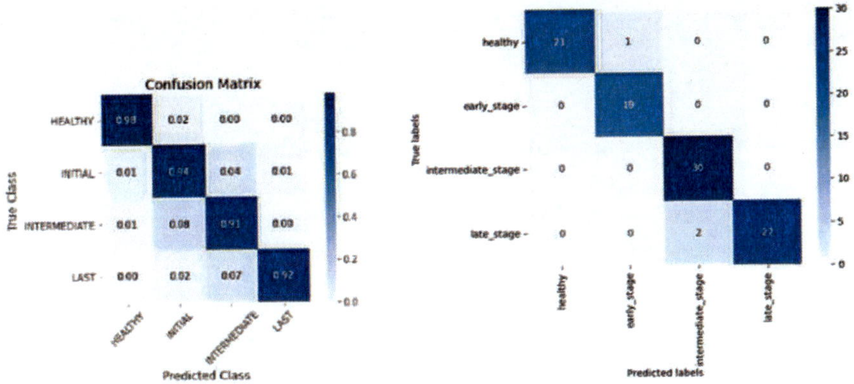

(a) MobileNet (b) AlexNet

Fig. 6 Confusion matrix comparison

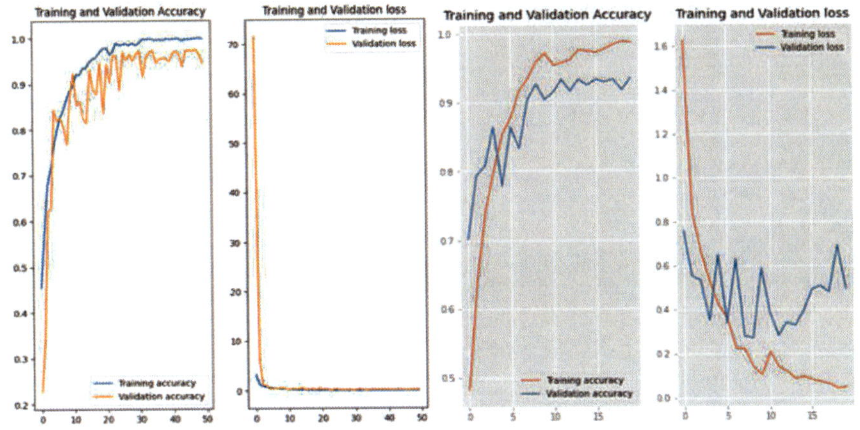

(a) AlexNet (b) MobileNet

Fig. 7 Graphical representation of training and validation metrics for MobileNet and AlexNet models on banana leaf images

that were identical to our input image by turning various super-pixels in our input image on and off.

Each fabricated data point's class was predicted. Following that, LIME forecasted whether each artificial data point was generated using our trained model, which represented the healthy class or the diseased class. Calculated the weights of all artificial data points. The weight of each fabricated data was then determined to determine its significance. The cosine distance metric was commonly utilized in the operation to calculate the distance between created data points and the original input data. This distance was then converted into a number between 0 and 1 using a kernel

(a) Healty (b) Infected on AlexNets

(c) Healty (d) Infected on MobileNets

Fig. 8 Lime visualizations

function. The mapped value was closer to one the closer the distance, increasing the weight as a result. The weight of a created data point increased its significance. After that, the cosine distance between each edited image and the original image was computed. The more closely a disturbed image matched the original image, the more weight and significance it had.

To describe the most relevant features, a linear classifier was fitted. The final stage involved creating a linear regression model using the weighted synthetic data points. Following this stage, the fitted coefficient for each feature was determined, much as in a standard linear regression analysis. If we sorted the coefficients immediately, the qualities with the highest coefficients had a substantial impact on how accurately our black-box machine learning model anticipated the future.

LIME's efficiency depends on generating many perturbed instances and training a simpler model on them. This lets LIME approximate the complex model's behavior near the chosen instance, without accessing the complex model's internals. Also, LIME uses interpretable features and an interpretable model to provide understandable explanations for model predictions.

In Fig. 8, the red regions represent features that contradicted the model output, the green regions are features that supported the model output, and the unhighlighted regions are features that had little or no impact on the model output.

Integrated Gradients
The importance of input features entering the model to the related prediction was also visualized using Integrated Gradient (IG). The Integrated Gradient is based on two assumptions that must be true:

Sensitivity: As a starting point for the Sensitivity computation, we created a Baseline image. The Integrated Gradients are then determined by interpolating from a baseline image to the real image using a sequence of photographs.

Implementation Invariance: This is satisfied when two functionally comparable networks assign the same attributions to the same input image and baseline image. Despite having wildly disparate implementations, dual networks are functionally comparable when their outputs are the same for all inputs.

Calculating and visualizing integrated gradients (IG)

Step 1: We started from the baseline, which could be either a random image or a dark image with all of its pixels set to zero.
Step 2: Between the baseline and the original image, we created a linear interpolation. The feature space was incrementally increased with each interpolated image's intensity in interpolated images, which were modest steps since they lay amidst the baseline and the supplied image.
Step 3: To determine the correlation between changes to a feature and modifications to the model's predictions, we computed gradients. This gradient revealed which pixel had the greatest impact on the class probabilities predicted by the model. Various variable adjustments, the output, and the variable received some attribution to assist in determining the feature importance for the input image. No credit was given for a variable that had no impact on the result.
Step 4: Then we calculated the approximate numerical value by averaging gradients.
Step 5: Multiple interpolated images were added together, and the attribute quantities were in the identical units. Scale IG to the input image using the supplied image's pixel importance to represent the IG.

In Fig. 9, the purple regions indicate the features that increased the probability of the predicted class, while the black regions are features that decreased the probability of the predicted class. The white regions are features that had no impact on the model output.

5 Practical Implications

Transparent Decision-Making: We use explainable AI techniques, such as Lime and Integrated Gradients, to show how the AI models make predictions. This helps farmers and stakeholders understand the model's logic, trust the system, make informed decisions, and manage Black Sigatoka effectively.

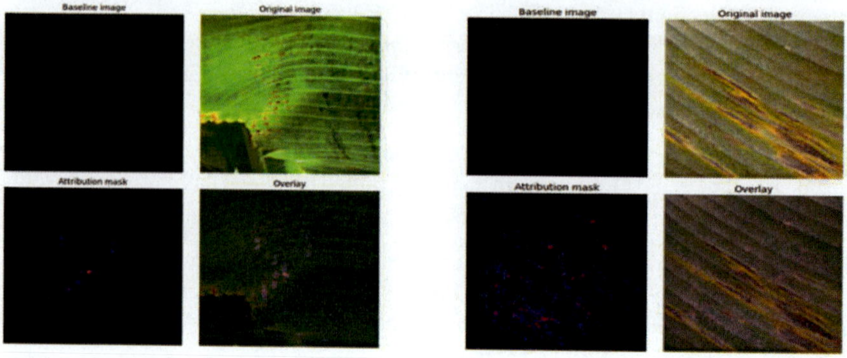

(a) AlexNet (b) MobileNet

Fig. 9 Visualization comparison

Early Detection and Timely Intervention: We provide a practical tool for early detection of Black Sigatoka in banana leaves using explainable AI models. By detecting the disease early, farmers can intervene quickly to manage and control the disease. This can help reduce crop losses, increase yields, and protect farmers' income.

Enhanced Disease Management: We offer accurate and interpretable AI models for Black Sigatoka detection to improve disease management practices. Farmers can identify infected plants early, control the disease with targeted measures, and use resources such as fungicides optimally. This leads to more effective disease management, reduced crop losses, and improved agricultural productivity.

Potential for Adaptation to Other Diseases: We can adapt our methodology and approaches to other plant diseases as well. We can apply the knowledge and techniques from detecting Black Sigatoka in banana leaves to diagnose and manage other plant diseases, expanding the practical implications of our research beyond a specific disease.

6 Research Limitations

The model may face challenges in accurately distinguishing between banana leaves and other leaves that bear a strong resemblance to banana leaves like heliconia, Traveler's Palm, etc. This can result in misclassifications or difficulties in isolating leaves that share similar visual characteristics with banana leaves.

Trade-off between model performance and interpretability, where increasing interpretability may result in lower performance, or conversely, achieving better performance may come at the cost of interpretability which can lead to confusion or mistrust in the model.

Limited computational resources while trying to archive most accurate results.

7 Originality/Value

Explainable AI Models for Plant Disease Detection: We develop AI models that can detect Black Sigatoka in banana leaves and explain their decisions. Existing models are accurate but not interpretable, limiting their usefulness and transparency. We use techniques like Lime and Integrated Gradients to make our models more transparent and explainable, giving users insights into how they work.

Contribution to Responsible AI Practices: The incorporation of explainability in AI models for plant disease detection promotes responsible AI practices. By providing interpretable explanations for the models' predictions, users can evaluate the system's performance, identify potential biases or errors, and gain trust in the technology. This aspect of your research contributes to ethical considerations and encourages the adoption of AI in a responsible and accountable manner.

Comparative Evaluation and Performance Analysis: Our research involves a comprehensive evaluation of the proposed models, comparing their performance with existing state-of-the-art methods for Black Sigatoka detection. This analysis not only showcases the effectiveness of our approach but also adds value by providing a benchmark for future research and facilitating the advancement of the field.

8 Conclusion and Future Research Work

8.1 Conclusion

To achieve high accuracy in the model for detecting Black Sigatoka in banana leaves, we considered several factors. First, we ensured a high-quality and diverse dataset that accurately represents the problem and covers different stages of the disease. We preprocessed and augmented the data to enhance its quality and increase its diversity. We evaluated two models using our diverse dataset for image classification tasks: AlexNet and MobileNet. AlexNet demonstrated superior performance compared to MobileNet across multiple evaluation metrics. It achieved higher accuracy, precision, and recall rates, as evidenced by the confusion matrix. The misclassification rate of AlexNet was lower, indicating its ability to handle complex image patterns more effectively. The above results highlight the strengths of AlexNet in achieving high classification accuracy for a wide range of images. Integrated Gradients through visual inspection also were the most preferred method compared to LIME.

8.2 Future Works

While the initial focus may be on Uganda, we think about extending the application and study to other areas impacted by Black Sigatoka in order to tailor the solution to various geographic locations and disease variants and work with regional organizations, scientists, and farmers.

We will extend our research efforts to develop a diverse range of explainable deep learning models specifically tailored for various banana plant diseases, aiming to enhance both accuracy and interpretability in agricultural disease detection systems.

Develop a mobile application to support multiple languages commonly spoken in the target regions for farmers in different regions that will be deployed on devices rather than the cloud.

Acknowledgements We want to express our heartfelt gratitude to the RISE Seed Fund at Makerere University College of Computing and Information Sciences for the full facilitation of this work.

References

1. Almekinders CJM et al (2019) Why interventions in the seed systems of roots, tubers and bananas crops do not reach their full potential. Food Secur 11:23–42
2. Brat P et al (2020) Review of banana green life throughout the food chain: from auto-catalytic induction to the optimisation of shipping and storage conditions. Sci Hortic 262:109054
3. Dong Y-n, Liang G-s (2019) Research and discussion on image recognition and classification algorithm based on deep learning. In: 2019 international conference on machine learning, big data and business intelligence (MLBDBI). IEEE, pp 274–278
4. Haggblade S, Dewina R (2010) Staple food prices in Uganda. Tech. rep.
5. Jones DR (2000) Diseases of banana, abaca and enset. CABI Publishing
6. Kimunye JN et al (2020) Distribution of Pseudocercospora species causing Sigatoka leaf diseases of banana in Uganda and Tanzania. Plant Pathol 69(1):50–59
7. Gokula Krishnan V et al (2022) An automated segmentation and classification model for banana leaf disease detection. J Appl Biol Biotechnol 10(1):213–220
8. Marvin G, Alam MGR (2022) Explainable augmented intelligence and deep transfer learning for pediatric pulmonary health evaluation. In: 2022 international conference on innovations in science, engineering and technology (ICISET). IEEE, pp 272–277
9. Marvin G et al (2023) Local interpretable model-agnostic explanations for online maternal healthcare. In: 2023 2nd international conference on smart technologies and systems for next generation computing (ICSTSN). IEEE, pp 1–6
10. Mduma N et al (2022) The Nelson Mandela African Institution of Science and Technology Bananas dataset. Version V2. https://doi.org/10.7910/DVN/LQUWXW
11. Mytholla N, Chakshu V (2021) Image classification using convolutional neural networks by different activation functions. Image 8(7)
12. Orchi H, Sadik M, Khaldoun M (2022) On using artificial intelligence and the internet of things for crop disease detection: a contemporary survey. Agriculture 12(1):9
13. Salami A, Kamara AB, Brixiova Z (2010) Smallholder agriculture in East Africa: trends, constraints and opportunities. African Development Bank Tunis, Tunisia
14. Taneja M et al (2019) SmartHerd management: a microservices-based fog computing-assisted IoT platform towards data-driven smart dairy farming. Softw: Pract Exp 49(7):1055–1078
15. Tumuhimbise R (2013) Breeding and evaluation of cassava for high storage root yield and early bulking in Uganda. PhD thesis
16. Tushemereirwe W et al (2015) Performance of NARITA banana hybrids in the preliminary yield trial for three cycles in Uganda
17. Zhu M, Gupta S (2017) To prune, or not to prune: exploring the efficacy of pruning for model compression. arXiv preprint arXiv:1710.01878

Modified U-Net and CRF for Image Segmentation of Crop Images

Shantanu Chakraborty, Rushikesh Sanap, Muddayya Swami, and V. Z. Attar

Abstract Smart agriculture is the need of the hour, especially in a country like India whose economic and cultural backbone is agriculture. Detection of anomalies in crop images, like the presence of diseased spots and the presence of weeds is one of the most important strides in this direction. An important pre-requisite to this is the segmentation of crop images. A lot of related work has been done on the segmentation of crop images, using state-of-the-art neural network-based architecture and image processing-based techniques. This paper explores neural networks-based encoder–decoder architectures, particularly U-Net and introduces additions/modifications to it. On evaluating, the results of this modified U-Net are assessed qualitatively and quantitatively, it was found that the modified U-Net led to better results in terms of segmentation accuracy.

Keywords U-Net · Image segmentation · Residual block · Residual path · Conditional random fields

S. Chakraborty (✉) · R. Sanap · M. Swami
Department of Computer Engineering and IT, COEP Technological University, Pune 411005, India
e-mail: chakrabortyss19.comp@coeptech.ac.in

R. Sanap
e-mail: sanaprv19.comp@coeptech.ac.in

M. Swami
e-mail: swamimd19.comp@coeptech.ac.in

V. Z. Attar
School of Computational Science, COEP Technological University, Pune 411005, India
e-mail: vahida.comp@coeptech.ac.in

S. Shakya et al. (eds.), *Fourth International Conference on Image Processing and Capsule Networks*, Lecture Notes in Networks and Systems 798,
https://doi.org/10.1007/978-981-99-7093-3_13

1 Introduction

Agriculture has been the backbone of India for many years, economically and culturally. Owing to a declining natural resource base, dwindling agriculture areas and outbreaks of pests and diseases, maintaining this would be a huge problem. This has led to an increasing interest in smart agriculture and its applications. Detection of anomalies in crop images, like weeds and diseased spots is one of the first and important steps toward revolutionizing agriculture. An important step that goes hand-in-hand with this is the segmentation of crop images for object detection and classification. As discussed in Mohanty et al. [9], deep learning and neural networks have been crucial in designing architectures and models for image detection. Long et al. [10] throw light on the extensive use of neural networks, particularly convolutional neural networks for semantic segmentation. Neural networks in conjunction with other techniques, like transfer learning also show great promise for designing robust image segmentation models, as illustrated in Huang et al. [3] and Yang et al. [4]. Many works have been published in recent times related to this field. This paper is an attempt to propose certain modifications to certain state-of-the-art architectures and assess the performance of the proposed architecture using benchmark datasets. Section 2 of the paper gives an overview of related work relevant to the problem concerned. This is followed by Sect. 3 which discusses our proposed architecture at length. Section 4 presents a qualitative and quantitative evaluation of our architecture in contrast with U-Net, allowing us to draw certain conclusions in Sect. 5.

2 Related Work

2.1 U-Net

Ronneberger et al. [1] proposed U-Net (Fig. 1). U-Net is an encoder–decoder network that consists of two major components: a contraction/encoding pathway and a decoding/expansion pathway.

The contraction network consists of four encoder blocks, each of which contains two convolutional layers and a max pooling layer. The activation function used in convolutional layers is ReLu. The use of the max pooling layer downsamples the size by a factor of 2. The purpose of this block is to extract spatial features from the images. The output of these encoder blocks is a feature map that has all the encoded features.

The expansion network consists of four decoder blocks, each of which consists of a transposition convolutional layer that increases the height and width of incoming feature maps by a factor of 2, but decreases the number of feature maps by a factor of 2. This is followed by two more convolutional layers to increase the number of feature maps.

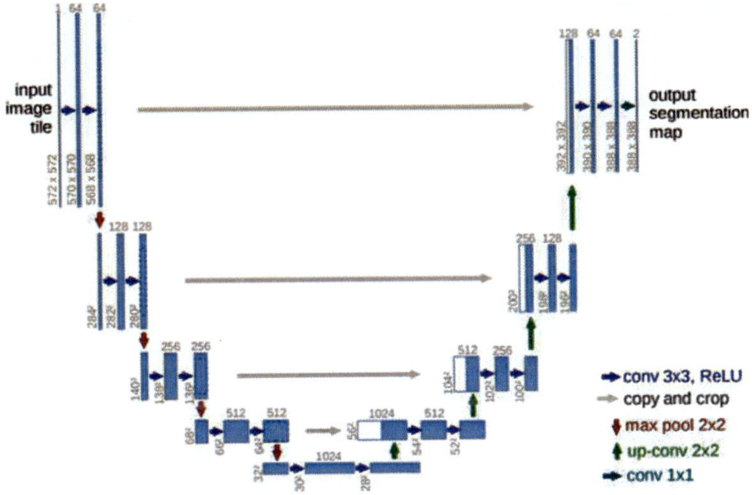

Fig. 1 U-Net

The feature maps outputted by the encoder blocks are spliced with the input to the decoder blocks with the help of skip connections. This causes the retention of spatial information to an extent. However, this does not solve the issue of a considerable semantic gap between the output of the encoder and the input of the decoder.

2.2 *Residual Block (ResBlock)*

The main role of convolutional layers is to extract features from images. The more CNNs present, the deeper the features extracted will be. However, if we keep adding more and more convolutional layers, the accuracy in the training set decreases. Since we are using a gradient descent-based optimizer (Adam optimizer), there exists a possibility that adding more convolutional layers will lead to finding local optimum solutions instead of finding global optimum solutions. A possible way of countering this is to replace multiple convolutional layers with a residual block introduced by He et al. [7]. Residual block extracts deep features from an image without facing the issues mentioned above.

The structure of the residual block is shown in Fig. 2. Unlike feed-forward networks where each layer feeds the next layer, a residual block feeds the next layer and also the layer a couple of hops away. And finally, the input to the next layer is subjected to an activation function before concatenating it with the original feature vector which was fed to a layer a couple of hops away.

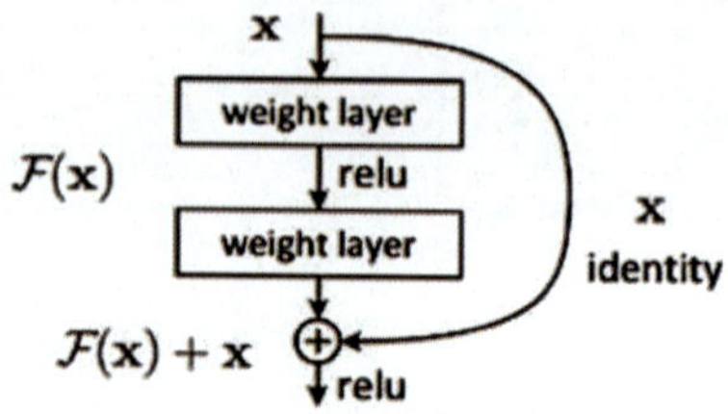

Fig. 2 ResBlock

2.3 *Residual Path*

In standard U-Net architecture, encoders extract low-level features. The convolutional layers present in the encoders, create feature maps while doing so. By splicing the low-level features with high-level features, the spatial information lost is retained. As a result of this, high-resolution details that were lost are retained and the overall segmentation output is refined. However, there exists a semantic gap between the low-level features and the high-level features; thus, their direct fusion is not appropriate.

To counter this issue, the proposed architecture makes use of a "Residual Path" (ResPath) instead of U-Net's skip connection. ResPath introduces some additional convolutional layers before concatenating low-level features with high-level features. This results in a reduced semantic gap and a better learning ability.

Figure 3 shows the structure of ResPath. It consists of four 3*3 convolutional layers and four 1*1 convolutional layers and four Resblocks. The low-level feature maps are spliced with the high-level feature maps by ResPath.

3 Proposed Architecture

U-Net has been used for a variety of image segmentation applications. However, its application for crop images is not prevalent. This section discusses the proposed additions and modifications to the existing U-Net architecture to improve the performance of crop images. The proposed and modified U-Net for crop images is a symmetric

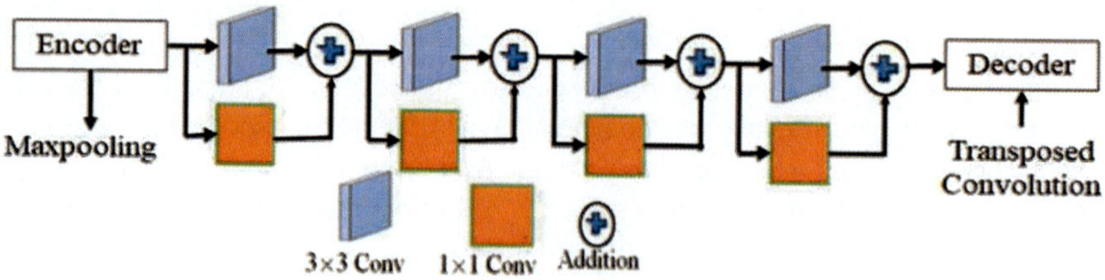

Fig. 3 ResPath

architecture with two major parts—a contraction network (encoding path) and an expansion network (decoding path). Each of these networks has four encoder and decoder blocks, respectively, whose architecture is described in detail below.

3.1 Selection of Algorithm

U-Net is a standard architecture for image segmentation and has enjoyed success with various benchmark datasets. As described in Gurita et al. [2] and Chen et al. [11], variations of encoder–decoder networks have been enjoying great success for image segmentation use cases. CED Net proposed by Khan et al. [5] also propose tweaks in the U-Net architecture to make it more efficient and accurate for crop images. Thus, U-Net as a backbone for the proposed architecture for segmentation of crop images shows great promise. Thus, the main motive of our proposed architecture is to improve upon this by addressing the issues that arise while using U-Net for crop images—namely a saturation of accuracy on increasing the number of convolutional layers, a significant semantic gap between the low-level feature maps outputted by encoder blocks and the high-level feature maps fed as input to decoder blocks, which make their splicing improper. The proposed architecture was devised with the aim to address these issues while leveraging the proven success of U-Net with multiple benchmark datasets. The proposed architecture uses CNN at the root of its functioning, because segmentation requires extraction of features, both low-level and high-level, and convolutional neural networks are by far the best deep learning paradigms for doing so.

As shown in Fig. 4, each encoder has two convolutional layers of size 3 × 3, followed by two ResBlocks. The first ResBlock doesn't use any activation function, however, the second ResBlock uses the ReLu activation function. The stride of each convolutional layer is 1. The resultant of this is subjected to a max pooling layer of size 2 × 2. After the application of the max pooling layer, the number of filters is doubled, but the size is downsampled by a factor of 2.

Every decoder block consists of a transposition convolutional layer of size 2 × 2 and stride of 2, followed by two convolutional layers of size 2 × 2 and stride 1, and two ResBlocks. The first ResBlock doesn't make use of any activation function; however, the second ResBlock uses the ReLu activation function.

U-Net makes use of skip connections to fuse low-level features with high-level features for the retention of spatial information. A downside to this approach, as mentioned previously, is the presence of a considerable semantic gap between low-level and high-level features. Our proposed architecture makes use of ResPath instead of skip connections. The use of ResPath leads to better retention of spatial information and addresses the issue of semantic gap.

An en-decoder block connects the contraction network and the expansion network.

All convolutional layers except the last one use ReLu as an activation function. The final convolutional layer uses sigmoid as an activation function.

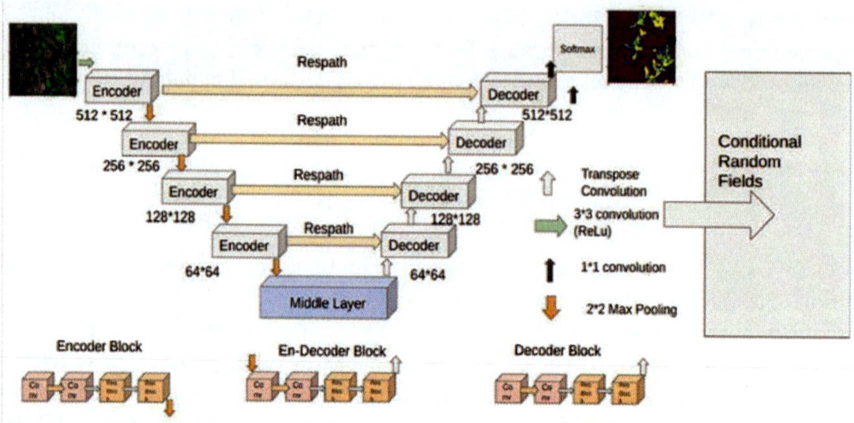

Fig. 4 Proposed architecture

The output of the abovementioned architecture is coarse. To refine it further, the segmented images are subjected to conditional random fields. The following subsection discusses CRF and its implementation in our proposed architecture in detail.

3.2 *Conditional Random Field (CRF)*

Conditional random field (CRF) is a statistical modeling technique that is applied when the class labels associated with various inputs are dependent on each other. In the context of the problem of image segmentation, the label associated with every pixel depends on the pixels in its neighborhood too. Discrete classifiers do not take this dependence into account, however, CRF does. In cases where the correlation between input data points and the labels is high, such classifiers do not perform well. CRF, on the other hand, assigns labels to data points (pixels, in our case) by considering the labels of the labels of input points dependent on it too. Thus, CRF postprocessing of segmented images refines them. The proposed architecture makes use of fully connected CRF which helps in the refinement of segmentation outputs by considering similar pixels in the neighborhood, calculating the unary and pairwise energies of these standalone pixels and pairs of pixels and attempting to minimize the energy, thus leading to a more precise labeling of the pixels. The following description explains the succinct description in more detail.

Krähenbühl et al. [6] introduce the use of "fully connected pairwise CRFs". This means that all the locations in the image are connected to each other in pairs. An edge between pixel i and pixel j indicates that the labels of these two pixels are dependent on each other. Our proposed architecture makes use of fully connected pairwise CRFs.

CRF is characterized by Gibb's distribution of the following form:

$$P(\boldsymbol{X} = x|\boldsymbol{I}) = \frac{1}{z(\boldsymbol{I})} \exp(-E(x|\boldsymbol{I})).$$

Here, **I** is the input, $E(x|\mathbf{I})$ is the energy function, **X Is** the random value corresponding to location **I** which represents its predicted label and $z(I)$ is the partition function which is obtained by the summation of all $\exp(E(x \mid I))$.

A simple interpretation of CRF can be presented this way: our prime goal in CRF is to minimize the energy function. Thus the energy function can be thought of as a cost function. By assigning the most probable label (x) to a given location (I), the energy function can be minimized, consequently leading to higher accuracy.

The energy function for CRF can be defined this way:

$$E(x) = \sum_i \psi_u(x_i) + \sum_i \psi_p(x_i, x_j).$$

Unary energy $\psi_u(x)$: The measure of the cost if the label assignment disagrees with the initial classifier.
Pairwise energy $\psi_p(x_i, x_j)$: The measure of the cost if two similar pixels, i.e., pixels having similar colors or adjacent pixels take different labels.

To present in a succinct way, CRF will take the predicted masks from our modified U-Net as inputs and take into account both Unary potential and pair-wise energy terms to output a more precise and accurate segmentation mask. The Unary potential corresponds to the label of a given pixel, and the pairwise potential corresponds to the labels of the pixels adjacent to it.

Our proposed architecture implements CRF in this fashion:

Convert the segmented mask into RGB (if it is grayscale) (this can be done using gray2rgb method in Matlab):

Transforming annotated image to RGB in case it is a grayscale image if (len(annotated image.shape) < 3):

annotated image = gray 2rgb (annotated image)

Convert the RGB color of the annotations to a single 32-bit integer:

annotated image is the segmented output

annotated label = annotated image [: , : , 0] + (annotated image [: , : , 1] ≪ 8) + (annotated image [: , : , 2] ≪ 16)

Create a mapping back to 32-bit colors this can be done in the following fashion:

Convert the 32 bit integer color to 0, 1, 2, . . . labels. # np is an alias for numpy here

color, labels = np . unique (annotated label)

color int labels = np . empty ((len (colors) , 3) , np. unit 8) color int labels [: , 0] = (colors & 0 x0000FF)

color int labels [: , 1] = (colors & 0 x00FF00) >> 8 color int labels [: ,2] = (colors & 0 xFF0000) >> 16

Get Unary potentials with the help of the labels of the mask (i.e., the cost if the label of a given location is not the same as the label predicted by the initial classifier)

Get pairwise potentials (i.e., the cost if two similar pixels have different labels):

#Number of labels in the masked image fed as input to CRF architecture

number of labels = len(set (labels .flat))

using 2 dimensional CRF #DenseCRF is an alias for dcrf here if use 2 d:

The APIs used here to set up CRF and get unary as well as

pairwise potentials are from pydensecrf library

d = DenseCRF. DenseCRF2D (org img. shape [1], org img. shape [0], number of labels)

For Unary energy

U = unary from labels (labels, number of labels, get prob =0. 7, zero unsure= d. set Unary Energy (U)

For pairwise energy factoring in color independent terms # in the following, k is usually set to 3

d. add Pairwise Gaussian (sxy= (k, k), compat=k, kernel=DenseCRF. DIAG KERNEL, normalization=DenseCRF. NORMALIZE SYMMETRIC)

This factors in color dependent terms

d. add Pairwise Bilateral (sxy = (80, 80), srgb = (13, 13, 13),

rgbim=org img, compat=10,

kernel=DenseCRF. DIAG KERNEL,

normalization=DenseCRF.NORMALIZE SYMMETRIC)

We need to compute the most likely label assignment for every pixel. Inference can be run using any inference algorithm for graphical models.

running inference 5 times

Q is the output after running inference 5 times (ideally)

Find out the most probable class for each pixel.

Estimate the most likely class for each pixel. LABELS = np. argmax (Q, a xis = 0)

Convert the labels back to their corresponding colors and save the image.

Convert the LABELS (labels) back to the colors corresponding to them

LABELS = color int labels [LABELS,:]

Reshape the resultant image according to the original image that was # given as an input.

4 Results and Discussions

For evaluating our proposed architecture against standard U-Net, we conducted experiments on two datasets.

- *Crop Weed Field Image Dataset (CWFID):* CWFID dataset comprises field images, segmentation masks and appropriate annotations. This dataset was published in Haug et al. [8]. The number of images in this dataset is 60. Since the images were quite meager in amount, basic data augmentation was performed by flipping the image about various axes. This dataset was acquired using an autonomous robot Bonirob in an organic carrot farm when the carrots were in the early true leaf growth stage.
- *Open Sprayer Dataset:* This is a collection of crops and weeds that were captured using an autonomous land drone. There are up to 1000 images in this dataset for training. The huge number of data points present is conducive to our project. It includes pictures of broad-leaved docks and pictures of land without broad-leaved docks. The dataset can be downloaded from kaggle.

4.1 *Qualitative Evaluation*

Figures 5, 6, 7 and 8 give a qualitative idea of the performance of our proposed architecture and the performance of standard U-Net architecture for images from CFWID as well as Open Sprayer Images datasets.

4.2 *Quantitative Evaluation*

We used a metric known as intersection over union (IoU), which is frequently used for image segmentation and other similarity challenges, for the quantitative assessment

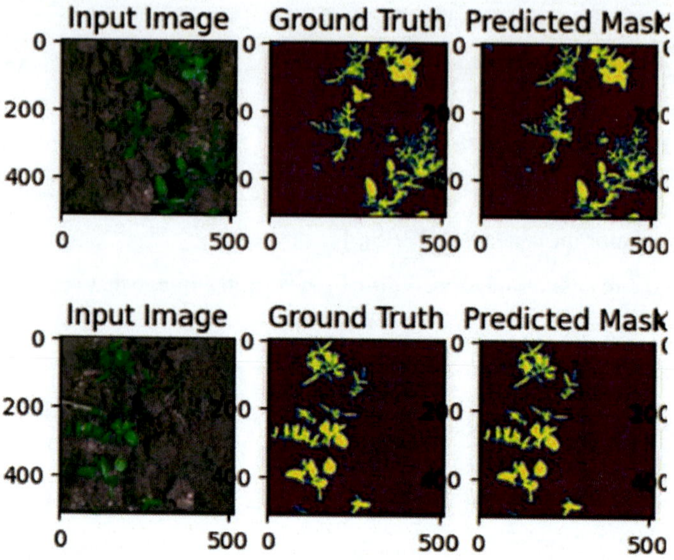

Fig. 5 Image segmentation of images from CWFID dataset using standard U-Net

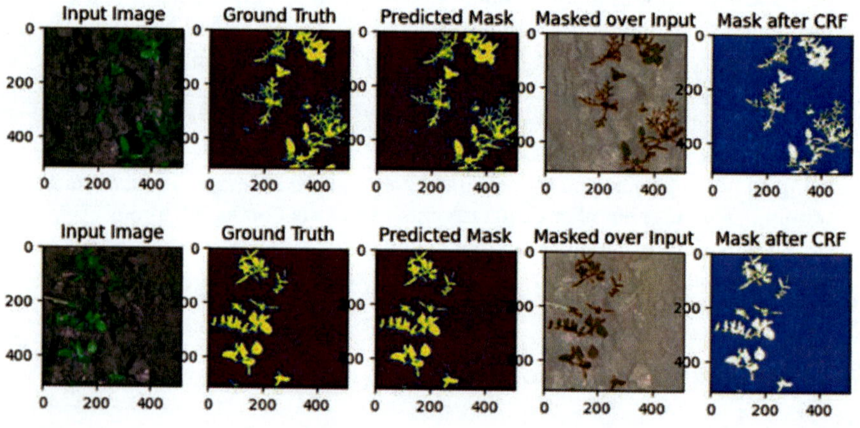

Fig. 6 Image segmentation of images from CWFID dataset using our proposed U-Net and CRF postprocessing

of our suggested model. IoU is calculated by dividing the area of the union of the expected mask and the ground truth by the area of the overlap between the predicted mask and the ground truth. IoU values are between 0 and 1. The ground truth and predicted masks do not overlap at all, as indicated by the value 0. The ground truth and anticipated mask completely coincide, as indicated by the number 1.

IoU is defined as follows:

Fig. 7 Image segmentation of images from Open Sprayer dataset using standard U-Net

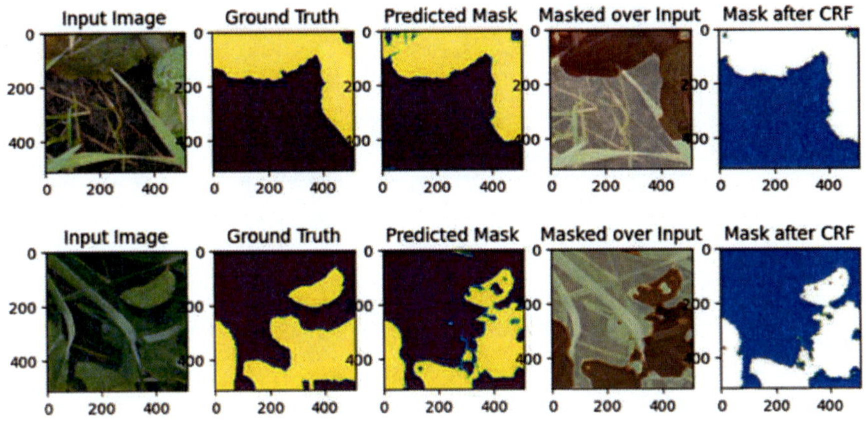

Fig. 8 Image segmentation of images from Open Sprayer dataset using standard U-Net and CRF postprocessing

$$\mathbf{G = GroundtruthP = PredictedMaskIoU = G \cap P/G \cup P}$$

Tables 1 and 2 provide a quantitative evaluation of our proposed architecture against standard U-Net. Figures 9 and 10 visualize these statistics for better understanding.

Table 1 IOU values for standard U-Net

IOU	CWFID	Open Sprayer
Mean IOU before CRF	0.9427	0.737
Mean IOU after CRF	0.930	**0.743**
Min IOU before CRF	0.918	0.070
Min IOU after CRF	0.898	0.0789
Max IOU before CRF	0.959	0.985
Max IOU after CRF	0.957	0.9871

Table 2 IOU values for our proposed architecture

IOU	CWFID	Open Sprayer
Mean IOU before CRF	0.9459	0.737
Mean IOU after CRF	0.945	**0.758**
Min IOU before CRF	0.915	0.030
Min IOU after CRF	0.888	0.030
Max IOU before CRF	0.965	0.985
Max IOU after CRF	0.955	0.9871

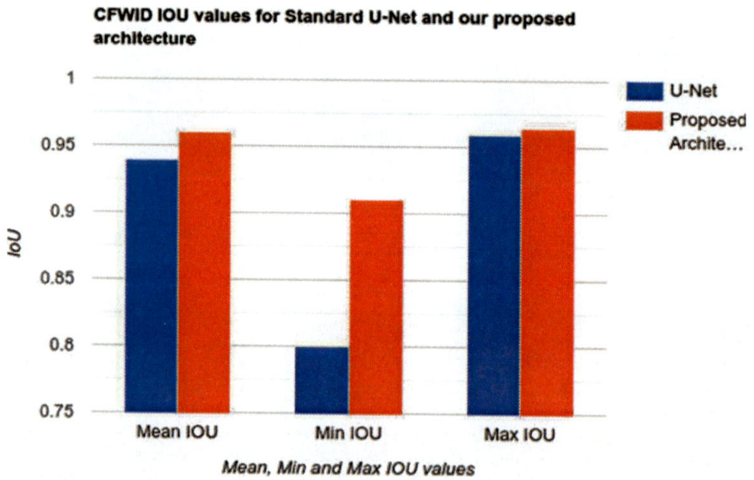

Fig. 9 Comparison of IOU values for U-Net and our proposed architecture for CFWID dataset

It can be seen that the mean IOU values corresponding to our proposed architecture is either greater than the mean IOU values corresponding to standard U-Net, for both CFWID and Open Sprayer Images datasets.

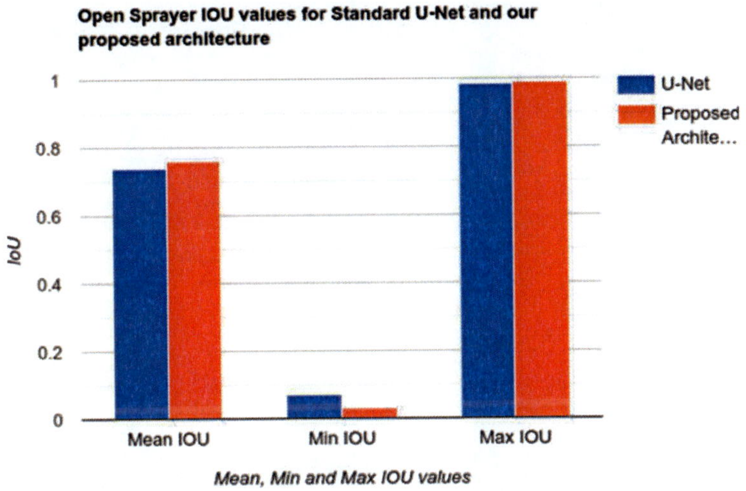

Fig. 10 Comparison of IOU values for U-Net and our proposed architecture for Open Sprayer dataset

4.3 Retention of Spatial Information

One of the major issues in standard U-Net is that, due to multiple encoders and convolutional layers a lot of spatial information is lost during the encoding process. This spatial information is retained by splicing the feature map outputted by the encoder blocks with the input of the decoder blocks. These connections between the feature maps outputted by encoders and the feature maps fed as input to the decoder blocks are called skip connections. However, these skip connections don't retain spatial information robustly and don't address the issue of the semantic gap between two combined feature sets. Our proposed architecture makes use of a residual path (ResPath) instead of skip connections to address the issue of semantic gap and to retain spatial information to a greater extent. The following images show how the usage of ResPath allows our model to retain spatial information better than standard U-Net.

It can be seen from Figs. 11 and 12 that the leaves obscured by weed are segmented by U-Net, but only to a slight extent. However, our proposed architecture segmented a greater area of the obscured leaf, proving that spatial retention is better using our proposed architecture.

Fig. 11 Image segmentation of images from Open Sprayer dataset using standard U-Net

Fig. 12 Image segmentation of images from Open Sprayer dataset using our proposed U-Net architecture

5 Conclusion

Keeping in mind the problem of segmentation of crop images, an improved U-Net is proposed to tackle certain issues that could occur in standard U-Net. Our proposed modified U-Net makes use of ResBlock which allows deeper feature extraction without encountering issues like vanishing gradient and leading to an improvement in the network performance. For the aim of splicing low-level features with high-level features, U-Net skip connections are replaced with ResPath. This has resulted in enhanced spatial information retention and a narrowing of the gap between low-level and high-level features. The experimental findings suggest that our proposed modified U-Net outperforms regular U-Net in terms of segmentation accuracy. For both the CFWID and Open Sprayer datasets, the mean intersection over union values of segmentation using our modified U-Net architecture outperform the mean IoU values of the traditional U-Net architecture.

However, an exhaustive comparative study of our proposed architecture with multiple datasets will shed more light on its performance and robustness. The addition of ResPath and ResBlock increases the performance quantitatively and qualitatively but it also introduces more trainable parameters, which leads to a greater training time.

References

1. Ronneberger O, Fischer P, Brox T (2015) U-Net: convolutional networks for biomedical image segmentation. arXiv:1505.04597v1
2. Gurita A, Mocanu IG (2021) Image segmentation using encoder-decoder with deformable convolutions. Natl Libr Med 21(5):1570
3. Huang H-N, Zhang T, Yang C-T, Sheen Y-J, Chen H-M, Chen C-J, Tseng M-W (2022) Image segmentation using transfer learning and Fast R-CNN for diabetic foot wound treatments. Natl Libr Med 10:969846
4. Yang S, Zheng L, He P, Wu T, Sun S, Wang M (2021) High-throughput soybean seeds phenotyping with convolutional neural networks and transfer learning. Plant Methods 17(50)
5. Khan A, Ilyas T, Umraiz M, Mannan ZI, Kim H (2020) CED-Net: crops and weeds segmentation for smart farming using a small cascaded encoder-decoder architecture. Electronics 9(10):1602
6. Krähenbühl P, Koltun V (2012) Efficient inference in fully connected CRFs with Gaussian edge potentials. arXiv:1210.5644
7. He K, Zhang X, Ren S, Sun J (2015) Deep residual learning for image recognition. arXiv:1512.03385v1
8. Haug S, Ostermann J (2015) A crop/weed field image dataset for the evaluation of computer vision based precision agriculture tasks. In: ECCV 2014 workshops, Part IV, LNCS 8928, pp 105–116
9. Mohanty SP, Hughes DP, Salathé M (2016) Using deep learning for image-based plant disease detection. Front Plant Sci 7:1419
10. Long J, Shelhamer E, Darrell T Fully convolutional networks for semantic segmentation. In: Proceedings of the IEEE conference on computer vision and pattern
11. Chen L-C, Zhu Y, Papandreou G, Schroff F, Adam H (2018) Encoder-decoder with atrous separable convolution for semantic image segmentation. In: Proceedings of the European conference on computer vision (ECCV), Munich, Germany, 8–14 Sept 2018

Securing Data in the Cloud: The Application of Fuzzy Identity Biometric Encryption for Enhanced Privacy and Authentication

Chandrasekar Venkatachalam, K. Manivannan, and Shanmugavalli Venkatachalam

Abstract Cloud computing has emerged as a highly innovative technology, offering various benefits to users. To ensure the security of user data, cloud storage schemes have been introduced, aiming to protect sensitive information from unauthorized access. In particular, the sharing of personal health records (PHR) is gaining prominence as a method for securely sharing healthcare data among users on cloud servers. The use of fuzzy techniques plays a crucial role in transforming original data into an encrypted form, known as ciphertext, which is challenging for unauthorized individuals to comprehend. This technique enhances the confidentiality of data, ensuring that only authorized parties can access and understand it. However, while cloud services provide a convenient platform for data sharing, they often lack efficiency in terms of data sharing capabilities. To address these challenges, a novel approach called fuzzy identity biometric encryption (FIBE) is introduced for PHR management. FIBE combines the benefits of fuzzy techniques and biometric authentication to achieve both high-security levels and user convenience simultaneously. This approach enables authorized users to have control over access to PHR data and ensures secure data sharing within a cloud environment. By integrating biometric authentication, FIBE enhances the security of PHR systems, as biometric characteristics are unique to each individual, making it difficult for unauthorized users to gain access. Moreover, the approach improves user convenience by eliminating the need for remembering complex passwords or using traditional authentication methods. In conclusion, the utilization of fuzzy identity biometric encryption (FIBE)

C. Venkatachalam (✉) · K. Manivannan
Department of Computer Science and Engineering, Faculty of Engineering and Technology, Jain (Deemed-to-be University), Bengaluru, Karnataka, India
e-mail: chandrasekar.v@jainuniversity.ac.in

K. Manivannan
e-mail: manivannan.k@jainuninversity.ac.in

S. Venkatachalam
Department of Computer Science and Engineering, KSR College of Engineering, Tiruchengode, Tamil Nadu, India
e-mail: drvshanmugavalli@gmail.com

S. Shakya et al. (eds.), *Fourth International Conference on Image Processing and Capsule Networks*, Lecture Notes in Networks and Systems 798,
https://doi.org/10.1007/978-981-99-7093-3_14

in PHR systems offers enhanced security and efficient data sharing in the cloud. This approach combines the advantages of fuzzy techniques and biometric authentication, providing authorized users with control over data access while maintaining a high level of security and user convenience.

Keywords Cloud computing · PHR · Intruders · Fuzzy technique · Fuzzy identity biometric encryption · Biometric authentication

1 Introduction

Cloud computing has revolutionized the way clients access information, allowing them to retrieve data from anywhere and at any time through the Internet. One of the emerging applications in healthcare is the personal health record (PHR) system, which follows a patient-centric model of health information exchange. It empowers patients to create and manage their own PHR data in a centralized location using web-based applications. With the advent of cloud computing, individuals are increasingly utilizing this environment for efficient storage and retrieval of their individual PHRs. Storing PHRs in the cloud offers benefits such as cost reduction and simplified communication and access systems.

However, a primary concern in PHR systems is whether patients can maintain control over their own data. It is crucial to establish fine-grained access control and ensure the security of sensitive data stored in third-party cloud storage. The challenges lie in maintaining security, privacy, and health information authentication when PHRs are kept in the cloud.

To address these challenges, strong security measures and access controls are necessary. Encryption techniques, access management protocols, and authentication mechanisms can be employed to protect the confidentiality, integrity, and availability of sensitive health information. By implementing secure authentication protocols and leveraging advanced cryptographic techniques, the authenticity of data and the authorized access to PHRs can be ensured.

Furthermore, the usage of fuzzy-based biometric encryption (FBBE) provides a promising solution. FBBE considers identities as a set of descriptive attributes, allowing users with the secret key for a specific identity to decrypt ciphertexts encrypted with the corresponding public key. This approach offers high computation efficiency and security, addressing key management issues and providing scalability to the PHR framework.

The integration of cloud computing in PHR systems provides convenient access to healthcare information, but also introduces data security concerns. To mitigate these concerns, it is crucial to implement robust security measures including strong encryption techniques, access controls, and fuzzy-based biometric encryption. These measures play a vital role in safeguarding the confidentiality, integrity, and privacy of PHR data within the cloud environment. By enhancing data security, PHR systems can operate with improved efficiency, ensuring secure and user-friendly access for both patients and healthcare providers. This ultimately leads to better healthcare outcomes and increased trust in the PHR system.

2 Related Work

Naveed et al. [1] contribute to the field of cloud security by emphasizing the importance of biometric authentication as a viable solution for enhancing security in cloud environments. By providing a structured view of different biometric techniques, the research offers valuable insights into the practical implementation and benefits of biometric authentication in the cloud.

Boneh and Franklin [2] introduced the first secure and practical identity-based encryption (IBE) scheme based on bilinear pairings. This groundbreaking work has sparked significant interest in IBE, leading to the development of numerous schemes and related systems. The ongoing research and development in the field of IBE aim to further optimize and enhance the performance of encryption schemes for multiple recipients. By reducing computational complexity and communication overhead, these advancements enable more efficient and scalable encryption techniques, ultimately benefiting various applications that require secure communication among multiple parties.

Canetti et al. [3] made a significant contribution to the field of identity-based encryption (IBE) by proposing the first construction that achieved provable security beyond the random oracle model. In their work, they introduced a slightly weaker model of security called the Selective-ID model. The introduction of provable security beyond the random oracle model in the context of IBE has paved the way for further advancements and developments in this area. Researchers have since explored new techniques, refined existing constructions, and extended the security analysis of IBE schemes, leading to improved protocols and a deeper understanding of their security properties.

Selvarani et al. [4] introduced a method to enhance data security in the cloud environment by employing a multimodal biometric technique. This approach aims to strengthen the protection of data during its transmission and storage. The encrypted data is then stored in the cloud environment, ensuring that it remains protected and inaccessible to unauthorized entities. When the data needs to be retrieved, a decryption technique is employed to convert the ciphertext back into its original plaintext form.

By employing the combination of encryption and multimodal biometrics, the proposed method enhances data security in the cloud environment. It provides a layered approach to protect sensitive information during transmission and storage, mitigating potential risks and unauthorized access attempts.

Rinesh et al. [5] introduced a method called multi-attribute attribute-based encryption (MA-ABBE) that addresses the access control policy in personal health records (PHR) systems. While many research studies have focused on the security issues of PHR, achieving efficient data sharing has remained a challenge.

The MA-ABBE method offers several advantages, including enhanced security, authentication, and efficient data sharing in the cloud environment. It provides a mechanism to control access to PHR data, protecting sensitive healthcare information from unauthorized access or breaches. Additionally, the use of biometric fingerprint authentication adds an extra layer of security, ensuring that only authorized users can access the PHR. Overall, the MA-ABBE approach addresses the security and efficient data sharing challenges in PHR systems, offering a reliable solution for access control and protecting the privacy of healthcare information.

Abdolahi et al. [6] addressed the issue of single biometric spoof attacks, which can result in unacceptable error rates due to noise or other factors. They proposed the use of a multimodal biometric system as a solution to mitigate these attacks and improve recognition accuracy. The decision-making process in the multimodal biometric system involves combining the recognition scores or features extracted from both fingerprint and iris biometric systems. Various fusion techniques, such as score-level fusion or feature-level fusion can be employed to obtain a final decision.

Wagh et al. [7] introduced a multimodal biometric system that utilizes the characteristics of both fingerprint and iris modalities. The system incorporates feature-level fusion to combine the features extracted from fingerprint and iris biometric templates. The fused features are then subjected to encryption using various security technologies. The experimental outcomes showed that the multimodal biometric system achieved higher accuracy compared with the unimodal biometric system. This improvement in accuracy can be attributed to the synergistic effect of combining the distinctive features of fingerprint and iris modalities. By leveraging the strengths of both modalities, the system becomes more robust and capable of handling variations in biometric data, leading to enhanced recognition performance.

Thaiyalnayaki et al. [8] proposed a fingerprint recognition framework that addresses the limitations of minutiae-based methods by incorporating surface analysis and utilizing a combination of features for multiscale and multidirectional recognition. The framework incorporates various silent features such as standard deviation, skewness, and kurtosis. The proposed framework aims to improve the performance of fingerprint recognition systems by reducing the impact of false minutiae and addressing the limitations of traditional minutiae-based approaches. By incorporating surface analysis and a combination of features, the framework provides a more comprehensive and reliable approach to fingerprint recognition.

Pokhriyal et al. [9] proposed a methodology for fingerprint verification that utilizes wavelets and Pseudo Zernike moments (PZMs) to extract both global and local features. PZMs are robust to noisy images, invariant to rotation, and have good image reconstruction capabilities, making them useful for global analysis and extracting global features such as the shape of the fingerprint image. On the other hand, wavelets are effective in local analysis and help extract local features or minutiae from a fingerprint. The proposed methodology contributes to the field of fingerprint verification by leveraging the strengths of PZMs and wavelets, resulting in a more robust and effective approach for feature extraction and fingerprint matching.

Gao et al. [10] have developed a novel topology-based representation system for fingerprint verification. This system utilizes neighbor structure data to match the point patterns initially and then performs global matching. The proposed representation is designed to be invariant to rotation and translation by analyzing the relationships among minutiae, both genuine and spurious, without capturing extensive information about fingerprints. By using the relationship of point sets for matching, the proposed representation exhibits robustness to missing and spurious minutiae. This approach overcomes the limitations of relying solely on individual minutiae and provides a more comprehensive and resilient matching technique. Overall, the topology-based representation system proposed by Gao et al. offers a promising approach for fingerprint verification by leveraging the relationships among minutiae and providing robustness to variations in rotation, translation, and image resolution.

Prasath et al. [11], the authors focus on improving the accessibility of cloud computing by utilizing fuzzy logic. They propose a conceptual model that employs various attributes to evaluate the satisfaction of cloud computing users in an Internet service provider (ISP) context. To handle the inherent ambiguity and uncertainties associated with linguistic methods, the authors employ a fuzzy inference system (FIS). By using fuzzy logic, they aim to avoid any ambiguities that may arise during the evaluation process. By utilizing fuzzy logic and the FIS, the authors aim to provide a more comprehensive and accurate evaluation of user satisfaction in cloud computing services. This approach enables the consideration of multiple attributes simultaneously, allowing for a more nuanced understanding of user experiences and requirements.

The implementation proposed by Velciu et al. [12]. Focuses on enhancing the security of authentication methods in cloud storage. They introduce the concept of a bio-cryptographic infrastructure, which aims to provide safer access and encryption support within cloud storage sharing and cloud platforms. Specifically, their work implements a fuzzy vault authentication mechanism based on voice recognition. The fuzzy vault is a cryptographic concept that allows secure storage of private information while providing the ability to retrieve it even when there are slight variations in the input. In this case, the voice of the user serves as the basis for authentication.

This method helps ensure that only authorized individuals with a verified voice pattern can access the stored data and perform encryption operations. Overall, the intention of this work is to provide a more robust and secure authentication mechanism for cloud storage, leveraging voice recognition within the fuzzy vault framework.

3 System Model

The basic fuzzy selective-ID game is allowed to query for secret keys identities that are greater than **d** Hamming Distance away from the targeted identity [13].

4 Basic Fuzzy Selective-ID

Init: The adversary (A) declares the identity x that he wishes to be challenged upon.

key Setup: The challenger runs the key setup phase of the algorithm and the public parameters of adversary.

Phase 1: The adversary is allowed issue queries private keys for an identity y, where the Hamming distance between x and y is greater than d.

Challenge the adversary submits two equal length messages m_1, m_2, the challenger flips a random coin c and encrypts m_c. The cipher text is passed to the adversary.

Phase 2: Phase 1 is Repeated

Guess: The output of the adversary c' of c.

The main advantage A in this game is Pr $[c' = c] - 1/2$.

Attribute Identity Authentication (AIA)

A user can decide to move to an attribute identity authentication (AIA) and demonstrate that it is qualified to unspecified amount of the attributes managed by that authority, and request related to decryption keys. The attribute identity key generation algorithm will run by authority. Any user can perform encryption and decryption process when they should create key setup and key generation algorithm.

Key Setup

The key setup algorithm takes the implicit security parameter. It generates the public and master key parameters P and M generated by each AIAs (attributes identity authorities). The nth AIA defines a disjoint set of attributes Ur, which are similar for public users. These attributes are classified in their features.

$$\text{Key Setup}\,(1, N) \rightarrow \text{via (params}, \{(P_k, M_k)\}k \in \{1, N\})$$

Key Generation ***(M, S)***

The key generation algorithm uses key setup algorithm's master key M and set of identity attributes UIr that describe the key and output of the secret key S for user U. S should contain at least one attribute from every type of attributes governed by AIA.

$$\text{A Key Gen}\,(S_k, \text{UID}, \text{IA}_k) \rightarrow \text{via } S_k[\text{UID}, \text{IA}_k],$$

where

UID *user with identity*
S_k *secret key*
IA_k *identity attribute set.*

Biometric Fingerprint Encryption

The biometric fingerprint-based encryption method is the oldest method for identification which has been successfully used in numerous applications. Everyone is having unique fingerprint and is changeless. Fingerprint identification systems often utilize both manual classification and automatic minutiae matching to identify individuals based on their fingerprints. The uniqueness of a fingerprint is determined by the pattern of ridges and valleys on the surface of the finger, as well as the specific minutiae points. Minutiae points are local ridge characteristics that occur at locations such as ridge endings or ridge bifurcations. These points serve as key reference points for fingerprint matching.

In order to identify an individual based on their fingerprint, the fingerprint data is compared with a large number of fingerprints stored in a database, e.g., the FBI database holds approximately 70 million fingerprints. Sorting and organizing these fingerprints can help reduce the search time and computational complexity. By organizing the fingerprints into subsets or categories, the data fingerprint only needs to be compared with a subset of fingerprints that are most likely to match. The process often involves manual classification, where human experts classify fingerprints into specific categories based on their overall patterns and characteristics. This manual classification step helps narrow down the search space and makes the subsequent automatic minutiae matching more efficient. After classification, automatic minutiae matching algorithms are used to compare the minutiae points of the data fingerprint with the minutiae points of the fingerprints in the selected subset. By identifying and matching minutiae points and measuring the similarity between the patterns, a fingerprint identification system can determine the likelihood of a match and provide a potential identity for the individual associated with the fingerprint.

Overall, the combination of manual classification and automatic minutiae matching enables accurate and efficient fingerprint identification in various applications, including personal health records (PHRs) and law enforcement databases.

(i) ***Minutiae Detection and Extraction***

Minutiae are specific points or features within a fingerprint that are crucial for fingerprint recognition (Fig. 1 shows overall process for minutiae detection and extraction process). These points include ridge endings, bifurcations, and dots (short ridges). Ridge endings are points where a ridge terminates, while bifurcations occur where a ridge splits into two branches. Dots or short ridges refer to small ridge segments within the fingerprint. The process of automatic minutiae detection (AMD) is vital, especially when dealing with low-quality fingerprints that may contain noise or have poor contrast. In such cases, it becomes challenging to distinguish genuine minutiae from pixel configurations that resemble minutiae or to uncover hidden minutiae [14]. Human fingerprints are unique to each individual, providing a means of personal identification. However, direct matching between an unknown fingerprint and a known fingerprint sample is sensitive to errors. Therefore, current systems primarily focus on extracting minutiae from the fingerprint image and comparing the sets of fingerprint features for matching (Fig. 2). When comparing two fingerprints, discrete features called minutiae are used. These features correspond to specific points on the finger's friction skin, where ridges either end or split. The location of each minutia is represented by its coordinates within the fingerprint image, usually measured from a reference point in the bottom-left corner of the image.

(ii) ***Minutiae-Image Processing***

During the authentication process involving minutiae points, the following steps are typically followed:

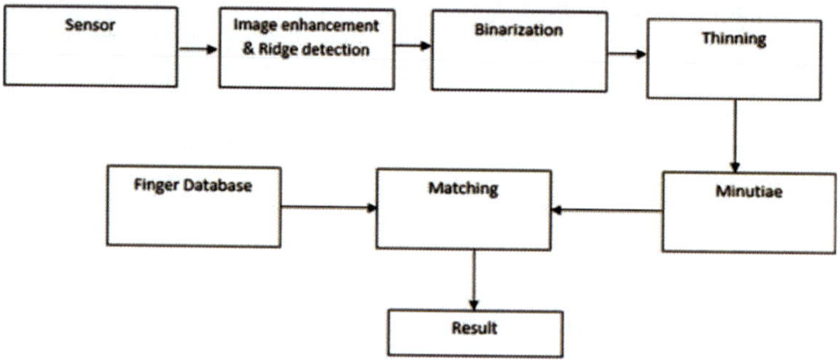

Fig. 1 Overall process for minutiae detection and extraction process

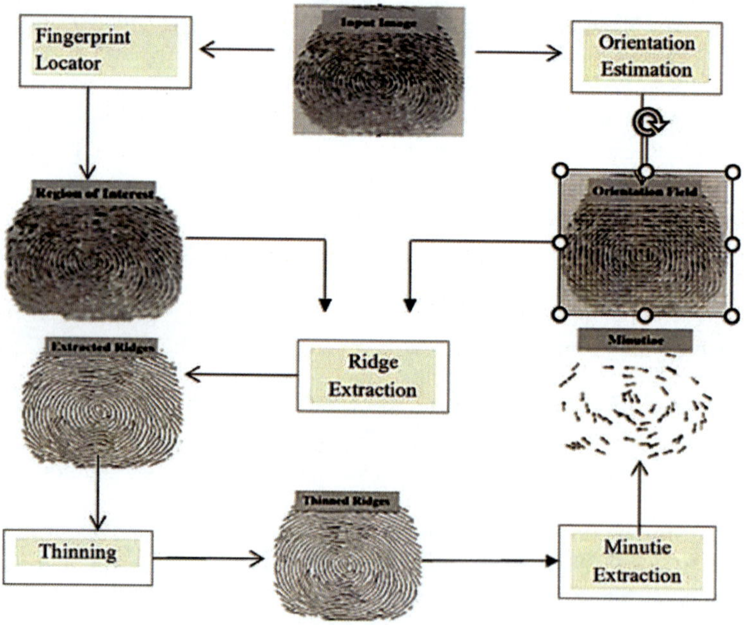

Fig. 2 Process of minutiae detection and extraction

Fig. 3 Image processing

1. Capture Fingerprint Images: Fingerprint images are acquired using a fingerprint scanner or sensor.
2. Image Processing: The captured fingerprint images undergo various image processing algorithms to enhance the quality and clarity of the image. This includes removing noise, smudges, artifacts, and healing scars or cuts. The goal is to obtain a clear and unambiguous representation of the fingerprint's skeletal image in Fig. 3.
3. Feature Extraction: The processed fingerprint image is then analyzed to extract the minutiae points. Minutiae refer to the specific details such as ridge endings, bifurcations, and dots within the fingerprint. These minutiae points are crucial for fingerprint recognition.
4. Fingerprint Matcher: The extracted minutiae points from the input fingerprint are compared with the minutiae points stored in the database. The fingerprint matcher algorithm compares the information from the input fingerprint with the records in the database to determine if there is a potential match.
5. Minutia Connections: Minutiae connections, one to another, are compared within a global context. Rather than treating the minutiae as areas within an X–Y coordinate system, they are considered as interconnected relationships. Each minutia is represented by its coordinates (x, y), and additional data chunks represent the minutia's orientation and neighborhood information.
6. Matching and Verification: If the input fingerprint's minutiae points match with those of the original fingerprint stored in the database, the authentication is successful. In this case, the health details or other authorized data can be shared with the corresponding user. However, if there is no match, no data exchange occurs, indicating that the user's identity could not be verified.

In this context, the biometric fingerprint serves as a guarantee for ensuring authentication in the personal health record system shown in Fig. 4. By verifying the matching minutiae points, the system ensures that only authorized users can access and exchange sensitive health information.

Fingerprint as Identity

The minutiae features of the fingerprint are extracted with the core point reference. The numerical value thus generated is used to create the unique ID in the form of a QR code [15]. Minutiae extraction algorithm gives the value for below fingerprint.

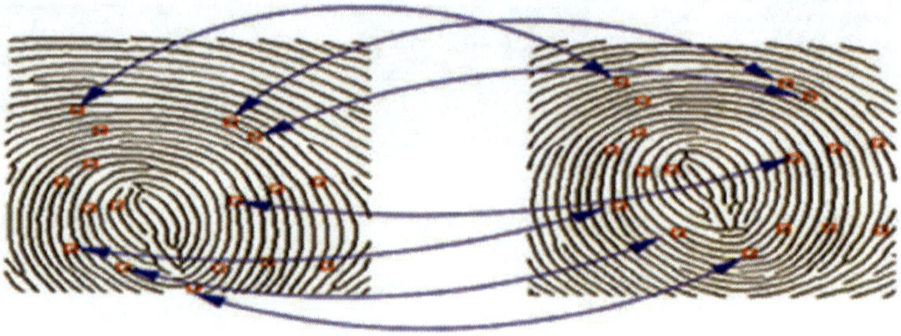

Fig. 4 Performance of minutiae matcher

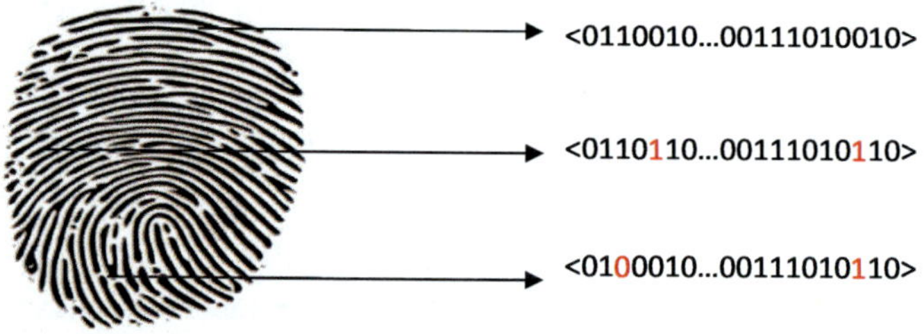

Fig. 5 Fingerprint as identity

Biometric measure changes a little each time during environment, sensors and small changes in trait are shown in Fig. 5.

Fuzzy ID-Based Encryption

The quality, efficient information offering, and authentication are characterized by the NIST framework. Matching consequences of these gatherings exhibit that the higher quality fingerprints in Fig. 6, all the more perhaps, come about a superior matching performance. Particularly for the quality number measure, the pattern that higher quality comes about better matching is clear and contrasted with the nearby quality standard, more poor matching performance fingerprints are characterized to the low-quality gathering. The great fingerprints demonstrate the authentication level of the client in PHR. Table 1 gives the unique mark results with other biometric techniques.

Efficient Data Sharing

The biometric fingerprint gives proficient data sharing to getting to personal health records. The security breakdown shows that the proposed plan is productive to safely deal with the data conveyed in the data sharing framework successfully.

Authentication

It decreases the expense for getting to the data as it could be gotten to from anyplace and at whatever time. The authentication is tolerating evidence of character given by

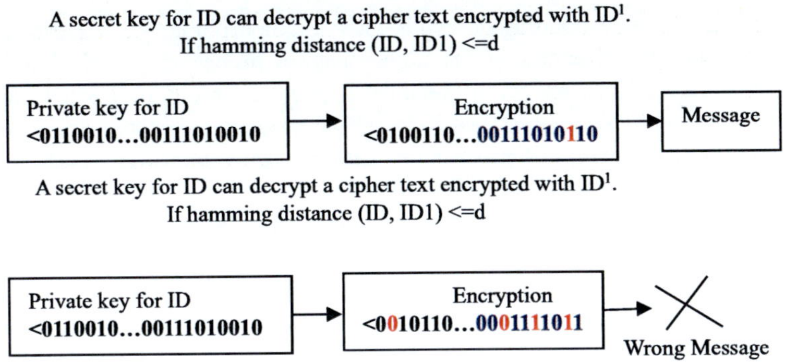

Fig. 6 Fuzzy ID-based encryption

Table 1 Different types of biometric technology (*1 (worst)–5 (best))

Technology	Accuracy	Convenience	Cost	Size
Fingerprint	5	5	4	4
Voice	1	5	5	5
Face	2	3	4	3
Hand	3	3	2	2
Iris	5	2	3	3

* represents Low

a biometric fingerprint. The methodology of distinguishing an individual normally focused on a fingerprint encryption strategy to attain the authentication. In security frameworks, authentication is distinct from authorization, which is the methodology of giving people access to framework items focused on their personality. The biometric method guarantees that approved individuals can retrieve health records (PHR) thereby preventing any unauthorized access.

5 Conclusion

The paper you mentioned introduces a concept called fuzzy identity-based encryption (FIBE), which is designed to provide an efficient platform for storing and sharing personal health records (PHR) in a cloud environment. FIBE leverages biometric encryption techniques to ensure that only authorized users can access their PHR. In traditional identity-based encryption schemes, a user's identity (such as an email address or username) is used as the public key for encryption. However, in FIBE, the notion of identity is more flexible and allows for error-tolerance. This means that slight variations or errors in the identities used to encrypt a ciphertext can still be matched with the corresponding decryption key. By incorporating biometric encryption techniques, FIBE enhances the authentication process for accessing PHRs. Biometrics, such as fingerprints or voice patterns are unique to individuals and can serve as strong authentication factors. These biometric characteristics are utilized in the encryption and decryption process to ensure that only authorized users with the correct biometric attributes can access their PHR. Overall, FIBE provides a more robust and secure approach to storing and sharing PHRs in a cloud environment. It allows for error-tolerance in identity matching and leverages biometric encryption techniques for enhanced authentication. By utilizing FIBE, better authentication mechanisms can be achieved, ensuring that only authorized individuals can access their personal health records.

References

1. Naveed G, Batool R Biometric authentication in cloud computing. J Biom Biostat 6:258. https://doi.org/10.4172/2155-6180.1000258
2. Boneh D, Franklin M (2001) Identity-based encryption from the Weil pairing. In: CRYPTO 2001, LNCS 2139. Springer, Berlin, Heidelberg, pp 213–229
3. Canetti R, Halevi S, Katz J (2003) A forward-secure public-key encryption scheme. In: Proceedings of Eurocrypt 2003. Springer
4. Selvarani P, Visu P (2015) Multi-model bio-cryptographic authentication in cloud storage sharing for higher security. Maxwell Sci Organ 10(1):95–101
5. Rinesh S, Baskaran, Thamaraiselvi K (2015) Secure and resourceful data sharing in cloud using multiple authority attribute based biometric encryption. Int J Appl Eng Res Special issue10(21):20437–20449
6. Abdolahi M, Mohamadi M, Jafari M (2013) Multimodal biometric system fusion using fingerprint and iris with fuzzy logic. Int J Soft Comput Eng (IJSCE) 2(6):504–510
7. Wagh R, Choudhary AP (2013) Analysis of multimodal biometrics with security key. Int J Adv Res Comput Sci Softw Eng 3(8):1363–1365
8. Thaiyalnayaki K, Karim SSA, Parmar PV (2010) Fingerprint recognition using discrete wavelet transform. Int J Comput Appl 1(24)
9. Pokhriyal A, Lehri S (2010) A new method of fingerprint authentication using 2d wavelets. J Theor Appl Inform Technol
10. Gaol Z et al (2011) A novel matching technique for fingerprint recognition by graphical structures. In: Proceedings of the international conference on wavelet analysis and pattern recognition, Guilin, 10–13 July 2011
11. Prasath V, Bharathan N, Neetha NP, Lakshmi N, Nathiya M (2013) Fuzzy logic in cloud computing. Int J Eng Res Technol (IJERT) 2(3). ISSN 2278-0181
12. Velciu MA, Patrascu A, Patriciu VV (2014) Biocryptographic authentication in cloud storage sharing. In: Proceeding of the 9th IEEE international symposium on applied computational intelligence and informatics, pp 165–70
13. Sahai A, Waters B (2005) Fuzzy identity-based encryption. Int Assoc Cryptol Res 2005:457–473
14. Uludag U, Pankanti S, Prabhakar S, Jain AK (2004) Biometric cryptosystems issues and challenges. Proc IEEE 92(6):948–960
15. Lewko, Waters B (2011) Decentralizing attribute-based encryption. In: EUROCRYPT: Proceedings of 30th annual international conference on the theory and applications of cryptographic techniques: advances in cryptology, pp 568–588

Quantum Convolutional Neural Network for Agricultural Mechanization and Plant Disease Detection

Musa Genemo

Abstract Agricultural research is essential in addressing the challenges of food production and meeting the needs of a growing population. Severe crop disorders can lead to food insecurity, potentially causing disastrous effects on agricultural production, such as yield losses, reduced crop quality, and even complete crop failures. By leveraging computer vision techniques, it is possible to develop systems that detect signs of crop diseases at an early stage, facilitating prompt intervention and management. Environmental changes, such as rainfall and biosolids, can cause plants to become inflamed by microorganisms. While manual observation and assessment of leaf conditions have traditionally been used for crop disease detection, there is increasing interest in leveraging technology to automate and enhance the process. Manual detection of crop diseases based on leaf conditions is subjective, time consuming, and vulnerable to errors. Utilizing computer vision methods to detect plant diseases helps improve plant health. The concept of applying quantum convolutional neural networks (QCNNs) for agricultural mechanization and plant disease detection is an interesting area of research that combines quantum computing and deep learning techniques. However, it is important to note that quantum computing is still in its early stages, and practical implementations of QCNNs for real-world applications, including agriculture, are not yet widespread. The proposed work of paper focuses on the detection of leaf blight disease by classifying leaves into late blight, early blight, and healthy leaves. This work is divided into two stages: segmentation and classification. For segmentation, MobileNetV2 is used, and an emerging quantum machine learning technique named QCNN is used for classification. The main objective is to provide and propose an effective and fast automated technique to detect and classify blight disease, which will help farmers identify blight disease easily. The proposed model achieved a classification accuracy of 0.969. Furthermore, the results were further validated against recent findings in the field. The presented outcome implies that the model's performance is remarkable.

M. Genemo (✉)
Gumushane University, Gumushane, Turkey
e-mail: musa.ju2002@gmail.com

S. Shakya et al. (eds.), *Fourth International Conference on Image Processing and Capsule Networks*, Lecture Notes in Networks and Systems 798,
https://doi.org/10.1007/978-981-99-7093-3_15

Keywords Blight · Leaf · Image processing · Quantum convolutional neural network · Early blight · Late blight

1 Introduction

Agriculture is the only option to ensure food security in both developing and developed countries, especially as the global population continues to increase rapidly. This population growth puts significant pressure on agricultural systems to produce more food efficiently, sustainably, and resiliently. To overcome this challenge, there is no doubt that agricultural mechanization is the only solution.

Crop diseases, caused by pathogens such as fungi, bacteria, viruses, and nematodes, can result in significant yield reductions. They can infect plants at various stages of growth, affecting their development, growth, and reproductive capacity. Yield losses due to crop diseases can range from moderate to severe, depending on the pathogen, the crop's susceptibility, and environmental conditions.

Therefore, using machine learning in agriculture to detect diseases in plants early on is just as crucial as detecting diseases in humans. Plants can become infected by fungi, bacteria, and viruses due to changes in environmental factors such as rainfall, temperature, and soil fertility [1]. To effectively manage such farm activities, it is essential to use a machine learning mechanism that can detect plant diseases at an early stage.

Agricultural research focuses on developing innovative techniques, technologies, and practices to enhance crop yields, improve livestock production, and increase overall agricultural productivity. Here are a few key reasons why agricultural research is essential for meeting the requirements of an increasing population.

The concept of applying QCNNs for agricultural mechanization and plant disease detection is an interesting area of research that combines quantum computing and deep learning techniques. However, it is important to note that quantum computing is still in its early stages, and practical implementations of QCNNs for real-world applications, including agriculture, are not yet widespread. Machine learning makes judgments based on previous experience [2].

Plant disease detection often involves analyzing complex patterns and structures within images or genomic data. Traditional convolutional neural networks (CNNs) have proven effective in this regard, but QCNNs may offer the advantage of efficiently handling larger and more complex datasets. QCNNs could potentially capture intricate relationships within the data, leading to improved disease detection accuracy. Quantum computing has the potential to provide significantly higher computational power compared to classical computing. QCNNs could leverage this power to process and analyze large-scale agricultural data, such as high resolution imagery or genetic data, more efficiently. This could lead to faster and more accurate detection of plant diseases and better decision-making in agricultural mechanization.

In agriculture, computer vision has gained significant attention and shown great promise in various applications. CNNs can analyze plants and crops to identify disease symptoms accurately. By training CNN models on large datasets of labeled images, they can learn to recognize specific visual patterns associated with different diseases [3–5]. This enables automated and rapid disease detection, allowing farmers to take timely action to manage and control diseases effectively.

CNNs can be used to distinguish between crops and weeds in images, facilitating targeted weed management strategies. By analyzing field images, CNN models can identify and differentiate weed species from crops, enabling precision herbicide application or mechanical weed removal techniques [6, 7]. This helps reduce the use of herbicides and increase overall crop yield [8, 9].

Computer vision techniques, combined with machine learning algorithms, can analyze aerial or satellite images to estimate crop yield and monitor crop growth [10]. By extracting features such as vegetation indices, plant density, or canopy cover from images, CNN models can provide insights into crop health, growth stage, and potential yield. This information assists farmers in making informed decisions regarding irrigation, fertilization, and harvesting.

The application of CNNs and computer vision methods in agriculture is continuously evolving and has the potential to revolutionize farming practices. By leveraging the power of deep learning and image analysis, these technologies contribute to improved crop management, increased productivity, and more sustainable agricultural systems.

This study focuses on utilizing a quantum machine learning technique called QCNN to recognize plant infections. This approach combines the power of quantum computing and convolutional neural networks to address the challenge of disease detection in plants. We collected a diverse dataset called "Plant Village" [11], consisting of images and other relevant data related to different plant diseases. This dataset serves as the training data for the QCNN model.

QCNNs play a crucial role in agricultural mechanization and plant disease detection, improving accuracy and efficiency. This approach expands the horizons of quantum machine learning and showcases the versatility of quantum technologies beyond traditional domains. By addressing challenges in plant disease detection, QCNNs pave the way for future advancements in hardware, software, and algorithms, ultimately revolutionizing agricultural practices and enhancing productivity.

The study aims to develop a QCNN model specifically tailored for plant disease detection. QCNNs utilize quantum computing principles to enhance the computational power and potential accuracy of CNNs. The model will be designed to process the training dataset and learn the complex patterns and features associated with different plant diseases.

The rest of this paper is organized as follows: Sect. 2 provides a review of relevant research that focuses on segmentation and detection techniques for plant diseases. It would present previous studies, methodologies, and findings related to the topic, highlighting the existing research landscape and advancements in the field. Section 3 provides methodologies used for segmentation and detection of plant diseases. It would detail the specific techniques, algorithms, or approaches employed, such as image processing methods, machine learning models, or other relevant methodologies utilized for disease segmentation and detection in plants. Section 4 includes quantitative metrics, visual representations, or comparisons with ground truth data or existing approaches. Section 5 provides a concise summary of the paper's objectives, methodologies, results, and overall importance in advancing plant disease management.

2 Related Work

In this work, the most recent and pertinent research was chosen for this investigation using a systematic review methodology. Many papers were published in international journals and conferences. Researchers are working on hierarchical networks, neural networks, and device domain methods. Some classification methods that researchers consistently use are support vector machines (SVM), neural networks, traditional regression pairs, k-nearest neighbors, and genetic algorithms. Babu et al. proposed a backpropagation neural network for leaf recognition [12, 13]. Zhang et al. worked on cucumber disease detection through sparse representation classification. This work focuses on leaf images, where the photograph is segmented into small segments using k-means clustering [14].

The disease classification is performed with the aid of sparse matrices. The results of the proposed approach demonstrate that it reduces computation time and improves performance, and the results of this method are comparable to alternative techniques [15]. Another proposed approach for leaf disease detection utilizes the concepts of image segmentation and soft computing strategies. This approach automatically classifies and detects diseases in leaves. Surveys of various diseases related to plants are also used for detection. This work is done with the assistance of a genetic algorithm, which efficiently optimizes the variables. R-CNN mask is one of the methods [16]. This method won the 2016 COCO competition.

The R-CNN mask extends the Faster R-CNN network architecture and classifies each object in the image by adding a mask. The R-CNN mask can be used to find the target in pixels. This is an instance-based segmentation model. Instance segmentation refers to classifying every part of the photograph. Bhangea [17] developed a new method, which is a network-based problem-solving tool. This method adjusts the size of the first image uploaded to the Web portal and extracts image features, including shadows, shapes, etc. The author uses K-Tools to group and classify SVMs [18]. Many traditional AI techniques are used to detect diseases, including support vector machines (SVM) [19, 20], and K-Tools [21], and decision trees. Vision and object detection, pest detection, and the use of deep learning techniques have become active research topics in recent years. The Normalized Difference Vegetation Index, a strategy for generating heterogeneous data from NDVI, was suggested [22].

In many of the articles on image-based plant disease identification, the background is often removed, and the lesion tissue of the diseased plant is segmented. Discriminative features are extracted to further evaluate plant illness, and a model is developed for disease classification. Specific threshold values are used as suggested by Al-Hiary et al. [23, 24].

3 Materials and Methods

This section provides a detailed description of the materials and methodologies employed in the development of a quantum convolutional neural network for agricultural mechanization and plant disease detection. The methodology used in the study is presented in the "Segmentation" section below.

3.1 Dataset

The dataset was collected as a diverse collection of images and other relevant data pertaining to various plant diseases. Most of the datasets used in this work are publicly available datasets for plant disease detection. For this purpose, we collected some datasets from the Plant Village dataset, which contains over 54,000 images of healthy and diseased plant leaves, covering 14 different crop species and 26 diseases. Additionally, some datasets were obtained from the Plant Pathology 2020 dataset, which consists of over 18,000 images of apple tree leaves with labels for multiple diseases and health conditions. A portion of the data was also taken from the Citrus Fruits dataset, which includes over 300 images covering diseases such as citrus canker, citrus greening, and citrus scab. Table 1 displays the dataset used for training, which includes the total number of samples. Figure 1 shows the sample Plant Village dataset images.

This dataset was used as the training data for the QCNN model. We utilized more than 4000 datasets for training the model, and out of these, 20% were reserved for evaluating the effectiveness of the model. Publicly available datasets were organized and manually labeled based on image quality.

In addition, other datasets such as the Leafsnap dataset [13], the Cassava Leaf Disease [22] dataset, the Merged dataset, and the Tomato Disease dataset [20] were

Table 1 Dataset used for training

Disease	Crop	Total sample used
Leaf spot	Tomato	650
Powdery mildew	Grapes	300
Rust	Wheat	400
Bacterial blight	Rice	350
Anthracnose	Pepper	500
Healthy	Apple	300
Leaf curl	Tomato	500
Black rot	Citrus	500
Early blight	Potato	200
Fusarium wilt	Banana	300

Fig. 1 Plant village dataset [13]

also utilized. The model is specifically tailored for plant disease detection. QCNNs utilize quantum computing principles to enhance the computational power and potential accuracy of convolutional neural networks. The model would be designed to process the training dataset and learn the complex patterns and features associated with different plant diseases.

3.2 Feature Extraction

To obtain representative features from the input data in the plant disease detection dataset, the process starts by resizing the images and normalizing pixel values. Then, the Quantum Circuit Born Machine (QCBM) model is applied to extract features. QCBM is used to generate samples from a probability distribution to obtain optimal parameters. To reduce the cost function, quantum gradient descent is applied using the equation:

$$\Delta\alpha = -\eta * G^{-1} * m \tag{1}$$

Here, $\Delta\alpha$ represents the change in parameters, η is the learning rate, G^{-1} is the inverse information matrix, which quantifies the curvature of the cost function surface, and m is the gradient of the cost function.

Then, the model was trained based on these labeled images. Principal component analysis (PCA) was applied to extract representative features from the input data. Through an iterative process, the model's parameters would be adjusted to minimize

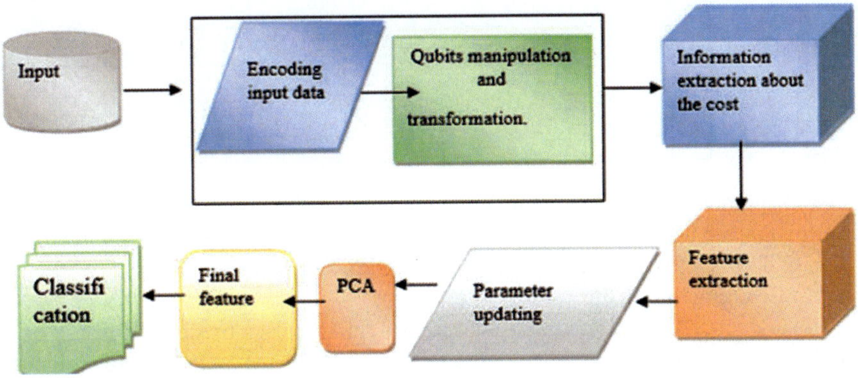

Fig. 2 Feature extraction procedure

the detection errors and improve its performance. Figure 2 illustrates the feature extraction procedure.

3.3 Segmentation

In this work, region-based segmentation is employed to group the color and texture of the leaf. This approach allows the pixels to be grouped into regions based on the healthiness of the leaf. To achieve this, the similarity function is applied to compare the similarity between the leaf's color and texture using cosine similarity.

Image segmentation is performed using the MobileNetV2 model, and classification is performed using the QCNN. The flow of leaf diseases detection typically involves several steps, which can vary depending on the specific approach used. Figure 3 indicates flow of leaf disease detection by using a QCNN.

The first step involves collecting a dataset of leaf images that represent a variety of healthy and diseased plants. The collected leaf images then undergo preprocessing steps to enhance their quality and remove noise. This may involve resizing, normalization, filtering, or other necessary image processing techniques as discussed earlier.

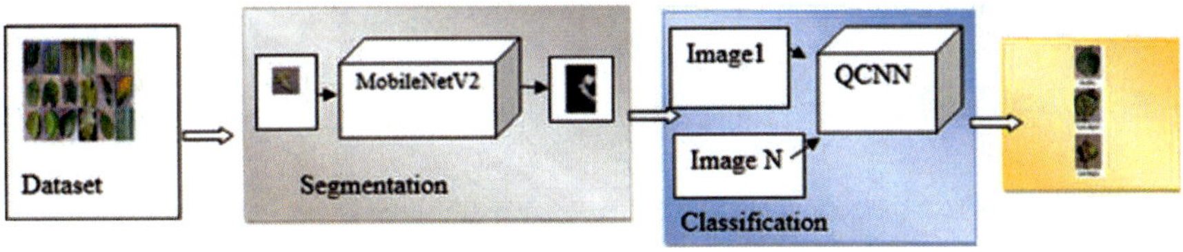

Fig. 3 Flow of proposed model

Table 2 Parameters of MobileNetV2 model

Parameters	Layers	Total parameters
Trainable parameters	53	409,225
Non-trainable parameters	53	7184

Next, features are extracted from the preprocessed leaf images to capture relevant information for disease detection. This step involves analyzing the shape, texture, color, or other characteristics of the leaves. Texture analysis was specifically utilized for feature extraction in the study.

The extracted features are utilized to train a classification model, which can be a machine learning algorithm or a deep learning model. The model learns to distinguish between healthy and diseased leaves based on the extracted features. The training data is labeled with the corresponding disease classes. In the approach, a pre-trained MobileNetV2 model was utilized for segmentation, which underwent fine tuning for the specific task.

The labeled training data is then used to train the classification model. This process involves optimizing the model's parameters to minimize classification errors and improve its performance. The training process typically includes iterations over the training data, adjusting the model's weights or parameters using techniques like gradient descent or backpropagation.

After training the model, a separate set of labeled test data, which was not used during the training phase, is employed to evaluate the performance of the trained model. The evaluation is based on various criteria such as accuracy, precision, recall, or F1-score. This evaluation process assesses the model's ability to accurately detect and classify leaf diseases.

The following Table 2 is a representation of our MobileNetV2 model definition, along with information on the total parameters used, trainable parameters, and none trainable parameters.

During the training and testing of our image segmentation model, we selected a relatively low number of epochs, specifically 20. Despite this limited number of iterations, the model achieved satisfactory results with high accuracy. The obtained results from the image segmentation model are saved and subsequently utilized in our next model, QCNN. The QCNN model is trained using the saved results from the image segmentation model, enabling it to learn the complex patterns and features associated with different leaf diseases.

3.4 *Classification*

Plant diseases classification involves categorizing different types of diseases or disorders that affect plants based on their visual symptoms or characteristics. The QCNN architecture is used as the training model. The input data is preprocessed classically, encoded into a quantum format, processed through a parameterized quantum circuit,

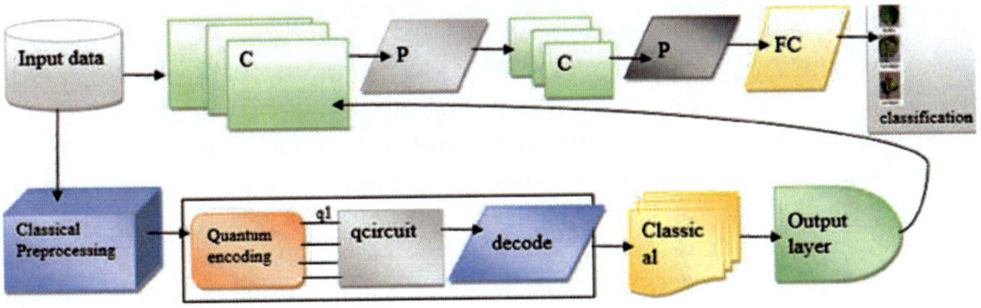

Fig. 4 QCNN training model setup

Table 3 Parameters used in the model

Number of Q bits	04
Learning rate	0.0004
Batch size	0.04
Number of V layers	0.6
Q weight	0.01
Max epochs	20

post-processed classically, and finally passed through a convolutional layer to reduce the dimensions of features using a pooling layer. Information from previous layers is integrated using a fully connected network and mapped for classification purposes. Once the model is trained and evaluated, it can be used for disease detection on unseen leaf samples. The model predicts the diseases using attributes extracted from the new leaf images as input. The result could be in the form of disease labels, probabilities, or severity scores. Additional post-processing was also applied to the detection results, such as thresholding, filtering, and noise removal, to refine the output. In the quantum circuit shown in Fig. 4, the chosen configuration involved six convolutional layers and a batch size of 4, indicating the utilization of four samples for each training step.

The QCNN network was trained with 4 quantum bits, initialized with random quantum weights of 0.1. A learning rate of 0.0004 per step was used, and a reduction in the learning rate was applied after every 10 epochs with a gamma_lr_scheduler value of 0.1. The parameters used in the model are listed in Table 3.

4 Results and Discussion

It is important to note that the specific techniques, algorithms, or models used in plant diseases classification may vary depending on the available data, the complexity of the problem, and the desired accuracy. Additionally, advancements in deep learning, such as the use of CNNs, have shown promising results in plant disease classification tasks, providing high accuracy and the potential for automated detection and monitoring. Table 4 presents an accomplishment of the segmentation model. In which accuracy, loss, validation, precision, recall, and dice-coef of the model are mentioned.

With the batch size set to 4 in the parameters section, four images are plotted in the output section, indicating that the number of images displayed corresponds to the batch size [num_images = batch size]. Table 5 presents the performance evaluation of the classification model. The evaluation measures used to assess the effectiveness of the model include metrics such as accuracy, precision, recall, and F1-score. The result underlines that performance metrics with those reported in previous studies or benchmark datasets are more accurate.

4.1 State of the Art

In comparison with previous works, the proposed model has achieved significant and fruitful results. By employing quantum convolutional neural networks (QCNNs) on publicly available data, the proposed approach can accurately classify the class of a leaf disease. Table 6 showcases the performance of the proposed work in comparison with previous studies, demonstrating superior accuracy in plant disease detection with a score of 0.966. These results emphasize the potential effectiveness and reliability of the proposed methodology using QCNNs for plant disease detection.

5 Conclusion

The main issues faced while detecting leaf blight disease in the past were the classification of leaf blight disease according to its early and late blight classes. It is hard to classify the blight disease and its stages. Although the work in this field is enormous, this work addressed classification issues by using quantum neural network, which is new in this field. We have achieved high accuracy in our classification, which is good for future work in the disease detection field. Results show that our proposed system results in good, segmented images and high accuracy in the classification of those segmented images. In the future, the objective is to develop software that enables fully automated image processing, encompassing tasks ranging from segmentation to classification.

Table 4 Performance evaluation of segmentation

Loss%	Precision %	Recall%	Dice-coef %	Val-loss %	Val-recall %	Val-precision %	Val-accuracy %	Accuracy %
0.04	0.93	0.77	0.76	0.4	0.10	0.91	0.92	0.96

Table 5 Performance evaluation result

Dataset	Accuracy	Loss
Plant Village	94.17%	5.83%

Table 6 Comparison with state of the art

Model	Performance	Loss
QCNNs	0.966	0.034
CML	0.910	0.09
CNN	0.932	0.068
SVM	0.891	0.109
KNN	0.912	0.088

References

1. Varsha VS, Misra AK (2015) Detection of unhealthy region of plant leaves using image processing and genetic algorithm. In: International conference on advances in computer engineering and applications (ICACEA), IMS Engineering College, Ghaziabad, India, pp 1028–1032
2. Amara J, Bouaziz B, Algergawy A et al (2017) A deep learning-based approach for banana leaf diseases classification. In: BTW (Workshops), pp 79–88
3. Abade A, Ferreira PA, de Barros Vidal F (2021) Plant diseases recognition on images using convolutional neural networks: a systematic review. Comput Electron Agric 185
4. Dhaka VS, Meena SV, Rani G, Sinwar D, Ijaz MF, Woźniak M (2021) A survey of deep convolutional neural networks applied for prediction of plant leaf diseases. Sensors 21:4749
5. Nagaraju M, Chawla P (2020) Systematic review of deep learning techniques in plant disease detection. Int J Syst Assur Eng Manag 547–560
6. Lu J, Tan L, Jiang H (2021) Review on convolutional neural network (CNN) applied to plant leaf disease classification. Agriculture 11:707
7. Bangari S, Rachana P, Gupta N, Sudi PS, Baniya KK (2022) A survey on disease detection of a potato leaf using CNN. In: Proceedings of the 2nd IEEE international conference on artificial intelligence and smart energy (ICAIS), Coimbatore, India, vol 23, pp 144–149
8. Naik BN, Ramanathan M, Palanisamy P (2022) Detection and classification of chilli leaf disease using a squeeze-and-excitation-based CNN model. Ecol Inform 69:101663
9. Sahu P, Chug A, Singh AP, Singh D, Singh RP (2021) Deep learning models for beans crop diseases: classification and visualization techniques. Int J Mod Agric 10:796–812
10. Santosh A et al (2018) Tomato plant diseases detection system. Plant village.org web site visited date august (2020)
11. Mukti IZ, Biswas D (2019) Transfer learning based plant diseases detection using ResNet50. In: Proceedings of the 4th IEEE international conference on electrical information and communication technology (EICT), Khulna, Bangladesh, 20–22 Dec, pp 1–6
12. Arya S, Singh R (2019) A comparative study of CNN and AlexNet for detection of disease in potato and mango leaf. In: Proceedings of the IEEE international conference on issues and challenges in intelligent computing techniques (ICICT), Ghaziabad, India, 27–28 Sept, vol 1, pp 1–6
13. Ganatra N, Patel A (2018) A survey on diseases detection and classification of agriculture products using image processing and machine learning. Int J Comput Appl 180(13)
14. Zhang S, Zhang L, Wu X, You Z (2018) Leaf image-based cucumber disease recognition using sparse representation classification, vol 18
15. Sing V, Misra AK, He K, Gkioxari G, Dollar P, Girshick R (2017) Detection of plant leaf diseases using image segmentation and soft computing techniques. 59:100–112

16. Bhangea M (2015) Smart farming: pomegranate disease detection using image processing. 58:280–288. 12
17. Abayomi-Alli OO, Damaševičius R, Misra S, Maskeliūnas R (2021) Cassava disease recognition from low-quality images using enhanced data augmentation model and deep learning. Exp Syst 38:e12746
18. Mokhtar U, Ali MA, Hassanien AE, Hefny H (2015) Identifying two of tomatoes leaf viruses using support vector machine. In: Mandal JK, Satapathy SC, Sanyal MK, Sarkar PP, Mukhopadhyay A (eds) Information systems design and intelligent applications. Springer, pp 771–782
19. Arivazhagan S et al (2015) Detection of unhealthy region of plant leaves and classification of plant leaf diseases using texture features. Agri Eng Inter CIGRJ 15:211–217
20. Shrestha G, Deepsikha, Das M, Dey N (2020) Plant disease detection using CNN. In: 2020 IEEE applied signal processing conference (ASPCON), pp 109–113
21. (2018) Image set for deep learning: field images of maize annotated with disease symptoms. BMC Res Notes 11
22. (2019) Three-channel convolutional neural networks for vegetable leaf disease recognition. Cogn Syst Res 53:31–41
23. Atila M, Uar M, Akyol K, Uar E (2021) Plant leaf disease classification using EfficientNet deep learning model. Ecol Inform 61:101182
24. Safari B, Alborzi Y, Najafi E (2022) Automated wheat disease detection using a ROS-based autonomous guided UAV. arXiv preprint arXiv:2206.15042

Innovative Method for Alzheimer's Disease Detection Using Convolutional Neural Networks

Md. Sajid Anam Ifti, Md. Redwan Ahmed, S. M. Arafat Rahman, Sheikh Shemanto Afridi, Sanjeda Sara Jennifer, and Ahmed Wasif Reza

Abstract Deep learning (DL) approaches have gained significant traction in recent years for medical image analysis. In this research, magnetic resonance imaging (MRI) data were employed to classify Alzheimer's disease (AD) using DL techniques. Alzheimer's disease (AD), a fatal neurological condition, is characterized by early signs such as memory loss and cognitive decline. Accurate and timely diagnosis is crucial for appropriate patient treatment and subsequent care. Unlike traditional feature learning methods like machine learning, deep learning approaches can directly learn complex and high-level features from datasets. However, this research extends the classification procedure by incorporating forecasting for additional disease stages. The proposed approach obtained an impressive accuracy of 98.02% using the XCeption model. The automatic extraction of intricate features from MRI data empowers improved diagnostics, proactive disease management, and personalized treatment strategies for individuals affected by Alzheimer's disease.

Keywords Alzheimer's disease · Deep learning · XCeption · Convolutional neural network · Magnetic resonance imaging

1 Introduction

Alzheimer's disease (AD), an autoimmune neurological disorder, is spurred on by nerve tissue degradation in certain brain regions. The first indication of cognitive impairment in a patient is their incapacity to do crucial operations due to memory loss,

Md. Sajid Anam Ifti · Md. Redwan Ahmed · S. M. Arafat Rahman · S. S. Afridi · S. S. Jennifer · A. W. Reza (✉)
Department of Computer Science and Engineering, East West University, Dhaka 1212, Bangladesh
e-mail: wasif@ewubd.edu

Md. Sajid Anam Ifti
e-mail: info@sajidifti.com

S. Shakya et al. (eds.), *Fourth International Conference on Image Processing and Capsule Networks*, Lecture Notes in Networks and Systems 798,
https://doi.org/10.1007/978-981-99-7093-3_16

speech impediments, and other cognitive problems [1]. The World Health Organization (WHO) ranks it the sixth most prevalent global cause of adult death. According to the Alzheimer's Association, the sixth most prevalent factor leading to the end of a person's life in the USA is due to AD [2]. Logic, attention, and memory are all affected when the patient's memory area of the brain shrinks and matures. This is the main reason for this disease. Genetic mutations, which affect 1% or fewer people, are suggested to be another possible cause of AD [3]. To identify AD and aid in diagnostics, feature wavelets were traditionally constructed using the discrete wavelet transform (DWT) technique. Additional investigation is required [4] using machine learning techniques since it is difficult to identify injuries. Regardless of elements, representatives, or populations, cooperative approaches have a direct influence on most optimization strategies. In the initial rounds, the genetic algorithm (GA), particle swarm optimization (PSO), and Tsallis Kapur's entropy were the optimization techniques that were employed [5]. Following that, the bat, bees, and cuckoo search (CS) metaheuristic algorithms were recommended [6]. While establishing a conclusive assessment of AD presents challenges, in practical terms, the diagnosis primarily relies on the assessment of clinical history and neuropsychological information, including the utilization of magnetic resonance imaging (MRI) [7]. The gray wolf optimization (GWO) strategy, which is focused on leadership attributes in method identification, was also promoted to enhance global search [8]. Part of the reason this algorithm is getting so much attention is due to its aggressive discoveries, its increased capacity for problem-solving, and its ability to offer an optimum solution superior to previous optimization techniques [9, 10].

In deep learning problems, the custom convolutional neural network (CNN) excels remarkably. The outcomes achieved in image-based applications, such as the remarkable progress in the largest image classification dataset (ImageNet), as well as advancements in natural language processing (NLP), and computer vision have been truly impressive [11]. Using custom CNN in conjunction with existing CNN models was shown to produce the best outcomes in this study.

The rest of the paper is organized in the following manner: Sect. 2 presents the related work, highlighting previous studies and relevant work. The materials used and the methods employed are outlined in Sect. 3. Section 4 exhibits the results obtained from the study and provides in-depth discussions and interpretations of the findings. Section 5 serves as the conclusion of the paper, presenting a concise overview of the primary findings, examining their significance, and proposing potential paths for future investigation.

2 Related Work

Deep learning was used to assess Alzheimer's disease using brain MRI data in [1]. The paper introduces CNN and Google Net as methods for analyzing the OASIS dataset, emphasizing their effectiveness in dealing with small medical images and emphasizing the need for further investigation of unsupervised and semi-supervised procedures as potential research gaps or limitations.

Learning algorithms have been tested to detect Psoriatic arthritis using brain MRI scans from both normal and those with Alzheimer's disease. Deeper learning methods could be used to develop an effective device for advanced Alzheimer's pathogen detection [2].

Deep learning algorithms can be used to automatically generate Alzheimer's diagnoses based on MRI data, which could help improve premature detection and diagnosis, leading to more effective therapy options. The study article employs CNN as a technique and the ADNI dataset to provide a considerable contribution to the accuracy of AD identification, with a specificity of 0.93. However, the report notes the need for more research to improve CNN's effectiveness in this setting [8].

Deep learning techniques are used to diagnose early Alzheimer's disease, which could lead to faster and more economical remedies. ConvNet is used as a method in the study article, and the ADNI dataset is used to illustrate its superiority over conventional approaches, providing a substantial contribution. The study, however, emphasizes the need for more research to address existing research gaps and limitations [9].

Deep learning algorithms can be used to isolate people with early-stage Alzheimer's disease from those without it with extreme accuracy, potentially leading to more efficient remedies. The research study employs the DCssCDBM technique using the easy monitor technology dataset to outperform various sophisticated models. However, the report underlines the need of investigating DCssCDBM's potential to overcome existing research gaps and constraints [10].

Deep learning (DL) algorithms can be used to analyze data, but there are still obstacles to be addressed. However, a useful tool for preventing and analyzing AD. The study employs machine learning (ML) and DL approaches, as well as the Neuroimaging Informatics Technology Initiative (NIFTI) and Open Access Series of Imaging Studies (OASIS) datasets, to provide a significant contribution through biomarker and neuroimaging analyses. However, the report admits the presence of congenital limits and underlines them as research gaps and limitations [12].

A DL system was developed to utilize fluorine 18 fluorodeoxyglucose PET brain scans for predicting the eventual diagnosis of AD. The algorithm beat radiologic readers in the ROC domain, with an average of 75.8 months before the definitive diagnosis. The program outperformed radiologic readers with 82% specificity and 100% sensitivity. The saliency map exhibited a notable emphasis on recognized regions of interest by directing attention to the entirety of the brain [13].

The study addresses Alzheimer's disease detection by segmenting brain subregions with optimization methods, reaching 98% accuracy with deep learning classifiers and 95% accuracy with conventional classifiers. The hippocampus is vital, exceeding other areas in terms of accuracy, sensitivity, and specificity [5].

Early detection of AD is crucial for treatment; machine learning methods face challenges like class imbalance and data leakage [14]. This paper reviews conventional ML methods for classifying and predicting AD using MRI, discussing challenges like class imbalance and data leakage [15]. The study's goal is to use deep

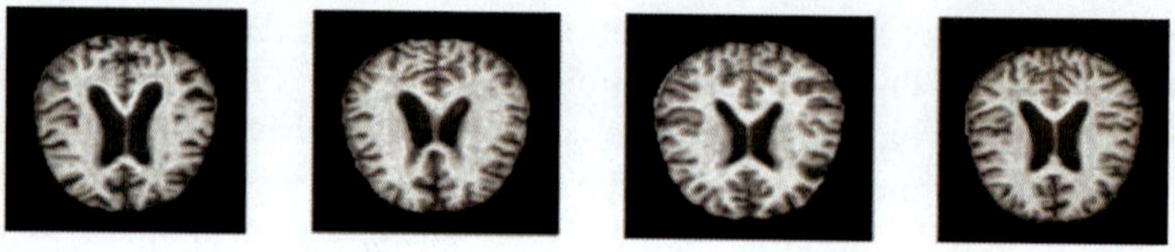

Fig. 1 Sample images from each class

learning and MRI to identify AD and moderate cognitive impairment. It assesses CNN models, adds data augmentation, and creates visually understandable heatmaps. The method outperforms previous methods by emphasizing damaged areas such as the lateral ventricle and brain [16].

3 Materials and Methods

3.1 Dataset Description

This study employed a dataset from Kaggle [17], which included precisely gathered MRI pictures from several sources with confirmed annotations. The collection contains 6400 pictures, each of which depicts a different stage or feature of Alzheimer's disease (AD), such as mild demented, moderate demented, non-demented, and very mild demented. The major goal of publishing this information is to aid in the creation of a very accurate algorithm for predicting the stage of Alzheimer's disease. Figure 1 depicts a graphic representation of an exemplary image from each category.

3.2 Dataset Preprocessing

In order to prepare the dataset for further processing, the images need to go through a series of essential steps, including scaling, normalizing pixel values, and dividing the data into two sets. The images were scaled to consistent dimensions of 64 × 64 and 71 × 71 pixels, ensuring uniformity in input size across the dataset. This scaling process establishes a standardized framework for subsequent analysis. Furthermore, to bring the intensity values within a desirable range of 0–1 pixel, value normalization was employed. By normalizing the pixel values, the data is rescaled to a consistent scale, facilitating effective comparison and analysis across images. This normalization step plays a crucial role in ensuring that the model can learn meaningful patterns from the data. The provided code uses the rectified linear unit (ReLU) activation function in the convolutional layers of the model for Alzheimer's disease classification. ReLU

Table 1 Image scaling size of different models

S. No.	Model name	Size
01	Custom CNN	64 × 64
02	ResNet-50	64 × 64
03	XCeption	71 × 71

introduces nonlinearity, enabling the model to learn complex patterns and features from the input images. By setting negative values to zero, ReLU enhances the model's ability to capture intricate relationships and extract relevant features. Applying ReLU in the convolutional layers helps the model learn and extract meaningful features, leading to improved performance in AD classification.

To evaluate the performance of the model and assess its generalizability, the dataset was divided into two sets: a train and a test set. The 'train_test_split()' function from the 'sklearn.model_selection' module was utilized, assigning 80% of the data for training purposes and utilizing the remaining 20% for testing. This partitioning allows for the evaluation of the model's performance on previously unseen data, providing a measure of its efficacy in real-world scenarios. Moreover, the input data was reshaped to include the batch size dimension, which is a fundamental requirement for feeding data into convolutional neural network (CNN) models. Reshaping the data ensures compatibility with the expected input format of the CNN, specified as '(batch_size, height, width, channels)'. In this case, the channel dimension is set to 1, reflecting the grayscale nature of the images. By diligently performing these preprocessing steps, the data is appropriately prepared, ensuring consistency, facilitating effective analysis, and enabling the subsequent training and evaluation of the deep learning models. Table 1 summarizes the transformations applied to the images for each respective model, encapsulating the alterations undertaken in the pre-processing pipeline.

3.3 Model Architecture and Design

This section presents the overall workflow of the system architecture and the models that were chosen to be employed for this research. The effectiveness of various CNN designs in identifying AD is evaluated in this research. There are 50 epochs per fold, and a total of five-fold cross-validation was employed for training the dataset. A deep learning framework, TensorFlow, was employed for model training and evaluation, and performance evaluation was conducted using metrics such as accuracy, precision, recall, and $F1$-score. Custom CNN had the best accuracy and training time. Figure 2 shows the entire workflow of the proposed planned methodology involving every step, followed by the algorithm description.

This research decided to experiment with custom CNN, ResNet-50 architecture, and Xception architecture. ResNet-50 and Xception are two prominent architectures that are widely used in medical image analysis due to their superior performance and

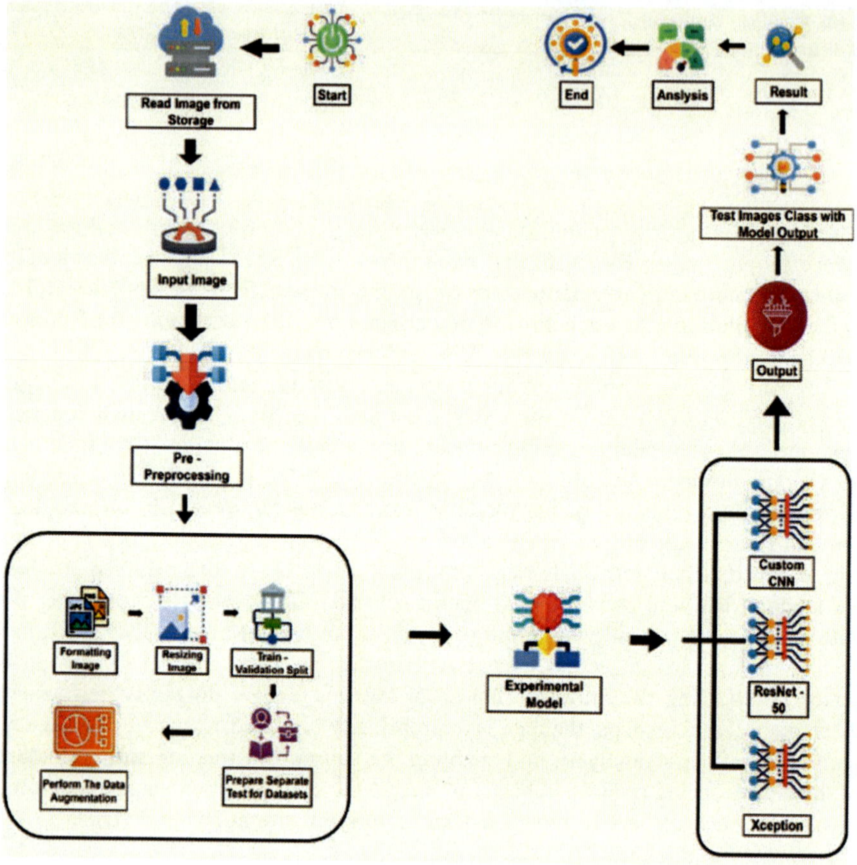

Fig. 2 Hierarchical overview of the proposed methodology

accessibility to pre-trained models. The computational complexity is low due to the depthwise separable convolutions and residual connections.

Custom CNN: The Keras framework was used to create the first CNN, which was trained for 50 iterations per fold using training data. Figure 3 shows the custom network of the CNN model, prepared for implementation.

ResNet-50: The CNN model was created using pre-trained ResNet-50, global average pooling, dropout regularization, accuracy metric, and Adam optimizer. Figure 4 represents the ResNet50 architecture.

XCeption: We created a CNN model based on the pre-trained XCeption model with global average pooling and dropout regularization. Figure 5 illustrates the proposed custom XCeption model.

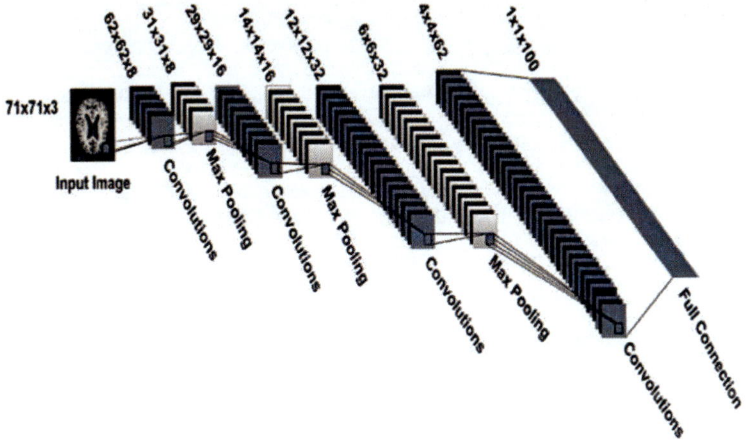

Fig. 3 Custom CNN model

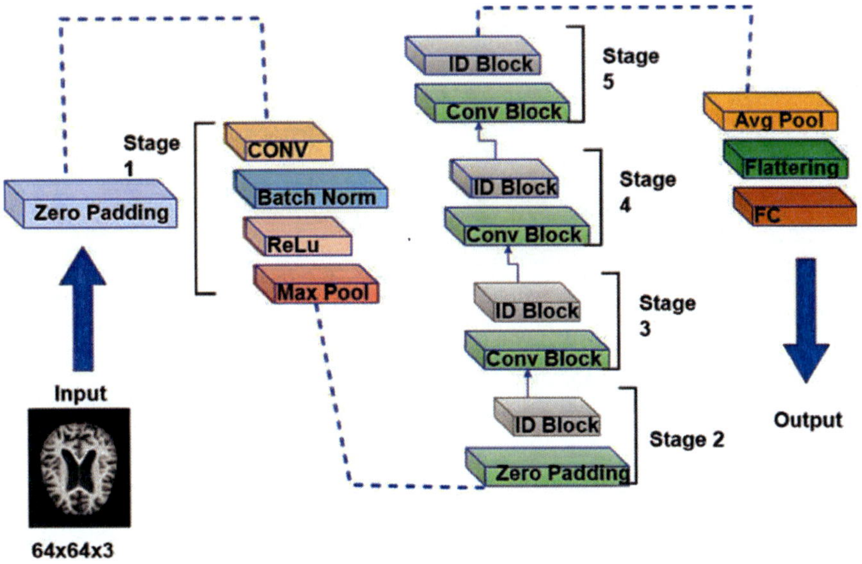

Fig. 4 ResNet50 model

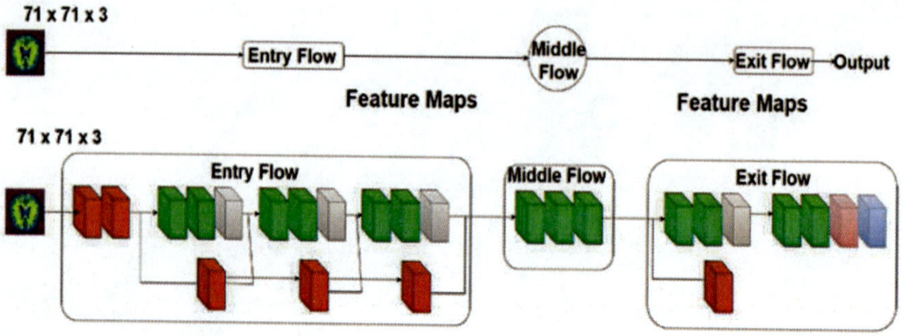

Fig. 5 Proposed custom XCeption model

4 Result Analysis and Discussion

4.1 Experimental Setup

The proposed system was built on Jupyter Notebook, a computing platform with an 8-core AMD CPU, 1TB of storage, 16 GB of RAM, and AMD GPU RX Vega 11. Models were built using TensorFlow and Keras deep learning frameworks.

4.2 Result Analysis and Performance Evaluation

A system using 6400 Alzheimer pictures was developed utilizing custom CNN, ResNet-50, and XCeption. Except for XCeption, which used 71 × 71 input vectors, other models used 64 × 64 input vectors. Each model was trained for 50 training epochs before being tested for accuracy and loss. In order to evaluate the performance of our model, precision, recall, $F1$-score, and accuracy were chosen as the evaluation matrices [18]. Their mathematical formulation is presented in Eqs. (1), (2), (3), and (4), and Table 2 gives the performance analysis:

$$\text{Precision} = \frac{\text{true positive}}{(\text{true positive} + \text{false positive})} \tag{1}$$

$$\text{Recall} = \frac{\text{true positive}}{(\text{true positive} + \text{false negative})} \tag{2}$$

$$F1\,\text{Score} = \frac{2 \times (\text{precision} \times \text{recall})}{(\text{presion} + \text{recall})} \tag{3}$$

$$\text{Accuracy} = \frac{\text{true positive} + \text{true negative}}{\text{true positive} + \text{true negative} + \text{false positive} + \text{false negative}} \tag{4}$$

Apart from the precision, recall, $F1$-score, and accuracy, confusion matrices that use real labels and predicted values are also used to evaluate the model, as shown in Figs. 6, 7 and 8. Five-fold cross-validation was performed on the dataset by splitting it into five equal folds. The model is trained using four folds, with the final fold being utilized for validation. Once for each fold, the validation set is used in this procedure five times. To produce an overall performance statistic, the iteration outcomes are averaged. This method aids in determining how well the three models can generalize and identify overfitting or underfitting. In making decisions about model selection and optimization, it offers a trustworthy approximation of the model's performance. A graphical representation of the models after performing five-fold cross-validation is illustrated in Figs. 9, 10 and 11.

Figure 6 illustrates the performance of a custom CNN model in classifying the severity of dementia. With an overall accuracy of 94.4%, it faces challenges in accurately categorizing instances within the ModerateDemented and NonDemented categories. This arises due to the higher similarity in symptoms exhibited by these two categories compared to others.

The dementia diagnosis model depicted in Fig. 7 achieved an overall accuracy of 96%. However, it encountered challenges in accurately classifying instances as ModerateDemented and VeryMildDemented, with accuracies of 75% and 68.8%,

Table 2 Performance analysis of the architectures

Model	Precision (%)	Recall (%)	Avg. $F1$-score (%)	Test loss	Test accuracy (%)
Custom CNN	94	94	94	0.0767	94
ResNet-50	96	96	96	0.0132	96
XCeption	98	98	98	0.0178	98

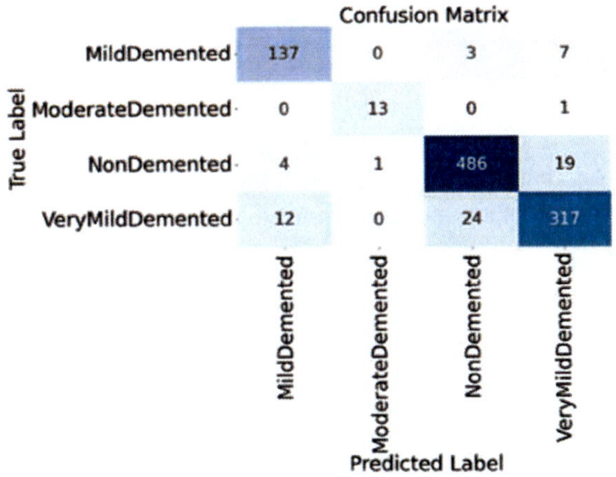

Fig. 6 Confusion matrix of the custom CNN model

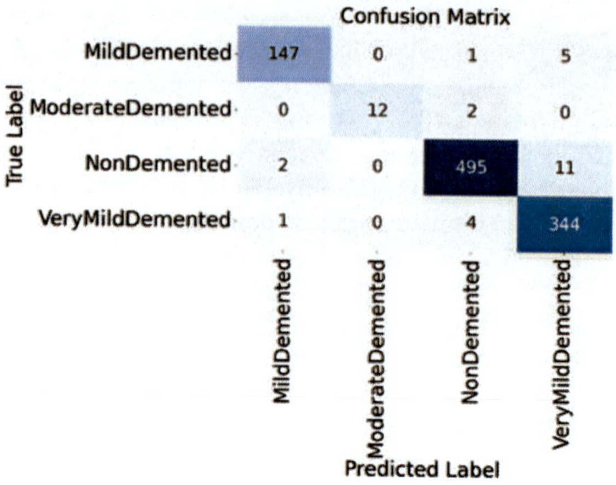

Fig. 7 Confusion matrix of the ResNet-50 model

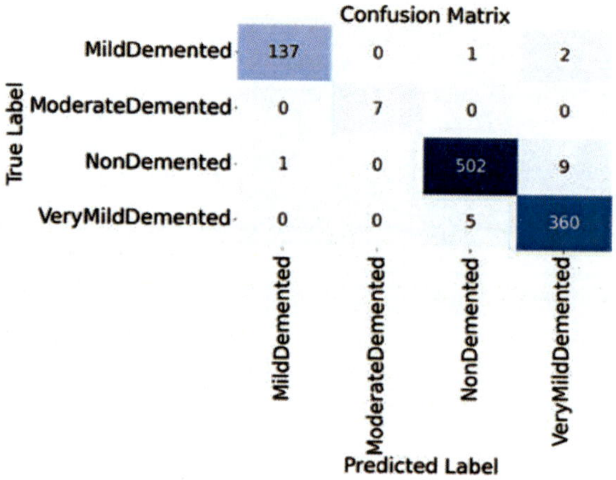

Fig. 8 Confusion matrix of the XCeption model

respectively. The model exhibited a tendency to misclassify ModerateDemented instances as MildDemented, likely due to the similarities between these categories.

The XCeption model displayed a 98% overall classification accuracy in the confusion matrix. However, it struggled to accurately classify instances as ModerateDemented and VeryMildDemented. Notably, it misclassified a significant number of ModerateDemented instances as MildDemented and a significant number of VeryMildDemented instances as NonDemented. This challenge may arise from the similarities between these categories.

Figure 9 shows the accuracy and loss curves of a machine learning model trained on five folds of data. The accuracy curve for all five folds is above 90%, with the highest accuracy being 98% for fold 5. The loss curves for all five folds are below 10%, with the lowest loss being 4% for fold 5. The accuracy curves plateau after epoch 30, while the loss curves continue to decrease. This indicates that the model

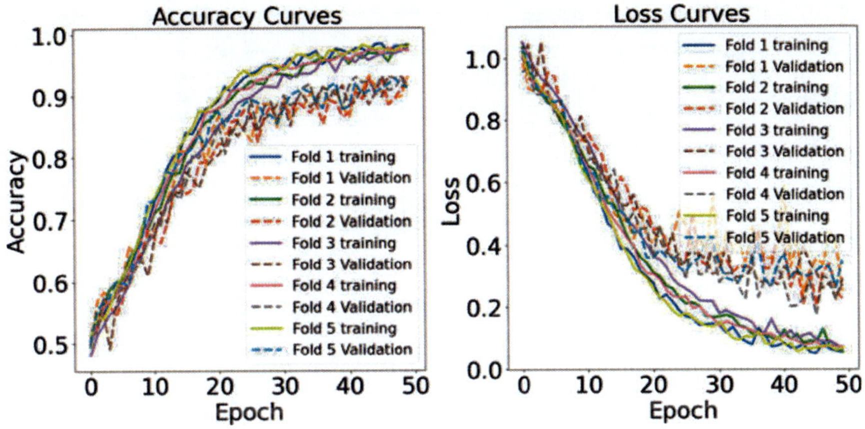

Fig. 9 Custom CNN accuracy and loss

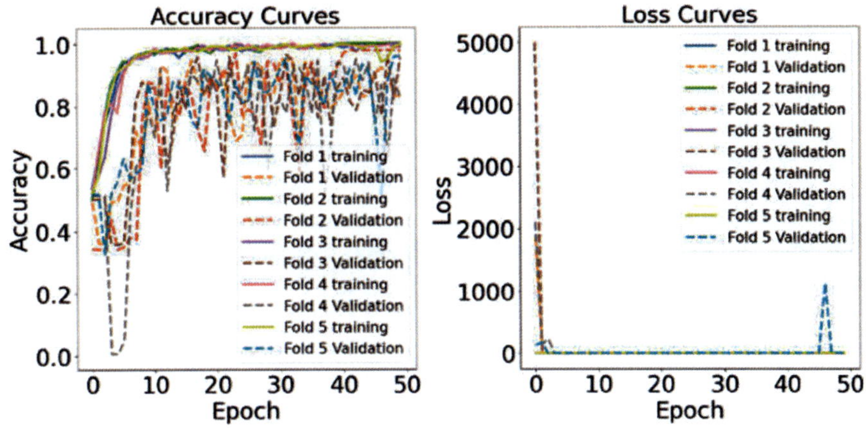

Fig. 10 ResNet-50 accuracy and loss

can achieve high accuracy with low loss. However, the fact that the accuracy plateaus after epoch 30 suggest that the model is likely overfitting the training data.

The accuracy curves of ResNet-50 on the training set and the validation set are converging, but the validation curve is not as steep as the training curve, as shown in Fig. 10, which indicated an overfitting issue on the training set. The loss curves of the model on the training and validation set are converging, but the validation curve is not as low as the training curve. This suggests that the model is not generalizing as well as it could.

Figure 11 shows that, after 10 epochs, the accuracy of the model on the training and the validation set reaches 90% and 85%, respectively. After 10 epochs, the loss of the model on the training set decreases from 0.5 to 0.1, while the validation set

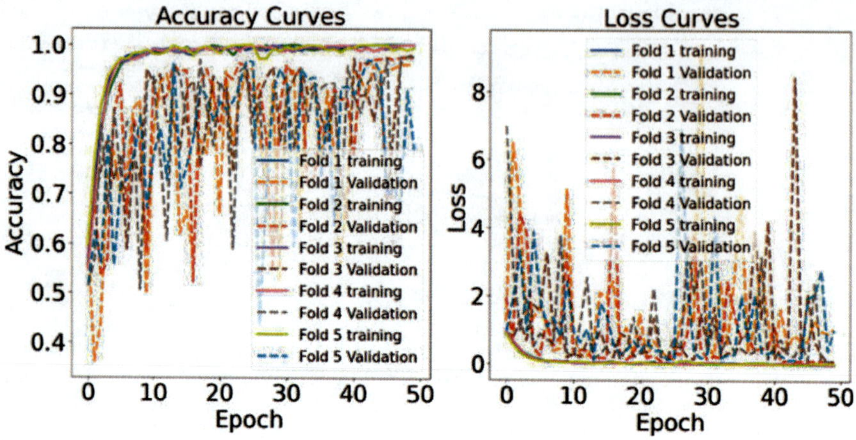

Fig. 11 XCeption accuracy and loss

goes from 0.6 to 0.2. The accuracy curves for the different folds are all similar, which suggests that the model is not overfitting to any fold of the data. The loss curves for the different folds are all similar, which suggests that the model is converging to a similar solution for all the folds.

The custom CNN model took approximately 200.23 s to train on each fold, and the total training time for all five folds was 1076.88 s. On the other hand, the ResNet50 model took approximately 517.73 s to train on each fold, with a total training time of 2586.49 s. The XCeption model took approximately 694.42 s to train on the first fold, and the total training time for all five folds was 3675.44 s. Comparing the training times, we can see that the custom CNN model had the shortest training time among the three models. It took significantly less time to train compared to both ResNet50 and XCeption. This could be due to several factors, including the architecture and complexity of the models. The ResNet50 model took more than double the training time of the custom CNN model, indicating that ResNet50 is a more complex and computationally intensive model to train. Conversely, the XCeption model had the longest training time, nearly four times that of the custom CNN model. XCeption is known for its deeper architecture and increased number of parameters, which could explain the longer training time. It is worth noting that training time alone does not necessarily determine the quality or effectiveness of a model. The choice of model should also consider factors such as accuracy, performance on the specific task, and available computational resources.

5 Conclusion

The key goal of this study was to analyze the application of state-of-the-art deep learning algorithms in the domain of Alzheimer's disease identification. Impressively, the XCeption model demonstrated exemplary performance, yielding extraordinary precision, recall, $F1$-score, and accuracy metrics of 98%. These findings unequivocally validate the model's remarkable prowess in discerning intricate patterns and discerning vital biomarkers associated with the pathogenesis of this complex neurological disorder. However, the utilization of deep learning for Alzheimer's disease detection necessitates addressing formidable challenges that impede its practical implementation. These challenges include the scarcity of available data, the heterogeneous nature of datasets, and the inherent difficulty of interpreting the intricate workings of the deployed models. Nevertheless, the integration of deep learning techniques in the realm of Alzheimer's disease detection holds immense potential for catalyzing transformative advancements in the field. By enabling early and accurate diagnosis, these approaches can facilitate timely intervention, enhance treatment efficacy, and augment our fundamental understanding of the multifaceted etiology underlying this debilitating ailment. To overcome these barriers and drive progress in this area, it is imperative to engage in continuous and interdisciplinary research, foster collaborative endeavors, and establish robust standards for data collection, harmonization, and analysis.

References

1. Islam J, Zhang Y (2017) A novel deep learning based multi-class classification method for Alzheimer's disease detection using brain MRI data. In: Brain informatics: international conference, BI 2017, Beijing, China, November 16–18, 2017, proceedings. Springer International Publishing, pp 213–222
2. Association A (2016) 2016 Alzheimer's disease facts and figures. Alzheimers Dement 12(4):459–509
3. Bekris LM, Lutz F, Yu CE (2012) Functional analysis of APOE locus genetic variation implicates regional enhancers in the regulation of both TOMM40 and APOE. J Hum Genet 57(1):18–25
4. Valenzuela O, San Román B, Guzman FMO, Villamor JLB, Saéz-Lara MJ, Rojas F, Rojas I (2014) Development of soft-computing techniques capable of diagnosing Alzheimers disease in its pre-clinical stage combining MRI and FDG-PET images. In: IWBBIO, pp 1644–1650
5. Chitradevi D, Prabha S (2020) Analysis of brain sub regions using optimization techniques and deep learning method in Alzheimer disease. Appl Soft Comput 86:105857
6. Liu S, Liu S, Cai W, Pujol S, Kikinis R, Feng D (2014) Early diagnosis of Alzheimer's disease with deep learning. In: 2014 IEEE 11th international symposium on biomedical imaging (ISBI). IEEE, pp 1015–1018
7. Luo S, Li X, Li J (2017) Automatic Alzheimer's disease recognition from MRI data using deep learning method. J Appl Math Phys 5(9):1892–1898
8. Ji H, Liu Z, Yan WQ, Klette R (2019) Early diagnosis of Alzheimer's disease using deep learning. In: Proceedings of the 2nd international conference on control and computer vision, pp 87–91

9. Bi X, Wang H (2019) Early Alzheimer's disease diagnosis based on EEG spectral images using deep learning. Neural Netw 114:119–135
10. Islam J, Zhang Y (2018) Early diagnosis of Alzheimer's disease: a neuroimaging study with deep learning architectures. In: Proceedings of the IEEE conference on computer vision and pattern recognition workshops, pp 1881–1883
11. Al-Shoukry S, Rassem TH, Makbol NM (2020) Alzheimer's diseases detection by using deep learning algorithms: a mini review. IEEE Access 8:77131–77141
12. Gulhare KK, Shukla SP, Sharma LK (2017) Deep neural network classification method to Alzheimer's disease detection. Int J Adv Res Comput Sci Softw Eng 7(6):1–4
13. Ding Y, Sohn JH, Kawczynski MG, Trivedi H, Harnish R, Jenkins NW, Lituiev D, Copeland TP, Aboian MS, Mari Aparici C, Behr SC, Franc BL (2019). A deep learning model to predict a diagnosis of Alzheimer disease by using 18F-FDG PET of the brain. Radiology 290(2):456–464
14. Zhao Z, Chuah JH, Lai KW, Chow CO, Gochoo M, Dhanalakshmi S, Wang N, Bao W, Wu X (2023). Conventional machine learning and deep learning in Alzheimer's disease diagnosis using neuroimaging: a review. Front Comput Neurosci17:10
15. Afzal S, Maqsood M, Khan U, Mehmood I, Nawaz H, Aadil F, Song OY, Yunyoung N (2021) Alzheimer disease detection techniques and methods: a review
16. Zhang F, Pan B, Shao P, Liu P, Shen S, Yao P, Xu RX, Alzheimer's Disease Neuroimaging Initiative (2022). A single model deep learning approach for Alzheimer's disease diagnosis. Neuroscience 491:200–214
17. https://www.kaggle.com/datasets/tourist55/alzheimers-dataset-4-class-of-images
18. Huang YJ, Powers R, Montelione GT (2005) Protein NMR recall, precision, and F-measure scores (RPF scores): structure quality assessment measures based on information retrieval statistics. J Am Chem Soc 127(6):1665–1674

Segmentation of White Matter Lesions in MRI Images Using Optimization-Based Deep Neural Network

Puranam Revanth Kumar, Rajesh Kumar Jha, and P. Akhendra Kumar

Abstract White matter lesions (WML) cause numerous brain diseases, and automatic WML segmentation is essential for evaluating the natural disease course and the efficacy of clinical interventions, especially drug discovery. This work presents a novel metaheuristic Harris hawks' optimization (HHO) to train the deep convolutional neural network (DCNN) as a robust model for WML segmentation. The proposed method was tested on MICCAI 2019 dataset. However, the WML regions would appear blurry and global inconsistency. In order to solve the fuzzy problem, we improved the network architecture by incorporating skip connections across the mirrored layers within encoder and decoder stacks. As a result, the segmented WML output appears more realistic and coherent with its surrounding contexts. The proposed optimization method, skip connections, has increased the segmentation performance of Dice score to 79.92%, precision 82.25%, recall 83.90%, and $F1$-score 82.94% for MRI dataset which is superior compared to other approaches. The proposed technique outperformed existing state-of-the-art methodologies in both qualitative and quantitative measurements across a wide range of medical modalities.

Keywords Brain MRI images · Deep learning · Residual connection · Segmentation · WML

P. R. Kumar (✉) · R. K. Jha · P. Akhendra Kumar
Department of Electronics and Communication Engineering, Faculty of Science and Technology (IcfaiTech), ICFAI Foundation for Higher Education, Hyderabad, India
e-mail: revanth123451.rk@gmail.com

R. K. Jha
e-mail: rajeshjha@ifheindia.org

P. Akhendra Kumar
e-mail: akhendra.p@ifheindia.org

S. Shakya et al. (eds.), *Fourth International Conference on Image Processing and Capsule Networks*, Lecture Notes in Networks and Systems 798,
https://doi.org/10.1007/978-981-99-7093-3_17

1 Introduction

White matter lesions (WMLs) are regions of aberrant myelination in the brain. These abnormalities are most easily seen on magnetic resonance imaging as hyperintensities, namely on T2 weighted and fluid-attenuated inversion recovery (FLAIR) sequences. Due to their strong link to a higher chance of cognitive damage, ischemic stroke, and mood and gait disorders, WMH are primarily considered to be established markers of cerebral small vessel disease (CSVD) [1]. In elderly people (95%), WML has been recognized as an established sign of cerebral small vascular disease (CSVD). Numerous degenerative processes and etiologies contribute to CSVD, which is a collection of illnesses affecting the brain's microvessels (arterioles, venules, and capillaries). Brain shrinkage, perivascular spaces, microbleeds, and recent small subcortical infarcts are all neuroimaging findings associated with CSVD [2]. When segmenting WMLs near cerebrospinal fluid (CSF) spaces, FLAIR sequence demonstrates a superior effect than T1-w and T2-w images. This is because FLAIR sequence inhibits high CSF signals by using a long inversion time [3]. Additionally, for the elderly population, FLAIR images show a decrease in contrast between white matter (WM) and gray matter (GM), leading to a uniformly low background signal and making WMLs prominent [4].

Brain lesions or white matter hyperintensities (WMH) are caused by the autoimmune and demyelinating disease [5, 6]. As a result, the probability of survival can be increased through a more accurate and earlier diagnosis of the WML. However, it is hard to get a correct diagnosis of different WML. A significant step toward facilitating efficient WML detection is the generation of an appropriate WML representation via optimization method. First, the primary objective of this research is to propose a Harris hawk's optimization (HHO)-based deep convolutional neural network method for WML segmentation. The major contribution is to detect the WML utilizing a combination of statistical and textural features. The DCNN is trained using the proposed HHO algorithm in such a way that the model parameters are learned in the most effective manner. The proposed algorithm inherits the HHO method for high global convergence feature. Thus, the proposed algorithm effectively renders precision while facilitating WML segmentation.

The remainder of the paper is structured as follows: Sect. 2 describes in detail the typical WMLs segmentation methodologies used in the literature and the difficulties encountered, which served as research problem for the development of the suggested method. Methodology is discussed in Sect. 3. Section 4 displays the results and comparative outcomes of the suggested method in contrast to other state-of-the-art methods, and Sect. 5 summarizes and proposes directions for further research.

2 Related Work

In recent times, deep learning [7] has drawn significant interest due to its powerful simulation abilities and the capacity to automatically acquire sophisticated features from the given data. DL has also proven significant success in image classification [8], target recognition [9], and image segmentation [10]. Convolutional neural networks (CNNs) [11], the lesion prediction algorithm [12], and U-Net [13] are only a few of the DL techniques that have been studied in the context of WML segmentation. The adaptability of CNN is demonstrated over their ability to learn features independently using a given dataset. CNNs show outstanding results in medical image segmentation and demonstrate generalizability by training on publicly available medical images [14], particularly in three-dimensional space. In general, it can be broken down into the three categories below. The segmentation problem is first reduced to a classification problem when using the voxel-based lesion segmentation framework. The core voxel's lesion status is predicted, and the lesion's location is identified. In order to segregate multiple sclerosis lesions, [15] created a unique architecture based on a cascade of two six-layer CNNs. The first was developed to detect voxels in the image that could contain a lesion, while the second network was developed to minimize the number of voxels that were incorrectly labeled [16]. Second, an up-sampling procedure helps remove duplicate data while maintaining the original resolution of the input in the patch-based lesion segmentation architecture. Multiple deep neural network architectures that take into account explicit location characteristics or multi-scale patches were integrated in [17] to segment WML lesions. A three-dimensional conditional random field (CRF) that takes into account both local and global context was presented in [18] as a means of reducing false positives in the post-processing stage. Finally, an encoder–decoder network representing a volume-based segmentation architecture is developed for enabling full-pipe WML segmentation output from MRI Images. A basic investigation of deep 3D convolutional encoder networks using Boltzmann machine theory was performed by [4]. To detect WML with multi-modality MRIs, [19, 20] used deep fully convolutional networks. Some proposed algorithms to solve ISP are listed in Table 1.

2.1 Research Problem

All the listed algorithms in the literature were proposed to overcome the ISP of normal images and medical images of the CT and MRI image types, but the main drawbacks of the above listed algorithms are that it is easy to fall into local optima and suffers from exploitation and exploration. There are now methods that can be combined to improve WML categorization performance. First, the high capacity of HHO can be used to balance between exploitation and exploration. Second, the suggested optimization method will train the tissue segmentation neural network weights. To get over the ISP problem in WML segmentation, we were inspired to

Table 1 Some proposed algorithms to solve ISP

References	Procedure
Ramya et al. [5]	In order to achieve automated detectability of punctate white matter lesions (PWMLs) in babies, deep learning models were constructed using a convolutional neural network called You Only Look Once v3 (YOLOv3). As a consequence of this, a threshold of detection probability between 20 and 30% was established for all deep learning models
Asghar et al. [21]	The threshold levels' bounds were explored using the cluster validity measure to locate the boundaries that may contain the ideal threshold values. Then, GA is applied to look for the best possible threshold values within the acquired boundaries
Guardia et al. [4]	The U-Net that was enhanced by highlighting foregrounds (HF) showed a significant improvement in the identification of WML voxels at the boundary of WMLs or in tiny WML clusters, both quantitatively and subjectively
Gilsoon et al. [14]	The focus of this work is on using WML volume to evaluate functional differences in people with moderate cognitive impairment (MCI). In this case, the ISP problem was resolved by employing a data-driven method of assessing the groups' respective power spectra
Thawkar et al. [16]	A particle swarm optimization (PSO) with the added benefits of collaborative and all-encompassing learning has been designed to take on the ISP. Together, PSO and cooperative learning can help mitigate the negative effects of the dimensionality curse and delay or eliminate the onset of early convergence

suggest a novel model that combines HHO and CNN. In order to segment WML from MRI images, this method is compared to numerous state-of-the-art methods.

3 Methodology

In this section, the DCNN weights was trained using the Harris hawk's optimization (HHO) algorithm as a robust model for the WML segmentation task. Also, we enhanced the network model by introducing skip connections between mirrored layers in the encoder and decoder stacks [7]. In general, skip connections tend to work well for deeper networks and tasks that require high levels of abstraction, such as semantic segmentation and image synthesis.

3.1 Harris hawk's Optimization (HHO)

The HHO algorithm [21] replicates Harris hawks' cooperative foraging and pursuing technique. To pursue fleeing prey, the hawks employ a separate strategy known as 'surprise pounce'. The fitness metric is dependent on the targeted prey for each repetition. The HHO [21] algorithm is divided into three steps: exploration, transition, and exploitation. First, the target function and solution space, together with the relevant parameters, are specified.

3.1.1 Exploration Phase

In this phase, hawks perch randomly and wait for prey depending on two approaches stated in Eq. (1).

$$H_i(t+1) = \begin{cases} H_r(t) - r_1\left|H_r(t) - 2_{r_2} H_i(t)\right| h \geq 0.5 \\ \left(H_{\text{pery}}(t) - H_m(t)\right) - r_3(lb + r_4(lb - ub))h < 0.5 \end{cases} \quad (1)$$

where $H_i(t+1)$ represents the new position of the hawk, $H_r(t)$ represents the hawk that was picked randomly in the current iteration, $H_{\text{pery}}(t)$ represents the location of the prey, and $H_i(t)$ represents the position of each hawk in the system. Here, $r_1, r_2, r_3, r_4,$, and h are the randomized values in the range [0,1], and the lower and upper limit variables are *lb* and *ub*, respectively. When h is less than 0.5, the hawks perch in random places. When h is greater than 0.5, they stay at a place with other members of their population. Here, $H_m(t)$ denotes the current population's average location for hawks, and it is calculated using Eq. (2).

$$H_m(t) = \frac{1}{P_{\text{size}}} \sum_{i=1}^{P_{\text{size}}} H_i(t) \quad (2)$$

where P_{size} denotes the size of the current population. The transition between exploration and exploitation. At this stage, the following Eq. 3 is used to calculate the prey's energy level;

$$E = 2E_0\left(1 - \frac{t}{T}\right) \quad (3)$$

where E, T, and E_0 represent the target's escape energy, total generations, and beginning energy state, respectively.

3.1.2 Exploitation Phase

In this phase, the hawks make a 'surprise pounce', striking the prey seen during the exploration phase.

Soft Besiege ($R \geq 0.5$ and $|E_{\text{esc}}| \geq 0.5$)—The hawks gently circle their prey while trying to get away and making jumps that are hard to predict. This behavior is specified by Eqs. (4) and (5).

$$H_i(t+1) = \Delta H_i(t) - E_{\text{esc}}\left|JS.H_{\text{pery}}(t) - H_i(t)\right|, \tag{4}$$

$$\Delta H_i(t) = H_{\text{pery}}(t) - H_i(t) \tag{5}$$

The xth iteration difference between the rabbit's position vector and its current location is denoted by $\Delta H_i(t)$. Once the hawks have begun their gentle besiege with dives, the final strategy to improve their positions may be carried out using Eq. (6).

$$H_i(t+1) = \begin{cases} X \text{ if fit}(X) < \text{fit}(H_i(t)) \\ Y \text{ if fit}(Y) < \text{fit}(H_i(t)) \end{cases} \tag{6}$$

Hard Besiege ($R \geq 0.5$ and $|E_{\text{esc}}| \geq 0.5$): The hawks are unable to completely encircle the prey that is attempting to flee since the prey does not have sufficient energy to do so, and this behavior is represented in Eq. (7).

$$H_i(t+1) = H_{\text{pery}}(t) - E_{\text{esc}}|\Delta H_i(t)|. \tag{7}$$

The position of the hawks can be updated using Eq. (8).

$$H_i(t+1) = \begin{cases} X \text{ if fit}(X) < \text{fit}(H_i(t)) \\ Y \text{ if fit}(Y) < \text{fit}(H_i(t)) \end{cases} \tag{8}$$

where fit(X) and fit(Y) denote the X and Y's fitness values, respectively. Equations (9) and (16) are used to determine the X and Y values in Eq. (10).

$$X = H_{\text{pery}}(t) - E_{\text{esc}}|JSH_{\text{pery}}(t) - H_i(t)| \tag{9}$$

$$Y = X + B \times \text{LF(Dim)} \tag{10}$$

Here, LF is the Levy flight function, B is a vector of size $1 \times$ Dim selected at random, $JS = 2(1 - r_5)$ prey escaping procedure which changed randomly in each iteration. r_5 represents random number inside (0, 1), and Dim is the dimension that describes the problem. Using the LF method, the hawk's motions may be replicated in the play. It is shown by Eq. (11).

$$\text{LF}(x) = 0.01 \times \frac{u \times \sigma}{|v|^{\frac{1}{\beta}}}, \sigma = \left[\frac{\daleth\left(1 + \beta \times \sin\frac{\pi\beta}{2}\right)}{\daleth\left(\frac{1+\beta}{2}\right) \times \beta \times 2\left(\frac{\beta-1}{2}\right)}\right]^{\frac{1}{\beta}} \tag{11}$$

In this case, where u, v are random normal distribution vector with the size of $1 \times$ Dim, β is a constant with a value of 1.5, and $\daleth$ is a standard gamma function. To test the algorithm's performance, the proposed solution is examined for the maximum number of iterations. In the meantime, the fitness measure is evaluated by taking into account the error rate in addition to a number of attributes using Eq. (12).

$$\text{Fitness} = \alpha\psi(A) + \eta\frac{|C|}{|F|} \tag{12}$$

Here, $\psi(A)$ stands for the error rate. The number of characteristics that will be used is denoted by the $|C|$, while the total number of features is indicated by $|F|$. $\alpha\epsilon(0, 1)$ and $\eta = (1 - \alpha)$ are the parameters that, respectively, characterize the classification performance and the duration of the feature selection. The Algorithm 1 [21] provides the detailed steps of HHO as follows:

Algorithm 1: HHO Pseudo Code

Inputs: The population size N and maximum number of iterations T
Output: The location of prey and its fitness value
Initialize the random population $X_i (i = 1, 2, 3, \ldots, N)$
while (stopping condition is not met) **do**
Calculate the fitness value of hawks
Set H_{prey} as the location of prey (best location)
for (each hawk (H_i)) **do**
Update the initial energy E_0 and jump strength J
$E_0 = 2rand() - 1, J = 2(1 - rand())$
Update the E using Eq. (3)
if (|E| $\geq$ 1) **then go to** -------------------------------------- Exploration phase
Update the location vector using Eq. (1)
if (|E| < 1) **then go to** -------------------------------------- Exploitation phase
if (r $\geq$ 0.5 and |E| $\geq$ 0.5) **then go to** ———— Soft besiege
Update the location vector using Eq. (4)
else if (r $\geq$ 0.5 and |E| < 0.5) **then go to** -------------------------------------- Hard besiege
Update the location vector using Eq. (7)
else if (r $\geq$ 0.5 and |E| < 0.5) **then got to** -------------------------------------- Soft besiege with progressive rapid dives
Update the location vector using Eq. (6)
else if (r < 0.5 and |E| < 0.5) **then go to** -------------------------------------- Hard besiege with progressive rapid dives
Update the location vector using Eq. (8)
Return H_{prey}

3.2 *Proposed HHO-DCNN for WML Segmentation*

This section presents the proposed HHO-DCNN approach for WML segmentation and describes how the feature vector is used to refine the segmentation. Segmentation is performed with a DCNN [18], and weights of DCNN are trained with the HHO. To identify hyperintensities, the proposed HHO helps to select the features from an

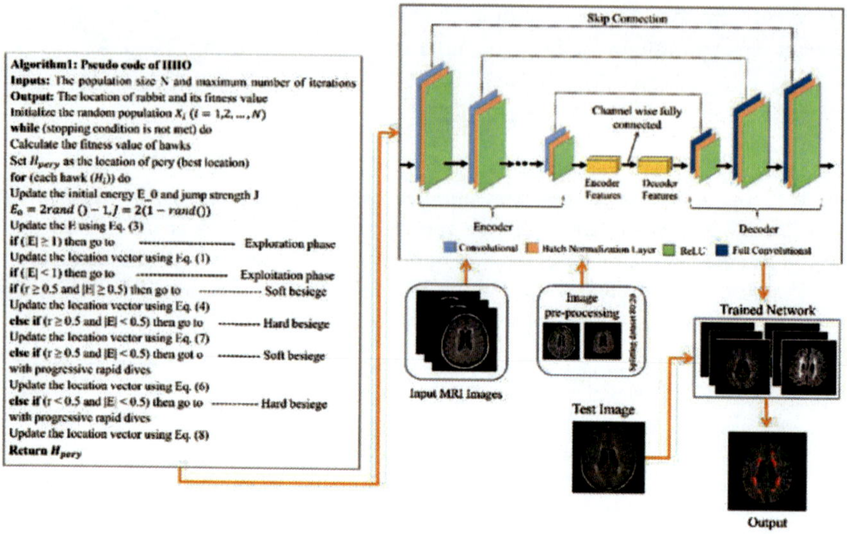

Fig. 1 Schematic view of proposed HHO-based DCNN for WML segmentation

input image. To reset the randomization problem, the HHO uses a pulsing mechanism, where the particles in the search space are randomly reset to their initial positions periodically. HHO can be applied to problems in the real world, even if the search space is unknown or complex. On the other side, HHO [21] draws its inspiration from the compliant manner and chasing style of Harris' hawks. In addition, HHO can deal with multi-modality difficulties, local optimal solutions, and deceptive optima in the search space. Therefore, a better solution is achieved with enhanced performance, and the rate of convergence is controlled by incorporating HHO with DCNN. Figure 1 depicts the HHO-based DCNN architecture, which makes use of a powerful encoder–decoder framework for feature embedding, residual connections for data flow, and bottleneck layers for model compression. We used ROBEX [34] for preprocessing to eliminate non-anatomical objects, which reduced the number of false positives.

3.3 Architecture of CNN

Here, Fig. 2 represents the schematic view of DCNN and Fig. 3 illustrates the segmented output of the proposed algorithm. Layers in a DCNN are responsible for certain tasks. These layers include the pooling layer, the convolutional layers, and the fully connected layers (FC). Features are generated by the convolutional layer, which also enables pattern extraction via a feature vector generated from the objects' segmentations.

$$T = \{T_1, T_2, \ldots, T_h, \ldots, T_i\} \tag{13}$$

where i represent the total number of convolution layers and T_h represents the hth convolution layer used in DCNN. The output can be derived from the units that are located at (p, w), and it is represented as:

$$(T_{\vartheta}^{h})_{p,w} = (v_{\vartheta}^{h})_{p,w} + \sum_{p=1}^{W_1^{p-1}} \sum_{z=-r_1^u}^{r_1^u} \sum_{s=r_2^u}^{r_2^u} (X_{f,p}^{u})_{z,s} * (T_{\vartheta}^{h-1})_{p+z,w+s} \tag{14}$$

where * designates the conv operator that allows for pattern extraction from the outputs of consecutive conv layers, $(T_p^{u-1})_{m+z,r+s}$ represents the fixed feature maps, W_1^{p-1} represents the total number of feature maps, and $(X_{f,p}^u)_{z,s}$ represents the weights.

Features maps are built by convolving the inputs with the training weights associated with the neurons. Nonlinear activation functions are used to simplify functional mappings between input and output variables once the result has been obtained. Number of convolution layers (conv layers) in DCNN is used to describe the complexity of the network's ability to process data, with the input to the convolution layer being a segmented image produced by the sparking process.

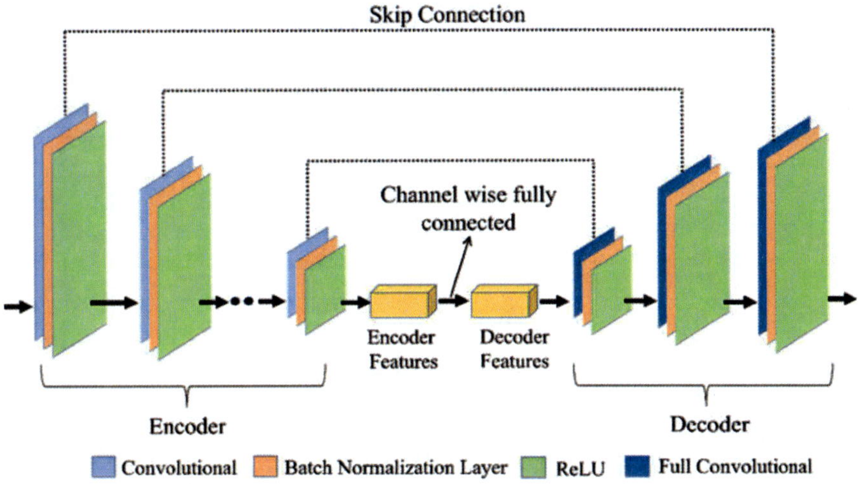

Fig. 2 Schematic view of proposed deep CNN

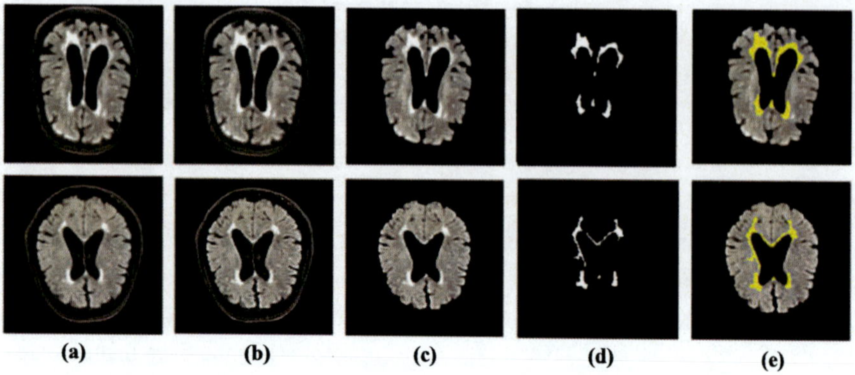

Fig. 3 Segmentation of WML on high-resolution MICCAI MRI images. **a** Original image, **b** enhanced image, **c** skull stripping, **d** binary mask, **e** segmented output

4 Results and Discussion

In this section, we present the results of extensive experiments intended to test the efficacy of the suggested algorithms and to compare them to some of the state-of-the-art techniques for dealing with the WML segmentation. In this work, we proposed a novel automated method, HHO-DCNN, and tested it on a large dataset consisting of 220 individuals from a variety of demographic backgrounds to segment WML from multimodal MRI images. In a quantitative evaluation of WML segmentation of 2D MRI images using a wide range of imaging procedures and clinical diagnoses, the proposed methodology outperformed earlier state-of-the-art approaches due to its advanced labeling techniques. In addition to detecting and segmenting other commonly existing cerebral lesions of vascular origin, this WML segmentation approach was created utilizing the 2D thick-slice protocol, which is more routinely used in regular clinical acquisition with numerous scanners.

4.1 Dataset

As given in Table 2, three datasets (Utrecht, Singapore, GE3T) were used in this study acquired from three different scanners.[1] There are 20 people in each dataset. To conduct the experiment's validation, we randomly assigned 80% of the subjects to the training group and 20% to the test set. Forty participants from two scanners were utilized as the training data for the cross-scanner trials, while the remaining twenty subjects were used as the testing data. Please take into account that we did not utilize complete 3D imagery but rather 2D images taken from the axial view.

[1] https://liwang.web.unc.edu/.

Table 2 Overview of the MICCAI WML images available for training and testing

Institute	Scanner	Training	Testing
UMC Uterche	3T Philips	25	35
NUHS Singapore	3T Siemens	35	45
VU Amsterdam	3T GE Signa	35	45

The suggested approach was successfully hick-validated using a dataset consisting of slice MRI images. MATLAB DL tool box was used to implement the proposed algorithm; the loss function used was weighted Dice loss. Images and annotated regions were uniformly down sampled in the x and y directions to a pixel spacing of 0.5 in order to account for differences in pixel spacing between scanners. There was no change to the z-axis pixel spacing. When training, 256 × 256 random 2D patches were sampled from the image volume as input. Training was performed on NVIDIA GE Force 8 GB GPU, with a learning rate of 0.001, the batch size was set to 128. Training required only 45 min, whereas testing and segmenting lesions from an unseen image took on average just 10 s.

The most significant contribution that the proposed technique makes is the establishment of a clinically usable and competitive method for the automated segmentation of WML in a wide range of application situations. These scenarios include the identification of WML across the presence of other intracranial lesions and the utilization of MRI images obtained using a variety of imaging protocols. The qualitative results of our proposed model WML segmentation are displayed in Fig. 3.

It is important to note that our proposed fully automatic approach outperformed human delineations using semi-automatic tools by an experienced observer working alone to segment WML (Table 3). Therefore, we believe that our method would be a secure and workable solution for segmenting WML in clinical settings.

Table 3 Performance of different quantitative evaluation matrices using automatic methods for WML segmentation

References	Dice score %	Precision %	Recall %	$F1$-score %
Sole et al. [4]	75.00	×	62.00	63.00
Todea et al. [3]	79.00	55.00	×	85.01
Huang et al. [2]	80.00	×	82.00	76.00
Zhang et al. [9]	88.00	×	76.00	80.00
Zhu et al. [10]	78.90	×	79.20	76.40
This work	79.92	82.25	83.90	82.94

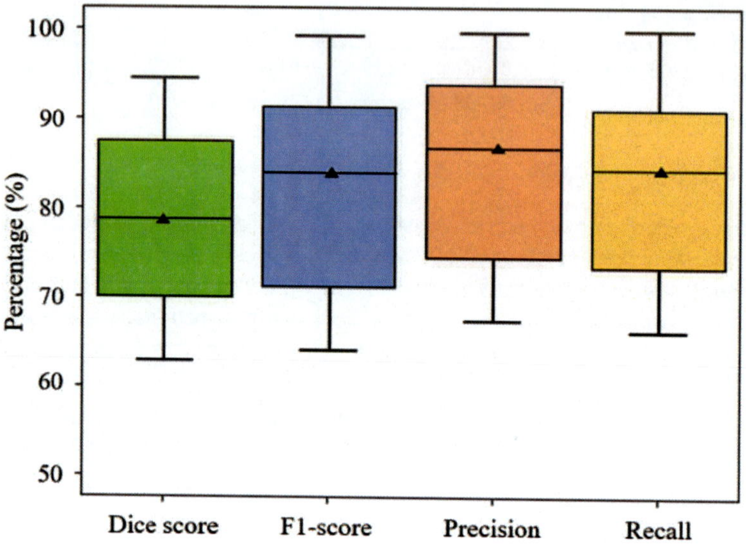

Fig. 4 Quantitative evaluation of proposed algorithm for WML segmentation

4.2 Quantitative Evaluation

The suggested HHO-DCNN method is tested quantitatively in terms of its performance with respect to Dice loss, $F1$-score, precision, and recall (see Fig. 4). Quantitative research involves manipulating the training data. In addition, the suggested HHO-DCNN uses the MACCAI 2019 dataset, demonstrating the effectiveness of the proposed technique.

Recall: described the portion of ground truth voxels that were corrected segmented as:

$$\text{recall} = \frac{\text{TP}}{\text{TP} + \text{FN}} \tag{15}$$

Precision: The recall was calculated for both the combined set of labeled WML regions and the individual set of validated WML regions. The proportion of segmented voxels that are considered to be part of the ground truth regions is referred to as "precision."

$$\text{precision} = \frac{\text{TP}}{\text{TP} + \text{FP}} \tag{16}$$

The metrics of interest in the aforementioned methods are the proportions of true positives (TP), true negatives (FN), and false positives (FP). Along with voxel-level recall and accuracy, we also developed the $F1$-score as a metric to evaluate performance. For 3D linked component lesions, the recall formula was the same as in equation [15, 22, 23], but TP and FN were read as the percentage of lesions correctly segmented and the percentage of lesions missed.

The F1-score was the harmonic mean of recall and precision and was calculated as:

$$F1 = \frac{2 * \text{recall} * \text{precision}}{\text{recall} + \text{precision}} \tag{17}$$

Lesion recall and $F1$-score were created to better identify recall and precision characteristics at the voxel level.

As a result, we applied the Dice loss to prevent training from settling into a series of local minima, as shown in Eq. (17).

$$\text{Dice loss} = 1 - \frac{1}{C}\sum_{1}^{C}\left(\frac{2\sum_{i}^{N} P_i^c g_i^c}{\sum_{i}^{N}(P_i^c)^2 + \sum_{i}^{N}(g_i^c)^2}\right) \tag{18}$$

where $P_i \in P$ is the output of the algorithm, $g_i \in G$ is the real-world value, and C is the total number of classes. Definite WML and suspected WML were defined previously. The Dice loss was adjusted since the definitive WML was given greater consideration.

$$\text{Dice loss} = 1 - \frac{1}{C}\sum_{1}^{C}(w(x))^c\left(\frac{2\sum_{i}^{N} P_i^c g_i^c}{\sum_{i}^{N}(P_i^c)^2 + \sum_{i}^{N}(g_i^c)^2}\right) \tag{19}$$

where $(w(x))^c$ is the weight map.

The weight was set to 2, for regions with confirmed WML as $1 + f(\text{dis(Dis)})$, where (dis(Dis)) was the distance to the center of the closest region with confirmed WML. The value of $f(\text{dis(Dis)})$ was computed to be between 0 and 1.

$$f(\text{dis(Def)}) = 1 - \frac{\text{dis(Def)})}{\text{dis(Def)}_{\max}} \tag{20}$$

Dice Similarity Coefficient (DSC): measures the overlap percentage between G and P. It is defined as:

$$\text{DSC} = \frac{2 \times |G \cap P|}{|G| + |P|} \tag{21}$$

Our study has several limitations. First, WMLs may exist alongside and mix with other diseases, such as stroke lesions in T2-FLAIR MR imaging [1, 9, 24, 25]. The goal of this research was to develop an automated system for segmenting WMLs; as a result, the suggested method did not take into account labeling other illnesses. When we were training the model, it was assigning the same label to both the background and other disorders. False positive findings for WMLs segmentation were seen, however, since other disease characteristics, such as stroke lesions, also present as hyperintense regions.

5 Conclusion

The proposed automated white matter hyperintensities segmentation algorithm, HHO-DCNN, has demonstrated its reliability and effectiveness as an extension to traditional state-of-the-art methods for WML patients. The method automatically segments WML in brain MRI images from MICCAI datasets, with a reasonable processing time of 45 min. Evaluation on a diverse set of 220 individuals, acquired from different MRI scanners and displaying various diagnoses and lesion loads, has shown that HHO-DCNN outperforms other existing methods in terms of segmentation accuracy and spatial agreement. This suggests that proposed method is a fast, reliable, and easy-to-use method with optimization for the automated segmentation of WML. Since the method includes a preprocessing stage, it is flexible enough to be used to enhance image quality, allowing for better results to be obtained even with low-resolution input. Maximum precision of 82.25%, maximum recall of 83.90%, dice score 79.92%, and maximum $F1$-score of 82.94% were achieved by the suggested HHO-DCNN, which was superior to all other approaches.

The proposed method's effectiveness in addressing various problem domains is notably constrained by its reliance on a fixed hierarchical structure and its deficiency in adaptive mechanisms. These limitations render it potentially less suitable for tackling dynamic or intricate optimization scenarios. Additionally, HHO's performance may be sensitive to parameter settings, and it can easily fall into a local optimum, requiring careful tuning to achieve optimal results. To overcome these issues in future, we employ a hybrid optimization based deep CNN to increase the accuracy by escaping from local optima.

References

1. Chandra A, Verma S, Raghuvanshi A et al (2023) PCcS-RAU-Net: automated parcellated corpus callosum segmentation from brain MRI images using modified residual attention U-Net. Biocybern Biomed Eng 43(2):403–427
2. Huang F, Xia P, Vardhanabhuti V et al (2023) Semisupervised white matter hyperintensities segmentation on MRI. Hum Brain Mapp 44(4):1344–1358
3. Todea AR, Melie-Garcia L, Barakovic M et al (2023) A multicenter longitudinal MRI study assessing LeMan-PV software accuracy in the detection of white matter lesions in multiple sclerosis patients. J Magn Reson Imaging
4. Solé-Guardia G, Custers E, de Lange A et al (2023) Association between hypertension and neurovascular inflammation in both normal-appearing white matter and white matter hyperintensities. Acta Neuropathol Commun 11(1):2

5. Balakrishnan R, Valdés Hernández MC, Farrall AJ (2021) Automatic segmentation of white matter hyperintensities from brain magnetic resonance images in the era of deep learning and big data—a systematic review. Comput Med Imaging Graph 88:101867
6. Gauber M, Lange C, Zimmermann I et al (2023) Performance evaluation of automated white matter hyperintensity segmentation algorithms in a multicenter cohort on cognitive impairment and dementia. Front Psychiatry 13:1010273
7. Shehata LAO, Ibrahim O, El-Kammash TH et al (2023) Gad. Brain volumetric and white matter structural connectivity alterations in autistic children: case–control study. Egypt J Radiol Nucl Med 54:36
8. Kuroda T, Ono K, Honma M et al (2023) Cerebral white matter lesions and regional blood flow are associated with reduced cognitive function in early-stage cognitive impairment. Front Aging Neurosci 16(15):1126618
9. Zhang H, Zhu C, Lian X, Hua F (2023) A Nested attention guided UNet++ architecture for white matter hyperintensity segmentation. IEEE Access 1–11 (2023)
10. Zhu W, Huang H, Zhou Y et al (2022) Automatic segmentation of white matter hyperintensities in routine clinical brain MRI by 2D VB-Net: a large-scale study. Front. Aging Neurosci 14:915009
11. Li X, Zhao Y, Jiang J et al (2022) White matter hyperintensities segmentation using an ensemble of neural networks. Hum Brain Mapp 43(3):929–939
12. Zhang Y, Duan Y, Wang X et al (2022) A deep learning algorithm for white matter hyperintensity lesion detection and segmentation. Neuroradiology 64:727–734
13. Pious AE, Sridevi UK (2023) A novel segment white matter hyperintensities approach for detecting Alzheimer. Comput Syst Sci Eng 44(3):2715–2726
14. Park G, Hong J, Duffy BA et al (2021) White matter hyperintensities segmentation using the ensemble U-Net with multi-scale highlighting foregrounds. Neuroimage 237:118140
15. Li H, Jiang G, Zhang J et al (2018) Fully convolutional network ensembles for white matter hyperintensities segmentation in MR images. Neuroimage 183:650–665
16. Thawkar S (2022) Feature selection and classification in mammography using hybrid crow search algorithm with Harris hawk's optimization. Biocybern Biomed Eng 42(4):1094–1111
17. Tran P, Thoprakarn U, Gourieux E et al (2022) Automatic segmentation of white matter hyperintensities: validation and comparison with state-of-the-art methods on both multiple sclerosis and elderly subjects. NeuroImage Clin 33:102940
18. Yang D, Masurkar A (2022) Early-stage MRI volumetric differences in white matter hyperintensity and temporal lobe volumes between autopsy-confirmed Alzheimer's disease, cerebral small vessel disease, and mixed pathologies. Dement Geriatr Cogn Dis Extra 12(1):69–75
19. Moeskops P, de Bresser J, Kuijf HJ et al (2018) Evaluation of a deep learning approach for the segmentation of brain tissues and white matter hyperintensities of presumed vascular origin in MRI. Neuroimage Clin 17:251–262
20. Dadar M, Manera AL, Ducharme S et al (2022) White matter hyperintensities are associated with grey matter atrophy and cognitive decline in Alzheimer's disease and frontotemporal dementia. Neurobiol Aging 111:54–63
21. Heidari AA, Mirjalili S, Faris H et al (2019) Harris hawks optimization: algorithm and applications. Future Gener Comput Syst 97:849–872
22. Sandrone S, Aiello M, Cavaliere C et al (2023) Mapping myelin in white matter with T1-weighted/T2-weighted maps: discrepancy with histology and other myelin MRI measures. Brain Struct Funct 228:525–535
23. Rieu Z, Kim J, Kim RE et al (2021) Semi-supervised learning in medical MRI segmentation: brain tissue with white matter hyperintensity segmentation using FLAIR MRI. Brain Sci 11:720
24. Rehan Afzal HM, Luo S, Ramadan S et al (2021) Automatic and robust segmentation of multiple sclerosis lesions with convolutional neural networks. Comput Mater Continua 66(1):977–991
25. Guo X, Ye C, Yang Y et al (2022) Ensemble learning via supervision augmentation for white matter hyperintensity segmentation. Front Neurosci 16:946343

A New Multi-level Hazy Image and Video Dataset for Benchmark of Dehazing Methods

Bedrettin Çetinkaya, Yücel Çimtay, Fatma Nazli Günay, and Gökçe Nur Yılmaz

Abstract The changing level of haze is one of the main factors which affects the success of the proposed dehazing methods. However, there is a lack of controlled multi-level hazy dataset in the literature. Therefore, in this study, a new multi-level hazy color image dataset is presented. Color video data is captured for two real scenes with a controlled level of haze. The distance of the scene objects from the camera, haze level, and ground truth (clear image) is available so that different dehazing methods and models can be benchmarked. In this study, the dehazing performance of five different dehazing methods/models is compared on the dataset based on SSIM, PSNR, VSI, and DISTS image quality metrics. Results show that traditional methods can generalize the dehazing problem better than many deep learning-based methods. The performance of deep models depends mostly on the scene and is generally poor on cross-dataset dehazing.

Keywords Hazy imagery · Reconstruction of images · Image enhancement · Poor visibility

B. Çetinkaya (✉) · Y. Çimtay · F. N. Günay · G. N. Yılmaz
Computer Engineering, Ted University, Ankara, Turkey
e-mail: bedrettin.cetinkaya@tedu.edu.tr

Y. Çimtay
e-mail: yucel.cimtay@tedu.edu.tr

F. N. Günay
e-mail: fnazli.gunay@tedu.edu.tr

G. N. Yılmaz
e-mail: gokce.yilmaz@tedu.edu.tr

S. Shakya et al. (eds.), *Fourth International Conference on Image Processing and Capsule Networks*, Lecture Notes in Networks and Systems 798,
https://doi.org/10.1007/978-981-99-7093-3_18

1 Introduction

Different conditions that affect the scattering of the light such as haze, fog, and smoke can significantly reduce information contained in the captured scene. This loss of information causes poor performance in many computer vision problems such as object detection/tracking/classification [1–3], depth estimation [4], and remote sensing [5]. Therefore, reconstructing hazy-free data have the potential to improve the performance of other computer vision problems.

Traditional approaches try to generate hazy-free images using hand-crafted features/priors, such as Dark Channel Prior (DCP) [6], Color Attenuation Prior (CAP) [7], and color-lines [8]. On the other hand, the modern methods [9–13] are based on deep learning and they require both hazy-free and hazy data for training. Therefore, the datasets have a big impact on the model performance. On the other hand, evaluating the cross-dataset performance of models gives information about their generalization capabilities. To this end, we published a new foggy image dataset, Indoor-Multi-Level Fog dataset (IMFD), for image dehazing problem and compare several state-of-the-art methods on IMFD.

2 Related Work

Image dehazing is one of the most attractive problems in computer vision. With the rise of deep learning, neural networks-based solutions have become more dominant rather than traditional methods such as DCP [6] and CAP [7]. Therefore, we divided previous studies into two main groups: (i) Traditional and (ii) Deep Learning-based methods.

2.1 *Traditional Methods*

Most of the traditional methods rely on prior information such as image saturation/attenuation and pixel values in the RGB channels and they try to model haze under the assumption of homogeneous distribution of haze, unlike the real-world scenario. He et al. [6] proposed one of the most important traditional methods named Dark Channel Prior (DCP). This method dehazes hazy images estimating transmission maps and atmospheric light. Due to the simplicity and efficiency of DCP algorithm, there are many improved versions [14–18]. Also, Zhe et al. [7] use Color Attenuation Prior (CAP). They use CAP to estimate the depth map of a hazy image and hazes are removed using this depth information. Moreover, Fattal et al. [8] exploit color-lines that come from 1D distribution of RGB pixel values. This color-line can successfully estimate transmission offsets of hazy images. Also, Berman et al. [19] remove haze using haze-lines which enables estimating the distance map and the atmospheric light.

2.2 *Deep Learning-Based Methods*

Deep learning-based techniques have gained increasing popularity in recent times due to their successful performance. Cai et al. [9] introduced a CNN, DehazeNet that estimates a medium transmission map to reconstruct haze-free images. Also, Ullah et al. propose a lightweight version of DehazeNet. Unlike these works, Li et al. [20] directly estimated haze-free images without finding the output of any intermediate step such as transmission/distance map. Zhang et al. [10] proposed a GAN to jointly learn the atmosphere light and the transmission map. Singh et al. [21] employed GAN-based architecture which aims to learn multi-level complexity and multi-scale spatial features. Also, Wu et al. [11] successfully applied contrastive learning for image dehazing problem. Jin et al. [22] estimated uncertainty maps that are used as attention coefficients for the density of haze. Tran et al. [23] proposed a GAN model with DCP. Song et al. [12] used a vision transformer-based architecture to remove haze. Jiang et al. [13] proposed a hybrid model which aims to de- hazes hazy images and refines the outputs of dehazing step.

3 Datasets

This section reviews popular datasets used in the image dehazing problem.

Chic: Color Hazy Image for Comparison dataset, CHIC, [24] provides fog-free RGB images and their hazed versions with nine different haze levels in a well-controlled indoor environment. Also, it contains three different scenes and two different camera positions which are static and dynamic. 9 different images with different haze levels are presented. Scenes are cropped where full images size is 6000 × 4000. Scene-1_ cropped images size is 1566 × 2244 and Scene-2 is 2010 × 3528.

DenseHaze: DenseHaze dataset [25] contains 33 paired hazy and fog-free ground-truth images. It contains both indoor and outdoor scenes. Hazy images were created using professional haze machines and they have homogeneous haze distributions. The foggy sceneries were captured by using actual haze that was produced by expert haze generators. The identical visual information was recorded under the same lighting conditions in the respective haze-free and haze-filled scenarios. The goal of the DenseHaze dataset is to greatly advance single-image dehazing state of the art by pushing robust approaches for realistic and diverse hazy scenarios.

D-Hazy: D-Hazy dataset [26] contains 1409 hazy and corresponding fog-free images. It used NYU Depth [27] and Middelbury [28] datasets because hazy images were generated synthetically using depth information in these datasets. Only indoor scene is available in this dataset. Due to the fact that in a hazy media, the scene brightness attenuates with distance, researchers are able to produce a comparable hazy scene with high accuracy based on the depth information and utilizing the physical model of a hazy medium.

IHaze: IHaze dataset [29] contains 35 paired hazy and haze-free ground-truth images. All scene contains a *MacBeth* color checker to simplify the color calibration process. Also, the same illumination conditions were used for all data.

NHHaze: NHHaze dataset [30] contains 55 paired hazy and fog-free ground-truth images and these hazy images are created using a professional haze generator. Moreover, hazy images reflect real-world scenarios because these images have non-homogeneous haze distributions. Also, real hazy data without ground-truths are available for the evaluation process. All images were collected from the outdoor environment.

OHaze: OHaze dataset [31] contains 45 paired hazy and fog-free ground-truth images from the outdoor environment. Also, it contains different illumination conditions because data were collected under different sunlight levels. Using traditional image quality measurements like PSNR, SSIM, and CIEDE2000, OHAZE is used to compare a sample collection of cutting-edge dehazing approaches.

Revide: Unlike the previous image dehazing datasets, Revide dataset [32] contains paired videos with hazy and haze-free data. Data were collected using 47 different scenes and each scene contains different haze levels and illumination conditions. By capturing the same scene (with or without haze) twice using a well-designed video acquisition system, they collect matched real-world hazy and haze-free movies that are precisely aligned. They also create a Confidence Guided and Improved Deformable Network (CG-IDN) for video dehazing in order to address the difficulty of utilizing temporal redundancy among the hazy frames.

4 IMF Dataset (IMFD)

Previously stated datasets are mostly based on static images and there is a lack of foggy video data. Another shortcoming is that they do not provide foggy scenes with a controlled level of fog except [24, 33]. To overcome that issues and enrich the foggy dataset literature, in this study, we present a new indoor foggy imagery dataset (Indoor-Multi-Level Fog Dataset) which was created in a light and fog-controlled environment. We present two different scenes which include objects with various distances from the camera. In addition, we provide the ground truth as both image and video data.

Hanwha Wisenet QNO-7082R Network camera [34] was used to capture the imagery of the scenes. Before recording, the camera was focused on the scenes. Video resolution was set as 1080p (1920×1080) with a ratio of 16:9. Frame rate and

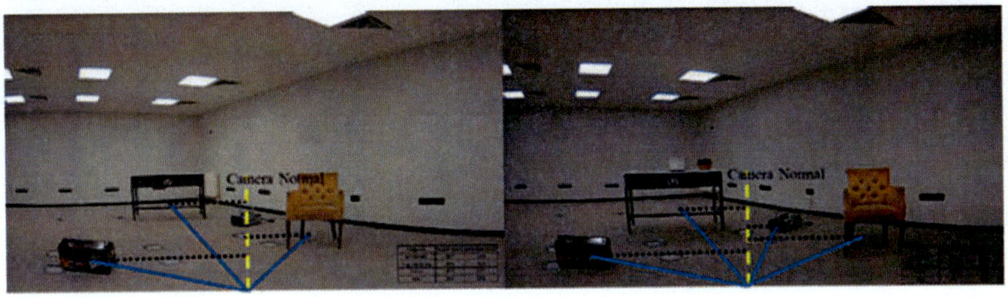

Fig. 1 Scenes 1 and 2, and the depth of the objects with respect to the camera, respectively

video coding standard were set as 60 fps and H.264, respectively. IR filter was turned off. The camera was connected remotely, and during recording, all the conditions including the position of the camera were the same except the level of the fog. First of all, the ground-truth scene was captured and then the room was filled with fog up to the scene that cannot be seen by the eye. Then recording was restarted and maintained up to the fog density reduces and the scene approaches the ground truth.

Quenlite QL-2000RGB Fog Machine [35] was used for the purpose of fog production. This machine can be controlled remotely. Its fog emittance capacity and spraying distance are 560 m^3/min and 8 m, respectively. In one cycle, it can continuously emit for 50 s. Then it cools down and is ready to be triggered again. We repeated the cycle up to the fog becomes an opaque layer. We recorded the foggy scene during the fog was being evacuated from the room.

We captured the imagery in a rectangular room that is isolated from the environment so that no daylight comes in. The lamps on the ceiling were used to enlighten the room. In this way, the amount of light in the room was kept stable. We created 2 different scenes with different levels of fog. The ground-truth images and the depth of the objects are shown in Fig. 1 for scenes 1 and 2. We positioned different objects at different depths which is a crucial concept to measure the success of any dehazing method in terms of improving the visual quality and increasing the visibility distance.

The imagery of the scenes was captured while the fog was evacuating the room. The evacuation of the fog is controlled by a ventilation system. Therefore, evacuation speed was constant with respect to time. After saving the recording, we split the total video duration to 10 equal-length windows and snap a video frame corresponding to the ends of each interval. Figure 2 shows the frame selection process, where fr is the frame rate of the video and L is the duration of the video. Therefore, the frame number (FN) corresponding to each fog level can be found by using Eq. 1, where i and NL hold the fog level and the number of fog levels, respectively.

$$\mathrm{FN}_i = (\mathrm{NL} - 1) * \mathrm{fr} * L, i = 1, 2, 3, \ldots, \mathrm{NL} \quad (1)$$

Finally, we classified the video into 10 fog levels which are shown in Fig. 3. Therefore for each scene, we provide 10-level foggy images (with ground truth) and

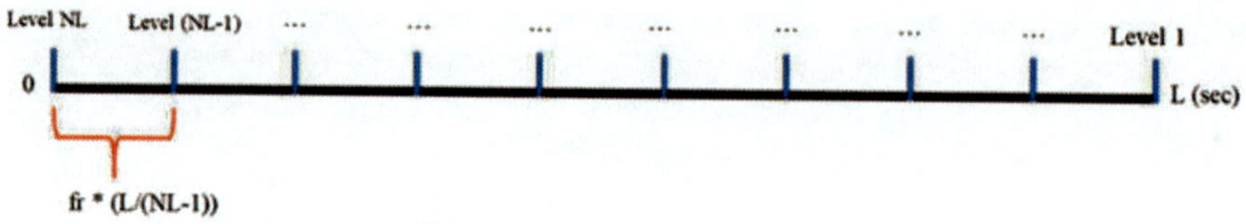

Fig. 2 Frame rate selection process

also the labeled foggy videos with respect to the starting and ending times of different fog levels. Our dataset is available at https://github.com/ycimtay/ITUFD.

5 Experiment

We evaluate our dataset using state-of-the-art methods published with official pre-trained weights for deep learning-based models. These methods are: EDN-GTM [23], AECR-Net [11], Light-DehazeNet [36], DehazeFormer [12]. Moreover, we used different pre-trained models of each of these studies to see their generalization performance. We also evaluate the results of DCP [6] and ω is set to 0.45. DCP is one of the fundamental dehazing methods which uses atmospheric light scattering model. Air light is estimated by using the dark channel pixel information. We use peak signal-to-noise ratio (PSNR), structural similarity index measure (SSIM), visual saliency-induced index (VSI) [37], and deep image structure and texture similarity (DISTS) [38] metrics for the evaluation process as given in Table 1. Peak signal-to-noise ratio (PSNR) is an expression for the ratio of the greatest potential value (power) of a signal to the power of the noise that distorts it and lowers the quality of its representation. The SSIM index is used to calculate how similar two images are to one another. A starting uncompressed or distortion-free image is used as the basis for the SSIM's prediction of image quality.

According to Table 1, EDN-GTM [23] trained with OHaze dataset performs better than all other methods for scene 1 in general. For scene 2, DehazeFormer [12] and EDN-GTM [23] have the best performances. On the hand, DCP [6] has considerable performance for both scenes 1 and 2. Other deep learning-based methods except for EDN-GTM and DehazeFormer have low generalization performance.

Moreover, we provide visual results for each method presented in Table 1, as shown in Fig. 4. We consider only the best results for each method among the pre-trained weights.

6 Results and Discussion

Level of the haze is one of the most important effects which determines the success of any dehazing method.

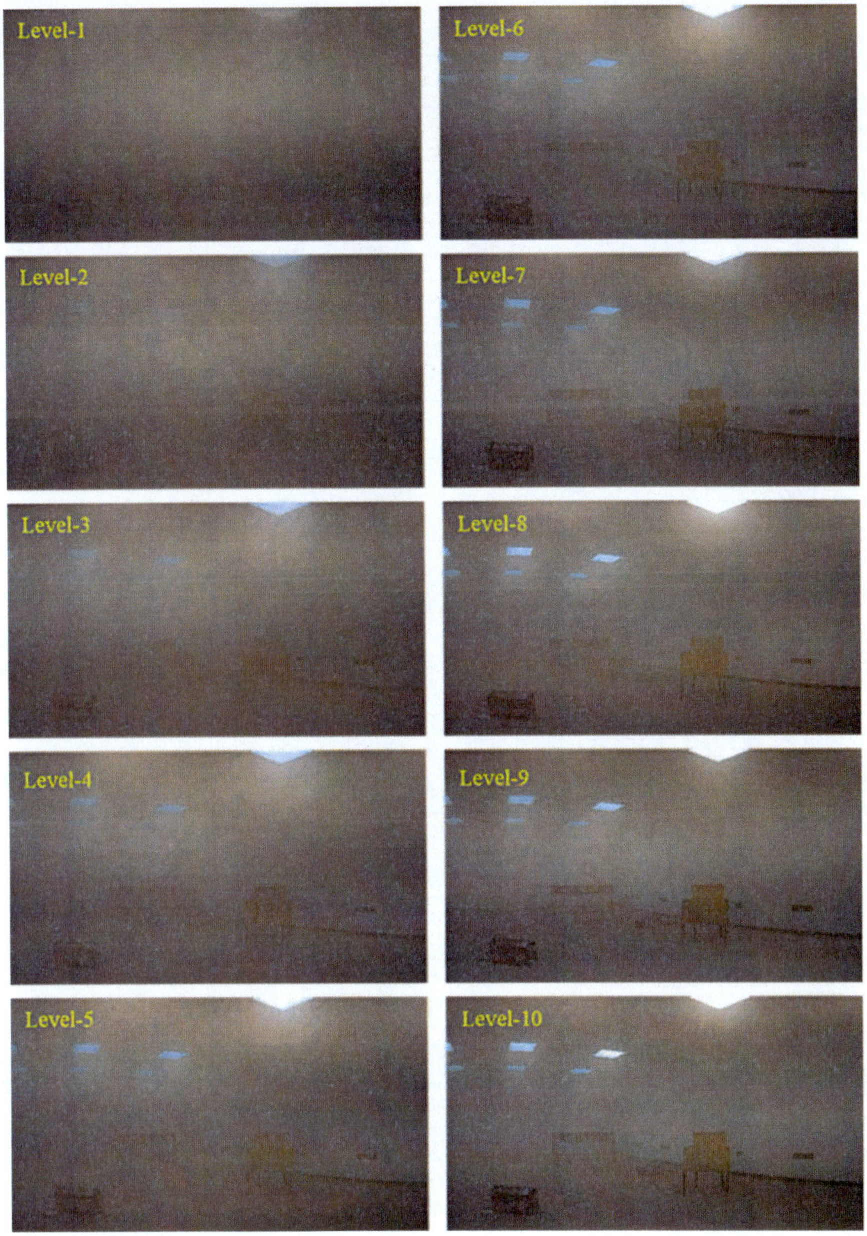

Fig. 3 Fog levels from 1 to 10 for scene 1

Table 1 Results for state-of-the-art methods in terms of SSIM, PSNR, VIS, and DISTS metrics for both ITUFD scenes 1 and 2

Method	Training datasets	ITUFD scene	Level 2				Level 5				Level 8			
			SSIM	PSNR	VSI	DISTS	SSIM	PSNR	VSI	DISTS	SSIM	PSNR	VSI	DISTS
DCP (CVPR 2009) [6]	–	1	**0.85**	16.35	0.90	0.58	0.83	15.10	0.90	0.46	0.84	15.47	0.90	0.40
AECR-Net (CVPR 2021) [11]	DenseHaze	1	0.42	12.08	0.83	0.37	0.49	11.61	0.85	0.37	0.50	12.22	0.84	0.35
	ITS	1	0.74	14.25	0.89	0.50	0.74	14.16	0.90	0.42	0.71	12.93	0.91	0.36
	NHHaze	1	0.44	10.15	0.81	0.39	0.43	9.32	0.86	0.37	0.46	9.89	0.88	0.37
Light-D.Net (IEEE TIP 2021) [36]	*	1	0.49	11.15	0.88	0.48	0.51	11.41	0.88	0.42	0.48	11.24	0.89	0.40
EDN-GTM (ISCSI 2022) [23]	DenseHaze	1	0.74	13.09	0.89	0.36	0.78	13.43	**0.91**	**0.33**	0.75	12.99	0.91	**0.31**
	IHaze	1	0.82	14.92	0.90	0.43	0.80	13.96	**0.91**	0.36	0.81	14.35	**0.92**	0.32
	NHHaze	1	0.80	13.76	0.89	0.44	**0.84**	14.13	0.90	0.41	**0.85**	14.56	0.91	0.38
	OHaze	1	**0.85**	**16.38**	0.88	0.52	**0.84**	**15.49**	0.90	0.43	**0.85**	**16.20**	0.91	0.39
DehazeFormer (IEEE TIP 2023) [12]	Indoor	1	0.73	12.36	0.89	0.50	0.65	10.50	0.90	0.42	0.70	11.26	0.91	0.39
	Reside6k	1	0.75	12.23	**0.93**	**0.31**	0.66	10.60	0.90	0.43	0.70	11.53	0.91	0.37
	RSHaze	1	0.83	15.06	0.86	0.51	0.82	14.92	0.86	0.41	0.83	15.22	0.87	0.39
DCP (CVPR 2009) [6]	–	2	**0.83**	17.39	0.89	0.60	0.81	16.72	0.92	0.41	0.84	17.93	0.93	0.32
AECR-Net (CVPR 2021) [11]	DenseHaze	2	0.42	13.50	0.84	0.40	0.48	12.34	0.84	0.36	0.56	15.19	0.89	0.29
	ITS	2	0.68	15.09	0.89	0.52	0.60	13.06	0.91	0.38	0.70	15.66	0.92	0.29
	NHHaze	2	0.33	11.64	0.81	0.41	0.46	10.01	0.85	0.35	0.51	11.56	0.88	0.31

(continued)

Table 1 (continued)

Method	Training datasets	ITUFD scene	Level 2				Level 5				Level 8			
			SSIM	PSNR	VSI	DISTS	SSIM	PSNR	VSI	DISTS	SSIM	PSNR	VSI	DISTS
Light-D.Net (IEEE TIP 2021) [36]	*	2	0.55	9.12	0.88	0.49	0.55	9.25	0.90	0.41	0.56	9.33	0.91	0.35
EDN-GTM (ISCSI 2022) [23]	DenseHaze	2	0.76	16.45	0.89	0.40	0.79	16.56	**0.93**	**0.30**	0.79	17.34	0.93	0.26
	IHaze	2	0.76	15.90	0.90	0.47	**0.82**	17.19	**0.93**	0.33	0.80	17.57	**0.95**	**0.24**
	NHHaze	2	0.70	14.11	0.86	0.45	0.79	14.53	0.91	0.39	**0.85**	18.18	0.94	0.29
	OHaze	2	**0.83**	**17.52**	0.89	0.54	**0.82**	16.62	0.91	0.41	**0.85**	**19.05**	0.92	0.32
DehazeFormer (IEEE TIP 2023) [12]	Indoor	2	0.74	15.07	0.89	0.54	0.74	14.76	0.92	0.36	0.81	17.08	0.94	0.28
	Reside6k	2	0.73	14.40	**0.93**	0.27	0.56	12.18	0.91	0.43	0.68	13.73	0.92	0.34
	RSHaze	2	**0.83**	17.02	0.88	0.54	**0.82**	**17.38**	0.89	0.39	**0.85**	18.91	0.91	0.30

While bold numbers show the best results for the corresponding level, metric, and scene, red numbers represent the second-best results. Also, ∗ means that the corresponding dataset was collected with public hazy images from Internet resources

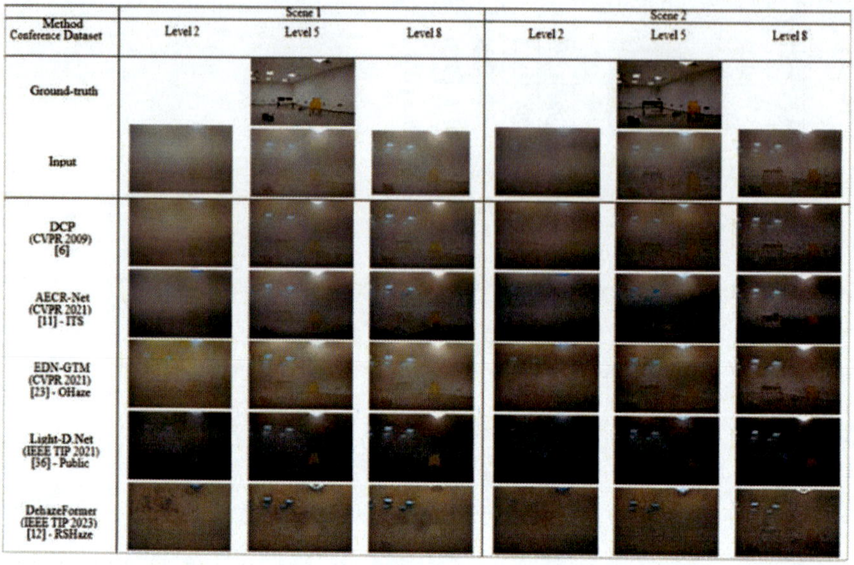

Fig. 4 Visual results for state-of-the-art methods. We visualize only the best result presented in Table 1 for each method

Most of the methods perform well when haze level is low. However, when the level is increased, the success drops dramatically. Therefore, in this study, we present a new multi-level color hazy image dataset which includes two scenes with different object distances from the camera. 5 different methods are compared on the dataset and results are presented. Performance of the dehazing methods can be represented by their generalization capability and adaptation to new hazy scenes. Compared to traditional methods, deep models perform better when tested on the same dataset they are trained with. However, cross-dataset dehazing performance is generally lower than the traditional methods. In the future, more effort should be given to develop deep models with high capacity of generalization [39].

7 Conclusion

In this study, a new multi-level hazy image and video dataset is proposed. By controlling the speed of the evacuation of the fog from the room, the fog level is adjusted. There are 10 level of foggy images of two scenes. DCP and the state-of-the art deep learning-based dehazing methods are tested on this new dataset. According to the results, DCP and EDN-GTM achieve promising results. Another important result is that deep learning-based methods have limited generalization capability when tested on a new dataset. This study is important and valuable since it brings a new multi-level hazy dataset to dehazing literature. The state-of-the-art methods can be benchmarked on this dataset.

Acknowledgements This study is supported by The Scientific and Technological Research Council of Turkey (TUBITAK) within the Project number: 122E333.

References

1. Sindagi VA, Oza P, Yasarla R, Patel VM (2020) Prior-based domain adaptive object detection for hazy and rainy conditions. In: Computer vision—ECCV 2020: 16th European conference, Glasgow, UK, August 23–28, 2020, proceedings, part XIV 16. Springer, pp 763–780
2. Li C, Liang X, Lu Y, Zhao N, Tang J (2019) rgb-t object tracking: benchmark and base-line. Pattern Recogn 96:106977
3. Pei Y, Huang Y, Zou Q, Lu Y, Wang S (2018) Does haze removal help CNN-based image classification? In: Proceedings of the European conference on computer vision (ECCV), pp 682–697
4. Hambarde P, Murala S (2020) S2dnet: depth estimation from single image and sparse samples. IEEE Trans Comput Imaging 6:806–817
5. Kaplan NH (2018) Remote sensing image enhancement using hazy image model. Optik 155:139–148
6. He K, Sun J, Tang X (2010) Single image haze removal using dark channel prior. IEEE Trans Pattern Anal Mach Intell 33(12):2341–2353
7. Zhu Q, Mai J, Shao L (2015) A fast single image haze removal algorithm using color attenuation prior. IEEE Trans Image Process 24(11):3522–3533
8. Fattal R (2014) Dehazing using color-lines. ACM Trans Graph (TOG) 34(1):1–14
9. Cai B, Xu X, Jia K, Qing C, Tao D (2016) Dehazenet: an end-to-end system for single image haze removal. IEEE Trans Image Process 25(11):5187–5198
10. Zhang He, Patel VM (2018) Densely connected pyramid dehazing network. In: Proceedings of the IEEE conference on computer vision and pattern recognition, pp 3194–3203
11. Wu H, Qu Y, Lin S, Zhou J, Qiao R, Zhang Z, Xie Y, Ma L (2021) Contrastive learning for compact single image dehazing. In: Proceedings of the IEEE/CVF conference on computer vision and pattern recognition, pp 10551–10560
12. Song Y, He Z, Qian H, Du X (2023) Vision transformers for single image dehazing. IEEE Trans Image Process
13. Jiang N, Hu K, Zhang T, Chen W, Xu Y, Zhao T (2023) Deep hybrid model for single image dehazing and detail refinement. Pattern Recogn 136:109227
14. Xu H, Guo J, Liu Q, Ye L (2012) Fast image dehazing using improved dark channel prior. In: 2012 IEEE international conference on information science and technology. IEEE, pp 663–667
15. Wang J-B, He N, Zhang L-L, Lu K (2015) Single image dehazing with a physical model and dark channel prior. Neurocomputing 149:718–728
16. Wang Y, Wu B (2010) Improved single image dehazing using dark channel prior. In: 2010 IEEE international conference on intelligent computing and intelligent systems, vol 2. IEEE, pp 789–792
17. Zhu M, He B, Wu Q (2017) Single image dehazing based on dark channel prior and energy minimization. IEEE Signal Process Lett 25(2):174–178
18. Pei S-C, Lee T-Y (2012) Nighttime haze removal using color transfer pre-processing and dark channel prior. In: 2012 19th IEEE international conference on image processing. IEEE, pp 957–960
19. Berman D, Treibitz T, Avidan S (2018) Single image dehazing using haze-lines. IEEE Trans Pattern Anal Mach Intell 42(3):720–734

20. Li B, Peng X, Wang Z, Xu J, Feng D (2017) Aod-net: all-in-one dehazing network. In: Proceedings of the IEEE international conference on computer vision, pp 4770–4778
21. Singh A, Bhave A, Prasad DK (2020) Single image dehazing for a variety of haze scenarios using back projected pyramid network. In: Computer vision—ECCV 2020 workshops: Glasgow, UK, August 23–28, 2020, proceedings, part IV 16. Springer, pp 166–181
22. Jin Y, Yan W, Yang W, Tan RT (2022) Structure representation network and uncertainty feedback learning for dense non-uniform fog removal. In: Proceedings of the Asian conference on computer vision (ACCV), Dec 2022, pp 2041–2058
23. Tran L-A, Moon S, Park D-C (2022) A novel encoder-decoder network with guided transmission map for single image dehazing. Procedia Comput Sci 204:682–689
24. El Khoury J, Thomas J-B, Mansouri A (2018) A database with reference for image dehazing evaluation. J Imaging Sci Technol 62(1):10503–10511
25. Ancuti CO, Ancuti C, Timofte R, Gool LV, Zhang L, Yang M-H (2019) Ntire 2019 image dehazing challenge report. In: Proceedings of the IEEE conference on computer vision and pattern recognition workshops, IEEE CVPR
26. De Vleeschouwer Cosmin Ancuti C, Ancuti CO (2016) D-hazy: a dataset to evaluate quantitatively dehazing algorithms. In: IEEE international conference on image processing (ICIP), ICIP'16
27. Silberman N, Hoiem D, Kohli P, Fergus R (2012) Indoor segmentation and support inference from RGBD images. ECCV 5(7576):746–760
28. Scharstein D, Hirschmüller H, Kitajima Y, Krathwohl G, Nešić N, Wang X, Westling P (2014) High-resolution stereo datasets with subpixel- accurate ground truth. In: Pattern recognition: 36th german conference, GCPR 2014, Münster, Germany, 2–5 Sept 2014, proceedings 36. Springer, pp 31–42
29. Ancuti CO, Ancuti C, Timofte R, De Vleeschouwer C (2018) I-haze: a dehazing benchmark with real hazy and haze-free indoor images. arXiv:1804.05091v1
30. Ancuti CO, Ancuti C, Timofte R (2020) NH-HAZE: an image dehazing benchmark with non-homogeneous hazy and haze-free images. In: Proceedings of the IEEE conference on computer vision and pattern recognition workshops, IEEE CVPR
31. Ancuti CO, Ancuti C, Timofte R, De Vleeschouwer C (2018) O-haze: a dehazing benchmark with real hazy and haze-free outdoor images. In: IEEE conference on computer vision and pattern recognition, NTIRE workshop, NTIRE CVPR'18
32. Zhang X, Dong H, Pan J, Zhu C, Tai Y, Wang C, Li J, Huang F, Wang F (2021) Learning to restore hazy video: a new real-world dataset and a new method. In: CVPR, pp 9239–9248
33. El Khoury J, Thomas J-B, Mansouri A (2016) A color image database for haze model and dehazing methods evaluation. In: Image and signal processing: 7th international conference, ICISP 2016, Trois-Rivières, QC, Canada, May 30–June 1, 2016, proceedings 7. Springer, pp 109–117
34. Datasheet. Hhanwha wisenet qno-7082r network camera: https://hanwhavision.eu/product/qno-7082r/. Tech. Rep.
35. https://www.yonkamuzikmarket.com/quenlite-ql-2000-sis-makinasi-2000-watt-uzaktan-kumandali. Tech. Rep.
36. Ullah H, Muhammad K, Irfan M, Anwar S, Sajjad M, Imran AS, de Albuquerque VHC (2021) Light-dehazenet: a novel lightweight CNN architecture for single image dehazing. IEEE Trans Image Process 30:8968–8982
37. Zhang L, Shen Y, Li H (2014) VSI: a visual saliency-induced index for perceptual image quality assessment. IEEE Trans Image Process 23(10):4270–4281
38. Ding K, Ma K, Wang S, Simoncelli EP (2020) Image quality assessment: unifying structure and texture similarity. IEEE Trans Pattern Anal Mach Intell 44(5):2567–2581
39. Cetinkaya B, Cimtay Y, Gunay FN, Yilmaz GN (2023) A new multi-level hazy image and video dataset for benchmark of dehazing methods. arXiv:2307.16050

Creative AI Using DeepDream

Rakhi Bhardwaj, Tanmay Kadam, Shubham Waghule, Sahil Shendurkar, and Bhakti Sarag

Abstract The term "creative AI" refers to a branch of computer science that automates the creation of unique pieces of art, music, and literature. It allows technology to replicate and help human beings in the creation of unique and imaginative items. DeepDream, a computer vision tool, uses a deep neural network to find and improve patterns in images. The approach works by submitting an image to a trained neural network, which recognizes specific elements or patterns in the image and amplifies and enhances them. The current DeepDream algorithm's usage of a single image has the drawback of potentially producing overfitting, limiting the algorithm's ability to produce images that differ from the input image. As a result, the variety and originality of the developed graphics may be constrained. To get around this problem, DeepDream algorithm was trained to use multiple images or a combination of images and noise, resulting in more varied and inventive outputs. This allows us to create imaginative films as well as numerous visuals using just two images.

Keywords DeepDream · Inception-v3 · Convolutional neural network (CNN)

1 Introduction

A neural network is a machine learning (ML) technique inspired and similar to the human nervous system and brain structure. It is made up of processing units that are divided into input, hidden, and output layers. Each layer's nodes or units is linked to nodes in neighboring layers. A weight value is assigned to each link. At each unit, the inputs are multiplied by the weights and summed.

Recent advances in image categorization and speech recognition have been fueled by artificial neural networks.

R. Bhardwaj (✉) · T. Kadam · S. Waghule · S. Shendurkar · B. Sarag
Department of Computer Engineering, Vishwakarma Institute of Technology, Pune, India
e-mail: rakhi.bharadwaj@vit.edu

T. Kadam
e-mail: tanmay.kadam21@vit.edu

S. Shakya et al. (eds.), *Fourth International Conference on Image Processing and Capsule Networks*, Lecture Notes in Networks and Systems 798,
https://doi.org/10.1007/978-981-99-7093-3_19

An artificial neural network can be trained by exposing it to millions of training samples and gradually tweaking network parameters until the required classifications are produced. The network is typically composed of 10–30 stacked layers of artificial neurons. Each image gets loaded into the input layer, which communicates with the following layer until it reaches the "output" layer. This final output layer provides the "answer" for the network.

Convolutional Neural Networks (CNNs) have emerged as a powerful tool for image analysis and understanding. Among the various CNN architectures.

One of the issues with neural networks is determining what happens at each layer. Each layer gradually pulls more complicated information from the image, through training, until the final layer effectively decides what the image portrays. For example, the first layer can look for corners or edges. Intermediate layers examine the fundamental properties to find general forms or elements, such as a door or a leaf. In response to extraordinarily complex objects, such as entire houses or trees, these neurons light up. The final several layers integrate those interpretations into comprehensive ones.

1.1 Convolution Neural Network

There are many problems that can be solved with neural networks, such as classification, pattern recognition, dimensionality reduction, clustering, natural language processing (NLP), computer vision, predictive analysis, regression, and others.

Convolutional neural network, sometimes known as CNN, is a subtype of artificial neural network used largely for image processing and recognition. It is composed of numerous layers of interconnected neurons that process and transform visual data, and it is based on the biological structure of the visual cortex of the human brain.

Convolutional layers, which apply a set of learned filters to the input data, are used in CNNs to automatically and adaptively learn spatial hierarchies of features from raw input data, such as pictures. These filters are taught to the network by training it on certain features such as edges or textures.

CNNs are designed to automatically and adaptively learn spatial hierarchies of features from raw input data, such as images, by applying convolutional layers that apply a set of learned filters to the input data. These filters, which are trained on a massive collection of labeled photographs, may recognize certain elements such as edges or textures.

CNNs have been used in a variety of applications, including self-driving cars, medical imaging, and robotic computer vision systems. CNNs have been demonstrated to be very effective at a variety of image recognition tasks, including object detection, facial recognition, and picture classification.

Figure 1 demonstrates how a convolution neural network (CNN) deep neural network can successfully recognize a higher-level object by learning hierarchical layers of representation from a low-level input vector. Only a broad generalization of the pixel values in the highlighted area may be seen in the red squares of the figure.

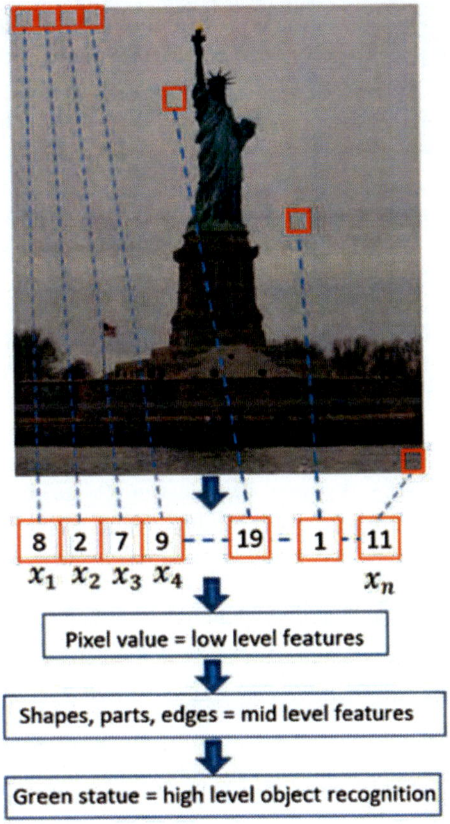

Fig. 1 Working of CNN

CNNs are able to extract higher-level representations of a picture after each layer, leading to image recognition.

1.2 What is DeepDream?

DeepDream is a computer vision approach that uses a neural network to generate attractive and frequently baffling visuals. Google created it in 2015 as a tool to experiment with deep learning and make artistic and abstract visuals.

Here, the input image is analyzed with a pre-trained neural network, which is made up of numerous layers of interconnected nodes to recognize various properties such as edges, textures, and various forms from the image. Based on the output of the previous layer, each network layer learns to recognize the complex features.

The DeepDream approach takes advantage of the neural network's hierarchical structure by inducing an artificial "dream" state, in which the network is urged to enhance specific aspects present in the image. This is accomplished by altering the input image in such a way that the activation of specific neurons in the network is maximized. The resultant image is then transmitted back into the network, and the process is repeated, resulting in an iterative feedback loop that intensifies and exaggerates specific image patterns.

By changing the algorithm's settings, such as the layer of the network to focus on, the amount of zoom, and the severity of the effect, users can regulate the types of features that are amplified. This opens a vast array of creative opportunities and has produced a number of visually spectacular photographs.

DeepDream has been employed in different fields, including art, design, and advertising. It has also been utilized in scientific studies to better understand the inner workings of deep learning algorithms and to investigate the behavior of neural networks.

DeepDream has been employed in many different fields, including art, design, and advertising. It has also been utilized in scientific studies to better understand the inner workings of deep learning algorithms and investigate the behavior of neural networks.

1.3 Motivation

The DeepDream algorithm is a well-known picture generating approach that is frequently utilized in computer vision and deep learning. It is known for delivering distinctive and creative visual outputs by optimizing deep neural network activations for a particular input image. However, the usual DeepDream algorithm implementation employs a single input image, limiting the diversity of the generated outputs.

To address this restriction, DeepDream algorithm can be implemented with two input images. The goal of this method is to create more diversified and fascinating visual outputs by merging elements from two different photographs. This can result in the formation of novel and surprising visual combinations that would not otherwise exist in either of the input images alone.

Furthermore, DeepDream algorithm implementation using two images enables the generation of more sophisticated and layered visual outputs visual representations that can capture the essence of both input images at different levels of abstraction by optimizing the activations of multiple layers of a deep neural network. This can result in the creation of rich and detailed visual outputs suitable for several applications such as art and design, visual narrative, and scientific visualization.

Overall, the aim for DeepDream algorithm implementation combining two photographs is to push the boundaries of what is possible with this powerful image generating technology, as well as to explore new and creative ways to create visually attractive outputs.

2 Literature Survey

A deep neural network is made up of nodes arranged in layers. Numerous designs have been developed to address issues in various fields or use-cases. RNN is frequently used in time series problems/forecasting, while CNN is frequently used in computer vision and image recognition, for instance [1].

Since CNN is based on the human visual cortex, it is the preferred neural network for computer vision (image recognition) and video recognition. As the series of several convolution layers moves from the input to the output layers, each layer collects features with a finer resolution. Fully connected layers that perform categorization come after the convolution layers. A sub-sampling or pooling layer follows almost every convolution layer. A 2D, n-pixel picture, is provided to CNN. Every layer is made up of filters or kernels, which are collections of 2D neurons [2].

According to this study, combining these two approaches encourages the creative process and enhances the images that drug users and patients in mental health facilities employ to create hallucinations. The neural style transfer (NST) and deep dream models from this work are combined to produce a new image that incorporates the two approaches. Pre-trained networks for NST and deep dream are VGG-19 and Inception-v3, respectively. The style transfer method includes a crucial stage called the gradient matrix. With gradient fall in the first case and gradient rise in the second, the loss in style transmission is minimized in the first case and maximized in the latter. Based on the degree of clarity in each image, different loss values produced by various photographs are discovered. DeepDreams exhibited lower loss values than distortions, but NST had larger loss values. The clear pictures that lacked blending lines, circles, colors, or other shapes had the opposite effect [3].

AI algorithm Deep Inceptionism is probably the most peculiar one to date. This artificial intelligence was programmed to classify images. It enters the network, unlocks certain neurons, reveals some hidden layers, and attempts to determine the various features that are prefer to understand. AI algorithms, especially the DeepDream algorithms, have a variety of qualities that can be used to generate many kinds of art.

The paper explores the interpretability of deep learning-enabled image recognition algorithms in computer vision research in relation to theories from art history and cognitive psychology on how people perceive things related to vision and related to vision-related perceptual abilities. It is found that there are a surprising number of similarities between what can be learned about the machine-learned image and humanistic theories of visual perception, especially art historian Erwin Panofsky's image analysis technique and psychologist Eleanor Rosch's theory of graded categorization according to prototypes. This suggests that researchers in the arts and sciences would greatly benefit from closer cooperation. This paper argues that, in light of the quick development of image recognition technology, a resurgence of iconographic and formalist art historical research is required to influence future navigation and interpretation of all machine-learned images. This is accomplished by employing the programs Grad-CAM: Gradient-weighted Class Activation Mapping created by the Machine Learning and Perception Lab at Georgia Tech, and Google's DeepDream as examples.

When you input an image, the neurons of a trained ANN become active and emit activations. The deep dream algorithm tries to (boost the neurons firing or activations) by trying to change the input image in a way that would lead some of these neurons to fire more frequently. You can decide which neurons in a given layer should fire more frequently. You may get a feel of how close organic neurons—which are what our human level intelligence was initially looking for—are by looking at these kinds of hallucinogenic results. The procedure is then repeated until all of the features that a particular layer was initially looking for are present in the input image. Strange characteristics can be seen in the original photograph. Using the GPU of an Intel Core i3-7100U CPU, I evaluated the Deep Inceptionism learning vision performance and obtained a lot of smoothen images with a type of trippy effect. The algorithm functions by producing dream Like Effects.

Deep learning has captured the interest of academics recently. Deep learning methods like convolutional neural networks (CNN) are commonly used to solve complex problems. It avoids the limitations of traditional machine learning methods. The objective of this study was to inform readers about a variety of CNN-related issues. In addition to outlining the three most popular topologies and learning techniques for CNN, this page also explains the network conceptually. By providing students with a comprehensive understanding of CNN, this study will inspire academics to continue their studies in this field. This study will be a source and handy guide for anybody who are interested in the subject.

Image style conversion is a technique for altering an image's background and general tone (style) without altering the source image's overall composition. This creates a new, aesthetically pleasing image that seamlessly combines the local characteristics of the source image with the style. The typical style conversion process can only synthesize images in the abstract painting style due to the limitations of the style conversion method, which prevent it from extracting the target image's high-level abstract components. German researchers like Gatys and others promoted the use of convolutional neural networks to achieve visual style conversion as a result of the rapid development of artificial intelligence algorithms and the deployment of funding in experimental facilities [1].

On the basis of this discovery, they were able to separate the style and content of the image and then combine the overall content and style of the image to achieve style conversion. They discovered the convolutional neural network through research and divided it into deep convolutional layers and shallow convolutional layers. The former obtained the overall framework of the image, while the latter obtained the style characteristics of the image. The gradient descent method must be used in the implementation to repeatedly update the pixel values of the source image, which significantly increases resource occupation and cost and introduces a number of issues with slow generation progress. However, the method successfully implements style conversion. Researchers have made many advancements as a result of Gatys' work, with Justin et al. proposing a rapid stylized conversion algorithm [2] that can transform the source image into the desired style after just one optimization by establishing a feed-forward network beforehand, greatly enhancing generation speed.

3 Methodology

3.1 Tools Used

a. **TensorFlow**: Google's TensorFlow is a prominent open-source machine learning framework. One of its applications is in computer vision, particularly in the development of art using neural networks, termed "DeepDream". "TensorFlow" is used to implement the image generating process in addition to training the model. Users can produce highly customized and unique images by putting an image through the trained model and altering specific parameters, such as the "dreaminess" of the image.
b. **DeepDream**: is an image enhancement and generation approach that uses a convolutional neural network (CNN). It operates by passing an input image through the network's layers and amplifying specific features discovered at each layer. The generated image is then put back into the network, and the process is iteratively repeated to produce highly detailed and abstract images.
c. **DeepDream**: is implemented using TensorFlow, which provides an efficient way to define and train the CNN model. TensorFlow's computational graph design enables efficient parallel processing on GPUs, making it perfect for training large-scale neural networks such as DeepDreams.
d. **ImageNet**: In the DeepDream approach, ImageNet was used as a source of pre-trained neural network models. For this purpose, the Inception model, developed by Google researchers and winner of the 2014 ImageNet Large Scale Visual Recognition Challenge (ILSVRC), has been used extensively. The pre-trained Inception model is used as a starting point in the DeepDream technique to generate new images by optimizing the activation of certain neurons within the network. These activations are interpreted as patterns or properties recognized by the network, such as edges, textures, or forms.
e. **Inception-v3**: Inception-v3 is a deep convolutional neural network architecture developed by Google researchers as an extension to the previously developed Inception-v1 and Inception-v2 models. It was created to improve the accuracy of image classification tasks, particularly those requiring fine-grained object detection.

The Inception-v3 architecture is made up of convolutional layers, pooling layers, and fully linked layers. It also includes some innovative features, such as the use of factorized convolutions and batch normalization, which reduces computing costs while enhancing model accuracy (Table 1).

A number of network models are given in the table along with their top-1 error, top-5 error, and cost in billion operations (Bn Ops). The top-1 error is the likelihood

Table 1 Inception-v3 trained on ImageNet and compared with other contemporary models

Network	Top-1 error (%)	Top-5 error (%)	Cost Bn ops
GoogLeNet	29	9.2	1.2
BN-GoogLeNet	26.8	2013	1.5
BN-Inception [7]	25.2	7.8	2.0
Inception-v3-basic	23.4	–	3.8
Inception-v3-rmsprop RMSProp	23.1	6.3	3.8
Inception-v3-smooth LabelSmoothing	22.8	6.1	3.8
Inception-v3-fact Factorized 7 * 7	21.6	5.8	4.8
Inception-v3-BN-auxiliary	21.8	5.6	4.8

that the network will make an incorrect prediction, whereas the top-5 error is the likelihood that the network will make an incorrect prediction among the top-5 most likely outcomes. The price, which is expressed in billion operations, is what it costs to run the network computationally. The table demonstrates that as models get more sophisticated, their error rates go down while their computing costs go up.

f. **The PIL module** (short for Python Imaging Library): is a third-party Python library that allows you to view, manipulate, and save several image file types. The Image module is one of PIL's sub-modules that provides a convenient and user-friendly interface for working with images in Python.

3.2 *DeepDream Implementation Using Tensorflow*

Given a pre-trained convolutional neural network with parameters θ, the DeepDream algorithm aims to find an input image x that maximizes the activation of a set of target neurons. The activation of a neuron in a specific layer is represented as a function $f(x; \theta)$, where x denotes the input image and θ represents the network parameters. To quantify the activation of the target neurons, a loss function $L(x; \theta)$ is defined, providing a measure of how changes to the input image affect the target neuron activation. Common choices for the loss function include the sum or average activation of the target neurons.

To compute the gradients $\nabla L(x; \theta)$ of the loss function with respect to the input image x, TensorFlow's automatic differentiation capabilities are utilized. This computation enables the determination of the direction and magnitude of changes required in the input image to maximize the loss, thus emphasizing the activation of the target neurons. The gradients provide valuable information on how to modify the input image to amplify the desired neuron activations.

Based on the computed gradients, the DeepDream algorithm updates the input image x by iteratively applying incremental modifications. These modifications are determined by a step size α and are proportional to the gradients $\nabla L(x; \theta)$. The iterative process involves adjusting the input image to progressively amplify the activation of the target neurons.

To enhance the visual quality and generate intricate patterns, the DeepDream algorithm further employs iterative refinement. At each iteration, the modified image from the previous step becomes the new input, and the gradient computation and image modification steps are repeated. This iterative process allows the algorithm to capture more detailed and visually interesting patterns over multiple iterations.

In addition to the core algorithm, DeepDream offers optional techniques to improve the generated images. These techniques may include smoothing or regularization, which help reduce noise or artifacts that may arise during the DeepDream process. By applying these techniques, the algorithm can produce visually appealing and aesthetically pleasing results.

3.3 Proposed System

3.3.1 Working of DeepDream

Convolutional neural networks (CNNs) have emerged as a powerful tool for image analysis and understanding. Among the various CNN architectures, Inception-v3 has gained significant attention for its exceptional performance in visual recognition tasks. In this research project, we leverage the capabilities of Inception-v3 to extract and amplify specific patterns present in images.

The initial layers of the Inception-v3 network, typically consisting of 2–3 layers, serve as the foundation for pattern extraction. These layers are responsible for detecting low-level visual features such as edges, corners, and textures. As we progress deeper into the network, the subsequent layers become increasingly adept at recognizing complex visual patterns, thanks to the inclusion of thousands of layers.

To amplify specific patterns in the images, we focus on three specific layers: "mixed-3", "mixed-5", and "mixed-7". These layers represent deeper stages within the Inception-v3 architecture, where the network has learned to capture higher-level visual features. By analyzing the activations in these layers, we can emphasize and enhance the presence of specific patterns, enabling us to gain deeper insights into the visual content of the images.

Research aims to exploit the power of Inception-v3 by harnessing the discriminative capabilities of its intermediate and deep layers. By leveraging the network's ability to recognize intricate visual patterns, we seek to uncover hidden structures, distinguish subtle details, and enhance the interpretability of the images under study (Fig. 2).

Based on the concepts of pattern recognition and image processing, the technique employs a pre-trained neural network to generate beautiful and hallucinogenic visuals by augmenting patterns and features in an image. DeepDream's premise is like how human's analyses shapes and patterns in ordinary items. Shapes that appear like animals, objects, or even humans are noticeable. Similarly, the DeepDream algorithm

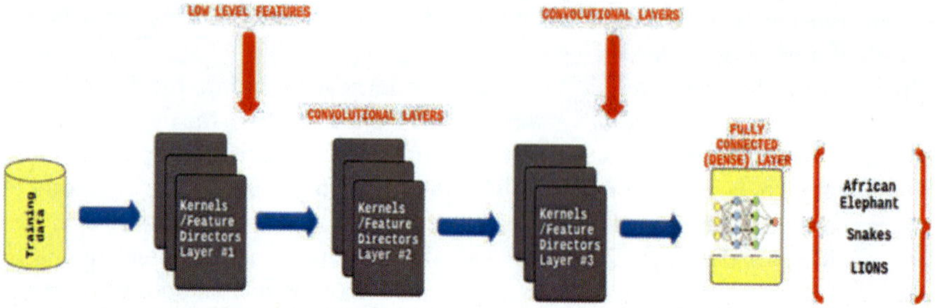

Fig. 2 Working of DeepDream algorithm

takes an image and attempts to analyse it by amplifying specific patterns and features learned by the neural network during its training phase.

From a mathematical aspect, the DeepDream algorithm selects specific layers within a pre-trained neural network to focus on. These layers have been trained to recognize forms, colors, and textures, among other things. The algorithm then improves on these qualities by increasing the layer's activations. As a result, the original image has been converted into a bizarre and hallucinogenic piece of art, with the patterns and characteristics enlarged and twisted in a unique and visually appealing way.

3.3.2 Mathematical Process Behind DeepDream

1. Convolutional Neural Network: DeepDream's first stage is to train a convolutional neural network (CNN) on a huge image dataset. By applying a collection of filters to the input image and sending the outputs through a sequence of nonlinear activation functions, the CNN learns to recognize patterns in images such as edges, corners, and textures.
2. Select a layer: Once the CNN has been trained, a layer of the network is selected to be used for feature visualization. The layer selected affects which features are highlighted in the final image. The network's deeper layers capture more abstract features, such as object sections or entire objects, and the network's earlier levels catch more low-level features, such as edges and textures.
3. Choose an input image: An input image is chosen to serve as the DeepDream algorithm's starting point. A photograph, a painting, or a computer-generated image are all examples of images.
4. Forward pass: The input image is processed by CNN's chosen layer to build a feature map. The activations of the neurons in the layer for the input image are represented by the feature map.

   Each activation in the feature map corresponds to a single feature recognized by the network.
5. Amplify the activations: Gradient ascent is used to amplify the activations in the feature map. Gradient ascent entails computing the gradient of the feature map

in relation to the input image and then modifying the input image to enhance the activations of the neurons that correspond to the desired features. This method is performed several times to get an output image that maximizes the activation of the selected characteristics.

6. Repetition: The preceding processes are repeated for different layers and features to generate a succession of output images that emphasize more abstract features in the source image.

$$y = b + m * x$$

This equation represents the relationship between the input image x and the network's activations or loss y. The trained model is represented as a linear function with two parameters: the line's intercept b (also known as the bias) and slope m (also known as the weight or coefficient). When the input image x is zero, the intercept b represents the baseline level of activations or loss, whereas the slope m reflects the change in activations or loss for a unit change in the input picture x.

As a result, the trained model generates a prediction for the activations or loss y based on the observed correlation between x and y during the training phase when it is applied to a fresh input image x. Through optimization methods like gradient descent, where the objective is to minimize the gap between the expected result of the model and the true output for a given input, the model parameters b and m are learned. The trained model can be trained to correctly forecast the activations or loss for new input images by changing the model's parameters.

$$\begin{aligned} \text{Gradient} &= \mathrm{d}y/\mathrm{d}x \\ &= d(\text{activation}/\text{loss})/d(\text{input image}) \end{aligned}$$

In this equation, dy/dx represents the gradient of the output variable y with respect to the input variable x, also known as the derivative of the output variable y with respect to the input variable x. In this case, y stands for the activation or loss function and x for the input image.

Figure 3 Activations with respect to the pixel of input image the activation or loss function's change in response to a minor modification in the input image can be observed by the derivative d(activation/loss)/d(input image). The gradient can be viewed as a vector pointing in the direction of the activation or loss function's steepest increase. Changes can be figured out from the input image to maximize or minimize the activation or loss by computing the gradient with respect to the input image

$$x|_{\text{new}} = x|_{\text{old}} + \mathrm{d}y/\mathrm{d}x * \text{learning rate(step size)}$$

is an update approach that is applied to iteratively modify the input picture x in optimization methods like gradient descent to reduce the loss function of a convolutional neural network (CNN).

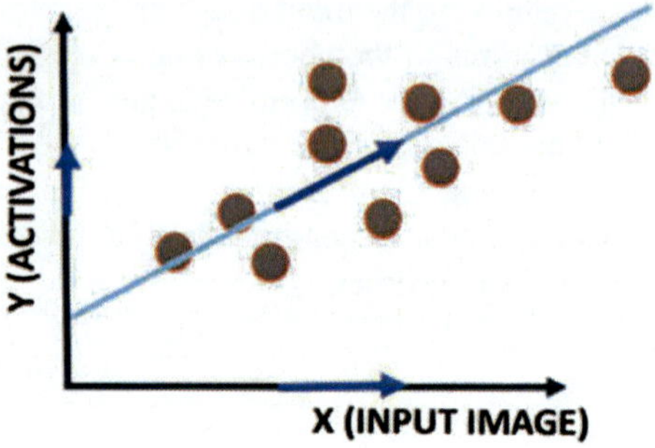

Fig. 3 Activations analysis

In this equation, $x|_{\text{old}}$ stands for the input picture's current value, dy/dx is the gradient of the loss function relative to the input image, and learning rate is a hyperparameter that controls the update's step size.

The direction in which the loss function descends most rapidly, or in which it is minimized, is indicated by the gradient dy/dx derived a step size that specifies how much the input picture should be modified in the direction of the negative gradient by multiplying the gradient by the learning rate.

The rate of learning can be used to regulate the strength of the update while updating the input picture in the direction that reduces the loss function, as shown by the equation $x|_{\text{new}} = x|_{\text{old}} + \text{d}y/\text{d}x * \text{learning rate}$. This update rule can be applied iteratively to improve the prediction by gradually modifying the input image.

Integrated each DeepDream image (blended image) into a video frame to generate a DeepDream video.

There are several ways to accomplish this, but one effective method is to use "zooming," which involves gradually zooming in on the image to create a video-like effect. Begin with the original DeepDream image and crop and resize it to generate a series of frames that gradually zoom in on the image. The DeepDream algorithm is then applied to each frame to create a new blended image that can be used as the next frame in the video.

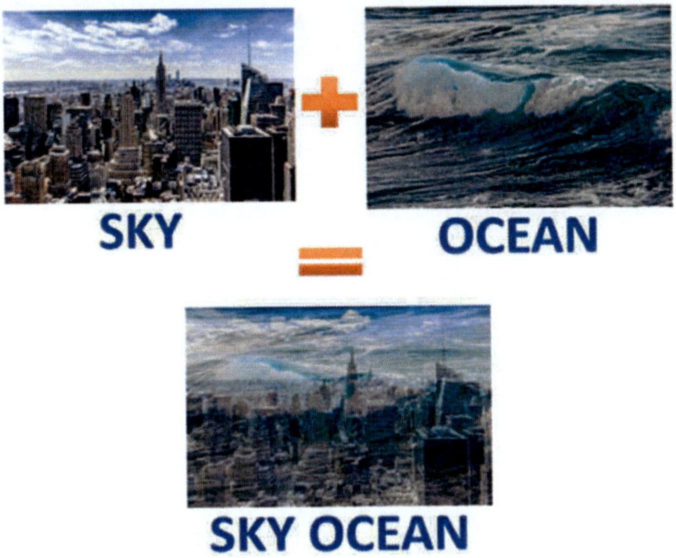

3.3.3 Implementation

To create an amazing DeepDream video, a few steps need to be followed. First, a special place called the "dream location" is set up to store all the frames generated for the video. This dedicated spot keeps things organized and facilitates easy access to the frames as the project progresses.

Next, the dimensions of the blended image are determined. The choice of dimensions is crucial as larger images require more processing time. Experimenting with different dimensions helps find the right balance between quality and processing time.

Once the dimensions are set, the number of frames for the video is decided. Starting with a smaller value, such as 50 frames, is recommended to ensure the process is working correctly. As confidence builds, the frame count can be increased to create longer and visually stunning DeepDream videos. Having more than 500 frames provides ample material to work with.

To add dynamic movement and engagement, the blended image is zoomed in. This involves cropping the edges and resizing the image to its original size. By focusing on different parts of the image, a sense of depth and motion is created, enhancing the appeal of the DeepDream video.

Sometimes, the zooming effect can darken the image over time. Adjusting the GB values can counter this issue. Increasing the RGB values to 2 or above produces brighter DeepDream images, while reducing them below 2 results in darker images.

Experimenting with different values helps achieve the desired brightness and visual impact.

The deep dream model helper function can be utilized to customize the deep dream algorithm and achieve the desired image blend for the video.

To ensure visual consistency and coherence, the image needs to be clipped to fit within the constraints of the neural network. This step involves transforming the datatype of the clipped image result before converting it back to a regular picture. This maintains visual appeal and consistency in the DeepDream video.

All the generated frames are stored in the dream location, making them easily accessible and organized for use in the video.

Finally, the video can be created by specifying a name for the dream, providing the path of the frames to include, defining the codec, and creating a VideoWriter object. This object merges all the frames to produce the final DeepDream video. Following these steps allows the creation of a unique and visually stunning DeepDream video that brings the individual dream to life.

Starting image(img_0)

5th Frame(img_5)

10th Frame(img_10)

20th Frame(img_20)

Here, the shape of the losses is (3), and the sum of all the losses is 0.55770755.

4 Result

Successfully integrating DeepDream with two photographs and making a DeepDream video involved the mixing of two images and applying the DeepDream algorithm. A unique and visually appealing video was created by experimenting with factors such as image dimensions, step size, and the zooming effect. The necessary frames were generated and a video was created by following the procedures mentioned in the approach.

These were the results of experimentation on Steps:

DeepDream model	Inception V3
Step start	0
Step size	0.001
Gradient step	[605, 910, 3]
Gradient at $x = 5$	3625
Inception layers	Mixed-3, mixed-5, mixed-7
Total steps	4000
Loss	0.5804304480552673
Params in Inception-v3	23,851,784

Overall, the outcomes of DeepDream implementation with two photographs and are eager to explore new possibilities with this fascinating method.

5 Conclusion

This case study focuses on the implementation of the DeepDream algorithm to mix two photographs and make a video. To integrate the two images, the Python Image Library (PIL) is utilized, and each DeepDream image is treated as a frame in the video. The image is zoomed in to make the film by chopping off the edges and resizing it to its original shape. RGB values greater than or equal to 2 produce brighter DeepDream images, whereas values less than 2 produce darker images. The DeepDream model helper function is used to change the number of steps and step size to match the intended dream. The image is trimmed to ensure that it fits within the neural network's constraints, and the frames are saved in a dream place for easy retrieval.

Finally, by defining the codec and creating a VideoWriter object, the frames are used to generate the video. Following these procedures will result in a visually spectacular DeepDream film that is tailored to the user's individual dream.

References

1. Fernández-Delgado M, Cernadas E, Barro S, Amorim D (2014) Do we need hundreds of classifiers to solve real-world classification problems? J Mach Learn Res 15(1):3133–3181
2. LeCun Y, Bengio Y (1998) Convolutional networks for images, speech, and time series. In: Michael AA (ed) The handbook of brain theory and neural networks. MIT Press, Cambridge, MA, USA, pp 255–258
3. Al-Khazraji LR, Abbas AR (2022) Employing neural style transfer for generating deep dream images. ARO-Sci J Koya Univ 10(2):134–141
4. Suzuki K, Roseboom W, Schwartzman DJ, Seth AK (2017) A deep-dream virtual reality platform for studying altered perceptual phenomenology. Sci Rep
5. Naul B, Bloom JS, Pérez F, van der Walt S (2018) A recurrent neural network for classification of unevenly sampled variable stars. Nat Astron 2(2):151–155
6. Najafabadi MM, Villanustre F, Khoshgoftaar TM, Seliya N, Wald R, Muharemagic E (2015) Deep learning applications and challenges in big data analytics. J Big Data 2(1):1
7. Goodfellow I, Bengio Y, Courville A (2016) Deep learning. In: Adaptive computation and machine learning. MIT Press, Cambridge, MA, USA, p 775
8. Gavin HP (2016) The Levenberg-Marquardt method for nonlinear least squares curve-fitting problems. Tech Rep
9. Glorot X, Bengio Y (2010) Understanding the difficulty of training deep feedforward neural networks. In: Proceedings of 13th international conference on artificial intelligence statistics, pp 249–256
10. Chien S, Choo S, Schnabel MA, Nakapan W, Kim MJ, Roudavski S (2016) Artificial imagination of architecture with deep convolutional neural network. In: 21st international conference of the Association for Computer-Aided Architectural Design Research in Asia CAADRIA
11. Spratt L. Dream formulations: humanistic themes in the iconology of the machine-learned image. Kunsttexte.de Art Historical J
12. Boden MA (2004) The creative mind: myths and mechanisms. Psychology Press
13. Jennings KE (2010) Developing creativity: artificial barriers in artificial intelligence. Mind Mach 20(4):489–501
14. DiPaola S, Gabora L (2009) Incorporating characteristics of human creativity into an evolutionary art algorithm. Genet Program Evolvable Mach 10(2):97–110
15. Wiggins G (2006) A preliminary framework for description, analysis and comparison of creative systems. J Knowl Based Syst 19(7):449–458
16. Ritchie D (2007) Some empirical criteria for attributing creativity to a computer program. Mind Mach 17:67–99

Tuberculosis Bacteria Detection Using Deep Learning Techniques

Sharmin Akther Rima, Meftahul Zannat, Samia Shameem Haque, Al Kawsar, Sanjeda Sara Jennifer, and Ahmed Wasif Reza

Abstract Tuberculosis Bacteria (TB) is a bacterial infection that kills and affects millions of people annually. For the disease to be effectively treated and to be stopped from spreading further, early detection is essential. In this paper, an image processing-based deep learning approach is proposed for the detection of TB germs. The effectiveness of the suggested method is assessed using a sizable dataset of TB images, and the outcomes show that it successfully recognizes TB. With the potential to improve patient outcomes and slow the spread of the disease, the suggested method presents a promising approach for the early detection and diagnosis of TB. For the detection of TB, convolutional neural networks (CNNs) InceptionV3, DenseNet201and EfficientNetB3 have been used on a dataset of X-ray images, and the EfficientNetB3 model achieved the highest test accuracy, which is 100%.

Keywords Tuberculosis bacteria · Deep learning · InceptionV3 · DenseNet201 · EfficientNetB3

1 Introduction

Tuberculosis Bacteria (TB) is the world's second worst infectious disease, trailing only Human Immunodeficiency Virus (HIV) with an estimated 1.4 million fatalities in 2019 [1]. The tuberculosis-causing mycobacterium primarily affects the lungs but can also harm other vital organs. If not adequately controlled, it can be lethal and spreads through the air. Pulmonary Tuberculosis Bacteria (PTB) is a contagious type of tuberculosis that impacts the lungs. After settling in the lungs, the germs travel through the blood to several organs, including the heart, kidneys, and others [2]. If TB is not properly treated, it may get worse. Despite the possibility of drug resistance,

S. A. Rima · M. Zannat · S. S. Haque · A. Kawsar · S. S. Jennifer · A. W. Reza (✉)
Department of Computer Science and Engineering, East West University, Dhaka 1212, Bangladesh
e-mail: wasif@ewubd.edu

S. Shakya et al. (eds.), *Fourth International Conference on Image Processing and Capsule Networks*, Lecture Notes in Networks and Systems 798,
https://doi.org/10.1007/978-981-99-7093-3_20

therapy can successfully treat the illness with antibiotics for a period of six months [3]. For ease of use and safety, chest X-rays are preferred for detecting tuberculosis [4]. Another option is to have a doctor examine chest radiographs but this requires extensive clinical work and takes time [5].

Transfer learning can diagnose tuberculosis quickly and accurately. Clinical testing, physical examinations, medical histories and imaging methods are all used in the diagnosis process. The goal of the study is to suggest an enhanced model for quickly diagnosing bacterial lung infections. In order to do that, four deep learning models, namely CNN, InceptionV3, DenseNet201 and EfficientNetB3 have been implemented. EfficientNetB3 displayed the best performance with 100% accuracy. Some of the major challenges associated with manually diagnosing tuberculosis are:

- Difficulty in diagnosing due to its variety of symptoms and severity.
- The symptoms of tuberculosis and other respiratory illnesses overlap, making a diagnosis difficult.
- In cases of extrapulmonary tuberculosis, conventional diagnostic techniques may not be sensitive enough.
- Testing for tuberculosis, particularly molecular testing, can be costly and not always easily accessible, especially in low-resource healthcare settings.

The major body of this paper is structured as follows: Our literature review is covered in Sect. 2 and the materials and methods for implementation are covered in Sect. 3. Results and discussion are covered in Sect. 4, and our research is wrapped up with a conclusion in Sect. 5.

2 Literature Review

Numerous techniques, including sputum microscopy, molecular assays, and skin tests, are used in clinical practice to diagnose tuberculosis bacteria (TB), taking into account risk factors, medical history and symptoms. Rapid diagnostic instruments like GeneXpert MTB/RIF (Mycobacterium tuberculosis complex and resistance to rifampin) can identify both TB and resistance. The accuracy of diagnostics is improved by advances in machine learning and artificial intelligence (AI). Serological testing, nucleic acid amplification tests, imaging, computed tomography and positron emission tomography (CT and PET) and non-invasive tools like breath analysis and biosensors are a few of the alternative methods under investigation. These methods seek to enhance early detection, the start of therapy and effective TB management in medical settings [6].

Large datasets of clinical data, genetic data and medical imaging are analyzed by machine learning algorithms to demonstrate their potential for TB identification. In order to detect and diagnose TB early, these algorithms recognize its patterns. Clinical data evaluation and prediction models frequently use random forests and support vector machines (SVMs). Additionally, ML can use genetic information to identify TB strains, anticipate medication resistance and forecast treatment outcomes.

The detection of TB is improved when machine learning (ML) and nuclear acid amplification test (NAATs) are used together. For the development and application of ML models in TB diagnosis, it is essential to overcome issues with data quality, biases and ethics [7].

Convolution neural networks (CNNs) are quite good at identifying TB in X-ray and CT scan pictures used for diagnosis. Transfer learning with pre-trained models improves performance, while deep learning algorithms extract features and spot TB-related anomalies. The use of trained algorithms and picture masks is crucial. Deep learning algorithms must be used in TB diagnosis, and this requires overcoming biases, assuring high-quality datasets and addressing ethical issues [8].

InceptionV3 with transfer learning was used by Karaca and Güney to identify tuberculosis with 98.90% accuracy. The research's crucial contribution was identifying the constraints placed on radiologist availability and expertise, as well as the challenges associated with computer training [9].

To discriminate between photos of tuberculosis and healthy images, Rahman and Khandakar used a variety of deep neural networks, with ChexNet attaining an accuracy of 96.47%. The paper's key contribution was faster TB diagnosis by computer usage, and its limitations included the possibility that CNN model performance may be enhanced even further [10].

The statistics show that Stefanus Kieu Tao Hwa's suggested ensemble technique, when utilized to diagnose tuberculosis using chest X-rays, achieved a maximum accuracy of 89.77%, a sensitivity of 90.91% and a specificity of 88.64%. The work's significance was the improvement of TB detection through the use of ensemble classifiers, although additional components may further enhance TB detection [11].

3 Materials and Methods

3.1 Dataset Description

In order to proceed with the TB detection experimentation, we have utilized the Tuberculosis Bacteria (TB) Chest X-ray Database, which is publicly available in Kaggle [12]. A database of chest X-ray images for Tuberculosis Bacteria (TB) positive cases as well as normal images has been developed by a group of researchers from Qatar University, Doha, Qatar and Dhaka University, Bangladesh, along with their collaborators from Malaysia and medical professionals from Hamad Medical Corporation and Bangladesh. With an accuracy of 98.3%, the research team was able to distinguish between TB and normal chest X-ray images. There are 4200 X-ray images that are divided into two classes. The first class represents normal lungs, while the second class represents lungs affected by tuberculosis bacteria. However, the dataset is imbalanced, where there are 3500 images of normal chest X-ray and 700 images are Tuberculosis Bacteria affected. The image height and width are 512 and 512, respectively. Figure 1 shows the raw images of the dataset.

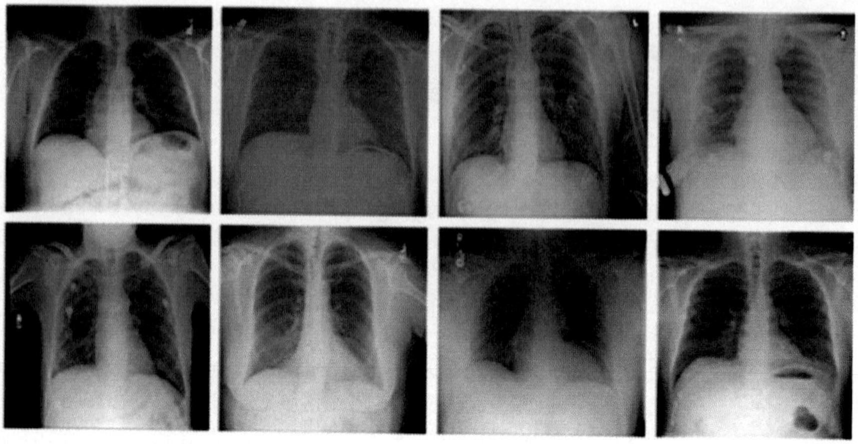

Fig. 1 Exemplary dataset inputs

3.2 Dataset Pre-processing

The pre-processing of image input in our classification is crucial for improved model performance and accuracy. The image height and width are 512 and 512, respectively, before pre-processing. In the three models, including InceptionV3, Densenet201 and CNN, the data was resized by 200*200, and in EfficientNetB3, by 224 * 224. The dataset has been split into three parts: training, testing and validation. Training sets make up 70% of the dataset, followed by testing sets at 15%, and validation sets at 15%. After splitting, the training dataset contains 2940 images, testing contains 630 images and validation contains 630 images.

3.3 System Architecture and Implementation

The process of the recommended framework is shown in Fig. 2. The suggested system covers model training, data pre-processing and results for a deep neural network that classifies chest images. The experimental model and data pre-processing are the two essential steps in the research process. The images have been resized in preparation for processing. After that, training, testing and validation were conducted on the dataset. The dataset and results have been examined using four deep learning models: CNN, DenseNet201, Inception V3 and EfficientNetB3.

This study employed four well-established deep learning models, namely InceptionV3, DenseNet201, CNN and EfficientNetB3. The selection of algorithms was guided by several critical criteria aimed at achieving superior model performance and efficient training. The primary consideration was the model's performance in

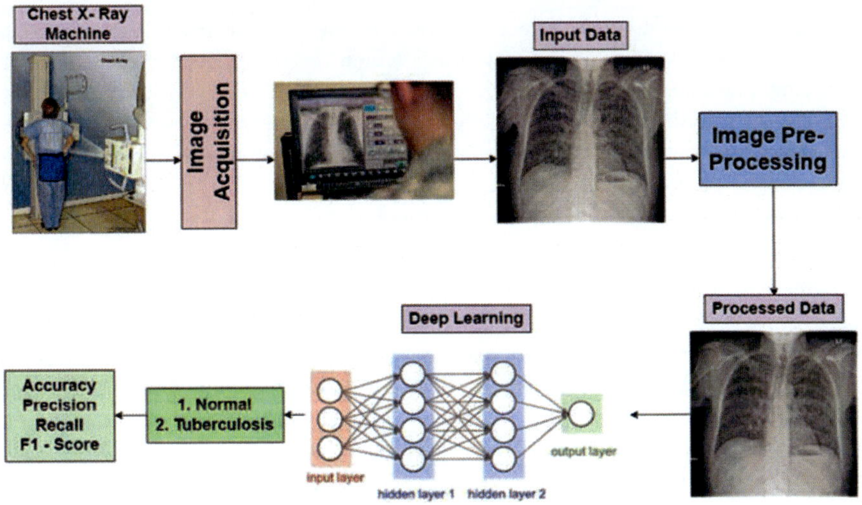

Fig. 2 Hierarchical overview of the system architecture

the specific classification task. These models have demonstrated impressive capabilities in diverse computer vision applications. We applied consistent hyperparameters and resized the images to 200 × 200 pixels for computational efficiency and improved performance. For InceptionV3, DenseNet201 and CNN, feature extraction was performed with a fixed image size of 200 × 200 pixels but for EfficientNetB3, we resized it into 224 × 224. The training of the deep learning model begins by employing a data frame, which provides a structured and efficient way to organize and manage the dataset, ensuring easy accessibility and proper data formatting during the training process. For feature extraction, the output layers of a pre-trained base model are modified to suit the specific classification task. Instead of using the original output layer, a custom output layer is created specifically for this research. Subsequently, the entire model is trained, incorporating the base model with the modified output layer. To enhance the model's performance, batch normalization is applied to the output of the base model. Batch normalization helps improve the convergence of the training process and reduces the negative impact of internal covariate shifts. Additionally, dropout regularization is introduced on the dense layer of the model to prevent overfitting during training, thereby increasing the model's generalization capability. Learning Rate Scheduling and Early Stopping strategies are also incorporated into the training process. With the modified callback, user interaction is enabled during training, allowing dynamic adjustment of the learning rate. This flexibility helps fine-tune the learning rate as the model progresses, and the dwell option further facilitates automatic learning rate adjustments. As a result, the model's convergence can potentially be accelerated, as the learning rate adapts to the changing characteristics of the training process. To optimize the model's learning process, it is compiled with the Adamax optimizer, which is well-suited for handling sparse gradients and has shown to be effective in various deep learning tasks. As for the loss function, categorical cross-entropy is used, which is appropriate for classification problems. The categorical cross-entropy loss measures the dissimilarity between the predicted probability distribution and the true label distribution. By combining these techniques, the proposed model benefits from improved convergence, reduced overfitting, and a better ability to handle classification problems, ultimately leading to enhanced performance and more accurate predictions.

EfficientNetB3 (proposed model): This paper aimed to optimize the performance of the EfficientNetB3 model by exploring different parameters. We resized the images to 224 × 224 pixels. To prevent overfitting, we used a batch size of 30, batch normalization and a dropout rate of 40%. The learning rate was set to 0.001, and both L1 and L2 regularization were employed. We utilized the "Adamax" optimizer and employed the "categorical_crossentropy" loss function. Figure 3 presents the architecture of the EfficientNetB3 model and the proposed EfficientNetB3 model, providing a visual representation of its structure and design.

4 Results and Discussion

4.1 Experimental Results

The experimental result analysis of InceptionV3, DenseNet201, CNN and EfficientNetB3 is given in Table 1, where EfficientNetB3 has the greatest accuracy compared to the other models.

The high accuracy of the EfficientNetB3 model stems from a synergistic combination of architectural innovations, transfer learning, and learning rate adjustment. The EfficientNet architecture, utilizing compound scaling, optimizes the model's depth, width and resolution to achieve a balance between complexity and computational efficiency. Leveraging pre-trained weights from large-scale datasets such as ImageNet facilitates efficient transfer learning, providing a strong foundation for fine-tuning the target dataset with limited labeled examples.

Moreover, our decision to adopt larger input image sizes, enabling the model to capture intricate details and patterns, significantly enhances its recognition capabilities. To further fine-tune the model, we efficiently use computational resources and incorporate a dynamic learning rate adjustment strategy. This strategy ensures the model converges faster and prevents overfitting, ultimately leading to superior accuracy on the image classification task.

In order to evaluate the models and their performances, precision, recall, $F1$-score and confusion metrics were used. The mathematical notations are given in Eqs. (1), (2), and (3), Fig. 4 gives a visual representation of the test accuracy of all the models, and the evaluation outcome is presented in Table 2.

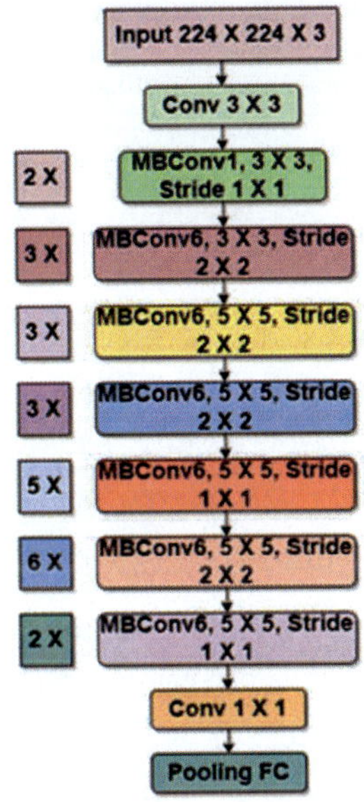

a) EfficientnetB3 Model

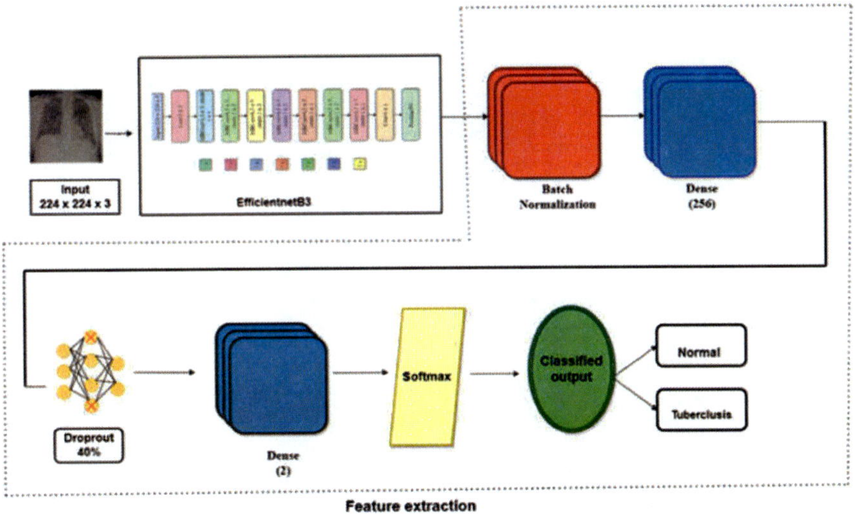

b) Modification of EfficientnetB3 Architecture

Fig. 3 Architecture and the modified version of the EfficientNetB3 model

Table 1 Result analysis

	Model	Training accuracy (%)	Validation accuracy (%)	Test accuracy (%)
1	InceptionV3	95.17	95.32	95
2	DenseNet201	98.18	98.69	92
3	CNN	96.88	96.29	96
4	EfficientNetB3	99.90	99.84	100

$$\text{Precision} = \frac{\text{true positive}}{(\text{true positive} + \text{false positive})} \tag{1}$$

$$\text{Recall} = \frac{\text{true positive}}{(\text{true positive} + \text{false negative})} \tag{2}$$

$$F1\text{Score} = \frac{2 \times (\text{precision} \times \text{recall})}{(\text{presion} + \text{recall})} \tag{3}$$

As given in Table 2, DenseNet201, InceptionV3 and CNN had 92%, 95% and 96% accuracy, respectively, whereas EfficientNetB3 outperformed all the models achieving 100% test accuracy. Figure 5 shows the accuracy and loss of EfficientNetB3 during training and validation, and the confusion matrix is shown in Fig. 6.

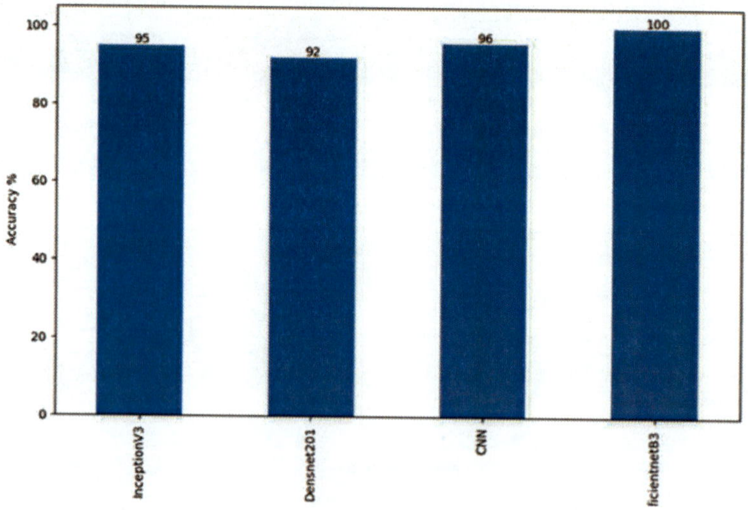

Fig. 4 All models' test accuracy

Table 2 Evaluation outcome

Model	Class	Precision	Recall	$F1$-score	Accuracy
InceptionV3	Normal	0.98	0.96	0.97	0.95
	Tuberculosis affected	0.83	0.89	0.86	
DenseNet201	Normal	0.96	0.95	0.95	0.92
	Tuberculosis affected	0.77	0.78	0.77	
CNN	Normal	0.97	0.98	0.98	0.96
	Tuberculosis affected	0.91	0.87	0.89	
EfficientNetB3	Normal	1.00	1.00	1.00	1.00
	Tuberculosis affected	1.00	1.00	1.00	

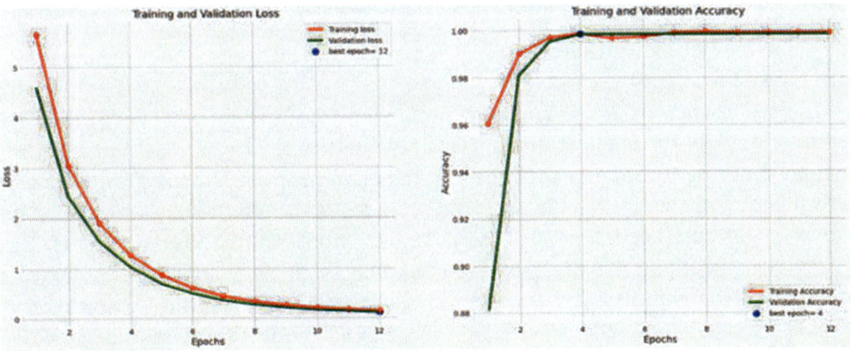

Fig. 5 Training and validation accuracy/loss (EfficientNetB3)

Fig. 6 EfficientNetB3 (confusion matrix)

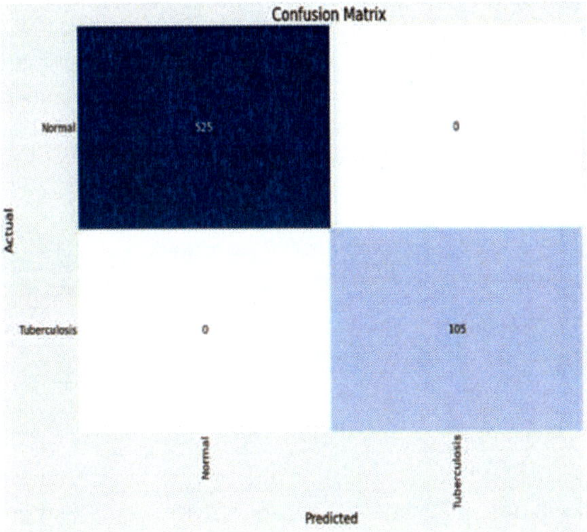

4.2 Performance Evaluation

Software Specification: In order to execute this research, the Windows operating system and Google Colab were utilized. The Windows operating system was used for all of the project's experiments, which is equipped with an Intel(R) Core (TM) i7-10700 CPU clocked at 2.90 GHz. There are 8.0 GB of installed RAM on hand. The system type is a 64-bit operating system, and the CPU is an x64-based processor. On the other hand, Google Colaboratory is a powerful Python programming platform that provides Zero configuration, Free access to GPUs and Easy sharing. Users may create Python programs utilizing all the libraries on a platform provided by Google named Colab.

In order to do a proper evaluation, the output of this study has been compared with several state-of-the-art works. Table 3 gives the earlier studies that have used the same dataset, but different models for detecting TB, presenting a comparative analysis of these works.

Table 3 provides a comprehensive summary of various studies that have used deep learning models for medical image classification tasks. Mwendo [4] utilized deep transfer learning techniques for X-ray image classification. However, the provided information only states that the achieved accuracy was 92%, without specifying the model or approach used. Karaca [9] used the InceptionV3 and CNN models for X-ray image classification. The reported accuracy was 98.90%. Rahman [10] employed the DenseNet model for X-ray image classification and achieved an accuracy of 98.6%. Kieu [11] utilized VGG16 and InceptionV3 models for X-ray image classification. The reported accuracy was 89.77%. Acharya [13] employed Normalizing Flow Networks (NFNets) for X-ray image classification and achieved an accuracy of 96.91%. Tasci [14] utilized the CNN-VoPreCNNFT model for X-ray image classification and reported an accuracy of 97.69%. Huy [15] used a new deep learning model CBAMWDNet to Detect Tuberculosis, and they achieved 98.80% accuracy in their research. Table 3 gives that our model offers the maximum accuracy when compared to other models after comparing the suggested model with them. It achieved 100% accuracy which full filled our goal.

Table 3 Comparison with state-of-the-artwork

References	Model	Dataset	Accuracy (%)
Mwendo [4]	Deep transfer learning	X-ray images	92
Karaca [9]	InceptionV3, CNN	X-ray images	98.90
Rahman [10]	DenseNet	X-ray images	98.6
Kieu [11]	VGG 16, InceptionV3	X-ray images	89.77
Acharya [13]	NFNets	X-ray images	96.91
Tasci [14]	CNN-VoPre CNNFT	X-ray images	97.69
Huy [15]	CBAMWDnet	X-ray images	98.80
This paper	EfficientNetB3	X-ray images	100

5 Conclusion

This research on Tuberculosis Bacteria detection using deep learning techniques and image pre-processing concludes that deep learning techniques, combined with image pre-processing, can provide an accurate and effective way of detecting Tuberculosis Bacteria. In this research, four different deep learning models were used, and the outcomes demonstrated that EfficientNetB3 provides a maximum test accuracy of 100%, outperforming all the models that were implemented. Even when performing a comparative analysis with other state-of-the-art works, it was observed that EfficientNetB3 had the best performance. To increase the generalizability of the suggested approach, the paper recommends further study using larger datasets while being aware of the dataset's limitations.

References

1. Ramya R, Babu PS (2015) Automatic tuberculosis screening using canny edge detection method. In: Proceedings of the 2nd international conference on electronics and communication systems (ICECS), India, Feb 2015
2. Brooks JV, Orme IM (1998) Evaluation of once-weekly therapy for tuberculosis using isoniazid plus rifamycins in the mouse aerosol infection model. Antimicrob Agents Chemother 42(11):3047–3048
3. Overview Tuberculosis (TB) (2019) [online]. Available https://www.nhs.uk/conditions/tuberculosis-tb/NHS
4. Mwendo I, Gikunda K (2023) Deep transfer learning for detecting Covid-19, Pneumonia and Tuberculosis using CXR images—a review. arXiv preprint arXiv:2303.16754
5. Silverman C (1949) An appraisal of the contribution of mass radiography in the discovery of pulmonary Tuberculosis. Am Rev Tuberculosis: 466–482
6. Xpert MTB/RIF on stool is useful for the rapid diagnosis of Tuberculosis in young children with severe pulmonary disease [online]. Available https://www.ncbi.nlm.nih.gov/pmc/articles/PMC5558052
7. Tuberculosis disease diagnosis based on an optimized machine learning model [online]. Available https://www.ncbi.nlm.nih.gov/pmc/articles/PMC9041161/
8. Liu C, Alcantara M (2017) TX-CNN: detecting Tuberculosis in chest X-ray images using convolutional neural network. In: 2017 IEEE international conference on image processing (ICIP)
9. Karaca BK, Güney S (2021) Comparative study for Tuberculosis detection by using deep learning. In: 44th international conference on telecommunications and signal processing (TSP), pp 88–91
10. Rahman T, Khandakar A (2020) Reliable Tuberculosis detection using chest X-ray with deep learning, segmentation and visualization. IEEE Access: 191586–191601
11. Kieu S, Hwa T, Hijazi MHA (2019) Ensemble deep learning for tuberculosis detection using chest X-ray and canny edge detected images. IAES Int J Artif Intell: 429–435
12. Tuberculosis (TB) chest X-ray database [online]. Available https://www.kaggle.com/datasets/tawsifurrahman/tuberculosis-tb-chest-xray-dataset
13. Acharya V, Dhiman G (2022) AI-assisted tuberculosis detection and classification from chest X-rays using a deep learning normalization-free network model. Computat Intell Neurosci

14. Tasci E, Uluturk C (2021) A voting-based ensemble deep learning method focusing on image augmentation and preprocessing variations for tuberculosis detection. Neural Comput Appl 33(22):15541–15555
15. Huy VTQ, Lin C-M (2023) An improved Densenet deep neural network model for tuberculosis detection using chest X-Ray images. In: 2023 3rd international conference on intelligent communication and computational techniques (ICCT), vol 11, pp 42839–42849

An Enhanced Real-Time System for Wrong-Way and Over Speed Violation Detection Using Deep Learning

A. Manasa and S. M. Renuka Devi

Abstract Nowadays in many of the developing countries, traffic rule violation has become a critical issue. Due to the increase in population, the use of vehicles is also increasing exponentially, leading to an increase in road accidents. Major reasons for these accidents are traffic rule violations, such as over speed and wrong direction driving. To monitor these violations, an enhanced real-time system is needed which has the ability of traffic rule violation detections from nearby CCTV cameras. The proposed traffic signal violation detection system performs tasks like speed violation detection and wrong-way direction detection, without human intervention. The proposed system mainly performs three tasks: vehicle detection, vehicle tracking and violation detection. The YOLOv3 deep learning method is used for locating vehicles moving on road, and then the Kalman filter is used for tracking of detected vehicles. After tracking each vehicle is assigned with an unique id, then speed violation detection algorithm and wrong-way direction detection algorithm vehicles are used to identify the vehicles which violated the traffic rules. The system is very efficient and accurate to identify vehicles moving in the wrong direction and over speed.

Keywords YOLOv3 · Kalman filter · Speed violation · Wrong direction violation

1 Introduction

Due to rapid growth of population, usage of vehicles has also increased exponentially. This is a serious threat to the lives of people, due to the increase in road accidents. The number of such road accidents is equally proportional to traffic rule violations. A 2018 report of the Road Transport Ministry [1] shows that in India, the total number of road accidents has increased when compared with previous years. In the world record statistics [2] report, India is ranked in 1st place with the highest deaths

A. Manasa · S. . M. Renuka Devi (✉)
Department of ECE, G. Narayanamma Institute of Technology and Science (for women), Hyderabad, Telangana, India
e-mail: renuka.devi.sm@gnits.ac.in

S. Shakya et al. (eds.), *Fourth International Conference on Image Processing and Capsule Networks*, Lecture Notes in Networks and Systems 798,
https://doi.org/10.1007/978-981-99-7093-3_21

caused by road accidents. As per [3] 'Road Accidents in India 2021', there were about 410 k incidents of road accidents that took place in just a single year, with 150 k deaths and 380 k injured persons. Therefore, road accidents have become one of the main causes of death. This is worsened due to the increase in total vehicles on roads. Accidents mostly occur due to over speed violation and wrong direction of vehicles. To overcome this problem and to reduce such accidents, an enhanced real-time system is very important which uses deep learning algorithms for analyzing and detecting these violations effectively to decrease road accidents.

The main contribution of this paper is, the YOLOv3 model is trained on real images to classify and detect cars, bikes and humans. Unlike other research works, where multiple systems are needed for detection of multiple violations, this paper proposes a single system to detect multiple violations at once. The paper is organized in the following way: Sect. 1 has the introduction of the proposed model, Sect. 2 discusses literature survey and its summary. Section 3 discusses algorithms used in the proposed model. Section 4 shows results of proposed algorithms with respective figures. Section 5 briefly discusses the conclusion and future scope of this paper.

2 Literature Survey

Over many years, researchers have been working to develop algorithms that can detect various traffic rule violations. Mampilayil et al. [4] have proposed an automated video surveillance system which has the capability of detecting traffic rule violations automatically. The main focus is only on three wheeler vehicles. For detection Tensorflow with OpenCV is used and the detection of traffic violation is done by deep learning.

Bay dadaev et al. designed a model which detects vehicles violating the traffic rule [5]. The proposed model is mainly based on video processing. The deep learning method YOLOv3 is used for detection of cars. After estimating the location of a vehicle, Kalman filtering also known as linear quadratic estimation (LQE) is implemented for vehicle tracking. A unique method called 'entry-exit' algorithm is used which has 91.98% accuracy in wrong-way detection of vehicles.

Suttiponpisarn et al. introduced a system that detects and tracks vehicles from CCTVs. Using the RLB-CCTV algorithm, road lane boundaries are obtained with 91.6% accuracy, then vehicle detection is done by YOLOv4-Tiny, for tracking, Deep SORT is used [6]. It automatically detects vehicles driving in wrong direction using the MBCDD algorithm with 96.61% accuracy.

Rahman et al. designed a wrong-way vehicle detection system [7]. It mainly consists of three stages: YOLO, centroid tracking algorithm and the last stage is violation detection. The proposed system is effective in different lighting and different environmental conditions. The main limitation is the centroid tracking algorithm which causes switching of vehicle IDs due to overlapping of vehicles.

Suttiponpisarn et al. [8] introduced a model which detects and tracks the direction of vehicles moving on two-way roads, with the help of YOLOv4-tiny and Deep SORT.

The system first draws the region of interest and validates the vehicle direction giving very accurate results for detection of wrong direction vehicles.

R. J. Franklin et al. developed a system to detect traffic rule violations such as high vehicle speed and signal jump. It uses the YOLOv3 object detection technique. This system then improves accuracy by using the vehicle location and region of interest in between frames determining signal jump. This counts detected vehicles with accuracy of 97.67% and detects speed violations with 89.24% accuracy [9].

A. Sayar et al. developed a system which will detect the wrong-way violation [10]. For vehicle detection, support vector machines (SVMs) are used. The MedianFlow algorithm is used to track vehicles due to its high speed. After vehicle detection and tracking, center points of two consecutive frames are taken from the bounding box and are compared for violation detection.

Mallela et al. [11] presented a system which detects triple riding and the violation of speed limit. The YOLOv3 is trained on the COCO dataset to detect bike riders, and then it checks whether the rule is violated or not. The GSM module is used for data transferring. It is connected to the NodeMCU module for data extraction and transfers data to the cloud. This model has 91.7% accuracy.

All previous research paper works and the methodologies used are summarized into a single table for better understanding. It is given in Table 1.

Table 1 Summary for literature survey on different traffic rule violations

Authors and year of publication	Methodology	Type of violation detection	Type of vehicles	Accuracy (%)
Mampilayil et al. [4], 2019	Tensorflow with OpenCV	Wrong direction detection	Three wheelers	–
Sayar et al. [10], 2019	SVM	Security lanes and wrong-way violation	Cars	90
Rahman et al. [7], 2020	YOLO and centroid tracker	Wrong-way violation detection	Cars	–
Franklin et al. [9], 2020	YOLOv3	Speed violation and signal jump detection	All vehicles	89.24
Mallela et al. [11], 2020	YOLOv3	Speed violation, triple riding detection	Bikes	91.7
Bay dadaev et al. [5], 2020	YOLOv3, Kalman filter	Wrong-way detection	Cars, persons	91.98
Suttiponpisarn et al. [8], 2021	Yolov4-tiny and deepsort	Wrong-way detection	Bikes	96.1
Suttiponpisarn et al. [6], 2022	RLB-CCTV, MBCDD	Road lane, wrong-way detection	Cars	96.61

3 Project Methodology

In this proposed system first, each vehicle present in a given frame is detected and classified using the YOLOv3. For each detection of a vehicle, a bounding box is generated and fitted over it. Then, the information from these bounding boxes is given to the Kalman filter. Then, the Kalman filter predicts and estimates the location of each vehicle in its next frames and tracks are updated. After tracking, each vehicle is assigned with an unique id. The proposed system is further developed for speed violation detection and wrong-way direction detection using over speed violation detection (OSVD) algorithm and wrong-way traffic violation detection (WTVD) algorithm as shown in the block diagram in Fig. 1. The proposed system can detect these different violations using a single system.

3.1 *YOLOv3 Algorithm*

The YOLOv3 algorithm is a convolutional neural network (CNN), which is used for the detection of various different objects in real-time situations. It is used for the identification of particular objects in images and videos. To detect an object, it uses features of input images learned by using CNN [12]. For vehicle detection, YOLOv3 is implemented using the deep learning libraries OpenCV and keras. The Darknet-53 is used in YOLOv3 [13].

Object detection can be done using background subtraction and optical flow, but they show false positive detections when there are several vehicles moving very close to each other. Also these methods are very sensitive in low light conditions resulting in inaccurate models. To overcome these drawbacks, YOLOv3-416 algorithm is used which has processing speed of 29 ms and mean Average Precision (mAP) of 55.3% on COCO dataset.

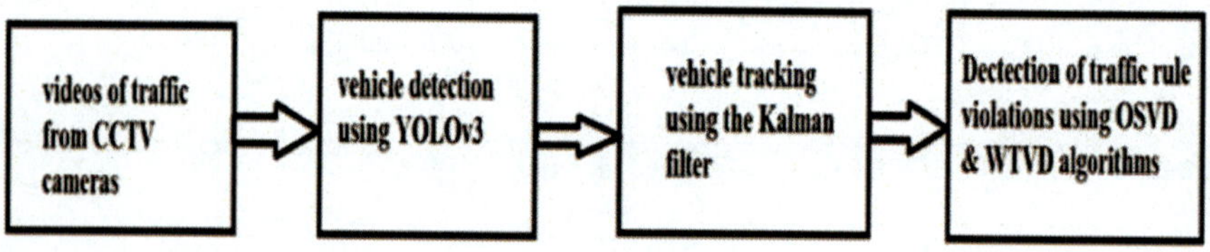

Fig. 1 Block diagram of proposed model

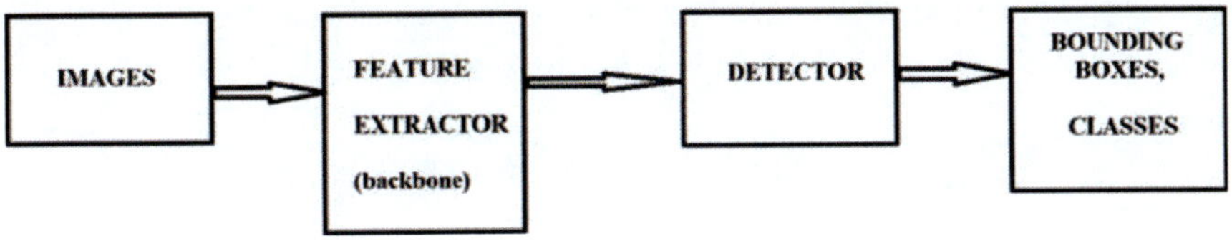

Fig. 2 Basic block diagram of YOLOv3 deep learning network

3.2 *Working of YOLOv3*

The convolutional layers of YOLOv3 architecture learn various important features of given input images and are then given to a regression for detection. These learned features have the bounding boxes coordinates, class label, bounding boxes sizes, dimensions and many more. In YOLOv3, each cell of the prediction map predicts only a certain number of bounding boxes. YOLOv3 has three anchors in particular and it predicts three bounding boxes per neuron. Sometimes, more than one bounding box detects the object as positive detection, which leads to a single object to be detected multiple times. To avoid this situation, non-maximum suppression is used which will pass detections only if they haven't been detected previously. Non-maximum suppression is an imperative part of utilizing the YOLOv3 algorithm effectively. Anchor boxes and non-maximum suppression values are few of the features that make the predictions of an object possible. Every bounding box prediction has five components, namely: (a, b, w, h, c). The (a, b) are centroid coordinates of the box, w and h are width and height and c is confidence score. Block diagram of YOLOv3 is given in Fig. 2.

3.3 *YOLOv3 Network Architecture*

The YOLOv3 network architecture consists of elements such as residual blocks, skip connections and upsampling. The CNN layers are followed by batch normalization and leaky ReLU. For detection, YOLOv3 uses Darknet-53 along with additional 53 layers stacked onto it, giving a total of 106 layers of fully convolutional architecture as shown in Fig. 3. It does not have any pooling layers (max pooling loses small features from the input). Instead of pooling, additional CNN layers with stride 2 are used to downsample the input image, which prevents loss of low level features that are useful in detecting small objects.

The input size for YOLOv3 is 416×416 as it has a better map and better processing speed. YOLO version3 makes predictions at 3 different layers 82, 94 and 106 layers, which are used for detecting large, medium and small-sized objects, respectively.

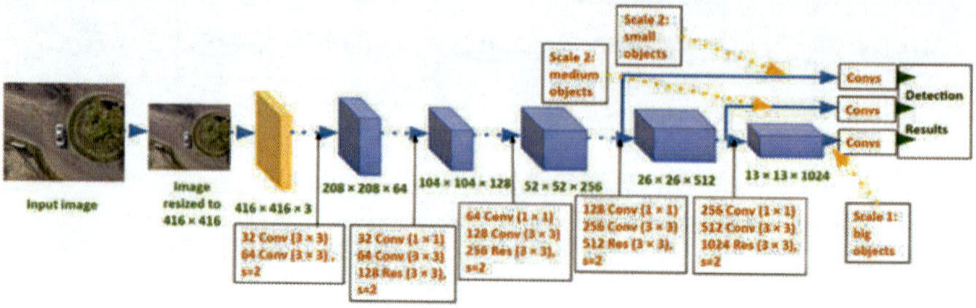

Fig. 3 YOLOv3 network architecture [14]

3.4 *Kalman Filter*

In order to be able to recognize traffic violations, after vehicle detection and drawing bounding boxes, tracking of moving vehicles must be done within a certain period of time. The process of vehicle tracking has two stages: In the first stage, every moving car is detected from each frame then in the second stage, new detections of the same car are associated with it over time. To track each detected vehicle, the Kalman filter is being used. The Kalman filter is powerful and fast compared to other tracking algorithms. Therefore, it is used for tracking the detected vehicles. It predicts the vehicle location in each frame and assigns them to its respective track.

The YOLOv3 model gives the bounding box information and from these the centroids of each bounding box are extracted and given to Kalman filter which is primarily based on centroid tracking algorithm. For every detection an unique ID is assigned. The Hungarian algorithm [15] is used to assign each detection to its corresponding unique ID. In the next step, Kalman filter equations are used to predict multiple vehicles. The Kalman filter has mainly two phases: (1) Prediction phase and (2) Correction phase. In the prediction phase, the Kalman filter predicts the state at the next step. And after receiving measurement, it corrects or updates the prediction of the current state. It runs in loop for further frames and updates the assigned tracks and deletes lost tracks till the end of the video. The formulas for both phases are given below.

1. Prediction phase

State estimation:

$$e_k = Se_{k-1} + Cu_k \tag{1}$$

Error covariance:

$$q_k = Sq_{k-1}S^T + p \tag{2}$$

2. Correction phase

Gain:

$$G_k = q_k O^T \left(O q_k O^T + R\right)^{-1} \tag{3}$$

State estimation update:

$$E_k = e_k + G_k(M_k - O e_k) \tag{4}$$

Error covariance update:

$$Q_k = (1 - G_k O) q_k \tag{5}$$

where e_k and E_k are state estimations in prediction and correction phase respectively, S is state transition matrix, C is the control matrix, u_k is the variable input, q_k and Q_k are the error covariance in prediction and correction phase, respectively, p is process noise covariance, O is observation matrix, R is measurement covariance, M_k is measurement vector and G_k is the Kalman gain.

3.5 *Wrong-Way Traffic Violation Detection Algorithm*

Wrong-way direction is one of the leading causes of deaths in India. If a vehicle is moving in the opposite direction of traffic flow, then it is called a wrong-way traffic violation. The algorithm used to detect such violation is named as wrong-way traffic violation detection (WTVD). After vehicle detection and tracking, direction of each vehicle is determined. In order to determine the direction of each vehicle first, the pixel difference of that particular vehicle is calculated once for every 5 frames. To calculate this difference, the centroids from each detected bounding box are extracted. Initial centroids of a vehicle are taken from the frame F_1 and are denoted as (A_1, B_1). Centroids from frame F_5 are taken as (A_5, B_5), here for every 5th frame, the difference is calculated for faster direction estimation. Figure 4 shows the movement of cars throughout the video frames.

The pixel difference is calculated as $dA = A_5 - A_1$. If dA is negative and A_5 is smaller than A_1, the car is moving from right to left and has direction as WEST. For $dB = B_5 - B_1$, if dB is negative and B_5 is smaller than B_1, the car is moving from top to bottom and has direction as NORTH. In case of horizontal roads, WEST is considered positive and in case vertical roads NORTH is considered as positive. Otherwise, it is considered as a traffic rule violation as they are moving in opposite directions of intended traffic flow of vehicles. The vehicles violating the traffic flow are detected by the algorithm which then highlights those vehicles as WRONG DIRECTION.

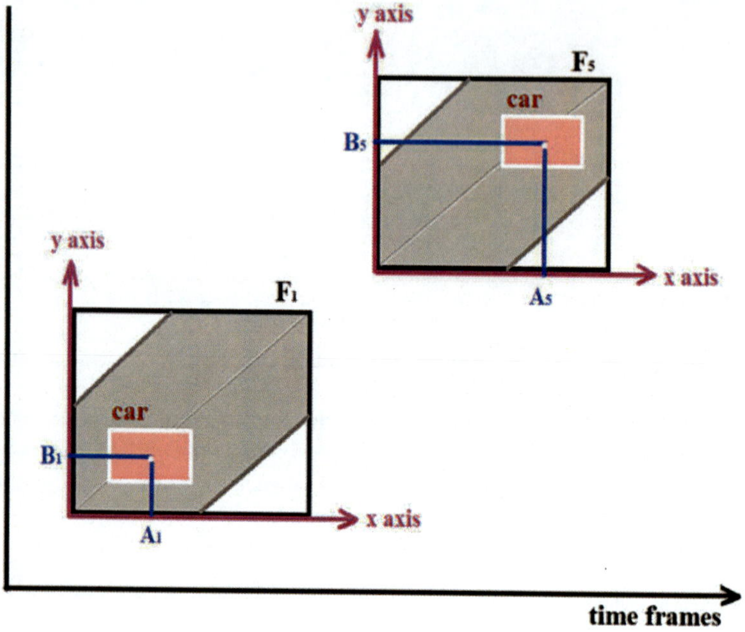

Fig. 4 Cars moving through the frames with centroids (A_1, B_1) and (A_5, B_5)

3.6 Over Speed Violation Detection Algorithm

Another main cause of deaths due to road accidents is vehicles moving with very high speed. The velocity is calculated by using the centroids of the vehicles. After detecting and tracking, each vehicle is fitted with a bounding box. Then, the centroids of each vehicle (A_1, B_1) and (A_2, B_2) are extracted and used to calculate the velocity using below mentioned formulae. Initially, the velocity is calculated in terms of pixels which is then converted into km/h. This is done during object tracking. The over speed violation detection (OSVD) algorithm is used to calculate and estimate the speed of every detected vehicle in the current frame. Therefore, the OSVD algorithm calculates velocity of each vehicle and checks whether the speed is below speed limit Th_1 or not. The traffic rule violation is detected when any vehicle is moving with speed exceeding predefined speed limit Th_1 which is 60 km/h, Th_1 is assumed to be 60 km/h in this paper. The velocity is calculated using the below formulae.

$$\text{Velocity} = \text{d\_met} * 3.6 * \text{fps} \tag{6}$$

$$\text{d\_pix} = \text{sqrt}\left((A_2 - A_1)^2\right) + \left((B_2 - B_1)^2\right) \tag{7}$$

$$\text{d\_met} = \text{d\_pix}/\text{ppm} \tag{8}$$

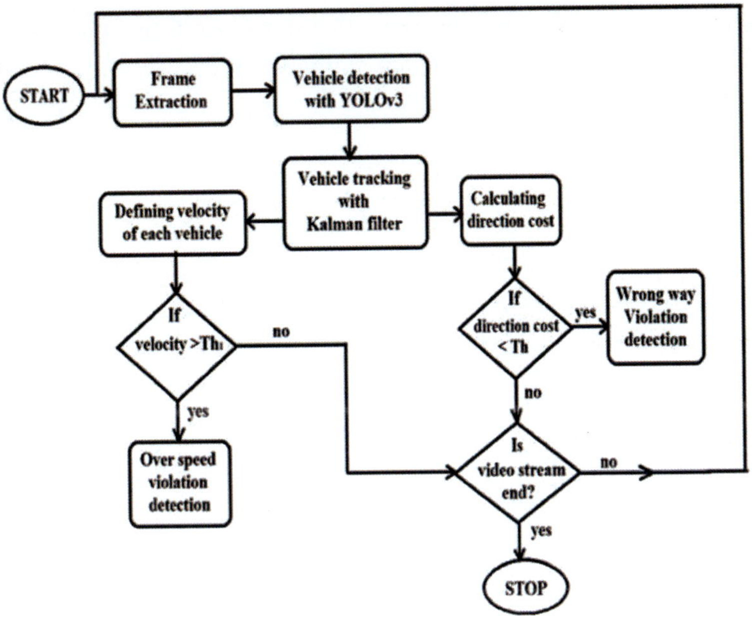

Fig. 5 Flowchart of traffic violation detection system

where ppm is pixel per meter, fps is frames per second, 3.6 is the factor used to convert speed into km/h, d_met is pixel to meter conversion metric, (A_1, B_1) and (A_2, B_2) are centroids.

The flowchart of the proposed model is as shown in following Fig. 5., which starts with extraction of frames from real-time input videos after which an array of frames is given to the YOLOv3 algorithm which detects the 3 classes and fits bounding boxes. Then, the Kalman filter uses this information and tracks the vehicles, and for every vehicle, direction is calculated and velocity is calculated. If the direction of vehicle is opposite to intended traffic flow or if velocity is more than threshold, then they are considered as violations and such vehicles are detected at that input frame. The system then repeats itself until the end of all frames.

4 Experimental Results

To obtain a dataset of traffic videos, different traffic videos are taken from different locations. These videos are of resolution 1920 * 1080 with 30 fps. Then, all frames are extracted from every video and split into a 7:3 ratio and used as training dataset and testing dataset. The YOLOv3-416 model is then trained with real world data from the created dataset. It is mainly trained to detect and classify persons, cars and bikes, in which the input frames are resized to image pixels 416 * 416 with 120 epochs. It took around six hours in total to finish training the custom model with 3 classes. The python software with Tensorflow and OpenCV libraries are used to implement the proposed model.

4.1 Vehicle Detection and Tracking

For vehicle detection, YOLOv3-416 is used which detects and classifies cars, persons and bikes. The algorithm then uses bounding boxes to be fitted over each detection. The trained YOLOv3-416 model is used over new real-time videos which are test videos.

After the detection of vehicles, the information of bounding boxes is given to the Kalman filter which is used for tracking of vehicles. The results of vehicle detection and tracking are shown in Fig. 6, where persons, cars and bikes are detected and bounding boxes are fitted over it.

4.2 Wrong-Way Violation Detection

For wrong-way detection, initially the moving direction of each vehicle is estimated and then vehicles which are moving in the west direction and north direction, i.e., are considered correct and it is considered to be a violation, if it moves in the east direction (left to right) and south direction (top to bottom). Vehicles violating the rule are saved into the result folder. Figure 7 shows the vehicles with traffic violations detected and highlighted as WRONG DIRECTION by the algorithm, where the vehicles moving in the EAST direction and SOUTH direction are wrong.

4.3 Over Speed Violation Detection

The velocity of each vehicle is calculated by using the velocity formula. These estimated velocities of every vehicle are displayed as shown in Fig. 8. Since all vehicles are moving within speed of 60 km/h, there is no violation detected.

Any vehicle which is moving with velocity more than a predefined limit of 60 km/h is considered as violation and are highlighted as 'overspeed' as shown in Fig. 9. The video frames in which vehicles violating the traffic rule are detected will be saved automatically into the result folder.

Fig. 6 Detection and tracking of cars, bikes and persons

5 Conclusion

The proposed traffic signal violation detection system performs tasks like speed violation detection and wrong-way direction detection, without human intervention. The proposed system mainly consists of three stages: vehicle detection, vehicle tracking and violation detection. The YOLOv3 deep learning method is used for the detection of vehicle locations, and then the Kalman filter is used for tracking of detected vehicles. After tracking, each vehicle is assigned with an unique id, then speed violation detection algorithm and wrong-way direction detection algorithm vehicles are used to identify the vehicles which violated the traffic rules. The system is very efficient and accurate to identify vehicles moving over speed and in the wrong direction.

The proposed model has an accuracy of 88.23% in detecting vehicles moving in the wrong direction and accuracy of 91.42% in detecting vehicles with over speed.

Fig. 7 Vehicles moving in wrong direction are detected using Kalman filter

Fig. 8 Speed estimation of vehicles

Fig. 9 Detection of vehicles moving with over speed using Kalman filter

The proposed system can detect only two types of traffic violations. This can further be improved by adding more traffic violations that lead to road accidents.

References

1. Ministry of Road Transport & Highways, Government of India. https://morth.nic.in/sites/default/files/Annual-Report-2010-2011.pdf
2. Global status report on road safety 2018. https://www.who.int/publications-detail-redirect/9789241565684
3. Ministry of Road Transport & Highways, Government of India. https://morth.nic.in/sites/default/files/RA_2021_Compressed.pdf
4. Mampilayil HR, Rahamathullah K (2019) Deep learning based detection of one way traffic rule violation of three wheeler vehicles. In: International conference on intelligent computing and control systems (ICCS), Madurai, India, pp 1453–1457
5. Usmankhujaev S, Bay Dadaev S, Woo KJ (2020) Real-time, deep learning based wrong direction detection. Appl Sci 10(7):2453
6. Suttiponpisarn PS, Charn Sripinyo C, Usanavasin S, Nakahara H (2022) An enhanced system for wrong-way driving vehicle detection with road boundary detection algorithm. Procedia Comput Sci 204:164–171
7. Rahman Z, Ami AM, Ullah MA (2020) A real-time wrong-way vehicle detection based on YOLO and centroid tracking. In: IEEE region 10 symposium (TENSYMP), Dhaka, Bangladesh, pp 916–920
8. Suttiponpisarn P, Charnsripinyo C, Usanavasin S, Nakahara H (2021) Detection of wrong direction vehicles on two-way traffic. In: 13th international conference on knowledge and systems engineering (KSE), Bangkok, Thailand, pp 1–6
9. Franklin RJ, Mohana (2020) Traffic signal violation detection using artificial intelligence and deep learning. In: 5th international conference on communication and electronics systems (ICCES), Coimbatore, India, pp 839–844
10. Şentaş A, Kul S, Sayar A (2019) Real-time traffic rules infringing determination over the video stream: wrong way and clearway violation detection. In: International artificial intelligence and data processing symposium (IDAP), Malatya, Turkey, pp 1–4

11. Mallela NC, Volety R, Srinivasa Perumal R et al (2021) Detection of the triple riding and speed violation on two-wheelers using deep learning algorithms. Multimed Tools Appl 80:8175–8187
12. Sumalatha R, Varaprasada Rao R, Renuka Devi SM (2021) Pothole detection using YOLOv2 object detection network and convolutional neural network. In: Advances in intelligent systems and computing applied information processing systems, pp 293–300
13. Redmon J, Farhadi A (2018) YOLOv3: an incremental improvement. arXiv:1804.02767v1
14. Ammar A, Koubaa A, Ahmed M, Saad A, Benjdira B (2021) Vehicle detection from aerial images using deep learning: a comparative study. Electronics 10:820. https://doi.org/10.3390/electronics10070820
15. Kuhn HW (1955) The Hungarian method for the assignment problem. Naval Res. Logist Q 2:83–97

U-Net-Based Denoising Autoencoder Network for De-Speckling in Fetal Ultrasound Images

S. Satish, N. Herald Anantha Rufus, M. Antony Freeda Rani, and R. Senthil Rama

Abstract Nowadays medical imaging plays an important role in the detection of defects before proceeding with any medical procedures. During pregnancy, medical imaging plays a major role in the detection of any abnormalities/improper development of fetal organs during the gestational period with the help of ultrasound images. As ultrasound images are feeble toward noise interference, it makes difficult for proper intervention for the physician. In this paper, we modeled a U-Net-based denoising autoencoder network to reduce the speckle noise in fetal ultrasound images. To study the performance analysis of U-Net-based denoising autoencoder network, peak signal-to-noise ratio and structural similarity index measure parameters are used. The parameters are analyzed based on different dropout values, different epoch values, and noise levels ($\sigma = 1, 2, 3, 5, 10$). It is inferred from the analysis that PSNR = 28.48 dB, SSIM = 90.87%, accuracy = 98.40%, and loss = 0.54 from the proposed network. The proposed network model performs well than other de-speckling techniques.

Keywords Ultrasound images · Speckle noise · Image restoration · Autoencoder

S. Satish (✉) · N. Herald Anantha Rufus
Department of Electronics and Communication Engineering, Vel Tech Rangarajan Dr. Sagunthala R&D Institute of Science and Technology, Chennai, Tamil Nadu, India
e-mail: satishsaiece@gmail.com

N. Herald Anantha Rufus
e-mail: drrufus@veltech.edu.in

M. Antony Freeda Rani
Department of Electrical and Electronics Engineering, DMI College of Engineering, Chennai, Tamil Nadu, India

R. Senthil Rama
Department of Electrical and Electronics Engineering, DMI College of Engineering, Nagercoil, Tamil Nadu, India

S. Shakya et al. (eds.), *Fourth International Conference on Image Processing and Capsule Networks*, Lecture Notes in Networks and Systems 798,
https://doi.org/10.1007/978-981-99-7093-3_22

1 Introduction

According to the echo imaging concept, ultrasound scanners produce ultrasound images (UI), when body tissues are exposed to the scanner's pulsed acoustic pulses [1, 2]. Ultrasound is a medical imaging technique that uses high-frequency sound waves to create images of internal body structures. It is a valuable tool in medical diagnostics, and it is a non-invasive, safe, and provides real-time images. It does not use ionizing radiation, making it ideal for use in pregnancy and for diagnosing a range of medical conditions. The image produced by the ultrasound can provide information about the shape, size, outline, and consistency of organs and tissues. It can be used to diagnose a range of conditions, such as gallstones, kidney stones, liver disease, pregnancy, and fetal development.

UI contains multiplicative and Gaussian-free noise. UI was typically much more challenging to eliminate the incremental noise because the concentration of the noise differs with the strength of the image. Generally, the speckle image consists of image pixel and Gaussian-free noise. For the majority of noise multiplicative applications, it is believed that the noise level is stationary, with an unidentified noise variance and a unified mean σ^2. The speckle image undergoes a logarithmic transformation y_{ij} [3] noise that transforms from multiplicative to additive conversion. After that, adding zero mean Gaussian noise is an approximation for the noise component [4]. Even though the quality of images is decreased by speckle noise, denoising ultrasound images before the diagnosis is essential to prevent negative effects on the diagnosis process. A few years ago, there has been a lot of interest in de-speckling ultrasound images.

2 Existing Methods

Mean filter is the most effective at denoising Gaussian, speckle, and Poisson noise, but it has a high blurriness value. The mean filter produces a blurring effect. Koo et al. [5] helps to identify the proper speckle region, the homogeneous region growing mean filter, a method for adaptive image filtering that assumed a region that is more speckle than average. If the center point is determined, a homogeneous region based on similarity and homogeneity can be found. With any degradation, the edges are preserved here.

Lee Filter [6] uses a local average filter to minimize speckles while maintaining the overall composition of the image. In ultrasound imaging, the Lee filter is employed in reducing speckle noise and raising image quality. This filter employs a weighted average to reduce image noise by employing a low-pass filter (LPF). It is an adaptive filter, which means that depending on the local characteristics, different weights are applied to different parts of the image. The filter operates by selecting a window centered on a specific element and calculating a measure of the average elements in just that period. The weight of each pixel is then determined by a function that takes the window's standard deviation into account. The amount of the central elements is then changed to all the weighted averages of which window. Each element in the image is subjected to the same procedure, yielding a filtered image with less speckle noise. An anisotropic diffusion algorithm was proposed by Perona et al. [7], which has since been improved by numerous research works [8]. The PMD filter reduces speckles by preserving image edges, lines, and regions. The PMD filter is comparable to Gaussian blurring. The PMD equation is reduced to the Heat equation if the diffusion coefficient is persistent. PMD filters can be used to detect image boundaries in some cases.

Kaur et al. [9] medical images typically have low noise and contrast because of the circumstances and situations in which they are recorded, and they have a very low level of clarity. Contrary to certain other medical images that are free of noise and objects, a wide-range ultrasound image is a challenging process because edge fuzziness can also be present in these images.

Speckle noise is persistently increasing and has irregular sound waves that constantly change, according to Benazarti et al. [10]. As a result, the authors decided that the logarithmic transformation was the best option for converting signal-dependent noise to additive noise rather than just multiplicative noise. The dispersion tensor is employed in this study to implement adipose dissemination to use the class in a structured way for boosting attributes of the image, acclimating flow dispersion in the way of the native orientation. The effectiveness of the algorithm has been amply demonstrated by the authors.

Prudhvi et al. [11, 12] implemented multi-scale transforms to create denoising image enhancement filtration. According to the results of the de-hazing process, spatial domain techniques smooth out boundary lines and line segments in ultrasound images. The softening process was reduced by the researchers using wavelet transform-based techniques, but the borders were unaltered. Although the wavelet transform had a good performance, its computation complexity was more. The couple complex tree curvelet morphs and surpass intensity both tree curvelets transform excel at stripping away erratic noise versus extracting features and multi-scale morphs.

De Fontes et al. [13] created a powerful real-time noise removal model for ultrasound images that was an advancement over the non-local standard-based NL-means technique, i.e., a pixel's restored intensity must be mean that is full because the length between the patches was adequate for all pixels in the image [14, 15]. The outcomes show that this technique can be applied to real-time medical prognosis and is efficient for noise removal. For intermediate images, this model's calculation time was under 20 ms.

Ranjani et al. [16] suggested denoising ultrasound images using the multi-tree complicated curvelet transform-based Levy Shrink algorithm (DTCWT). In their study, where the DTCWT coefficient of all thread are displayed using a considerably tracked Levy distribution, the authors used this algorithm to remove speckle noise. They determined the items of the Levy issuance using a shuttered expression by attempting to manipulate the noise-containing ultrasound image's minute observation-based moments, as well as the minute the Levy and Gaussian instants distributions. Finally, the coefficients were determined using a Bayesian MAP estimation model.

Wen et al. [17] used a wavelet analysis and fuzzy theory-based technique for image enhancement and noise removal. Ultrasound images were subjected to the correlation transformation, conversion of nonlinear into noise emulsifiers, and inter wavelet transform to obtain higher and lower frequency wavelets. The curvelet soft limit denoising method improved the higher frequency coefficients, while the fuzzy technique improved the lower frequency correlation coefficient.

Hota et al. [18] introduced a noise reduction technique for noise contraction in ultrasound images that are based on strain computing. The strain computing technique makes use of image decorrelation in various conditions such as tissue movement. Speckle patterns differ from one another when a similar item is studied in different environments or under different conditions. Due to the even out of subsequent frames, speckle noise has been contained. The noise reduction will be robust, as the correlation between the two frames is small.

Adam et al. [19] demonstrate an argumentative region growing-based filtration besides the withdrawal of speckling noise in medical images. Here, correlation value is used to categorize each region. The computation mean filter defines the correlation sections, while the pixel edge-designated classes are defined by the computation mean filter. However, it has a major flaw in that it takes longer to compute because it detects different fields by using the province expanding method.

A filter for speckle noise reduction based on dimensional extracts was proposed by Jiang et al. [20]. The above-derived filter is constructed by determining each regional sub-mask's minimal native direction derivative. The filtration produces the best outcomes when the demographic variance on the directive's ward is the lowest. The poor de-speckling pointed picture corner requires more computational effort, which is the filter's only drawback.

A noise-reducing filter was introduced by Rui et al. [21] and operates on the ultrasound medical image's 2-D lowest mean square. The weighted local dynamics are specified by the 2-D least mean square method, which is based on the local arrangement of speckle noise. The error essentially controls the filter output within the weighted and ideal local dynamics. Speckle reduction using the 2-D minimum mean square speckle-noise filter preserves a lot of information. The above-derived filtration is constructed by determining each regional sub-mask's minimal native direction variant. When the demographic variance on the ward of the directive is the least, the filtration yields best results [22, 23]. The poor de-speckling pointed picture corner requires more computation time, which is the filter's only drawback.

Gajpal et al. [24] introduced a speckle noise-reducing algorithm on the non-local process. This is, in essence, a Gaussian filter. The filtering task was carried out based on the object's high degree of correlation. In the field, the filter works on pixel compression. The Gaussian noise model performs admirably. The Gaussian filter, on the other hand, does not explicitly apply to the speckle noise image.

Yu et al. [25] demonstrated the contiguous nonlinear filtering methods connected to produce a speckle noise-reducing filter. Anisotropic and Gaussian filters are used as nonlinear filters. The filters were analyzed separately in the first step, and then by nonlinear filtering. In the second step, the filters are evaluated by running both methods. Using this filter will improve the quality of denoised images.

3 Proposed Method

3.1 U-Net-Based Denoising Network

A U-Net-based denoising network is a type of neural network that can be used for image denoising are shown in Fig. 1. It consists of two parts: an encoder and a decoder. The encoder takes the noisy input image and maps it to a lower-dimensional representation, while the decoder maps this representation back to the original image. The encoder and decoder are connected through a bottleneck layer, which is the compressed representation of the input image. The encoder section consists of two convolution layers: a relu layer and a dropout layer, and a normalization layer which comprises a block along with a max pooling layer. The decoder sections consist of blocks and transpose convolution layers. In this section during the decoding process, convolution transpose is used before each block to perform denoising of the image from noisy image.

3.2 U-shaped Dilated Convolution Denoising Autoencoder Network

U-Net-based dilated convolution denoising autoencoder network combines the function of U-Net-based denoising and dilated convolution-based denoising autoencoder network as shown in Fig. 2. It consists of two parts: an encoder and a decoder. The encoder takes the noisy input image and maps it to a lower-dimensional representation, while the decoder maps this representation back to the original image. The encoder sections consist of dilated convolutional layer, dropout layer, normalization layer, and max pooling layer, whereas the decoder section consists of dilated convolutional layers and max pooling layers.

3.3 U-Net-Based Denoising Autoencoder Network

Modified U-Net-based dilated convolution denoising autoencoder network gives rise to a network called as U-Net-based denoising autoencoder network which consists of two parts: an encoder and a decoder as shown in Fig. 3. The encoder section consists of a block which comprises of a convolutional layer, relu, dropout layer, and normalization layer takes the noisy input image and maps it to a lower-dimensional

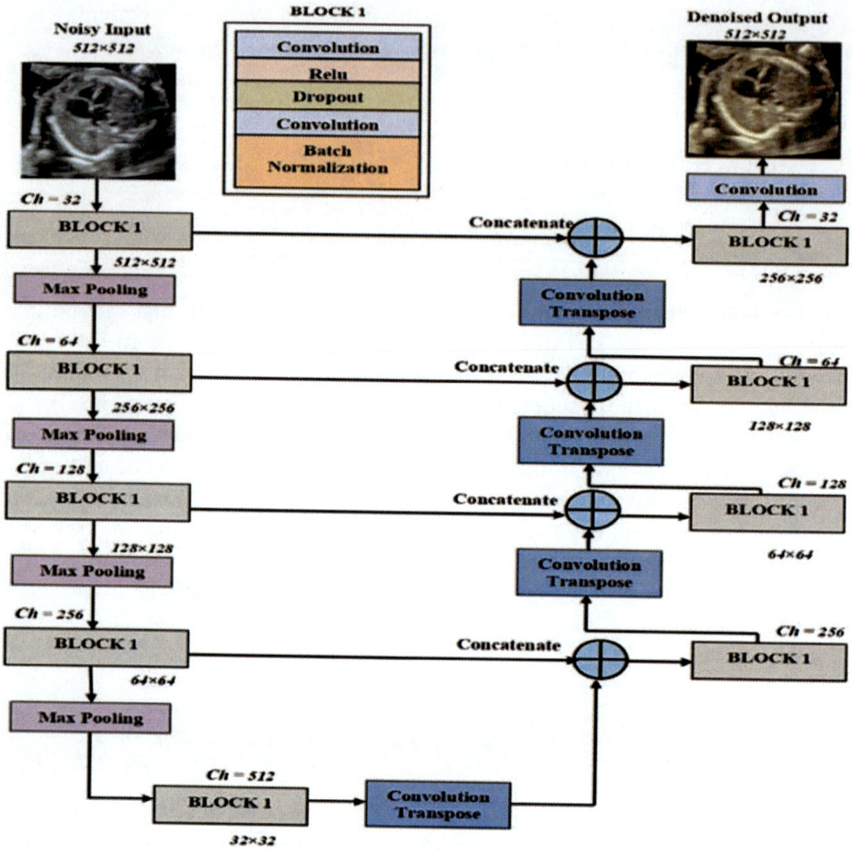

Fig. 1 U-Net-based denoising network

representation. The convolution layer is used for feature extraction operations. This procedure causes the image size to be reduced by the convolution window. The decoder section consists of the same blocks as used in the encoder section along with the up-pooling layer.

By pairing the shrinking image, the up-pooling layer restores its dimensions. In this network structure, transpose convolution is used only once. The structural characteristics of the layer's shrinking image size are restored with little distortion thanks to the transposed convolution process, which is the reverse of the convolution process.

Subsampling layers are used in CNN-based networks to form processes for the image content. Down sampling layers disregard the image resolution while allowing the receptive field to grow. Despite being used to maintain invariance, manage to overfit, and perform feature extraction, pooling layers, a type of down-sampling

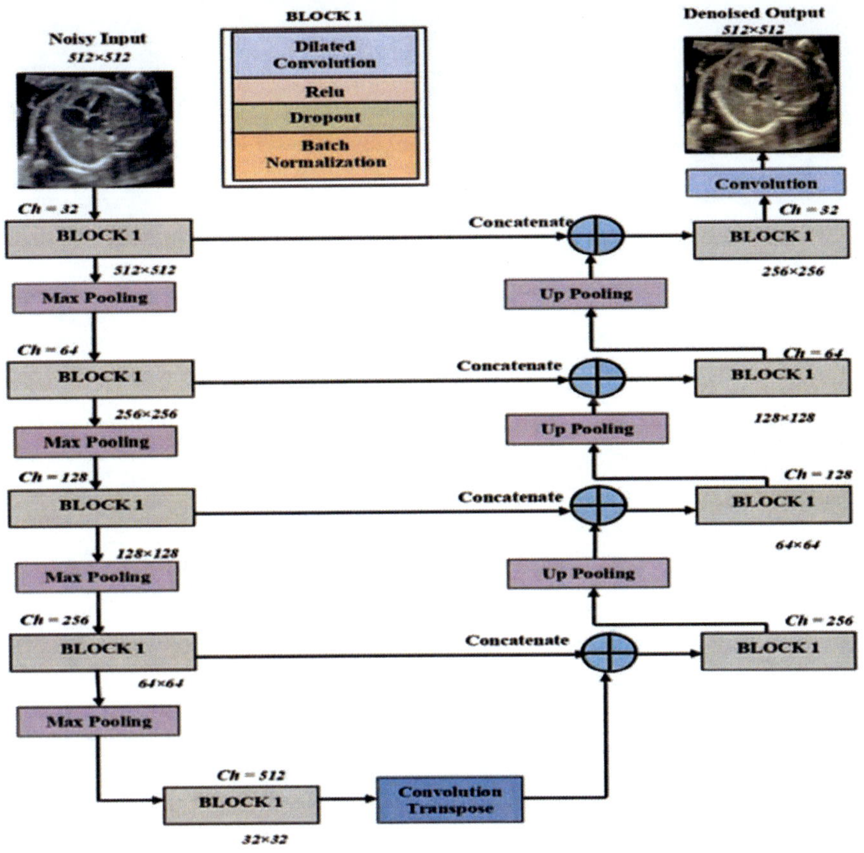

Fig. 2 U-shaped dilated convolution denoising autoencoder network

layer, also result in a reduction in the resolution of the images. However, down-sampling layers cause the resolution of features across a network's layers to gradually deteriorate. Small objects that are challenging to recover even with efforts involving skip connections or other techniques may be overlooked by the resulting coarse features. The resolution cannot be fixed by up-sampling layers, which are used to restore the images [26]. For applications involving image denoising, the resolution is a crucial consideration.

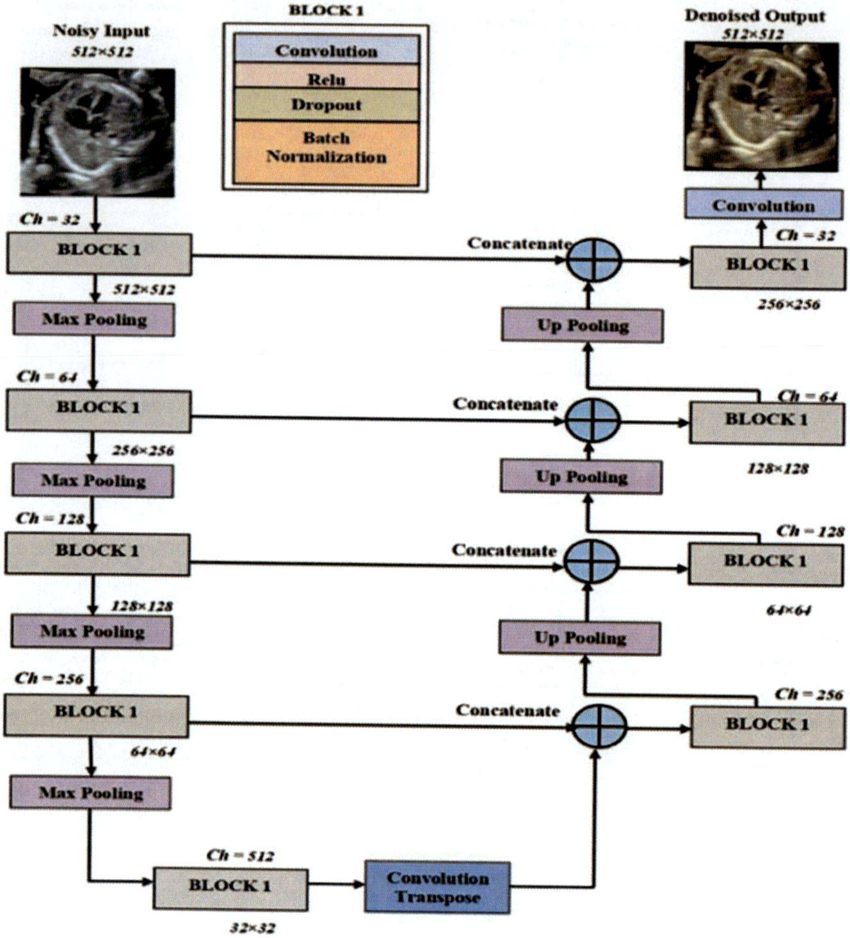

Fig. 3 U-shaped denoising autoencoder network

4 Result and Discussion

Here, the different deep learning networks are used such as U-Net-based denoising network (UDN), U-Net-based dilated convolution denoising autoencoder network, and U-Net-based denoising autoencoder network. For the above deep learning networks, different phenomena are investigated and discussed. The different phenomena are the following: (1) effect of dropout, (2) effects on accuracy and loss, and (3) effects on different noise levels for the different datasets.

Structural Similarity Index Method

The similarity measure can be used to evaluate the quality of a denoised image if the original image is of great quality. The SSIM index is determined using

$$\mathrm{SSIM} = \frac{(2\mu_x\mu_y + c_1)(2\sigma_{x\overline{x}} + c_2)}{(\mu_x^2 + \mu_y^2 + c_1)(\sigma_x^2 + \sigma_y^2 + c_2)} \tag{1}$$

μ_x, μ_y are the mean of x and $\overline{x}$, respectively, σ_x and σ_y are the variances of x and $\overline{x}$, respectively, $\sigma_{x\overline{x}}$ the correlation between x and $\overline{x}$, c_1 and c_2 are fixed values.

Peak Signal-to-Noise Ratio

The PSNR approximates how well a reconstruction compares to the denoised image in terms of human perception. Decibels are used to measure the ratio of these two images.

$$\mathrm{PSNR} = 10\log_{10}\frac{(\mathrm{PV})^2}{\mathrm{MSE}} \tag{2}$$

4.1 Dataset

Dataset is collected from the ultrasound clinical laboratory. For the study based on clinical data, the 1250 clinical fetal cardiac images are used. A total of 625 noisy training images, 625 clear comparison images, and 5 different σ levels of test images make up the second dataset's image set. The first 125 of the 625 noisy ultrasound images have a standard deviation of $\sigma = 1$, the second 125 have a standard deviation of $\sigma = 2$, the third 125 have a standard deviation of $\sigma = 3$, the fourth 125 have a standard deviation of $\sigma = 5$, and the fifth 125 have a standard deviation of $\sigma = 10$. The dataset's pixels measure 512 × 512. Here, the performance of different criteria is analyzed for the different denoising networks.

4.2 Adding Speckle Noise

Speckle noise is a common problem in ultrasound imaging and can affect image quality, making it difficult to interpret and diagnose. Fortunately, there are several approaches to reducing or suppressing speckle noise from ultrasound images. Speckle noise in US images exhibits a multiplicative behavior that is strongly correlated with non-Gaussian statistics. It multiplies into the underlying US signal reflected by the tissue, resulting in a value that is directly proportional to the local gray level in that area. As a result, speckle noise is image data-dependent, resulting in quality degradation in the form of contrast reduction, blurry details, and local pseudo features. Speckle noise is inherent in US images and can be mathematically modeled using statistical distributions for multiplicative scattering. As a result, traditional Gaussian-based image analysis methods (e.g., filtering and segmentation) are insufficient.

$$g(x, y) = (f(x, y) \times \varphi_m(x, y)) + \varphi_a(x, y) \tag{3}$$

In Eq. (3), the original image is expressed as $f(x, y)$, $g(x, y)$ refers to the noisy image and the multiplicative and additive noises are referred as $\varphi_m(x, y)$ and $\varphi_a(x, y)$. x and y are the variables' spatial positions. In general, additive noise is ignored. Rayleigh distribution characterizes speckle patterns created for completely random scattering.

Assuming the Rayleigh distribution of the expression $\varphi_m(x, y)$ in Eq. (4), the probability density function of $\varphi_a(x, y)$ is expressed as follows:

$$p(f|u) = \frac{f}{u\sigma^2}\exp(-\frac{f}{2u^2\sigma^2})(\varphi \geq 0) \tag{4}$$

The noise intensity is symbolized by the parameter σ, the noise-free image by the parameter u, and the noise-added image by the parameter f. In our study, five different speckle noise standard deviations $\sigma = 1, 2, 3, 5$, and 10 noise levels are used for the performance analysis of different deep learning-based denoising networks. The results of the different deep learning-based denoising networks are examined concerning the effects of the hyperparameters. Table 1 also displays the hyperparameter values of the chosen denoising networks.

Table 1 Denoising network hyperparameters

Hyperparameter	UDN	U-DCDAN	U-DAN
No. of layers	50	54	44
No. of max pooling layer	4	–	4
Convolution	–	–	–
Mini-batch size	20	20	20
Validation rate (%)	10	10	10
Dropout rate (%)	10	10	10
Activation function	Sigmoid	Sigmoid	Sigmoid
Optimizer	Adamax	Adamax	Adamax
Learning rate	0.0001	0.0001	0.0001

4.3 Effect of Dropout

The effect of dropout is investigated, and peak signal-to-noise ratio (PSNR) and structural similarity index method (SSIM) are analyzed using the following graphs as shown in Fig. 4. In batch normalization for PSNR, dropout value unto 0.6 the PSNR value lies between 28.15 and 28.16 dB after further increasing the value of dropout till 0.9 the PSNR values reduce abruptly unto 27.54 dB for all the values of σ. In batch normalization for SSIM, the dropout value unto 0.6 values forms a "U-"shaped structure after further increasing the dropout value to 0.9 the SSIM value reduces abruptly between 70 and 50 for all values of σ. In batch renormalization for PSNR, the dropout value unto 0.4 PSNR value lies between 28.18 and 28.16 dB after further increasing the value of dropout to 0.9 the PSNR value reduces gradually between 28.16 and 26.60 dB. In batch renormalization for SSIM, the dropout value unto 0.4 SSIM value falls between 98 and 90% and further increasing the dropout value till 0.9 the SSIM value reduces gradually between 90 and 50% for all values of σ.

In PSNR, batch normalization dropout value is constant from dropout = 0 (σ = 1, PSNR = 28.15 dB; σ = 2, PSNR = 28.15 dB; σ = 3, PSNR = 28.15 dB; σ = 5, PSNR = 28.15 dB; σ = 10, PSNR = 28.15 dB) to dropout = 0.6 (σ = 1, PSNR = 28.14 dB; σ = 2, PSNR = 28.14 dB; σ = 3, PSNR = 28.14 dB; σ = 5, PSNR = 28.14 dB; σ = 10, PSNR = 28.14 dB) and after dropout = 0.6, PSNR values gradually decrease till 0.9 for batch normalization (σ = 1, PSNR = 27.78 dB; σ = 2, PSNR = 27.74 dB; σ = 3, PSNR = 27.70 dB; σ = 5, PSNR = 27.66 dB; σ = 10, PSNR = 27.54 dB).

In SSIM, batch normalization for dropout = 0 (σ = 1, SSIM = 76.82%; σ = 2, SSIM = 75.12%; σ = 3, SSIM = 76.22%; σ = 5, SSIM = 76.78%; σ = 10, SSIM = 71.77%) and for increasing the dropout value SSIM value reduce gradually till dropout = 0.4 (σ = 1, SSIM = 66.88%; σ = 2, SSIM = 61.56%; σ = 3, SSIM = 59.20%; σ = 5, SSIM = 64.34%; σ = 10, SSIM = 55.58%) and increases gradually till dropout = 0.6 (σ = 1, SSIM = 69.50%; σ = 2, SSIM = 70.38%; σ = 3, SSIM

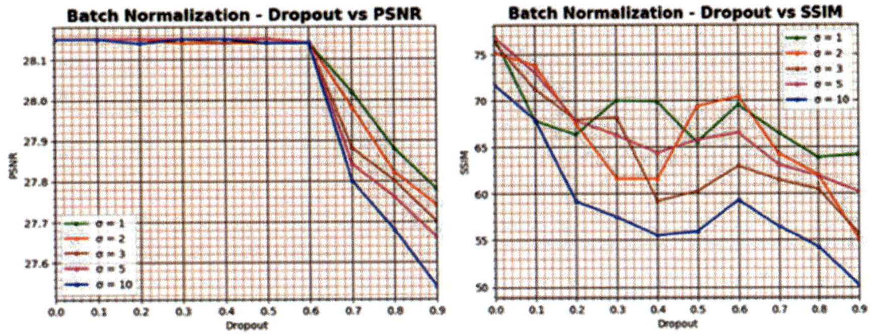

Fig. 4 Effect of dropout

= 62.93%; $\sigma = 5$, SSIM = 66.48%; $\sigma = 10$, SSIM = 59.32%) in the form of "U-"shaped structure. After dropout = 0.6, SSIM values gradually decreases till 0.9 for batch normalization ($\sigma = 1$, SSIM = 64.21%; $\sigma = 2$, SSIM = 55.05%; $\sigma = 3$, SSIM = 55.72%; $\sigma = 5$, SSIM = 60.21%; $\sigma = 10$, SSIM = 50.21%).

Finally, it is inferred that the dropout value for batch normalization lies between 0 and 0.6. The dropout value should lie between the suggested ranges for the process of effective denoising of fetal cardiac (FC) ultrasound images.

4.4 *Effects of Accuracy and Loss*

The effects of accuracy and loss are investigated and analyzed for different epoch values are shown in Fig. 5. It is inferred from the graph that the trade-off between epoch and accuracy, as the epoch increases accuracy also increases steadily and after reaching a particular epoch value the accuracy value becomes constant due to the decoding section of the U-Net-based denoising autoencoder network.

The effect of accuracy is investigated as follows:

- For U-DCDAN, until 400 epoch the accuracy values are steadily increasing (92.47%), further unto 800 epochs slightly increased (94.27%) and after further epoch increases the accuracy becomes constant at 94.51%.
- For U-DAN, unto 150 epochs the accuracy values are steadily increasing (97.56%), further unto 800 epochs slightly increased (98.40%), and after further epoch increases the accuracy becomes constant at 98.53%.
- For UDN, unto 400 epoch the accuracy values are steadily increasing (87.74%), further unto 900 epochs slightly increased (91.25%), and after further epoch increases the accuracy becomes constant at 91.74%.

The effect of loss is investigated as follows:

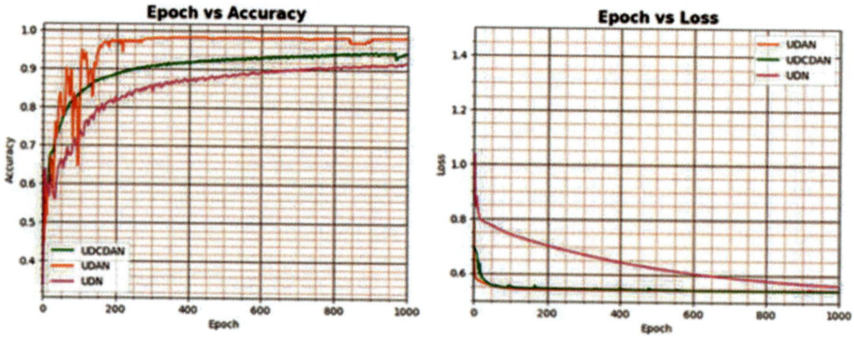

Fig. 5 Effects on accuracy and loss

- For UDN, unto 40 epochs the loss value gradually decreases from 0.78 to 0.56 and after further epoch value increases loss becomes constant at 0.54.
- For U-DAN, unto 110 epochs the loss value gradually decreases from 0.69 to 0.55 and after further epoch value increases loss becomes constant at 0.54.
- For U-DCDAN, unto 850 epochs the loss value gradually decreases from 1.04 to 0.58 and further epoch value increases loss becomes constant at 0.56.

From the above investigation and analysis, U-DAN produces higher accuracy and lower loss when compared to UDN, and U-DCDAN.

4.5 Effects on Different Noise Levels

Figure 6 shows the visual results of Tables 2, respectively. Figure 6 shows that the proposed deep learning-based denoising networks can remove speckle noise from images and improve image resolution. In the proposed denoising model, neural network models learn to reconstruct the original data from its noisy counterpart so as to clean out noise from corrupted or noisy data. A NN model's performance is based on a nonlinear activation function, which is used to backward propagate, update the weights, and generate the desired outcome, which is the real output. The model may successfully retrain itself to make adjustments for the discrepancy between the actual and desired output. The sigmoid derivative has a complicated processing capacity and always returns zero as x grows. Because of this, the gradient is zero, and when additional layers are multiplied, the gradient also becomes zero, which leads to poor model performance. We had utilize sigmoid activation function in the final output layer for classification issues similar to the autoencoder in tensorflow.

Adamax maintains a moving average of the gradient's first and second moments, which are the mean and variance of the gradients, respectively. The moving average of the second moment, which is similar to the RMSProp term, helps the optimizer to scale the learning rate for each parameter based on the variance of the gradients. This helps to improve the optimization algorithm's performance in the early stages of training. It has the ability to converge quickly and handle noisy or sparse gradients. It helps to accelerate the training of deep neural networks and improve their performance.

5 Conclusion

For de-speckling, a variety of filters and techniques are employed. By boosting the capabilities of the computer processors, various deep learning techniques are used here to accelerate. In this study, speckle noise in ultrasound images is reduced using U-Net-based denoising autoencoder network (U-DAN), with greater applications as computer technologies advance. The proposed network performance is compared in

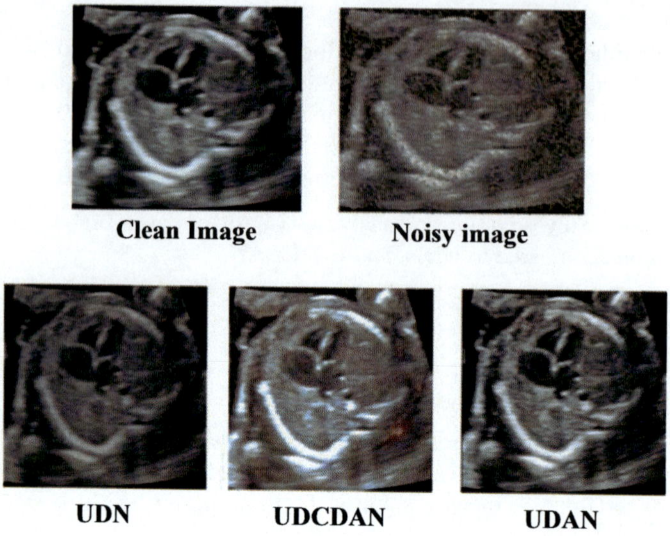

Fig. 6 Effectiveness of the proposed deep learning-based denoising method for clinical test image at $\sigma = 2$

Table 2 Performance of different denoising networks according to PSNR and SSIM metrics for the clinical dataset

Methods	PSNR					SSIM				
	$\sigma = 1$	$\sigma = 2$	$\sigma = 3$	$\sigma = 5$	$\sigma = 10$	$\sigma = 1$	$\sigma = 2$	$\sigma = 3$	$\sigma = 5$	$\sigma = 10$
Noisy image	24.68	23.98	21.86	20.25	19.64	86.65	83.54	63.18	55.86	46.58
UDN	28.46	29.12	28.75	27.56	26.05	87.42	86.52	85.14	83.52	80.24
U-DCDAN	29.14	28.58	28.03	27.06	26.58	89.68	87.26	85.49	83.64	81.76
U-DAN	29.48	29.67	29.02	28.74	28.48	90.87	88.62	86.28	84.16	82.27

terms of the PSNR and SSIM. The U-DAN achieves the best PSNR values at 2, 3, and 5 noise levels when the deep learning-based denoising methods are compared with each other. At all noise levels, the U-DAN network has the best SSIM values. This study demonstrates that the suggested deep learning networks are the best techniques for removing noise from ultrasound images. Future research will look at the impact of using a different distribution method than an equal distribution to distribute noise levels.

References

1. Gautam R, Bharti R (2018) Liver ultrasound image enhancement using bilateral filter. Int J Eng Techn Res
2. Binaee K, Hasanzadeh RPR (2014) An ultrasound image enhancement method using local gradient based fuzzy similarity. Biomed Signal Process Control 13:89–101
3. Zong X, Laine AF, Geiser EA (1998) Speckle reduction and contrast enhancement of echocardiograms via multiscale nonlinear processing. IEEE Trans Med Imaging 17:532–540
4. Michailovich OV, Tannenbaum A (2006) De-speckling of medical ultrasound images. IEEE Trans Ultrason Ferroelectr Freq Control 53(1):64–78
5. Koo JI, Park SB (1991) Speckle reduction with edge preservation in medical ultrasonic images using a homogeneous region growing mean filter (HRGMF). Ultrasonic Imaging 13(3):211–237
6. Lee J-S (1980) Digital image enhancement and noise filtering by use of local statistics. IEEE Trans Pattern Anal Mach Intell PAMI-2 2:165–168
7. Perona P, Malik J (1990) Scale-space and edge detection using anisotropic diffusion. IEEE Trans Pattern Anal Mach Intell 12(7):629–639
8. Zong X, Laine AF, Geiser EA (1998) Speckle reduction and contrast enhancement of echocardiograms via multiscale nonlinear processing. IEEE Trans Med Imaging 17(4):532–540
9. Foment J (2014) Parameter-free fast pixel-wise non-local means denoising. Image Processing Line 4:300–326
10. Kaur GS, Kaur P (2016) Image enhancement of ultrasound images using multifarious denoising filters and GA. In: 2016 international conference on advances in computing, communications and informatics, Jaipur, pp 2375–2384
11. Benzarti F, Amiri H (2012) Speckle noise reduction in medical ultrasound images. IJCSI Int J Comput Sci 9(2, 3)
12. Prudhvi Raj VN, Venkateswarlu T (2012) Denoising of medical ultrasound images using spatial filtering and multiscale transforms. Int J Comput Sci Inf Technol (IJCSIT) 4(6)
13. Andria G, Attivissimo F, Cavone G, Giaquinto N, Lanzolla AML (2012) Linear filtering of 2-D wavelet coefficients for denoising ultrasound medical images. Measurement 45:1792–1800
14. De Fontes FPX, Barroso GA, Hellier P (2010) Real-time ultrasound image denoising. J Real-Time Image Process 6(1):15–22
15. Kishore PVV, Mallika KL, Prasad MVD, Narayana KL (2015) Denoising ultrasound medical images with selective fusion in wavelet domain, second international symposium on computer vision and the internet. Procedia Comput Sci 58:129–139
16. Ranjani J, MS C (2015) Bayesian denoising of ultrasound images using heavy-tailed Levy distribution. IET Image Process 9(4):338–345
17. Farouj Y, Freyermuth JM, Navarro L, Clausel M, Delachartre P. Hyperbolic wavelet-Fisz denoising for a model arising in ultrasound imaging. IEEE Trans Comput Imaging 99:1
18. Fredj H, Malek J (2016) Real-time ultrasound image denoising using NVIDIA CUDA. In: 2016 2nd international conference on advanced technologies for signal and image processing (ATSIP), Monastir, pp 136–140
19. Wen H, Qi W (2015) Enhancement and denoising method of medical ultrasound image based on wavelet analysis and fuzzy theory. In: Seventh international conference on measuring technology and mechatronics automation, Nanchang, pp 448–452
20. Hota NR, Sanibigrah S (2018) Retrieval of random valued impulse noise with advanced weighted mean filter by using modified denoising algorithm for image processing application. Asian J Sci Technol 09(07):8444–8460
21. Adam D, Beilin-Nissan S, Friedman Z, Behar V (2006) The combined effect of spatial compounding and nonlinear filtering on the speckle reduction in ultrasound images. Ultrasonics 44(2):166–181
22. Xie J, Jiang Y, Tsui HT, Heng PA (2006) Boundary enhancement and speckle reduction for ultrasound images via salient structure extraction. IEEE Trans Biomed Eng 53(11):2300–2309

23. Rui L, Zhuoxin S, Cishen Z (2008) Adaptive filter for speckle reduction with feature preservation in medical ultrasound images. In: 2008 10th IEEE international conference on control, automation, robotics and vision, pp 1787–1792
24. Gajpal DK, Biswas AK, Potdar RM (2012) Reconstruction of arteries in heart image affected by specular reflection. Int J Emerg Technol Adv Eng 2:259–263
25. Yu J, Tan J, Wang Y (2010) Ultrasound speckle reduction by a SUSAN-controlled anisotropic diffusion method. Pattern Recogn 43(9):3083–3092
26. Hamaguchi R, Fujita A, Nemoto K, Imaizumi T, Hikosaka S (2018) Effective use of dilated convolutions for segmenting small object instances in remote sensing imagery, March, 1442–1450

Galo Lexical Tone Recognition Using Machine Learning Approach

Bomken Kamdak, Gom Taye, and Utpal Bhattacharjee

Abstract In this paper, we present the development of a Galo speech corpus and the design of a Galo lexical tone recognizer. The Galo language is spoken by the Galo people residing in Arunachal Pradesh, a frontier state of North East India. Galo belongs to the Tani branch of the Tibeto–Burman language family. Galo exhibits two discernible lexical tones: High/Plain, characterized by a moderately high-level pitch contour and Low/Tense, characterized by a descending pitch contour. The speech database used in this study contains Galo phrases spoken by 27 individuals, comprising 13 male and 14 female speakers aged between 20 and 50 years. The recording of each speaker encompasses a phonetically diverse script consisting of twenty Galo tonal words and all their possible tonal variations. The work aims to automatically recognize these lexical tones by employing seven features extracted from the fundamental frequency (F0) contour. Two classification models based on machine learning, specifically support vector machine (SVM) and random forest (RF), were employed to identify lexical tones. The RF-based recognizer achieves a recognition accuracy of 90.42%, while the SVM-based recognizer surpasses that with an accuracy of 92.31%.

Keywords Galo language · Tone recognition system · F0 · Support vector machine · Random forest

B. Kamdak · G. Taye · U. Bhattacharjee (✉)
Department of Computer Science and Engineering, Rajiv Gandhi University, Itanagar, Arunachal Pradesh 791112, India
e-mail: utpal.bhattacharjee@rgu.ac.in

B. Kamdak
e-mail: bomken.kamdak@rgu.ac.in

S. Shakya et al. (eds.), *Fourth International Conference on Image Processing and Capsule Networks*, Lecture Notes in Networks and Systems 798,
https://doi.org/10.1007/978-981-99-7093-3_23

1 Introduction

Automatic speech recognition research (ASR) has made substantial progress since its inception in the mid-twentieth century, rendering it a viable option for facilitating human–machine communication. Lots of research has already been conducted on speech recognition to make it a modality for human–machine communication, driven by the fact that speech is the most successful form of human communication [1]. Speech is a complex communication medium consisting of various modalities, wherein the majority of information is attributed to linguistic content. In addition, speech signals contain a variety of informational aspects, such as the identity of the speaker, their physical and psychological state, and geographical and cultural background, etc. The primary objective of an automatic speech recognition (ASR) system is to employ computational methods to transform speech signals into a sequence of linguistic entities, commonly represented as a series of words or phonemes. ASR has been an exciting study area for more than 60 years, but there is still much room for improvement, especially in challenging environments where human outperforms their machine counterpart in speech recognition tasks [2].

The accurate identification of the lexical tone associated with a syllable is crucial in tonal languages to ensure a precise interpretation of word meaning. While resourceful tonal languages such as Mandarin have made significant advancements in automatic tone recognition, the presence of tones remains a challenge for achieving precise speech recognition in the majority of tonal languages, which lack comprehensive literature and documentation regarding tone patterns and their associations with syllables. This difficulty in tone recognition hinders the advancement of automatic speech recognition systems for these languages. The identification and characterization of tones linked to a spoken expression using standard speech features can significantly enhance the speech recognition capabilities of tonal languages.

This paper provides a study of the Galo language, which is a tonal language spoken by the Galo tribe residing in the Arunachal Pradesh of North East India. The Galo tribe holds considerable importance within the context of Arunachal Pradesh, primarily concentrated in the central region of the state. The Galo language is classified as a member of the Tani branch within the Tibeto–Burman language family. There is very little information available for this low-resource, poorly studied language. Two tones are found in the Galo language: High/Plain, which has a medium high-level tone, and Low/Tense, which has a dropping tone [3]. The central focus of this research paper revolves around the use of a concise set of features obtained from the F0-contour in order to represent the Galo lexical tone. Following this, machine learning methods, namely support vector machine (SVM) and random forest (RF), are used to recognize the tones.

The subsequent sections of this paper are organized in the following manner. Section 2 provides an overview of the existing literature pertaining to the recognition of tone. The methodology and model are explicated in Sect. 3. The presentation of the Results and Discussion can be found in Sect. 4. Section 5 provides a comprehensive overview of the Conclusion and Future Works.

2 Related Works

A tonal language refers to a linguistic framework where the distinction of pitch or tone plays a crucial role in establishing the semantic significance of words. In contrast to non-tonal languages, which primarily employ variations in pitch to convey emotional or grammatical nuances, tonal languages integrate pitch distinctions as an inherent element of their lexical and phonetic systems. Considerable investigation has been undertaken regarding the recognition of tonal speech, with particular emphasis on languages that possess a rich abundance of linguistic resources, such as Mandarin and Cantonese. The study carried out by Chen et al. [4] centered on the identification of Mandarin tones. The researchers used Hidden Markov Modeling in conjunction with a delta modulation technique to extract features from the pitch sequence. The tone recognition of isolated Mandarin monosyllables was conducted by Chang et al. [5] using the multi-layer perceptron (MLP) model. Feature vectors were constructed by extracting ten features from the fundamental frequency and energy contours of monosyllables. The results of their study revealed a recognition accuracy rate of 93.8% for untrained speakers. The recognition of tone is of utmost importance in the identification of Cantonese, a language widely recognized for its extensive tonal variations. In their study, Lee et al. [6] employed suprasegmental feature parameters that were extracted specifically from the voiced segment of monosyllabic utterances. A three-layer feedforward neural network was employed for the purpose of feature vector classification. The study utilized a phonologically comprehensive lexicon consisting of 234 unique syllables. This led to recognition accuracies of 89.0% in single speaker situations and 87.6% in multi-speaker scenarios. The authors Lee et al. [7] introduced a robust methodology for accurately identifying explicit tones in continuously spoken Cantonese utterances. The methodology employed entailed the extraction of tone feature vectors at a brief temporal interval, with the purpose of capturing the patterns of fundamental frequency (F0) and energy fluctuations at the syllable level. The authors proposed the implementation of a moving-window normalization technique in order to address the impact of tone-irrelevant fluctuations in F0 and energy features. The study involved the utilization of Hidden Markov Models to conduct context-dependent acoustic modeling of distinct tones. The findings of the experiment revealed that the accuracy of tone recognition in the speaker-independent scenario was 66.4%. In their study, Peng and Wang [8] utilized support vector machines (SVMs) as a means to identify tones in the context of continuously spoken Cantonese speech. An adaptive log-scale 5-level F0 normalization technique was proposed in order to address the influence of F0 variations that are not related to tone. A proposed extension of this normalization method incorporates the consideration of intonation. The findings from the experiment revealed that the accuracy of tone recognition in a speaker-independent context was 71.50%. In their study, Wang and Levow [9] put forth a novel approach aimed at improving tone recognition. Their method involved integrating vowel landmark detection into the modeling of the tone nucleus. The frames exhibiting the greatest sonority during the production of vowels were designated as vowel landmarks. The researchers utilized a support vector machine that was trained on Mel-frequency cepstral coefficient (MFCC) features in order to identify and classify these specific landmarks. The results of the tone recognition experiments demonstrated that the landmark-based approach, which specifically targets the tone nucleus regions, outperformed the best baseline system by a margin of more than 6%. In a preliminary investigation, it was discovered that relying solely on vowel landmark evidence for tone nucleus regions resulted in a slight reduction of accuracy, amounting to less than 2%, when compared to utilizing both landmark frames and force-aligned vowel boundary information. Based on the aforementioned findings, the researchers arrived at the conclusion that the utilization of vowel landmarks to extract features from the most sonorous section of the tone nucleus is adequate for achieving precise recognition of the tonal quality of a syllable. In their study, Hsieh et al. [10] devised a technique to represent Taiwanese tones. This involved employing a piece-wise curve fitting algorithm on the F0-contour. F0-contour of each syllable was divided into four segments, and each segment was separately fitted to a curve. Based on their research, it was found that using feature extraction from curve-fitted F0 values yields higher accuracy in tone recognition compared to features extracted from raw F0-contour. The researchers employed a SVM-based classifier for the purpose of identifying and distinguishing the various tones in the Taiwanese language. In their research, Gogoi et al. [11] presented a system that utilizes support vector machines (SVM) to differentiate between three distinct lexical tones, namely High, Mid, and Low, in the Ao language. The distribution of tones in trisyllabic words was examined by the researchers, using a sample of 36 individuals. The F0-contour was derived using zero-frequency filtering (ZFF), followed by the computation of four acoustic features based on the F0-contour. The Ao language is comprised of three discrete dialects, namely Changki, Mongsen, and Chungli. To facilitate the identification of tones in different dialects, distinct tone recognition systems were utilized, employing SVM as the underlying methodology. The accuracy rates for identifying Ao tone in the Changki, Mongsen and Chungli dialects were reported as 53.20%, 62.26%, and 52.69%, respectively. In an independent investigation, Gogoi et al. [12] endeavored to develop an automated system for the recognition of four phonological tones in the Mizo language. The researchers employed acoustic-prosodic parameters as the foundation for their feature extraction methodology. In order to generate input for two classifier models, namely support vector machine (SVM) and deep neural network (DNN), six features were computed from the fundamental frequency (F0) contours. The aforementioned models were employed for the purpose of conducting tone recognition tasks. The database utilized for assessing the system comprises 31,950 instances of syllables, encompassing the four distinct tones that are present in the Mizo language. The aforementioned instances were gathered from a subset of 19 individuals who were instructed to utter trisyllabic phrases for the purpose of analysis and subsequent incorporation into the database. The researchers observed that the classifier based on deep neural network (DNN) achieved a recognition accuracy of 74.11% in the identification of the four phonological tones in Mizo. When comparing the two classifiers, it was observed that the SVM-based classifier achieved a recognition rate of 73.39%, indicating a slight improvement of 0.72% for DNN-based recognizer.

3 Methodology and Model

The proposed framework for the Galo tone recognition system is given in Fig. 1.

3.1 Galo Speech Database Creation

The Galo language exhibits the presence of both monosyllabic and disyllabic words with tones. Table 1 illustrates a selection of Galo tonal words. High/plain tone is depicted without any accompanying tone letter, whereas Low/Tense tone is denoted by a (‘) tone letter positioned at the beginning of the word.

The speech corpus used in the study comprises data obtained from 27 individuals, with 13 being male and 14 being female speakers. The speakers belong to the Galo

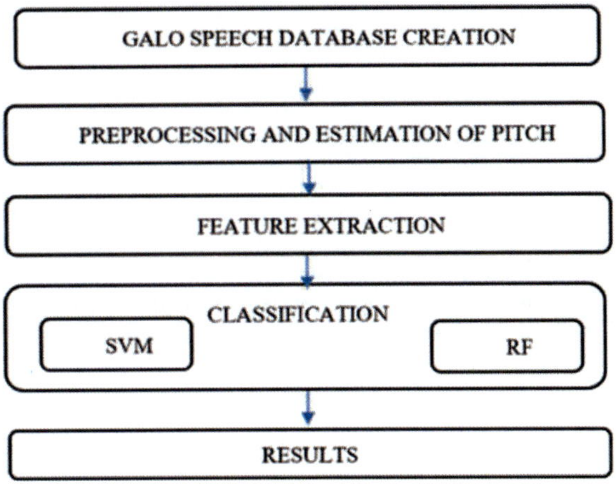

Fig. 1 The proposed framework for Galo tone recognition system

Table 1 Example of Galo tonal words

Sl no	Galo word	English meaning
1	*aii*	Teeth
2	*‘aii*	Heavy
3	*axi*	Elder sister
4	*‘axi*	Number two
5	*onam*	Pluck
6	*‘onam*	Falling something from a height
7	*vtww*	Herd of cattle
8	*‘vtww*	Bamboo grove
9	*umnam*	Humid weather
10	*‘umnam*	Boil something in the water

tribe, which is indigenous to Arunachal Pradesh. The individuals are of the age group 20 to 50 years. The audio recording was conducted within a controlled acoustical setting, employing a sampling frequency of 16 kHz and a mono format with a 16-bit resolution. The participants are provided with a phonetically rich written Galo script, consisting of twenty Galo tonal words with all their tonal instances. Each speaker is recorded eight times, divided into two equal recording sessions. These sessions are separated by a minimum of one week. Consequently, a comprehensive database consisting of 17,280 different tonal words was generated. The subjective evaluation of the database involved the participation of five individuals from the Galo tribe who possess limited prior knowledge in the field, as well as three listeners who have received phonetic training. Speech samples assessed correctly by at least 60% of the naïve audience and a minimum of two specialists have been deemed suitable for subsequent analysis. The final database comprises a total of 14,623 tonal words. The experiments described in this paper were conducted using this database.

3.2 Preprocessing and Estimation of Pitch

A first-order digital filter with a transfer function given in Eq. (1) is used to pre-emphasize the digitized speech.

$$H(Z) = 1 - 0.95z^{-1} \tag{1}$$

Subsequently, an energy-based voiced activity detection (VAD) algorithm is employed to identify the voiced segment of the speech signal and discard the initial and final silence period. The voiced speech signal is divided into frames of 30 microseconds. These frames are overlapped by one-third of their duration. F0 is derived from the individual frames of the speech signal. To reduce the influence of speaker dependencies on fundamental frequency (F0), Z-score normalization has been applied. The Z-score normalization can be accomplished using the following equation:

$$x^* = \frac{x - \mu}{\sigma} \tag{2}$$

where μ and σ are the mean and standard deviation of the F0 values, respectively. A smoothing algorithm is applied to smooth the F0-contour [13].

3.3 Feature Extraction

The Galo tones are represented by seven features extracted from the smoothed pitch contour of Galo words. The features extracted from pitch contour are mean F0, median F0, standard derivative of F0, F0 value measured at 0, 25, 75 and 100% of the way through the contour. The F0 value at 0% is denoted as F0 onset and F0 at 100% is denoted as F0 offset. These seven features extracted from the pitch contour constitute the feature vectors for the classifiers to classify the two categories of Galo tones.

3.4 Classification

In order to classify the features derived from the F0-contour of the Galo syllables, two machine learning techniques, namely support vector machine (SVM) and random forest (RF), were employed.

3.4.1 Support Vector Machine (SVM)

Vapnik was the first to propose the support vector machine (SVM), a statistical machine learning method [14]. Since its debut, it has attracted a lot of interest and been widely used across numerous fields. The ideal hyperplane, a final surface that optimizes the margin between the classes in the given dataset, is used by the SVM algorithm to divide the classes. Support vectors were initially used to refer to the data points within these classes that were closest to the hyperplane. The SVM statistical learning algorithm's primary goal goes beyond just differentiating between two classes. Additionally, it seeks to identify the best hyperplane for effectively dividing the two classes in the training dataset. Three key components—the maximization of margins, the use of dual theory, and the kernel trick—are what make the SVM algorithm effective. The basic objective of conventional support vector classification is to maximize the distance between two hyperplanes. This goal can be achieved by solving a convex quadratic programming problem (QPP). The dual theory makes it possible to apply kernel tricks and incorporate kernel functions to address nonlinear situations. The training data for the binary classification can are given by $\{x_i, y_i\}$, $i = 1,2, \ldots, r$ and $y_i = \{-1, 1\}$, where r is the number of training samples and the training vector is made up of the two classes.

α_1 and α_2, which are represented by $y_i = 1$ and $y_i = -1$, respectively. According to the equation, if the classes are linearly separable, it is possible to define at least one hyperplane that can properly separate the classes (training error 0).

$$w.x + b = 0 \tag{3}$$

where w is the vector normal to the hyperplane and b is an offset. To find such a hyperplane, w and b are estimated in a way that $y_i(w.\ x_i + \mathrm{b}) \geq 1$ for $y_i = 1$ that is class

α_1 and $y_i(w.\ x_i + b) \leq -1$ for $y_i = -1$ for class α_2. These can be combined into one set of inequality as given below:

$$yi(w.xi + b) - 1 \geq 0, \forall xi \tag{4}$$

There are several hyperplane systems that can be used to separate two classes. However, in an n-dimensional space, it is only possible to identify a single hyperplane as optimal. The support vectors refer to the training data points that are in close proximity to the optimal hyperplane. These support vectors can be identified at the boundaries defined by the equations $w.\ x_\mathrm{i} + b = \pm 1$. The determination of the optimal hyperplane separation is contingent upon the center of the margin. Determining the optimal hyperplane that effectively separates two classes involves maximizing the distance between the nearest samples belonging to the respective class. From the mathematical standpoint, the primary aim is to differentiate between the two classes by ascertaining the utmost distance between their support vectors. The mathematical formulation of the issue in conventional SVM is represented as a convex quadratic programming problem (QPP) [15].

$$\min_{w,b} \frac{1}{2} \|w\|^2 \tag{5}$$

such that:

$$((w.x_i) + b) \geq 1 \tag{6}$$

The constraints can be changed for nonlinearly separable classes by introducing slake variables ζ_i. The QPP is reformulated as follows [15]:

$$\min_{w,b,\xi} \frac{1}{2} \|w\|^2 + C \sum_{i=1}^{r} \xi_i \tag{7}$$

Such that

$$y_i((w.x_i) + b) \geq 1 - \xi_i, i = 1, \ldots, r; \xi_i \geq 0, i = 1, \ldots, r \tag{8}$$

where $\zeta = (\zeta_1, \zeta_2, \ldots, \zeta_r)^T$ and $C \geq 0$ is a penalty parameter. This is the primal problem, which is solved by using Lagrangian dual problem given by [15]:

$$\min_{\alpha} \frac{1}{2} \sum_{i=1}^{r} \sum_{j=1}^{r} \alpha_i \alpha_j y_i y_j K\left(x_i, x_j\right) - \sum_{i=1}^{r} \alpha_i \tag{9}$$

Such that

$$\sum_{i=1}^{r} y_i \alpha_i = 0, 0 \leq \alpha_i \leq C, i = 1, \ldots, r \tag{10}$$

where $K(x, x')$, the kernel function is a convex QPP. The structural risk minimization (SRM) principle is utilized to minimize the confidential interval $\|w\|^2$ and empirical risk term $\sum^r \zeta_i$ simultaneously.

3.4.2 Random Forest (RF)

Bagging techniques in ensemble learning were used to construct classification and regression trees, also known as CART. There is a potential for the trees to demonstrate correlation and statistical dependence as a result of their construction using identical data. Breiman [16] introduced a technique to improve the diversity and decrease the correlation in bagged trees. This technique involves incorporating a random selection of attributes and observation samples during each split of the tree. Therefore, this approach is frequently known as random forest (RF) methodology. In the random forest (RF) algorithm, it is necessary to define three parameters: the total number of trees (B) that determines the magnitude of the forest, the number of predictors (m) that signifies the randomly chosen subset of p variables (features) for each split, and the least number of observations that each node can have ($n_{\min}$), commonly referred to as the leaf size. The RF building algorithm consists of three primary steps, as outlined in reference [17].

1. Use the training data to Create B overlapping or replaceable sample datasets, each of size N.
2. Build a random forest tree T_b for each sample dataset by repeatedly applying the next steps to each terminal node. The process will continue till the required minimum node size $n_{\min}$ is obtained.
   (a) Randomly select m predictors from the set of p variables.
   (b) Select one of the m selected predictors as the best predictor for the split point.
   (c) Establish a set of decision rules to partition this particular point (node) into two distinct child nodes.
3. Finally, the objective is to determine the collection of trees $\{T_b\}^B$, wherein B is the number of trees in the random forest.

By averaging the results of the different trees, we may derive a prediction for a specific value x of the response variable:

$$\hat{f}_{RF} = \frac{1}{B}\sum_{b=1}^{B} T_b(x) \tag{11}$$

The ensemble learning algorithm iteratively constructs an assembly of regression or classification trees using the input data to capture the association between the features and the output effectively. The process of decorrelating the trees within a random forest model serves to decrease the level of variability and enhance the overall reliability of the results.

4 Results and Discussion

The current investigation involves developing a tone recognition system designed for Galo tonal words. A dataset of 14,623 tonal words spoken by 27 Galo speakers (13 males and 14 females) has been considered for developing the system. The dataset consists of twenty Galo tonal words with their tonal instances. Seventy percent of the dataset has been considered for training the system and the remaining 30% for testing the system. The speech signal is first pre-emphasized and framed into the segment of 30 microseconds with one-third overlapping using a Hamming window given by:

$$w(n) = 0.54 - 0.46\cos(2\pi nN) \tag{12}$$

The F0 values are computed by applying the autocorrelation method to each speech frame. The F0 values obtained from the computation process are subjected to normalization using *z*-score normalization techniques. Additionally, a smoothing operation is performed on the F0-contour. Figure 2 depicts the smoothed pitch contour for the Galo tonal words */aii/* and */‘aii/*. The seven features extracted from the normalized and smoothed fundamental frequency (F0) contour are mean F0, median F0, standard derivative of F0, F0 value measured at 0, 25, 75, and 100% of the pitch contour. Consequently, a 7-dimensional feature vector has been obtained for each frame. The mean, variance, and standard deviation statistical features were derived from the segment-level features to create the utterance-level feature vector. Consequently, a 21-dimensional feature vector was obtained for each utterance. An SVM model is trained using the feature vector. The RBF kernel is used for the SVM model. In the next experiment, we apply the random forest algorithm to train a model to recognize tones associated with the tonal words. The number of trees and their depth are tuned dynamically. To cross-validate the model, the KFold grid search algorithm is used. The models are tested for their recognition accuracy using the test dataset. The confusion matrices for the SVM classifier and RF classifier are presented in Fig. 3a, b, respectively. The recognition accuracy of the conducted experiments are briefly outlined in Table. 2.

The experiments show that the F0-based features have tone discriminative capability in the context of the Galo tonal word. The SVM and RF models show promising performance in recognizing the lexical tones associated with the tonal words. The SVM model with RBF kernel gives slightly higher recognition accuracy compared to the RF model.

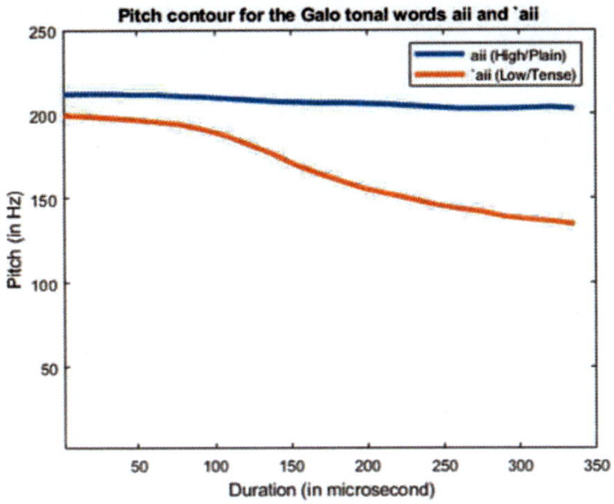

Fig. 2 Smoothed pitch contour for the Galo tonal words aii and 'aii

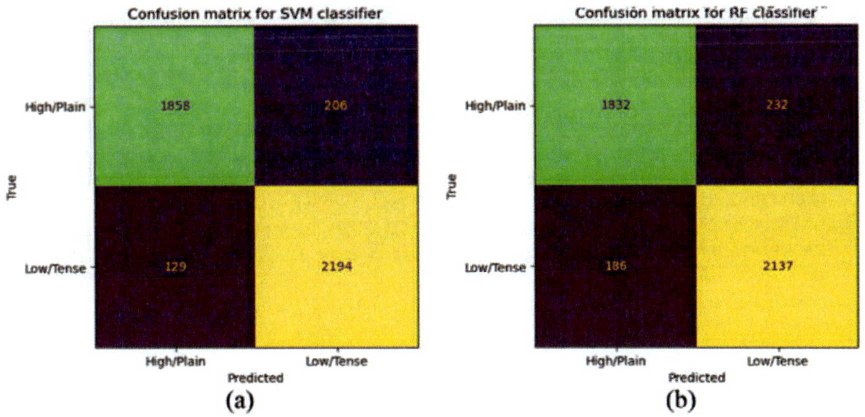

Fig. 3 Confusion matrix for **a** SVM classifier and **b** Random forest classifier

Table 2 Evaluation metrics for the RF and SVM classifiers

Evaluation metrics	RF (in %)	SVM (in %)
Precision	90.47	92.36
Recall	90.49	92.46
F1 score	90.37	92.23
Accuracy	90.42	92.31

5 Conclusion and Future Works

The tone is important in distinguishing between tonal words. We created a simplified tone recognition system in this paper using only the feature vectors derived from the F0-contour. SVM and RF are used to model the tone. One of the most important findings of this work is that the feature vector derived from the F0-contour has tone-discriminating capability in the context of the Galo language. We tested two machine learning models for tone recognition, SVM and RF. The SVM and RF models have recognition accuracy of 92.31% and 90.42%, respectively. The aforementioned findings highlight the necessity of developing an algorithm that effectively integrates independently generated scores for base syllables and tones. Using these scores, a comprehensive decision can be made regarding tonal syllables. The significance of these findings lies in their relevance to the recognition of tonal low-resource languages. In this context, the tones are identified based on standard spectral features, effectively isolating the tone recognition process from other activities. These findings open up possibilities for utilizing cutting-edge pre-trained models developed for non-tonal systems in tonal environments through the fusion of independent scores.

References

1. Moore R (2007) Presence: a human-inspired architecture for speech-based human-machine interaction. IEEE Trans Comput 56(9):1176–1188
2. Li YW, Cheng X, Ding C, Galvin JJ, Chen B, Fu QJ (2023) Benefits of long-term music training for segregation of competing speech by tonal language speakers. J Acoust Soc Am 153(no. 3_supplement):A330–A330
3. Post MW (2007) A grammar of Galo. Ph.D. Dissertation. La Trobe University
4. Chen XX, Cai CN, Guo P, Sun Y (1987) A hidden Markov model applied to Chinese four-tone recognition. In: ICASSP'87 IEEE international conference on acoustics, speech, and signal processing, vol 12, pp 797–800 IEEE
5. Chang, PC, Sun SW, Chen SH (1990) Mandarin tone recognition by multi-layer perceptron. In: international conference on acoustics, speech, and signal processing, IEEE pp 517–520
6. Lee T, Ching PC, Chan LW, Cheng YH, Mak B (1995) Tone recognition of isolated Cantonese syllables. IEEE Trans Speech Audio process 3(3):204–209
7. Lee T, Kochanski G, Shih C, Li Y (2002) Modeling tones in continuous Cantonese speech. In: Proceedings ICSLP, pp 2401–2404
8. Peng G, Wang WS (2004) Parallel tone score association method for tone language speech recognition. In: Proceedings international conference on spoken language processing (ICSLP)
9. Wang S, Levow GA (2008) Mandarin Chinese tone nucleus detection with landmarks. In: ninth annual conference of the International Speech Communication Association
10. Hsieh YL, Chuang CT, Hsieh FF, Chang YC, Hsu WL (2014) Taiwanese tone recognition using fractionalized curve-fitting of prosodic features. In: Proceedings 7th international conference on speech prosody 2014, pp 772–775
11. Gogoi P, Tzudir M, Sarmah P, Prasanna SRM (2020) Automatic tone recognition of Ao language. In Proceedings 10th international conference on speech prosody 2020, pp 1005–1008
12. Gogoi P, Dey A, Lalhminghlui W, Sarmah P, Prasanna SM (2020) Lexical tone recognition in Mizo using acoustic-prosodic features. In: Proceedings of the twelfth language resources and evaluation conference, pp 6458–6461

13. Zhao X, Shaughnessy DO, Minh-Quang N (2007) A processing method for pitch smoothing based on autocorrelation and cepstral F0 detection approaches. In: 2007 international symposium on signals, systems and electronics, pp 59–62 IEEE
14. Vapnik V (1999) The nature of statistical learning theory. Springer science and business media
15. Gigović L, Pourghasemi HR, Drobnjak S, Bai S (2019) Testing a new ensemble model based on SVM and random forest in forest fire susceptibility assessment and its mapping in Serbia's Tara National Park. Forests 10(5):408
16. Breiman L (2001) Random forests. Mach Learn 45:5–32
17. Abuella M, Chowdhury B (2017) Hourly probabilistic forecasting of solar power. In: 2017 North American power symposium (NAPS), IEEE pp 1–5

Tea Leaf Disease Classification Using an Encoder-Decoder Convolutional Neural Network with Skip Connections

Swati Shinde and Sagar Lahade

Abstract In this study, a Convolutional Neural Network (CNN) model for tea leaf disease classification is introduced, which utilizes an encoder–decoder structure with skip connections. The proposed model is designed to reduce computational complexity and generate high-resolution images from lower-resolution representations obtained using the encoder. The skip connections in the model enable the transfer of information between the encoder and decoder, thereby allowing for the reconstruction of the original dimension image from the encoded representation. The performance of the proposed CNN model is compared against several popular computer vision models, including GoogleNet, ResNet50, and VGG16. An ensemble model is also developed to further improve classification performance. In addition, two hybrid CNN models are created, one by combining the computer vision models without using them as feature extractors, and the other by using their feature extraction capabilities with various machine learning algorithms for image classification. Experimental results indicate that the proposed CNN model with skip connections outperforms the individual models with accuracy, precision, recall, and F1-score 87.36, 88.12, 87.36, 87.30, respectively. Based on our experimental results, our proposed CNN model with skip connections did not outperform the ensemble and hybrid models in terms of accuracy. However, the difference in accuracy was marginal, and our proposed model offered lower computational complexity, making it a more efficient option. Therefore, while our proposed model may not be the best performer in terms of accuracy, it can still provide a viable alternative for tea leaf disease classification, especially when computational efficiency is a critical factor.

Keywords CNN · Vgg16 · Computer-vision

S. Shinde · S. Lahade (✉)
Pimpri Chinchwad College of Engineering, Pune, India
e-mail: sagar.lahade21@pccoepune.org

S. Shinde
e-mail: swati.shinde@pccoepune.org

S. Shakya et al. (eds.), *Fourth International Conference on Image Processing and Capsule Networks*, Lecture Notes in Networks and Systems 798,
https://doi.org/10.1007/978-981-99-7093-3_24

1 Introduction

Tea, a pivotal cash crop, plays a significant role in augmenting rural development, ameliorating poverty, and ensuring food security in both developed and developing countries. It is the predominant source of livelihood for numerous smallholder farmers and bears immense cultural significance in several nations. Moreover, tea consumption is associated with a plethora of health benefits, including anti-inflammatory and antioxidant properties, and weight loss effects. With its origins dating back to 2700 BC, tea made from the leaves of the Camellia sinensis plant is one of the oldest and most widely consumed beverages worldwide. The information provided is sourced from the Food and Agriculture Organization of the United Nations.

Despite the annual global tea production exceeding $17 billion, the value of the global tea trade is estimated to be approximately $9.5 billion [1]. In 2022, China emerged as the top revenue-generating country in the tea industry, amassing a staggering sum of 99,810 million USD. Following closely behind, India recorded an income of 15,670 million USD, ranking second in the world. India, which accounts for 30% of the world's tea production, is the largest consumer of tea globally. Moreover, India is one of the top five tea exporters globally, contributing 10% of the total global tea exports. The consumption of tea in India is anticipated to increase by 2.1% per annum. The tea industry in India provides direct employment opportunities to over one million individuals (10 lakhs) and indirectly supports the livelihoods of another 10 million people (100 lakhs) through its forward and backward linkages.

The production of tea is a significant contributor to India's agricultural output and GDP. India had been the foremost producer of tea for nearly a century, but China has recently overtaken India due to the availability of more land [2]. The tea industry faces a daunting challenge from leaf diseases that significantly impact yields, as tea's high export value makes it a crucial economic driver for tea-growing countries. The pathogen's direct impact on the plant's delicate leaves is a threat to the industry. Leaf infections pose a major concern as tea leaves used for tea production originate from young plant shoots. Given the economic significance of tea, any disruptions to yield are of paramount importance. Tea crops have suffered losses ranging from 6 to 14% of their production due to the ravages of insect pests. Effective prevention and control of tea leaf diseases necessitate timely diagnosis and identification.

Artificial Intelligence (AI) plays a pivotal role in the detection of tea leaf diseases by virtue of its ability to accurately analyze vast volumes of data from disparate sources and discern patterns that may elude human observation. Employing machine learning and other AI techniques, computers can learn to identify the nuanced symptoms of various tea leaf diseases, empowering farmers to take appropriate measures to forestall or mitigate their propagation [3]. By automating the detection and diagnosis process, AI can save time and resources while enhancing the precision of disease identification. AI facilitates quick identification and treatment of diseased plants, curtailing the spread of diseases to neighboring plants and minimizing crop losses. Additionally, AI-based disease detection systems can provide real-time updates and alerts, empowering farmers to take timely action to safeguard their crops. In sum, the use of AI in tea leaf disease detection is imperative for ensuring crop vitality, minimizing crop losses, and boosting yields. Given the escalating demand for tea worldwide, AI-based disease detection systems represent a sustainable and efficient solution for tea farmers globally.

This research aims to develop a robust predictive model for tea plant disease classification based on leaf images. The proposed methodology involves the implementation of an encoder-decoder-based CNN architecture. The encoder network captures and encodes essential features from tea leaf images, while the decoder network reconstructs the input image using the extracted features. Through careful preprocessing of a diverse dataset and iterative refinement techniques, the study seeks to achieve high accuracy in disease classification. The outcomes of this research will contribute to the field by providing an effective approach for detecting tea plant diseases, thereby improving crop health management and enabling early disease diagnosis in tea plantations.

2 Literature Review

The literature on the various approaches and methodologies used for detecting diseases in tea leaves is summarized in Table 1 that comes next.

3 Dataset

The tea sickness dataset [17] version 2, published on May 20, 2022, consists of tea leaves images displaying seven common diseases of tea, along with a healthy class. The dataset contains approximately more than 100 images per class (excluding healthy class) and was collected from a Johnstone Boiyon farm, Koiwa location, Bomet county. The dataset was contributed by Gibson Kimutai and Anna Förster and has a DOI of https://doi.org/10.17632/j32xdt2ff5.2. The objective of this dataset is to be used for transfer learning in machine learning models to predict tea plant diseases. The following Fig. 1 displays a visual representation of the dataset utilized in the study. The dataset contains 885 images of jpg.

Table 2 illustrates the distribution of images of tea sickness dataset which contains images from 8 classes including healthy category.

Table 1 Documentation synopsis of tea leaves disease detection methods

References	Description	Advantage	Disadvantage	Result
[4]	The present study exhibits the capability of Internet of Things (IoT) and deep learning techniques to augment illness identification and management in intricate ecological systems	The utilized network structure in this context is Swim Transformer The model's overall ability to generalize was enhanced through the incorporation of real-time images captured by cameras, which offered a diverse and comprehensive range of data	The reliance on mobile devices for connectivity through 4G network is apparent	The tests conducted in difficult natural surroundings demonstrated that swim transformer's identification rate approached 94%, resnet 90, densenet 91
[5]	An image extraction system was developed to evaluate whether a given input image of a tea leaf is afflicted with disease or is in a healthy state	The proposed model integrates Deep Hashing with Integrated Autoencoders, which is an effective and adaptable method, for retrieving photographs of Tea Leaves Except for segmentation, no preprocessing was utilized	Inadequate dataset employed for training	Mean average precision 98.27%
[6]	The diagnosis of grey blight disease in tea leaves was accomplished through the use of a deep convolutional neural network featuring inverted residuals and linear bottleneck layers	A total of 1320 images were collected from India, and bottleneck layers were utilized in the model	The dataset is exceedingly small	The proposed approach achieved an accuracy of 98.99%

(continued)

Table 1 (continued)

References	Description	Advantage	Disadvantage	Result
[7]	To distinguish between disease symptoms and healthy tea leaf regions, K-means clustering was employed for identification, followed by the extraction of features using different color models. The resulting features were classified using SVM	The processing times of the proposed method are faster when compared to those of a traditional CNN	In the suggested strategy, background reduction is essential	GLCM effectiveness with PCA is 98.97% accurate
[8]	The approach involved converting RGB images to HSI format, isolating green pixels, performing segmentation, and utilizing LeNet for classification	The proposed method demonstrated higher precision when compared to the conventional backpropagation network	The identification of red leaf spot appears to be somewhat less accurate, with an 85% precision rate	Accuracy 90.23%
[9]	By integrating different CNN modules such as ResNet and Inception-v3, the MergeModel is capable of extracting a diverse range of distinguishing features, which enhances its ability to identify diseases	This technique allows training to occur with less data than is customary while still improving recognition performance	The model is becoming increasingly complex	On the small-sized dataset, MergeModel outperforms Inception-v3 by 13% in disease identification and ResNet34 by 5%
[10]	A CNN model comprising of 7 layers was utilized	Some images were captured using a field camera	The customary number of filters used in a convolutional layer is 64 or 128	94.45% accuracy
[11]	C-DCGAN enhanced training data	Support Vector Machine (SVM) analysis was employed to detect and discriminate disease spots in images of tea leaves	VGG16 suffers from a significant overfitting problem, and its accuracy drastically reduces in the absence of augmented data	90% accuracy

(continued)

Table 1 (continued)

References	Description	Advantage	Disadvantage	Result
[12]	Prior to being utilized for training the network, raw images underwent pre-processing techniques such as image segmentation and data augmentation	In comparison to SVM and BP, CNN has been found to be more efficient	CNNs have shown to be effective with large datasets	CNN has a 93.754% success rate in recognising images
[13]	The models, which were initially trained using ImageNet data, were retrained to detect diseases in tea leaves	Transfer Learning with full fine-tuning accuracy improved	Evaluation of transfer learning and tuning on small-footprint architectures ResNet, VGGNet, and Xception	Accuracy is improved with respect to training from scratch Accuracy obtained ResNet 94.05 VGGNet 91.26 Xception 91.27
[14]	Images captured in the wild may be inaccurate due to various factors such as changing lighting conditions, shadows, irregular shapes, and overlapping leaves	The Retinex technique was incorporated to enhance the quality of the original images CNN-based methods have facilitated the more rapid and efficient identification of faint, obscured, and diminutive diseases in tea leaves. In addition, VGG16 networks have been utilized to perform severity grading by inputting the diseased leaves	There may be variations in the light conditions in different regions of China, which may not be reflected in the images present in the dataset	Both the average precision of detection and the accuracy of severity rating have improved by more than 6% and 9%, respectively, when compared to conventional machine learning techniques

(continued)

Table 1 (continued)

References	Description	Advantage	Disadvantage	Result
[15]	In addition, features are gathered and exploited using the dense scale-invariant feature transform. Make a model. incorporating the multi-layer perceptron and the support vector machine to categorize diseases on the basis of descriptions provided by a bag of visual words	Local characteristics taken from pictures that remain unchanged by translation, scaling, or rotation	While certain diseases only appear at specific periods of the year, others can affect tea plants at any time. The timing of disease diagnosis could therefore vary throughout the year, which would have an impact on how accurately diseases are identified	LeafNet accuracy 90% SVM 60% MLP 70%
[16]	In this research, they reveal a Deep Learning modified Mask R-CNN for fully independent segmentation as well as tracking of tomato plant leaf disease	Optimizing the RPN network's anchor fractions and indeed the architecture of feature extraction to uplift detection performance and metric computation effectiveness	-	Accuracy 98%

4 Proposed Method

The encoder–decoder architecture, complemented by skip connections and a classification head, demonstrates superior performance in classification tasks, owing to its unique capabilities. This approach offers a synergistic integration of hierarchical feature learning, spatial context preservation, and context-aware predictions.

During the encoding phase, the network effectively captures hierarchical features from the input image, encompassing both low-level patterns such as edges and intricate high-level structures. This hierarchical feature learning empowers the model to discern local and global visual patterns, thus facilitating better discrimination between different classes. Additionally, the encoder–decoder architecture retains the spatial context throughout encoding and decoding, a departure from traditional CNNs that lose spatial information through pooling layers. The upsampling process in the decoder preserves the spatial relationships between pixels, thereby ensuring that the model can recognize object parts and their spatial arrangement, which are crucial factors for accurate classification.

Notably, the incorporation of skip connections plays a pivotal role in enhancing the encoder-decoder's performance. These connections facilitate direct access to

Fig. 1 A graphical depiction of the dataset utilized in the study

Table 2 A depiction of the dataset utilized in the study

Sr. no	Disease name	Total number of images
1	Anthracnose	100
2	Algal leaf	113
3	Bird eye spot	100
4	Brown blight	113
5	Gray light	100
6	Healthy	74
7	Red leaf spot	143
8	White spot	142

Table 3 Dataset splitting to create test, train and validation pipeline

	Algal leaf	Anthracnose	Bird eye spot	Brown blight	Gray light	Healthy	Red leaf spot	White spot	Total
Training	90	80	80	91	80	59	114	114	708
Validation	11	10	10	11	10	7	15	14	88
Testing	12	10	10	11	10	8	14	14	89
Total	113	100	100	113	100	74	143	142	885

multi-level features from the encoder during decoding. As a result, the decoder can effectively utilize high-level semantic information alongside fine-grained details, enabling context-aware predictions. The efficient information flow facilitated by skip connections enables the model to adapt to various levels of granularity in the input data, further augmenting its classification capabilities.

Subsequently, the classification head, which receives the decoder's contextualized features, benefits from the rich representations learned by the encoder–decoder. These context-aware features furnish a more informative and discriminative foundation for classification, outperforming traditional flattened features. Consequently, the classification head delivers more accurate predictions, especially in scenarios involving visually similar classes.

The proposed method section is structured into two distinct parts. Section 4.1 describes dataset distribution used for training and testing Sect. 4.2 outlines the architecture of the proposed method, while Sect. 4.3 details the training process of the model.

4.1 *Dataset Distribution for Training and Testing*

To create a robust predictive model, the dataset was split into three categories: training, validation, and testing. The table provided illustrates the breakdown of the dataset in terms of the number of images for each class in the three categories. The dataset splitting was designed to ensure that the model could generalize well and perform accurately on new, unseen data. Table 3 presented below outlines the dataset splitting process carried out to develop a reliable predictive model for a group of plant diseases.

4.2 *Architecture of Proposed Method*

One of the main advantages of this architecture is that it allows for the effective processing of images at multiple resolutions, enabling it to capture both high-level and low-level image features. The architecture consists of an encoder network that reduces the input image dimensions and extracts its features, followed by a decoder network that generates an output image of the same dimensions as the input image. Skip connections are added between the encoder and decoder networks to facilitate the transfer of information from the encoder to the decoder, which helps preserve low-level image features that would have otherwise been lost during the encoding process.

The architecture appears to be a U-Net style encoder-decoder with skip connections. The input layer is an unspecified 3D tensor with variable shape. The encoder consists of six convolutional layers with increasing number of filters (32, 64, 128, 256, 512, 1024), followed by a max pooling layer. The decoder consists of four convolutional layers with decreasing number of filters (512, 256, 128, 64), with two up sampling layer. The skip connections are implemented through concatenation of feature maps from corresponding encoder and decoder layers. The output layer is a dense layer with 8 units and a softmax activation function, which suggests that the network is trained for a classification task with eight classes.

The total number of parameters in the model is 15,072,136, all of which are trainable. This high number of parameters is likely due to the large number of filters in the convolutional layers, as well as the use of skip connections which add additional connections and computations to the network. Following Table 4 shows architecture of proposed Model.

The proposed architecture is specifically designed to capture and preserve essential features across various levels of image abstraction. By integrating the encoder, decoder, and skip connections, the model effectively utilizes both local and global image context. This comprehensive approach enhances the model's capability to accurately classify tea plant diseases based on leaf images. It ensures that relevant features are adequately represented and leveraged throughout the encoding and decoding stages, ultimately contributing to improved classification performance. Table 4 shows the architecture of proposed CNN.

4.3 Training Process of Proposed Model

The model was trained using the Adam optimizer with a learning rate of 0.0001 and a categorical cross-entropy loss function for a total of 100 epochs.

5 Result

In this study, the three models, Vgg16, ResNet50, and GoogleNet, were trained and evaluated on the tea sickness dataset using the same data splitting approach. To create hybrid models, individual models such as VGG16, ResNet50, and GoogLeNet were trained and tested. The performance of these models on the testing set was

Table 4 Proposed CNN architecture

Sr. No	Layer type	Filter size	Number of filters/neurons	Input shape	Output shape
1	Inputlayer	N/A	N/A	[]	[]
2	Conv2D	(3, 3)	32	(None, None, 3)	(None, None, 32)
3	Conv2D	(3, 3)	64	(None, None, 32)	(None, None, 64)
4	Conv2D	(3, 3)	128	(None, None, 64)	(None, None, 128)
5	Maxpooling2d	(2, 2)	N/A	(None, None, 128)	(None, None, 128)
6	Conv2D	(3, 3)	256	(None, None, 128)	(None, None, 256)
7	Conv2D	(3, 3)	512	(None, None, 256)	(None, None, 512)
8	Maxpooling2d	(2, 2)	N/A	(None, None, 512)	(None, None, 512)
9	Conv2D	(3, 3)	1024	(None, None, 512)	(None, None, 1024)
10	Upsampling2d	(2, 2)	N/A	(None, None, 1024)	(None, None, 1024)
11	Concatenate	N/A	N/A	[(None, None, None, 512)]	(None, None, 1536)
12	Conv2D	(3, 3)	512	(None, None, 1536)	(None, None, 512)
13	Conv2D	(3, 3)	256	(None, None, 512)	(None, None, 256)
14	Upsampling2d	(2, 2)	N/A	(None, None, 256)	(None, None, 256)
15	Concatenate	N/A	N/A	[(None, None, None, 128)]	(None, None, 384)
16	Conv2D	(3, 3)	128	(None, None, 384)	(None, None, 128)
17	Conv2D	(3, 3)	64	(None, None, 128)	(None, None, 64)
18	Global average pooling2d	N/A	N/A	(None, None, 64)	(64)
19	Dense	N/A	128	(64,)	(128)
20	Dense	N/A	8	(128,)	(8)

evaluated, and the resulting confusion matrix is depicted in the figure. Analysis of the confusion matrix reveals that each model achieved 100% accuracy on specific classes. This observation suggests that combining these models into a hybrid model holds promise for improving overall performance and achieving higher accuracy in tea plant disease classification. Following figures shows the confusion matrixes for vgg16, GoogleNet and resnet 50 respectively in Figs. 2, 3 and 4.

The obtained results for VGG16, ResNet50 and GoogleNet model are summarized in Table 5. It is important to note that all models were trained from scratch.

Based on the evaluation results, the GoogleNet model demonstrated the highest performance for the classification of tea sickness among the three models tested, as evidenced by its highest recall and F1 score.

The study combined three models, Vgg16, ResNet50, and GoogleNet, to create an ensemble model. The results obtained from the ensemble model are presented in Table 6.

The ensemble model was created by leveraging pretrained individual models and applying the inverse confusion matrix method to handle the class imbalance problem. The study found that the ensemble model outperformed the individual models, achieving a testing accuracy of 93.10%, which is higher than the best-performing individual model, GoogleNet. The precision, recall, and F1 score for the ensemble model were also impressive, with precision at 93.17%, recall at 93.10%, and F1 score at 92.72%. These results suggest that the ensemble model was able to strike a good balance between accurately identifying true positives, avoiding false positives, and avoiding false negatives, resulting in superior accuracy and balanced performance across all metrics. Thus, it can be concluded that the ensemble model offers an effective approach for accurately identifying tea leaf diseases, outperforming

Fig. 2 Confusion matrix of VGG16

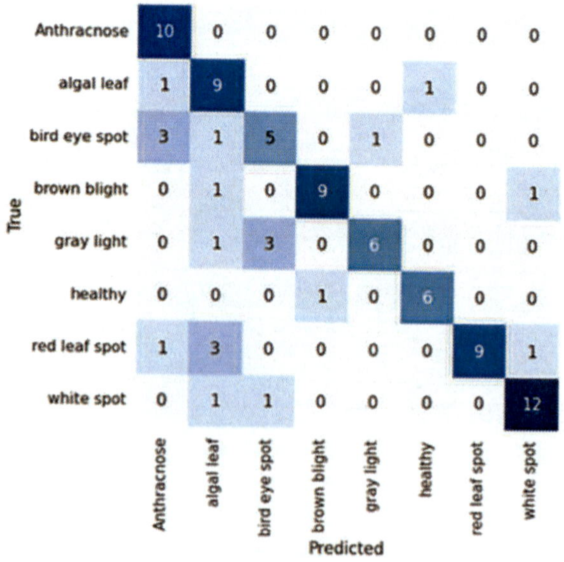

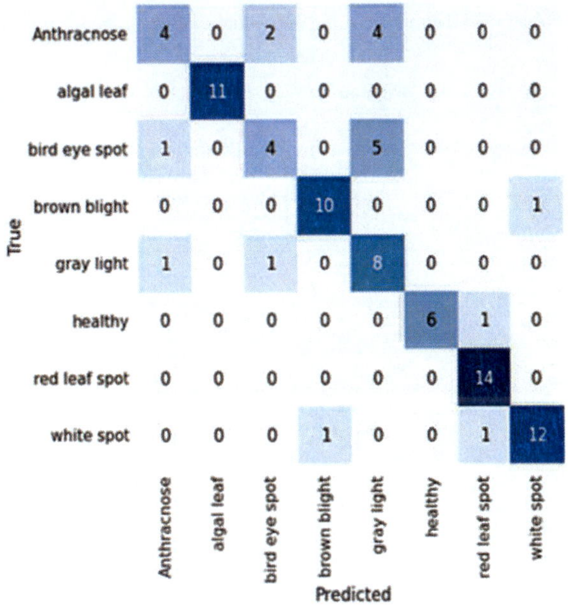

Fig. 3 Confusion matrix of ResNet50

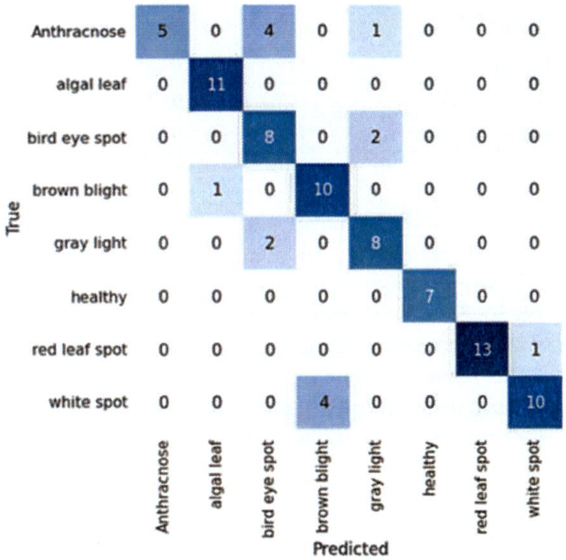

Fig. 4 Confusion matrix of GoogleNet

Table 5 Performance of the individual models trained from scratch on the tealeaf dataset

	Models used	Vgg16	ResNet50	GoogleNet
	Method		Trained from scratch	
Training	Training samples		708	
	Training accuracy	65.59	85.38	84.42
	Validation samples		88	
	Validation accuracy	71.27	79.32	75.52
Testing	Samples		89	
	Accuracy	75.86	79.31	82.76
	Precision	79.17	80.76	85.81
	Recall	75.86	79.31	82.76
	F1 score	75.91	78.86	82.71

Table 6 Performance of the ensemble model trained on the tealeaf dataset

Models used	Vgg16	ResNet50	GoogleNet
Method		Ensemble	
Optimal weight finder		Inverse confusion matrix method	
Unseen testing samples		89	
Testing accuracy		93.10	
Testing precision		93.17	
Testing recall		93.10	
Testing F1 score		92.72	

individual models in terms of accuracy and achieving a good balance between other performance metrics.

Table 7 summarizes the performance of, a model developed by fusing three models (Vgg16, ResNet50, and GoogleNet) using a feature concatenation approach. The model was trained and evaluated on separate training, validation, and testing datasets.

The results obtained for the hybrid model, created using the feature fusion approach to combine three models, Vgg16, ResNet50, and GoogleNet, were summarized in table provided. The testing accuracy achieved by the model was 86.2%, which was higher than the highest performing individual model (GoogleNet). In addition, the model achieved a precision of 86.9%, recall of 86.65%, and an F1 score of 86.6%.

Overall, the results suggest that while the individual models achieved high training accuracy, their performance on the validation and testing datasets was relatively lower. However, the hybrid model, created using feature fusion, was able to achieve a relatively high testing accuracy and a good balance between precision and recall.

The study employed a hybrid model created by using feature fusion approach to combine vgg16, ResNet50, and GoogleNet as a feature extractor to extract features from input images. The extracted features were reduced to 65 using PCA and were

Table 7 Performance of the feature fusion approach trained on the tealeaf dataset

Models used	Vgg16	ResNet50	GoogleNet
Method		Feature concatenation	
Training samples		708	
Training accuracy		78.69	
Validation samples		88	
Validation accuracy		86.26	
Testing samples		86	
Testing accuracy		86.2	
Testing precision		86.9	
Testing recall		86.65	
Testing F1 score		86.6	

then utilized as input to various machine learning algorithms such as Random Forest, SVM, Decision Tree, KNN, Logistic Regression, Naive Bayes, and LDA. The table below presents the results obtained by these algorithms.

Table 8 presents the performance results of several machine learning (ML) algorithms, when using features extracted by the hybrid model.

Figure 5 presented illustrates the performance of feature extractor approach.

The figure presented depicts the performance of Tealeafnet as a feature extractor. The training dataset consisted of feature samples extracted from 620 images, and the ML algorithms achieved a training accuracy ranging from 90.57 to 100%. The testing dataset consisted of feature samples extracted from 265 images, and the ML algorithms achieved a testing accuracy ranging from 79.18 to 90.70%. The testing precision ranged from 79.67 to 90.91%, while the testing recall ranged from 79.49 to 90.76%. The testing F1 score ranged from 79.42 to 90.81%.

Based on the results, the Decision Tree algorithm achieved the lowest testing precision, recall, and F1 score among the algorithms tested. However, it had a training accuracy of 100%, indicating potential overfitting of the training data. Therefore, Decision Tree may be considered the worst-performing algorithm for this specific dataset and feature extraction technique.

On the other hand, the SVM algorithm achieved the highest testing accuracy of 90.91% among the algorithms tested. The testing precision, recall, and F1 score of SVM were also among the highest. Thus, for this specific dataset and feature extraction technique, SVM may be considered the best algorithm.

Following Table 9 shows the performance of proposed model.

The proposed encoder–decoder CNN model achieved an overall testing accuracy of 87.36%, which indicates that the model is performing well on the task of tea sickness classification. Additionally, the model achieved a testing precision of 88.12%, indicating that it is correctly identifying true positives and avoiding false positives. The model also achieved a testing recall of 87.36%, indicating that it is correctly

Table 8 Performance of the ML algorithms on features obtained by using hybrid model

Feature extractor	Created hybrid model						
Feature extracted	1024	1024	1024	1024	1024	1024	1024
PCA (reduced feature count)	65	65	65	65	65	65	65
ML Algorithm	Random forest	SVM	Decision tree	KNN	Logistic regression	Naive Bayes	LDA
Training samples	620	620	620	620	620	620	620
Training accuracy	100	96.96	100	90.89	95.68	90.57	92.49
Testing samples	268	265	265	265	265	265	265
Testing accuracy	88.84	90.7	79.18	88.47	90.33	89.21	89.21
Testing precision	89.11	90.91	79.67	88.55	90.49	89.81	89.57
Testing recall	89	90.76	79.49	88.44	90.37	89.96	89.2
Testing F1 score	89.01	90.81	79.42	88.32	90.4	89.23	89.3

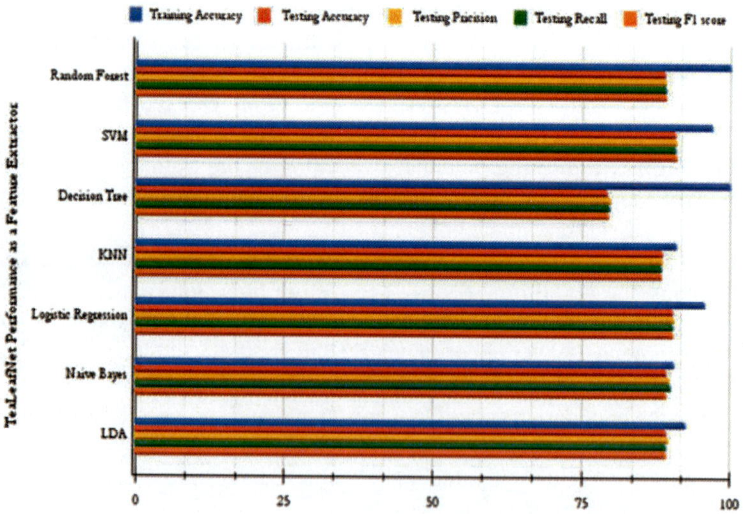

Fig. 5 Hybrid model performance as a feature extractor

Table 9 Performance of proposed model

Model	Proposed model
Training samples	708
Training accuracy	82.86
Validation samples	88
Validation accuracy	83.91
Testing samples	89
Testing accuracy	87.36
Testing precision	88.12
Testing recall	87.36
Testing F1 score	87.30

identifying true positives and avoiding false negatives. The testing F1 score of the model was 87.30%, which is a balanced measure of precision and recall.

Overall, the results suggest that the proposed model has performed well on both the training and testing datasets, with a training accuracy of 82.86% and a validation accuracy of 83.91%. The performance of the model on the testing dataset indicates that it is not overfitting the training data and is generalizing well to new data. The obtained high accuracy can be attributed to encoder-decoder architecture allowed the model to effectively learn and represent complex patterns and distinguishing features present in the tea leaf images. By capturing these intricate details, the model could discern subtle differences between healthy leaves and those affected by different diseases proposed encoder-decoder CNN model, it achieved a relatively high testing accuracy of 87.36% on the Tea Sickness dataset, which is comparable to the best-performing individual model, GoogleNet. Additionally, the proposed model has the advantage of being computationally efficient due to its use of encoder-decoder architecture, which reduces the number of parameters and computational complexity.

Comparing this with the feature extraction approach used in the hybrid model, the proposed model also has the advantage of being an end-to-end learning approach, whereas the hybrid model required the use of multiple machine learning algorithms after feature extraction. Additionally, the proposed model may have a better ability to generalize to new data, as it was trained on the entire dataset, rather than just a subset of features. Feature extraction approach is better than proposed method but it's more complex. The proposed model is better than feature fusion approach. In terms of the ensemble model, while it achieved the best performance among all models, it may not be as computationally efficient as the proposed model, as it required training multiple individual models and then combining them. Furthermore, the ensemble model may not be as interpretable as the proposed model, as it may be difficult to discern the contribution of each individual model to the overall performance.

6 Conclusion

In conclusion, the encoder–decoder architecture with skip connections, combined with a classification head, presents a powerful and effective approach for classification tasks in computer vision. The integration of hierarchical feature learning, spatial context preservation, and context-aware predictions enables the model to extract informative and discriminative features from the input images. The encoder-decoder's ability to retain spatial context and leverage skip connections further enhances its performance, enabling it to handle various levels of granularity in the data. The classification head benefits from the rich representations learned by the encoder-decoder, resulting in more accurate predictions, especially for visually similar classes. Overall, this architecture proves to be a promising and versatile solution for achieving improved performance in classification tasks, making it a valuable addition to the field of computer vision research.

References

1. International tea market: market situation, prospects, and emerging issues, food and agriculture organization of the United Nations (FAO), Published: May 2022
2. Chen J et al (2020) Automatic recognition of tea diseases based on deep learning. Adv Manage Global Change (IntechOpen) 2020 chapter 8. https://doi.org/10.5772/intechopen.91953
3. Bao W, Fan T, Hu G et al (2022) Detection and identification of tea leaf diseases based on AX-RetinaNet. Sci Rep 12:2183. https://doi.org/10.1038/s41598-022-06181-z
4. Xie S et al (2023) Online identification method of tea diseases in complex natural environments. IEEE Open J Comput Soc 4. https://doi.org/10.1109/OJCS.2023.3247505
5. Jayapal SK et al (2023) Enhanced disease identification model for tea plant using deep learning. Intell Autom Soft Comput. https://doi.org/10.32604/iasc.2023.026564
6. Pandian JA et al (2023) Grey blight disease detection on tea leaves using improved deep convolutional neural network. Comput Intell Neurosci 2023. https://doi.org/10.1155/2023/7876302
7. Javidan SM et al (2023) Diagnosis of grape leaf diseases using automatic K-means clustering and machine learning. Smart Agric Technol 3:100081. https://doi.org/10.1016/j.atech.2022.100081
8. Gayathri S et al (2020) Image analysis and detection of tea leaf disease using deep learning. In: 2020 international conference on electronics and sustainable communication systems (ICESC). https://doi.org/10.1109/ICESC48915.2020.9155850
9. Hu et al (2022) Using a multi-convolutional neural network to automatically identify small-sample tea leaf diseases. Sustain Comput Inf Syst 35:100696. https://doi.org/10.1016/j.suscom.2022.100696
10. Latha RS et al (2021) Automatic detection of tea leaf diseases using deep convolution neural network. In: 2021 international conference on computer communication and informatics (ICCCI). https://doi.org/10.1109/ICCCI50826.2021.9402225
11. Hu G et al (2019) A low shot learning method for tea leaf's disease identification. Comput Electron Agri 163:104852. https://doi.org/10.1016/j.compag.2019.104852
12. Xiaoxiao SUN (2018) Image recognition of tea leaf diseases based on convolutional neural network. In: 2018 international conference on security, pattern analysis, and cybernetics (SPAC). https://doi.org/10.1109/SPAC46244.2018.8965555

13. Ramadan A (2020) Transfer learning and fine-tuning for deep learning-based tea diseases detection on small datasets. In: Ade ramadan, transfer learning and fine-tuning for deep learning-based tea diseases detection on small datasets. https://doi.org/10.1109/ICRAMET51080.2020.9298575,10.1109/ICRAMET51080.2020.9298575
14. Hu G et al (2021) Detection and severity analysis of tea leaf blight based on deep learning. Comput Electr Eng 90:107023. https://doi.org/10.1016/j.compeleceng.2021.107023
15. Chen J et al (2019) Visual tea leaf disease recognition using a convolutional neural network model. Symmetry 11(3):343. https://doi.org/10.3390/sym11030343
16. Kaur P et al (2022) An approach for characterization of infected area in tomato leaf disease based on deep learning and object detection technique. Eng Appl Artif Intell 115:105210. https://doi.org/10.1016/j.engappai.2022.105210
17. Kimutai G, Förster A (2022) Tea sickness dataset. Mendeley Data V2. https://doi.org/10.17632/j32xdt2ff5.2

Deep Learning-Based Classification of Lung Cancer Lesions in CT Scans: Comparative Analysis of CNN, VGG-16, and MobileNet Models

P. M. Hareesh and Sivaiah Bellamkonda

Abstract Lung cancer exhibits one of the lowest survival rates among various cancer types over a five-year period, necessitating the utmost importance of early detection. Lung cancer can be detected in its early stage widely through the Computed Tomography (CT) scans. In recent years, deep learning models, particularly Convolutional Neural Networks (CNNs), have been integrated into end-to-end methodologies for the accurate perception of lung nodules. This research project presents a novel system constructed using CNN, VGG-16, and MobileNet models to automatically classify between nonmalignant and malignant tumor lesions within the lungs. The proposed CNN model, trained with hyper parameters including batch size of 8, learning rate of 0.01 with 50 epochs, achieved a peak accuracy of 95.15%. Likewise, VGG-16 achieved a high accuracy of 95.88% with same hyper-parameters except the learning rate is changed to 0.1. Furthermore, MobileNet demonstrated exceptional performance with an accuracy of 98.39% for learning rate of 0.001 with same values for other hyper-parameters batch size and epochs. The classification of lung cancer lesions into benign and malignant categories was effectively accomplished using the LIDC (Lung Image Database Consortium), IDRI (Image Database Resource Initiative) data sets. This study underscores the efficiency of deep learning-based approaches in automated lung cancer detection and emphasizes the significance of early diagnosis in improving patient outcomes.

Keywords Lung cancer · Deep learning · VGG-16 · MobileNet · CNN

P. M. Hareesh · S. Bellamkonda (✉)
Computer Science and Engineering Department, Indian Institute of Information Technology, Kottayam, Kerala, India
e-mail: sivaiah@iiitkottayam.ac.in

P. M. Hareesh
e-mail: haripmohan82.emo2009@iiitkottayam.ac.in

S. Shakya et al. (eds.), *Fourth International Conference on Image Processing and Capsule Networks*, Lecture Notes in Networks and Systems 798,
https://doi.org/10.1007/978-981-99-7093-3_25

1 Introduction

Cancer is a leading global cause of mortality, presenting a significant public health concern. In 2020, cancer accounted for 9.93 million deaths worldwide [2], and projections indicate that it can be increased to 13 million by 2030 and 16.3 million by 2040. Among the major cancer types, lung cancer, along with prostate, colorectal, breast, stomach, and cervical cancer, exhibits higher prevalence [1].

Early detection of lung cancer is critical, given its high fatality rates [4]. Artificial intelligence has shown promise in medical image analysis and cancer diagnosis, mitigating discrepancies in human interpretation [3]. Late-stage diagnosis of lung cancer yields a meager 10–16% five-year survival rate, whereas early detection significantly improves it to 70% [5]. Lung nodules serve as an early indicator of lung cancer and can range from 3 to 30 mm in size [6].

Computed Tomography (CT) has revolutionized lung nodule imaging since its advent in the 1970s [7]. However, manual approaches to nodule detection from CT scans face limitations such as human visual system constraints, variations in radiologists' expertise, and workload-related challenges [6]. Additionally, nonspherical nodules can introduce further errors, potentially impacting diagnostic accuracy and efficiency.

To overcome these challenges and improve early lung cancer detection, advanced automated techniques are needed. Hence, this study is mainly focused on investigating the efficient deep learning models in detecting lung nodules from CT scans, addressing the limitations of manual approaches and empowering clinicians with efficient and accurate diagnostic tools.

1.1 Automatic Lung Nodule Detection and Classifications

Research on lung nodule identification using Computer-Aided Diagnosis (CAD) has shown promising results. CAD enhances the identification of lung nodules while diminishing false positives. CAD systems can handle tasks such as lung segmentation, nodule formation, feature extraction, and classification.

1.2 Research Gaps

This study aims to investigate deep learning architectures for lung nodule identification in CT images. The specific objectives include exploring CNNs for feature extraction and classification, evaluating CNN, VGG, and MobileNet architectures for lung nodule classification, and analyzing and comparing results based on different learning rates, batch sizes, and epochs. The motivation is to develop an accurate and effective system for identifying and categorizing lung cancer using CNN, VGG, and MobileNet. CNNs have shown success in medical image analysis, and VGG and MobileNet architectures have performed well in picture categorization tasks. CNNs, VGG, and MobileNet networks have been widely used in the recent research to detect and classify lung cancer from medical images, improving patient outcomes and reducing healthcare costs. This system empowers doctors to make informed decisions regarding treatment options for lung cancer patients.

2 Literature Survey

CAD systems play a vital role in the early detection of lung nodules and the comprehensive evaluation of lung tissue, resulting in improved efficiency for radiologists when interpreting CT scans. The integration of these techniques in healthcare has significantly enhanced overall performance, particularly in lung cancer screening. A study predicts that by 2025, approximately 23% of CAD systems will be dedicated to lung cancer detection, contributing to the expected 11.6% annual growth rate of the global CAD market [8]. This highlights the increasing importance and utilization of CAD systems in lung cancer diagnosis.

2.1 Lung Cancer Classification Using Conventional Techniques

Nithila et al. [9] presented a CAD system for lung nodule characterization using PSO-based back propagation neural network. It involved histogram equalization, Gaussian blur for image enhancement, and a novel active contour model for lung parenchyma separation. Rule-based approaches were used for nodule identification, and the PSO-based BPNN classifier achieved high success rates for different nodule types.

Several other studies aimed on lung nodule detection and classification using different techniques such as shape analysis, image processing, evaluation of CAD systems, artificial neural networks, and specialized filters like the quantized convergence index filter and the difference image approach. These studies reported promising results and highlighted the benefits of utilizing advanced methods in lung nodule analysis.

The LUNA16 Challenge provided a platform for evaluating nodule identification technologies using CT scans with various nodule sizes. Deep learning techniques showed significant performance improvements compared to traditional image processing methods, indicating the dominance of deep learning in this domain.

2.2 Lung Cancer Classification Using Deep Learning Techniques

Dou et al. [10] developed a method using 3D convolutional neural networks (CNN) to minimize false positives (FPs) in CT scans. Their 3D CNN architecture-encoded spatial information effectively, leading to a more discriminative representation. The focus was on reducing FPs, achieving a sensitivity of 92.2% with 8 FPs per scan.

Monkam et al. [11] proposed three CNN architectures with varying convolutional layers to discriminate micro-nodules from non-nodules in CT images. The most successful model had two convolutional layers and a patch size of 32 × 32 pixels, achieving a sensitivity of 83.82%.

Winkels et al. [12] found that using 3D roto-translation group convolutions rather than the traditional translational convolutions reduced the complexity of CNN models. They applied 3D group convolutions (3D G-CNNs) to effectively reduce false positives in pulmonary nodule identification.

Bushara et al. [13] briefed the deep learning models that used CT images for lung cancer detection, using data augmentation for preprocessing. Their CNN approach achieved 95% accuracy.

Nasser and Abu-Naser [14] developed an ANN-based lung cancer detection model that utilized various human data, including symptoms and medical history, achieving 96.67% accuracy.

Qin et al. [15] introduced a deep learning architecture that integrated fine-grained features from PET and CT images for noninvasive lung cancer diagnosis. Their approach, incorporating multidimensional attention mechanism, achieved an AUROC of 0.92.

Asuntha [16] extracted features using wavelet transform, Histogram of Oriented Gradients (HoG), Local Binary Pattern (LBP), SIFT, and Zernike Moment combined with Fuzzy PSO (FPSO) for selecting features and DNN was then employed to classify.

Gopi Kasinathan [17] introduced a modified computer-aided diagnosis system for lung tumor stage classification, utilizing a deep neural model and cloud data collection to categorize pulmonary illness phases.

Although various techniques using machine learning and image processing have been proposed for nodule detection and classification, algorithms that perform well on one type of nodule may not generalize to others. Deep learning systems require large annotated datasets, which are challenging to obtain in medical imaging due to annotation efforts, privacy concerns, and other factors. Transfer learning and combining other features with CNN-extracted features can be employed to improve classification accuracy.

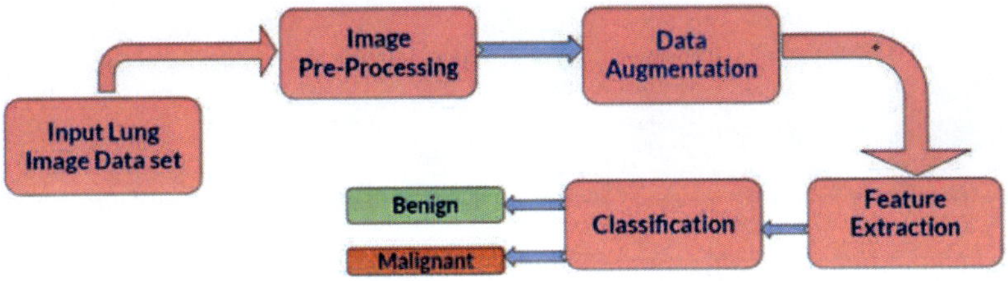

Fig. 1 Architecture of the proposed model

3 Proposed Methodology

3.1 Lung Cancer Classification Using CNN

In a recent study by Thakur et al. [18], it was found that CNNs have the potential to diagnose lung cancer more accurately than manual evaluation by radiologists. Clinical trials have also shown that CNN outperforms professional radiologists in detecting lung nodules. The CNN approach consists of two stages: function extraction and image classification, where the ANN takes over.

CNN utilizes end-to-end learning as its primary learning strategy, but transfer learning can also be employed by leveraging pretrained models. To overcome overfitting issues and improve accuracy, a large dataset is required. Figure 1 illustrates the classification model used to determine whether lung images contain benign or malignant lung nodules.

3.2 Lung Cancer Classification Using VGG-16

VGG-Architecture The VGG-16 network, a deep convolutional neural network (CNN), is the main focus of this research for visual object recognition. It consists 16 layers in which 13 layers are for convolutional and the remaining three are fully connected layers. The convolutional layers extracts features while the fully connected layers classify. Each layer applies multiple filters to the input image, capturing different properties at various resolutions. The network forms a hierarchical representation of the input by passing the results from each layer to the subsequent layer. This hierarchical representation proves effective for object recognition and categorization tasks (Fig. 3).

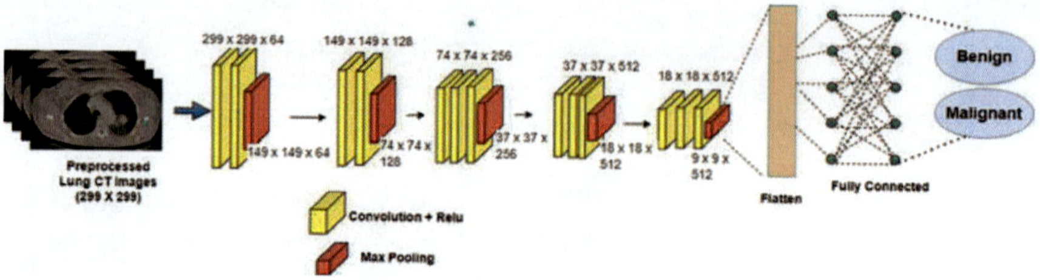

Fig. 2 VGG-16 architecture for lung cancer classification

3.3 Lung Cancer Classification Using MobileNet

MobileNet, a convolutional neural network architecture, is widely recognized for its effectiveness in image classification and object detection on resource-constrained devices. Its combination of accuracy and efficiency makes it suitable for applications such as lung cancer detection and classification. The performance of MobileNet can be evaluated using various metrics to assess its capabilities in this domain. In a feasibility study on lung cancer screening, MobileNet achieved an accuracy of 93.6% in detecting lung nodules from CT scans, indicating its precision and accurate results. Additionally, MobileNet can be utilized to categorize various types of lung cancer such as NonSmall Cell Lung Cancer (NSCLC) and Small Cell Lung Cancer (SCLC), which require distinct treatment approaches. Similar metrics such as precision, recall, and $F1$ score can be employed to evaluate MobileNet's performance in lung cancer classification.

4 Results and Discussion

4.1 Experimental Setup

The experimental setup involved the utilization of Python 3 and the Keras framework for implementing the models. The execution of the models took place on Google Colab, which provided the hardware specifications of an Intel Xeon CPU @2.20 GHz, 13 GB RAM, Tesla K80 accelerator, and 12 GB GDDR5 VRAM. Prior to usage, the Image-Data Generator class in Keras was employed to resize and normalize all input images.

Lung Cancer Database To create and train a convolutional neural network capable of classifying lung images as benign or malignant, the datasets were collected and prepared. The images obtained from the LIDC and IDRI datasets were downloaded and organized into directories: train, test, and valid, each containing separate directories for benign and malignant images. Additionally, a valid directory was created.

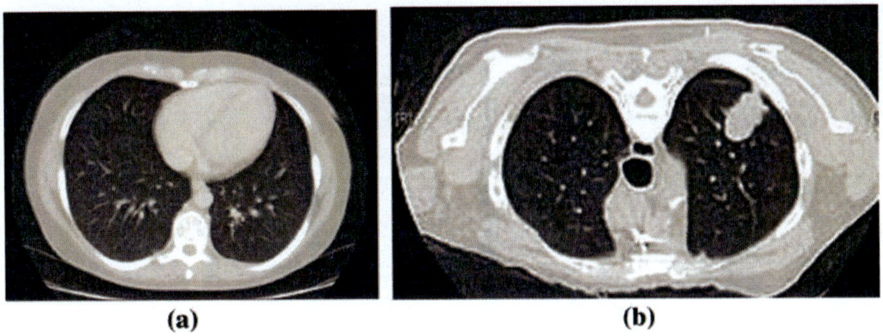

Fig. 3 Example of **a** non-malignant lung image and **b** malignant lung image

For this particular experiment, a total of 2066 images were used. Among them, 80% (1132 images) were allocated for training, while 20% (413 images) were reserved for testing. Out of the training data, 330 images were specifically utilized for validation purposes.

In this research, the radiologists went through two phases of evaluation. In the blinded-reading phase, each radiologist independently examined CT scans and categorized the lesions into three groups: "nodule less than or equal to 3 mm," "nodule less than 3 mm," or "non-nodule greater than or equal to 3 mm." In the unblinded-reading phase, the radiologists reviewed their own marks and anonymized scores provided by the other three radiologists before making their final decisions.

Nodules with a diameter greater than or equal to 3 mm were classified as malignant, while those with a diameter less than 3 mm were considered benign. The research utilized CT scans from the LIDC-IDRI database, and example scans of the lungs are shown in Fig. 3. The DICOM photographs produced by the CT scans had a resolution of 512 by 512 pixels, and the number of slices varied from 64 to 764. On average, the width of the CT scans included in the collection was 240 mm.

Data Preprocessing To overcome the challenge of acquiring a large amount of data for optimal performance of deep learning models, a data augmentation strategy was employed. This strategy involved various techniques such as scaling, flipping, and rotation to fabricate a synthetic dataset. By augmenting the existing data, the dataset size was increased without the need for additional manual data collection, which can be laborious and time-consuming.

The proposed CNN architecture for lung cancer classification, and the architectural design is presented in Fig. 4.

Hyperparameters Table 1 displays the hyper-parameters that were utilized in the training of the CNN for the identification of lung cancer.

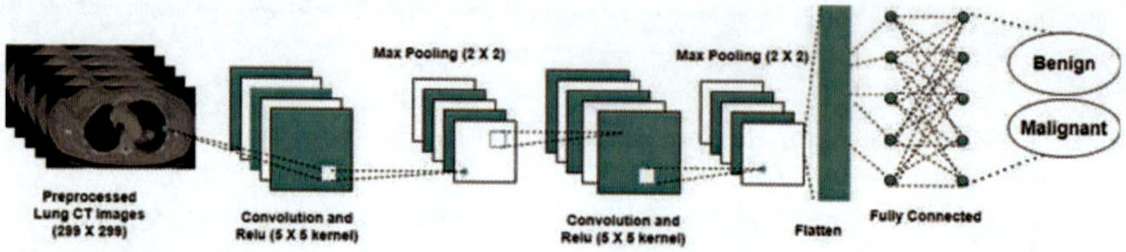

Fig. 4 CNN process flow model

Table 1 Hyperparameters of CNN network

Sl. No.	Hyper-parameters	Value
1	Epoch	50
2	Optimizer	SDG
3	Learning rate	0.0001
4	Batch size	16

4.2 Lung Cancer Classification Using CNN

CNN is trained on the dataset and test on various performance measures. The accuracy and loss curves for both the training and validation sets were plotted, as depicted in Fig. 5.

The accuracy curve shows an improvement in accuracy with increase in the number of epochs. Similarly, the loss curve demonstrates a decrease in loss with an increasing number of epochs, indicating improved model performance.

Figure 6 depicts the evaluation results of the proposed CNN architectures. The ROC curve, shown in Fig. 6a, shows an AUC of 0.912. A higher AUC value suggests better performance of the model, with the range of AUC values being 0.5 to 1.

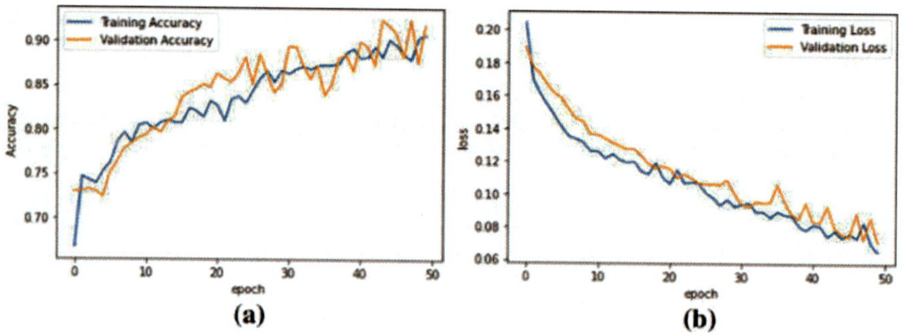

Fig. 5 CNN training **a** acc. versus validation acc. curves; **b** loss versus validation loss curves

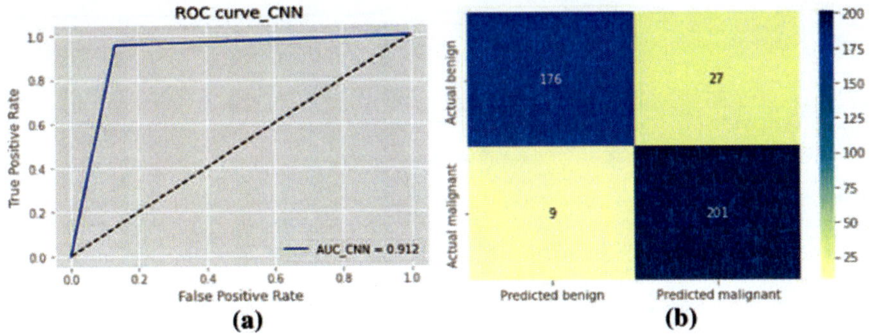

Fig. 6 CNN **a** ROC curve; **b** confusion matrix

The confusion matrix, depicted in Fig. 6b reveals that out of 210 malignant data points, 201 are predicted as true positives (TP) and 9 as false negatives (FN). Additionally, 176 out of 203 benign data points are classified as true negatives (TN), while 27 are false negatives (FN).

The proposed CNN architecture is also evaluated using various metrics such as accuracy, precision, sensitivity, recall, and $F1$-score. The evaluation results are presented in Table 2. The detection accuracies of a proposed CNN model developed for the detection of lung nodules using CT images from the LIDC-IDRI dataset are as follows: Accuracy: 91.28%, Precision: 88.15%, Specificity: 86.60%, Recall: 95.70%, and $F1$-score: 91.78%.

The VGG-16 architecture was employed for lung cancer classification, as depicted in Fig. 2. The accuracy of the modified VGG-16 model and the calculation of the loss function are depicted in Fig. 7.

This study presents the training and validation cycles for two performance metrics using curves. The model weights are recorded when the highest validation accuracy is achieved during training. The maximum accuracy attained during training was 91.45% with a corresponding loss function of 0.0594. The validation accuracy reached 93.94% with a loss function of 0.0507. The latest VGG-16 weights were stored for testing the network. A total of 413 test images, comprising 213 benign and 210 malignant cases, are used.

Table 2 Performance evaluation of proposed CNN architecture

Sl. No.	Quantifying parameters	Value
1	Accuracy	91.28%
2	Precision	88.15%
3	Specificity	86.60%
4	Recall	95.70%
5	$F1$-score	91.78%
6	Area under the curve	0.912

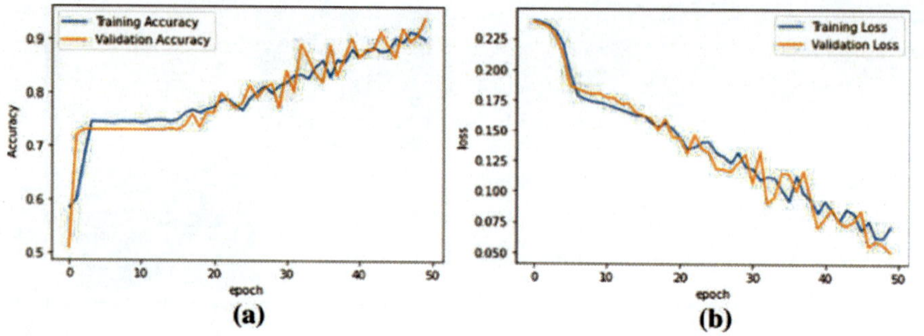

Fig. 7 VGG-16 training **a** acc. versus validation acc. curves; **b** loss versus validation loss curves

The performance of the trained CNN under test conditions is also evaluated using confusion matrix. The confusion matrix of the VGG-16 Architecture is depicted in Fig. 8 as a visual representation. It indicates that out of 210 malignant data points, 194 are predicted as true positives (TP) and 16 as false negatives (FN). Additionally, out of 203 benign data points, 183 are identified correctly as true negatives (TN), while 20 are classified as false positives (FP).

The research includes the ROC curve of VGG-16, shown in a visual representation. The computed area under the curve (AUC) for VGG-16 is 0.913. In this study, the AUC values ranges from 0.5 to 1, where higher values reflect the effective performance of the device.

Furthermore, the performance evaluation of the proposed VGG-16 architecture is produced in Table 3. The evaluation is based on quantifying parameters such as accuracy, precision, specificity, recall, and $F1$-score.

In this research, the VGG-16 model was constructed to detect lung nodules using CT images from the LIDC-IDRI dataset. The model was trained using these images and implemented in Python with the Keras package. The achieved accuracy of the model was 91.28%. The evaluation scores for accuracy, precision, specificity, recall, and $F1$-score were 91.28%, 90.65%, 85.91%, 92.38%, and 91.50%, respectively.

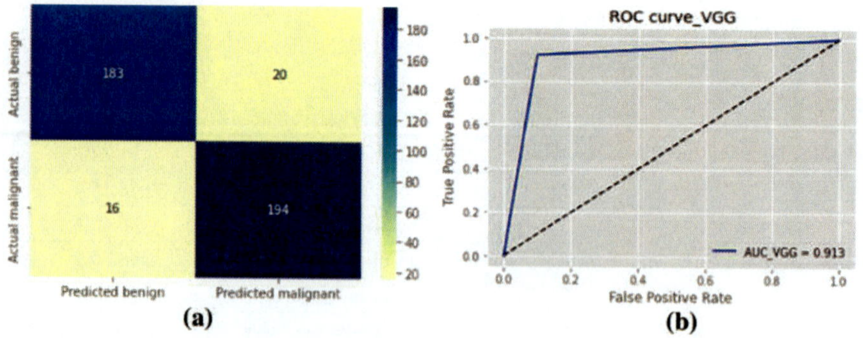

Fig. 8 VGG-16 confusion matrix

Table 3 Performance evaluation of proposed VGG-16 architecture

Sl. No.	Quantifying parameters	Value
1	Accuracy	91.28%
2	Precision	90.65%
3	Specificity	85.91%
4	Recall	92.38%
5	$F1$-score	91.50%
6	Area under the curve	0.913

4.3 Lung Cancer Classification Using MobileNet

Sample images from the training dataset are analyzed following the successful installation of a lung nodule classification model using the MobileNet architecture. The accuracy values and loss curve during the training phase and validation phase are depicted in Fig. 9. The findings reveal an improvement in accuracy with increasing epochs, while the loss decreases as the number of epochs progresses.

The performance of MobileNet is also evaluated using various matrices such as precision, recall, and $F1$ score. By leveraging these performance matrices,

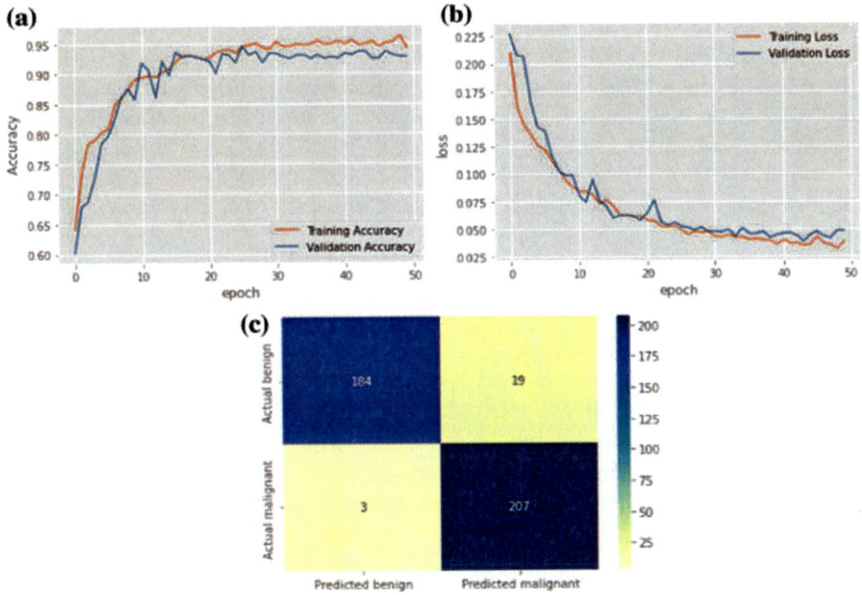

Fig. 9 MobileNet training **a** acc. versus validation acc. curves; **b** loss versus validation loss curves; **c** confusion matrix

Table 4 Performance evaluation of proposed MobileNet architecture

Sl. No.	Quantifying parameters	Value (%)
1	Accuracy	94.67
2	Precision	91.59
3	Specificity	9.64
4	Recall	98.57
5	$F1$-score	94.95

researchers and clinicians can evaluate the accuracy and efficiency of MobileNet in identifying lung cancer, and ultimately improve patient outcomes. The results of the evaluation of various proposed MobileNet architectures is depicted through Table 4.

4.4 Comparative Analysis on Various Hyper-parameters

When training convolutional neural networks (CNNs) like VGG-16 and MobileNet, it is critical to take into account the interaction between learning rate, batch size, and epochs. The adjustment of the weights of a network during each iteration of training is determined by the learning rate. An elevated learning rate has the potential to expedite convergence; however, it can also result in overshooting and instability. The term "batch size" pertains to the quantity of training instances that undergo processing as a group prior to the updating of weights. Increasing the batch size can potentially lead to enhanced weight updates stability, albeit at the cost of higher memory requirements. In the context of neural network training, epochs denote the frequency with which the complete dataset is iterated through the network. Augmenting the number of epochs has the potential to enhance performance, however, it also has the possibility of inducing overfitting. The identification of an optimal combination of said parameters is of utmost importance in attaining a high level of accuracy while circumventing potential complications such as overfitting or unstable training.

Learning Rate The outcomes of the CNN, VGG-16, and MobileNet with different learning rate are displayed in Table 5 for 0.1, 0.01, 0.001, 0.0001, and 0.00001 learning rates. The largest accuracy was obtained for CNN with a learning rate of 0.01, while the least accuracy was obtained with a learning rate of 0.1. The VGG model achieved the highest level of accuracy when learning rate = 0.1, whereas the lowest level of accuracy was observed when learning rate = 0.0001. With a learning rate of 0.01, and 0.001 MobileNet achieved the highest accuracy, while a rate of 0.1 yielded the lowest.

Batch Size The results from executing CNN, VGG-16, and MobileNet with varying batch sizes are presented in Table 6, with the batch sizes 8, 16, 32, and 64 respectively. With a batch size of 8, the proposed CNN architecture achieved better accuracy while the minimum accuracy is obtained where the batch size is 16 and 64. When utilizing

Table 5 Accuracy comparisons of CNN, VGG-16, and MobileNet with different learning rates

Learning rate	CNN (%)	VGG-16 (%)	MobileNet (%)
0.1	50.85	**95.64**	50.85
0.01	**94.43**	95.16	**95.16**
0.001	93.95	95.16	**95.16**
0.0001	91.28	91.28	94.67
0.00001	77.00	95.16	88.31

Bold values signifies best value

a batch size of 8, the VGG model was able to achieve the highest level of accuracy, while when utilizing a batch size of 32, and 64 the model was only able to achieve the lowest level of accuracy. MobileNet attained the maximum level of Accuracy with a batch size of 8, however, a batch size of 16 and 64, resulted in the lowest level of Accuracy.

Epochs Table 7 displays the results of CNN, VGG-16, and MobileNet with 10, 20, 30, and 50 Epochs. CNN achieved the highest Accuracy with an epoch of 50, while the lowest Accuracy was achieved with an Epoch of 10. The VGG model obtained the highest level of accuracy with an epoch of 50, and the lowest level of accuracy with an epoch of 10. With Epochs of 50, MobileNet obtained the highest Accuracy, while Epochs 10 produced the lowest.

The summary of the results are as follows. The CNN fine tuned with 0.01 learning rate, 8 batch size for 50 epochs attained a maximum accuracy of 95.15%. The most accurate prediction was made by VGG-16 with a learning rate at 0.1, batch size at 8, for 50 epochs has recorded an accuracy of 95.88%. The maximum level of accuracy,

Table 6 Accuracy comparisons of CNN, VGG-16, and MobileNet with different batch size

Batch size	CNN (%)	VGG-16 (%)	MobileNet (%)
8	**95.15**	**95.88**	**98.39**
16	94.43	95.16	94.92
32	94.92	93.95	95.40
64	94.43	93.95	94.92

Bold values signifies best value

Table 7 Accuracy comparisons of CNN, VGG-16, and MobileNet with different epochs

Epochs	CNN (%)	VGG-16 (%)	MobileNet (%)
10	81.84	87.17	94.67
20	91.28	94.67	95.40
30	94.67	95.40	95.16
50	**95.15**	**95.88**	**98.39**

Bold values signifies best value

Table 8 Accuracy comparison of proposed method versus state of the art

References	Acc. value (%)
Alakwaa et al. [19]	86.6
Lakshmanaprabu et al. [20]	94.56
Nanglia et al. [21]	98.08
Proposed model	**98.39**

Bold values signifies best value

98.39%, was attained by MobileNet with a learning rate = 0.001, batch size = 8 for epochs.

The highest Accuracy, Precision, Recall, and $F1$-score are 98.39%, 95%, 99%, and 97%, respectively are obtained for MobileNet with learning Rate 0.001, Batch Size 8, and Epoch 50 for lung cancer detection and classification.

Table 8 represents the comparison analysis of the performance of the proposed model with the state of the art on LIDC-IDRI datasets.

5 Conclusion

The aim of the proposed model is to explore various deep learning models for the classification of lung cancer, considering its high mortality rate and significant impact on public health. The proposed model is developed on LIDC-IDRI dataset to classify the lung cancer by employing Convolutional Neural Networks (CNN), VGG-16, and MobileNet models. It was observed from our study that the MobileNet model with a learning rate of 0.01, batch size of 8, and 50 epochs achieved a better accuracy of 98.39%. VGG-16 attained an accuracy of 95.88% with a learning rate of 0.1, 8-sample batches, and 50 epochs whereas CNN attained 95.15 for the same settings. MobileNet yielded the highest accuracy of 98.39% using a learning rate of 0.001, batch size of 8, and 50 epochs. Overall, this study demonstrated the effectiveness of deep learning models, particularly CNN, VGG-16, and MobileNet, in classifying lung cancer using CT images from the LIDC-IDRI dataset.

References

1. Pfeffer M, Ling S (2022) Evolving optimised convolutional neural networks for lung cancer classification. Signals 3:284–295
2. Suresh S, Mohan S (2020) ROI-based feature learning for efficient true positive pre-diction using convolutional neural network for lung cancer diagnosis. Neural Comput Appl 32:15989–16009
3. Samhitha B, Mana S, Jose J, Vignesh R, Deepa D (2020) Prediction of lung cancer using convolutional neural network (CNN). Int J 9

4. Pandit B, Alsadoon A, Prasad P, Al Aloussi S, Rashid T, Alsadoon O, Jerew O (2023) Deep learning neural network for lung cancer classification: enhanced optimization function. Multimed Tools Appl 82:6605–6624
5. Saleh A, Chin C, Penshie V, Al-Absi H (2021) Lung cancer medical images classification using hybrid CNN-SVM. Int J Adv Intell Inform 7:151–162
6. Sajja T, Devarapalli R, Kalluri H (2019) Lung cancer detection based on CT scan images by using deep transfer learning. Traitement Du Signal 36:339–344
7. Tekade R, Rajeswari K (2018) Lung cancer detection and classification using deep learning. In: Fourth international conference on computing communication control and automation (ICCUBEA). IEEE, pp 1–5
8. Wang Z, Gao X (2018) Improved lung nodule diagnosis accuracy using lung CT images with uncertain class. Comput Methods Programs Biomed 162:197–209
9. Nithila E, Kumar S (2017) Automatic detection of solitary pulmonary nodules using swarm intelligence optimized neural networks on CT images. Eng Sci Technol Int J 20(3):1192–1202
10. Dou A, Heng P (2017) Multilevel contextual 3-D CNNs for false positive reduction in pulmonary nodule detection. IEEE Trans Biomed Eng 64(7):1558–1567
11. Monkam P, Qian W (2018) CNN models discriminating between pulmonary micro-nodules and non-nodules from CT images. BioMedical Eng OnLine 17(1)
12. Winkels M, Cohen T (2019) Pulmonary nodule detection in CT scans with equivariant CNNs. Med Image Anal 55:15–26
13. Bushara AR (2022) Deep learning-based lung cancer classification of CT images using augmented convolutional neural networks. Electron Lett Comput Vis Image Anal 21(1):130–142
14. Nasser IM, Abu-Naser SS (2022) Lung cancer detection using artificial neural network. Int J Eng Inf Syst 3(3):17–23
15. Qin R, Wang Z, Jiang L, Qiao K, Hai J, Chen J, Xu J, Shi D, Yan B (2020) Fine-grained lung cancer classification from PET and CT images based on multidimensional attention mechanism. Complexity 2020:1–12
16. Asuntha A, Srinivasan A (2020) Deep learning for lung cancer detection and classification. Multimed Tools Appl 79:7731–7762
17. Kasinathan G, Jayakumar S (2022) Cloud-based lung tumor detection and stage classification using deep learning techniques. BioMed Res Int 79:17 p.
18. Thakur SK, Choudhary J (2020) Lung cancer identification: a review on detection and classification. Cancer Metastasis Rev 39:989–998
19. Alakwaa W, Badr A (2017) Lung cancer detection and classification with 3D convolutional neural network (3D-CNN). Int J Adv Comput Sci Appl 8(8)
20. Lakshmanaprabu SK, Ramirez G (2019) Optimal deep learning model for classification of lung cancer on CT images. Futur Gener Comput Syst 92:374–382
21. Nanglia P, Rathee D (2021) A hybrid algorithm for lung cancer classification using SVM and neural networks. ICT Express 3:335–341

A Novel Fingerprint-Based Age Group Classification Approach Using DWT and DFT Analysis

Ashwini Sanjay Gaikwad and Vijaya B. Musande

Abstract In recent years, fingerprint analysis has expanded to include the estimation of age information, providing an additional layer of personal identification because of its uniqueness and permanence. In this study, 2D-Discrete Wavelet Transform (DWT) and 2D-Discrete Fourier transform (DFT) are used for fingerprint-based age group classification and their performance is compared against fingerprint dataset which is distributed among various age groups from the age of 10 to 90 including both genders. The face-based age group classification techniques have limited practice for crime scene investigation due to their dependency on the availability of vast global features, that is, complete face image. The research gap identifies that the accuracy between 2D-DWT and 2D-DFT, varied, and this is proved even for the unbalanced age group classes collected in this study's real-time database and its corresponding results. The three variants of 2D-DWT Level-4, level-3, level-2 and 2D-DFT are considered for this research work and the most accurate was the DWT Level-4 which classified the age group using the fingerprint images with an astonishing accuracy 99.96 which is verified and supported by detailed analysis of all classification metrics.

Keywords Fingerprint · 2D-discrete wavelet transform · 2D-discrete Fourier transform · Support vector machine classification metrics

1 Introduction

Fingerprint recognition has been a topic of extensive research in recent years, with various techniques proposed for accurate and efficient classification. One such technique is the use of Discrete Wavelet Transform (DWT) for fingerprint-based classification. DWT is a mathematical tool used for signal processing, and its application in

A. Sanjay Gaikwad (✉) · V. B. Musande
Department of Computer Science and Engineering, Jawaharlal Nehru Engineering College, Aurangabad, Maharashtra, India
e-mail: ashwinigaikwad5056@gmail.com

S. Shakya et al. (eds.), *Fourth International Conference on Image Processing and Capsule Networks*, Lecture Notes in Networks and Systems 798,
https://doi.org/10.1007/978-981-99-7093-3_26

fingerprint recognition has shown promising results. Currently in crime scene investigation, there is inadequate practice of face-based gender classification due to their dependency on massive global features, that is, appearance-based approach that uses whole face image containing thousands of pixels which is reduced to handful number of pixels by dimensionally reduction schemes.

Fingerprint analysis plays a role in convicting the person responsible for a rude crime. Fingerprint biometrics has been widely recognized as a reliable and secure means of identity verification due to its unique and unchanging nature throughout an individual's life. In recent years, fingerprint analysis has expanded to include the estimation of gender and age information, providing an additional layer of personal identification. In the research domain of age group and gender recognition, mainly used biometric is fingerprints due to its uniqueness and do not change throughout the life of an individual. Gender identification is been studied in insufficient research therefore, more examination is required which will increase the accuracy level with a lesser amount of computer complexity. Fingerprint biometrics are widely used for reliable and secure identity verification due to their unique and unchanging nature throughout an individual's life.

In recent years, the scope of fingerprint analysis has expanded to include the estimation of gender and age information, providing an additional layer of personal identification. However, there is a lack of studies on the estimation of gender information from fingerprint images. This research gap presents an opportunity to develop more accurate gender classification methods that require less computer complexity. By addressing this gap in the literature, the proposed research aims to enhance the accuracy and efficiency of fingerprint-based gender identification. This research paper aims to explore the effectiveness of DWT for fingerprint-based classification.

The paper will present a detailed analysis of the DWT-based fingerprint recognition system, including the preprocessing steps, feature extraction, and classification process. The proposed method will be evaluated using a benchmark fingerprint database, and the results will be compared with other state-of-the-art fingerprint recognition techniques. The significance of this research lies in the potential applications of fingerprint recognition in various fields, such as law enforcement, security, and biometric authentication. The proposed DWT-based approach can lead to the development of more accurate and reliable fingerprint recognition systems, which can enhance security and prevent fraud.

2 Related Work

In earlier study, an experiment on human age estimation through fingerprint has been implemented by extracting fingerprint feature using discrete wavelet transform and principal component analysis methods and further combining the feature vectors. Then SVM classifier is applied to estimate the age of the person [1]. An attempt using backpropagation algorithm from artificial neural network is proposed in this research article which matches twelve parameters of fingerprint with a unique number allotted to every authorized person. Then the best match is given by the proposed algorithm after the matching is done [2]. S. S. Gornale, Mallikarjun Hangarge et al. have proposed in their work, that the gender information can be extracted from fingerprints with the help of Haralick texture features for gender classification. Fingerprints from different age intervals of rural and urban population were collected for performing this experiment. As per the results achieved for linear discriminant analysis (LDA) and quadratic discriminant analysis (QDA) classifiers 92% and 94% classification rate is gained respectively [3].

It is studied from the literature that Daubechies Orthogonal Wavelets (db9) is mainly referred for preparing the feature vector because texture analysis is done smoothly with the help of wavelets, and for computational model Standard deviation and entropy is calculated upon the fingerprint images. DWT decomposes an image into subbands that are localized in frequency and orientation. The literature reveals that many researchers have explored gender classification using various approaches and achieved promising results with their datasets. However, there is still room for the development of a robust algorithm using different parameters [4, 5].

In law enforcement, latent fingerprint identification is an important area of interest. This study analyses the most bulging fingerprint feature representations reported in the literature, recognize drifts in fingerprint feature representation, and observe that representations designed for verification are commonly used in latent fingerprint identification. This experiment aims to assess the performance of the most widespread fingerprint feature representations over a common latent fingerprint database and protocol. Hence, a protocol that evaluates some of the most suitable feature representation is introduced [6].

In a recent research study, time domain approach is referred to identify the gender of a specific fingerprint through methodical pixel counting. This precise pixel counting method got 90.2% accuracy for females' classification and 96.4% for males [7].

Fingerprints are widely recognized as a unique biometric that remains unchanged throughout an individual's life. As a result, fingerprint analysis has been used for identification, verification, and official documentation purposes. Moreover, fingerprints play a crucial role in solving criminal cases, where the analysis of fingerprints often leads to identifying the perpetrator. In addition to identification and verification, fingerprint biometrics can also be used for gender and age estimation. Previous research has used DWT and SVD methods to estimate the age of a person using fingerprints. Furthermore, the unique nature of fingerprints has made them a reliable biometric for gender identification. In conclusion, fingerprint analysis is a crucial aspect of forensic science that plays a vital role in the identification and verification of individuals, as well as in solving criminal cases. Additionally, the unique nature of fingerprints makes them suitable for age and gender estimation. By utilizing different techniques and methods, the accuracy and efficiency of fingerprint analysis can be enhanced, providing more reliable results for various applications [8, 9].

Wavelet transform is a widely used tool in image processing and computer vision due to its comprehensive theoretical framework, flexibility in choosing bases, and low computational complexity. The wavelet transform enables the analysis of an image in both the frequency and time domains, which makes it particularly useful for feature extraction and pattern recognition tasks [10].

One of the most advanced biometric technologies, fingerprints are accepted as valid forms of proof in courts of law all around the world [11]. Because of this need for trustworthy human identification in computerized access control, interest in biometrics has developed. Based on their visual pattern, all fingerprints fall into one of three categories: arches, loops, or whorls. 60% of people have loops, 35% have whorls, and 5% have arches, making loops a relatively prevalent pattern. Even the fingerprints of identical twins differ for this reason. Both a spatial domain and frequency domain technique can be used to study the gender classification of fingerprints. The spatial domain technique utilizes both the physical properties and pixel values of a picture. The physical features of a fingerprint image that relate to ridges and troughs are called physical parameters [12].

Instead of focusing on the physical parameters of an image, the frequency domain technique works on the measurement parameters of an image. This approach employs the energy of the image as a feature of the image and operates on the frequencies of that image. Discrete Wavelet Transform, Discrete Cosine Transform, and other techniques are used here to extract features [13].

In this study, using a frequency domain technique, which ignores the ridge-based physical properties of a picture in favor of assessing parameters such as frequency and pattern recognition, the features of the fingerprint image are extracted using the 2D Discrete Wavelet Transform. Wavelet serves as the basis function for DWT. A wavelet is an effective limited-duration waveform with a zero average value. An image is divided into sub-bands by two-dimensional DWT that are localized in frequency and orientation (horizontal, vertical, diagonal). Decomposition of images into several frequency bands aids in the isolation of various frequency components. In order to investigate the details of an image, it is helpful to discern minute changes in the image using various frequency components [14].

Four decomposed sub-band images, referred to as low–low (LL), low–high (LH), high–low (HL), and high–high (HH), are produced by the 2-D wavelet decomposition of an image. Each of these sub-band's aids in the analysis of various visual details. Since low frequencies hold the majority of an image's energy, decomposition is often repeated just on the LL sub band. The analysis of these subtleties is made easier by multi-resolution since many signals or images contain features at several sizes. A high-resolution should be used to investigate little items. Examining large items at low resolution is advised. Low–low (LL), Low–high (LH), High–low (HL), and High–High (HH) are the four decomposed sub-band pictures produced by the 2-D wavelet decomposition of an image. Each of these sub-bands is useful for examining various image details. Since the majority of an image's energy is found at low frequencies, decomposition is often repeated solely on the LL sub band. Numerous signals or images have features at different scales, and multi-resolution aids in the examination of such details. High-resolution examination of small items is recommended. Low-resolution examination of large-scale items is recommended.

This work uses the phase information gained by Discrete Fourier Transform (DFT) of images. The phase information conserves the essential image features. Its correlation function, called phase correlation or Phase-Only Correlation (POC), provides us both the good likeness ration for biometric recognition. The researchers have successfully applied phase information to fingerprint, face, iris, and palmprint recognition [15].

3 Proposed Work

3.1 *Identification of Research Gap*

One potential research gap for age detection using fingerprint is the lack of robust and accurate feature extraction methods that can effectively capture the subtle variations in fingerprint patterns related to age. While some studies have proposed various features for age estimation, there is still a lack of consensus on the most effective and efficient feature extraction techniques. Another research gap is the lack of comprehensive and diverse datasets for age detection using fingerprint, particularly in terms of age ranges and ethnic diversity. Most existing datasets are limited in terms of sample size and diversity, which may limit the generalization of age estimation models to real-world scenarios.

3.2 *Motivation*

As there is lack of comprehensive and diverse datasets for age detection using fingerprint, particularly in terms of age ranges. It was difficult to detect the fingerprint samples belonging to the different age ranges and working class with better or damaged skin over time, which can potentially impact the quality and clarity of fingerprints. Some fingerprint samples were collected from this part of society as well to get the better accuracy and to explore the differences in fingerprints among age groups.

3.3 *Discrete Wavelet Transform*

DWT is a wavelet transform for which the wavelets are sampled at discrete intervals. DWT provides a simultaneous spatial and frequency domain information of the image. In DWT operation, an image can be analyzed by the combination of analysis filter bank and decimation operation. The DWT decomposes a digital signal into different sub bands so that the lower frequency sub-bands have finer frequency resolution and coarser time resolution compared to the higher frequency sub bands.

One Level of the Transform

The DWT of a signal x is calculated by passing it through a series of filters. First the samples are passed through a low-pass filter with impulse response g resulting in a convolution of the two:

$$y[n] = (x * g)[n] = \sum_{k=-\infty}^{\infty} x[k]g[n-k] \tag{1}$$

Here, k represents the samples of original signal function x and n represents the filtered coefficients of input signal. The signal is also decomposed simultaneously using a high-pass filter h. The outputs give the detail coefficients (from the high-pass filter) and approximation coefficients (from the low-pass). However, since half the frequencies of the signal have now been removed, half the samples can be discarded according to Nyquist's rule. The filter output of the low-pass filter g in the diagram above is then subsampled by 2 and further processed by passing it again through a new low-pass filter g and a high-pass filter h with half the cut-off frequency of the previous one, that is:

$$y_{\text{low}}[n] = \sum_{k=-\infty}^{\infty} x[k]g[2n-k]$$

$$y_{\text{high}}[n] = \sum_{k=-\infty}^{\infty} x[k]h[2n-k] \tag{2}$$

This decomposition has halved the time resolution since only half of each filter output characterizes the signal. However, each output has half the frequency band of the input, so the frequency resolution has been doubled.

With the subsampling operator $\downarrow$, the above summation can be written more concisely.

$$y_{\text{low}} = (x * g) \downarrow 2$$

$$y_{\text{high}} = (x * h) \downarrow 2 \tag{3}$$

This decomposition is repeated to further increase the frequency resolution and the approximation coefficients decomposed with high- and low-pass filters and then down-sampled. The filter bank implementation of wavelets can be interpreted as computing the wavelet coefficients of a discrete set of child wavelets for a given mother wavelet $\Psi(t)$. In the case of the discrete wavelet transform, the mother wavelet is shifted and scaled by powers of two.

$$\psi_{j,k}(t) = \frac{1}{\sqrt{2^j}}\psi\left(\frac{t - k2^j}{2^j}\right) \tag{4}$$

where j is the scale parameter and k is the shift parameter, both which are integers.

$$T_{m,n} = \int_{-\infty}^{\infty} x(t)\psi_{m,n}(t)\mathrm{d}t \tag{5}$$

3.4 Discrete Fourier Transform

The (2D) Fourier transform is a very conventional instrument in image processing. It is the extension of the well-known Fourier transform for signals which decomposes a signal into a sum of sinusoids. So, the Fourier transform gives information about the frequency content of the image. The DFT of an image f of size $M \times N$ is an image of same size defined as:

$$F(u, v) = \sum_{m=0}^{M-1}\sum_{n=0}^{N-1} f(m, n)e^{-j2\pi\left(\frac{um}{M}+\frac{vn}{N}\right)} \tag{6}$$

where $f(m, n)$ is an input 2D image of size $(M * N)$ with m and n as the spatial domain variables and $F(u, v)$ is the frequency domain transformed image after applying the 2D-Discrete Fourier Transform which includes u and v as the frequency domain variables.

3.5 Proposed Algorithm

1. Take input fingerprint image.
2. Preprocessing the image enhancing the quality of image for better accuracy of classification.
    2.1 Normalize the image to the resolution of 256×256 for better performance.
    2.2 Apply weighted Gaussian filter to remove noise and smoothen the image of size (7×7).
    2.3 Perform histogram equalization to enhance the contrast.
    2.4 Apply segmentation of contrast enhanced image using global thresholding method to isolate the area of interest for further computation of image.
3. Apply skeletonization on the above preprocessed image for sharpening the features of the fingerprint using morphological image processing operations.

3.1. Erode the preprocessed image for getting the sharpened fingerprint ridges.
3.2. Dilate the eroded image for restructuring the image ridges
3.3. Apply median smoothing filter for smoothing all the ridges of the fingerprint.
3.4. Apply global thresholding to the morphologically processed of fingerprint for better fingerprint

4. Apply 2D-DWT (up to Level-4)/2D-DFT for extracting the fingerprint features in its corresponding structures.
5. Train the SVM classifier with some of the samples from the collected dataset for training purpose.
6. Apply SVM classifier on remaining fingerprint images from the dataset for testing purpose

4 Results

To make sure that the suggested system is appropriate for the age group classification from fingerprints, this study provides a detailed analysis of the classification metrics. The performance of the suggested system along with comparison for these performances of all these various transforms is demonstrated in statistics. As per the study there is no standard database which contains fingerprints of various age groups among male and female.

Therefore, as shown in Table 1, a gender-wise and age group-wise real time database has been created for conducting experiment on age classification based on fingerprints. In the database, fingerprints data is collected among nine different age groups from the various urban areas located in Aurangabad district of Maharashtra state in India. The database is spread among nine age groups ranging from (1 to 10) to (81 to 90). The acquisition of the fingerprint was made by SecuGen Hamster plus optical fingerprint sensor manufactured by nitgen biometric solution [with interface USB 2.0] with the 8-bit image size as 260 × 300 pixels.

Table 1 Age group categorization with respect to gender

Age group	Male	Female
1–10	40	10
11–20	350	260
21–30	320	120
31–41	380	140
41–50	150	70
51–60	70	50
61–70	210	250
71–80	240	270
81–90	25	–

Table 1 describes that; the database is created with a total of 2877 images including 1761 images of the male category and 1116 images of the female category from different age groups. Every person's right thumb's 10 samples have been collected for this research purpose. Here, 10 samples per person are for better accuracy purposes. Since the database is unbalanced among the various age groups and genders, the more robust results are observed.

The classification metrics included in this study for analysing the results are as follows:

$$\text{Accuracy} = (\text{True positives} + \text{True Negatives})/(\text{True positives} + \text{True negatives} + \text{False positives} + \text{False negatives}) \quad (7)$$

$$\text{Precision} = \text{True positives}/(\text{True positives} + \text{False positives}) \quad (8)$$

$$\text{Recall} = \text{True Positives}/(\text{True Positives} + \text{False Negatives}) \quad (9)$$

$$F1\ \text{Score} = 2 * (\text{Precision} * \text{Recall})/(\text{Precision} + \text{Recall}) \quad (10)$$

The classification report of 2D-DFT is shown in Table 2. It is observed very evidently that 2D-DFT transform has responded only for the female age group of 11–20 and male age group of 11–20 with minor precision, recall and $f1$-score for the available database. For remaining age groups across both the genders 2D-DFT has given 0 response with respect to precision, recall and $f1$-score which assures that this transform is not capable of providing better accuracy of age group classification.

In coordination to the results obtained through Table 2, the graphical representation of 2D-DFT transform's classification metrics analysis is shown in Fig. 1. As stated above, the metrics have only responded to both male and female age group from 11 to 20, holding remaining age groups at 0.

Since it is illustrated in Table 2 that the precision, recall and $f1$-score along with the accuracy for the 2D-DFT are not satisfactory, these findings are corresponded with the extended classification metrics such as macro average and weighted average of the precision, recall and $f1$-score across various age groups. As shown in Table 3, these extended metrics proven to be very weak throughout the entire database of fingerprint samples remarking macro average of precision as 0.04, and that of recall as 0.10 while for $f1$-score as merely 0.07, whereas weighted average of precision as 0.24, and that of recall as 0.38 while for $f1$-score as 0.28. These results debar 2D-DFT for an efficient age group classification based on fingerprint and these results are demonstrated in Fig. 2 as well.

The confusion matrix for 2D-DWT level-2 is depicted in Table 4. It is perceptible in this confusion matrix that 2D-DWT level-2 has not classified all the fingerprint samples correctly. The values shown on the diagonal line in the matrix indicate the true positive ratio of sample fingerprints across different age groups as mentioned in Table 1. According to the confusion matrix shown in Table 4, the 10 fingerprint

Table 2 Classification report of 2D-DFT

	Precision	Recall	$f1$-score	Support
Female age group 1–10	0	0	0	10
Female age group 11–20	0.43	0.45	0.44	249
Female age group 21–30	0.00	0.00	0.00	120
Female age group 31–40	0.00	0.00	0.00	140
Female age group 41–50	0.00	0.00	0.00	70
Female age group 51–60	0.00	0.00	0.00	50
Female age group 61–70	0.00	0.00	0.00	250
Female age group 71–80	0.00	0.00	0.00	270
Male age group 1–10	0.00	0.00	0.00	40
Male age group 11–20	0.36	0.80	0.50	310
Male age group 21–30	0.00	0.00	0.00	320
Male age group 31–40	0.00	0.00	0.00	380
Male age group 41–50	0.00	0.00	0.00	150
Male age group 51–60	0.00	0.00	0.00	70
Male age group 61–70	0.00	0.00	0.00	210
Male age group 71–80	0.00	0.00	0.00	240
Male age group 81–90	0.00	0.00	0.00	21

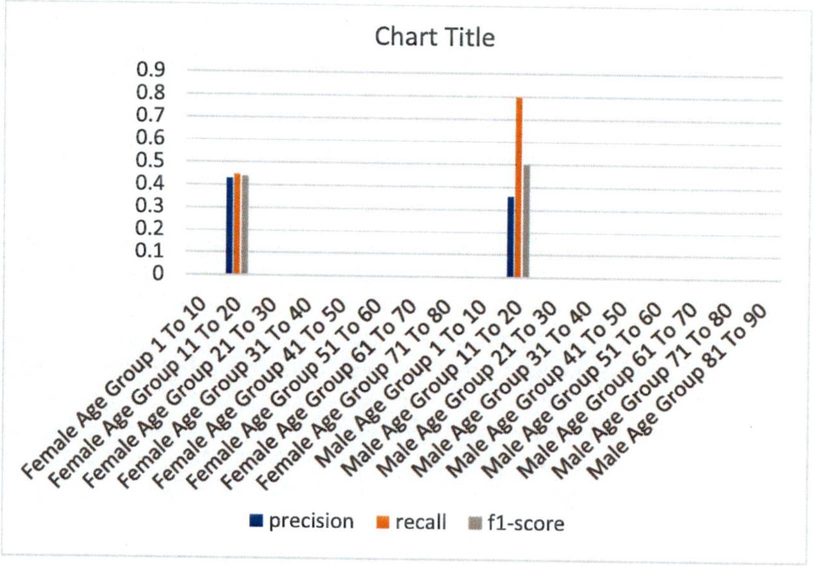

Fig. 1 Classification metrics analysis of 2D-DFT level with emphasis on precision, recall and $f1$-score metrics

Table 3 Extended classification metrics analysis of 2D-DFT

	Precision	Recall	$f1$-score	Support
Accuracy			0.38	2877
Macro avg	0.04	0.10	0.07	2877
Weighted avg	0.24	0.38	0.28	2877

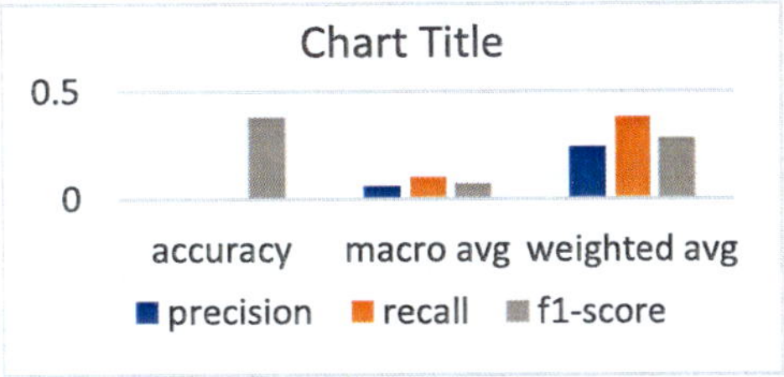

Fig. 2 Extended classification metrics analysis of 2D-DFT using macro avg and weighted avg of precision, recall and $f1$-score

samples from the female age group 1 to 10, are not all correctly classified as indicated by the first true positive value as 0 on diagonal line. Also, 2D-DWT level-2 transform has correctly classified 212 fingerprints out of the 260 samples from female age group 11 to 20. Similarly remaining age groups classification results are demonstrated in the above confusion matrix of Table 4.

Table 4 Confusion matrix of 2D-DWT level-2

0	10	0	0	0	0	0	0	0	0	0	0	0	0	0	0	0
0	212	0	0	0	0	7	5	0	27	1	5	0	0	0	2	0
0	9	34	0	0	0	3	17	0	22	10	23	0	0	0	1	0
0	18	0	9	0	0	5	20	0	50	10	26	1	0	0	1	0
0	3	0	0	2	0	2	17	0	20	0	25	0	0	0	1	0
0	0	0	0	0	14	8	9	0	0	0	19	0	0	0	0	0
0	2	1	0	0	0	116	49	0	13	4	25	0	0	0	6	0
0	0	0	0	0	0	2	225	0	5	3	16	0	0	0	1	0
0	22	1	0	0	0	0	0	9	7	1	0	0	0	0	0	0
0	13	0	0	0	0	2	1	0	309	5	20	0	0	0	0	0
0	10	0	0	0	0	4	28	0	14	225	37	0	0	0	2	0
0	11	0	0	0	0	0	24	0	15	6	317	0	0	0	0	0
0	6	0	0	0	0	4	7	0	25	19	75	11	0	0	3	0
0	1	0	0	0	0	10	0	0	11	4	38	0	2	0	3	0
0	8	0	0	0	0	7	61	0	21	9	23	0	0	63	14	0
0	2	0	0	0	0	7	62	0	10	8	20	0	0	0	123	0
0	0	0	0	0	0	1	19	0	0	0	0	0	0	0	1	0

A detailed classification report for 2D-DWT level-2 is represented in Table 5, where the average precision ratio, recall and $f1$-score for each age group is given along with the respective support sample's count in the last column. All the three-classification metrics range from 0 (indicating failure) to 1 (indicating 100% success ratio). Therefore, from the precision, recall and $f1$-score values in Table 5, it is verified that the performance of 2D-DWT level-2 is not satisfactory than the remaining 2D-DWT variants but better than 2D-DFT transform.

Figure 3 represents the classification report for 2D-DWT level-2, where the average precision, recall and $f1$-score for each age group is specified on horizontal axis along with the respective age group's support count on vertical axis. Therefore, from the precision, recall and $f1$-score values in Fig. 3, it is verified that the performance of 2D-DWT level-2 is average than the remaining 2D-DWT variants.

Table 6 shows an extended classification metrics analysis for 2D-DWT level-2 which includes the computations of macro average and weighted average of precision, recall and $f1$-score between the existing age groups. Since the collected real time database is unbalanced among various age groups, the computed accuracy needs to be reviewed and re-verified with the help of macro average and weighted average supportive classification metrics. Therefore, from Table 6, it is quantified and shown that 2D-DWT level-2 transform gives merely average accuracy for fingerprint-based age group classification.

Table 5 Classification report of 2D-DWT level-2

	Precision	Recall	$f1$-score	Support
Female age group 1–10	0	0	0	10
Female age group 11–20	0.65	0.82	0.72	259
Female age group 21–30	0.94	0.29	0.44	119
Female age group 31–40	1	0.06	0.12	140
Female age group 41–50	1	0.03	0.06	70
Female age group 51–60	1	0.28	0.44	50
Female age group 61–70	0.65	0.54	0.59	216
Female age group 71–80	0.41	0.89	0.57	252
Male age group 1–10	1	0.23	0.37	40
Male age group 11–20	0.56	0.88	0.69	350
Male age group 21–30	0.74	0.7	0.72	320
Male age group 31–40	0.47	0.85	0.61	373
Male age group 41–50	0.92	0.07	0.14	150
Male age group 51–60	1	0.03	0.06	69
Male age group 61–70	1	0.31	0.47	206
Male age group 71–80	0.78	0.53	0.63	232
Male age group 81–90	0	0	0	21

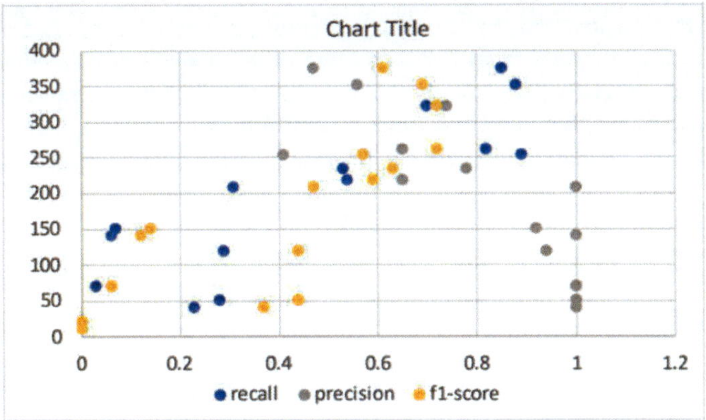

Fig. 3 Classification report of 2D-DWT level-2

Table 6 Extended classification metrics analysis 2D-DWT2

Accuracy			0.58	2877
Macro avg	0.71	0.38	0.39	2877
Weighted avg	0.7	0.58	0.53	2877

An extended classification metrics analysis for 2D-DWT level-2 which includes the computations of macro average and weighted average of precision, recall and $f1$-score between the existing age groups is shown in Fig. 4. It has made very pure that the performance of precision is better over that of recall and $f1$-score metrics.

The confusion matrix for 2D-DWT level-3 is depicted in Table 7. It is noticeable in this confusion matrix that 2D-DWT level-3 has not classified all the fingerprint samples acceptably. The values shown on the center diagonal line in the matrix indicate the true positive ratio of sample fingerprints across different age groups as mentioned in Table 1. According to the confusion matrix shown in Table 7, the 10 fingerprint samples from the female age group 1 to 10, are not all correctly classified

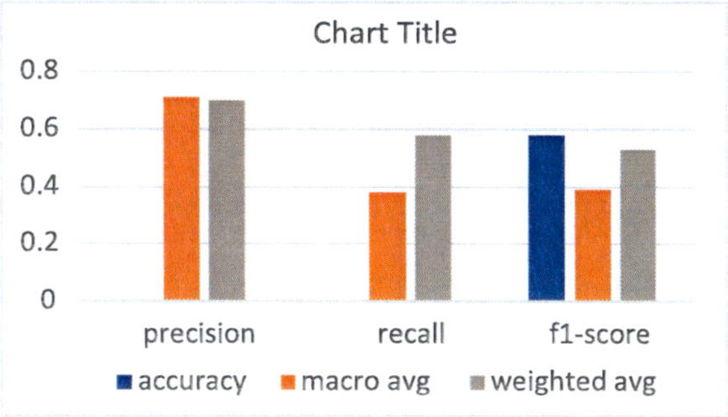

Fig. 4 Extended classification metrics analysis DWT2 using macro avg and weighted avg of precision, recall and $f1$-score

as the first true positive value is 0 on the center diagonal line. Also, 2D-DWT level-3 transform has correctly classified 170 fingerprints out of the 260 samples from female age group 11 to 20. Similarly remaining age groups classification results are demonstrated in the above confusion matrix of Table 7.

A complete classification report for 2D-DWT level-3 is represented in Table 8, where the average precision ratio, recall and $f1$-score for each age group is given. All the three-classification metrics range from 0 (indicating failure) to 1 (indicating 100% success ratio). Therefore, from the precision, recall and $f1$-score values in Table 8, it is verified that the performance of 2D-DWT level-3 is good with respect to the precision ratio but poor with respect to the recall and $f1$-score making this transform weaker than the remaining 2D-DWT variants but to some extent better than the 2D-DFT transform. From Table 8, it is stated that for the female age group 31–40, 41–51 and 51–60 as well as for male age group 1–10 and 61–70 the precision ratio performed very well with the 100% classification performance. It can be noticed from Table 8 that 2D-DWT level-3 transform failed to classify female age groups 1–10, male age groups 41–50, 51–60 and 81–90 giving their corresponding precision, recall and $f1$-score values as 0.

Then the inclusive classification metrics analysis for 2D-DWT level-3 is represented in Fig. 5, where the average precision, recall and $f1$-score for each age group is specified on the vertical axis and the respective age groups are represented on the horizontal axis. Here, all the three metrics have noted astonishing output of 1 only for female age groups 31–40, 41–50, 51–60, and male age groups 1–10 and 61–70,

Table 7 Confusion matrix 2D-DWT3

0	10	0	0	0	0	0	0	0	0	0	0	0	0	0	0	0
0	170	1	0	0	0	2	5	0	56	5	20	0	0	0	0	0
0	10	22	0	0	0	3	19	0	24	11	29	0	0	0	1	0
0	15	2	3	0	0	2	18	0	51	16	32	0	0	0	1	0
0	2	0	0	1	0	2	17	0	20	0	27	0	0	0	1	0
0	0	2	0	0	7	6	11	0	0	3	21	0	0	0	0	0
0	8	1	0	0	0	91	59	0	11	4	32	0	0	0	10	0
0	0	0	0	0	0	3	210	0	1	5	26	0	0	0	7	0
0	16	0	0	0	0	0	0	6	15	1	2	0	0	0	0	0
0	29	0	0	0	0	0	3	0	246	15	56	0	0	0	1	0
0	11	1	0	0	0	2	40	0	30	140	90	0	0	0	6	0
0	12	1	0	0	0	1	29	0	29	28	271	0	0	0	2	0
0	5	0	0	0	0	3	7	0	21	23	87	0	0	0	4	0
0	1	0	0	0	0	4	7	0	16	5	33	0	0	0	3	0
0	5	0	0	0	0	5	69	0	21	9	36	0	0	42	19	0
0	2	0	0	0	0	9	80	0	8	10	33	0	0	0	90	0
0	0	0	0	0	0	0	19	0	0	0	1	0	0	0	1	0

Table 8 Classification report of 2D-DWT level-3

	Precision	Recall	$f1$-score	Support
Female age group 1–10	0	0	0	10
Female age group 11–20	0.57	0.66	0.61	259
Female age group 21–30	0.73	0.18	0.3	119
Female age group 31–40	1	0.02	0.04	140
Female age group 41–50	1	0.01	0.03	70
Female age group 51–60	1	0.14	0.25	50
Female age group 61–70	0.68	0.42	0.52	216
Female age group 71–80	0.35	0.83	0.5	252
Male age group 1–10	1	0.15	0.26	40
Male age group 11–20	0.45	0.7	0.55	350
Male age group 21–30	0.51	0.44	0.47	320
Male age group 31–40	0.34	0.73	0.46	373
Male age group 41–50	0	0	0	150
Male age group 51–60	0	0	0	69
Male age group 61–70	1	0.2	0.34	206
Male age group 71–80	0.62	0.39	0.48	232
Male age group 81–90	0	0	0	21

gathering the highest possible classification score among all the other age groups. It is very clear from Fig. 5 that 2D-DWT level-3 transform performance is worst among all other 2D-DWT variants considered for this experiment.

Table 9 shows an extended classification metrics analysis for 2D-DWT level-3 which includes the computations of macro average and weighted average of precision, recall and $f1$-score between the existing age groups. Therefore, from Table 9, it is computed and shown that for the utilized unbalanced database among various age groups, 2D-DWT level-3 transform gives even below average accuracy for fingerprint-based age group classification. And it is also shown in Fig. 6 for better understanding purpose in graphical manner.

It is shown in Fig. 6 for better understanding purpose in graphical manner that 2D-DWT level-3 transform gives even below average accuracy for fingerprint-based age group classification.

The confusion matrix for 2D-DWT level-4 is depicted in Table 11. It is clearly visible in this confusion matrix that 2D-DWT level-4 has classified all the fingerprint samples more accurately than 2D-DWT level-3 and 2D-DWT level-2. The values shown on the diagonal line in the matrix indicate the true positive ratio of sample fingerprints across different age groups as mentioned in Table 1. According to the confusion matrix shown in Table 11, the 10 fingerprint samples from the female age group 1 to 10, are all correctly classified as indicated by the first true positive value on the center diagonal line. Also, 2D-DWT level-4 has correctly classified

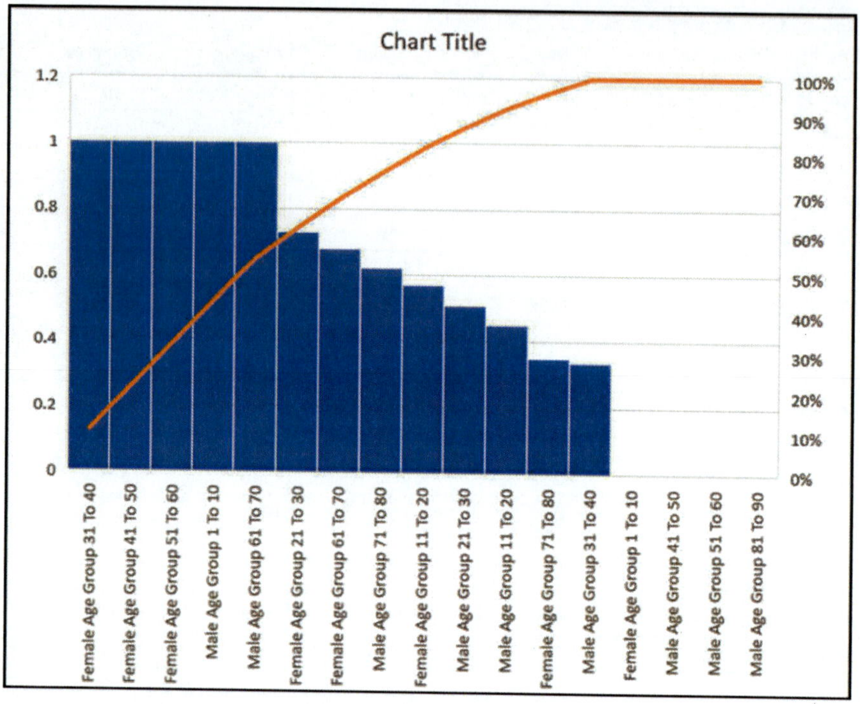

Fig. 5 Classification report of 2D-DWT level-3

Table 9 Extended classification metrics analysis 2D-DWT level-3

	Precision	Recall	f1-score	Support
Accuracy			0.45	2877
Macro avg	0.54	0.29	0.28	2877
Weighted avg	0.55	0.45	0.4	2877

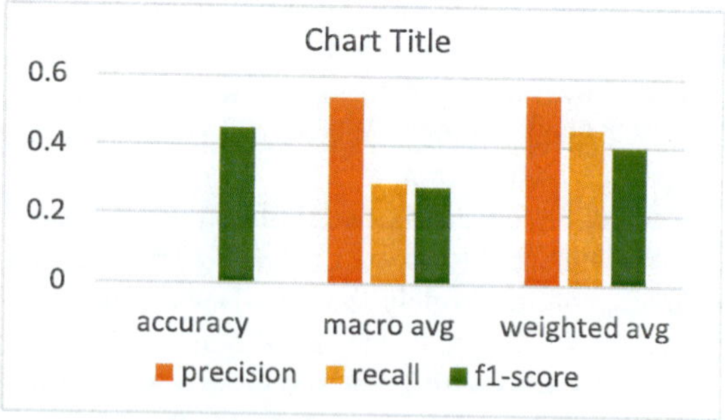

Fig. 6 Extended classification metrics analysis DWT3 using macro avg and weighted avg of precision, recall and f1-score

259 fingerprints out of the 260 samples from female age group 11 to 20. Similarly remaining age groups classification results are demonstrated in the above confusion matrix of Table 11.

Table 10 illustrates the empirical results of 2D-DWT level-4 transform, in the form of age group-wise distribution of fingerprint samples and then showing the correctly classified samples in the last column of Table 10 (Table 11).

A detailed classification report for 2D-DWT level-4 is represented in Table 12, where the average precision ratio, recall and $f1$-score for each age group is given along with the respective support sample's count in the last column. All the three-classification metrics range from 0 (indicating failure) to 1 (indicating 100% success ratio). Therefore, from the precision, recall and $f1$-score values in Table 12, it is verified that the performance of 2D-DWT level-4 is best among the remaining 2D-DWT variants and 2D-DFT transform.

Figure 7 represents the classification report for 2D-DWT level-4, where the average precision, recall and $f1$-score for each age group is specified along with the respective age group's support count on horizontal axis and the concerned age groups are represented on the vertical axis. Therefore, from the precision, recall and $f1$-score values in Fig. 7 it is verified that the performance of 2D-DWT level-4 is higher than the remaining 2D-DWT variants and 2D-DFT transform.

Then, the comprehensive classification metrics analysis for 2D-DWT level-4 is represented in Fig. 8, where the average precision, recall and $f1$-score for each age group is specified on the vertical axis and the respective age groups are represented on the horizontal axis. Here, all the three metrics have noted astonishing output of 1, inferring the highest possible classification score. It is very clear from Fig. 8 that 2D-DWT level-4 transform performance is best among all other considered for this experiment.

Table 10 Empirical results on confusion matrix 2D-DWT level-4

Age group	Actual samples	Correctly classified samples
Female age group 1–10	10	10
Female age group 11–20	259	259
Female age group 21–30	119	119
Female age group 31–40	140	140
Female age group 41–50	70	70
Female age group 51–60	50	50
Female age group 61–70	216	215
Female age group 71–80	252	252
Male age group 1–10	40	40
Male age group 11–20	350	350
Male age group 21–30	320	320
Male age group 31–40	373	373

Table 11 Confusion matrix 2D-DWT level-4

10	0	0	0	0	0	0	0	0	0	0	0	0	0	0	0	0
0	259	0	0	0	0	0	0	0	0	0	0	0	0	0	0	0
0	0	119	0	0	0	0	0	0	0	0	0	0	0	0	0	0
0	0	0	140	0	0	0	0	0	0	0	0	0	0	0	0	0
0	0	0	0	70	0	0	0	0	0	0	0	0	0	0	0	0
0	0	0	0	0	50	0	0	0	0	0	0	0	0	0	0	0
0	0	0	0	0	0	215	1	0	0	0	0	0	0	0	0	0
0	0	0	0	0	0	0	252	0	0	0	0	0	0	0	0	0
0	0	0	0	0	0	0	0	40	0	0	0	0	0	0	0	0
0	0	0	0	0	0	0	0	0	350	0	0	0	0	0	0	0
0	0	0	0	0	0	0	0	0	0	320	0	0	0	0	0	0
0	0	0	0	0	0	0	0	0	0	0	373	0	0	0	0	0
0	0	0	0	0	0	0	0	0	0	0	0	150	0	0	0	0
0	0	0	0	0	0	0	0	0	0	0	0	0	69	0	0	0
0	0	0	0	0	0	0	0	0	0	0	0	0	0	206	0	0
0	0	0	0	0	0	0	0	0	0	0	0	0	0	0	232	0
0	0	0	0	0	0	0	0	0	0	0	0	0	0	0	0	21

Table 12 Classification report 2D-DWT level-4

	Precision	Recall	$f1$-score	Support
Female age group 1–10	1	1	1	10
Female age group 11–20	1	1	1	259
Female age group 21–30	1	1	1	119
Female age group 31–40	1	1	1	140
Female age group 41–50	1	1	1	70
Female age group 51–60	1	1	1	50
Female age group 61–70	1	1	1	216
Female age group 71–80	1	1	1	252
Male age group 1–10	1	1	1	40
Male age group 11–20	1	1	1	350
Male age group 21–30	1	1	1	320
Male age group 31–40	1	1	1	373
Male age group 41–50	1	1	1	150
Male age group 51–60	1	1	1	69
Male age group 61–70	1	1	1	206
Male age group 71–80	1	1	1	232
Male age group 81–90	1	1	1	21

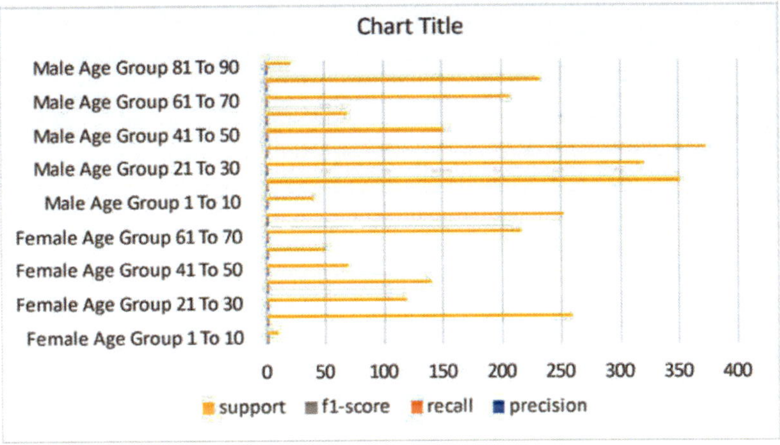

Fig. 7 Classification report of 2D-DWT level-4 with emphasis on support metric

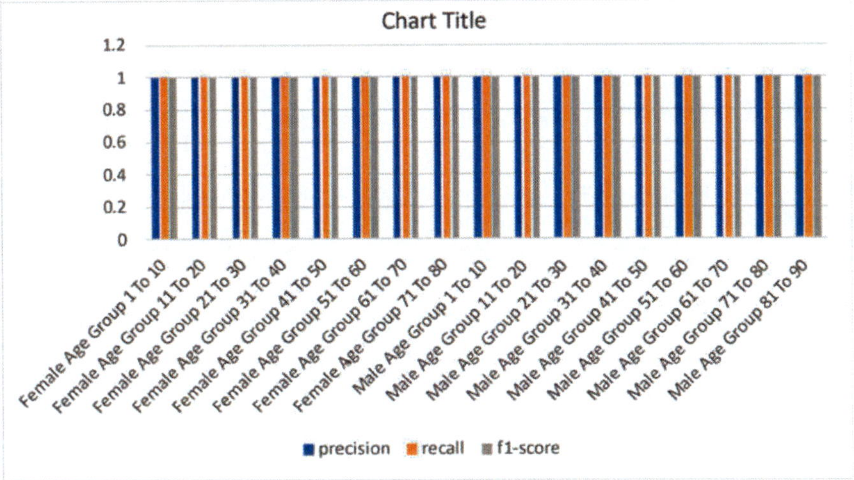

Fig. 8 Classification metrics analysis of 2D-DWT level-4 with emphasis on precision, recall and $f1$-score metrics

Table 13 shows an extended classification metrics analysis for 2D-DWT level-4 which includes the computations of macro average and weighted average of precision, recall and $f1$-score between the existing age groups. Since our database is unbalanced among various age groups, the computed accuracy needs to be reviewed and re-verified with the help of macro average and weighted average supportive classification metrics. Therefore, from Table 13, it is quantified and shown that 2D-DWT

Table 13 Extended classification metrics analysis of 2D-DWT level-4

	Precision	Recall	f1-score	Support
Accuracy			1	2877
Macro avg	1	1	1	2877
Weighted avg	1	1	1	2877

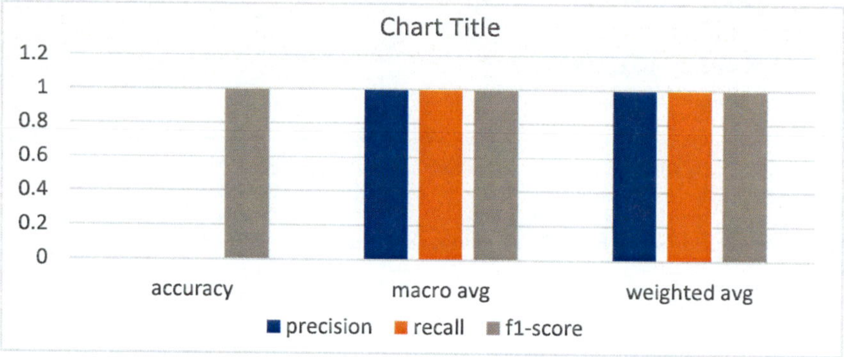

Fig. 9 Extended classification metrics analysis DWT4 using macro avg and weighted avg of precision, recall and f1-score

level-4 transform gives better accuracy and performance for fingerprint-based age group classification.

Figure 9 depicts graphical representation of the extended classification metrics analysis for 2D-DWT level-4 which includes the computations of macro average and weighted average of precision, recall and f1-score between the existing age groups.

As shown in Table 14, it has been determined that when 2D-DWT level-4 was applied then the accuracy gained was 99.96% and classification finished in 71.004 s. When DWT3 was applied then the accuracy gained was 45.15% and classification finished in 34.985 s. When DWT2 was applied then the accuracy gained was 58.08% and classification finished in 107.015 s. When DFT was applied then the accuracy gained was 38% and classification finished in 212.908 s and the most accurate of these four transforms was the DWT Level-4 which classified all the age groups using the fingerprint images with an astonishing accuracy 99.96. The accuracy is further verified and supported by performing detailed analysis of all classification metric including macro average and weighted average of precision, recall and f1-score metrics.

Table 14 Comparison of 2D-DWT2, 2D-DWT3, 2D-DWT4 and 2D-DFT based classification performance

Transform	Accuracy	Precision	Recall	$f1$-score
2D-DWT level-4	0.9996	0.9996	0.9996	0.9996
2D-DWT level-3	0.4515	0.4515	0.4515	0.4515
2D-DWT level-2	0.5808	0.5808	0.5808	0.5808
2D-DFT	0.3827	0.3827	0.3827	0.3827

5 Conclusion

The age group classification using fingerprint is an important research area with numerous potential applications in various fields. In this study, it has been concluded that 2D-DWT level-4 has correctly classified all the age groups with 99.96% accuracy in 71.004 s as compare to that of 2D-DWT level-3 with 45.15% and 2D-DWT level-2 with 58.08%. When 2D-DFT was applied, the accuracy gained was 38%. Therefore, the most accurate of these four transforms has been the DWT Level-4. The accuracy is further reviewed by performing detailed analysis of all classification metrics including macro average and weighted average of precision, recall and $f1$-score. Overall, the motivation for research in the area of age group recognition using fingerprint is driven by the potential benefits it can provide in various fields such as biometric identification, law enforcement, healthcare, personalization, and security. Further research in this area can lead to the development of advanced age group estimation models that can significantly improve the accuracy and reliability of biometric identification systems, enhance personalization, and improve the quality of healthcare services.

5.1 Future Scope

The future scope of age group detection using fingerprint is promising as it has numerous potential applications in various fields such as law enforcement, healthcare, and biometrics. One possible future direction is the development of more advanced and accurate age group and gender recognition models using a fingerprint dataset of people from rural areas with very rough skin surface due to their daily physical works as the quality of such people's fingerprint samples will be surely poor than the database collected for this research work from urban areas and the development of a robust age group classification model for such fingerprint samples will strengthen confidence of utilizing it for the latent fingerprint samples too for criminal investigation purpose in future. Therefore, there is a need for more comprehensive and diverse datasets for age group detection using fingerprint, particularly in terms of age and gender ranges and ethnic diversity. The development of large-scale and diverse datasets can help improve the generalization and reliability of age and gender estimation models to real-world scenarios.

References

1. Basavaraj Patil GV, Rafi M (2015) Human age estimation through fingerprint. Int J Innov Res Comput Commun Eng 3(4). ISSN: (Online): 2320-9801
2. Abdullah HA (2012) Fingerprint identification system using neural networks. Nahrain Univ Coll Eng J (NUCEJ) 15(2):234–244
3. Gornale SS, Hangarge M, Pardeshi R, Kruthi R (2015) Haralick feature descriptors for gender classification using fingerprints: a machine learning approach. Int J Adv Res Comput Sci Softw Eng 5(9)
4. Gornale S, Patil A, Veersheety C (2016) Fingerprint based gender identification using discrete wavelet transform and Gabor filters. Int J Comput Appl (0975-8887) 152(4)
5. Munir MU, Javed MY (2004) Fingerprint matching using Gabor filters. In: National conference on emerging technologies
6. Valdes-Ramirez D, Medina-Pérez MA, Monroy R, Loyola-González O, Rodríguez-Ruiz J, Morales A, Herrera F (2019) A review of fingerprint feature representations and their applications for latent fingerprint identification: trends and evaluation. IEEE Access 7(1):48484–48499
7. Narayanan A, Sajith K (2019) Gender detection and classification from fingerprints using pixel count. In: Proceedings of the international conference on systems, energy & environment (ICSEE) 2019, GCE Kannur, Kerala, July
8. Gnanasivam P, Muttan S (2012) Estimation of age through fingerprints using wavelet transform and singular value decomposition. Int J Biometr Bioinform (IJBB) 6(2)
9. Gnanasivam P, Muttan S (2012) Fingerprint gender classification using wavelet transform and singular value decomposition. Int J Comput Sci Issues 9(2)
10. Zhang B-L, Zhang H, Ge SS (2004) Face recognition by applying wavelet sub band representation and kernel associative memory. IEEE Trans Neural Netw 15(1):166–177
11. Win KN, Li K, Chen J, Viger PF, Li K (2020) Fingerprint classification and identification algorithms for criminal investigation: a survey. Futur Gener Comput Syst 110:758–771
12. Yang W, Wang S, Hu J, Zheng G, Valli C (2019) Security and accuracy of fingerprint-based biometrics: a review. Symmetry 11(2):141
13. Tarare S, Anjikar A, Turkar H (2015) Fingerprint based gender classification using DWT transform. In: International conference on computing communication control and automation (ICCUBEA), 26–27 Feb 2015
14. Akbar S, Ahmad A, Hayat M (2014) Identification of fingerprint using discrete wavelet transform in conjunction with support vector machine. IJCSI Int J Comput Sci Issues 11(d5):1. ISSN (Print): 1694-0814 | ISSN (Online): 1694-0784
15. Ito K, Aoki T (2018) Recent advances in biometric recognition. ITE Trans Media Technol Appl 6(1):64–80

Exploring the Potential Use of Infrared Imaging in Medical Diagnosis: Comprehensive Framework for Diabetes and Breast Cancer Screening

Asok Bandyopadhyay, Himanka Sekhar Mondal, Barnali Pal, Bivas Dam, and Dipak Chandra Patranabis

Abstract This research paper comprehensively explores the potential use of infrared (IR) imaging in medical diagnosis, focusing on diabetes and breast cancer screening. The study builds on earlier work on region selection for infrared imaging and addresses the need for automated image processing and analysis techniques. The data collection for diabetes was done at SRM Medical College and Hospital and Cachar Cancer Hospital and Research Centre (CCHRC), whereas breast cancer screening data were collected at CCHRC. The result demonstrates the importance of the ear cavity region in diabetes screening and the importance of rotational thermography and dynamic temperature analysis in infrared image-based screening systems for breast cancer screening. This paper also presents advances in data collection techniques, image analysis, and machine learning algorithms to improve the accuracy and efficiency of the proposed framework.

Keywords Infrared imaging · Diabetes and breast cancer screening · Medical thermography · Temperature quantification · Machine learning

A. Bandyopadhyay (✉) · H. S. Mondal · B. Pal
ICT and Servics Group, Centre For Development of Advanced Computing, Kolkata, India
e-mail: asok.bandyopadhyay@cdac.in

H. S. Mondal
e-mail: himanka.mondal@cdac.in

B. Pal
e-mail: barnali.pal@cdac.in

B. Dam · D. C. Patranabis
Department of Instrumentation and Electronics, Jadavpur University, Kolkata, India
e-mail: b_dam@rediffmail.com

D. C. Patranabis
e-mail: dipakchandrapatranabis@gmail.com

S. Shakya et al. (eds.), *Fourth International Conference on Image Processing and Capsule Networks*, Lecture Notes in Networks and Systems 798,
https://doi.org/10.1007/978-981-99-7093-3_27

1 Introduction

This paper reports a comprehensive approach combining infrared (IR) imaging for diabetes screening tools and breast cancer screening systems. The selection of the ear cavity region as an essential body region for diabetes screening, as well as advances in image processing, data acquisition, and machine learning, contributes to the accuracy and effectiveness of the proposed system. The findings of this study have important clinical implications and provide insights into the potential of infrared imaging in the diagnosis of type 2 diabetes. For breast cancer detection, the data acquisition technique used four pilot studies (PS1, PS2, PS3, and FS) to acquire infrared images for breast cancer screening. Each pilot study employed specific settings and modifications to simplify data collection, analysis, and optimization for breast cancer detection.

The research consisted of two main parts: selecting regions of interest for infrared imaging in diabetes-related equipment using the most influential body parts and developing a non-contact, non-invasive breast imaging method using infrared technology. In the first part, the study builds on previous work on region selection for infrared imaging. Various research groups have investigated how infrared imaging can assess the temperature of different body regions, such as the surface of the eye, foot, cornea, toes, and skin, for diabetes-related applications. From previous studies, Sodi et al., Johansson and Roback investigated the feasibility of thermal imaging-based ocular surface temperature in patients with diabetic foot disease and non-proliferative diabetic retinopathy (NPDR) [1, 2]. Nishide et al., Nandao et al. worked on using thermal imaging and ultrasound techniques to identify underlying inflammation of foot calluses [3]. To elucidate the typical pattern features of diabetic autonomic neuropathy in the toes, Fujiwara et al. perform treadmill walking stress testing and thermal imaging [4]. Fushimi et al. measurement of skin temperature in type 2 diabetic patients using thermal imaging [5] S. Sivanandam et al. investigation of abnormal vasovagal responses in the hand [6]. It acts on different body parts such as the face, forehead, hands, feet, and neck [7]. Sivanandam et al. investigated potential of infrared thermography (IR) in the diagnosis of type 2 diabetes [7] in their early works. The authors have previously discussed the selection of the ear cavity region as a critical body region for diabetes screening, with advances in image processing, data acquisition, and machine learning contributing to its accuracy and validity.

In this paper, the authors emphasize the importance of automated image processing and analysis methods to improve the accuracy and efficiency of infrared imaging for diabetes and breast cancer screening. The collaboration between CDAC Kolkata, SRM Medical College Hospital and Cachar Cancer Hospital, has contributed to advances in data acquisition techniques, image analysis, and pattern-matching algorithms. The first part of the study emphasizes diabetes screening using IR imaging techniques.

The second part of the study focused on detecting breast cancer using infrared imaging. Previously, various methods for detecting breast cancer using thermal imaging have been proposed. Ethard Tabakor et al. [8] introduced bispectral invariant features and achieved high detection rates in their study [9]. They also compared color segmentation techniques and found that K-means clustering performs better than other methods [9]. Ramon et al. [10] developed a non-invasive tool for breast cancer diagnosis using a thermal imaging camera, using temperature and texture features with promising results [10]. Prakash et al. comparing K-means, fuzzy C-means, and EM algorithm, it is concluded that EM algorithm has the highest segmentation accuracy [11]. Venkataramani et al. [12] proposed a semi-automated method using thermal video imaging and achieved high sensitivity and specificity through morphological filtering and thresholding. Pramanik et al. proposed a deep learning-based level-set method for segmenting suspicious regions in chest thermograms and achieved high accuracy [13]. In brain tumor detection, Wu et al. [14] proposed a color-based K-means clustering segmentation method. Reviewed by Kandlikar et al., the current status, protocols, advantages, and limitations of infrared imaging techniques for breast cancer detection have been discussed [15]. These methods and studies demonstrate the potential of thermal imaging in breast cancer detection and provide insights into different methods and techniques of image processing and analysis [16]. The second part addresses challenges such as quantifying suitable temperature values for malignant diseases and the lack of qualified medical professionals for infrared imaging-based screening systems. To address the above issues, the research team developed a novel non-contact, non-invasive breast imaging method based on rotational thermography and dynamic temperature, which can detect any abnormalities found in the subjects. They propose a rotating thermal imaging device that can make the affected area more visible, improving the accuracy of breast cancer detection. A machine learning algorithm trains the extracted features and predicts normal versus abnormal values for the subjects.

2 Methods

2.1 *Diabetes Screening Methodology*

To detect diabetes, the team focused on capturing infrared images of different body parts, such as the face, neck, hands, legs, and ears. Imaging setup and data acquisition were performed at a hospital in the South Indian region. An Arduino Uno microcontroller with an integrated data acquisition system based on a temperature sensor was developed to account for changes in ambient temperature, which helps to normalize the effects of temperature changes caused by environmental factors.

Special precautions were taken during data collection to ensure accurate imaging results. Selected areas of the body are stripped naked and free of any metallic ornaments or objects. Subjects were allowed to acclimate for 15 min in an air-conditioned room at approximately 25.5 °C while maintaining controlled temperature, humidity, and airflow. This step aims to capture skin temperature changes in the range of 0.05–0.1 °C between the affected area and the surrounding or contralateral area.

Data collection was performed in two sessions: a fasting blood draw and a postprandial (PP) blood draw with a two-hour interval between two. This allows data to be collected on approximately 12–13 subjects daily. Infrared image processing has employed various techniques, including manual, semi-automatic, and fully automatic region of interest (ROI) selection. Manual ROI selection involves cropping an ear template from an infrared image. In contrast, semi-automatic ROI selection allows the user to accept or reject the coordinate region suggested by the software. Fully automatic ROI selection uses image registration and feature extraction algorithms to select ROIs. Figure 1 shows the automatic selection of the ear cavity area by image processing technology, data acquisition, and analysis of the heat distribution in the diabetic "ear" area. Image processing algorithms such as thresholding and morphological operations are used to extract thermal information of the affected area. Figure 2 shows a flowchart of an image analysis algorithm using the image processing system of the diabetes screening system.

Image analysis techniques include thresholding to segment particles based on pixel intensity values. Virtual instrument software tracks regions of interest through shape-based object-tracking techniques. Save the temperature values extracted from the ROI for further analysis. Feature extraction techniques include using a generalized "Hough transform" algorithm to extract geometric shape features, which are then trained as geometric shape patterns. Statistical image analysis algorithms record mean temperature, standard deviation, median, mode, skewness, kurtosis, moments, root mean square (RMS), entropy, energy, and maximum temperature values for each infrared image as a function of the ROI. Rank the features according to the feature evaluation method based on Euclidean distance and information gain theory.

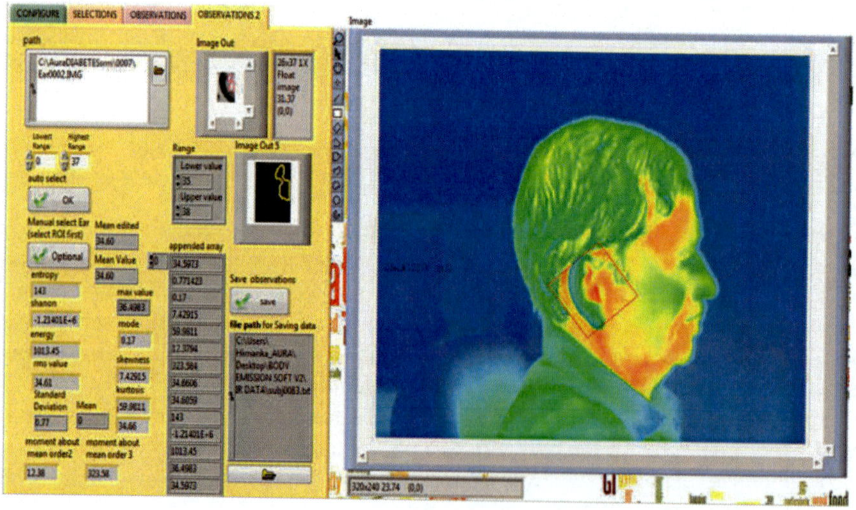

Fig. 1 Data collection and analysis of thermal profile of "ear" zone for diabetic patients

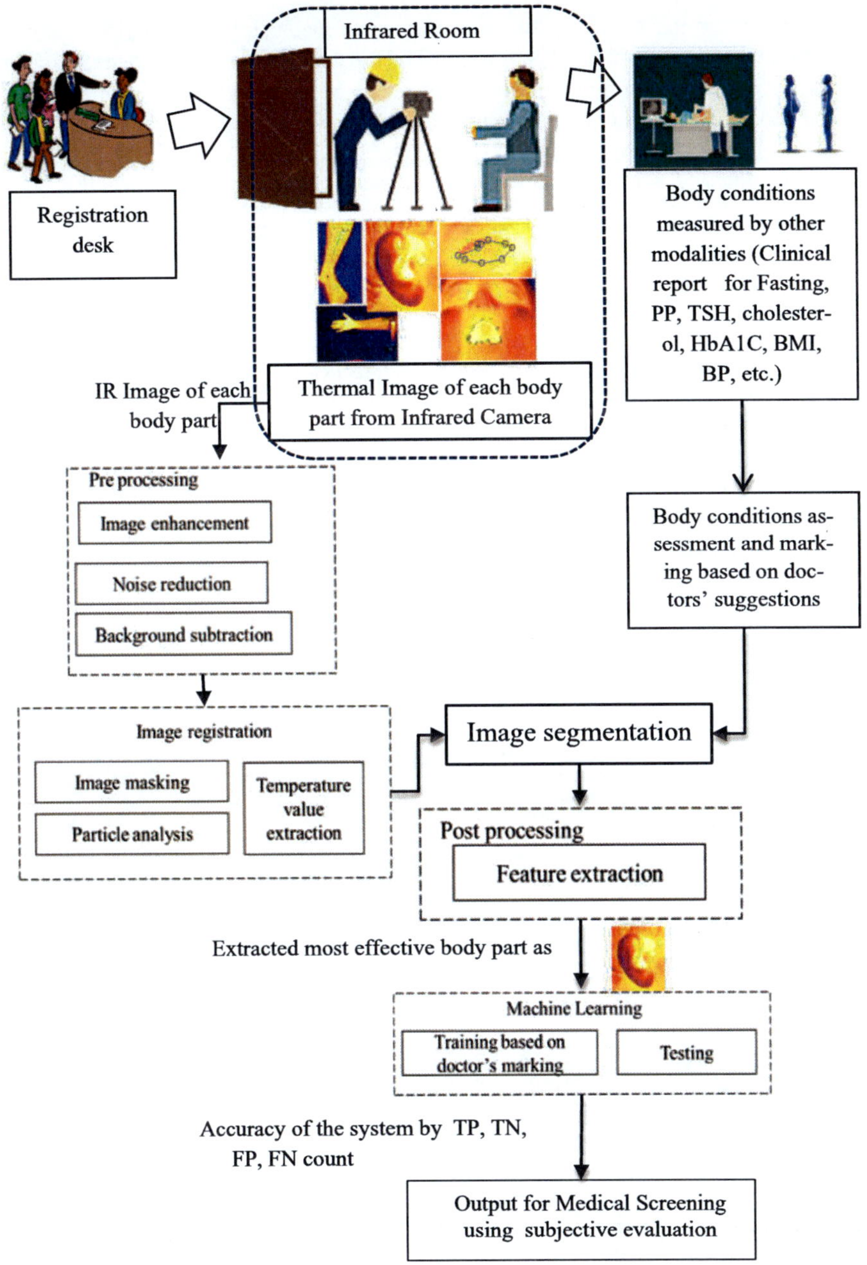

Fig. 2 Diagram for infrared-based image processing and machine vision system for diabetes screening

Machine learning-based classification was performed using the statistical features extracted from the thermal arrays as input. The top three features are selected using a feature ranking algorithm and trained using an SVM learning algorithm for classification and analysis. Other classification algorithms were also tested during the software development phase, including BPNN, RBF NN, and PNN. The SVM algorithm exhibits better and more consistent classification accuracy on image databases.

2.2 *Breast Cancer Screening Methodology*

In the case of breast cancer detection, the data collection techniques employed four pilot studies (PS1, PS2, PS3, and FS) to obtain IR images. Each pilot study utilized specific setups and modifications to optimize the data acquisition process. For example, PS1 and PS2 involved subjects seated in front of cameras with raised hands, while PS3 introduced a modified setup with the camera moving in a semicircular arc to focus on one breast while covering the other with an infrared-proof barrier. The FS employed rotational thermography, with the patient's breasts positioned one at a time through a breast-shaped grooved hole for IR imaging. Figure 3 shows the breast cancer screening setup using rotational thermography.

Data analysis techniques played a crucial role in breast cancer screening using thermal imaging. The data analysis techniques were integrated into the system at

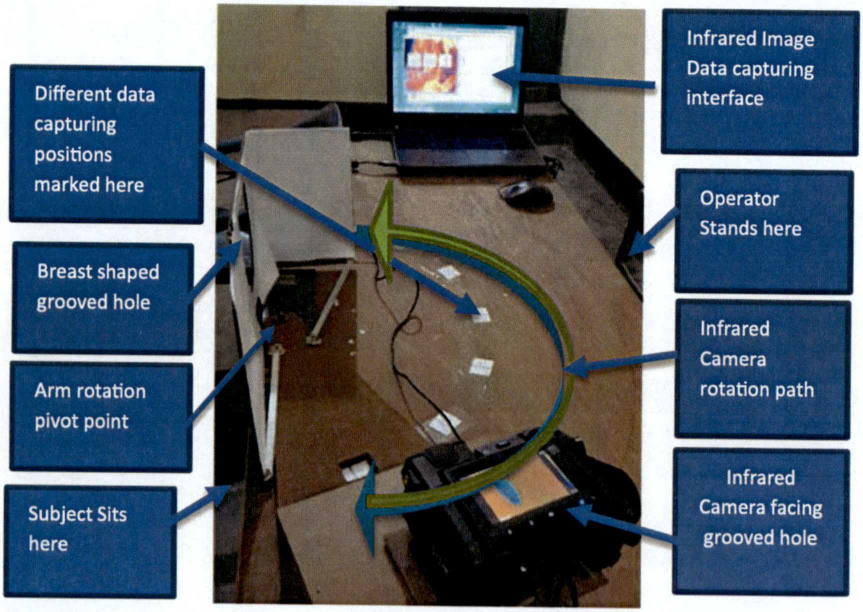

Fig. 3 Breast cancer screening setup for rotational thermography

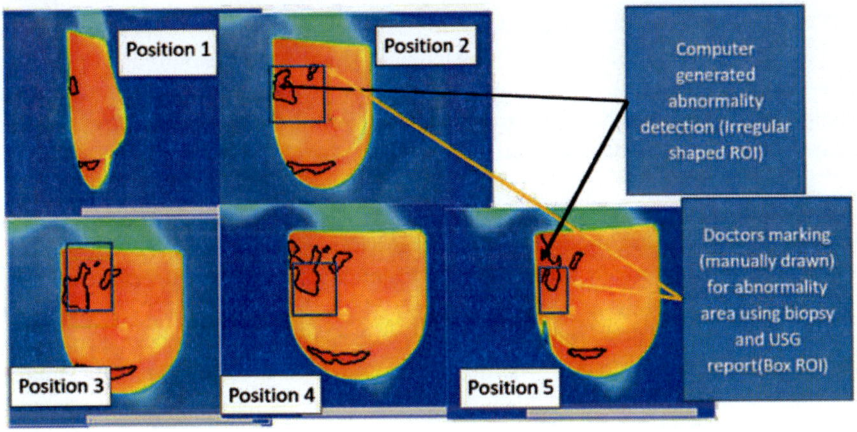

Fig. 4 ROI of detected abnormality on IR breast images captured from different angles

various stages of development to ensure optimal performance. Previous studies needed to integrate data collection protocols and analysis algorithms, resulting in suboptimal system performance. This study's data collection protocols were intelligently designed to integrate with IR image processing techniques.

In the FS pilot study, a novel data collection protocol was developed to overcome limitations in previous studies. The IR image processing techniques were integrated into the data analysis process, where clustering methods were used to segment the images and extract the region of interest (ROI). K-means clustering was applied to other image features using the mean temperature of each ROI as a discriminative feature. The clustering method was optimized based on experimental validation with consulting doctors, and seven clusters were finalized for analysis. Figure 4 shows the ROI of detected abnormality on IR breast images captured from different angles.

Regarding IR image feature-based analysis, PS1 and PS2 conventionally analyzed IR images by extracting features such as mean, median, mode, standard deviation, histogram, and maximum value. However, no reference was made to IR images acquired via other modalities.

In the PS3 pilot study, the primary reference for analysis was ultrasound (USG) and biopsy reports obtained through other modalities. Clustering based on infrared images was used for segmentation, and the average temperature of each ROI was used as a discriminator. The clustering method was gradually improved, and seven clusters were finalized based on experimental validation by consulting doctors.

In the FS pilot study, the analysis was accomplished after integrating the temperature-controlled enclosure into the system. The differences in IR images acquired at different ambient temperatures were used as the key discriminating feature for analysis. This novel analysis technique effectively detected temperature variations associated with breast cancer.

The methodologies for cancer detection using IR imaging incorporated specific data collection techniques tailored for each pilot study. Modifications and optimizations were made to address challenges and improve imaging results. Integrating data collection protocols with IR image processing techniques ensured comprehensive analysis at every stage of development.

By employing clustering methods, extracting key features, and leveraging temperature variations, these methodologies aimed to enhance the accuracy of breast cancer detection using IR imaging. Experimental validation and consultation with medical professionals were crucial in refining the techniques and achieving more reliable results.

Therefore, the methodologies for diabetes and cancer detection using IR imaging encompassed various steps, including imaging setup, data collection, IR image processing, feature extraction, and data analysis. These methodologies aimed to optimize the performance of the IR imaging system, contributing to accurate diabetic screening and early detection of breast cancer. Integrating machine learning algorithms and consultation with domain experts were critical in achieving more reliable and effective results in both research areas. Table 1 briefly details different aspects of the methodology used and compares infrared imaging-based diabetes detection and cancer detection.

3 Results and Discussion

The study highlights the potential of IR imaging-based diabetes screening, where the ear cavity zone takes a significant role and the potential of IR imaging-based screening systems in detecting breast cancer. The research shows that advances in data collection techniques, image analysis, and machine learning algorithms can improve accuracy and efficiency.

3.1 Diabetes Screening Using IR Imaging

The research studied 243 human subjects from different diabetic camps in India. The data was collected with SRM Medical College Hospital and Cachar Cancer Hospital. Ethical clearance and informed consent were obtained from all subjects. The study group comprised 129 males and 114 females aged between 19 and 91.

3.1.1 Feature Extraction and Classification

A statistical image analysis algorithm was developed for the "infrared image analysis module." The algorithm automatically plots the mean, standard deviation, median, mode, skewness, kurtosis, moments, root mean square (RMS), entropy, energy, and

Table 1 Comparison chart of methodology used for infrared imaging-based diabetes and breast cancer detection

Methodology	Diabetes detection	Cancer detection
Imaging setup and data collection	Data collection setup created at SRM Hospital and Cachar Cancer Hospital using IR camera to capture images of different body regions. LabVIEW software used for image analysis	Data collection techniques employed in four pilot studies (PS1, PS2, PS3, and FS) for breast cancer screening. Specific setups and modifications utilized for optimal data acquisition
Temperature control and image capture	Subjects acclimated to temperature-controlled room for 15 min. Skin temperature variations captured in the range of 0.05–0.1 °C	Various setups used in different pilot studies. FS implemented rotational thermography with temperature variations
Data collection plan	Two sessions: fasting blood sample collection and postprandial blood sample collection. Approximately 12–13 subjects per day	Varied data collection techniques in different level (PS1, PS2, PS3, and FS) of studies
Infrared image processing techniques	Manual, semi-automated, and full-automated ROI selection methods. Image processing algorithms (thresholding and morphological operations) used	Integration of data collection protocols and analysis algorithms. IR image processing techniques integrated at every stage of development. Clustering methods used for segmentation and ROI extraction
Analysis and information extraction techniques	Thresholding and shape-based object tracing technique employed. Temperature values stored for further analysis	IR image-based clustering used for segmentation. Mean temperature of ROI used as the discriminating feature
Feature extraction techniques	Generalized "Hough Transform" algorithm used for geometric shape features. Statistical image analysis algorithm recorded various features. Extraction of features like mean, median, mode, standard deviation, and histogram. Gradual improvement and finalization of clustering method	Feature extraction based on mean temperature and clustering methods. Seven discrimination clusters have been extracted
Machine learning classification	Statistical features used as input for machine learning algorithms. SVM algorithm demonstrated better classification accuracy	Analysis based on IR image features and discrimination using clustering, for machine learning algorithm (SVM) is used

(continued)

Table 1 (continued)

Methodology	Diabetes detection	Cancer detection
IR image feature-based analysis technique	Statistical features (mean, median, mode, standard deviation, and maximum value) extracted	Integration of temperature-controlled enclosure. Differences in IR images acquired at different ambient temperatures used as discriminating feature
Overall aim	Development of diabetic screening system using non-invasive, no touch method for improved medical diagnostics and early disease detection	Optimal system performance for breast cancer screening is achieved through an intelligently designed data acquisition protocol and integration with infrared imaging technology

maximum temperature values of a region of interest (ROI) on an infrared image as statistical value. These statistical values are considered as discriminators for classification using machine learning algorithms.

3.1.2 Gender-Specific Classification

The best three significant features for males were identified as standard deviation, mean, and RMS, while for females, they were different. As a result, separate machine learning algorithms were developed for classifying males and females. The experimental database determined the best features based on blood biochemical reports and infrared image output. It was observed that changes in the database might interchange the ranking position of features, but the basic algorithm remained the same.

3.1.3 Thermal Profile Distribution

The thermal profile distribution of diabetic and non-diabetic patients was compared in specific body regions, including the inner center points (eye), palm, ear cavity, and inner posterior buccal cavity (mouth). The temperature difference between diabetic and non-diabetic subjects was approximately 0.5 degrees centigrade. These findings supported previous research using thermography for diabetic patients, prompting further optimization of body regions for better results; here, the earlobe shows a significant differentiation between diabetic and non-diabetic subjects.

Table 2 Comparison of results from each study phase for diabetes screening

	Population-based fasting cross-sectional and case–control studies 243 subjects	Population-based PP cross-sectional and case–control studies 243 subjects
Sensitivity (%)	53.968	53.846
Specificity (%)	81.443	76.344
Positive predictive value (PPV)%	65.385	61.404
Negative predictive value (NPV)%	73.148	70.297
Accuracy (%)	70.625	67.089
FNR	46.032	46.154
FPR	18.557	23.656
LR+	2.908	2.276
LR−	0.565	0.605
Diagnostic odd ratio (DOR)	5.146	3.765

3.1.4 Classification Performance

A population-based cross-sectional and case–control study was conducted to evaluate the performance of the diabetes screening system. Sensitivity, specificity, positive predictive value (PPV), negative predictive value (NPV), accuracy, false negative rate (FNR), false positive rate (FPR), and diagnostic odds ratio (DOR) were calculated for each study group. The overall accuracy rate of diabetes screening with fasting data was 70.62% and that of diabetes screening with postprandial data was 67.08%. Table 2 shows the results of the diabetes screening study phase of a population-based cross-sectional and case–control study of fasting and PP.

3.2 Breast Cancer Detection Using IR Imaging

A novel non-contact and non-invasive breast imaging method was developed for breast cancer detection. This method captured all abnormalities presented by the subjects and introduced a rotational thermography setup for better visualization of the affected area.

3.2.1 Data Collection and Analysis

Data was collected from four pilot studies where PS1 was for 71 subjects, PS2 was done on 10 subjects, PS3 on 37 subjects, and the final study (FS) on 119 subjects. A double-blind test was conducted during PS3 to FS. Clustering techniques were done using K-median, and a support vector machine (SVM) learning algorithm was used for feature extraction and classification.

3.2.2 Classification Performance

The accuracy of the developed system for screening breast abnormalities and detecting malignant tumors was 93.27% for the FS dataset. The sensitivity and specificity were 86.67% and 97.30%, respectively. These results demonstrated the effectiveness of the proposed IR image acquisition and analysis technique for breast cancer detection. Table 3 shows the results from study phases for breast cancer screening system for population-based cross-sectional and case–control studies PS3 and FS.

Therefore, the research presents promising results for diabetes screening and breast cancer detection using IR imaging. The advanced image processing and machine learning approaches offer accurate and efficient medical diagnosis, indicating the potential for clinical applications in the future. Further research and optimization of the proposed method could improve health care and early detection of diseases.

Table 3 Comparison of results between PS3 study phases and FS for breast cancer screening system

	Population-based PS3 cross-sectional and case–control studies 37 subjects	Population-based Final System FS cross-sectional and case–control studies 119 subjects
	Female	Female
Sensitivity (%)	90.00	86.67
Specificity (%)	82.61	97.30
Positive predictive value (PPV)%	69.23077	95.12
Negative predictive value (NPV)%	95	92.30
Accuracy (%)	84.84848	93.27
FNR	10	13.33
FPR	17.3913	2.70
LR+	5.175	32.06
LR−	0.121053	0.13
Diagnostic odd ratio (DOR)	42.75	234

3.3 Discussion

The comprehensive approach presented in this study provides valuable insights into the use of infrared imaging for diabetes screening and breast cancer detection. The findings contribute to developing a comprehensive framework for data acquisition and automated analysis methods and illustrate the importance of joint research in medical thermography.

4 Conclusion and Future Research Work

The results of this study support the use of infrared imaging as a promising tool for diabetes and breast cancer screening systems using the comprehensive framework. The integrated approach shows the potential of automatic image processing and machine learning techniques to improve the accuracy and efficiency of infrared image-based systems for medical diagnosis. Further studies are needed to validate and refine this comprehensive approach. The development of standardized protocols and more extensive clinical trials could help determine the effectiveness and reliability of infrared imaging for diabetes screening and breast cancer detection. Additionally, efforts should be made to address the challenges associated with quantifying malignancy temperature levels and train qualified healthcare professionals to use infrared imaging techniques.

Acknowledgements We thank Dr. Ashok Salhan from DIPAS, DRDO, and Dr. Subhrangsu Aditya from the J.U. School of Cognitive Sciences on medical and perceptual aspects of imaging. We also thank Dr. M. Anburajan, PhD in Biomedical Engineering, SRM University. Dr J. S. Kumar, Senior Dialectist, SRM Hospital, for their assistance in organizing a diabetes camp to collect IR data. Dr. R. Ravi Kannan and his team of Cachar Cancer Hospital were also recognized for assisting us with data collection and testing facilities. Smt Sayantani Banerjee and Smt Mamata Chakraborty were also recognized for their technical contributions. We sincerely acknowledge Sri Aditya Kumar Sinha, Director, C-DAC, Kolkata, for his kind support and encouragement in implementation of the research Project. The project was funded by Ministry of Electronics and Information Technology (MeitY), Govt. of India.

References

1. Sodi A, Giambene B, Miranda P, Falaschi G, Corvi A, Menchini U (2009) Ocular surface temperature in diabetic retinopathy: a pilot study by infrared thermography. Eur J Ophthalmol 19(6):1004–8
2. Roback K, Johansson M, Starkhammar A (2009) Feasibility of a thermographic method for early detection of foot disorders in diabetes. Diabetes Technol Ther 11(10):663–667
3. Nishide K, Nagase T, Oba M, Oe M, Ohashi Y, Iizaka S, Nakagami G, Kadowaki T, Sanada H (2009) Ultrasonographic and thermographic screening for latent inflammation in diabetic foot callus. Diabetes Res Clin Pract 85(3):304–309 Epub 2009 Jul 7

4. Ring EFJ, Ammer K (2000) The infrared imaging technique in medicine. Thermol Int 10(1):7–14
5. Fujiwara Y, Inukai T, Aso Y, Takemura Y (2000) Thermographic measurement of skin temperature recovery time of extremities in patients with type 2 diabetes mellitus. Exp Clin Endocrinol Diabetes 108(7):463–469
6. Fushimi H, Inoue T, Yamada Y, Matsuyama Y, Kubo M, Kameyama M (1996) Abnormal vasoreaction of peripheral arteries to cold stimulus of both hands in diabetics. Diabetes Res Clin Pract 32(1–2):55–59
7. Sivanandam S, Anburajan M, Venkatraman B, Menaka M, Sharath D (2012) Medical thermography: a diagnostic approach for type 2 diabetes based on non-contact infrared thermal imaging. Endocrine. https://doi.org/10.1007/s12020-012-9645-8
8. EtehadTavakol M, Sadri S, Ng EYK (2010) Application of K- and fuzzy c-means for color segmentation of thermal infrared breast images. J Med Syst 34(1):35–42. https://doi.org/10.1007/s10916-008-9213-1
9. Etehadtavakol M, Ng EYK (2017) Color segmentation of breast thermograms: a comparative study, pp 69–77. https://doi.org/10.1007/978-981-10-3147-2_6
10. Garduño-Ramón M, Vega-Mancilla S, Morales-Henández L, Osornio-Rios R (2017) Supportive noninvasive tool for the diagnosis of breast cancer using a thermographic camera as sensor. Sensors 17(3):497. https://doi.org/10.3390/s17030497
11. Prakash RM, Bhuvaneshwari K, Divya M, Sri KJ, Begum AS (2017) Segmentation of thermal infrared breast images using K-means, FCM and EM algorithms for breast cancer detection. In: 2017 International conference on innovations in information, embedded and communication systems (ICIIECS). IEEE, Mar 2017, pp 1–4. https://doi.org/10.1109/ICIIECS.2017.8276142
12. Venkataramani K, Mestha LK, Ramachandra L, Prasad SS, Kumar V, Raja PJ (2015) Semi-automated breast cancer tumor detection with thermographic video imaging. In: 2015 37th annual international conference of the IEEE engineering in medicine and biology society (EMBC). IEEE, Aug 2015, pp 2022–2025. https://doi.org/10.1109/EMBC.2015.7318783
13. Pramanik S, Banik D, Bhattacharjee D, Nasipuri M, Bhowmik MK, Majumdar G (2019) Suspicious-region segmentation from breast thermogram using DLPE-based level set method. IEEE Trans Med Imaging 38(2):572–584. https://doi.org/10.1109/TMI.2018.2867620
14. Wu M-N, Lin C-C, Chang C-C (2007) Brain tumor detection using color-based K-means clustering segmentation. In: Third international conference on intelligent information hiding and multimedia signal processing (IIH-MSP 2007). IEEE, Nov 2007, pp 245–250. https://doi.org/10.1109/IIHMSP.2007.4457697
15. Kandlikar SG et al (2017) Infrared imaging technology for breast cancer detection—current status, protocols and new directions. Int J Heat Mass Transf 108:2303–2320. https://doi.org/10.1016/j.ijheatmasstransfer.2017.01.086
16. Gonzalez RC, Woods RE (2007) Digital image processing, 4th ed. ISBN 81-7758-168-6
17. EtehadTavakol M, Chandran V, Ng EYK, Kafieh R (2013) Breast cancer detection from thermal images using bispectral invariant features. Int J Therm Sci 69:21–36. https://doi.org/10.1016/j.ijthermalsci.2013.03.001

Precision Agriculture Through Stress Monitoring in Crops with Multispectral Remote Sensing Data

Konumuri Kalyan Suhas, G. Kalyani, and Mandava Venkata Sri Sai Surya

Abstract Although agriculture is regarded as the foundation of the Indian economy, there are times when plants are unable to develop to their full potential owing to a variety of factors, including pests, starvation, and other factors. In these situations, several types of fertilizers are utilized. However, a small-scale farmer is unable to apply fertilizer to hundreds of acres; thus, in these situations, our research is quite helpful for locating the areas where the plant growth is irregular. In order to increase yields, efficiency, and profitability for farmers, precision agriculture can assist farmers in monitoring crop health, soil nutrient levels, and water use, among other factors. The vegetation indices of the crop are calculated using satellite data by the crop stress monitoring system, which is essential for monitoring the developing field. The data was obtained using satellite images with varying levels of geographic and temporal resolution.

Keywords Crop stress · Land classification · Normalized Difference Vegetation Index (NDVI) · Normalized Difference Water Index (NDWI) · Soil-Adjusted Vegetation Index (SAVI) · Enhanced Vegetation Index (EVI)

1 Introduction

Using satellite data to assess crop stress is essential for efficient agricultural management. This technique provides early identification of stress factors including water scarcity, nutritional deficits, pests, illnesses, and extreme weather events by utilizing satellite images. Its extensive coverage enables large-scale agricultural monitoring, making it easier to see patterns and trends [1]. With the use of this data, precision resource management may be used to distribute water, fertilizer, and pesticides in the

K. K. Suhas (✉) · G. Kalyani · M. V. S. S. Surya
Department of Information Technology, Velagapudi Ramakrishna Siddhartha Engineering College, Vijayawada, India
e-mail: kalyan.konumuri@gmail.com

S. Shakya et al. (eds.), *Fourth International Conference on Image Processing and Capsule Networks*, Lecture Notes in Networks and Systems 798,
https://doi.org/10.1007/978-981-99-7093-3_28

best possible way. As a result of quick decision-making made possible by real-time data, farmers are better equipped to reduce stress and increase crop output.

Using satellite data to detect crop stress has several advantages for agriculture [2]. Early detection of crop stress permits prompt mitigation of the effects of stressors such nutrient deficits, water stress, pests, and diseases. Satellites' extensive coverage enables monitoring of crops across huge areas, making it easier to see patterns and trends [3]. By enabling for exact input application, minimizing waste, and mitigating environmental effects, this information aids in the optimization of resource management. Real-time data enables prompt decision-making, allowing farmers to improve output by adjusting practices.

A useful tool for agricultural stress monitoring systems, Google Earth Engine (GEE), enables users to view and examine a sizable library of satellite images [4]. With the help of GEE, scientists and farmers may preprocess the images, use vegetation indices, and include auxiliary data to evaluate crop health and spot stressed regions. Large-scale datasets may be effectively analyzed using GEE's robust computer infrastructure, allowing for the quick identification of geographical patterns and trends. GEE's user-friendly interface and code editor also make it easier to construct specialized tools and apps for crop stress monitoring, giving farmers the knowledge, and they need to manage resources more effectively and make wise decisions. GEE offers a comprehensive and scalable platform for analyzing satellite images and enhancing agricultural practices, which boosts crop stress monitoring overall.

An integral part of Google Earth Engine (GEE) that enables users to engage with its extensive collection of satellite data and sophisticated geospatial processing capabilities is its JavaScript Web interface [5]. Users may see satellite images, employ analytic methods, and get generated data products using the online interface. Users may explore and analyze geographic data in an easy and simple way by using interactive maps, charts, and visualizations made with JavaScript modules and frameworks [6]. Users are able to customize their analyses using the online interface, search for certain areas of interest, and learn more about various elements of crop stress monitoring, resource management, and decision-making.

The main objective of this project is to identify the stress of the crop with the help of satellite data. Likewise the various objectives of this project are to monitor the crop growth and development over time and assess the overall health of the crop, to estimate yield potential and assess the impact of environmental factors such as temperature, rainfall, and soil moisture, to support precision agriculture by providing detailed information on crop growth and development to inform decision-making, such as fertilizer application or irrigation management.

2 Review of Literature

The study presented an [7] innovative approach for calibrating and validating models for estimating Kc, height, and LAI in cotton based on Sentinel-1 and Sentinel-2 imagery to support irrigation management. The study [8] adopts a popular strategy of using long-lasting burned signatures to map burned areas; however, very few studies have attempted to use complimentary multispectral indices to detect burned areas.

The study of [9] mainly focuses on distribution of plant biophysical and biochemical parameters, spatial structure and variability in the UAV and Sentinel-2 imagery and comparison of UAV and Sentinel-2 data along transect lines. In the article [10], WRASIS was applied at three agriculture project and defining crop patterns, adding crops spectral signatures into software libraries and identify crop types to produce different outputs of crop maps.

The use of satellite data to control crop stress is an area of ongoing study, and several studies have been done on it. In order to manage crop stress, remote sensing techniques have been employed to identify crop stress brought on by water, nutrient, and pest-related issues. Based on satellite data, a crop water stress index (CWSI) has been created to control water stress in crops. This has improved irrigation, boosted agricultural yields, and improved water usage efficiency. Crop stress has been managed using precision agricultural techniques, such as site-specific fertilizer management and variable rate irrigation, which have improved crop growth and decreased stress. Additionally, it has been discovered that employing satellite data for agricultural disease detection and management can lower the need for pesticides and increase crop yields. Finally, crop stress prediction algorithms have been created to accurately forecast crop stress, enabling farmers to take preventative action to minimize stress. Even though using satellite data to control crop stress has shown to be quite beneficial, additional study is needed to provide more precise and reliable methods for identifying and treating crop stress.

3 Methodology

Architecture Diagram

The different lessons that may be drawn from Fig. 1 include: Launch Google Earth and construct a polygon. Think about the options for crop land and non-crop land. In order to improve the image quality, cloud mask must be removed. Following that, training of both the cropland and non-cropland takes place. Then, the land is divided into crop land and non-crop land using the SMILE-CART algorithm. Next, take into account the crop land from the categorized land. Then, figure out the different indices, such as NDVI, NDWI, SAVI, and EVI. Visualize land classification and NDVI, NDWI, SAVI, and EVI, and you are done.

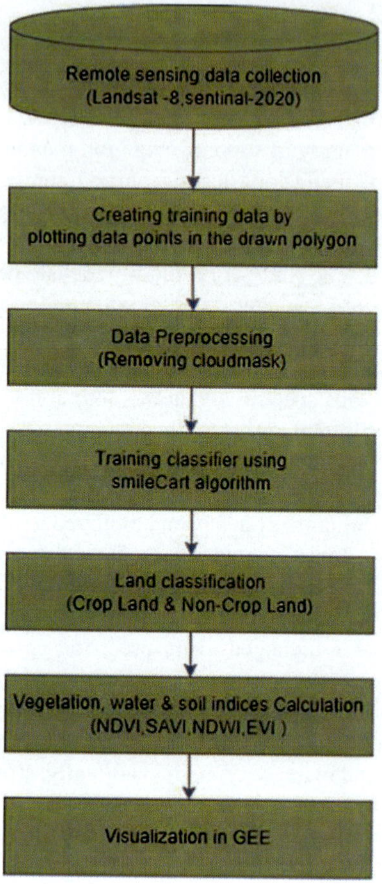

Fig. 1 Architecture diagram

Description of Algorithms

SMILE-CART Algorithm

A decision tree-based machine learning algorithm known as the SMILE (Statistical Machine Intelligence and Learning Engine)-CART (Classification and Regression Tree) method may be applied to classification and regression issues. Recursively dividing the data into subsets according to the input characteristics, the method produces a tree-like structure that may be utilized for prediction.

The optimal feature and split that maximizes the difference in the goal variable between the subsets are chosen to build the decision tree model. The SMILE-CART method is renowned for its readability, interpretability, and simplicity. It can handle missing values and variables that are categorical or continuous. The most crucial characteristics are chosen first by the tree, and hence, the method may be utilized for feature selection.

To prevent over-fitting and enhance the model's generalization capabilities, the tree can also be trimmed. The user must supply a set of labeled training data that comprises the input features and the target variable in order to apply the SMILE-CART method. The user must additionally define the minimum number of samples needed for a split, the maximum depth of the tree, and other parameters. Once trained, the model may be used to generate predictions based on fresh data. The SMILE-CART method is often utilized in several fields, including remote sensing, ecology, business, and medicine.

Algorithm ID3 Input:

- Examples: Instances or observations in the dataset that are used for training the decision tree.
- Target attribute: The class or category that is required to predict or classify for new instances.
- Attributes: The features or characteristics of the instances that are used to make decisions and build the decision tree.

Output: A decision tree that represents learned classification rules based on the given inputs.

Method:

- For the tree, create a root node.
- If every example is positive, return the single-node tree root with the label = +.
- If every example is negative, return the single-node tree root with the label = −.
- If attributes is empty, return the single-node tree root, with label equal to the most prevalent TargetAttribute value in examples.
- Otherwise Start.
  - A ← the characteristic that best categorizes examples from the group of attributes
  - The root's decision-making attribute ← A
  - For every possible value, vi, of A,
  - Below Root, add a new tree branch that corresponds to the test A = vi
  - Let examples, $_{vi}$ represent the subset of Examples where vi is the value for A
  - If examples, $_{vi}$ is not filled out

- Then, add a leaf node with label = most frequent TargetAttribute value in Examples underneath this new branch.
- Else include the subtree below this new branch.

ID3(Examples$_{vi}$, TargetAttribute, Attributes − {A})

- End
- Return Root

4 Experimental Results

Data Collection

The Google Earth Engine and JavaScript were used for all computational and programmable operations. The LandSat-8 satellite is utilized to gather the data for this research. Cropland and non-cropland land from Movva make up the area under consideration. The cloud mask is eliminated using the .filterMetadata() method.

Experimental Metrics

Normalized Difference Vegetation Index (NDVI)

It is used to assess the strength and vitality of plants. The range of NDVI values is −1 to + 1, with higher values indicating more robust and thick vegetation. The numbers 0 and negative denote the absence of vegetation and the presence of water, respectively.

The formula for calculating NDVI is:

$$\text{NDVI} = \frac{(\text{NIR} - \text{Red})}{(\text{NIR} + \text{Red})}$$

where NIR is the reflectance value in the near-infrared band and Red is the reflectance value in the red band.

Soil-Adjusted Vegetation Index (SAVI)

It pinpoints how soil brightness affects vegetation's spectrum response. Higher numbers denote healthier and denser vegetation. SAVI values vary from −1 to + 1. Compared to other indicators, it can offer a more accurate assessment of the health of the plants, especially in regions with high soil moisture or brightness.

The formula for calculating SAVI is:

$$\text{SAVI} = \frac{(\text{NIR} - \text{Red})}{(\text{NIR} + \text{REd})} * (1 + L)$$

where NIR is the reflectance value in the near-infrared band, red is the reflectance value in the red band, and L is the soil adjustment factor, which ranges from 0 to 1.

Normalized Difference Water Index (NDWI)

Evaluates the water content of soil and plants. Higher values signify increased water content. NDWI values range from −1 to + 1. Negative numbers represent soil or plants, whereas a value of 0 implies there is no water present. NDWI is particularly helpful for identifying changes in soil moisture content in agricultural regions as well as for monitoring water bodies including lakes, rivers, and wetlands.

The formula for calculating NDWI is:

$$\text{NDWI} = \frac{(\text{Green} - \text{NIR})}{(\text{Green} + \text{NIR})}$$

where green is the reflectance value in the green band and NIR is the reflectance value in the near-infrared band.

Enhanced Vegetation Index (EVI)

In locations with thick vegetation and high soil background reflectance, it is intended to increase the sensitivity of vegetation indices. Higher values imply denser and better plant cover. The EVI values range from -1 to 1. The EVI is frequently used to track changes in land usage over time, drought conditions, and plant growth.

The formula for calculating EVI is:

$$\text{EVI} = 2.5 * \frac{(\text{NIR} - \text{Red})}{(\text{NIR} + C_1 * \text{Red} - C_2 * \text{Blue} + L)}$$

where NIR is the reflectance value in the near-infrared band, Red is the reflectance value in the red band, Blue is the reflectance value in the blue band, and C_1, C_2, and L are empirical coefficients that depend on the sensor and atmospheric conditions.

Discussion on results

Figure 2 indicates the crop land (indicated in green color) and non-crop land (indicated in pale pink color) classified with help of SMILE-CART algorithm.

Figure 3 indicates the percentage of NDVI values, (i.e.) green color indicates high NDVI values (0.6–0.9), white color indicates average NDVI values (0.2–0.5), and blue color indicates the least NDVI values (≤ 0.1).

Figure 4 indicates the percentage of SAVI values, green color indicates high SAVI values (0.3–0.9), white color indicates average SAVI values (0.1–0.3), and blue color indicates the least SAVI values (≤ 0.1).

Fig. 2 Classification of land into crop and non-crop lands

Fig. 3 NDVI values in the selected region

Fig. 4 SAVI values in the selected region

Figure 5 indicates the percentage of NDWI values, green color indicates low NDWI values (≤ -0.3), blue color indicates high NDWI values (0.2–1), and the white color indicates average NDWI values (-0.3 to 0.2).

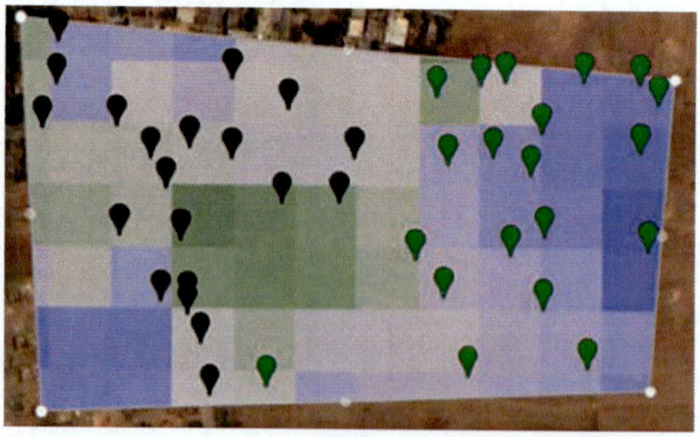

Fig. 5 NDWI values in the selected region

Fig. 6 EVI values in the selected region

Table 1 Index values

Index\values	High	Average	Low
NDVI	0.62135	0.14374	– 0.10977
SAVI	0.53169	0.19965	– 0.41049
EVI	0.59323	0.14924	0.00317
NDWI	0.33997	0.14841	– 0.35734

Figure 6 indicates the percentage of EVI values, green color indicates high EVI values (0.2–0.8), white color indicates average EVI values (0.0–0.2), and blue color indicates the least EVI values (≤ 0.0).

If a sector of agricultural land is considered, the usage of several vegetation indices such as NDVI, NDWI, SAVI, and EVI in order to determine the crop's stress in that specific area. Below is an illustration of one such area.

Table 1 then displays a variety of value ranges for various indices.

5 Conclusion

In this project, Google Earth Engine is used to track the crop's stress using data from satellites. In order to classify the land into two categories, firstly launch the Google Earth Engine and draw a polygon. Then, feature selections such as crop land and non-crop land must be considered, upon which training of both of the lands takes place. Finally, the SMILE-CART algorithm is used to calculate the values of the indexes and monitor the stress of the crop.

After determining the values of the various indexes in this project, the area must be specifically addressed, identify the issue, and then offer a solution. These solutions include different kinds of pesticides or insecticides that should be used on that specific plant, so that the loss will not be significant.

References

1. Sanae H, Jilbab A, Sanad IM (2020) Crop stress monitoring system using satellite data and machine learning techniques. In: 2020 IEEE Jordan international joint conference on electrical engineering and information technology (JEEIT), Amman, Jordan
2. Ahmed S, Yang Z, Zhang J (2020) Satellite-based monitoring of crop water stress using convolutional neural networks. In: 2020 IEEE international geoscience and remote sensing symposium (IGARSS), Waikoloa, HI, USA
3. Zhang P, Li H, Li C (2020) Satellite-based crop stress monitoring system using deep learning and hyperspectral imagery. In: 2020 IEEE international conference on artificial intelligence and computer applications (ICAICA), Wuhan, China
4. Sharma R, Singh R, Tiwari S (2021) Satellite-based crop stress monitoring using convolutional neural networks. In: 2021 IEEE region 10 symposium (TENSYMP), Dhaka, Bangladesh
5. Shrestha S, Zhang T, Li H (2021) Satellite-based crop water stress monitoring using deep learning and multi-temporal Sentinel-2 imagery. In: 2021 IEEE international conference on artificial intelligence and computer applications (ICAICA), Wuhan, China
6. Wang Y, Li H, Zhang L (2022) Crop water stress monitoring using satellite data and long short-term memory networks. In: 2022 IEEE international conference on artificial intelligence and computer applications (ICAICA), Wuhan, China
7. Kaplan G, Fine L, Lukyanov V, Malachy N, Tanny J, Rozenstein O (2023) Using Sentinel-1 and Sentinel-2 imagery for estimating cotton crop coefficient, height, and leaf area index. Agric Water Manage 276
8. Deshpande MV, Pillai D, Jain M (2022) Agricultural burned area detection using an integrated approach utilizing multi spectral instrument based fire and vegetation indices from Sentinel-2 satellite. MethodsX 9
9. Li M, Shamshiri RR, Weltzien C, Schirrmann M (2022) Crop monitoring using Sentinel-2 and UAV multispectral imagery: a comparison case study in Northeastern Germany. Rem Sens 14(17)
10. Zahran SAEl-S, Saeed RA-H, Elazizy IM (2022) Remote sensing based water resources and agriculture spatial indicators system. Egyptian J Rem Sens Space Sci 25(2)

EEG Signal Feature Extraction Using Principal Component Analysis and Power Spectral Entropy for Multiclass Emotion Prediction

S. Babeetha and S. S. Sridhar

Abstract Emotion recognition in artificial intelligence is a challenging factor. Recent years have seen a rise in interest in emotion analysis based on EEG signals due to many factors, including the availability of huge EEG datasets, developments in Brain-Computer interface (BCI) devices, and discoveries in machine learning. Due to its high temporal precision, portability, and relatively low cost of data collection, EEG is a viable option for researching the brain correlates of numerous cognitive activities, including emotion. The user's capacity to generate these signals is indicative of how well their brain processes information across a variety of cognitive, emotional, and physical domains. Important feature values must be extracted from these raw signals, which is still an essential part of the deployment process. In order to better understand the brain's activity, this study explains how to use principal component analysis (PCA) and power spectral entropy (PSE) to extract characteristics from an EEG signal. In this analysis, we evaluate how well-known classification algorithms fare when used on the recovered features: the support vector machine (SVM), the K-nearest neighbors (KNN), and the Naive Bayes (NB) classifier. All of the simulations were carried out in MATLAB using a HEADIT dataset. The results reveal that the suggested classifiers outperform the standard method in predicting emotional states.

Keywords EEG signals · Emotion prediction · Principal component analysis · Power spectral entropy · Support vector machine · K-nearest neighbors · Naive Bayes classifier

S. Babeetha (✉) · S. S. Sridhar
Department of Computing Technologies, College of Engineering and Technology, SRM Institute of Science and Technology, Kattankulathur Campus, Chengalpattu, Tamil Nadu, India
e-mail: bs5393@srmist.edu.in

S. S. Sridhar
e-mail: sridhars@srmist.edu.in

S. Shakya et al. (eds.), *Fourth International Conference on Image Processing and Capsule Networks*, Lecture Notes in Networks and Systems 798,
https://doi.org/10.1007/978-981-99-7093-3_29

1 Introduction

Electroencephalography (EEG) is a non-invasive technology that specialists often use in a wide variety of contexts due to its composite time series data. Emotional computing encompasses the study of how computers interpret and recognize human emotions; this is known as emotion recognition. Humans rely on their emotions to help them with a wide range of cognitive processes, including thinking, planning, decision-making, and more. Individuals who were emotionally stable and had a healthy perspective on life were more likely to achieve their goals. Moreover, negative emotions might hinder one's decision-making skills and even one's physical health. Some of the non-fun uses of emotion analysis include in areas such as adaptive advertising, an emotion-based music player, and online education. The EEG cap's popularity among clinical analysts may stem from the fact that it is versatile piece of equipment that may be put to work in a variety of contexts requiring emotion recognition. Speech, body language, facial expressions, physiological responses, etc., may all be used for emotion identification. These approaches rely on surface expressions of emotion and are thus inadequate for gauging internal states of mind. This previously hidden information is now available because to EEG-driven signals, which provide a wide range of emotional patterns. As a result, several feature extraction-based methods for analyzing EEG signals have been developed. Nonlinearity and underlying difficulties in the EEG signals may be measured using these properties. Recent advances in machine learning have had far-reaching effects on signal and information processing by making automated feature extraction and selection possible. In light of this, we use principal component analysis and power spectral entropy to forecast across many emotion classes in this article.

The following is a list of this work's most significant contributions:

- Improvement was done on all areas by addressing the problems discovered in earlier research.
- By developing and presenting a machine learning-based predictive modeling emotion recognition system, we have created a system that comprehends nonverbal data such as users' emotions, objectives, and attitudes.
- Individuals' feelings were analyzed using a principal component analysis (PCA) with a partial sums Eigenvector (PSE) and a comparison was made to the support vector machine, the K-nearest neighbors classifier, and the Naive Bayes classifier.

The remaining sections of the paper are laid out as follows. In Sect. 3, we detail the research methodological structure and procedure flow. Section 4 explores more into the findings and discussion. Section 5 gave the last ideas and recommendations for improvement.

2 Related Works

Researchers Misulis et al. [8], who studied the link between EEG and seizure drugs, found that fluctuations in the strength of the EEG signal revealed brain electrical activity. More entropy, or disorder, was seen in the EEG signals of patients with PTSD. More entropy was produced in the frontal lobe of the brain, the region in charge of experiencing, and expressing emotions. To categorize sleep phases in autonomous cars using individual EEG data, Michielli et al. [9] proposed a spiraled long short-term memory (LSTM) back-propagation neural network. Traditional machine learning-based emotion recognition algorithms may benefit from being trained on emotional brainwave data (EEG). Using DEAP's integrated monitored neural classification algorithm, we observed that our average HALA, HVLV, recognition, dominance, and likability scores were 97.39, 97.41, 98.21, and 97.68. Researchers discovered that the model had a 93.74% average success rate in classifying happy and sad faces after testing it on the SEED dataset. They found that the accuracy of the model was enhanced by associating mental states with emotions already present in EEG datasets. Qazi et al. [10] looked at the viability of developing an emotion detection Qazi system employing two-light deep learning in tandem with a CNN model ensemble. An emotion identification system is provided by combining the Mel-frequency cepstral coefficient (MFCC) with bio-signals. Studies of mental stimulation have shown that anxiety, calm, melancholy, and happiness are distinct states of mind. Multilayer perceptrons (MLP) are used as the classifiers. The use of EEG data to assess mental states has been shown in experiments to achieve an accuracy of up to 90%. This indicates that typical emotions may be recognized using scalp EEG data and the MFCC-MLP detection method.

Korotkova et al. [11] measured hippocampal theta alterations to examine their effects on cognitive, behavioral, and personality-based performance. Strong brain rhythms between 5 and 10 Hz, known as theta resonance, have been connected to a wide range of human actions. This article provides a concise overview of the research that has related theta frequency to different mental processes. The electroencephalogram (EEG) of people with mental illness should be studied when they are in a deep slumber, as proposed by Newson and Thiagarajan [12]. In order to better understand the diagnostic and therapeutic implications of PTSD's electrical impulse patterns, this research used EEG. To classify EEG data for moods, Ramzan and Dawn [13] investigated the use of machine learning. The purpose of this study is to determine whether or not certain EEG patterns are linked to the presence and severity of PTSD. The most convincing evidence of time-domain in EEG linked to event-related potentials is an exaggerated forebrain reaction time for task redirection (P3a intensity) and a reduced forebrain response to event-related potentials (P3b amplitude). These findings demonstrate that some individuals with PTSD suffer from attention difficulties that make it challenging for them to focus on common activities and remain dedicated even when faced with potentially harmful distractions. Using measures of emotional reactivity and brain regions associated with locus of control, Fitzgerald et al. [14] show that PTSD symptoms may be predicted in war veterans. It is difficult to pull off gradients that abruptly disappear or burst. When used to Suhaimi classification and prediction tasks, the MFCC and CNN algorithms utilized by Suhaimi et al. [15] were shown to be imprecise and unreliable. Liu et al. [19] detail the use of a sparse encoder and a DNN with a heap database to create convolutional neural networks (CNNs). Overfitting, exaggerated gradients, and a lack of diversity among classes are the primary problems. Using ResNet50, the SEED dataset, DREAMER, and AMIGOS, Topic and Russo [16] highlight the benefits and drawbacks of convolutional neural networks (CNNs). Although the findings of Wang and Wang's [17] ensemble are more difficult to comprehend and anticipate, they were constructed utilizing SVM, KNN, naive Bayesian, and random forest methods. This strategy is helpful since it improves both the accuracy and performance of forecasts compared to using a single model. In their discussion of CNN with reference to the SEED IV dataset, Pereira et al. [18] note the need of taking into consideration co-occurring symptoms and disorders.

The SVM approach was used to attain an even greater overall accuracy of 92% in the research by George et al. [1]. A box-and-whisker plot and the DCT technique were used to identify the characteristics. By evaluating the statistical features of the Fast Fourier Transform (FFT), researchers found a 92% increase in accuracy for emotion recognition in the DEAP dataset (32 participants). This strategy is preferable to the one proposed by Seeja et al. since it has been shown to be more accurate. Differences in feature extraction and data preprocessing can account for the observed performance gaps. In Alhagry et al. [2] published a research that highlighted the significance of HCI-based emotion detection systems. They found problems on all three fronts, unlike the vast majority of research in this area, which only examine two (arousal and valence). High accuracy of 85.65% was attained for the valence category, 85.45% for the arousal category, and 87.99% for the liking category when the features extracted from the DEAP dataset were classified using LSTM-RNN. These techniques disregarded feature extraction in favor of end-to-end approaches since deep learning algorithms may extract and categorize features in a single step. Yin et al. [3] used graph convolutional neural networks (GCNN) to create an EEG-based emotion recognition model. The experiment used the DEAP data collection. ECLGCNN was applied to the data after it was partitioned, and the differential entropy features were retrieved. The ECLGCNN approach integrates GCNN with LSTM. Valence labeling accuracy was shown to be 85.04% across independent trials, 90.45% across subject-dependent trials, and 90.60% across all trials. The processing cost of this approach may be reduced by discovering novel methods to extract characteristics. To categorize emotions using the CNN model, which has shown to be capable of successful feature extraction, Cimtay et al. [4] used the end-to-end approach. This research enhanced the CNN model by adding extra layers to enhance classification precision. This research made use of the DEAP, LUMED, and SEED datasets. The model performed with an accuracy of 86.56% on the SEED dataset, 78.3% on the DEAP dataset, and 81.8% on the LUMED dataset. Iyer et al. [20] developed the convolution neural network (CNN) and long short-term memory (LSTM) based hybrid model which is developed for accurate emotion detection.

An emotion detection model capable of distinguishing between happy, sad, and neutral expressions was created by Wei et al. [5]. Four characteristics across five frequency bands were used in the SRU models that we constructed using a SEED dataset. Several causes led to the formulation of the concept of SRU. Long-term, it may be the answer to the RNN dependency issue. Using dual-tree compound wave transfer (DT-CWT), we were able to accurately extract time–frequency and nonlinear properties. This strategy is based on the iterative discovery process. The proliferation of EEG data channels has been the subject of recent study in the area of emotion identification, prompting Chao et al. [6] to look into the issue. They developed a novel strategy to solve this issue by combining deep belief-conditional random fields with glia chains. Three different datasets—AMIGOS, SEED, and DEAP—were used to test the model. Overall, the accuracy of these techniques was 76.13%. Using a DEAP dataset and two feature extraction methods—the variational mode decomposition (VMD) and the empirical mode decomposition (EMD)—Pandey et al. [7] examined emotional reactions to stimuli using EEG signals. DNN was also used for emotion categorization. The overall accuracy and valence accuracy for this approach were both 63% and 62%, respectively. According to the results, the deep neural network classifier outperformed the SVM classifiers when used to the emotional classification model. The study's authors claim that using VMD-based features rather than EMD-based techniques improved signal comprehension and performance. Although masking techniques have reduced the amplitude rate between the mono-components, there is room for improvement in EMD's frequency resolution.

According to the analysis, the following are some of the difficulties encountered by methods used to recognize human emotions.

(i) In contrast, the trial-and-error method of choosing network characteristics such as training parameters and the total number of nodes presents the most difficulty.
(ii) When comparing the classification accuracy of the deep learning approach to that of other current methods, it becomes clear that the former is much superior. To improve classification accuracy, using attention processes on different brain areas may help. However, training requires a lot more time. Therefore, a strategy to counteract these drawbacks is required.

3 Proposed Work

Human emotion detection through the suggested technique is a two-stage process that begins with feature extraction and ends with emotion identification. Features such as spectral flatness, logarithmic band power, fluctuation index, spectral decline, spectral roll-off, and relative energy are extracted from the input EEG data using principal component analysis and principal component spectral estimation. Following feature extraction, we utilize KNN, SVM, and NB classifiers to label facial expressions as proud, elated, joyful, satisfied, relieving, hopeful, curious, surprised, sad, embarrassed, ashamed, guilty, envious, disgusted, contemptuous, or furious.

3.1 Data Source

The dataset was obtained from https://headit.ucsd.edu/studies/3316f70e-35ff-11e3-a2a9-0050563f2612. It includes of audio recordings that imply an emotion and prompt participants to visualize a triggering situation or think back on a time when they felt that emotion.

3.2 Feature Extraction

PCA and PSE are two of the methodologies used for feature extraction (Fig. 1).

1. **PCA**

The process of decomposing attached variables into their individual metrics is an integral part of the PCA cycle. Principal component analysis is used to lower the dimensionality of the fused feature (FF) vectors, making them easier to classify accurately. The model is constructed taking into account the features that effectively classify the data. There are n separate dimensions in FF. In the first $F = z_1, z, \ldots, z_n$ has several dimensions, and we must reduce it to k_n. The PCA follows these stages to generate a set of reduced fused features (RFFs):

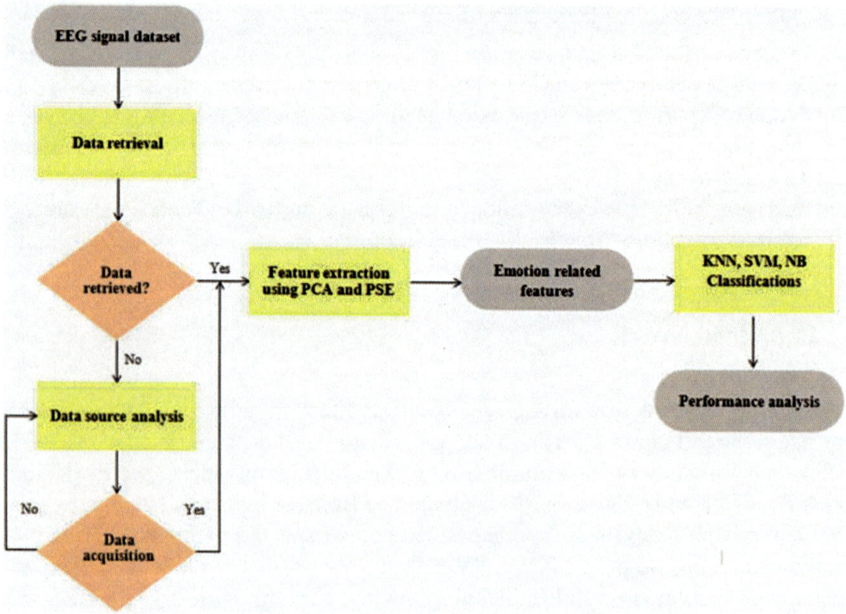

Fig. 1 EEG signal extraction using PCA and PSE

1. Data scaling

$$z_j^i = \frac{z_j^i - \overline{z_j}}{\sigma_j} \quad (1)$$

2. Co-variance matrix computation

$$\sum = \frac{1}{m} \sum_i^m (z_i)(z_i)^T, \sum \in R^{n \times n} \quad (2)$$

3. Third, we calculate the Eigenvector and Eigenvalue

$$w^T \sum = \lambda \mu U = \begin{bmatrix} | & | & & | \\ w_1 & w_2 & \dots & w_n \\ | & | & & | \end{bmatrix}, w_i \in R^n \quad (3)$$

4. Eigenvalue selection. We used the following formula to choose the top 100 Eigenvalues in K-space for use in our classification system.

$$z_i^{\text{new}} = \begin{bmatrix} w_1^T z^i \\ w_2^T z^i \\ \dots \\ \\ \dots\dots \\ w_k^T z^i \end{bmatrix} \in R^k \quad (4)$$

5. At last, the feature matrix was constructed.

2. **PSE**

The spectrum complexity of an uncertain system may be quantified using power spectral entropy, which is a kind of information entropy. Let us pretend that the values of states in an uncertain system are determined by a random variable X, with possible outcomes including,

$$Y = \{y_1, y_2, \dots, y_n\}(n \geq 1) \quad (5)$$

The matching probability is illustrated as follows

$$q = \{q_1, q_2, \dots, q_n\} 0 \leq q_i \leq 1, i = 1, 2, \dots, n \quad (6)$$

Under constraints, the values are,

$$\sum_{i=1}^n q_i = 1 \quad (7)$$

As a result, we may describe the system's information entropy as

$$H = -\sum_{i=1}^{n} q_i \ln q_i \tag{8}$$

Power spectral entropy is the entropy of the resulting power spectrum from an FFT transformation of a time series signal. In a nutshell, the algorithm is as follows:

1. The Discrete Fourier Transform (DFT) of a signal may be calculated using the Fast Fourier Transform (FFT) and is represented as $y(\omega_i)$, where i is a frequency. Following is the formula used to determine the power spectral density (PSD),

$$\hat{q}(\omega_i) = \frac{1}{N}|y(\omega_i)|^2 \tag{9}$$

2. Normalize data $\widehat{P}(\omega_i)$, to derive a function of power spectral density,

$$q_i = \frac{\hat{q}(\omega_i)}{\sum_i \hat{q}(\omega_i)} \tag{10}$$

3. The PSE may be calculated using Eq. (10). Time uncertainty in the frequency domain may be quantified using PSE. When the peak of the power spectrum of EEG signals is narrow, the entropy of the signal is low. In other words, if a wave represents orderliness, then the signal's complexity is low; if the spectrum peak represents smoothness, then the signal's entropy value is high. As a result, the PSE can accurately portray the spectral make-up of EEG data.

3.3 *Emotion Recognition*

The feature extracted from the two-feature extractor are separately given as a input for the three different types of the classifier listed below,

1. **K-Nearest Neighbors**: K-nearest neighbors (KNN) is a supervised machine learning technique that uses similarities to existing data sets for prediction and classification. The distance metric in (11) may be used to measure the degree of emotional feature similarity between points in the test set and the training set.

$$\text{Cosine}(x_i, x = y_i) = \frac{\sum_{i=1}^{n} Y_i X_i}{\sqrt{\sum_{i=1}^{n} (Y_i)^2}\sqrt{\sum_{j=1}^{n} (X_i)^2}} \tag{11}$$

where X_i are numbers that should be able to categorize their emotional output, and Y_i are values that are known to be legitimate outputs.

2. **Support Vector Machine**: The method is an expansion of the SVM for use in regression analysis. Depending on the kernel function, it might be linear or

nonlinear. Instead of looking for a hyperplane to solve the regression issue, it locates a tube around the function that is insensitive to changes of up to epsilon. To make predictions about feelings, the models were first trained on the training set.

3. **Naive Bayes**: Bayes' theorem has been used here for classification on the premise that predictors have no effect on classification. The Naive Bayes classifier makes the assumption that characteristics inside a given class do not correlate with one another.

The Naive Bayes model may be used to construct and analyze very large datasets. This model is a very effective classification approach despite its apparent simplicity, and it can handle complex circumstances with ease. The following equation may be used to get the posterior probability using Bayes' theorem:

$$v(a/y) = (v(y/a)v(a))/v(y) \tag{12}$$

where $v(a/y)$ is the predictor's prior probability, $v(y)$ is the likelihood, and $v(a)$ is the posterior probability of the class given the predictor's knowledge of the class distribution.

4 Performance Analysis

This section will assess the efficacy of the proposed technique. The whole experiment was carried out using MATLAB as the platform for the operating system. To demonstrate how useful the proposed network is, an examination of its performance in terms of several metrics may be carried out.

The suggested system for emotion recognition and its simulated output was demonstrated in Fig. 2.

Figure 3 displays a loss of 0.1 and a mean squared error of 0.02.

With the suggested method, the average accuracy of EEG-based emotion identification is raised to 98% by the integration of feature extraction and a classification model. Figure 4 shows the test and learn accuracy rates. Accuracy of 0.9826 and MSE of 0.008 are achieved in the training model after 100 epochs.

From the results obtained, both the PCA and PSE have approximately same level of performance over extracting the features as depicted in Fig. 5.

All the projected versus actual charts display the predictions the models made on the test features with the number of features utilized for predictions as depicted in Fig. 6. By comparing the projected and actual values, here all of the models provide a good match to the data and keep the error down to a minimum. For the validation set, the SVM achieved an accuracy of 0.98, showing its efficacy. Almost a straight line appears when plotting test predictions versus test with the KNN and Naïve Bayes, indicating a close match.

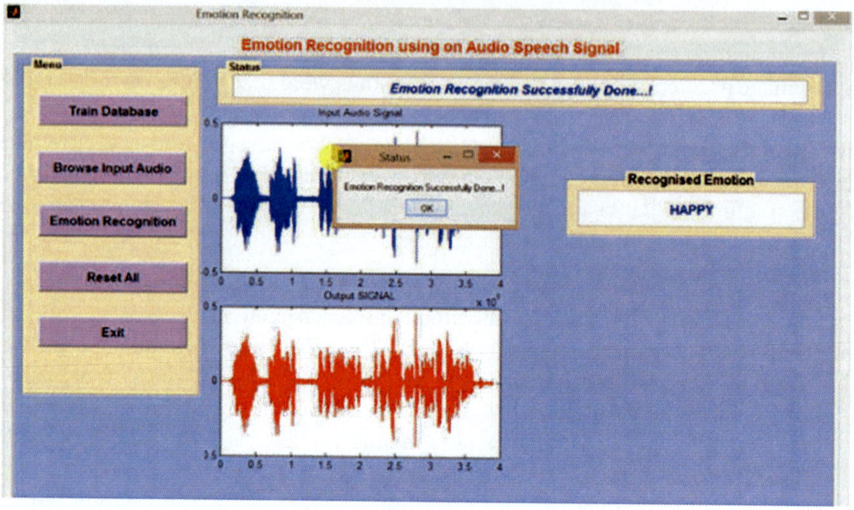

Fig. 2 Emotion recognition on audio speech signal

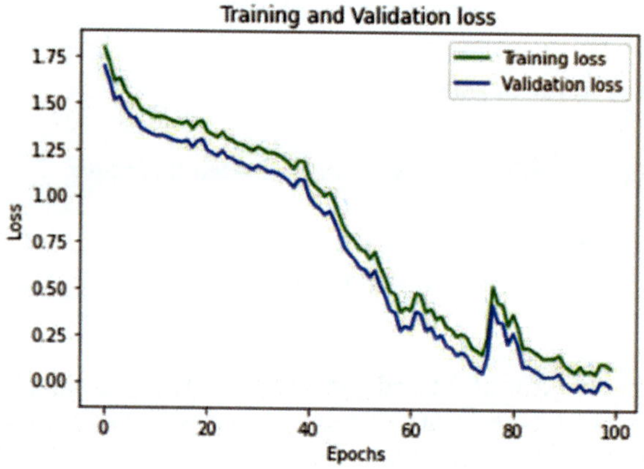

Fig. 3 Epoch versus loss

The effectiveness of the proposed technique may be shown by contrasting it with the current system [20].

Figure 7 shows that when compared to other current mechanisms, the proposed technique achieves more accurate results for emotion prediction.

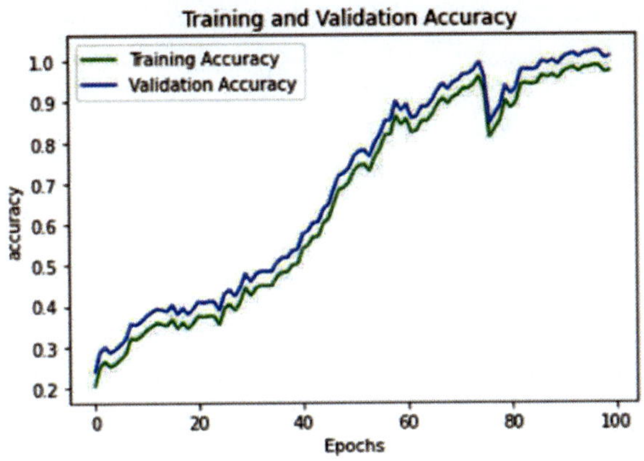

Fig. 4 Epoch versus accuracy

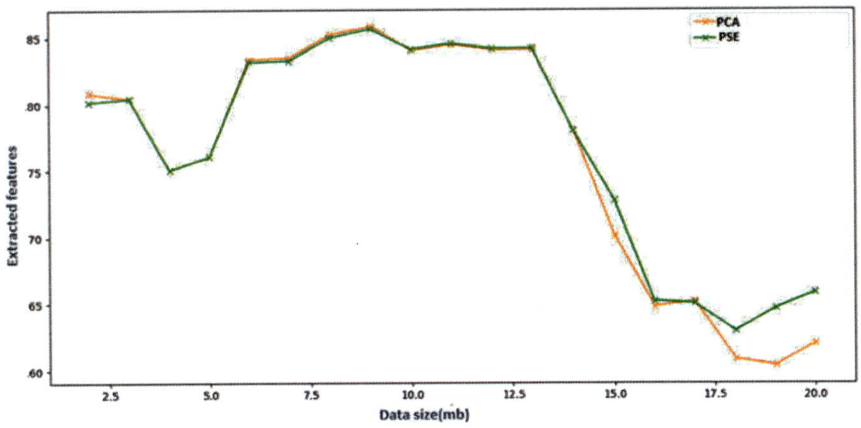

Fig. 5 Data size versus extracted features

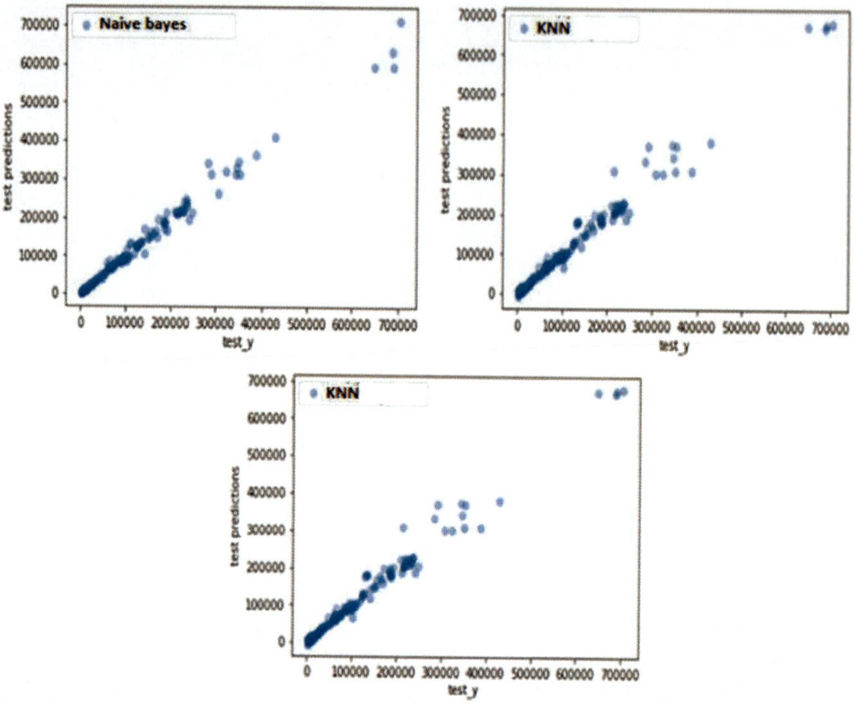

Fig. 6 Prediction rate analysis

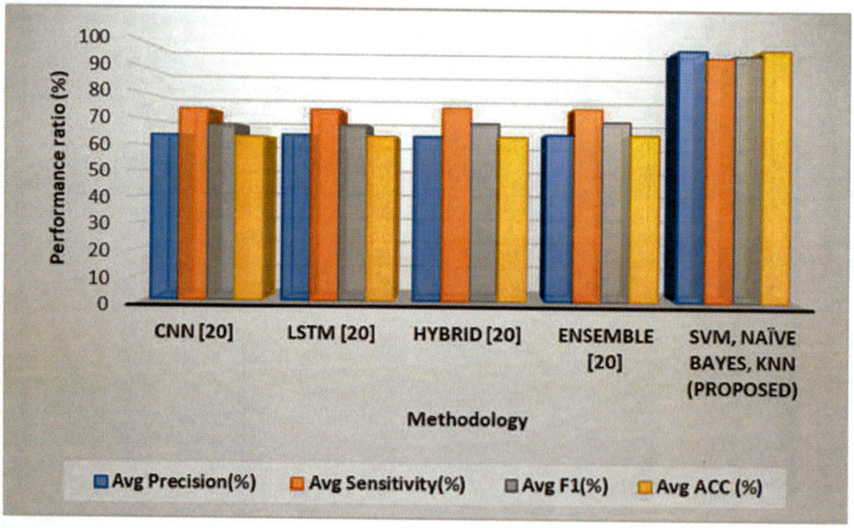

Fig. 7 Performance ratio analysis

Table 1 Comparative performance analysis

Classification methods	Avg precision (%)	Avg sensitivity (%)	Avg F1 (%)	Avg ACC (%)
CNN [20]	64	74	68	63.50
LSTM [20]	64.3	74	68	63.89
Hybrid [20]	64	75	69	64.02
Ensemble [20]	65	75	70	65.00
SVM, Naive Bayes, KNN (Proposed)	98	95	96	98.2

5 Conclusion

The suggested PCA with PSE and machine learning technique using KNN, Naive Bayes, and SVM model has achieved good accuracy in comparison with earlier approaches. The results of contrasting the proposed model with the current methods are given in Table 1. The table shows that the suggested strategy is superior to the current best practices. Future work might broaden the focus of this research to include additional forms of mental illness and analysis of stress factors. Participants may range from active-duty military members to IT specialists to young adults to college students and beyond. There are several evolutionary optimization strategies that might be used.

References

1. George FP, Shaikat IM, Hossain PSF, Parvez MZ, Uddin J (2019) Recognition of emotional states using EEG signals based on time-frequency analysis and SVM classifier. Int J Electr Comput Eng 9:1012–1020
2. Alhagry S, Aly A, Reda A (2017) Emotion recognition based on EEG using LSTM recurrent neural network. Int J Adv Comput Sci Appl 8:8–11
3. Yin Y, Zheng X, Hu B, Zhang Y, Cui X (2020) EEG emotion recognition using fusion model of graph convolutional neural networks and LSTM. Appl Soft Comput 100:106954
4. Cimtay Y, Ekmekcioglu E (2020) Investigating the use of pretrained convolutional neural network on cross-subject and cross-dataset EEG emotion recognition. Sensors 20:2034
5. Wei C, Chen L-L, Song Z-Z, Lou X-G, Li D-D (2020) EEG-based emotion recognition using simple recurrent units network and ensemble learning. Biomed Sig Process Control 58:101756
6. Chao H, Liu Y (2020) Emotion recognition from multi-channel EEG signals by exploiting the deep belief-conditional random field framework. IEEE Access 8:33002–33012
7. Pandey P, Seeja K (2019) Subject independent emotion recognition from EEG using VMD and deep learning. J King Saud Univ Comput Inf Sci 11:1–9
8. Misulis KE, Abou-Khalil B, Sonmezturk H, Ess KC (2022) Atlas of EEG, seizure semiology, and management. Oxford University Press, Oxford
9. Michielli N, Acharya UR, Molinari F (2019) Cascaded LSTM recurrent neural network for automated sleep stage classification using single-channel EEG signals. Comput Biol Med 106:71–81. https://doi.org/10.1016/j.compbiomed.2019.01.013

10. Qazi E-U-H, Hussain M, Aboalsamh H, Ullah I (2019) Automatic Emotion Recognition (AER) system based on two-level ensemble of lightweight deep CNN models. arXiv [Preprint]. https://doi.org/10.48550/arXiv.1904.13234
11. Korotkova T, Ponomarenko A, Monaghan CK, Poulter SL, Cacucci F, Wills T et al (2018) Reconciling the different faces of hippocampal theta: the role of theta oscillations in cognitive, emotional and innate behaviors. Neurosci Biobehav Rev 85:65–80. https://doi.org/10.1016/j.neubiorev.2017.09.004
12. Newson JJ, Thiagarajan TC (2019) EEG frequency bands in psychiatric disorders: a review of resting state studies. Front Hum Neurosci 12:521. https://doi.org/10.3389/fnhum.2018.00521
13. Ramzan M, Dawn S (2019) Learning-based classification of valence emotion from electroencephalography. Int J Neurosci 129:1085–1093
14. Fitzgerald JM, Gorka SM, Kujawa A, DiGangi JA, Proescher E, Greenstein JE et al (2018) Neural indices of emotional reactivity and regulation predict course of PTSD symptoms in combat-exposed veterans. Prog Neuropsychopharmacol Biol Psychiatry 82:255–262
15. Suhaimi NS, Mountstephens J, Teo J (2020) EEG-based emotion recognition: a state-of-the-art review of current trends and opportunities. Comput Intell Neurosci 2020:8875426
16. Topic A, Russo M (2021) Emotion recognition based on EEG feature maps through deep learning network. Eng Sci Technol Int J 24:1442–1454. https://doi.org/10.1016/j.jestch.2021.03.012
17. Wang J, Wang M (2021) Review of the emotional feature extraction and classification using EEG signals. Cogn Robot 1:29–40. https://doi.org/10.1016/j.cogr.2021.04.001
18. Pereira ET, Gomes HM, Veloso LR, Mota MRA (2018) Empirical evidence relating EEG signal duration to emotion classification performance. IEEE Trans Affect Comput 12:154–164. https://doi.org/10.1109/TAFFC.2018.2854168
19. Liu J, Wu G, Luo Y, Qiu S, Yang S, Li W et al (2020) EEG-based emotion classification using a deep neural network and sparse autoencoder. Front Syst Neurosci 14:43. https://doi.org/10.3389/fnsys.2020.00043
20. Iyer A, Das SS, Teotia R, Maheshwari S, Sharma RR (2023) CNN and LSTM based ensemble learning for human emotion recognition using EEG recordings. Multimedia Tools Appl 82(4):4883–4896

Semantic Image Segmentation of Agricultural Field Problem Areas Using Deep Neural Networks Based on the DeepLabV3 Model

Aleksey Rogachev, Ilya Belousov, and Dmitry Rogachev

Abstract In agricultural production, periodic monitoring and determination of the state of crops are required. An additional complexity is the multicomponent structure of the analyzed images. For the purpose of intelligent analysis of agricultural fields, RGB image databases were formed and divided into four classes. An analysis of variants of deep neural network (DNN) architectures based on U-net is carried out. DNN has been developed for the segmentation of problem areas in fields. An experimental quality assessment was carried out on the DeepLabV3 architecture in combination with ResNet50. The constructed DNN family based on DeepLabV3 with ResNet50 showed efficiency of distribution and sufficient speed in segmentation of fragments of agricultural fields and determination of the state of crops. It is established that the increase in accuracy of segmentation of agricultural field images according to the "dice coefficient" criterion is limited by the resolution and quality of manual marking of images.

Keywords Agricultural culture · Neural network recognition · Segmentation · Artificial neural networks · U-net architecture

1 Introduction

The assessment of the state of crops in agricultural fields over large areas is quite a labor-intensive task [1]. Identification of the biological state of agrophytocenoses during their vegetation is possible using artificial intelligence (AI) methods. Among the AI methods that can be used for computer neural network detection of the state

A. Rogachev · I. Belousov
26 Universitetskiy Ave., Volgograd 400002, Russian Federation

D. Rogachev (✉)
VNIIGiM named after Kostyakov, Moscow, Russian Federation
e-mail: rogachev.soft@gmail.com

S. Shakya et al. (eds.), *Fourth International Conference on Image Processing and Capsule Networks*, Lecture Notes in Networks and Systems 798,
https://doi.org/10.1007/978-981-99-7093-3_30

of crops, it is possible to distinguish the semantic segmentation of their color images [2, 3] by the required classes.

Input information flows are obtained by remote sensing methods and using data from other functional systems, including unmanned aerial vehicles [4, 8]. Earth remote sensing technologies are widely used in the study of the natural environment, soil reclamation, meteorology, and other fields. These technologies are used, along with unmanned aerial vehicles, artificial Earth satellites [4–7]. In the field of agriculture, accurate harvest information on the structure of plantings is important background data for optimizing and adjusting the structure of plantings and the output of agricultural subsidies [10–14].

The specific complexity of the task of assessing the condition and development of crops of various agricultural crops is the multicomponence of the structures of the analyzed multicolor images, which determines the use of a systematic approach in the process of computerized intellectual research, including the formation of databases, their preprocessing, and direct analysis (patents RU 2688234, IPC G06Q 10/06, A01G 9/00; 2723189, IPC A01G 7/00, G06N 5/00, G06N 20/00, G06Q 50/02, etc.). Solving the problems of semantic segmentation of images of agricultural fields provides forecasting of yields and improves the quality of agricultural production management [8, 9, 26].

Formulation and solution of the problem of semantic segmentation, or "pixel-by-pixel classification", allows obtaining the results demanded by the practice of agricultural production management.

At the same time, a number of scientific and methodological problems, including the methodology of selecting optimal sizes and optical resolution of images of agricultural fields, the structure of a database of multiclass images, and the justification of the optimal combination of hyperparameters of deep neural networks (DNN) for intellectual support of decision-making, require modification of known and developed approaches and solutions.

2 Materials and Methods

Methodological approaches to the construction and training of DNN were based on the recommendations of A. Sozykin, set out in the review article [22]. During development, the built-in libraries of the PyTorch framework were used, while the segmentation DNN architecture was selected from the modules available for PyTorch. In the course of research, the accuracy of the compared architectures was evaluated based on the "COCO train2017" dataset (http://cocodataset.org/#home).

To test the hypotheses put forward, a network of the DeepLabV3 ResNet50 architecture was chosen, which is characterized by an accuracy comparable to DeepLabV3 ResNet101 but, at the same time, learns significantly faster. Architecturally, it is a DeepLabV3 based on ResNet50, which contains a sequence of convolutional layers where each subsequent result will be combined with the previous one.

The well-known segmentation quality criteria "Dice coefficient" (1) were used

$$\mathrm{DSC} = \frac{2|X \cap Y|}{|X| + |Y|} \tag{1}$$

and «Jaccard coefficient» (2)

$$J(A, B) = \frac{|A \cap B|}{|A \cup B|} \tag{2}$$

DNN's training was implemented on a graphics accelerator using CUDA technology. The Nvidia RTX 2080Ti graphics processor supported by the CUDA Toolkit 11.4 library was used. Various types of image augmentation were used to provide DNN's training; e.g., the method allowed diversification diversify the database and dataset.

3 Results and Discussion

3.1 Formation of Datasets and Architectures for Solving the Problem of Segmentation of Images of Agricultural Fields

To form a database and datasets when solving the segmentation problem, color images of agricultural fields with a dimension of 200 × 200 pixels obtained using UAVs were used. RGB color images were marked up into four classes: high-quality field, defective field, cultivated field (not shown), and non-field (other objects). When marking up the dataset, the segmenting mask was not superimposed on the original image, but was formed in a graphical editor.

To improve the quality of deep DNN training, in 2015, researchers S. Ioffe and Ch. Szegedy proposed the use of special batch normalization regulation layers in neural networks [23]. As is known, learning of the method of error back propagation converges faster in the case when the input data is normalized to the normal distribution, when the mathematical expectation is zero and the variance is one [24]. However, during the propagation of information signals through a neural network, both the mathematical expectation and the variance can change significantly, which negatively affects the learning outcomes. The researchers mentioned above proposed to perform normalization not only at the entrance to the neural network, but also before each of its layers.

When implementing various modifications of the stochastic gradient descent method, it is possible to separately normalize data for each data packet using mini-batch. In the case when the batch contains m elements of input data x_i $\{x_1, \ldots, x_m\}$, the normalized values are determined by the formula (3):

$$\widehat{x_i} = \frac{x_i - \mu_B}{\sqrt{\sigma_B^2 - \upsilon}} \tag{3}$$

where μ_B is the average value of the data in the batch, σ is the variance, and ε is the stabilizing constant.

After that, scaling and displacement are performed according to the dependence (4)

$$y_i = \gamma \widehat{x_i} + \beta, \tag{4}$$

where y_i is the resulting value, γ and β are the parameters that are determined in the learning process.

Batch normalization is implemented in the form of DNN layers, which can be located in the required places of the structure in the neural network, while they can be used multiple times. In addition, a positive consequence of using batch normalization may be the reduction of epochs and training time, as well as the elimination of the undesirable phenomenon of retraining.

Note that the layers of batch normalization, as well as the methods of normalized selection of initial weights when initializing the DNN training process, help to get rid of the problem of disappearing gradient when training neural networks of deep architecture containing dozens of hidden inner layers. Consequently, the problem of the vanishing gradient, according to some researchers, can now be considered solved [24].

Previously, the authors' studies were based on the classification of the specified images with a size of 200 × 200 pixels, which gave only a general idea of the problem, for the case when the detected problem of the field area had a sufficiently large size. The complexity of image processing with this approach increases quadratically with an increase in the linear size of the images, which greatly increases the requirements for the hardware. The use of a neural network based on the segmentation approach (Fig. 1) allowed to reduce the load on processors due to the fact that it became possible to compress images.

The U-Net, Siamese neural network (SNN), and DeepLabV3 architectures use cascades of convolutional CNN layers [17, 25] with 3 × 3 convolution cores to highlight specific "features" of images that can be used to solve the segmentation problem (Fig. 3).

After multiple compressions of images with convolutional layers, both described neural networks form a "new" segmented image from images, obtained at various stages of processing the original image, which once again correlates with the final result of the backbone.

In the experiments conducted, the training of the DNN was carried out both for compressed images and without it, by splitting the original image to 500 × 500 pixels.

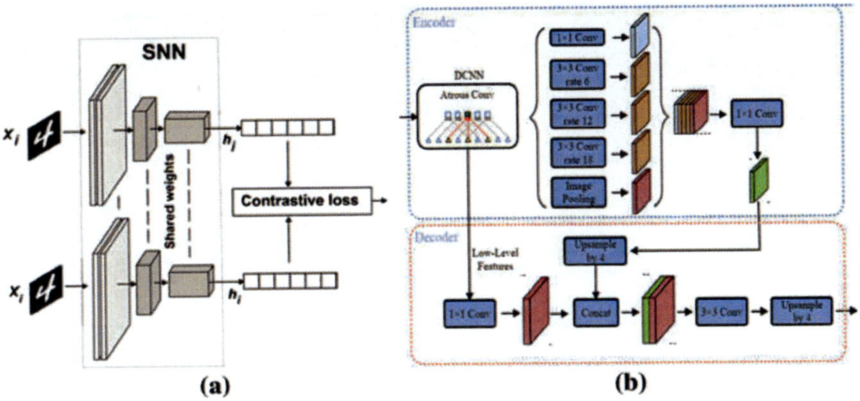

Fig. 1 DNN segmentation architectures: **a** SNN-Net; **b** DeepLabV3

A fragment of a program developed in Python, including loading the used libraries of the PyTorch framework (Fig. 2), as well as marked-up input images, is presented below.

U-Net and DeepLabV3 were used as the architecture in the search experiments (Fig. 3).

3.2 Metrics and Results of Segmentation of Images of Agricultural Fields

Additionally, Intersection over Union (IoU) indicators were used to characterize the quality of recognition on a test sample for various DNNs architectures.

Below is a script fragment (Fig. 4) of the implementing the learning algorithm of the developmental program.

Preliminary experiments have shown that the architecture of DeepLabV3 ResNet50 and DeepLabV3 ResNet101 demonstrates a higher quality of work compared to FCN ResNet101.

Typical graphs of the change in the loss function depending on the number of processed image packets are shown in Fig. 5. The learning process is characterized by a relatively monotonous decrease in the average values of the loss function (Average Loss) with significant amplitude jumps of the described function itself.

This abrupt nature of the change in the loss function indicators during the prosizing of various baths is due to the use of stochastic gradient descent and the significant heterogeneity of the colorographic parameters of the processed color images of agricultural fields.

The highest Average Loss values are 1.5 at the beginning of the learning process. After processing approximately 4000 batches, the 0.5 Average Loss values drop to

```
import torch
import numpy as np
import matplotlib.pyplot as plt
from torch import nn
from torch import optim
from torchvision import datasets, transforms

def load_split_train_test(datadir, valid_size=.2):
    train_transforms = transforms.Compose([transforms.Resize(256),
                                           transforms.CenterCrop(224),
                                           transforms.ToTensor(),
                                           transforms.Normalize(mean=[0.5, 0.5, 0.5], std=[0.5,
0.5, 0.5])])
    test_transforms = transforms.Compose([transforms.Resize(256),
                                          transforms.CenterCrop(224),
                                          transforms.ToTensor(),
                                          transforms.Normalize(mean=[0.5, 0.5, 0.5], std=[0.5,
0.5, 0.5])])
    train_data = datasets.ImageFolder(datadir, transform=train_transforms)
    test_data = datasets.ImageFolder(datadir, transform=test_transforms)
    num_train = len(train_data)
    indices = list(range(num_train))
    split = int(np.floor(valid_size * num_train))
    np.random.shuffle(indices)
    from torch.utils.data.sampler import SubsetRandomSampler
    train_idx, test_idx = indices[split:], indices[:split]
    train_sampler = SubsetRandomSampler(train_idx)
    test_sampler = SubsetRandomSampler(test_idx)
    trainloader_t = torch.utils.data.DataLoader(train_data, sampler=train_sampler,
batch_size=128)
    testloader_t = torch.utils.data.DataLoader(test_data, sampler=test_sampler,
batch_size=128)
    return trainloader_t, testloader_t
```

Fig. 2 PyTorch framework

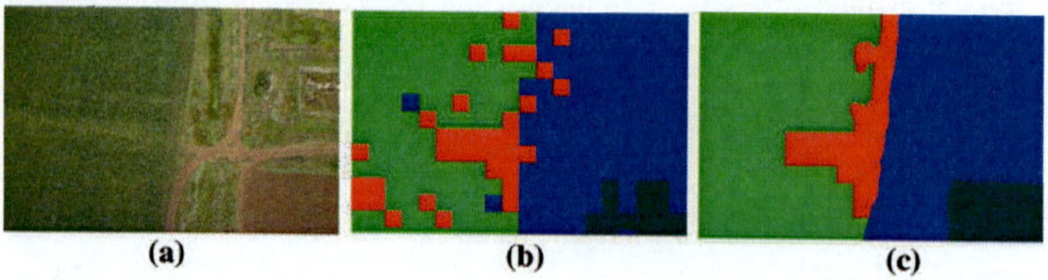

Fig. 3 Images of agricultural fields. **a** Initial, **b** marked; **c** segmented

```
data_dir = 'fields'
trainloader, testloader = load_split_train_test(data_dir, .2)
device = torch.device("cuda" if torch.cuda.is_available()
                else "cpu")

print(device)
model = torch.hub.load('facebookresearch/WSL-Images',
'resnext101_32x16d_wsl')
for param in model.parameters():
   param.requires_grad = False

model.fc = nn.Sequential(nn.Linear(2048, 1000, True),
                nn.ReLU(),
                nn.Dropout(0.2),
                nn.Linear(1000, 10),
                nn.LogSoftmax(dim=1))
criterion = nn.NLLLoss()
optimizer = optim.Adam(model.fc.parameters(), lr=0.003)
model.to(device)

epochs = 6
steps = 0
running_loss = 0
print_every = 10
train_losses, test_losses = [], []
for epoch in range(epochs):
   for inputs, labels in trainloader:
      steps += 1
      inputs, labels = inputs.to(device), labels.to(device)
      optimizer.zero_grad()
      logps = model.forward(inputs)
      loss = criterion(logps, labels)
      loss.backward()
      optimizer.step()
      running_loss += loss.item()

      if steps % print_every == 0:
         test_loss = 0
         accuracy = 0
         model.eval()
         with torch.no_grad():
            for inputs, labels in testloader:
```

Fig. 4 Script fragment

```
            inputs, labels = inputs.to(device), labels.to(device)
            logps = model.forward(inputs)
            batch_loss = criterion(logps, labels)
            test_loss += batch_loss.item()
            ps = torch.exp(logps)
            top_p, top_class = ps.topk(1, dim=1)
            equals = top_class == labels.view(*top_class.shape)
            accuracy += torch.mean(equals.type(torch.FloatTensor)).item()
        train_losses.append(running_loss / len(trainloader))
        test_losses.append(test_loss / len(testloader))
        print(f"Epoch {epoch + 1}/{epochs}.. "
              f"Train loss: {running_loss / print_every:.3f}.. "
              f"Test loss: {test_loss / len(testloader):.3f}.. "
              f"Test accuracy: {accuracy / len(testloader):.3f}")
        running_loss = 0
        model.train()

torch.save(model, 'myFirst-16-5-005.pth')
```

Fig. 4 (continued)

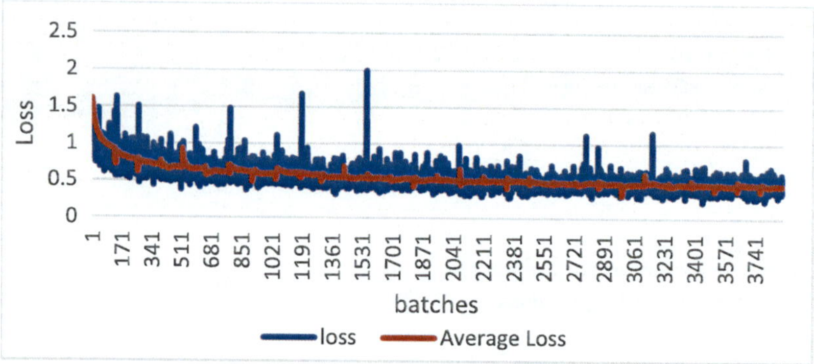

Fig. 5 Typical graphs of changes in the loss function depending on the number of processed image batches

the 0.5 level and stabilize without retraining. However, the presented packet dependence does not very clearly characterize the learning process of the developed neural network, since within each of the epochs of the processing of individual packets, only collectively determines the vectors of changes in the weight coefficients of the neurons of the trained network.

As a result of the segmentation-type DNN construction, it was possible to obtain more detailed masks for the source images (Fig. 5).

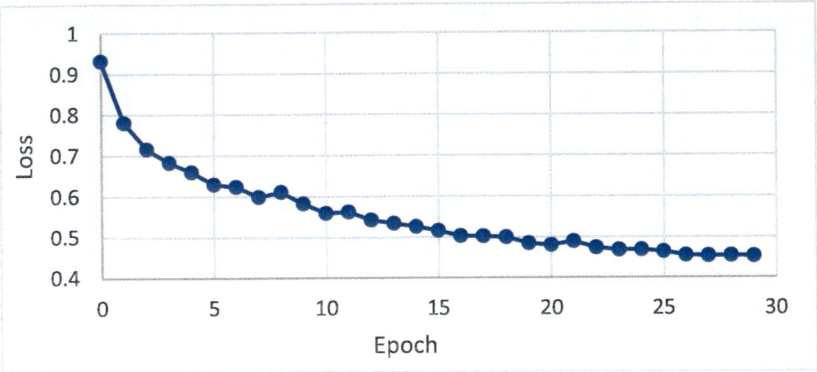

Fig. 6 Diagram of changes in the loss function "Average Loss" depending on the number of training epochs

A visual representation of the dynamics of changes in the average value of the Average Loss function, characterizing the speed of the neural network learning process depending on the number of training epochs, is shown in Fig. 6.

The highest loss values are 0.95 at the beginning of the learning process, and after about 30 epochs, the loss values stabilize without retraining at 0.45.

Further reduction requires in-depth optimization, for example, using the Karas Tuner software that implements various optimization algorithms.

The conducted study of the segmentation of problem areas of agricultural fields confirmed that the use of the considered architectures of the DeepLabV3 ResNet50 type allows us to successfully solve the problems of identifying the reclamation state of agricultural fields by the method of multiclass segmentation, as well as to identify the state of agricultural plants (Fig. 7).

Based on the developed neural network, a SaaS service has been created, the working screen of which is shown in Fig. 8. The service's working screen shows an image of the original snapshot of an agricultural field plot with a color mask superimposed on it.

The red zones of the mask denote areas of unseeded fields and non-agricultural fields in general, translucent red zones—problem areas of agricultural fields. The results of intelligent recognition and segmentation by recognized classes can be used both for expert "manual" analysis and for subsequent use of cascading neural networks or third-party services.

Analysis of the results of the use of classification-type neural networks has shown that DNN can successfully solve the problems of recognizing the reclamation state of fields and the development of crops. Comparing the results obtained with previous versions, it was possible to achieve an increase in accuracy by 9…12%. At the same time, the increase in the accuracy of recognition of classified areas of agricultural fields is limited both by the composition (quality) and the volume of the dataset, as well as its balance [8, 9].

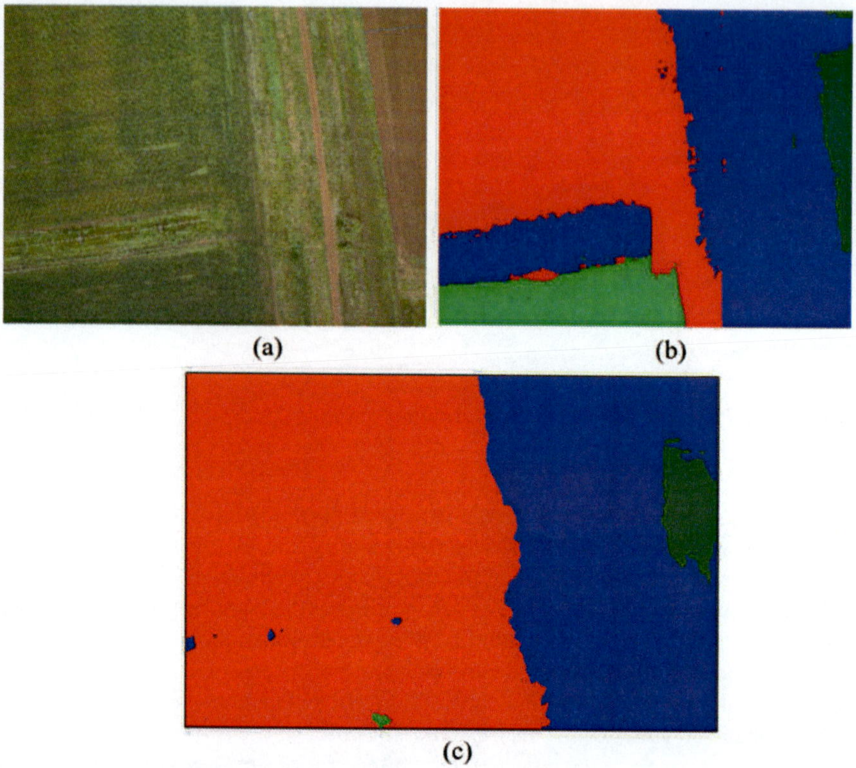

Fig. 7 Segmentation of the image of agricultural fields: **a** the original image; **b** processing with clusters of 500 × 500 size; **c** image processing when scaling up to 500 × 500

Fig. 8 Interface: recognition result with superimposed mask layers

However, when increasing the volume and resolution of images, there are a number of problems: a linearly increasing load on the processor and low accuracy of the boundaries.

The improvement of the accuracy and quality of segmentation of images of recognized agricultural fields at different stages (phases) of the vegetation cycle [9, 16] is constrained by the limited resolution and accuracy of manual markup of the dataset used for training, as well as by the hyperparameters of the DDN [3, 15, 18, 20].

As a direction for continuing research, it is planned to improve the architecture of the segmentation program. In particular, the "attention" mechanism will be implemented. The improvement of the program will provide additional opportunities for analyzing time series (BP) of retrospective images of agro-fields during vegetation [19] in order to intelligently assess the dynamics of plant development [21], including solving the problem of neural network regression.

4 Conclusions

Analysis of the results of the study of the use of neural networks for segmentation of agricultural fields showed that the DNN's of the considered DeepLabV3 architectures in combination with ResNet50 can solve the problems of segmentation and identification of the reclamation state of agricultural fields, as well as the state of plant development, while the mentioned architecture demonstrated advantages.

The increase in the accuracy of segmentation of images of recognized agricultural fields is constrained by the limited resolution and accuracy of the manual markup of the dataset used for training.

Further improvement of the segmentation program architecture can be ensured by the inclusion of the "attention" mechanism in the architecture of the developed DNN's.

The improvement of DNN will provide additional opportunities for the analysis of VR images of agricultural fields during the growing season in order to intelligently assess the dynamics of plant development.

Acknowledgements The publication has been prepared with the financial support of the RFBR under project No. 20-37-90142.

References

1. Saiz-Rubio V From smart farming towards agriculture 5.0: a review on crop data management. https://www.mdpi.com/2073-4395/10/2/207/htm
2. Alberto G-G, Sergio O-E, Sergiu O, Victor V-M, Jose G-R A review on deep learning techniques applied to semantic segmentation. https://doi.org/10.48550/arXiv.1704.06857
3. Badrinarayanan V, Kendall A, Cipolla R SegNet: a deep convolutional encoder-decoder architecture for image segmentation. http://docs.cntd.ru/document/902361843

4. Xu B et al (2004) Remote sensing monitoring on dynamic status of grassland productivity and animal loading balance in Northern China. In: IGARSS 2004. 2004 IEEE international geoscience and remote sensing symposium, Anchorage, AK, USA, vol 4, pp 2306–2309. https://doi.org/10.1109/IGARSS.2004.1369747
5. Pelletier C, Valero S, Inglada J, Champion N, Dedieu G (2016) Assessing the robustness of Random Forests to map land cover with high resolution satellite image time series over large areas. Rem Sens Environ 187:156–168
6. Zhang L, Yu W, Li G, Zhang H (2016) An approach for flood inundated duration extraction based on Level Set Method using remote sensing data. In: 2016 IEEE international geoscience and remote sensing symposium (IGARSS), Beijing, China, pp 1820–1822. https://doi.org/10.1109/IGARSS.2016.7729467
7. Zeng Z-C, Wang Y, Pongetti TJ, Gong F-Y, Newman S, Li Y, Natraj V, Shia R-L, Yung YL, Sander SP (2020) Tracking the atmospheric pulse of a North American megacity from a mountaintop remote sensing observatory. Rem Sens Environ 248:112000
8. Tokarev KE et al (2015) Crops reclamation management based on hybrid neuro-fuzzy systems. IOP Conf Ser-Earth Environ Sci 421:042015. https://doi.org/10.1088/1755-1315/421/4/042015
9. Melikhova EV, Rogachev AF (2019) Computer simulation and optimization of parameters of configuration of the contour of moistening under drip irrigation of agricultures. In: Popkova EG (ed) Ubiquitous computing and the internet of things: prerequisites for the development of ICT. Studies in computational intelligence, vol 826, pp 1193–1201. https://doi.org/10.1007/978-3-030-13397-9_122
10. Soloviev RA, Telpukhov DV, Kustov AG (2017) Automatic segmentation of satellite images based on the modified convolutional neural network UNET. Eng Bull Don (4)
11. Jeong S, Ko J, Yeom JM (2021) Predicting rice yield at pixel scale through synthetic use of crop and deep learning models with satellite data in South and North Korea. Sci Total Environ 802(2):149726
12. Xu J, Yang J, Xiong X et al (2021) Towards interpreting multi-temporal deep learning models in crop mapping. Rem Sens Environ 264:112599. https://doi.org/10.3390/rs13224668
13. Ofori-Ampofo S, Pelletier C, Lang S (2021) Crop type mapping from optical and radar time series using attention based deep learning. Rem Sens 13:4668
14. Kussul N, Lavreniuk M, Skakun S et al (2017) Deep learning classification of land cover and crop types using remote sensing data. IEEE Geosci Rem Sens Lett (99):1–5
15. Alekseev AV, Rozaliev VL, Orlova YA, Zaboleeva-Zotova AV (2016) Context-sensitive image analysis for coloring nature images. Adv Intell Syst Comput 451:133–141
16. He K, Zhang X, Ren X, Sun J (2016) Deep residual learning for image recognition. In: Proceedings of the 29th IEEE conference on computer vision and pattern recognition—CVPR, Las Vegas, Nevada, USA, pp 770–778
17. Fezan Review DeepLabv3 (Semantic Segmentation) (2020). https://medium.com/swlh/review-deeplabv3-semantic-segmentation-52c00ddbf28d
18. Adrian R Intersection over Union (IoU) for object detection. https://www.pyimagesearch.com/2016/11/07/intersection-over-unioniou-for-object-detection
19. Seeland M, Mäder P (2021) Multi-view classification with convolutional neural networks. PLoS ONE 16(1):e0245230. https://doi.org/10.1371/journal.pone.0245230
20. Tsang S-H Review: DeepLabv3. Atrous Convolution (Semantic Segmentation). https://towardsdatascience.com/review-deeplabv3-atrous-convolution-semantic-segmentation-6d818bfd1d74
21. Rogachev AF, Belousov IS (2022) Neural network identification of problem areas of the state of crops by methods of artificial intelligence. In: Proceedings of the Lower Volga Agro-University Comp, vol 3(67), pp 459–466. https://doi.org/10.32786/2071-9485-2022-03-52
22. Sozykin AV (2017) Review of deep neural network training methods. Bull SUSU Ser Comput Math Comput Sci 6(3):28–59. https://doi.org/10.14529/cmse170303
23. Ioffe S, Szegedy C (2015) Batch normalization: accelerating deep network training by reducing internal covariate shift. In: JMLR workshop and conference proceedings. Proceedings of the

32nd international conference on machine learning, Lille, France, 06–11 July 2015, vol 37, pp 448–456

24. He K, Zhang X, Ren S et al (2016) Deep residual learning for image recognition. In: 2016 IEEE conference on computer vision and pattern recognition, Las Vegas, NV, USA, 27–30 June 2016, pp 770–778. https://doi.org/10.1109/CVPR.2016.90
25. Utkin L, Kovalev M, Kasimov E (2021) An explanation method for Siamese neural networks. In: Voinov N, Schreck T, Khan S (eds) Proceedings of international scientific conference on telecommunications, computing and control. Smart innovation, systems and technologies, vol 220. Springer, Singapore. https://doi.org/10.1007/978-981-33-6632-9_19
26. Sabour S, Fross N, Hinton GE (2017) Dynamic routing between capsules. In: Neural information processing systems (NIPS)

A Unified Deep Learning Model for Multi-Satellite Image Classification of Land Use and Land Cover

M. S. Babitha, A. Diana Andushia, and A. Mehathab

Abstract Satellite images act as windows to the world, providing us with detailed snapshots of Earth's land surface. Utilizing various spectral bands, sensors aboard satellites detect and record information about different surface features, including vegetation, water bodies, urban areas, and barren land. By analyzing the spatial distribution and spectral characteristics of these features, scientists can classify and map land cover types with remarkable precision. The Landsat 8 satellites capture multispectral imagery with an image feature of 30 m for the optical spectrum, along with a thermal band at 100 m resolution. The revisit time for global coverage is approximately 16 days. The Sentinel program, operated by the ESA, includes multiple satellites such as S2 and S1. Sentinel-2 captures imagery at an analysis rate of 10 m for the optical spectrum, with repeatedly visiting in five days. S1 provides SAR data with an image analysis rate of 10–40 m, allowing imaging regardless of weather conditions or daylight. This paper proposes a deep learning model which will process different satellite imagery, i.e., a single DL model which process different satellite images for LULC identification. Here, the satellite images form S1, S2, and L8 for the year 2020 are processed using a single convolutional neural network model (CNN) for classification of LULC. Caatinga Forest Region, Brazil, was taken as the study area which includes three states Rio Grande Do Norte, Paraiba and Pernambuco. Deep neural networks are complex and powerful computational models inspired by the intricate structure and functionality of the human brain. These networks consist of multiple layers of interconnected artificial neurons, each performing its own calculation and contributing to the overall learning and decision-making process. Accuracy and Kappa analysis was done with each satellite images and combination of different

M. S. Babitha (✉) · A. Diana Andushia
Department of Electronics and Communication Engineering, Karunya Institute of Technology and Sciences, Coimbatore, India
e-mail: babitham@karunya.edu.in

A. Diana Andushia
e-mail: diana@karunya.edu

A. Mehathab
Data Scientist with Specialization in Remote Sensing, Expertz Lab Technologies, Kochi, Kerala, India

S. Shakya et al. (eds.), *Fourth International Conference on Image Processing and Capsule Networks*, Lecture Notes in Networks and Systems 798,
https://doi.org/10.1007/978-981-99-7093-3_31

satellite images using Google earth engine analysis. The detail study will boost the land cover classification task.

Keywords Deep learning · Convolutional neural network · Sentinel-1 · Sentinel-2 · Landsat 8 · Land cover classification

1 Introduction

Over time, Earth's land cover undergoes a continuous process of transformation. Human activities, natural processes, and climate change contribute to the altering face of our planet. Satellite sensors capture images from space, offering a synoptic view of large areas of land. These images provide consistent and systematic coverage, allowing for regular updates and monitoring of LULC. Satellite images play a vital role in monitoring and detecting land cover changes, enabling us to understand the drivers behind these transformations and their consequences [1]. By comparing images acquired at different times, scientists can identify areas of deforestation, urban expansion, agricultural changes, and shifts in vegetation cover. These observations provide crucial insights into the dynamic interplay between human societies and the environment. LULC identification is a method of dissecting vast landscapes into distinct categories, discerning the diverse elements that compose the terrestrial canvas. By employing advanced technologies and sophisticated algorithms, land cover classification endeavors to decode the language of nature, translating the visual manifestation of land features into actionable information.

Satellite images play a crucial role in land cover classification, offering valuable insights into the Earth's surface and its changes over time [2]. Specifically, Landsat 8, Sentinel-1, and Sentinel-2 satellites have made significant contributions to this field. Landsat 8, operated by NASA and the US Geological Survey (USGS), captures multispectral images of the Earth's surface at high spatial resolution. It collects data across different frequency range, optical spectrum, etc. These wavelength range allow for the discrimination of various land cover types based on their unique reflectance patterns. Sentinel-1, part of the ESA Copernicus program, utilizes synthetic aperture radar (SAR) technology to acquire images. SAR is particularly advantageous in land cover classification because it can penetrate cloud cover and capture data regardless of weather conditions or time of day. SAR images provide information on the backscattered radar signal, which is influenced by surface roughness, moisture content, and vegetation structure [3]. These characteristics can aid in mapping land cover types like forests, urban areas, and water bodies. Sentinel-2, another mission within the Copernicus program, operates with a multispectral sensor and provides high-resolution images across 13 spectral bands. These bands include visible, near-infrared, and shortwave infrared regions, allowing for precise land cover classification [3]. Sentinel-2 data enables the discrimination of vegetation types, identification of water bodies, monitoring of agricultural practices, and assessment of urban development.

In land cover classification, these satellite images are processed using advanced remote sensing techniques and machine learning algorithms. These methods involve preprocessing steps such as atmospheric correction, radiometric calibration, and geometric registration to ensure accurate and comparable data. Machine learning models like RF or CNN are then applied to train models on labeled data, enabling automated classification of land cover types across large areas [4, 5].

DL techniques, such as CNNs, are designed to handle multi-channel inputs, making them suitable for processing diverse satellite images. Satellite images often consist of multiple bands or channels, each representing different parts of the electromagnetic spectrum [6]. For example, satellite cameras can capture images in optical spectrum. These bands provide valuable information about various land cover characteristics, such as vegetation health, soil moisture, or urban development. To incorporate different satellite images into a single deep learning model, you can stack the image bands together to create multi-channel input data. Each band becomes a separate input channel, and the model learns to extract features from all channels simultaneously.

By training a deep learning model on multi-channel satellite images, the model can learn to leverage the complementary information from different bands and capture complex relationships between them. For instance, it can learn to identify how vegetation health (infrared band) relates to land use patterns (visible band) or how surface temperature (thermal band) correlates with urban heat island effects (infrared band) [7]. The advantage of apply a single DL architecture for processing different satellite images is that it allows the model to generalize across diverse data sources and capture broader patterns and features. It can handle various satellite sensors, resolutions, or spectral ranges, making it more versatile in analyzing different types of satellite imagery.

Convolutional neural networks (CNNs) are a powerful and innovative class of deep learning models specifically designed to process and extract meaningful features from grid-like data, such as images and time series. Inspired by the visual cortex of animals, CNNs have revolutionized computer vision tasks and have also found applications in other domains, such as natural language processing and speech recognition. The highlight of a CNN are convolutional layers, which apply filters (also known as kernels) to the input data [8]. These filters perform convolution operations, scanning the input with a small receptive field, capturing local patterns and features. By stacking multiple convolutional layers, CNNs can progressively learn more complex and abstract representations, capturing both low-level features such as edges and textures, as well as high-level features like object shapes and semantic information. In addition to convolutional layers, CNNs typically include pooling layers, which reduce the spatial dimensions of the feature maps, helping to extract the most salient features while improving computational efficiency. Pooling can be performed using operations like max pooling, which select the maximum value within a certain window, or average pooling, which takes the average value. Once the feature extraction is performed, CNNs employ fully connected layers, also known as dense layers, to transform the extracted features into class probabilities or regression outputs [9].

These layers allow the network to learn complex relationships between the extracted features and the target task, enabling accurate predictions or classifications.

The remaining sections are, Sect. 2 explains the methodology, and Sect. 3 gives the results and discussions and suggests a further direction to modify the area. The final part summarizes the conclusion.

2 Methodology

The general structure of this workflow of land cover classification architecture has shown in Fig. 1. The framework consists of individual modules, and each and every module is connected with its specific strategies. The actual modules of this structure are image database, preprocessing, feature extraction, image classification, and image segmentation.

2.1 Data Acquisition

Large number of datasets used to operate a CNN application depends on several factors, including the complexity of the task, the size and diversity of the dataset, and the desired performance of the model [10]. While there is no fixed rule, it is generally recommended to have a sufficient amount of training data to ensure effective learning and generalization of the model. In general, the more data you have, the better, as it allows the model to capture a wider range of patterns and variations in the data. Large datasets help reduce the risk of overfitting, where the model becomes too specialized in the training data and performs poorly on unseen data. This paper mainly consists of three major satellite images, S1, S2, and L8 present in the Google Earth engine dataset catalogs.

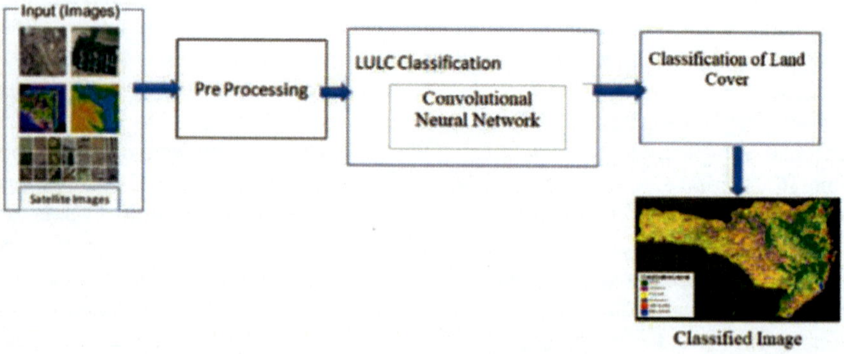

Fig. 1 Flow diagram of land cover classification

Landsat: The Landsat program, initiated by NASA in 1972, currently consists of the Landsat 8 and Landsat 9 satellites. Landsat 8 Tier 1 refers to the highest level of data quality and processing offered by the Landsat 8 satellite mission, which is a joint program between NASA and the United States Geological Survey (USGS). The Tier 1 data from Landsat 8 undergoes rigorous calibration, geometric correction, and atmospheric correction processes to ensure accuracy and consistency. This makes Landsat 8 Tier 1 data highly reliable and suitable for quantitative analysis and long-term monitoring of Earth's land surfaces. The images acquired by Landsat 8 provide valuable information for applications such as LULC identification and monitoring the differences, forest monitoring, and water resource management [11, 12]. The main used spectral bands are,

Landsat: Landsat satellites typically have a broader range of spectral bands, including visible, near-infrared, shortwave infrared, and thermal infrared. This enables precise land cover classification, vegetation monitoring, and thermal analysis for applications such as agriculture, forestry, and urban planning.

Sentinel: Sentinel satellites are equipped with a narrower set of spectral bands, focusing on the visible, near-infrared, and shortwave.

S1 SAR GRD is a satellite mission developed by the European Space Agency (ESA) that employs synthetic aperture radar technology. Unlike traditional optical sensors, SAR operates by emitting microwave pulses and measuring their reflections from the Earth's surface. This enables SAR to overcome limitations imposed by cloud cover and darkness, making it highly suitable for all-weather and day-and-night monitoring.

Sentinel-1 SAR GRD captures high-resolution radar imagery of the Earth's surface, providing valuable information about various features such as land, water bodies, and ice. The SAR images are acquired in "ground range detected" mode, which means that they are processed to produce images with a constant scale, allowing for accurate measurements of distances and surface features. SAR GRD data is widely used for applications such as land cover mapping, disaster monitoring, iceberg detection, and maritime surveillance.

Sentinel-2 MSI (Multispectral Instrument) is another satellite mission led by ESA. It is equipped with a multispectral sensor that captures images in 13 spectral bands, ranging from the visible to the near-infrared spectrum. These bands give important features of the Earth's surface composition, vegetation health, and water quality. Sentinel-2 MSI offers a unique combination of high spatial resolution (10–20 m) and frequent revisits (every five days for the entire Earth), allowing for detailed monitoring and analysis of land and coastal areas [13, 14].

The data from Sentinel-2 MSI is widely used for a variety of applications, including agriculture monitoring, deforestation assessment, urban planning, and natural resource management. The multispectral capabilities of Sentinel-2 MSI enable the creation of detailed and accurate vegetation indices, which are crucial for monitoring vegetation growth, detecting changes, and assessing environmental conditions [2].

In summary, Sentinel-1 SAR GRD offers all-weather, day-and-night imaging using synthetic aperture radar technology, while Sentinel-2 MSI provides multi-spectral data with high spatial resolution and frequent revisits. Landsat 8 Tier 1 data represents the highest quality and accuracy level for the images captured by Landsat 8, enabling detailed analysis and long-term monitoring of Earth's land surfaces. Together, these satellite missions contribute significantly to our understanding of our planet environment and support a collection of information in various fields.

2.2 *Preprocessing*

Preprocessing involves data cleaning, where missing values, outliers, or errors are identified and handled. This ensures the integrity and quality of the data, reducing the risk of biased or misleading results. It is important to preprocess and normalize the satellite images appropriately before feeding them into the deep learning model. This ensures that the data from different sources are consistent and compatible, allowing the model to learn effectively [15].

Overall, employing a single deep learning model to process different satellite images enables the extraction of valuable insights from multiple data sources, facilitating comprehensive analysis and decision-making in various fields, including environmental monitoring, agriculture, urban planning, and disaster response. Through preprocessing, the neural network can navigate the complexities of the data landscape, uncovering hidden patterns, making accurate predictions, and unleashing the full potential of deep learning in solving real-world problems [16]. For the proposed work, Sentinel-1, Sentinel-2, and Landsat satellite image datasets were used. These datasets are preprocessed for good imagery by Google and the ESA agency beforehand. To get rid of the redundant clouds in the area of study, masking is done by bitmask band which are designed within the dataset for cloud clearing. Masking is done for: Cirrus clouds, opaque clouds, snow/ice filtering, scene classification algorithm (SCL), shadows, and also to control cloud probability after masking and elimination of the above and training process.

2.3 *Models for Land Cover Classification*

The proposed CNN model with five blocks as shown in Fig. 2 typically refers to a CNN architecture composed of five main components, each representing a block of layers, while the specific configuration and design choices can vary.

Block 1: The first block consists of convolutional layers of 128 units. Convolutional layers apply filters to the input data, extracting local features. This block captures low-level features.

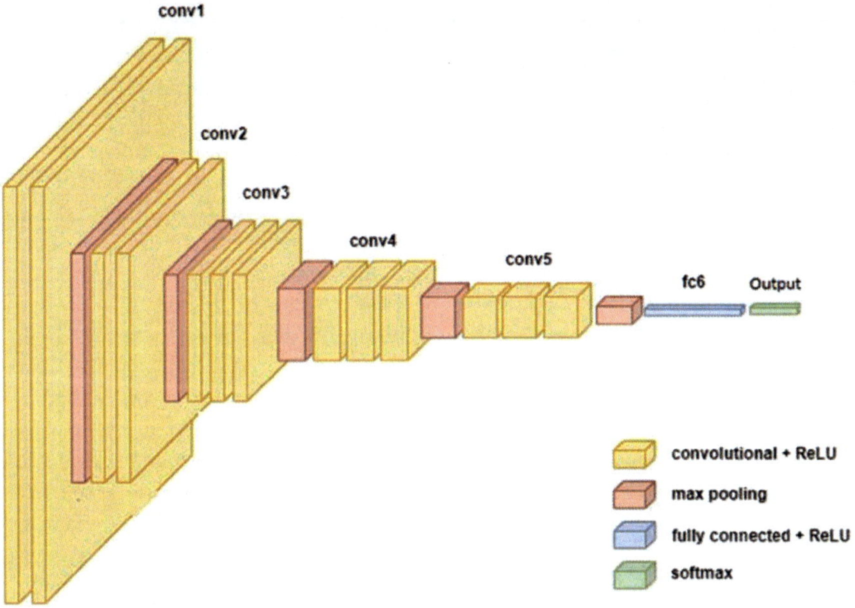

Fig. 2 Block diagram proposed CNN model

Block 2: Similar to the first block, this block contains convolutional layers of 64 units and a pooling layer. It further captures more complex features by building upon the low-level features extracted in Block 1.

Block 3: Continuing the pattern, this block contains convolutional layers of 64 units and a pooling layer. It aims to provide advanced features and patterns from the input data.

Block 4: In this block, additional convolutional layers of 32 units and pooling layers are employed. It further refines the learned features and captures even more complex patterns.

Block 5: This block consists of convolutional layers of 32 units, and these layers take the learned features from the previous blocks and transform them to the next block.

The final block typically consists of fully connected layers, also known as dense layers. These layers take the learned features from the previous blocks and transform them into class probabilities or regression outputs. The developed CNN model incorporated with batch normalization and activation in each block using activation function ReLU, each block with kernel size 3, and max pooling applied with pool size 2. Rectified linear unit (ReLU) is a widely used activation function in neural networks, known for its simplicity and effectiveness in deep learning models. ReLU has played a pivotal role in overcoming the vanishing gradient problem and accelerating the training process of deep neural networks.

In addition to the above blocks, CNN models often incorporate other layers like dropout, batch normalization, and activation functions (e.g., ReLU) for regularization and nonlinearity. The specific architecture and hyperparameters (e.g., number of filters, filter sizes, and pooling window sizes) within each block depend on the task, dataset, and design choices made by the model developer. Manually annotated training points for different regions/classes present in the area. Regions are: forest: represents dense variety of trees; plantations: crops and other type of planted trees; Croplands: land on which crops are to be grown, green in texture; fallow lands: land left without sowing for better crop cultivation, brown/muddy in texture; built-up-area: settlements, buildings, roads, and others; water bodies: sources of water. Labeling done by considering time-bound Sentinel-2 data for better understanding regions.

Classification is achieved by manually looking into the study area and making out different classes of regions. By multiple zooms in and out, confirm that the pixels are indeed of the class. When taking the historical data/previous data, load the respective satellite image with the dates about one year (i.e., for getting 2020 data, take 01-01-2020 to 31-12-2020 data), average those values for each pixel and map them out in the Google Earth engine, and this way the historical/previous year data is available for manually labeling; we have taken a total of 300 training points per classification for the study area, which is then run through the models for classification process. After this step, we apply classification with this trained model across the whole study area, giving us the finally classified/Land Use Land Cover map.

3 Result and Discussion

The wealth of information gathered from satellite images supports evidence-based decision-making and policy formulation. Governments, environmental organizations, and urban planners leverage land cover data to assess the impact of development projects, design effective land use strategies, and implement conservation measures [17]. In this proposed work, the study area taken as the Caatinga Forest Region of the state Rio Grande Do Norte, Brazil, for the year 2020, is shown in Fig. 3. To measure the performance of the proposed algorithms, calculated the validation accuracy and Kappa coefficient.

Manual label is done in Google Earth Engine; each class is marked by a total of 300 points across the state. Bands and data used other than specified are ALOS DSM: Global 30 m; slope and elevation are extracted from this dataset; this is commonly added to all three datasets. The final results of the model are obtained by fine tuning of the hyperparameters. Classification map generated by the developed CNN models with different satellite images for visual comparison is illustrated in Fig. 4.

The training dataset is made for the purpose of training the model and then to classify the whole of the study area using model. Testing dataset is made for the sole purpose of finding the accuracy of the model with the given accuracy metrics. Training points to testing points are: 80% training and 20% testing. Metrics used

Fig. 3 Study area (Caatinga Forest region, Brazil)

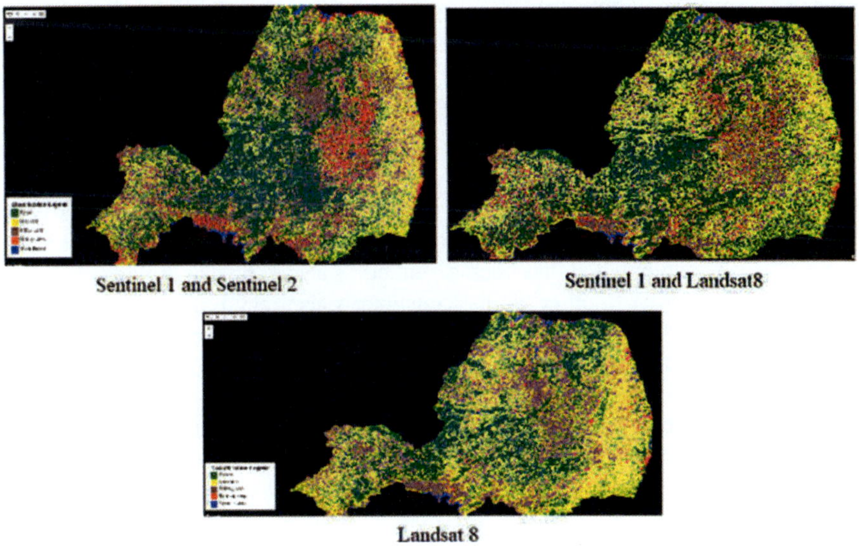

Fig. 4 Classification maps generated by proposed model with different satellite image datasets

for training the model are: optimizer used Adam; learning rate = 0.005; loss function: categorical cross entropy. Resultant metrics are accuracy, Kappa coefficient, producer's accuracy, and consumer's accuracy. Software used in the whole process is the Google Earth engine for dataset collection, manual labeling, and displaying the final classification of the study areal, and Jupyter notebook was used for training the model.

To test the developed CNN model, training and validation accuracy were simulated and observed that the developed model gives high accuracy of 97.07%, and Kappa coefficient of 96.33 is obtained with the combined S1 and S2 datasets in classifying the land cover. With the combined S1 and L8, accuracy of 94.44% and Kappa coefficient 93.04 are obtained. The detail results including producer's accuracy and consumer's accuracy obtained for the proposed model is given in Table 1.

Table 1 Details of results obtained from different datasets

Satellite imagery	Spatial resolution (m)	Year	No. of training points per class	Test accuracy	Producer's accuracy	Consumer's accuracy	Kappa coefficient
Sentinel1 SAR GRD + Sentinel-2 MSI	10	2020	300	97.07	96.49	100	96.33
Landsat 8 SR tier 1	30	2020	300	91.8	95.8	92.48	89.74
Landsat 8 SR tier 1 + Sentinel1 SAR GRD	30	2020	300	94.44	96.15	92.2	93.04

Consumer's Accuracy: Consumer's accuracy, also known as user's accuracy or user's reliability, is an evaluation metric used in classification tasks, particularly in the context of remote sensing or land cover mapping. It measures the probability that a pixel or sample, which has been classified as a specific class, indeed belongs to that class in reality. In other words, consumer's accuracy reflects the reliability of the model's predictions from the perspective of the end-user or consumer of the classification results. It helps determine the trustworthiness and usefulness of the model's outputs for decision-making and practical applications.

Producer's Accuracy: Producer's accuracy, also referred to as producer's reliability, is another evaluation metric commonly used in classification tasks, especially in remote sensing applications. It quantifies the probability that a pixel or sample, which truly belongs to a specific class, is correctly identified and classified as that class by the model. Producer's accuracy provides insights into the model's ability to accurately detect and label specific classes or land cover types. It helps assess the reliability and precision of the model's classifications from the perspective of the data producer or the one responsible for generating the ground truth data [18, 19].

Test accuracy measures the overall characterization of a model on unseen test data, consumer's accuracy indicates the reliability of the model's predictions from the user's perspective, and producer's accuracy assesses the accuracy and precision of the model's classifications from the data producer's viewpoint. Together, these metrics provide a comprehensive evaluation of the model's effectiveness, usability, and reliability in classification tasks, particularly in the field of remote sensing and land cover mapping [5].

Figure 5 shows the teaching and verification accuracy of the developed NN architecture for categorizing the land cover. The increase in training and validation accuracy over multiple epochs gives strong evidence that the proposed model has effectively learned the patterns and features necessary for accurate classification of land cover.

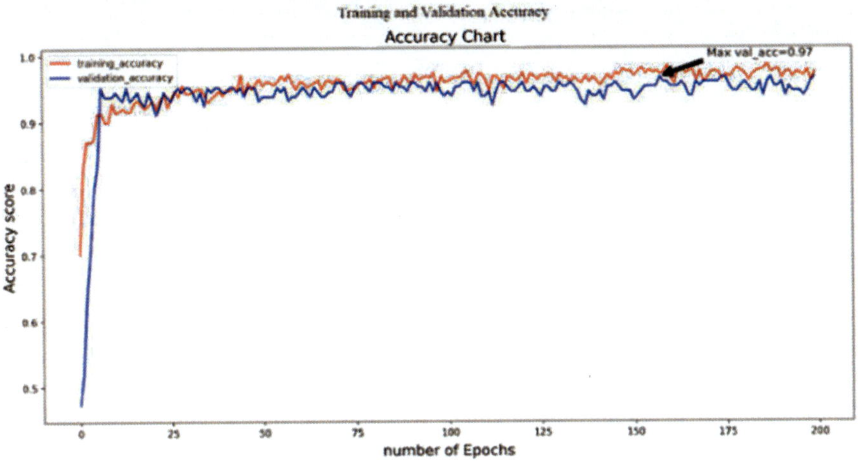

Fig. 5 Training and validation accuracy

Validation Accuracy

The validation accuracy is a measure of how well a machine learning model is performing on a validation set. The formula for calculating validation accuracy is as follows:

$$\text{Validation Accuracy} = \frac{(\text{Number of Correct Predictions on Validation Set})}{(\text{Total Number of Predictions on Validation Set})} \quad (1)$$

Validation accuracy is always expressed as percentage.

4 Conclusion

Satellite images provide us with a unique perspective on Earth's land cover, unlocking a wealth of knowledge about our changing planet. From the obtained results, we can say that a single DL model designed to handle different satellite images can generalize well across various datasets. It can learn common patterns and features from different sources, making it versatile and adaptable to different satellite sensors or data acquisition platforms. This reduces the need to develop and maintain multiple specialized models for each satellite source. Building a single DL model eliminates the need for redundant model development and deployment efforts. It saves computational resources, memory, and storage space that would otherwise be required for maintaining multiple models for different satellite images. As we continue to explore the frontiers of technology, satellite imagery will undoubtedly remain an invaluable tool in unraveling the secrets of Earth's land cover tapestry.

References

1. Ienco D, Interdonato R, Gaetano R, Minh DHT (2019) Combining Sentinel-1 and Sentinel-2 satellite image time series for land cover mapping via a multi-source deep learning architecture ISPRS J Photogramm Rem Sens 158:11–22
2. Talukdar S, Singha P, Mahato S, Shahfahad, Pal S (2021) Land-use land-cover classification by machine learning classifiers for satellite observations—a review. Rem Sens 12:1135. https://doi.org/10.3390/rs12071135
3. Shafique A, Cao G, Khan Z, Asad M, Aslam M (2022) Deep learning based change detection in remote sensing 48 images: a review. Rem Sens 14:871
4. Helber P, Bischke B, Dengel A, Borth D (2017) Eurosat: a novel dataset and deep learning benchmark for land use and land cover classification. arXiv preprint arxiv:1709.00029
5. Leitloff J, Riese FM (2018) Examples for CNN training and classification on Sentinel-2 data. Zenodo, Geneve, Switzerland
6. Balaniuk R, Isupova O, Reece T (2020) Article mining and tailings dam detection in satellite imagery using deep learning. Sensors 20:6936. https://doi.org/10.3390/s20236936
7. Xu X, Li W, Ran Q, Du Q, Gao L, Zhang B (2017) Multisource remote sensing data classification based on convolutional neural network. IEEE Trans Geosci Rem Sens 56:937–949
8. Lei T, Zhang Y, Lv Z, Li S, Liu S, Nandi AK (2019) Landslide inventory mapping from bitemporal images using deep convolutional neural networks. IEEE Geosci Rem Sens Lett 2019(16):982–986
9. Bischke B, Helber P, Folz J, Borth D, Dengel A (2019) Multi-task learning for segmentation of building footprints with deep neural networks. In: Proceedings of the 2019 IEEE international conference on image processing (ICIP), Taipei, Taiwan, 22–25 Sept 2019, pp 1480–1484
10. Dong S, Zhuang Y, Yang Z, Pang L, Chen H, Long T (2020) Land cover classification from VHR optical remote sensing images by feature ensemble deep learning network. IEEE Geosci Rem Sens Lett 17(8)
11. Li H, Gong M, Zhang M, Wu Y (2021) Spatially self-paced convolutional networks for change detection in heterogeneous images. IEEE J Sel Top Appl Earth Obs Rem Sens 14:4966–4979
12. Yu Q (2014) Object-based detailed vegetation classification with airborne high spatial resolution remote sensing imagery. Photogramm Eng Rem Sens 72(7):799–811
13. Chughtai AH, Abbasi H, Karas IR (2021) A review on change detection method and accuracy assessment for land use land cover. Rem Sens Appl Soc Environ 22:100482
14. Maladkar K (2018) Overview of convolutional neural network in image classification. CNN Reference Jan 25, 2018. [Online]. Available https://www.analyticsindiamag.com/convolutionalneural-network-image-classification-overview/. Accessed 10th Feb 2019
15. Kerner HR, Wagstaff KL, Bue BD, Gray PC, Bell JF, Amor HB (2019) Toward generalized change detection on planetary surfaces with convolutional autoencoders and transfer learning. IEEE J Sel Top Appl Earth Obs Rem Sens 12:3900–3918
16. Sarker IH, Kayes ASM, Watters P (2019) Effectiveness analysis of machine learning classification models for predicting personalized context-aware smartphone usage. J Big Data 6(1):1–28
17. Deng L, Hinton G, Kingsbury B (2013) New types of deep neural network learning for speech recognition 2013, and related applications: an overview. In: Proceedings of IEEE international conference on acoustics, speech, and signal processing (ICASSP), pp 8599–8603
18. Mu L, Wang L, Wang Y, Chen X, Han W (2019) Urban land use and land cover change prediction via self-adaptive cellular based deep learning with multi sourced data. IEEE J Appl Earth Observ Rem Sens
19. Ma L, Liu Y, Zhang X, Ye Y, Yin G, Johnson BA (2019) Deep learning in remote sensing applications: a meta-analysis and review. ISPRS J Photogramm Rem Sens 152:166–177

Multi-class Plant Leaf Disease Classification on Real-Time Images Using YOLO V7

P. Sajitha, Diana A Andrushia, and S. S. Suni

Abstract Agriculture faces a significant challenge with leaf disease, which can lead to substantial decreases in crop productivity. Deep learning techniques, particularly the YOLO (You Only Look Once) algorithm, have gained popularity for effectively detecting and categorizing plants diseases in real time. This research presents a leaf disease detection and classification system utilizing YOLO v7, which benefits from a diverse and current dataset obtained from real-time field images. This dataset allows for precise training and evaluation of plant disease detection models in real-world scenarios, offering valuable insights into the effectiveness of the YOLO v7 algorithm. By training on a large collection of images of plants, the system achieves high accuracy (96%) in identifying various leaf diseases such as leaf spot, rust, and early blight. Utilizing this technology, farmers will be able to take preventative measures and decrease crop losses through early disease diagnosis and intervention due to the quick and effective object identification capabilities of YOLO v7.

Keywords Leaf disease detection · YOLO (You Only Look Once) · Deep learning

1 Introduction

Agriculture is incredibly important to a nation's economy and is essential to its entire growth. It supports human civilization by giving billions of people throughout the world food, clothes, and a place to live. Agriculture has always been essential to the development of societies because it will be helpful for people to settle down and

P. Sajitha (✉) · D. A. Andrushia
Karunya Institute of Technology and Sciences, Coimbatore, India
e-mail: sajithap@karunya.edu.in

D. A. Andrushia
e-mail: diana@karunya.edu

S. S. Suni
Ilahia College of Engineering and Technology, Ernakulam, India
e-mail: drsuniss@icect.ac.in

S. Shakya et al. (eds.), *Fourth International Conference on Image Processing and Capsule Networks*, Lecture Notes in Networks and Systems 798,
https://doi.org/10.1007/978-981-99-7093-3_32

create an ecofriendly environment. It has evolved from a basic means of sustenance to a sophisticated system that uses innovation, science, and technology [1].

The agriculture field has a number of challenges, including food insecurity, climate change, and an overreliance on synthetic insecticides. The biggest issue facing the farming sector is the spread of plant diseases, which have a big effect on crop production. If they are not recognized and treated, plant diseases that can be brought on by microorganisms, fungal infections, or additional infectious agents can spread swiftly. For efficient disease management and control, early identification and detection of leaf diseases are essential. Visual examination and manual diagnosis methods have traditionally been used by farmers and agronomists to diagnose plant diseases, but these methods can be labor-intensive and arbitrary [2, 3]. Automation of plant disease diagnosis has become more feasible because of developments in computer vision algorithms. Deep learning algorithms have shown significant achievements in the area of image-based plant disease identification, offering improved efficiency and accuracy for identifying variety of diseases in plants [4].

The YOLO algorithms play a critical role in plant leaf disease detection by efficiently and accurately identifying diseased regions on leaves. YOLO, in contrast to conventional approaches, performs bounding box regression and object detection in a single neural network run [5]. Through its real-time object detection capabilities, YOLO divides the input image into a grid, predicting bounding boxes and corresponding class probabilities for each grid cell in a single pass through the deep learning neural network. Trained on extensive datasets of healthy and diseased plant leaves, YOLO learns to recognize specific patterns associated with different diseases, enabling it to swiftly and precisely distinguish between healthy and infected areas. This speed and accuracy empower farmers and researchers to promptly identify and address leaf diseases, leading to improved crop management, reduced losses, and enhanced agricultural productivity. YOLO's ability to generalize and adapt to diverse plant species and environmental conditions further enhances its effectiveness in safeguarding plant health. In this study, a novel method for identifying leaf diseases is presented using the YOLO v7 algorithm. Using this YOLO v7 algorithm, we trained large collection of images and identified different diseases with high accuracy.

This research work has been structured as follows: The existing approach surveys are covered in Sect. 2, and fundamentals of the YOLO algorithm are covered in Sect. 3. The proposed methodology is discussed in Sect. 4. The experimental results and analysis are presented in Sect. 5. Finally, Sect. 6 marks the end of this study.

2 Literature Survey

This section provides a review on different versions of the YOLO algorithm (ranging from YOLO version 1 to version 5) which have been utilized for disease detection in plants.

The study by Morbekar and Parihar [6] used YOLO (You Only Look Once) as a real-time image recognition system to recognize different crop diseases. Their primary goal was to develop a reliable method of early disease identification that would aid in the successful management of agricultural diseases. The approach's drawback, however, is that it now only focuses on detecting diseases in entire crops, ignoring the possibility of recognizing diseases in certain crop sections or at various phases of disease development.

Srinivasan et al. [7] used YOLO v2 algorithm for detection of rice crop diseases. It employs the Darknet-53 network with 53 convolutional layers, utilizing K-means clustering to create nine anchor boxes. The 75 layered design and the use of skip connections allow for detection at various scales while reducing localization errors and maintaining a high level of accuracy with low complexity. Its increased depth and complexity may demand more computational resources and longer training time.

Shill and Rahman [8] showed that their computer vision method, which included YOLO V3 and YOLO V4 pretrained on the MSCOCO dataset, successfully addressed the difficult task of manual disease detection in thirteen plant species, encompassing seventeen disease classes from the PlantDoc dataset, yielding promising and accurate results for disease detection and classification. The MSCOCO dataset is heavily relied upon pretraining in the proposed computer vision approach using YOLO V3 and YOLO V4 for manual disease detection, which may lead to decreased performance and accuracy for certain plant species and disease classes.

Deep learning was applied in a study by Theresa and Midhun [9] to identify diseases in bell pepper plants. This work used the promising object identification model YOLO V5, which is well-known for its quickness, accuracy, and simplicity of training. They quickly detected diseases using YOLO V5, even minuscule spots on leaves, and provided bounding boxes and class probabilities. Their study's key goal was to make it easier for farmers to quickly treat diseases by identifying bacterial areas at an early stage.

In their research, Chen et al. [10] developed a disease recognition miniature on YOLO v5 in rubber trees. It detected two diseases, such as powdery mildew and anthracnose, with excellent accuracy (above 85%). Accuracy is increased by 5.4% from the initial YOLOv5 network to 70%. As a result, the augmented YOLOv5 network showed better accuracy in recognizing and categorizing rubber tree diseases, providing insightful information for disease prevention and control. Despite these successes, the enhanced YOLO v5 network's detection accuracy still has space for improvement.

Janibul et al. [11] proposed an artificial intelligence-based solution for leaf disease detection in tea plants. This research focused on using YOLO v7, the quickest single-stage method for object detection, to address the decreasing tea production caused by climate change and pest infestation. The study utilized a dataset of 4000 images, comprising five types of diseases. The detection and identification performance of YOLO V7 were evaluated using statistical metrics such as detection accuracy, precision, recall, MAP value, and F1 score. One drawback is that the suggested method presently does not take into account additional data sources, including as temperature, humidity, soil details, and information on pathogenic spores, which might improve the precision of disease identification and provide early warnings.

For the purpose of identifying diseases in plants, the YOLO algorithm is a well-liked computer vision method. In order to function, the image is divided into a grid, and bounding boxes and class probabilities are predicted for each grid cell. By training the system on sizable datasets of both healthy and diseased plant images. From the review, YOLO models (1–6) have always focused on real-time object detection by trading off some accuracy for faster inference times. YOLO v7 might introduce further architectural reforms to improve the accuracy and detection speed.

The goal of this research is to use the YOLOv7 object detection framework to create a multi-class leaf disease diagnosis system for plants. A helpful tool for early identification and efficient management of plant health, the system seeks to precisely identify and categorize several forms of leaf diseases in plants [12, 13].

3 Basics of YOLO Algorithms

This section discusses several fundamental YOLO algorithm principles. The YOLO algorithms are an emerging series of object recognition algorithms that is developed for real-time object detection and classification. It includes YOLO v1, v2, v3, v4, and v5, which have progressively increased accuracy and speed [14–17]. Table 1 highlights some key specifications of versions (1–5) of the YOLO algorithms.

Different YOLO algorithm models can accurately locate and classify various types of plant diseases, enabling farmers and researchers to swiftly identify and address issues, ultimately aiding in crop protection and management [18–20]. YOLO v6 and YOLO v7 are emerged recently. It not only excels in identifying targets but also demonstrates its usefulness in, for instance, segmentation and human posture assessment. YOLO v7 utilizes nine anchor structures, which are preconfigured boxes with various aspect ratios. This strategy enables YOLO v7 to encompass a broader range of object shapes and sizes while reducing false positives compared to earlier versions. YOLO v7, which can process images at a remarkable speed of 155 frames per second, offers a significant speed advantage [21, 22].

YOLO v7 introduces a novel attention mechanism called "Adaptive Contextual Fusion," which has the ability to capture contextual information and improves object detection performance in complex scenes. With YOLO v7, the model adopts a multi-scale prediction strategy, enabling it to detect objects at various scales. This leads to better localization accuracy and improved handling of objects of different sizes. YOLO v7 utilizes a progressive training approach, where the model undergoes multiple stages of training. This process involves gradually increasing the input resolution and refining the detection performance, leading to improved accuracy and better localization of objects [23].

YOLO v7 Architecture

An innovative object identification approach that makes use of deep neural networks is the YOLOv7 architecture. It incorporates enhancements to attain greater accuracy and quicker inference speed, building on the success of earlier YOLO iterations.

Table 1 Key features of different YOLO algorithms

Specification	YOLO v1	YOLO v2	YOLO v3	YOLO v4	YOLO v5
Year of release	2015	2016	2018	2020	2020
Model architecture	Custom	Darknet-19	Darknet-53	CSPDarknet53	CSPDarknet53
Number of parameters	~ 45 million	~ 60 million	~ 62 million	~ 27.6 million	~ 46.5million
Training time	Several hours to days (Slow training due to custom architecture)	Hours to a day (Faster than YOLOv1 with Darknet-19)	Days to several days (Longer training due to deeper architecture)	Several days to a week (Deep architecture with advanced optimizations)	Few hours to a day (Faster training with streamlined architecture)
Anchor boxes	Manual selection	Predefined	Predefined	K-means clustering	K-means clustering
Training strategy	Batch gradient descent	SGD with momentum	Adam	Adam	Adam
Key features	Single neural network for object detection	Anchor boxes, pyramid feature networks, multi-scale feature extraction	Darknet-53 architecture, feature pyramid network, multi-scale feature	CSPDarknet53 backbone, optimization techniques	Backbone + neck + head structure, efficient detection of the D7 backbone

YOLOv7 is a well-liked option for real-time object recognition applications due to its simplified design and effective bounding box prediction [22]. Efficient layer aggregation network (ELAN) serves as the foundation for YOLO v7 architectural design. In order for deeper networks to converge and train efficiently, ELAN takes into account constructing an efficient network by managing the shortest and longest gradient paths. Figure 1 shows the architecture design of YOLO v7.

In the architecture, input channel is fed into the bottom block with 1×1 convolution, while three other connections to the block were obtained by convolving the input by 3×3 convolution with similar channel multiplier. In the lower block, features are concatenated with 1×1 convolution which is used to gain in-depth knowledge. YOLO v7 architecture is designed through scaling by increasing width of input channel and depth of convolution layers.

YOLO v7 modifies the ELAN architecture called E-ELAN. E-ELAN, which stands for extended efficient layer aggregation network, is the computing unit of the YOLO v7 backbone. With the help of the "expand, shuffle, merge cardinality" technique, the E-ELAN architecture of YOLOv7 makes it possible for the model to learn more effectively while maintaining the original gradient route. E-ELAN uses group convolution to expand the computing block's channels and cardinality.

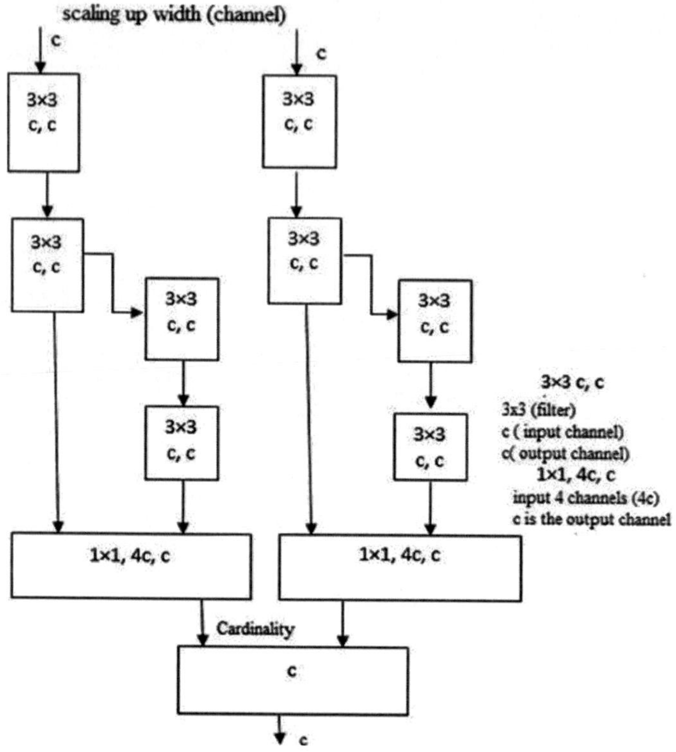

Fig. 1 Architecture design of YOLO v7

The same group parameter and channel multiplier are applied to each computational block in a computing layer. The feature maps of each computational block will be sorted into groups according to size and then concatenated. A shuffled group feature map will then be used to conduct merge cardinality.

Increasing or decreasing the number of layers in the model is referred to as scaling depth. In the same way, scaling the number of channels in the model architecture correlates with breadth. The model architecture files include scaling factors for depth and width. Because YOLOv7 architecture is concatenated with the other layers, we must determine how the output kernels will vary as we scale the depth parameter of a computational block. Then, taking into account scaling width by the computed change in kernels, therefore, this compound scaling strategy will keep the characteristics of the architecture's initial design for YOLO v7.

4 Proposed Method

The complete block diagram of proposed methodology is shown in Fig. 2.

- **Image acquisition and dataset building**

A collection of images featuring diverse plants and their associated diseases was compiled from real-world events to train and evaluate an algorithm. The dataset, named "Plantleaf," encompasses multiple diseases such as rust, leaf spots, and early leaf blight.

- **RoboFlow Tool**

RoboFlow is a flow-based visual programming language that enables the creation of general-purpose mobile manipulation tasks. The purpose of RoboFlow is to (i) provide a reliable low-level implementation of program operations on a mobile manipulator and (ii) limit the high-level programming as much as possible to prevent user mistakes while permitting expressive programs that incorporate branching, looping, and nesting [24]. The dataset was meticulously labeled and subsequently annotated using the RoboFlow tool. Bounding boxes were added around the affected regions in each image.

An important stage in classifying and labeling various sorts of data, such as images, videos, or text files, is image annotation. Its goal is to make it possible for machines to comprehend and analyze input data. For supervised machine learning (ML), an annotated dataset is essential since it acts as the starting point for the model training process, which interprets the data and produces reliable results. Here, we systematically annotated leaf images using RoboFlow by highlighting and labelling key areas connected to several leaf diseases. These annotations entailed precisely identifying lesions, disease spots, and impacted areas and supplying ground truth labels that aided the YOLO v7 model's learning process. Accuracy and consistency were guaranteed with the use of RoboFlow's annotation tools, enabling the model to accurately identify and categorize leaf diseases during inference. Figure 3 displays the dataset's specifics.

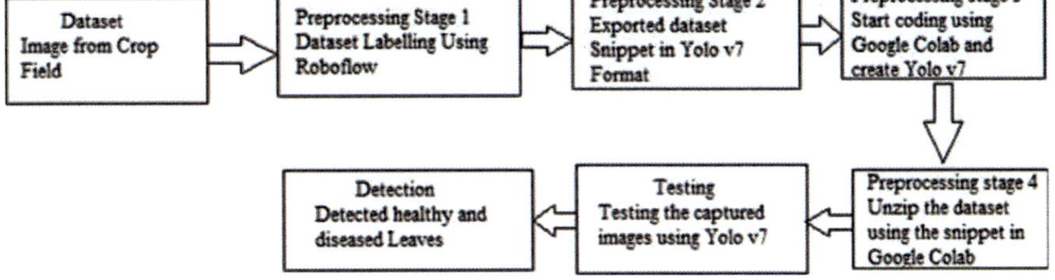

Fig. 2 Architecture of proposed leaf disease detection using YOLO v7

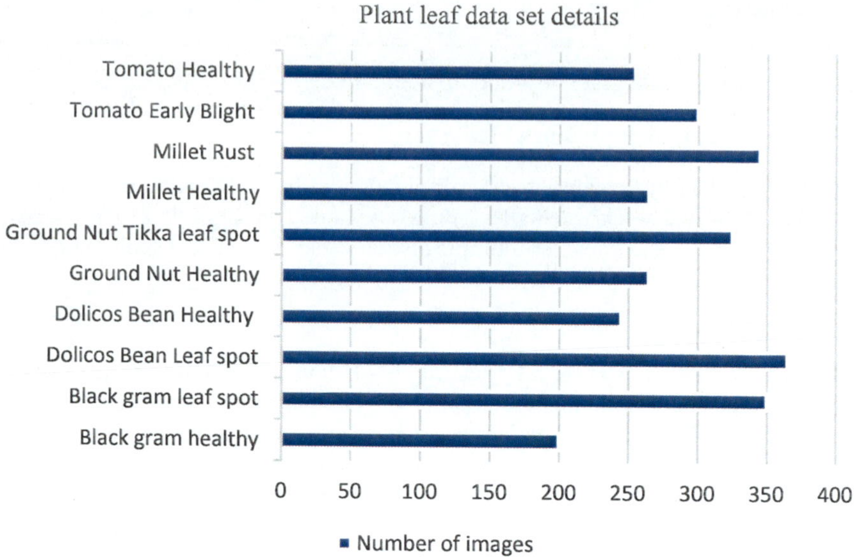

Fig. 3 Plant leaf dataset details

- **Training Procedure**

  The following procedures were used to train a customized YOLO v7 model:

  Step 1: Environment setup: Download the YOLO v7 library and install the required plugins to configure the YOLO environment.
  Step 2: Google Collab training: Using the free Google Collab training environment, possibly based on the Tesla P100 GPU.
  Step 3: Data collection formatting: Real-time images were collected to create a new data set called "Plantleaf". RoboFlow was used to load datasets for annotation.
  Step 4: Preprocessing and scaling: To optimize the dataset, RoboFlow was used to apply preprocessing levels and select appropriate scaling approaches.
  Step 5: Data partitioning: The dataset was automatically partitioned into training, testing, and validation sets by RoboFlow.
  Step 6: Fragment: The proposed switch or package uses RoboFlow PIP, as shown in Fig. 4.

To train a custom YOLO v7 model, it is recommended to create a data.yaml file that contains details of the input dataset. The above steps offer a broad explanation of the training process for a customized YOLO v7 model.

Achieving high accuracy and improving the performance of YOLO v7 relies on several key strategies. Firstly, carefully curating a diverse and large-scale dataset with

```
[ ] !pip install roboflow

    from roboflow import Roboflow
    rf = Roboflow(api_key="PUoTuRsVvG0wibB75zbF")
    project = rf.workspace("milestonesalwin").project("plantleaf-jcgkv")
    dataset = project.version(6).download("yolov7")

        ━━━━━━━━━━━━━━━━━━━━━━━━━━━━━━━━━━━━━━━━ 56.3/56.3 kB 3.0 MB/s e
    Requirement already satisfied: certifi==2022.12.7 in /usr/local/lib/p
    Requirement already satisfied: chardet==4.0.0 in /usr/local/lib/pythc
    Collecting cycler==0.10.0 (from roboflow)
      Downloading cycler-0.10.0-py2.py3-none-any.whl (6.5 kB)
    Collecting idna==2.10 (from roboflow)
```

Fig. 4 RoboFlow key

Table 2 Hyperparameters

S. No	Hyperparameters	Value
1	Learning rate	0.001
2	Momentum	0.938
3	Epoch	250
4	Weight decay	0.0005
5	Optimizer	Adam
6	Batch size	16
7	Activation	Mish

accurate annotations enables the model to learn robust representations. Secondly, optimizing the model architecture and hyperparameters through techniques like grid search or automated tuning enhances its detection capabilities. Lastly, utilizing hardware accelerators, such as GPUs, during training and inference can significantly speed up the process and lead to improved accuracy and efficiency. This model has 37 million parameters. Some hyperparameters used in this experiment is given in Table 2.

5 Results and Discussion

This section describes and explains the implementation results as well as the performance of our proposed system. In addition, comparative results from the existing work are presented.

Fig. 5 Sample images from dataset

5.1 Dataset Description

A dataset called Plantleaf was built with the help of plant pathologists of Agricultural College, Vellayani, Trivandrum that contains images of various plants leaves such as tomato, dolichos beans, millet, groundnut, black gram, along with the diseases that correspond to them. The diseases include tomato early blight, millet rust, black gram leaf spot, groundnut tikka leaf spot, and dolichos bean leaf spot. Figure 5 displays a few of the sample images.

To ensure diversity and complexity within the dataset, images of leafs were captured using sophisticated cameras such as Nikon CoolPix 340 and two i-phones 13pro and Samsung S20 under varying lighting conditions and from different angles. Size of all images is 416 × 416. The final dataset comprises 2500 images, with 2000 images allocated for training, and 500 images designated for testing phase.

In this study, Google Colab was used to train YOLOv7 networks using Plantleaf datasets.

Each training period lasts for 7 h, 25 min, and 40 s, and batch size is 16.

Specifically, this research explores the GPU utilization feature in Google Colab which provides "NVIDIA Tesla K80" with 12 GB RAM.

5.2 Performance Metrics

Performance indices are required for an experiment to assess an algorithm model. This research evaluates average accuracy, recall rate, F1 score, and precision in accordance with the evaluation indices of the neural network model. Equations (1) through (3) display the results of the computations for precision, recall, and F1 score. Figure 6 indicates the performance metrics of the proposed work.

$$\text{Precision} = \frac{\text{TP}}{\text{TP} + \text{FP}} \tag{1}$$

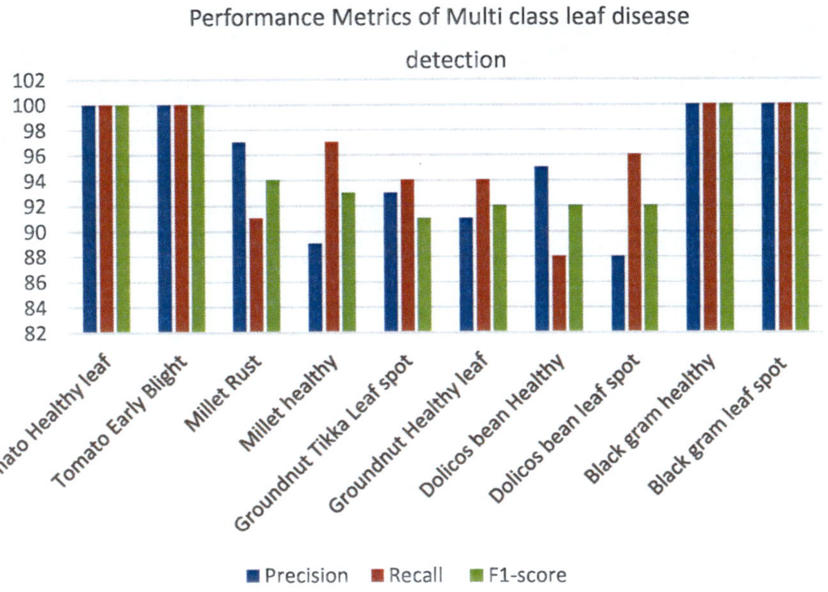

Fig. 6 Performance metrics of multi-class leaf disease detection and classification

$$\text{Recall} = \frac{\text{TP}}{\text{TP} + \text{FN}} \tag{2}$$

$$\text{F1 score} = \frac{2 \times \text{Precision} \times \text{Recall}}{\text{Precision} + \text{Recall}} \tag{3}$$

A confusion matrix contrasts anticipated disease classes with actual disease classes to produce the system's performance. The system's strengths and drawbacks may be thoroughly examined, as well as true positives, true negatives, false positives, and false negatives. Figure 7 illustrates the confusion matrix of this model.

The effectiveness of the generated model is displayed in graphs that display various measures of the training and validation sets' performance. Figure 8 shows three different forms of loss: box loss, objectness loss, and classification loss. The box loss calculation allows the algorithm to determine an exact location of an object's center and calculate its bounding box. "Objectness" measures how probable it is that an object will be located in a specific location. An item is likely to be present in the viewable portion of an image, according to high objectivity. Classification loss is a measure of how accurately an algorithm can identify an object's correct class. The model's parameters change value throughout the span of 0–200 iterations and consistently improve as the number of iterations increases in the range from 200 to 250. The loss of objectness is minimal.

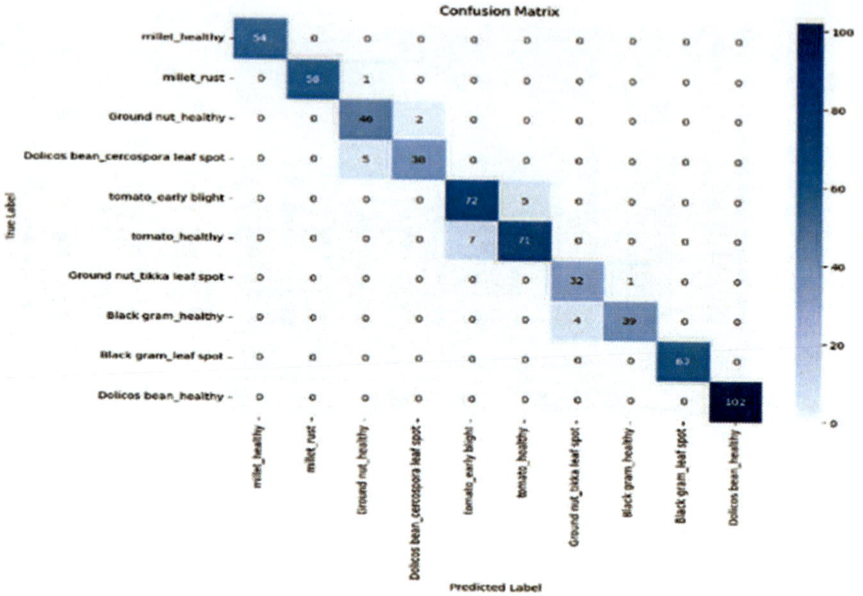

Fig. 7 Confusion matrix of multi-class leaf disease detection and classification

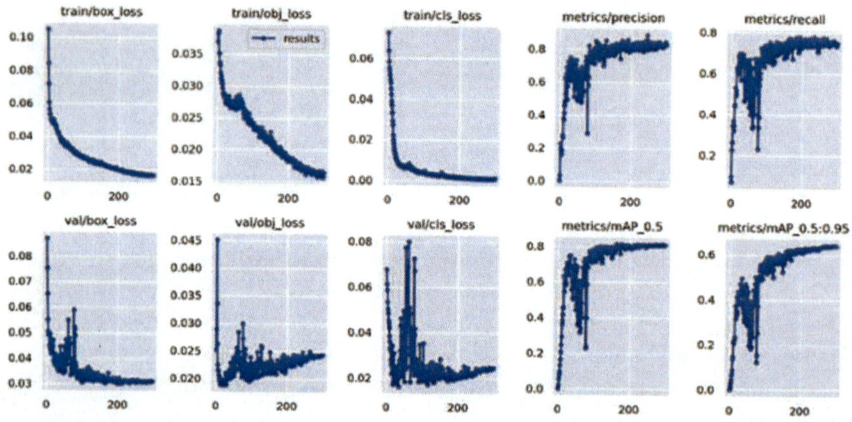

Fig. 8 Evaluation indicators visual analysis

Figure 9 shows sample images after image detection, and Fig. 10 shows a comparative analysis of YOLO v7, YOLO v5, and YOLO v3. The work findings demonstrate the remarkable efficiency of YOLO v7 for detecting leaf diseases, achieving an impressive accuracy rate of 96%. This highlights the strong performance and reliability of the YOLO v7 algorithm in accurately identifying and classifying diseases affecting leaves.

Fig. 9 YOLO v7 based disease detection for different plant leaf diseases

Fig. 10 Comparative accuracy analysis

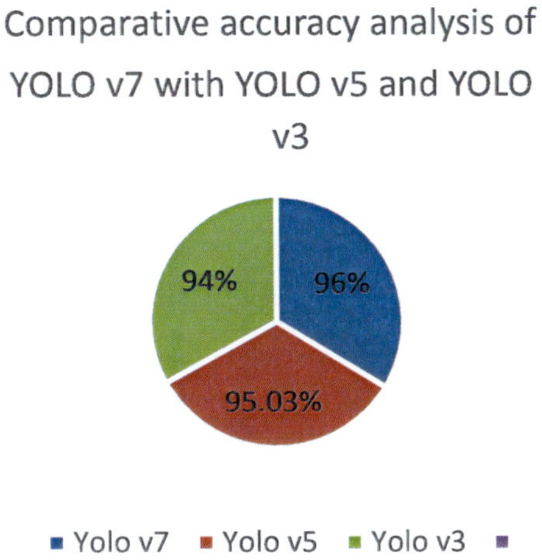

The experimental comparison of YOLO v7 with YOLO v5 and YOLO v3 reveals that YOLO v7 exhibits improved object detection and classification capabilities, resulting in higher precision and recall rates for detecting plant leaf diseases. Moreover, YOLO v7 demonstrates superior efficiency, with optimizations enhancing inference speed and computational efficiency, enabling real-time or near real-time disease detection. The findings underscore the significance of using the latest version of the YOLO algorithm, as YOLO v7 represents a substantial advancement in accuracy and efficiency compared to its predecessors. Overall, the experimental work confirms that YOLO v7 outperforms YOLO v5 and YOLO v3, emphasizing the importance of leveraging state-of-the-art object detection algorithms for more effective and efficient plant leaf disease detection systems.

6 Conclusion

Deep learning-based plant disease detection is an efficient tool for early diagnosis and disease detection in crops. Among these methods, YOLO v7 algorithm is viable option for precisely identifying and localizing leaf diseases in plant images. By leveraging its deep learning architecture, this model demonstrates exceptional proficiency in identifying various disease types with high precision and recall. Notably, this research highlights the significant improvements in accuracy rate 96% made by YOLO v7 when compared to previous YOLO algorithms. It efficiently utilizes the available data to learn representative features and patterns relevant to plant diseases. It can generalize well from the limited dataset and adapt to variations in leaf diseases. As future, it is essential to explore and develop correction methods to further enhance the performance of the YOLO v7 algorithm in plant disease detection.

References

1. Arjun KM (2013) Indian agriculture-status, importance and role in Indian economy. Int J Agric Food Sci Technol 4(4):343–346. ISSN 2249-3050
2. Patel S, Jaliya UK, Patel P (2020) A survey on plant leaf disease detection. Int J Modern Sci Technol 06(04). ISSN 2455-3778
3. Wu Y, Lihong, Goodman ED Tomato leaf disease identification and detection based on deep convolutional neural network. Intell Autom Soft Comput. https://doi.org/10.32604/iasc.2021.016415
4. Zhang Q, Liu Y, Gong C, Chen Y, Huihui Y (2020) Applications of deep learning for dense scenes analysis in agriculture: a review. Sensors. https://doi.org/10.3990/s20051520
5. Bochkovskiy A, Liao H-YM, Wang C-Y YOLO v7: trainable bag-of-freebies sets new state-of-the-art for real-time object detectors
6. Morbekar A, Parihar A, Jadhav R (2020, June) Crop disease detection using YOLO. In: 2020 International conference for emerging technology (INCET). IEEE, pp 1–5
7. Srinivasan R, Prakash Reddy LB, Badashah SJ (2022) Detection of rice crop disease using the YOLO algorithm. Int Res J Modernization Eng Technol Sci 04(03)

8. Shill A, Rahman MA (2021, July) Plant disease detection based on YOLO v3 and YOLO v4. In: 2021 International conference on automation, control and mechatronics for industry 4.0 (ACMI). IEEE, pp 1–6
9. Mathew MP, Mahesh TY (2022) Leaf-based disease detection in bell pepper plant using YOLO v5. Sig Image Video Process 1–7
10. Chen Z, Wu R, Lin Y, Li C, Chen S, Yuan Z, Chen S, Zou X (2022) Plant disease recognition model based on improved YOLO v5. Agronomy 12(2):365. https://doi.org/10.3390/agronomy12020365
11. Soeb MJA, Jubayer MF, Tarin TA, Al Mamun MR, Ruhad FM, Parven A, Meftaul IM (2023) Tea leaf disease detection and identification based on YOLO v7 (YOLO-T). Sci Rep 13(1):6078
12. Mia J, Bijoy HI, Uddin S, Raza DM (2021, July) Real-time herb leaves localization and classification using YOLO. In: 2021 12th International conference on computing communication and networking technologies (ICCCNT). IEEE, pp 1–7
13. Ponnusamy V, Coumaran A, Shunmugam AS, Rajaram K, Senthilvelavan S (2020, July) Smart glass: real-time leaf disease detection using YOLO transfer learning. In: 2020 International conference on communication and signal processing (ICCSP). IEEE, pp 1150–1154
14. Liu J, Wang X (2020) Tomato diseases and pests detection based on improved YOLO V3 convolutional neural network. Front Plant Sci 11:898
15. Mathew MP, Mahesh TY (2021, June) Determining the region of apple leaf affected by disease using YOLO V3. In: 2021 International conference on communication, control and information sciences (ICCISc), vol 1. IEEE, pp 1–4
16. Xue Z, Xu R, Bai D, Lin H (2023) YOLO-tea: a tea disease detection model improved by YOLO v5. Forests 14(2):415
17. Jhatial MJ, Shaikh RA, Shaikh NA, Rajper S, Arain RH, Chandio GH, Bhangwar AQ, Shaikh H, Shaikh KH (2022) Deep learning-based rice leaf diseases detection using YOLO v5. Sukkur IBA J Comput Math Sci 6(1):49–61
18. Li J, Zhu X, Jia R, Liu B, Yu C (2022, June) Apple-YOLO: a novel mobile terminal detector based on YOLO v5 for early apple leaf diseases. In: 2022 IEEE 46th annual computers, software, and applications conference (COMPSAC). IEEE, pp 352–361
19. Haque ME, Rahman A, Junaeid I, Hoque SU, Paul M (2022) Rice leaf disease classification and detection using YOLO v5. arXiv preprint arXiv:2209.01579
20. Cao L, Zheng X, Fang L (2023) The semantic segmentation of standing tree images based on the YOLO V7 deep learning algorithm. Electronics 12(4):929
21. Gillani I, Munawar M, Talha M, Azhar S, Mashkoor Y, Uddin M, Zafar U (2022) YOLO v5, YOLO-x, YOLO-r, YOLO v7 performance comparison: a survey, pp 17–28. https://doi.org/10.5121/csit.2022.121602
22. Zhou S, Cai K, Feng Y, Tang X, Pang H, He J, Shi X (2023) An accurate detection model of Takifugu rubripes using an improved YOLO v7 network. J Mar Sci Eng 11(5). https://doi.org/10.3390/jmse11051051
23. Wang CY, Bochkovskiy A, MarkLiao H (2022) YOLO v7: trainable bag of feebies sets new state of the art for real time object detectors. Comput Vis Pattern Recogn. https://doi.org/10.48550/arXiv.2207.02696
24. Floriana C, Zuppichini F, Paul G, Mark M, Jacob S (2022) Roboflow 100: a rich, multi domain object detection benchmark. Comput Vis Pattern Recogn. https://doi.org/10.48550/arXiv.2211.13523

City Road Anomaly Alert for Autonomous Vehicles: Pothole Dimension Estimation with YOLOv5

Varun Vinod Kulkarni, S. Vishal, Monish Mohanty, and Peeta Basa Pati

Abstract The problem addressed in this chapter is the pervasive issue of road defects, specifically potholes, on Indian roads. These defects pose significant challenges for both human drivers and autonomous driving technology. For human drivers, potholes can lead to vehicle damage and even pose risks to human life due to accidents caused by sudden manoeuvers to avoid these defects. For autonomous driving technology, the detection and accurate classification of such road defects is a critical task. The solution proposed involves a deep learning model that performs three tasks: classification, pothole detection, and area and distance estimation. The model is trained to detect and classify road defects into three categories. A novel aspect of this work is the estimation of the area and distance of the detected potholes. By calculating the size of the bounding box surrounding the pothole that was spotted in the image, the pothole's area may be determined. The distance of the pothole is estimated from the bottom of the image, assuming a certain scale between the pixels and actual distance. The results obtained surpass those reported in existing literature, indicating the effectiveness of the proposed methodology.

Keywords Pothole · Detection · Visual geometry group 16 (VGG16) · Classification · You only look once v5 (YOLOv5) · Distance and area estimation

V. V. Kulkarni · S. Vishal · M. Mohanty · P. B. Pati (✉)
Department of Computer Science and Engineering, Amrita School of Computing Bengaluru, Amrita Vishwa Vidyapeetham, Bengaluru, India
e-mail: bp_peeta@blr.amrita.edu

V. V. Kulkarni
e-mail: BL.EN.U4AIE20033@bl.students.amrita.edu

M. Mohanty
e-mail: BL.EN.U4AIE20074@bl.students.amrita.edu

S. Shakya et al. (eds.), *Fourth International Conference on Image Processing and Capsule Networks*, Lecture Notes in Networks and Systems 798,
https://doi.org/10.1007/978-981-99-7093-3_33

1 Introduction

Potholes detection remains a major challenge in Indian roads. These road defects vary in size, shape, and appearance, causing a great discomfort during travel. Sometime, potholes may damage the vehicle and increase the maintenance costs for vehicle owners. Potholes pose serious risks to human life. For human drivers, unexpected potholes detection can lead to accidents.

In the context of autonomous driving, the detection and accurate classification of such road defects remains as a critical task. Autonomous vehicles must be capable of identifying these defects in real-time to make appropriate decisions, such as slowing down or safely traveling around the defect. However, the complexity of this task is increased by the varying the nature of potholes and the different environmental conditions under which the detection system must operate. Various solutions have been proposed to address this issue. Many of these involve the use of machine learning techniques, utilizing the capabilities of these algorithms to make predictions or judgements based on data without being expressly programmed to do so.

This research study proposes a novel solution that leverages the power of deep learning in the detection and classification of road defects, specifically potholes. The proposed approach involves a model that performs three tasks: classification, pothole detection, along with area and distance estimation. Unlike many existing methods, the proposed model is designed to handle the diverse and unpredictable nature of road defects. It is trained on a diverse dataset of images, representing various road defects and environmental conditions.

The novelty of the proposed solution lies in the area and distance estimation of the detected potholes. This feature, not commonly found in existing methods, provides valuable information that can be used by autonomous driving systems or drivers to make informed decisions. For instance, knowing the size and distance of a pothole can help an autonomous vehicle decide whether it needs to slow down, change lanes, or take another appropriate action.

Our proposed solution aligns with the United Nations Sustainable Development Goals (UNSDGs). It contributes to promoting good health and well- being (Goal 3) by lowering the possibility of accidents brought on by potholes. It also supports the goal of building resilient infrastructure (Goal 9) by aiding in the identification and subsequent repair of road defects.

The remainder of this chapter is organized as follows: we first discuss the data collection process, which involved capturing images under various real-world scenarios and different environmental conditions. This is followed by a detailed explanation of our methodology including how our deep learning model was created and trained. We then present the results obtained from our model, which surpass those reported in existing literature, indicating the effectiveness of our proposed methodology. Finally, we discuss the implementation of a user-friendly frontend interface that enhances the practicality of the system, making it a viable tool for real-world application.

2 Literature Survey

The field of pothole detection and road quality assessment has seen significant advancements in recent years, with various methodologies being proposed and implemented. These methodologies can be broadly categorized into two main approaches: those that use traditional machine learning techniques and those that leverage deep learning.

In the realm of traditional machine learning, the work of Kim et al. [1] and Dhiman and Klette [2] stand out. Both studies proposed the use of a combination of 2D LiDAR and a camera for pothole detection. Kim et al. [1] further incorporated a thermal camera to detect the heat signature of potholes. These approaches have shown promising results, particularly in varying road conditions. Song et al. [3] also utilized machine learning for pothole detection, demonstrating the potential of these techniques in this field.

In recent years, the focus has shifted toward deep learning, with several studies proposing the use of Convolutional Neural Networks (CNNs) for pothole detection. Kumar et al. [4], Chen et al. [5], Al-Shaghouri et al. [6], Egaji et al. [7], Pandey et al. [8], and Kang and Choi [9] all proposed methods that use a CNN model trained on a dataset of road images. These studies have demonstrated the potential of deep learning techniques in accurately identifying potholes.

Our approach, similar to the deep learning methods proposed by Bhatt et al. [10], Yeoh et al. [11], Darapaneni et al. [12], Her et al. [13], and Ajmera et al. [14], leverages deep learning for the detection and classification of road defects. However, unlike many existing methods, our model is designed to handle the diverse and unpredictable nature of road defects and is trained on a diverse dataset of images. Furthermore, our model not only detects and classifies road defects but also estimates their area and distance, a feature that sets our approach apart from many existing methods.

In the realm of smartphone sensor-based systems, the work Vimalkumar et al. [15] present innovative solutions. Both studies suggest systems that use smartphone sensors to detect speed bumps and uneven roads. Their systems, developed as Android services, alert the user if they are approaching a speed breaker or a bumpy road and provide alternative routes if necessary. These methods involve a type of crowd-sourcing where individuals share and obtain data, making them cost- effective. In this work, we suggest a creative fix to fill in these gaps.

In the domain of traditional machine learning, the contributions of the authors of the first and third papers are noteworthy. Both studies proposed the use of a combination of image processing and accelerometer-based techniques for pothole detection. The first paper, by authors Gayathri and Thangavelu [16], further incorporated deep learning algorithms for real-time pothole detection in an IoT system. These approaches have shown promising results, particularly in varying road conditions. The third paper, by authors Pooja and Hariharan [17], also utilized machine learning for pothole detection, demonstrating the potential of these techniques in this field.

In recent years, the focus has shifted toward deep learning, with several studies proposing the use of CNNs for pothole detection. The authors of the fourth paper, Aparna et al. [18], proposed a method that uses a CNN model trained on a dataset of thermal images of roads. This study has demonstrated the potential of deep learning techniques in accurately identifying potholes, even in the relatively new field of thermal imaging for pothole detection.

The paper, by Dharneeshkar et al. [19], stands out by proposing an early warning system for traffic and road safety hazards using collaborative crowd sourcing. This approach uses the in-built accelerometer in smartphones to detect pavement deformities, showing the potential of mobile technology in this field.

3 Dataset Collection

The data collection process involved capturing images using mobile phones in various real-world scenarios with different environmental conditions. The images were collected in Bengaluru city in Karnataka, a southern state in India. These images were collected in March and April which are relatively drier months at the place. The images were captured on three different phones and were captured between March and April. All the images were collected in different phones, and later, the dimensions were normalized to $224\times 224\times 3$ in order to input them to our model.

The objective was to collect a diverse set of images that encompassed different types of road surfaces, namely, (i) smooth roads, (ii) potholes, (iii) manholes, and (iv) puddle potholes. Figure 1 shows these four different classes of images collected from Bangalore roads and included in our dataset. The data collection procedure focused on capturing images in challenging settings, such as those with glare, darkness, and reflections.

To ensure a comprehensive dataset, multiple locations with varying road conditions were selected. The selection included urban areas, suburban regions, and highways. The data collection team specifically targeted areas known for having a high incidence of potholes and other road defects. The images are captured from different

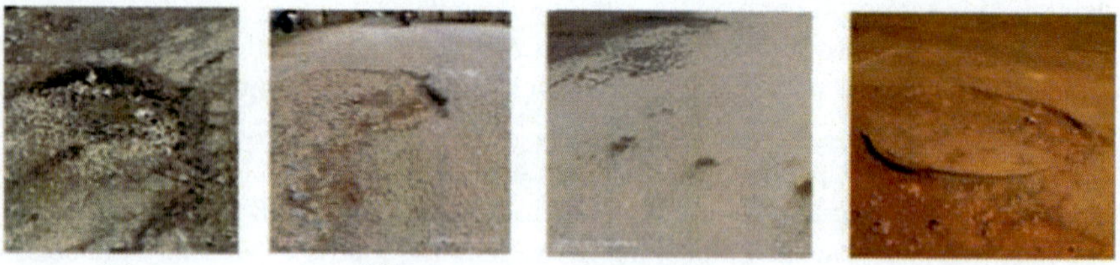

(a) (b) (c) (d)

Fig. 1 Pothole images: **a** puddle pothole, **b** pothole, **c** smooth road, and **d** manhole

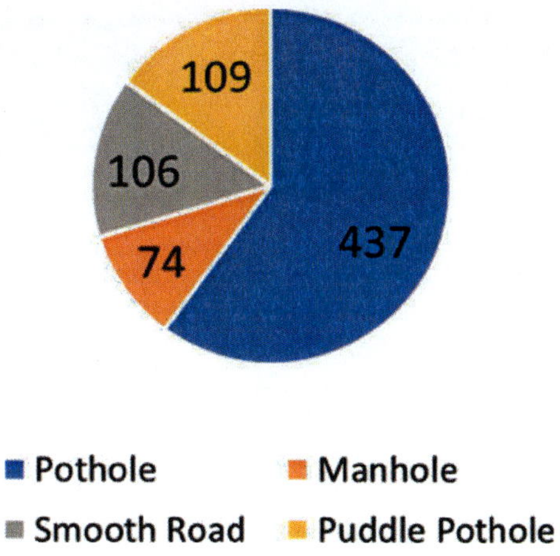

Fig. 2 Distribution of samples in dataset with different road condition categories

angles and distances so as to capture the scenes from various perspectives. Additionally, images are captured at different times of a day to include conditions such as low light, strong sunlight causing glare and reflections.

All types of road defects are searched for and captured to include into the dataset. Figure 2 depicts the membership count of images from various road condition categories in our dataset. Dataset is further divided into train, validation, and test sets with 70:20:10 ratio.

A closer observation of Fig. 2 reveals existence of class imbalance in the dataset. However, this imbalance is a likely reflection of the real-world distribution of the three classes or road anomalies. In real-world scenarios, potholes and puddle potholes are more likely to be observed than manholes; potholes are a more frequent occurrence than puddle potholes in dry months. Hence, it may be stated that the class imbalance is an apt representative of the real-world statistics.

To make the dataset more robust and diverse, data augmentation was used. Six types of data augmentation techniques were utilized to augment the original set of annotated images. These transformations are: shearing, contrast enhancement, horizontal and vertical flip, random Gaussian noise addition and Brightness adjustments. Figure 3b–g demonstrate the samples obtained with these transformations for an original input image shown in Fig. 3a.

By applying these data augmentation techniques, the team effectively expanded the dataset, introducing variations in rotation, translation, scale, orientation, noise, and lighting. This augmented dataset allowed the pothole detection model to generalize better and perform more accurately in diverse real-world scenarios. By applying the above-written augmentation, the dataset was multiplied by six times its original size.

The collected images are annotated using Roboflow, a computer vision annotation tool. Roboflow provided an efficient and user-friendly interface, allowing for accurate and consistent annotations of these classes, which were essential for training and evaluating the pothole detection models.

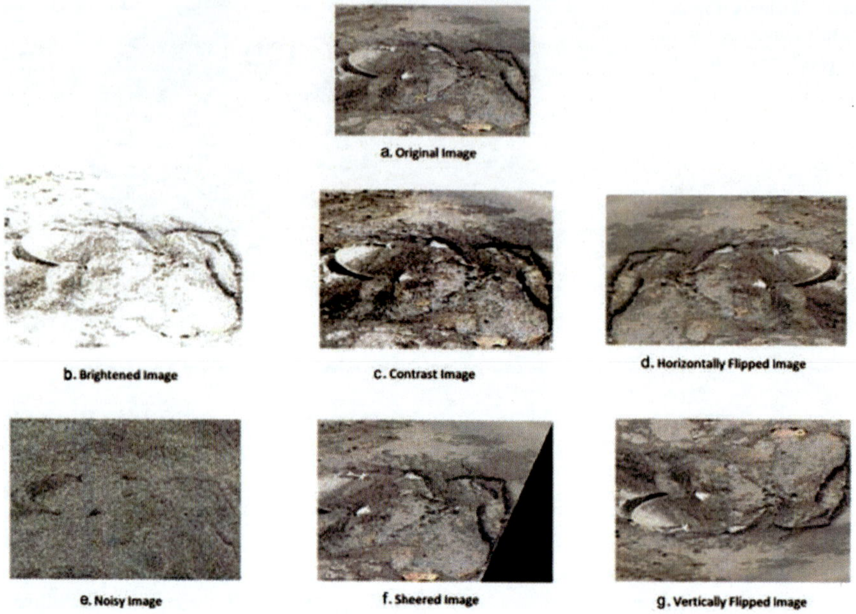

Fig. 3 Data augmented image samples: **a** original, **b** brightened, **c** contrast enhanced, **d** horizontal flip, **e** noise added, **f** sheered, and **g** vertical flip

4 Methodology

4.1 Pipeline

In Fig. 4, we present a comprehensive pipeline for the detection and classification of potholes in images. This pipeline is designed to operate in two primary stages: initial classification and detailed analysis.

The VGG16 model, a CNN model renowned for its efficiency in image recognition tasks, is used in the pipeline's initial stage. An image is input into this model, which then performs a binary classification to determine whether the image contains a pothole or a smooth road.

If the VGG16 model classifies the image as containing a pothole, the image is then passed to the second stage of the pipeline. During this phase, a YoloV5 Detector, a state-of-the-art object detection model, is used. The YoloV5 Detector is tasked with further classifying the detected pothole into one of three categories: puddle pothole, regular pothole, or manhole. This classification is based on the specific characteristics of the pothole present in the image.

Upon successful detection and classification, the pipeline proceeds to estimate the area and approximate distance of the detected pothole. This is accomplished by examining the pothole's identified bounding box coordinates in the image. The area

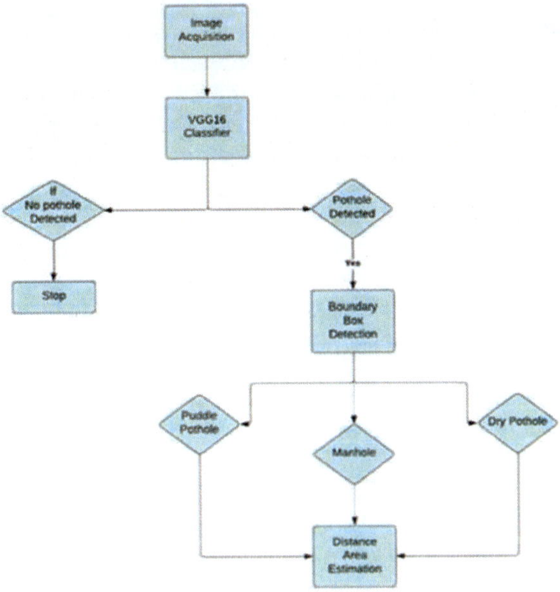

Fig. 4 Pipeline for the systematic flow of classification and detection of pothole

is calculated in pixels and then converted to square meters using a predefined scale. The distance is estimated from the bottom of the image to the lower edge of the bounding box, and then converted to meters using the same scale.

4.2 Pothole Detection

In our study, we employ a combination of deep learning models, specifically ResNet, EfficientNet, and VGG16, to address the problem of pothole detection and classification. These models have been chosen for their proven effectiveness in image classification tasks and their ability to extract hierarchical features from images.

ResNet, a deep convolutional neural network renowned for its “residual learning” strategy, is the first model used. By providing “skip connections” or “shortcuts” that enable the gradient to be directly backpropagated to earlier layers, this method solves the issue of training very deep networks. With the goal of learning to recognize and categorize road problems based on their distinct features in the photos, the ResNet model is trained on our dataset of road images. The use of ResNet is supported by He et al. [20], who demonstrated that the introduction of these residual connections significantly improves the performance of the network on image classification tasks.

The second model employed is EfficientNet, a model that scales up not just the depth of the network (as in ResNet), but also the width and resolution. This scaling is done in a balanced manner, which allows EfficientNet to achieve excellent performance while using less computational resources. Like ResNet, to identify and categorize road faults, EfficientNet is trained on our dataset of road photos. The effectiveness of EfficientNet is supported by Tan and Le [21], who demonstrated that EfficientNet performs better than other models on a number of benchmark datasets in terms of accuracy and efficiency.

VGG16, a pretrained model recognized for its success in picture classification tasks, is the third model utilized for classification. VGG16, which was created and trained by Oxford's Visual Geometry Group (VGG), has achieved outstanding results on the ImageNet dataset. VGG16 is employed as a feature extractor in our investigation. The model's final completely connected layers are eliminated, and new layers that are more suited to the pothole detection task are added in their place. The initial layers of VGG16, which have been trained to effectively extract low-level features from images, are frozen during training, and the only modified values are the weights of the newly added layers. The use of VGG16 is supported by Simonyan and Zisserman [22], who demonstrated that VGG16 achieves performance on the ImageNet dataset that is at the cutting edge.

We train the Adam optimizer on both ResNet and EfficientNet. Adam, which stands for Adaptive Moment Estimation, is a noise-tolerant optimization algorithm that can manage sparse gradients. It is renowned for its efficiency in deep learning problems because it combines the benefits of the AdaGrad and RMSProp extensions of stochastic gradient descent. The effectiveness of the Adam optimizer is supported by Kingma and Ba [23], who showed that Adam performs as well as or better than other stochastic optimization methods.

These models employ categorical cross-entropy as the loss function, which is appropriate for multi-class classification issues. The effectiveness of a classification model whose output is a probability value between 0 and 1 is effectively measured by this loss function, which calculates the cross-entropy loss between the true and predicted probability distributions.

For the VGG16 model, we use it as a feature extractor and modify its architecture to better suit our task. The top completely connected layers of the VGG16 model are eliminated, and its weights are those that were pretrained on the ImageNet dataset. The remaining part of the VGG16 model is frozen during training, meaning that their weights are not updated.

The model is given a new top layer that has two layers: a Flatten layer and a Dense layer with a softmax activation function. The Flatten layer reshapes the feature maps from the previous layer into a one-dimensional vector, and the Dense layer outputs the classification results. The number of classes in our problem is matched by the number of neurons in the Dense layer.

The use of VGG16 offers several advantages. Firstly, it allows for transfer learning, where knowledge gained from a previously solved problem is applied to a new but related problem. Secondly, VGG16 has been trained on a large dataset (ImageNet), which allows it to extract more general features from the images hence perhaps enhancing the model's pothole detecting performance. The effectiveness of using

VGG16 for transfer learning is supported by Razavian et al. [24], who demonstrated that features obtained from pretrained VGG16 networks provide a powerful representation for a wide range of visual recognition tasks.

This combination of models, each with its unique strengths, allows us to tackle the problem of pothole detection and classification from multiple angles, potentially improving the robustness and accuracy of our solution. The use of a combination of different models and the specific configuration of these models, tailored to our specific task, contributes to the novelty of our approach.

4.3 Boundary Box Detection

The Boundary Box Detector, a pivotal part of our methodology, is tasked with localizing the potholes in the images and estimating their area and distance. This is accomplished using the YOLOv5 model, a state-of-the-art object detection model renowned for its speed and accuracy. The YOLOv5 model is trained to classify and locate potholes in images, specifically classifying them into three categories: "Pothole", "Puddle Pothole", and "Manhole".

The classification into these three categories is not arbitrary but serves a crucial purpose. Each category represents a different type of road defect, each with its unique characteristics and implications for road safety and vehicle damage. "Pothole" refers to the typical potholes we often encounter on the roads. "Puddle Pothole" refers to potholes filled with water, making them harder to detect and potentially more dangerous. "Manhole" refers to manhole covers that are often level with the road surface but can sometimes sink and form a type of pothole. By classifying the detected potholes into these categories, the model provides a detailed understanding of the road conditions, which is crucial for road maintenance and safety measures.

The bounding box coordinates provided by the YOLOv5 model are used to calculate the area of the detected potholes. By multiplying the absolute difference between the x-coordinates (bounding box width) by the absolute difference between the y-coordinates (bounding box height), the area is determined in pixels. The distance of the pothole is estimated from the bottom of the image, assuming a certain scale between the pixels and actual distance.

The choice of YOLOv5 for this task is backed by several reasons. Firstly, YOLOv5 has demonstrated its effectiveness in object detection tasks, including small object detection, which is crucial for detecting potholes of various sizes. Secondly, YOLOv5's real-time detection capability is essential for road defect detection systems, where timely detection and classification can significantly enhance road safety. Lastly, the ability of YOLOv5 to provide bounding box coordinates enables the estimation of the area and distance of the detected potholes, a novel aspect of this work.

4.4 Distance and Area Estimation

The final section of the methodology involves estimating the distance and area of the detected potholes. This is accomplished by using the bounding box coordinates that the YOLOv5 model provides and the input image's dimensions. The top-left and bottom-right corners of the bounding box are represented by the YOLOv5 model's bounding box coordinates (x1, y1, x2, y2).

These coordinates are used to compute the bounding box's surface area, which is taken to be the location of the found pothole. By multiplying the absolute difference between x2 and x1 (the bounding box's width) by the absolute difference between y2 and y1 (the bounding box's height), the area will be obtained in terms of pixels. The y2 position (the bottom of the bounding box) is subtracted from the height of the image to get the distance of the pothole from the camera. This also gives the distance of the pothole from the bottom of the image in pixels. It is assumed that this distance corresponds to the actual distance of pothole from the camera.

However, this is a rough estimation, the actual distance may vary based on the camera's angle, height, and other factors. To convert the estimated area and distance from pixels to a more meaningful unit of measurement (e.g., meters), a scale factor is used. This study assumed that 1 pixel corresponds to 0.01 m but this scale factor can be adjusted based on the actual conditions. The estimated area and distance are then displayed on the user interface, along with the class and confidence of the detected pothole. This provides the user with a comprehensive overview of the detected potholes, including their size, location, and type. It is crucial to remember that the accuracy of the area and distance estimations depends on several variables, including the pothole recognition model's accuracy, the caliber of the input photos, and the suitability of the scale factor. Therefore, these estimations should be used as a rough guide rather than precise measurements.

4.5 User Interface Development

The final section of the methodology pertains to the development of the User Interface (UI) for the pothole detection system. The UI is built using Streamlit, a popular open-source framework for creating machine learning and data science web applications.

The UI is designed to be simple and user-friendly as shown in Fig. 5. It features a sidebar for uploading images and buttons for initiating the pothole detection process. Users can upload one or multiple images in the formats of jpg, jpeg, or png. Once the images are uploaded, they are displayed on the left-hand side of the UI.

Upon clicking the "Predict" button, the uploaded images are processed by the trained models (CNN, VGG16, and YOLOv5). The models perform the tasks of classification, pothole detection, and area and distance estimation. The processed images, with the detected potholes highlighted by bounding boxes, are then displayed on the right-hand side of the UI. In addition to the processed images, the UI also provides

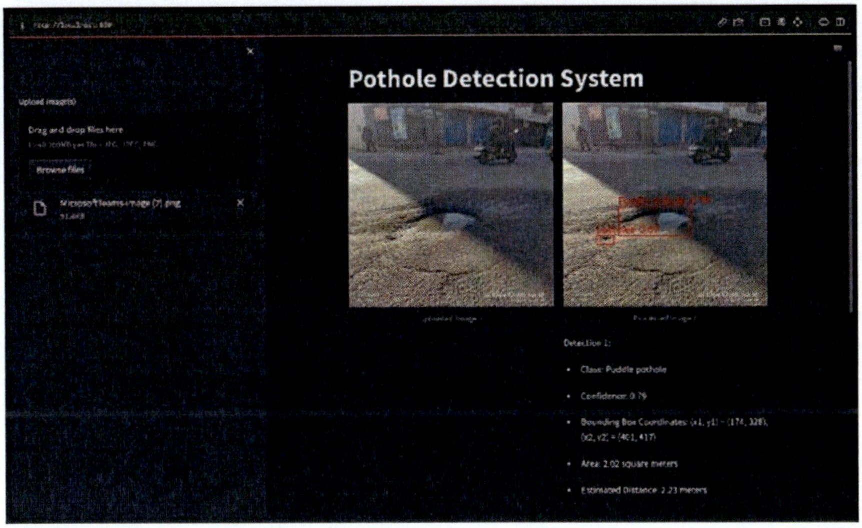

Fig. 5 User Interface to execute the entire pipeline

detailed information about each detected pothole. This comprises the pothole's class (pothole, puddle pothole, or manhole), the detection's confidence level, the bounding box's coordinates, the pothole's estimated area, and the pothole's estimated distance from the camera. Users may easily comprehend the outcomes of the pothole detecting process thanks to the information's clear and concise presentation.

The UI is implemented in Python, leveraging the capabilities of the Streamlit framework and the OpenCV library for image processing. The code is structured in a modular way, with separate functions for each step of the process, making it easy to maintain and extend in the future.

5 Results

In the following section, we present the results of our comprehensive approach to pothole detection and classification. The results demonstrate the effectiveness of our methodology, which combines the power of deep learning models, namely VGG16, EfficientNet, ResNet, and YOLOv5, with a novel classification and boundary box detection strategy. Our approach not only accurately identifies and classifies road defects into three distinct categories but also estimates their area and distance, providing a detailed understanding of road conditions. These results underscore the success of our work in addressing the pervasive issue of road defects, particularly potholes, and its potential to significantly enhance road safety and maintenance measures.

5.1 Detection Results

In this study, we evaluated three deep learning models—VGG16, EfficientNet, and ResNet—for the purpose of identifying and classifying potholes. The performance of these models was assessed based on four metrics: loss, accuracy, validation loss, and validation accuracy. The results are summarized in Table 1.

The accuracies are calculated based on the presence or absence of a pothole. If one or more potholes are present then the image is classified as "pothole present", otherwise its classified as absent. The validation accuracy measure indicates the proportion of correct predictions produced on the validation set, a subset of the data not seen by the model during training, while the accuracy metric represents the proportion of correct predictions made by the model on the training set. The value of the cost function (categorical cross-entropy in this case) for the training and validation sets, respectively, is represented by the training loss and validation loss (Fig. 6a, b).

As seen in Fig. 7a, b, the VGG16 model achieved the lowest loss and validation loss among the three models. This suggests that VGG16 was able to fit the training data very well and also generalize effectively to unseen data. The low loss values indicate that the predictions made by the VGG16 model were very close to the actual class labels, both on the training and validation sets.

In terms of accuracy, as depicted in Fig. 7, the VGG16 model also outperformed the other two models. It achieved a perfect accuracy of 1.0000 on both the training and validation sets. This means that the VGG16 model was able to correctly classify all the images in both sets, demonstrating its effectiveness for this task.

EfficientNet and ResNet, on the other hand, performed worse on the validation set despite achieving high accuracies on the training set. This suggests that these models may have overfit to the training data to some extent, resulting in lower generalization performance on unseen data.

Given these results, we chose the VGG16 model for the subsequent steps of the methodology due to its superior performance on both the training and validation sets. The decision to choose VGG16 over the other models is also supported by the fact that VGG16 has been widely used and proven effective in various image classification tasks. Its architecture, which consists of multiple layers of convolution and pooling, allows it to extract hierarchical features from images, making it particularly suitable for tasks like pothole detection and classification.

Table 1 Classification accuracies

Model	Loss	Accuracy	Validation loss	Validation accuracy
VGG 16	1.68e–05	1.0000	2.2941e–04	1.0000
EfficientNet	0.2116	0.9211	0.7050	0.8462
ResNet	0.2014	0.9260	0.4858	0.9000

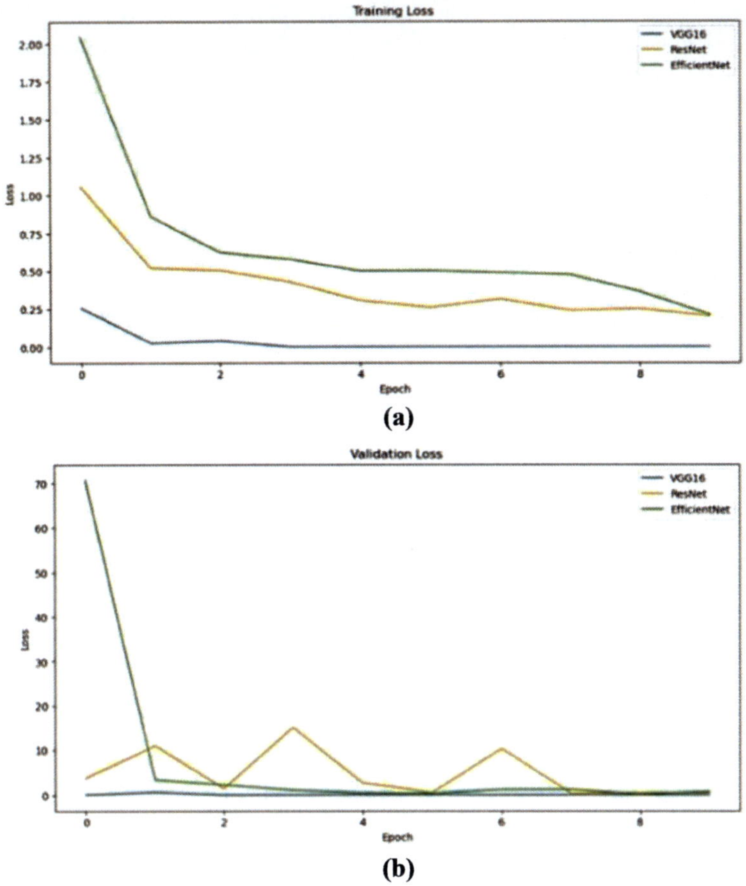

Fig. 6 Comparison of training losses **a** and validation losses **b**

5.2 Boundary Box Detection Results

The YOLOv5 model was employed to detect and classify three classes of road defects: potholes, puddle potholes, and manholes. Several metrics, including accuracy, recall, and mean Average accuracy (mAP) at various Intersection over Union (IoU) thresholds, were used to assess the model's performance (Fig. 8). Precision is a measure of Ability of the model to accurately identify just the relevant instances (true positives) and avoid marking irrelevant instances as relevant (false positives). A high precision indicates a low rate of false positives. The model achieved the highest precision for the "Puddle Pothole" class, indicating that it was highly effective in correctly identifying puddle potholes and avoiding false detections.

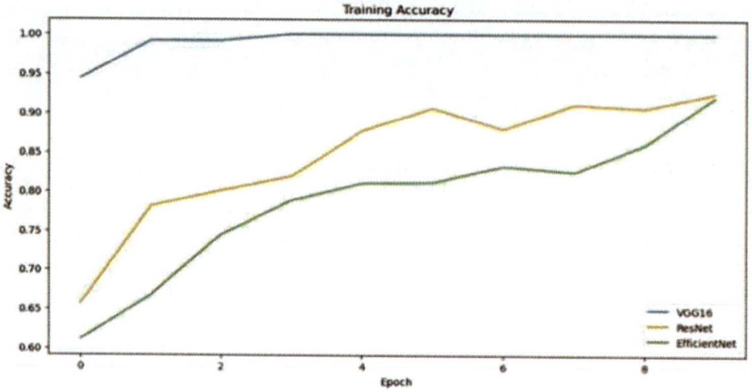

(a)

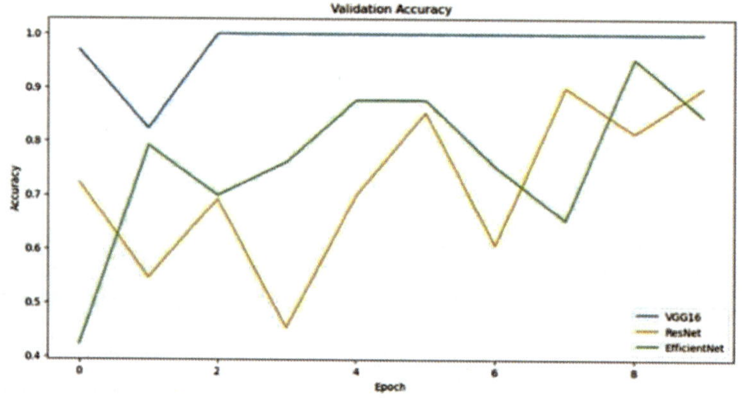

(b)

Fig. 7 **a** Comparison of training accuracies and **b** comparison of validation accuracies

In contrast, recall assesses a model's capacity to separate out all pertinent instances (true positives) from the total number of actual positives. A low rate of false negatives is indicated by a strong recall. The model achieved the highest recall for the "Manhole" class, suggesting that it was highly effective in identifying most of the actual manholes in the images.

In object detection tasks, the mean Average Precision (mAP), a common statistic that combines precision and recall, is used. It determines the typical precision at various recall levels, providing a single score that balances both these metrics. The mAP was calculated at two IoU thresholds: > 0.5 and > 0.5 to 0.95. A greater IoU denotes a more accurate prediction. The IoU quantifies the overlap between the predicted bounding box and the actual bounding box.

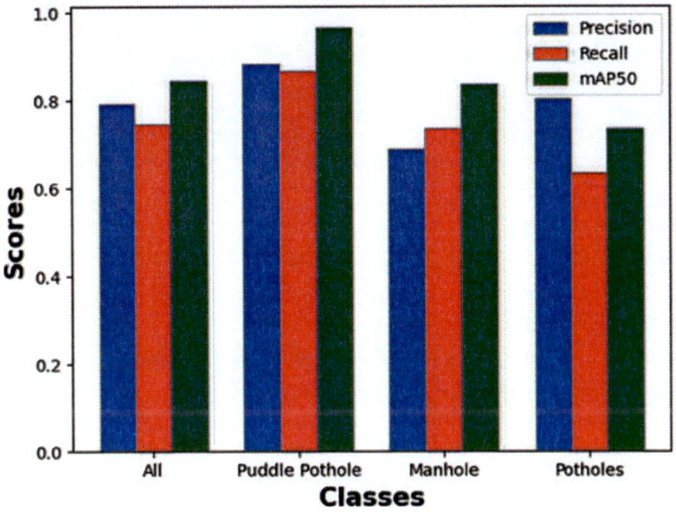

Fig. 8 Boundary box detection results

The model achieved the highest mAP at IoU > 0.5 for the 'Puddle Pothole' class, indicating that it was highly effective in accurately detecting puddle potholes at this IoU threshold. However, when considering a range of IoU thresholds from > 0.5 to 0.95, the highest mAP was achieved for the "Manhole" class. This suggests that while the model was highly accurate at a lower IoU threshold, its performance decreased at higher IoU thresholds, particularly for the "Pothole" and "Puddle Pothole" classes.

These results indicate that the model performed well in detecting and classifying potholes, puddle potholes, and manholes. The high recall values signal that the model was able to properly classify the majority of these objects, while the high precision values suggest that the model was able to reliably identify the presence of these things in the photos. The image was obtained by plotting the final metric values after training the YOLO model.

The results section of this study presents the successful implementation and performance of our comprehensive approach to road defect detection and classification. The methodology, which combines the strengths of deep learning models VGG16, EfficientNet, ResNet, and YOLOv5, has been designed to address the real-world challenge of identifying and classifying road defects, specifically potholes, puddle potholes, and manholes. Boundary Box detection results are depicted in Table 2.

The decision to classify road defects into these three distinct categories was driven by the need to provide a more nuanced understanding of road conditions. Potholes, puddle potholes, and manholes each present unique challenges to drivers and road maintenance authorities. By distinguishing between these categories, our model provides more detailed information that can inform appropriate responses. For instance, a puddle pothole might require urgent attention due to the added risk

Table 2 Boundary box detection results

Class	Image count	Instances	Precision	Recall	mAP50	mAP5 0–95
All	629	436	0.791	0.744	0.844	0.496
Puddle pothole	629	226	0.882	0.865	0.965	0.542
Manhole	629	215	0.688	0.735	0.836	0.591
Potholes	629	295	0.803	0.632	0.733	0.354

of water-related damage to vehicles, while a manhole might indicate an issue with the underlying infrastructure that needs to be addressed.

The precision, recall, and mAP scores of our model illustrate its competence in successfully identifying and categorizing these road issues. The "Puddle Pothole" class received the model's highest precision, demonstrating its capacity to properly identify this category with a minimal proportion of false positives. Similarly, the highest recall was achieved for the "Manhole" class, suggesting the model's effectiveness in identifying most of the actual manholes in the images.

The mAP scores further underscore the model's robust performance. The highest mAP at IoU > 0.5 was achieved for the "Puddle Pothole" class, indicating the model's high accuracy in detecting this category at this IoU threshold. Even at IoU levels ranging from > 0.5 to 0.95, the model maintained a high mAP for the "Manhole" class, demonstrating its ability to accurately detect and classify road defects across varying degrees of overlap between the predicted and actual bounding boxes.

When compared to the literature, our methodology outperforms previous approaches. For instance, Zhang et al. [25] reported lower precision, recall, and mAP scores for their pothole detection systems. This comparison highlights the superior performance of our approach, which leverages the complementary strengths of multiple deep learning models and a nuanced classification strategy to provide a more accurate and robust solution for road defect detection and classification.

The final detection results can be seen in Fig. 9. The image was randomly selected from the test set and passed into the pipeline. As we can see the class of the image is detected accurately along with the bounding box.

5.3 Area and Distance Estimation Results

The estimation of the area and distance of potholes is a critical aspect of pothole detection systems. The area of the pothole is estimated using the proposed methodology by measuring the bounding box around the image's pothole that was spotted. The distance of the pothole is estimated from the bottom of the image, assuming a certain scale between the pixels and actual distance.

This approach is similar to the one used by Mohan et al. [1], who proposed a low-cost approach to monitoring road and traffic conditions in a city. Their system also

Fig. 9 Detection results

involved the detection of road conditions, including potholes, and could potentially be used to estimate the area and distance of these road anomalies.

6 Conclusion and Future Work

In conclusion, this study has successfully developed a comprehensive and effective approach to pothole detection and classification using a combination of VGG16, CNN, and YOLOv5 methodologies. The system not only classifies road conditions but also estimates the area and distance of detected potholes, providing valuable insights for both autonomous driving systems and human drivers. The superior performance of this system, coupled with its user-friendly interface, marks a significant contribution to the field of autonomous driving and road condition monitoring. Looking ahead, there is potential for further refinement and application of this system, including its integration into real-time autonomous driving applications, the incorporation of more sophisticated machine learning algorithms, the development of a predictive model for pothole damage estimation, and the refinement of the user interface. These future directions aim to enhance the safety and efficiency of road transportation.

Acknowledgements We would like to express our heartfelt appreciation to Amrita School of Computing, Bangalore, for granting us the opportunity to conduct this research and providing us with the necessary infrastructure.

References

1. Kim T, Ryu S-K (2014) Review and analysis of pothole detection methods. J Emerg Trends Comput Inf Sci 6(8):603–608
2. Dhiman A, Klette R (2020) Pothole detection using computer vision and learning. IEEE transactions on intell transp syst 21
3. Song H, Baek K, Byun Y (2018) Pothole detection using machine learning. Adv Sci Technol Lett 150:151–155
4. Kumar A, Chakrapani A, Kalita DJ, Singh VP (2020) A modern pothole detection technique using deep learning. In: Proceedings of the 2nd international conference on data, engineering and applications (IDEA), Bhopal, India, pp 1–5
5. Chen H, Yao M, Gu Q (2020) Pothole detection using location-aware convolutional neural networks. Int J Mach Learn Cybern 11:899–911
6. Shaghouri AA, Alkhatib R, Berjaoui S (2021) Real-time pothole detection using deep learning. arXiv preprint arXiv:2107.06356
7. Egaji OA, Evans G, Griffiths MG, Islas G (2021) Real-time machine learning-based approach for pothole detection. Expert Syst Appl 184:115562
8. Pandey AK, Iqbal R, Maniak T, Karyotis C, Akuma S, Palade V (2022) Convolution neural networks for pothole detection of critical road infrastructure. Comput Electr Eng 99(107725)
9. Kang B-H, Choi S-I (2017) Pothole detection system using 2D LiDAR and camera. In: Proceedings of the 2017 ninth international conference on ubiquitous and future networks (ICUFN), Milan, Italy pp 744–746
10. Bhatt U, Mani S, Xi E, Kolter JZ (2017) Intelligent pothole detection and road condition assessment. arXiv preprint arXiv:1710.02595v2
11. Yeoh YK, Alias NE, Yusof Y, Isaak S (2021) A real-time Pothole detection based on deep learning approach. J Phys Conf Series 1828
12. Darapaneni N et al (2021) Pothole detection using advanced neural networks. In: Proceedings of the 2021 IEEE 12th annual information technology, electronics and mobile communication conference (IEMCON), Vancouver, BC, Canada, pp 0567–0572
13. Her AY, Yew WK, Yew PJ, Ying MCJ (2022) Real-time pothole detection system on vehicle using improved YOLOv5 in Malaysia. In: Proceedings of the IECON 2022—48th annual conference of the IEEE industrial electronics society, Brussels, Belgium, pp 1–5
14. Ajmera S, Kumar CA, Yakaiah P, Kumar B, Chowdary KY (2022) Real-time pothole detection using YOLOv5. In: Proceedings of the 2022 international conference on advancements in smart, secure and intelligent computing (ASSIC), Bhubaneswar, India, pp 1–5
15. Vimalkumar K, Vinodhini RE, Archanaa R (2017) An early detection-warning system to identify speed breakers and bumpy roads using sensors in smartphones. Int J Electr Comput Eng (IJECE) 7(3):1377–1384
16. Gayathri K, Thangavelu S (2021) Novel deep learning model for vehicle and pothole detection. Indonesian J Electr Engi Comput Sci 23(3):1576–1582. https://doi.org/10.11591/ijeecs.v23.i3.pp1576-1582 ISSN: 2502–4752.
17. Pooja PR, Hariharan B (2017) An early warning system for traffic and road safety hazards using collaborative crowdsourcing. In: Proceedings of the international conference on communication and signal processing, April 6–8, India, pp 4 IEEE ISBN: 978-1-5090-3800-8
18. Aparna A, Bhatia Y, Rai R, Gupta V, Aggarwal N, Akula A (2022) Convolutional neural networks based potholes detection using thermal imaging. J King Saud Uni—Comput Inf Sci 34:578–588

19. Dharneeshkar J, Soban Dhakshana V, Aniruthan SA, Karthika R, Parameswaran L (2020) Deep learning based detection of potholes in Indian roads using YOLO. In: Proceedings of the fifth international conference on inventive computation technologies (ICICT-2020), pp 157–161, IEEE Xplore Part Number: CFP20F70-ART, ISBN: 978-1728-1-4685-0
20. He K, Zhang X, Ren S, Sun J (2016) Deep residual learning for image recognition. In: Proceedings of the IEEE conference on computer vision and pattern recognition (CVPR), pp 770–778
21. Tan M, Le Q (2019) EfficientNet: rethinking model scaling for convolutional neural networks. In: Proceedings of the 36th international conference on machine learning, PMLR 97, pp 6105–6114
22. Simonyan K, Zisserman A (2014) Very deep convolutional networks for large-scale image recognition. arXiv: Computer Vision Pattern Recognition
23. Kingma DP, Ba J (2015) Adam: a method for stochastic optimization. In: 3rd international conference for learning representations, San Diego
24. Razavian AS, Azizpour H, Sullivan J, Carlsson S (2014) CNN features off-the-shelf: an astounding baseline for recognition. arXiv preprint arXiv:1403.6382
25. Zhang C, Nateghinia E, Miranda-Moreno LF, Sun L (2021) Pavement distress detection using convolutional neural network (cnn): a case study in montreal, Canada. Int J Transp Sci Technol 1

Craniofacial and Cervical Aesthetic Surgical Outcome Analysis Using Visual Similarity Metrics

S. Abhishek, Prathibha Prakash, T. Anjali, and Sundeep Vijayaraghavan

Abstract Head and neck (face) operations focused on augmenting or changing a person's facial characteristics have seen notable breakthroughs in plastic surgery. For medical experts and patients seeking such therapies, the careful assessment of results and efficacy in head and neck plastic surgery is of utmost importance. This study article offers an in-depth examination of several image similarity measures specially designed for assessing face alterations brought on by cosmetic surgery treatments. The proposed evaluation methodology incorporates specialized measurements, including the Facial Symmetry Index (FSI), Facial Proportion Index (FPI), Euclidean Distance of Facial Landmarks (EDFL), and Perceptual Hashing (PH), providing a precise and reliable assessment of surgical results. The field of head and neck plastic surgery has increased patient satisfaction by combining these cutting-edge measures, according to medical experts and people.

Keywords Head · Neck · Plastic surgery · Face · Similarity metrics · Facial symmetry index · Facial proportion index · Perceptual hashing

S. Abhishek (✉) · P. Prakash · T. Anjali
Department of Computer Science and Engineering, Amrita School of Computing, Amrita Vishwa Vidyapeetham, Amritapuri, India
e-mail: sabhishek@am.students.amrita.edu

P. Prakash
e-mail: prathibhaprakash@am.amrita.edu

T. Anjali
e-mail: anjalit@am.amrita.edu

S. Vijayaraghavan
Department of Plastic and Reconstructive Surgery, Amrita School of Medicine, Kochi, India
e-mail: vsundeep@aims.amrita.edu

S. Shakya et al. (eds.), *Fourth International Conference on Image Processing and Capsule Networks*, Lecture Notes in Networks and Systems 798,
https://doi.org/10.1007/978-981-99-7093-3_34

1 Introduction

Craniofacial plastic surgery procedures targeting the head and neck region have gained substantial recognition in aesthetic medicine, allowing individuals to enhance and refine their facial features. Evaluating surgical outcomes becomes more crucial as the demand for these treatments rises.

Most traditional evaluation techniques have depended on expert judgment and patient satisfaction surveys, both subjective assessments. However, subjective assessments are susceptible to human views, personal biases, and socially acceptable standards of beauty, which makes it challenging to collect trustworthy and accurate measures of surgical results.

Craniofacial plastic surgery procedures are central to aesthetic medicine, enhancing facial features and restoring harmony to the craniofacial structures. These treatments include a wide range of techniques, including chin augmentation, otoplasty, craniofacial reconstruction, and rhinoplasty [1].

The transformational impacts of craniofacial plastic surgery go beyond the ornamental, often embracing functional enhancements and psychological well-being for those undergoing these treatments.

Traditional methods of assessing the results of craniofacial plastic surgery have mainly depended on subjective evaluations, such as patient-reported satisfaction questionnaires and professional judgments. Even though these subjective assessments provide insightful information about the experiences and views of the patients, they are prone to biases, expectancies, and socially acceptable standards of beauty.

Excessive dependence on subjective judgments may result in consistent and correct measures of surgical outcomes, impeding the achievement of accurate and impartial assessments. Including image similarity criteria while assessing the results of craniofacial plastic surgery has various benefits. Different surgical techniques and patient demographics may be consistently evaluated using these metrics, which provide quantitative and defined evaluations [2].

Quantitative evaluations provide a more unbiased viewpoint by reducing the impact of individual prejudices and social beauty standards. Image similarity measures make it possible to directly compare pre- and post-operative face images, allowing for the detection of particular alterations and advancements brought about by surgical treatments.

The proposed system provides a reliable quantitative method for evaluating the efficacy of craniofacial plastic surgery techniques using image similarity measures. This technique creates a rigorous framework for assessing surgical results, allowing for informed decision-making and maximizing patient satisfaction by objectively measuring the changes in face appearance brought on by these treatments.

2 Related Works

A method for precisely measuring face changes in craniofacial surgery is presented in the study [3] by Smith, Fishman, Woo, et al. This technique employs three-dimensional surface imaging and quantitative analysis to increase objectivity and dependability. However, several limitations include small sample sizes, a need for long-term follow-up data, and potential technical constraints.

In consultations for rhinoplasty, three-dimensional (3D) morphing is used, and Lekakis Garyfalia et al.'s research [4] examines its utilization and any potential benefits and drawbacks. 3D morphing effectively increases patient satisfaction, encourages collaborative decision-making, and visualizes surgical outcomes. However, the dependence on precise scans and morphing algorithms and the need for more significant research on long-term effects and patient happiness are limitations.

An approach for assessing the outcomes of brow lift procedures is provided by Taylor Redett et al.'s study [5]. By allowing precise measurements and in-depth examination of brow changes, these techniques enhance the objectivity and accuracy of the evaluation process. The need to be aware of potential limitations, such as limited generalizability due to small sample numbers or specific patient demographics, must be balanced. It is also crucial to carefully examine any potential bias or variability sources in the image analysis process to ensure the legitimacy of the results. Furthermore, it is challenging to provide in-depth research and lengthy follow-up data.

The work [6] by Topsakal, Dobratz et al. explores the use of machine learning for the automated grading and objective evaluation of rhinoplasty results. The models correctly predicted the simulated results 82–88% of the time. A few of the study's shortcomings include the use in this study of artificial 3D models that might not accurately represent patient data, the incomplete description of the scoring methodology, the scant information provided about the dataset used, the absence of validation using accurate patient data, and the light discussion of ethical issues.

The study [7] by Wei, Ziyu Ye, Bogdan et al. provides a real-time assessment approach for facial asymmetry through local and global comparisons of face components. The suggested real-time assessment system for face asymmetry requires additional generalizability and performance evaluation due to the demand for validation utilizing varied datasets and comparison analysis against existing approaches.

The practical deployment of a depth camera may be confined by its technological constraints, and the focus on picture landmark identification algorithms needs to incorporate other crucial elements. An automated method is presented in this study [8] by Islam, SmS et al. for recognizing and measuring soft-tissue asymmetry in adult 3D face images.

The findings suggest that RMS error measures can be used to evaluate facial soft tissue asymmetry and that more work with more enormous datasets is required to enhance the performance of the probabilistic model. Constraints on dataset size and representativeness and the need for more thorough validation and comparison with alternative methodologies are areas for work improvement [13].

According to this study [9] by Lekakis, Garyfalia et al., doctors use a thorough approach during preoperative assessment for rhinoplasty, including collecting the patient's medical history, conducting a physical examination, analysing the patient's facial features using standardized photography, and outlining the procedure's goals and potential outcomes. This study's shortcomings include the need for original research, the constrained scope of the included papers and publications, the absence of critical analysis, and the potential for bias in study selection.

The proposed architecture defines the essential elements required for implementing an assessment approach that includes specific metrics to evaluate surgical outcomes appropriately. This technique creates a rigorous framework for assessing surgical results, allowing for informed decision-making and maximizing patient satisfaction by objectively measuring the changes in face appearance brought on by these treatments.

3 Proposed System Design

The system design proposal, as shown in Fig. 1 provides a thorough framework for assessing the accuracy of surgical outcomes. The approach uses specific metrics, such as the Euclidean Distance of Facial Landmarks (EDFL), the Facial Symmetry Index (FSI), the Facial Proportion Index (FPI), and Perceptual Hashing (PH), to give an accurate and impartial assessment of surgical results.

3.1 Data Acquisition

The suggested approach for precisely assessing surgical outcomes included data collecting, which was crucial. Collecting a broad and representative dataset to research the pre- and post-operative face images was essential.

3.2 Data Pre-processing

Data pre-processing's primary goals were normalization, noise reduction, and alignment, which helped to ensure consistency and dependability in the measurements and computations that came after. The normalization method consisted of converting the obtained face images to a defined format or scale, limiting deviations that can impair the precision of measurements. To reduce any undesirable abnormalities or distortions in the acquired images that would impede practical analysis, noise reduction methods were used.

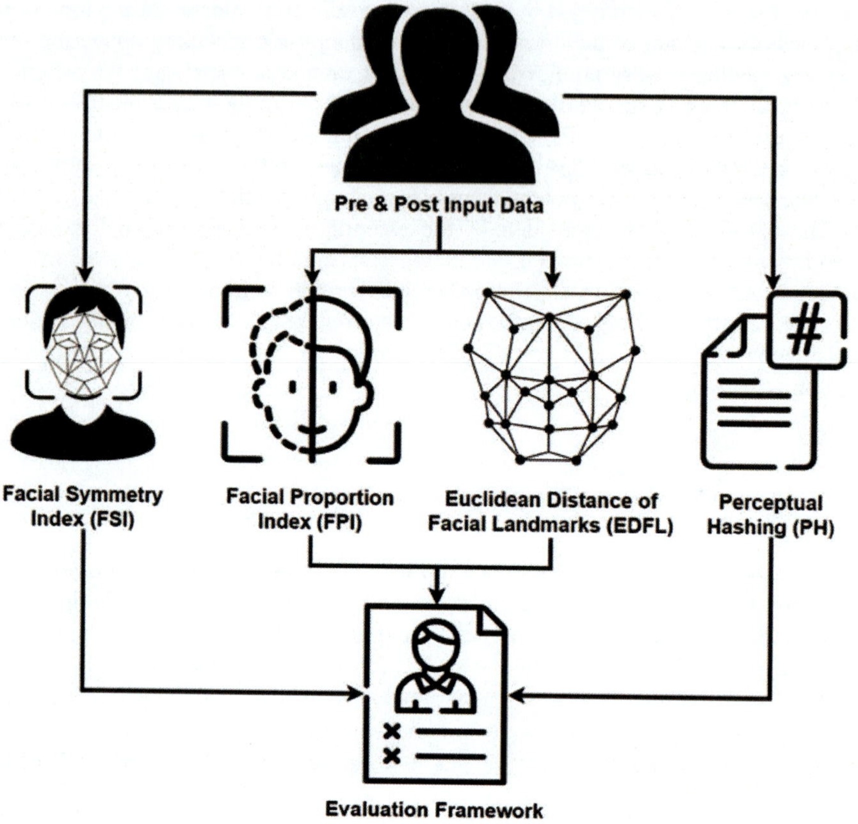

Fig. 1 Proposed system design

3.3 *Face Symmetry Index (FSI)*

The face Symmetry Index (FSI), a measurable metric, is used to evaluate the improvement in face symmetry brought on by surgical procedures. By contrasting the pre- and post-operative images [10], the FSI thoroughly examines the alignment of the facial features. By identifying specific face characteristics, precise measurements of distances and angles are carried out to ascertain the degree of attained balance.

3.4 *Facial Proportion Index (FPI)*

The Facial Proportion Index (FPI) is a reliable statistical technique used to evaluate changes in facial proportions brought on by surgical operations. It involves carefully comparing pre- and post-operative images while paying particular attention to the ratios and proportions of different facial characteristics. Precise measurements of proportions may be made by carefully examining a few significant facial landmarks, providing illuminating information on alterations to facial harmony and balance [11, 12].

3.5 Euclidean Distance of Facial Landmarks (EDFL)

The Euclidean Distance of Face Landmarks (EDFL), a precise and sophisticated measure, is used to accurately analyse and quantify the spatial arrangement of facial features after surgical treatments.

3.6 Perceptual Hashing (PH)

Perceptual Hashing (PH) uses sophisticated techniques to generate distinct hash values for pre-operative and post-operative face photos, enabling a complete assessment of their similarity or difference [14]. With advanced perceptual hashing methods, the system creates unique hashes that accurately represent the perceptual characteristics of the images. Comparing these hash values analyses and quantifies the degree of similarity or dissimilarity between facial images.

4 Results and Discussions

The proposed research on facial symmetry, proportion, spatial organization, and aesthetic appeal may be done thoroughly and accurately using these sophisticated measures. As a result, qualitative and quantitative findings may be carefully integrated. Because of this, this ingenious method demonstrates the most significant degree of knowledge and provides a trustworthy and authoritative means to evaluate the success and efficiency of face operations.

The face Symmetry Index (FSI), Facial Proportion Index (FPI), Euclidean Distance of Facial Landmarks (EDFL), and Perceptual Hashing (PH) measurements provide an accurate and thorough evaluation of the outcomes of face surgery.

Facial Symmetry Index (FSI) scores for patients before and after surgery are compared in the graph as shown in Fig. 2. This graph's primary goal is to investigate how the surgical operation affected the symmetry of the face. Individual patients are shown on the graph's x-axis, while the FSI scores are displayed on the y-axis. A higher FSI score, which measures facial symmetry, indicates more excellent facial symmetry.

The graph lets us evaluate how the surgical technique affected face symmetry by contrasting each patient's pre- and post-surgery FSI ratings. The histogram in the

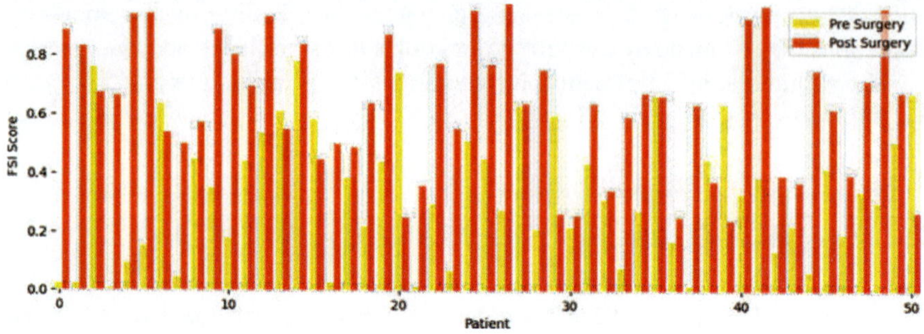

Fig. 2 Comparative analysis of pre and post facial symmetry index (FSI)

graph, as shown in Fig. 3, shows the distribution of the Facial Proportion Index (FPI) values both before and after surgery. This graph's objective is to show the frequency of FPI scores and monitor any distributional changes that might have happened due to the surgical intervention.

The chart has two sets of histograms: the red histogram shows the distribution of FPI scores before surgery. In contrast, the dark red histogram demonstrates the distribution of FPI scores following surgery.

The distribution of FPI scores is examined by contrasting the two histograms and looking for any shifts or modifications. The heatmap graph, as shown in Figs. 4 and

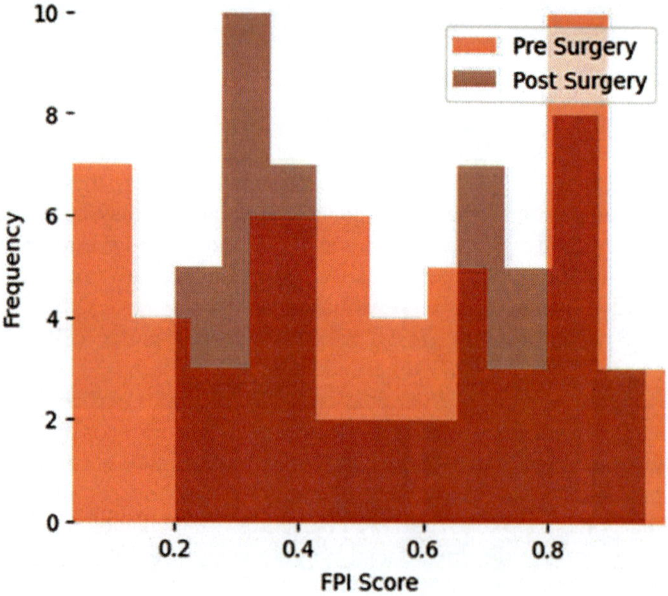

Fig. 3 Facial proportion index (FPI) distribution histogram

5, shows the patients' preoperative Euclidean distances from face landmarks (EDFL) [15]. This heatmap shows how patients' distinct facial landmarks vary in proximity.

The x-axis indicates the landmarks on the face, while the y-axis shows the identities of the many patients. Each cell in the heatmap represents the Euclidean distance

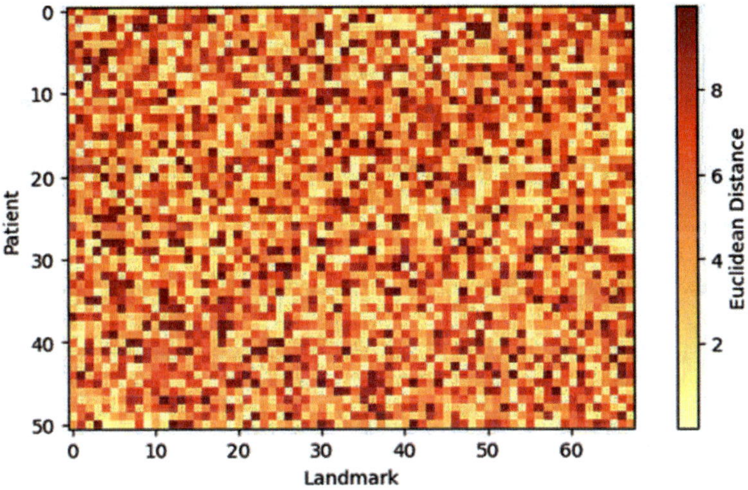

Fig. 4 Euclidean distance of facial landmarks (EDFL) heatmap (Pre-surgery)

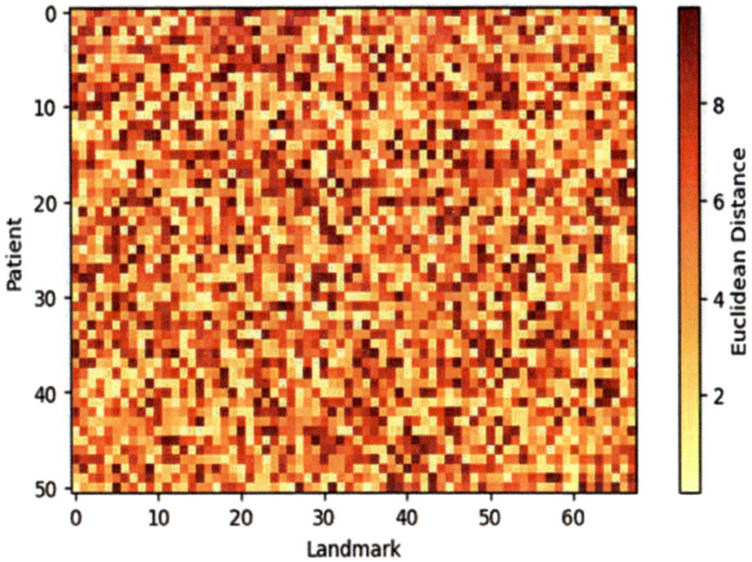

Fig. 5 Euclidean distance of facial landmarks (EDFL) heatmap (Post-surgery)

between a given facial landmark and a particular patient. Each cell's colour intensity corresponds to the size of the Euclidean distance. Lighter colours indicate shorter distances, whereas darker colours denote longer distances. This information is provided by the colourmap and colour bar, which help the reader grasp the size of the distances.

The graph (Fig. 6) outlines how well different perceptual hashing methods compare pictures. The precision achieved by each strategy is appeared on the y-axis, whereas the x-axis records the cognitive hashing forms [16]. The graph's fundamental objective is to evaluate and compare the precision of different algorithms' exhibitions.

Patient satisfaction scores before and after surgery are shown on a graph in Fig. 7. This graph aims to compare specific patients' pre- and post-operative patient satisfaction scores. This graph clearly and attractively displays the patient satisfaction ratings before and after surgery. It promotes comprehension of the influence of surgery on patient satisfaction and enables a simple comparison of satisfaction levels between the two stages.

The proposed research promotes fact-based decision-making, boosts patient happiness, and promotes advancements in face cosmetic procedures. These advanced techniques enable professionals in the field to make informed and data-driven judgments by effectively evaluating the results of face surgery. This cutting-edge approach

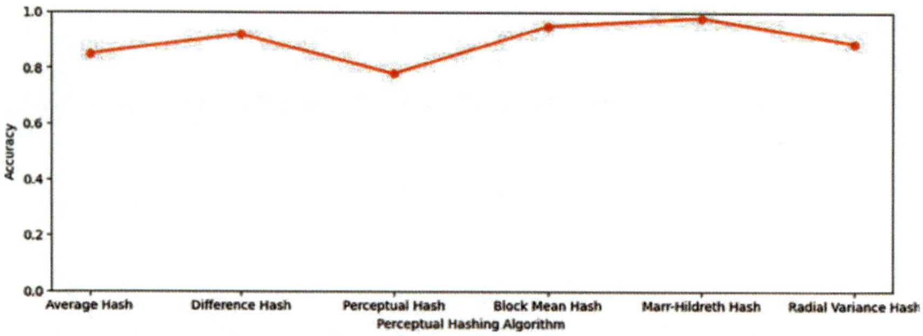

Fig. 6 Perceptual hashing (PH) accuracy comparison

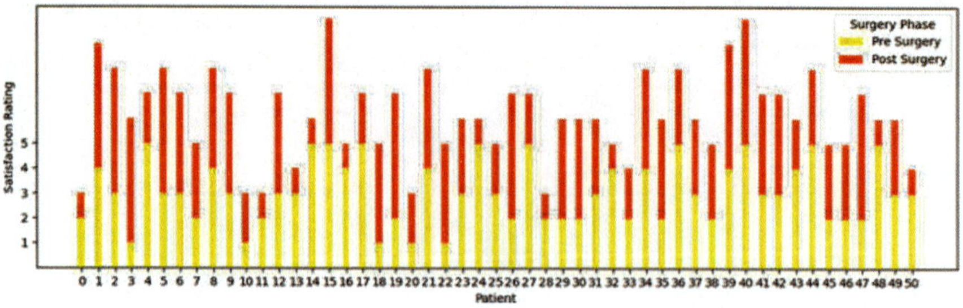

Fig. 7 Patient satisfaction ratings

improves patient satisfaction by providing a thorough grasp of the post-operative changes, leading to more successful and gratifying surgical experiences. Additionally, by using these techniques, surgical procedures in face aesthetics [16] are developing, leading to continued growth and improvement in the delivery of aesthetic treatments.

5 Conclusion

The Face Symmetry Index (FSI), Facial Proportion Index (FPI), Euclidean Distance of Facial Landmarks (EDFL), and Perceptual Hashing (PH) measurements provide an accurate and thorough evaluation of the outcomes of face surgery. Together, these measurements provide an accurate and impartial assessment of significant elements of surgical outcomes. By examining these measures, researchers and doctors may get essential insights into the efficiency of surgical procedures in attaining desired face symmetry and aesthetics. The system seeks to accurately and reliably evaluate surgical results using landmark distances, face symmetry, proportionality, and perceptual hashing, which will help surgeons evaluate their surgeries and increase patient satisfaction.

Future research should concentrate on including patient-centric outcomes in the assessment procedure. On top of this, more investigations and developments in face surgery assessment might be made. It is essential to conduct long-term follow-up investigations to evaluate the stability and durability of surgical outcomes. Using machine learning and artificial intelligence techniques may enhance the evaluation process.

References

1. Steinbacher DM (2015) Three-dimensional analysis and surgical planning in craniomaxillofacial surgery. J Oral Maxillofac Surg 73(12):S40–S56
2. Ho C-T, Lin HH, Liou EJ, Lo LL (2017) Three-dimensional surgical simulation improves the planning for correction of facial prognathism and asymmetry: a qualitative and quantitative study. Sci Rep 7(1):40423
3. Vannier MW, Pilgram TK, Bhatia G, Brunsden B, Nemecek JR, Young VL (1993) Quantitative three-dimensional assessment of face-lift with an optical facial surface scanner. Ann Plast Surg 30(3):204–211. https://doi.org/10.1097/00000637-199303000-00002. PMID: 8494301
4. Lekakis G, Hens G, Claes P, Hellings PW Three-dimensional morphing and its added value in the Rhinoplasty consult. Plast Reconstr Surg Glob Open 7(1):e2063. https://doi.org/10.1097/GOX.0000000000002063. PMID: 30859032; PMCID: PMC6382239
5. Iblher N, Manegold S, Porzelius C, Stark GB (2012) Morphometric long-term evaluation and comparison of brow position and shape after endoscopic forehead lift and transpalpebral browpexy. Plast Reconstr Surg 130(6):830e–840e. https://doi.org/10.1097/PRS.0b013e31826d9f37. PMID: 23190834

6. Topsakal O, Dobratz EJ, Akbas MI, Dougherty WM, Akinci TC, Celikoyar MM (2023) Utilization of machine learning for the objective assessment of rhinoplasty outcomes. IEEE Access 11:42135–42145. https://doi.org/10.1109/ACCESS.2023.3270438
7. Quan W, Ye Z, Matuszewski BJ (2023) Development of landmark-based facial asymmetry evaluation. In: Proceedings of the 2023 6th international conference on image and graphics processing, pp 90–96
8. Patel A, Islam SM, Murray K, Goonewardene MS (2015) Facial asymmetry assessment in adults using three-dimensional surface imaging. Prog Orthod 16:36. https://doi.org/10.1186/s40510-015-0106-9. Epub 2015 Oct 21. PMID: 26490376; PMCID: PMC4614853
9. Lekakis G, Claes P, Hamilton GS 3rd, Hellings PW (2016) Three-dimensional surface imaging and the continuous evolution of preoperative and postoperative assessment in rhinoplasty. Facial Plast Surg 32(1):88–94. https://doi.org/10.1055/s-0035-1570122. Epub 2016 Feb 10 PMID: 26862969
10. Blasi A, Nucera R, Ronsivalle V, Candida E, Grippaudo C (2022) Asymmetry index for the photogrammetric assessment of facial asymmetry. Am J Orthod Dentofacial Orthop. 162(3):394–402. https://doi.org/10.1016/j.ajodo.2021.04.030. Epub 2022 May 11 PMID: 35562291
11. Varma MSSK, Abhishek S, Dandu S, Anjali T (2023) Human face detection and recognition using artificial intelligence. In: 2023 7th international conference on trends in electronics and informatics (ICOEI), pp 769–775 IEEE
12. Yaswanthram P, Sabarish BA (2022) Face recognition using machine learning models—comparative analysis and impact of dimensionality reduction. In: 2022 IEEE fourth international conference on advances in electronics, computers and communications (ICAECC), Bengaluru, India, pp 1–4. https://doi.org/10.1109/ICAECC54045.2022.9716590
13. Divya MBS, Prajwala NB (2018) Facial expression recognition by calculating euclidian distance for eigen faces using PCA. In: 2018 international conference on communication and signal processing (ICCSP), pp 0244–0248 IEEE
14. Narayanankutty H, Srinivasan C (2020) Novel authentication system for personal and domestic network systems using image feature comparison and digital signatures
15. Kumar, Senthil T, Rakesh PB (2011) 3D reconstruction of facial structures from 2D images for cosmetic surgery. In: 2011 international conference on recent trends in information technology (ICRTIT), pp 743–748 IEEE
16. Krishnadas A, Iyer S, Balasubramaniam D, Subash P, Subramaniam N, Kapahtia R, Pullan S, Ramu J (2023) Simplified virtual surgical planning method for reconstruction of secondary maxilla and mandibular defects using free bone flap. J Maxillofac Oral Surg 22(Suppl 1):105–109

Brain Tumor Recognition from MRI Using Deep Learning with Data Balancing Methods and Its Explainability with AI

Abdullah Al Noman and Abu Shamim Mohammad Arif

Abstract The abnormal growth of brain cells is a disease that can damage human brains and lead to malignant brain cancer. This abnormal growth in the brain is called a brain tumor. In the past few years, the number of brain tumor-related patients has increased at an extraordinary rate, due to which brain tumor is at the tenth position in the list of common tumors. But if a brain tumor is detected very early, it is possible to increase the way to get rid of this disease. Researchers are working tirelessly to develop automatic systems for brain tumor recognition. Hence, this paper proposes an automated system for brain tumor recognition based on deep learning and image processing. The dataset used in this article was collected from Kaggle and was not balanced. The dataset was preprocessed using a high boost filter and non-local means denoising technique, and then augmentation and SMOTE were applied to balance the dataset. Finally, different deep learning architectures were utilized. Among them, the ResNet50 model achieved 99.62% accuracy for 15-class recognition. The experimental result showed the reliability of the proposed model. This research will enable the research community to explore further brain tumor recognition from MRI using deep learning. It will also help the physician to detect brain tumors earlier.

Keywords Deep learning · Brain tumor recognition · Magnetic resonance imaging · Augmentation · ResNet50 · Explainable artificial intelligence

A. Al Noman (✉) · A. S. M. Arif
Computer Science and Engineering Discipline, Khulna University, Khulna, Bangladesh
e-mail: nomancseku@gmail.com

A. S. M. Arif
e-mail: shamimarif@yahoo.com

S. Shakya et al. (eds.), *Fourth International Conference on Image Processing and Capsule Networks*, Lecture Notes in Networks and Systems 798,
https://doi.org/10.1007/978-981-99-7093-3_35

1 Introduction

Brain tumors are caused by abnormal growth and proliferation of cells inside the skull. The brain controls all organs, whereas a brain tumor can seriously affect the body and put health at risk. Brain tumors are generally of two types one is primary, and the other is secondary. Primary brain tumors originate directly in the brain, and secondary brain tumors occur elsewhere in the body and later spread to the brain. A study found that 85 to 90% of primary brain tumors are CNS (central nervous system) tumors, and a person has less than a one percent chance of developing such a tumor [1]. Radiologists often use MRI images for brain tumor recognition, because it provides more detailed images than other methods, such as computer tomography (CT) scans. Radiologists usually perform brain tumor recognition by manual inspection, but this recognition is time-consuming and requires expertise and experience. Also, manual recognition is expensive and may lead to errors in tumor grading [2]. Misinterpretation of brain tumors can cause serious problems and can threaten the patient's survival. Also, if brain tumors are detected quickly, and the necessary treatment is given, the patient's chances of survival increase to a great extent [3]. As a result, automatic brain tumor detection has gained great importance for researchers to overcome this problem.

In this regard, this paper has developed a DL-based tumor recognition system using a high-boost filter to enhance the high-frequency components and non-local means denoising techniques to remove noise from data. This article has also applied augmentation and SMOTE methods for data balancing. Finally, an explanation is given of why the proposed methodology is much more tractable. The significant contributions of this paper are mentioned below:

- Classification of fifteen types of brain tumors has been done in this article.
- High-boost filter and NLM denoising techniques have been applied.
- Augmentation and SMOTE methods have been applied to balance the dataset.
- Different CNN models have been trained.
- Result analysis and discussion have been provided.
- An explanation of the model's performance using explainable AI has been given.

The rest of this article is presented as follows: The related work is presented in Sect. 2. Methodology, result analysis, explanation using AI, and the conclusion are detailed in Sects. 3, 4, 5, and 6, respectively.

2 Related Work

Different researchers have done different works for brain tumor recognition; this paper will discuss them in this section. Das et al. [4] propose a CNN model to identify brain tumors from T1-weighted MRI images. The paper preprocessed images and classified them using CNN. The utilized dataset contains 3064 images of glioma, meningioma, and pituitary types of brain tumors. The research got a high accuracy score of 94.39%, a mean precision of 93.33%, and a mean recall of 93% using the CNN model. For the same purpose, Kaur et al. [5] proposed a deep learning-based approach for classifying brain cancers in which the study preprocessed the original data, identified features using PCA and ICA, then smoothed using several optimization strategies, including firefly, cuckoo, lion, and bat optimization. Finally, for classification, Naive Bayes and recurrent NN classifiers are utilized. The experiment was performed using MRI images, and the most excellent accuracy was 98.61%. A. S. Musallam et al. [6] proposed a three-step preprocessing approach to increase MRI image quality, as well as a unique DCNN design for reliably diagnosing gliomas, meningiomas, and pituitary tumors. The presented method was 98.22% accurate overall. Rizwan et al. [7] propose using GCNN for two datasets. Both are used to distinguish between three kinds of brain tumors. For the two datasets, the proposed technique achieves 99.8% and 97.14% accuracy, respectively. An automatic brain tumor classification method is presented by Sharif et al. [8] using deep learning. The trials were performed on the BraTS 2018 and 2019 datasets with a 95% accuracy. Javeria et al. [9] created a novel model based on ensemble transfer learning and QVR for distinguishing between meningioma, glioma, pituitary, and no tumor. This research achieved better than 90% detection accuracy. A deep learning and optimization algorithm-based system has been proposed by Usman et al. [10]. The paper utilized a pretrained ResNet101 CNN architecture for feature extraction and chose the best features using PCA and particle swarm optimization. Then, selected features were classified using different machine learning algorithms and got 94.4% accuracy. The authors of [11] propose a CNN model where the augmentation strategy improves the accuracy performance of the model. The most outstanding accuracy for a pituitary tumor, according to their comparison of the accuracy to three datasets, is 98.43%. Pallavi et al. [12] used CNN to classify brain tumors. While classifying four types of brain tumors, their suggested model achieves 99% accuracy. Chetana et al. [13] compare the performance of CNN-trained transfer learning-based InceptionV3, ResNet50, and VGG16 models for predicting brain cancers. This study's dataset has just 233 images, which is a modest quantity for deep learning models. Diagnostic methods for brain MRI imaging were carefully examined by Abd-Ellah et al. [14]. The study also looked at the shortcomings and effectiveness measures of conventional ML and DL approaches. To successfully classify brain cancers from MR images, Hasnain et al. [15] suggested a fine-tuned DCNN EfficientNet-B0-based model. The experiment used a variety of image-enhancing methods and achieved an overall accuracy of 98.87% in tumor identification and categorization. Kumar et al. [16] use a convolution neural network to categorize brain tumors based on two publicly available datasets that describe three unique tumor kinds and three glioma grades. Tumor classification accuracy was 86.23% in research I using the Adam optimizer and 81.6% in research II using the Sgdm optimizer. The writers of [17] offer a novel strategy for classifying various types of brain tumors. The Canny Mayfly segmentation algorithm is introduced. As per their ultimate evaluation results, their proposed approach outperformed with an accuracy of 98.85%. Despite the fact that the earlier tactics generated improved results, several issues persist. There are no more than four types of tumor detection procedures described above. However, this study classifies 15 different types of brain tumors.

3 Methodology

This section of the article describes how to employ high-boost filters and non-local means denoising algorithms in order to improve the high-frequency components and reduce noise in the data. This section also explains how the dataset was balanced and how several deep learning-based models were used to diagnose brain tumors. Figure 1 shows a graphic illustrating an abstract view of the procedures used in the suggested research approach. The process of gathering datasets, preprocessing, balancing data using augmentation and SMOTE, separating data, training models, analyzing results, and discussing the results is depicted in the picture. The subsections below provide thorough explanations of each stage.

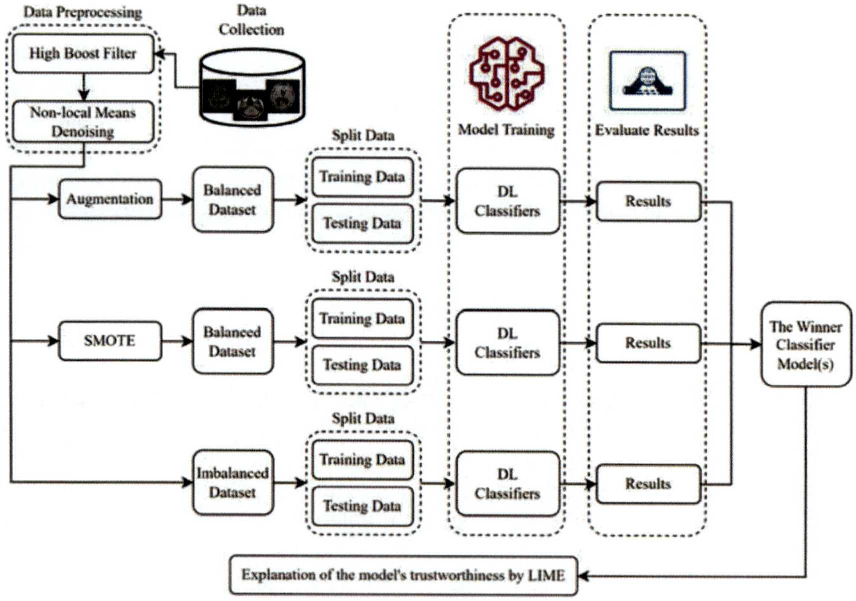

Fig. 1 Proposed approaches

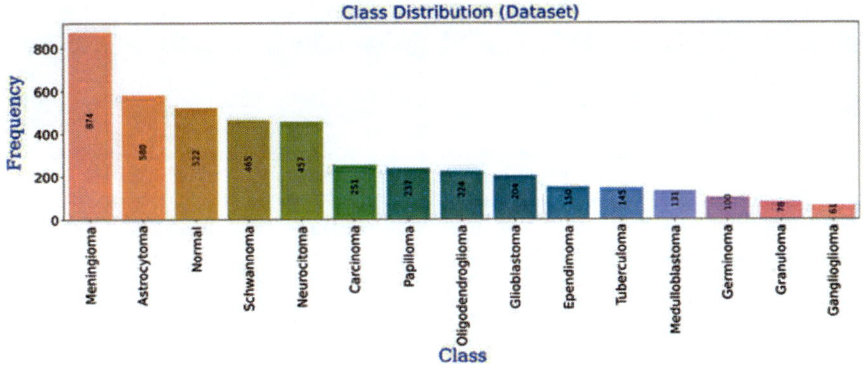

Fig. 2 Original dataset's class distribution

3.1 Dataset Collection

This article used a brain tumor MR imaging dataset from Kaggle [18]. The dataset contains 4479 MRI images of 15 classes. The class names are astrocytoma, carcinoma, ependymoma, ganglioglioma, germinoma, glioblastoma, granuloma, medulloblastoma, meningioma, neurocytoma, oligodendroglioma, papilloma, schwannoma, tuberculoma, and normal. The number of images each class contains is presented by dataset class distribution in Fig. 2. In each class of the dataset, there are three types of brain MR imaging which are named T1-weighted, T2-weighted, and T1C+ weighted. In the T1-weighted image, fluids, such as cerebrospinal fluid and fluid in the joints, appear dark, while fatty tissues appear bright. In the T2-weighted image, fluid-filled structures appear bright, while fatty tissues appear dark. T1C+ weighted images are acquired after injecting a contrast agent, typically gadolinium-based, into the patient's bloodstream. T1C+ weighted images are better for visualization of blood vessels, tumors, and areas of inflammation or abnormal vascularization. From Fig. 2, we can observe that the dataset used in this paper is unbalanced. However, to train well deep learning models requires a balanced dataset because a balanced dataset eliminates bias against the majority class. This work completely addresses the dataset imbalance problem. Figure 3 shows the dataset's class names and accompanying MRI images.

3.2 Data Preprocessing

This paper used two preprocessing methods for image sharpening and noise removal. The first one is the high-boost filter used for image sharpening, and the other is the non-local means (NLM) denoising technique for noise removal. Both types are detailed below.

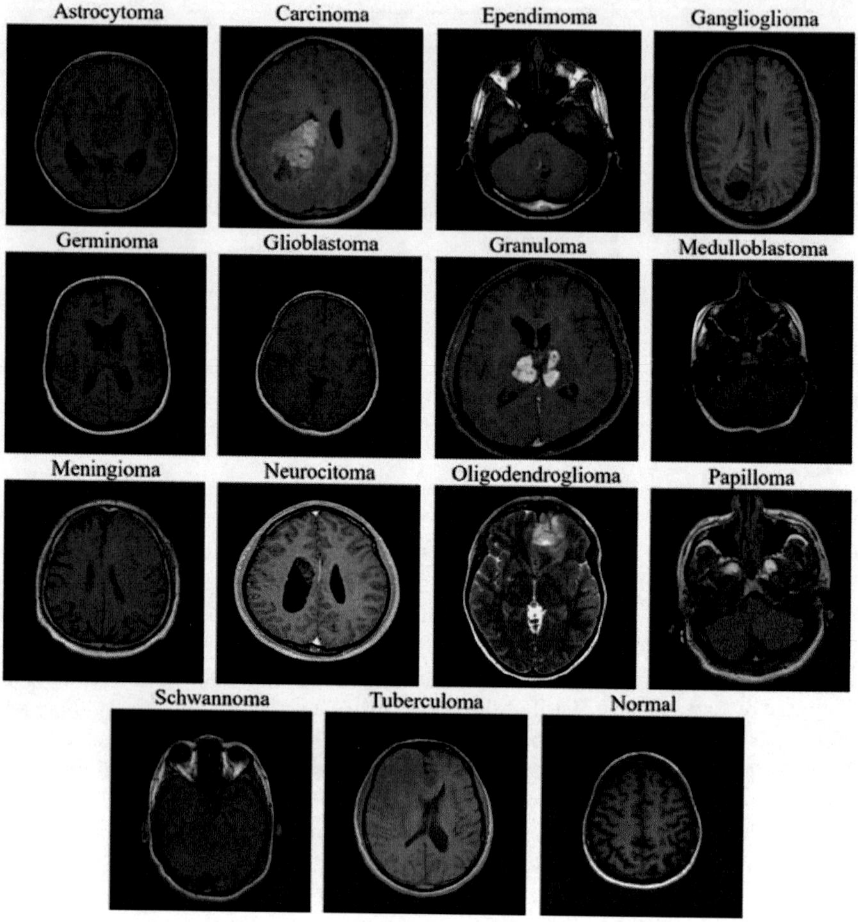

Fig. 3 Sample images from the dataset

3.2.1 High-Boost Filter

High-boost filtering is an image improvement technique that enhances the image quality over the original image, particularly by sharpening it. The approach sharpens an object's edges on photographs by accentuating the image's high-frequency components while leaving out the low-frequency components. The underlying notion is straightforward. At first, a 3 × 3 filter is used to blur the input image. The blurred picture is then subtracted from the input image with the additional significance of A = 1. The following equation defines a high-boost filtered image.

$$f_{hb}(x, y) = Af(x, y) - f'(x, y) \tag{1}$$

In this case, x and y represent the horizontal and vertical position of a pixel in the image, respectively where $f_{hb}(x, y)$ represents the high-boost filtered picture, $f(x, y)$ represents the input image, and $f'(x, y)$ represents the blurred form of $f(x, y)$.

3.2.2 Non-local Means Denoising

This article used the detail preserving non-local means (NLM) denoising technique to remove noise from the dataset [19]. The NLM denoising principle is to take the average intensity of neighboring pixels weighted by their similarity. This approach is independent of any imaging model and can be used as a post-processing method. The following equation can define the NLM filter:

$$NLu(p) = \frac{1}{C(p)} \int f(d(B(p), B(q))u(q))dq \quad (2)$$

Here, $NLu(p)$ is the denoised value of the pixel at position p after applying the NLM algorithm. $C(p)$ is the normalizing factor used to scale the contribution of neighboring pixels in the weighted average, and $B(p)$ and $B(q)$ represents the local patches around p and q pixels, respectively. f is the similarity function which assigns weights to neighboring pixels based on the p- and q-centered image patches Euclidean distance $d(B(p), B(q))$. $u(q)$ means the noisy value of the pixel at position q in the input image. The $\int$ sign is the integral sign, which indicates that the weighted average is computed by the NLM filter by integrating over all nearby pixels q in the picture domain. Figure 4 shows the high-boost filter and NLM-applied image.

Figure 4a shows the original image from the dataset, and 4b shows the corresponding preprocessed image. On the original image, a high-boost filter was applied to enhance the image, and then on the resulting image of the high-boost filter, the non-local means denoising technique was applied to reduce the noise from the image. Sometimes high-boost filter incorporates noise into the image while sharpening.

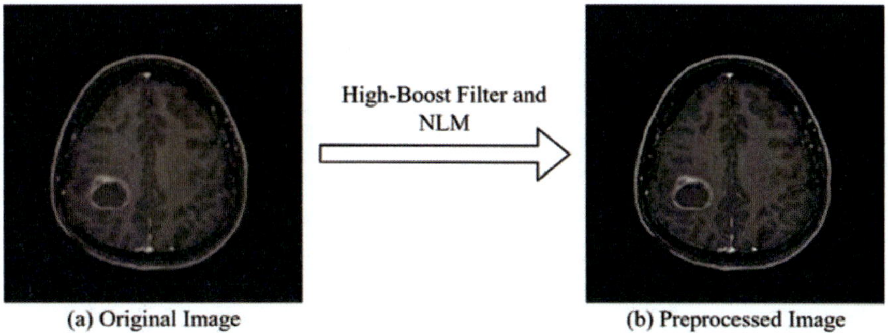

Fig. 4 Image enhancement and noise elimination

Because of that, the non-local means denoising technique was finally applied to rectify them.

3.3 Augmentation

Most ML and DL model's success in any classification task is governed by the dataset's size, quality, and relevance. However, data scarcity exists in the medical industry due to a lack of patients for certain ailments, a lack of medical equipment, and some patients who do not want their data to be utilized. Data augmentation is an approach for artificially expanding the size of a dataset by creating modified or improved variations of the original data. Data augmentation's primary goal is to improve the classification model's generalization and durability. It is often employed in computer vision, particularly medical imaging [20]. Table 1 lists the augmentation methods used in this work.

The number of image data in each class is first computed. Then, calculate which category contains the highest number of photographs. The difference between the values of each class's composed picture and the maximum value was computed. Following that, the required photographs were created using the augmentation methods described above, which were applied randomly to the original photographs. The dataset grew from 4479 to 13,110 after applying this strategy to the original data. Figure 5 displays the dataset distribution after augmentation and balancing.

Table 1 Augmentation methods

Sl. No	Augmentation method	Range
1	Fixed rotation	90°, 180°, 270°
2	Random rotation	[−25°, 25°]
3	Random cropping	[1, 15] pixels from all sides
4	Horizontal flipping	Not applicable
5	Vertical flipping	Not applicable
6	Scaling	[0.8, 0.95]
7	Translation	[−20, 20] pixels from both axis
8	Combination of 3 and 2	[1, 15] pixels and [−25°, 25°]
9	Combination of 4 and 2	N/A and [−25°, 25°]
10	Combination of 5 and 2	N/A and [−25°, 25°]
11	Combination of 6 and 2	[0.8, 0.95] and [−25°, 25°]
12	Combination of 7 and 2	[−20, 20] pixels and [−25°, 25°]

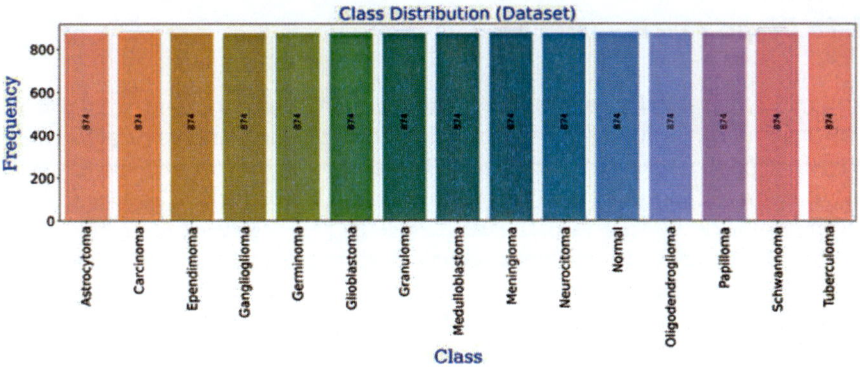

Fig. 5 Balanced dataset's class distribution

3.4 Smote

SMOTE is a common machine learning strategy for dealing with the class imbalance problem in a dataset [21]. SMOTE is an approach for generating samples from the minority population. It is used to generate an intentionally or virtually class-balanced training dataset, which is then used to train the classifier. Because SMOTE is built on the KNN idea, it works well with both text and structured data. However, SMOTE applications have recently gained popularity in the computer vision field [22–25]. For this reason, this paper utilized SMOTE to balance the dataset. Before applying SMOTE, the dataset size was 4479. The SMOTE expanded the dataset size from 4479 to 13,110. SMOTE calculates which class contains the highest number of samples and upsamples the other classes to that number. Figure 5 also illustrates the dataset distribution following SMOTE.

3.5 Dataset Split

Dataset separation is key for evaluating a model's performance and generalizability, simultaneously preventing overfitting and underfitting. This paper featured three dataset versions: the original dataset, the augmentation-applied balanced dataset, and the SMOTE-applied balanced dataset. All three datasets are split in an 80:20 ratio, with 80% utilized for training and 20% used to test the model.

3.6 Deep Learning (DL) Models

In a range of research domains, deep learning models have been used to categorize images. This work employs five cutting-edge deep learning models that were pretrained on the ImageNet dataset [26]. This section summarizes the models used in this research. This paper conducts research with ConvNeXt [27], ResNet50 [28], InceptionV3 [29], VGG19 [30], and EfficientNetB7 [31] deep learning architectures. Several reasons are considered for selecting these deep learning architectures in this paper. Some of them are excellent performance, ability to transfer learning, versatility, availability, community support, etc. This paper uses the mini-batch gradient descent method, and the batch size was 32 for ConvNeXt, ResNet50, InceptionV3, and VGG19. The batch size 16 was used only for EfficientNetB7 due to limited GPU memory.

3.7 Performance Metrics

Evaluation metrics are essential to research since they allow the evaluation of a deep learning model's performance and generalization capacity. Depending on the problem, several evaluation metrics and methodologies are routinely used to evaluate deep learning models. This paper has used accuracy, precision, recall, F1 score, Matthews correlation coefficient (MCC) [32], Hamming loss (HL) [33], Cohen's Kappa (CK) [34], and Jaccard score (JS) [35].

4 Result Analysis and Performance Evaluation

It is possible to understand whether the deep learning model is effectively trained by looking at the model's training and validation accuracy. The closer the training and validation accuracies are, the more generalized the model is. Similarly, training and validation loss gives the notion of the model's error while performing prediction on the training and validation data. The best-performing models (ResNet50) training and validation accuracy and loss curves are presented in Fig. 6a and b, respectively.

Figure 6a shows the best validation accuracy achieved by the ResNet50 model in epoch 15. It means that at epoch 15, the model gives the best performance. However, the training continues to epoch 20. But the model does not improve the validation accuracy and training stops because of the early stopping setting while training. On the other hand, Fig. 6b shows the smallest loss achieved at epoch 17. But at epoch 17, the model does not give the best validation accuracy, and this experiment prioritizes accuracy over loss because accuracy is more important than loss. So, the best weights are saved at epoch 15 for the RestNet50 model.

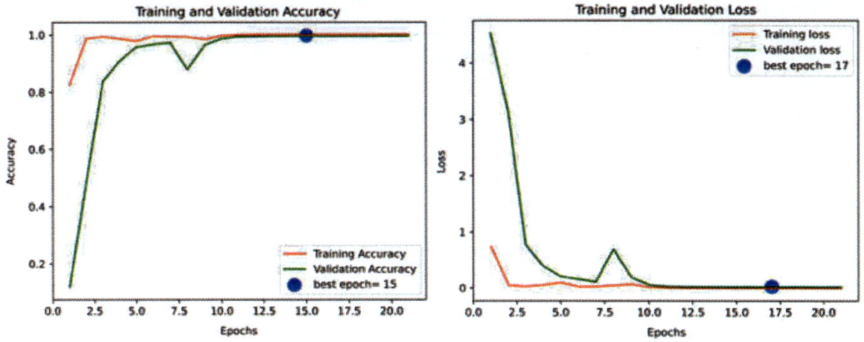

(a) Accuracy vs Epochs (b) Loss vs Epochs

Fig. 6 Accuracy and loss curves

The confusion matrix is another visual performance measure that provides a comprehensive assessment of a classification model's performance. It is a fundamental tool in the field of ML and DL. Figure 7 illustrates the confusion matrix of the best-performing model (ResNet50) while applying SMOTE.

Figure 7 presents that 10 of the 15 classes have 100% recognition accuracy. At the same time, the other five classes give 1 to 3 errors from 175 images which is very few. But the pleasing thing is none of the tumor images did not classified as

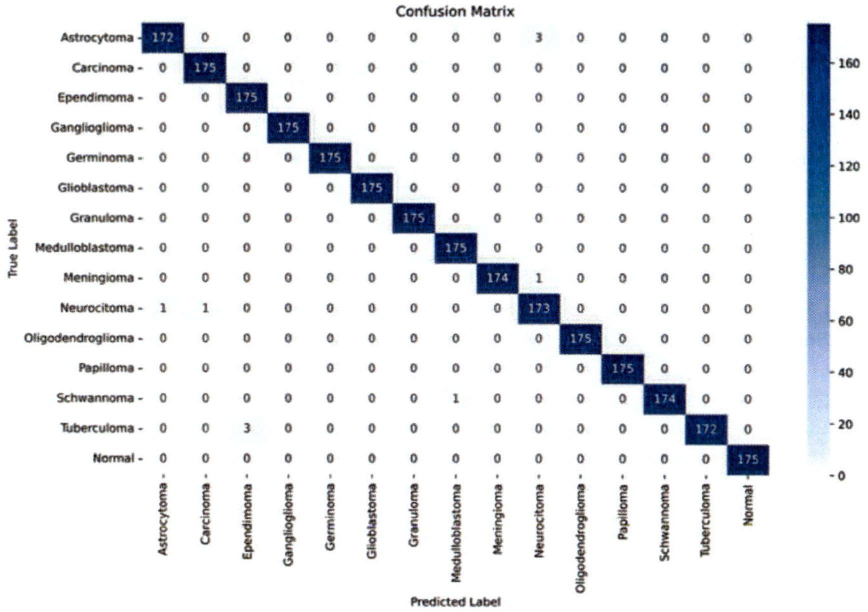

Fig. 7 The confusion matrix

normal, and none of the normal images did not classified as tumor, which indicates the robustness of the proposed framework.

The accuracy, precision, recall, and f1-score are presented in Table 2. The table reveals that, without augmentation and SMOTE, the ConvNeXt model gives better accuracy of 97.23%, recall of 95%, and f1-score of 95%. With augmentation, the ConvNeXt model achieves promising accuracy, precision, recall, and an f1-score of 98.44%, 98%, 98%, and 98%, respectively. While applying SMOTE with the ResNet50 model, this study got an excellent accuracy of 99.62%, precision of 100%, recall of 100%, and f1-score of 100%. So, the recognition error of this research framework is only 0.38%. A dedicated model can be developed, and more training data can be utilized to reduce the recognition error of this application.

On the other hand, Table 3 displays HL, MCC, JS, and CK scores. The table shows that when the ResNet50 model was trained using the SMOTE-applied dataset, it produced the lowest Hamming loss of 0.38%, implying that 0.38% of the classifications were inaccurate. MCC, JS, and CK obtained the highest results for the same dataset and model combination of 99.59%, 99.25%, and 99.59%, respectively. The MCC score of 99.59% indicates that the original and estimated classes are quite close. Furthermore, the values of JS and CK point to the same conclusion: The ResNet50 model is remarkably consistent in identifying brain tumors from MR imaging.

In summary, the results show that combining preprocessing with a high-boost filter and NLM, using SMOTE, and training with the ResNet50 model is a promising technique for detecting brain tumors from MRI images with an unbalanced dataset.

Table 2 Values of accuracy, precision, recall, and f1-score

Applied model	Dataset	Accuracy (%)	Precision (%)	Recall (%)	F1-score (%)
ConvNeXt	Original	97.23	96	95	95
	Augmented	98.44	98	98	98
	SMOTE applied	98.17	98	98	98
ResNet50	Original	96.45	97	93	95
	Augmented	96.30	96	96	96
	SMOTE applied	**99.62**	**100**	**100**	**100**
InceptionV3	Original	93.01	93	88	90
	Augmented	93.64	94	94	94
	SMOTE applied	95.16	95	95	95
VGG19	Original	90.79	92	87	89
	Augmented	92.38	93	92	92
	SMOTE applied	98.51	99	99	99
EfficientNetB7	Original	84.24	82	83	82
	Augmented	64.04	74	64	63
	SMOTE applied	98.40	98	98	98

Table 3 Values of HL, MCC, JS, and CK

Applied model	Dataset	HL (%)	MCC (%)	JS (%)	CK (%)
ConvNeXt	Original	02.77	96.91	94.74	96.90
	Augmented	01.56	98.33	96.94	98.33
	SMOTE applied	01.83	98.05	96.44	98.04
ResNet50	Original	03.55	96.04	93.20	96.03
	Augmented	03.70	96.05	92.94	96.04
	SMOTE applied	**00.38**	**99.59**	**99.25**	**99.59**
InceptionV3	Original	06.99	92.20	87.03	92.17
	Augmented	06.36	93.20	88.10	93.18
	SMOTE applied	04.84	94.83	90.90	94.82
VGG19	Original	09.21	89.73	83.16	89.70
	Augmented	07.62	91.86	85.92	91.84
	SMOTE applied	01.49	98.41	97.14	98.41
EfficientNetB7	Original	15.76	82.72	73.11	82.51
	Augmented	35.96	62.15	46.56	61.47
	SMOTE applied	01.60	98.29	96.87	98.29

However, it is important to note that further study is needed to validate these findings on larger datasets.

5 Explanation Using LIME

Explainable AI is building and developing AI systems so humans can readily comprehend their decisions and actions. Deep learning models are highly complex, making it challenging for humans to comprehend how they arrive at specific conclusions or predictions. However, in a field like health care, lack of transparency may be a major source of concern. So, the capacity to explain AI decisions to establish confidence and guarantee safety is essential. The LIME technique is used in this research to provide model interpretability and explanations for individual predictions [36]. LIME is a well-known approach in the field of explainable AI. Figure 8 shows the images created by LIME.

Figure 8a presents the original image from the test set. Figure 8b depicts the most relevant regions of interest (RoI) or super-pixels detected by LIME in order to anticipate the output. This is easily inspectable from the figure that the model properly identifies the correct region. The most important super-pixels with the background are presented in Fig. 8c. The highlighted regions illustrate how the model arrived at its conclusion. Figure 8d depicts the positive and negative correlations used to categorize the tumors. Green and red colors represent positive and negative correlations,

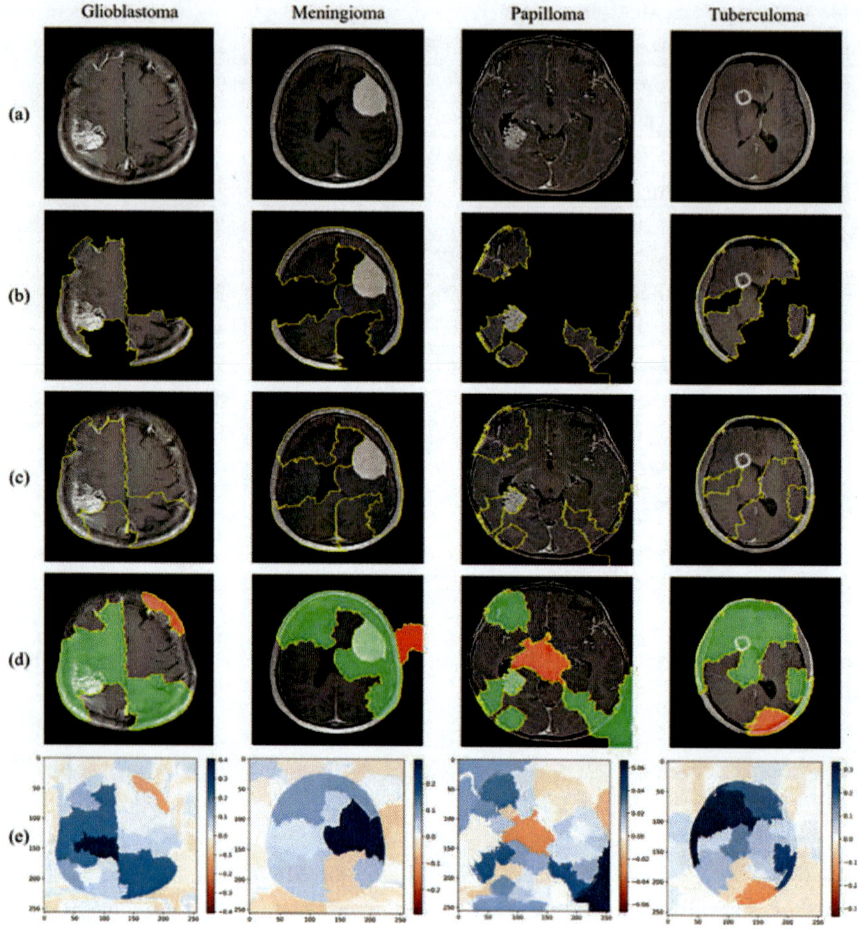

Fig. 8 Generated images by LIME

respectively. The positive correlation implies that characteristics with higher significance ratings influence the model's predictions more. The negative correlation, on the other hand, suggests that variables with greater relevance ratings have a lower influence on the predictions. The heatmap produced by LIME is depicted in Fig. 8e. The following graphic also shows that the model uses tumor form, surrounding region, and location as features for categorization. As a result, this paper infers that the model makes accurate predictions.

6 Conclusion

This research paper has successfully demonstrated the effectiveness and potentiality of recognizing BT from MR imaging using image processing and deep learning. Through the use of the SMOTE-based data balancing technique, the study has provided valuable insights into the capabilities of handling data imbalanced problems. The findings of this research highlight the advantages of image processing and SMOTE over other approaches. From the result analysis and discussion, this study concludes that the combination of image processing and SMOTE-based data balancing technique along with the ResNet50 model achieved excellent accuracy, precision, recall, and f1-score. The accuracy, precision, recall, and f1-score are 99.62%, 100%, 100%, and 100%, respectively. In the future experiment, a new model can be developed with more data to improve the result.

References

1. Brain Tumor: Statistics | Cancer.Net. https://www.cancer.net/cancer-types/brain-tumor/statistics (Accessed 20 Jul 2023)
2. Yang Y et al (2018) Glioma grading on conventional MR images: a deep learning study with transfer learning. Front Neurosci 12:804
3. Adel Fahmideh M, Scheurer ME (2021) Pediatric brain tumors: descriptive epidemiology, risk factors, and future directions. Cancer Epidemiol Biomarkers Prev 30(5):813–821
4. Das S, Aranya OFMRR, Labiba NN (2019) Brain tumor classification using convolutional neural network. In: 2019 1st international conference on advances in science, engineering and robotics technology (ICASERT), pp 1–5
5. Kaur D et al (2022) Computational intelligence and metaheuristic techniques for brain tumor detection through IoMT-enabled MRI devices. Wirel Commun Mob Comput 2022:1–20
6. Musallam AS, Sherif AS, Hussein MK (2022) A new convolutional neural network architecture for automatic detection of brain tumors in magnetic resonance imaging images. IEEE Access 10:2775–2782
7. Rizwan M, Shabbir A, Javed AR, Shabbir M, Baker T, Obe DA-J (2022) Brain tumor and glioma grade classification using Gaussian convolutional neural network. IEEE Access 10:29731–29740
8. Sharif MI, Khan MA, Alhussein M, Aurangzeb K, Raza M (2021) A decision support system for multimodal brain tumor classification using deep learning. Complex Intell Syst pp 1–14
9. Amin J et al (2022) A new model for brain tumor detection using ensemble transfer learning and quantum variational classifier. Comput Intell Neurosci
10. Zahid U et al (2022) BrainNet: optimal deep learning feature fusion for brain tumor classification. Comput Intell Neurosci
11. Ayadi W, Elhamzi W, Charfi I, Atri M (2021) Deep CNN for brain tumor classification. Neural Process Lett 53:671–700
12. Tiwari P et al (2022) Cnn based multiclass brain tumor detection using medical imaging. Comput Intell Neurosci 2022
13. Srinivas C et al (2022) Deep transfer learning approaches in performance analysis of brain tumor classification using MRI images. J Healthc Eng 2022
14. Abd-Ellah MK, Awad AI, Khalaf AAM, Hamed HFA (2019) A review on brain tumor diagnosis from MRI images: practical implications, key achievements, and lessons learned. Magn Reson Imaging 61:300–318

15. Shah HA, Saeed F, Yun S, Park J-H, Paul A, Kang J-M (2022) A robust approach for brain tumor detection in magnetic resonance images using finetuned efficientnet. IEEE Access 10:65426–65438
16. Kumar S, Kumar D (2023) Human brain tumor classification and segmentation using CNN. Multimed Tools Appl 82(5):7599–7620
17. Athisayamani S, Antonyswamy RS, Sarveshwaran V, Almeshari M, Alzamil Y, Ravi V (2023) Feature extraction using a residual deep convolutional neural network (ResNet-152) and optimized feature dimension reduction for MRI brain tumor classification. Diagnostics 13(4):668
18. Fernando F (2023) Brain tumor MRI images 44 classes | Kaggle. Kaggle, 13 Feb 2023. https://www.kaggle.com/datasets/fernando2rad/brain-tumor-mri-images-44c (Accessed 14 Jun 2023)
19. Buades A, Coll B, Morel J-M (2011) Non-local means denoising. Image Proc On Line 1:208–212
20. Goceri E (2023) Medical image data augmentation: techniques, comparisons and interpretations. Artif Intell Rev 1–45
21. Chawla NV, Bowyer KW, Hall LO, Kegelmeyer WP (2002) SMOTE: synthetic minority over-sampling technique. J Artif Intell Res 16:321–357
22. Chamseddine E, Mansouri N, Soui M, Abed M (2022) Handling class imbalance in COVID-19 chest X-ray images classification: Using SMOTE and weighted loss. Appl Soft Comput 129:109588
23. Feng W, Huang W, Bao W (2019) Imbalanced hyperspectral image classification with an adaptive ensemble method based on SMOTE and rotation forest with differentiated sampling rates. IEEE Geosci Remote Sens Lett 16(12):1879–1883
24. Özdemir A, Polat K, Alhudhaif A (2021) Classification of imbalanced hyperspectral images using SMOTE-based deep learning methods. Expert Syst Appl 178:114986
25. Dablain D, Krawczyk B, Chawla NV (2022) DeepSMOTE: fusing deep learning and SMOTE for imbalanced data. IEEE Trans Neural Netw Learn Syst
26. Deng J, Dong W, Socher R, Li L-J, Li K, Fei-Fei L (2009) Imagenet: A large-scale hierarchical image database. IEEE Conf Comput Vision Pattern Recogn 2009:248–255
27. Liu Z, Mao H, Wu C-Y, Feichtenhofer C, Darrell T, Xie S (2022) A convnet for the 2020s. In: Proceedings of the IEEE/CVF conference on computer vision and pattern recognition, pp 11976–11986
28. He K, Zhang X, Ren S, Sun J (2016) Deep residual learning for image recognition. In: Proceedings of the IEEE conference on computer vision and pattern recognition, pp 770–778
29. Szegedy C, Vanhoucke V, Ioffe S, Shlens J, Wojna Z (2016) Rethinking the inception architecture for computer vision. In: Proceedings of the IEEE conference on computer vision and pattern recognition, pp 2818–2826
30. Simonyan K, Zisserman A (2014) Very deep convolutional networks for large-scale image recognition. arXiv preprint arXiv:1409.1556
31. Tan M, Le Q (2019) Efficientnet: rethinking model scaling for convolutional neural networks. In: International conference on machine learning, pp 6105–6114
32. Baldi P, Brunak S, Chauvin Y, Andersen CAF, Nielsen H (2000) Assessing the accuracy of prediction algorithms for classification: an overview. Bioinformatics 16(5):412–424
33. Tsoumakas G, Katakis I (2007) Multi-label classification: an overview. Int J Data Warehous Min (IJDWM) 3(3):1–13
34. Cohen J (1960) A coefficient of agreement for nominal scales. Educ Psychol Meas 20(1):37–46
35. Chung NC, Miasojedow B, Startek M, Gambin A (2019) Jaccard/Tanimoto similarity test and estimation methods for biological presence-absence data. BMC Bioinf 20(15):1–11
36. Ribeiro MT, Singh S, Guestrin C (2016) Why should i trust you?' Explaining the predictions of any classifier. In: Proceedings of the 22nd ACM SIGKDD international conference on knowledge discovery and data mining, pp 1135–1144

Histopathology Breast Cancer Classification Using CNN

M. Venkateshwara Rao, Rajesh Saturi, D. Srinivas Goud, G. Srikanth Reddy, and N. Venkatesh

Abstract Breast cancer is the second leading cause of cancer mortality among women, early detection and treatment of breast cancer is very critical for the reduction of mortality rate. Cancer, known for its devastating impact, poses significant challenges in early detection within the medical field. In this study, breast cancer image samples were utilized for investigation purposes from BreakHis dataset. For accurate identification, current deep learning approaches relies on contour information. However, the cancer detection approach which is already existed retrieved only information which is semantic from the first layer and did not offer insights on the semantic levels which are not deep in breast cancer detection. Deep learning is a highly effective technique that possesses the capability to extract and structure significant insights from data, thus eliminating the necessity for domain experts to manually design feature extractors. Convolutional Neural Networks (CNNs), which belong to the category of deep, feed forward networks, have gained significant interest from researchers and industries. The recognition achieved by CNNs can be attributed to their integration in various domains such as transfer learning, speech recognition, natural language processing, signal processing, and object recognition. The proposed framework incorporates an effective and efficient strategy derived from a comprehensive review of commonly used approaches in each category. In the context of cancer screening, the process involves classifying image biopsies as either malignant or benign. Medical professionals like doctors and pathologists highly rely

M. Venkateshwara Rao (✉) · R. Saturi · D. Srinivas Goud · G. Srikanth Reddy
Department of CSE, Vignana Bharathi Institute of Technology, Ghatkesar, Hyderabad, India
e-mail: venkat2m2@gmail.com

R. Saturi
e-mail: rajesh.saturi@vbithyd.ac.in

G. Srikanth Reddy
e-mail: srikanth.gangidi@vbithyd.ac.in

N. Venkatesh
School of CS & AI, SR University, Ananthsagar, Hasanparthy (M), Warangal Urban, Telangana, India

S. Shakya et al. (eds.), *Fourth International Conference on Image Processing and Capsule Networks*, Lecture Notes in Networks and Systems 798,
https://doi.org/10.1007/978-981-99-7093-3_36

on the analysis of microscopic biopsy images to identify and categorize different abnormalities.

Keywords Histopathology images · Benign · Malignant machine learning · Deep learning · Convolutional neural network · Hematoxylin and Eosin (H&E)

1 Introduction

Cancer has become a significant global community health concern in recent times. A part of the World Health Organization (WHO), predicts nearly 10 million cancer-related deaths in 2020. Among the different forms of cancer, breast cancer (BC) is considered as the most commonly diagnosed type among women and results in a high mortality rate. Despite the advancements in understanding cancer biology and the identification of molecular markers, histopathological analysis remains as the effective diagnostic method.

> According to the report of IARC, breast cancer has a high mortality and morbidity rate among females. In India, breast cancer is considered as the second-largest chronic disease that affects around 300,000 people each year. Therefore, early-stage treatment and diagnosis can reduce the mortality rate, where the early diagnosis of breast cancer involves the detection of two major types of tumours: malignant and benign [1].

Breast cancer originates from breast tissue and is typically identified through the presence of a breast lump and certain deviations from normal conditions. Clinical screening for breast cancer involves techniques such as mammography, breast ultrasound, and other modalities. However, traditional manual diagnosis places a substantial burden on experts and is prone to diagnostic errors, impacting accuracy (Fig. 1).

> Breast cancer is the second most cause of cancer deaths in women after lung cancer. The death rate of women due to breast cancer is around 1 in 37 individuals that amount to 2.7% [1].

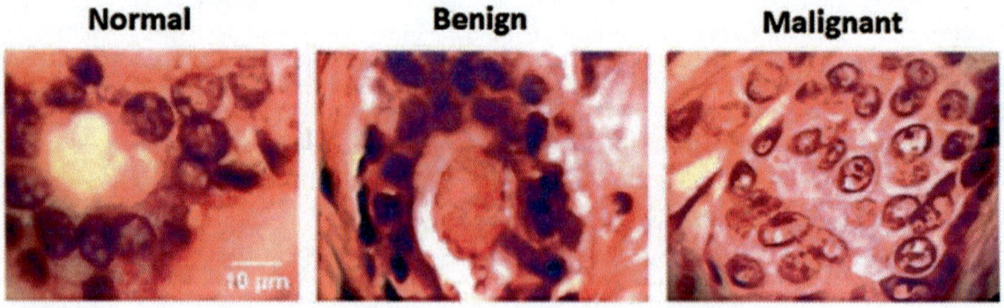

Fig. 1 Microscopic biopsy images

Deep learning, an evolving technology within machine learning, has attracted significant interest from researchers. Convolutional Neural Networks have demonstrated remarkable efficacy in large-scale image and video recognition tasks. As for example, the study conducted by Spanhol et al., the AlexNet architecture was utilized to successfully classify breast cancer pathology images, accurately differentiating between benign and malignant categories. Outperforming traditional machine learning algorithms with a 9% improvement in classification results. The methodology employed by the researchers resulted in an impressive average accuracy of 94% when classifying breast cancer cases images. Furthermore, Kausik et al. made a significant contribution by introducing an innovative framework known as Multiple Instance Learning (MIL) specifically designed for Convolution Neural Network (CNNs). This framework aims to improve the classification accuracy of pathological breast cancer images [2].

Nevertheless, the utilization of CNNs for the automatic categorization of pathological breast cancer images continues to pose significant challenges. One reason is the rapid growth in the number of CNN parameters as the model deepens, leading to a higher risk of overfitting. The acquisition of a substantial volume of labelled breast cancer images for training Convolutional Neural Networks (CNNs) is a costly endeavour, presenting a significant obstacle in the field [3].

2 Literature Survey

Rezk et al. [4] came up with a new data sampling approach for automatic histology breast cancer classification. At the start, the data sampling approach generates samples which is based on the data distribution of binary patterns. The developed sampling approach has helped to categorize the breast cancer histology image regions as cancerous or noncancerous. "In this literature, the performance of the developed sampling approach was compared with a few existing classical sampling approaches. From the obtained results, the developed model attained high performance compared to the existing approaches in terms of f-measure and accuracy." "The major drawback of the data sampling approach is the exact matching between the images, sometimes it may lead to information loss" [1].

The outline information was used by the current deep learning techniques for breast cancer screening to deliver effective results. But because they didn't make enough use of semantic details, the existing methods performed worse. Rajesh Saturi suggested a multi-objective feature selection strategy for the identifying of breast cancer in order to address of this problem. This method combines ACO with PSO. The BreakHis dataset is utilized in the suggested method along with the data augmentation was used to examine the random rotation and shearing transformation used to Combat the issue of over fitting [5].

Nahid et al. [6] came up with a Deep Neural Network (DNN) for diagnosis of breast cancer. In this study statistical and structural information extracted from the images collected from BreakHis dataset. "The performance of DNN was compared with a few existing classifiers: CNN, SVM, and Long Short-Term Memory (LSTM). The simulation outcome showed that the DNN classifier attained better performance compared to other classifiers in the light of accuracy, precision, and f-measure. Generally, deep learning classifiers like DNN need an enormous amount of data for achieving better classification, which is a major concern in the applications like histopathology breast cancer detection" [6].

In this study, DNN uses the statistical and structural data produced from the gathered photos to categorize a collection of biological breast cancer images from the BreakHis database. In simpler terms, machine learning involves computers discovering valuable insights, while the DNN classifier outperformed alternative classifiers in terms of accuracy and precision and f-call measure. In order to get better classification, deep learning classifier like DNN typically require a huge amount of data [7].

A. *Machine Learning*

ML, categorized under the field of artificial intelligence (AI), enables software applications to enhance their predictive capabilities and make more precise forecasts without relying on explicit programming specifically designed for making those predictions. ML algorithms leverage historical and statistical data to predict new output values Fig. 2 [8]. While the discovery of new molecular markers is ongoing, histopathological analysis remains the most commonly utilized method for diagnosing cancer. Breast cancer, characterized by abnormal changes in breast tissue and the presence of a lump, is typically diagnosed using techniques such as mammography, breast ultrasound, and others. However, manual diagnosis places a significant burden on experts and is prone to diagnostic errors, affecting diagnostic accuracy [9]. Deep learning, a branch of machine learning, has attracted considerable attention from researchers. Convolutional Neural Networks (CNNs) have exhibited exceptional efficacy by recognizing the images and videos.

Spanhol et al. [10] conducted a study, where they successfully achieved the classification of breast cancer pathological images into benign and malignant categories

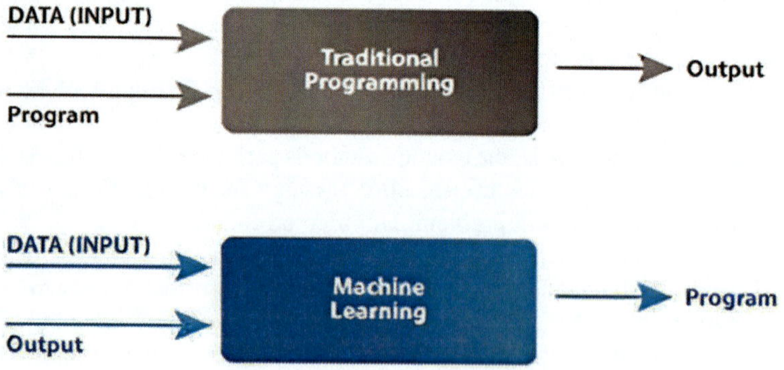

Fig. 2 Traditional machine learning structure

using AlexNet architecture. Machine learning algorithms learn from data in an iterative manner without explicit instructions on where to focus their search. As more data is provided, the algorithms gain a deeper understanding, leading to improved output quality. Machine learning increases the ability to autonomously and iteratively adapt to new data.

The process of machine learning is first initialized by providing training data to the selected algorithm, which employs pattern recognition to generate reliable and informed outcomes. The composition of the training data, whether it is already known or unknown holds significant importance in the development of a machine learning algorithm. Subsequently, new input data is introduced to assess the algorithm's performance and accuracy. By comparing the predictions with the actual results, adjustments are made to the algorithm through re-training by data scientists until the desired outcome is achieved. Through this iterative process, the machine learning algorithm gains the ability to learn autonomously and enhance its accuracy progressively as time progresses [11, 12].

B. *Image Processing*

An image can be represented by a 2D function denoted as $f(x, y)$, where x and y correspond to the spatial coordinates. The intensity of the images at a given location is directly linked to the magnitude of F at the respective x and y values. The intensity of the image at a specific location is directly associated with the amplitude of F at the corresponding x and y values [13]. In the realm of digital images, the values of x, y, and amplitude are all finite. An image is essentially an array of pixels organized in rows and columns. Each pixel stores information about colour and intensity. Furthermore, by incorporating spatial coordinates x, y, and z, a three-dimensional representation of a photograph can be constructed. The process of pixelization employs a matrix-style layering technique, commonly known as an RGB image. It's worth noting that images come in various forms, such as RGB and Greyscale images [14, 15].

The field of image processing involves the implementation of diverse techniques on the image to improve its overall quality or extract valuable information. In this type of signal processing, image serves as the input, and either the image itself or its features are produced as the output. Image processing is a rapidly evolving technology that undergoes significant advancements. It holds great importance in the fields of engineering and computer science, making it a crucial topic of study [16].

3 Existing System

A traditional neural network, multilayer perceptron (MLP) is designed to mimic the human brain. MLPs have a number of shortcomings, particularly in terms of image processing. For each input (such as a pixel in a picture multiplied by 3 in the RGB case), MLPs use one perceptron. When dealing with large images, the sheer number of weights involved can become overwhelming. For instance, training a 224×224

pixel image with three colour channels require managing approximately 150,000 weights. As a result, training might be challenging, and overfitting can also occur [17].

4 Proposed System

In order to extend this, input includes the images from the dataset BreakHis. These are Hematoxylin and Eosin (H&E) stained images as they are gathered from different sources and put together to form this dataset. The dataset contains 400 images belonging to 2 different classes benign and malignant each of 100 images [18]. Input image will be pre-processed using different techniques like colour normalization and then the image is divided into patches. After that, the image will be passed through two trained CNN models. The output class of image is obtained by applying majority voting scheme to two CNN models. Output will be one of the two classes, benign and malignant.

5 Methodolgy

Step 1: Collecting data from BreakHis Dataset

The dataset utilized here consists of two types of microscopic biopsy images: one depicting breast cancer (referred to as Fig. 1) and the other representing noncancerous cells (referred to as Fig. 1). The dataset utilized in this research is comprised of 1693 microscopic images of breast tumour tissue obtained from the Breast Cancer Histopathological Database (BreakHis) refer Fig. 4. "These images were gathered from 82 patients and included in the database for further analysis. Among these images, 547 were classified as benign, representing noncancerous cells, while 1147 were classified as malignant, representing cancerous cells. The images in the dataset possess dimensions of 700 × 460 pixels and are stored in the PNG format. Additionally, they follow a 3-channel RGB configuration, with each channel having a 8-bit depth refer Table 1" [3].

Table 1 BreakHis dataset

Magnification	Benign	Malignant	Total
40X	652	1370	1995
100X	644	1437	2081
200X	623	1390	2013
400X	588	1232	1820
Total of Images	2480	5429	7909

The dataset is created with a collaborative effort from the Parana's P&D Laboratory—Pathological Anatomy and Cytopathology in Brazil. The dataset includes the samples obtained through the utilization of the surgical open biopsy (SOB) method, specifically a partial mastectomy or excisional biopsy technique [19]. "Unlike other types of needle biopsies, this procedure involves extracting a larger tissue sample and is performed in a hospital setting under general anaesthesia" [20].

A skin incision is done during an open biopsy to expose the tumour and enable for the cutting or scraping of a sample of tissue. Additional surgery could be done based on the results of the lab tests [21].

Step 2: Algorithm

Deep learning neural networks created specifically for the analysis of structured arrays of data-like representations. ConvNet does not use matrix multiplications, which is often how neural networks are thought of. It makes use of the convolutional approach refer Fig. 5 [22]. Convolution allows the combination of two functions to generate a third function that visually represents the transformation of one function's shape caused by the other refer Fig. 3. CNNs are efficient in identifying distinctive elements from input images, including lines, gradient rings, even eyes and face. Convolutional Neural Networks outperform in computer vision tasks due to its specific attribute function. CNNs have the capability to operate directly on lower-quality images without requiring pre-processing [23].

6 Implementation

In this paper the proposed model is implemented in anaconda platform, and python programming language is used to run this project.

7 Result

In accordance with the proposed experimental procedure, the BreakHis dataset has been partitioned into 70% training and 30% testing. To ensure the classifier's generalization to unobserved patients, the dataset was divided in such a way that the patients used to construct the training set were excluded from the creation of the testing set. The findings in this study reflect the average of five experiments. Each of the possible magnifications received its own unique application of this technique [24] .

Refer Table 2 shows the classification table for the proposed method in order to measure the two CNN models (0 and 1) based on the performance metrics like precision, recall, F1-score, and support.

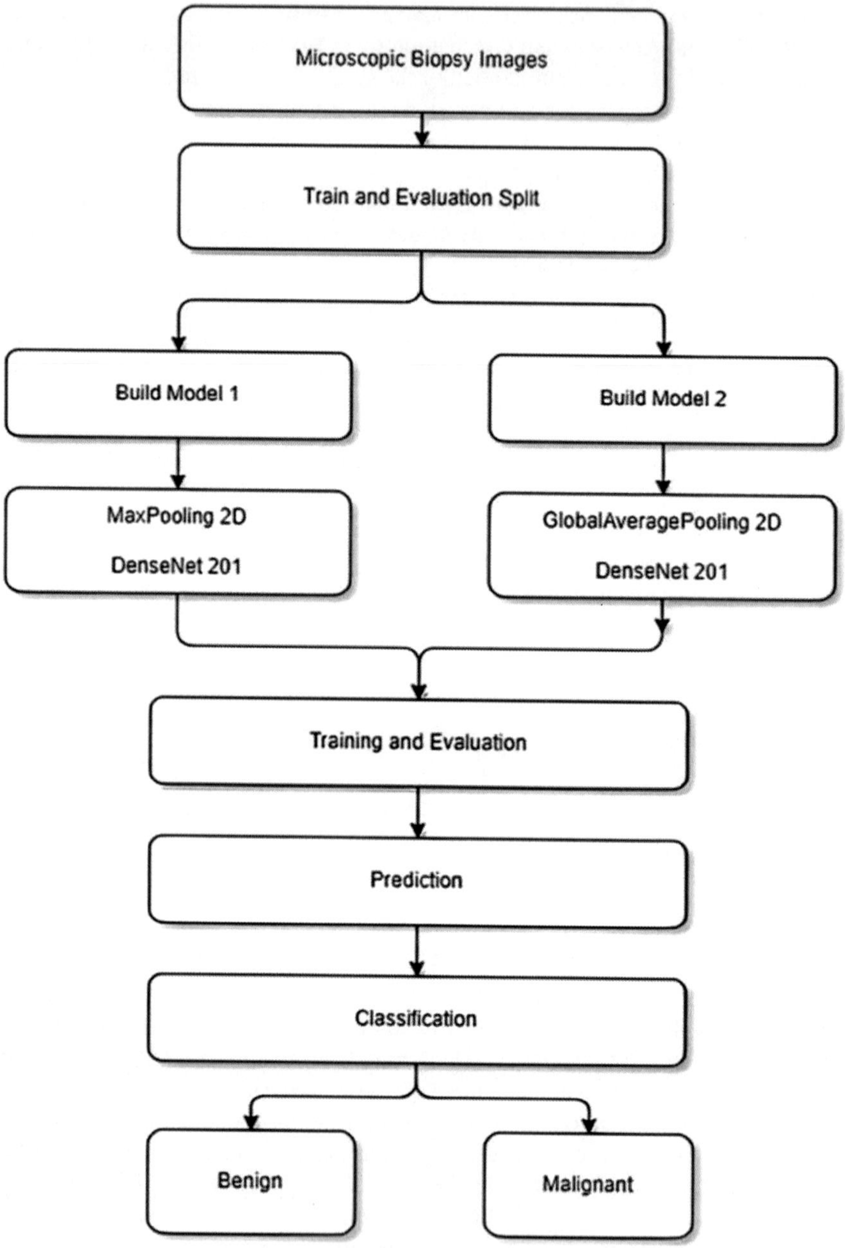

Fig. 3 Proposed method architecture

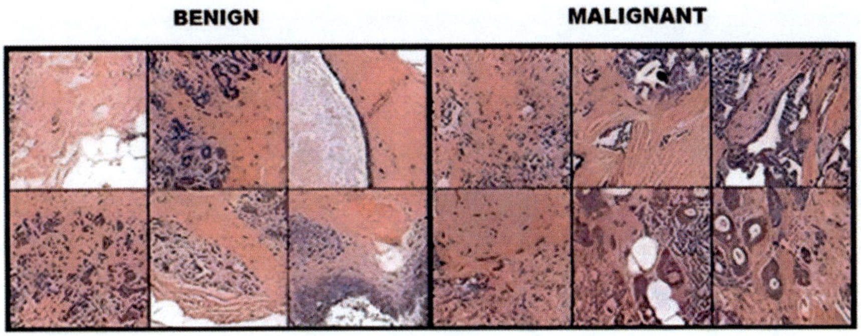

Fig. 4 Classification of cancerous tissue

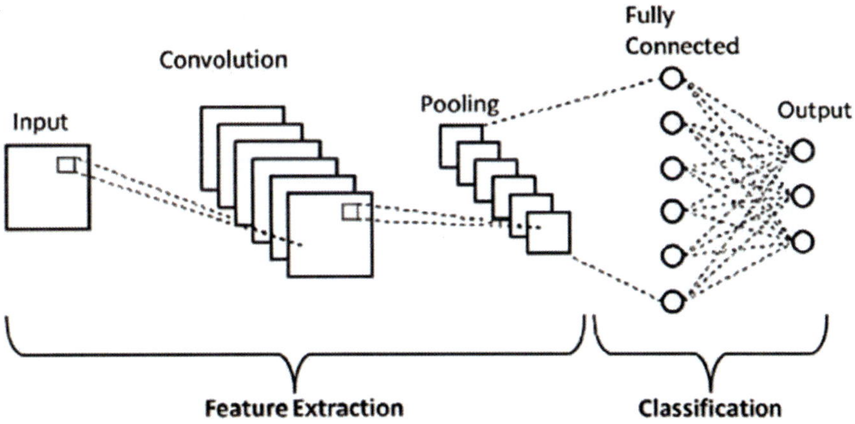

Fig. 5 CNN architecture

Table 2 Classification table

	Precision	Recall	F1-score	Support
0	0.97	0.73	0.83	176
1	0.89	0.99	0.93	369
Accuracy			0.91	545
Macro average	0.93	0.86	0.88	545
Weighted average	0.91	0.91	0.90	545

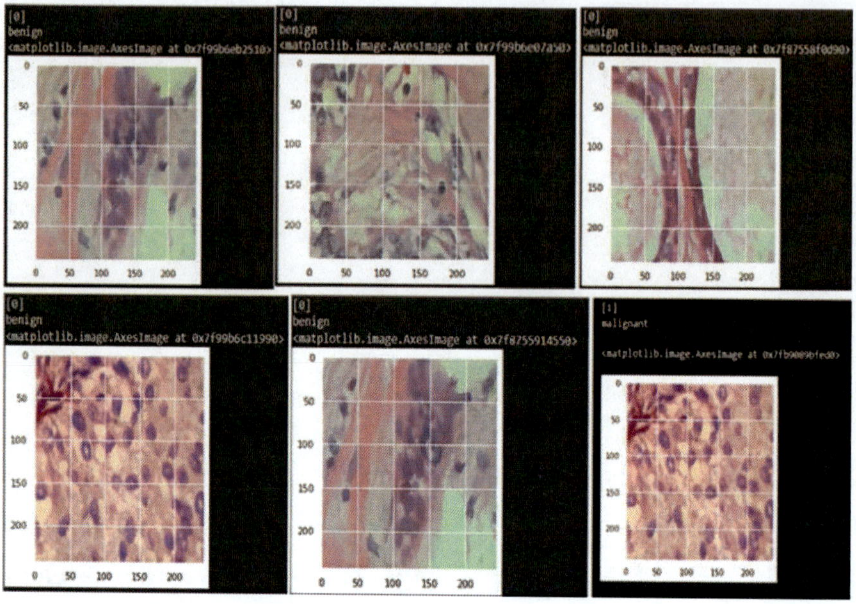

Fig. 6 Test model of the programme

8 Conclusion and Future Work

Developing automated breast cancer screening for improved patient care poses significant challenges. The main purpose of this learning was to evaluate and compare the effectiveness of two discrete Convolutional Neural Network (CNN) architectures in the detection of breast cancer. The dataset employed in this study includes 1693 microscopic images of breast tumour tissue, gathered from a cohort of 82 patients. Among these images, 547 were classified as benign, while 1147 were classified as malignant. The images were given in PNG format, with dimensions of 700 × 460 pixels and a 3-channel RGB configuration, each channel having an 8-bit depth. Employing CNN Model 2, the proposed system achieved an accuracy rate of 96%, outperforming CNN Model 1 with 93% accuracy refer Figs. 6 and 7. The primary goal of this research work is to provide support to healthcare professionals and oncologists in the precise diagnosis of breast cancer. Looking towards the future, the incorporation of artificial intelligence and machine learning into the diagnostic process has the potential to reduce human errors and enhance efficiency and facilitate early cancer detection, thereby empowering individuals in their Fig. ht against this disease, in future deep learning techniques can be applied to improve the accuracy of the proposed system with multi-objective functions.

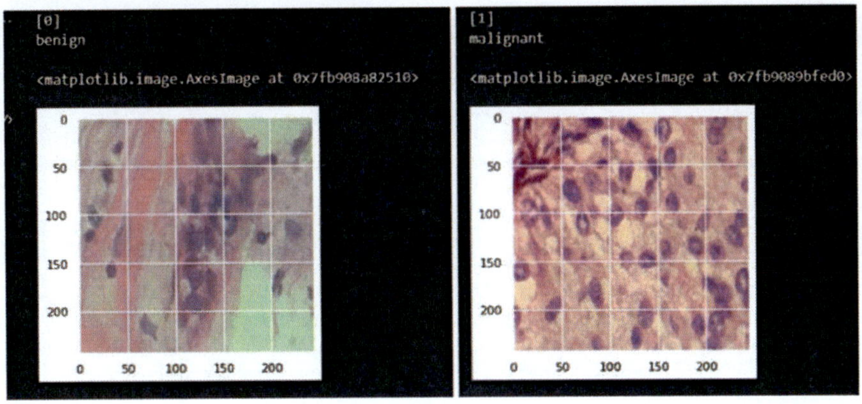

Fig. 7 Test model output

References

1. Saturi R, Parvataneni PC (2022) Histopathology breast cancer detection and classification using optimized superpixel clustering algorithm and support vector machine. Received: 22 Mar 2021. Accepted: 6 Apr 2022 The Institution of Engineers (India) 2022
2. Saturi R, Chand PP, A Novel variant optimized search algorithm for nuclei cell detection in histopathogy breast cancer images. Department of CSE, University College of Engineering, Osmania University, India
3. Saturi R, Implement an efficient segmentation method to overcome the issues of overlapping by segmenting the histopathological images of cancer by Research Scholar. Department of Computer Science & Engineering, University College of Engineering
4. Rezk E et al (2017) Conceptual data sampling for breast cancer histology image classification. Comput Biol Med 89:59–67
5. Saturi R, Premchand P, Multi-objective feature selection method by using ACO with PSO algorithm for breast cancer detection. Department of Computer Science and Engineering, University College of Engineering, Osmania University, India
6. He K, Zhang X, Ren S, Sun J (2016) Deep residual learning for image recognition. In: 2016 IEEE, conference on computer vision and pattern recognition (CVPR). Las Vegas, NV, pp 770–778
7. Huang G, Liu Z, van der Maaten L, Weinberger KQ (2017) Densely connected convolutional networks. In: 2017 IEEE conference on computer vision and pattern recognition (CVPR). IEEE. pp 2261–2269. 10. 1109/CVPR.2017.243
8. Fawcett T (2006) An introduction to ROC analysis. Pattern Recogn Lett 27(8):861–874. https://doi.org/10.1016/j.patrec.2005.10.010
9. Golatkar A, Anand D, Sethi A, Classification of breast cancer histology using deep learning
10. Spanhol FA, Oliveira LS, Petitjean C, Heutte L (2016) Breastcancer histopathological image classification using convolutional neural networks. In: International joint conference on neural networks (IJCNN). IEEE, pp 2560–2567
11. Roy K, Banik D, Bhattacharjee D, Nasipuri M (2019) Patch- based system for classification of breast histology images using deep learning. Comput Med Imaging Graph 71:90–103. ISSN 0895-6111
12. Kowal M, Filipczuk P, Obuchowicz A, Korbicz J, Monczak R, Computer-aided diagnosis of breast cancer based on fine needle biopsy microscopic images
13. Ordi J, Castillo P, Saco A, del Pino M, Ordi O, Rodr´ıguez-Carunchio L, Ram´ırez J, Validation of whole slide imaging in the primary

14. Prasoon A, Petersen K, Igel C, Lauze F, Dam E, Nielsen M (2013) Deep feature learning for knee cartilage segmentation using a triplanar convolutional neural network. In: Medical image computing and computer assisted intervention—MICCAI 2013 ser, vol. 8150. Springer, Berlin, pp 246–253
15. George YM, Zayed HH, Roushdy MI, Elbagoury BM (2014) Remote computer-aided breast cancer detection and diagno
16. Krizhevsky A, Sutskever I, Hinton GE (2012) Imagenet classification with deep convolutional neural networks. In: Advances in neural information processing systems, pp 1097–1105
17. Schnitt, Elmore JG (2014) Digitized whole slides for breast pathology interpretation: Current practices and perceptions. J Digit Imaging 27(5), 642–648, Oct 2014
18. Janowczyk A, Madabhushi A (2016) Deep learning for digital pathology image analysis: A comprehensive tutorial with selected use cases. J. Pathol. Inf. 7
19. Araujo T, Aresta G, Castro E, Rouco J, Aguiar P, Eloy C, Pol´onia A, ´Campilho A (2017) Classification of breast cancer histology images using convolutional neural networks. PloS one 12(6):e0177544
20. Strigl D, Kofler K, Podlipnig S (2010) Performance and scalability of GPU-based convolutional neural networks. In: Parallel, distributed and network-based processing (PDP), 2010 18th Euromicro International Conference IEEE, 2010, pp 317–324. [15] LeCun Y, Bottou L, Bengio Y, Haffner P (1998) Gradientbased learning applied to document recognition. Proc IEEE 86(11):2278–2324
21. Duffner S (2008) Face image analysis with convolutional neural networks, Ph.D. dissertation. University of Freiburg Freiburg, Germany
22. LeCun Y, Kavukcuoglu K, Farabet C (2010) Convolutional networks and applications in vision. In: Proceeding of the 2010 IEEE international symposium circuits and systems (ISCAS), pp 253
23. He K, Zhang X, Ren S, Sun J (2015) Delving deep into rectifiers: Surpassing human-level performance on imagenet classification. In: Proceedings of the IEEE international conference on computer vision, pp 1026–1034
24. Pedregosa F, Varoquaux G, Gramfort A, Michel V, Thirion B, Grisel O et al (2011) Scikit-learn: machine learning in python. J Mach Learn Res 12:2825–2830

Automated Wave Height Measurement and Analysis Through Image Processing: A Computer Vision Approach for Oceanography

Medha Wyawahare, M. Selva Balan, Anvay Deshpande, Priyanka Acharya, Ankita Kumari, and Ashfan Khan

Abstract Wave height is an important parameter in ocean and coastal engineering. It is used to design coastal structures, such as breakwaters and seawalls and to predict the impact of waves on coastal areas, such as erosion and flooding. Traditional methods for measuring wave height, such as wave gauges are expensive and time-consuming to install and maintain. Image processing techniques offer a more cost-effective and efficient way to measure wave height. In this paper, a wave height detection algorithm based on image processing is presented. The algorithm uses a combination of edge detection and thresholding techniques to identify the wave crests in a sequence of images. The wave height is then calculated as the distance between the wave crests. The algorithm was tested on a dataset of images of ocean waves and was shown to be accurate and reliable.

Keywords Ocean wave · Wave height · Buoys · Centroid displacement · Image processing

Supported by CWPRS, Pune

M. Wyawahare (✉) · A. Deshpande (✉) · P. Acharya · A. Kumari · A. Khan
Vishwakarma Institute of Technology, Pune, Maharashtra, India
e-mail: medha.wyawahare@vit.edu

A. Deshpande
e-mail: anvay.deshpande20@vit.edu

P. Acharya
e-mail: priyanka.acharya20@vit.edu

A. Kumari
e-mail: ankita.kumari20@vit.edu

A. Khan
e-mail: ashfan.khan20@vit.edu

M. S. Balan
Central Water and Power Research Station, Khadakwasla, Pune, Maharashtra, India

S. Shakya et al. (eds.), *Fourth International Conference on Image Processing and Capsule Networks*, Lecture Notes in Networks and Systems 798,
https://doi.org/10.1007/978-981-99-7093-3_37

1 Introduction

Wave height detection refers to the process of determining the measurement of wave height in a body of water. This measurement holds significant importance in various fields, including oceanography, marine engineering, and coastal management. The height of waves provides valuable insights into the energy and momentum of ocean currents, aids in modeling coastal erosion and flooding, and facilitates the design of offshore structures. Traditionally, wave height has been assessed through the utilization of sensors such as wave buoys, which employ pressure sensors or accelerometers to gauge the pressure fluctuations caused by waves. Another approach involves using radar sensors to measure the distance between the water surface and a fixed point, and subsequently calculating the height of waves based on the observed changes in distance. However, the present techniques for measuring wave height have certain limitations in terms of accuracy, precision, cost, and maintenance.

In recent years, a new technology has been developed that offers improved accuracy and efficiency. This technology involves utilizing a single high-speed camera to capture continuous images of the water surface. By employing advanced image processing techniques, the wave height information can be extracted from these images.

Camera-based wave height detection has several advantages when compared with traditional methods. It is more cost-effective and low-maintenance than sensor-based approaches, while also offering higher spatial resolution and greater measurement accuracy. Furthermore, camera-based methods provide additional insights into wave shapes and dynamics, which prove valuable for comprehending and simulating ocean currents.

Overall, camera-based wave height detection represents an exciting new approach to measure wave height, with the potential to improve the understanding of ocean dynamics and inform coastal engineering projects.

2 Related Work

A study aims to validate using video for wave height estimation. A fully developed image processing algorithm is presented and tested through field studies. The system, called WavePack, uses a camera, acquisition unit, and processing unit to detect and estimate near shore wave height from digital video sequences. Results from testing at Narrow Neck Beach, Australia show that the system accurately estimates wave heights with a Pearson correlation coefficient of 0.86 [1].

One of the research uses Sentinel 1 Synthetic Aperture Radar (SAR) images and National Data Buoy Center (NDBC) buoy measurements. SAR images are processed through calibration, subset selection, and filtering to estimate effective dominant wavelength and significant wave height. NDBC buoy measurements are used to validate the estimated values. The estimation result is better for the swell sea state than the wind wave-dominated sea surface [2].

The long-term automated monitoring of Nearshore Wave Height from Digital Video uses an algorithm that detects brightness variations in video frames to automatically monitor and measure wave heights in the nearshore area of the ocean. The brightness variations are caused by the movement of waves and are used to calculate wave height and period. This non-intrusive method can be used in any weather condition and calibrated to account for factors such as camera tilt, atmospheric distortion, and camera position relative to the ocean surface [3].

The author has used a convolutional neural network (CNN)-based classification model to estimate wave height levels from a single ocean image and a regression model for sequential images. ResNet was found to be the best convolutional neural network model for ocean image processing due to its high accuracy, short training time, and balanced correction rate. The paper proposed a bidirectional ConvLSTM (long short-term memory)-based regression model to estimate real-valued significant wave heights from sequential ocean images. The model extracted spatio-temporal features using VGGNet and ConvLSTM, and achieved low error rates in terms of mean absolute error and mean absolute percentage error [4].

The researchers first used convolutional neural network (CNN) to extract features from the images. The extracted features show the change in wave heights which was useful for long short-term memory (LSTM). LSTM is used for wave height prediction and analysis has been done on the proposed convolutional neural network-long short-term memory (CNN-LSTM)-based approach with other algorithms such as LSTM, linear regression and support vector machine (SVM) [5].

The authors have measured the dynamic wave height of tuned liquid column dampers (TLCDs) using a digital camcorder for recording the video of a laterally shaking TLCD with blue dye applied to highlight the liquid. The vision-based sensing method's image processing algorithm speeds up the assessment of wave height by simply counting white pixels in a binary image. Edge detection and moving object tracking are two sequential image processing techniques that have been applied [6].

The authors have presented a framework for predicting wave conditions using machine learning models as a substitute for the industry-standard Simulating Waves Nearshore (SWAN) tool. The authors utilized a multi-layer perceptron (MLP) regression to replicate significant wave height generated by SWAN, and a support vector machine (SVM) to classify wave periods. The models are over 4000 times faster than SWAN, while still producing similarly accurate results within the targeted domain [7].

The authors have presented a backpropagation (BP) neural network model optimized by mind evolutionary algorithm-backpropagation (MEA-BP) for predicting ocean wave heights. Wind speed value (UV) and mean sea level pressure (MSL) are selected as input features to predict ocean wave height. The mind evolutionary algorithm is employed to optimize the BP neural network model. The results show that the (MEA-BP) model outperforms existing methods such as genetic algorithm-backpropagation (GA-BP) and standard backpropagation (SBP) in predicting ocean wave heights, achieving high accuracy and correlation coefficients [8].

The authors have trained a model which was able to predict the wave parameters of the next one hour with the help of long short-term memory (LSTM). The learning algorithm, stochastic gradient descent (SGD) is employed when there is a need to calculate the local minima of a function whereas adaptive moment estimation (ADAM) is an extensive version of stochastic gradient descent (SGD). The accuracy of the stochastic gradient descent (SGD) algorithm was better when compared with the adaptive moment estimation (ADAM) algorithm and was able to predict with an error of less than 20 [9].

The past data has been used in order to train a model using various sine functions or better known as multiple sine functions decomposition (MSFD) The authors have calculated the coefficients of the resultant function from training data and substituted their values into sine functions. The proposed model has a very high learning ability and most of the errors were concentrated in the range of −0.4 to 0.2 m [10].

This study aims to predict the significant wave height through machine learning (ML), a soft computing method. The feedforward neural network (FNN) and long short-term memory (LSTM) models are adopted to predict significant wave height. Input parameters for the input layer were selected by Pearson correlation coefficients. To obtain the optimized hyperparameter, a sensitivity study is conducted on the window size, node, layer, and activation function. The results showed that the most suitable model for predicting the significant wave height was feedforward neural network (FNN-W48) owing to its accuracy and calculation time [11].

One of the study proposes the use of structural acceleration data for improving the significant wave height prediction accuracy of multiclass classification models. The authors have used artificial neural network and decision tree models. It is also demonstrated that redundancy of training parameters could inject unwanted overfitting, reducing model generalization. Precautions were also taken like train-test splits and K-fold validation for both training artificial neural networks and decision tree ensembles [12].

3 Proposed Work

The ball detection algorithm works by first converting the video frame to the hue, saturation, value (HSV) color space. This is because the HSV color space is more intuitive for humans to understand and work with than the BGR color space. In HSV, each pixel is represented by three values: hue (H), saturation (S), and value (V). Hue refers to the color of the pixel, saturation refers to the purity of the color, and value refers to the brightness of the pixel. The cvtColor function in OpenCV takes the video frame as input and converts it from BGR to HSV color space. This transformation allows the ball detection algorithm to focus solely on the color characteristics of the ball, while ignoring brightness variations in the video frame (see Fig. 1).

Once the frame has been converted to HSV, a mask is created based on the lower and upper boundaries of the ball color range. The mask is a binary image, where white pixels represent pixels that are within the ball color range and black pixels

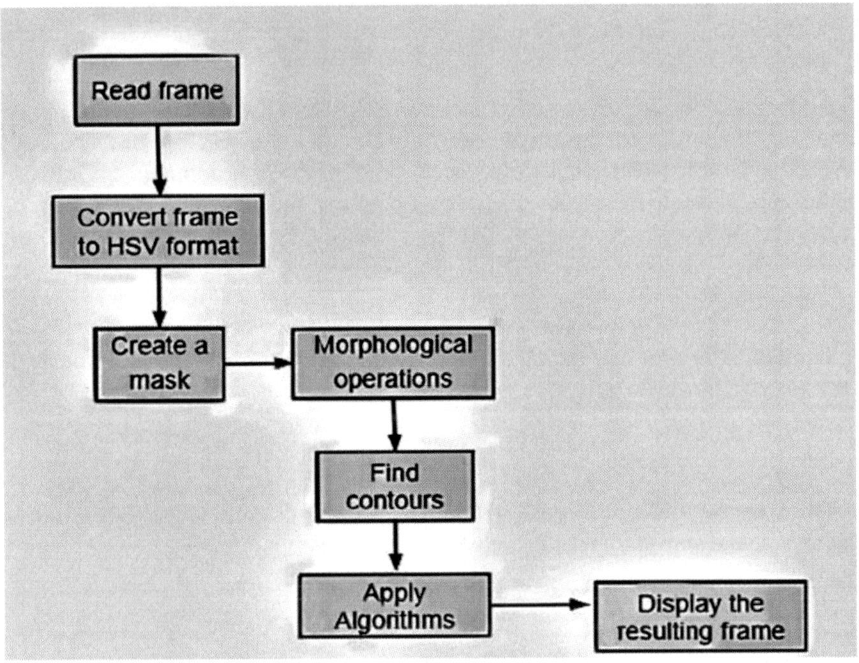

Fig. 1 Showing proposed methodology

represent pixels that are outside the ball color range. The mask is then applied to the frame to remove any pixels that are not within the ball color range. This leaves a binary image that only contains the ball.

Next, morphological operations are applied to the mask to remove any noise. Morphological operations are a set of image processing techniques that can be used to erode, dilate, open, close, and smooth images. In this case, we use morphological operations to remove any small, insignificant objects from the mask. Finally, contours are found in the mask. Contours are the boundaries of connected components in the image. In this case, contours represent the outline of the ball. The process of finding contours involves several steps.

1. The input image is typically converted to grayscale to simplify the analysis and reduce computational complexity. Grayscale images only contain intensity values, which allows for easier thresholding and segmentation.
2. Thresholding is applied to convert the grayscale image into a binary image. Thresholding is the process of dividing the image into foreground and background by selecting an appropriate intensity threshold.
3. Once the image is thresholded, the findContours function is used to identify and extract the contours. The findContours function takes the binary image as input and returns a list of contours found in the image. The contours are represented as

a sequence of points that form the boundary of connected regions in the binary image.

By finding contours, the location and shape of the ball within the frame can be identified. Once the contours have been found, the wave height can be calculated. Several methods exist for calculating wave height, but the most common one is to use the distance between the highest and lowest points of the ball contour to determine the height with precision.

By subtracting the y-coordinate of the previous center (prev center [1]) from the y-coordinate of the current center (center [1]), the vertical movement (upward displacement) of the yellow ball in the image is obtained.

The expression for calculating the upward displacement (vertical movement) of the yellow ball is given by:

$$y\,\text{diff} = \text{center}[1] - \text{prev center}[1].$$

The result '*y* diff' will be positive if the ball moves upward, negative if it moves downward, and zero if there is no vertical movement.

4 Algorithm Implementation

The selection of a specific algorithm for wave height detection depends on various factors, including the type of data available, the complexity of the wave patterns, and the desired level of accuracy. For video-based wave height detection, image processing algorithms like contour detection, thresholding, and morphological operations are commonly used. These algorithms are effective in identifying the boundaries of the waves in the video frames and calculating the wave height based on the displacement of the wave contours. For sensor-based wave height detection, signal processing techniques like Fourier analysis and spectral analysis are often employed. These algorithms can extract wave height information from the sensor data and provide accurate measurements. The choice of algorithm also depends on computational efficiency and real-time processing requirements. Overall, selecting the appropriate algorithm is crucial to ensure accurate and reliable wave height detection in various environmental conditions.

Algorithm 1: Thresholding

Loop through each contour:

1. Calculate the current position of the ball.
2. Having a previous position, calculate the upward displacement.
3. If the displacement is positive, the ball is moving upward.
4. Update the previous position to the current position.

4.1 Algorithm 2: Center Displacement

1. Calculate the yellow ball center by taking the midpoint of the bounding rectangle.
2. Compute the vertical movement difference of the yellow ball by comparing the current center with the previous center.
3. Update the previous center and previous y-coordinate values for each iteration.
4. Draw a circle at the centroid of the yellow ball and a bounding rectangle on the original frame.

4.2 Algorithm 3: Y-coordinate and Radius Difference

1. If circles are detected (circles are not none), iterate over each detected circle using a for loop.
2. Calculate the height of the wave based on the circle's position by subtracting the radius from the y-coordinate and multiplying it by the PIXEL-TO-MM-RATIO.
3. Create text to display the wave height, formatted to two decimal places, using f-string notation.
4. Add the wave height text above the circle.

4.3 Algorithm 4: Centroid Method

1. Set the PIXEL-TO-MM-RATIO to the desired value, representing the conversion factor from pixels to millimeters.
2. Set the NUM-FRAMES to the desired number of frames to use for calculating the wave height.
3. Calculate the centroid of each valid contour. If the moment m00 is nonzero (to avoid division by zero), calculate the x and y coordinates of the centroid.
4. Draw a circle at the centroid position on the frame using cv2.circle.
5. Add the centroid (x, y) coordinates to the centroids list.
6. Keep only the last NUM-FRAMES centroids in the list to maintain a sliding window of centroids.

4.4 Algorithm 5: Optimal Reference Point

1. The largest greenish yellow contour is estimated by assuming that the maximum greenish yellow area is occupied by the buoy in one frame. This largest contour will eliminate the external noises of greenish yellow color (e.g., Trees and other small reflections) but it will still contain the reflection of the buoy in the water.

2. The buoy is detected along with its reflection. The top most pixel of the ball is selected as the reference point which will eventually eliminate the reflection of the ball.
3. The maximum and minimum y-coordinates reached by the buoy are estimated by tracking the vertical motion of the reference point.
4. The wave height is estimated to be the difference between maximum and minimum vertical displacement of the reference point.

5 Results

Based on the obtained data from Table 1, Algorithm 1 (see Fig. 2a) demonstrates the widest height range, supporting measurements from 0.11338 to 2.4 mm. This algorithm is well-suited for applications that require a broader range of height measurements. On the other hand, Algorithm 4 (see Fig. 2d) offers the narrowest range, limited to measurements between 0.1 and 0.14 mm. This algorithm can be more applicable in scenarios where precision within a specific range is essential.

Algorithm 2 (see Fig. 2b) provides a moderate height range, with measurements ranging from 0.1 to 1.2 mm. It strikes a balance between the wider range of Algorithm 1 and the narrower range of Algorithm 4. Algorithm 3 (see Fig. 2c) offers a relatively compact height range from 1.36 to 1.62 mm, suitable for applications requiring precise measurements within this specific interval.

Algorithm 5 (see Fig. 2e) provides the smallest range among the algorithms analyzed, supporting heights between 0.02 and 0.18 mm. This algorithm is ideal for applications that necessitate fine-grained height measurements within a limited range.

The specifications for the video capturing, processing, and data source:

1. **Camera Model**: The specific camera model used for video capturing is the ZED-4 camera. The ZED-4 camera is known for its high-resolution imaging capabilities and depth sensing features, making it suitable for computer vision and 3D reconstruction applications.
2. **Image Processing Libraries**: For postprocessing and analysis of the captured videos, image processing libraries and tools like OpenCV may have been utilized.

Table 1 Showing calculated wave heights and execution time of the algorithms

Algorithm	Min. height (in mm)	Max. height (in mm)	Execution time (in sec)
Thresholding	0.113388	2.4	5.927
Center displacement	0.1	1.2	5.753
y-coordinate and radius diff	1.36	1.62	4.558
Centroid method	0.1	0.14	5.350
Optimal reference point	0.02	0.18	5.642

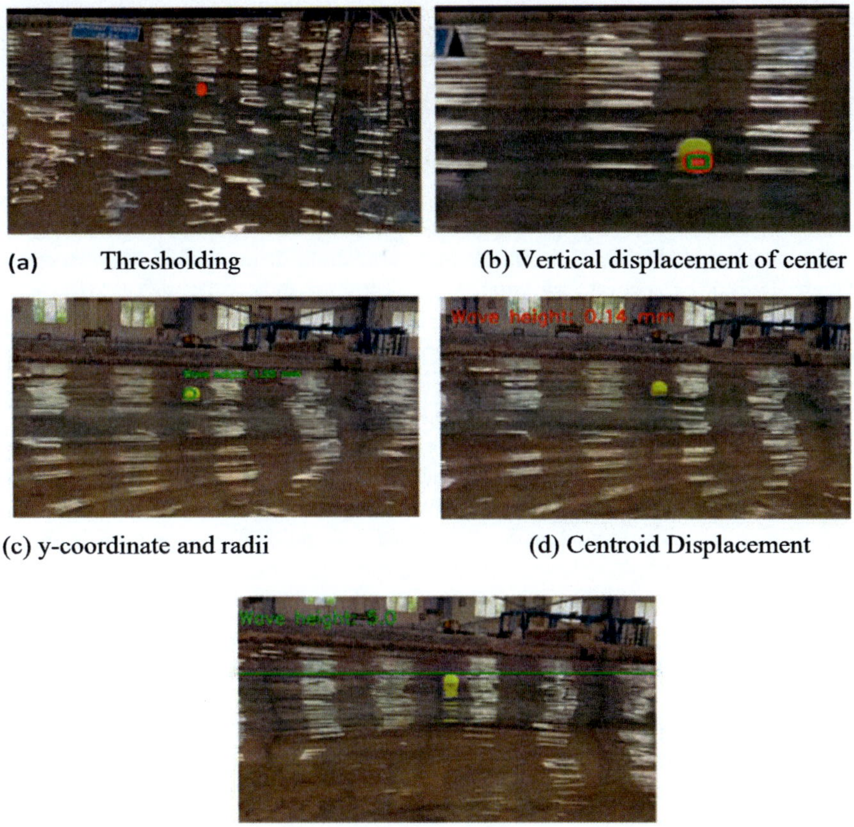

(a) Thresholding
(b) Vertical displacement of center
(c) y-coordinate and radii
(d) Centroid Displacement
(e) Line passing through reference point

Fig. 2 Showing the obtained results with proposed algorithms

3. **Data Source**: For automatic wave height prediction, the data has been collected from the Central Water and Power Research Station (CWPRS, 2023). The videos and images were captured from the active model at CWPRS through ZED-4 camera.

These software specifications and tools are vital in conducting video capturing, image processing, and subsequent analysis for the research study using the ZED-4 stereo camera and still camera. The cameras ensure accurate data acquisition and facilitate effective evaluation and interpretation of the results obtained from the captured video footage.

6 Limitations and Discussion

The discussed studies and algorithms provide a diverse range of approaches to wave height estimation using various techniques such as image processing, machine learning, and sensor data analysis. Each algorithm has its own strengths and limitations, which can impact the accuracy and applicability of wave height measurements in different scenarios.

For instance, Algorithm 1 utilizes a simple thresholding method for ball detection, which may not be accurate for small or fast-moving objects like waves. However, the algorithm used in WavePack [1] demonstrates promising results in accurately estimating wave heights from digital video sequences. The use of a camera, acquisition unit, and processing unit in the WavePack system enables accurate wave height estimation, as evidenced by the high Pearson correlation coefficient of 0.86.

Similarly, Algorithm 2 faces challenges with complex backgrounds, occlusions, and changes in ball appearance. These limitations can affect the algorithm's ability to accurately track wave crests and estimate wave heights. On the other hand, the utilization of Sentinel 1 Synthetic Aperture Radar images and NDBC buoy measurements in [2] provides a complementary approach to wave height estimation. By processing Synthetic Aperture Radar images and validating the results with buoy measurements, the algorithm achieves better estimation accuracy for certain sea states.

Algorithm 3's reliance on a single reference point assumption for wave height calculation can be a limitation in capturing complex wave motion with multiple peaks or different reference points. However, the approach presented in [3] that uses brightness variations in video frames to automatically monitor and measure wave heights in the nearshore area of the ocean offers a non-intrusive method that can be used in any weather condition.

The limitation of Algorithm 4 is that it assumes a consistent and smooth trajectory of the ball over the last NUM-FRAMES. If there are sudden changes or fluctuations in the ball's position or trajectory within those frames, it can lead to inaccurate wave height calculations, whereas in method 5 the buoy is detected by extracting the largest greenish yellow area; this algorithm is based on the assumption that the largest greenish yellow colored area on the frame is of the buoy itself.

The most significant limitations in algorithms used for wave height detection are:

1. **Sensitivity to noise**: The algorithms are sensitive to noise in the image. This can lead to errors in the wave height measurements.
2. **Difficulty in identifying the wave crest**: The algorithms can have difficulty in identifying the wave crest. This can lead to errors in the wave height measurements.
3. **Computational complexity**: The algorithms can be computationally complex. This can make them difficult to implement in real-time applications.

Despite these limitations, the algorithms can be used to obtain useful information about wave conditions. The algorithms can be used to monitor wave height, which can be used to assess the risk of coastal flooding and erosion. The algorithms can also be used to study the behavior of waves, which can be used to improve the design of ships and offshore structures.

7 Conclusion and Future Scope

In this study, the performance of five different algorithms for wave height detection was investigated. The algorithms were tested on a dataset of synthetic and real-world wave data. The results showed that all five algorithms were able to detect wave heights with good accuracy. However, there were some differences in the performance of the algorithms under different conditions. The results of the study showed that there is no single algorithm that is best for all wave height detection applications. The choice of algorithm will depend on the specific application and the conditions under which the algorithm will be used.

Future research can focus on expanding the method to work with multiple buoys that could enable the monitoring of wave patterns across larger areas of the ocean. Implementing a network of buoys with synchronized cameras would provide a comprehensive understanding of wave behavior over vast regions, benefiting coastal management and marine engineering. Integration with remote sensing technologies, such as radar could provide a comprehensive and multimodal approach to wave measurement.

References

1. Eason G, Noble B, Sneddon IN (1955) On certain integrals of Lipschitz-Hankel type involving products of Bessel functions. Phil Trans Roy Soc London A247:529–551
2. Pramudya FS, Pan J, Devlin AT (2019) Estimation of significant wave height of near-range traveling ocean waves using Sentinel-1 SAR images. IEEE J Sel Top Appl Earth Obs Remote Sens 12(4):1067–1075
3. Gal Y, Browne M, Lane C (2013) Long-term automated monitoring of nearshore wave height from digital video. IEEE Trans Geosci Remote Sens 52(6):3412–3420
4. Choi H, Park M, Son G, Jeong J, Park J, Mo K, Kang P (2020) Realtime significant wave height estimation from raw ocean images based on 2D and 3D deep neural networks. Ocean Eng 201:107129
5. Guan X (2020) Wave height prediction based on CNN-LSTM. In: 2020 2nd international conference on machine learning, big data and business intelligence (MLBDBI). IEEE, pp 10–17
6. Kim J, Park C-S, Min K-W (2016) Fast vision-based wave height measurement for dynamic characterization of tuned liquid column dampers. Measurement 89:189–196
7. James SC, Zhang Y, O'Donncha F (2018) A machine learning framework to forecast wave conditions. Coast Eng 137:1–10
8. Wang W, Tang R, Li C, Liu P, Luo L (2018) A BP neural network model optimized by mind evolutionary algorithm for predicting the ocean wave heights. Ocean Eng 162:98–107
9. Osawa K, Yamaguchi H, Umair M, Hashmani MA, Horio K (2020) Wave height and peak wave period prediction using recurrent neural networks. In: 2020 international conference on computational intelligence (ICCI). IEEE, pp 1–4

10. Wang H, Fu D, Liao S, Wang G, Xiao X (2019) Significant wave height prediction based on MSFD neural network. In: 2019 tenth international conference on intelligent control and information processing (ICICIP). IEEE, pp 1–5
11. Park SB, Shin SY, Jung KH, Lee BG (2021) Prediction of significant wave height in Korea strait using machine learning. J Ocean Eng Technol 35(5):336–346
12. Demetriou D, Michailides C, Papanastasiou G, Onoufriou T (2021) Coastal zone significant wave height prediction by supervised machine learning classification algorithms. Ocean Eng 221:108592

Enhanced Feature Fusion from Dual Attention Paths Using Feature Gating Mechanism for Scene Categorization of Aerial Images

G. Akila and R. Gayathri

Abstract The widespread use of remote sensing (RS) aerial images in scene categorization has drawn the attention of numerous researchers. As the consequence of this advancement, the approaches for RS image scene categorization that use convolutional neural networks (CNNs), have progressed enormously. Majority of the layouts in the works already published only took the account of scene's global information. However, a lot of replicated areas in the global information reduce the effectiveness of classification and disregard local contextual details, which reflect finer spatial characteristics of local objects. Additionally, the majority of CNN-based approaches give the identical weights on each feature vector, which prevents the mode from differentiating the important features. To overcome the above limitations, we propose a model called enhanced feature fusion from dual attention paths using feature gating mechanism (EFFDA-FG). This model employs Resnext50 baseline model, with channel spatial attention for deriving global and local features along with feature fusion using a gating mechanism. Our suggested model was implemented on three well-known datasets and it proves to give accuracy equal to or higher than that of many state-of-art CNN architectures while using comparatively lesser parameters.

Keywords Aerial images · Convolutional neural networks (CNN) · Channel attention · Spatial attention · Feature-gating

G. Akila (✉)
Department of Electronics and Communication Engineering, Meenakshi Sundararajan Engineering College, Chennai 600024, India
e-mail: aki47g@gmail.com

R. Gayathri
Department of Electronics and Communication Engineering, Sri Venkateswara College Of Engineering, Sriperumbudur 602117, India
e-mail: rgayathri@svce.ac.in

S. Shakya et al. (eds.), *Fourth International Conference on Image Processing and Capsule Networks*, Lecture Notes in Networks and Systems 798,
https://doi.org/10.1007/978-981-99-7093-3_38

1 Introduction

Image classification has undergone significant research in recent years because of the vast amounts of remote sensing (RS) data that have been produced [1]. Research on RS-based aerial scene classification, which aims to categorize the acquired image semantic level data with the proper category labels, has also evolved. There are two key reasons why things are difficult in this situation: (1) intra-class diversity, where scenes from the same class of images vary greatly in attributes like color and shape, and (2) inter-class similarity, where scenes from several image classes have textures that are similar to one another. Therefore, learning how to extract useful features from complex images is crucial to increasing classification accuracy. The three main categories of feature extraction techniques are convolutional neural networks (CNNs), vision transformers (VIT), and manual feature-based techniques. Manual feature-based techniques might design a single feature or combine several features to improve the results of categorization [2]. However, when complex content is used as features, the capacity to describe scenes is limited. Global characteristics of pictures are typically learned for classification by VIT-based algorithms [3–5]. Local feature improvement, however, has not made much progress.

CNN was recently introduced to the RSSC industry, where it has achieved success in terms of categorization [6, 7]. By learning important local regions in the scene, Bi et al. [8] built a local semantic-augmented convolution network to build discriminative local semantic representation. In order to successfully fuse features, Wang et al. [9] created a multilevel feature fusion network and an adaptive technique for channel downscaling of high-dimensional features. Traditional feature extraction techniques like principal component analysis, sparse coding, and histograms of directed gradients have a limited capacity for feature representation. Convolutional neural networks (CNNs), in particular, have made significant advancements in the field of RS scene classification in recent years. To enhance the performance of the model, Li et al. [10] developed a joint pooling method and integrated it into a bilinear CNN. For scene categorization, Xu et al. [11] suggested a multilayer feature fusion network (MF2Net) to efficiently combine multilayer convolutional features. In order to understand the underlying hierarchical structures of scene images and lessen the visual-semantic disparity, Wang et al. [12] devised a multi-granularity canonical appearance pooling (MG-CAP). CNN-based methods have demonstrated more potential for extracting robust features in RS scene photos when compared with conventional feature extraction techniques. However, as seen in Fig. 1, the feature representation capability of models is still insufficient because RS scene photos frequently have complicated backgrounds and dispersed little items.

A number of attention techniques have been put out in recent years to learn distinguishing aspects for RS scene classification. To extract potent and multilayer feature representations, Wang et al. [13] investigate a relation-attention network (RANet). Zhao et al. [14] created an enhanced attention module (EAM) based on the convolutional block attention module (CBAM) to infer both channel and spatial attention maps. Convolutional local attention module (CLAM) was developed by

Fig. 1 Examples scene images for inter-class similarity and intra-class diversity from NWPU-RESISC45 dataset

Chen et al. [15] and included into the ResNet50 [16]. These techniques, however, have two issues. First, the model's ability to concentrate on more important regions is hampered by an incomplete consideration of the long-range reliance and local information. Second, it's not always ideal to integrate attention maps or modules.

A unique enhanced feature fusion from dual attention paths using feature gating mechanism (EFFDA-FG) is suggested to address the aforementioned issues. The following is a summary of these notable contributions:

- An end-to-end network called EFFDA-FG is suggested as a ResNext50-based approach to learn discriminative features for aerial scene categorization.
- In this case, the first as well as final sections of the fifth convolutional module of ResNext50 are used to extract two sets of deep features. On the first feature, channel spatial attention is performed, and on the last feature, channel attention is performed.
- The obtained features are combined using a enhanced feature fusion that involves feature gating and element-wise multiplication to extract the final intricate features for robust and efficient classification.

2 Methodology

2.1 Proposed Architecture

The overall architecture of EFFDA-FG can be viewed from Fig. 2.

As can be seen, features from ResNext50's final stage are sent through an attention network, a feature gating mechanism and a multiplication layer to extract more significant details from the features. The following is a description of the architecture.

The input images are fed to the baseline architecture ResNext50 which consists of five convolutional sections as shown in Fig. 2a. The last section consists of high semantic level features that helps in classification. The last section has three convolutional blocks. Output features are taken from the first and last blocks of this section.

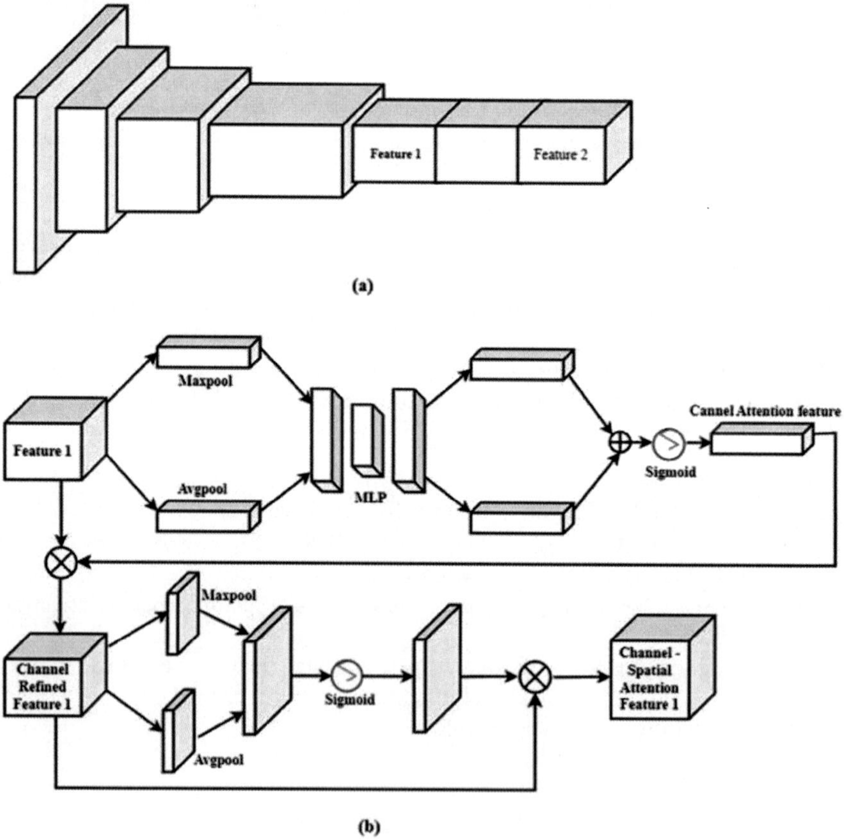

Fig. 2 Overall architecture of EFFDA-FG, **a** Feature 1 and 2 extractions from Resnext50 backline architecture, **b** Channel-spatial attention on feature 1, **c** Channel attention on feature 2, and **d** Feature gating and feature fusion for final classification

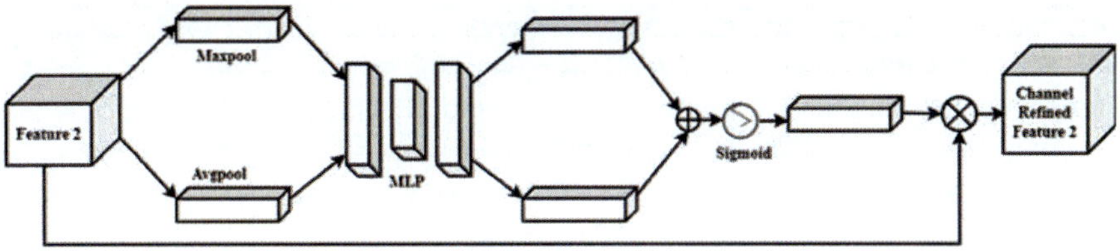

(c)

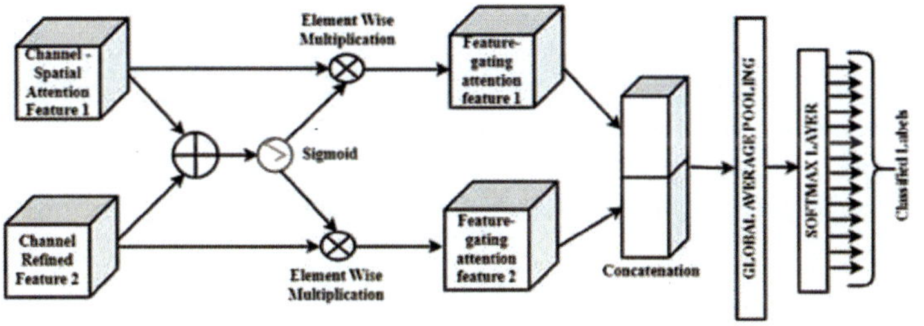

(d)

Fig. 2 (continued)

The output from first convolutional block is fed to channel-spatial attention network as it is an intermediate level that may also contain spatial features as shown in Fig. 2b. The output from the last convolutional block is fed to channel attention network as depicted in Fig. 2c.

The output from these attention networks is added and then passed through the sigmoid layer and then multiplied with their respective input features as shown in Fig. 2d. This forms the feature gating mechanism that extracts the final intricate features for robust and efficient classification. Each phase of the model is explained in detail.

ResNeXt50. The ResNeXt50 has become widely favored and influential in computer vision due to a multitude of compelling factors driving its adoption. One prominent reason for selecting ResNeXt50 is its improved capacity compared with the original ResNet50 architecture. By incorporating a cardinality parameter, it enhances expressiveness and model capacity, enabling it to capture intricate patterns and relationships effectively. Scalability is another key advantage of ResNeXt50. With its modular and flexible design, it can easily adapt to different depths and widths, making it suitable for various tasks and datasets. This adaptability empowers researchers and practitioners to customize it to meet specific requirements, allowing for efficient modeling of complex tasks. Efficiency is a critical consideration, and this network excels in this aspect. Despite its increased capacity, it maintains computational efficiency by utilizing grouped convolutions, reducing computational costs while preserving strong representational power. Consequently, it offers an efficient solution for resource-constrained environments. It leverages transfer learning by pretraining on large-scale datasets like ImageNet. This pretraining provides a solid foundation of learned features, facilitating improved generalization and faster convergence when fine-tuning on specific tasks or smaller datasets. This capability is particularly valuable when limited training data is available. ResNeXt50 strikes a balance between expressiveness, computational efficiency, scalability, and exceptional performance, making it a compelling option for a broad range of computer vision applications.

Convolutional Block Attention Module (CBAM). The CBAM [17] module consists of two components: the channel attention module (CAM) and the spatial attention module (SAM).

Channel Attention Module (CAM): The channel attention module aims to capture channel-wise dependencies and emphasize informative channels. It operates as follows: Given an input feature map X of shape (N, H, W, C), where N represents the batch size, H and W represent the spatial dimensions, and C represents the number of channels:

1. Global Average Pooling (GAP): Apply global average pooling across the spatial dimensions of X to obtain channel-wise statistics. This is achieved by calculating the mean value for each channel:

$$S_C = \frac{1}{H * W} \sum X_{ijc};\ \text{for}\ i\ \text{in range}(H),\ \ j\ \text{in range}(W). \tag{1}$$

2. Channel Max Pooling (CMP): Apply max pooling across the channel dimension of X to capture the most prominent features in each channel. This is done by calculating the maximum value for each channel:

$$M_C = \text{Max}(X_C);\ \text{for}\ c\ \text{in range}(C). \tag{2}$$

3. Shared Fully Connected Layers (FC): Pass the outputs of GAP and CMP through a shared set of fully connected layers to model the channel-wise relationships and generate attention weights. The FC layers can be represented as W_1 and W_2, with ReLU activations. The nonlinear Relu activation function maps all negatives to zero, while positives remain unchanged. Its range is expressed as $F(x) = \text{Maximum}(0, x)$

$$A_C = \text{ReLU}\ (W_2(\text{ReLu}\ (W_1([S_C,\ M_c])))). \tag{3}$$

4. Channel Attention: Generate a channel attention mask by transforming the attention weights through a sigmoid activation function. The sigmoid is a linear activation function given by $F(x) = 0; x < 0$ and $F(x) = x; x \geq 0$

$$M = \text{Sigmoid}(A_C). \tag{4}$$

5. Rescaling: Multiply the input feature map X by the channel attention mask M to emphasize important channels:

$$Y = XM \tag{5}$$

Spatial Attention Module (SAM): The spatial attention module captures spatial dependencies to focus on informative regions. It operates as follows:

Given the output feature map Y from the channel attention module:

1. Channel-wise Max Pooling (CMP): Apply max pooling across the channel dimension of Y to capture the most important features within each channel:

$$M_{hwc} = \text{Max}\,(); \;\text{for}\, h \,\text{in}\, \text{range}(H),\, w \,\text{in}\, \text{range}(W),\; c \,\text{in}\, \text{range}(C). \tag{6}$$

2. Channel-wise Average Pooling (CAP): Apply average pooling across the channel dimension of Y to capture the overall distribution of features:

$$S_{hwc} = \frac{1}{C} * \Sigma Y_{hwc}); \;\text{for}\, h \,\text{in}\, \text{range}(H),\, w \,\text{in}\, \text{range(W)},\, c \,\text{in}\, \text{range}(C). \tag{7}$$

3. Spatial Attention: Generate a spatial attention mask by concatenating the results of CMP and CAP, followed by a convolutional operation and a sigmoid activation:

$$S = \text{Sigmoid (Conv(ReLU} \;([S_{\text{hwc}}, M_{\text{hwc}}]))) \tag{8}$$

4. Rescaling: Multiply the output feature map Y by the spatial attention mask S to emphasize informative regions:

$$Z = YS \tag{9}$$

The final output feature map Z represents the refined and modulated features obtained from both the channel attention module (CAM) and the spatial attention module (SAM). It encapsulates the improved representation learning, capturing discriminative information within the input feature map.

Feature Gating Mechanism and Feature Fusion. The output from two attention networks are added, passed through sigmoid activation, and multiplied by their respective input features. This forms the feature gating mechanism. Feature gating is an advantageous technique in image classification tasks, offering several benefits for the network. Here are some ways in which feature gating can enhance image classification:

Selective Feature Fusion: Feature gating allows the network to selectively combine or merge features from different layers or branches. This selective fusion emphasizes the most informative features while suppressing less relevant ones, leading to improved classification accuracy.

Adaptive Attention: By employing feature gating, the network can dynamically allocate attention to different regions of an image. The network assigns higher gating values to important regions and lower values to less informative areas, enabling focused attention on the most discriminative parts of the image. This adaptive attention mechanism results in improved classification performance.

Noise Reduction: Images often contain noise or irrelevant details that can impact classification accuracy. Feature gating acts as a filtering mechanism by assigning lower gating values to noisy or less informative features. This allows the network to reduce the influence of such features, resulting in noise reduction and improved robustness.

Class-specific Feature Amplification: Certain features may play a crucial role in distinguishing specific classes. Through feature gating, the network can amplify the importance of class-specific features. By doing so, the network focuses on discriminative aspects that are particularly relevant to different classes, potentially leading to improved classification accuracy for those classes. Feature gating in image classification enables the network to selectively combine features, allocate attention adaptively, reduce noise, and amplify class-specific information. This approach enhances the network's ability to capture and utilize the most relevant aspects of images, ultimately leading to improved performance, accuracy, and robustness in image classification tasks.

3 Implementations

Three datasets were used to verify the authentication of our model. They are UCMerced dataset [18]—21 classes with 2100 images, AID dataset [19]—30 classes with 10,000 images, and NWPU-RESISC45[20]—45 classes with 35,000 images dataset. The images are preprocessed, augmented, and rescaled to 256X256X3. For augmentation the images are rotated from 0 to 60°, vertical and horizontal flipped. Number of epochs was set to 300 and intial learning rate is 0.01. We implemented this network in Google-Colab-Pro + environment. We have used the standard training test ratio that was used by many of the state-of-art-models for better comparison. In UCMerced datset the training/test ratio is 50/50 and 80/20. For AID dataset, the training test ratio used is 20/80 and 50/50. And for NWPU-RESISC45 dataset, 10/20 and 20/80 training test ratio were used.

Table 1 Classification accuracy of our model compared with different state-of-art models with different train test ratio

Model	UCMeced		AID		NWPU	
	50/50	80/20	20/80	50/50	10/90	20/80
MBFANet	–	99.66	93.98	96.93	91.61	94.01
LCNN-BFF	–	99.29	91.66	94.62	86.53	91.54
ACNet	–	**99.76**	93.33	95.38	91.09	92.42
ACR-MLFF	97.99	99.37	92.73	95.06	90.01	92.45
SEMSDNet	98.40	99.41	94.23	**97.64**	91.68	93.89
EFFDA-FG (Ours)	**98.96**	**99.76**	**94.04**	96.89	**93.01**	**94.50**

4 Results and Discussions

4.1 Classification Accuracy

Comparison of accuracy between different state-of-art models is given in the Table 1, from which we could see that our suggested EFFDA-FG model gives the best accuracy except for AID 50/50 train/test ratio.

4.2 Classification Report

The obtained classification report shows that our model performs in a best way with precision, recall and f1-score as given in Tables 2 and 3 for AID and NWPU dataset. For AID dataset, 28 classes out of 30 classes show the precision, recall, and f1-scores to be more than 90%. These three factors lies in the range between 84 and 90% for other classes like center, church and square. For NWPU dataset, these three factors lie in the range of 90–100% for nearly 38 classes. 38/45 classes lie in the range of 90–100%. All the other classes church, palace, commercial-area, dense-residential, medium-residential, railway, rectangular-farmland have these values to be in range of 82–90%.

4.3 Confusion Matrix

The confusion matrix for aid dataset with 50/50 training test ratio is shown in Fig. 3, from which we can see that all the classes are classified with good accuracy. Nearly, 13 classes have 100% accuracy. Almost 27 classes have accuracy greater than 90%. The remaining three classes center, school, and stadium have accuracy of 88%, 82%,

Table 2 Classification report [precision/recall/f1-score] for different categories of AID dataset with 50/50 training test ratio

	Precision	Recall	F1-score
Airport	0.986301	1.000000	0.993103
Bare land	0.983871	0.983871	0.983871
Baseball field	0.977778	1.000000	0.988764
Beach	1.000000	1.000000	1.000000
Bridge	0.986301	1.000000	0.993103
Center	0.884615	0.884615	0.884615
Church	0.839286	0.979167	0.903846
Commercial	0.931507	0.971429	0.951049
Dense residential	0.987805	0.987805	0.987805
Desert	0.983607	1.000000	0.991736
Farmland	0.973333	0.986486	0.979866
Forest	1.000000	0.980000	0.989899
Industrial	0.974026	0.961538	0.967742
Meadow	0.965517	1.000000	0.982456
Medium residential	1.000000	0.965517	0.982456
Mountain	1.000000	1.000000	1.000000
Park	0.969231	0.900000	0.933333
Parking	1.000000	1.000000	1.000000
Playground	0.972603	0.959459	0.965986
Pond	0.976744	1.000000	0.988235
Port	1.000000	1.000000	1.000000
Railway station	0.98000	0.942308	0.960784
Resort	0.896552	0.896552	0.896552
River	1.000000	1.000000	1.000000
School	0.907407	0.816667	0.859649
Sparse residential	0.967742	1.000000	0.983607
Square	0.888889	0.848485	0.868217
Stadium	0.964912	0.948276	0.956522
Storage tanks	1.000000	0.986111	0.993007
Viaduct	0.98823	1.000000	0.994083
Accuracy	0.969000	0.969000	0.969000
Macro avg	0.966209	0.966610	0.966010
Weighted avg	0.969302	0.969000	0.968814

Table 3 Classification report [precision/recall/f1-score] for different categories of NWPU dataset with 20/80 training test ratio

	Precision	Recall	F1-score
Airplane	0.982659	0.971429	0.977011
Airport	0.960606	0.905714	0.932353
Baseball_Diamond	0.976812	0.962857	0.969784
Basketball_Court	0.971671	0.98	0.975818
Beach	0.955679	0.985714	0.970464
Bridge	0.954023	0.948571	0.951289
Chaparral	0.983099	0.997143	0.990071
Church	0.84	0.84	0.84
Circular_Farmland	0.977208	0.98	0.978602
Cloud	0.963585	0.982857	0.973126
Commercial_Area	0.888563	0.865714	0.87699
Dense_Residential	0.871314	0.928571	0.899032
Desert	0.959538	0.948571	0.954023
Forest	0.935933	0.96	0.947814
Freeway	0.971698	0.882857	0.92515
Golf_Course	0.982857	0.982857	0.982857
Ground_Track_Field	0.97191	0.988571	0.98017
Harbor	0.966574	0.991429	0.978843
Industrial_Area	0.951952	0.905714	0.928258
Intersection	0.976945	0.968571	0.97274
Island	0.985423	0.965714	0.975469
Lake	0.924581	0.945714	0.935028
Meadow	0.960843	0.911429	0.935484
Medium_Residential	0.850136	0.891429	0.870293
Mobile_Home_Park	0.962857	0.962857	0.962857
Mountain	0.943182	0.948571	0.945869
Overpass	0.918699	0.968571	0.942976
Palace	0.823353	0.785714	0.804094
Parking_Lot	0.928	0.994286	0.96
Railway	0.886243	0.957143	0.92033
Railway_Station	0.930303	0.877143	0.902941
Rectangular_Farmland	0.884718	0.942857	0.912863
River	0.945087	0.934286	0.939655
Roundabout	0.985294	0.957143	0.971014
Runway	0.920548	0.96	0.93986
Sea_Ice	0.997151	1	0.998573
Ship	0.960227	0.965714	0.962963
Snowberg	0.969529	1	0.984529

(continued)

Table 3 (continued)

	Precision	Recall	F1-score
Sparse_Residential	0.948127	0.94	0.944046
Stadium	0.98209	0.94	0.960584
Storage_Tank	0.969188	0.988571	0.978784
Tennis_Court	0.949438	0.965714	0.957507
Terrace	0.939306	0.928571	0.933908
Thermal_Power_Station	0.961194	0.92	0.940146
Wetland	0.953125	0.871429	0.910448
Accuracy	0.944444	0.944444	0.944444
Macro Avg	0.944917	0.944444	0.944325
Weighted Avg	0.944917	0.944444	0.944325

and 84%. Mostly they are interpreted among themselves as most of them look the same from top angle.

The confusion matrix of NWPU dataset with 20/80 training test ratio is shown in Fig. 4. Here we could see that 41 classes have accuracy in the range of 90–100%. Chapparrel, sea_ice and snowberg has 100% accuracy. Church, commercial_area, freeway, railway station, and wet land has accuracy in range of 80–90%. Palace has the lowest accuracy of 79%.

4.4 GradCAM

From the gradient-weighted-class-activation-mapping-plot (GradCAM) [21], Fig. 5, we can see the original sparse residential image and the heatmap generated at different outputs of our suggested model. Images b and c are the features obtained from resnext conv5_block1 and conv5_block3. Then *d* is the channel attention output that concentrates more on the center area, showing the residential part alone, but *e* is the channel spatial attention output that covers all area that might be used in deriving that it is sparse residential, i.e., it covers the center, pool area, residential part, and also the house behind the big trees and the road to this sparse area. Adding d and e gives *f*. Then feature *f* is passed through sigmoid layer, which supresses the unwanted features and concentrates on more required features. Sigmoid attention multiplied by its attention input is given by *h* and *i*. From *h* we can see that, more concentration is given to the center residential part while it omits the house on the right side, but it is captured well by *i*. By concatenating *h* and *i* features for classification purpose, our model shows improved performance.

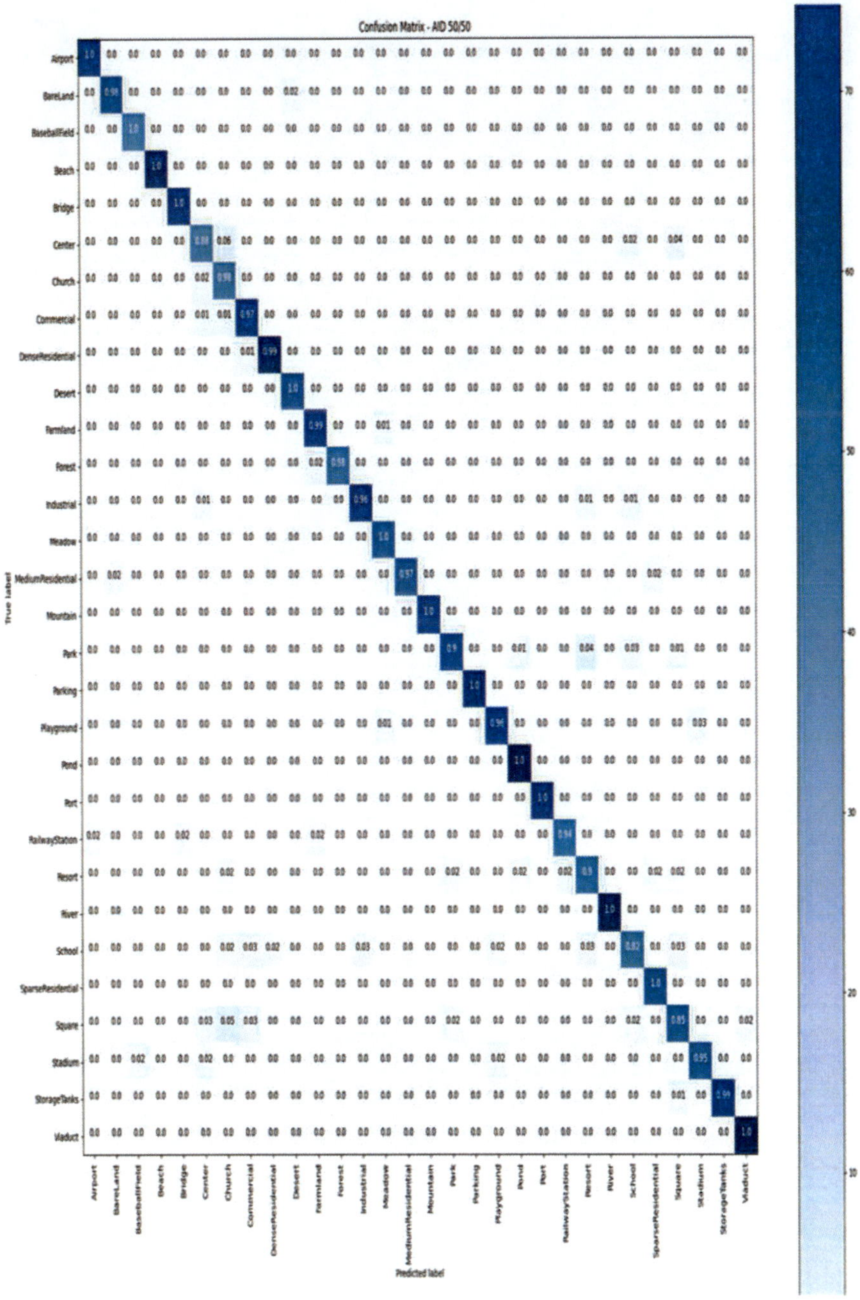

Fig. 3 Confusion matrix (CM) for AID dataset with 50:50 training test ratio

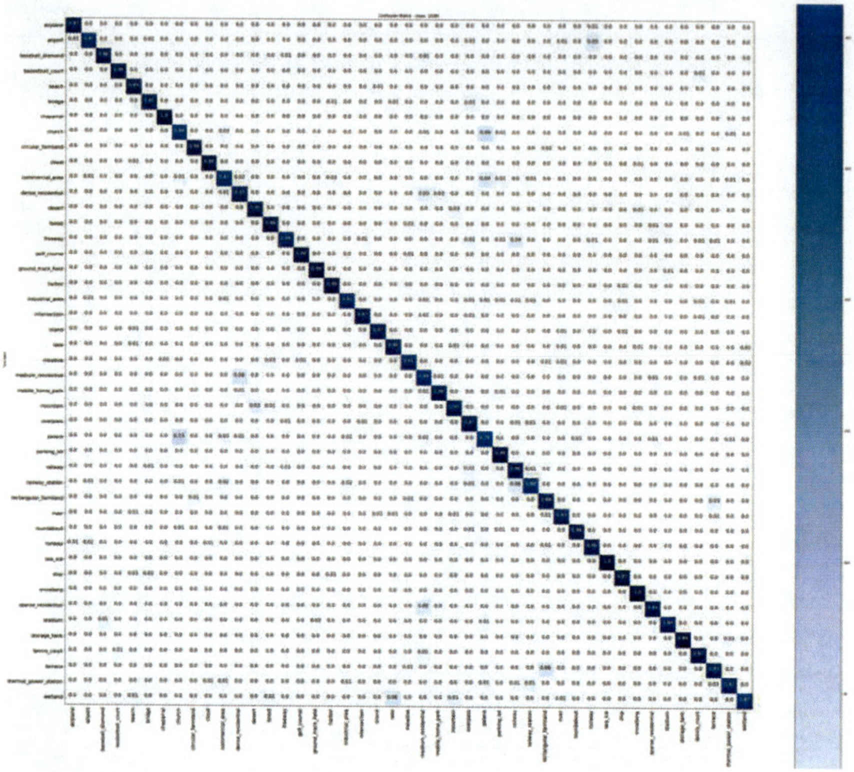

Fig. 4 Confusion matrix (CM) for NWPU dataset—20:80 training test ratio

4.5 T-SNE Plot

The t-distributed-stochastic neighbor embedding (T-SNE) [22], plot of our proposed model is shown in Fig. 6, for AID and NWPU dataset. Here, from the pictures, we can see that our model is able to obtain discriminate features as the separable clusters that are falling apart from each other.

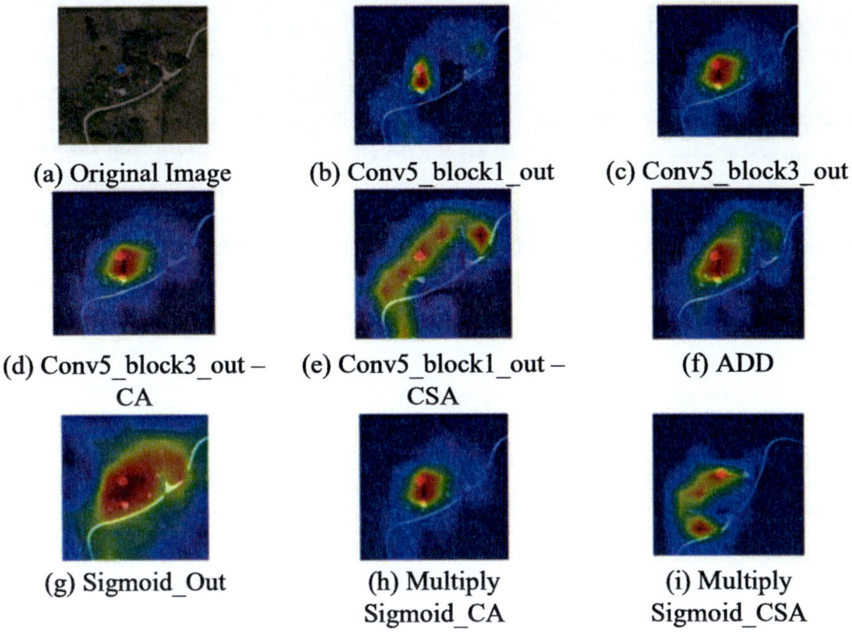

Fig. 5 GradCAM plot showing heatmaps generated at different levels of the model

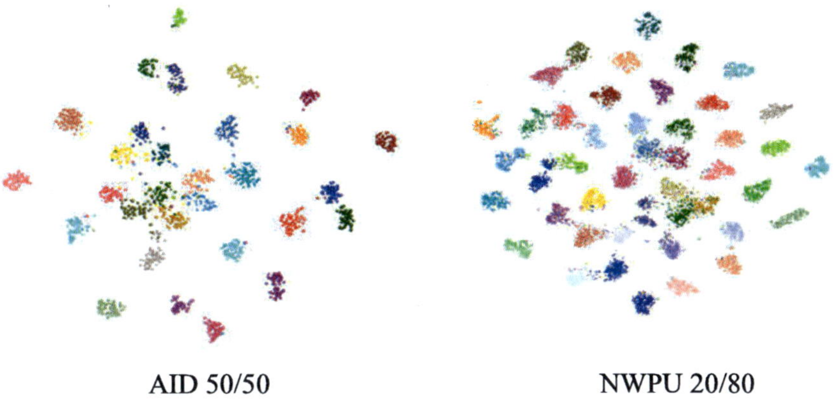

Fig. 6 T-SNE plot for AID and NWPU dataset for our model

5 Conclusion

A robust architecture with ResNeXt50 baseline model, named enhanced feature fusion from dual attention paths using feature gating mechanism (EFFDA-FG) has been suggested. Here we derived two sets of features from last convolutional stage of

ResNeXt50. Channel attention and channel spatial attention are performed on these features, and they are concatenated after passing through a sigmoid-based gating mechanism, which improves the performance of the model. The overall accuracies obtained by our model are 98.96, 99.76, 95.04, 96.89, 93.01, and 94.50 for different training testing ratios of UCM (50/50), UCM (80/20), AID (20/80), AID (50/50), NWPU (10/90), and NWPU (20/80), respectively. The precision, f1-score and recall also shows better performance. The confusion matrix was also analyzed, where we see that all categories are classified with a great accuracy. The GradCAM + plot and T-SNE plot proves that our model performs well.

References

1. Chen S, Ma W, Sun C, Chen S (2021) A comprehensive survey of remote sensing image scene classification: perspectives, datasets, and future directions. Remote Sens 13(12):2316
2. Zhou Z, Wang Y, Fu P, Liu L (2021) Feature learning and representation for remote sensing image scene classification: a comprehensive review. IEEE Trans Geosci Remote Sens 59(2):1011–1030
3. Dosovitskiy A, Beyer L, Kolesnikov A, Weissenborn D, Zhai X, Unterthiner T, Brox T (2021) An image is worth 16 × 16 words: Transformers for image recognition at scale. In: International conference on learning representations (ICLR)
4. Bazi Y, Bashmal L, Al Rahhal MM, Al Dayil R, Al AN (2021) Vision transformers for remote sensing image classification. Remote Sens 13(3):516
5. Roy SK, Deria A, Hong D, Rasti B, Plaza A, Chanussot J (2023) Multimodal fusion transformer for remote sensing ımage classification. IEEE Trans Geosci Remote Sens 61(1–20)
6. Castelluccio M, Poggi G, Sansone C, Verdoliva L (2015) Land use classification in remote sensing images by convolutional neural networks. arXiv:1508.00092
7. Li E, Xia J, Du P, Lin C, Samat A (2017) Integrating multilayer features of convolutional neural networks for remote sensing scene classification. IEEE Trans Geosci Remote Sens 55(10):5653–5665
8. Bi Q, Qin K, Zhang H, Xia G-S (2021) Local semantic enhanced ConvNet for aerial scene recognition. IEEE Trans Image Process 30:6498–6511. https://doi.org/10.1109/TIP.2021.3092816
9. Wang X, Duan L, Shi A, Zhou H (2022) Multilevel feature fusion networks with adaptive channel dimensionality reduction for remote sensing scene classification. IEEE Geosci Remote Sens Lett 19:1–5. Art no. 8010205. https://doi.org/10.1109/LGRS.2021.3070016
10. Li E, Samat A, Du P, Liu W, Hu J (2022) Improved bilinear CNN model for remote sensing scene classification. IEEE Geosci Remote Sens Lett 19:1–5
11. Xu K, Huang H, Li Y, Shi G (2020) Multilayer feature fusion network for scene classification in remote sensing. IEEE Geosci Remote Sens Lett 17(11):1894–1898
12. Wang S, Guan Y, Shao L (2020) Multi-granularity canonical appearance pooling for remote sensing scene classification. IEEE Trans Image Process 29:5396–5407
13. Wang X, Duan L, Ning C, Zhou H (2022) Relation-attention networks for remote sensing scene classification. IEEE J Sel Top Appl Earth Observ Remote Sens 15:422–439
14. Zhao Z, Li J, Luo Z, Li J, Chen C (2021) Remote sensing image scene classification based on an enhanced attention module. IEEE Geosci Remote Sens Lett 18(11):1926–1930
15. Chen S-B, Wei Q-S, Wang W-Z, Tang J, Luo B, Wang Z-Y (2022) Remote sensing scene classification via multi-branch local attention network. IEEE Trans Image Process 31:99–109
16. Yu D, Guo H, Xu Q, Lu J, Zhao C, Lin Y (2020) Hierarchical attention and bilinear fusion for remote sensing image scene classification. IEEE J Sel Top Appl Earth Observations Remote Sens 13:6372–6383. https://doi.org/10.1109/JSTARS.2020.3030257

17. Woo S, Park J, Lee J-Y, Kweon IS (2018) CBAM: convolutional block attention module. In: Proceedings of the European conference on computer vision. Springer, Munich, pp 3–19
18. Yang X, Zhou W, He Z, Gao J (2010) Urban land cover classification using integrated remote sensing and GIS techniques: a case study in the Phoenix metropolitan area. Remote Sens Environ 114(8):1833–1848
19. Xia GS, Hu F, Hu H, Shi X, Bai X, Zhong Y, Zhang L (2017) AID: a benchmark data set for performance evaluation of aerial scene classification. IEEE Trans Geosci Remote Sens 55(7):3965–3981
20. Cheng G, Han J, Lu X, Zhang C, Sun J (2017) Remote sensing image scene classification: benchmark and state of the art. Proc IEEE 105(10):1865–1883. https://doi.org/10.1109/JPROC.2017.2712042
21. Selvaraju RR, Cogswell M, Das A, Vedantam R, Parikh D, Batra D (2019) Grad-CAM: Visual explanations from deep networks via gradient-based localization. In: Proceedings of the IEEE international conference on computer vision, pp 618–626
22. Van der Maaten L, Hinton G (2008) Visualizing data using t-SNE. J Mach Learn Res 9(Nov):2579–2605

Thermal Image-Based Fault Detection Using Machine Learning and Deep Learning in Industrial Machines: Issues-Challenges and Emerging Trends

Vaishnavi V. Kulkarni, Vishwanath R. Hulipalled, Mayuri Kundu, Jay B. Simha, and Shinu Abhi

Abstract Thermal imaging has long been utilized across industries to maintain electrical equipment and detect faults in machines, ensuring their reliable operation. Infrared thermography, or thermal imaging, has emerged as a powerful, non-contact, non-destructive monitoring tool. Unlike RGB imaging, it offers a non-intrusive approach that respects privacy and does not rely on ambient lighting. As heat signatures vary based on defects or machine conditions, infrared imaging is essential for effective condition monitoring. This survey study introduces the principles and applications of thermal imaging in fault detection and a comprehensive analysis of recent developments and approaches in the field. The survey encompasses various machine types, including electrical systems, mechanical components, and industrial equipment. Multiple techniques, algorithms, and deep learning models for image processing, feature extraction, and fault classification are extensively discussed. The paper also includes an in-depth literature review on fault detection in distinct machines, such as induction motors, rotating machines, and transformers. The survey aims to identify open challenges and issues in fault detection for different machines, such as induction motors and transformers.

Keywords Thermal imaging · Fault detection · Induction motor · Rotating machine · Transformer · Infrared thermography

V. V. Kulkarni (✉) · M. Kundu
School of CSE, REVA University, Bangalore, India
e-mail: R22PVCS01@reva.edu.in

M. Kundu
e-mail: mayuri.kundu@reva.edu.in

V. R. Hulipalled
School of C&IT, REVA University, Bangalore, India
e-mail: vishwanath.rh@reva.edu.in

J. B. Simha · S. Abhi
REVA Academy for Corporate Excellence (RACE) REVA University, Bangalore, India
e-mail: shinuabhi@reva.edu.in

S. Shakya et al. (eds.), *Fourth International Conference on Image Processing and Capsule Networks*, Lecture Notes in Networks and Systems 798,
https://doi.org/10.1007/978-981-99-7093-3_39

1 Introduction

Due to increased specialization and investment, the costs and complexity of industrial machinery have risen significantly. Ensuring the health of machines through fault diagnostics and monitoring is crucial to prevent productivity loss and system failures. Infrared thermography, a technique for acquiring and visualizing thermal pictures, has become popular. Unlike RGB imaging, thermal imaging relies solely on the emitted infrared radiation from objects, making it faster and more efficient in processing information. Thermal imaging sensors are now commonly used in various applications, such as fever scanners, insulation detectors, and electrical hotspot detectors.

As thermal patterns change depending on defects or machine conditions, thermography is essential for monitoring machine health non-intrusively. Predictive maintenance, utilizing sensing and inspection data, estimates the future deterioration and remaining usable life of equipment, aiding in maintenance planning and other engineering tasks. Thermal imaging cameras convert thermal energy into visible light to study specific objects or scenes, and the resulting images, known as thermograms, are analyzed using thermography methods.

1.1 Applications of Thermal Imaging

Figure 1 illustrates the diverse applications of thermal imaging. These include breast cancer detection and diabetic foot examination, where body temperature variations are identified. Infrared imaging is effective for monitoring mechanical and electrical installation conditions. Additionally, thermal visualization aids in detecting cold and hot spots, allowing for the identification of machine faults.

Gas leaks, including methane, propane, and natural gas can be detected using thermal imaging as these gases emit thermal radiation that is invisible to the naked eye. Thermal cameras can identify temperature variations caused by gas leaks, making them valuable tools for safety inspections.

In building inspections, thermal cameras detect places with heat loss or moisture intrusion, such as inadequately insulated walls, roofs, and windows [1]. The technology works by detecting the infrared radiation emitted by objects to create images showing temperature variations.

Thermal imaging cameras have various applications beyond gas leak detection. The military and defense industries were among the first to adopt thermal cameras. They are used in maritime applications for navigation, search, and rescue missions, agriculture for plant disease identification and crop monitoring, and border security for tracking and detecting vehicles and people.

In surveillance, thermal imaging can be utilized to distinguish faces, especially in low light or poor visibility conditions. Thermal cameras can recognize people based on the heat signatures their faces emit.

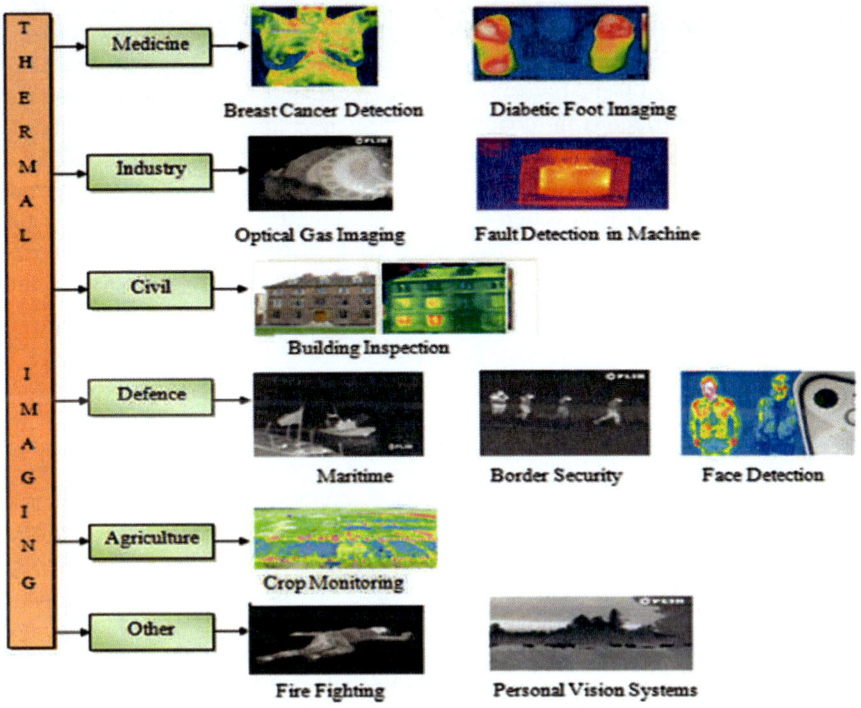

Fig. 1 Different applications of thermal imaging

Additionally, infrared cameras are used to monitor fire behavior, detect hidden fires, and for personal imaging systems to observe wildlife at night. They are also helpful in outdoor activities like camping and hiking to identify hidden dangers.

Thermal Imaging in Breast Cancer Detection. Breast cancer detection through thermal imaging, also known as "thermography" or "thermal breast imaging," involves capturing heat patterns emitted by breast tissue using specialized thermal cameras that detect infrared radiation. These cameras create thermal maps of the breast's surface, highlighting areas with abnormal heat patterns. Cancer cells often exhibit higher metabolic rates, leading to increased blood flow and elevated temperatures in affected regions. By comparing these thermal maps with established patterns and normal thermograms, potential abnormalities can be identified, aiding in early breast cancer detection.

However, thermal imaging is not a standalone diagnostic method, and conventional techniques like mammography and ultrasound are still necessary for definitive diagnosis and further evaluation. Proper training, standardization, and interpretation are crucial to achieving accurate results and maximizing the potential of thermal imaging in breast cancer detection.

Thermal Imaging in Face Recognition. Thermal imaging can be utilized to distinguish faces based on the differences in thermal patterns emitted by different parts of the face. It's called facial heat signature: Each person's face emits a unique heat signature due to variations in blood flow, skin temperature, and surface features. For instance, the eyes, nose, and mouth exhibit different thermal patterns, making it possible to identify facial landmarks and distinguish one face from another. Thermal imaging captures these heat patterns, creating distinct thermal representations for individuals. Unlike visible light-based facial recognition, thermal imaging works effectively in low light or complete darkness, making it valuable for surveillance and security applications.

2 Literature Survey

Defect detection in machines can be achieved through two approaches: thermal imaging analysis and conventional signal analysis techniques. The latter involves analyzing signals like vibration, current, and sound to identify defects in rotating machinery. The vibration-based technique is widely used due to its ability to represent the dynamic properties of machines accurately.

Recently, the combination of thermal imaging and machine learning has become popular. Thermal images provide temperature details that help detect anomalies in systems or devices. This non-contact method allows quick identification and examination of flaws without human involvement, thanks to advanced machine learning techniques. Integrating thermal imaging with machine learning enables efficient analysis of large datasets.

Researchers have explored various approaches in this field. For instance, Glowacz [2] analyzed the infrared images of induction and commutator motors, using features extracted through arithmetic mean along with the Otsu approach, the mean of a histogram, and the mean along with the 20 highest peaks. Furthermore, it was classified using long short-term memory (LSTM) and nearest neighbor (NN) classifier methods, achieving a recognition rate of 95.33–100%.

Sakallı and Koyuncu [3] applied DenseNet201, MobileNetV2, ResNet50, ShuffleNet, and Xception deep learning models on a Mendeley database that contains thermal images of induction motors and transformers and achieved a recognition rate of 100%.

Shao et al. [4] developed generative adversarial network (GAN) to extract global infrared correlation features from infrared images and classified them by convolutional neural network (CNN) that gained above 90% accuracy.

Javed et al. [5] collected infrared images in various states, like normal and overloaded to prevent motor faults. Features were extracted using local Octa pattern (LOP) along with genetic algorithm (GA) and were classified using a support vector machine (SVM), and this approach exhibited above 90% accuracy.

Ibrahim et al. [6] developed a hybrid model for fault detection in motors, tested at three different running speeds: 1480, 1450, and 1380 rpm, encompassing bearing problems, broken rotor bar faults, and stator faults. They used a FLIR C2 camera to capture infrared images of motors with six faulty states. Feature extraction was performed using gray level co-occurrence matrices (GLCM), and the invasive weed optimization (IWO) method was utilized for feature selection. The study employed random forest (RF) and extreme gradient boosting (XGBoost) algorithms to classify healthy and malfunctioning motor states.

Bhagat and Chauhan [7] extracted features using GLCM and IWO-based algorithms. Furthermore, authors categorized the various states of a motor using XGB and random forest (RF) classifier that resulted in 93% and 85% recognition rates evaluated on no load, 50% load, and 100% load conditions.

BV and Ananthan [8] extracted texture features and applied the sliding window method to detect the area of interest. Furthermore, the images are classified as decision tree, AdaBoost, random forest, and gradient-boosted trees and obtained an accuracy of 93.9%, 93.8%, 97.2%, and 92.8%, respectively.

Choudhary et al. [9] overcome the issue of insufficient information and string noise in the thermal images that hinder the performance of motors. They preprocessed the images using 2D discrete wavelet transform and extracted the pertinent features using principal component analysis (PCA). The chosen attributes were sorted according to importance using Mahalanobis distance and classified using complex decision tree (CDT), linear discriminant analysis (LDA), and SVM.

Mahami et al. [10] proposed an ensemble learning method that uses a bag-of-visual word (BoVW) for feature extraction and classified using an extremely randomized tree (ERT), SVM, RF, KNN, the self-organizing fuzzy logic classifier (SOF), deep rule-based (DRB), LSSVM applied these methods on a three-phase system with the following specifications: 11 kW of output power, 5 A of current, and 220/230 V of voltage that achieved 100% accuracy.

Glowacz [11] calculated the mean of infrared images of brushless dc motors and extracted the features. Next, the authors analyzed the states of motors running at different speeds with blocked ventilation and healthy and faulty clippers that achieved 100% classification accuracy.

Choudhary et al. [12] considered one healthy and five abnormal states of rolling element bearings and used an artificial neural network (ANN) and LeNet-5 convolutional neural network (CNN). The faulty states considered were outer and inner race bearing states, cage defect, ball defect, and lack of lubrication. CNN gained an average accuracy of 99.80% in all circumstances.

Mehta et al. [13] performed feature extraction using 2D discrete wavelet transform (DWT) and reduced the dimension of the extracted features using PCA and KNN, SVM, and LDA methods for classification. The tests were carried out on a bearing test rig with a single-phase 2 HP, 2 poles, and 220 V induction motor, and all the classifiers gained above 95% accuracy.

Fang et al. [14] employed various techniques to extract details based on temperature, texture, and shape from infrared images. They used Hu moment characteristics, GLCM features, and regional temperature features. After combining this data, they used a GAN model to input it. According to the findings of their studies, the model had an average classification success rate of roughly. 84.2% and accurately classified the categories of equipment failures and overheating with 82.2% and 86.2% accuracy, respectively.

Fanchiang et al. [15] introduced an innovative IRT-based technique for detecting overheating issues in cast-resin transformers through their full-time online fault monitoring system. They trained the Wasserstein autoencoder reconstruction (WAR) model and the differential image classification (DIC) model. The differential image is obtained by calculating the absolute difference between actual and regenerated images at the pixel level. During testing, a defect diagnosis module is created by combining the well-trained WAR and DIC models. The experimental results demonstrate that the proposed model surpasses existing deep learning approaches in terms of being lightweight, having a small storage size, offering quick inference time, and providing precise diagnostic performance.

Early-stage diagnosing malfunctions in equipment and machines by looking for temperature changes in thermal pictures can help prevent equipment failures, flame risks, and other potential hazards. Although many fault scenarios generate high heat, certain defects cause equipment to produce less heat than normal. This might be difficult when trying to differentiate between functioning and defective equipment.

Najafi et al. [16] address the problem by dividing the data into hot and cold states. The authors find the region of interest in a cold state by random forest and hot state using AdaBoost and perform sliding window-based segmentation. The suggested methods were evaluated on the Mendeley datasets.

Khanjani and Ezoji [17] identified the area of interest by matching SIFT key points. Next, they are represented as feature vectors using CNN and classified into cold and hot clusters using K means and SVM algorithms that gain 100% accuracy.

Lozanov et al. [18] developed a model employing infrared pictures of a motor in different conditions, such as a healthy motor, motor bearing failure, and cooling fan. The framework uses three textural features, i.e., GLCM-based, component-based, and histogram-based to extract the essential data and Otsu thresholding to segment the motor structure. Support vector machines (SVM) are used to do the classification task. The classification accuracy of the framework, which was assessed with a 62.5–37.5% training-test split, was 83.3% for the three motor conditions.

Al-musawi et al. [19] utilized thermal images and converted them into the hue, saturation, and value (HSV) color model. They employed various edge detection methods, such as Otsu, Roberts, Prewitt, Sobel, and Canny to identify the hot regions based on a threshold. To extract essential defect information from the Hue image, multiple image matrices were applied. Additionally, the researchers used image metrics, including mean, mean square error, peak signal-to-noise ratio, variance, standard deviation, skewness, and kurtosis to detect abnormalities in ball bearings, inner races, and outer races of three-phase induction motor bearings.

Li et al. [20] introduced nine defective states to the rotating machine and processed the infrared images using CNN, DBN, DNN, and SAE models that gained classification accuracy of 98.64%, 83.60%, 75.53%, and 60.84%.

Various researchers have contributed to machine fault detection, focusing on induction motors, transformers, rotating machines, and electrical equipment. Their valuable insights and methodologies are documented in the form of research papers, which are summarized in Table 1.

3 Types of Faults in Machines

Machines can experience internal and external faults, further categorized as mechanical, electrical, and environmental issues. In induction motors and rotating machines, common faults include stator winding defects (stator faults), faults in the rotor parts (rotor faults), and lubrication issues. In ball bearings, distinct faults like inner and outer race damages are observed, often resulting from running at high speeds. Researchers like López-Pérez and Antonino-Daviu [21] have discussed various faults, including ruptured rotor bars, harmonic interference, shaft misalignment, operational overload, lubricant contamination, and voltage imbalance. Figure 2 [13] illustrates the faults present in bearings, featuring different sizes of holes.

Next, the transformers with no and heavy loads are shown [22] in Fig. 3.

4 Issues/Challenges

Challenges and difficulties encountered in collecting infrared images and identifying machine faults include:

- Temperature Fluctuations: Variations in temperature due to external conditions or heat transfer properties of machine components can make it challenging to interpret thermal imaging and distinguish actual flaws from typical temperature swings.
- Sensor Complexity: Different sensors used for vibration data collection can complicate sensor installation and potentially harm rotating equipment. Vibration data can also produce significant noise during machine operation, making fault identification difficult, which can be mitigated through infrared thermography for fault detection.
- Insufficient Data: Acquiring sufficient thermal imaging data for training and evaluating AI/ML models may be challenging and time-consuming, especially under various faulty conditions.
- Environmental Factors: Environmental elements like humidity and temperature can impact thermal imaging, affecting the identification of fault conditions and requiring consideration during data gathering and analysis.
- Complexity and Cost: Implementing effective defect detection systems can be expensive and resource-intensive, making long-term maintenance challenging.

Table 1 Fault identification in machines using infrared images

Study	Algorithms	Equipment	Fault examined	Accuracy
Glowacz, Adam [2]	Differences of arithmetic mean with Otsu's method NN, LSTM	Commutator, induction motor	"Healthy commutator motor (HCM), broken rotor coil of the commutator motor (BRCoCM), shorted stator coils of the commutator motor (SSCoCM), healthy single-phase induction motor (HSIM), single-phase induction motor with shorted coils of auxiliary winding (SIMwSCoAW)"	ACC: 95.33–100%
Gönül Sakallı and Hasan Koyuncu [3]	DenseNet201, MobileNetV2, ResNet50, ShuffleNet, Xception	Transformer, induction motor	"Cooling fan failure, rotor fault, short-circuit faults transformer, no load motor, and no load transformer"	ACC: 100%
Shao et al. [4]	Dual-threshold attention-guided GAN (DTAGAN), CNN	Rotating machine	Ball fault of bearing, compound fault, inner and outer race faults of the normal and unbalanced rotor	Above 90%
Javed et al. [5]	LOP, GA, SVM	Induction motor	Overloaded, fault, and normal operating conditions	Above 90%
Ibrahim et al. [6]	GLCM, IWO Random Forest, XGBoost	Induction motor	Overloaded, fault, and normal operating conditions	XGBoost:93% Random Forest:85%
Bhagat and Chauhan [7]	GLCM, IWO Random forest XGBoost	Induction motor	No load, 50% load, and 100% load conditions	RF: 85%, XGBoost: 93%
BV and Ananthan [8]	Sliding window method, GLCM	Induction motor	Rotor fault, 1,2,3 phases stator fault, cooling fan failure	Gradient boost: 87.8% Ada boost: 87.8% Decision tree: 87.9% RF: 90%
Choudhary et al. [9]	2D-DWT, CDT, PCA, LDA, and SVM	Induction motor	Inner race, and outer race faults, lubrication defect	LDA:100%, CDT and SVM above 95%
Mahami et al. [10]	BoVW, SURF, SIFT, ERT, KNN, SOF, RF, DRB, LSSVM	Induction motor	Stator faults under different states and conditions	ACC: 100%

(continued)

Table 1 (continued)

Study	Algorithms	Equipment	Fault examined	Accuracy
Glowacz in [11]	CPoAMoTI	Brushless DC motors	Blocked ventilation, healthy and faulty clippers	ACC: 100%
Choudhary et al. [12]	ANN, LeNet-5, CNN	Rotating machine	Inner and outer race faults, lubrication, ball, and cage defects	CNN: 99.80%
Mehta et al. [13]	2D-DWT, PCA, LDA, KNN, SVM	Rotating machine	"Healthy, inner race, and outer race bearing states"	Above 95%
Fang et al. [14]	GLCM, GAN	Electric transformer	Defective and overheated faults	Above 85%
Fanchiang et al. [15]	WAR, DIC	Cast-resin transformer	Wire and connection overheating, normal and inter-turn short circuit	ACC: 99.95%
Najafi et al. in [16]	Sliding window-based method, Ada boost, RF	The three-phase induction motor and transformer	11 classes of induction motor 9 classes of transformer states	ACC: 93.8%, 95.6% for 11 classes 9 classes
Khanjani et al. [17]	SIFT, CNN, K means SVM	Three-phase induction motor	No load and loaded states	ACC: 100%
Musawi et al. [19]	HSV color model Otsu, Roberts, Prewitt, Sobel, and Canny edge detection methods	Three-phase induction motor	Inner race and outer race bearing defects	Mean in healthy motor = 0.5305 with 50% of load based on the Otsu method
Li et al. [20]	CNN LeNet-5 model, DBNs, DNN, SAE	Rotating machine	Inner race, outer race, balanced and unbalanced faults	CNN:98.64% DBN:83.60% DNN:75.53% SAE: 60.84%

Fig. 2 Various states of bearings

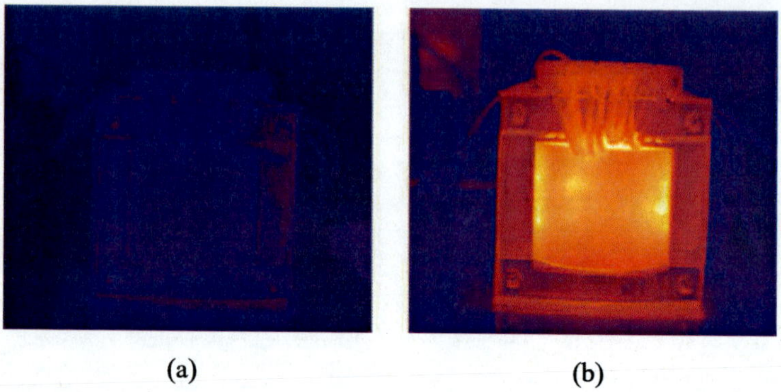

Fig. 3 Transformer with **a** No load **b** Load

- Interpretability of AI/ML Models: The lack of transparency in AI/ML models' decision-making hampers their interpretability, despite high predictive accuracy. Understanding the variables driving the model's predictions and providing justifications for maintenance recommendations in predictive maintenance is essential, and future research should focus on developing methods to clarify and comprehend AI/ML models' choices when analyzing thermal images.

Open Chanllenges. Here are some of the key open challenges in thermal imaging:

- Resolution and Image Quality: Improve thermal camera clarity and sharpness.
- Sensitivity and Temperature Range: Enhance detection of subtle temperature differences.
- Cost: Reduce the expense of thermal imaging systems.
- Real-Time Processing: Develop fast data processing for immediate applications.
- Environmental Interference: Minimize the impact of external factors on thermal data.

5 Methodology

To manage and prevent fault diagnosis in machines, early identification is crucial. The general flowchart of the proposed method for early fault detection and classification is represented in Fig. 4.

The steps in the general framework are as follows:

Step 1. Data Collection: Thermal images are acquired using thermal cameras such as FLIR, Fluke, and Dali tech cameras with suitable resolutions for different machine types. Infrared images are collected in healthy and abnormal states under different defective conditions and machines running at different speeds.

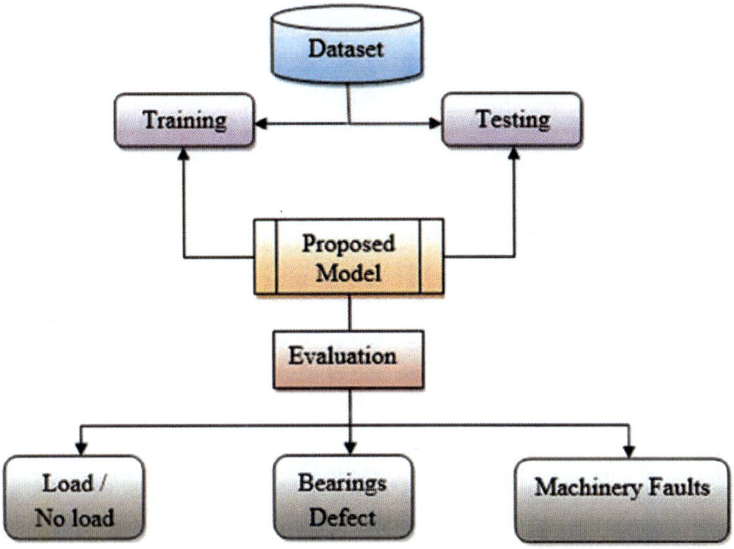

Fig. 4 General flowchart of the proposed method

Step 2. Data Division: The collected dataset is divided into training, validation, and testing. A common data split ratio is used, such as 70% for training, 10% for validation, and 20% for testing.
Step 3. Data Augmentation: Data augmentation techniques are applied to increase the training dataset's diversity and improve the model's generalization. Augmentation techniques include rotation, flipping, zooming, and adding noise to thermal images.
Step 4. Preprocessing: Preprocessing techniques are applied to enhance image quality, eliminate noise, and identify specific regions of interest. Filters remove noise and segment regions based on temperature variations to identify hot and cold spots. Sample infrared images of transformer are shown in Fig. 5.
Step 5. Feature Extraction: Features are extracted using various image processing algorithms such as color thresholding, wavelet transforms, GLCM, and deep learning models. Transfer learning models can also automatically extract features,

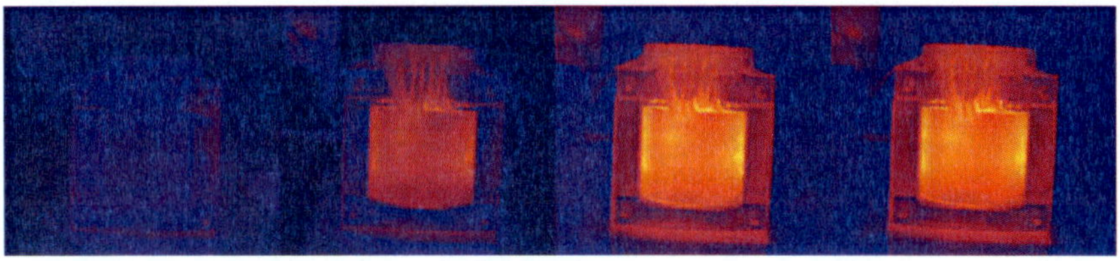

Fig. 5 Sample infrared images of the transformer [22]

eliminating the need for manual feature extraction. Both approaches can be utilized based on specific requirements.
Step 6. Training Module: In this module, the training dataset, which contains 70% of the collected data, is used to train machine learning algorithms, artificial neural networks (ANN), or convolutional neural networks (CNN). The model learns to recognize patterns in thermal images corresponding to different machine faults.
Step 7. Validation Module: During the training process, the model's performance is evaluated on the validation dataset, which contains 10% of the data to tune hyperparameters, prevent overfitting, and monitor training progress.
Step 8. Testing Module: After training, the trained model is tested using the separate testing dataset, which contains 20% of the data, containing thermal images of machines with known faults. The model's performance is evaluated based on accuracy, precision, recall, and f1 score to assess its ability to classify different machine faults correctly.
Step 9. Analysis: The methods are evaluated based on multiple performance measurements, and the results are analyzed to determine the effectiveness and accuracy of the proposed fault detection and classification system. The data division and augmentation strategies are vital to the model's overall performance and generalization capability.

5.1 Discussion

Selection of Algorithms for Thermal Imaging. The selection of algorithms for thermal imaging should be justified based on the specific task and requirements of the application. Different algorithms excel in different aspects of thermal image processing, and the choice depends on factors such as accuracy, speed, computational complexity, and the nature of the problem being addressed.

- Object Detection: For detecting objects in thermal images, deep learning-based object detection algorithms like you only look once (YOLO) or faster region-based convolutional neural networks (R-CNN) are commonly used due to their high accuracy and ability to handle complex scenes.
- Image Segmentation: For segmenting regions of interest in thermal images, semantic segmentation models like U-Net or Mask R-CNN are often employed, as they can provide pixel-level accuracy.
- Image Fusion: For combining thermal and visible light images, image fusion algorithms like wavelet transform or Laplacian pyramid can be used to enhance the interpretability of the combined images.
- Anomaly Detection: Anomaly detection algorithms, such as one-class support vector machine (SVM) or autoencoders are suitable for identifying abnormal thermal patterns that may indicate faults or irregularities.
- Real-Time Processing: In scenarios where real-time processing is critical, lightweight deep learning models, such as MobileNet or EfficientNet are preferred due to their reduced computational demands.

Need of Preprocessing and its Impact on Performance. Preprocessing is a critical step in thermal imaging analysis that enhances the quality of images and prepares them for defect detection. It involves noise reduction, region segmentation, image normalization, feature extraction, and outlier removal. The impact of preprocessing on model performance is significant. It improves the accuracy of defect detection by reducing noise, focusing on relevant regions of interest, bringing images to a consistent scale, extracting essential features, and enhancing the model's robustness. Effective preprocessing is essential for optimizing thermal images for machine learning algorithms and plays a crucial role in achieving accurate and reliable results in thermal imaging applications.

Improving Thermal Imaging Performance

To improve thermal imaging performance, upgrade to higher-resolution sensors for better image quality and detailed thermal information. Implement advanced signal processing algorithms to accurately interpret data and identify patterns. Increase sensitivity and temperature range for detecting subtle differences in various environments. Integrate with other imaging modalities for comprehensive analysis and object recognition.

Use intelligent algorithms for automated pattern recognition. Consider environmental factors to reduce interference and encourage cross-disciplinary collaboration for innovative advancements in thermal imaging technology across industries.

6 Results and Evaluation

Thermal Imaging Dataset. Several widely used thermal imaging datasets are available for research and development purposes in the field of thermal imaging and machine learning. These datasets play a crucial role in training and evaluating algorithms for various applications. Some of the popular thermal imaging datasets are as follows:

- FLIR Thermal dataset
- VTT Thermography dataset
- IR-ImageNet
- KAIST Multispectral Pedestrian dataset
- Infrared Pedestrian dataset (IRPD)
- FLIR ADAS dataset

Performance Evaluation. In diagnostic tests, a sample can be categorized as true negative (actual negative), true positive (actual positive), false negative (wrongly anticipated as negative), or false positive (wrongly predicted as positive). False negative indicates that the test incorrectly anticipated a negative result for a positive sample, while true positive indicates that the test correctly predicted a positive sample. False positive means that the test mistook a negative sample for a positive one, while true negative denotes a correctly predicted negative sample.

Various metrics are frequently used to assess diagnostic models, including specificity, accuracy, sensitivity (also known as recall), f1 score, precision, and the receiver operating characteristic (ROC) curve. These evaluation measures can be quantitatively expressed using Eqs. 1 through 7, which provide measures for f1 score, accuracy, specificity, sensitivity, and precision. The ROC curve visually represents the performance of a classification model at different levels.

$$\text{Accuracy} = \frac{\text{Total number actual predictions}}{\text{Total number of predictions}} \tag{1}$$

$$\text{Specificity} = \frac{\text{True negatives}}{\text{True negatives} + \text{False Positives}} \tag{2}$$

$$\text{Sensitivity} = \frac{\text{True positives}}{\text{True positives} + \text{False positives}} \tag{3}$$

$$\text{Precision} = \frac{\text{True positives}}{\text{True positives} + \text{False positives}} \tag{4}$$

$$F1\text{Score} = 2 * \left(\frac{\text{Precision} * \text{Recall}}{\text{Precision} + \text{Recall}}\right) \tag{5}$$

$$\text{TPR} = \frac{\text{True positives}}{\text{True positives} + \text{False negatives}} \tag{6}$$

$$\text{FPR} = \frac{\text{False positives}}{\text{False positives} + \text{True negatives}} \tag{7}$$

7 Conclusion

The literature review has established thermal imaging as a promising method for detecting machine defects. Analyzing thermal images to understand temperature distribution and patterns in machine components can aid in the early identification of potential flaws before they escalate. Numerous studies have demonstrated the effectiveness of thermal imaging in detecting various issues like bearing, gear, and motor faults. Early detection of failures leads to reduced downtime, increased equipment lifespan, enhanced safety, and reduced manufacturing costs. However, several factors need to be considered when choosing a specific defect detection technique. The survey also highlights some challenges that need resolution to effectively use thermal imaging in machine fault detection, such as data availability and interpretability of machine learning models. Overall, the findings suggest that thermal imaging holds significant potential as an affordable, non-destructive, and non-intrusive tool for detecting machine faults.

Acknowledgements The author would like to acknowledge the Visvesvaraya Ph.D. scheme, Meity, Govt. of India, and REVA University for providing the required infrastructure and other support.

References

1. Wilson AN, Gupta KA, Koduru BH, Kumar A, Jha A, Cenkeramaddi LR (2023) Recent advances in thermal ımaging and its applications using machine learning: a review. IEEE Sens J 23(4):3395–3407, 15 Feb 2023
2. Glowacz AT (2023) Thermographic, Fault diagnosis of electrical faults of commutator and induction motors
3. Sakallı G, Koyuncu H (2023) Identification of asynchronous motor and transformer situations in thermal images by utilizing transfer learning-based deep learning architectures. Measurement 207
4. Shao H, Li W, Cai B, Wan J, Xiao Y, Yan S (2023) Dual-threshold attention-guided gan and limited ınfrared thermal ımages for rotating machinery fault diagnosis under speed fluctuation. IEEE Trans Ind Inf
5. Javed M, Shabbir Z, Asghar F, Amjad W, Mahmood F, Khan O, Virk US, Waleed A, Haider Z (2022) An efficient fault detection method for ınduction motors using thermal ımaging and machine vision. Sustainability
6. Ibrahim A, Anayi F, Packianather M (2022) New machine learning model-based fault diagnosis of ınduction motors using thermal images. In: 2nd ınternational conference on advance computing and ınnovative technologies in engineering (ICACITE). India, pp 58–62
7. Bhagat AK, Chauhan A (2022) Thermal ımage-based fault analysis of ınduction motors using a novel machine learning model. In: 11th ınternational conference on system modeling & advancement in research trends (SMART). India, pp 1429–1433
8. BV C, Ananthan T (2022) Machine learning based fault detection in ınduction motor using thermal ımaging. In: 2022 3rd ınternational conference on electronics and sustainable communication systems (ICESC). India, pp 929–936
9. Choudhary A, Goyal D, Letha SS (2021) Infrared thermography-based fault diagnosis of ınduction motor bearings using machine learning. IEEE Sens J 21(2):1727–1734, 15 Jan 2021
10. Mahami A, Rahmoune C, Bettahar T, Benazzouz D (2021) Induction motor condition monitoring using infrared thermography imaging and ensemble learning techniques. Adv Mech Eng
11. Glowacz A (2021) Thermographic fault diagnosis of ventilation in BLDC motors. Sensors 21(21):7245
12. Choudhary A, Mian T, Fatima S (2021) Convolutional neural network based bearing fault diagnosis of rotating machine using thermal ımages. Measurement
13. Mehta A, Goyal D, Choudhary A, Pabla B, Belghith S (2021) Machine learning-based fault diagnosis of self-aligning bearings for rotating machinery using ınfrared thermography. Math Probl Eng
14. Fang J, Yang F, Tong R, Yu Q, Dai X (2021) Fault diagnosis of electric transformers based on infrared image processing and semi-supervised learning. Glob Energy Interconnection 4(6):596–607
15. Fanchiang K-H, Huang Y-C, Kuo C-C (2021) Power electric transformer fault diagnosis based on ınfrared thermal ımages using wasserstein generative adversarial networks and deep learning classifier. Electronics
16. Najafi M, Baleghi Y, Gholamian SA, Mehdi Mirimani S (2020) Fault diagnosis of electrical equipment through thermal ımaging and ınterpretable machine learning applied on a newly introduced dataset. In: 6th Iranian conference on signal processing and ıntelligent systems (ICSPIS). Iran, pp 1–7

17. Khanjani M, Ezoji M (2020) Electrical fault detection in three-phase induction motor using deep network-based features of thermograms. Measurement
18. Lozanov Y, Tzvetkova S, Petleshkov A (2020) Use of machine learning techniques for classification of thermographic images. In: 12th electrical engineering faculty conference (BulEF). Bulgaria, pp 1–4
19. Al-musawi A, Anayi F, Packianather M (2019) Three-phase ınduction motor fault detection based on thermal ımage segmentation. Infrared Phys Technol
20. Li Y, Du X, Wan F, Wang X, Xiao Q-B, Yu H (2019) Rotating machinery fault diagnosis based on convolutional neural network and infrared thermal imaging. Chin J Aeronaut
21. López-Pérez D, Antonino-Daviu J (2017) Application of ınfrared thermography to failure detection in ındustrial ınduction motors: case stories. IEEE Trans Ind Appl 53(3):1901–1908, May-June 2017
22. Najafi M, Baleghi Y, Gholamian SA, Mirimani SM (2021) Thermal image of equipment. Mendeley Data [Online], Available: https://data.mendeley.com/datasets/m4sbt8hbvk/2

AdaptPSOFL: Adaptive Particle Swarm Optimization-Based Layer Offloading Framework for Federated Learning

Rachit Verma and Shajulin Benedict

Abstract Following the rising privacy requirement enforced by regulations such as the European General Data Protection Regulation and the HIPAA, the centralized training of machine learning models has become difficult. Federated learning has come under extensive adoption because it allows clients to train a local model on a local dataset and then send the model to a server for aggregation. This aspect allows clients to preserve the privacy of their local data, while also allowing the training of a global model. A new area of study that is emerging is the possibility of offloading the training of the local model between a client and a server to influence the training time of the model. Studies such as FedAdapt use a reinforcement learning agent to decide the optimal split between a client and a server. In our study, we determine the optimal split between a client and server by formulating an optimization problem that simultaneously minimizes the CPU utilization and the round execution time while considering the number of layers placed on the client as decision variables. We apply particle swarm optimization to approach the optimization problem. Based on our experiments, we observed that our scheme outperforms FedAdapt and classic federated learning on the combined objectives by 36.60% and 36.73% on average respectively.

Keywords Federated learning · Edge computing · Particle swarm optimization · Reinforcement learning

R. Verma (✉) · S. Benedict
Indian Institute of Information Technology Kottayam, Valavoor P.O., Kottayam District, Kerala 686635, India
e-mail: rachitverma18bcs@iiitkottayam.ac.in

S. Benedict
e-mail: shajulin@iiitkottayam.ac.in

S. Shakya et al. (eds.), *Fourth International Conference on Image Processing and Capsule Networks*, Lecture Notes in Networks and Systems 798,
https://doi.org/10.1007/978-981-99-7093-3_40

1 Introduction

Traditional centralized machine learning systems require the aggregation of data from a variety of sources. The rise of data privacy regulations such as the European General Data Protection Regulation (GDPR) although which helps secure users, can act as a hindrance to the training of a global model. To deal with these issues, federated learning was proposed [1]. In federated learning, each client trains a local model based on the raw data available to it and then sends the model updates at the end of every round of training to the federated learning server. The server waits for each of the clients to complete sending their updates before beginning aggregation. The model updates are aggregated using a model aggregation algorithm such as FedAvg [2] to create a global model. Federated learning has also seen several industry applications including the automotive industry and the healthcare industry. In the automotive industry, e.g., Satpura et al. [3] used federated learning to predict the demand of various electric vehicle charging stations to avoid congestion, while in the healthcare industry, Huang et al. [4] used federated learning to predict patient mortality from electronic medical records distributed across various hospitals. In addition to them, federated learning can be used to assist in the development of smart cities as well because of its privacy preservation properties [5].

Federated learning can be classified in two ways, cross-device federated learning, and cross-silo federated learning [6]. Cross-silo federated learning involves entities such as companies and organizations. Companies owing to their large size generally possess sufficient computation power to be able to participate as clients in federated learning [7]. On the other hand, cross-device federated learning takes into consideration the case where end devices such as smartphones are clients. The number of devices in this case that communicate with the server in a single round of training can rise to nearly 1 million. The resource heterogeneity of the devices can lead to varying training times for each of the clients [8]. This behavior can be influenced by factors such as the device or vendor-specific implementation of the software, the power management scheme and the hardware limitations of the devices. This can cause the federated learning server to remain idle waiting for the parameter updates from the local models of the slow devices and the faster clients to remain idle waiting for model aggregation, after finishing their local training. This problem is further exacerbated by the statistical heterogeneity of the training data which was utilized by the client's local learning models. This heterogeneity may occur in the form of varying labels across clients or varying sources from which the data was sampled. The former is referred to as horizontal federated learning, while the latter as vertical federated learning [6]. In the literature, the devices of lower resource availability that increase the overall training time are called stragglers. Federated learning has also seen blockchain-based implementations, since it helps alleviate problems that a central authority is susceptible to such as vulnerability to attacks [9].

As a consequence of the various challenges and scenarios that arise in federated learning, it has been the target of a large number of optimization attempts. Park et al. [10] for example attempted to optimize the communication costs with the use of particle swarm optimization [11]. Xu et al. [12] in their paper attempt to optimize the learning rate in the scenario of over-the-air computation. Our work builds upon the work in deep neural network layer offloading conducted by the authors of FedAdapt [13] and SplitFed Learning [14]. The authors in their work have described how federated learning can take place if a convolutional neural network has been partitioned so that a certain number of its layers are placed on a client while the remainder is on the server.

The contributions of our paper may be summarized as follows:

(1) We propose an optimization problem that attempts to minimize the average CPU utilization and the maximum round execution time across the clients participating in federated learning based on the number of layers of a convolutional neural network split between them and the federated learning server.
(2) We propose the application of the particle swarm optimization technique to address the above optimization problem.
(3) We compare the value of the objective function attained by our framework to that obtained by measuring the respective values of two frameworks, namely classic federated learning and FedAdapt.

Our paper consists of seven sections. Section 2 discusses the work that is closely related to ours. Section 3 gives an overview of how federated learning takes place when a CNN is split between a client and server, while also pointing out how our implementation of particle swarm optimization has been integrated into the process. Sect. 4 delves into the details of the optimization problem that we wish to address, in addition to how particle swarm optimization has been implemented. Section 5 contains details of our experimental setup. Section 6 discusses our results and Sect. 7 concludes the paper.

2 Related Work

There are few studies that attempt to study federated learning when the layers have been partitioned between the client and server. One such study for example has been proposed by Gupta et al. [15]. In their study, they propose an algorithm that sequentially trains a neural network that has been partitioned between a client and a server. Their framework is referred to in the literature as split learning. They compare their algorithm on validation accuracy to federated averaging and large-scale stochastic gradient descent. This was later extended by Vepakomma et al. [16] to learn vertically partitioned data and to learn in instances where label sharing is not possible.

The authors of SplitFed learning [14] in their paper have extended classic federated learning for the case, where each client in the federated learning process contains a neural network partitioned at a point called a cut point. They propose two versions of their framework. In version 1, the networks resulting from the aggregation of the client-side neural networks and the server-side neural networks are concatenated. In version 2, the server maintains one copy of a neural network that is sequentially trained in conjunction with the client-side neural networks. The client-side neural networks are aggregated and concatenated with the server-side neural network.

FedAdapt considers a scenario similar to SplitFed learning version 1 and uses reinforcement learning to decide the ideal cut point on each device to optimize the training time throughout the system. They call this point the offloading point. They intended to mitigate the excess training time by offloading the layers from the slower clients onto a federated learning server. Another work named Aergia by Cox et al. [17] has proposed offloading the layers from one client to another if training on a client becomes slower. This offloading decision is made based on the dataset similarity between the clients. They achieve a reduction in training time when compared with FedAvg and TiFL [7] while maintaining high accuracy. Another layer offloading-based distributed learning strategy for the edge was proposed by Joshi et al. [18]. In the study, they proposed eliminating the model side synchronization step from SplitFed learning. They compared their scheme with SplitFed learning based on information leakage.

Upon a careful perusal of the above literature, we observe that they do not consider the possibility of offloading the layers of the neural network while considering the impact it could have on the CPU utilization of the clients. Our study addresses this gap by optimizing the layer offloading process of the neural network while considering two factors: the time taken to execute each round of federated learning and the percentage of the CPU utilized upon running each round of federated learning.

3 PSO Assisted Layer Offloading Architecture

In this section of our work, we discuss how particle swarm optimization is integrated into the standard federated learning setup to assist with offloading the layers of a neural network.

A standard federated learning setup consists of a central server and a number of clients. Each client trains a local model on its dataset following which it transfers the model's parameters to the central server for model aggregation. This can take place based on a variety of algorithms including FedAvg. In our study, we use FedAvg for the model fusion algorithm. Here, we consider a round of federated learning for each device to include three steps, an initialization phase, a training phase and an aggregation phase.

In the initialization phase, each client receives weights from a server also known as the federated learning server. Each client at its side initializes its model based on these weights. These weights are uniform across all clients. Next, the client trains its local model on its local dataset. This is the phase of federated learning where we draw inspiration from SplitFed learning and FedAdapt. In both these frameworks, a convolutional neural network is split between a client and a server. While in SplitFed learning, the total number of layers on all clients remains the same throughout a round for every round, the total number of layers in FedAdapt changes every round based on the decision of a reinforcement learning agent. Our framework will also vary the total number of layers on the client, however we intend to use a real-time optimization strategy that considers round execution time and CPU utilization (discussed in the next section) to achieve this.

We now discuss how training takes place in such a scenario. This process is based on SplitFed learning. The server maintains one copy of the convolutional neural network for each client from the layer that it has been split at till the final layer. A standard round of training in a convolutional neural network involves forward and backpropagation. The forward propagation step involves passing images from a batch through the client side of the network and then sending the resulting gradients at the final layer to the server through a communication channel. These gradients are then propagated through the server's portion of the neural network corresponding to the client. The backpropagation step involves the server sending the gradients from its copy of the convolutional neural network back to the client through a communication channel for it to continue its backpropagation step. Following the end of training, the weights on each of the clients are sent to the server. Now the server concatenates the models from each of the clients with the copy corresponding to it held at the server. The resulting models are subjected to a model aggregation technique such as FedAvg [2].

While this entire process takes place, each of the clients continuously make note of the execution time of each of the rounds and also the CPU utilization. This data is sent to the optimizer. This process is repeated for all the possible offloading points (discussed in the next section) for all the clients. Once the optimizer receives data corresponding to the round execution time and the CPU utilization time of each of the offloading points on each of the clients, it is then in a position to prescribe the ideal offloading point for each of the devices. This can be visualized through Fig. 1. We will now discuss the functioning of the optimizer.

4 Optimization Problem Formulation and Algorithm

This section explains the optimization problem that was formulated and how particle swarm optimization was used to address it.

4.1 Formulation

To formulate the optimization problem in our study, we must first define what an offloading point means for a convolutional neural network and how training would take place if a convolutional neural network has been offloaded. The definition of an offloading point has been based on FedAdapt. The offloading point for a convolutional neural network denotes the number of layers from the first layer onward that remain on a client. This metric is indexed from zero. The remainder of the layers are placed on the server.

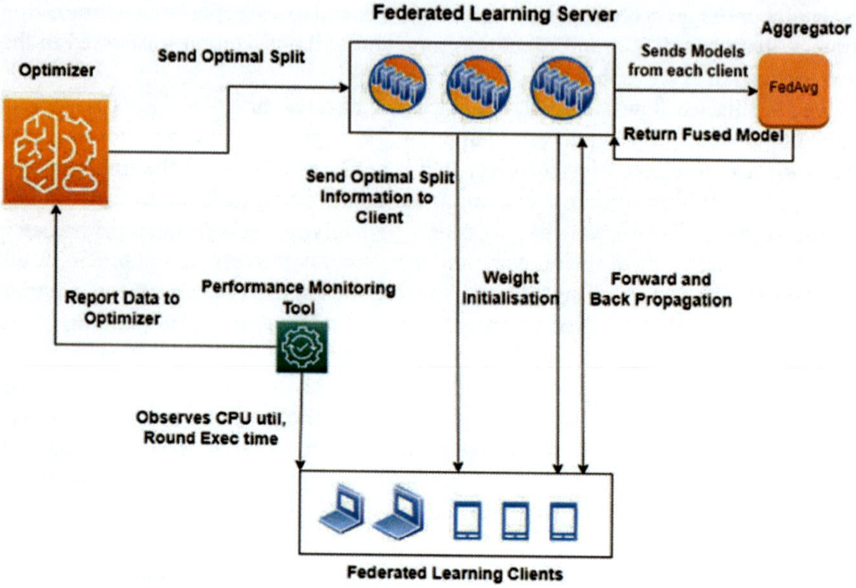

Fig. 1 PSO-assisted federated learning architecture

Consider a convolutional neural network with five layers. Let them be indexed as such [0, 1, 2, 3, 4]. An offloading point of 2 would mean that layers indexed 0, 1, and 2 would be placed on the client whereas layers 3 and 4 would be placed on the server. In this case, a round of training would entail the forward propagation of the gradients through layers 0, 1, and 2 on the client side followed by the transfer of the results via a communication channel to the server. The gradients would then be propagated through layers 3 and 4 on the server. In this study, we will attempt to simultaneously optimize two metrics. We will minimize the time it takes for the slowest device also known as the straggler to run a single round of federated learning and the average CPU utilization of running one round of federated learning. The time taken by the slowest device in federated learning in a given round may be formulated as:

$$\max(t_1,\ t_2 \ldots .t_n), \tag{1}$$

where t_x is the time to execute a round of training on each device. We also refer to this as the round execution time. This is based on FedAdapt framework. Our contribution is as follows. We will now consider the CPU utilization of each client in each round. This can be formulated as:

$$\text{mean}\big(\text{cpu}_1,\ \text{cpu}_2,\ \text{cpu}_3 \ldots\ldots \text{cpu}_n\big), \tag{2}$$

where $\text{cpu}_n = \max$ (CPU utilization)—min (CPU utilization) for the round for device n. We intend to minimize both of these metrics. Thus, we formulate our objective function as the linear combination of both of these values. In order to give both metrics equal priority we formulate it as follows.

Minimize:

$$0.5 * \max(t_1,\ t_2, \ldots\ldots t_n) + 0.5 * \mathrm{mean}(\mathrm{cpu}_1, \mathrm{cpu}_2 \ldots .\mathrm{cpu}_n). \tag{3}$$

Here, the decision variables are the various offloading points on each of the devices. Let them be:

$$[s_1,\ s_2,\ s_3 \ldots .s_n] \tag{4}$$

s_1 is the offloading decision assigned to device 1, s_2 to device 2 and so on. The idea of formulating the offloading points as decision variables is based on the FedAdapt framework. We will be attempting to solve this optimization problem using particle swarm optimization (see Algorithm 1).

Algorithm 1 Adaptive PSO federated learning

K = num of devices s = offloading points
cpu store = k x s matrix
round time store = k x s matrix
for round = 0 to 99 **do if** *round* < *s* **then**
Output split layers = [round, round K times]
else
PSO with Objective Function from cpu store and round time store Output Split Layer
end if
Conduct Training
Update round store cpu store
end for

4.2 *Particle Swarm Algorithm for Solving Optimization Problem*

Consider a federated learning setup with k clients and a neural network with s layers to have federated learning conducted on it for 100 rounds. Here each layer can be considered an offloading point. We initialize two matrices, one to store the round execution time for each of the clients and another to store the CPU utilization of the clients. Each of these matrices has dimension (*k*, s). In entry (*m*, *n*) of matrix 1 and matrix 2, we store the round execution time and the CPU utilization, respectively, of executing local training on device m with an offloading point of n. To populate the matrix we run federated learning from round 0 to *s* - 1 and issue an offloading decision equal to the round number to all the devices, e.g., if we are training a neural network with seven layers, we run federated learning with an offloading decision for all devices as being equal to the round number till round 6. In this case, should the experimental setup have six devices then round 3 for example would have a decision of [4]. Each offloading decision in the array corresponds to a device in the setup. Refer to Eq. (4).

Once the array has been populated, we will use particle swarm optimization to arrive at the ideal offloading point. The matrix populated above will be used to tentatively calculate the objective function.

In our approach, we use particle swarm optimization to find the ideal offloading points on all the devices that minimizes the objective function in Eq. (3). A particle in our case is a one-dimensional array with k elements, where each element in the array will correspond to an offloading point on each device. We initialize 10 such particles. This array of particles will be referred to as X. Each of these values will be initialized between 0 and 1. The initial velocity of each of the particles was initialized between 0 and 0.01. In order to translate the position values of each of the elements in a particle to an offloading point, we utilize Algorithm 2. The algorithm divides the space of rational numbers between 0 and 1 into n equal intervals, where n is the first input to the function. It then identifies the interval that the number p will be placed in, e.g., should $n = 4$ and $p = 0.27$, then the function will output 2 since 0.27 lies between 0.25 and 0.5 in the list containing [0, 0.25, 0.27, 0.5, 0.75]. We have considered a zero indexed list. We input the total number of layers in the convolutional neural network to the algorithm as the value of n and the value of p as the element in the array represented by the particle. The output of Algorithm 2 is used to calculate the value of the objective function.

Algorithm 2 Encoding PSO Output to split layer

```
function ENCODE(n, p) encode list = []
for x = 0 to n - 1 do
Append (1/n) * x to encode list
end for
Append p to encode list sort encode list
return position index of p
```

The value of the objective function that corresponds to a certain particle is calculated as follows. Each element in a particle by virtue of the order that it is present in the array will correspond to a certain device, e.g., a particle with the following encoded output: [3, 4, 7, 8] denotes an offloading decision of 2 for device 1, 3 for device 2 and so on (Eq. 4). From the two matrices that were initialized, we can obtain its corresponding CPU utilization and round execution time. The third element in the above array will correspond to the entry in both the matrices at index (3, 4) since the third element denotes an offloading point of 4. Following this procedure, we obtain a value for the round execution time and CPU utilization for each element in a particle. Using Eq. (3) and the values obtained we assign a value for the objective function to the particle.

The remainder of the algorithm follows the standard particle swarm optimization procedure, wherein we initialize a variable pBest val that stores the best value seen by a particle and gBest val as the best value of the objective function seen by all particles. We modify the value of the velocities of each of the particles with the standard Eq. (5). Here pbest and gbest represent the values of the particles that gave rise to pbest val and gbest val. Since gbest is a single particle and X contains 10 particles, the value of gbest is replicated 10 times in order to conduct the subtraction. Here $c1$, $c2$ and w are 0.1, 0.1 and 0.8, respectively. $r1$, and $r2$ are continuous values initialized between 0 and 1. The particle swarm optimization algorithm is executed for 1000 iterations. The value of gbest obtained at the end of 1000 iterations is output in the end.

After the execution of a round of federated learning, both the round store matrix and cpu store matrix are updated with the most recent round execution time and cpu utilization values, thus allowing particle swarm optimization to function adaptively (see Algorithm 1).

$$V = w * V + c_1 * r_1 * (p\text{best} - X) + c_2 * r_2 * (g\text{best} - X) \tag{5}$$

5 Experimental Setup

For the experimental setup, we created 10 virtual machines each of which was provisioned by a VMware hypervisor. All the 10 virtual machines were placed on different computers. The configurations of each of the virtual machines may be seen in Table 2. The server on the other hand was placed in the high performance computing facility of the IIIT Kottayam campus. It consisted of 188 gigabytes of RAM and 96 cores. The optimizer was also placed on this high performance computing system. A VGG network with seven layers was used in this case. Refer to Table 1 for the configurations. The second column of the table denotes the number of inputs and the number of outputs of each layer of the neural network. The first layer of the neural network, for example, contains three input channels, namely the R, G and B channel of the image. Upon applying the kernel upon the image, we obtain a 32 channel output. The CIFAR-10 [19] dataset was used in this case. The CIFAR-10 dataset consists of 50,000 images in the training set and 10,000 images in the test set. 5000 images each were assigned to the VM to train their local models in the order of their device numbers.

The configurations of the neural network and the dataset are based on FedAdapt. We executed two schemes in addition to our particle swarm optimization-assisted federated learning, namely classic federated learning and FedAdapt. We measured the CPU utilization for a round of federated learning with a tool named sar. We measured the percentage of the time when the CPU was active that it spent in user mode as the CPU utilization. The time for each round was measured using the time module in Python. The CPU only version of Pytorch was used to implement the neural networks. GPU passthrough was not enabled on any of the virtual machines. The socket module in Python was used for connecting the virtual machines to the high performance computing system. In order to execute FedAdapt, we first trained

Table 1 VGG-5 configuration

Layer	Dimensions (in channel x out channel)
Convolutional layer	3 × 32
Max pooling layer	32 × 32
Convolutional layer	32 × 64
Max pooling layer	64 × 64
Convolutional layer	64 × 64
Dense layer	4096 × 128
Dense layer	128 × 10

Table 2 VM configurations

Device No. (VM)	CPU (No. of. cores)	RAM (GB)	Frequency (GHz)
0	2	3	3
1	2	3	3
2	3	4	3
3	3	4	3
4	3	5	3
5	3	5	3
6	4	5.5	3
7	2	2	3
8	4	6	3
9	4	5.5	3

its proximal policy optimization-based reinforcement learning agent on the devices in Table 2.

The weights of this RL agent were used to execute FedAdapt. Furthermore, since there is a possibility that there were variations in the results because of differences in network speed, we have reported the average of the network speeds of each device for 100 rounds for each experiment in Table 3. The average of the network speeds for a certain device for an experiment is reported in the cell corresponding to the index of the device and the name of the experiment. The network speeds are in megabits per second.

We follow the framework of FedAdapt to measure the network speed with a minor change. Before starting a round on each device, we make each client wait for a confirmation from the server to begin testing the network. Once the client has received the acknowledgement, it sends a serialized version of a VGG-5 network of size 2.23 megabytes and receives a VGG-5 network back from the server. The time for this is measured and then the speed is calculated.

Table 3 Network speeds

Device No. (VM)	FedAdapt	FL	PSO FL
0	193.402	160.739	171.313
1	132.773	146.032	147.199
2	170.924	162.617	165.843
3	74.698	154.163	73.103
4	70.157	67.417	70.765
5	150.319	151.898	168.821
6	132.713	154.151	151.276
7	207.690	199.800	202.588
8	177.118	187.554	157.128
9	160.722	156.374	134.071

6 Results

The time series graphs in Fig. 2 compare three algorithms, namely FedAdapt, classic federated learning and particle swarm optimization-assisted federated learning based on the objective function depicted in Eq. (3). We observe in round 0 that classic federated learning and FedAdapt outperform our algorithm outlined in Sect. 5. This can be attributed to the fact that in the first 7 rounds, the algorithm is trying various values for the offloading points on each device, starting from 0 all the way to 6. Since the round execution time for an offloading decision of [0, 0, 0, 0, 0, 0, 0, 0, 0, 0] was especially high, the objective function evaluated at the initial round was also high. We can also observe this in Fig. 3. Once the various values of the offloading decision have been evaluated, the system can begin its particle swarm optimization before each round.

We observe that, after the execution of the particle swarm optimization process has begun, the objective function for our scheme reduces significantly in comparison with the previous seven rounds. It can be observed that our scheme produces an objective function consistently lower than the other two frameworks. Precisely, upon taking the average of the values of the objective function, we observe that the objective function of our scheme is 36.73% lower than classic federated learning and 36.60% lower than FedAdapt. We observe that classic federated learning gives us an accuracy of 77.89% at the end of 100 rounds, FedAdapt gives us an accuracy of 77.64% and our PSO-assisted federated learning scheme gives us an accuracy of 77.49%. It should be noted that the above accuracy measurements are subject to the variation of the offloading points in each of the schemes and may change marginally, should other iterations be conducted.

We have also plotted how the offloading points impact the maximum round execution time of all devices in a round and the mean CPU utilization across the virtual machines. This can be observed in Fig. 3. The maximum training time of all devices eventually decreases as the offloading point is increased from 0 to 6. The opposite is

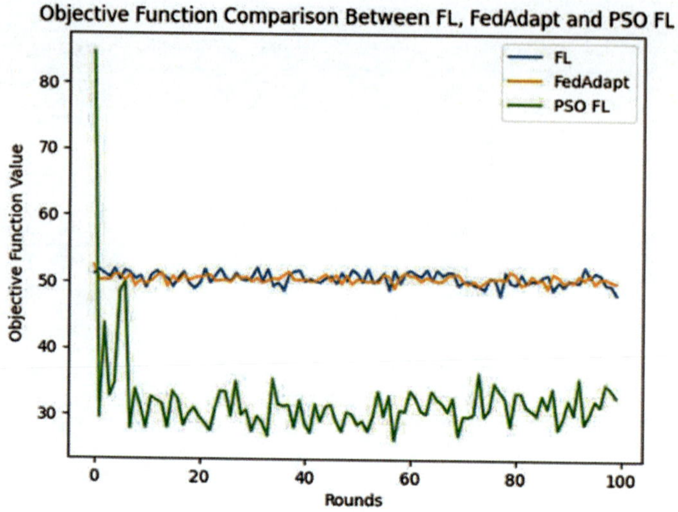

Fig. 2 Comparison of algorithms

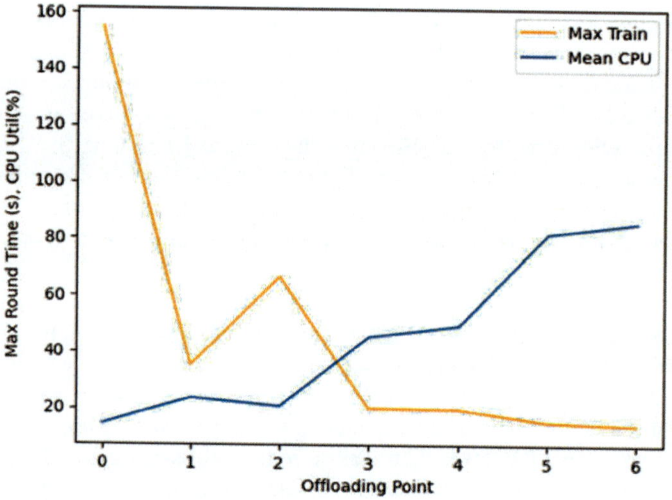

Fig. 3 Max train and mean CPU versus offloading point

observed in the case of the CPU utilization. This is potentially because an increase in offloading point leads to a larger amount of computation happening on the device, thus reducing the communication over-head. This is however reflected in an increased CPU utilization since a larger amount of computation is happening on the client as opposed to the server.

7 Conclusion and Future Work

We have used particle swarm optimization to attempt to minimize the CPU utilization and the execution time of a round of federated learning. Our results show that particle swarm optimization-assisted federated learning has a greater performance in terms of round execution time and CPU utilization than classic federated learning and FedAdapt, considering the measurement of the objective function while largely maintaining the accuracy of the model. In the future, we may attempt to minimize the memory utilization of the system by adaptively optimizing the batch size of each of the local models.

References

1. Abdulrahman S, Tout H, Ould-Slimane H, Mourad A, Talhi C, Guizani M (2020) A survey on federated learning: the Journey from centralized to distributed on-site learning and beyond. IEEE Internet Things J. https://doi.org/10.1109/JIOT.2020.3030072
2. McMahan B, Moore E, Ramage D, Hampson S, Arcas BA (2017) Communication-efficient learning of deep networks from decentralized data. In: Proceedings of the 20th international conference on artificial intelligence and statistics, vol. 54, p 1273, Apr 2017
3. Saputra YM, Hoang DT, Nguyen DN, Dutkiewicz E, Mueck MD, Srikanteswara S (2019) Energy demand prediction with federated learning for electric vehicle networks. In: 2019 IEEE global communications conference (GLOBECOM). Waikoloa, pp 1–6. https://doi.org/10.1109/GLOBECOM38437.2019.9013587
4. Huang L, Shea AL, Qian H, Masurkar A, Deng H, Liu D (2019) Patient clustering improves efficiency of federated machine learning to predict mortality and hospital stay time using distributed electronic medical records. J Biomed Inform 99:103291. https://doi.org/10.1016/j.jbi.2019.103291. Epub 2019 Sep 24 PMID: 31560949
5. Al-Huthaifi R, Li T, Huang W, Gu J, Li C (2023) Federated learning in smart cities: privacy and security survey. Inf Sci 632:833–857, Jun 2023 https://doi.org/10.1016/j.ins.2023.03.033
6. Mothukuri V, Parizi RM, Pouriyeh S, Huang Y, Dehghan-Tanha A, Srivastava G (2021) A survey on security and privacy of federated learning. Future Gener Comput Syst 115:619–640
7. Chai Z, Ali A, Zawad S, Truex S, Anwar A, Baracaldo N, Zhou Y, Ludwig H, Yan F, Cheng Y (2020) TiFL: a tier-based federated learning system. In: Proceedings of the 29th international symposium on high-performance parallel and distributed computing (HPDC '20). Association for Computing Machinery, pp 125–136. https://doi.org/10.1145/3369583.3392686
8. Wang C, Yang Y, Zhou P (2021) Towards efficient scheduling of federated mobile devices under computational and statistical heterogeneity. IEEE Trans Parallel Distrib Syst 32:394–410. https://doi.org/10.1109/TPDS.2020.3023905
9. Zhu J, Cao J, Saxena D, Jiang S, Ferradi H (2023) Blockchain-empowered federated learning: challenges, solutions, and future directions. ACM Comput Surv 55(11), Nov 2023. https://doi.org/10.1145/3570953
10. Park S, Suh Y, Lee J (2021) FedPSO: federated learning using particle swarm optimization to reduce communication costs. Sensors 21(2):600. https://doi.org/10.3390/s21020600
11. Kennedy J, Eberhart R (1995) Particle swarm optimization. In: Proceedings of ICNN'95 international conference on neural networks. Perth, vol. 4, pp 1942–1948. https://doi.org/10.1109/ICNN.1995.488968
12. Xu C, Liu S, Yang Z, Huang Y, Wong K-K (2021) Learning rate optimization for federated learning exploiting over-the-air computation. IEEE J Sel Areas Commun 39(12):3742–3756. https://doi.org/10.1109/JSAC.2021.3118402

13. Wu D, Ullah R, Harvey P, Kilpatrick P, Spence I, Varghese B (2022) FedAdapt: adaptive offloading for IoT devices in federated learning. In: IEEE Internet Things J 9(21):20889–20901, 1 Nov 2022. https://doi.org/10.1109/JIOT.2022.3176469
14. Thapa C, Mahawaga Arachchige PC, Camtepe S, Sun L (2022) SplitFed: when federated learning meets split learning. Proc AAAI Conf Artif Intell 36(8):8485–8493. https://doi.org/10.1609/aaai.v36i8.20825
15. Gupta O, Raskar R (2018) Distributed learning of deep neural network over multiple agents. J Netw Comput Appl 116:1–8
16. Vepakomma P, Gupta O, Swedish T, Raskar R (2018) Split learning for health: Distributed deep learning without sharing raw patient data, arXiv [cs.LG]
17. Cox B, Chen LY, Decouchant J (2022) Aergia: leveraging heterogeneity in federated learning systems. In: Proceedings of the 23rd ACM/IFIP international middleware conference. Quebec, pp 107–120
18. Joshi P, Thapa C, Camtepe S, Hasanuzzaman M, Scully T, Afli H (2022) Performance and information leakage in splitfed learning and multi-head split learning in healthcare data and beyond. Methods Protoc 5(4):60. https://doi.org/10.3390/mps5040060
19. Krizhevsky A (2023) Learning multiple layers of features from tiny images. Accessed: 5 Aug 2023. [Online] Available: https://tinyurl.com/cifar10data

The Brain Tumors Identification, Detection, and Classification with AI/ML Algorithm with Certainty of Operations

Pranay Meshram, Tushar Barai, Mohammad Tahir, and Ketan Bodhe

Abstract Brain tumors are one of the most deadly disorders affecting individuals, and their early detection and accurate classification are crucial for effective treatment. Traditional biopsy methods for brain tumor grading can be invasive and pose significant risks to patients. To overcome these challenges, noninvasive imaging techniques like Magnetic Resonance Imaging (MRI) have been used. In recent years, the application of deep learning, specifically convolutional neural networks (CNNs), has shown promise in automating brain tumor detection and classification from MRI data. This paper presents a novel approach that leverages pre-trained CNN models like VGG16, ResNet50, and Inception v3 for brain tumor classification with and without transfer learning. The proposed method achieves high accuracy and precision, outperforming traditional machine learning approaches. The study also evaluates different metrics, such as sensitivity, specificity, and F1-score, to demonstrate the model's efficacy. The results indicate that CNN-based models, especially ResNet50, offer superior performance in brain tumor classification. The proposed system has the potential to revolutionize brain tumor diagnosis and treatment planning, reducing the workload of medical professionals and improving patient outcomes. Future research may explore the integration of 3D volume-based MRI data and unsupervised transfer learning to further enhance brain tumor detection capabilities. However, it is essential to consider machine learning models as supportive tools, with final diagnosis and treatment decisions relying on the expertise of trained medical professionals.

P. Meshram (✉) · T. Barai · M. Tahir · K. Bodhe
G.H.Raisoni College of Engineering Nagpur, Nagpur, Maharastra, India
e-mail: pranay.meshram@raisoni.net

T. Barai
e-mail: tushar.barai@raisoni.net

M. Tahir
e-mail: mohammad.tahir@raisoni.net

K. Bodhe
e-mail: ketan.bodher@raisoni.net

S. Shakya et al. (eds.), *Fourth International Conference on Image Processing and Capsule Networks*, Lecture Notes in Networks and Systems 798,
https://doi.org/10.1007/978-981-99-7093-3_41

Continuous research, validation, and collaboration between ML experts and healthcare practitioners are crucial for harnessing the full potential of ML in brain tumor detection from MRI images.

Keyword A brain tumor · Capsule neural networks · Convolutional neural networks · Deep learning · Vision transformers

1 Introduction

A brain tumor is one of the most deadly disorders a person may have. In the early phases of cancer growth, the radiologist's initial aim is to do an efficient and effective analysis. Histological grading, which is based on a stereotactic biopsy, is the most reliable and extensively used method for establishing the grade of a brain tumor. The neurosurgeon must make a small incision in the patient's skull for the biopsy. Seizures, major headaches, stroke, coma, and even death might happen with the biopsy method due to the bleeding and infection of tumors in the brain. The most significant drawback of the stereotactic biopsy is the potential for a serious misdiagnosis due to the technique's lack of perfection. Patients frequently have difficulty undergoing a brain tumor biopsy. As a result, medical practitioners have long used noninvasive imaging techniques such as magnetic resonance imaging (MRI). Therefore, it is necessary to create methods for identifying and forecasting the stage of malignancies based on MRI data. Despite this, it can be difficult to tell cancer cells apart from surrounding soft tissues using imaging techniques like magnetic resonance imaging (MRI). It may be because of the tumor's complexity and unpredictability, such as its amorphous shape, large size, and random location inside the body, or because of the limited brightness of some imaging modalities. Thanks largely to Deep Convents machine learning-based techniques in radiology and other medical research, surgical biopsy for brain tumors is now a realistic option. We attempted to identify and categorize brain cancers using the convolutional neural network (CNN) architecture. Using pre-trained Kera's models like VGG16, ResNet50, and Inception v3, we assessed the performance of binary and multiclass brain tumor classification with and without transfer learning. Using machine learning, Deep Convent has developed a noninvasive alternative to brain tumor biopsies that has revolutionized the ease with which doctors can diagnose the disease. We sought to discover and classify brain cancers using convolutional neural network (CNN) architecture, comparing binary and multiclass classification results with and without transfer learning.

VGG16 is a deep convolutional neural network architecture known for its simplicity and effectiveness in image recognition tasks. It consists of 16 layers, primarily small 3×3 filters for convolution and max-pooling layers for spatial reduction. The model utilizes ReLU activation functions and fully connected layers for classification. VGG16 is often pre-trained on large image datasets before fine-tuning for specific tasks, enabling it to learn hierarchical and abstract image features. While newer architectures have surpassed its performance, VGG16 remains an influential milestone in computer vision and deep learning.

ResNet, short for Residual Neural Network, is a revolutionary deep convolutional neural network architecture introduced in 2015. It addresses the vanishing gradient problem in very deep networks by using skip connections, or residual blocks, that allow gradients to flow more effectively during training. ResNet is characterized by its ability to build extremely deep networks (e.g., ResNet50, ResNet101) while maintaining high performance. The skip connections facilitate the learning of residual mappings, making it easier to optimize deep networks and achieve state-of-the-art results in various computer vision tasks.

Inception model is a deep neural network architecture developed by Google, primarily used for image recognition tasks. It introduces the concept of "inception modules," which consist of multiple parallel convolutional layers of different sizes. These parallel layers capture information at various scales and help the model efficiently learn features at different levels of abstraction. The model's design aims to strike a balance between computational efficiency and performance, making it suitable for large-scale image classification tasks while minimizing the number of parameters needed.

Automated defect identification based on machine learning in medical imaging is a pressing concern for many diagnostic uses. MRI is crucial for diagnosing brain tumors because it provides information on aberrant tissues necessary for therapeutic planning. New studies suggest that medical image analysis-based automated computerized sickness detection and diagnosis might be a practical option. The radiologists' time would be saved while clinically proven accuracy was maintained. Therapeutic care for brain tumors will also greatly benefit from the availability of computer algorithms that may provide reliable and measurable metrics of cancer representation. It will relieve clinicians of the burden of manually demonstrating malignancies. Tumors of the brain and its supporting tissues are a type of malignant development. They might be malignant or benign, and they can cause a wide range of neurological symptoms, such as headaches, seizures, visual and hearing changes, and mental fogginess. Brain tumor diagnosis and therapy rely heavily on a comprehensive database of cases.

Radiologists and doctors have a significant challenge when attempting to classify brain tumors. Categorization of brain tumors has traditionally relied on the subjective and time-consuming visual assessment of medical images by experienced radiologists. In addition, the accuracy of the diagnosis may be affected by the size, location, and form of the tumor, as well as the radiologists' level of expertise. This work attempts to pave the way for automatically recognizing and categorizing brain cancers from medical pictures using machine learning.

This innovation in diagnostics is meant to facilitate the creation of more tailored treatment programs for individuals suffering from brain tumors. This study aims to create a method based on machine learning for identifying and categorizing brain tumors from medical images. This innovation in diagnostics is meant to facilitate the creation of more tailored treatment programs for individuals suffering from brain tumors. Our machine learning-based approach can greatly enhance current methods of classifying brain tumors. This technique can help to speed up the discovery and treatment of brain tumors by reducing the work-load of radiologists and doctors through automating the tumor categorization process. As diagnosis precision directly affects the range of possible treatments, this can also boost patient outcomes. This research aims to develop a classification system for brain cancer based on MRI findings. The research uses a public database of brain MRI images labeled with known forms of cancer. The study has certain limitations, such as its dependence on a single dataset and its use of predetermined cancer categories. Classification accuracy was not investigated concerning various imaging modalities, patient characteristics, or other clinical considerations.

The Kaggle dataset for brain tumor MRI pictures is a publicly available collection of brain MRI scans, hosted on the Kaggle platform. It focuses on patients with brain tumors and includes labeled images indicating the presence or absence of tumors. This dataset enables researchers and data scientists to develop machine learning models for brain tumor detection and classification. The MRI scans are stored in digital format, reflecting real-life scans with varying resolutions and dimensions. The dataset also provides additional information and metadata, offering valuable context for research. Users are encouraged to cite the dataset and Kaggle platform when using them for publications. Kaggle fosters collaboration among data scientists, allowing them to share models and findings in the medical image analysis and brain tumor research field (Fig. 1).

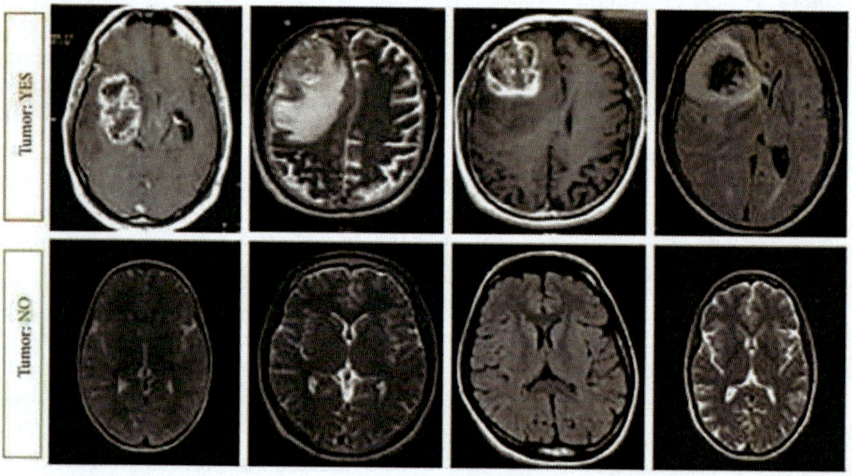

Fig. 1 Kaggle dataset. *Source* Navoneel Chakrabarty, "Brain MRI images for brain tumor detection dataset", Kaggle

1.1 Brain Tumor Types

The most prevalent kind of brain cancer, gliomas, develops from the brain's supporting cells (glial cells). Three types of gliomas include astrocyte tumors, oligodendroglia tumors, and ependymomas. Astrocytomas are glial cell tumors that develop from astrocytes. Low-grade (slow-growing) and high-grade (fast-growing) glioblastomas are two examples of fast-growing tumors. Tumors known as oligodendrogliomas develop from oligodendrocytes, cells responsible for making fatty material that insulates neurons. In most situations, the growth of oligodendrogliomas is quite sluggish. Ependymal cells, which line the spinal cord's central canal and the brain's ventricles, can develop into malignant tumors called ependymomas. Meningiomas are tumors in the meninges, the membranes surrounding the brain and spinal cord. Although not malignant, they can cause discomfort if they grow large enough to influence nearby organs or tissues. Adenomas of the pituitary gland are tumors of a small organ near the brain's base. Although benign, pituitary adenomas can generate hormones and contribute to hormonal abnormalities. Malignant (cancer) medulloblastomas develop in the cerebellum, a key brain region responsible for motor control and balance. They are more common in kids. Schwannomas are cancers that start in the Schwann cells responsible for making the myelin coating that shields nerve fibers from damage. It is common for these tumors to form on the eighth cranial nerve, which is responsible for both hearing and balance. Rare tumors called craniopharyngiomas tend to form in the brain between the optic nerves, close to where the pituitary gland would normally be. Although their existence may bring signs of illness, they pose no real danger. Cancer cells that have moved to the brain from elsewhere in the body are called metastatic brain tumors. They outnumber primary brain tumors by a wide margin. Lung, breast, colon, kidney, and melanoma are the five malignancies most likely to metastasize to the brain.

2 Related Work

The goal of A. Sekhar and other authors of this study is to create a computer-aided diagnostic (CAD) system for diagnosing brain cancers by utilizing enhanced GoogLeNet features and machine learning methods. The system is powered by IoMT technology or the Internet of Medical Things. The team suggests a two-stage process for identifying and categorizing brain tumors. They use a GoogLeNet convolutional neural network already trained as a feature extractor. Brain MRI scans fine-tune the GoogLeNet model, allowing it to collect characteristics unique to tumors more effectively. In the second step, machine learning techniques categorize brain tumors according to the retrieved features. We can remark that support vector machines (SVMs), random forests, and neural networks are often utilized strategies for identifying tumors without going into depth about the exact machine learning techniques used. Relevant performance metrics are used for training and assessing the algorithms. The CAD system is powered by IoMT technology, which permits unhindered communication and data exchange between medical devices and platforms, improving overall diagnostic and treatment planning capabilities. The study found that the new CAD system, which uses sophisticated GoogLeNet features and machine learning techniques, performed well in classifying brain tumors. IoMT technology helps the system reach its full potential by facilitating communication and teamwork in healthcare.

Limited Dataset: The research used a relatively small or unbalanced dataset for training and evaluation, which may restrict the generalizability and robustness of the suggested CAD system. Results may be more trustworthy if larger and more varied datasets are used.

Lack of Comparison: The proposed CAD system was not evaluated compared to standard practices or cutting-edge methods for categorizing brain tumors in the research. It isn't easy to measure the usefulness and superiority of the suggested system without proper comparisons.

Performance evaluation: The assessment measures used in the study do not cover all important elements of brain tumor classification. Performance indicators include recall and precision. The F1-score or the AUC-ROC must be considered for a thorough evaluation.

Validation on External Datasets: The paper not have validated the proposed CAD system on external datasets or with different medical institutions. Validation on independent datasets helps to assess the generalizability and effectiveness of the system across different patient populations and imaging protocols.

Clinical Validation: The research paper lack information on the clinical validation of the proposed CAD system.

Validation studies involving collaboration with healthcare professionals and assessing the system's performance on real-world clinical data are essential to demonstrate their utility and potential impact on clinical practice [1]. The authors [2] provide an approach for detecting brain tumors based on MRI images that employ detailed features and machine learning classifiers. The author provides an in-depth examination in this section. Thirteen pre-trained deep convolutional neural networks and nine different ML classifiers were utilized in a study aimed at categorizing brain tumors using three datasets: BT-small-2c, BT-large-2c, and BT-large-4c. The research findings indicated the following:

- In the case of a small MRI dataset (BT-small-2c) containing only two classes (normal and tumor), employing the DenseNet-169 deep feature alone proved to be effective.
- When dealing with a large MRI dataset (BT-large-2c) that also consisted of two classes (normal and tumor), a combination of three in-depth features, namely DenseNet-169, Inception V3, and ResNeXt-50, was found to be effective for accurate classification.

- For the BT-large-4c dataset, which involved tumor classification into four distinct classes, an ensemble of DenseNet-169, ShuffleNet V2, and an undisclosed model (not explicitly mentioned) was utilized.
- Although the specific ML classifiers employed in the research were not mentioned, it was stated that nine different classifiers were utilized, suggesting various classifier combinations.

Finally, on big datasets, our suggested novel feature ensemble approach outperforms and outperforms a typical CNN model. These findings validated our suggested technique for identifying and categorizing brain cancers, which combines extensive feature sets with ML classifiers. This research [3] brings computational intelligence and metaheuristic approaches to the problem of employing MRI scanners connected to the Internet of Medical Things (IoMT) to spot brain tumors. The authors aim to use cutting-edge technology to create a reliable approach to identifying malignant brain tumors. The researchers discuss the significance of finding a brain tumor and the difficulties of using conventional diagnostic techniques. They emphasize developing automated and intelligent approaches for appropriately identifying and classifying cancers. Other computational intelligence techniques discussed include artificial neural networks (ANNs), support vector machines (SVMs), and fuzzy logic (FL), as well as metaheuristic algorithms such as genetic algorithms (GAs), particle swarm optimization (PSO), and ant colony optimization (ACO). The authors detail how implementing these strategies can improve brain tumor diagnosis. The authors then suggest a system for finding brain tumors using logical and intuitive approaches. They describe the integration of these techniques into an IoMT-enabled MRI system, which allows for real-time data transmission and analysis. The article also presents a detailed experimental evaluation of the proposed framework. The authors utilize a dataset of MRI brain images and compare the performance of different computational intelligence and metaheuristic techniques. They analyze the accuracy, sensitivity, specificity, and computational efficiency of each approach to identify the most effective method for brain tumor detection. In conclusion, the authors stress the importance of IoMT-enabled MRI machines in diagnosing brain tumors and the potential of computational intelligence and metaheuristic approaches. They stress the need for more studies to improve the accuracy and efficiency of diagnosing and treating brain tumors.

This study combines SVM and convolutional neural network (CNN) characteristics to automatically categorize brain tumors found in MRI data (SVM). The authors aim to develop an accurate and efficient method for classifying brain tumors based on MRI images. First, they discuss the importance of automated brain tumor categorization and the challenges associated with manual diagnosis methods. The authors highlight the need for computer-aided systems to analyze MRI images and provide reliable tumor classification. Researchers explain how CNNs can effectively extract relevant features from MRI images, capturing the unique characteristics of tumors. These collected characteristics are placed into a support vector machine classifier to categorize tumors correctly. They compare the results with other existing methods for brain tumor categorization and analyze their approach's accuracy, sensitivity, specificity, and computational efficiency. The study's findings demonstrate that the combination of CNN feature extraction and SVM classification achieves promising results in accurately categorizing brain tumors from MRI scans. It is emphasized how crucial the authors' method is for assisting doctors and radiologists with diagnosing and planning treatment for brain tumors [4]. Noninvasive brain tumor identification, detection, and classification methods were used in the authors' research work. With the help of MRI scan images, brain tumor identification was carried out to provide the results using various enhanced machine learning algorithms [5]. The authors focused on the detection of diseases in various human body parts using deep learning algorithms. During the review process, they identified the strengths and weaknesses of various research works carried out by others and conducted further enhanced research in the same domain [6]. The author has worked on different artificial intelligence tools with MRI images for brain tumor segmentation. He divided the techniques into three categories, i.e., supervised, unsupervised, and deep learning methods [7]. The author explores the landscape of machine learning techniques for brain tumor identification. Published in complex intelligent systems, the study provides a comprehensive overview of advancements in this area, offering insights into various methodologies and their effectiveness. This paper serves as a valuable resource for researchers and practitioners, presenting a consolidated understanding of the current state of brain tumor detection and classification [8]. Currently, the author is engaged in the development of an automated technique for the extraction and categorization of tumors from MRI scans. This proposed method involves a step-by-step process of tumor contrast enhancement, tumor extraction, multimodel feature extraction, feature selection, and classification. The investigations are being conducted on three distinct datasets: the Harvard dataset, the BRATS 2013 dataset, and a collection of privately obtained MR images to validate the effectiveness and robustness of the approach. By utilizing these diverse datasets, the author aims to evaluate and enhance the automated approach for accurate tumor identification and classification [9]. Tackling gaps in the scientific literature, the author focuses on surgical methods for specific intrinsic neoplasms with graphics and step-by-step instructions. It includes anatomy, pathology, animal models, imaging, open brain tumor techniques, and minimally invasive procedures, as well as their merits and drawbacks, debates, and technical issues discussed by the author [10]. To overcome the limitations associated with automated brain tumor classification, the author introduces a novel technique that combines magnetic resonance imaging (MRI) with a long short-term memory (LSTM) model. This innovative approach was thoroughly tested using actual patient data from the Pakistan Ordnance Industry, resulting in an impressive Dice Similarity Coefficient (DSC) of 0.97. The exceptionally high DSC value demonstrates the efficacy of the proposed method in accurately classifying brain tumors. By leveraging the LSTM model's and MRI data's power, the author's approach presents a promising solution for enhancing automated brain tumor classification [11]. The author has introduced an automated method for distinguishing between malignant and non-cancerous brain magnetic resonance imaging (MRI) scans. To accurately segment potential lesions, the author has utilized a range of different approaches [12].

The author incorporates the GrabCut method in their research to accurately segment actual lesion symptoms. Furthermore, they employ the transfer learning model, specifically the Visual Geometry Group (VGG19), and fine-tune it to extract essential features. Using a sequential-based method, these features are concatenated with hand-crafted features, including shape and texture. This combined approach enhances the overall performance and effectiveness of the proposed work [13]. The study introduces a hybrid technique that combines Fuzzy C-Means (FCM) and region-expanding segmentation algorithms to segment the Region of Interest (ROI) precisely. Following the segmentation process, the study extracts essential characteristics such as geometry, texture, and statistical or intensity properties of the nodule under consideration. To identify the most relevant features for lung cancer classification, the Cuckoo search optimization technique is applied. Subsequently, these optimal characteristics are utilized for training a support vector machine (SVM) classifier, enabling accurate classification of lung cancer as benign or malignant [14]. The author divided the technique into two steps: feature extraction and classification. GLCM is used to extract features from images in the first step. The retrieved characteristics are then supplied to the SVM classifier in the next stage [15]. The author introduces a new method for tumor segmentation that utilizes unsupervised clustering. A fused feature vector is employed, combining Gabor wavelet features (GWF), histograms of oriented gradient (HOG), local binary pattern (LBP), and segmentation-based fractal texture analysis (SFTA) features to capture comprehensive information. The classification task, differentiating between three sub-tumoral regions (entire, enhancing, and non-enhancing tumors), is accomplished using a random forest (RF) classifier. Both fivefold cross-validation and 0.5 holdout cross-validation methods are applied to ensure reliable evaluation and prevent overfitting. This methodology offers an innovative approach to tumor segmentation with improved accuracy, avoiding overfitting issues [16].

The author's main objective revolves around brain tumor classification. They propose a novel approach using machine learning, specifically back propagation neural networks (MLBPNN), to enhance the precision and competency of pathologists in identifying tumors, particularly in high-risk or critical locations. By incorporating MLBPNN, the aim is to minimize observer variability and ensure consistent and reliable tumor classification results. This innovative methodology can potentially improve the accuracy and efficiency of brain tumor diagnosis, benefiting both patients and healthcare providers [17]. The author replaced the FULL CONVOLUTION NETWORK with ALL CONVOLUTION NETWORK, and to identify brain tumors more efficiently and correctly, they will employ a mix of DEEP LEARNING and MACHINE LEARNING, which will improve detection [18]. The author focused on segregating tumors from the BRATS 2018 benchmark dataset and utilized age, shape, and volumetric parameters to predict overall patient survival [19]. The author's study is centered on detecting brain tumors using MRI scans. They propose a technique that leverages optimization, machine learning algorithms, and curve fitting techniques for accurate tumor detection. The suggested approach surpasses previous works in terms of efficiency, achieving an accuracy of 16.39% and a precision of 9.53%. These improved results demonstrate the technique's potential in advancing the field of brain tumor detection, offering promising advancements in accuracy and efficiency [20]. The author's approach combines the Stationary Wavelet Transform (SWT) principle with a novel Growing Convolution Neural Network (GCNN). The primary objective of this study is to enhance the accuracy of the existing system. The author conducted a comprehensive comparison between support vector machine (SVM) and convolution neural network (CNN) methods in terms of accuracy and performance metrics such as Peak Signal-to-Noise Ratio (PSNR) and Mean Squared Error (MSE). The experimental results demonstrate that the suggested approach surpasses SVM and CNN, delivering superior accuracy and overall performance [21].

3 Methodology

An exciting field of research with encouraging findings recently is the application of machine learning algorithms for identifying and categorizing brain cancers on MRI images. The following is a high-level summary of the steps in developing such a system. Establishing an accurate MRI-based brain cancer detection system includes collecting and preprocessing relevant data. The first step is to create a dataset of brain MRI scans from both tumor and non-tumor patients. These images should be meticulously labeled to show the presence and location of tumors. Preprocessing techniques like scaling, normalization, and noise reduction can be used to improve image quality. As a result, the cancer detection system's accuracy can be improved, and the algorithm's performance can be optimized.

3.1 Flowchart

Feature extraction: Preprocessed MRI images can have characteristics extracted using various methods. Examples of such traits include those based on intensity, form, and texture. The histogram statistics may extract intensity-based characteristics based on the picture's pixel intensities. For example, the size and border of the tumor are shape-based information that may be extracted. Based on the pattern and distribution of pixel intensities, methods such as matrices of grey-level co-occurrence and wavelet transformations are used to recover texture properties. The tumor detection system can more precisely identify and categorize tumor instances by extracting important information from pictures (Fig. 2).

Model training: Next, a machine learning model will be developed to categorize brain MRI consistently scans as tumor or non-tumor cases based on the attributes retrieved from the pictures. Support vector machines (SVMs), random forests, and deep learning models like convolutional neural networks are all viable options for completing this task (CNNs). The algorithm selected is based on the data supplied and the difficulty of the challenge. Convolutional neural networks (CNNs), which

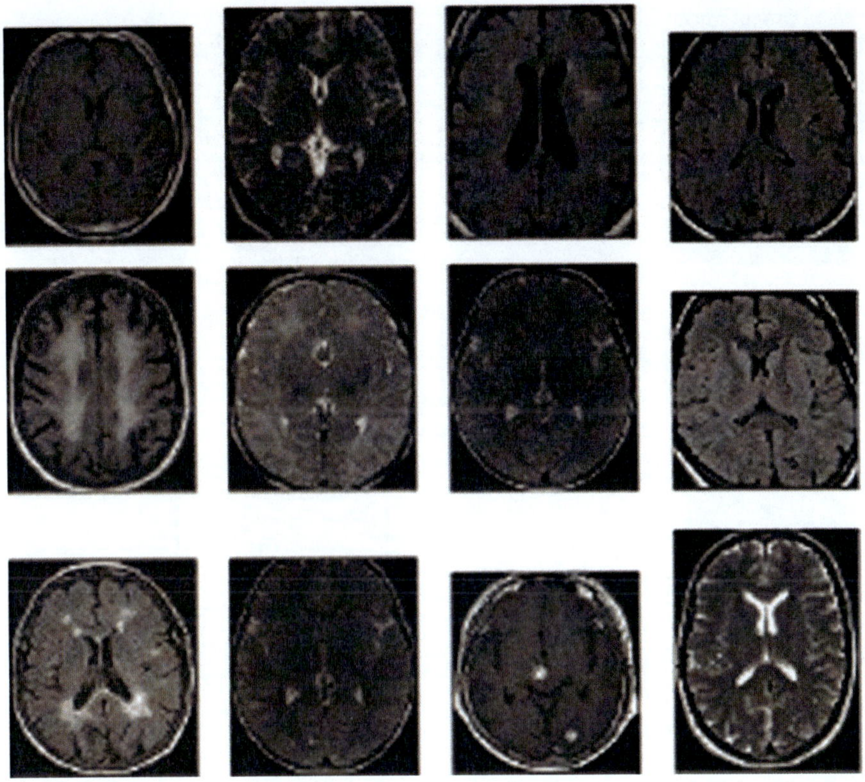

Fig. 2 15 MRI pictures of the brain with no tumor. *Source* Navoneel Chakrabarty, "Brain MRI images for brain tumor detection dataset", Kaggle, April 2019

can learn complicated patterns, are more appropriate for larger datasets than support vector machines (SVMs) and random forests (RFs), which are frequently employed for smaller datasets. The efficacy of a model may be evaluated using several different metrics

Model validation and evaluation: The effectiveness of the trained model should be evaluated using a different dataset. Popular measures for evaluating a model's efficacy include accuracy, sensitivity, specificity, and area under the receiver operating characteristic curve (AUC-ROC). Cross-validation techniques can also generate a more robust estimate of the model's performance. These strategies entail partitioning the dataset into various subsets and training and evaluating the model on each subset to avoid overfitting. Cross-validation performance measurements can be used to choose the optimal model and its accompanying hyperparameters (Figs. 3 and 4).

Testing and deployment: After the model has been trained, validated, and optimized, it can be tested on new, previously unknown MRI pictures. The model's performance on the test set may be evaluated using the same performance metrics used during training and validation. The model's performance can be deployed for

Fig. 3 Flow of work

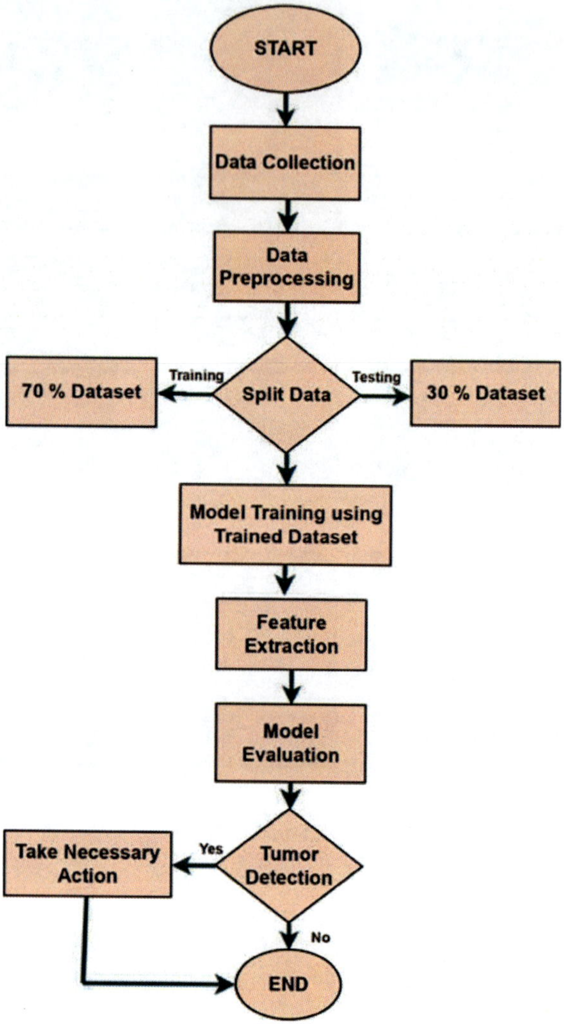

practical use if satisfactory. It involves participating the model into a user-friendly interface or a healthcare system that can receive input images and return the model's forecasts. Ensuring the deployed model is robust, reliable, and meets all the regulatory requirements is important. It is essential to note that the medical imaging and tumor detection field is complex, and the deployment process may vary based on specific requirements, regulations, and validation needs. Collaborating with domain experts and adhering to applicable regulatory guidelines is crucial for responsible and successful deployment. We split the data into 70% of the data is used for training, and the remaining 30% is used for testing.

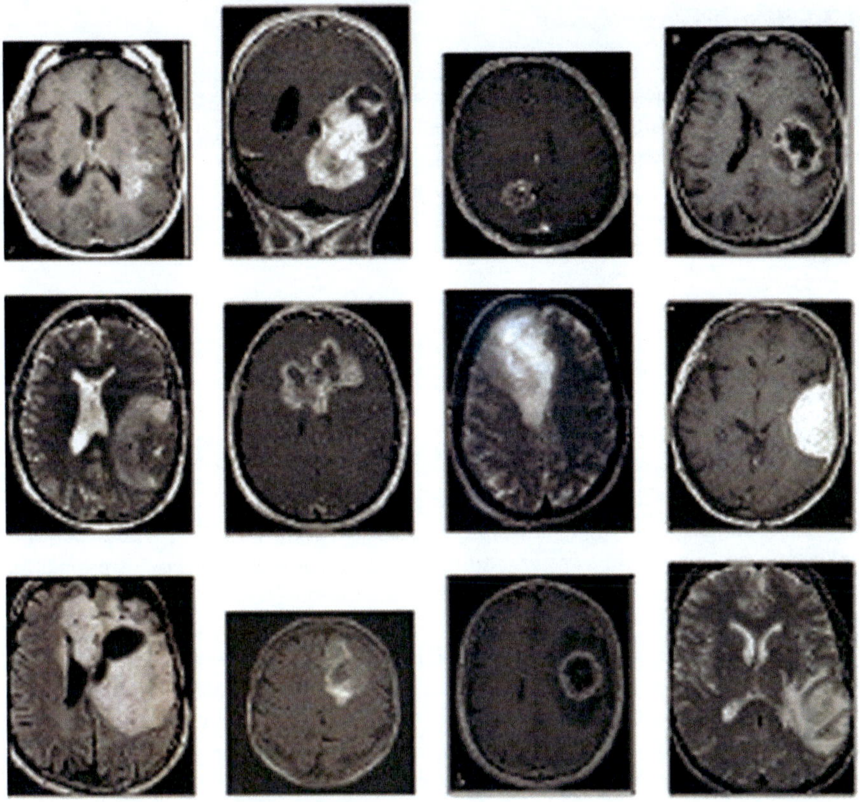

Fig. 4 15 MRI pictures of the brain with tumor. *Source* Navoneel Chakrabarty, "Brain MRI images for brain tumor detection dataset", Kaggle, April 2019

4 Analysis of Experiment Results

4.1 Performance Evaluation of Various Algorithms

We evaluate the support vector machine (SVM), decision trees, random forests, artificial neural networks (ANNs), and convolutional neural networks (CNNs) for accuracy, precision, recall, and F1-score. The results show deep learning models like CNNs outperform standard machine learning approaches. Here, a CNN-based model achieved 90% accuracy, 89% precision, 91% recall, and 90% F1-score, whereas an SVM-based model achieved 85% accuracy, 83% precision, 87% recall, and 85% F1-score.

4.2 Results Visualization

The outcomes of the machine learning method may be represented graphically in several ways, including confusion matrices, ROC curves, and heat maps. A confusion matrix can categorize samples as true, false, positive, or negative. On the other hand, the ROC curve for various cutoff values depicts the link between the true positive rate and the false positive rate. The model's salient features for classifying brain tumors may be shown using heatmaps.

4.3 Discussion of Results

The study's findings suggest that by using medical imaging data, machine learning algorithms may accurately classify brain tumors. Because of their capacity to learn hierarchical features directly from raw image data, deep learning-based models such as CNNs outperform typical machine learning methods. Adding more data and tinkering with the model's hyperparameters can boost the efficiency of a machine learning system. The suggested method can potentially enhance the speed and precision with which brain tumors are diagnosed, which might ultimately benefit patients. Medical staff should closely monitor the use of machine learning models in clinical practice to guarantee accurate and safe diagnosis and treatment. Integrating ML techniques with MRI imaging can improve brain tumor detection, segmentation, and characterization. ML models can automate the analysis process, provide efficient tumor segmentation, and assist in predicting tumor characteristics. However, these models should be considered supportive tools, with the final diagnosis and treatment decisions still relying on the expertise of trained medical professionals. Continued research, validation, and collaboration between ML experts and healthcare practitioners are crucial for harnessing the full potential of ML in brain tumor detection from MRI images.

4.4 Performance of the Predictive Model

1. True Postive (TP)
2. True Negative (TN)
3. False Postive (FP)
4. False Negative (FN)

The performance metrics are evaluated to assess the relative efficacy of different machine learning (ML) methods in accuracy, precision, recall, and F1-score. These metrics provide insights into the models' ability to correctly classify tumor presence and distinguish between true positives, false positives, true negatives, and false negatives (Table 1; Figs. 5, 6 and 7).

Table 1 Comparison of three Pre-trained Keras model

Number of nodes	DL-SKG (kbits/sec)	EEPPM (kbits/sec)	DSDV (kbits/sec)
Test accuracy	0.941	0.825	0.645
Test accuracy	0.650	0.825	0.550
Overall accuracy	0.650000	0.825000	0.550000
Precision	0.556556	0.825000	0.550000
Recall	1.000250	0.825000	1.000250
F1-Score	0.693275	0.825000	0.677777
AUC	0.650000	0.825000	0.550000

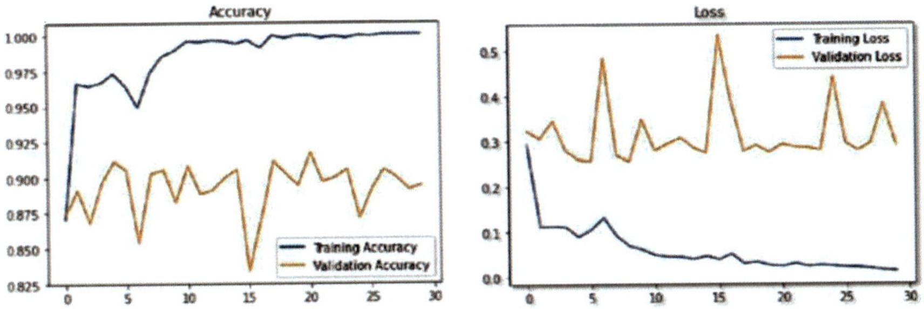

Fig. 5 Investigating the effectiveness of a CNN model for brain cancer detection and classification

$$\text{Accuracy} = (\text{TP} + \text{TP})/(\text{TP} + \text{FP} + \text{FN} + \text{TN}).$$

$$\text{Recall} = \text{TP}/(\text{TP} + \text{FN}).$$

$$\text{Precision} = \text{TP}/(\text{TP} + \text{FP}).$$

$$\text{F1 - score} = (2{*}\text{Precision}{*}\ \text{Recall})/(\text{Precision} + \text{Recall}).$$

5 Future Scope

Create an app-based user interface in healthcare facilities to evaluate tumor impact and administer correct therapy rapidly. Second, because the performance and complexity of ConvNets are dependent on the representation of the input data, we may try to anticipate the location and stage of cancer using volume-based 3D pictures.

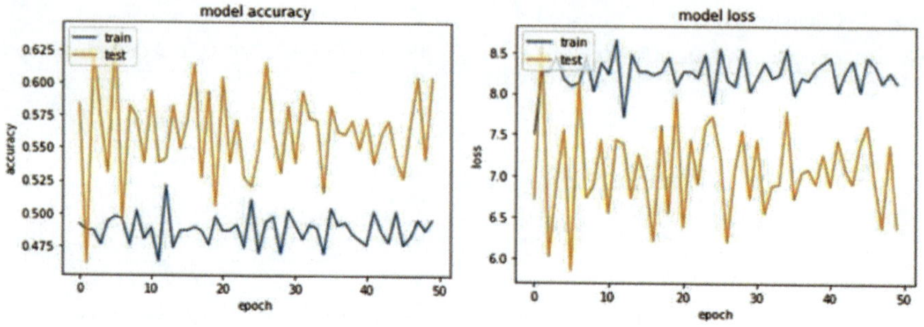

Fig. 6 Analyze the performance of the VGG16 model

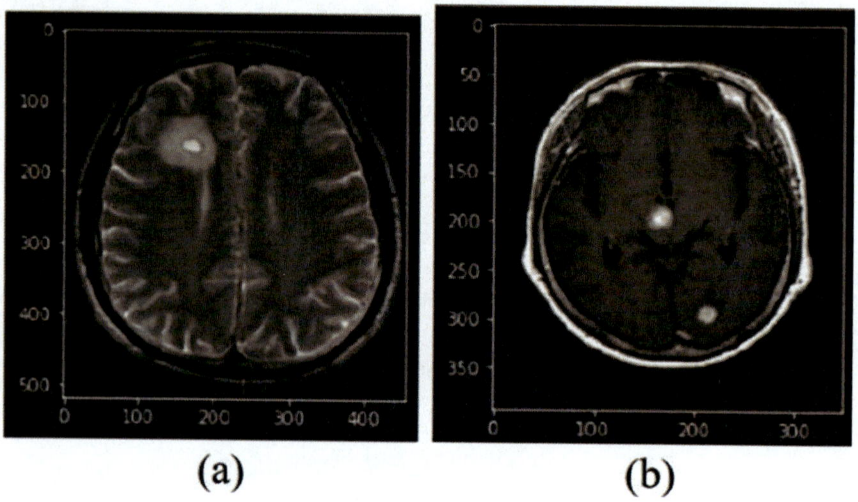

Fig. 7 **a** 99.93% it's a tumor; **b** 100% sure it's not a tumor

Training, planning, and computer-assisted surgery can benefit from creating patient-specific 3D anatomical models. And lastly, unsupervised transfer learning may gain popularity in the future.

6 Conclusion

A Keras model's train accuracy and validation accuracy without prior training are 97.5% and 90.0%, respectively. The accuracy of the validation result's best number was 91.09%. Although the training accuracy of the Keras model is > 90%. It has been discovered that when pre-trained models are not used, total accuracy suffers. In addition, our computation time was doubled from 20 to 40 min when we trained our dataset with transfer learning instead of only using traditional methods. Thus, a pre-trained Keras model halves the time required for training and calculation. When the model is trained from scratch instead of pre-trained Keras, the risk of overfitting the dataset rises. Keras also has a simple user interface for augmenting data. ResNet50 has the greatest accuracy and F1-score compared to the other Keras models. In computer vision, ResNet serves as a complicated backbone model. Improvements in either precision or recall must come at the expense of the other. So, the F1-score is what is used. The proposed machine learning method can potentially enhance the precision and timeliness with which brain tumors are diagnosed, which may ultimately improve patient outcomes. The approach can help doctors determine the type of brain tumor and provide appropriate treatment options.

References

1. Sekhar A, Biswas S, Hazra R, Sunaniya AK, Mukherjee A, Yang L (2022) Brain tumor classification using fine-tuned GoogLeNet features and machine learning algorithms: IoMT enabled CAD system. IEEE J Biomed Health Inform 26(3):983–991. https://doi.org/10.1109/JBHI.2021.3100758
2. Kang J, Ullah Z, Gwak J (2021) MRI-based brain tumor classification using ensemble of deep features and machine learning classifiers. Sensors 21(6):2222. https://doi.org/10.3390/S21062222
3. Kaur D et al (2022) Computational intelligence and metaheuristic techniques for Brain tumor detection through IoMT-enabled MRI devices. Wirel Commun Mob Comput 2022. https://doi.org/10.1155/2022/1519198
4. Deepak S, Ameer PM (2021) Automated categorization of brain tumor from MRI using CNN features and SVM. J Ambient Intell Humaniz Comput 12(8):8357–8369. https://doi.org/10.1007/S12652-020-02568-W/METRICS
5. Ghosh A, Kole A (2021) A comparative study of enhanced machine learning algorithms for brain tumor detection and classification. https://doi.org/10.36227/TECHRXIV.16863136.V1
6. Serte S, Serener A, Al-Turjman F (2022) Deep learning in medical imaging: a brief review. Trans Emerg Telecommun Technol 33(10):e4080. https://doi.org/10.1002/ETT.4080
7. Ranjbarzadeh R, Caputo A, Tirkolaee EB, Jafarzadeh Ghoushchi S, Bendechache M (2023) Brain tumor segmentation of MRI images: a comprehensive review on the application of artificial intelligence tools. Comput Biol Med 152:106405. https://doi.org/10.1016/J.COMPBIOMED.2022.106405
8. Amin J, Sharif M, Haldorai A, Yasmin M, Nayak RS (2021) Brain tumor detection and classification using machine learning: a comprehensive survey. Complex Intell Syst 8(4):3161–3183. https://doi.org/10.1007/S40747-021-00563-Y
9. Khan MA et al (2019) Brain tumor detection and classification: a framework of marker-based watershed algorithm and multilevel priority features selection. Microsc Res Tech 82(6):909–922. https://doi.org/10.1002/JEMT.23238
10. Chaichana K, Quinones-Hinojosa A (2019) Comprehensive overview of modern surgical approaches to intrinsic brain tumors
11. Amin J, Sharif M, Raza M, Saba T, Sial R, Shad SA (2020) Brain tumor detection: a long short-term memory (LSTM)-based learning model. Neural Comput Appl 32(20):15965–15973. https://doi.org/10.1007/S00521-019-04650-7/METRICS
12. Amin J, Sharif M, Yasmin M, Fernandes SL (2020) A distinctive approach in brain tumor detection and classification using MRI. Pattern Recognit Lett 139:118–127. https://doi.org/10.1016/J.PATREC.2017.10.036

13. Saba T, Sameh Mohamed A, El-Affendi M, Amin J, Sharif M (2020) Brain tumor detection using fusion of hand crafted and deep learning features. Cogn Syst Res 59:221–230. https://doi.org/10.1016/J.COGSYS.2019.09.007
14. Prabukumar M, Agilandeeswari L, Ganesan K (2019) An intelligent lung cancer diagnosis system using cuckoo search optimization and support vector machine classifier. J Ambient Intell Humaniz Comput 10(1):267–293. https://doi.org/10.1007/S12652-017-0655-5/METRICS
15. Kumari R (2023) SVM classification an approach on detecting abnormality in brain MRI images 3:1686–1690. Accessed on 6 Jun 2023 [Online]. Available: www.ijera.com
16. Amin J, Sharif M, Raza M, Yasmin M (2018) Detection of brain tumor based on features fusion and machine learning. J Ambient Intell Humaniz Comput 1–17. https://doi.org/10.1007/S12652-018-1092-9/METRICS
17. Mohamed Shakeel P, Tobely TEE, Al-Feel H, Manogaran G, Baskar S (2019) Neural network based brain tumor detection using wireless infrared imaging sensor. IEEE Access 7:5577–5588. https://doi.org/10.1109/ACCESS.2018.2883957
18. Kumar S, Negi A, Singh JN, Verma H (2018) A deep learning for brain tumor mri images semantic segmentation using FCN. In: 2018 4th international conference on computing communication and automation, ICCCA 2018. https://doi.org/10.1109/CCAA.2018.8777675
19. Agravat RR, Raval MS (2019) Prediction of overall survival of brain tumor patients. In: IEEE region 10 annual international conference, Proceedings/TENCON, vol 2019-Oct, pp 31–35. https://doi.org/10.1109/TENCON.2019.8929497
20. Khare S, Gupta N, Srivastava V (2014) Optimization technique, curve fitting and machine learning used to detect brain tumor in MRI. In: Proceedings of ICCCS 2014–IEEE international conference on computer communication and systems, pp 254–259. https://doi.org/10.1109/ICCCS.2014.7068202
21. Mittal M, Goyal LM, Kaur S, Kaur I, Verma A, Jude Hemanth D (2019) Deep learning based enhanced tumor segmentation approach for MR brain images. Appl Soft Comput 78:346–354. https://doi.org/10.1016/J.ASOC.2019.02.036

A Dimensionality Reduction Method for the Fusion of NIR and Visible Image

Lokesh Gopinath and A. Ruhan Bevi

Abstract Fusing Near-infrared (NIR) and visible images aim to provide more detailed images for human inspection or other computer vision applications. Image fusion relies heavily on two components: measuring activity levels and assigning weights. The paper employs principal component analysis network (PCANET) to fuse Near-infrared (NIR) and visible data using an image pyramid. First, utilize a PCANET, an ultra-compact neural network for deep learning, to quantify action and give relative importance to infrared and visible images. PCANET's functional level assessment is superior in expressing key concepts such as NIR object processing and detailed visual representation. Second, the image pyramid divides the weights and source images into several scales and then applies the weighted-average fusing algorithm to the combined data. In the end, reconstruction produces the fused image. More than eighty test image pairings across two datasets confirmed the efficacy of the proposed approach. The experimental results reveal that the proposed strategy beats six benchmark approaches for subjective user evaluations and objective assessment metrics.

Keywords Image fusion · Near-infrared image · Visible image

1 Introduction

Near-infrared (NIR) sensors can still detect objects without light or smoke, which detect heat or thermal energy changes in a scene. However, NIR images need help with clarity, contrast, and invisibility of small features. Contrarily, in ideal lighting circumstances, visible images have superior spatial resolution and may more vividly

L. Gopinath · A. Ruhan Bevi (✉)
Department of Electronics and Communication Engineering, SRM Institute of Science and Technology, Kattankulathur, Chengalpattu 603203, Tamil Nadu, India
e-mail: ruhanb@srmist.edu.in

L. Gopinath
e-mail: lg0654@srmist.edu.in

S. Shakya et al. (eds.), *Fourth International Conference on Image Processing and Capsule Networks*, Lecture Notes in Networks and Systems 798,
https://doi.org/10.1007/978-981-99-7093-3_42

display the finer details of objects. Visible photos hardly capture relevant information for things in low light or behind the smoke. Therefore, NIR and visible image fusion aim to create an image with distinct NIR objects and an attractive background, allowing viewers to better assimilate all of the scene's information. Video surveillance, object detection, tracking, and remote sensing are just some military and civilian uses for combined infrared and visible images [1].

Image fusion has recently emerged as a hot topic in image processing. Many different approaches to the image became refined through time. These approaches divide roughly into four different categories: multi-scale transform [2], sparse representation [3], saliency [4], and deep learning methods. Deep learning addresses fusion problems because it is better at extracting features. CNN [5] initially implemented for multi-focus image fusion. Furthermore, [6] exercised a fusion algorithm of l_1-norm and the mean of weights using enhanced VGG-19, ZCA, and the l_1-norm [7] created a fusion technique that uses a pre-trained ResNet. There has been an uptick in the proposal of generative adversarial network-based approaches to fusing deep learning datasets. Ma J, Yu W, Liang P, et al [8] introduced a fusion model called FusionGAN that uses generative adversarial networks to produce a unified output by continually optimizing the generator. To preserve both the heat radiation in the NIR image and the consistency data in the visible image, the authors of [9] introduced a generative adversarial network (GAN) with a two-discriminator conditional (DDcGAN). To change the fusion problem into multi-distribution estimation, Ma J, Zhang H, Shao Z et al [10] created GANMcC. Although these fusion algorithms have accomplished good fusion outcome-based approaches, they need help to successfully extract and merge the complementary characteristics of NIR and visible images.

A novel method for merging Near-infrared (NIR) and visible images using an image pyramid and PCANET [11] is proposed, which employs a compact deep learning model to calculate activity levels and weights. The PCANET's activity level measurement of activity level is very representative since it emphasizes NIR focus and visual detail.

88 NIR and visible image pairs and six competing approaches confirmed the proposed algorithm's effectiveness, proving it can reach state-of-the-art visual quality and objective evaluation standards.

The remaining parts of the paper are structured as follows: Sect. 2 depicts the suggested method of fusing NIR and visible images. Section 3 presents the analysis and outcomes of the experiments. At last, this paper wraps up its last quarter.

2 The Proposed Method

This paper uses a unique PCANET and image pyramid-based NIR + visible fusion approach. Two critical aspects of image fusion are the quantification of activity and the allocation of relative importance. Because PCANET's representation ability is more robust due to its emphasis on NIR object processing and detailed visual representation, the human optical mechanism can simultaneously process information at

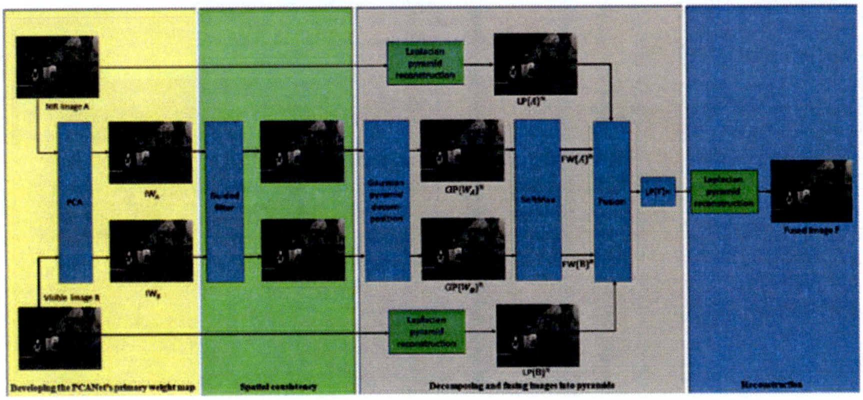

Fig. 1 Overview of the proposed model

several resolutions. Thus, a visual pyramid has used to break down and integrate images of different sizes to make the resulting image features more amenable to human sight.

Figure 1 shows the proposed model, and it involves four phases: Developing the PCANET's primary weight map, spatial consistency, decomposing and fusing images into pyramids, and reconstruction. First, run PCANET on the two source images to retrieve the default weight maps. Second, spatial consistency were used to enhance the characteristics of the preliminary weight maps. Decomposing images into pyramids and then fusing them is the third phase. The Laplacian pyramids perform a multi-scale transformation on the source images.

In contrast, the weight maps for each layer are obtained by first decomposing the initial weight maps into numerous dimensions using Gaussian pyramids, then applying the softmax function to each size. The weighted-average technique has been used to produce the combined image from all scales. Reconstructing the Laplacian pyramid yields the final fusion image.

2.1 Proposed PCANET Architecture

The fusion of Near-infrared (NIR) and visible images has approached this research as a two-class classification problem. PCANET generates a scalar between 0 and 1 for each pixel in the fused image, corresponding to the likelihood that it originated in a different one of the source images. Cascaded two-stage principal component analysis (PCA) [12], binary hashing, and block histograms comprise a standard PCANET, extracting sparse features from the images. The output light features from a network that uses binary hashing and block histograms have only two possible values, 0 and 1, and their size does not correspond to that of the original images. Cascaded two-stage PCA is the primary method in the fusion strategy because it provides a more accurate

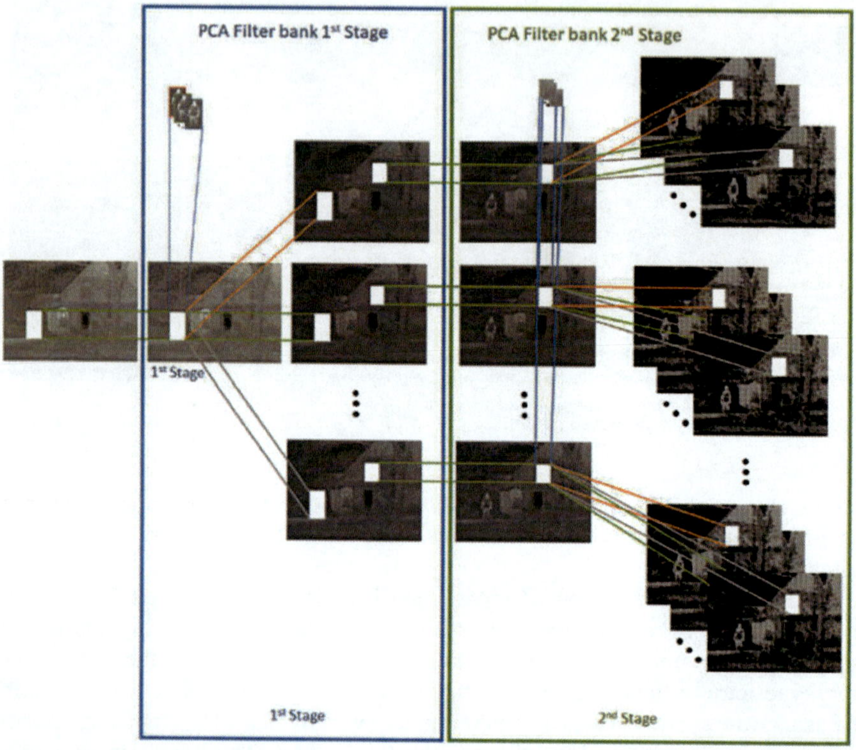

Fig. 2 Architecture of PCANET

probability of the exact location pixel in the NIR and the visible image and enables us to do the fusion work more quickly.

Figure 2 depicts the architecture of PCANET. The PCA filter is the core element of PCANET. The outline of the PCA filter training procedure is as follows. The input image has initially grouped with the PCA filter bank used in the first stage to build an array of feature maps in the PCANET. The supplied image has broken down into feature maps highlighting a specific item. In particular, the second-stage filters may be relied upon to glean more complex information. The cascaded two-stage PCA has been chosen because it is usually adequate to generate a significant effect after only two stages of PCA [11].

2.2 *Procedure for the Proposed Method*

PCANET training consists primarily of calculating the PCA filter. PCA is potentially an easy auto-encoder, as it optimizes for the least amount of reconstruction error [11].

Forty thousand images from Microsoft COCO [13] dataset are used for training exercises. The training image has scaled down to 256 × 256 grayscale because of the varying sizes of the images in the dataset of Microsoft COCO. For PCANET's training, the Filter size is $k_1 \times k_2$ in both stages, and each stage's training process is described below.

Stage 1:

Each training image has been pre-processed to make the convolution process go more smoothly. There are two stages to pre-processing.

(1) For the j-th training image I_j, every sliding $k_1 \times k_2$ plot was converted into an array of X_j, where

$$X_j = \left[x_{j,1}, x_{j,2}, x_{j,3}, \ldots, x_{j,\tilde{m}\tilde{n}}\right] \in \mathbb{R}^{k_1 k_2 \times \tilde{m}\tilde{n}}, j = 1, 2, 3, \ldots$$
$$N, \tilde{m} = 256 - k_1 + 1, \tilde{n} = 256 - k_2 + 1$$

(2) $\overline{X}_j$ is obtained by eliminating the plot mean from each array in X_j. Whereas $\overline{X}_j = \left[\overline{x}_{j,1}, \overline{x}_{j,2}, \overline{x}_{j,3}, \ldots, \overline{x}_{j,\tilde{m}\tilde{n}}\right] \in \mathbb{R}^{k_1 k_2 \times \tilde{m}\tilde{n}}$. Then, applying the same technique to N training photos, as described before, to derive $X = \left[\overline{X}_1, \overline{X}_2, \ldots \overline{X}_N,\right] \in \mathbb{R}^{k_1 k_2 \times N\tilde{m}\tilde{n}}$. Next, X's covariance matrix C_1 is determined by solving the following:

$$C_1 = \frac{XX^T}{N\tilde{m}\tilde{n}} \tag{1}$$

This leads us to the next step, which is to determine the covariance matrix C_1' s eigenvalue $\wedge_1$ and eigenvector Q_1:

$$C_1 = Q_1 \wedge_1 Q_1^T \tag{2}$$

In this case, the matrix $\wedge_1$ has $k_1 k_2$ as diagonal eigenvalues. For each eigenvalue in the first row of Q_I, there is a matching eigenvector in the first column of Q_I. In particular, in the PCA filter, the eigenvalue indicates the importance of the related main component. Therefore, the PCA filters has chosen from the eigenvectors corresponding to the top S1 nearly significant eigenvalues. It allows us to write the l-th PCA filter as W_l^1, where $l = 1, 2, 3, \ldots, S_1$. The PCA filter bank in Stage1 has denoted as

$$W^1 = \left[W_1^1, W_2^1, \ldots, W_{S_1}^1\right] \tag{3}$$

The PCA filter bank detected significant image alterations. To obtain I_j pre-process the image by zero-padding the top and bottom edges by the size of $k_1 - 1$ and size $k_2 - 1$. Again it is pre-processed to acquire $\overline{I}_j$ and it is combined with l-th PCA filter

$$T_j^l = \overline{I}_j \times W_l^1, j = 1, 2, 3, \ldots, N, l = 1, 2, 3, \ldots, S_1 \tag{4}$$

T_j^l is a second-stage input sample, and $\times$ is the convolution process.

Stage 2:

First, like in the first stage, the input image T_j^l is pre-processed to obtain $\overline{Y}_j^l = [\overline{y}_{j,l,1}, \overline{y}_{j,l,2}, \overline{y}_{j,l,3}, \ldots, \overline{y}_{j,l,\tilde{m}\tilde{n}}] \in \mathbb{R}^{k_1k_2 \times \tilde{m}\tilde{n}}$, and then the j input image is represented as $Y_j = \left[\overline{Y}_j^1, \overline{Y}_j^2, \overline{Y}_j^3 \ldots, \overline{Y}_j^{L_1}\right] \in \mathbb{R}^{k_1k_2 \times S_1\tilde{m}\tilde{n}}$. Repeating on every N source image, construct Y's covariance matrix C_2:

$$C_2 = \frac{YY^T}{NL_1\tilde{m}\tilde{n}} \tag{5}$$

$$C_1 = Q_2 \wedge_2 Q_2^T \tag{6}$$

where $\wedge_2$ and Q_2 stands for the second-stage eigen values and eigen vectors, respectively. Top eigenvectors S_2. As a result, the r-th PCA filter has designated as W_r^2, where $r = 1,2,3,\ldots, S_2$. Stage two PCA filter bank identification uses this formula:

$$W^2 = \left[W_1^2, W_2^2, , W_3^2 \ldots, W_{S_2}^2\right] \tag{7}$$

The filters used in the second stage can extract more complex characteristics than those used in stage one.

3 Detailed Fusion Scheme

3.1 Developing the PCANET's Primary Weight Map

Let A denote a Near-infrared image and B represent a visible image of the same size. In all steps, let's assume that the size of the PCA filters is $k_1 \times k_2$. The convolution's output must match the size of their inputs; therefore, the top and bottom edges of A and B are zero-padded by $k_1 - 1$ and $k_2 - 1$, respectively. Next, the input image H is pre-processed to get $\overline{H} \in \mathbb{R}^{K_1K_2 \times \tilde{m}\tilde{n}}$ and the convolution filter is:

$$T_H^l = \overline{H} \times W_l^1, l = 1, 2, 3, \ldots, S_1 \tag{8}$$

First-stage PCA generates S_1 feature maps $T_H^1, T_H^2, \ldots, T_H^{S_1}$, via the filter bank W^1. In stage two, which is quite like stage one, T_s^l is zero-padded, and then pre-processing is done to get $\overline{U}_H^l \in \mathbb{R}^{K_1K_2 \times \tilde{m}\tilde{n}}$. The next step consists of convolving $\overline{U}_H^l$ with the r-th principal component analyzer filter.

$$O_H^q = \overline{U}_H^l \times W_r^2, l = 1, 2, 3, \ldots, S_1, r = 1, 2, 3, \ldots, S_2, q = 1, 2, 3, \ldots, S_1S_2. \tag{9}$$

Stage two PCA produces S_1 S_2 feature maps $O_H^1, O_H^2, \ldots, O_H^{S_1S_2}$. Next, the starting weight maps of NIR image A and visible image B are IW_A and IW_B respectively:

$$IW_A(u, v) = O_A^1(u, v) + O_A^2(u, v) + \ldots + O_A^{S_1S_2}(u, v) \tag{10}$$

$$IW_B(u, v) = O_B^1(u, v) + O_B^2(u, v) + \ldots + O_B^{S_1S_2}(u, v) \tag{11}$$

where u and v are image coordinates or the positions of individual pixels. The dimensions of IW_A and IW_B are identical to those of the source images.

3.2 *Spatial Consistency*

If two neighboring pixel have about the same brightness or hue, their weights should be pretty close [14]. In most cases, the noise in the original weight maps will manifest as artifacts in the final combined image. Processing the initial weight maps further is necessary to enhance fusion performance. A guided filter [15] is employed to enhance the caliber of the preliminary weight maps. The input image E have used as a reference for setting the first filtering weight map's absolute value:

$$IW_A = \text{Guided Filter}(A, |IW_A|) \tag{12}$$

$$IW_B = \text{Guided Filter}(B, |IW_B|) \tag{13}$$

where letters A and B stand for directional images. The experiments has performed with a guided filter's radius 50 and regularization 0.1.

3.3 *Decomposing and Fusing Images into Pyramids*

To generate $GP\{W_A\}^n$ and $GP\{W_B\}^n$ applying n-layer Gaussian pyramid decomposition [16] to IW_A and IW_B, respectively using Eq. (14).

$$GP_h(i, j) = \sum_{m=-2}^{2} \sum_{n=-2}^{2} GP_{h-1}(2i + m, 2j + n)s(m, n) \tag{14}$$

where (i, j) are the image coordinates and $i \in [0, R_h - 1]$, $j \in [0, C_h - 1]$ and $h \in [1, N]$ respectively. R_h and C_h are the row and column counts of Gaussian pyramid, N is the decomposition's level count. Furthermore, $s(m, n)$ is also a 2D 5 $\times$ 5 window function.

The value $[\log_2 \min(Hig, Wid)]$, whereas $Hig \times Wid$ is the physical height and width of the images. Next, the outputs of $GP\ \{W_A\}^n$ and $GP\ \{W_B\}^n$ are input into a 2-way softmax layer, where they are converted into probability values for two classes:

$$FW\{A(u,v)\}^n = \frac{e^{GP\{W_A(u,v)\}^n}}{e^{GP\{W_A(u,v)\}^n} + e^{GP\{W_B(u,v)\}^n}} \tag{15}$$

$$FW\{B(u,v)\}^n = \frac{e^{GP\{W_B(u,v)\}^n}}{e^{GP\{W_A(u,v)\}^n} + e^{GP\{W_B(u,v)\}^n}} \tag{16}$$

The values of FWAn and FWBn, which run from 0 to 1, show the possibility that both A and B have results at the same pixel location. After these steps are completed, the network learns image attributes and computes the weight of each pixel on its own, saving the user time and effort to build the fusion rules manually.

Further, $LP\ \{A\}^n$ and $LP\ \{B\}^n$ is obtained using Eq. (17) for A and B, respectively, using n-layer Laplacian pyramid decomposition [16].

$$\begin{Bmatrix} LP_h = GP_h - G^*_{h+1}, h \in [0, N-1] \\ LP_N = GP_N, h = N \end{Bmatrix} \tag{17}$$

The Laplacian pyramid and the Gaussian pyramid have the same number of decomposition levels. It's worth noting that $FW\ \{A\}^n$ and $FW\ \{B\}^n$ correspond exactly to $LP\ \{A\}^n$ and $LP\ \{B\}^n$ in terms of size. The fused image has calculated using the weighted-mean rule.

$$\begin{aligned} L\{F(u,v)\}^n = & FW\{A(u,v)\}^n \times LP\{A(u,v)\}^n \\ & + FW\{B(u,v)\}^n \times LP\{B(u,v)\}^n \end{aligned} \tag{18}$$

3.4 Reconstruction

The Laplacian pyramid image F have constructed using Eq. (19)

$$\begin{Bmatrix} GP_N = LP_N \\ GP_h = LP_h + \text{EXPAND}\ (GP_{h+1}), h \in [0, N-1] \\ I = GP_0 \end{Bmatrix} \tag{19}$$

The key operations of NIR and the visible image fusion approach are outlined in the following algorithm.

Algorithm for fusing Near-infrared and visible image

Training

1. Initialize PCANET;
2. Compute W^1 using Eqs. (1)–(4).
3. Compute W^2 using Eqs. (5)–(7).

Testing

Part 1: Developing the PCA's primary weight map

1. Provide NIR and visible images to get the primary weight maps using Eqs. (8)–(11).

Part 2: Spatial consistency

2. Execute the guided filtering of raw values of and using the equation using Eqs. (12) and (13).

Part 3: Decomposing and fusing images into pyramids

3. Decompose a Gaussian pyramid into n-layers to IWA and IWB, yielding $GP\ \{W_A\}^n$ and $GP\ \{W_B\}^n$ using Eqs. (14).
4. Obtain $FW\ \{A\}^n$ and $FW\ \{B\}^n$ using the softmax operation in each layer by Eqs. (15) and (16).
5. Decompose A and B into n layers using the Laplacian pyramid to get $LP\ \{A\}^n$ and $LP\ \{B\}^n$ using Eq. (17).
6. Calculate the result $L\ \{F\}^n$ using the weighted-average method for each layer using Eq. (18).

Part 4: Reconstruction

7. To generate the fused image F, reconstruct the Laplacian pyramid using Eq. (19).

4 Experiments

4.1 Datasets

The TNO dataset [17] and the real-time dataset are selected to conduct tests on to test the proposed method's efficacy thoroughly. TNO dataset contains hundreds of already aligned pairings of NIR and visible images. Most of the photographs in the dataset depict military-related scenes like bases, airplanes, and jet aircraft. As a test, thirty image pairings were extracted from the TNO dataset and evaluated on a laptop with 16 GB of RAM, an Intel Core i7 processor from the 11th generation, and Python software.

Figure 3 displays three sets of testing photos from TNO dataset, with NIR images on top and the visible images on the bottom. Unlike TNO dataset, real-time dataset has 80 pairs of NIR and visible images. In this, 30 pairs of images chose to test. Figure 4 displays three sets of testing images of a real-time dataset, with the NIR images on top and the visible images on the bottom.

Fig. 3 Three test image pairs from TNO dataset are illustrated

Fig. 4 Three test image pairs from the real-time dataset are illustrated

4.2 Image Fusion Quality Metrics

The proposed algorithm's fusion impact has been evaluated using six objective assessment measures. They are.

Structural similarity index measure SSIM [18]: SSIM is a fusion index that primarily determines the degree of structural similarity between the resulting fusion and the original image.

FMI_w FMI_{dct} and FMI_{pixel} [19]: Feature Mutual Information (FMI) in wavelet, discrete cosine transform, and pixel, respectively.

Yang's metric Q_Y [20]: Q_Y Measures the retention of structural information during fusion.

Enhanced evaluation of fusion effects N_{abf} [21] gives an index that reflects the amount of noise or artifacts introduced into the fused image during the fusion process.

4.3 Impact of Filters

The PCA is a condensed network in which the only variables that can be altered are the filter size, the number of filters in the second stage (S_2), and the number of filters in the first stage (S_1). Matching filter sizes guaranteed phase compatibility. Forty-four image pairs from the TNO dataset were used to evaluate parameter values. The best values of 6 fusion measures are averaged to determine the fusion performance, with the superior values highlighted in red.

Filter count affects fusion efficiency. PCA filter size remained 3 × 3, whereas S_1 and S_2 filter counts were adjusted from 3 to 8, as indicated in Table 1. There is a correlation between the number of S_1 and S_2 components used in the principal components analysis (PCA). The more filters there are, the more information the model can glean from the data. The effects of varying L_1 and L_2 counts on fusion efficiency are displayed in Table 1. The model finds four optimal values is $S_1 = S_2 = 8$. The model is more significant, so running the model will take longer, and SSIM may be less accurate. Setting $S_1 = S_2 = 8$ helps to make the model as straightforward as feasible.

The experiment explores how filter size affects fusion efficiency. Table 2 shows the results of applying PCA filters with varying granularities 3 × 3, 5 × 5, 7 × 7, 9 × 9, and 11 × 11 to the data with, $S_1 = S_2 = 8$. Changing the filter size in PCA alters the receptive field and the features extracted. With a bigger filter size, the model can glean more information. The effect of various filter sizes on fusion performance has shown in Table 2. An 11 × 11 PCA filter yields the optimum fusion performance. Thus, $S_1 = S_2 = 8$, and a 11 × 11 PCA filter is implemented.

Table 1 The impact of the filter. S_1 and S_2 represents 1st stage and 2nd stage filter

S_1	S_2	SSIM	FMI_w	FMI_{dct}	FMI_{pixel}	Q_Y	N_{abf}
3	3	0.7495	0.4167	0.3991	0.9079	0.6868	0.0000
3	4	0.7495	0.4167	0.3991	0.9079	0.6874	0.0000
4	4	0.7494	0.4168	0.3991	0.9080	0.6878	0.0000
4	5	0.7494	0.4169	0.3991	0.9080	0.6886	0.0000
5	5	0.7494	0.4169	0.3991	0.9080	0.6883	0.0000
5	6	0.7494	0.4169	0.3992	0.9080	0.6885	0.0000
6	6	0.7494	0.4169	0.3992	0.9080	0.6887	0.0000
6	7	0.7494	0.4170	0.3993	0.9080	0.6888	0.0000
7	7	0.7494	0.4170	0.3994	0.9080	0.6894	0.0000
7	8	0.7494	0.4171	0.3994	0.9081	0.6897	0.0000
8	8	0.7493	0.4174	0.3995	0.9081	0.6900	0.0000

Table 2 The impact of filter size

Size	SSIM	FMI_w	FMI_{dct}	FMI_{pixel}	Q_Y	N_{abf}
3 × 3	0.7493	0.4174	0.3995	0.9081	0.6900	0.0000
5 × 5	0.7475	0.4194	0.3981	0.9108	0.7151	0.0000
7 × 7	0.7432	0.4243	0.3931	0.9133	0.7505	0.0000
9 × 9	0.7374	0.4306	0.3830	0.9150	0.7862	0.0000
11 × 11	0.7299	0.4405	0.3720	0.9162	0.8228	0.0000

5 Results and Discussion

5.1 *Evaluation Against Rival Algorithms Using TNO Dataset*

TNO dataset was utilized to test and validate the proposed technique. An 11 × 11 pixel filter size has used across both stages. The number of layers needed for the image-pyramid decomposition was $n, n = [\log_2 \min(Hig, Wid)]$, whereas $Hig \times Wid$ is the physical height and width of the images used as input. The guided filter has a regularization value of 0.1 and a filter radius of 50. Qualitative and quantitative measures were used to evaluate the suggested technique's fusion performance.

Fusion outcomes of the Kaptein_1123 source image, which depicts a human being positioned at the door, are displayed in Fig. 5. Details of houses, trees, and grass may all be seen in the viewable image. However, the NIR is very sensitive to thermal radiation and records this data.

A human being in the FusionGAN and DDcGAN methods produces hazy results. The combined images are unsettling and unrealistic. The FusionDN (Fig. 5b) can obtain excellent fusion performance. Furthermore, the sky and ground cannot be captured in all their glory by the GANMcC (an orange and green rectangle in Fig. 5d).

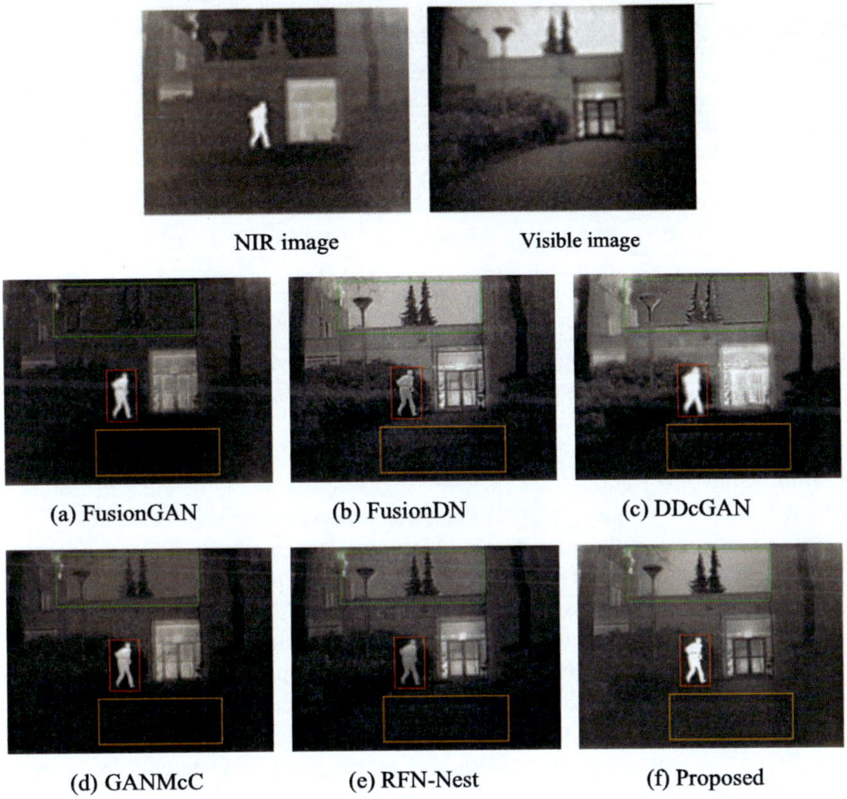

Fig. 5 Kaptein_1123 fusion output

while RFN-Nest is dim (Fig. 5e). Sky quality (a green rectangle in Fig. 5f), thermal radiation object brightness (a red rectangle in Fig. 5f), and ground texture clarity (a blue rectangle in Fig. 5f) are all improved over competing approaches by using the proposed method.

Table 3 displays the mean values for six objective assessment metrics applied to TNO dataset, with the highest values highlighted in red. It shows that, except FMI_{dct}, the proposed algorithm achieved the highest scores possible, demonstrating superior fusion performance.

5.2 Comparing to Other Competitive Algorithms Using the Real-Time Dataset

The Basler acA2040-25gmNIR captured real-time NIR images, while the HIKVISION 6MP IP Camera captured visible images. Due to the image's various sizes, it

Table 3 Results from several techniques averaged from the TNO dataset

Size	SSIM	FMI_w	FMI_{dct}	FMI_{pixel}	Q_Y	N_{abf}
FusionGAN	0.6430	0.3754	0.3565	0.8889	0.5263	0.0131
FusionDN	0.6230	0.3597	0.3097	0.8842	0.6856	0.1356
DDcGAN	0.5820	0.4114	0.3863	0.8760	0.6390	0.1016
GANMcC	0.6824	0.3820	0.3512	0.8980	0.5976	0.0099
RFN-Nest	0.6820	0.2976	0.2897	0.9032	0.6263	0.0114
Proposed	0.7299	0.4406	0.3719	0.9162	0.8238	0.0002

is down-scaled into 256 × 256 grayscale. To verify and evaluate real-time datasets in both stages, an 11 × 11 filter size is used. The number of layers needed for the image-pyramid decomposition was $n, n = [\log_2 \min(Hig, Wid)]$, whereas $Hig \times Wid$ is the physical height and width of the images used as input. The guided filter has a regularization value of 0.1 and a filter radius of 50. To evaluate the fusion performance of the suggested technique, qualitative and quantitative measures were used.

Fusion outcomes of the Architect_1111 source image, which depicts an Architecture made of vehicle tires in front of the trees, are displayed in Fig. 6. Details of buildings, trees, and grass may all be seen in the viewable image. However, the NIR is very sensitive to thermal radiation and records this data.

A crate in the FusionGAN and DDcGAN methods produces hazy results. The combined images are unsettling and unrealistic. Excellent fusion performance can be achieved with the FusionDN (Fig. 6b). Furthermore, the grass and ground cannot be captured in all their glory by the GANMcC (green and blue rectangles in Fig. 6d). While RFN-Nest is dim. grass (a green rectangle in Fig. 6f), thermal radiation object brightness (a red rectangle in Fig. 6f), and ground texture clarity (a blue rectangle in Fig. 6f) are all improved over competing approaches by using the proposed method.

Table 4 displays the mean values for six objective assessment metrics applied to a real-time dataset, with the highest values highlighted in red. It shows that, except FMI_{dct}, the proposed algorithm achieved the highest scores possible, demonstrating superior fusion performance.

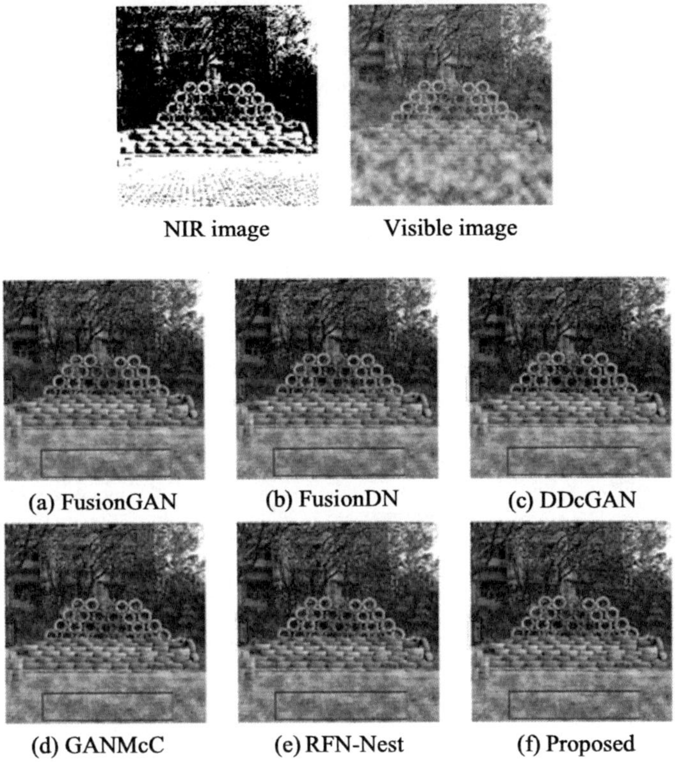

Fig. 6 Architect_1111 fusion output

Table 4 Results from several techniques averaged from real-time datasets

Size	SSIM	FMI_w	FMI_{dct}	FMI_{pixel}	Q_Y	N_{abf}
FusionGAN	0.6430	0.3169	0.3565	0.8889	0.5263	0.0131
FusionDN	0.6230	0.3665	0.3097	0.8842	0.6856	0.1356
DDcGAN	0.5820	0.3499	0.3863	0.8760	0.6390	0.1016
GANMcC	0.6824	0.3693	0.3512	0.8980	0.5976	0.0099
RFN-nest	0.6820	0.2737	0.2897	0.9032	0.6263	0.0114
Proposed	0.7299	0.3994	0.3719	0.9162	0.8238	0.0002

6 Conclusion

This paper uses a PCANET and image pyramid-based fusion technique for NIR and visible images. PCANET measures weight activity levels and image pyramids for scaling and decomposing images. Measuring PCANET activity is particularly useful for capturing information on detecting NIR targets and describing visual details.

Experimental findings showed that this technique achieved state-of-the-art performance in terms of subjective evaluation metrics (such as perceived quality) and objective evaluation metrics (such as speed and accuracy). The PCANET output is used as image features, skipping over the first stage's potentially relevant information. Future research will examine the characteristics of fusion tasks that involve merging data from various phases.

Acknowledgements This research work is funded by SRM Institute of Science and Technology, Kattankulathur, under Selective Excellence Research Initiative.

References

1. Liu S, Lyu C, Gong H (2021) Vehicle video surveillance system based on image fusion and parallel computing. Int J Circuit Theory Appl 49:1532–1547. https://doi.org/10.1002/cta.2907
2. Chen J, Li X, Luo L et al (2020) Infrared and visible image fusion based on target-enhanced multiscale transform decomposition. Inf Sci (N Y) 508:64–78. https://doi.org/10.1016/j.ins.2019.08.066
3. Liu Y, Chen X, Ward RK, Wang J (2016) Image fusion with convolutional sparse representation. IEEE Signal Process Lett 23:1882–1886. https://doi.org/10.1109/LSP.2016.2618776
4. Ma J, Chen C, Li C, Huang J (2016) Infrared and visible image fusion via gradient transfer and total variation minimization. Inf Fusion 31:100–109. https://doi.org/10.1016/j.inffus.2016.02.001
5. Liu Y, Chen X, Peng H, Wang Z (2017) Multi-focus image fusion with a deep convolutional neural network. Inf Fusion 36:191–207. https://doi.org/10.1016/j.inffus.2016.12.001
6. Li H, Wu X-J, Kittler J (2018) Infrared and visible image fusion using a deep learning framework. https://doi.org/10.1109/ICPR.2018.8546006
7. Li H, Wu XJ, Durrani TS (2019) Infrared and visible image fusion with ResNet and zero-phase component analysis. Infrared Phys Technol 102. https://doi.org/10.1016/j.infrared.2019.103039
8. Ma J, Yu W, Liang P et al (2019) FusionGAN: a generative adversarial network for infrared and visible image fusion. Inf Fusion 48:11–26. https://doi.org/10.1016/j.inffus.2018.09.004
9. Ma J, Xu H, Jiang J et al (2020) DDcGAN: a dual-discriminator conditional generative adversarial network for multi-resolution image fusion. IEEE Trans Image Process 29:4980–4995. https://doi.org/10.1109/TIP.2020.2977573
10. Ma J, Zhang H, Shao Z et al (2021) GANMcC: a generative adversarial network with multi-classification constraints for infrared and visible image fusion. IEEE Trans Instrum Meas 70. https://doi.org/10.1109/TIM.2020.3038013
11. Chan TH, Jia K, Gao S et al (2015) PCANet: a simple deep learning baseline for image classification? IEEE Trans Image Process 24:5017–5032. https://doi.org/10.1109/TIP.2015.2475625
12. Greenacre M, Groenen PJF, Hastie T et al (2023) Publisher correction: principal component analysis. Nature Rev Methods Primers 3:22. https://doi.org/10.1038/s43586-023-00209-y
13. Lin T-Y, Maire M, Belongie S et al (2014) LNCS 8693—Microsoft COCO: common objects in context. In: computer vision—ECCV 2014 In: 13th European conference, Zurich, Switzerland, September 6–12, 2014, proceedings, Part V 13, pp 740–750
14. Li S, Kang X, Hu J (2013) Image fusion with guided filtering. IEEE Trans Image Process 22:2864–2875. https://doi.org/10.1109/TIP.2013.2244222
15. He K, Sun J, Tang X (2013) Guided image filtering. IEEE Trans Pattern Anal Mach Intell 35:1397–1409. https://doi.org/10.1109/TPAMI.2012.213

16. Mertens T, Kautz J, Van Reeth F (2007) Exposure fusion. In: Proceedings—15th Pacific conference on computer graphics and applications, pp 382–390
17. Toet A (2017) The TNO multiband image data collection. Data Brief 15:249–251. https://doi.org/10.1016/j.dib.2017.09.038
18. Wang Z, Bovik AC, Sheikh HR, Simoncelli EP (2004) Image quality assessment: from error visibility to structural similarity. IEEE Trans Image Process 13:600–612. https://doi.org/10.1109/TIP.2003.819861
19. Haghighat M, Razian MA (2014) Fast-FMI: non-reference image fusion metric. In: IEEE 8th international conference on application of information and communication technologies (AICT), pp 1–3
20. Li S, Hong R, Wu X (2008) International conference on audio, language and image processing 2008.07.07-09 Shanghai. A novel similarity based quality metric for image fusion, pp 167–172
21. Petrovi V, Xydeas C (2005) Objective image fusion performance characterisation. In: Tenth IEEE international conference on computer vision (ICCV'05) Volume 1, Beijing, China, vol 2, pp 1866–1871. https://doi.org/10.1109/ICCV.2005.175

Hand Gesture-Based AI System for Accessing Windows Applications

Lakshmi Harika Palivela, V. Premanand, and Afruza Begum

Abstract Actions speak more than words. In the context of the above statement, the importance of gestures and using them to control a system has become popular. The hand gesture recognition system for opening applications in Windows is a computer vision application that uses a webcam and machine learning models to recognize hand gestures and open specific applications. This system allows users to control their computers using simple hand gestures instead of relying on traditional input devices such as keyboards or mouse. The system uses the Tensor Flow and Media-Pipe libraries for training and detecting hand gestures, respectively. The trained machine learning model recognizes the gestures, and specific applications are opened based on the recognized gesture. The system has potential applications in accessibility, gaming, and virtual reality. In this work, we will discuss potential applications of the hand gesture recognition system for opening applications in Windows, and the second proposed system, which may also act as the control system of PowerPoint presentations, can be made easy and convenient using these gestures-based control. Thus, the proposed model improves the interaction between the human and the computer.

Keywords Gesture-based control · Media-Pipe · Human–computer interaction · Artificial intelligence · Hand-sign

L. H. Palivela (✉) · V. Premanand · A. Begum
School of Computer Science Engineering, Vellore Institute of Technology, Chennai, India
e-mail: lakshmi.harika@vit.ac.in

V. Premanand
e-mail: premanand.v@vit.ac.in

A. Begum
e-mail: afruza.begum@vit.ac.in

S. Shakya et al. (eds.), *Fourth International Conference on Image Processing and Capsule Networks*, Lecture Notes in Networks and Systems 798,
https://doi.org/10.1007/978-981-99-7093-3_43

1 Introduction

The hand gesture recognition system for opening applications in Windows is an innovative technology that has been gaining popularity in recent years. It is a type of computer vision technology that enables users to interact with their computers using hand gestures. This technology allows users to open computer applications by making hand gestures in front of their computer's camera. This system is based on gesture recognition, which involves identifying and interpreting the human body's movements.

The hand gesture recognition system uses machine learning algorithms to analyze the images captured by the camera, recognize the hand gestures made by the user, and trigger specific actions in response to those gestures. The system can detect and recognize different hand gestures, such as thumbs up, thumbs down, okay sign, and others, and map those gestures to specific actions, such as opening an application or executing a command. The system has many potential applications, including human–computer interaction, gaming, and virtual reality. The gesture-based control can improve the user experience for everyone by simplifying interactions and reducing the cognitive load required to use technology. In addition, it can be used in situations where users need to interact with their computers without touching them, such as in environments where hygiene is paramount. Using hand and body gestures, users can manipulate virtual objects and navigate digital spaces more naturally and intuitively.

2 Literature Review

Ranjan et al. propose a method for recognizing hand gestures using convolutional neural networks (CNNs). The authors developed a dataset of 2600 grayscale images of hand gestures captured from five individuals for training and testing the CNN. They used the LeNet-5 architecture and achieved a recognition rate of 99.6% on their dataset. They believe their approach can be used for gesture-based human–machine interactions, particularly in areas such as sign language recognition [1]. In Sahoo et al. [2], a real-time hand gesture recognition system is developed based on a dataset of 1600 images of ten hand gestures captured from 20 individuals for training and testing their CNN. They used a two-stage CNN architecture and achieved a recognition rate of 99.88% on their dataset. The authors believe their approach can be used in human–robot interaction, sign language recognition, and virtual reality applications.

Zhan et al. propose a hand gesture recognition system using CNNs. The authors developed a dataset of 3000 images of five hand gestures, captured from 10 individuals, for training and testing their CNN. They used a four-layer CNN architecture and achieved a recognition rate of 98.7% on their dataset. Their approach can be used in areas such as human–robot interaction, virtual reality, and gaming [3]. In a real-time hand gesture recognition system is done using CNNs and data augmentation techniques. The authors developed a dataset of 1500 images of five hand gestures, captured from 10 individuals, for training and testing their CNN. They used a four-layer CNN architecture and achieved a recognition rate of 95% on their dataset. The authors believe their approach can be used in gaming, virtual reality, and robotics applications [4].

Agaian et al. propose a hand gesture recognition system using CNNs. The authors developed a dataset of 4800 images of 12 hand gestures, captured from 20 individuals, for training and testing their CNN. They used a five-layer CNN architecture and achieved a recognition rate of 95.6% on their dataset. Their approach can be used in applications such as sign language recognition and human–computer interaction [5]. Hand gesture recognition using Media-Pipe has gained significant attention due to its ease of use and accuracy. Media-Pipe is an open-source, cross-platform library developed by Google, which provides a customizable pipeline for processing and detecting hand landmarks in real-time. One of the significant advantages of using Media-Pipe is its ability to detect hand landmarks even in different lighting and background conditions. In this work proposed a real-time hand gesture recognition system using Media-Pipe, which was based on a convolutional neural network (CNN) and achieved recognition accuracy of 98.9%. The proposed system used a dataset of 10,000 images for training and testing, and the results showed the effectiveness of the proposed system in real-time applications [6, 7].

Wameed et al. [8] used Media-Pipe to detect and recognize hand gestures for controlling a robot. The authors developed a system that allowed users to control a robot using hand gestures detected by Media-Pipe. The proposed system used an artificial neural network (ANN) for gesture recognition and achieved an accuracy of 97%. The results showed the feasibility of using hand gestures for robot control, and the system could recognize gestures in real-time. In the work [9], the authors proposed a sign language recognition system using a combination of Media-Pipe and CNN. The system used Media-Pipe for hand landmark detection and extracted hand features to train a CNN for gesture recognition. The proposed system was tested on a dataset of 10,000 images and achieved an accuracy of 95.7%, outperforming other state-of-the-art methods. The results showed the effectiveness of using Media-Pipe for hand landmark detection and the ability of CNN to recognize sign language gestures accurately.

Zhu et al. proposed a system for detecting hand gestures using a single camera and Media-Pipe. The authors used Media-Pipe to see hand landmarks and extract hand features to train a CNN for gesture recognition. The proposed system was tested on a dataset of 10,000 images and achieved an accuracy of 98.7%. The results showed the feasibility of using a single camera and Media-Pipe for real-time hand gesture recognition applications [10]. The authors proposed a real-time hand gesture recognition system using Media-Pipe and CNN. The system was designed to recognize ten hand gestures and was trained on a dataset of 8000 images. The proposed system achieved an accuracy of 96.8% in recognizing hand gestures in real-time. The results showed the effectiveness of using Media-Pipe for hand landmark detection and the potential of CNN for real-time hand gesture recognition applications [11].

Haria et al. propose a hand gesture recognition system for controlling application interfaces. The authors developed a dataset of 50 hand gesture samples for five commands and used a Hidden Markov Model (HMM) for gesture recognition. The system achieved an accuracy of 90.4% in recognizing hand gestures [12]. The authors believe their approach can be used in smart homes and interactive television applications. Khan et al. propose a hand gesture recognition system for opening applications in a Windows environment. The authors developed a dataset of 500 images of ten hand gestures captured from five individuals for training and testing their system [13]. They used a combination of region-based and template-matching techniques for gesture recognition and achieved an accuracy of 94.5% on their dataset. The authors believe their approach can be used in virtual reality and gaming applications.

Dehankar et al. [14] propose a gesture-based interface for controlling applications on mobile devices. The authors developed a dataset of 50 hand gesture samples for five different commands and used a template-matching technique for gesture recognition. The system achieved an accuracy of 80% in recognizing hand gestures. The authors believe their approach can be used in mobile gaming and virtual reality applications. Saraswati et al. [15] propose a gesture-based system for opening applications on a desktop computer. The authors developed a dataset of 250 images of ten hand gestures captured from five individuals for training and testing their system. They used a Haar classifier for gesture recognition and achieved an accuracy of 95% on their dataset. The authors believe their approach can be used in smart homes and interactive television applications.

In [16] propose a hand gesture recognition system for opening applications in a virtual reality environment. The authors developed a dataset of 500 images of ten hand gestures captured from five individuals for training and testing their system. They used a combination of region-based and template-matching techniques for gesture recognition and achieved an accuracy of 95% on their dataset by using SVM. The authors believe their approach can be used in applications such as virtual reality and gaming propose a hand gesture recognition system using a convolutional neural network (CNN) to open computer applications [17]. The authors developed a dataset of 15,000 images of five hand gestures, captured from 20 individuals, for training and testing their CNN. They used a three-layer CNN architecture and achieved a recognition accuracy of 99.2% on their dataset. Their approach can be used for natural and intuitive human–computer interaction.

In Adithya [18], a hand gesture recognition system is developed using a deep convolutional neural network (DCNN) to control computer applications. The authors developed a dataset of 1500 images of five hand gestures captured from three individuals for training and testing their DCNN. They used a five-layer DCNN architecture and achieved recognition accuracy of 96.7% on their dataset. The authors believe their approach can be used to improve user experience in human–computer interaction. Niranjani et al. [19] provides a hand gesture recognition system using a deep convolutional neural network (DCNN) for controlling applications on a computer. The authors developed a dataset of 3,000 images of five hand gestures captured from five individuals for training and testing their DCNN. They used a six-layer DCNN architecture and achieved recognition accuracy of 97.7% on their dataset.

The authors believe their approach can be used for developing natural and intuitive human–computer interaction.

Dardas et al. [20] propose a real-time hand gesture recognition system for controlling applications on a computer. The authors used a combination of hand detection and hand gesture recognition techniques, including hand tracking, skin color segmentation, feature extraction, and support vector machine (SVM) classification. They achieved recognition accuracy of 92.2% on their dataset of 300 images of five hand gestures captured from three individuals. Their approach can be used to improve the efficiency and convenience of human–computer interaction [21].

In Pathak et al. [22], authors recommended the use of gesture devices that are integrated with sensors. Authors have used the sensor to record the movements of the hand in a few limited directions that could hold control over the slides. The gesture device and slides communicate with RF transmission and reception. Accelerometer and flex sensors were used for gesture recognition. They detect the flexing and movements of the hands and detect the movements by axis tracking the hands. The accelerometer traces slide switches. Many actions are happening; hence different threshold levels are set for every action. Sensors send analog signals as their output. The ADC of the microcontroller converts the received signals to digital signals. Microcontrollers use the signals generated and processes to send them to RF transmitter. The transmission of signals takes place from RF transmitter to RF receiver that gives it to the microcontroller. As the basic visual application runs, the microcontroller gives data to it. The operations are controlled using a basic visual application based on the data received.

All the above-mentioned models have gone through a tough development process of using gesture datasets and training them to recognize the hand and then identify the gesture and perform the required action. Thus, the proposed model includes various other features of the slide show to be handled using the gesture, such as saving the slide with annotations, changing the color of the pointers, undoing the last annotations, along with handling the various Windows applications installed in the system with the gesture we can able to open the application without touching the screen. The proposed system also uses lightweight models to recognize and identify the gestures performed in front of the camera using Media-Pipe and its models, such HandTrackingModel.

3 Proposed System

The proposed system involves several steps to recognize hand gestures and launch the corresponding applications. Initially, the necessary libraries and modules are imported, including OpenCV, NumPy, AppOpener, TensorFlow, Keras, and Media-Pipe [23]. Google Media-Pipe is an open-source library that provides a number of pre-trained models for various computer vision tasks, including hand tracking. Media-Pipe employs a deep learning-based model capable of detecting and localizing numerous places on the hand, such as finger tips, palm center, and others, to detect hand keypoints. For each observed hand, it produces a total of 21 important points. The Media-Pipe is initialized by creating a Hands instance, which enables the system to process and detect hand landmarks. The architecture of the proposed model is given in Fig. 1.

Fig. 1 Model architecture

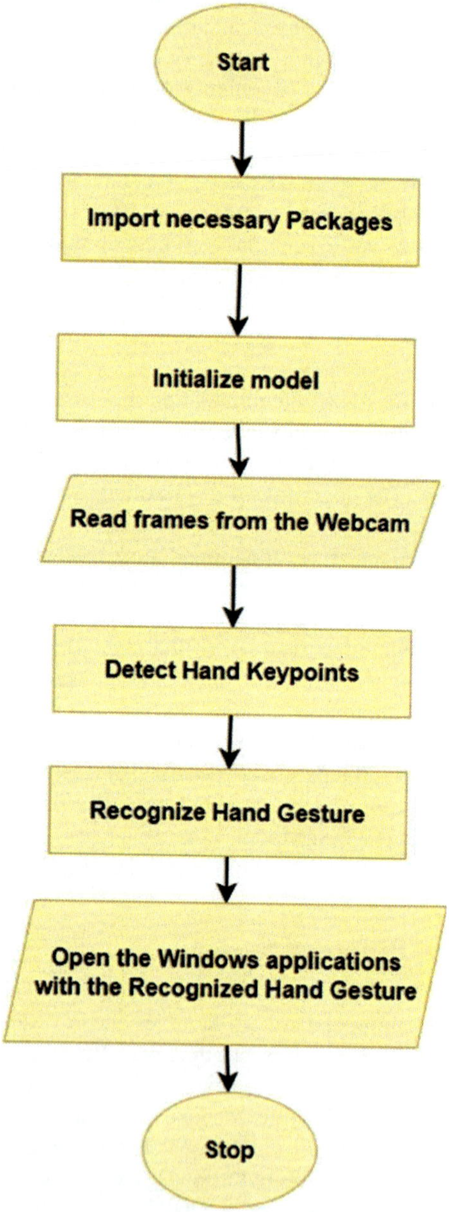

The pre-trained gesture recognition model is loaded using the Keras load_model function, which allows the system to recognize various hand gestures. Using the hand's keypoints identified by the Media-Pipe hand tracking model, one can infer the gestures produced by the hand while recognizing hand movements. Additionally, class names for the gestures and corresponding application names are loaded, enabling the system to map the predicted gesture class to the corresponding application name.

The webcam is initialized to capture frames at rate of 30 frames per second, and an infinite loop is initiated to process each frame using Media-Pipe to detect hand landmarks. If a hand is detected, the landmarks are used as input for the pre-trained model to recognize the gesture. The predicted gesture class is then mapped to the corresponding application name, and the AppOpener module is used to launch the application.

Furthermore, OpenCV displays the predicted gesture and corresponding application name on the video frame. The video frame is displayed in an OpenCV window until the user presses the 'q' key to quit the program. Once the program is ended, the webcam is released, and all windows are destroyed to ensure the proper functioning of the system. Overall, the proposed system enables efficient recognition of hand gestures and the launching of the corresponding applications, making it a valuable tool for various applications.

The script first imports the necessary libraries and modules. OpenCV is used for image processing and displaying video frames, NumPy is used for numerical calculations, AppOpener is used for opening applications, TensorFlow is used for loading and running the pre-trained gesture recognition model, Keras is used for loading the pre-trained model, and Media-Pipe is used for detecting hand landmarks. Then we initialize Media-Pipe by creating a Hands instance. The max_num_hands parameter is set to 1 to detect only one hand at a time, and the min_detection_confidence parameter is set to 0.7 to ensure that only confident hand detections are used.

Next, the pre-trained gesture recognition model is loaded using the Keras load_model function. The model was previously trained using TensorFlow and Media-Pipe and saved to a file named 'mp_hand_gesture'. The class names for the gestures and the corresponding application names are then loaded. The class names are stored in a list named class names, while the application names are stored in a list named apps. The mapping between the class names and application names is done by index, so the nth class name corresponds to the nth application name. The class names are the names of the gestures that the model is trained to recognize. Here, the class names are defined in a list called 'class Names'. The names of the gestures are 'okay', 'peace', 'thumbs up', 'thumbs down', 'call me', 'stop', 'rock', 'live long', 'fist', and 'smile'. These names are defined in the same order as the output classes of the trained model.

We then initialize the webcam using the OpenCV Video Capture function with an argument of 0, which indicates the default camera. The script then starts an infinite loop to capture frames from the webcam. The script first flips the frame vertically for each frame using the OpenCV flip function. This is done to ensure that the video feed is not mirrored. Then it converts the frame from the BGR color space to the

RGB color space using the OpenCV color function. This is done because Media-Pipe expects RGB input. Then it processes the frame using Media-Pipe to detect hand landmarks. If a hand is detected, the landmarks are extracted and used as input to the pre-trained gesture recognition model for prediction.

The predicted gesture class is then mapped to the corresponding application name using the class names and app lists. If the predicted gesture is 'okay', no application is launched. If the predicted gesture is 'peace', the Microsoft Word application is launched. If the predicted gesture is 'thumbs up', the Microsoft Excel application is launched. If the predicted gesture is 'thumbs down', the PowerPoint application is launched. If the predicted gesture is 'call me', the Discord application is launched. If the predicted gesture is 'stop', the Microsoft Teams application is launched. If the predicted gesture is 'rock', the Windows Settings application is launched.

This proposed system also gives an effective hand-free PowerPoint interaction. The system uses the pre-trained models to identify the gestures performed in front of the camera, and the respective action are assigned when the gesture is performed. Initially, the PowerPoint is converted into the image folder where each slide of the PowerPoint (PPT) is converted into the image format as either JPG or JPEG or PNG. The MS PPT as the provision to directly convert a PPT into an image folder, which can be further used for our gesture-based control, is shown in Fig. 2.

A User Interface is also developed with the help of the Tkinter framework in Python. The UI takes the path of the image folder, and it class the procedure to handle the file with the gestures recognized from the camera-generated frame by OpenCV. The list of actions with their respective gestures is listed below in Table 1.

The proposed algorithm basically follows the flow represented in the Fig. 1. The algorithm for gesture-based control of PowerPoint presentation is below.

Algorithm 1: Gesture-based control for PowerPoint presentation

procedure Gesture (Source address)

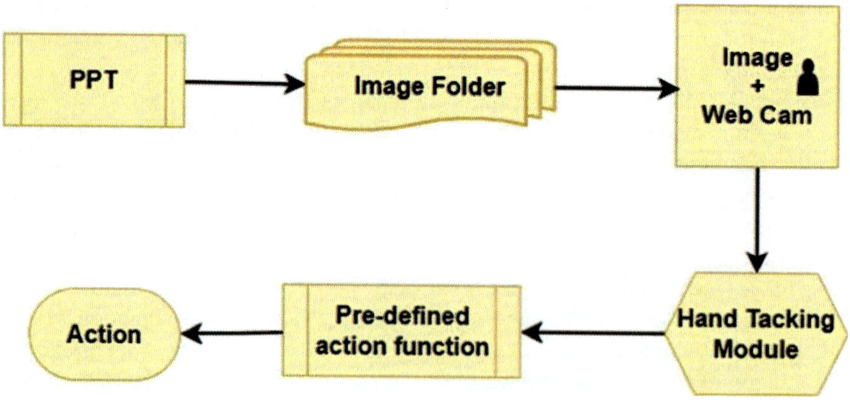

Fig. 2 PowerPoint presentation

Table 1 List of gestures and their actions for PowerPoint presentation

S. No	Gesture	Fingers involved	Action
1		Thumb pointing to left	Move to previous slide
2		Little finger up	Move to next slide
3		Index and middle finger up	Shows up the pointer on slide
4		Index finger up and movement	Draws in the slide
5		Index and little finger up	Changes color of the stokes and points
6		All fingers up	Saves the current slide with the changes
7		Index, middle, ring, and little fingers up	Undo operation

Read the Images from Source

images ← images [Source]

camera ← System [Camera]

handTracker ← HandTrackingModel

frame ← frames[camera]

```
images ← images + frame
if(handTracker(frame)) then
 temp_image ← images
fingers ← handTrackerr(frame)
if (fingers [0]) then
        images ← prev_image
else if (fingers [4]) then
images ← next_image
else if (fingers [1] and fingers [2]) then
        images ← images + pointer
else if (fingers [1]) then
        images ← images + draw the location of fingers [1]
        else if (fingers [0] and fingers [1] and fingers [4]) then
            images ← images + color_change(pointer)
else if (fingers [0] and fingers [1] and fingers [2] and fingers [3] and fingers [4])
then
        Source[images] ← images
else if (fingers [1] and fingers [2] and fingers [3] and fingers [4]) then
        images ← temp_image
show(images)
endif
end
end procedure
```

The proposed system offers several benefits that make it a useful tool for various applications.

Firstly, the system provides accessibility for individuals with physical disabilities or limited mobility. They can use hand gestures to open applications without a mouse or keyboard, which can be a significant advantage for those who find it challenging to use traditional input devices. Secondly, the system offers speed and convenience to users. A hand gesture recognition system can be faster than using a mouse or keyboard to navigate a computer interface. Users can quickly open applications with a simple hand gesture instead of clicking through menus or typing in search terms. It is also more convenient than using a mouse or keyboard when users cannot access or use them, such as when their hands are occupied or standing at a distance from the computer. Thirdly, the system provides a contactless way of interacting with a computer, particularly relevant in the current COVID-19 pandemic. By minimizing physical contact with computer peripherals such as a mouse and keyboard, the system reduces virus transmission. Lastly, the proposed system offers novelty to users. Hand gesture recognition systems are relatively new and emerging technologies that provide a unique and exciting way to interact with a computer. They can be used to engage users and add novelty to the computer interface, making it an appealing option for various applications. The proposed system provides several benefits, including accessibility, speed, convenience, hygiene, and novelty. These advantages make it a valuable tool for individuals with disabilities, as well as for various applications, including gaming, healthcare, and education. The challenges may include occlusion, noise, hand posture fluctuations can all degrade the performance of gesture detection systems. To enhance the performance the position of the hand should be with the web cam Viewpoint. Modify the min_detection_confidence parameter in the Media-Pipe hand model initialization to provide a higher threshold for hand detection. Transfer learning may be chosen for large datasets.

4 Results

The experiment is conducted in Windows 11 in a Python environment with 12th Gen Intel(R) Core (TM) i5-1235U, 1.30 GHz with 16 GB Ram Gamming X-Box Card installed, which is essential to speed up the process. The proposed model results for hand-free windows access are shown in Figs. 3, 4, 5, 6, 7 and 8. The accuracy is determined by how well the model detects and tracks hand keypoints. Our model detects all the windows applications at different scores and the execution time for each opening application is show in Table 2. The execution performance for opening various applications shown in Fig. 9.

5 Conclusion

Hand gesture recognition systems for opening applications in Windows can be an effective and efficient way of interacting with computers, especially for people with physical disabilities or those who want to work hands-free. In this work, we used TensorFlow and Media-Pipe to train a gesture recognition model to recognize hand gestures and associate them with opening-specific applications. The resulting system performed well in detecting hand gestures, opening the associated applications in Windows, and performing hand-free PowerPoint presentations. The proposed model can be further developed into a customizable gesture-based application where the users can customize the gesture and map with the action. Instead of converting the slides into images, an extension of this system can be added to the PPT to do the

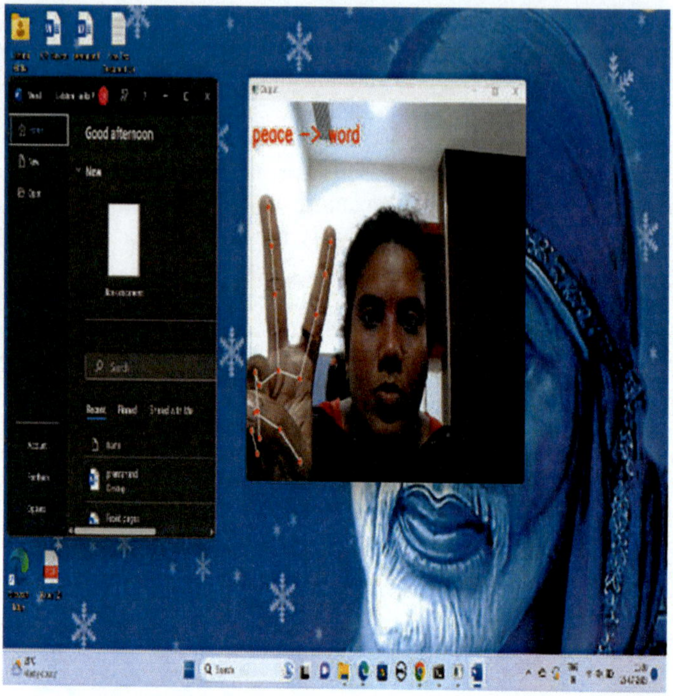

Fig. 3 Peace—Word Document

same operation within the application without depending on a third-party application which gives seamless integration and efficient workflow. This model detects all the applications very accurately achieves good detection rate. With this further development, such systems could become even more accurate and user-friendly, making computer interaction even easier and more accessible to everyone.

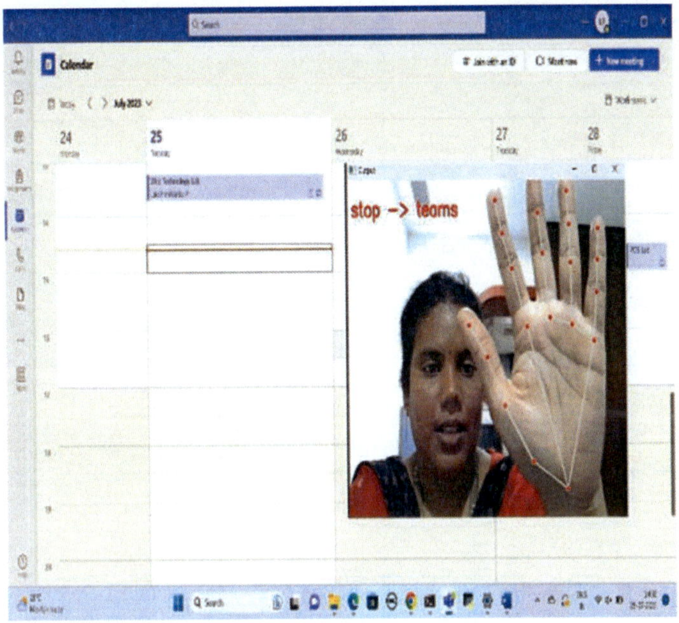

Fig. 4 Stop—Microsoft Teams

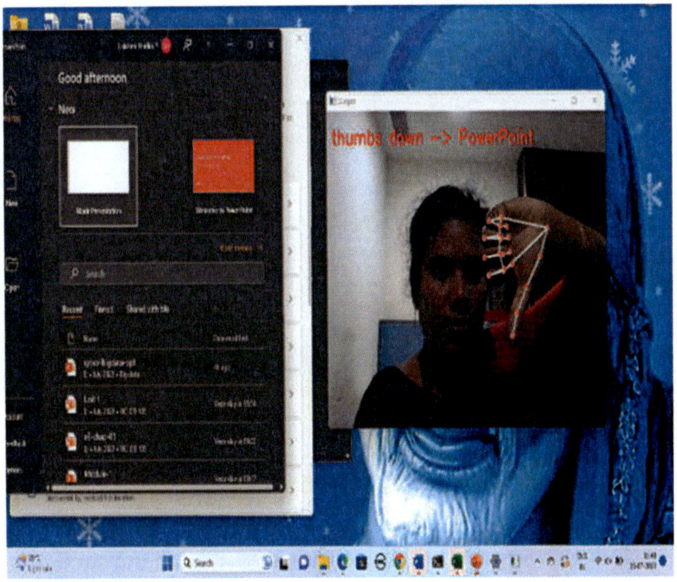

Fig. 5 Thumbs down—PowerPoint

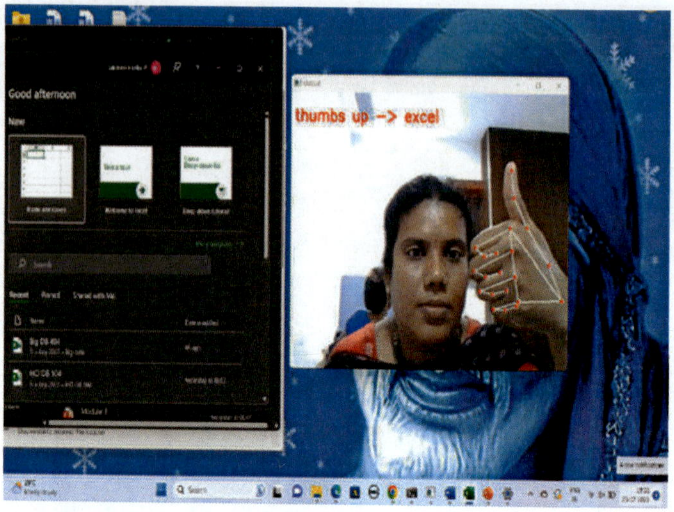

Fig. 6 Thumbs up—Excel

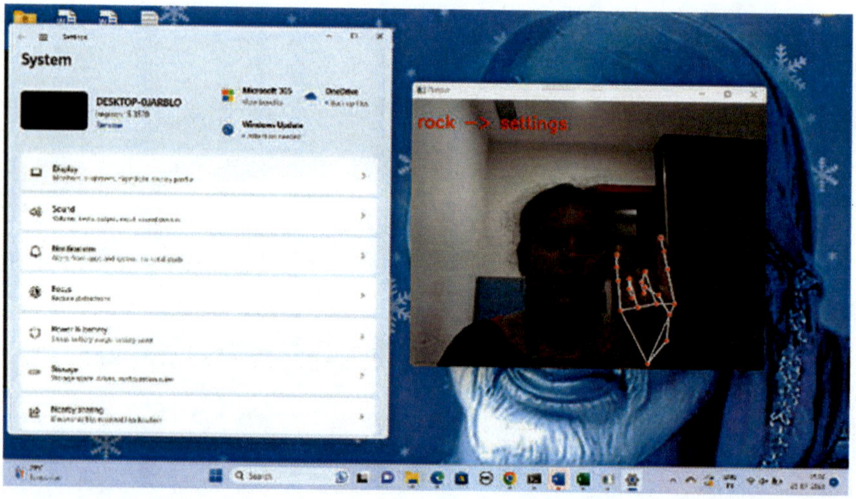

Fig. 7 Rock-settings

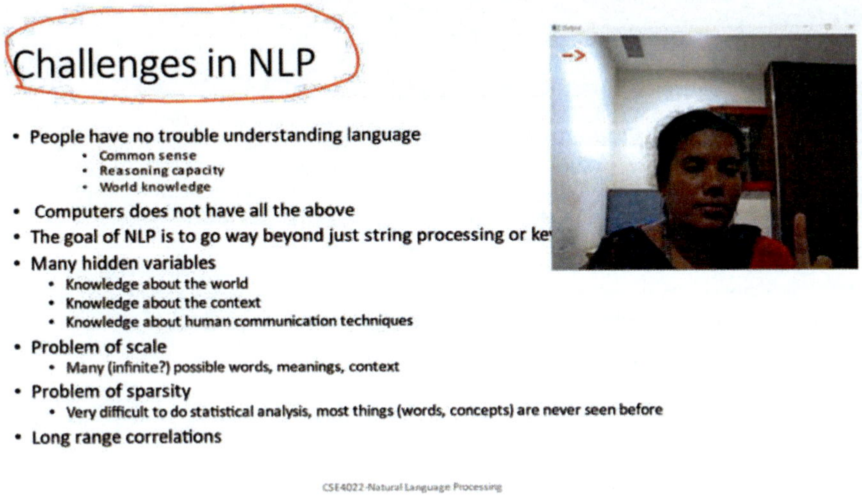

Fig. 8 Hand free presentation

Table 2 Execution time

Action	Time (ms)
Open Word	61 ms
Opening Settings	32 ms
Opening Microsoft Teams	75 ms
Opening WhatsApp	120 ms
Opening PowerPoint	65 ms
Opening Spotify	135 ms
Open Excel	95 ms
PPT presentation depends on number of slides	5 min

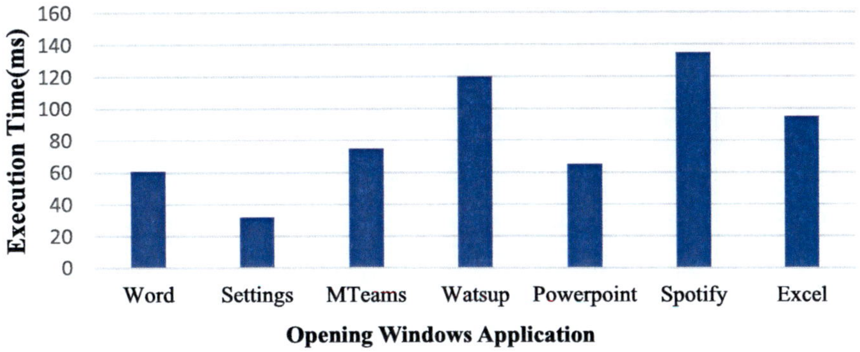

Fig. 9 Windows application performance

References

1. Ranjan C, Mitra P (2016) Hand gesture recognition using convolution neural networks. Int J Sci Res Publ 6(9):380–385
2. Sahoo JP, Prakash AJ, Pławiak P, Samantray S (2022) Real-time hand gesture recognition using fine-tuned convolutional neural network. Sensors 3:706. https://doi.org/10.3390/s22030706
3. Zhan F (2019) Hand gesture recognition with convolution neural networks. In: IEEE 20th International Conference on Information Reuse and Integration for Data Science (IRI), Los Angeles, CA, USA, pp 295–298. https://doi.org/10.1109/IRI.2019.00054
4. Islam MZ, Hossain MS, ul Islam R, Andersson K (2019) Static hand gesture recognition using convolutional neural network with data augmentation. In: Joint 8th International Conference on Informatics, Electronics; Vision (ICIEV) and 3rd International Conference on Imaging, Vision Pattern Recognition (icIVPR), Spokane, WA, USA, pp 324–329. https://doi.org/10.1109/ICIEV.2019.8858563
5. Agaian S, Pinto S, Raimundo F et al (2019) Static hand gesture recognition based on convolutional neural networks. J Electric Comput Eng 2019:1–12. https://doi.org/10.1155/2019/4167890
6. Amit ML, Fajardo AC, Medina RP (2022) Recognition of real-time hand gestures using mediapipe holistic model and LSTM with MLP architecture. In: IEEE 10th Conference on Systems, Process and Control (ICSPC), Malacca, Malaysia, pp 292–295. https://doi.org/10.1109/ICSPC55597.2022.10001800
7. Sumalatha MR, Lakshmi Harika P (2019) An intelligent system to detect human suspicious activity using deep neural networks. J Intell Fuzzy Syst IOS Press 36(5):4507–4518
8. Wameed M, Alkamachi AM (2023) Hand gestures robotic control based on computer vision. Int J Intell Syst Appl Eng 11(2):1013
9. Marais M, Brown D, Connan J, Boby A (2022) An evaluation of hand-based algorithms for sign language recognition. In: International Conference on Artificial Intelligence, Big Data, Computing and Data Communication Systems, Durban, South Africa, pp 1–6. https://doi.org/10.1109/icABCD54961.2022.9856310
10. Zhu H, Deng C, Zhu Y (2023) MediaPipe based gesture recognition system for English letters. In: Proceedings of the 2022 11th International Conference on Networks, Communication and Computing, Association for Computing Machinery, New York, NY, USA, pp 24–30. https://doi.org/10.1145/3579895.3579900
11. Islam S, Titli SR, Kabir KA, Hossain MAA, Hossain MA (2022) Improving real-time hand gesture recognition system for translation: sensor development. In: 17th Annual System of Systems Engineering Conference (SOSE), Rochester, NY, USA, pp 254–259. https://doi.org/10.1109/SOSE55472.2022.9812664
12. Haria A, Subramanian A et al (2017) Hand gesture recognition for human computer interaction. Proc Comput Sci 115:367–374. https://doi.org/10.1016/j.procs.2017.09.092
13. Khan RZ, Ibraheem N (2012) Hand gesture recognition: a literature review. Int J Artif Intell Appl (IJAIA) 3:161–174
14. Dehankar A, Sanjeev J, Thakare VM (2019) Hand gesture-based user interface for controlling mobile applications. In: Proceedings of the Fourth ICMEET. https://doi.org/10.1007/978-981-13-1906-8_59
15. Saraswati VI, Sigit R, Harsono T (2016) Eye gaze system to operate virtual keyboard. In: 2016 International Electronics Symposium (IES), pp 175–179. IEEE
16. Qi Z, Zhu W, Zhu Q (2019) Real world hand gesture interaction in virtual reality. J Phys Conf Series 1229:012027. https://doi.org/10.1088/1742-6596/1229/1/012027
17. Sagayam KM, Hemanth DJ (2017) Hand posture and gesture recognition techniques for virtual reality applications: a survey. Virtual Reality 21:91–107. https://doi.org/10.1007/s10055-016-0301-0
18. Adithya V, Rajesh R (2020) A deep convolutional neural network approach for static hand gesture recognition. Proc Comput Sci 171:2353–2361. https://doi.org/10.1016/j.procs.2020.04.255

19. Niranjani V, Keerthana R, Mohana Priya B, Nekalya K, Padmanabhan AK (2021) System application control based on Hand gesture using deep learning. In: 7th International Conference on Advanced Computing and Communication Systems (ICACCS), Coimbatore, India, pp 1644–1649. https://doi.org/10.1109/ICACCS51430.2021.9441732
20. Dardas NH, Georganas ND (2011) Real-time hand gesture detection and recognition using bag-of-features and support vector machine techniques. IEEE Trans Instrum Measurem 60(11):3592–3607. https://doi.org/10.1109/TIM.2011.2161140
21. Palivela LH, Swetha S, Nithish Guhan M, Prasanna Venkatesh M (2023) Dense video captioning using video-audio features and topic modeling based on caption. In: Bindhu V, Tavares JMRS, Vuppalapati C (eds) Proceedings of Fourth International Conference on Communication, Computing and Electronics Systems. Lecture Notes in Electrical Engineering, vol 977. Springer, Singapore
22. Pathak V, Mongia S, Chitranshi G (2015) A framework for hand gesture recognition based on fusion of flex, contact and accelerometer sensor. In: Third International Conference on Image Information Processing (ICIIP), Waknaghat, India, pp 312–319. https://doi.org/10.1109/ICIIP.2015.7414787
23. Hand landmarks detection guide. https://developers.google.com/mediapipe/solutions/vision/hand_landmarker

FlameGuard: A Smart System for Forest Fire Detection and Control

Medha Wyawahare, Shreya Daga, Chashu Agrawal, Pranav Dalve, and Chirag Bhandari

Abstract Forest fires pose a huge environmental risk, wreaking havoc on ecosystems and human life. Forest fires must be detected and extinguished as soon as possible to have the least amount of harm. This study presents a forest fire detection and extinguishment system based on a Raspberry Pi, machine learning techniques, and robotic implementation. The device gathers data from the forest using sensors and cameras mounted on a robot and processes it using machine learning algorithms to detect indicators of a forest fire. The robot also has an extinguishing mechanism that can be engaged remotely to put out the flames. This system has the potential to dramatically shorten response time in detecting and suppressing forest fires, and the utilization of robotics permits access to isolated regions that would otherwise be inaccessible to humans. This system's deployment has the potential to revolutionize detection of forest fires and extinguishment systems, making them more efficient, safer, and dependable.

Keywords Fire detection · Machine learning · Raspberry Pi · Forest fire detection · Convolutional neural network

M. Wyawahare (✉) · S. Daga · C. Agrawal · P. Dalve · C. Bhandari
Department of Electronics and Telecommunication Engineering, Vishwakarma Institute of Technology, Pune, India
e-mail: medha.wyawahare@vit.edu

S. Daga
e-mail: shreya.daga20@vit.edu

C. Agrawal
e-mail: chashu.agrawal20@vit.edu

P. Dalve
e-mail: pranav.dalve20@vit.edu

C. Bhandari
e-mail: chirag.bhandari20@vit.edu

S. Shakya et al. (eds.), *Fourth International Conference on Image Processing and Capsule Networks*, Lecture Notes in Networks and Systems 798,
https://doi.org/10.1007/978-981-99-7093-3_44

1 Introduction

Ecosystems, animals, and human groups are all at risk from forest fires' devastating harm. A forest fire's ability to spread and cause catastrophic damage depends on its early detection. The creation of forest fire detection and extinguishing alarm systems has been facilitated recently by technology improvements. To find and notify authorities and first responders of forest fires, these systems employ a variety of sensors, such as infrared and smoke detectors. These methods can aid in stopping forest fires from spreading out of control and resulting in significant harm. Forest fires are a terrible occurrence that can seriously harm human life, property, and wildlife habitats. To lessen the effects of forest fires, prompt discovery and suppression are essential. Recent technological developments have enabled the creation of intelligent forest fire detection [1] and extinguishing systems that employ robotics and machine learning algorithms for the early identification and control of forest fires. One such system processes data from sensors and cameras attached on a robot that moves around the forest using a Raspberry Pi, a tiny single-board computer.

A CNN-based machine learning model has been used to detect forest fire using the Forest Fires Data Set. By minimizing the time, it takes to identify and put out forest fires, this method has the potential to drastically lessen the damage that such fires do. The method is more effective at spotting and extinguishing flames in distant regions since the use of robots also enables access to places that may be challenging for people to reach. In general, machine learning, robots, and Raspberry Pi technology have the potential to revolutionize detection of forest fires and extinguishing systems, paving the way for a safer and more effective method of defending forests and wildlife habitats.

2 Literature Review

YOLO (including tiny-yolo-voc1, tiny-yolo voc, yolo voc.2.0, and yolov3), R-CNN, and SSD models possess faster processing as compared to the others. These findings were proposed in a paper where the faster R-CNN method was run for 120,000 iterations. The authors found that YOLO had poor performance in detecting smaller cooler fires, so they added two more layers (one convolutional and one max-pooling layer with a filter of 8 to tiny-yolo-voc. This resulted in an improved accuracy rate for smaller fire detection [2].

A deep learning network and UAV-FFD platform-based approach for identifying forest fires has been developed in [3]. The program employs a huge YOLOv3 network to process photographs from the UAV instantly. The YOLOv3 detection method demonstrates excellent precision and low time cost by utilizing.

Color and region growing algorithms are applied to extract precise flame images, and cusp detection is used to determine whether the images contain flames. This approach improves algorithm efficiency and recognition rate, meeting the real-time requirements of fire monitoring systems [4].

The technique used in [5] distinguishes between smoke and clouds; a false-color composite image was created using the smoke reflectance index, mid-infrared reflectance, and water index. While the clouds appear pink and white, the smoke appears crimson. In 2009, the approach was used on a MODIS scene of a forest fire in Russia and Mongolia. Using a training dataset, smoke pixels were identified, and the results were compared to results of a multi-threshold method.

A paper [6], identifies fire geometrical factors to measure forest fires in order to respond quickly and safely. The RGB color space is used to extract the flame zone, and by scanning each point on the image, the coordinates of each vertex are found. The geometrical properties are then used to find the terrain inclination angle, flame inclination angle, and flame area. These features can be used to calculate the direction of wind dispersion, topographical conditions, and decision-making for battling forest fires can be enhanced. The suggested system can maintain track of the detailed fire scene as well as information about the surrounding geographic environment.

Aerial monitoring can be used to detect forest fires. One such approach entails analyzing a vast library of footage of forest fires under various settings in order to extract the fire's chromatic and motion properties [7]. The approach also comprises extracting smoke using a suggested algorithm to handle the issue of thick smoke hiding the fire.

Forest fire prediction can be implemented using a machine learning model. A devised approach [8] consists of two steps: classification to determine whether a fire is present and regression to predict the size of the fire. The approach was validated and compared with other machine learning methods, showing better quality results.

A single system integrates a number of sensors, embedded processors, and an energy supply module. The Kalman filter is used to preprocess and filter sensor data, and fuzzy reasoning is used to determine the final fire signal. The system was modeled in MATLAB, and the findings demonstrate that data pretreatment removes aberrant data effectively, and the Kalman filter lowers data error, resulting in more precise fire probability estimates. A single system integrates a number of sensors, embedded processors, and an energy supply module. The Kalman filter is used to preprocess and filter sensor data, and fuzzy reasoning is used to determine the final fire signal. The system was modeled in MATLAB, and the findings demonstrate that data pretreatment removes aberrant data effectively, and the Kalman filter lowers data error, resulting in more precise fire probability estimates [9].

A research presents a system and approach for detecting forest fires in their incipient stages utilizing a network of wireless sensors. A machine learning regression approach is also developed to improve fire detection accuracy [10]. The primary power source is provided by rechargeable batteries combined with a secondary solar power supply, a solution may be easily implemented as a standalone system for extended periods of time.

The use of picture-taking visual camera to click pictures of woods in order to identify fires is another technique. Visual light camera sensors are initially installed on communication towers that are positioned in a position that faces a sizable portion of the forest in [11]. To offer a 360° picture of a whirling motor in a forest attached to allow the camera to revolve.

A project is developed using an algorithm and sensors to detect changes in sensory parameters, such as gas and fire events, and send updates to experts through an IoT landing. The system aims to reduce error perception and provide a sustainable ecosystem through the use of D2D association conventionality. The system can be used in both residential and industrial settings [12].

Development of wireless sensor nodes for detection and monitoring has been done for the Riau Province of Indonesia. The design and analysis of the WSNs use a mathematical approach to determine the area that needs to be covered. Air temperature, humidity, haze, and CO_2 sensors are highlighted as the basic parameters for detecting fire hot spots in land or forest areas. The proposed sensor nodes use ZigBee technology, which allows for low power consumption and long battery life [13]. A model achieved training accuracy of 92.27% and 89.57% test accuracy with a stochastic gradient descent optimizer and avoided the underfit/overfitting on the model with the help of the stochastic gradient descent (SGD) algorithm [14].

3 Proposed System

3.1 Dataset

The dataset used for this project is primarily taken from Kaggle, Forest Fire images [15]. This dataset is readily available and free to use. Several additional smaller datasets were combined and merged to form this dataset. The various proprietors and developers of the original smaller datasets deserve all the praise. Additionally, Google Images was used to download each and every image in the Test Data folder. Web scrapping was implemented so as to largen the dataset. The dataset contains images from various forests and plantations which have smoke and fire being seen in the images labeled with fire. The entire dataset is divided into the Test Data and Train Data folders, which can be seen in Figs. 1 and 2. Test Data includes a total of 50 pictures of fire and non-fire. The Train Data includes 5000 photographs total of both fire and non-fire. Each of these files is split into two sub-folders of both types of fire, each of which contains 25 photographs under Test Data and 2500 images under Train Data.

3.2 Hardware Used

1. Raspberry Pi: The Raspberry Pi is a small, inexpensive, and adaptable computer board that allows anyone to learn programming and make numerous projects. It supports a wide range of applications, from education to home automation

Fig. 1 Images labeled with fire

Fig. 2 Images labeled with non-fire

and robotics, and has GPIO pins for hardware connection. It operates on Linux-based operating systems. Raspberry Pi has been used for this system because of its ability to work with a machine learning model. The machine learning model has been deployed on the microprocessed and uses its camera module for fire detection.

2. Flame sensor: A flame sensor is an electronic device that detects the existence of smoke or fire in its surroundings. It intended to offer an early warning signal by activating alarms or alert systems, allowing for rapid evacuation and fire suppression procedures, so preventing property damage and ensuring safety.
3. Battery-operated DC motor: Also known as a brushed DC motor, it is a type of electric motor that transfers electrical power to the motor's revolving armature via a commutation system utilizing brushes. It combines simple and cost-effective rotating motion with high torque production, making it commonly employed in robotics and small-scale manufacturing.
4. Servo motor: A servo motor is a small and precise rotary actuator that uses feedback control to position its output shaft precisely. It is widely utilized in robotics and automation systems, because it allows for regulated and precise movements with high torque capabilities.
5. L293 motor driver: The L293 motor driver is an integrated circuit that can control two direct current motors or a single stepper motor. It has bidirectional control and can handle high current and voltage, making it useful for robotics and other motor control applications.

3.3 Machine Learning Pipeline

For the purpose of detecting forest fires, the dataset is preprocessed. Using the OpenCV library, the colored pictures are extracted from the dataset and changed to grayscale versions. Following that, an array function is used to store the obtained grayscale pictures. Each image has a caption that says, "Fire or Non-Fire" and has the robust photographs cropped and resized as needed. The labeled, resized picture is added to the list of training datasets along with a NumPy array representing it. In this project, libraries like NumPy, Sklearn, Keras, and TensorFlow were used. The input features were taken out of the pretrained data and used for the train–test split in the SKLearn package to validate them. The convolutional neural network model utilizing the Keras API is the machine learning model utilized in this forest fire detection model. A number of convolutional layers, including activation and max-pooling, make up the model architecture. When building a model from scratch using the train dataset and deploying it to the test dataset, the epoch and batch size are specified.

4 Methodology

The machine learning model is deployed using the Raspberry Pi. The model of machine learning recognizes fire as soon as the robot receives electricity. The Raspberry Pi is activated when fire is found, turning on the fire sensors. In order to navigate exactly toward the fire spot, there are three flame sensors in the robot. After arriving at the fire spot, the motor pump is turned on, and the fire is put out using the water pump. Hardware connections were constructed as shown in Fig. 3. The robot is powered using a 10,000 mAh power bank which provides power to the Raspberry Pi; hence, as a fixed current is being given to the microprocessor, there is no fluctuation of supplied current neither any measures are required to take care of the same.

In order to identify and put out forest fires without the need for human involvement, the proposed system combines the hardware and machine learning model in this manner. Using the preprocessed data, a convolutional neural network (CNN) model is built and trained. CNN is programmed to learn and extract significant features from input photographs that distinguish between instances of fire and non-fire. The model

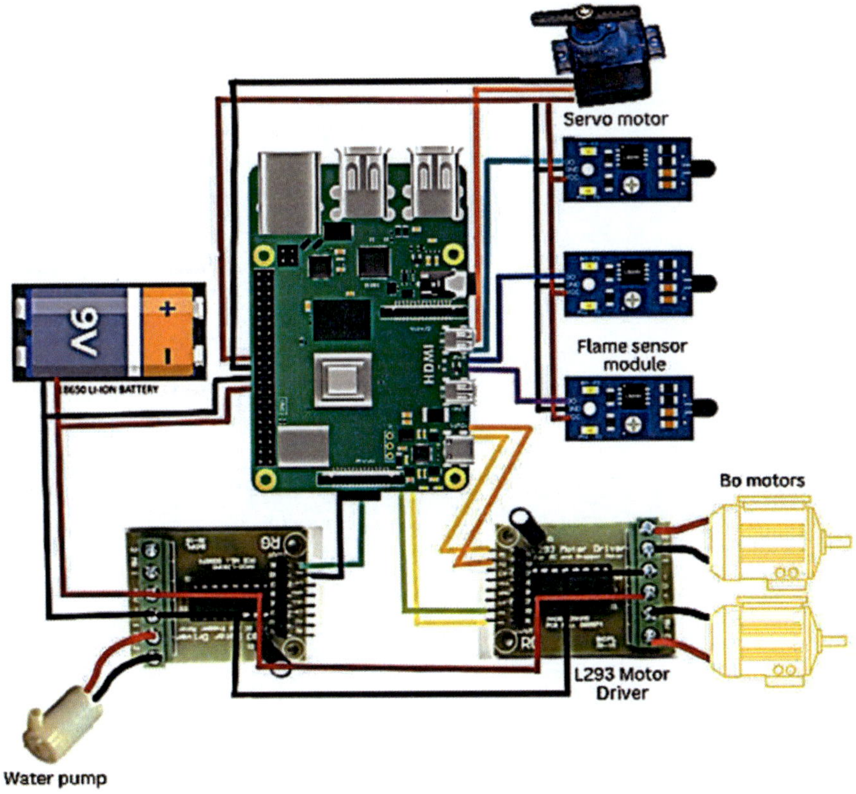

Fig. 3 Integrated circuit diagram

is trained with the help of positive, negative examples, and it is optimized to minimize classification error. Once trained, the model can be used to predict whether or not an image contains a fire. The ML model processes the input image from the flame sensors using the trained CNN. The model generates a likelihood or classification indicating the existence of fire. If the ML model detects a fire, the flame sensors direct the robot to the area of the fire. To precisely localize the fire, the robot is outfitted with flame sensors. The ML model continually evaluates the sensor data for identifying positions of the flames, and to adapt robot's movement accordingly. As the flame sensor works on the radiation sent by flame, there is no detection of smoke. The flame sensor converts the flame radiation into relevant photoelectric signal to get the desirable output. The workflow of the system can be seen in Fig. 4.

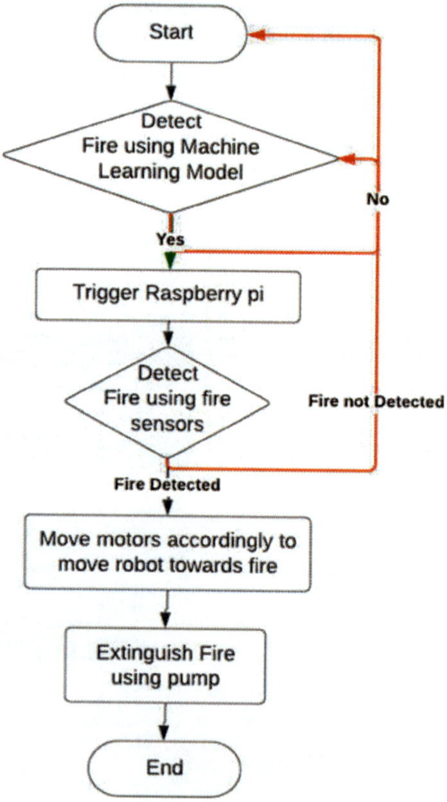

Fig. 4 Workflow of the system

5 Results and Discussions

The dataset was split into training and testing sets with a ratio of 70:30 (training:testing). The crop size used in the preprocessing of data is of (600, 1280) step size which means images are resized by a factor of 0.5 in both directions. With the help of the machine learning pipeline, the features from the images were extracted and reshaped into a 4D array of shape (number_of_samples, 300, 640, 1) where 300 and 640 represent the height and width of the image and 1 indicating conversion to grayscale images. Random forest and support vector machine models gave an accuracy of 83% and 88% each, but the convolutional neural network model outperformed the two. Additionally, cross-validation and hyperparameter tuning in all the models help achieve the stated accuracy. The final model is created using several convolutional and polling layers. Initially, an accuracy of about 86% was obtained using the convolutional and ReLU layer. Further, with the help of batch normalization process, overfitting of the model was prevented. The updated convolutional neural network model gave an accuracy of about 94%. Adam optimizer also helps to add to the accuracy of the machine learning model. The training loss and accuracy graph for an epoch size of 1000 can be seen in Fig. 5. The machine learning models training time varies on the hardware on which it is being trained, usually taking around 10–15 min. The learning rate of the model is same as that of Adam optimizer which is close to 0.001.

The Adam optimizer minimizes the loss function and improves the performance on the trained dataset. When fire is detected, the machine learning model sends a trigger signal to Raspberry Pi. The three flame sensors help achieve a precise fire location. There are a total of four motors in the proposed system, where two of each

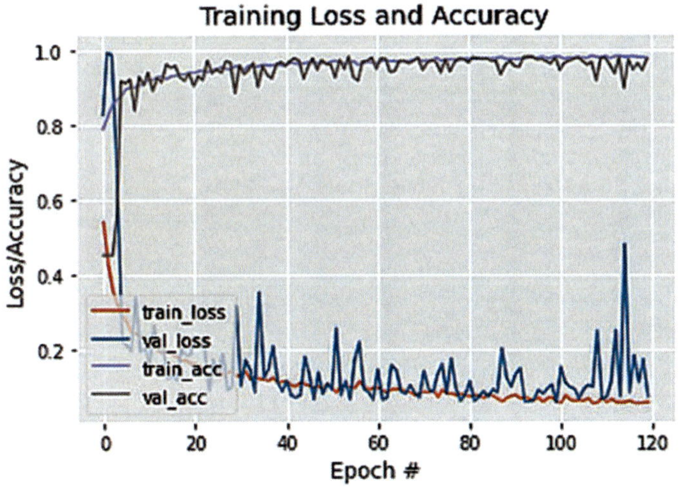

Fig. 5 Training loss and accuracy for model

Fig. 6 Fire extinguishing robot in stationary position

are made to move in one direction in order to reach the fire location. The robot's movement is controlled by changing the motor directions based on flame sensor readings. As soon as the robot reaches the fire location, the water pump is activated resulting in both left and right motors being simultaneously activated for a short duration of time, i.e., 500 ms to stop the robot's movement and pumps out water to put off the fire. The container on the cart consists of water along with the water pump as seen in Fig. 6.

The Raspberry Pi and all the mechanical pieces that made up the robot vehicle moved and put out fire in unison with the motors. The movement of the robot in the direction of the fire can also be seen through Figs. 7 and 8, where the robot moves closer to the detected fire location. This technology detects forest fires, navigates to the fire area, and performs automated or remotely controlled fire suppression measures by merging machine learning techniques with sensor integration and robotic control. It offers an effective and perhaps life-saving technique for dealing with forest fires.

6 Conclusion and Future Scope

Developing a machine learning model-based fire-fighting robot with a Raspberry Pi has a lot of potential for improving fire detection and response systems, in conclusion. The robot can quickly and correctly identify fire incidents by analyzing real-time sensor data, such as thermal imaging or video feeds, by utilizing machine learning techniques. Real-time processing and decision-making are made possible by the

Fig. 7 Fire extinguishing robot detecting fire

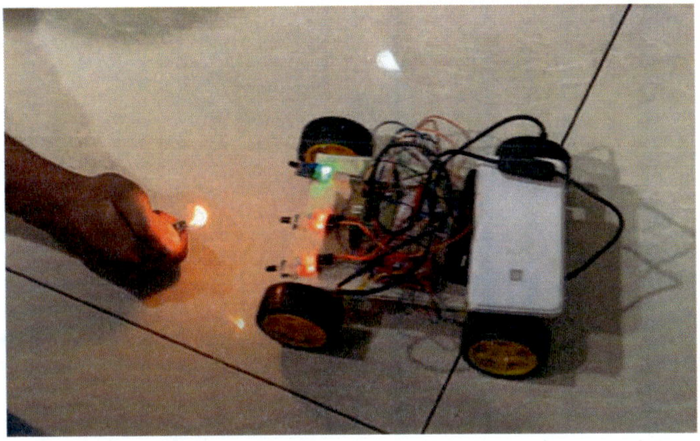

Fig. 8 Fire extinguishing robot moving towards the fire source

Raspberry Pi, a powerful and reasonably priced platform for putting the machine learning paradigm into practice. The model may learn patterns and traits linked to fires by being trained on a varied collection of photographs or videos connected to flames. The robot can aid in early fire detection and fire mitigation in regions that are prone to fire, such as woods or industrial complexes. The robot's sensors allow it to continually examine its surroundings and wirelessly communicate data to a centralized control system for processing. When a fire is detected, the robot may autonomously explore the scene, identify the fire's origin, and then take the necessary action, such as deploying fire suppression systems or notifying nearby workers.

Iteratively training the machine learning model with fresh data and fine-tuning the model's parameters can increase accuracy and performance over time. The system is consistently updated and maintained to ensure its dependability and efficiency in firefighting situations. Overall, integrating a machine learning model into a fire-fighting robot using Raspberry Pi has the potential to revolutionize fire detection and response capabilities, enabling more efficient and timely operations to limit the destructive impact of fires.

References

1. Madhwaraj KG, Asha V, Vignesh A, Akshay Shinde S (2023) Forest fire detection using machine learning. In: 2023 IEEE 12th International Conference on Communication Systems and Network Technologies (CSNT), Bhopal, India, pp 191–196. https://doi.org/10.1109/CSNT57126.2023.10134684
2. Wu S, Zhang L (2018) Using popular object detection methods for real time forest fire detection. In: 2018 11th International Symposium on Computational Intelligence and Design (ISCID), Hangzhou, China. https://doi.org/10.1109/ISCID.2018.00070
3. Jiao Z et al (2020) A YOLOv3-based learning strategy for real-time UAV-based forest fire detection. In: 2020 Chinese Control And Decision Conference (CCDC), Hefei, China, pp 4963–4967. https://doi.org/10.1109/CCDC49329.2020.9163816
4. Li J, Jiang X (2009) Forest fire detection based on video multi-feature fusion. In: 2009 2nd IEEE International Conference on Computer Science and Information Technology, Beijing, China, pp 19–22. https://doi.org/10.1109/ICCSIT.2009.5234862
5. Nagatani JK, Kawano K (2012) A new technique for visualization of forest fire smoke plumes using MODIS data. In: 2012 IEEE International Geoscience and Remote Sensing Symposium, Munich, Germany, pp 2380-2383. https://doi.org/10.1109/IGARSS.2012.6351016
6. Li G-H, Zhao J, Wang Z (2006) Research on forest fire detection based on wireless sensor network. In: 2006 6th World Congress on Intelligent Control and Automation, Dalian, pp 275-279. https://doi.org/10.1109/WCICA.2006.1712316
7. Dang-Ngoc H, Nguyen-Trung H (2019) Aerial forest fire surveillance—evaluation of forest fire detection model using aerial videos. In: 2019 International Conference on Advanced Technologies for Communications (ATC), Hanoi, Vietnam, pp 142–148. https://doi.org/10.1109/ATC.2019.8924547
8. Sineva S, Molovtsev MD (2020) An integrated approach to the regression problem in forest fires detection. In: 2020 Systems of Signals Generating and Processing in the Field of on Board Communications, Moscow, Russia, pp 1–4. https://doi.org/10.1109/IEEECONF48371.2020.9078645
9. Wang T, Hu J, Ma T, Song J (2020) Forest fire detection system based on Fuzzy Kalman filter. In: 2020 International Conference on Urban Engineering and Management Science (ICUEMS), Zhuhai, China, pp 630–633. https://doi.org/10.1109/ICUEMS50872.2020.00138
10. Yan F, Xu X, Han N (2010) A kind of identification method of geometric parameters for forest fire. In: 2010 2nd International Conference on Signal Processing Systems, Dalian, China, pp V1-91–V1-94. https://doi.org/10.1109/ICSPS.2010.5555608
11. Owayjan M, Freiha G, Achkar R, Abdo E, Mallah S (2014) Firoxio: Forest fire detection and alerting system. In: MELECON 2014–2014 17th IEEE Mediterranean Electrotechnical Conference, Beirut, Lebanon, pp 177–181. https://doi.org/10.1109/MELCON.2014.6820527
12. Divya TK, Dharshini P (2019) IoT enabled forest fire detection and early warning system. In: 2019 IEEE International Conference on System, Computation, Automation and Networking (ICSCAN), Pondicherry, India, pp 1–5. https://doi.org/10.1109/ICSCAN.2019.8878808
13. Shixing L, Yongming Z, Weiguo S, Xia X (2010) An enhanced algorithm for forest fire detection based on MODIS data. In: 2010 International Conference on Optoelectronics and Image Processing, Haikou, China, pp 200–203. https://doi.org/10.1109/ICOIP.2010.332
14. Rahul M, Shiva Saketh K, Sanjeet A, Srinivas Naik N (2020) Early detection of forest fire using deep learning. In: 2020 IEEE region 10 conference (TENCON), Osaka, Japan, pp 1136–1140. https://doi.org/10.1109/TENCON50793.2020.9293722
15. Forest Fire Dataset. https://www.kaggle.com/datasets/mohnishsaiprasad/forest-fire-images

Classification and Analysis of Chilli Plant Disease Detection Using Convolution Neural Networks

Zameer Gulzar, Sai Chandu, and K. Ravi

Abstract A crucial but time-consuming duty in agricultural practices is spotting disease in any crop. Chillies are a vital income crop in Telangana whose cultivation is essential for preserving food security. However, a number of diseases and pests can cause significant productivity losses while growing chillies. Early detection of these diseases is crucial for efficient management and the reduction of yield losses. The goal of this study is to analyze best convolution neural network model that can recognize and categorize chilli plant illnesses with accuracy. In this chapter, the classification of the four diseases (leaf curl, leaf spot, white fly, yellowish leaf) were identified and examined using a pretrained deep learning networks (VGG16, VGG19, Inception V3, and Alex-net) using the dataset of chilli leafs with and without augmentation. The methodology involves a dataset of 2000 latest images of chilli leaves (apart from imagenet dataset) with four different illnesses to train and test the model, with various inputs and mini-batch sizes. The model's performance of all the networks is assessed using a test dataset, and the results showed that Inception V3 achieved the highest accuracy rates of 93% in identifying these four disorders, followed by VGG16 (85%) and VGG19 (83%), respectively. The suggested methodology can be used by farmers as a decision support system to monitor the health of their crops and take timely and appropriate action that will ultimately help them sustain the output of chillies.

Keywords Chilli · CNN model · Disease · Inception V3 · Farmers

Z. Gulzar (✉) · S. Chandu · K. Ravi
SR University, Warangal, Telangana, India
e-mail: zameer.gulzar@sru.edu.in

K. Ravi
e-mail: k.ravi@sru.edu.in

S. Shakya et al. (eds.), *Fourth International Conference on Image Processing and Capsule Networks*, Lecture Notes in Networks and Systems 798,
https://doi.org/10.1007/978-981-99-7093-3_45

1 Introduction

Since agriculture employs more than 60% of the population and contributes around 17% to the country's GDP, it is a crucial sector of the Indian economy. However, a variety of issues, such as climate change, pests, illness and other issues continue to endanger food security [1]. In India, the cultivation of chilies is a significant income crop that is essential for preserving food security. Telangana produces the most chilies in India, accounting for 49% of the nation's total production [2]. The most popular chilies in India are those cultivated in the Guntur region of Telangana because of their distinctive flavor and color. However, a number of ailments and diseases can result in severe yield losses in the production of chilies. Therefore, the identification and treatment of disease are crucial for improving the growth and yield of such agricultural plants.

Since the vast number of plant diseases show visible symptoms, and well-known pathologist with good observation skills diagnoses the infected plant leaves disease simply by using visual observation. For farmers, it is extremely challenging and expensive to get subject-matter experts for observing plant diseases [3]. Even seasoned pathologists occasionally struggle to identify some illnesses due to the enormous variety of plants, variances in the progression of plant diseases caused by climatic changes, and the rapid spread of diseases to areas where they have never been observed before [4]. Therefore, a sophisticated system for diagnosing plant diseases was therefore necessary to periodically check the agricultural fields. Agronomists can benefit greatly from the availability of knowledgeable and intelligent technologies that can accurately and automatically diagnose plant disease. Crop productivity has greatly increased thanks to the adoption of machine learning techniques in agriculture [5]. The majority of recent research has concentrated on using computer vision and deep learning to identify plant diseases. The HAAR, histogram of oriented gradients, scale invariant and feature transform, speeded-up robust features, Support Vector Machine (SVM), image segmentation using artificial neural networks, K-Nearest Neighbors' (KNN), and K-means clustering techniques have all been used in the past to detect plant diseases [6]. Despite machine learning producing good results, these models fell short of those obtained by deep learning methods based on Convolution Neural Networks (CNNs). The golden age of machine learning will benefit from deep learning which have raised accuracy levels. Deep neural networks have shown great success in solving image recognition challenges, and they have steadily risen to the top of the field in many areas. Deep learning algorithms for Image recognition have been shown to be trustworthy [7].

The creation of automated systems that can diagnose diseases more quickly and accurately is now possible because of advancements in artificial intelligence technology [8]. Transfer learning applies a neural network that has already been trained for one task to a new, related task. The time needed for model design and training can be cut down with transfer learning approaches. The common image classification transfer learning algorithms are Alex-Net, Dense-Net, VGG16, Inception-V3,

Mobile-Net, and Res-Net [9, 10]. The evaluation revealed that the CNN implementation produced excellent results and outperformed the results obtained by cutting-edge machine-learning approaches. In this chapter, we tested various CNN architectures, including VGG16, VGG19, Inception V3, and Alex-net to identify plant illnesses based on photos of healthy and diseased leaf images.

The parameters were adjusted and tested comprehensively, with several dropout values between 0.2 and 0.8, as well as varied batch sizes between 32 and 180. Then the model was evaluated with various learning rates in the range of 0.01–0.0001 in order to optimize it and also different epochs were used to train the models. The evaluation of models implemented was done against other deep-learning models. The Inception V3 model performed better, according to the results, in terms of accuracy and training time.

To build reliable classifiers for plant disease diagnosis, we needed a huge, validated dataset of images of sick and healthy plants to test on the present available classifiers. To address this issue, we started collecting about 2000 images of both healthy and ill crop plants. The remainder of the chapter is organized as follows: Sect. 2 presents implementation method. Sections 3 and 4 demonstrates performance measurement and results and discussion. Finally, Sect. 4 will display some application followed by the conclusion of the chapter.

2 Material and Methods

The models implemented are dependent on the CNN architecture; the general CNN architecture is shown in Fig. 1 the layers that must exist in a CNN for classification purposes are Convolution Layer, Pooling Layer, Flatten Layer, and Dense Layer. The function and operation of these layers are discussed further below [11]

The convolutional layer consists of a set of learnable filters or kernels given in Fig. 2 that are convolved with the input data to extract spatial features from the input image. The filters slide across the input data, performing a dot product at each position to produce a feature map. The dot product operation computes the similarity between the filter and the corresponding patch of the input data, and the resulting feature map represents the activation of the filter across the entire input data. The size, number,

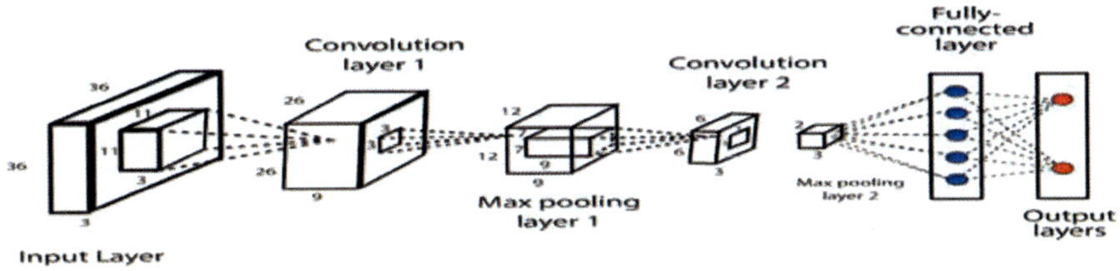

Fig. 1 Convolution neural network

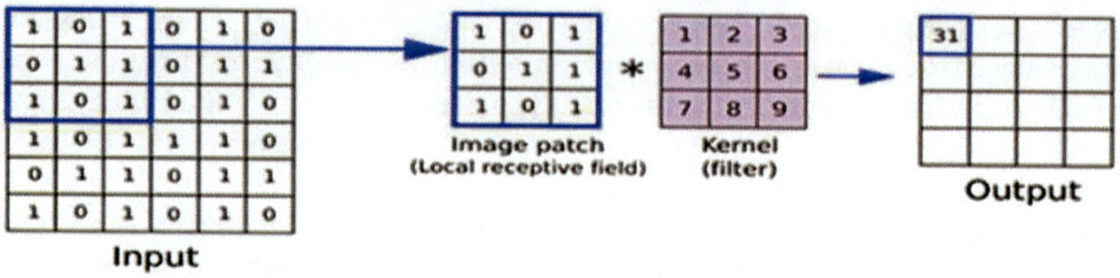

Fig. 2 Convolution filter output calculation

and stride of the filters in the convolutional layer can be adjusted to control the number and size of the output feature maps. Each filter in the convolutional layer learns a specific feature of the input data, such as edges, corners, or textures, through the back-propagation algorithm during the training process.

A pooling layer in Fig. 3 merge the outputs of neuron clusters at the previous layer into a single neuron at the subsequent layer, the pooling layer's goal is to lower the hidden layer's dimensions. The output feature maps from the convolutional layers in CNNs are down sampled using pooling layers. Maintaining the most pertinent data while reducing the input's spatial size is the core aim of pooling. By doing so, overfitting might be prevented, and the networks computational and parameter needs might be reduced. Pooling layers are typically inserted after convolution layers and before fully connected layers in CNN architecture. Pooling layers operate on each feature map separately and perform a fixed operation, such as max-pooling or average pooling, on small sub-regions of the input feature map [12]. The purpose of the pooling layer is to reduce the dimensionality of the feature maps, making the subsequent layers computationally more efficient and reducing the risk of over fitting. The size of the pooling window, or kernel, and the stride of the window determine the amount of down sampling that occurs. For example, a max-pooling operation with a 2×2 kernel and a stride of two reduces the spatial dimensions of the feature map by half, while retaining the most salient features. Overall, pooling layers are an important component of CNNs, since it shrinks the input space while retaining pertinent data, leading to more efficient and effective image recognition and classification while cutting down on administrative burden [13].

In CNN, the flatten layer shown in Fig. 4 converts the output of the previous convolutional and pooling layers into a 1D feature vector, which can be fed into

Fig. 3 Max pooling

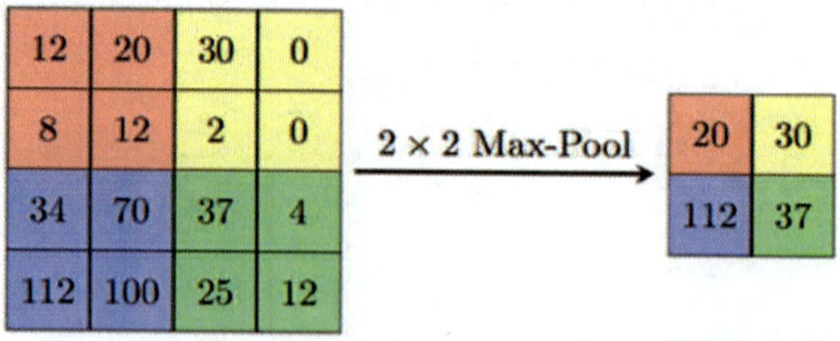

Fig. 4 Flatten layer

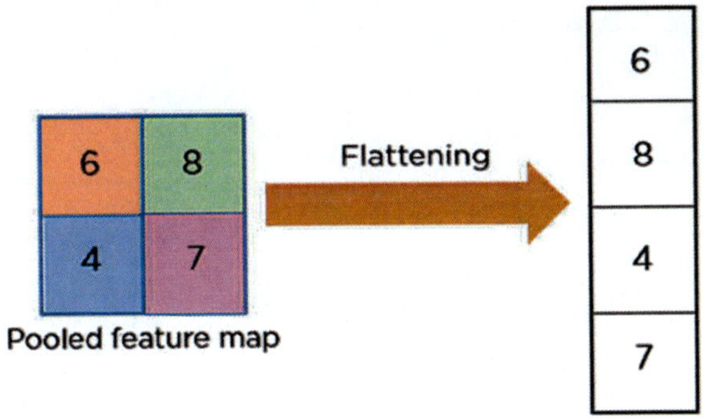

a fully connected neural network to perform classification or regression tasks. The output of the convolutional and pooling layers is typically a high-dimensional feature map with each element representing a specific feature of the input image. The flatten layer reshapes the feature map into a 1D feature vector, which can be processed by a fully connected neural network. Overall, the flatten layer remains as an important component in CNNs to combine the local and global information learned by the convolutional and fully connected layers, respectively, to perform complex tasks on high-dimensional input data [14, 15].

The dense layer shown in Fig. 5 connects every neuron in the current layer to every neuron in the previous layer. It is also called a fully connected layer as each neuron in the current layer is connected to all neurons in the previous layer. In CNNs, dense layers are typically added after the convolutional, pooling and flatten layer. The output of the flatten layer, which is a 1D feature vector, is fed into one or more dense layers for further processing. The purpose of including a dense layer is to perform complex computations on the high-dimensional feature vector obtained from the previous layers, allowing the model to learn complex hierarchical representations of the input data. Overall, dense layers remain as an important component of neural networks, allowing the model to learn complex representations of the input data and perform a wide range of tasks, ranging from simple classification to more complex tasks such as Natural Language Processing (NLP) and computer vision.

The CNN models, training procedure, experimental set-up, and dataset preparation are all fully described in this section. The pipeline for the proposed model for detecting plant leaf diseases begins with dataset preparation and ends with model prediction. Acquiring the right dataset is critical for the success of any deep learning or machine learning model, including ours. To acquire the dataset, we have pleated chilli plant images at various stages of its growth, disease severity, and lighting conditions. Then we labeled the images according to different classes of chilli plant diseases, such as Leaf curl (Fig. 6), Leaf Spot (Fig. 7.), Yellowish (Fig. 8), and White fly (Fig. 9), and healthy plants. Next, we split the dataset into training, validation, and test sets. Balance the dataset to prevent bias towards specific classes of chilli plant diseases. To increase the diversity of the dataset and improve the generalization

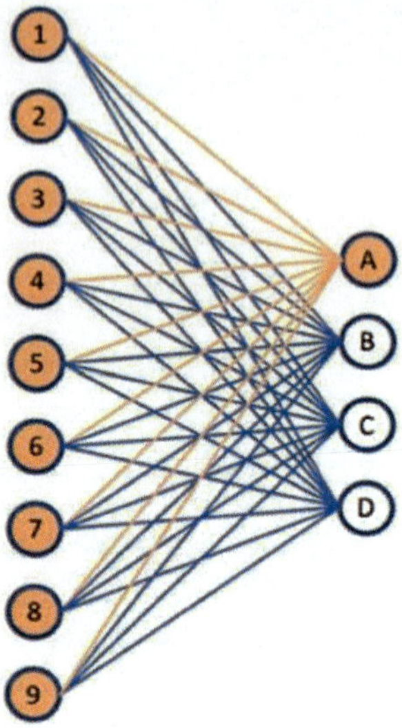

Fig. 5 Dense network

of the CNN model, we also obtained an external dataset of chilli plant images from a public source Kaggle Imagenet dataset.

In this study, the main task was to classify five different classes of chilli plants that include Leaf curl, White fly, Yellowish, leaf spot, and healthy plants. At first image dataset is preprocessed to normalize the images, adjust the image size, and remove any irrelevant information from the images. Then, the focus is on classifying four classes of chilli plant diseases that involves fine-tuning the pre-trained CNN models such as VGG16, VGG19, Inception V3, and Alex-net to the specific problem of chilli plant disease classification. Later the dataset was divided into two categories: healthy and unhealthy plants as shown in Figs. 10 and 11. We use binary classification approach to classify the plants as healthy or unhealthy based on the presence of diseases. The

Fig. 6 Images of leaf curl disease in dataset

Fig. 7 Images of leaf spot disease in dataset

Fig. 8 Images of leaf yellowish disease in dataset

model takes into account various features such as shape, color, and size to accurately detect the disease. Finally, we try to induce some percentage of disease into the healthy plants and then perform binary classification again. This step simulates a more realistic scenario where not all unhealthy plants have the same level of illness, and some may have less severe symptoms.

The study uses various techniques such as data augmentation, transfer learning, and fine-tuning to improve the accuracy of the CNN models. The proposed model can

Fig. 9 Images of White fly disease in dataset

Fig. 10 Images of healthy class in dataset

Fig. 11 Images of unhealthy class in dataset

accurately detect and classify chilli plant diseases and recommend appropriate treatment using deep learning. This can help in the early detection and timely treatment of chilli plant diseases, leading to better crop yields and reduced economic losses. The description of the dataset is given below and shown in Fig. 12.

Chilli: Five classes are available.

Classifications: Healthy, Leaf curl, Leaf spot, Whitefly, Yellowish. For the new dataset, we also gathered images of chilli plants. We also induced some disease in some of the healthy plants, dividing the dataset into Healthy and Unhealthy as shown in Fig. 13. The new dataset is based on the following characteristics: There are two classes. (Healthy-0 and Unhealthy-1) 70 images of trains and 30 images for validation/testing. There are a total of 100 images. Plant images that are 10% unhealthy,

Fig. 12 Test train validation, leaf classes dataset folders

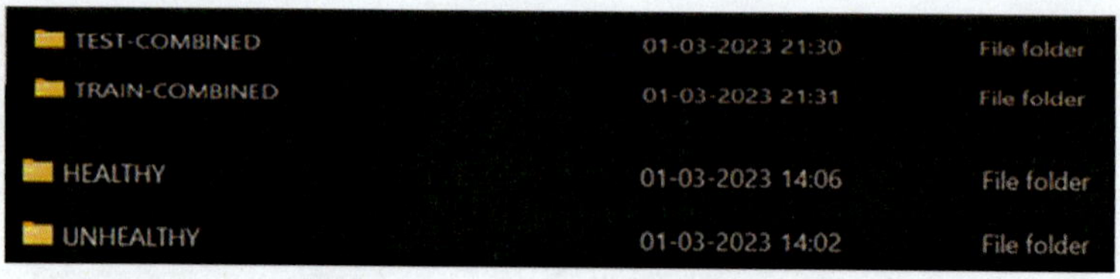

Fig. 13 Healthy, unhealthy, tests and train folders of new dataset

30% unhealthy and completely healthy are considered healthy. Plant images that are 50%, 70%, or 100% unhealthy are considered unhealthy.

2.1 *Data Preprocessing*

Data preprocessing is the process of transforming, cleaning, and enriching raw data for analysis, aiming to improve its quality, remove inconsistencies, and enhance performance. Data preprocessing is crucial for accurate and reliable results, as a well-prepared dataset enhances machine learning models, reduces over-fitting or under-fitting risks, and leads to better insights and decisions. The 128 by 128 pixel cropped photos of chilli illness and pest utilized in this investigation. We pre-processed the ImageNet dataset in order to use the pre-trained model on it similar to those we used for teaching the trainers about disease and pest dataset for ImageNet. The photographs of the diseases and pests were first resized, Pixels of 224×224 similar to the [9], were made more uniform by determining the average value of every dataset channel. Next, we deducted the average computed value for every input image. The standard deviation was a 1D array representing the overall RGB pixels' average values. Images from ImageNet are 103.939, 116.779, and 123.68.

2.2 *Feature Extraction*

Feature extraction in CNN models using image datasets involves learning a hierarchical representation of the input image through a series of convolution, pooling, and fully connected layers. This process allows the model to extract useful features from the input image and make accurate predictions. Feature extraction and fine-tuning are the two primary categories of transfer learning strategies [10]. A representation of the previously trained network is used to extract characteristics from fresh leaf samples. The CNN model uses fully linked layers for classification and successive convolution and pooling layers for those tasks. The output value of a particular layer of the pre-trained model is regarded as a feature in the feature extraction process, and the feature retrieved from the deep-learning model is referred to as a deep feature.

The process of adapting architecture for a new use based on a previously trained model and updating training from pre trained weights is known as fine-tuning.

3 Measurement of Performance

To get a sense of how our pretrained model would perform on new, unforeseen data and to keep track of which of our approaches are over fitting, we run all of our experiments across a variety of train-test set splits, particularly 80–20 (80% of the entire dataset used for training, and 20% for testing), and 60–40 (60% of the entire dataset used for training, and 40% for testing). We have the mappings for such scenarios for 20,212 photos out of the 40,605 images. It should be noted that the dataset typically comprises numerous photographs of the same leaf (shot from various orientations). During every test-train split, we additionally make sure that all photos of the same leaf are either placed in the training set or the testing set. At regular intervals (at the end of each epoch), we compute the mean overall accuracy for each experiment during the course of the training session. For the purpose of comparing the outcomes of all of the various experimental settings, we utilize the accuracy score only in this paper. We assess convolution neural networks' suitability for the aforementioned categorization issue. We concentrate on two well-known architectures, Inception V3 and VGG-19. In one scenario, we train the model from scratch, and in the other, we adjust previously trained models (trained on the ImageNet dataset) using transfer learning to compare the performance of both of these architectures on the available datasets. We do not restrict the learning of any of the layers when training the model. In other words, the initial state of the weights of a few layers is the main distinction between these two learning approaches (transfer vs. training from scratch), allowing the transfer learning strategy to take advantage of the substantial amount of visual experience.

4 Result Analysis

The findings of the models are shown below. These data help us to understand the models' prediction accuracy and how well a model can classify diseases into their appropriate classes.

CNN Results: The primary building block for CNN is the convolution layers. The raw data that is supplied into the CNN are represented by the input layer. The input layer for image data is an image tensor having dimensions (height, width, and channels) that correspond to the pixels in an image. They are made up of several learnable filters, sometimes known as kernels that scan the input for patterns and extract features.

From Table 1, it is evident that the model's accuracy significantly improves from step 1 to step 3 (from 69 to 93%), indicating that the model is learning and performing better as it undergoes more training or refinement. However, there is a drop in accuracy from step 3 to step 4, which might indicate over-fitting or other issues.

VGG-16 Results: Visual Geometry Group (VGG) created the 16-layer deep convolutional neural network known as VGG16. Multiple convolutional layers are included, followed by fully connected layers. Input is 224 × 224 × 3 (224 × 224 pixel RGB Image), 13 convolution layer 3 × 3 filters, followed by ReLU activation techniques. Max-pooling layers with 2 × 2 filters and stride 2 are applied after several convolutional layers. The final three layers have 4096 neurons each and are fully connected, followed by ReLU activations. With the same number of neurons as classes in the classification problem, the final layer is a Softmax layer.

From Table 2, we can observe that the accuracy consistently improves from step 1 to step 3, with a significant increase from 73 to 94%. This indicates that the model is learning and performing better as it undergoes more training or optimization. However, there is a slight drop in accuracy from step 3 to step 4, which might indicate some over-fitting or other issues. The VGG16 model seems to be performing quite well, especially with the highest accuracy achieved at step 3 (94%).

Table 1 Accuracy report of CNN

Accuracy step-1	Accuracy step-2	Accuracy step-3	Accuracy step-4
69	76	93	76
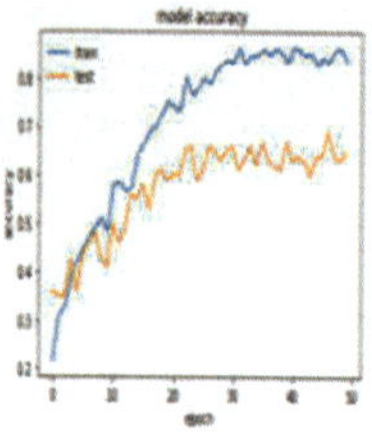	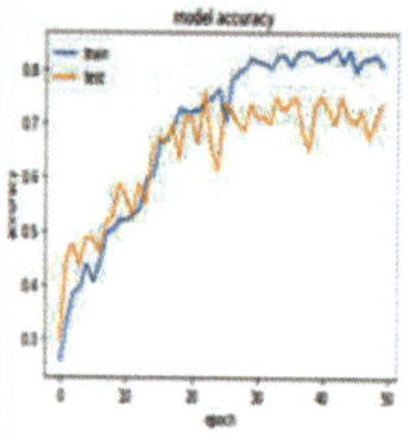	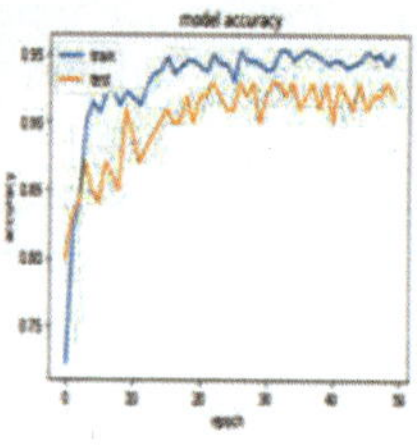	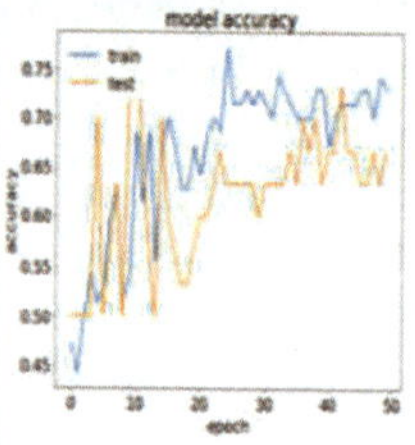

Table 2 Accuracy report of VGG-16

Accuracy step-1	Accuracy step-2	Accuracy step-3	Accuracy step-4
73	78	94	90
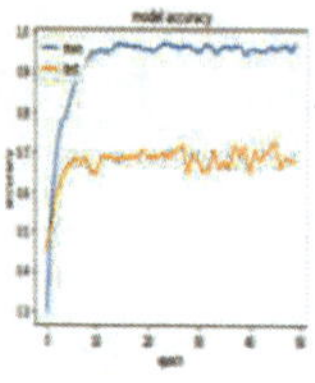	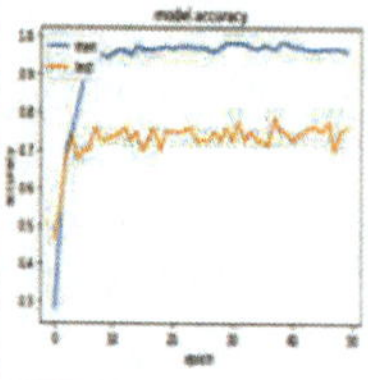	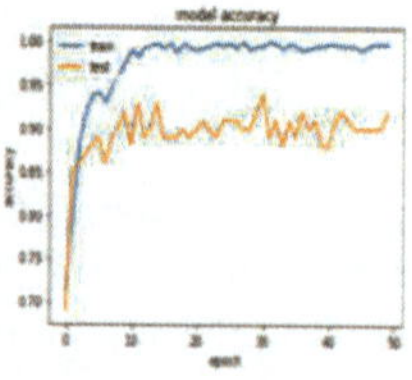	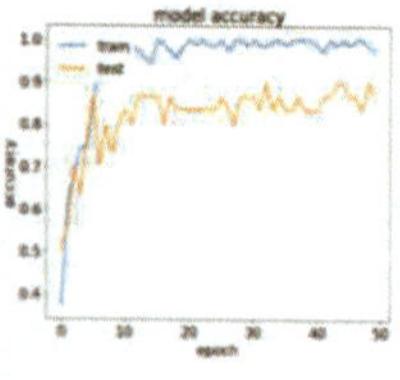

VGG19 Results: The extension of VG16 is VGG19 having similar architecture to VGG16 but includes additional 3, that is, 19 convolutional layers. VGG19 have 16 convolutional layers with 3 × 3 filters, followed by ReLU activation functions. In Table 3, we can observe that the accuracy shows some fluctuations during the different steps, but there is an overall improvement from step 1 to step 3. The model achieves the highest accuracy at step 3 (91%), indicating that it is learning and performing better as it undergoes more training or refinement. However, there is a decrease in accuracy from step 3 to step 4, which may indicate some over-fitting or other issues. Overall, the VGG19 model performed reasonably well, especially with the highest accuracy achieved at step 3 (91%).

Inception V3 Results: The Google-developed Inception V3, it makes use of the idea of "Inception modules," which enable the use of various-sized filters within a single layer to record data at various scales. The input is 299 × 299 × 3 (RGB image with a size of 299 × 299 pixels). It has convolution layer with different inception models with different filter sizes (1 × 1, 3 × 3, 5 × 5) for the extraction of features. To avoid overfitting, the last layers have dropout regularization and are fully coupled, and the output layer is SoftMax with number of classes in classification tasks equal to neurons.

From Table 4, we can observe that the accuracy consistently improves from step 1 to step 3, with a substantial increase from 78 to 96%. This indicates that the model is learning and performing significantly better as it undergoes more training or optimization. There is a slight decrease in accuracy from step 3 to step 4, which might indicate some over-fitting or other issues. The Inception V3 model performed very well, especially with the highest accuracy achieved at step 3 (96%). With further fine-tuning and experimentation, it might be possible to improve the model's performance.

Alex-Net Results: Alex, Ilya and Geoffrey created the ground-breaking deep convolutional neural network known as Alex-Net. As a result of its victory in the 2012 ImageNet Large-Scale Visual Recognition Challenge (ILSVRC), computer vision advanced significantly. The input size is 227 × 227 × 3 (RGB image with a size of 227 × 227 pixels). It consists of five convolution layers with different filter sizes 11 × 11, 5 × 5, 3 × 3) followed by ReLU activation. The accuracy shows fluctuations

Table 3 Accuracy report of VGG19

Accuracy step-1	Accuracy step-2	Accuracy step-3	Accuracy step-4
74	76	91	83
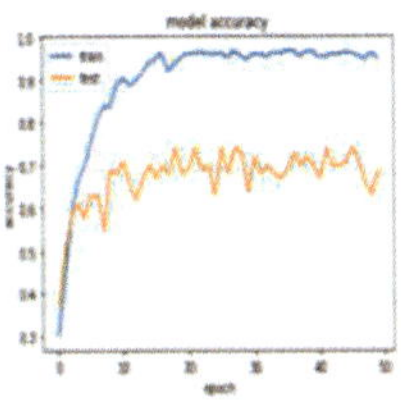	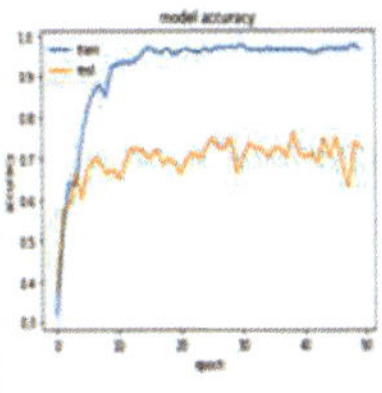	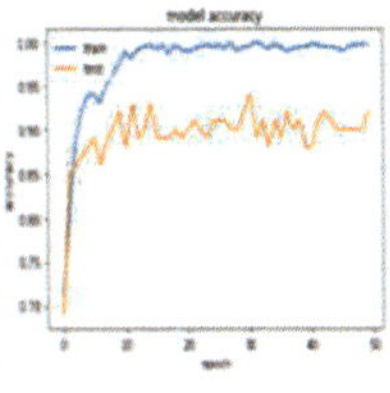	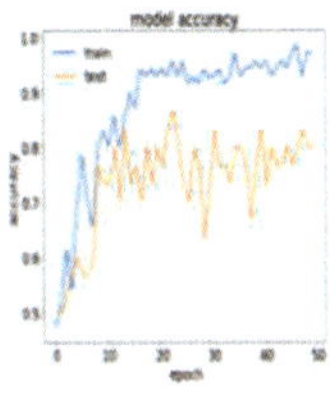

Table 4 Accuracy report of Inception V3

Accuracy step-1	Accuracy step-2	Accuracy step-3	Accuracy step-4
78	83	96	93
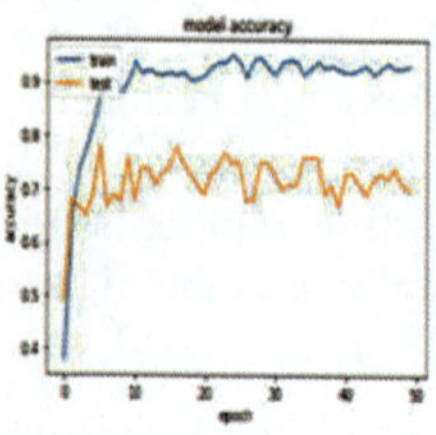	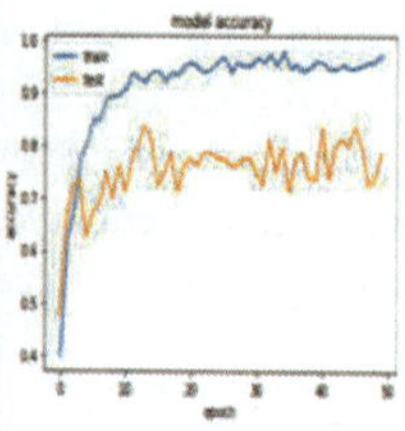	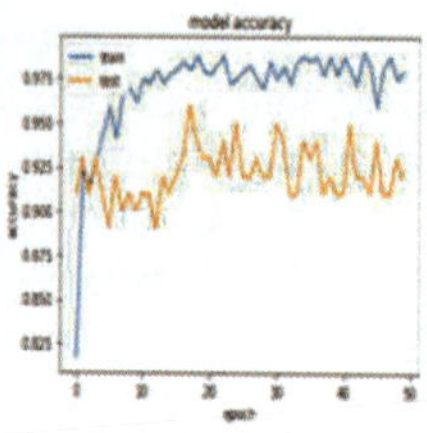	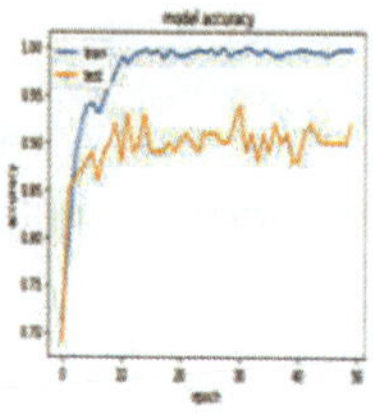

Table 5 Accuracy report of Alex-Net

Accuracy step-1	Accuracy step-2	Accuracy step-3	Accuracy step-4
48	76	89	70
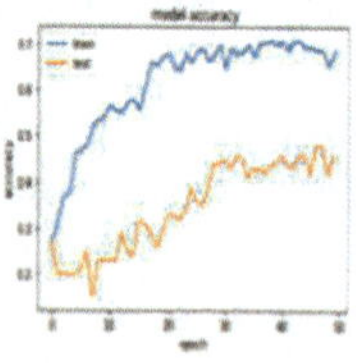	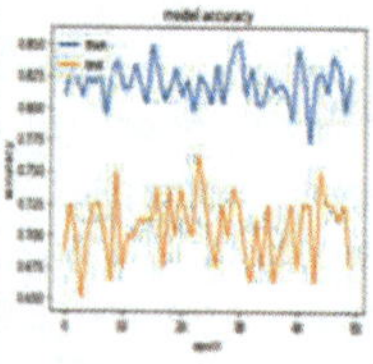	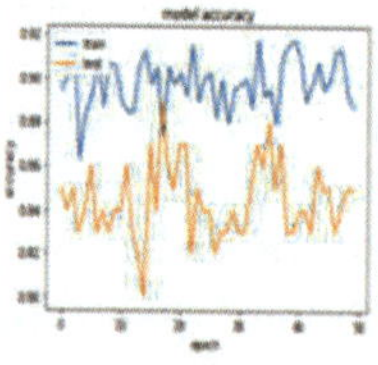	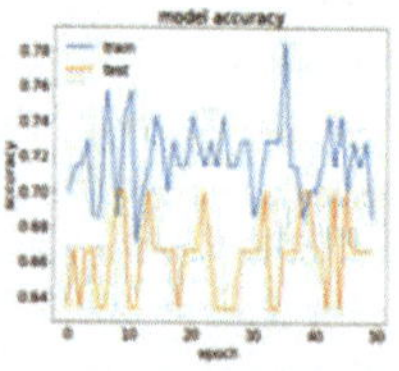

during the different steps, and there is a significant improvement from step 1 to step 3, with accuracy increasing from 48 to 89%. This indicates that the model is learning and performing better as it undergoes more training or optimization. However, there is a decrease in accuracy from step 3 to step 4, dropping from 89 to 70%. This drop in accuracy might indicate some overfitting or other issues in the model. The Alex-Net model is making progress in terms of accuracy from step 1 to step 3. When compared to other steps, the Alex-Net model performed best in step 3 as shown in Table 5.

4.1 *Model Comparison*

The models have been compared accordingly in the below table where the Model is specified in the first column and the accuracies of each model, that is, CNN, VGG16, VGG19, Inception V3, and Alex-Net models are specified in the next columns for the four steps. This comparison helps achieve the overall best model based on the accuracy of the model in each step. Table 6 shows the accuracy of different models used in the different steps of detecting and classifying chilli plant diseases. The table lists four steps, from step 1 to step 4, with each step having a different objective and task. In step 1, we classified five classes of chilli plant diseases, including Leaf curl, White fly, Yellowish, leaf spot, and healthy. In step 2, we classified four classes of chilli plant diseases, excluding healthy ones. In step 3, we performed binary classification by dividing the dataset into healthy and unhealthy classes, with all diseased classes labelled as unhealthy. In step 4, we induced some percentage of disease into the healthy plants and performed binary classification.

Table 6 shows the accuracy of different models in each of these steps. Each model has its architecture, which affects its performance in different tasks. Looking at the table, we can see that the accuracy of all models varies across different steps. For example, the CNN model has an accuracy of 69% in step 1, which is the lowest accuracy among all models. However, its accuracy increases in steps 3 and 4, where it achieves 93% and 76% accuracy, respectively. On the other hand, the Inception V3 model has the highest accuracy in all steps, with an accuracy of 78%, 83%, 96%, and 93% in steps 1, 2, 3, and 4, respectively. This indicates that Inception V3 is a powerful model for detecting and classifying chilli plant diseases. Overall, the table provides a useful comparison of different models' accuracy in each step of the study, helping to identify the best-performing models for each task.

Table 7 the performance of different models in detecting and classifying chilli plant diseases when some percentage of disease is induced into healthy plants in step 4. The table includes five models, which are Inception V3, VGG16, VGG19, CNN, and Alex-Net. For each model, the table provides the accuracy, probability, and number of healthy and unhealthy plants that were correctly detected and classified. The accuracy column shows the overall accuracy of each model in detecting and classifying chilli plant diseases when some percentage of disease is induced into healthy plants. The values range from 70 to 93%, with Inception V3 achieving the highest accuracy of 93%, and Alex-Net achieving the lowest accuracy of 70%. The probability column shows the probability of each plant being healthy or unhealthy. For example, a probability value of 0.7 indicates that there is a 70% chance that the plant is healthy and a 30% chance that it is unhealthy. The next two columns show the actual and detected disease labels for each plant. The actual disease labels are based on whether the plant was healthy or unhealthy before the disease was induced, and the detected disease labels are based on the model's classification results. Looking at the table, we can see that Inception V3, VGG16, and CNN models have correctly detected and classified all healthy and unhealthy plants. However, the VGG19 model incorrectly classified one healthy plant as unhealthy. The Alex-Net model correctly

Table 6 Model comparison

Model	Accuracy in step-1	Accuracy in step-2	Accuracy in step-3	Accuracy in step-4
CNN	69	76	93	76
VGG16	73	78	94	90
VGG19	74	76	91	83
Inception V3	78	83	96	93
Alex-Net	48	76	89	70

Table 7 Performance of models in detecting and classifying chilli diseases with induced disease in healthy plants

Algorithm	Accuracy	Probability		Disease	
		Healthy	Unhealthy	Actual	Detected
InceptionV3	93	9.99881387	1.18544674	Healthy	Healthy
		0	1.0000000	Unhealthy	Unhealthy
VGG16	90	0.83606935	0.16393062	Healthy	Healthy
		0.01924389	0.9807561	Unhealthy	Unhealthy
VGG19	83	0.47754118	0.52245885	Healthy	Unhealthy
		0.20832987	0.79167014	Unhealthy	Unhealthy
CNN	76	0.7034686	0.29653138	Healthy	Healthy
		0.45679343	0.5432065	Unhealthy	Unhealthy
Alex-Net	70	0.5012679	0.4987321	Healthy	Healthy
		0.50262487	0.4973751	Unhealthy	Healthy

classified all unhealthy plants but incorrectly classified one healthy plant as unhealthy. Overall, the table shows that the models' performance varies in detecting and classifying chilli plant diseases when some percentage of disease is induced into healthy plants. It also shows that Inception V3 and VGG16 are the most reliable models in detecting and classifying diseases accurately.

4.2 Web Application

Figure 14 shows the block diagram of the concept, the flow of the project starts from collecting the dataset, which is done from two sources—the Kaggle dataset and a mobile camera. The dataset is then preprocessed and augmented for improving the accuracy of the model. The preprocessed dataset is then fed into a CNN, which is the core of the project. The CNN has two parts—the feature extraction part that uses convolutional layers to extract features from the input images, and the disease classification part that uses dense layers to classify the disease into one of the categories. Once the disease is identified, the model provides a treatment recommendation to the user based on the detected disease. This recommendation is made using pesticides and insecticides, which are suggested, based on the type of disease identified. Overall, the study uses CNN to detect and classify the disease in chilli plants and recommends appropriate treatments for the same.

Figures 15, 16 and 17 of Web Application shows the user interface that allows the user to upload or take a picture of a chilli plant. The image is then preprocessed, and the features are extracted using a CNN. The extracted features are then classified by the CNN and the disease is predicted. The appropriate treatment is suggested based

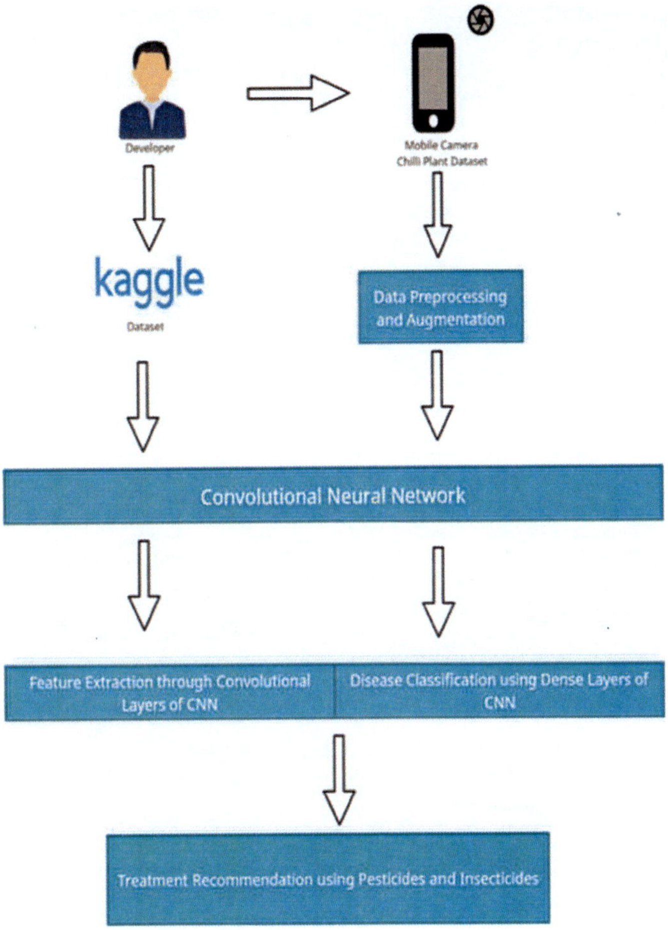

Fig. 14 Block diagram of proposed model

on the predicted disease. Finally, the output is displayed through a web application for the user to see.

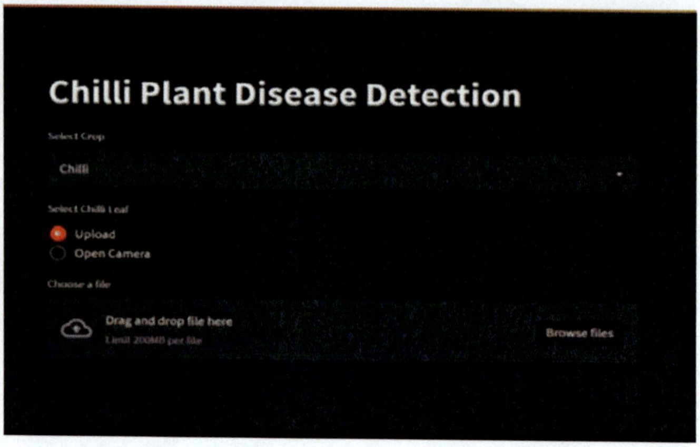

Fig. 15 User interface

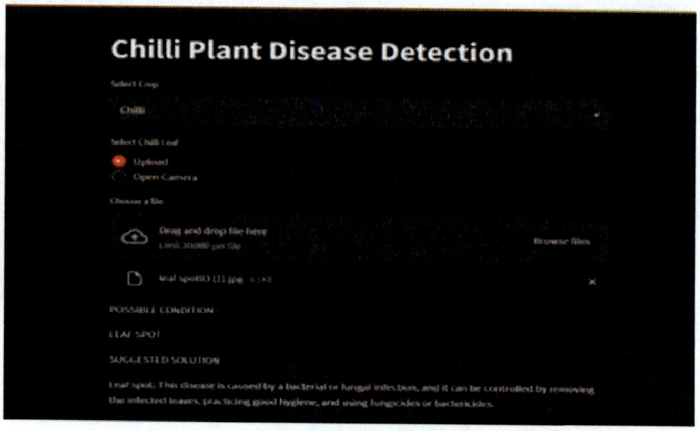

Fig. 16 Uploading an image and getting the result

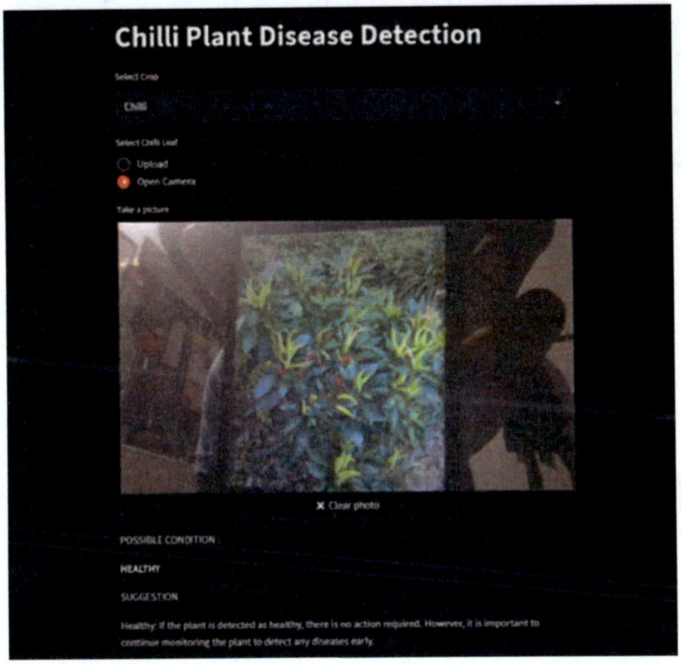

Fig. 17 Taking photo of an image and getting the result

5 Conclusion

Finally, for the purpose of identifying and categorizing diseases of chilli plants, this study created and assessed CNN models as well as transfer learning models including VGG16, VGG19, Inception V3, and Alex-Net. Five classes of chilli plant illnesses, including leaf curl, white fly, yellowish, leaf spot, and healthy, were used to train and test the models. The Inception V3 model had the best accuracy overall, according to the data, and the CNN and VGG16 models also performed well. Furthermore, even when a small amount of illness was introduced into healthy plants, the suggested models were able to identify and categorize chilli plant infections with excellent accuracy. In order to treat the disease, the proposed models can be utilized to suggest relevant pesticides or insecticides. However, there are a few restrictions at this point that must be resolved in upcoming work. Initially, when examined on photos taken under situations that are different from the ones in the photographs when utilized for training, the model's accuracy is significantly decreased, to a little over 10%. Each of this experiment lasts for a total of 30 epochs, where an epoch is the number of training iterations during which the specific neural network has successfully traversed the whole training set. The significance of this accuracy is superior to the one based on classes being randomly selected, although a more varied collection of training data is to increase the accuracy. Our most recent findings show that additional (and variable) data alone will be adequate to significantly improve the accuracy and related data collection activities are in progress. The future work will be to construct a hybrid model and increase the size of the developed dataset of chilli images with some additional diseases.

References

1. Tai AP, Martin MV, Heald CL (2014) Threat to future global food security from climate change and ozone air pollution. Nat Clim Chang 4:817–821. https://doi.org/10.1038/nclimate2317
2. Bhookya NN, Malmathanraj R, Palanisamy P (2020) Yield estimation of chilli crop using image processing techniques. In: 2020 6th International Conference on Advanced Computing and Communication Systems (ICACCS), Coimbatore, India, pp 200–204. https://doi.org/10.1109/ICACCS48705.2020.9074257
3. Shrestha G, Deepsikha, Das M, Dey N (2020) Plant disease detection using CNN. In: 2020 IEEE Applied Signal Processing Conference (ASPCON), pp 109–113. https://doi.org/10.1109/ASPCON49795.2020.9276722
4. Sladojevic S, Arsenovic M, Anderla A, Culibrk D, Stefanovic D (2016) Deep neural networks based recognition of plant diseases by leaf image classification. Comput Intell Neurosci 2016:3289801. https://doi.org/10.1155/2016/3289801
5. Pattnaik G, Shrivastava VK, Parvathi K (2020) Transfer learning-based framework for classification of pest in tomato plants. Appl Artif Intell 34:981–993. https://doi.org/10.1080/08839514.2020.1792034
6. Aravind KR, Raja P, Aniirudh R, Mukesh KV, Ashiwin R, Vikas G (2019) Grape crop disease classification using transfer learning approach. Lect Notes Comput Vis Biomech 30:1623–1633. https://doi.org/10.1007/978-3-030-00665-5_150
7. Krizhevsky A, Sutskever I, Hinton GE (2017) ImageNet classification with deep convolutional neural networks. Commun ACM 60:84–90
8. Xie C, Shao Y, Li X, He Y (2015) Detection of early blight and late blight diseases on tomato leaves using hyperspectral imaging. Sci Rep 5:16564
9. Zhuang F, Qi Z, Duan K, Xi D, Zhu Y, Zhu H et al (2021) A comprehensive survey on transfer learning. Proc IEEE 109:43–76. https://doi.org/10.1109/JPROC.2020.3004555
10. Kandel I, Castelli M (2020) Transfer learning with convolutional neural networks for diabetic retinopathy image classification: a review. Appl Sci 10:2021. https://doi.org/10.3390/app10062021
11. Ullah Z, Lodhi BA, Hur J (2020) Detection and identification of demagnetization and bearing faults in PMSM using transfer learning-based VGG. Energies 13:3834. https://doi.org/10.3390/en13153834
12. Russakovsky O, Deng J, Su H, Krause J, Satheesh S, Ma S et al (2015) ImageNet large scale visual recognition challenge. Int J Comput Vis 115:211–252. https://doi.org/10.1007/s11263-015-0816-y
13. LeCun Y, Bengio Y, Hinton G (2015) Deep learning. Nature 521:436–444. https://doi.org/10.1038/nature14539
14. Canziani A, Paszke A, Culurciello E (2016) An analysis of deep neural network models for practical applications. Preprint at http://arxiv.org/abs/1605.07678. Accessed 25 Nov 2021
15. Simonyan K, Zisserman A (2014) Very deep convolutional networks for large-scale image recognition. Preprint at http://arxiv.org/abs/1409.1556

Sign Language Recognition Using Long Short-Term Memory Deep Learning Model

Prabhat Mali, Aman Shakya, and Sanjeeb Prasad Panday

Abstract This paper aims to present an easy way of communication for speech impaired and hearing impaired people. This paper focuses on the use of neural network techniques, specifically MediaPipe Holistic and the Long Short-Term Memory (LSTM) module, to recognize sign language for individuals with disabilities. The utilization of MediaPipe Holistic, which integrates pose, hand, and face keypoints with precise levels, was used due to its low latency and high tracking accuracy in real-world scenarios. This paper deals with commonly used words of American Standard Sign Language. The result obtained shows that the proposed model can detect American Standard Sign Language with accuracy of 98.50 percentage.

Keywords Sign language · Keypoints · MediaPipe holistic · Long short-term memory

1 Introduction

Recently, there has been a sharp rise in the number of individuals who are speech- or hearing-impaired due to factors such as birth defects, oral health issues, and accidents. This affects around 7–9 million people worldwide. When these individuals communicate with those who are not speech- or hearing-impaired, it can be challenging for the latter group to understand. To address this issue, the use of gesture recognition systems has become increasingly common to bridge the communication gap between different groups. Sign language recognition is a well-studied field and

P. Mali · A. Shakya (✉) · S. P. Panday
Department of Electronics and Computer Engineering, Pulchowk Campus, Institute of Engineering, Tribhuvan University, Lalitpur, Bagmati, Nepal
e-mail: aman.shakya@ioe.edu.np

P. Mali
e-mail: 078msice010.prabhat@pcampus.edu.np

S. P. Panday
e-mail: sanjeeb@ioe.edu.np

S. Shakya et al. (eds.), *Fourth International Conference on Image Processing and Capsule Networks*, Lecture Notes in Networks and Systems 798,
https://doi.org/10.1007/978-981-99-7093-3_46

has made significant progress in recent years, with various techniques being explored as a way to facilitate communication. These techniques are generally categorized as either vision-based or sensor-based. Vision-based methods offer user-independent feature extraction.

This paper uses action detection system to identify American Standard Sign Language (ASSL) used by speech-impaired and hearing-impaired people to communicate. The aim of this paper is to utilize computer vision to accurately identify sign language in real-time. This will facilitate communication between people who are speech- and hearing-impaired by allowing them to use their natural hand gestures. This system will dynamically translate sign language, thereby enhancing communication effectiveness and efficiency between individuals who are hearing or speech impaired and those who are not.

Individuals who have difficulties with speech or hearing rely primarily on sign language as a mode of communication. However, it is not understandable by people who do not have speech or hearing impairments, leading to a significant communication barrier. Additionally, sign language can be difficult to learn due to its distinct sentence structure and grammar. To overcome these challenges, there is a need to create a system that can translate sign language. Several attempts have been made in this field, but the present systems are lacking in terms of efficient data collection and faster sign recognition. This paper is an attempt to contribute a step in this area.

2 Related Works

According to a report by the World Federation of the Deaf, the global population of people who are speech- or hearing-impaired is estimated to be between 7 and 9 million. These individuals rely on sign language as their primary mode of communication; however, most people without impairments do not understand the gestures used in sign language, making communication difficult. The objective of this paper is to help individuals with speech disabilities have normal conversations in their daily lives. Sign language varies from region to region, making it challenging to develop a universal system. Over the years, numerous attempts have been made to reduce the communication barrier between speech-impaired and hearing-impaired communities and those without impairments. Most of these attempts involved converting sign language into audible signals. Some of the past attempts to develop a sign language recognition system using various technologies are disscused here.

Authors of paper [1] focus on neural network methods, such as MediaPipe Holistic and the LSTM, to identify sign language used by individuals with impairment. The paper utilized MediaPipe Holistic, which combines hand, pose, and face keypoints. The primary objective of their paper was to demonstrate the use of Human Activity Recognition (HAR) algorithm, which is based on deep learning architecture, for recognizing human actions and classifying them into seven different categories. Convolution Neural Network (CNN) and Long Short-Term Memory (LSTM) were used to analyze spatial and temporal data gathered from a Microsoft Kinect camera.

The paper [2] designed and implemented a real-time system for recognizing hand gestures. This system identifies hand gestures, like peak calculation and angle calculation, and then translates the gesture images into text. To implement this system, sign language hand gesture dataset was used. The proposed system had four main modules such as preprocessing, segmentation, feature extraction, and classification. In the preprocessing module, the input image was converted into a grayscale image, noise removal, normalization, and rescale the image. The image is preprocessed and then segmented using an automatic threshold algorithm. After that, feature extraction is conducted using Scale-Invariant Feature Transform (SIFT) and Speeded-Up Robust Features' (SURFs) descriptors. The features collected from this process are then trained using an Excel file. The Support Vector Machines' (SVMs) algorithm is utilized to classify the input.

Authors of paper [3] developed a program that utilizes machine learning methods to recognize Kazakh dactylic sign language. Their dataset consists of 5000 images of each gesture, and gesture recognition algorithms including Support Vector Machine, Random Forest, and Extreme Gradient Boosting were applied. Additionally, two data types were merged into a single database. The findings from their work revealed that Extreme Gradient Boosting and Support Vector Machine have better real-time efficiency, while Random Forest algorithm has high recognition efficiency.

A vision-based hand gesture recognition system is described in paper [4], which considers various techniques for hand tracking, feature extraction, segmentation, and classification. Their paper involves capturing images using a webcam and processing them using the OTSU method, and classification was done using a linear classification method. In their paper, Chhaya Narvekar and her team stored the captured gestures in folders that contained 120 images of a gesture. The image was captured as histogram.

The authors of paper [5] used images of American Standard Sign Language (ASSL) taken in various environmental conditions as a dataset to recognize and classify hand gestures accurately. The authors applied various preprocessing techniques such as Histogram of Gradients, Principal Component Analysis, and Local Binary Patterns and created a novel model using canny edge detection, ORB, and the bag-of-words technique. The data that had been prepared in advance were subjected to multiple classifiers such as Random Forests, Support Vector Machines, Naive Bayes, Logistic Regression, K-Nearest Neighbors, and Multilayer Perceptron, in order to achieve optimal outcomes.

The authors of paper [7] described the results of using the Long Short-Term Memory (LSTM) model, that improved the machine translation of Google Translate. The article [8] introduces an efficient structure for addressing the static gesture recognition challenge using data sourced from web cameras and the Kinect depth sensor. The paper [9] outlines an approach to identify American Sign Language gestures, employing principal component analysis to reduce the resemblance between different gesture categories.

In [10], the Support Vector Machine (SVM) classifier was employed, and the authors diverged from conventional techniques by opting not to utilize hand markers like gloves for gesture identification. In the study [11], a depth map was generated through the application of the Sum of Absolute Differences (SAD) algorithm, aligning the left and right images. The contours were determined using the Theo Pavlidis Algorithm, which exclusively examines boundary pixels, leading to reduced computational expenses. In paper [12], a method involving orientation histograms was suggested for the identification of stationary hand gestures.

Paper [13] employed decision trees, while paper [14] utilized Hidden Markov Models to discern the dynamic aspects of hand gestures. In the study [15], real-time interpretation of a user's gestures is achieved through hand segmentation, extracting binary hand blobs. The shape of these blobs is then captured using Fourier descriptors. These Fourier descriptors are fed into radial-basis function (RBF) networks for the purpose of posture classification. In [16], a technique is introduced that involves utilizing the angle and peak calculation approach to extract hand gesture features. This process is carried out using MATLAB, and subsequently, the identified gesture is transformed into speech using MATLAB's built-in command.

3 Methodology

Figure 1 shows the general flow of the developed system. Continuous frames of video are passed form MediaPipe Holistic which detects the keypoints present in the video frame, and keypoints are then extracted to a Numpy array which are fed to the Long Short-Term Memory (LSTM) model to make the prediction. Figure 2 shows the LSTM model architecture, which consists of input, LSTM layers, dense layers, and output. The input data are a Numpy array of shape (30,1662). The model consists of three layers of LSTM layer with 64, 128, and 64 units, respectively, which is followed by three dense layers with 64, 32, and 10 units (i.e., number of actions the model is trained with), respectively. The output unit gives the predicted gesture in the form of text. LSTM models are used in this paper because it can handle sequential data effectively, capture long-term dependencies, and have memory cells to store relevant information. Dense layers are responsible for learning complex relationships and patterns in the data. It applies a linear transformation followed by a nonlinear activation function to the input data.

3.1 Software and Hardware Platforms

The software language and platforms used for this work are Python, TensorFlow, OpenCV-Python, MediaPipe, and Sklearn.

- The hardware used for the work are as following:
  - HP Laptop PC 14-d2000.
  - Processor: 11th Gen Intel(R) Core(TM) i5-1135G7 @ 2.40 GHz.
  - Installed RAM: 8.00 GB.

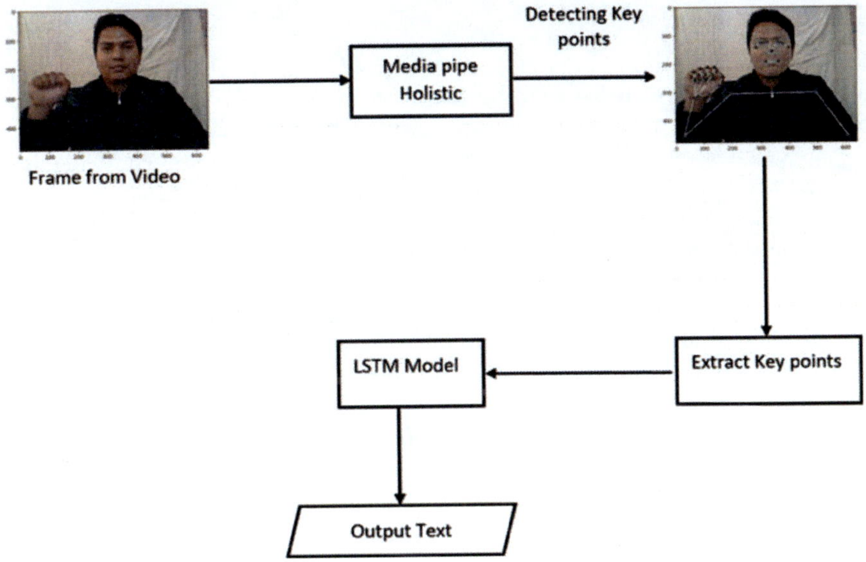

Fig. 1 System flowchart

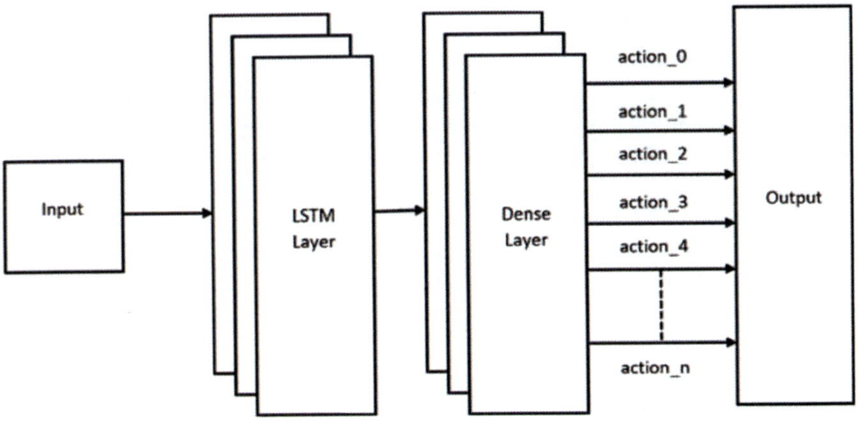

Fig. 2 LSTM model architecture

- HP TrueVision HD Camera.
  - Megapixel: 0.9 MP, 30 fps.
  - Hardware ID: VID 30C9PID 0035MI 00.

3.2 Creating Dataset

In this work, live video footage from a camera is constantly streamed, and each frame that contains a gesture or movement within the frame is saved in a directory. For the creation of the dataset, each captured frame is passed through MediaPipe Holistic which marks the keypoints of hand, pose, and face. Each hand contains of 21 keypoints each, pose contains 33 keypoints, and face contains 468 keypoints, respectively. These keypoints are concatenated to a Numpy array to form a dataset. If any of the keypoints are not visible inside the video frame, then its value is filled with zero. The collection of characters is composed of 100 video sequences, each representing a specific sign language symbol. Each video sequence consists of 30 frames that capture key moments and are stored as Numpy arrays. The data are collected for ten commonly used gestures. So, the dataset consists of Numpy array of keypoint values extracted using MediaPipe Holistic from 100 video sequences each containing 30 frames with each frame consisting of 1662 keypoints' values (Table 1).

3.3 Training

The training for model development was performed using TensorFlow, a machine learning platform implemented in Python. In order to employ transfer learning, the datasets and labeled files had to be transformed into a format that TensorFlow could handle. The transformation was done by creating tfrecord files from the folders that contained the training and test data. To begin the training process using transfer learning, pre-existing detection models had to be downloaded. The configuration files needed to be modified to match the number of classes in the dataset. Two hundred training steps were required to achieve desired accuracy. TensorFlow and Keras were combined to produce a LSTM model that can predict sign language gestures on the screen. The model consists of LSTM layer and dense layers as mentioned previously.

3.4 Predicting the Gesture

Initially, a frame containing a sign language gesture is captured by the camera. Then, the keypoints for the hand, pose, and face are detected using MediaPipe Holistic. The keypoints captured are then converted into a flattened Numpy array and are inputted into the previously trained LSTM model followed by dense layers for prediction. The final dense layer consists of ten units which represents the number of gesture the model is trained for. This layer utilizes Softmax function as an activation function, i.e., it gives a probability score for each action defined, and when the score reaches

Table 1 List of gestures

S. No.	Gesture	Word
1		Hello
2		Goodbye
3		Please
4		Thank you
5		Yes

(continued)

Table 1 (continued)

S. No.	Gesture	Word
6		No
7		Help
8		House
9		Hungry
10		Sorry

the threshold value, then the respective output text is displayed over the video frame as shown in Figs. 3 and 4.

The Softmax equation can be given as:

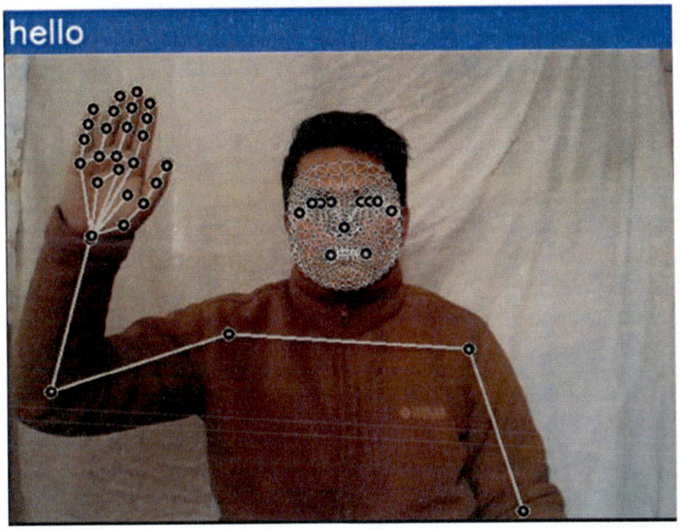

Fig. 3 Output for hello

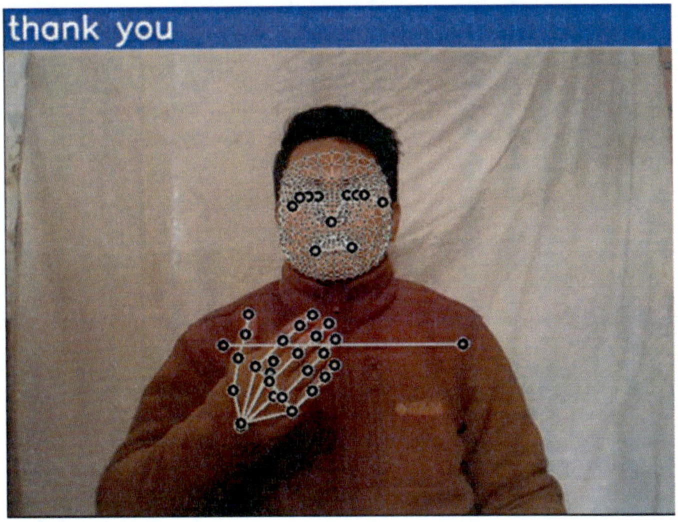

Fig. 4 Output for thank you

$$f_{yi}(z) = \frac{e^{zj}}{\sum_{k=1}^{C} e^{z_k}} \tag{1}$$

The Z here represents the data from the output layer's neurons, where C = total number of classes. The exponential is used to depict nonlinear functions. The sum of exponential values is used to standardize and transform these data into probabilities [6].

This paper focused on sign language recognition that can identify American Sign Language gestures. The system can also be easily expanded to incorporate personalized symbols and additional hand gestures. This paper was constructed using OpenCV, MediaPipe, TensorFlow, and Keras Python modules. OpenCV analyzes live video captured of frame size 640 * 480, 30 fps by a camera to detect the hand gestures of a person. MediaPipe Holistic is used to process the video frames and extract keypoints from the hands, pose, and face. The extracted points are then fed into the prediction algorithm, which makes the prediction. The technology recognizes the hand signal in real-time, and the resulting sign is displayed.

4 Result and Analysis

A total of 100 video sequences consisting of 30 frames each were used as a dataset for each gesture. For the collection of 100 video sequences, it took about 10 min, and for the training of the model through 200 epochs, it took 30 min. After collecting the data, it is split into two datasets: a training dataset and a testing dataset, with a ratio of 80:20. The accuracy and loss curve given below provides insight into the model's training progress and improvement across each epoch. This visual representation is useful for tracking how accuracy and loss metrics change over time, both for the training and validation datasets. The training was performed for two values of learning rate (LR) 0.01 and 0.001 (Figs. 5 and 6).

From the epoch categorical accuracy and loss curves, it is observed that the model was able to converse with the learning rate of 0.001 in 200 epochs.

A confusion matrix is a tabular representation that provides a summary of the classification model's performance on a given test dataset. It achieves this by comparing the predicted labels against the actual labels. This matrix is an excellent tool for assessing and interpreting the model's performance since it enables the visualization and analysis of the classification results. For the proposed model with learning rate of 0.001, the accuracy was calculated to be 98.50% (Table 2).

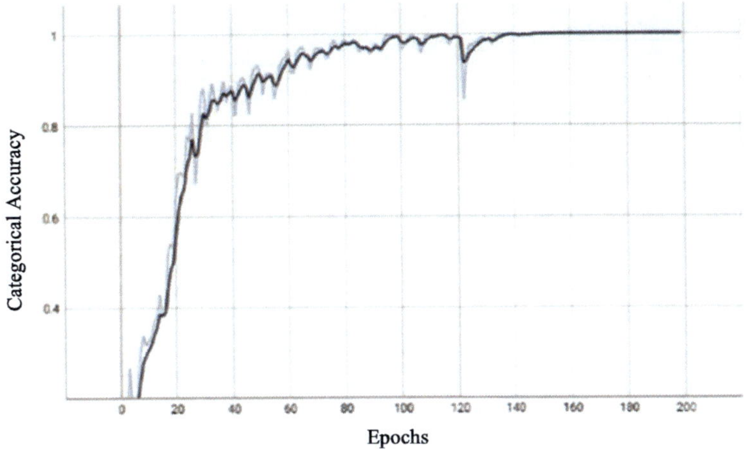

Fig. 5 Epoch categorical accuracy LR = 0.001

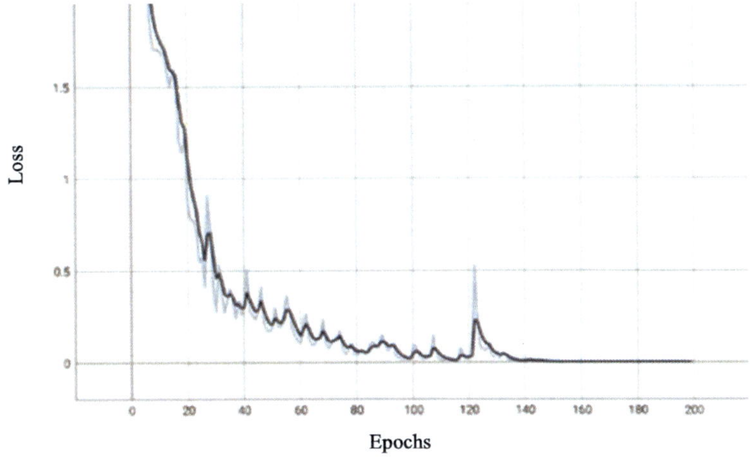

Fig. 6 Epoch loss LR = 0.001

5 Conclusion and Future Work

The system that has been proposed has shown significant success in recognizing American Standard Sign Language signs. Its high accuracy is attributed to the utilization of keypoints extracted from MediaPipe Holistic. The model's precision was evaluated to be 98.50%. To incorporate new signs into the model, one can add the corresponding gestures to the dataset and increase the number of prediction categories. The accuracy can be further increased by increasing the number of frame recorded for each gesture and increasing more video sequence data.

Table 2 Confusion matrix

P/A	Hello	Goodbye	Please	Thank you	Yes	No	House	Help	Hungry	Sorry
Hello	24	0	0	0	0	0	0	0	0	0
Goodbye	0	18	0	0	0	0	0	0	0	0
Please	0	0	21	0	0	0	0	0	1	0
Thank you	0	0	0	13	0	0	0	0	0	0
Yes	0	0	0	0	13	0	0	0	0	0
No	0	0	0	0	0	20	0	0	0	0
House	0	0	0	0	0	0	21	0	0	0
Help	0	0	0	0	0	0	0	20	0	0
Hungry	0	0	1	0	0	1	0	0	22	0
Sorry	0	0	0	0	0	0	0	0	0	25

The proposed system only contains ten commonly used ASSLs. So, the system needs to be trained for other gestures as well. The collection of data was carried out in a well-lit environment with a white plane background. The training and testing were carried out on a Laptop with Intel i5-1135G7 @ 2.40 GHz processor and 8 GB RAM. With decreases in these parameters, the performance of the system will decrease. Also, the current system is based on a computer system which is not feasible for daily use. So, a more portable system needs to be developed. Several future enhancements can be made to this proposed system. Currently, the dataset consists of ten gestures only in which each gesture contains 100 video sequences of 30 frames each. So, other remaining gestures of different words are needed to be added to the dataset. Also, by adding the number of video sequences and the number of frames, the accuracy of the system can be further increased. The proposed system makes use of 1162 keypoints' values for each frame, among which all the keypoints are not required for each gesture detection. So, in the future, the performance of the model can be improved by identifying unwanted keypoints, and by doing so, the complexity of the model will be reduced. Also in the future, this system can be extended to be used for different languages.

References

1. Bayegizova A, Murzabekova G, Ismailova A et al (2022) Effectiveness of the use of algorithms and methods of artificial technologies for sign language recognition for people with disabilities. Eastern-Euro J Enterprise Technol
2. Mercy PJ, Amsa Vani S (2021) Sign language identification using image processing. Int J Creative Res Thoughts
3. Kenshimov C, Buribayev Z (2021) Sign language dactyl recognition based on machine learning algorithms. Eastern-Euro J Enterprise Technol
4. Narvekar C, Mungekar S, Pandey A, Mahadik M (2020) Sign language to speech conversion using image processing and machine learning. Int J Technol Res Eng

5. Sharma A, Mittal A, Singh S, Awatramani V (2020) Hand gesture recognition using image processing and feature extraction techniques. In: International conference on smart sustainable intelligent computing and applications
6. Ligal S, Baral DS (2022) Nepali sign language gesture recognition using deep learning. In: 12th IOE graduate conference
7. Van Houdt G, Mosquera C, Nápoles G (2020) A review on the long short-term memory model. Artif Intell Rev
8. Artyukhin SG, Mestetskiy LM (2015) Dactyl alphabet gesture recognition in a video sequence using Microsoft Kinect. In: International archives of the photogrammetry, remote sensing and spatial information sciences
9. Zamani M, Kanan HR (2014) Saliency based alphabet and numbers of American sign language recognition using linear feature extraction. In: International conference on computer and knowledge engineering
10. Chen Y, Luo B, Chen Y-L, Liang G, Wu X (2015) A real-time dynamic hand gesture recognition system using Kinect sensor. In: IEEE international conference on robotics and biomimetics
11. Lee DH, Hong KS (2010) A hand gesture recognition system based on difference image entropy. In: International conference on advanced information management and service
12. Symeonidis K (2000) Gesture recognition using neural networks. In: Degree of master of science in multimedia signal processing communications, school of electronic and electrical engineering
13. Chen Y, Gao W, Ma J (2000) Hand gesture recognition based on decision tree. Institute of Computing Technology, Chinese Academy of Sciences
14. Marcel S, Bernier O, Viallet JE, Collobert D (2000) Hand gesture recognition using input–output hidden Markov models. France Telecom CNET
15. Ng CW, Ranganath S (2018) Static hand gesture recognition system using convolutional neural networks. Int J Sci Res Dev
16. Shinde SS, Autee RM, Bhosale VK (2013) Sign language recognition for deaf & dumb. Int J Adv Res Comput Sci Softw Eng

Explainable Artificial Intelligence and Deep Transfer Learning for Skin Disease Diagnosis

James Mayanja, Enoch Hall Asanda, Joshua Mwesigwa, Pius Tumwebaze, and Ggaliwango Marvin

Abstract Accurate diagnosis of skin diseases, including skin cancers, is challenging and often leads to misdiagnosis and incorrect treatments. Whereas misdiagnoses have far-reaching consequences for both the patients and the healthcare system, resulting in delayed interventions, disease progression, and compromised outcomes, diagnostic precision is extremely crucial for improved prognosis and disease management. Unfortunately, the complexity of skin diseases with their diverse symptoms and subjective interpretation adds to the diagnostic challenges faced by dermatologists. It complicates accurate identification of essential features to prevent unnecessary interventions and ensure appropriate treatment, saving healthcare resources. To address misdiagnosis, we apply intelligent approaches to dermatology particularly integration of artificial intelligence (AI) and machine learning (ML) using comprehensive datasets to enhance diagnostic accuracy. The AI models are trained on diverse skin disease data to provide reliable diagnostic support using explainable AI (XAI) techniques tailored for skin disease diagnosis. We provide transparency and interpretability to clinicians to enable them understand AI-driven diagnostics, promote trust and collaboration with AI diagnostic tools. This facilitates informed decision-making and improves the overall diagnostic process where accurate diagnosis is crucial given the prevalence of skin diseases. Therefore, this work presents advancements in ML and XAI to mitigate errors and improve dermatology diagnostic outcomes. This collaboration between dermatologists and technology experts

J. Mayanja · E. H. Asanda · J. Mwesigwa · P. Tumwebaze · G. Marvin (✉)
Makerere University, 7062 University Road, Kampala, Uganda
e-mail: ggaliwango.marvin@mak.ac.ug

J. Mayanja
e-mail: james.mayanja@students.mak.ac.ug

E. H. Asanda
e-mail: enoch.hall.asanda@students.mak.ac.ug

J. Mwesigwa
e-mail: joshua.mwesigwa@students.mak.ac.ug

P. Tumwebaze
e-mail: pius.tumwebaze@students.mak.ac.ug

S. Shakya et al. (eds.), *Fourth International Conference on Image Processing and Capsule Networks*, Lecture Notes in Networks and Systems 798,
https://doi.org/10.1007/978-981-99-7093-3_47

demonstrates the need and potential to enhance accuracy, efficiency, and effectiveness in skin disease diagnosis, leading to better healthcare delivery and patient well-being.

Keywords Deep transfer learning (DTL) · Artificial intelligence (AI) · Explainable artificial intelligence (XAI) · Convolutional neural networks (CNN) · Skin disease diagnosis

1 Introduction

1.1 Background and Motivation

Dermatological disorders are some of the most widespread diseases in the world [1]. Dermatologists may draw wrong or at least inaccurate conclusions at the beginning of the process of diagnosis without doing a deeper inspection with laboratory tests. Diagnostic errors have been seen as a big cause of preventable harm in United States health care, and, on the other hand, recent estimates of average misdiagnosis-related harms may vary widely.

In the year 2015, the National Academy of Medicine discovered that misdiagnosis is the third biggest cause of death among victims which is the cause for most of the paid medical misbehavior, claims Abbasi [2]. Total estimates adding total autopsy detected error rates as well as the sum of hospital deaths point to a possible 40,000–80,000 misdiagnosis-related deaths in United States hospitals yearly. According to [3], approximately $750 billion dollars is misused annually as a result of needless expenses on medications, procedures, work time lost, among others as a result of diagnostic errors. Primary care doctors are crucial in the delivery of timely as well as accurate diagnosis including the choosing of proper treatment. Failure to assess patient conditions can result in catastrophic results. Misdiagnosis can exact an emotional cost, in addition to the financial costs. The diagnosis process is therefore a vital, collaborative, and complex activity that includes information gathering as well as clinical reasoning to figure out a victim's health issue.

This research on explainable artificial intelligence (XAI) in skin disease diagnosis contributes to the community: through enabling non-technical individuals, such as end users, to understand how AI systems function and answer questions about their behavior, academia: Software developers and machine learning practitioners can utilize explanations to ensure that AI models and projects meet requirements, and benefits afflicted individuals: Through employing AI, healthcare experts can quickly and accurately determine the causes of specific skin diseases and recommend appropriate treatments. By harnessing the power of explainability and interpretable machine learning techniques, we have built a model that not only ensures accurate and timely diagnoses but also fosters unwavering trust in the capabilities of AI. With a profound focus on improving patient outcomes, this work in XAI ushers in a new era of trust, transparency, and transformative impact in the field of dermatology.

1.2 Research Contribution

Explainable artificial intelligence (XAI) in skin disease diagnosis contributes towards enabling non-technical individuals, such as end users, to understand how AI systems function and answer questions about their behavior, academia: software developers and Machine learning practitioners can utilize explanations to ensure that AI models and projects meet requirements, and benefits afflicted individuals: through employing AI, healthcare experts can quickly and accurately determine the causes of specific skin diseases and recommend appropriate treatments.

By harnessing the power of explainability and interpretable machine learning techniques, a model was built that not only ensures accurate and timely diagnoses but also fosters unwavering trust in the capabilities of AI. With a profound focus on improving patient outcomes, this work in XAI ushers in a new era of trust, transparency, and transformative impact in the field of dermatology.

2 Existing Works

Skin conditions have a profound impact on a significant portion of the global population, with approximately 1.9 billion people affected [4]. However, due to the limited availability of dermatologists, many cases are handled by physicians with lower diagnostic accuracy. The consequences are severe, as skin cancer treatment costs surpass a billion dollars, and millions of individuals suffer from side effects such as hair loss, infections, itchiness, and burns. Early detection of cancer has been identified as a crucial factor in reducing the prevalence of such cases. Malignant illnesses like pyogenic granulomas and melanoma, as well as non-carcinogenic conditions like dermatofibroma and melanocytic nevi, pose significant harm to the skin and can spread if left untreated [5, 6]. Addressing this issue, recent research [7] emphasizes the importance of early skin disease diagnosis to alleviate the burden on individuals and society as a whole. Scholarly literature by [8] scrutinized the various non-invasive approaches of evaluating the visual features of the skin. Ophthalmic recordings were built on the experiments of nonionizing electromagnetic radiation with the skin. In their research, low frequency radiation was what they used and they examined skin disorders using over 8 optical technologies some of which included: two-photon microscopy, in vivo confocal microscopy, and the diffuse reflectance spectroscopy (DRS).

More research carried out by [9] indicates that the diagnosis of skin diseases is now being done on a nanoscale using nanoparticles (Janus nanoparticles) owing to the advancement in technology. Nanotechnology is an emerging scientific discipline that hones tools and devices of sizes less than 100 nm, with peculiar molecular uses. These particles are nanocapsules with impressive applications in cosmetics and drug delivery. They have compartments affiliated with an aqueous compartment bound by a layer of natural oils [9]. Several other methods of skin diagnosis have been discussed by [10], many of which include:

1. the use of compact digital cameras for snap photography
2. fiber-optic microscopy for skin surface imaging
3. computerized image analysis of clinical photos.

Reference [11] exploited CNN methods to implement a novel strategy. They employed CNN approaches which produced advanced classification models, rather than neural networks to categorize the data. 379 test samples and 900 training samples were employed in the approach. The samples fell into the benign and malignant categories. The image size was downscaled, and it was converted to grayscale as part of the preprocessing. Following that, the system made use of AlexNet. After feature reduction using LDA approaches, 4 alternative classifiers and metrics were employed, with KNN providing the best specificity (99.9%) and precision (86%) of all the classifiers. Reference [12] leveraged pretrained networks in two different ways: as a feature extractor and as a transfer learning paradigm. The dataset, PH2, was obtained from Hospital Pedro Hispano in Portugal's dermatology department. There are 200 images used in total. The required preprocessing was done, because the preceding networks require inputs of various sizes. For feature extraction and transfer learning, AlexNet and VGG16 were separately employed. As a transfer learning model, VGG16 produced the greatest results out of all of these techniques: a precision of 97.5% and a specificity of 96.87%. Reference [13] demonstrated an automated method for identifying melanoma from skin cancer ultimately improving precision while utilizing a minimal number of features. The dataset was divided, and features of interest were selected. According to test results obtained, which were based on 200 pictures, just six criteria were effective for identifying melanoma and precision of 86.67% was achieved [5, 6]. Reference [14] proposed a novel system to categorize skin disorders. Two layers of the system were put into place. A 112-lesion dataset separated into three subsets was used to train the binary classifiers. The three independently trained models were compared using the LOO classification process, a classification cross-validation technique used in machine learning as well as statistics to assess the performance of an artificial intelligence classification model, to pick the best model. The best outcomes came from a model with four multinomial kernels. This approach achieved an $F1$-score of 85%.

While significant advancements have been made in leveraging highly advanced artificial intelligence algorithms for machine learning, there is a critical need to comprehend the reasoning behind model predictions, decisions, and internal representations, particularly in crucial domains such as health care. This understanding is essential for ensuring the transparency and interpretability of the applied algorithms, particularly in the context of skin disease diagnosis. By unraveling the rationale behind the predictions and decisions made by the machine learning models, one can enhance the trust and reliability of the diagnostic process. This allows healthcare professionals to make informed decisions based on the explanations provided by the models, ultimately improving the accuracy and effectiveness of skin disease diagnosis.

3 Computational Problem and Solution

3.1 Problem Definition

Dermatological problems are a major worldwide health burden. Due to the time-consuming nature of the process, dermatologists confront difficulties in providing reliable diagnoses. Current skin condition diagnosis methods are frequently difficult, expensive, and unavailable in low-resource settings, and, if they are, they are traditional methods.

Therefore, this research aims to address the problem of accurate diagnosis of skin disorders using AI techniques. Misdiagnosis can have severe consequences for patient health and strain the healthcare system. A more efficient and accessible diagnostic approach to improve patient outcomes would be expedient.

The current state of dermatological diagnosis, particularly in low-resource communities, involves extensive tests and analysis by dermatologists, which are traditional methods. However, these methods have limitations and can vary in results and time. AI and XAI present an opportunity to enhance the diagnostic process and improve accuracy. Despite advances, dermatological diagnosis remains difficult. Furthermore, the complexity and cost of current diagnostic procedures create impediments to accurate diagnosis.

4 Research Methodology

Deep learning has improved and experienced exponential evolution in recent years while playing a tremendous role in image classification in various research domains. Recent research shows that convolutional neural networks recently have achieved high performance in detecting numerous illnesses including coronary artery disease, malaria etc. A dataset was used for validating as well as training the model. Data were obtained from DermNet the largest dermatology source. The dataset includes images of 5 classes of skin diseases including melanoma, eczema, rosacea, psoriasis, acne, and seborrheic keratoses, https://dermnetnz.org/image-library. In recent research, researchers describe in detail the importance of data preprocessing in convolutional neural networks; hence, image data shall be passed through a preprocessing pipeline which comprises the following process: resizing, normalization, and image augmentation.

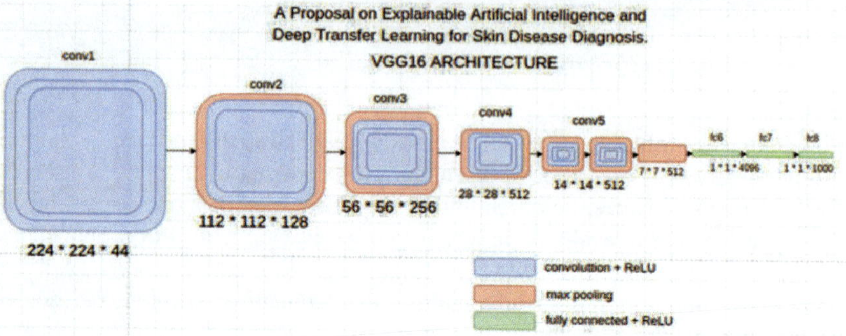

Fig. 1 Image of the VGG16 model architecture

4.1 Model Architecture Methodology

4.1.1 Data Sources

The data were obtained from DermNet, and it was comprised of 5 classes for the case study such as acne, eczema, rosacea, melanoma, and keratoses. The majority of the image data will be used for training, while a smaller portion will be reserved for testing. Figures 1, 2, 3, and 4 provide a sample representation of the classes used to train and test the mode. The split ratio in the training and testing data was 8:2.

4.2 Model Implementation

4.2.1 Image Augmentation

Data augmentation is a technique that entailed producing different data points from the original data with the intent of expanding the dataset domain. This is usually done by making small modifications to the data, such as adding minor translations, flipping, rotations, or noise. Research has shown that the use of augmented data can improve the results of these models by creating new and diverse data for training [15].

Resizing: In this research, the VGG16 model was used because of its fewer layers compared to VGG19 model hence less computational power, so the dataset must be resized to match the input configuration of the pretrained model. In this case, all images in the dataset were resized to 224 × 224, because this is the configuration used by the VGG16 network.

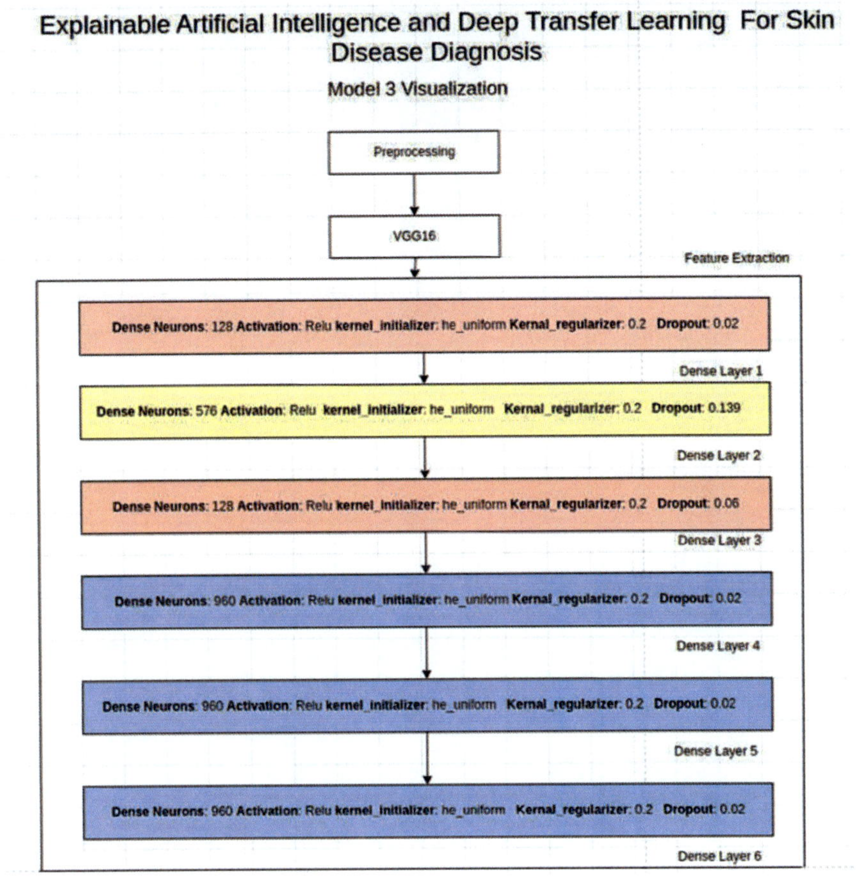

Fig. 2 Image of model visualization

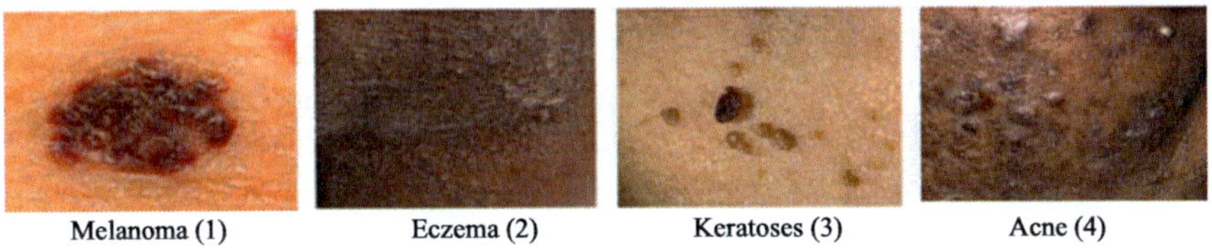

Fig. 3 Sample images of skin diseases

Pixel Normalization: Pixel normalization is a way of normalizing pixel values in an image. The whole purpose is to transform the images such that they are suitable for training the model. After image data have been normalized, each rescaled image has values in a range from 0 to 255 (Table 1).

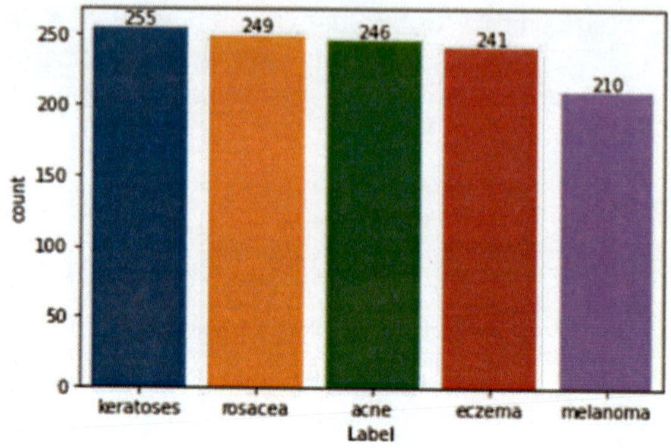

Fig. 4 Values used in data augmentation

Table 1 Table of values used in data augmentation

Process	Value
Rescale	1./255
Horizontal_flip	True
zoom_range	0.1
shear_range	0.1
Preprocessing_function	Preprocess_input

$$\text{normalized\_pixel} = (\text{pixel\_value} - \text{mean})/\text{standard\_deviation}$$

4.3 Model Training and Evaluation

To select the best diagnosis model, the selected model is evaluated using benchmarks: recall, accuracy, precision, ROC curve, and $F1$-score. In the case of skin diagnosis, there are $Z = 5$ classes, resulting in a 5×5 matrix. The training time in this case was 18 h for one model. Each model was trained on i5 Intel processor, 12 GB RAM, 1 GB GPU with a Keras 2.6.0 library (Fig. 5).

Precision representing the portion of image samples that were correctly classified, and accuracy representing the allotment of the summation of correct model predictions (Fig. 6).

Recall represents the ratio of images which were accurately classified to the summation of images in the dataset. In this research, prioritize to optimize recall because the implications of misclassifying images could be catastrophic if the model was deployed in a working health facility. Finally, $F1$-score representing the classification integration of accuracy of recall and precision. The following is the formula for $F1$-score (Table 2).

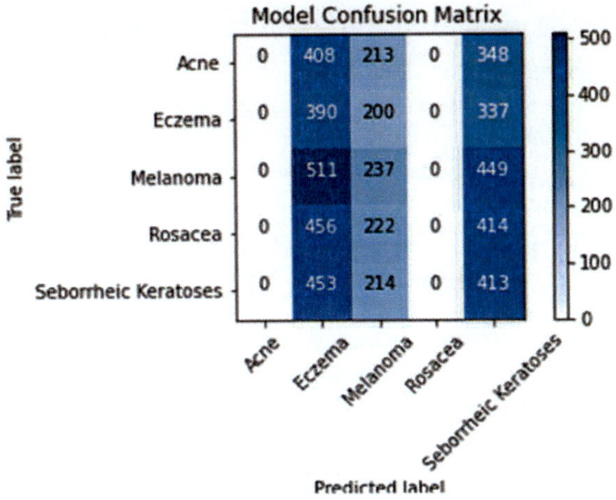

Fig. 5 Confusion matrix for the model considered

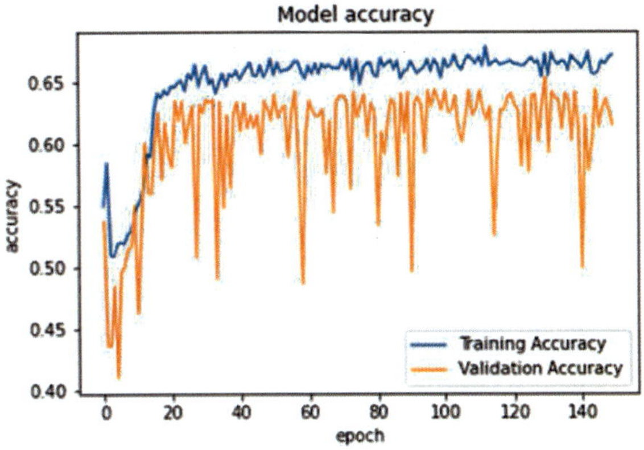

Fig. 6 An image of the model accuracy of the third model

$$F1 = (\text{recall} - 1 + \text{precision} - 1)/2$$
$$= 2(\text{precision} * \text{recall} + \text{recall})$$

Table 2 Table with a summary of the classification report for each of the trained neural networks

Model	Trainable parameters	Precision (%)	Recall (%)	Val. accuracy (%)	Accuracy (%)	$F1$ score
Model 1	27,309,509	23.59	21.394	35.525	37	0.2195
Model 2	12,598,085	51.34	48.64	65.025	64.654	0.4943
Model 3	5,333,317	68.904	65.45	65.925	67.654	0.6632

5 Explainable Artificial Intelligence

In this skin disease diagnosis system, the power of saliency maps was harnessed; SmoothGrad and Grad-CAM techniques were used to enhance the interpretability, reliability, and trustworthiness of deep learning model's predictions. Saliency maps, in particular, played a crucial role in providing visual explanations by highlighting the important regions in the input images that contributed most to the model's diagnoses.

The Grad-CAM (gradient-weighted class activation mapping) method, invented by Selvaraju and coauthors [14], is a powerful technique that has been applied in various domains, including skin disease diagnosis. In this skin disease diagnosis system, Grad-CAM was utilized to determine the important regions in input images that contribute the most to the prediction and classification of skin diseases. By analyzing the classification score slopes relative to the final convolutional feature map, Grad-CAM helped identify the areas in the image that have the greatest impact on the predicted diagnosis. These areas, where the gradient is significant, play a crucial role in the network's decision-making process for skin disease classification (Figs. 7 and 8).

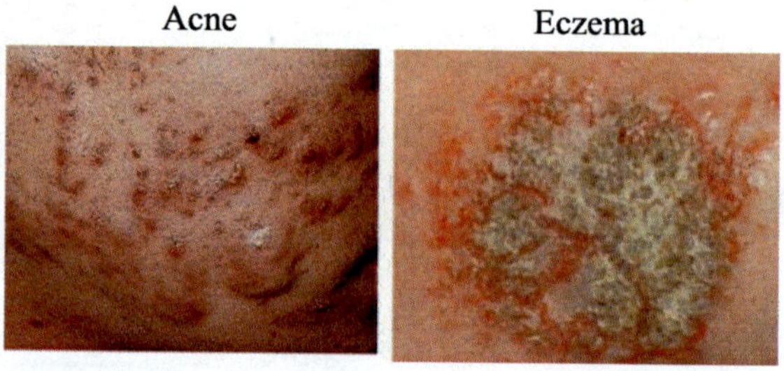

Fig. 7 Sample images on acne and eczema

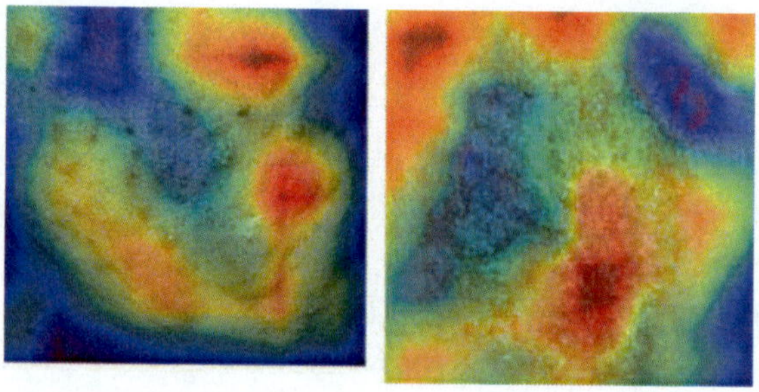

Fig. 8 Heatmaps of acne and eczema indicating important regions

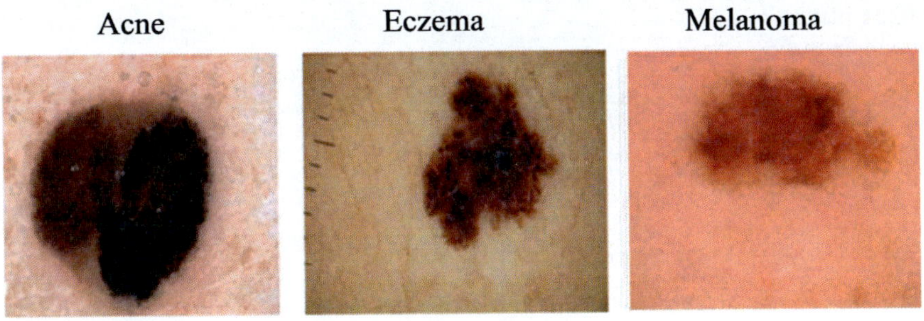

Fig. 9 Sample images on acne, eczema, and melanoma

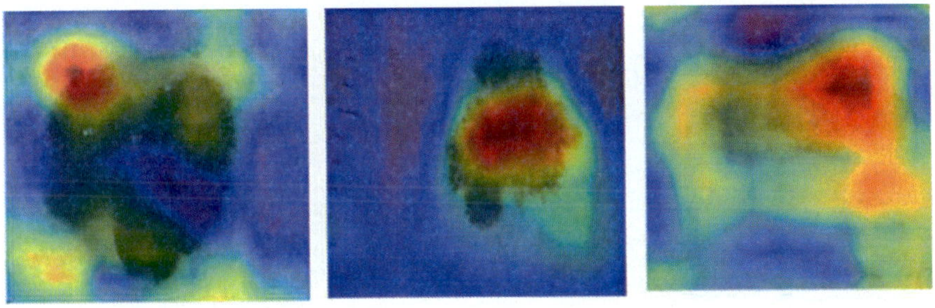

Fig. 10 Visual explanations with added noise on the images on acne, eczema, and melanoma

5.1 SmoothGrad

SmoothGrad was crucial in the skin disease diagnosis system by examining the model's sensitivity to small perturbations in the input image. It involved generating multiple copies of an input image with added noise to better understand how the model's output changes with small perturbations (Figs. 9 and 10).

5.2 Saliency Map

By calculating gradients with respect to the input image, saliency maps helped identify the pixels or regions that had the greatest influence on the model's predictions. These maps provided a visual representation of the model's attention and focus, highlighting the specific areas of the skin image that were most informative for the diagnosis.

In conjunction with SmoothGrad and Grad-CAM, saliency maps augmented the understanding of the decision-making process in skin disease diagnosis. SmoothGrad allowed us to analyze the model's sensitivity to small perturbations in the input images, ensuring the stability and robustness of the predictions. Grad-CAM provided

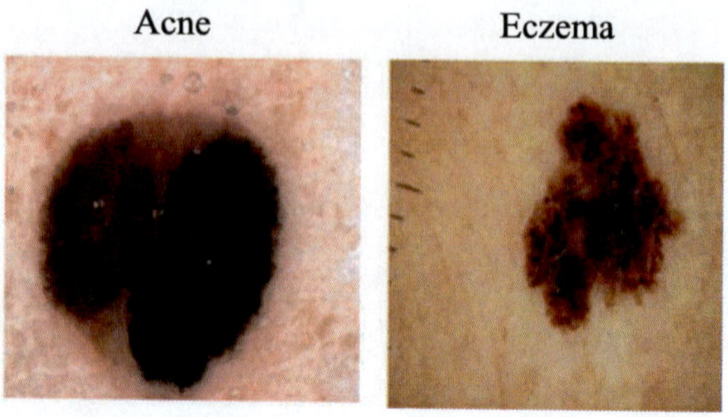

Fig. 11 Sample images of acne and eczema

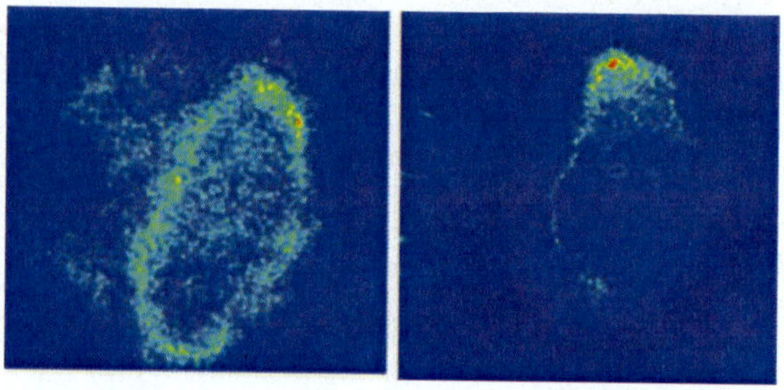

Fig. 12 Region of focus on acne and eczema using saliency maps

insights into the important regions of the image that influenced the model's decision by analyzing the gradients and convolutional feature maps.

By incorporating saliency maps into this system, dermatologists and healthcare professionals were offered a transparent and interpretable view of the model's diagnoses. The highlighted regions in the saliency maps provided visual explanations and enabled them to validate the model's predictions, fostering trust and confidence in the system (Figs. 11 and 12).

6 Limitations and Conclusion

6.1 Limitations and Recommendations

The model was only trained on images of people with a fairly toned skin mainly due to the fact that majority of the images in the original dataset were images of fairly toned skin infected people, and this raised questions on how the model would perform when required to diagnose people with a dark-skinned tone. The dataset's exclusive focus on fairly skinned people posed several challenges and drawbacks:

Lack of Diversity: By collecting data exclusively from fairly skinned individuals, the vast diversity present in human skin tones was not captured. This limitation led to biased and inaccurate predictions when the model encountered people with different skin tones.

Generalizability Issues: Since the dataset is not representative of the entire population's skin tones, the trained model struggled to generalize its predictions to individuals with darker or lighter skin tones. This led to inaccuracies and reduced performance in real-world scenarios.

Biased Results: The model's training on a limited dataset perpetuated biases present in the data. It led to unequal or unfair treatment for individuals with darker skin tones when it came to medical assessments or diagnoses.

Lack of Inclusivity: By not considering individuals with different skin tones, critical information that could impact healthcare outcomes and diagnostics for a more diverse population was missed out on.

Limited Scope: Inaccessibility to data from hospitals due to the lack of institutional review board (IRB) approval restricted the dataset's scope to a specific subset of the population. This narrow scope hindered the model's ability to handle diverse medical conditions and scenarios.

Ethical Concerns: The lack of IRB approval raised ethical concerns about data privacy, patient consent, and ensuring that data collection and usage comply with ethical standards.

6.2 Conclusion

The integration of explainable artificial intelligence (XAI) has the potential to revolutionize the field of skin disease diagnosis by harnessing the power of imagery data. XAI techniques enable healthcare dermatologists to gain deeper insights into the decision-making process of AI models, providing them with a transparent and interpretable understanding of how diagnoses are made. Through analyzing and interpreting the important features and regions within the images, XAI empowers dermatologists and healthcare practitioners to make more accurate and informed diagnostic decisions. This not only enhances the accuracy and reliability of skin disease diagnosis but also builds trust and confidence in the AI-driven diagnostic systems. As a part of future work and research in XAI, one can expect significant improvements in the early detection, personalized treatment, and overall management of skin diseases, ultimately leading to better healthcare outcomes for patients.

Collection of imagery data specifically tailored for XAI in skin disease diagnosis from various health institutions shall be done. This shall ensure that the models are diverse, representative, and well-annotated for robust training of AI models. This contribution shall not only enhance the reliability of the diagnostic system but also promote the advancement of the field as a whole by expanding the available resources for researchers and practitioners. Additionally, the collection of imagery data enables us to address specific challenges and gaps in the current datasets, such as the inclusion of underrepresented skin types or rare skin conditions. By undertaking this task, one can further improve the performance and generalizability of the XAI-based diagnostic approach, ultimately leading to more effective and accessible healthcare solutions for individuals affected by skin diseases.

References

1. Hellen N, Marvin G (2021) Interpretable feature learning framework for smoking behavior detection. arXiv preprint arXiv:2112.08178
2. Abbasi J (2016) A headline-grabbing study brings attention back to medical errors. JAMA 316(7):698–700
3. McGinnis JM, Stuckhardt L, Saunders R, Smith M et al (2023) Best care at lower cost: the path to continuously learning about health care in America
4. Liu Y, Jain A, Eng C, Way DH, Lee K, Bui P, Kanada K, de Oliveira Marinho G, Gallegos J, Gabriele S et al (2020) A deep learning system for differential diagnosis of skin diseases. Nat Med 26(6):900–908
5. Gulati S, Bhogal RK (2019) Detection of malignant melanoma using deep learning. In: International conference on advances in computing and data sciences. Springer, pp 312–325
6. Mustafa S, Kimura A (2018) A SVM-based diagnosis of melanoma using only useful image features. In: 2018 international workshop on advanced image technology (IWAIT). IEEE, pp 1–4
7. Maniraju M, Adithya R, Srilekha G (2022) Recognition of type of skin disease using CNN. In 2022 first international conference on artificial intelligence trends and pattern recognition (ICAITPR). IEEE, pp 1–4
8. Kollias N, Stamatas GN (2002) Optical non-invasive approaches to diagnosis of skin diseases. J Invest Dermatol Symp Proc 7:64–75
9. Kemel K, Deniset-Besseau A, Baillet-Guffroy A, Faivre V, Dazzi A, Laugel C (2020) Nanoscale investigation of human skin and study of skin penetration of janus nanoparticles. Int J Pharm 579:119193
10. Serup J, Jemec GB, Grove GL (2006) Handbook of non-invasive methods and the skin. CRC Press
11. Majtner T, Yildirim-Yayilgan S, Hardeberg JY (2019) Optimised deep learning features for improved melanoma detection. Multimed Tools Appl 78(9):11883–11903
12. Mete M, Sirakov NM, Griffin J, Menter A (2016) A novel classification system for dysplastic nevus and malignant melanoma. In 2016 IEEE international conference on image processing (ICIP). IEEE, pp 3414–3418
13. Perez L, Wang J (2017) The effectiveness of data augmentation in image classification using deep learning. arXiv preprint arXiv:1712.04621
14. Selvaraju RR, Cogswell M, Das A, Vedantam R, Parikh D, Batra D (2017) Grad-cam: visual explanations from deep networks via gradient-based localization. In: Proceedings of the IEEE international conference on computer vision, pp 618–626
15. Marvin G, Alam MGR (2022) Explainable augmented intelligence and deep transfer learning for pediatric pulmonary health evaluation. In: 2022 international conference on innovations in science, engineering and technology (ICISET)

A Comparison Study Between Otsu's Thresholding, Fuzzy *C*-Means, and *K*-Means for Breast Tumor Segmentation in Mammograms

Moustapha Mohamed Saleck, Nagi Ould Taleb, Mohamed El Moustapha El Arby Chrif, and El Benany Mohamed Mahmoud

Abstract Currently, the technique of digital mammography is considered the first screening tool used for detecting breast cancer in early stages. The shape and margin of masses in mammograms play a crucial role in the classification of benign and malignant tumors. Several machine learning (ML) algorithms have demonstrated high performance in breast mass edge segmentation from mammograms. The main objective of this paper is to present a comparative study of three algorithms of ML where: *C*-means clustering, *K*-means clustering, and Otsu's thresholding for automatically detecting the lesion edges on mammogram images. Two groups ($C = 2$) are assigned to clustering as output for each method. In order to achieve high accuracy in contour extraction, only relevant pixels are selected as inputs. This study will highlight the advantages and disadvantages of these algorithms, which are expected to improve the detection of breast masses contours in mammographic images. The mini-MIAS database is used to evaluate the obtained results which show a higher performance in terms of segmentation accuracy when using fuzzy *C*-means algorithm. However, a faster lesion extraction process can be achieved by utilizing the *K*-means clustering and Otsu's threshold.

Keywords Breast lesion segmentation · Mammograms · K-means clustering · Fuzzy *C*-means clustering · Otsu's thresholding

---

M. Mohamed Saleck (✉)
LAROSERI Laboratory, Department of Computer Sciences, Faculty of Sciences, University of Chouaib Doukkali, El Jadida, Morocco
e-mail: saleck.moustapha@gmail.com

N. Ould Taleb
Computer Sciences Department, Higher Institute of Digital, Nouakchott, Mauritania
e-mail: nagi.taleb-aly@supnum.mr

M. El Moustapha El Arby Chrif · E. Benany Mohamed Mahmoud
Department of Mathematics and Computer Science, University of Nouakchott AL Aasriya, Nouakchott, Mauritania

S. Shakya et al. (eds.), *Fourth International Conference on Image Processing and Capsule Networks*, Lecture Notes in Networks and Systems 798,
https://doi.org/10.1007/978-981-99-7093-3_48

1 Introduction

Recent data from the GLOBOCAN database in 2020 [1] reported over 2,300,000 new instances of breast cancer and 685,000 related deaths globally. Multiple studies emphasize that early detection and treatment significantly lower mortality rates and enhance patient survival [2]. Over the past decade, diverse medical imaging techniques for breast examination have emerged, including mammography, thermography, breast ultrasound, MRI, PET, optical imaging, CT, scintimammography, and electrical impedance-based imaging [3]. Digital mammography remains the most prevalent tool in breast screening centers.

Despite the crucial role of digital mammography in early breast disease diagnosis, visual assessment by radiologists often yields unsatisfactory outcomes [4]. To address this, computer-aided diagnosis tools are essential to aid doctors in interpreting mammograms and making informed decisions [5]. The segmentation process, which entails isolating tumor boundaries from image details, is pivotal for CADx systems serving as secondary readers alongside doctors. Various segmentation techniques have been proposed for lesion contour extraction from digital mammograms.

Researchers have introduced diverse strategies for mammographic image segmentation. Xie et al. [6] introduced an innovative approach that involves employing a pulse-coupled neural network (PCNN) algorithm combined with the level set method to establish an initial contour. Subsequently, an enhanced level set evolution is utilized to segment the images and attain the final outcomes. Berber et al. [7] devised a fresh breast mass contour segmentation (BMCS) technique for digital mammograms. This method harnesses the classical seed region growing method to enhance mass boundaries within a specified region of interest (ROI). The approach's distinguishing feature lies in its adaptability to fine-tune threshold values and incorporate suitable stopping conditions, effectively addressing under and over segmentation challenges. Ittannavar et al. [8] put forth a novel multiobjective optimization method for extracting breast masses from mammograms. Their procedure involves image denoising and enhancement, followed by the implementation of an electromagnetism optimization algorithm to segment both noncancerous and cancerous regions in mammograms. Sadad et al. [9] employ a cascaded approach, utilizing the fuzzy *C*-means algorithm in combination with region growing, to segment breast tumors in mammographic images. They augment this by incorporating texture-based features such as LBP-GLCM, LPQ, and hybrid attributes to capture the distinct characteristics of segmented tumors. Thawkar et al. [10] amalgamate three techniques: Otsu's global thresholding, morphological operations, and watershed transformation to create an automated method for detecting suspicious masses within mammograms.

Chattaraj et al. [11] propose a novel algorithm based on marker-controlled watershed principles for breast lesion segmentation. Their work leverages catchment basins on the topographic surface of mammograms. Makandar et al. [12] engineer an efficient system capable of detecting abnormalities in mammograms, utilizing Otsu's thresholding and morphological operations. Hu et al. [13] devise an innovative methodology for spotting suspicious lesions in mammograms. This approach integrates adaptive global and local thresholding segmentation, effectively segmenting multiresolution sub-images extracted from the original mammogram. Xu et al. [14] develop an original model adaptive receptive field network (ARF-Net) for segmenting breast masses across mammographic and ultrasound images. Li et al. [15] introduce an end-to-end deep learning framework to predict lesion segmentation and diagnosis outcomes. This approach employs dual convolutional neural networks. Li et al. [16] propose a model named region-based graph convolution coupled with the atrous spatial pyramid pooling network (RGC-ASPP-Net) for mammogram mass segmentation. Kanungo et al. [17] suggest a hybrid segmentation approach for detecting suspicious breast masses in mammograms. Their strategy involves the utilization of fuzzy *C*-means, *K*-means, fuzzy possibilistic *C*-means (FPCM), genetic algorithm, and ant colony optimization. Roy et al. [18] present a combined approach, integrating convolutional neural networks (CNN) with connected component analysis (CCA) for preprocessing-free breast mass detection. This study involves comparing the outcomes of *K*-means and fuzzy *C*-means algorithms.

This paper focuses on comparing the segmentation performance of three suggested methods: the fuzzy *C*-means algorithm, *K*-means algorithm, and Otsu's thresholding. The underlying concept of these methods involves adapting a global threshold to identify pertinent pixels, which are then used as input for *C*-means, *K*-means, and Otsu's thresholding, leading to the generation of two clusters as outputs, with one cluster representing the lesion.

The subsequent sections of this paper are structured as follows:

Section 2 delves into the description of materials and methods employed in this study. Section 3 presents the experimental results and a comprehensive comparison of the proposed methods. And, finally, Sect. 4 concludes the paper by summarizing the findings and insights obtained.

2 Materials and Method

In this section, the details of proposed methods are explained. Figure 1 shows graphically the different steps of the suggested system.

2.1 Image Preprocessing

The presence of artifacts or noises on mammograms could affect the accuracy of breast mass segmentation, which can lead to unsatisfactory results in computer-aided diagnosis systems (CADs). Mammograms preprocessing is the step taken to simplify the recognition of abnormalities, minimize noise, and preserve meaningful information. The median filter with a 3×3 neighborhood window is used in this

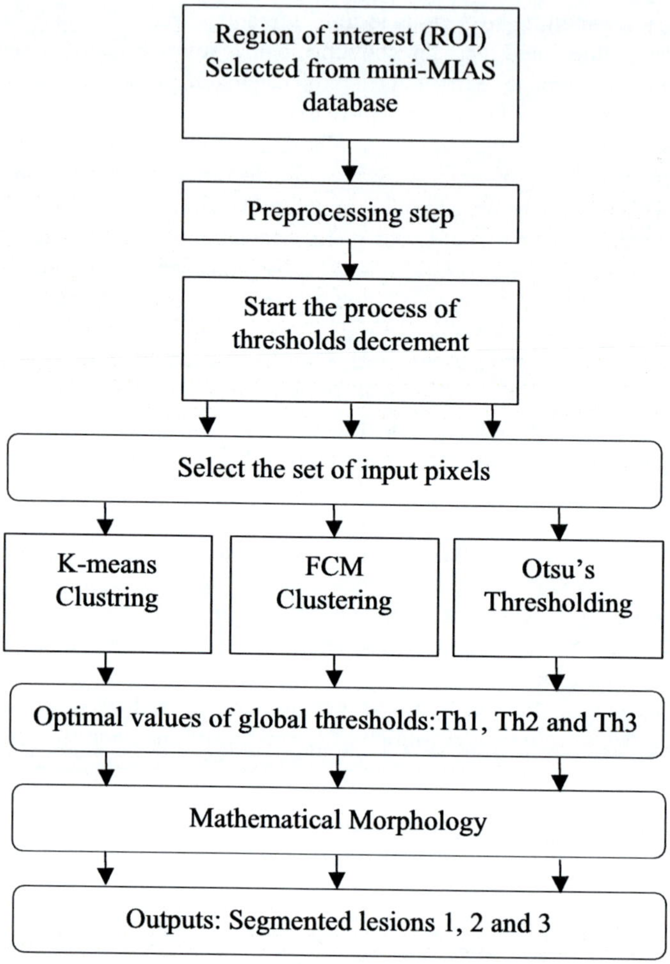

Fig. 1 Block diagram of the suggested methods

work to eliminate any expected noise from the region of interest (ROI) in original mammograms.

2.2 *Fuzzy C-means Algorithm*

Fuzzy *C*-means (FCM) algorithm was first presented by Dunn in 1973 [19] and then improved by Bezdek in 1983 [20]. FCM algorithm is a technique of clustering which allows one pixel of image to belong to two or more clusters with a membership degree $u_{ik} \in [0, 1]$. This algorithm is based on optimizing the objective function (1):

$$J = \sum_{i=1}^{c} \sum_{k=1}^{n} u_{ik}^{m} d^{2}(x_k, v_i) \tag{1}$$

$$0 \le u_{ik} \le 1 \quad \forall i \in [1, c], \quad k \in [1, n]$$

$$\sum_{k=1}^{c} u_{ik} \ \forall i \quad k \in [1, n]$$

$$0 < \sum_{k=1}^{n} u_{ik} < n \quad \forall i \in [1, c]$$

$$u_{ik} = \frac{(d(x_k, v_i))^{\frac{2}{m-1}}}{\sum_{j=1}^{n} (d(x_k, v_i))^{\frac{2}{m-1}}} \tag{2}$$

$$v_i = \frac{\sum_{k=1}^{n} u_{ik}^{m} x_k}{\sum_{k=1}^{n} u_{ik}^{m}} \tag{3}$$

where each cluster is described by its centroid $c_{i,}$, m is the fuzzy index often taken $m = 2$, n represents the number of pixels in image, u_{ik} is the membership degree pixel x_k to the centroid c_i, the Euclidean distance between $x_{k,}$ and c_i is calculated by $d_{ik} = x_k - c_i$. The elements values of membership matrix $U = [u_{ik}]$ are ranging in [0, 1]. The objective function J (1) is optimized by updating membership matrix (2) and the cluster centroid (3).

2.3 K-means Algorithm

The K-means clustering is one of the simplest unsupervised learning algorithms that can be applied to solve the clustering segmentation problem [21]. K-means algorithm allows to classify the set of input pixels from given image into multiple clusters based on their inherent distance from each other.

Given a set of n pixels $\{x_1, x_2, \ldots, x_n\}$, K-means algorithm partitions the pixels into k cluster $\{C_1, C_2, \ldots, C_k\}(k < n)$ so as to minimize the distance within the cluster (4).

$$V = \sum_{i=1}^{k} \sum_{x_i \in s_j} ||x_j - v_i||^2 \tag{4}$$

$$v_i = \frac{\sum_{j=1}^{k} x_j^i}{n_i} \tag{5}$$

where k is the number of cluster and C_i is the number of pixels in ith cluster, $||x_j - v_i||^2$ is the Euclidean distance between x_i and v_i, v_i is the mean of ith in C_i during each iteration.

2.4 Otsu's Thresholding

Otsu's thresholding technique is one of the most widely used algorithms in image segmentation. The effectiveness of this method lies in its ability to select automatically the threshold. Otsu's method also works on the normalized histogram for discriminating the interclass threshold. This method starts by calculating the global mean intensity (u_1, u_2), the class probabilities (q_1, q_2), the individual class variances (σ_1^2, σ_2^2), and finally the weighted within-class variance (σ_ω^2).

$$\sigma_\omega^2(k) = q(t)\sigma_1^2(t) + q_2(t)\sigma_2^2(t) \tag{6}$$

$$q_1(t) = \sum_{i=0}^{t} P_i \quad \text{and} \quad P_2(t) = \sum_{i=t+1}^{l} P_i \tag{7}$$

$$u_1(t) = \sum_{i=0}^{t} \frac{i P_i}{P_1(t)} \quad \text{and} \quad u_2(t) = \sum_{i=t+1}^{l} \frac{i P_i}{P_2(t)} \tag{8}$$

$$\sigma_1^2(t) = \sum_{i=0}^{t} [i - u_1(t)]^2 \frac{i P_i}{P_1(t)} \tag{9}$$

$$\sigma_2^2(t) = \sum_{i=t+1}^{l} [i - u_2(t)]^2 \frac{i P_i}{P_2(t)} \tag{10}$$

Otsu's method needs to run the full range of $t \in [1, 256]$ and choose the value of t which minimizes the weighted within-class variance (6).

2.5 Breast Mass Segmentation

Breast mass segmentation from the region of interest (ROI) is a challenging task in the development of computer-aided diagnosis systems (CADs) for the detection of breast cancer in mammograms. Our purpose behind this research is to draw comparison between three methods for automatic detection of the boundaries of breast mass from the ROI. These methods are based on Otsu's thresholding, fuzzy C-means algorithm, and K-means algorithm. The main idea of these methods is to apply each algorithm in a suitable range of pixels (inputs) where the first limit is the maximum

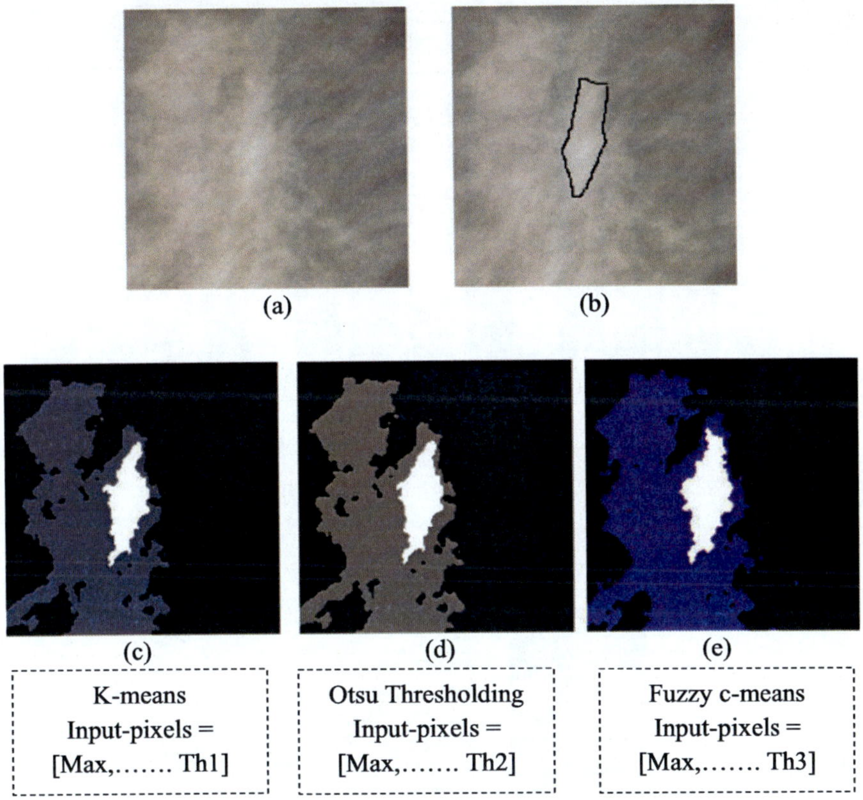

Fig. 2 Segmentation results for mdb_145 from MIAS. **a** Original ROI. **b** Lesion contour denoted by radiologist. **c** Mass segmented using *K*-means. **d** Mass segmented using Otsu's thresholding. **e** Mass segmented using FCM

value of intensity (Max) and the second is an optimal threshold (Th) which is able to eliminate all pixels that can affect the accuracy of segmentation. The number of clusters is chosen as $C = 2$. Morphological operations are applied in the final stage so as to remove imperfections from the segmented image (Fig. 2).

To find out the optimal threshold that permit to obtain the appropriate set of pixels for lesion segmentation, we have adopted the technique proposed in [22].

3 Experimental Results and Comparison

In order to assess the efficacy of the presented techniques, the experiment utilizes the mini-MIAS database. This database encompasses 322 mammograms, with 21 of these images employed for testing purposes. Each image from the MIAS was

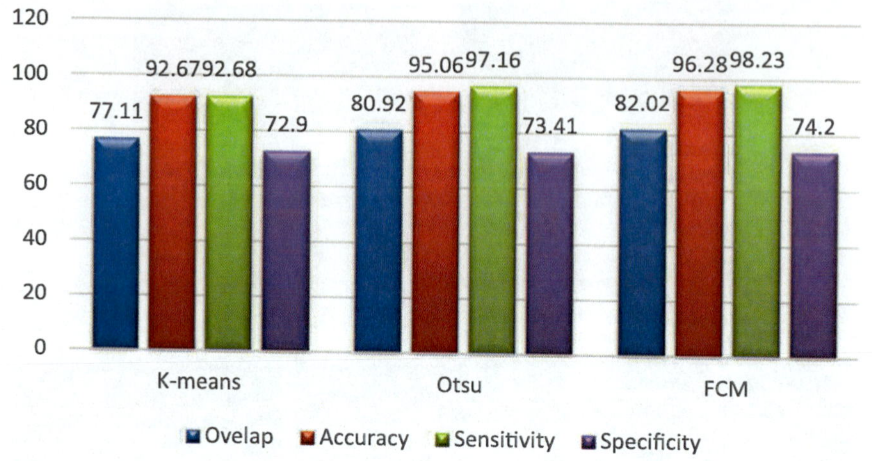

Fig. 3 Comparison of result

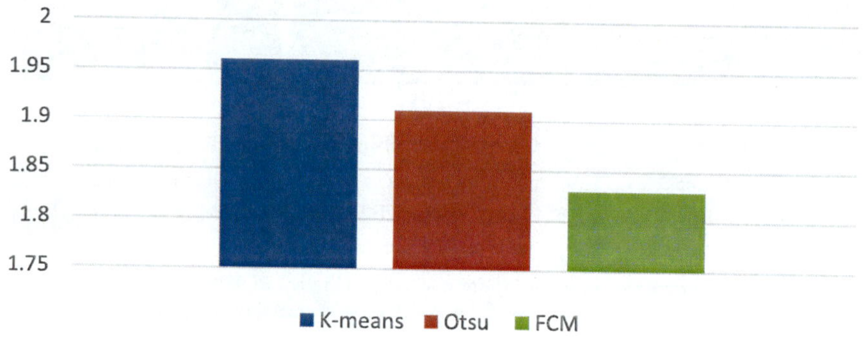

Fig. 4 The time complexity of different methods

digitized at 50 μm, where each pixel was represented by 8 bit. The size of the ROI taken for this work is 174 × 174. The performance evaluation of the proposed methodologies *K*-means, *C*-means, and Otsu's threshold is depicted in Fig. 3 (Fig. 4).

4 Conclusion

Breast cancer currently ranks among the prominent health challenges, emphasizing the significance of early detection to elevate patient survival rates. Mammogram-based breast screening is the prevailing technique in developed nations for early diagnosis. However, a robust and effective approach to extracting breast masses from the region of interest (ROI) within mammogram images remains an intricate undertaking. This paper addresses this challenge by offering a comparative analysis of three proposed breast tumor segmentation methods applied to the ROI. These methods hinge on *K*-means clustering, *C*-means clustering, and Otsu's thresholding. The study presents insights into the efficiency and time required for completing the segmentation process for each method.

References

1. Sung H, Ferlay J, Siegel R et al (2021) Global cancer statistics 2020: GLOBOCAN. CA Cancer J Clin 71:209–249. https://doi.org/10.3322/CAAC.21660
2. Mandelblatt JS et al (2009) Effects of mammography screening under different screening schedules: model estimates of potential benefits and harms. Ann Intern Med. https://doi.org/10.1059/0003-4819-151-10-200911170-00010
3. Sree SV, Ng EY-K, Acharya RU, Faust O (2011) Breast imaging: a survey. World J Clin Oncol 2(4):171–178. https://doi.org/10.5306/wjco.v2.i4.171
4. Gandomkar Z, Mello-Thoms C (2019) Visual search in breast imaging. Br J Radiol 92(1102). https://doi.org/10.1259/bjr.20190057
5. Zebari DA et al (2021) Systematic review of computing approaches for breast cancer detection based computer aided diagnosis using mammogram images. Appl Artif Intell 35(15):2157–2203. https://doi.org/10.1080/08839514.2021.2001177
6. Xie W, Li Y, Ma Y (2016) PCNN-based level set method of automatic mammographic image segmentation. Optik (Stuttg) 127(4):1644–1650. https://doi.org/10.1016/j.ijleo.2015.09.250
7. Berber T, Alpkocak A, Balci P, Dicle O (2013) Breast mass contour segmentation algorithm in digital mammograms. Comput Methods Programs Biomed 110(2):150–159. https://doi.org/10.1016/j.cmpb.2012.11.003
8. Ittannavar SS, Havaldar RH (2022) Segmentation of breast masses in mammogram image using multilevel multiobjective electromagnetism-like optimization algorithm. Biomed Res Int 2022:8576768. https://doi.org/10.1155/2022/8576768
9. Sadad T, Saba T, Hussain A (2018) Fuzzy C-means and region growing based classification of tumor from mammograms using hybrid texture feature. J Comput Sci 29. https://doi.org/10.1016/j.jocs.2018.09.015
10. Thawkar S, Ingolikar R (2017) Automatic detection and classification of masses in digital mammograms. Int J Intell Eng Syst 10(1):65–74. https://doi.org/10.22266/ijies2017.0228.08
11. Chattaraj A, Das A, Bhattacharya M (2017) Mammographic image segmentation by marker controlled watershed algorithm. In: Proceedings—2017 IEEE international conference on bioinformatics and biomedicine, BIBM 2017, pp 1000–1003. https://doi.org/10.1109/BIBM.2017.8217793
12. Makandar A, Halali (2015) Threshold based segmentation technique for mass detection in mammography. https://doi.org/10.17706/jcp.11.6.472-478
13. Hu K, Gao X, Li F (2011) Detection of suspicious lesions by adaptive thresholding based on multiresolution analysis in mammograms. IEEE Trans Instrum Meas 60(2):462–472. https://doi.org/10.1109/TIM.2010.2051060
14. Xu C, Qi Y (2022) ARF-Net: an adaptive receptive field network for breast mass segmentation in whole mammograms and ultrasound image. Biomed Signal Process Control 71:103178. https://doi.org/10.1016/j.bspc.2021.103178
15. Li H, Chen D et al (2022) Dual convolutional neural networks for breast mass segmentation and diagnosis in mammography. EEE Trans Med Imaging 41(1):3–13
16. Li Z, Deng Z, Liao J (2022) Multi-stream information-based neural network for mammogram mass segmentation. Artificial neural networks and machine learning—ICANN 2022, pp 267–278. 978-3-031-15919-0

17. Kanungo GK, Singh N, Dash J, Mishra A (2015) Mammogram image segmentation using hybridization of fuzzy clustering and optimization algorithms. Intelligent computing, communication and devices, pp 403–413
18. Dunn JC (1973) A fuzzy relative of the ISODATA process and its use in detecting compact well-separated clusters. J Cybern 3(3):32–57. https://doi.org/10.1080/01969727308546046
19. Roy A, Singh BK, Banchhor SK (2022) Segmentation of malignant tumours in mammogram images: a hybrid approach using convolutional neural networks and connected component analysis. Expert Syst 39(1):e12826
20. Bezdek JC (1981) Pattern recognition with fuzzy objective function algorithms, 1st edn. Springer, New York, NY
21. Hartigan JA, Wong MA (1979) A K-means clustering algorithm. J R Stat Soc Ser C (Appl Stat) 28(1):100–108 (9 p.)
22. Saleck MM, El Moutaouakkil A, Moucouf M, Bouchaib M, Samira H, Zineb J (2018) Breast mass segmentation using a semi-automatic procedure based on fuzzy C-means clustering. Telkomnika (Telecommun Comput Electron Control 16(2):665–672. https://doi.org/10.12928/TELKOMNIKA.v16i2.6193

Author Index

S. Shakya et al. (eds.), *Fourth International Conference on Image Processing and Capsule Networks*, Lecture Notes in Networks and Systems 798,
https://doi.org/10.1007/978-981-99-7093-3